ASTRONOMY AND ASTROPHYSICS ABSTRACTS

A Publication of the Astronomisches Rechen-Institut Heidelberg
Member of the Abstracting Board of the International
Council of Scientific Unions

Volume 9
Literature 1973, Part 1

Edited by
S. Böhme · W. Fricke · U. Güntzel-Lingner
F. Henn · D. Krahn · U. Scheffer · G. Zech

Springer-Verlag Berlin Heidelberg GmbH 1973

Astronomisches Rechen-Institut
Heidelberg
Director: Prof. Dr. W. Fricke

Astronomy and Astrophysics Abstracts
Editor-in-Chief: F. Henn

Astronomy and Astrophysics Abstracts
is prepared under the auspices
of the International Astronomical Union

ISBN 978-3-662-12289-1   ISBN 978-3-662-12287-7 (eBook)
DOI 10.1007/978-3-662-12287-7

Library of Congress
Catalog Card Number 72-104650.

Preface

Astronomy and Astrophysics Abstracts, which has appeared in semi-annual volumes since 1969, is devoted to the recording, summarizing and indexing of astronomical publications throughout the world. It is prepared under the auspices of the International Astronomical Union (according to a resolution adopted at the 14th General Assembly in 1970).

Astronomy and Astrophysics Abstracts aims to present a comprehensive documentation of literature in all fields of astronomy and astrophysics. Every effort will be made to ensure that the average time interval between the date of receipt of the original literature and publication of the abstracts will not exceed eight months. This time interval is near to that achieved by monthly abstracting journals, compared to which our system of accumulating abstracts for about six months offers the advantage of greater convenience for the user.

Volume 9 contains literature published in 1973 and received before August 15, 1973; some older literature which was received late and which is not recorded in earlier volumes is also included.

We acknowledge with thanks contributions to this volume by Dr. J. Bouška, who surveyed journals and publications in the Czech language and supplied us with abstracts in English, and by the Commonwealth Scientific and Industrial Research Organization (C.S.I.R.O.), Sydney, for providing titles and abstracts of papers on radio astronomy.

We also extend our warmest thanks to Miss Helga Ballmann, Mrs Monika Betz, Mrs Karola Gudé, and Mrs Ingrid Wolf, who typed the text of this volume on IBM 72 Composers and compiled the pages from abstract slips in a perfect form for offset reproduction, to Miss Gisela Nollert, for punching material for the author index and the subject index.

Heidelberg, September 1973

Siegfried Böhme
Walter Fricke
Ulrich Güntzel-Lingner
Frieda Henn
Dietlinde Krahn
Ute Scheffer
Gert Zech

Contents

Planetary System

Stellar Systems

Introduction

Astronomical bibliographies

Astronomy and Astrophysics Abstracts begins documentation and abstracting as from the year 1969. For information on astronomical literature before this date consultation of one of the following bibliographies is suggested:

(1) J. J. de Lalande, Bibliographie Astronomique, Paris 1803 (this work covers the time from 480 B. C. to the year 1803, VIII + 966 pages).

(2) J. C. Houzeau, A. Lancaster, Bibliographie générale de l'astronomie, Volume I (in two parts), Bruxelles 1882, 1887, Volume II, Bruxelles 1889. The complete title of Volume II is "Bibliographie générale de l'astronomie ou catalogue méthodique des ouvrages, des mémoires et des observations astronomiques, publiés depuis l'origine de l'imprimerie jusqu'en 1880". A new edition of these volumes was prepared by D. W. Dewhirst in 1964.

(3) Bibliography of Astronomy, 1881 - 1898. The literature of this period was recorded on standard slips by the Observatoire Royal de Belgique. From the material (some 52.000 items) a microfilm version was produced by University Microfilms Limited, Tylers Green, High Wycombe, Buckinghamshire, England, in 1970.

(4) Astronomischer Jahresbericht, 1899 gegründet von Walter Wislicenus, herausgegeben vom Astronomischen Rechen-Institut in Heidelberg (formerly in Berlin), Verlag W. de Gruyter, Berlin. For the period from 1899 to 1968 sixty-eight volumes were published, each of which, in general, covers the literature of one year.

(5) Bulletin Signalétique – Section, Astronomie, Physique Spatiale, Géophysique. Published by Centre de Documentation du Centre National de la Recherche Scientifique, Paris. This publication is a continuation of "Bibliographie Mensuelle de l'Astronomie" founded in 1933 by the Société Astronomique de France. The publication is continued.

(6) Referativnyj Zhurnal. Founded in 1953 and published by Vsesoyuznyj Institut Nauchnoj i Tekhnicheskoj Informatsii, Akademiya Nauk, Moskva. The publication is continued.

Concept of Astronomy and Astrophysics Abstracts

This abstracting service aims to present a comprehensive documentation of the literature in all fields of astronomy and astrophysics. It appears in semi-annual volumes, two of which cover the literature of a calendar year. The half-yearly period of issue is regarded as an optimal period of time for summarizing papers into subject categories and for the presentation of abstracts as quickly as possible after the publication of the original literature. The time limits at which the documentation begins and ends for a volume are not sharply defined, except in the sense that all literature will be covered which was received by the editors within these limits.

Vol. 9 is devoted to the recording, summarizing and indexing of astronomical publications of the year 1973 received from January 1, 1973 to August 15, 1973; it also records a number of papers issued before 1973 but received within the given period of time.

The main characteristics of the concept of Astronomy and Astrophysics Abstracts may be summarized briefly.

(1) Titles of papers are given in the language of their authors whenever possible. If they are not in English but supplied with English translations they will be given in English. Abstracts are presented in English, French or German. Titles of papers in Russian are given in English.

(2) Authors' abstracts are used whenever possible. As a rule, popular articles were not abstracted; however their titles are usually given with the notation "Popular article".

(3) As a rule, each paper has been classified into one of 108 numbered subject categories and allocated a serial number within the category. In this way each item is numbered by six figures, the first three of which indicate the number of the category. Three further figures indicate the serial number within the category, which was allocated in the order of the receipt of the abstract. Reference to an abstract in Volume 1 is indicated by "01" before the number of the category; for example, 01.074.028, denotes Volume 1, category 074, abstract 028. Vol. 2 is indicated by "02", etc., Vol. 9 by "09".

A paper may have been classified into more than one category. Then its abstract has been allocated a number in one of the categories involved, and in the other category (or categories) the paper has been indicated by the title and a reference to the abstract number.

Papers whose authors are not named were treated like those with authors' names, with one exception: reports from correspondents of journals whose names were unknown were not numbered.

(4) There are categories which suggest the presentation of the material in subject groups. For instance, a subject group may be formed by all information received on the same solar eclipse, comet, nova, etc. The unsorted presentation of such material in a subject category would be inconvenient for the user, even if the individual comet, etc. were included in the subject index.

The following subject categories are subdivided into subject groups:

008 Observatories, Institutes. The publications of observatories and astronomical institutes are listed in alphabetical order of the towns of the institutions, each town forming a numbered subject group. For each publication a reference to an abstract number is made.

010 Societies, Associations, Organizations. The publications of each one form a subject group. The groups are presented in alphabetical order.

079 Solar eclipses. All publications related to one solar eclipse form a subject group.

103 Comets: Listed Objects. All publications related to the same comet form a numbered group.

124 Novae. All publications related to one nova form a subject group.

125 Supernovae. All publications related to one supernova form a subject group.

(5) Border fields of astronomy and astrophysics have been taken into account by presenting titles of papers occasionally without abstracts. The selection of papers for inclusion has been made according to the degree of relevance to astronomical research.

Transliteration of the Russian alphabet

The transliteration of the Russian alphabet in use in Astronomy and Astrophysics Abstracts is presented here.

А	а	a	Р	р	r
Б	б	b	С	с	s
В	в	v	Т	т	t
Г	г	g	У	у	u
Д	д	d	Ф	ф	f
Е	е	e	Х	х	kh
Ё	ё	e	Ц	ц	ts
Ж	ж	zh	Ч	ч	ch
З	з	z	Ш	ш	sh
И	и	i	Щ	щ	shch
Й	й	j	Ъ	ъ	''
К	к	k	Ы	ы	y
Л	л	l	Ь	ь	'
М	м	m	Э	э	eh
Н	н	n	Ю	ю	yu
О	о	o	Я	я	ya
П	п	p			

This transliteration was recommended by the Abstracting Board of the International Council of Scientific Unions in 1969. It is essentially the same as the transliteration proposed by the Academy of Sciences, Moscow, and used by the Referativnyj Zhurnal. It may be noted that the letters can be read and printed by usual data processing machines.
If the names of Russian authors in the literature are transliterated very different from this scheme we present the names in the form in which they are given in the references cited and in addition in round brackets according to our transliteration table.

Sources of information

The majority of sources of information for this volume are given in section **001 Periodicals** and in section **008 Observatories, Institutes.** The term "periodical" has been used in its widest sense for publications in a sequence of undetermined duration, even if the intervals of appearance are not regular. Section 001 records 291 periodicals with their full titles and with abbreviations which are in use in Astronomy and Astrophysics Abstracts. It may be noted that the titles of the periodicals are given in their original languages, and that Russian titles have been transliterated applying the transliteration given above. Section 008 records 130 periodicals; these are publication series of observatories and astronomical institutes which have not been included in section 001. The abbreviations of the titles of the periodicals have been given so that in most cases they permit recognition of the full title without recourse to the key in section 001. The steadily growing number of periodicals makes it necessary to use more extensive abbreviations and to abandon the use of very condensed ones.
Other abstracting journals have been consulted in order to examine the degree of completeness of our service. Occasionally, in particular in Physics Abstracts, Referativnyj Zhurnal, and Bulletin Signalétique abstracts of papers were found which had not come to our attention. In such cases Astronomy and Astrophysics Abstracts cites these papers, but also gives reference to the abstracting service which acted as the source.

Classification into a scheme of subject categories

The subdivision of astronomy and its border fields into sub-ject categories is facilitated by the fact that the astronomical objects appear to be particularly well suited for the formation of categories. Sun, moon, earth, planets, comets, and meteorites, the various kinds of stars, galaxies, radio sources, quasars, and pulsars etc. suggest natural subdivisions. It may be assumed that such subdivisions can be maintained for long periods of time. Experience shows, however, that progress in research may imply changes in the classification scheme, in particular, in fields where the expansion of knowledge is explosive.
A few explanatory remarks may be in order on some of the subject categories. Section 002 includes short news notes whose titles and authors are given, but the authors of the notes have not been included in the author index. In section 003 books on astronomy and astrophysics and its border fields are listed which came to our notice from January 1973 to August 1973. References to book reviews are given if the review appeared quickly.
For completeness of documentation, personal notes (section 006) and obituaries (section 007) are listed. In section 012 (Proceedings of Colloquia, Congresses, Meetings, and Symposia) the proceedings etc. are listed with titles and editors. The individual papers are classified into their corresponding subject categories, but not included in the subject index. The main subjects of these symposia are cited in the index under section 012.
Errata to papers communicated by the authors are listed at the end of the corresponding subject categories.

Author index and subject index

The subject category and the serial number forming six figures for each abstract have been used as a means of reference in the author index and the subject index. These references are more precise than page references. They offer considerable advantages in indexing by means of data processing machines, and they are more convenient for the user.
The author index of this volume contains 8285 names. A complete reference comprises six figures, three for the subject category and three for the serial number within the category. In the case of more than one reference to abstracts in one category, the number of the category is given only once and not repeated in the immediately following references. The total number of papers (some do not give names of authors) recorded in this volume is about 7200.
We consider the subject index as only a first approximation to an optimal index covering all fields of astronomy and astrophysics and their border fields. Several iterative steps appear to be necessary until an index has been compiled for one of the subsequent volumes which may then serve as a kind of standard for the near future. The assigning of one or more key words to a paper is undoubtedly a difficult task. Some journals have started giving key words together with the titles of papers. These key words are chosen by the authors themselves and are in many cases identical with our designations of subject categories with no additional specification. In fact, in some cases it may be more useful to refer to a subject category as a whole than to an item number, in particular, if the total number of abstracts in a category is very small, and if more specific key words do not provide a proper description of the paper.
While each volume is scheduled to contain an author index and a subject index, the magnetic tapes containing the index information will be used to produce separate index volumes (authors and subjects) at intervals of a few years.
The text of the publication was typed on IBM 72 Composers in the editorial office, and it was given to the printer in a form ready for offset reproduction. The author index and the subject index were compiled and printed by means of electronic computer (Siemens 2002).

Abbreviations

AAS	American Astronomical Society	Geogr.	Geography, etc.
AAVSO	American Association of Variable Star Observers	Geophys.	Geophysics, etc.
		Ges.	Gesellschaft
Abh.	Abhandlungen	Glav.	Glavnyj (Main)
Abstr.	Abstract	Gos.	Gosudarstvennyj (State)
Abt.	Abteilung	HRD	Herzsprung-Russell diagram
Acad.	Academy, etc.	Hydrogr.	Hydrography, etc.
Accad.	Accademia	IAF	International Astronautical Federation
Adv.	Advances	IAU	International Astronomical Union
AG	Astronomische Gesellschaft	ICSU	International Council of Scientific Unions
AIAA	American Institute of Aeronautics and Astronautics	IEEE	Institute of Electrical and Electronics Engineers
AJB	Astronomischer Jahresbericht	Industr.	Industry, etc.
Akad.	Akademie	Inform.	Information
An.	Anales, etc.	Inst.	Institute, etc.
Ann.	Annals, etc.	Instn.	Institution
Arch.	Archiv, etc.	Ionosph.	Ionosphere, etc.
Ark.	Arkiv	Issled.	Issledovaniya (Research)
ASA	Astronomical Society of Australia	Ist.	Istituto
Asoc.	Asociación	Izv.	Izvestiya (News)
ASP	Astronomical Society of the Pacific	Jb.	Jahrbuch
Ass.	Association	JO	Journal des Observateurs
ASSA	Astronomical Society of Southern Africa	Journ.	Journal
Astrofis.	Astrofisica, etc.	Kl.	Klasse
Astrofiz.	Astrofizika, etc.	Lab.	Laboratory
Astron.	Astronomy, etc.	Mag.	Magazine
Astronaut.	Astronautics, etc.	Mat.	Matematica, etc.
Astrophys.	Astrophysics, etc.	Math.	Mathematics, etc.
ASV	Astronomical Society of Victoria	Mech.	Mechanics, etc.
ASWA	Astronomical Society of Western Australia	Med.	Mededelingen
Atmosph.	Atmosphere, etc.	Medd.	Meddelande, Meddelser
BA	Bulletin Astronomique	Mekhan.	Mekhanika, etc.
BAA	British Astronomical Association	Mém.	Mémoires
BAN	Bulletin of the Astronomical Institutes of the Netherlands	Mem.	Memoirs, Memorandum, etc.
		Meteorol.	Meteorology, etc.
Ber.	Berichte	MIT	Massachusetts Institute of Technology
BIH	Bureau International de l'Heure (Paris)	Mitt.	Mitteilungen
Bol.	Boletin	MVS Sonneberg	Mitteilungen über Veränderliche Sterne, Sonneberg
Boll.	Bolletino		
Bull.	Bulletin	Nachr.	Nachrichten
Byull.	Byulleten' (Bulletin)	NASA	National Aeronautics and Space Administration
Circ.	Circular		
Cl.	Classe	Nat.	Naturwissenschaftlich, etc.
Coll.	Collection	Naut.	Nautics, etc.
Commun.	Communication	NBS	National Bureau of Standards
Comun.	Comunicazioni	NRAO	National Radio Astronomy Observatory (Green Bank)
Contr.	Contributions, etc.		
COSPAR	Committee on Space Research	NRL	Naval Research Laboratory (Washington)
C.S.I.R.O.	Commonwealth Scientific Industrial Research Organization	Obs.	Observatory, etc.
		OSA	Optical Society of America
Dep.	Department	Oss.	Osservatorio, Osservazioni, etc.
Diss.	Dissertation	Ped.	Pedagogika, etc. (Pedagogics)
Div.	Division	Phil.	Philosophical
Dokl.	Doklady (Reports)	Phys.	Physics, etc.
ESO	European Southern Observatory	Planet.	Planetary
ESRO	European Space Research Organization	Priklad.	Prikladnoj (Applied)
Fis.	Fisica, etc.	Proc.	Proceedings
Fiz.	Fizika, etc.	Progr.	Progress, etc.
Fys.	Fysica, etc.	Pubbl.	Pubblicazioni
Géod.	Géodésie, etc.	Publ.	Publications
Geod.	Geodesy, etc.	Rap.	Raportoj
Geofis.	Geofisica, etc.	RAS	Royal Astronomical Society
Geofiz.	Geofizika, etc.	RAS Canada	Royal Astronomical Society of Canada
Geofys.	Geofysik, etc.	Rech.	Recherches
Geol.	Geology, etc.	Rend.	Rendiconti

Abbreviations

Rep.	Report	Techn.	Technics, etc.
Repr.	Reprint	Tekhn.	Tekhnika, etc.
Res.	Research	Teor.	Teoreticheskij
Rev.	Review, etc.	Terr.	Terrestrial, etc.
Ric.	Ricerche	TH	Technische Hochschule
Roy.	Royal, etc.	Theor.	Theoretical
SAF	Société Astronomique de France	Tidssk.	Tidsskrift
SAI	Società Astronomica Italiana	Trans.	Transactions
SAO	Smithsonian Astrophysical Observatory	Trudy	Trudy (Publications)
SAS	Société Astronomique de Suisse	Tsentr.	Tsentral'nyj (Central)
Sci.	Science, etc.	Tsirk.	Tsirkulyar (Circular)
Sect.	Section	TU	Technical University
Ser.	Series, etc.	Uch. Zap.	Uchenye Zapiski (Treatise)
S. I. R.	Service International Rapide des Latitudes	Univ.	University, etc.
Sitz.-Ber.	Sitzungsberichte	URSI	Union Radio Scientifique Internationale
Soc.	Society	Verh.	Verhandlungen
Soobshch.	Soobshcheniya (Communications)	Veröff.	Veröffentlichungen
Sternw.	Sternwarte	Wet.	Wetenschappen
Stud. Cerc.	Studii şi Cercetari	Wiss.	Wissenschaften, etc.
Supl.	Suplemento	Zeitschr.	Zeitschrift
Suppl.	Supplement	ZfA	Zeitschrift für Astrophysik
SuW	Sterne und Weltraum	Zhurn.	Zhurnal (Journal)

Periodicals, Proceedings, Books, Activities

001 Periodicals

AAS Photo-Bull.
AAS (American Astronomical Society) Photo-Bulletin.
Published by the Working Group on Photographic Materials. Produced by Eastman Kodak Co., Rochester, N. Y.

Abh. Deutsch. Akad. Wiss. Berlin
Abhandlungen der Deutschen Akademie der Wissenschaften zu Berlin. Klasse für Mathematik, Physik und Technik. Publisher: Akademie-Verlag, Berlin.

Acad. Roy. Belgique, Bull. Cl. Sci.
Académie Royale de Belgique, Bulletin de la Classe des Sciences (Koninklijke Academie van België, Mededelingen van de Klasse der Wetenschappen). 5ᵉ Série. Palais des Académies, Bruxelles.

Acta Astron.
Acta Astronomica. Publisher: Komitet Astronomii, Polskiej Akademii Nauk, Warszawa - Kraków.

Acta Phys. Austriaca
Acta Physica Austriaca. Publisher: Springer-Verlag, Wien.

Acta Univ. Carolinae Math. Phys.
Acta Universitatis Carolinae, Mathematica et Physica. Administrace: Matematicko-fyzikální fakulta University Karlovy, Praha.

Actas Acad. Nacional Cienc. Lima
Actas de la Academia Nacional de Ciencias Exactas, Fisicas y Naturales de Lima. Lima - Peru.

Adv. Astron. Astrophys.
Advances in Astronomy and Astrophysics. Publisher: Academic Press, New York − London.

AIAA Journ.
AIAA Journal. A Publication of the American Institute of Aeronautics and Astronautics devoted to Aerospace Research and Development. Published by the American Institute of Aeronautics and Astronautics, New York, N.Y.

American Scient.
American Scientist. Society of Sigma Xi, New Haven, Conn.

.Ann. d'Astrophys.
Annales d'Astrophysique. Revue internationale bimestrielle publiée par le Centre National de la Recherche Scientifique et éditée par son Service d'Astrophysique, Paris. After Vol. 31 replaced by "Astronomy and Astrophysics".

Ann. Françaises Chronométrie Micromécanique
Annales Françaises de Chronométrie et de Micromécanique, publication annuelle de l'Observatoire de Besançon, du Centre Technique de l'Industrie Horlogère et de la Société Française de Chronométrie et de Micromécanique. Rédaction et administration: Observatoire de Besançon. Publiées avec le concours du Centre National de la Recherche Scientifique et des organismes corporatifs.

Ann. Géophys.
Annales de Géophysique. Revue Internationale trimestrielle, publiée par le Centre National de la Recherche Scientifique, Paris.

Ann. Obs. Astron. Météorol. Toulouse
Annales de l'Observatoire Astronomique et Météorologique de Toulouse. Publisher: Gauthier-Villars, Paris.

Ann. Physics
Annals of Physics. Publisher: Academic Press Inc., New York, N.Y.

Ann. Physik
Annalen der Physik. 7. Folge. Publisher: Johann Ambrosius Barth, Leipzig.

Ann. Physique
Annales de Physique. Publisher: Masson et Cie., Paris.

Ann. Soc. Sci. Bruxelles
Annales de la Société Scientifique de Bruxelles. Série I: Sciences Mathématiques, Astronomiques et Physiques. Published by Institut de Physique, Heverlé-Louvain.

Annual Rep. Astron. Inst. Greece
Annual Reports of the Astronomical Institutes of Greece. Published by the Greek National Committee for Astronomy. Academy of Athens, Research Center for Astronomy and Applied Mathematics.

Annual Rev. Astron. Astrophys.
Annual Review of Astronomy and Astrophysics. Publisher: Annual Reviews Inc., Palo Alto, California.

Ann. Univ.-Sternw. Wien
Annalen der Universitäts-Sternwarte Wien. In Kommission bei Ferd. Dümmlers Verlag, Bonn.

Anzeiger. Österreich. Akad. Wiss. Math.-Nat. Kl.
Anzeiger. Österreichische Akademie der Wissenschaften. Mathematisch-Naturwissenschaftliche Klasse. Publisher: Springer-Verlag, Wien.

Applied Optics
Applied Optics. A monthly publication of the Optical Society of America. Published for the Optical Society of America by the American Institute of Physics, New York, N. Y.

Arch. Sci. Genève
Archives des Sciences, éditées par la Société de Physique et d'Histoire Naturelle de Genève. Publisher: Imprimerie Kundig, Genève. Subscription address: Librairie Payot, Genève.

Ark. Astron.
Arkiv för Astronomi. Utgivet av Kungliga Svenska Vetens-

kapsakademien, Stockholm. Printed by Almqvist & Wiksell, Stockholm.

Ark. Geofys.
Arkiv för Geofysik. Kungliga Svenska Vetenskapsakademien, Stockholm.Printed by Almqvist & Wiksell, Stockholm.

Artificial Satellites
Artificial Satellites. Publication of Polish Scientific Institutions. Polish Academy of Sciences, National Committee of Geophysics and Geodesy, National Committee for Space Research, Warsaw. Publishing Office: Palac Kultury i Nauki, Warszawa.

Asoc. Argentina Astron. Bol.
Asociación Argentina de Astronomía. Boletin. Editor: Instituto Argentino de Radioastronomía, Provincia de Buenos Aires, Argentina. Printer: Talleres Gráficos "Renovación", La Plata, República Argentina.

Astrofizika
Astrofizika. Izdatel'stvo Akademii Nauk Armyanskoj SSR, Erevan. [An English translation is published in "Astrophysics".]

Astrofiz. Issled. Izv. Spets. Astrofiz. Obs.
Astrofizicheskie Issledovaniya. Izvestiya Spetsial'noj Astrofizicheskoj Observatorii. Akademiya Nauk SSSR. Publishers: Izdatel'stvo "Nauka", Leningradskoe Otdelenie, Leningrad.

Astron. Astrophys.
Astronomy and Astrophysics. A European Journal. Published by Springer-Verlag, Berlin — Heidelberg— New York.

Astron. Astrophys. Suppl. Ser.
Astronomy and Astrophysics. Supplement Series. A European Journal. Published by the Astronomical Institute Lausanne and Geneva Observatory, Switzerland, on behalf of the Board of Directors.

Astronaut. Acta
Astronautica Acta. An Archive Journal of the International Academy of Astronautics. Published by Pergamon Press, New York — Oxford.

Astronaut. Aeronaut.
Astronautics & Aeronautics. A Publication of the American Institute of Aeronautics and Astronautics. Published monthly by the American Institute of Aeronautics and Astronautics, Easton, Pennsylvania.

Astron. in der Schule
Astronomie in der Schule. Zeitschrift für die Hand des Astronomielehrers. Herausgegeben vom Verlag Volk und Wissen, Berlin. Redaktion: Sternwarte Bautzen.

Astron. Journ.
The Astronomical Journal. Published for the American Astronomical Society by the American Institute of Physics, New York, N. Y. Editorial Office: Department of Astronomy, Columbia University, New York, N. Y.

Astron. Nachr.
Astronomische Nachrichten. Publisher: Akademie-Verlag, Berlin.

Astron. Soc. Pacific Leaflet
Astronomical Society of the Pacific. Leaflet. Edited by the Astronomical Society of the Pacific, San Francisco, California.

Astron. Tidssk.
Astronomisk Tidsskrift. Edited by Astronomisk Selskab, København; Norsk Astronomisk Selskap, Oslo; Svenska Astronomiska Sällskapet, Stockholm. Printed by John Griegs Boktrykkeri, Bergen.

Astron. Tsirk.
Astronomicheskij Tsirkulyar, izdavaemyj Byuro Astronomicheskikh Soobshchenij Akademii Nauk SSSR. Tipografiya Astrosoveta AN SSSR, Moskva.

Astron. Vestn.
Astronomicheskij Vestnik. Publishers: Izdatel'stvo "Nauka", Moskva.

Astron. Zhurn. Akad. Nauk SSSR
Astronomicheskij Zhurnal. Akademiya Nauk SSSR. Publishers: Izdatel'stvo "Nauka", Moskva. [An English translation is published in "Soviet Astronomy AJ"].

Astrophysics
Astrophysics. The Faraday Press cover-to-cover translation of Astrofizika. The Faraday Press, Inc., New York, N. Y.

Astrophys. Journ.
The Astrophysical Journal. Published in collaboration with the American Astronomical Society by the University of Chicago Press, Chicago, Illinois.

Astrophys. Journ. Suppl. Ser.
The Astrophysical Journal. Supplement Series. Published in collaboration with the American Astronomical Society by the University of Chicago Press, Chicago, Illinois.

Astrophys. Letters
Astrophysical Letters. An International *EXPRESS* Journal. Published monthly by Gordon and Breach Science Publishers Ltd., New York — London — Paris.

Astrophys. Norvegica
Astrophysica Norvegica. Edited by The Institute of Theoretical Astrophysics, University of Oslo (Det Norske Videnskaps-Akademi i Oslo). Universitets-forlaget, Oslo.

Astrophys. Space Sci.
Astrophysics and Space Science. An International Journal of Cosmic Physics. Published by D. Reidel Publishing Company, Dordrecht — Holland.

Atti Accad. Nazionale Lincei. Mem.
Atti della Accademia Nazionale dei Lincei. Serie Ottava. Memorie. Classe di Scienze fisiche, matematiche e naturali. Sezione I: Matematica, Meccanica, Astronomia, Geodesia e Geofisica. Published by Accademia Nazionale dei Lincei, Roma.

Atti Accad. Nazionale Lincei. Rend.
Atti della Accademia Nazionale dei Lincei. Serie Ottava. Rendiconti. Classe di Scienze fisiche, matematiche e naturali. Published by Accademia Nazionale dei Lincei, Roma.

Australian Journ. Phys.
Australian Journal of Physics. Published by the Commonwealth Scientific and Industrial Research Organization, East Melbourne, Victoria.

Australian Journ. Phys. Astrophys. Suppl.
Australian Journal of Physics, Astrophysical Supplement.
Published by Commonwealth Scientific and Industrial
Research Organization, East Melbourne, Victoria.

BAV Rundbrief
BAV Rundbrief. Mitteilungsblatt der Berliner Arbeitsge-
meinschaft für Veränderliche Sterne. Editor: BAV Berliner Arbeitsgemeinschaft für Veränderliche Sterne eV.,
Berlin.

BBSAG Bull.
Bedeckungsveränderlichen Beobachter der Schweizerischen Astronomischen Gesellschaft, [Swiss Astronomical
Society's Eclipsing Variable Observers], Bulletin. To be
obtained from R. Diethelm, Winterthur, Switzerland.

Bild der Wiss.
Bild der Wissenschaft. Zeitschrift über die Naturwissen-
schaften und die Technik in unserer Zeit. Publisher:
Deutsche Verlagsanstalt, Stuttgart.

Bol. Inst. Mat., Astron., Fis. Univ. Nacional Córdoba
Boletin del Instituto de Matematica, Astronomia y
Fisica, Universidad Nacional de Córdoba (R. A.).Direc-
ción General de Publicaciones, Córdoba (Argentina).

Bol. Liga Latinoamericana Astron.
Boletin de la Liga Latinoamericana de Astronomia. Publicado por la Asociacion Argentina Amigos de la Astronomia, Buenos Aires, Argentina.

Boll. Geod. Sci. Affini
Bolletino di Geodesia e Scienze Affini. Pubblicazione
dell'Istituto Geografico Militare, Firenze.

Boundary-Layer Meteorology
Boundary-Layer Meteorology. An International Journal
of Physical and Biological Processes in the Atmospheric
Boundary Layer. Published by D. Reidel Publishing Company, Dordrecht–Holland.

British Astron. Ass. Circ.
British Astronomical Association, Circular. Editorial
Office: 97 Hawkswood Drive, Hailsham, Sussex.

Bull. American Astron. Soc.
Bulletin of the American Astronomical Society. Published
for the American Astronomical Society by the American
Institute of Physics Inc., New York, N. Y.

Bull. Astron. Inst. Czechoslovakia (BAC)
Bulletin of the Astronomical Institutes of Czechoslovakia.
Published under the auspices of the Czechoslovak Acade-
my of Sciences by Academia, Praha. Editor: Astronomic-
al Institutes of the Czechoslovak Academy of Sciences,
Praha.

Bull. Astron. Inst. Netherlands (BAN)
Bulletin of the Astronomical Institutes of the Nether-
lands. Publisher: North-Holland Publishing Company,
Amsterdam. After Vol. 20 replaced by "Astronomy and
Astrophysics".

Bull. Géod.
Bulletin Géodésique, being the Journal of the International Association of Geodesy. Nouvelle Série. Publié
par le Bureau Central de l'Association Internationale
de Géodésie, Paris.

Bull. Geograph. Survey Inst.
Bulletin of the Geographical Survey Institute. Published
by the Geographical Survey Institute, Ministry of Construction, Tokyo, Japan.

Bull. Obs. Astron. Beograd
Bulletin de l'Observatoire Astronomique de Béograd.
Editor: Observatoire Astronomique de Béograd. Printed
by Naucna delo, Béograd.

Bull. Sci. Yougoslavie
Bulletin Scientifique. Conseil des Academies des Sciences
et des Arts de la RSF de Yougoslavie. Section A: Sciences
Naturelles, Techniques et Médicales. Redaction et Administration: Opatička ul. 18/II, Zagreb (Yougoslavie).

Bull. Signal.
Bulletin Signalétique. Section 120: Astronomie, Physique
spatiale, Géophysique. Centre de Documentation du
Centre Nationale de la Recherche Scientifique, Paris.

Bull. Signal.
Bulletin Signalétique. Bibliographie des Sciences de la
Terre. Section 220, Cahier A: Minéralogie, Géochimie,
Géologie extraterrestre. Centre de Documentation du
C.N.R.S., Paris; Département Documentation du B.R.
G.M., Orléans.

Bull. Soc. Roy. Sci. Liège
Bulletin de la Société Royale des Sciences de Liège.
L'Université, Liège.

Byull. Abastuman. Astrofiz. Obs.
Abastumanskaya Astrofizicheskaya Observatoriya, Gora
Kanobili. Byulleten'. Akademiya Nauk Gruzinskoj SSR.
Publishers: Izdatel'stvo "Metsniereba", Tbilisi.

Byull. Stantsij Optichesk. Nablyud. Iskusstv. Sputnikov Zemli
Byulleten' Stantsij Opticheskogo Nablyudeniya Iskusstvennykh Sputnikov Zemli. Published by Astronomicheskij Sovet Akademii Nauk SSSR, Moskva.
Beginning with number 60 (1971) the title of the publication changed in Nablyudeniya Iskusstvennykh Nebesnykh Tel.

Canadian Journ. Phys.
Canadian Journal of Physics. Published by the National
Research Council of Canada, Ottawa. Printed in Canada
by the University of Toronto Press, Toronto, Ont.

Celestial Mechanics
Celestial Mechanics. An International Journal of Space
Dynamics. Publishers: D. Reidel Publishing Company,
Dordrecht–Holland.

Ciel et Terre
Ciel et Terre. Bulletin de la Société Belge d'Astronomie,
de Météorologie et de Physique du Globe. Administration: Avenue Circulaire, 3, Bruxelles. Printed by Imprimerie R. Louis, Bruxelles.

Circ. d'Information
Circulaire d'Information. Union Astronomique Internationale. Commission des Etoiles Doubles. Address: Observatoire de Meudon, Meudon, France.

Coelum
Coelum. Periodico bimestrale per la Divulgazione dell'
Astronomia. Editor: Osservatorio Astronomico Univer-
sitario di Bologna.

Comments Astrophys. Space Phys.
Comments on Astrophysics and Space Physics. A Journal of Critical Discussion of the Current Literature. Comments on Modern Physics: Part C. Publishers: Gordon and Breach Science Publishers, Inc., New York — London

Comptes Rendus Acad. Bulg. Sci.
Comptes Rendus de l'Académie bulgare des Sciences. (Doklady Bolgarskoj Akademii Nauk). Sofia.

Comptes Rendus Acad. Sci. Paris
Comptes Rendus hebdomadaires des Séances de l'Académie des Sciences, publié avec le concours du Centre National de la Recherche Scientifique. Imprimerie: Gauthier-Villars, Paris.

Contr. Atmosph. Phys.
Contributions to Atmospheric Physics — Beiträge zur Physik der Atmosphäre. Publisher: Friedrich Vieweg & Sohn, Braunschweig.

Cosmic Electrodynamics
Cosmic Electrodynamics. An International Journal devoted to Geophysical and Astrophysical Plasmas. Printed in The Netherlands by D. Reidel Publishing Company, Dordrecht—Holland.

COSPAR Inform. Bull.
COSPAR. Information Bulletin. Address: COSPAR Secretariat, Paris.

Deutsche Geod. Kommission Bayer. Akad. Wiss.
Deutsche Geodätische Kommission bei der Bayerischen Akademie der Wissenschaften. Reihe A: Höhere Geodäsie; Reihe B: Angewandte Geodäsie; Reihe C: Dissertationen; Reihe D: Tafelwerke; Reihe E: Geschichte und Entwicklung der Geodäsie. Published by Verlag der Bayerischen Akademie der Wissenschaften, München.

Documentat. Observateurs
Documentation des Observateurs. Rédaction: Station d'Astrophysique de Forcalquier.

Documentat. Observateurs Circ.
Documentation des Observateurs. Circulaire. Rédaction: Station d'Astrophysique de Forcalquier.

Dokl. Akad. Nauk
Doklady Akademii Nauk SSSR. Seriya Matematika, Fizika. Publishers: Izdatel'stvo "Nauka", Moskva.

Dunsink Obs. Publ.
Dunsink Observatory Publications. The Observatory of the School of Cosmic Physics, Dublin Institute for Advanced Studies, Dublin.

Earth Extraterr. Sci.
Earth and Extraterrestrial Sciences. Published by Gordon and Breach Science Publishers, London.

Earth Planet. Sci. Letters
Earth and Planetary Science Letters. A Letter Journal devoted to the Development in Time of the Earth and Planetary System. Publisher: North-Holland Publishing Company, Amsterdam.

El Universo
El Universo. Organo de la Sociedad Astronomica de Mexico, Mexico, D. F.

Endeavour
Endeavour. A review of the progress of science, published in four languages by Imperial Chemical Industries Limited, London.

ESO Bull.
European Southern Observatory, Bulletin. Edited by European Southern Observatory. Office of the Director: Hamburg.

Fortschritte Phys.
Fortschritte der Physik. Publisher: Akademie-Verlag, Berlin.

Gaz. Astron. Mém.
Gazette Astronomique. Mémoires van het Sterrenkundig Genootschap van Antwerpen, (de la Société d'Astronomie d'Anvers), Antwerpen. Printer: «De Voorzorg», A. Van Leuvenhaege, Antwerpen.

Geochim. Cosmochim. Acta
Geochimica et Cosmochimica Acta. Journal of the Geochemical Society. Publishing House: Pergamon Press, Ltd., Oxford.

Geodezja Kartografia
Geodezja i Kartografia. Komitet Geodezji Polskiej Akademii Nauk. Publisher: Państwowe Wydawnictwo Naukowe, Warszawa.

Geomagn. Aeronom.
Geomagnetizm i Aehronomiya. Akademiya Nauk SSSR. Izdatel'stvo "Nauka", Moskva [An English translation is published in "Geomagnetism and Aeronomy".]

Geophys. Journ.
The Geophysical Journal of the Royal Astronomical Society. Published for the Royal Astronomical Society by Blackwell Scientific Publications, Oxford — Edinburgh.

Gerlands Beiträge Geophys.
Gerlands Beiträge zur Geophysik. Publisher: Akademische Verlagsgesellschaft Geest & Portig K.-G., Leipzig.

Glasnik Mat.
Glasnik Matematicki. Published by the Society of Mathematicians and Physicists of the S. R. of Croatia. Publisher: Drustvo Matematicara i Fizicara S. R. Hrvatske, Zagreb.

Helvetica Phys. Acta
Helvetica Physica Acta. Schweizerische Physikalische Gesellschaft. Publisher: E. Birkhäuser, Basel.

Hemel en Dampkring
Maandblad van de Nederlandse Vereniging voor Weer-en Sterrenkunde en van de Vereniging voor Sterrenkunde, Meteorologie, Geophysica en Aanverwante Wetenschappen in Belgie. Publisher: Wolters-Noordhoff N. V., Groningen.

IAU Circ.
International Astronomical Union, Circular. Central Bureau for Astronomical Telegrams, Smithsonian Astrophysical Observatory, Cambridge, Mass.

IBM Journ. Res. Development
IBM Journal of Research and Development. Published bimonthly by International Business Machines Corporation, Armonk, New York.

Icarus
Icarus. International Journal of Solar System Studies. Publisher: Academic Press, New York — London.

ICSU Bull.
ICSU Bulletin. International Council of Scientific Unions. Secretariat: 7, Via Cornelio Celso, Rome, Italy.

IEEE Spectrum
IEEE Spectrum. Published monthly by the Institute of Electrical and Electronics Engineers, Inc., New York, N. Y.

Inform. Bull. Southern Hemisphere
Information Bulletin of the Southern Hemisphere. Editorial Office: Observatorio Astronómico, La Plata, Argentina.

Inform. Bull. Variable Stars
Commission 27 of the I.A.U. Information Bulletin on Variable Stars. Konkoly Observatory, Budapest.

Infrared Physics
An International Research Journal. Publisher: Pergamon Press Ltd., Oxford — London — New York.

International Journ. Theor. Phys.
International Journal of Theoretical Physics. Publisher: Plenum Publishing Company, Donington House, London.

Irish Astron. Journ.
The Irish Astronomical Journal. A Quarterly Publication under the auspices of the Observatories of Armagh and Dunsink. Subscription address: Managing Editor, Irish Astronomical Journal, Armagh Observatory, Northern Ireland.

Izv. Akad. Nauk Armyan. SSR
Izvestiya Akademii Nauk Armyanskoj SSR. Fizika. Publisher: Izdatel'stvo AN Armyanskoj SSR, Erevan.

Izv. Glav. Astron. Obs. Pulkovo
Izvestiya Glavnoj Astronomicheskoj Observatorii v Pulkove. Akademiya Nauk SSSR. Izdanie Glavnoj astronomicheskoj observatorii v Pulkove, Leningrad.

Izv. Komissii Fiz. Planet
Izvestiya Komissii po Fizike Planet. Akademiya Nauk SSSR. Astronomicheskij Sovet. Moskva.

Izv. Krymskoj Astrofiz. Obs.
Izvestiya Krymskoj Astrofizicheskoj Observatorii. Akademiya Nauk SSR. Publishers: Izdatel'stvo "Nauka", Moskva.

Jenaer Rundschau (Jena Review)
Jenaer Rundschau (Jena Review). Publisher: VEB Verlag Technik, Berlin.

JETP Letters
JETP Letters. A translation of JETP Pis'ma v Redaktsiyu of the Academy of Sciences in the USSR. Published semimonthly by the American Institute of Physics, Lancaster, Pennsylvania.

Journ. Astronaut. Sci.
The Journal of the Astronautical Sciences. Published by the American Astronautical Society Inc., Baltimore, Md.

Journ. Astron. Soc. Victoria
The Journal of the Astronomical Society of Victoria.

Printed by D. Buscombe Printers, Glen Waverley, Victoria

Journ. Astron. Soc. Western Australia
The Journal of the Astronomical Society of Western Australia. Edited by the Astronomical Society of Western Australia, Perth, W. A.

Journ. Atmosph. Sci.
Journal of the Atmospheric Sciences. Published by the American Meteorological Society, Boston, Mass.

Journ. Atmosph. Terr. Phys.
Journal of Atmospheric and Terrestrial Physics. Publishers: Pergamon Press, Oxford — London — New York.

Journ. British Astron. Ass.
Journal of the British Astronomical Association. Subscription address: British Astronomical Association, Burlington House, Piccadilly, London.

Journ. British Interplanet. Soc.
Journal of the British Interplanetary Society. Printed in Great Britain by Unwin Brothers Ltd., The Gresham Press, Old Woking, Surrey, and published by The British Interplanetary Society, London.

Journ. Fluid Mechanics
Journal of Fluid Mechanics. Published by Cambridge University Press, London — New York.

Journ. Geophys. Res.
Journal of Geophysical Research. An International Scientific Publication. Published three times a month by the American Geophysical Union, Washington, D. C. First section: Space Physics; Second section: Physics and chemistry of the solid earth, planetology, geodesy; Third section: Oceans and atmospheres.

Journ. History Astron.
Journal for the History of Astronomy. Publisher: Science History Publications Ltd., Cambridge, England. American Representative: Neale Watson Academic Publications, Inc., New York City, U.S.A.

Journ. Math. Phys.
Journal of Mathematical Physics. Published by the American Institute of Physics, New York, N. Y.

Journ. Navigation
The Journal of Navigation. Published quarterly by The Royal Institute of Navigation at the Royal Geographical Society, London.

Journ. Optical Soc. America
Journal of the Optical Society of America. Publisher: American Institute of Physics, New York.

Journ. Phys. A. General Phys.
Journal of Physics A. General Physics. Europhysics Journal. Published by the Institute of Physics and the Physical Society, London, England, in association with the American Institute of Physics, New York.

Journ. Physique
Journal de Physique. Publication de la Société Française de Physique, Paris.

Journ. Plasma Phys.
Journal of Plasma Physics. Publishers: Cambridge University Press, London.

Journ. Proc. Roy. Soc. New South Wales
Journal and Proceedings of the Royal Society of New South Wales. Published by the Society, Science House, Sydney.

Journ. Quant. Spectrosc. Radiat. Transfer
Journal of Quantitative Spectroscopy & Radiative Transfer. Publisher: Pergamon Press, Oxford – New York.

Journ. Roy. Astron. Soc. Canada
The Journal of the Royal Astronomical Society of Canada, devoted to the advancement of astronomy and allied sciences. Printed by the University of Toronto Press, Toronto, Ontario, Canada.

Kometn. Tsirk. *Kiev*
Kometnyj Tsirkulyar. Gruppa po Issledovaniyu Komet Astrosoveta i Mezhduvedomstvennyj Geofizicheskij Komitet. Akademii Nauk SSSR. Kievskij Universitet im. T. G. Shevchenko.

Komety i Meteory
Komety i Meteory. Akademiya Nauk Tadzhikskoj SSR. Astronomicheskij Sovet Akademii Nauk SSSR. Publishers: Izdatel'stvo "Donish", Dushanbe.

Kosmich. Issled.
Kosmicheskie Issledovaniya. Akademiya Nauk SSSR. Publishers: Izdatel'stvo "Nauka", Moskva.

Kozmos
Kozmos. Popular Astronomical Journal of the Slovak Central Observatory in Hurbanovo. Publisher: Slovenská ústredná hvezdáreň v Hurbanove.

L'Astronomie
L'Astronomie et Bulletin de la Société Astronomique de France. Revue mensuelle. Rédaction: Société Astronomique de France, Paris.

L'Universo
L'Universo. Rivista dell'Instituto Geografico Militare. Direzione, Redazione e Amministrazione: Istituto Geografico Militare, Firenze.

Magnitnye Polya Solnech. Pyaten
Magnitnye Polya Solnechnykh Pyaten. (Supplements to Solnechnye Dannye. Byulleten' (*Solar Data*)). Publishers: Izdatel'stvo "Nauka", Leningrad.

Math. Rev.
Mathematical Reviews. Published by the American Mathematical Society, Providence, R. I.

Mem. Fac. Sci. Kyoto Univ.
Memoirs of the Faculty of Science, Kyoto University. Series of Physics, Astrophysics, Geophysics, and Chemistry. Printed by Yamashiro Printing Publishing Co. Ltd., Kamigyo, Kyoto.

Mem. Roy. Astron. Soc.
Memoirs of the Royal Astronomical Society. Published for the Royal Astronomical Society by Blackwell Scientific Publications, Oxford – Edinburgh.

Mem. Soc. Astron. Italiana
Memorie della Società Astronomica Italiana. Nuova Serie. Pubblicate sotto gli auspici del Consiglio Nazionale dell Ricerche. Publisher: Tipografia Baccini & Chiappi, Firenze.

Mercury
Mercury. The Journal of the Astronomical Society of the Pacific. Published by the Astronomical Society of the Pacific, San Francisco, California.

Messtechnik
Messtechnik (Zeitschrift für Instrumentenkunde). Publishers: Verlag Friedrich Vieweg & Sohn GmbH, Braunschweig.

Meteoritics
Meteoritics. The Journal of the Meteoritical Society. Published quarterly by The Meteoritical Society and Arizona State University Bureau of Publications. Editorial address: Center for Meteorite Studies, The Arizona State University, Tempe, Arizona.

Meteoritika
Akademiya Nauk SSSR. Komitet po Meteoritam. Publishers: Izdatel'stvo "Nauka", Moskva.

Mitt. Astron. Ges.
Mitteilungen der Astronomischen Gesellschaft, Hamburg. Printed by G. Braun, GmbH, Karlsruhe.

Monatsber. Deutsch. Akad. Wiss. Berlin
Monatsberichte der Deutschen Akademie der Wissenschaften zu Berlin. Mitteilungen aus Mathematik, Naturwissenschaft, Medizin und Technik. Publisher: Akademie-Verlag, Berlin.

Monthly Notes Astron. Soc. Southern Africa
Monthly Notes of the Royal Astronomical Society of Southern Africa. Published by the Astronomical Society of Southern Africa, Royal Observatory, Cape Province, South Africa.

Monthly Notes International Polar Motion Service
Monthly Notes of the International Polar Motion Service. Published by the Central Bureau, International Latitude Observatory of Mizusawa, Mizusawa-shi, Iwate-ken, Japan.

Monthly Notices Roy. Astron. Soc.
Monthly Notices of the Royal Astronomical Society. Published for the Royal Astronomical Society by Blackwell Scientific Publications, Oxford – Edinburgh.

Moon
The Moon. An International Journal of Lunar Studies. Publisher: D. Reidel Publishing Company, Dordrecht – Holland.

MVS Sonneberg
Mitteilungen über Veränderliche Sterne. Edited by Sternwarte Sonneberg (Zentralinstitut für Astrophysik, Bereich Sternphysik) der Deutschen Akademie der Wissenschaften zu Berlin.

Nablyud. Iskusstv. Nebesn. Tel
Nablyudeniya Iskusstvennykh Nebesnykh Tel. Published by Astronomicheskij Sovet Akademii Nauk SSSR, Moskva.

Nachr. Akad. Wiss. Göttingen
Nachrichten der Akademie der Wissenschaften in Göttingen. II. Mathematisch-Physikalische Klasse. Vandenhoeck & Ruprecht, Göttingen.

Nachr. Karten-, Vermessungswesen
Nachrichten aus dem Karten- und Vermessungswesen. Editor: Institut für Angewandte Geodäsie (Abt. II des

Deutschen Geodätischen Forschungsinstituts). Published by Verlag des Instituts für Angewandte Geodäsie, Frankfurt a. M.

Nature
Nature. Editorial and Publishing Offices: Macmillan Journals Limited, 4 Little Essex Street, London; 711 National Press Building, Washington, D. C.

Nature, Phys. Sci.
Nature, Physical Science. Editorial and Publishing Offices: Macmillan Journals Limited, London – Washington.

Naturwissenschaften
Die Naturwissenschaften. Publisher: Springer-Verlag, Berlin – Heidelberg – New York.

Nauchn. Informatsii
Nauchnye Informatsii. Astronomicheskij Sovet Akademii Nauk SSSR, Moskva.

Nuovo Cimento
Il Nuovo Cimento. Rivista Internazionale e Organo della Società Italiana di Fisica, Series A, B. Publisher: Nicola Zanichelli, Editore, Bologna.

Nuovo Cimento Lettere
Lettere al Nuovo Cimento, a Cura della Società Italiana di Fisica. Editrice Compositori, Bologna.

Nuovo Cimento Rivista
Rivista del Nuovo Cimento a cura della Società Italiana di Fisica. Editrice Compositori, Bologna.

Nuovo Cimento Suppl.
Supplemento al Nuovo Cimento. Publisher: Nicola Zanichelli, Editore, Bologna.

Observations Artificial Earth Satellites
Observations of Artificial Satellites of the Earth (Nablyudeniya Iskusstvennykh Sputnikov Zemli). Magyar Tudományos Akadémia Csillagvizsgáló Intézete, Budapest.

Observatory
The Observatory. A Review of Astronomy. Publishers: The Editors of "The Observatory", Royal Greenwich Observatory, Herstmonceaux Castle, Hailsham, Sussex, England.

Optik
Optik. Zeitschrift für das gesamte Gebiet der Licht- und Elektronenoptik. Publishers: Wissenschaftliche Verlagsgesellschaft mbH., Stuttgart.

Orion Schaffhausen
Orion. Zeitschrift der Schweizerischen Astronomischen Gesellschaft (SAG). Bulletin de la Société Astronomique de Suisse (SAS). Administration: Generalsekretariat der SAG, Schaffhausen.

Österreich. Zeitschr. Vermessungswesen
Österreichische Zeitschrift für Vermessungswesen. Editor and Publisher: Österreichischer Verein für Vermessungswesen, Wien.

Peremennye Zvezdy, Byull.
Peremennye Zvezdy, Byulleten', izdavaemyj Astronomicheskim Sovetom Akademii Nauk SSSR. Published by Astronomicheskij Sovet Akademii Nauk SSSR, Moskva.

Peremennye Zvezdy, Prilozhenie
Peremennye Zvezdy, Prilozhenie (The Variable Stars, Supplement). Astronomicheskij Sovet Akademii Nauk SSSR, Moskva.

Phil. Mag.
The Philosophical Magazine. A Journal of Theoretical, Experimental and Applied Physics. Eighth Series. Publisher: Taylor & Francis, Ltd., London.

Phil. Trans. Roy. Soc. London
Philosophical Transactions of the Royal Society of London. Series A, Mathematical and Physical Sciences. Published by the Royal Society, London.

Phys. Abstr.
Physics Abstracts. Science Abstracts, Series A. An INSPEC Publication, published by The Institution of Electrical Engineers, London.

Phys. Ber.
Physikalische Berichte. Herausgegeben von der Deutschen Physikalischen Gesellschaft e. V.und von der Deutschen Akademie der Wissenschaften zu Berlin. Friedrich Vieweg & Sohn, Braunschweig.

Phys. Blätter
Physikalische Blätter. Physik-Verlag, Mosbach/Baden.

Phys. Bull.
Physics Bulletin. Published by the Institute of Physics and the Physical Society, London, England.

Phys. Earth Planet. Interiors
Physics of the Earth and Planetary Interiors. A journal devoted to observational and experimental studies of the Earth and Planetary interiors and their theoretical interpretation by the physical sciences. Publisher: North-Holland Publishing Company, Amsterdam, Netherlands.

Phys. Fluids
The Physics of Fluids. Published by the American Institute of Physics, New York, N.Y.

Phys. Letters
Physics Letters. Volumes A and B. Publisher: North-Holland Publishing Company, Amsterdam.

Phys. Rev. A
Physical Review A, General Physics. Published for the American Physical Society by the American Institute of Physics, Lancaster, Pa., and New York, N.Y.

Phys. Rev. B
Physical Review B, Solid State. Published for the American Physical Society by the American Institute of Physics, Lancaster, Pa., and New York, N. Y.

Phys. Rev. C
Physical Review C, Nuclear Physics. Published for the American Physical Society by the American Institute of Physics, Lancaster, Pa., and New York, N.Y.

Phys. Rev. D
Physical Review D, Particles and Fields. Published for the American Physical Society by the American Institute of Physics, Lancaster, Pa., and New York, N.Y.

Phys. Rev. Letters
Physical Review Letters. Published weekly by The Amer-

ican Physical Society, New York, N. Y.

Phys. Today
Physics Today. Published by the American Institute of Physics, New York, N.Y.

Physica
Physica. Publishers: North-Holland Publishing Company, Amsterdam, The Netherlands, on request of the Foundation "Physica", Utrecht.

Physica Scripta
Physica Scripta. (Formerly Arkiv för Fysik). Published by the Royal Swedish Academy of Sciences, Stockholm.

Planet. Space Sci.
Planetary and Space Science. Pergamon Press, Oxford – London – New York.

Plasma Physics
Plasma Physics. Publisher: Pergamon Press, Oxford, England.

Pokroky
Pokroky matematiky, fyziky a astronomie. Editor: Jednota čs. matematiků a fyziků. Publisher: Academia, Praha.

Postępy Astron.
Postępy Astronomii. Czasopismo Poświecone Upowszechnianiu Wiedzy Astronomicznej. Polskie Towarzystwo Astronomiczne, Warszawa. Printed in Poland by Pánstwowe Wydawnictwo Naukowe, Lódź.

Priroda
Priroda. Publishers: Izdatel'stvo "Nauka", Moskva.

Proc. Astron. Soc. Australia
Proceedings of the Astronomical Society of Australia. Published for the Society by Sydney University Press, Sydney.

Proc. Cambridge Phil. Soc.
Proceedings of the Cambridge Philosophical Society (Mathematical and Physical Sciences). Publishers: Cambridge University Press, London.

Proc. IEEE
Proceedings of the IEEE. Published monthly by the Institute of Electrical and Electronics Engineers, Inc., New York, N. Y.

Proc. Koninkl. Nederl. Akad. Wet.
Koninklijke Nederlandse Akademie van Wetenschappen. Proceedings. Series B, Physical Sciences. Publishers: North-Holland Publishing Company, Amsterdam.

Proc. National Acad. Sci. U.S.A.
Proceedings of the National Academy of Sciences of the United States of America. Published monthly by the National Academy of Sciences, Washington, D.C.

Proc. Roy. Soc. London
Proceedings of the Royal Society of London. Series A: Mathematical and Physical Sciences. Published by the Royal Society, London.

Progr. Theor. Phys. Japan
Progress of Theoretical Physics. Published for the Research Institute for Fundamental Physics and the Physic-

al Society of Japan. Publication Office: Progress of Theoretical Physics, Yukawa Hall, Kyoto University, Kyoto, Japan.

Progr. Theor. Phys. Suppl.
Supplement of the Progress of Theoretical Physics. Published for the Research Institute for Fundamental Physics and The Physical Society of Japan. Publication Office: Progress of Theoretical Physics, Yukawa Hall, Kyoto University, Kyoto, Japan.

PTB Mitt.
PTB Mitteilungen. Amts- und Mitteilungsblatt der Physikalisch-Technischen Bundesanstalt, Braunschweig – Berlin.

Publ. Astron. Soc. Japan
Publications of the Astronomical Society of Japan. Published by the Astronomical Society of Japan. Office of the Society: Tokyo Astronomical Observatory, Mitaka, Tokyo. Agent: Maruzen Co. Ltd. (Export Department), Nihonbashi, Tokyo, Japan.

Publ. Astron. Soc. Pacific
Publications of the Astronomical Society of the Pacific. Published in Provo, Utah, by the Astronomical Society of the Pacific, San Francisco, California. Printed by Brigham Young University Press, Provo, Utah.

Publ. Roy. Obs. Edinburgh
Publications of the Royal Observatory, Edinburgh. Published by The Royal Observatory, Edinburgh, Scotland.

Publ. Tartu Astrofiz. Obs.
W. Struve nimelise Tartu Astrofüüsika Observatooriumi, Publikatsioonid. Eesti NSV Teaduste Akadeemia, Tartu.

Quarterly Journ. Roy. Astron. Soc.
Quarterly Journal of the Royal Astronomical Society. Published for the Royal Astronomical Society by Blackwell Scientific Publications, Oxford.

Radio Sci.
Radio Science. Published by the American Geophysical Union, Richmond, Virginia.

Referativ. Zhurn. 51. Astron.
Referativnyj Zhurnal. 51. Astronomiya. Vsesoyuznyj Institut Nachnoj i Tekhnicheskoj Informatsii. Moskva.

Referativ. Zhurn. 52. Geod. i Aehros"emka
Referativnyj Zhurnal. 52. Geodeziya i Aehros"emka. Vsesoyuznyj Institut Nauchnoj i Tekhnicheskoj Informatsii. Moskva.

Referativ. Zhurn. 62. Issled. kosm. prostranstva
Referativnyj Zhurnal. 62. Issledovanie Kosmicheskogo Prostranstva. Vsesoyuznyj Institut Nauchnoj i Tekhnicheskoj Informatsii. Moskva.

Rep. Progr. Phys.
Reports on Progress in Physics. Published by The Institute of Physics and the Physical Society, London.

Rev. Geophys. Space Phys.
Reviews of Geophysics and Space Physics (formerly Reviews of Geophysics). Published by the American Geophysical Union, Richmond, Virginia.

Revista Astron.
Revista Astronomica. Organo de la Asociación Argentina Amigos de la Astronomia, Buenos Aires.

Rev. Modern Phys.
Reviews of Modern Physics. Published for The American Physical Society by the American Institute of Physics, Lancaster, Pa., and New York, N.Y.

Rev. Sci. Instruments
Reviews of Scientific Instruments. Published by the American Institute of Physics, Lancaster, Pa., and New York, N.Y.

Rezul'taty Nablyud. Sovet. Iskusstv. Sputnikov Zemli
Rezul'taty Nablyudenij Sovetskikh Iskusstvennykh Sputnikov Zemli. Published by Astronomicheskij Sovet Akademii Nauk SSSR, Moskva. Replaced after No. 140 by Rezul'taty Nablyudenij Iskusstvennykh Sputnikov Zemli.

Rezul'taty Nablyud. Iskusstv. Sputnikov Zemli
Rezul'taty Nablyudenij Iskusstvennykh Sputnikov Zemli. Published by Astronomicheskij Sovet Akademii Nauk SSSR, Ryazanskij Gosudarstvennyj Pedagogicheskij Institut, Ryazan'.

Ric. Sci.
La Ricerca Scientifica. Serie Seconda. Rivista del Consiglio Nazionale delle Ricerche. Consiglio Nazionale delle Ricerche, Roma.

Říše hvězd
Říše hvězd. Czechoslovak popular astronomical journal. Publisher: Orbis, Praha.

Roy. Astron. Soc. New Zealand Circ.
Royal Astronomical Society of New Zealand, Variable Star Section, Circular. Publication Office: Greerton, Tauranga, New Zealand.

Roy. Astron. Soc. New Zealand Variable Star Sect. Repr.
Royal Astronomical Society of New Zealand, Variable Star Section, Reprint. Publication Office: Greerton, Tauranga, New Zealand.

Rumanian Sci. Abstr.
Rumanian Scientific Abstracts. Natural Sciences. Publishers: The Scientific Documentation Centre of the Academy of the Socialist Republic of Romania, Bucureşti.

Sci. American
Scientific American. Published monthly by Scientific American, Inc., New York, N.Y.

Science
Science. American Association for the Advancement of Science, Washington, D.C.

Sci. Progr. Découverte
Science Progrès Découverte (formerly Science Progrès, La Nature). Revue publiée avec la participation du Palais de la Découverte. Published by Dunod, Editeur, Paris. Imprimerie Bayeusaine, Bayeux.

Sci. Rep. Tôhuku Univ.
The Science Reports of the Tôhuku University. First Series (Physics, Chemistry, Astronomy). Published by the Faculty of Science, Tôhuku University, Sendai, Japan.

Sitz.-Ber. Bayer. Akad. Wiss.
Bayerische Akademie der Wissenschaften. Mathematisch-Naturwissenschaftliche Klasse. Sitzungsberichte. Publisher: Verlag der Bayerischen Akademie der Wissenschaften, München.

Sitz.-Ber. Deutsch. Akad. Wiss. Berlin
Sitzungsberichte der Deutschen Akademie der Wissenschaften zu Berlin. Klasse für Mathematik, Physik und Technik. Publisher: Akademie-Verlag, Berlin.

Sitz.-Ber. Heidelberger Akad. Wiss.
Sitzungsberichte der Heidelberger Akademie der Wissenschaften. Mathematisch-Naturwissenschaftliche Klasse. Publisher: Springer-Verlag, Heidelberg.

Sitz.-Ber. Österreich. Akad. Wiss.
Sitzungsberichte. Österreichische Akademie der Wissenschaften. Mathematisch-Naturwissenschaftliche Klasse. Abteilung II: Mathematik, Astronomie, Meteorologie und Technik. Publisher: Springer-Verlag, Wien.

Sky Telescope
Sky and Telescope. Published by Sky Publishing Corporation, Cambridge, Mass.

Smithsonian Contr. Astrophys.
Smithsonian Contributions to Astrophysics. Smithsonian Institution Astrophysical Observatory, Cambridge, Mass. Printed by Smithsonian Institution Press, City of Washington. For sale by the Superintendent of Documents, U. S. Government Printing Office, Washington, D. C.

Smithsonian Year
Smithsonian Year. Annual Report of the Smithsonian Institution, including the financial report of the Executive Committee of the Boards of Regents. Published by the Smithsonian Institution, Washington, D.C.

Solar Physics
Solar Physics. A Journal for Solar Research and the Study of Solar Terrestrial Physics. Publishers: D. Reidel Publishing Company, Dordrecht–Holland.

Solnechnye Dannye Byull.
Solnechnye Dannye. Byulleten'. *(Solar Data)*. Publishers: Izdatel'stvo "Nauka", Leningradskoe Otdelenie, Leningrad.

Soobshch. Byurakan. Obs.
Soobshcheniya Byurakanskoj Observatorii. Akademiya Nauk Armyanskoj SSR, Erevan.

Soobshch. Gos. Astron. Inst. Shternberg
Soobshcheniya Gosudarstvennogo Astronomicheskogo Instituta im P.K. Shternberga. Publishers: Izdatel'stvo Moskovskogo Universiteta, Moskva.

Southern Stars
Southern Stars. The Journal of the Royal Astronomical Society of New Zealand (Inc.). Address of the Society: P.O. Box 3181, Wellington C1, New Zealand.

Soviet Astron. AJ
Soviet Astronomy AJ. A translation of the Astronomical Journal of the Academy of Sciences of the USSR. Published by the American Institute of Physics, Inc., New York, N.Y.

Spaceflight
Spaceflight. A Publication of the British Interplanetary

Society. Printed by Eyre & Spottiswoode Limited at Grosvenor Press, Portsmouth, and published by the British Interplanetary Society, London.

Space Science Rev.
Space Science Reviews. Publishers: D. Reidel Publishing Company, Dordrecht—Holland.

Springer Tracts Modern Phys.
Springer Tracts in Modern Physics. (Ergebnisse der exakten Naturwissenschaften). Springer-Verlag, Berlin—Heidelberg—New York.

Sterne
Die Sterne. Zeitschrift für alle Gebiete der Himmelskunde. Johann Ambrosius Barth, Leipzig.

Sternenbote
Sternenbote. Monatsschrift für Österreichs·Amateurastronomen. Publisher: Astronomisches Büro, Hermann Mucke, Wien.

Stockholms Obs. Ann.
Stockholms Observatoriums Annaler. Printed by Almquist & Wiksell, Stockholm.

Strolling Astronomer
The Strolling Astronomer. The Journal of The Association of Lunar and Planetary Observers, Publication Office: The Strolling Astronomer, Box 3 AZ, University Park, New Mexico.

Stud. Cerc. Astron.
Studii şi Cercetări de Astronomie. Editura Academiei Republicii Socialiste România. Editorial Office: Observatorul Astronomic, Bucureşti.

Stud. Geophys. Geod.
Studia geophysica et geodaetica. Published for the Geophysical Institute of the Czechoslovak Academy of Sciences by Academia, Praha.

Stud. Soc. Sci. Torunensis
Studia Societatis Scientiarum Torunensis, Toruń – Polonia. Sectio F (Astronomia).

Stud. Univ. Babeş-Bolyai
Studia Universitatis Babeş-Bolyai. Series Mathematica-Physica. Publishers: Intreprinderea Poligrafica, Cluj.

SuW
Sterne und Weltraum. Astronomische Monatsschrift. Publisher: Verlag Sterne und Weltraum Dr. Vehrenberg, Düsseldorf, Germany.

Tellus
Tellus, a bi-monthly Journal of Geophysics. Svenska Geofysiska Foreningen. Printed in Sweden by Almqvist & Wiksells Boktryckeri AB, Uppsala.

Trans. Astron. Obs. Yale Univ.
Transactions of the Astronomical Observatory of Yale University. Published by the Observatory, New Haven.

Trans. Roy. Soc. Canada
Transactions of the Royal Society of Canada. Published by the Royal Society of Canada, National Research Building, Ottawa.

Trudy Astrofiz. Inst. Alma-Ata
Trudy Astrofizicheskogo Instituta, Alma-Ata. Akademiya

Nauk Kazakhskoj SSR. Publishers: Izdatel'stvo "Nauka" Kazakhskoj SSR, Alma-Ata.

Trudy Glav. Astron. Obs. Pulkovo
Trudy Glavnoj Astronomicheskoj Observatorii v Pulkove. Akademiya Nauk SSSR. Izdanie Glavnoj astronomicheskoj observatorii v Pulkove, Leningrad.

Trudy Inst. Teor. Astron.,Leningrad
Trudy Instituta Teoreticheskoj Astronomii. Akademiya Nauk SSSR. Publishers: Izdatel'stvo "Nauka", Leningrad.

Trudy Tashkent. Astron. Obs.
Trudy Tashkentskoj Astronomicheskoj Observatorii. Akademiya Nauk Uzbekskoj SSR. Publishers: Izdatel'stvo "FAN" Uzbekskoj SSR, Tashkent.

Tsirk. Astron. Inst. Tashkent
Tsirkulyar Astronomicheskogo Instituta. Akademiya Nauk Uzbekskoj SSR. Izdatel'stvo "FAN" Uzbekskoj SSR, Tashkent.

Tsirk. Astron. Obs. L'vov
Tsirkulyar. Astronomicheskaya Observatoriya. L'vovskij Ordena Lenina Gosudarstvennyj Universitet imeni Ivana Franko. Publisher: Izdatel'stvo L'vovskogo Universiteta, L'vov.

Umschau
Umschau in Wissenschaft und Technik. Umschau-Verlag Frankfurt a. M.

Urania Barcelona
Urania. Revista de Astronomia y Ciencias Afines. Organo de la Sociedad Astronómica de España y América, Barcelona; Unión Nacional de Astronomia y Ciencias Afines, Madrid.

Urania Kraków
Urania. Miesiecznik Polskiego Towarzystwa Milośników Astronomii, Kraków. Publisher: Krakowska Drukarnia Prasowa, Kraków.

Vasiona
Vasiona. Revue d'Astronomie et d'Astronautique. Bulletin de la Société Astronomique "R. Bosković", Beograd.

VdS Nachrichtenblatt
Nachrichtenblatt der Vereinigung der Sternfreunde e.V. After Vol. 18, No. 3 published in combination with "Sterne und Weltraum". Verlag Sterne und Weltraum Dr. Vehrenberg, Düsseldorf, Germany.

Veröff. Astron. Rechen-Inst. Heidelberg
Veröffentlichungen des Astronomischen Rechen-Instituts Heidelberg. Verlag G. Braun, Karlsruhe.

Veröff. Sternw. Sonneberg
Deutsche Akademie der Wissenschaften zu Berlin. Institut für Sternphysik. Veröffentlichungen der Sternwarte in Sonneberg. Publisher: Akademie-Verlag, Berlin.

Vesmír
Vesmír. Přírodovědecky časopis Čs. akadmie věd. Publisher: Academia, Praha.

Vestn. Khar'kov. Univ.
Vestnik Khar'kovskogo Universiteta. Seriya Astronomicheskaya. Publishers: Izdatel'stvo Khar'kovskogo Universiteta, Khar'kov.

Vestn. Kiev. Univ.
Vestnik Kievskogo Universiteta. Seriya Astronomii.
Publishers: Izdatel'stvo Kievskogo Universiteta, Kiev.

VJS Naturforsch. Ges. Zürich
Vierteljahresschrift der Naturforschenden Gesellschaft
in Zürich. Printer and Publisher: Leeman AG, Zürich.

Weltraumfahrt
Weltraumfahrt, Raketentechnik. Publisher: Umschau-
Verlag, Frankfurt a/Main.

Wiss. Zeitschr. Friedrich-Schiller Univ. Jena
Wissenschaftliche Zeitschrift der Friedrich-Schiller-Uni-
versität. Jena. Mathematisch-Naturwissenschaftliche
Reihe. Edited by the Rektor der Friedrich-Schiller-Uni-
versität Jena.

Wiss. Zeitschr. Humboldt-Univ. Berlin
Wissenschaftliche Zeitschrift der Humboldt-Universität
zu Berlin. Mathematisch-Naturwissenschaftliche Reihe.
Edited by the Rektor der Humboldt-Universität, Berlin.

Yamamoto Circ.
Yamamoto Circular. Published by the Yamamoto Observatory, Kamitanakami − Kiryutyo, Otu, Siga-ken, Japan.

Zeitschr. Angew. Physik
Zeitschrift für Angewandte Physik. Publisher: Springer-
Verlag, Berlin−Heidelberg−New York.

Zeitschr. Astrophys. (ZfA)

Zeitschrift für Astrophysik. Publisher: Springer-Verlag,
Berlin−Heidelberg−New York. After Vol. 69 (1968)
replaced by "Astronomy and Astrophysics".

Zeitschr. Geophys.
Zeitschrift für Geophysik. Publisher: Physica-Verlag,
Würzburg, Germany.

Zeitschr. Naturforschung
Zeitschrift für Naturforschung. Europhysics Journal.
Teil a: Astrophysik, Physik, Physikalische Chemie.
Published by Verlag der Zeitschrift für Naturforschung,
Tübingen, Germany.

Zeitschr. Physik
Zeitschrift für Physik. Publisher: Springer-Verlag, Berlin-
Heidelberg−New York.

Zemlya i Vselennaya
Zemlya i Vselennaya. Astronomiya, Geofizika, Issledovaniya Kosmicheskogo Prostranstva. Nauchno-Populyar-
nyj Zhurnal Akademii Nauk SSSR. Publishers: Izdatel'-
stvo "Nauka", Moskva.

Zentralblatt Math. Grenzgebiete
Zentralblatt für Mathematik und ihre Grenzgebiete. Publisher: Springer-Verlag, Berlin−Heidelberg−New York.

Zvaigžņota Debess
Latvijas PSR Zinātņu Akadēmijas Radioastrofizikas
Observatorijas Populārzinatnisks Gadalaiku Izdevums.
Izdevnieciba "Zinātne", Riga.

002 Bibliographical Publications

002.001 Science news.
Priroda, No. 1.73, p. 97 - 106 (1973). In Russian.
Tidal interactions of galaxies, p. 99; Powerful explosion in the Galaxy, p. 99; Water in the Martian atmosphere, p. 99; News of the moon's structure, p. 99; Satellite ERTS-1, p. 100.

002.002 News from science and other informations.
Zemlya i Vselennaya, 1973, No. 1. In Russian.
Lunokhod 2 on the moon, p. 11; Nuclear reactions on the solar surface, p. 20; "Thermal" Saturn, p. 79.

002.003 Science news.
Priroda, No. 2.73, p. 102 - 110 (1973). In Russian.
Starts of space apparatus in the USSR (September - October 1972), p. 102 - 103; Satellite Copernicus, p. 103; Velocity of motion of the Martian satellites, p. 103; Observation of cosmic radio radiation sources, p. 103 - 104.

002.004 News from science and other informations.
Zemlya i Vselennaya, 1973, No. 2. In Russian.
Final lunar expedition according to the "Apollo" program (*D. Yu. Gol'dovskij*), p. 15; Raging sun (*M. A. Livshits*), p. 36 - 37; Search for a transplutonian planet, p. 37; Photographs of the geocorona, p. 38.

002.005 Science news.
Priroda, No. 3.73, p. 103 - 115 (1973). In Russian.
(Landing of) Lunokhod 2 near Taurus Mountains, p. 103 - 104; Intercosmos 8 (*S. A. Nikitin*), p. 104 - 105; Last Apollo expeditions, p. 105 - 106; Hypothesis on the lunar nucleus, p. 106; Radio radiation of Callisto, p. 106.

002.006 Table of contents to "Astronomicheskij Tsirkulyar" Nos. 151 - 400 (1954 - 1967).
Compiled by N. B. Lavrova, N. D. Petrova.
Published by Byuro astronomicheskikh soobshchenij Akademii nauk SSSR, Moskva. 135 pp. (1972). In Russian.

002.007 Centre de Données Stellaires, Inform. Bull. No. 4.
J. Jung (Editor).
Published by Stellar Data Center, Obs. Strasbourg, Strasbourg, France. 2 + 34 pp. (1973). — Contents: The dissemination of data by the Stellar Data Center, *J. Jung*, p. 1 - 2; 2nd supplement to the list of catalogues available at the CDS, *J. Jung*, p. 3 - 5; Astronomical catalogs available from SAO in machine readable form, *K. Haramundanis*, p. 6 - 10; Bibliographical catalogue of variable stars, *W. Wenzel*, p. 11 - 13; 2nd complement to the list of spectroscopic and photometric catalogues lately published or to be published, *B. Hauck*, p. 14 - 19; A new catalogue of stellar UBV photoelectric photometry II, *J.-C. Mermilliod*, p. 20 - 21; First supplement to the list of transit tables for star numberings in open clusters, *J.-C. Mermilliod*, p. 22 - 26; The catalogue of stellar identifications; progress report II, *J. Jung, M. Bischoff, F. Ochsenbein*, p. 27 - 30.

002.008 Science news.
Priroda, No. 4.73, p. 19, 99 - 110 (1973).
In Russian.
A new method of calculating meteorite orbits, p. 19; Starts of space vehicles in the USSR (November–December 1972), p. 99; Radio waves investigate the Martian atmosphere, p. 100; How many solar neutrinos are likely? p. 100.

002.009 Hydrographische Bibliographie. (Ozeanographie, Erdmagnetismus, Nautik).
Selected and arranged by F. Model, H. Mädler.

Separate prints from Deutsche Hydrographische Zeitschr. [edited by Deutsches Hydrographisches Institut, Hamburg], Jahrgang 24 (1971), 88 pp (1972). — The hydrographic bibliography is a generalized extract from the "Hydrographische Dokumentation" (hydrographic documentation). It is available on international catalogue cards.

002.010 Doctoral theses.
Journ. History Astron., Vol. 4, 143 (1973). — Concerning three theses from 1971.

002.011 Annotations on papers on geomagnetism and aeronomy published in "News of Universities.
Radiophysics", 1971, Vol. 14, Nos. 11–12; 1972, Vol. 15, Nos. 1, 4–6.
Geomagn. Aeronom., Vol. 13, 550 - 554 (1973). In Russian.

002.012 Science news.
Priroda, No. 5.73, p. 102 - 112 (1973). In Russian.
Pioneer 10 passed through the belt of asteroids, p. 103; Outburst of a supernova, p. 103; Discovery of interstellar formaldimine, p. 103 - 104; Interstellar hydrogen sulfide, p. 104; Search for gravitational radiation, p. 104 - 105; The atmosphere of the earth older than 4×10^9 years, p. 109.

002.013 Science news.
Priroda, No. 6.73, p. 105 - 114 (1973). In Russian.
Starts of space vehicles in the USSR (January–February 1973), p. 105 - 106; Prognoz 3, p. 106; American space vehicles investigate the sun, p. 106 - 107; Anomalies of magnetism on the moon, p. 107; Analogue of the nucleic acid in meteorites, p. 107 - 108; The age of the Sikhote-Alin meteorite was established, p. 108; Coriolis force and continental drift, p. 110 - 111.

002.014 Chronicle.
Urania Kraków, Vol. 44, 16 - 20, 81 - 85, 114 - 116, 175 - 180 (1973). In Polish.
(1) Orbital Astronomical Observatory "Copernicus", (*Z. Paprotny*); Cosmochemical symposium in Cambridge, (*B. Kuchowicz*); Explosions of stars in the pre-telescopic era, (*T. Szymczak*). (3) Gamma radiation from the Crab nebula, (*B. Kuchowicz*); A proposed experimental verification of the general relativity (*M. Pańków*); Radio stars finally discovered!, (*A. Marks*); Orbiting telescope, (*M. Pańków*); Paleoselenomorphical charts, (*S. R. Brzostkiewicz*); Eclipses of stars by Jupiter, (*M. Pańków*); Do the intramercurial planetoids exist? , (*S. R. Brzostkiewicz*); (4) Does the X-ray pulsar Cen X3 belong to a double system? , (*B. Kuchowicz*); Again on X-ray source Sco X1, (*B. Kuchowicz*); Detection of oxygen molecules on Mars, (*B. Kuchowicz*); Hydrogen and methane in the atmosphere of Titan, (*B. Kuchowicz*). (6) Another pulsar in the vicinity of supernova remnant, (*B. Kuchowicz*); Five new radiosources in the neighbourhood of the Crab nebula, (*B. Kuchowicz*); Infrared observations of interstellar clouds, (*B. Kuchowicz*); British program of search for superheavy elements in cosmic rays, (*B. Kuchowicz*); Three new double systems, (*T. Szymczak*); Winds can blow on the moon, (*B. Kuchowicz*); On the eclipse of Mercury by Venus in 1737, (*M. Zawilski*).

002.015 AFCRL bibliography for the third and fourth quarter of 1971. J. W. Salisbury (Editor).
Icarus, Vol. 19, 247 - 286, 287 - 317 (1973).
Presented is a bibliography on lunar and planetary subjects furnished by the Air Force Cambridge Research Laboratory, Laurence G. Hanscom Field, Bedford, Mass.

002.016 Rassegna delle riviste e notizie brevi. P. Maffei.
Coelum, Vol. 41, 24 - 30, 64 - 69, 106 - 110 (1973).

002.017 News from science and other informations.
Zemlya i Vselennaya, 1973. No. 3. In Russian.
Neutrinos and models of the sun, p. 11 - 12; Spiral arms and
activity of the nuclei of galaxies, p. 17; 3rd day of Lunokhod,
p. 39; Decameter radio radiation of Jupiter (*S. P. Bozhich*),
p. 73 - 74.

002.018 Astronomical notebook. J. S. Griffith.
Journ. British Interplanet. Soc., Vol. 26, 51 - 61,
114 - 123, 180 - 187, 246 - 252, 305 - 309, 369 - 373, 427 -
432 (1973).
(1) The earth's acceleration; Mars; Mercury; Halley's comet;
Atmospheric entry probes; Additional planets; Faculae and
the solar oblateness; Interstellar clouds and stellar formation;
Contact binaries; Stellar instabilities; Circumstellar material;
K and M stars and the solar neighbourhood; Deformed galaxies;
Black holes; Galactic evolution and cosmology; Pulsars, quasars
and neutron stars; The oldest supernova remnant; The solar corona and pulsar occultations. (2) An oscillating sun? ; Time; The
internal structure of the planets; The airglow of the night sky;
The atmosphere and meteoroids; Planetary formation; A supergiant in the Small Magellanic Cloud; Pulsar radiation scintillation; Quasars and Seyfert nuclei; The cooling of pulsars;
GALAXY and parallaxes; RY Sagittarii, a pulsating star; The
Galaxy and the scale of distance; Non-spherical nebulae from
novae eruptions; Clusters of galaxies; Antimatter search. (3)
The history of the earth; Galactic cosmic rays and Pioneer 10;
The origin of interstellar grains; Main sequences for model
stellar clusters; Binary dwarfs; Mass and magnetic fields of
pulsars; Galactic hydrogen profiles; Quasars, radio galaxies and
cosmology; Maffei 2; Collapsed objects, X-ray stars and spectroscopic binaries; Gamma rays and OSO-3; Cygnus X-1 a
black hole? ; Black hole at the centre of our Galaxy? . (4)
Minor planets in the inner solar system; Jupiter's decametric
radio emission; The oldest disk stars; X-ray binary with neutron star component? ; Stellar hot spots; Galactic formation;
The Hubble red shift; Rotating black holes; The universe as a
black hole; Superlight velocities? ; Missing cosmological mass;
Interstellar alcohol and isocyanic acid. (5) The course of the
solar cycle; Future lunar research; Deuterium in Jupiter;
Uranus and Neptune; Starquakes on Algol? ; Star formation
and evolution in galaxies; Galactic bridges and tails. (6) Formation of the solar system; Very young stars; Supernovae;
Neutron-star accretion in a stellar wind; Quasar-events in galactic nuclei? ; An 'exploding galaxy'; Synthesis of elements
during the big-bang; The future of the universe. (7) Jupiter
and its satellites; Young stars; The origin of the light elements;
Outflow velocities in stars, Seyfert nuclei and quasars; An
attack on the density-wave; Shakhbazian I (a cluster of
galaxies); Cosmology and quasars; Olbers' paradox revisited;
Radio emission from binary stars; Cygnus X-1 a black hole?

002.019 Nouvelles de la science.
L'Astronomie, 87ᵉ année, p. 31 (1973). – Les
conjonctions quintuples de Neptune (*J. Meeus*).

002.020 Noted in the current journals.
D. Morrison, N. D. Morrison.
Mercury, (Journ. Astron. Soc. Pacific), Vol. 2, No. 1, p. 11 -
12, No. 2, p. 9 - 10 (1973). – (1) Is there star formation in the
galactic nucleus? ; Infrared observations of galaxies; The five-
minute oscillations in the solar atmosphere; The atmosphere of
Uranus. (2) Possible discovery of a black hole; The submillimeter: a new spectral region; Evidence against planet X.

002.021 Forschung und Technik.
Phys. Blätter, 29. Jahrgang, p. 44, 181 - 186, 236 -
237 (1973). – Schwefelwasserstoff in interstellaren Wolken;

seltene Isotope in Cyanwasserstoff, p. 44; Ergebnisse neuer
Beobachtungen des Jupitermondes Io, p. 181 - 182; Neuer
Satellit für Erforschung der kosmischen Gammastrahlung, p.
182 - 183; Bestätigung des Uhrenparadoxons bei Flügen um
die Erde, p. 183 - 184; Erste Meßwerte vom deutschen For-
schungssatelliten AEROS, p. 236; Lichtgeschwindigkeit nun-
mehr auf ± 1.1 m/s genau bestimmt! p. 237.

002.022 Science and the citizen.
Sci. American, Vol. 228, No. 1, p. 44 - 46; No. 2, p.
46 - 49; No. 6, p. 38 - 40 (1973). – Blowup (Cygnus X-3), No.
1, p. 45 - 46; Missing pulses, No. 2, p. 48; Looking backward,
No. 6, p. 38.

002.023 News notes.
Sky Telescope, Vol. 45, 22 - 23, 91 - 92, 159 - 160,
214 - 216, 277 - 278, 353 - 354, 358 (1973). – (1) Does planet
X exist? ; New findings about Jupiter's and Saturn's satellites;
Lowell Observatory leader; Measurer of star distances; Total
eclipses in the 1970's. (2) Very young galaxies? ; Surface composition of Mercury; Diameters of two red stars; More about
Stephan's Quintet. (3) Big Australian Schmidt; Three astrometric binaries; Did a comet strike the moon? ; Unusual variable in Lacerta. (4) Interstellar hydrogen map; Calendar reformer; Radar echoes from Saturn; A well-observed fireball; Canter-
bury sky atlas; Orange soil very old; Asteroids 1797 to 1813;
Nova Carinae 1970; Pleione's new shell. (5) A bright comet
next winter!; Pluto's rotation; A strange star in Aquila; Radioactivity of Aristarchus; Salute to E. J. Öpik; Mass of the moon.
(6) Largest red shift; More about a nearby galaxy; Velocity of
light 300 years ago; American Meteor Society moves; San Juan
Capistrano meteorite; Sirius not triple; Nearby star; Diamonds
in impactite; Mars map; Polish comet expert.

002.024 News notes.
Sky Telescope, Vol. 46, 26 (1973). – (1) Radio detection of a giant dust cloud; Meteorite crater in India; New
society in India; Center for Astrophysics; An astronomical
coincidence.

002.025 Mitteilungen aus Wissenschaft und Literatur.
Sterne, 49. Jahrgang, p. 53 - 54, 121 - 122 (1973).
(1) Formaldehyd in Meteoriten (*H. Lambrecht*); (2) Die Leucht-
kraft der RR-Lyrae-Sterne (*F. Schmeidler*); Förderung der
Astronomie (*H. Lambrecht*).

002.026 Kurzberichte aus der Forschung
SuW, Vol. 12, 17 - 20, 47 - 49, 82 - 84, 113 - 115,
142 - 144, 174 - 176 (1973).
(1) Voraussage für den Sonnenfleckenzyklus Nr. 21; Mehr
von und über Cyg X-3; 5-km-Radioteleskop fertiggestellt; HD
33579 – der hellste Stern der LMC; Nichtkosmologische Rot-
verschiebung? ; Geplante Kooperation der AAT (*AAT =
Anglo - Australian Telescope projects*)- und ESO-Schmidt-
Teleskope. (2) Röntgen-Leuchtkraft und Geschwindigkeits-
dispersion in Galaxienhaufen; Multipolschwingungen in
Weißen Zwergen; Interstellare Absorptionslinien im Spektrum
der Supernova in NGC 5253; Zahl der M-Hauptreihen-Sterne
vervielfacht? . (3) Eine Bedeckung von β Scorpii durch Jupiter; Radiobeobachtungen der Sonnenkorona bei mehreren
Frequenzen; Nachweis eines ultralangen Zyklus der Sonnen-
tätigkeit; Gravitationswellen vom Pulsar CP 1133; Ein neues
Alter für die Welt. (4) Radiostrahlung von Galaxien und die
Entstehung der Spiralarme; Spiralstruktur in Kometen:
Uranusforschung mit dem Stratoscope II. (5) Nochmals der
hypothetische Transpluto; Apollo 15: Krater Aristarchus eine
geologisch 'aktive' Zone? ; Infrarotstrahlung von extragalak-
tischen Objekten; 'Einzel-Wolken'-Modell für R CrB ; Cyano-
acetylenwolke in Sgr B2. (6) Erzeugen alle Supernovae Pul-
sare? ; Shakhbazian I, ein entfernter kompakter Haufen von
kompakten Galaxien; Interstellares Deuterium; Entdeckung

neuer Radiosterne; Die Saturn-Ringe.

002.027 **News and views.**
Nature, Vol. 241, 9 - 18, 85 - 94, 161 - 170, 237 - 246, 307 - 316, 364 - 371, 423 - 432, 497 - 506 (1973). – Discoveries of new radio stars, p. 9; BL Lac objects: Link with galaxies?, p. 16 - 17; Variations of Cygnus X-3, p. 18; Surveyor spacecraft results verified, p. 93 - 94; Soft X-ray source towards the galactic centre, p. 94; Problems of understanding pulsars, p. 161; Background radiation no longer anomalous, p. 167; Methane and carbide on the moon, p. 168 - 169; Progress in gamma-ray astronomy, p. 169; Relativistic astrophysics: State of the art, p. 169 - 170; Mariner 1977: Experiments chosen, p. 170; Fixes on radio sources, p. 237 - 238; Explaining the spectra of heavy cosmic rays, p. 245; X-ray star problems defined, p. 246; Record solar activity in August 1972, p. 307; Two new infrared sources discovered, p. 312; Infrared observations of the stratosphere, p. 314 - 315; Celebrating the birth of Copernicus, p. 316; Explaining the deuterium in the universe, p. 364; Gamma-ray statistics, p. 366; Hubble constant: New determination, p. 370 - 371; Optical studies of Cyg X-1 (HDE 226868), p. 370; Relativistic forms, p. 423 - 424; Orbiting systems, mass loss and gravity pulses, p. 430 - 431; Microtektites; Bottle green variety, p. 431; Origin of bridges and tails in galaxies, p. 432; Does polar wandering occur?, p. 497 - 498; Rapid and strong variations in BL Lac, p. 502 - 503; QSOs: Redshifts and distances, p. 506.

002.028 **News and views.**
Nature, Vol. 242, 9 - 18, 83 - 92, 155 - 164, 225 - 234, 295 - 304, 365 - 374, 431 - 438, 493 - 500, 551 - 558 (1973). – Lunar seismology, p. 17; Are all QOSs in the nuclei of galaxies? p. 18; Spin, torsion and gravitational singularities, p. 18; Ubiquitous neutrinos, p. 83 - 84; Drift in the ionosphere, p. 84 - 85; Solar system: Primordial field, p. 90 - 91; Soft X-ray structure of Cassiopeia A, p. 91 - 92; Circular polarization in compact radio sources, p. 92; New oceanic model, p. 162; Structure in solar radio bursts, p. 162; Moon matters, p. 231 - 232; Amino-acids in lunar samples, p. 232; Wolf-Rayet systems and X-ray binaries, p. 303; Universal isotropy: Essential requirement for life?, p. 304; Is Cyg X-3 similar to Sco X-1?, p. 304; Record QSO redshift observed, p. 365 - 366; Mars: Up hill down dale, p. 372 - 373; Radio emission from nebulae around stars, p. 374; Mars: Polar wandering? p. 436; Stellar nucleosynthesis and the s-process, p. 438; SMC X-1 observed at energies > 7 keV, p. 498; Singularities and the C field, p. 556 - 557; QSOs and galaxies: Nature of Ton 256, p. 557; Origin of cosmic X-ray background, p. 557 - 558.

002.029 **News and views.**
Nature, Vol. 243, 7 - 14, 55 - 62, 114 - 122, 184 - 192, 258 - 265, 318 - 324, 434 - 440, 490 - 496 (1973). Puppis A mapped by Copernicus, p. 12; Reddening of stars near the galactic centre, p. 61; Gravitational radiation: Waving goodbye to Weber's waves, p. 61 - 62; Invoking black holes, p. 114 - 115; Radio emission from supernova 1970g, p. 120; Continuum radio emission from vicinity of pulsars, p. 190; From quantum crystals to neutron stars, p. 192; Radio source associated with OH source ON-1, p. 264; Lunar conductivity, p. 265; Mariner results, p. 322 - 323; Microcrater in lunar meteorite, p. 323; Link between coronal and interplanetary shocks, p. 323; Astronomy: Goddard symposia, p. 324; Galac-

tic escape and cosmic electrons, p. 438; Gravitational radiation: Magnetic correlation?, p. 439; Spiral structure and nuclear activity in galaxies, p. 439; Mars: Are ice ages temporary?, p. 439 - 440; Solar eclipse of a lifetime, June 30, 1973, p. 490; QSOs and intergalactic gas, p. 495; COSPAR: Plenty in prospect, p. 496; Candidate for LMC X-1, p. 496.

002.030 **News and views.**
Nature, Vol. 244, 6 - 12, 68 - 74 (1973). – Very long baseline interferometry, p. 6; Selenology: Stepwise velocity, p. 10 - 11; Formation of earth craters, p. 68 - 69; Blowing bubbles, p. 69; Gravitational radiation: The ten year test, p. 72 - 73.

002.031 **News.**
Nature, Phys. Sci., Vol. 241, 1, 25, 49, 73, 89, 105, 121, 137, 153 (1973). – Selenology: Deficient oxides, p. 1; Dynamics and coordinates of the moon, p. 105; Martian magnetism, p. 121; Hot and cold on Mars, p. 137.

002.032 **News.**
Nature, Phys. Sci., Vol. 242, 1, 17, 33, 49, 65, 81, 97, 113, 129; Vol. 243, 1, 17, 41, 57, 73, 105 (1973).– An unstable galaxy pair, Vol. 242, p. 113; The earth: Is gravity drifting? p. 113; Four-day weather cycle on Venus?, p. 129. Infrared connexions, Vol. 243, p. 1; Martian water vapour, p. 17; Cosmic rays: Two mechanisms, p. 41; Was the early moon a permanent magnet? p. 57; X-rays: Origin of soft background p. 73; Cosmic gamma rays come in bursts, p. 105.

002.033 **News.**
Nature, Phys. Sci., Vol. 244, 1, 17, 33 (1973).– Circular polarization in astronomy, p. 1; Dark asteroids and regoliths, p. 1; Winds in the Venusian atmosphere, p. 17; Selenology: Age of orange soil from Taurus-Littrow, p. 17; Copernicus at the Royal Society, p. 33.

002.034 **Astronomy and Astrophysics Abstracts. Vol. 8, Literature 1972, Part II.**
S. Böhme, W. Fricke, U. Güntzel-Lingner, F. Henn, D. Krahn, U. Scheffer, G. Zech (Editors).
Published for Astronomisches Rechen-Institut, Heidelberg by Springer-Verlag, Berlin – Heidelberg – New York. 10 + 594 pp. Price DM 78.00; (US $ 28.90) [Subscription price DM 62.40; (US $ 23.10)] (1973).

002.035 **Nouvelles brèves.**
Ciel et Terre, Vol. 89, 63, 147 - 150 (1973).
(1) Observation de l'occultation des Pléiades par la lune le 19 mars 1972; (2) Une brillante supernova dans NGC 5253; La distance des pulsars; La petite planète 1036 Ganymed; Un catalogue d'éphémérides cométaires; Un nouvel Astronome Royal.

002.036 **Bibliography.**
Z. Kopal, M. Moutsoulas, J. W. Salisbury (Editors). The Moon, Vol. 6, 212 - 228 (1973). – Current critical bibliography of the entire field of lunar studies.

Yearbook of the books published in the USSR in 1969. Systematic index, Vol. 2. Books on natural sciences, technique and other sciences. See Abstr. 003.146.

003 Books (Astronomy and Astrophysics)

**003.001 Luna 20. A study of samples from the lunar high-
lands returned by the unmanned Luna 20 spacecraft.** E. Anders, A. L. Albee (Editors).
Geochim. Cosmochim. Acta, Vol. 37, 719 - 1109 (1973).
The individual contributions are included in their corresponding subject categories – see abstracts 094.071 - 094.106.
Some further papers published in Earth Planet. Sci. Letters,
Vol. 17, have been included in Vol. 08 of these Abstracts, see
08.094.191 - 08.094.198.

**003.002 Radio astronomy observations of the solar eclipse
on 20 May, 1966.**
AN SSSR. Nauch. sovet sluzhby "Solntse – Zemlya". Nauka,
Moskva, 142 pp. Price 86 Kop. (1972). In Russian. – Review in Referativ. Zhurn. 51. Astron., 3.51.492 (1973).
The individual contributions are included in their corresponding subject categories – see abstracts 072.019, 072.020,
077.016 - 077.034.

**003.003 Special Supplement to the third edition of the Gen-
eral Catalogue of Variable stars, containing the list**
of 32731 stars arranged in the order of right ascensions for the
equinox 1950.0.
B. V. Kukarkin, P. N. Kholopov, Yu. N. Efremov, N. P. Kukar-
kina, N. E. Kurochkin, G. I. Medvedeva, N. B. Perova, V. P.
Fedorovich, M. S. Frolov.
Academy of Sciences of the USSR, Astronomical Council,
Sternberg State Astronomical Institute of the Moscow State
University, Moscow. 234 pp. (1972). In Russian and English.

**003.004 Nicolaus Copernicus – 1473–1973. Das Bild vom
Kosmos und die Copernicanische Revolution in den**
gesellschaftlichen und geistigen Auseinandersetzungen.
J. Herrmann (Editor).
(Akademie der Wissenschaften der DDR. Studien zum Coper-
nicus-Jahr 1973). Akademie-Verlag, Berlin. 215 pp. Price
DM 18.00 (1973). – The individual papers are included in
their corresponding subject categories – see abstracts 004.024
- 004.032, 013.004.

**003.005 Astronomy. Volume 8. Physics and evolution of
stars.** A. G. Masevich, B. M. Shustov.
Itogi nauki i tekhniki. Seriya Astronomiya, tom 8, Moskva.
116 pp. Price 77 Kop. (1972). In Russian. – Results of investigations on the internal structure and evolution of stars
published in the USSR and abroad from 1967 to 1971 are
generalized. – Review in Referativ. Zhurn. 51. Astron.
4.51.651 (1973).

003.006 Astrometriya i Astrofizika, Vypusk 16.
E. P. Fedorov (Editor).
Respublikanskij Mezhvedomstvennyj Sbornik. Akademiya
Nauk Ukrainskoj SSR, Glavnaya Astronomicheskaya Observatoriya. Izdatel'stvo "Naukova Dumka", Kiev. 124 pp. Price
1 Rbl. 6 Kop. (1972). In Russian. – The papers included are
abstracted in their corresponding subject categories – see
abstracts 031.016, 041.006, 041.007, 071.023, 071.024,
073.028, 091.037, 094.131 - 094.134, 103.104, 122.028,
158.087.

003.007 Atomic physics and astrophysics. Brandeis Univer-
sity Summer Institute in Theoretical Physics, 1969.
Vol. 1. M. Chrétien, E. Lipworth (Editors).
Gordon and Breach Science Publishers, New York – London –
Paris. 12 + 216 pp. Price DM 53.00 (1971). – Review in Space
Sci. Rev., Vol. 14, 344; 1973 (R. Mewe). – For the individual
contributions – see abstracts 022.066, 022.067.

003.008 Thermal characteristics of the moon.
J. W. Lucas (Editor).
Progress in Astronautics and Aeronautics, Vol. 28. The MIT
Press, Cambridge, Mass. – London, England. 14 + 340 pp.
Price $ 14.95 (1972). – The individual contributions are in-
cluded in their corresponding subject categories – see abstracts
094.538 - 094.548.

003.009 Solar activity observations and predictions.
P. S. McIntosh, M. Dryer (Editors).
Progress in Astronautics and Aeronautics, Vol. 30.
The MIT Press, Cambridge, Mass. 15 + 442 pp. Price $ 17.50
(1972). – The individual contributions are included in their
corresponding subject categories – see abstracts 072.046,
072.047, 073.044 - 073.050, 074.049 - 074.051, 075.002,
075.003, 076.016, 077.051, 078.029, 078.030, 080.019 -
080.021, 082.072, 083.061, 084.247, 084.248.

003.010 Zur Geschichte der Erde und des Kosmos.
H.-J. Treder (Editor).
Akademie-Verlag, Berlin = Veröff. Forschungsbereich Kosm.
Phys., Akad. Wiss. DDR, No. 1. 74 pp. Price MDM 10.00
(1973).

003.011 Interstellar matter. Swiss Society of Astronomy
and Astrophysics second advanced course, Saas-Fee,
1972 March 20–25.
N. C. Wickramasinghe, F. D. Kahn, P. G. Mezger.
Edited by Astron. Inst. Univ. Basel, Binningen; published by
Geneva Obs., Sauverny, Switzerland. 11 + 437 pp. Price
sf. 35.00 (1972). – For the individual contributions – see abstracts 131.153 - 131.155.

003.012 Young stellar groups. Astroclimate.
V. S. Shevchenko (Editor).
Akademiya Nauk Uzbekskoj SSR, Astronomicheskij Institut;
Izdatel'stvo "FAN" Uzbekskoj SSR, Tashkent. 180 pp. Price
1 Rbl. 2 Kop. (1972). In Russian. – The individual contributions are included in their corresponding subject categories –
see abstracts 082.103 - 082.107, 113.052, 122.117, 152.011,
152.012, 153.001.

**003.013 Cosmic rays. Articles No. 13. Results of researches
on the International Geophysical Projects.**
S. N. Vernov, L. I. Dorman (Editors).
Publishing House "Nauka", Moscow. 260 pp. Price 2 Rbl.
21 Kop. (1972). In Russian. – The individual contributions
within the subject scope of Astronomy and Astrophysics
Abstracts are included in their corresponding subject categories
– see abstracts 034.095, 078.044, 078.045, 085.011, 143.060 -
143.066.

003.014 Nicolaus Copernicus zum 500. Geburtstag.
F. Kaulbach, U. W. Bargenda, J. Blühdorn (Editors).
Böhlau Verlag, Köln–Wien. 15 + 270 pp. Price DM 48.00
(1973).
Contents: Die Astronomie des Copernicus (*F. Schmeidler*);
Zur Frage der Tragweite der Copernicanischen Wende (*P. Jor-
dan*); Die Copernicanische Wende als philosophisches Prinzip.
Nachgewiesen bei Kant und Nietzsche (*F. Kaulbach*); Nicolaus
Copernicus als ermländischer Domherr (*B. Stasiewski*); Die
Vorschläge des Nicolaus Copernicus zu einer Reform des
preußischen Münzwesens (*H. Kellenbenz*); Das ärztliche Wirken
des Frauenburger Domherrn Nicolaus Copernicus (*B.-M.
Rosenberg*); Das westliche Preußen und das Ermland zur Zeit
des Copernicus (*W. Hubatsch*); Copernicus und die deutschen
Theologen des 16. Jahrhunderts (*Z. Wardęska*); Die Schreib-

weise des Namens Copernicus. Betrachtungen zur Schreibung des Namens des großen Astronomen, ausgehend von der Kontroverse im Dritten Reich (*H. Koeppen*); Dietrich von Reden und Nikolaus von Schönberg. Zwei Freunde von Copernicus (*K. Forstreuter*); Lebensdaten des Copernicus (*W. Thimm*); Kleine Copernicus-Bibliographie (*H.-J. Schuch, W. Thimm*).

003.015 Atomic physics and astrophysics.
Brandeis University Summer Institute in Theoretical Physics, 1969. Vol. 2. M. Chrétien, E. Lipworth (Editors). Gordon and Breach Science Publishers, New York – London – Paris. 12 + 337 pp. (1973). – For the individual contributions within the subject scope of Astronomy and Astrophysics Abstracts – see abstracts 061.064, 061.065.

003.016 Astrometriya i Astrofizika, Vypusk (No.) 18.
E. P. Fedorov (Editor).
Respublikanskij Mezhvedomstvennyj Sbornik. Akademiya Nauk Ukrainskoj SSR. "Naukova Dumka", Kiev. 104 pp. Price 87 Kop. (1973). In Russian. – The papers included are abstracted in their corresponding subject categories – see abstracts 065.177, 071.056, 081.028, 082.127, 094.919, 094.920, 097.120, 100.056, 103.102, 103.113, 122.140, 122.141.

003.017 J. C. Poggendorff, Biographisch-literarisches Handwörterbuch der exakten Naturwissenschaften.
Vol. VIIb, Part 4, 4th number.
H. Salié (Editor).
Published by Akademie-Verlag, Berlin. 236 pp. Price DM 36.00 (1973).

003.018 Hipótesis del sum. F. Soler Batlle.
Caracas, 2 + 147 pp. (1973).

003.019 Optical transforms. H. Lipson.
Academic Press, London – New York. 436 pp. (1972). – Applications of optical transform techniques.
DNC

003.020 Theory and analysis of phased array antennas.
N. Amitay, V. Galindo, Chen Pang Wu.
Wiley Interscience Publishers, New York. 443 pp. (1972).
The book presents a mathematical approach for analysis and design of a broad class of phased array antennas and examines their electromagnetic properties. – *ACM*

003.021 Polskie Miasta Kopernika. J. Adamczewski.
Wydawnictwo Interpress, Warszawa. 191 pp. Price zł 75.00 (1972). – Review in Urania Kraków, Vol. 44, 92 - 93; 1973 (*Z. Maślakiewicz*).

003.022 Atom, man, universe. H. Alfvén.
Translated from the English edition. Znanie, seriya "Kosmonavtika, astronomiya", 1973, No. 1, Moskva. 64 pp. Price 10 Kop. In Russian. – Review in Priroda, No. 5.73, p. 120 - 121 (1973).

003.023 Other worlds, other beings. S. W. Angrist.
Thomas Y. Crowell Company, New York. 119 pp. Price $ 4.95 (1973). – Review in Sky Telescope, Vol. 46, 44 (1973).

003.024 Emission, absorption and transfer of radiation in related atmospheres.
B. H. Armstrong, R. W. Nicholls.
Pergamon Press Ltd., Oxford. 320 pp. Price £ 7.00 (1972). Review in Journ. British Interplanet. Soc., Vol. 26, 127 - 128 (1973).

003.025 Elementare Plasmaphysik. L. A. Arzimowitsch.

Akademie-Verlag GmbH, Berlin. 178 pp. Price MDM 15.00 (1972). – Review in Phys. Blätter, 29. Jahrgang, p. 240; 1973 (*H. Zwicker*).

003.026 Comets and meteors. I. Asimov.
Follett Publishing Co., Chicago, Ill. 32 pp. Price $ 1.25 (1972).

003.027 The sun. I. Asimov.
Follett Publishing Co., Chicago, Ill. 31 pp. Price $ 1.25 (1972).

003.028 Collection of problems of celestial mechanics and cosmodynamics. Manual for college students.
M. B. Balk, V. G. Demin, A. L. Kunitsyn.
Nauka, Moskva. 336 pp. Price 95 Kop. (1972). In Russian. Review in Referativ. Zhurn. 51. Astron., 5.51.38 (1973).

003.029 Binary and multiple systems of stars. (International Series of Monographs in Natural Philosophy, Vol. 51). A. H. Batten.
Pergamon Press Ltd., Oxford – New York – Toronto – Sydney – Braunschweig. 12 + 278 pp. Price £ 4.00 (1973). – Review in SuW, Vol. 12, 188; 1973 (*L. D. Schmadel*).

003.030 The sun and the amateur astronomer.
W. M. Baxter.
David & Charles Ltd., Newton Abbot. 156 pp. Price £ 2.95 (1973). – Review in Journ. British Astron. Ass., Vol. 83, 232 (1973).

003.031 Exploring tomorrow in space. T. W. Becker.
Sterling Publishing Co., New York. 160 pp. Price $ 6.95 (1972). – Review in Sky Telescope, Vol. 45, 48 (1973).

003.032 Introductory Fourier transform spectroscopy.
R. J. Bell.
Academic Press, New York – London. 398 pp. Price £ 9.10 (1972). – Review in Journ. British Interplanet. Soc., Vol. 26, 64 (1973).

003.033 Einstein. J. Bernstein.
The Viking Press Inc., New York. 12 + 242 pp. Price $ 6.95 (1973). – Review in Science, Vol. 180, 620 - 623; 1973 (*S. Goldberg*).

003.034 Świat Kopernika. H. Bietkowski, W. Zonn.
Wydawnictwo Arkady, Warszawa. 167 pp. Price zł 120.00 (1972). – Review in Urania Kraków, Vol. 44, 93; 1973 (*Z. Maślakiewicz*).

003.035 Zagadka rozszerzającego się Wszechświata.
W. Bonnor.
Państwowe Wydawnictwo Naukowe, Warszawa – Biblioteka Problemów. Translated from English into Polish by M. Kubiak. 249 pp. Price zł 30.00 (1972). – Review in Urania Kraków, Vol. 44, 92; 1973 (*K. Ziołkowski*).

003.036 The new astronomies. B. Bova.
St. Martin's Press,Inc., New York. 214 pp. Price $ 7.95 (1972). – Reviews in Sky Telescope, Vol. 45, 47, 380 - 381; 1973 (*M. R. Chartrand III*).

003.037 Introduction to the solar wind. J. Brandt.
Translated from the English edition. Mir, Moskva. 207 pp. Price 1 Rbl. 30 Kop. (1973). In Russian. – Reviews in Referativ. Zhurn. 51. Astron., 7.51.392; 62. Issled. kosmich. prostranstva, 7.62.182 (1973).

003.038 Pieces of another world. F. M. Branley.

Thomas Y. Crowell Company, New York. 58 pp. Price $ 4.50 (1972). – Review in Sky Telescope, Vol. 45, 312 (1973).

003.039 Environmental space science.
D. G. Carpenter (Editor).
Whitehall Co., Northbrook, Ill. 719 pp. Price $ 10.95 (1972). Review in Sky Telescope, Vol. 45, 248 - 249 (1973).

003.040 Ellipsoidal figures of equilibrium.
S. Chandrasekhar.
Translated from the English edition. Mir, Moskva. 288 pp. Price 1 Rbl. 96 Kop. (1973). In Russian. – Review in Referativ. Zhurn. 51. Astron., 7.51.713 (1973).

003.041 Zonnewijzers aan en bij Gebouwen in Nederland, benevens een korte beschrijving van twee astronomische torenuurwerken. J. G. van Cittert-Eymers.
Thieme & Cie, Zutphen. 114 pp. Price f 14.50 (1972). Review in Hemel en Dampkring, Vol. 71, 75; 1973 (*T. de Vries*).

003.042 The interplanetary pioneers. W. R. Corliss.
National Aeronautics and Space Administration, NASA SP-278. [Available from Superintendent of Documents U.S. Government Printing Office, Washington, D.C.], Vol. 1: 130 pp. Price $ 1.25; Vol. 2: System design and development, 304 pp. Price $ 2.50; Vol. 3: Operations, 152 pp. Price $ 1.75 (1972). – Reviews in Sky Telescope, Vol. 45, 249; Vol. 46, 44 (1973).

003.043 The pulse of the planet.
J. Cornell, J. Surowiecki (Editors).
Crown Publishers, Inc., New York. 129 pp. Price $ 2.95 (1972). – Review in Sky Telescope, Vol. 45, 114 (1973).

003.044 Celescope catalog of ultraviolet stellar observations. 5068 objects measured by the Smithsonian experiment aboard the Orbiting Astronomical Observatory (OAO-2).
R. J. Davis, W. A. Deutschman, K. L. Haramundanis, with a preface by F. L. Whipple.
Smithsonian Astrophysical Observatory, Cambridge, Mass.; Smithsonian Institution, Washington, D. C. 5 + 248 pp. For sale by the Superintendent of Documents, U. S. Government Printing Office, Washington, D. C. Price $ 4.85, $ 4.50 respectively (1973). – This Catalog contains the results of photometric measurements of 5068 stars between 1200 and 3000 Angstroms. The catalog is based on more than 8000 television pictures obtained by the Smithsonian's experiment aboard the Orbiting Astronomical Observatory (OAO-2) launched by the National Aeronautics and Space Administration on December 7, 1968. Besides this printed book the data are available on magnetic tape, together with utility programs for reading and printing the contents of the tape. This magnetic-tape version is available from the National Space Sciences Data Center, Code 601, National Aeronautics and Space Administration, Goddard Space Flight Center, Greenbelt, Maryland 20771. The NSSDC identifies our data by the number "68-110A-01 (Smithsonian OAO data)." Scientists who are not U. S. citizens have access to these data by writing to: World Data Center A for Rockets and Satellites, Code 601, NASA Goddard Space Flight Center, Greenbelt, Maryland 20771, USA.

003.045 Enciclopedia dello spazio. G. De Fiore.
Edited by La Scuola, Brescia, Italy, 334 pp. Price L. 3200. – Review in Coelum, Vol. 41, 113 - 114; 1973 (*A. Betti*).

003.046 Układ Słoneczny.
W. G. Demin (*V. G. Demin*). Translated from Russian into Polish by C. Krepski.

Państwowe Wydawnictwo Naukowe – Biblioteka Problemów, Warszawa. 301 pp. Price zł 28.00 (1972). – Review in Urania Krakow, Vol. 44, 93 - 94; 1973 (*K. Ziołkowski*).

003.047 The new outline of science. D. Dietz.
Dodd, Mead & Company, New York. 495 pp. Price $ 17.50 (1972). – Review in Sky Telescope, Vol. 45, 114 (1973).

003.048 Cosmic rays in the earth's magnetic field.
L. I. Dorman, V. S. Smirnov, M. I. Tyasto.
Nauka, Moskva. 400 pp. Price 1 Rbl. 90 Kop. (1971). In Russian. – Review in Vestn. AN SSSR, No. 10, p. 142 - 143; 1972 (*V. P. Shabanskij*).

003.049 Mechanik, Relativität, Gravitation.
G. Falk, W. Ruppel.
Springer-Verlag, Berlin – Heidelberg – New York. 16 + 442 pp. Price DM 38.00 (1973). – Reviews in Orion Schaffhausen, 31. Jahrgang, p. 71; 1973 (*E. Wiedemann*); SuW, Vol. 12, 156; 1973 (*C. Thum*).

003.050 Magnetosphere–ionosphere interactions. Proceedings of a Study Institute, Dalseter, Norway, April 1971. K. Folkestad (Editor).
Universitetsforlaget, Oslo, Norway. 254 pp. Price $ 28.00 (1972).

003.051 Leben unter fernen Sonnen? W. R. Fuchs.
Droemersche Verlagsanstalt Th. Knaur Nachf., München. 256 pp. Price DM 25.00 (1973). – Review in Sky Telescope, Vol. 45, 313 (1973).

003.052 Astrofizica centemporană. V. L. Ginzburg.
EB-Enciclopedia de Buzunar. Editura Enciclopedia Romănă, Bucureşti. Translated from the Russian Edition (1970). 218 pp. Price Lei 7.50 (1973).

003.053 Al–Bituji: On the principles of astronomy.
An edition of the Arabic and Hebrew versions with translation, analysis and an Arabic–Hebrew–English glossary. Vol. 1, 2. B. R. Goldstein (Editor).
Yale University Press, New Haven – London. Price $ 35.00 (1971). – Essay review by E. S. Kennedy in Journ. History Astron., Vol. 4, 134 - 136 (1973).

003.054 New and full moons: 1001 B. C. to A. D. 1651.
H. H. Goldstine.
American Philosophical Society, Philadelphia, Pa. 221 pp. Price $ 5.00 (1973). – Review in Sky Telescope, Vol. 45, 312 (1973).

003.055 Nicolas Copernic. To the 500th anniversary of his birthday. E. A. Grebenikov.
Nauka, Moskva. 96 pp. Price 20 Kop. (1973). In Russian. Review in Referativ. Zhurn. 51. Astron., 7.51.11 (1973).

003.056 Cosmonauts in orbit.
G. Gurney, C. Gurney.
Franklin Watts, Inc., New York. 192 pp. Price $ 7.95 (1972). Review in Sky Telescope, Vol. 45, 114 (1973).

003.057 Doppelplanet Erde–Mond. E. Hantzsche.
BSB B. G. Teubner Verlagsgesellschaft, Leipzig. 256 pp. Price MDM 9.90 (1973). – Review in Astron. in der Schule, 10. Jahrgang, p. 69 (1973).

003.058 The large scale structure of space–time.
S. W. Hawking, G. F. R. Ellis.
At the University Press, Cambridge, England. 11 + 391 pp. Price £ 10.00 (1973). – Contents: (1) The role of gravity;

(2) Differential geometry; (3) General relativity; (4) The physical significance of curvature; (5) Exact solutions; (6) Causal structure; (7) The Cauchy problem in general relativity; (8) Space—time singularities; (9) Gravitational collapse and black holes; (10) The initial singularity in the universe.

003.059 **The comet of 1577. Its place in the history of astronomy.**   C. D. Hellman.
AMS Press, New York. 488 pp. Price £ 5.65 (1971).' – Review in Journ. History Astron., Vol. 4, 61 - 62; 1973 (*W. H. Donahue*).

003.060 **DTV–Atlas zur Astronomie.**   J. Herrmann.
Deutscher Taschenbuch Verlag GmbH und Co., München. 287 pp. Price DM 9.80 (1973). – Review in Sky Telescope, Vol. 45, 387 (1973).

003.061 **City of the stargazers.**   K. Heuer.
Charles Scribner's Sons, New York. 170 pp. Price $ 7.95 (1972). – Reviews in Sky Telescope, Vol. 45, 113, 244 - 247 (1973).

003.062 **Albert Einstein. Creator and rebel.**
B. Hoffmann, with the collaboration of H. Dukas. The Viking Press, Inc., New York. 16 + 272 pp. Price $ 8.95 (1972). – Reviews in Science, Vol. 180, 620 - 623; 1973 (*S. Goldberg*); Sky Telescope, Vol. 45, 313 (1973).

003.063 **Optical production technology.**   D. F. Horne.
Crane, Russak and Co., Inc., New York. 567 pp. Price $ 42.50 (1972). – Review in Sky Telescope, Vol. 45, 114 (1973).

003.064 **From Stonehenge to modern cosmology.**
F. Hoyle.
W. H. Freeman and Company Ltd., San Francisco. 8 + 96 pp. Price $ 4.95, £ 2.10 respectively (1972). – Contents: (1) Science and society in modern times; (2) Stonehenge; (3) Recent developments in cosmology I; (4) Recent developments in cosmology II.

003.065 **Norwood Russell Hanson: Constellations and conjectures.**   W. C. Humphreys, Jr. (Editor).
Synthese Library. D. Reidel Publishing Company, Dordrecht – Holland/Boston – U. S. A. 10 + 282 pp. Price $ 22.25 (1973).

003.066 **Astronomie Mikuláše Kopernika.**
C. Iwaniszewska.
Orbis, Praha. 92 pp. Price Kčs 9.00 (1972). Translated from the Polish edition. In Czech. – Review in Říše hvězd, Vol. 54, 141 - 142 (1973).

003.067 **Ontstaan en levensloop van sterren.**
C. de Jager, E. P. J. van den Heuvel.
Thieme & Cie., Zutphen. 2nd edition. 245 pp. Price f 24.50 (1973). – Review in Hemel en Dampkring, Vol. 71, 229 (1973).

003.068 **The Milky Way.** An elusive road for science.
S. L. Jaki.
Science History (Neale Watson Academic) Publications, New York. 10 + 352 pp. Price $ 14.95 (1973). – Review in Science, Vol. 180, 1269 - 1270; 1973 (*R. Berendzen*).

003.069 **Stellar chromospheres.**
S. D. Jordan, E. H. Avrett (Editors).
National Aeronautics and Space Administration, NASA SP-317. [Available from Superintendent of Documents, U. S. Government Printing Office, Washington, D. C.], 318 pp. Price $ 2.00 (1973). – Review in Sky Telescope, Vol. 45, 386 - 387 (1973).

003.070 **Elementy astronomii dla geografów.**   M. Karpowicz
Wydawnictwa Uniwersytetu Warszawskiego, Warszawa. 340 pp. Price zł 23.00 (1972). – Review in Urania Kraków, Vol. 44, 125 - 126; 1973 (*L. Zajdler*).

003.071 **Copernicus und seine Welt. Biographie.**
H. Kesten.
Deutscher Taschenbuch Verlag GmbH & Co., München. 328 pp. Price DM 6.80 (1973).

003.072 **Le roman de la lune.**   P. Kohler.
Éditions France-Empire, Paris. 303 pp. Price F 24.85 (1973). – Review in Sky Telescope, Vol. 46, 44 (1973).

003.073 **The solar system.**   Z. Kopal.
Oxford University: Oxford, London and New York, 8 + 152 pp. Price £ 2.25, $ 6.00 respectively, cloth, £ 1.00, $ 1.95 respectively, paper (1973). – Review in Nature, Vol. 242, 484; 1973 (*D. W. Hughes*).

003.074 **Introduction to the physics of stellar interiors.**
V. Kourganoff.
First published in 1969 by Dunod, Paris. Translated from the French by J. R. Lesh. Astrophysics and Space Science Library, Vol. 34. D. Reidel Publishing Company, Dordrecht – Holland/Boston – U.S.A. 7 + 115 pp. Price hfl. 50.00 (1973). Contents: (1) General considerations concerning the energy radiated by stars; (2) Mechanical equilibrium: the equilibrium between the gravitational force per unit volume and the gradient of the total pressure; (3) The determination of the internal structure by the density distribution $\rho(r)$; (4) Energy equilibrium and nuclear reactions; (5) Evolutionary models. The actual determination of structure.

003.075 **Principles of plasma physics.**
N. A. Krall, A. W. Trivelpiece.
International Series in Pure and Applied Physics. McGraw-Hill Book Company, New York. 14 + 674 pp. Price $ 25.00 (1973).

003.076 **Nuclear reactions in cosmic bodies.**
A. K. Lavrukhina.
Nauka, Moskva. 256 pp. Price 2 Rbl. 62 Kop. (1972). In Russian. – Review in Referativ. Zhurn. 51. Astron., 3.51.52 (1973).

003.077 **Origins of astrology.**   J. Lindsay.
Frederick Muller, London. 480 pp. Price £ 4.00 (1971). – Review in Journ. History Astron., Vol. 4, 59; 1973 (*D. Pingree*).

003.078 **Al di là della luna. (Beyond the moon).**
P. Maffei.
Edizioni Scientifiche e Tecniche Arnoldo Mondadori, Milano, Italy. 315 pp. Price L 4000 (1973). – Reviews in Coelum, Vol. 41, 112 - 113 (1973); Sky Telescope, Vol. 45, 386 (1973).

003.079 **Energy distribution in the solar spectrum and the solar constant.**   E. A. Makarova, A. V. Kharitonov.
Nauka, Moskva. 288 pp. Price 1 Rbl. 78 Kop. (1972). In Russian. – Review in Referativ. Zhurn. 51. Astron., 7.51.361 (1973).

003.080 **Dall'astronomia alla astronautica.**   R. Masini.
Edit. Industrie Grafiche Nistri Lischi, Pisa. 175 pp. Price L 3000. – Review in Coelum, Vol. 41, 72 - 73; 1973 (*A. Betti*).

003.081 **The lunar rocks.**   B. Mason, W. G. Melson.
Translated from the English edition.

Mir. Moskva. 165 pp. Price 1 Rbl. 47 Kop. (1973). In Russian. – Reviews in Priroda, No. 6. 73, p. 122 (1973); Referativ. Zhurn. 51. Astron., 6.51.313 (1973).

003.082 L'univers relativiste. S. Mavridès.
Masson et Cie. S.A., Paris. 19 + 383 pp. Price F 198.00 (1973). – Review in Sky Telescope, Vol. 45, 386 (1973).

003.083 Meteorites and their origins. G. J. H. McCall.
John Wiley & Sons Inc., New York. 352 pp. Price $ 12.95 (1973). – Reviews in Science, Vol. 180, 1165; 1973 (*B. Mason*); Sky Telescope, Vol. 45, 387 (1973).

003.084 Science and controversy. A biography of Sir Norman Lockyer. A. J. Meadows.
The MIT Press, Cambridge, Mass. 9 + 331 pp. Price $ 16.95, £ 6.95 respectively (1972). – Essay review by H. Dingle in Journ. History Astron., Vol. 4, 131 - 133 (1973). – Reviews in Science, Vol. 179, 556 - 557; 1973 (*D. S. Evans*); Sky Telescope, Vol. 46, 39 - 40; 1973 (*J. A. Eddy*).

003.085 Physique et dynamique planétaires. Géodynamique. Vol. 4. P. Melchior.
Vander – éditeur, Louvain – Bruxelles. 8 + 257 pp. (1973). Contents: (1) Le freinage séculaire de la rotation de la terre; (2) Théorie de la précession – nutation et potentiel des marées; (3) Oscillations d'une enveloppe ellipsoidale en rotation contenant un fluide homogène; (4) Les oscillations libres de la terre; (5) Introduction à l'étude du noyau liquide; (6) Rotation et marées de la lune.

003.086 Zu neuen Horizonten. H. Mielke.
Transpress VEB Verlag für Verkehrswesen, Berlin. 3rd revised edition. 368 pp. Price MDM 22.00 (1972). – Review in Astron. in der Schule, 10. Jahrgang, p. 70; 1973 (*W. König*).

003.087 Earth – cosmos – moon.
S. N. Minchin, A. T. Ulubekov.
Mashinostroenie, Moskva. 244 pp. Price 46 Kop. (1972). In Russian. – Review in Referativ. Zhurn. 62. Issled. kosmich. prostranstva, 6.62.64 (1973).

003.088 Johannes Kepler, 1571 - 1630.
V. V. Mišković.
Beograd Srpska Akad. nauka i umet. 48 pp. (1972). In Serbo-Croatian. – Review in Referativ. Zhurn. 51. Astron., 4.51.12 (1973).

003.089 Can you speak Venusian? A guide to the independent thinkers. P. Moore.
David and Charles Ltd., Newton Abbot. 176 pp. Price £ 2.75 (1972). – Reviews in Journ. British Astron. Ass., Vol. 83, 232; 1973 (*C. A. Ronan*); Observatory, Vol. 93, 125; 1973 (*M. V. Penston*).

003.090 The sky at night 4. P. Moore.
British Broadcasting Corporation, (Publications Management), London, 184 pp. Price £ 2.25 (1972). – Review in Journ. British Astron. Ass., Vol. 83, 228; 1973 (*D. Howse*).

003.091 The southern stars. P. Moore.
Whitcombe and Tombs, Christchurch. 159 pp. Price $ 6.50 (1972). – Review in Southern Stars, Vol. 24, 153 - 155 (1973).

003.092 The story of astronomy. P. Moore.
Macdonald, London. 253 pp. Price £ 2.95 (1972). Review in Spaceflight, Vol. 15, 119; 1973 (*M. J. Anslow*).

003.093 How to recognize the stars. P. Moore, L. Clarke.
Corgi, London. 96 pp. Price 20p. (1972). – Review in Journ. British Astron. Ass., Vol. 83, 149; 1973 (*C. A. Ronan*).

003.094 Geology of the moon. T. A. Mutch.
Princeton University Press, Princeton, NJ. 391 pp. Price $ 22.50 (1972). – Review in Sky Telescope, Vol. 45, 386 (1973).

003.095 Asteroizi şi comete. V. Nadolschi.
Editura Albatros, Bucureşti. 289 pp. Price Lei 6.25 (1971).

003.096 Il libro dello spazio. H. E. Newell.
Edit. La Scuola, Brescia, 150 pp. Price L 1300. Review in Coelum, Vol. 41, 113; 1973 (*A. Betti*).

003.097 Find a falling star. H. H. Nininger.
P. S. Eriksson, Inc., New York; George J. McLeod Ltd., Toronto, Ont. 254 pp. Price $ 8.95 (1972). – Review in Journ. Roy. Astron. Soc. Canada, Vol. 67, 95 - 96; 1973 (*P. M. Millman*).

003.098 Introduction to stellar atmospheres & interiors. E. C. Novotny.
National Aeronautics and Space Administration, Manned Spacecraft Center Greenbelt, Maryland. 576 pp. Price $ 19.50 (1973). – Review in Sci. American, Vol. 227, No. 5, p. 128 (1973).

003.099 Some peculiarities of the Fourier spectra of basalts returned by Luna 16 and Apollo 11 according to results of investigations in the USSR and materials of the conference in Houston (USA). M. Omuraliev.
Redkollegiya zhurn. "Izv. AN Kirg SSR", Frunze. 11 pp. (1973). In Russian. – Review in Referativ. Zhurn. 62. Issled. kosmich. prostranstva, 6.62.167 (1973).

003.100 High-energy astrophysics.
L. M. Ozernoj, O. F. Prilutskij, I. L. Rozental'.
Atomizdat, Moskva. 1973. In Russian. – Review in Zemlya i Vselennaya, 1973, No. 3, p. 33.

003.101 Radio astrophysics. Non-thermal processes in galactic and extragalactic sources. A. G. Pacholczyk.
Translated from the English edition. Mir. Moskva. 252 pp. Price 1 Rbl. 71 Kop. (1973). In Russian. – Review in Referativ. Zhurn. 51. Astron., 6.51.42 (1973).

003.102 Introductory astronomy. N. A. Pananides.
Addison-Wesley Publishing Co. Inc., Reading, MA – London – Amsterdam. 344 pp. Price $ 12.50 (1973). – Review in Sky Telescope, Vol. 46, 44 (1973).

003.103 Problems of lunar geology. A. V. Peyve (Editor).
National Aeronautics and Space Administration. Superintendent of Documents, U. S. Government Printing Office, Washington, D. C. 421 pp. Price $ 6.00 (1973). – Review in Sky Telescope, Vol. 45, 387 (1973).

003.104 The opaque minerals in stony meteorites.
P. Ramdohr.
Elsevier Publishing Company, Amsterdam – London – New York. 245 pp. Price DM 68.75 (1973).

003.105 Astronomy data book. J. H. Robinson.
John Wiley & Sons Inc., New York; David & Charles Ltd., Newton Abbot. 271 pp. Price $ 10.95, £ 4.20 respectively (1972). – Review in Sky Telescope, Vol. 45, 249 (1973).

003.106 Das Fernrohr für Jedermann. H. Rohr.

Orell Füssli Verlag, Zürich. 5th edition. 265 pp.
Price DM 24.50 (1972). To be obtained from Treugesell-
Verlag, Düsseldorf. — Review in SuW, Vol. 12, 125; 1973
(*H. Vehrenberg*).

003.107 **Illustrated sources in history: Astronomy.**
C. A. Ronan.
David & Charles Ltd., Newton Abbot; Barnes & Noble (Harper
and Row) Inc., New York, NY. 112 pp. Price £ 3,25, $ 10.50
respectively (1973). — Reviews in Journ. British Astron. Ass.,
Vol. 83, 309; 1973 (*J. L. White*); Sky Telescope, Vol. 45, 313
(1973).

003.108 **Sterne und Planeten.** Sterne erkennen — Sterne
beobachten.
G. D. Roth; paintings: B. Damnitz.
BLV Verlagsgesellschaft, München — Bern — Wien. 232 pp.
Price DM 22.00 (1972). — Reviews in Sky Telescope, Vol. 45,
248 (1973); SuW, Vol. 12, 126; 1973 (*G. Klare*).

003.109 **The earth in the doctrine of Copernic.**
M. Rushinek.
Translated from the Polish edition. Nauka i izkustvo, Sofiya.
83 pp. Price Lv. 0.78 (1972). In Bulgarian.

003.110 **The background to Copernicus. The Copernican
Revolution.** C. A. Russell.
The Open University Arts Course: Renaissance and Reforma-
tion, Units 15-16. Open University Press and Eyre and Spottis-
woode, Walton Hall — London. 72 pp. Price $ 1.10 (1972).
Review in Journ. History Astron., Vol. 4, 139 - 140; 1973
(*O. Pedersen*).

003.111 **Ibn al-Haytham, al-Shukūk 'alā Batlamyūs (Dubita-
tiones in Ptolemaeum).**
A. Sabra, N. Shehaby (Editors), Prologue by I. Madkour.
The National Library Press, Cairo. 10 pp. English; 23 + 94 pp.
Arabic (1971). — Review in Journ. History Astron., Vol. 4,
138 - 139; 1973 (*B. R. Goldstein*).

003.112 **Geochemistry and cosmic chemistry of isotopes of
rare gases.** Yu. A. Shukolyukov, L. K. Levskij.
Atomizdat, Moskva. 336 pp. Price 2 Rbl. 4 Kop. (1972).
In Russian. — Review in Referativ. Zhurn. 51. Astron., 3.51.51
(1973).

003.113 **Physics of stellar interiors.** T. L. Swihart.
Intermediate short texts in astrophysics. Pachart
Publishing House, Tucson, Arizona. 10 + 119 pp. Price
$ 7.95 (1972). — Contents: (1) Introduction; (2) Radiation
theory; (3) Gas in thermodynamic equilibrium; (4) Polytropes; (5) Stellar energies; (6) Structure and evolution of the
stars; Appendixes.

003.114 **Meyers Handbuch über das Weltall.**
Compiled by K. Schaifers, G. Traving.
5th revised and enlarged edition. Bibliographisches Institut
(Meyers Lexikonverlag), Mannheim — Wien — Zürich. 780 pp.
Price DM 39.00 (1973). — Review in SuW, Vol. 12, 187 - 188;
1973 (*H. H. Voigt*).

003.115 **General astrophysics with elements of geophysics.**
J. S. Stodółkiewicz.
American Elsevier Publishing Company, Inc., New York —
London — Amsterdam. Translated from Polish by E. Lepa.
11 + 218 pp. Price Dfl. 36.00 (1973). — Contents: (1) Astronomical observations; (2) The motions of the earth; (3) The
physics of the solar system; (4) Stellar structure; (5) Galactic
structure; (6) Interstellar matter; (7) Extra-galactic astronomy.

003.116 **Variable stars.**

W. Strohmeier, edited by A. J. Meadows.
Pergamon Press, Oxford — New York — Braunschweig. 8 + 279
pp. Price DM 72.30 (1972). — Contents: (1) Variability and
high-energy astrophysics; (2) Variability in the form of lower-
energy outbursts; (3) Variability in young stars; (4) Variability
due to pulsation; (5) Semi-regular and irregular variability; (6)
Variability with extensive convection; (7) Variability due to
geometrical and physical factors; (8) Variability and magnetism;
(9) Variability of an entire galaxy; Appendices A — F.

003.117 **An introduction to the theory of plasma turbulence.**
V. N. Tsytovich.
Pergamon Press Ltd., Oxford. 160 pp. Price £ 3.75 (1972).
Review in Journ. British Interplanet. Soc., Vol. 26, 319 (1973).

003.118 **History of chronology.**
B. E. Tumanyan.
Yerevan State University Publishing House. 230 pp.(1972).
In Armenian with Russian Summary. — Review in Journ.
History Astron., Vol. 4, 137 - 138; 1973 (*K. A. Kazarian*).

003.119 **Nicolaus Copernicus.**
W. Thimm.
Verlag Gerhard Rautenberg, Leer (Ostfriesland, Germany).
80 pp. (1972).

003.120 **Einführung in die Astronomie.**
K. Thoene.
Hallwag, Bern — Stuttgart. 95 pp. (1971).

003.121 **Handbuch der Sternbilder. (Handbook of the con-
stellations).** H. Vehrenberg, D. Blank.
Treugesell-Verlag KG, Düsseldorf; Sky Publishing Corporation,
Cambridge, Mass. Second edition. 197 pp. Price DM 48.50,
$ 17.50 respectively. — Reviews in Orion Schaffhausen, 31.
Jahrgang, p. 71; 1973 (*E. Wiedemann*); Sky Telescope, Vol. 45,
313, 384 (1973); SuW, Vol. 12, 155 - 156; 1973 (*A. Kunert*).

003.122 **Erschröckliche und warhafftige Wunderzeichen
1543 - 1586.** B. Weber (Editor).
Urs Graf-Verlag, Zürich; American distributor: B. M. Rosenthal,
Inc. Price $ 336.00 (1972). — Review in Journ. History
Astron., Vol. 4, 143 (1973).

003.123 **Magic without magic.**
J. A. Wheeler: A collection of essays in honor of his
sixtieth birthday. J. R. Klauder (Editor).
Freeman & Sons, San Francisco. 14 + 492 pp. Price $ 19.50
(1972). — Review in Science, Vol. 180, 176; 1973 (*S. Weinberg*).

003.124 **Light scattering functions for small particles with
applications in astronomy.** N. C. Wickramasinghe.
Adam Hilger Ltd., London. 8 + 506 pp. Price £ 12.00 (1973).
Contents: Part I. Astrophysical background and light scattering theory (Prologue; Solid particles in space; Light scattering
by spherical particles; Light scattering by cylinders and ellip-
soids); Part II. Numerical results (Light scattering functions
for homogeneous spheres; Light scattering functions for infinite cylinders; Light scattering functions for spheres of ice,
iron and graphite; Extinction and scattering efficiencies for
composite grains). — Review in Nature, Vol. 243, 552; 1973
(*H. Seddon*).

003.125 **Physics of the magnetosphere.**
D. Williams, J. Mid.
Translated from the English edition. Mir, Moskva. 592 pp.
Price 3 Rbl. 66 Kop. (1972). In Russian. — Review in Refe-
rativ. Zhurn. 62. Issled. kosmich. prostranstva, 7.62.213
(1973).

003.126 **Progress in optics. Vol. 10.** E. Wolf (Editor).
North-Holland Publishing Company, Amsterdam –
London; Elsevier Publishing Company, Amsterdam – London
– New York. 16 + 394 pp. Price $ 30.00 (1973).

003.127 **Nicolaus Copernicus.** H. Wussing.
Urania-Verlag Leipzig–Jena–Berlin, Leipzig. 117 pp.
Price MDM 6.80 (1973).

003.128 **What is time?** G. J. Whitrow.
Thames and Hudson Ltd., London. Price £ 2.00
(1972). – Review in Nature, Vol. 242, 280; 1973 (*D. Layzer*).

003.129 **Von nun an bis in Ewigkeit. Die ewige Sekunde.**
G. J. Whitrow.
Translated from the English edition 'What is time?'. [Thames
and Hudson, London (1972)]. Translated by J. Knust, T.
Knust. Econ-Verlag, Düsseldorf – Wien. 222 pp. Price
DM 24.80 (1973).

003.130 **Nicolaus Copernicus.**
J. Dobrzycki, M. Biskup.
BSB B. G. Teubner Verlagsgesellschaft, Leipzig. 100 pp.
Price MDM 5.00 (1973).

003.131 **Sun and the amateur astronomer.** W. M. Baxter.
Newton Abbot, Devon, England: David & Charles.
165 pp. (1973). – Review in Phys. Abstracts, Vol. 76,
No. 31574 (1973).

003.132 **Gravitation waves in Einstein's theory of gravitation.** V. D. Zakharov.
Translated from the Russian edition (1972). John Wiley &
Sons Ltd., Chichester. 180 pp. Price £ 5.85 (1973). – Review in Journ. British Interplanet. Soc., Vol. 26, 446 (1973).

003.133 **Laser conquers the sky.** V. E. Zuev.
Zap-Sib. kn. izdatel'stvo, Novosibirsk. 191 pp.
Price 50 Kop. (1972). In Russian. – Review in Referativ.
Zhurn. 51. Astron., 5.51.46 (1973).

003.134 **Populated cosmos.**
Nauka, Moskva. 371 pp. Price 2 Rbl. 45 Kop.
(1972). In Russian. – Reviews in Referativ. Zhurn. 51. Astron.
7.51.5; 62. Issled. kosmich. prostranstva, 4.62.62 (1973).

003.135 **Interplanetary medium and physics of the magnetosphere.**
In-t kosmich. issled. AN SSSR. Nauka, Moskva. 212 pp.
Price 1 Rbl. 67 Kop. (1972). In Russian. – Review in Referativ. Zhurn. 62. Issled. kosmich. prostranstva, 5.62.282
(1973).

003.136 **Geomorphology. Abstracts of papers.**
Materialy Mosk. fil. Geogr. o-va SSSR, Moskva.
79 pp. Price 45 Kop. (1973). In Russian. – Review in Referativ. Zhurn. 51. Astron., 6.51.39 (1973).

003.137 **Problems of the history of mathematics and astronomy.**
Trudy Samarkand. un-ta, vyp. (No.) 229. Samarkand. 128 pp.
Price 42 Kop. (1972). In Russian.

003.138 **Stellar atmospheres and interplanetary plasma.**
Technical details of radio astronomical reception.
Trudy Fiz. in-ta. AN SSSR, Vol. 62, Nauka, Moskva. 199 pp.
Price 1 Rbl. 4 Kop. (1972). In Russian.

003.139 **The interplanetary medium and physics of the**
magnetosphere.
In-t kosmich. issled. AN SSSR. Nauka, Moskva. 212 pp.

Price 1 Rbl. 67 Kop. (1972). In Russian. – Review in Referativ. Zhurn. 51. Astron., 5.51.508 (1973).

003.140 **Problems of satellite astrometry.**
Sbornik statej. Nauch. inform. Astron. sovet. AN
SSSR, vyp. (No.) 22, 116 pp. Price 65 Kop. (1972).
In Russian.

003.141 **Essays on modern geochemistry and analytic**
chemistry. On the occasion of the seventy-fifth
anniversary of academician A. P. Vinogradov.
Institute of Geochemistry and Analytic Chemistry of the
USSR Academy of Sciences. Nauka, Moskva. 642 pp. Price
4 Rbl. 44 Kop. (1972). In Russian.

003.142 **Tables of minor planets.**
F. Pilcher, J. Meeus.
To be obtained from F. Pilcher, Illinois College, Jacksonville,
Ill. 104 pp. Price $ 4.00 (1973).

003.143 **Theoretical micro- and macrophysics.**
Sbornik nauch. rabot. Leningr. gos. ped. in-t im.
A. I. Gertsena. Leningrad. 152 pp. Price 67 Kop. (1972).
In Russian.

003.144 **Gravitation. Its problems and vistas. Dedicated to**
the memory of Aleksej Zinov'evich Petrov.
Naukova dumka, Kiev. 359 pp. Price 2 Rbl. 39 Kop. (1972).
In Russian.

003.145 **Automation of a transit instrument.**
Uch. zap. Latv. un-t, 169, Riga. 80 pp. Price
42 Kop. (1972). In Russian. – Review in Referativ. Zhurn.
51. Astron., 3.51.233 (1973).

003.146 **Yearbook of the books published in the USSR in**
1969. Systematic index, Vol. 2. Books on natural
sciences, technique and other sciences.
Kniga, Moskva. 1104 pp. Price 5 Rbl. 27 Kop. (1972).
In Russian.

003.147 **Science year – The world book science annual**
1973.
Field Enterprises Educational Corp., Chicago, Ill. 443 pp.
Price $ 7.95 (1972). – Review in Sky Telescope, Vol. 45, 48
(1973).

003.148 **Weigert/Zimmermann: Brockhaus ABC Astronomie.**
VEB F. A. Brockhaus-Verlag, Leipzig. 4th edition, revised by
H. Zimmermann. 464 pp. Price MDM 18.00 (1973).

003.149 **Eclipsing variable stars.**
V. P. Tsesevich.
Translated from the Russian edition (1972). John Wiley &
Sons Ltd., Chichester. 350 pp. (1973). – Review in Journ.
British Interplanet. Soc., Vol. 26, 446 (1973).

003.150 **Atlas of finding charts of variable stars.**
V. P. Tsesevich, M. S. Kazanasmas.
Nauka, Moskva. 350 pp. Price DM 57.50 (1971). – To be obtained from Treugesell-Verlag KG, Düsseldorf. – Reviews in
Orion Schaffhausen, 31. Jahrgang, p. 102; 1973 (*K. Locher*);
SuW, Vol. 12, 157; 1973 (*R. Lukas*).

003.151 **General bibliography of solar prominence research**
1880 - 1970. J. Kleczek, J.-L. Leroy, F. Q. Orrall.
Czechoslovak Academy of Sciences, Praha, Czechoslovakia.
2 + 151 pp. (1972). – Presents a bibliography of nearly 100
years of research with over 1300 references arranged in alpha-

betic order for author and according to a simple classification scheme.

003.152 **Astronomie für Sternfreunde.**
        W. Schroeder.
Franckh'sche Verlagshandlung/Kosmos-Verlag, Stuttgart.
180 pp. (1973).

003.153 **Die Erreichbarkeit der Himmelskörper.**
        W. Hohmann.
Sändig, Walluf (near Wiesbaden, Germany). 88 pp. (1973).

003.154 **Welt und Weltraum.**   R. Rietschel-Kluge.
        Delphin-Verlag, Stuttgart – Zürich. 93 pp. (1972).

003.155 **Forschen mit Kopernikus.**
        D. Guerrier.
Franckh'sche Verlagshandlung/Kosmos-Verlag, Stuttgart.
64 pp. (1973).

004 History of Astronomy, Chronology

004.001 **The Mesopotamian origin of early Indian mathematical astronomy.** D. Pingree.
Journ. History Astron., Vol. 4, 1 - 12 (1973).

004.002 **Megalithic yard or Megalithic myth?**
H. L. Porteous.
Journ. History Astron., Vol. 4, 22 - 24 (1973).

004.003 **New light on Tycho's instruments.**
V. E. Thoren.
Journ. History Astron., Vol. 4, 25 - 45 (1973).

004.004 **Astronomy archives in Australia (1).**
S. Mourot, D. J. Cross.
Journ. History Astron., Vol. 4, 66 - 68 (1973).

004.005 **Copernicus' place in the history of astronomy.**
E. Rosen.
Sky Telescope, Vol. 45, 72 - 75 (1973).

004.006 **Newton and the fudge factor.** R. S. Westfall.
Science, Vol. 179, 751 - 758 (1973).

004.007 **La genèse des lois de Kepler.** P. Russo.
L'Astronomie, 87ᵉ année, p. 1 - 17 (1973).

004.008 **Kepler et l'astronomie.** B. Morando.
L'Astronomie, 87ᵉ année, p. 33 - 34 (1973).

004.009 **L'aube de l'astronomie moderne: Nicolas Copernic.**
M. Tellier.
L'Astronomie, 87ᵉ année, p. 57 - 75 (1973).

004.010 **Copernicus und die Bezugssysteme in Physik und Himmelsmechanik.** H.-J. Treder.
Sterne, 49. Jahrgang, p. 15 - 22 (1973).

004.011 **Die Planetentheorie des Nicolaus Copernicus. Vorgeschichte und Inhalt.** J. Hoppe.
Sterne, 49. Jahrgang, p. 23 - 33 (1973).

004.012 **Die Astronomie in Wittenberg zur Zeit des Copernicus.** D. Wattenberg.
Sterne, 49. Jahrgang, p. 33 - 43 (1973).

004.013 **Kepler und die Begründung der Dynamik.**
H.-J. Treder.
Sterne, 49. Jahrgang, p. 44 - 48 (1973).

004.014 **De geschiedenis van de sterrenkunde (7).**
G. W. E. Beekman.
Hemel en Dampkring, Vol. 71, 10 - 20 (1973).

004.015 **De geschiedenis van de sterrenkunde (8).**
G. W. E. Beekman.
Hemel en Dampkring, Vol. 71, 55 - 59 (1973).

004.016 **Détermination astronomique du temps d'Hésiode.**
M. E. Dehousse.
Ciel et Terre, Vol. 89, 38 - 44 (1973).

004.017 **A note on Copernicus' instruments.** R. W. Tanner.
Journ. Roy. Astron. Soc. Canada, Vol. 67, 40 - 41 (1973).

004.018 **De astrolabe.** W. Kastelein.
Hemel en Dampkring, Vol. 71, 104 - 111 (1973).

004.019 **De geschiedenis van de sterrenkunde (9).**
G. W. E. Beekman.
Hemel en Dampkring, Vol. 71, 111 - 116 (1973).

004.020 **Die Entdeckungen des Copernicus. Zur 500. Wiederkehr seines Geburtstages.** H.-B. Brenske.
SuW, Vol. 12, 36 - 39 (1973).

004.021 **Das astrologische Mittelalter. Über die Wissenschaft im Jahrhundert des Copernicus.** G. D. Roth.
SuW, Vol. 12, 40 - 42 (1973).

004.022 **From ancient ideas to the Copernican heliocentric system of the world.** I. N. Veselovskij.
Zemlya i Vselennaya, 1973, No. 1, p. 47 - 51. In Russian.

004.023 **Astronomy at the Cracow University of the 15th century.** E. Rybka.
Zemlya i Vselennaya, 1973, No. 1, p. 58 - 62. In Russian.

004.024 **Die Entwicklung des Weltbildes in der Antike.**
F. Jürss.
Nicolaus Copernicus − 1473−1973, (see 003.004), p. 21 - 51 (1973).

004.025 **Das Weltbild der arabischen Astronomie.**
G. Strohmaier.
Nicolaus Copernicus − 1473−1973, (see 003.004), p. 53 - 68 (1973).

004.026 **Die gesellschaftlichen Bedingungen für das Wirken von Nicolaus Copernicus und die philosophisch-weltanschaulichen Voraussetzungen und Wesenszüge seiner Lehre.** H. Mielke.
Nicolaus Copernicus − 1473−1973, (see 003.004), p. 69 - 116 (1973).

004.027 **Zu den weltanschaulichen Auslassungen in dem Hauptwerk von Nicolaus Copernicus.** H. Ley.
Nicolaus Copernicus − 1473−1973, (see 003.004), p. 117 - 131 (1973).

004.028 **Die Weiterentwicklung des Copernicanischen Weltbildes und seine Stellung in den Auseinandersetzungen am Ende des 16. und während des 17. Jahrhunderts. Johannes Kepler, Galileo Galilei und René Descartes.**
O. Günther.
Nicolaus Copernicus − 1473−1973, (see 003.004), p. 133 - 146 (1973).

004.029 **Das Copernicanische Weltbild und die Arbeiten Otto von Guerickes.** A. Kauffeldt.
Nicolaus Copernicus − 1473−1973, (see 003.004), p. 147 - 160 (1973).

004.030 **Die experimentelle Bestätigung des Copernicanischen Weltbildes.** K.-H. Hintze.
Nicolaus Copernicus − 1473−1973, (see 003.004), p. 161 - 165 (1973).

004.031 **Einige wissenschaftliche und erkenntnistheoretische Aspekte der Zeit Isaak Newtons und ihre Auswirkungen bis in die Gegenwart.** G. Jackisch.
Nicolaus Copernicus − 1473−1973, (see 003.004), p. 167 - 178 (1973).

004.032 **Bibliographisches zu "De revolutionibus".**

H. Mielke.
Nicolaus Copernicus − 1473−1973, (see 003.004), p. 199 - 205 (1973).

004.033 Nicolaus Copernicus' minor commentary on the hypotheses of celestial motions established by him.
Translated from Latin by I. N. Veselovskij.
Priroda, No. 2.73, p. 10 - 11 (1973). In Russian.

004.034 Copernicus and planetary astronomy.
I. N. Veselovskij.
Priroda, No. 2.73, p. 12 - 21 (1973). In Russian.

004.035 Studies on Genkareki or Yüan-chia-li. M. Utida.
Tokyo Astron. Obs., Report No. 61, Vol. 16, 416 - 423 (1973). In Japanese.

004.036 Galileo Gleanings. XXI: On the probable order of Galileo's notes on motion. S. Drake.
Physis, Vol. 14, 55 - 68 (1972).
The extent of Galileo's work on motion completed at Padua before 1610 is revealed by a study of watermarks on papers used for his notes over about 30 years, developing theorems published in the "Discorsi" of 1638. The parabolic trajectory is assignable to 1608−09. Some evidence of experiments appears in unpublished notes. Galileo planned to publish a treatise on motion in mid-1609, interrupted by the telescope, and again in 1617−18, interrupted by the comets of 1618. Watermark descriptions are given for most of Galileo's dated letters.

004.037 Datos sobre historia de la astronomía.
A. E. Olivares.
Bol. Acad. Ci. Fis., Mat., Nat., Vol. 31, No. 93, p. 29 - 38 (1971/73).

004.038 Die astronomische Datierung von Kunstwerken. II.
A. Beer.
Sterne, 49. Jahrgang, p. 84 - 104 (1973).

004.039 Zur Geschichte der Astronomie in Berlin im 16. bis 18. Jahrhundert. II. Eine Quellenübersicht.
D. Wattenberg.
Sterne, 49. Jahrgang, p. 104 - 116 (1973).

004.040 Early sundials and the discovery of the conic sections. W. W. Dolan.
Math. Mag., Vol. 45, 8 - 12 (1972). − Abstr. in Zentralblatt Math. Grenzgebiete, Vol. 242, No. 01003 (1973).

004.041 Copernicus' revolution. W. Zonn.
Simpozion Copernic, (see 012.009), p. 15 - 17 (1973). In Romanian.

004.042 Nicolaus Copernicus and his time. C. Drâmbă.
Simpozion Copernic, (see 012.009), p. 19 - 27 (1973). In Romanian.

004.043 Copernicus and Renaissance philosophy.
C. Ionescu-Gulian.
Simpozion Copernic, (see 012.009), p. 29 - 32 (1973). In Romanian.

004.044 Copernicus and the astronomy in Transylvania in the XVIth century. A. Dankanits.
Simpozion Copernic, (see 012.009), p. 33 - 34 (1973). In Romanian.

004.045 Nicolaus Copernicus, precursor of a new mechanics.
C. Iacob.

Simpozion Copernic, (see 012.009), p. 35 - 48 (1973). In Romanian.

004.046 Levels of enlightenment. From Tycho Brahe to Einstein. D. Andrews.
Spaceflight, Vol. 15, 222 - 227 (1973).

004.047 The spheres of Eudoxus. E. G. Forbes.
Journ. British Astron. Ass., Vol. 83, 196 - 198 (1973).

004.048 Nicolaus Copernicus und seine Beobachtungen. Zum 500. Geburtstag des großen Astronomen am 19. Februar 1973. D. Wattenberg.
Blick in das Weltall, Archenhold-Sternw. Berlin-Treptow, 21. Jahrgang, p. 13 - 29 (1973).

004.049 Eudoxus encircled. E. Maula.
Yearbook Philos. Soc. Finland, Vol. 33, 201 - 253 (1971). − Abstr. in Zentralblatt Math. Grenzgebiete, Vol. 245, No. 01001 (1973).

004.050 Adriaan van Maanen's influence on the island universe theory: Part 2.
R. Berendzen, R. Hart.
Journ. History Astron., Vol. 4, 73 - 98 (1973).

004.051 al-Khalīlī's auxiliary tables for solving problems of spherical astronomy. D. A. King.
Journ. History Astron., Vol. 4, 99 - 110 (1973).

004.052 A megalithic lunar observatory in Orkney: The Ring of Brogar and its cairns. A. Thom, A. S. Thom.
Journ. History Astron., Vol. 4, 111 - 123 (1973).

004.053 The Aberdeen copy of Copernicus's Commentariolus. J. Dobrzycki.
Journ. History Astron., Vol. 4, 124 - 127 (1973).

004.054 Copernicus and Naṣīr al-Dīn al-Ṭūsī.
I. N. Veselovsky.
Journ. History Astron., Vol. 4, 128 - 130 (1973). − Note.

004.055 Copernico e l'eliocentrismo. R. Migliavacca.
Coelum, Vol. 41, 45 - 57 (1973).

004.056 Duizend Paasdata kritisch bekeken (van 1583 tot 2582). S. Verhezen.
Hemel en Dampkring, Vol. 71, 130 - 132 (1973).

004.057 De frequentie van de Paasdata. J. Meeus.
Hemel en Dampkring, Vol. 71, 132 - 135 (1973).

004.058 De geschiedenis van de sterrenkunde (10).
G. W. E. Beekman.
Hemel en Dampkring, Vol. 71, 135 - 141 (1973).

004.059 De comput. J. Meeus.
Hemel en Dampkring, Vol. 71, 141 - 144 (1973).

004.060 L'importance des textes cunéiformes antiques pour la mécanique céleste moderne et les recherches historiques de G. V. Schiaparelli. J. O. Fleckenstein.
Mem. Soc. Astron. Italiana, Nuova Ser., Vol. 43, 745 - 750 (1973).
La mécanique céleste determine les élements orbitaux perturbés sous la forme d'une série de Taylor. Depuis un longue intervalle du temps (4×10^3 ans) sont encore utilisables mêmes les observations de l'antiquité d'une précision de $\pm 0°.1$. G. V. Schiaparelli était un des précurseurs qui ont souligné l'importance des observations assyro-babylonéennes

dans l'écriture cunéiforme (élongations de Venus) pour l'astronomie moderne.

004.061 Copernican truth. N. I. Idel'son.
Astron. vestn., Vol. 7, 3 - 8 (1973). In Russian.

004.062 Wann sind seit der Geburt des Nicolaus Copernicus genau 500 Jahre verflossen? J. Wempe.
Blick in das Weltall, Archenhold-Sternw., Berlin-Treptow, 21. Jahrgang, p. 47 - 48 (1973).

004.063 Nicolaus Copernicus, astronomer and scholar, 1473 - 1543. J. L. Perdrix.
Journ. Astron. Soc. Victoria, Vol. 26, 18 - 27 (1973). – The third Philipp Simon lecture delivered at the general meeting of the Society, 1973 March 15.

004.064 Nicolaus Copernicus and modern science. W. Iwanowska.
Journ. Roy. Astron. Soc. Canada, Vol. 67, 105 - 114 (1973). An address delivered to several Centres of the R. A. S. C. in January, 1973 as part of the celebrations of the 500th anniversary of the birth of Copernicus.

004.065 Nicolaus Copernicus who set the earth whirling. G. Rodionova.
Nauka i zhizn', 1973, No. 1, p. 10 - 15. In Russian. – Abstr. in Referativ. Zhurn. 51. Astron., 5.51.8 (1973).

004.066 Zodiaco clasico, incaico y azteca; similitud y diferencias. L. E. Arochi.
El Universo, No. 102, Vol. 27, 9 - 10 (1973).

004.067 La astronomía en el México Antiguo, primera parte. J. G. Hernández.
El Universo, No. 102, Vol. 27, 21 - 23 (1973).

004.068 Astronomical photography. A brief history. D. Calder.
Southern Stars, Vol. 24, 142 - 152 (1973).

004.069 Astronomy and astronomers at the mountain observatories. A. E. Whitford.
Ann. New York Acad. Sci., Vol. 198, (see 012.020), 202 - 210 (1972).

004.070 The early history of radio astronomy. G. Westerhout.
Ann. New York Acad. Sci., Vol. 198, (see 012.020), 211 - 218 (1972).

004.071 On the Mishima–Goyomi for the 13th year of Kan'yei. S. Kanda.
Mem. Japan Astron. Study Ass., No. 19, Vol. 5, 181 - 196 (1972). In Japanese.

004.072 Has Copernicus had an astronomical observatory? J. Classen.
Urania Kraków, Vol. 44, 9 - 12 (1973). In Polish.

004.073 Before Copernicus came. T. Z. Dworak.
Urania Kraków, Vol. 44, 13 - 16 (1973). In Polish.

004.074 At the source of the great Copernicus ideas. E. Rybka.
Urania Kraków, Vol. 44, 34 - 39 (1973). In Polish.

004.075 Where did Nicolaus Copernicus live and observe? J. Pagaczewski.
Urania Kraków, Vol. 44, 46 - 51, 73 - 81, 104 - 110, 144 - 150 (1973). In Polish.

004.076 Historical chronicle.
Urania Kraków, Vol. 44, 91 - 92, 122 - 125, 152 - 156, 185 - 187 (1973). In Polish.

004.077 Concerning the philosophical estimation of Newton's theoretical heritage. G. U. Likhosherstnykh.
Nauch. dokl. vyssh. shkoly. Filos. n., 1973, No. 1, p. 97 - 107. In Russian. – Abstr. in Referativ. Zhurn. 51. Astron., 6.51.1 (1973).

004.078 Short survey on the history of development of astronomy in the medieval East before the Ulugh Begh epoch. A. U. Usmanov.
Trudy Samarkand. un-ta, 1972, vyp. (No.) 229, p. 60 - 97. In Russian. – Abstr. in Referativ. Zhurn. 51. Astron., 6.51.6 (1973).

004.079 al-Ferghani's proof of the fundamental theorem on the stereographic projection.
N. D. Sergeeva, L. M. Karpova.
Vopr. istorii estesvozn. i tekhn. Vyp. (No.) 3 (40). Moskva, Nauka, 1972, p. 50 - 53. In Russian. – Abstr. in Referativ. Zhurn. 51. Astron., 6.51.7 (1973).

004.080 About the astronomical treatise by Dzhagmini. (Brief essay). Z. A. Pashaev.
Trudy Samarkand. un-ta, 1972, vyp. (No.) 229, p. 24 - 26. In Russian. – Abstr. in Referativ. Zhurn. 51. Astron., 6.51.8 (1973).

004.081 On a comment of al-Birdgandi to astronomical tables of Ulugh Begh. (Preliminary communication).
A. E.-A. Khatipov.
Trudy Samarkand. un-ta, 1972, vyp. (No.) 229, p. 119. In Russian. – Abstr. in Referativ. Zhurn. 51. Astron., 6.51.11 (1973).

004.082 The role of scientists of Ulugh Begh's Samarkand astronomical school in the development of astronomy. A. U. Usmanov.
Trudy Samarkand. un-ta, 1972, vyp. (No.) 229, p. 101 - 118. In Russian. – Abstr. in Referativ. Zhurn. 51. Astron., 6.51.12 (1973).

004.083 Johannes Kepler: from "Mysterium" to "Harmonia". Yu. A. Danilov, Ya. A. Smorodinskij.
Uspekhi fiz. nauk, Vol. 109, 175 - 209 (1973). In Russian. Abstr. in Referativ. Zhurn. 51. Astron., 6.51.22 (1973).

004.084 Kepler's works in optics. (To the 400th birthday). V. P. Linnik.
Uspekhi fiz. nauk, Vol. 109, 167 - 174 (1973). In Russian. Abstr. in Referativ. Zhurn. 51. Astron., 6.51.25 (1973).

004.085 Dissemination of the heliocentric doctrine in Armenia. B. E. Tumanyan.
Vest. Erevan. un-ta. Obshchestv. n., 1972, No. 3 (18), p. 120 - 127. In Armenian. – Abstr. in Referativ. Zhurn. 51. Astron., 6.51.26 (1973).

004.086 Copernicus' heliocentric system. V. Vanýsek.
Říše hvězd, Vol. 54, 105 - 109 (1973). In Czech.

004.087 L'astronomia araba e la sua diffusione. M. Cimino.
Reprinted from 13° Convegno Volta, Accad. Nazionale Lincei, Roma – Firenze 1969, p. 647 - 674 = Oss. Astron. Roma, Contr. Sci., Ser. III, No. 100 (1971).

004.088 Astronomische Uhrgenauigkeit vor 3000 Jahren. K. Ferrari d'Occhieppo.
Sternenbote, 16. Jahrgang, p. 2 - 8 (1973).

004.089 Zum Kopernikus-Jahr 1973. Kometenbewegung und nichtgravitationelle Effekte. H. Mucke.
Sternenbote, 16. Jahrgang, p. 22 - 29 (1973).

004.090 Die internationalen Sternwarten vor 100 Jahren. J. Classen.
Veröff. Sternw. Pulsnitz (Sachsen), No. 9, 23 pp. (1972). – Reprints of some papers in 'Sterne', Vols. 47, 48 (1971/72).

004.091 Die Astronomie des Copernicus. F. Schmeidler.
Separate print from 'Nicolaus Copernicus zum 500. Geburtstag' [Böhlau Verlag, Köln – Wien], 21 pp. (1973).

004.092 Die Ausbreitung neuer Erfindungen in der neuzeitlichen Astronomie. F. Schmeidler.
Separate print from 'RETE – Strukturgeschichte der Naturwissenschaften' [Verlag Dr. H. A. Gerstenberg, Hildesheim], Vol. 1, 117 - 124 (1972).

004.093 Entstehung und Wirkung der heliozentrischen Lehre von Copernicus. F. Schmeidler.
Naturwiss. Rundschau, [Wiss. Verlagsgesellschaft, Stuttgart], Vol. 26, 53 - 57 (1973).

004.094 Search for direct evidence for the motions of the earth. A. A. Mikhajlov.
Zemlya i Vselennaya, 1973, No. 3, p. 50 - 53. In Russian.

004.095 Notes on the Maya calendar. Year of 365 days. Yu. V. Knorozov.
Sov. ehtnografiya, 1973, No. 1, p. 70 - 80. In Russian.
Abstr. in Referativ. Zhurn. 51. Astron., 7.51.7 (1973).

004.096 Nicolas Copernic – to the 500th anniversary of his birthday. H. Leśniok.
Geod. i kartografiya, 1973, No. 2, p. 61 - 66. In Russian.
Abstr. in Referativ. Zhurn. 51. Astron., 7.51.13 (1973).

004.097 Nicolas Copernic, 1473 – 1543. A. Markushevich.
Nar. obrazovanie, 1973, No. 3, p. 82 - 86. In Russian.
Abstr. in Referativ. Zhurn. 51. Astron., 7.51.14 (1973).

004.098 On the genesis of "De revolutionibus" of Copernic. I. N. Veselovskij.
Vopr. istorii estestvozn. i tekhn. Vyp. (No.) 1 (42). Moskva, Nauka, 1973, p. 9 - 16, 94. In Russian. – Abstr. in Referativ. Zhurn. 51. Astron., 7.51.16 (1973).

004.099 On the essential features of the Copernican revolution in the history of human thought. V. Vuaze.
Vopr. istorii estestvozn. i tekhn. Vyp. (No.) 1 (42). Moskva, Nauka, 1973, p. 16 - 22, 94. In Russian. – Abstr. in Referativ. Zhurn. 51. Astron., 7.51.17 (1973).

004.100 N. Copernic's doctrine in Lithuania. P. V. Slavenas.

Vopr. istorii estestvozn. i tekhn. Vyp. (No.) 1 (42). Moskva, Nauka, 1973, p. 23 - 25, 94. In Russian. – Abstr. in Referativ. Zhurn. 51. Astron., 7.51.20 (1973).

004.101 D. I. Mendeleev about N. Copernic. A. A. Makarenya.
Vopr. istorii estestvozn. i tekhn. Vyp. (No.) 1 (42). Moskva, Nauka, 1973, p. 26 - 27, 94. In Russian. – Abstr. in Referativ. Zhurn. 51. Astron., 7.51.26 (1973).

004.102 The Nicolas Copernic-year.
Vopr. istorii estestvozn. i tekhn. Vyp. (No.) 1 (42). Moskva, Nauka, 1973, p. 3 - 8, 94. In Russian. – Abstr. in Referativ. Zhurn. 51. Astron., 7.51.61 (1973).

Nicolaus Copernicus zum 500. Geburtstag. See Abstr. 003.014.

Polskie Miasta Kopernika. See Abstr. 003.021.

Świat Kopernika. See Abstr. 003.034.

New and full moons: 1001 B. C. to A. D. 1651. See Abstr. 003.054.

From Stonehenge to modern cosmology. See Abstr. 003.064.

Astronomie Mikuláše Kopernika. See Abstr. 003.066.

Origins of astrology. See Abstr. 003.077.

Dall'astronomia alla astronautica. See Abstr. 003.080.

The story of astronomy. See Abstr. 003.092.

Illustrated sources in history: Astronomy. See Abstr. 003.107.

The background to Copernicus. The Copernican Revolution. See Abstr. 003.110.

History of chronology. See Abstr. 003.118.

Erschröckliche und warhafftige Wunderzeichen 1543 - 1586. See Abstr. 003.122.

Problems of the history of mathematics and astronomy. See Abstr. 003.137.

On the near approach of Mars in 1877 and the so-called "Saigo" star. See Abstr. 097.038.

005 Biography

005.001 The American Kepler: Daniel Kirkwood and his analogy. R. L. Numbers.
Journ. History Astron., Vol. 4, 13 - 21 (1973).

005.002 Adriaan van Maanen's influence on the island universe theory: Part 1. R. Berendzen, R. Hart.
Journ. History Astron., Vol. 4, 46 - 56 (1973).

005.003 The adventures of C. H. F. Peters – I, II.
J. Ashbrook.
Sky Telescope, Vol. 45, 90 - 91, 152 - 153 (1973).

005.004 Ira Sprague Bowen (1898 - 1973). O. C. Wilson.
Sky Telescope, Vol. 45, 212 - 214 (1973).

005.005 Copernicus published as he perished.
E. Rosen.
Nature, Vol. 241, 433 - 434 (1973).
Nicholas Copernicus was born on February 19, 1473, yet his great work *Revolutions* was only published in March 1543, two months before he died.

005.006 Leben und Persönlichkeit des Nicolaus Copernicus.
F. Schmeidler.
Sterne, 49. Jahrgang, p. 1 - 14 (1973).

005.007 M. L. Humason – Some personal recollections.
N. U. Mayall.
Mercury, (Journ. Astron. Soc. Pacific), Vol. 2, No. 1, p. 3 - 8, 22 (1973).

005.008 Copernicus 1473–1543. A. V. Douglas.
Journ. Roy. Astron. Soc. Canada, Vol. 67, 1 - 7 (1973).

005.009 Mikołaj Kopernik. W. Iwanowska.
Postępy Astron., Vol. 21, 3 - 7 (1973).

005.010 H. C. Urey – Octogenarius. Special issues, dedicated to Professor Harold C. Urey on the occasion of his 80th birthday.
The Moon, Vol. 7, Nos. 1/2, 3/4, 4 + 504 pp. (1973). – The individual contributions are included in their corresponding subject categories – see abstracts 034.007, 081.005, 091.008, 094.041 - 094.053, 095.001; 091.060, 094.575 - 094.587, 105.137, 107.015.

005.011 Copernicus and his long revolution. J. R. Ravetz.
Endeavour, No. 116, Vol. 32, 57 - 59 (1973).

005.012 Johannes Kepler (1571–1630). On the occasion of the 400th birthday. Yu. Belij, V. Metev.
Fiz.-matem. spisanie, Vol. 15, No. 1, p. 34 - 58 (1972). In Bulgarian. – Abstr. in Referativ. Zhurn. 51. Astron., 3.51.4 (1973).

005.013 Great Copernicus. A. A. Gurshtejn.
Zemlya i Vselennaya, 1973, No. 1, p. 53 - 57. In Russian.

005.014 Polish country – country of Copernicus.
G. V. Cheremushkin.
Zemlya i Vselennaya, 1973, No. 1, p. 63 - 64. In Russian.

005.015 Nikolaus Kopernikus. Zur 500. Wiederkehr seines Geburtstages am 19. Februar 1973. E. Krug.
Orion Schaffhausen, 31. Jahrgang, p. 39 - 44 (1973).

005.016 A great reformer of astronomy. On the occasion of the 500th anniversary of Nicolaus Copernicus' birthday. A. A. Mikhailov.
Priroda, No. 2.73, p. 2 - 9 (1973). In Russian.

005.017 Nicolaus Copernicus' portraits. F. K. Velichko.
Priroda, No. 2.73, p. 126 - 128 (1973). In Russian.

005.018 Richard C. Carrington. M. Barnes.
Journ. British Astron. Ass., Vol. 83, 122 - 124 (1973).

005.019 Johannes Kepler and Bohemia before the battle of White Mountain. J. Hanzal.
Dějiny Věd Techn., 1971 (4), p. 1 - 12 (1971). In Czech. Abstr. in Zentralblatt Math. Grenzgebiete, Vol. 244, No. 01004 (1973).

005.020 Johann Bayer and his star nomenclature.
J. Ashbrook.
Sky Telescope, Vol. 45, 292 - 294 (1973).

005.021 Shapley und das Sternsystem.
O. Heckmann.
SuW, Vol. 12, 100 - 103 (1973).

005.022 Harlow Shapley – A tribute to a great man.
F. W. Wright.
Mercury, (Journ. Astron. Soc. Pacific), Vol. 2, No. 2, p. 3 - 4 (1973).

005.023 Franz Timerman and the astronomical determination of Moscow's longitude. Yu. Kh. Kopelevich.
Priroda, No. 4.73, p. 90 - 93 (1973). In Russian.

005.024 Semnificaţia generală a operei lui N. Copernic.
C. Popovici.
Stud. Cerc. Astron., Vol. 18, 3 - 5 (1973).

005.025 Nicolae Copernic. 500 de ani de la naştere.
C. Drâmbă.
Stud. Cerc. Astron., Vol. 18, 7 - 15 (1973).

005.026 Un grand astronome: Harlow Shapley (1885 - 1972). A. Brun.
L'Astronomie, 87ᵉ année, p. 209 - 212 (1973).

005.027 Portraet af en schweizisk amatørastronom, R. A. Naef. P. Darnell.
Astron. Tidssk., Årg. 6, p. 78 - 80 (1973).

005.028 The eagle eye of William Rutter Dawes.
J. Ashbrook.
Sky Telescope, Vol. 46, 27 - 28 (1973).

005.029 Thomas Young, 1773–1829.
A. V. Douglas.
Journ. Roy. Astron. Soc. Canada, Vol. 67, 150 - 151 (1973).

005.030 Second profession of a great astronomer.
O. Korottsev.
Nauka i zhizn', 1973, No. 1, p. 14 - 15. In Russian. – Abstr. in Referativ. Zhurn. 51. Astron., 5.51.10 (1973).

005.031 Einblick in das Leben und Werk von Friedrich Wilhelm Herschel. D. B. Herrmann.
Astron. in der Schule, 10. Jahrgang, p. 55 - 57 (1973).

005.032 Copernicus, how did he look like?
S. R. Brzostkiewicz.
Urania Kraków, Vol. 44, 51 - 56 (1973). In Polish.

**005.033 The mathematician and astronomer Nasir at-Din
at-Tusi.   U. Ataev.**
Trudy Samarkand. un-ta, 1972, vyp. (No.) 229, p. 119 - 123.
In Russian. — Abstr. in Referativ. Zhurn. 51. Astron., 6.51.9
(1973).

**005.034 Kazi-Zade at-Rumi's comment on the astronomical
treatise of Nari at-Din at-Tusi.   U. Ataev.**
Trudy Samarkand. un-ta, 1972, vyp. (No.) 229, p. 124 - 127.
In Russian. — Abstr. in Referativ. Zhurn. 51. Astron., 6.51.10
(1973).

005.035 N. Copernicus – half a millenium in retrospect.
J. M. Mohr.
Říše hvězd, Vol. 54, 25 - 30 (1973). In Czech.

005.036 M. Kopernik.   Ľ. Pajdušáková.
Kozmos, Vol. 4, 1 - 4 (1973). In Slovak.

**005.037 Professor M. G. Pereira de Barros and the Astrono-
mical Observatory of the University of Porto.**
J. Osório.
Anais Faculdade Ciências do Porto, Vol. 55, Fasc. 1/2, 20 pp.
= Publ. Obs. Astron. Porto, No. 26 (1972).

005.038 E. J. Delporte, astronome, directeur de l'Observa-
toire royal de Belgique.   S. Arend.
Biographie Nationale publiée par l'Academie Royale des
Sciences, des Lettres et des Beaux-Arts de Belgique, Vol. 37,
Fasc. 1, p. 205 - 211 (1971).

Nicolaus Copernicus zum 500. Geburtstag.
See Abstr. 003.014.

Albert Einstein. Creator and rebel.
See Abstr. 003.062.

Copernicus und seine Welt. Biographie.
See Abstr. 003.071.

**Science and controversy. A biography of Sir Nor-
man Lockyer.   See Abstr. 003.084.**

Johannes Kepler, 1571 - 1630.
See Abstr. 003.088.

Nicolaus Copernicus.
See Abstr. 003.119.

Nicolaus Copernicus.   See Abstr. 003.127.

Nicolaus Copernicus.   See Abstr. 003.130.

Nicolaus Copernicus und seine Beobachtungen.
Zum 500. Geburtstag des großen Astronomen am 19. Februar
1973.   See Abstr. 004.048.

006 Personal Notes

W. Becker received the honory degree of D. Sc.
G. A. Tammann.
Orion Schaffhausen, 31. Jahrgang, p. 27 (1973).

L. Biermann received the Emil-Wiechert-Medal.
Phys. Blätter, 29. Jahrgang, p. 194 (1973).

H. C. Freiesleben was elected into honorary membership of the Royal Institute of Navigation, London.
Journ. Navigation, Vol. 26, 125 - 126 (1973).

H. Keres, 60th birthday.
Izv. AN EhstSSR. Fiz., mat., Vol. 21, 458 - 459 (1972). In
Russian. — Abstr. in Referativ. Zhurn. 51. Astron., 5.51.17
(1973).

K. A. Kulikov, 70th birthday.
Astron. vestn., Vol. 7, 54 - 55 (1973). In Russian.

B. Yu. Levin received the Kepler Gold Medal of the
USA.
Priroda, No. 4.73, p. 109 (1973). In Russian.

B. Mason received the Leonard Medal of the Meteoritical Society.
Meteoritics, Vol. 7, 611 - 612 (1972).

M. J. McCutcheon received the Gold Medal of the
Roy. Astron. Soc. Canada.
Journ. Roy. Astron. Soc. Canada, Vol. 67, 100 (1973).

J.-C. Pecker received la médaille de l'Université
de Nice.
Bull. d'Information, Ass. Développement International Obs.
Nice, No. 9, p. 11 - 19 (1972).

L. Perek received the 10th annual medal of the
A.D.I.O.N.
Bull. d'Information, Ass. Développement International Obs.
Nice, No. 9, p. 41 - 43 (1972).

J. Reynolds received the Leonard Medal of the
Meteoritical Society.
Meteoritics. Vol. 8, 90 (1973).

H. C. Urey, 80th birthday.   Z. Kopal.
The Moon, Vol. 7, Nos. 1/2, p. II - IV (1973). In Latin.

007 Obituaries

R. L. Aikens died 1972 September 20.
M. W. Burke-Gaffney.
Journ. Roy. Astron. Soc. Canada, Vol. 67, L2 (1973).

G. B. van Albada, 1911 March 28 - 1972 December 18.   J. H. Oort, T. de Groot.
Hemel en Dampkring, Vol. 71, 47 - 48 (1973).

G. B. van Albada died on December 18, 1972.
Sky Telescope, Vol. 45, 160 (1973).

J. (G.) Alter died 1972 October 30.
Říše hvězd, Vol. 54, 11 (1973).   In Czech.

I. S. Bowen died 1973, Februar 6.
Publ. Astron. Soc. Pacific, Vol. 85, 174 (1973).

I. S. Bowen died 1973 February 6.
Science, Vol. 180, 1158 (1973).

I. S. Bowen, 1898 - 1973 February 6th.
O. C. Wilson.
Sky Telescope, Vol. 45, 212 - 214 (1973).

R. C. Cameron died on December 13, 1972.
Sky Telescope, Vol. 45, 158 (1973).

J. Cox, 1898 August 16 - 1972 October 20.
P. Melchior.
Ciel et Terre, Vol. 89, 69 - 79 (1973).

V. A. Dombrovskij, 1914 - 1972 February 1.
Trudy Astron. Obs., *Leningrad,* Vol. 29 (= Uchenye Zapiski Leningr. Un-ta, No. 363 = Seriya Matem. Nauk, vyp. (No.) 48), p. 3 - 5 (1973). In Russian.

P. W. Gast died 1973 May 16.
Sky Telescope, Vol. 46, 9 (1973).

K. A. Grigoryan, 1928 - 1970 August 29.
Soobshch. Byurakan. Obs., vyp. (No.) 44, p. 137 - 138 (1972). In Russian.

O. Günther, 1911 June 8 - 1973 January 8.
Astron. in der Schule, 10. Jahrgang, p. 3 (1973).

C. Doris Hellman, 1910 - 1973.
Journ. History Astron., Vol. 4, 136 (1973).

C. D. Hellman died 1973 March 3.
Sky Telescope, Vol. 45, 278 (1973).

M. L. Humason, 1892 - 1972.   E. Wiedemann.
Orion Schaffhausen, 31. Jahrgang, p. 15 (1973).

A. H. Joy 1882 September 23 – 1973 April 18.
Sky Telescope, Vol. 46, 3,9 (1973).

A. Kahrstedt, 1897 August 24 - 1971 January 11.
F. Gondolatsch.
Astron. Nachr., Vol. 294, 147 - 148 (1973).

M. Kamieński, 1879 - 1973.   L. Zajdler.
Urania Kraków, Vol. 44, 162 - 166 (1973). In Polish.

K. Ledersteger, 1900 Nov. 11 - 1972 Sept. 24.
M. Kneissl.
Bull. Géod., Nouvelle Sér., Année 1973, No. 108, p. 114.

K. Ledersteger died 1972 September 24.
Österreich. Zeitschr. Vermessungswesen, 60. Jahrgang, p. 77 (1972).

K. Ledersteger, 1900 November 11 – 1972 September 24.   F. Hauer.
Österreich. Zeitschr. Vermessungswesen, 60. Jahrgang, p. 109 - 111 (1973).

W. M. Lindley, 1891 July 27 - 1972 September 2.
Journ. British Astron. Ass., Vol. 83, 201 - 203 (1973).

J. B. Ohlsson, 1894 October 7 - 1972 December 30.
N. Hansson.
Astron. Tidssk., Årg. 6, p. 88 (1973).

R. Prinz died 1973 February 5.
H. Oberndorfer.
SuW, Vol. 12, 152 (1973).

H. Shapley, 1885 November 2 - 1972 October 20.
Z. Kopal.
Astrophys. Space Sci., Vol. 18, 258 - 266 (1972).

H. Shapley, 1885 - 1972 October 20.
L. Rosino.
Coelum, Vol. 41, 13 - 15 (1973).

H. Shapley, 1885 November 2 - 1972 October 20
H. S. Hogg.
Journ. Roy. Astron. Soc. Canada, Vol. 67, 31 - 33 (1973).

H. Shapley died 1972 October 20.
Říše hvězd, Vol. 54, 36 (1973). In Czech.

C. L. Stearns died 1972 November 28.
Science, Vol. 179, 1212 (1973).

P. Tardi, 1897 - 1972.   J. Kovalevsky.
L'Astronomie, 87e année, p. 128 - 130 (1973).

P. Tardi, 1897 June 4 - 1972 August 5.
Bull. d'Information, Ass. Développement International Obs. Nice, No. 9, p. 3 - 4 (1972).

P. Tardi, 1897 June 4 – 1972 August 5. J. Mitter.
Österreich. Zeitschr. Vermessungswesen, 60. Jahrgang, p. 102 - 103 (1972).

A. Wilke died on July 24, 1972.
E. Scholz.
Sterne, 49. Jahrgang, p. 52 - 53 (1973).

008 Observatories, Institutes

Reports, communications and publications of observatories and astronomical institutes are recorded in this section; included are numbered series of reprints. Whenever possible, the numbers of the abstracts referring to the publications are given. Observatories and institutes are listed in alphabetical order of their towns. In some cases observatory publications do not give the name of the town; the following list which gives names and towns of some institutions may serve as an aid in such cases.

Aarne Karjalainen Observatory	Oulu, Finland
Algonquin Radio Observatory	Lake Traverse, Ontario, Canada
Allegheny Observatory	Pittsburgh, Pennsylvania
Archenhold-Sternwarte	Berlin-Treptow, Germany
Arthur J. Dyer Observatory	Nashville, Tennessee
Astronomical Latitude Station, Polish Academy of Sciences	Borowiec, Poland
Bosscha Observatory	Lembang, Indonesia
Boyden Observatory	Bloemfontein, South Africa
Bureau International de l'Heure	Paris, France
Cajigal Observatory	Caracas, Venezuela
California Institute of Technology	Pasadena, California
Cape of Good Hope	Cape Town, South Africa
Carter Observatory	Wellington, New Zealand
Catalina Station	Tucson, Arizona
Cavendish Laboratory	Cambridge, England
Ceskoslovenská Akademie Ved Astronomický Ustav	Praha, Czechoslovakia
Chamberlin Observatory, University of Denver	Denver, Colorado
Commonwealth Observatory	Canberra, Australia
Corralitos Observatory	Las Cruces, New Mexico
David Dunlap Observatory, University of Toronto	Richmond Hill, Ontario
Dearborn Observatory	Evanston, Illinois
Department of Astronomy and Observatory, Univ. California	Los Angeles, California
Department of Astronomy, University of Texas	Austin, Texas
Division Radiophysics, C.S.I.R.O. University Grounds	Sydney, N.S.W., Australia
Dominion Astrophysical Observatory	Victoria, British Columbia
Dominion Observatory	Ottawa, Ontario
Dominion Radio Astrophysical Observatory	Penticton, British Columbia
Dudley Observatory	Albany, New York
Dunsink Observatory	Dublin, Ireland
Engelhardt Observatory	Kazan, U.S.S.R.
European Southern Observatory	Hamburg, Federal German Republic
Five College Observatories	Amherst, Massachusetts
Florida State University Radio Observatory	Tallahassee, Florida
Flower and Cook Observatories, University of Pennsylvania	Philadelphia, Pennsylvania
Fraunhofer Institut	Freiburg, Federal German Republic
Georgetown Observatory	Washington, D.C.
Goddard Space Flight Center	Greenbelt, Maryland
Goethe Link Observatory, University of Indiana	Bloomington, Indiana
Hale Observatories	Pasadena, California
Harvard College Observatory	Cambridge, Massachusetts
Harvard Radio Astronomy Station	Cambridge, Massachusetts
Haystack Observatory	Westford, Massachusetts
Heinrich-Hertz-Institut	Berlin, Germany
High Altitude Observatory, University of Colorado	Boulder, Colorado
Institute for Astronomy, University of Hawaii	Honolulu, Hawaii
Institute for Theoretical Astronomy (Institut Teoreticheskoj Astronomii)	Leningrad, U.S.S.R.
Institute of Theoretical Astrophysics, Blindern	Oslo, Norway
Inter-American Observatory	Cerro-Tololo, (La Serena), Chile
International Latitude Observatory	Mizusawa, Japan
Joint Institute for Laboratory Astrophysics (JILA)	Boulder, Colorado
Kandilli Observatory	Istanbul, Turkey
Kansas University Observatory	Lawrence, Kansas
Kapteyn Astronomical Laboratory	Groningen, Netherlands
Karl-Schwarzschild-Observatorium	Tautenburg, German Democratic Republic
Kenneth Mees Observatory	Rochester, New York
Kwasan Observatory	Kyoto, Japan
Lamont-Hussey Observatory	Bloemfontein, South Africa
Leander McCormick Observatory University of Virginia	Charlottesville, Virginia
Lee Observatory	Beirut, Lebanon
Leopold-Figl-Observatorium	Wien, Austria
Leuschner Observatory	Berkeley, California
Lick Observatory	Santa Cruz, (Mount Hamilton), California
Lindheimer Astronomical Research Center	Evanston, Illinois
Lockheed Solar Observatory	Saugus, California
Lohrmann-Observatorium für Geodätische Astronomie	Dresden, German Democratic Republic
Louisiana State University Observatory	Baton Rouge, Louisiana
Lowell Observatory	Flagstaff, Arizona
Lunar and Planetary Laboratory	Tucson, Arizona
Max-Planck-Institut für Astronomie	Heidelberg, Federal German Republic
Max-Planck-Institut für Phyik und Astrophysik	München, Federal German Republic
Max-Planck-Institut für Radioastronomie	Bonn, Federal German Republic
McDonald Observatory	Fort Davis, Texas
McMath Hulbert Observatory	Pontiac, Michigan
Michigan State University Observatory	East Lansing, Michigan
Molonglo Radio Observatory, University of Sydney	Sydney, New South Wales
Mount Cuba Observatory	Wilmington, Delaware
Mount John Observatory	Lake Tekapo, New Zealand
Mount Palomar Observatory	Pasadena, California
Mount Wilson Observatory	Pasadena, California
Mullard Radio Astronomy Observatory	Cambridge, England
Narrabri Observatory, University of Sydney	Sydney, New South Wales

National Bureau of Standards	Washington, D. C.	Sagamore Hill Radio Observatory	Bedford, Massachusetts
National Observatory,USA	Kitt Peak, Arizona	Saint-Michel, l'Observatoire	Haute Provence, France
National Radio Astronomy	Charlottesville, Virginia	San Fernando Observatory	El Segundo, California
Observatory	Green Bank, West Virginia	Smithsonian Astrophysical	
	Tucson, Arizona	Observatory	Cambridge, Massachusetts
New Mexico State		Specola Astronomica Vaticana	Castel Gandolfo, Italy
University Observatory	Las Cruces, New Mexico	Specola di Padova	Asiago, Italy
Nizamiah Observatory	Hyderabad, India	Sproul Observatory	Swarthmore, Pennsylvania
Nuffield Radio Astronomy		Sternberg Observatory	Moscow, U.S.S.R.
Laboratories, Jodrell Bank		Steward Observatory,	
University of Manchester	Manchester, England	University of Arizona	Tucson, Arizona
Observatoire Royal de Belgique	Uccle, Belgium	United States Naval Observatory	Washington, D.C.
Observatorio de Cartuja	Granada, Spain	University of Florida,	
Observatorio del Ebro	Tortosa, Spain	Radio Observatory	Gainesville, Florida
Observatorio Fabra	Barcelona, Spain	University of Illinois Observatory	Urbana, Illinois
Observatory, University of		University of Michigan	
Michigan	Ann Arbor, Michigan	Observatories	Ann Arbor, Michigan
Ohio State University		University of South Florida	
Radio Observatory	Columbus, Ohio	Observatory	Tampa, Florida
Ole Roemer-Observatoriet	Aarhus, Denmark	Uttar Pradesh State Observatory	Naini Tal, India
Owens Valley Radio	Big Pine, California	Van Vleck Observatory	Middletown, Connecticut
Observatory		Wallace Observatory	Cambridge, Massachusetts
Perkins Observatory, Ohio State		Warner and Swasey Observatory	Cleveland, Ohio
and Wesleyan Universities	Delaware, Ohio	Washburn Observatory	Madison, Wisconsin
Purple Mountain Observatory	Nanking, China	West Melton Observatory	Christchurch, New Zealand
Radcliffe Observatory	Pretoria, South Africa	Yale University Observatory	New Haven, Connecticut
Remeis-Sternwarte	Bamberg,	Yerkes Observatory	Williams Bay, Wisconsin
	Federal German Republic	Zentralinstitut für Astrophysik,	
Republic Observatory	Johannesburg, South Africa	Sternwarte Babelsberg, (Fach-	
Rosemary Hill Observatory	Gainesville, Florida	bereich Kosmische Physik)	Potsdam-Babelsberg, German
Royal Radar Establishment,			Democratic Republic
Radio Astronomy Division	Malvern, England		

008.001 Albany

Dudley Observatory, *Albany, New York,* Reports, No. 7 (R. H. Giese, 09.106.034).

Dudley Observatory, *Albany, New York.* Reprint No. B43 (C. L. Hemenway, D. S. Hallgren, D. C. Schmalberger, 08.082.001).

Dudley Observatory, *Albany, New York.* Reprint Nos. C30 (A. G. D. Philip, 07.154.029), C44 (A. G. D. Philip, J. Stock, 08.114.173), C45 (J. M. Greenberg, 09.063.044).

008.002 Ames

Erwin W. Fick Observatory, Iowa State University, Ames, Iowa. – Observatory report. W. I. Beavers.
Bull. American Astron. Soc., Vol. 5, 101 - 103 (1973).

008.003 Amherst

Five College Astronomy Department: Amherst College, Amherst, Massachusetts; Hampshire College, Amherst, Massachusetts; Mount Holyoke College, South Hadley, Massachusetts; Smith College, Northampton, Massachusetts; University of Massachusetts, Amherst, Massachusetts. – Observatory report. W. M. Irvine.
Bull. American Astron. Soc., Vol. 5, 104 - 108 (1973).

Contributions from the Five College Observatories, *Amherst,* Nos. 135 (T. F. Tascione, 08.061.050), 139 (R. N. Manchester, G. R. Huguenin, J. H. Taylor, 07.141.542).

008.004 Ann Arbor

Department of Astronomy, University of Michigan, Ann Arbor, Michigan, 1971–1972. – Observatory report. W. A. Hiltner.
Bull. American Astron. Soc., Vol. 5, 183 - 186 (1973).

008.005 Arcetri

Elenco dei lavori eseguiti dal personale dell'Osservatorio Astrofisico di Arcetri durante il 1971.
Boll. Geod. Sci. Affini, Anno 32, p. 74 - 78 (1973).

008.006 Arecibo

Arecibo Observatory report, Arecibo, Puerto Rico. Observatory report. F. D. Drake, T. Hagfors.
Bull. American Astron. Soc., Vol. 5, 58 - 62 (1973).

008.007 Armagh

Armagh Observatory, Leaflet, Nos. 109 (E. Öpik, 06.105.158), 110 (D. J. Mullan, 06.131.157), 111 (D. J. Mullan, 06.126.024), 112 (D. J. Mullan, 07.071.015), 113 (E. M. Lindsay, J. McFarland, 07.159.012), 114 (D. Crowe, 07.005.004), 115 (A. D. Andrews, 07.122.158).

Contributions from the Armagh Observatory, Nos. 68 (E. M. Lindsay, 05.159.007), 69 (D. J. Mullan, 05.141.059), 70 (T. W. Rackham, 06.105.157), 71 (D. J. Mullan, 06.064.

058), 72 (E. M. Lindsay, 06.122.140), 73 (D. J. Mullan, 06. 062.001), 74 (D. J. Mullan, 06.071.072), 75 (E. J. Öpik, 06. 102.025), 76 (D. J. Mullan, 06.064.028), 77 (E. M. Lindsay, P. A. Wayman, 08.122.088), 78 (A. D. Andrews, D. J. Mullan, 08.123.038), 80 (E. M. Lindsay, 08.005.027).

Contributions from the Armagh Observatory, Quarto Series, No. 5 (D. J. Mullan, 08.064.012).

008.008 Asiago

Programmi astrometrici in corso presso gli osservatori astronomici di Padova e Asiago. C. Barbieri, M. Capaccioli, R. Ganz, G. Pinto. Mem. Soc. Astron. Italiana, Nuova Ser., Vol. 43, 635 - 636 (1973).

008.009 Atlanta

Fernbank Observatory, Fernbank Science Center, Atlanta, Georgia. – Observatory report. P. H. Knappenberger. Bull. American Astron. Soc., Vol. 5, 103 - 104 (1973). This is the first report from the Observatory and contains a brief description of the facilities and philosophy of operation.

008.010 Austin

Publications of the Department of Astronomy, The University of Texas, *Austin,* Series I, Vol. 2, Nos. 12 (B. Warner, AJB 68, 16.130), 13 (B. Warner, AJB 68, 64.136), 17 (B. Warner, 01.114.012), 18 (B. Warner, AJB 68, 16.131), 19 (D. L. Lambert, E. A. Mallia, B. Warner, 01.071.021), 20 (B. Warner, 01.071.032), 21 (E. F. Montgomery, P. Connes J. Connes, F. N. Edmonds, Jr., 02.114.016), 22 (F. N. Edmonds, Jr., 02.064.021), 23 (B. Warner, 02.115.002), 24 (B. Warner, R. C. Kirkpatrick, 02.022.017), 25 (B. Warner, 02.114.090), 26 (B. W. Bopp, F. N. Edmonds, Jr., 03.112. 009), 27 (B. Warner, R. C. Kirkpatrick, 03.022.002), 28 (R. W. Day, 02.151.062), 29 (T. J. Deeming, 03.141.054), 30 (R. R. Robbins, 03.132.020).

Publications of the Department of Astronomy, The University of Texas, *Austin,* Series II, Vol. 2, Nos. 13 (F. N. Edmonds, Jr., 01.034.013), 15 (G. de Vaucouleurs, W. L. Peters, 02.151.067), 16 (D. S. Evans, 02.082.029); Vol. 3, No. 6 (W. H. Jefferys, 09.042.049).

008.011 Baton Rouge

Louisiana State University Observatory, Baton Rouge, Louisiana. – Observatory report.   A. U. Landolt. Bull. American Astron. Soc., Vol. 5, 167 - 169 (1973).

Contributions of the Louisiana State University Observatory, Nos. 58 (J. S. Drilling, 06.113.048), 59 (J. S. Drilling, 07.114.059), 60 (D. L. Crawford, J. V. Barnes, G. Hill, C. L. Perry, 06.153.025), 61 (J. G. Peters, W. A. Fowler, D. D. Clayton, 07.061.025), 62 (H. E. Bond, 08.114.004), 63 (A. U. Landolt. K. L. Blondeau, 08.121.008), 64 (A. U. Landolt, 08.121.011), 65 (H. E. Bond, 08.114.077), 66 (H. E. Bond, 07.158.164), 67 (J. S. Drilling, 08.155.004),

68 (J. B. Irwin, A. U. Landolt, 08.121.076), 69 (J. B. Irwin, 09.119.001), 70 (P. Lee, P. Daigle, 08.114.166).

008.012 Bedford

Sagamore Hill Radio Observatory, Air Force Cambridge Research Laboratories, Bedford, Massachusetts. Observatory report.   J. P. Castelli. Bull. American Astron. Soc., Vol. 5, 229 - 230 (1973).

008.013 Beirut

Lee Observatory, American University of Beirut, Lebanon. Monthly Bulletin, Astronomical Section, 1972 October - 1973 April (F. Bruin, H. Hourani, N. G. Bustati, 09.075.010).

008.014 Beograd

University of Beograd, Faculty of Sciences. **Publications of the Department of Astronomy** (Publications de la Chaire d'Astronomie), No. 4 (G. Teleki, 09.082.122; T. Angelov, 09.065.168; R. Dejaiffe, 09.032.030).

008.015 Berkeley

Research Units and Academic Departments, University of California: Berkeley, Los Angeles, San Diego, and Santa Cruz. – **I. Berkeley Campus.** – Observatory report. Bull. American Astron. Soc., Vol. 5, 64 - 72 (1973).

008.016 Berlin

Heinrich-Hertz-Institut. Solare Beobachtungsergebnisse. Deutsche Akademie der Wissenschaften zu Berlin, Zentralinstitut für Solar-Terrestrische Physik, Berlin-Adlershof. HHI Solar Data, Vol. 23, 1972 September – December; Vol. 24, 1973 January – March (C.-U. Wagner, A. Böhme, F. Fürstenberg, D. Scholz, S. Böhm, 09.075.011).

Heinrich-Hertz-Institut. Supplement Series of Solar Data. Deutsche Akademie der Wissenschaften zu Berlin, Zentralinstitut für Solar-Terrestrische Physik, Berlin-Adlershof. HHI Suppl. Ser. Solar Data, Vol. 3, Nos. 1 (F. Fürstenberg, A. Krüger, 09.077.065), 2 (H. Künzel, 09.072.072).

008.017 Big Pine

Owens Valley Radio Observatory, California Institute of Technology, Big Pine, California. – Observatory report.   G. J. Stanley. Bull. American Astron. Soc., Vol. 5, 222 - 225 (1973).

008.018 Bloomington

Goethe Link Observatory, Indiana University, Bloomington, Indiana. – Observatory report.

F. K. Edmondson.
Bull. American Astron. Soc., Vol. 5, 114 - 116 (1973).

008.019 **Bologna**

Laboratorio di Radioastronomia, Consiglio Nazionale delle Ricerche, Istituto di Fisica, Bologna (Italy). Separate prints (G. Colla, C. Fanti, R. Fanti, A. Ficarra, L. Formiggini, E. Gandolfi, I. Gioia, C. Lari, B. Marano, L. Padrielli, P. Tomasi, 09.141.132; G. Grueff, M. Vigotti, 09.141.133).

008.020 **Bonn**

Max-Planck-Institut für Radioastronomie, Bonn. Sonderdrucke, Nos. 56 (Y. K. Minn, J. M. Greenberg, 08.131. 008), 63 (J. E. Wink, W. J. Altenhoff, W. J. Webster, Jr., 09.131.011), 64 (F. F. Gardner, H. R. Dickel, J. B. Whiteoak, 09.131.024), 67 (E. Churchwell, C. M. Walmsley, 09.132.004), 74 (K. Rohlfs, E. Braunsfurth, U. Mebold, 08.131.080).

008.021 **Bordeaux**

Observatoire de l'Université de Bordeaux, Floirac (Gironde). Rapport présenté au Conseil de l'Université, année scolaire 1971–1972. J. Delannoy.
Imprimerie Centrale, Bordeaux. 19 pp. (1972).

008.022 **Borowiec**

Polish Academy of Sciences, Astronomical Latitude Station, Borowiec, Circular Nos. 124, 125 (09.044.022).

008.023 **Boulder**

Joint Institute for Laboratory Astrophysics of the National Bureau of Standards and the University of Colorado, Boulder, Colorado. – Observatory report. J. I. Castor.
Bull. American Astron. Soc., Vol. 5, 134 - 141 (1973).

National Oceanic and Atmospheric Administration, Boulder, Colorado. – Observatory report. H. Leinbach.
Bull. American Astron. Soc., Vol. 5, 204 - 205 (1973). – This report covers activities of the Solar Physics Group of the Space Environment Laboratory, plus related activities of the Space Environment Services Center of the Laboratory. These programs are part of the larger mission of the Laboratory, concerned with solar-terrestrial physics.

008.024 **Buenos Aires**

Plans for the Naval Observatory of Buenos Aires. J. Marpegán.
IAU Colloquium No. 1, (see 012.019), p. 23 - 25 (1972).

008.025 **Byurakan**

Chronicle.

Soobshch. Byurakan. Obs., vyp. (No.) 44, p. 139 - 145 (1972). In Russian.

Byurakan Astrophysical Observatory, Armenia, USSR, Reprints, Nos. 83 (B. E. Markarian, V. A. Lipovetsky, 07.158.063), 84 (A. T. Kalloghlian, 07.158.064), 85 (E. S. Parsamian, 07.122.062), 86 (V. A. Ambartsumian, 07.122.063), 87 (N. B. Yengibarian, 07.063.018), 88 (Yu. L. Vartanian, A. V. Hovsepian, G. S. Hajian, 07.065.074), 89 (V. V. Papoyan, D. M. Sedrakian, E. V. Chubarian, 07.126.016), 90 (M. A. Kazarian, E. Ye. Khachikian, 08.132.001), 91 (M. A. Arakelian, E. A. Dibay, V. F. Yesipov, 08.158.001), 92 (A. T. Kalloghlian, 08.160.001), 93 (N. B. Yengibarian, A. G. Nicoghossian, 08.063.002), 94 (Yu. L. Vartanian, 08.061.001), 95 (G. S. Sahakian, R. M. Avakian, 08.061.002), 96 (N. B. Yengibarian, 08.063.003), 97 (B. E. Markarian, V. A. Lipovetsky, 09.158.030), 98 (B. E. Markarian, 09.158.031), 99 (M. A. Arakelian, E. A. Dibay, V. F. Yesipov, 09.158.032), 100 (N. B. Yengibarian, A. G. Nikogosian, 09.063.012), 101 (G. S. Saakian, D. M. Sedrakian, 09.064.017), 102 (M. A. Arakelian, E. A. Dibay, V. F. Yesipov, 09.158.049), 103 (V. V. Papoyan, D. M. Sedrakian, E. V. Chubarian, 09.061. 014), 104 (Yu. L. Vartanian, 09.126.007), 105 (G. G. Harutyunian, D. M. Sedrakian, 09.065.037), 106 (M. A. Arakelian, E. A. Dibay, V. M. Lyutij, 09.158.051; R. M. Avakian, G. G. Harutyunian, G. S. Saakian, 09.065.040; G. A. Gurzadyan, 09.126.008).

Soobshcheniya Byurakanskoj Observatorii, vyp. (No.) 44 (E. S. Parsamian, 09.122.127; E. S. Parsamian, 09. 122.128; E. S. Parsamian, H. S. Chavushian, 09.122.129; H. S. Badalian, L. K. Erastova, 09.122.130; L. K. Erastova, 09.122.131; R. G. Mnatsakanian, K. A. Sahakian, 09.113.053; N. L. Ivanova, 09.114.166; N. L. Ivanova, N. K. Andreasian, 09.114.167; E. D. Arsenievich, 09.131.195; K. A. Grigorian, M. A. Eritsian, 09.122.132; R. A. Vardanian, Yu. K. Melik-Alaverdian, 09.141.127; E. H. Harutjunian, Yu. K. Melik-Alaverdian, 09.141.557; R. M. Avakian, 09.126.028; K. A. Grigorian, H. V. Abrahamian, G. Lelievre, M. A. Eritsian, 09.034.096; V. A. Malarev, E. M. Neplokhov, I. K. Pavlov, G. A. Tambovski, 09.034.097; 09.007.000; 09.008.025).

008.026 **Cambridge, Mass.**

Harvard College Observatory, Cambridge, Massachusetts. – Observatory report. A. Dalgarno.
Bull. American Astron. Soc., Vol. 5, 116 - 120 (1973).

George R. Wallace, Jr. Astrophysical Observatory, Massachusetts Institute of Technology, Cambridge, Massachusetts. – Observatory report. T. B. McCord.
Bull. American Astron. Soc., Vol. 5, 240 - 242 (1973).

Smithsonian Institution. Astrophysical Observatory, Research in Space Science. SAO Special Reports, Nos. 348 (L. G. Jacchia, J. W. Slowey, 09.082.120), 349 (Y. Kozai, 09.052.032), 351 (R. L. Kurucz, 09.022.080).

008.027 **Cape Town**

Royal Observatory, Cape of Good Hope. – Report for the year ending 1971 December 31. G. A. Harding.
Quarterly Journ. Roy. Astron. Soc., Vol. 14, 103 - 106 (1973).

Royal Observatory Bulletins, (Joint Publications of the Royal Greenwich Observatory, Herstmonceux, Royal Observatory, Cape of Good Hope), No. 177 (D. J. Stickland, 09. 119.016).

008.028   Castel Gandolfo

Specola Vaticana. Annual report 1972: Report of the Astronomical Observatory; Report of the Astrophysical Laboratory.   P. J. Treanor, J. Junkes. Printed in Vatican City, 19 pp. (1973).

Ricerche Astronomiche, Specola Vaticana, Città del Vaticano, Vol. 8 Nos. 17 (P. Smeyers, 09.065.167), 18 (W. J. Miller, A. A. Wachmann, 09.123.035).

Specola Vaticana, *Castel Gandolfo*, Comunicazione, Nos. 56 (D. J. K. O'Connell, 07.121.082), 57 (D. J. K. O'Connell, 08.005.027).

Vatican Observatory Publications, Specola Vaticana, Città del Vaticano, Vol. 1, No. 4 (F. C. Bertiau, E. de Graeve, P. J. Treanor, 09.082.121).

008.029   Catania

Osservatorio Astrofisico di Catania, Pubblicazione, No. 150 (G. Godoli, V. Sciuto, R. A. Zappalà E. Catinoto, G. Domina, G. Celeani, G. Sapienza, S. Sciuto, 09.075.009).

008.030   Cerro Tololo

Kitt Peak National Observatory, Tucson, Arizona and Cerro Tololo Inter-American Observatory, La Serena, Chile. – Observatory reports.   L. Goldberg. Bull. American Astron. Soc., Vol. 5, 142 - 162 (1973).

Geodetic and astronomical coordinates of the Cerro Tololo Inter-American Observatory. See Abstr. 046.020.

Cerro Tololo Inter-American Observatory, Contributions, No. 126 (R. S. Harrington, B. M. Blanco, V. M. Blanco, 09.046.020).

008.031   Charlottesville

Leander McCormick Observatory, University of Virginia, Charlottesville. – Observatory report. C. R. Tolbert. Bull. American Astron. Soc., Vol. 5, 162 - 164 (1973).

Publications of the Leander McCormick Observatory of the University of Virginia, Vol. 11, Part 27 (T. E. Corbin, S. J. Goldstein, Jr., 09.046.018).

National Radio Astronomy Observatory, Charlottesville, Virginia, Green Bank, West Virginia, and Tucson, Arizona. Observatory reports.   D. S. Heeschen. Bull. American Astron. Soc., Vol. 5, 206 - 214 (1973). – This report covers the period July 1971 through June 1972.

008.032   Cincinnati

Minor Planet Circulars (MPC), Nos. 3407 - 3534 (P. Herget, 09.098.064).

008.033   Cleveland

Warner and Swasey Observatory, Case Western Reserve University, Cleveland, Ohio. – Observatory report. W. P. Bidelman. Bull. American Astron. Soc., Vol. 5, 242 - 247 (1973).

Publications of the Warner and Swasey Observatory, Case Western Reserve University, Vol. 1, No. 2 (W. H. Wooden II, 09.155.047).

008.034   College Park

Astronomy Program, University of Maryland, College Park, Maryland. – Observatory report.   G. Westerhout. Bull. American Astron. Soc., Vol. 5, 173 - 180 (1973).

008.035   Columbus

The Observatories of the Ohio State and Ohio Wesleyan Universities, Columbus and Delaware, Ohio. – Observatory reports.   A. Slettebak. Bull. American Astron. Soc., Vol. 5, 216 - 221 (1973).

Ohio State University Radio Observatory, Columbus, Ohio. – Observatory report.   J. D. Kraus. Bull. American Astron. Soc., Vol. 5, 221 (1973).

008.036   Copenhagen

Copenhagen University Observatory, Reprint Nos. 214 (O. H. Einicke, S. Laustsen, H. Schnedler Nielsen, 05.031.001), 215 (K. T. Johansen, H. E. Jørgensen, V. Bohr, 05.121.004), 216 (K. T. Johansen, 05.121.028), 217 (H. E. Jørgensen, K. T. Johansen, E. H. Olsen, 05.122.044), 219 (J. Andersen, 05.032.027), 221 (L. Hansen, P. Kjaergaard, 06.064.021), 222 (E. H. Olsen, 06.113.029), 223 (B. Strömgren, 06.159.032), 224 (K. T. Johansen, 06.122.070), 225 (R. F. Nielsen, 05.032.056), 226 (J. O. Petersen, H. E. Jørgensen, 07.122.034), 228 (J. O. Petersen, 07.065.113), 229 (E. H. Olsen, 08.114.017), 231 (H. Stub, 08.121.014), 232 (O. C. Wilson, E. H. Olsen, P. Kjaergaard, 07.117.040), 233 (B. Strömgren, 07.131.151), 234 (K. Gyldenkerne, 05.121. 076), 235 (B. Strömgren, 09.061.046), 236 (H. J. Fogh Olsen, P. Jensen, T. Knudsen, 09.041.001).

008.037   Delaware

The Observatories of the Ohio State and Ohio Wesleyan Universities, Columbus and Delaware, Ohio. – Observatory reports.   A. Slettebak. Bull. American Astron. Soc., Vol. 5, 216 - 221 (1973).

Contributions from the Perkins Observatory, Ohio State – Ohio Wesleyan Universities. Series I, Nos. 133 (G. W. Collins II, 07.064.047), 134 (I. Marenin, A. E. Greene, 08. 064.042), 135 (E. R. Capriotti, 09.133.002), 136 (R. F. Wing, J. W. Warner, M. G. Smith, 09.114.015), 137 (R. F. Wing, 09.114.080).

Contributions from the Perkins Observatory, Ohio State University – Ohio Wesleyan Universities. Series II, No. 31 (A. E. Greene, 09.064.081).

008.038 Dublin

Dunsink Observatory. – Report for the year ending 1972 March 31.   P. A. Wayman.
Quarterly Journ. Roy. Astron. Soc., Vol. 14, 67 - 72 (1973).

008.039 Dunedin

Astrophysics at the University of Otago (N.Z.).
P. J. Edwards.
Proc. Astron. Soc. Australia, Vol. 2, 138 - 139 (1972).

008.040 East Lansing

Michigan State University, East Lansing, Michigan.
Observatory report.   A. P. Linnell.
Bull. American Astron. Soc., Vol. 5, 186 - 188 (1973).

008.041 Edinburgh

Report of the Astronomer Royal for Scotland for the year ending 31st March 1973.   H. A. Brück.
Science Research Council. The Royal Observatory, Edinburgh, 14 pp. (1973).

Communications from the Royal Observatory, Edinburgh, Nos. 114 (K. Nandy, H. Seddon, 06.131.039), 115 (G. E. Bromage, M. T. Brück, K. Nandy, 06.131.047), 128 (M. J. Smyth, 09.114.076), 134 (K. Nandy, A. Kelly, 09.131.100), 135 (R. J. Dodd, W. McD. Napier, A. A. Preece, 08.061.049), 136 (B. N. G. Guthrie, 08.142.058), 137 (B. N. G. Guthrie, 09.131.003),138(G. E. Bromage, 09.132.001), 139 (K. Nandy, N. Pratt, 09.114.165), 140 (G. E. Bromage, K. Nandy, 09.114.047), 143 (P. Barker, A. Boksenberg, H. E. Butler, S. Gardier, L. Houziaux, C. Humphries, C. Jamar, D. Macau-Hercot, D. Malaise, A. Monfils, K. Nandy, G. I. Thompson, R.Wilson, H. Wroe, 09.114.039), 144 (R. Q. Twiss, W. T. Welford, 09.034.100), 145 (W. McD. Napier, R. J. Dodd, 09.101.004).

008.042 El Segundo

The Aerospace Corporation, El Segundo, California: (I)Electronics Research Laboratory; (II) Space Physics Laboratory. – Observatory report.   G. A. Paulikas.
Bull. American Astron. Soc., Vol. 5, 53 - 57 (1973).
Research in astronomy at Aerospace is carried out by groups in the Electronics Research Laboratory and the Space Physics Laboratory. The Space Physics Laboratory operates the Aerospace Corporation's San Fernando Observatory, and also conducts astronomical observations at X-ray and *EUV* wavelengths utilizing satellites. The activities of these two groups are described below.

008.043 Evanston

Lindheimer Astronomical Research Center, and Dearborn Observatory, Evanston, Illinois; Corralitos Observatory, Las Cruces, New Mexico. – Observatory reports.
J. A. Hynek.
Bull. American Astron. Soc., Vol. 5, 164 - 166 (1973).

008.044 Flagstaff

Lowell Observatory, Flagstaff, Arizona. – Observatory report.   J. S. Hall.
Bull. American Astron. Soc., Vol. 5, 169 - 173 (1973).

Lowell Observatory Bulletin, *Flagstaff, Arizona,* No. 160, Vol. 7, No. 23 (H. L. Giclas, R. Burnham, Jr., N. G. Thomas, 09.112.005).

008.045 Fort Davis

The University of Texas. Contributions from the McDonald Observatory, Fort Davis, Texas, Nos. 434 (K. Ishida, 01.153.016), 437 (R. E. Nather, B. Warner, M. MacFarlane, 01.141.021), 438 (R. E. Nather, B. Warner, 01.124.106), 440 (B. Warner, R. E. Nather, 01.141.074), 441 (B. Warner, R. E. Nather, M. MacFarlane, 01.141.072), 442 (G. de Vaucouleurs, H. D. Ables, 03.158.029), 443 (G. de Vaucouleurs, 03.158.030), 444 (J. T. Bergstralh, 04.034.021), 448 (A. P. Fairall, R. J. Angione, 02.158.051), 449 (D. W. Weedman, 03.158.027), 450 (P. R. Jordahl, 04.064.028).

The University of Texas. Contributions from the McDonald Observatory, Fort Davis, Texas, Series II, Nos. 24 (B. Warner, R. E. Nather, 01.022.029), 25 (R. G. Tull, 02.034.031), 27 (G. de Vaucouleurs, 01.158.065), 28 (R. G. Tull, 02.034.005), 30 (B. Warner, R. E. Nather, 03.126.001).

008.046 Freiburg

Fraunhofer Institut, Map of the Sun. 1973 January 1 - June 30 (09.075.008).

Mitteilungen aus dem Fraunhofer Institut, *Freiburg,* Nos. 103 (A. Bruzek, 07.072.038), 105 (U. Grossmann-Doerth, M. von Uexküll, 06.073.058).

008.047 Gainesville

University of Florida Observatories, Gainesville, Florida: Rosemary Hill Observatory, F. B. Wood; **University of Florida Radio Observatory,** A. G. Smith. – Observatory reports.
Bull. American Astron. Soc., Vol. 5, 108 - 112 (1973).

Rosemary Hill Observatory, Department of Physics and Astronomy, University of Florida, Gainesville, Florida, Contributions, Nos 16 (H. L. Cohen, 06.121.063), 21 (G. H. Folsom, A. G. Smith, R. L. Hackney, K. R. Hackney, 05.141.076), 22 (S. M. Ruciński, 06.117.027), 27 (R. Bloomer, 06.123.022), 28 (R. Bloomer, 06.123.023), 29 (G. H. Folsom, A. G. Smith, H. W. Schrader, 09.141.131), 30 (F. B. Wood, R. H. Bloomer, 07.123.051).

008.048 Genève

Publications de l'Observatoire de Genève, Série A, Fasc. 79 (E. Lindemann, B. Hauck, 09.113.055; P. Bouvier, 09.151.049).

008.049 Gothenburg

Research Laboratory of Electronics and Onsala Space Observatory, Chalmers University of Technology, Gothenburg, Sweden. **Research Report,** No. 108 (B. T. Cato, 08.131.106).

008.050 Green Bank

National Radio Astronomy Observatory, Charlottesville, Virginia, Green Bank, West Virginia, and Tucson, Arizona. Observatory reports. D. S. Heeschen.
Bull. American Astron. Soc., Vol. 5, 206 - 214 (1973). — This report covers the period July 1971 through June 1972.

National Radio Astronomy Observatory, *Green Bank,* **Reprints,** Series A, Nos. 261 (I. I. K. Pauliny-Toth, K. I. Kellermann, 08.141.056), 262 (K. I. Kellermann, 08.158.063), 263 (R. M. Hjellming, M. Hermann, E. Webster, 07.142.104), 264 (M. R. Kundu, T. Velusamy, 08.125.006), 265 (J. S. Gallagher III, 08.158.064), 266 (F. J. Kerr, G. R. Knapp 08.154.006), 267 (B. Balick, 08.141.025), 268 (G. R. Knapp, F. J. Kerr, 08.131.059), 269 (P. Lantos, M. R. Kundu, 08.077.016), 270 (E. B. Fomalont, 08.141.085), 271 (G. R. Knapp, G. L. Verschuur, 08.131.081), 272 (H. M. Tovmassian, 08.158.097), 273 (I. I. K. Pauliny-Toth, K. I. Kellermann, 08.141.112), 274 (Y. K. Minn, J. M. Greenberg, 09.131.010), 275 (G. L. Verschuur, 09.155.005), 276 (M. Reinhardt, M. S. Roberts, 08.160.017), 277 (R. H. T. Bates, P. T. Gough, P. J. Napier, 09.034.002), 278 (P. L. Baker, 09.157.002), 279 (S. Weinreb, A. R. Kerr, 09.033.045), 280 (D. S. Heeschen, 09.008.000), 281 (R. M. Hjellming, B. Balick, 08.142.057), 282 (T. Velusamy, M. R. Kundu, 09.131.036), 283 (J. W. Findlay, J. Payne, 09.033.046).

National Radio Astronomy Observatory, *Green Bank,* **Reprints,** Series B, Nos. 331 (M. R. Kundu, 08.073.018), 332 (B. Balick, 08.131.028), 333 (M. A. Gordon, T. Cato, 08.157.001), 334 (W. C. Erickson, T. B. H. Kuiper, T. A. Clark, S. H. Knowles, J. J. Broderick, 08.141.073), 335 (B. Zuckerman, J. L. Yen, C. A. Gottlieb, P. Palmer, 08.131.065), 336 (P. R. Schwartz, W. J. Wilson, 08.131.091), 337 (W. J. Wilson, P. R. Schwartz, G. Neugebauer, P. M. Harvey, E. E. Becklin, 08.114.099), 338 (H. M. Tovmassian, R. Šramek, 08.158.102), 339 (E. K. Conklin, B. H. Andrew, B. J. Wills, J. D. Kraus, 08.141.076), 340 (R. M. Hjellming, E. Webster, B. Balick, 08.141.111), 341 (R. M. Hjellming, 08.121.107), 342 (E. E. Epstein, W. G. Fogarty, K. R. Hackney, R. L. Hackney, R. J. Leacock, R. B. Pomphrey, R. L. Scott, A. G. Smith, R. W. Hawkins, R. C. Roeder, B. L. Gary, M. V. Penston, K. P. Tritton, C. Bertaud, M. P. Véron, G. Wlérick, A. Bernard, J. H. Bigay, P. Merlin, A. Durand, G. Sause, E. E. Becklin, G. Neugebauer, C. G. Wynn-Williams, 08.141.105), 343 (A. A. Penzias, K. B. Jefferts, R. W. Wilson, H. S. Liszt, P. M. Solomon, 08.131.098), 344 (D. S. Heeschen, 09.158.019), 345 (H. M. Tovmassian, 08.158.110), 346 (F. H. Briggs, F. D. Drake, 08.097.098), 347 (A. H. Barrett, R. N. Martin, P. C. Myers, P. R. Schwartz, 08.141.103), 348 (R. N. Manchester, J. H. Taylor, G. R. Huguenin, 09.141.501), 349 (J. Pfleiderer, 08.158.154), 350 (R. M. Hjellming, B. Balick, 08.142.077), 351 (B. E. Turner, 08.142.073), 352 (R. N. Manchester, J. H. Taylor, G. R. Huguenin, 08.141.525), 353 (S. T. Gottesman, M. A. Gordon, 08.155.048), 354 (W. W. Warnock, J. R. Dickel, 08.093.038), 355 (M. A. Gordon, R. L. Brown, S. T. Gottesman, 08.157.004), 356 (B. Margon, H. Spinrad, C. Heiles, H. Tovmassian, E. Harlan, S. Bowyer, M. Lampton, 08.158.111), 357 (K. B. Jefferts, A. A. Penzias, R. W. Wilson, 09.132.003), 358 (K. I. Kellermann, B. G. Clark, M. H. Cohen, D. B. Shaffer, J. J. Broderick,

D. L. Jauncey, 09.158.029), 359 (N. R. Vandenberg, T. A. Clark, W. C. Erickson, G. M. Resch, J. J. Broderick, R. R. Payne, S. H. Knowles, A. B. Youmans, 09.141.506), 360 (Y. N. Parijskij, 09.162.016), 361 (R. L. Brown, 09.160.012), 362 (D. S. De Young, D. E. Hogg, 09.141.046), 363 (D. S. De Young, 08.141.099), 364 (B. Zuckerman, B.E. Turner, D. R. Johnson, P. Palmer, M. Morris, 08.131.087), 365 (B. E. Turner, M. A. Gordon, G. T. Wrixon, 08.131.088), 366 (L. E. Snyder, D. Buhl, 08.131.089), 367 (D. Buhl, L. E. Snyder, J. Edrich, 08.131.090), 368 (M. C. H. Wright, 09.158.014), 369 (W. D. Gwinn, B. E. Turner, W. M. Goss, G. L. Blackman, 09.131.027), 370 (N. Z. Scoville, P. M. Solomon, 09.131.032), 371 (N. Z. Scoville, P. M. Solomon, 09.155.013), 372 (M. A. Gordon, S. T. Gottesman, 09.131.097), 373 (E. B. Fomalont, L. Weliachew, 09.131.102), 374 (D. Buhl, L. E. Snyder, 09.131.106), 375 (D. Buhl, L. E. Snyder, 09.131.049), 376 (S. H. Knowles, K. J. Johnston, J. M. Moran, J. A. Ball, 09. 131.052), 377 (R. L. Brown, J. J. Broderick, 09.114.062), 378 (R. L. Brown, 09.131.001), 379 (R. M. Hjellming, 09. 142.065).

008.051 Greenbelt

Goddard Space Flight Center, Greenbelt, Maryland. **Preprints,** X-592-73-11 (R. S. Mather, 09.081.025), X-661-73-79 (P. J. Serlemitsos, E. A. Boldt, S. S. Holt, R. Ramaty, A. F. Brisken, 09.125.041), X-592-73-105 (M. A. Khan, 09. 081.026), X-592-73-130 (C. A. Wagner, 09.081.027), X-592-73-162 (H.-S. Liu, L. Carpenter, R. W. Agreen, 09.045. 026), X-592-73-164 (R. S. Mather, 09.046.024).

008.052 Greenwich

Royal Greenwich Observatory. — Report for the year ending 1971 December 31. R. Woolley. Quarterly Journ. Roy. Astron. Soc., Vol. 14, 81 - 102 (1973).

Das Royal Greenwich Observatory in Herstmonceux. Eindrücke von einer Besichtigung. M. Lammerer. Orion Schaffhausen, 31. Jahrgang, p. 3 - 7 (1973).

Royal Observatory Bulletins, (Joint Publications of the Royal Greenwich Observatory, Herstmonceux, Royal Observatory, Cape of Good Hope), No. 177 (D. J. Stickland, 09. 119.016).

008.053 Groningen

Nederlandse Vereniging voor Weer- en Sterrenkunde. **Observations of Variable Stars. Report** (Kapteyn Astronomical Laboratory, Groningen–Netherlands), No. 23 (L. Plaut, H. Feijth, 09.123.057).

008.054 Hamburg

Deutsches Hydrographisches Institut, Hamburg. **Astronomische Zeit- und Breitenbestimmungen, Empfangszeiten von Zeitsignalen,** 1972 October - December (09.044. 029).

008.055 Haute Provence

Le grand Schmidt de l'Observatoire de Haute-Provence. A. Heck.
L'Astronomie, 87ᵉ année, p. 241 - 250 (1973).

008.056 Helsinki

Publications of the Finnish Geodetic Institute, *Helsinki,* Nos. 73 (V. R. Ölander, 09.045.025), 74 (J. Kakkuri, K. Kalliomäki, 09.034.108).

008.057 Holmdel

Bell Telephone Laboratories, Incorporated, Crawford Hill Laboratory, Holmdel, New Jersey. – Observatory report. A. A. Penzias.
Bull. American Astron. Soc., Vol. 5, 62 - 64 (1973).

008.058 Honolulu

University of Hawaii, Institute for Astronomy, Honolulu, Hawaii. – Observatory report. J. T. Jefferies.
Bull. American Astron. Soc., Vol. 5, 120 - 126 (1973). – This report covers progress at the Institute for Astronomy over the twelve month period July 1971 through June 1972.

008.059 Ioannina

Department of Astronomy, University of Ioannina. Annual report 1971. S. N. Svolopoulos.
Annual Rep. Astron. Inst. Greece 1971, p. 20 (1972).

University of Ioannina, Contributions from the Laboratory of Astronomy, Ioannina, Greece, No. 7 (S. N. Svolopoulos, 09.114.110).

008.060 Iowa City

The University of Iowa, Iowa City, Iowa. – Observatory report. J. S. Neff.
Bull. American Astron. Soc., Vol. 5, 131 - 134 (1973). – This report describes the astronomical activities and facilities of the Department of Physics and Astronomy for the period 31 August 1971 to 31 August 1972.

008.061 Kazan

Trudy Kazanskoj Gorodskoj Astronomicheskoj Observatorii, *Kazan',* No. 37 (M. I. Lavrov, 09.121.078; Sh. T. Khabibullin, Yu. A. Chikanov, 09.094.921; S. S. Peruanskij, 09.094.922; N. A. Sakhibullin, 09.133.039; S. I. Petrusevich, 09.065.178; V. I. Stebnev, 09.064.085; V. I. Stebnev, 09.064.086; V. P. Merezhin, V. I. Stebnev, 09.064.087; M. A. Vajsov, 09.031.057; M. A. Vajsov, 09.031.058; Yu. V. Evdokimov, 09.103.105; E. D. Kondrat'eva, 09.102.023).

Trudy Kazanskoj Gorodskoj Astronomicheskoj

Observatorii, *Kazan',* No. 38 (V. P. Merezhin, 09.117.020; V. P. Merezhin, 09.121.025; E. E. Belyaeva, L. L. Shishkina, 09.064.039; E. E. Belyaeva, L. L. Shishkina, 09.064.040; M. I. Lavrov, N. V. Lavrova, 09.122.053; Yu. V. Evdokimov, 09.103.105; Yu. V. Evdokimov, E. D. Kondrat'eva, 09.104.010; A. A. Nemo, 09.042.023; D. D. Dzyaman, 09.033.010; D. D. Dzyaman, 09.033.011; M. A. Vajsov, 09.046.009; M. A. Vajsov, 09.046.010).

008.062 Kiel

Sonderdrucke der Sternwarte Kiel, Nos. 180 (D. Koester, 07.126.007), 181 (I. Bues, 06.126.014), 182 (W. Wolfram, 07.064.007), 183 (R. Wehrse, 08.064.005), 184 (A. Unsöld, 08.061.010), 185 (H. Holweger, 08.071.010), 186 (I.-J. Sackmann, V. Weidemann, 08.125.025).

008.063 Kitt Peak

Kitt Peak National Observatory, Contributions, Nos. 552 (C. R. Lynds, 07.158.099), 553 (C. R. Lynds, 07.141.082), 555 (L. H. Auer, J. N. Heasley, R. W. Milkey, 09.064.082), 556 (D. N. B. Hall, 09.071.051).

Photographic report from Kitt Peak. See Abstr. 032.001.

008.064 Krim

Chronicle. Izv. Krymskoj Astrofiz. Obs., Vol. 46, 167 - 168 (1972). In Russian.

Izvestiya Krymskoj Astrofizicheskoj Observatorii, Akademiya Nauk SSSR, Tom (Vol.) 46 (Yu. S. Efimov, N. M. Shakhovskoy, 09.141.085; P. F. Chugainov, 09.122.082; P. P. Petrov, 09.113.042; T. M. Rachkovskaya, 09.121.034; M. E. Boyarchuk, 09.064.053; M. E. Boyarchuk, 09.064.054; R. E. Gershberg, 09.122.083; O. P. Hollandskij, 09.064.055; S. I. Gopasyuk, T. T. Tsap, 09.073.042; S. I. Gopasyuk, T. T. Tsap, 09.073.043; E. A. Baranovsky, N. N. Stepanyan, 09.072.044; V. A. Kotov, 09.072.045; N. Ya. Nikolaev, Yu. F. Yurovsky, 09.077.049; Yu. F. Yurovsky, L. I. Yurovskaya, 09.077.050; E. I. Terez, 09.034.046; A. P. Kulčizky, 09.034.047; G. M. Popov, 09.031.026; 09.008.064).

008.065 Las Cruces

Lindheimer Astronomical Research Center, and Dearborn Observatory, Evanston, Illinois; Corralitos Observatory, Las Cruces, New Mexico. – Observatory reports. J. A. Hynek.
Bull. American Astron. Soc., Vol. 5, 164 - 166 (1973).

New Mexico State University, Department of Astronomy, Las Cruces, New Mexico. – Observatory report. W. L. Reitmeyer.
Bull. American Astron. Soc., Vol. 5, 214 - 216 (1973).

008.066 Lawrence

Kansas University Observatory, Lawrence, Kansas.

Observatory report.   P. A. Wehinger.
Bull. American Astron. Soc., Vol. 5, 141 - 142 (1973).

008.067 **Leningrad**

Ephemerides of minor planets for 1974.
(G. A. Chebotarev, 09.098.019).

Trudy Astronomicheskoj Observatorii, (Transactions of the Astronomical Observatory), *Leningrad,* Vol. 29 (A. K. Kolesov, 09.063.049; D. I. Nagirner, A. B. Schneeweiss, 09.063.050; V. M. Loskutov, 09.063.051; A. A. Nikitin, T. Kh. Feklistova, 09.114.176; V. A. Dombrovskij, T. A. Polyakhova, V. A. Yakovleva, 09.122.142; V. A. Yakovleva, 09.122.143; T. E. Derviz, V. A. Dombrovskij, 09.122.144; M. K. Babadzhanyants, V. A. Hagen-Thorn, E. N. Kopatskaya, V. B. Nebelitskij, E. L. Polyanskaya, 09.141.135; G. V. Khozov, V. V. Shalberova, L. V. Danilova, 09.113.058; O. S. Shulov, 09.121.079; V. G. Khristich, 09.031.059; L. P. Osipkov, 09.151.050; M. A. Belozerova, 09.151.051; V. A. Antonov, E. M. Nezhinskij, 09.151.052; S. P. Yakimov, 09.151.053; V. A. Antonov, 09.042.051; E. N. Polyakhova, 09.052.038; E. I. Timoshkova, K. V. Kholshevnikov, 09.052.039; G. V. Ufimtsev, 09.052.040; M. S. Zverev, D. D. Polozhentsev, 09.041.047; M. P. Mishchenko, A. V. Shiryaev, 09.044.031; S. V. Gromov, 09.031.060).

008.068 **London, Canada**

The Observatories of the University of Western Ontario, London, Canada. – Observatory report.
W. H. Wehlau.
Bull. American Astron. Soc., Vol. 5, 254 - 256 (1973).

008.069 **Los Angeles**

Research Units and Academic Departments, University of California: Berkeley, Los Angeles, San Diego, and Santa Cruz. – **II. Los Angeles Campus.** – Observatory report.
G. O. Abell.
Bull. American Astron. Soc., Vol. 5, 72 - 76 (1973).

008.070 **Louvain**

Institut d'Astronomie et de Géophysique, Georges Lemaître, Université Catholique de Louvain, Heverle-Louvain, Belgique. Contribution Nos. 11 (O. Godart, G. Schayes, J.-P. Peters, 09.082.126), 12 (O. Godart, 09.052.034).

Publications de l'Institut d'Astronomie et de Géophysique, Georges Lemaître, Louvain, Vol. 5, Nos. 2 (P. Melchior, 06.043.008), 3 (P. Melchior, 08.044.002), 4 (P. Melchior, 08.045.038), 6 (R. Dejaiffe, 07.051.003).

008.071 **Madison**

Washburn Observatory, University of Wisconsin, Madison, Wisconsin. – Observatory report.
D. E. Osterbrock.
Bull. American Astron. Soc., Vol. 5, 247 - 251 (1973).

008.072 **Manchester**

University of Manchester, Nuffield Radio Astronomy Laboratories, Jodrell Bank. – Report for the year ending 1972 August 31.   B. Lovell.
Quarterly Journ. Roy. Astron. Soc., Vol. 14, 73 - 80 (1973).

Astronomical Contributions from the University of Manchester, Series II, Jodrell Bank Reprints, Nos. 455 (P. N. Wilkinson, 06.158.022), 465 (A. G. Lyne, 08.141.005), 468 (A. Pedlar, R. D. Davies, 08.131.002), 470 (R. G. Conway, R. E. B. Munro, 08.141.034), 471 (R. D. Davies, H. E. Matthews, A. Pedlar, 08.155.002), 472 (J. Critchley, H. P. Palmer, B. Rowson, 08.141.088), 473 (R. G. Conway, D. Stannard, 08.099.065), 474 (R. G. Conway, D. Stannard, 08.141.068), 475 (P. K. Wraith, 08.141.089), 476 (P. N. Wilkinson, 08.141.090), 477 (R. D. Davies, 08.155.054), 478 (I. W. A. Browne, N. J. McEwan, 08.141.118), 479 (R. G. Conway, D. Stannard, 08.122.099), 481 (J. E. B. Ponsonby, I. Morison, A. R. Birks, J. K. Landon, 08.094.143), 482 (R. J. Peckham, H. P. Palmer, 08.141.083), 483 (B. Anderson, R. G. Conway, R. J. Davis, R. J. Peckham, P. J. Richards, R. E. Spencer, P. N. Wilkinson, 08.142.041), 485 (J. G. Davies, A. Lyne, J. H. Seiradakis, 08.141.524), 486 (J. Milogradov-Turin, F. G. Smith, 09.157.001).

008.073 **Middletown**

Van Vleck Observatory, Wesleyan University, Middletown, Conn. – Observatory report.   A. R. Upgren.
Bull. American Astron. Soc., Vol. 5, 237 - 240 (1973).

008.074 **Milano-Merate**

Contributi dell'Osservatorio Astronomico di Milano-Merate, Nuova Serie, Nos. 330 (P. Galeotti, 03.121.022), 331 (G. Beltrami, P. Galeotti, 03.121.051), 332 (A. Manara, 09.055.002), 333 (R. Gallino, A. Ferrari, A. Masani, 09.162.078), 334 (R. Gallino, A. Ferrari, A. Masani, 05.155.002), 335 (A. Manara, I. Almàr, E. I. Almàr, 09.082.125), 336 (M. Fracassini, L. E. Pasinetti, 04.121.015), 337 (F. Chlistovsky, E. Proverbio, 04.035.002), 338 (F. Zagar, 04.005.008), 339 (C. Casini, L. E. Pasinetti, 09.114.168), 340 (A. Manara, 09.055.003), 341 (A. Kranjc, 07.042.066), 342 (M. Fracassini, L. E. Pasinetti, 05.121.043), 343 (F. Zagar, 05.010.027), 344 (A. Kranjc, 09.052.033), 345 (R. Gallino, A. Masani, G. Silvestro, 06.065.154), 346 (A. Masani, 06.065.090), 347 (C. Casini, L. E. Pasinetti, 09.114.169), 348 (P. L. Battistini, M. Fracassini, L. E. Pasinetti, 06.121.075), 349 (F. Zagar, 08.004.026), 350 (L. Buffoni, A. Manara, 07.042.023), 351 (F. Mazzoleni, 08.032.044).

008.075 **Minneapolis**

University of Minnesota, Minneapolis, Minnesota, Separate print (W. J. Luyten, 09.112.015).

008.076 **Mizusawa**

Annual report of geophysical observations made at the International Latitude Observatory of Mizusawa for the year 1970.   T. Okuda.
Published by the International Latitude Observatory of Mizusawa, Japan. 44 pp. (1972).

This annual report contains the values read every hour in the records of gravity changes at our Observatory and tilting motion of the ground at Akagane Geodetic Station, about 23 km east from Mizusawa. It also contains the results of seismological observations at our Observatory. Outlines of the observations and the data processings are described.

Bulletins, Time Service of the Mizusawa Observatory, Vol. 15, No. 1 - 12, 1970 (S. Takagi, I. Okamoto, M. Aihara, K. Yokoyama, T. Hara, G. Murakami, 09.044.035).

Monthly Notes of the International Polar Motion Service, 1972 Nos. 11 - 12, 1973 Nos. 1 - 4 (09.045.027).

Proceedings of the International Latitude Observatory of Mizusawa, No. 12 (T. Hara, K. Horiai, K. Iwadate, 09.035.003; M. Ooe, S. Kuji, 09.034.039; C. Sugawa, C. Kakuta, 09.045.010; H. Okawa, Y. Goto, 09.032.016; K. Iwadate, 09.034.040; T. Gotō, N. Kikuchi, 09.045.011; N. Kikuchi, 09.044.011; E. Onodera, 09.044.012; T. Gotō, 09.045.012; C. Sugawa, H. Ishii, G. Teleki, 09.045.013).

Publications of the International Latitude Observatory of Mizusawa, Vol. 8, No. 1 (S. Takagi, 09.045.008; S. Takagi, 09.044.010).

008.077   Mons

Communications du Département d'Astrophysique de la Faculté des Sciences de Mons. Mons Astrophysical Papers, Nos. 29 (R. Wilson, S. Gardier, C. Jamar, J. P. Macau, D. Malaise, A. Monfils, H. E. Butler, C. M. Humphries, K. Nandy, G. I. Thompson, P. J. Barker, H. Wroe, L. Houziaux, A. Boksenberg, 08.114.029), 32 (A. Delcroix, S. Volonte, 09.062.063), 33 (P. Barker, A. Boksenberg, H. E. Butler, S. Gardier, L. Houziaux, C. Humphries, C. Jamar, D. Macau-Hercot, D. Malaise, A. Monfils, K. Nandy, G. I. Thompson, R. Wilson, H. Wroe, 09.114.039), 34 (A. Delcroix, 09.064. 083), 35 (Y. Andrillat, L. Houziaux, 09.114.001).

008.078   Montevideo

Departamento de Astronomía y Física, Facultad de Humanidades y Ciencias, Universidad de la República, Montevideo, Uruguay. **Publicación,** Nos. 40 (R. E. Caligaris, J. C. Grangel, 09.022.085), 41 (G. Rossel, R. Donangelo, 09.022. 086), 42 (F. Cernuschi, F. R. Marsicano, 09.131.209).

008.079   Moskva

Trudy Gosudarstvennogo Astronomicheskogo Instituta im. P. K. Shternberga. Izdatel'stvo Moskovskogo Universiteta, Vol. 43, vyp. (No.) 1 (P. A. Stroev, Ju. A. Pavlov, V. L. Panteleev, V. O. Bagramjants, 09.081.012).

008.080   München

Max-Planck-Institut für Physik und Astrophysik, Institut für Extraterrestrische Physik, Garching bei München, Separate prints (J. Trümper, V. Schönfelder, 09.142.094; D. Hovestadt, O. Vollmer, G. Gloeckler, C. Y. Fan, 09.143. 095).

Mitteilungen aus dem Institut für Astronomische und Physikalische Geodäsie der Technischen Universität München, Nos. 99 (M. Schneider, 09.021.006), 100 (E. Nagel, C. Reigber, 09.052.042), 102 (K. H. Ilk, 09.052.043).

008.081   Nashville

Dyer Observatory, Vanderbilt University, Nashville, Tennessee. – Observatory report.   A. M. Heiser. Bull. American Astron. Soc., Vol. 5, 99 - 101 (1973). – Report 1971 September 15 – 1972 September 15.

008.082   Neuchâtel

Rapport d'activité pour l'exercice 1972 et Rapport sur le Concours chronométrique 1972.   J. Bonanomi. Observatoire Cantonal de Neuchâtel, 23 pp. (1973).

Observatoire de Neuchâtel, Bulletin. Série B, 1972 January - December (09.044.036); Série D, 1972 January - December (09.044.037).

008.083   New Haven

Yale University Observatory, New Haven, Connecticut. – Observatory report.   P. Demarque. Bull. American Astron. Soc., Vol. 5, 258 - 262 (1973).

Transactions of the Astronomical Observatory of Yale University, Vol. 32, Part 1 (D. Hoffleit, 09.034.101).

008.084   Nice

Rapport d'activité de l'Observatoire de Nice pour 1971.   P. Delache. Bull. d'Information, Ass. Développement International Obs. Nice, No. 9, p. 45 - 85, 101 - 107 (1972).

Les opérations "portes ouvertes" à l'Observatoire de Nice.   P. Franck. Bull. d'Information, Ass. Développement International Obs. Nice, No. 9, p. 95 - 98 (1972).

008.085   Ondřejov

Seventy-five years of the observatory Ondřejov. Říše hvězd, Vol. 54, 57 - 58 (1973). In Czech.

008.086   Ottawa

Contributions Astrophysics Branch, National Research Council of Canada, Ottawa, Ontario, NCR 12824 (W. J. Medd, B. H. Andrew, G. A. Harvey, J. L. Locke, 08. 141.070).

Contributions from the Earth Physics Branch, Ottawa, Canada, Nos. 393 (M. R. Dence, 09.105.159), 417 (J. C. Gupta, S. R. C. Malin, 08.084.304), 433 (E. G. Woolsey, 09.045.028).

008.087 Oxford

Department of Astrophysics, University Observatory, Oxford, Publication Nos. 26 (R. A. Lyttleton, 04.102. 023), 29 (A. D. Petford, D. E. Blackwell, B. S. Collins, P. A. Ibbetson, E. A. Mallia, G. Smith, D. Emerson, 06.034.006), 30 (D. L. Lambert, E. A. Mallia, A. D. Petford, 06.071.043), 33 (M. G. Adam, 06.072.052), 37 (D. L. Lambert, E. A. Mallia, 07.072.018), 38 (E. A. Mallia, D. E. Blackwell, A. D. Petford, 06.072.057), 39 (E. A. Mallia, A. D. Petford, 07.072. 022), 40 (R. Hunt, D. W. Sciama, 07.161.011), 41 (M. F. Ingham, 07.082.009), 43 (D. E. Blackwell, B. S. Collins, 07.022.088), 45 (J. R. Sternberg, 08.082.006), 46 (D. E. Blackwell, B. S. Collins, A. D. Petford, 07.071.027), 48 (W. D. Evans, G. T. Bath, 07.117.027), 49 (R. Hunt, D. W. Sciama, 08.161.001), 50 (N. L. Balazs, 08.062.032), 51 (D. E. Blackwell, G. Calamai, R. B. Willis, 08.071.039), 52 (D. E. Blackwell, J. H. Kirby, G. Smith, 08.071.049), 53 (S. Chandrasekhar, J. L. Friedman, 08.066.082), 54 (G. Wegner, 08.126.016), 55 (D. W. Sciama, 08.155.049), 56 (S. Chandrasekhar, 08.066. 051), 57 (G. Smith, 09.022.081).

008.088 Padova

Programmi astrometrici in corso presso gli osservatori astronomici di Padova e Asiago.
C. Barbieri, M. Capaccioli, R. Ganz, G. Pinto.
Mem. Soc. Astron. Italiana, Nuova Ser., Vol. 43, 635 - 636 (1973).

008.089 Palo Alto

Lockheed Palo Alto Research Laboratory, Palo Alto, California. – Observatory report.   B. McCormac.
Bull. American Astron. Soc., Vol. 5, 166 - 167 (1973).

008.090 Paris

Bureau International de l'Heure. Annual report for 1972.   R. Michard (Editor).
Printing Office: Observatoire de Paris. 6 + A15 +B43 + C15 pp. (1973).
Contents: Methods of computation; Tables and figures; Time signals.

Bureau International de l'Heure, (B. I. H.), Circulaires B/C, Nos. 201 - 207 (09.045.032).

Bureau International de l'Heure, (B. I. H.), Circular D 74 - D 80 (09.044.039).

008.091 Pasadena

Hale Observatories, operated by Carnegie Institution of Washington and California Institute of Technology, Pasadena, California. Annual report of the director, July 1, 1971 - June 30, 1972.   H. W. Babcock.
Reprinted from Carnegie Institution, Washington, Year Book, Vol. 71, 649 - 716 (1972).

008.092 Philadelphia

Flower and Cook Observatory, University of Pennsylvania, Philadelphia, Pennsylvania. – Observatory report.
R. H. Koch.
Bull. American Astron. Soc., Vol. 5, 112 - 114 (1973).

008.093 Pittsburgh

Allegheny Observatory, University of Pittsburgh, Pittsburgh, Pennsylvania. – Observatory report.
J. Kiewiet de Jonge.
Bull. American Astron. Soc., Vol. 5, 57 (1973).

008.094 Porto

Observatorio Astronómico da Universidade do Porto, Monte da Virgem – Vila Nova de Gaia (Portugal). Informações do Observatório Astronómico da Faculdade de Ciências do Porto, Nos. 3 (T. Torrão, J. Osório, 09.041.049), 4 (A. J. Pascoal, 09.031.061).

Observatorio Astronómico da Universidade do Porto, Monte da Virgem – Vila Nova de Gaia (Portugal). Publicações do Observatório Astronomico («Prof. Manuel de Barros») da Faculdade de Ciências do Porto, Nos. 25 (J. Osório, 09.096.029), 26 (J. Osório, 09.005.037), 27 (J. Osório, T. Torrão, 09.046.023), 28 (J. Osório, N. Rego, 09.096.030).

Professor M. G. Pereira de Barros and the Astronomical Observatory of the University of Porto.
See Abstr. 005.037.

008.095 Praha

Académie Tchécoslovaque des Sciences, Institut Astronomique, Station de l'Heure à Prague, Série 6, Nos. 3 - 4 (L. Webrová, V. Ptáček, 09.044.033).

Contributions and Observations from the People's Observatory of Prague, Vol. 8, Ser. 5, No. 1 (M. Procházková, 09.098.065).

008.096 Princeton

Princeton University Observatory, Princeton, New Jersey. – Observatory report.   L. Spitzer, Jr.
Bull. American Astron. Soc., Vol. 5, 225 - 229 (1973).

008.097 Pulkovo

Trudy Glavnoj Astronomicheskoj Observatorii v Pulkove, Seriya 2, Vol. 77 (O. T. Markina, G. M. Petrov, 09. 041.037; L. F. Gorel, 09.041.038; G. S. Kosin, 09.041.039; L. F. Gorel, 09.041.040; G. D. Baturina, 09.041.041; V. N. Boyko, S. V. Shilova, 09.041.042; O. T. Markina, G. M. Petrov, 09.041.043; I. I. Bozhko, G. K. Zimmerman, 09.041.044; G. S. Kosin, 09.041.045; G. S. Kosin, 09.041.046; V. A. Sokolova, 09.124.109; Zh. P. Anosova, 09.117.039; T. S. Belorossova, D. D. Maksutov, N. V. Merman, M. A. Sosnina, 09.031.055).

Trudy Glavnoj Astronomicheskoj Observatorii v Pulkove. Seriya 2, Vol. 79 (V. I. Sakharov, 09.045.006; G. S. Sheptunov, 09.045.007).

008.098 Pulsnitz

Veröffentlichungen der Sternwarte Pulsnitz (Sachsen) No. 9 (J. Classen, 09.004.090).

008.099 Richmond Hill

David Dunlap Observatory, University of Toronto, Richmond Hill, Ontario, Canada. – Observatory report. D. A. MacRae.
Bull. American Astron. Soc., Vol. 5, 92 - 99 (1973).
The following report of the work of the Observatory is for the period 1 July 1971 to 30 June 1972.

Communications from the David Dunlap Observatory, University of Toronto, Richmond Hill, Ontario, Canada, Nos. 335 (R. G. Conway, J. A. Gilbert, P. P. Kronberg, R. G. Strom, 07.141.167), 336 (K. M. Yoss, T. E. Lutz, 06.112.010), 337 (S. van den Bergh, 08.158.071), 338 (J. D. Fernie, 08.082.090), 339 (S. P. S. Anand, M. M. Shara, 07.141.157), 340 (N. R. Walborn, 08.121.049).

008.100 Rochester

C. E. Kenneth Mees Observatory, Rochester, New York. – Observatory report. S. Sharpless.
Bull. American Astron. Soc., Vol. 5, 180 - 183 (1973).

C. E. Kenneth Mees Observatory, University of Rochester, Rochester, N. Y., Reprints, Nos. 36 (C. R. Sturch, H. L. Helfer, 08.113.034), 39 (L. G. Taff, J. E. Littleton, 08.155.046), 40 (S. Sharpless, 09.008.100), 42 (L. G. Taff, J. E. Littleton, 09.153.012).

008.101 Roma

Osservatorio Astronomico di Roma, Monte Mario – Monte Porzio – Stazione Astrofisica sul Gran Sasso. Contributi scientifici, Serie III, Nos. 100 (M. Cimino, 09.004.087), 119 (P. Giannone, M. A. Giannuzzi, 07.117.005), 120 (P. Giannone, M. A. Giannuzzi, 07.117.032), 121 (N. Virgopia, 06.065.133), 122 (K. Nandy, F. Smriglio, 06.114.073), 123 (G. A. De Biase, F. Sacchetti, D. Trevese, 09.034.109).

Monthly Bulletin. Osservatorio Astronomico di Roma, Nos. 176, 177, 179 (M. Cimino, M. Torelli, V. Croce, R. Flamini, F. Casamassima, 09.075.022).

Photographic Journal of the Sun, Osservatorio Astronomico di Roma, Nos. 57, 58, 66, 68 (M. Cimino, 09. 075.023).

008.102 Rosario

Observatorio Astronómico Municipal, Rosario – Argentina. Departamento de Fisica Solar, Contribuciones, Serie 1, No. 6 (R. Barbarroja, V. Capolongo, J. A. Gutierrez, O. F. Liesche, L. A. Mansilla, 09.075.012).

008.103 San Diego

Research Units and Academic Departments, University of California: Berkeley, Los Angeles, San Diego, and Santa Cruz. – III. San Diego Campus. – Observatory report.
Bull. American Astron. Soc., Vol. 5, 76 - 78 (1973).

008.104 San Fernando

Memoria de las actividades en 1972.
Separate print Inst. y Obs. de Marina, San Fernando (Cadiz), 12 pp. (1973).

008.105 Santa Cruz

Research Units and Academic Departments, University of California: Berkeley, Los Angeles, San Diego, and Santa Cruz. – IV. Santa Cruz Campus: A.) Board of Studies in Astronomy and Astrophysics, J. Faulkner; B.) Lick Observatory, R. P. Kraft. – Observatory Report.
Bull. American Astron. Soc., Vol. 5, 78 - 86 (1973).

008.106 Seattle

University of Washington, Astronomy Department, Seattle, Washington. – Observatory report. G. Wallerstein.
Bull. American Astron. Soc., Vol. 5, 251 - 254 (1973).

008.107 Sonneberg

Zentralinstitut für Astrophysik. Mitteilungen über Veränderliche Sterne, Sonneberg, Vol. 6, No. 4 (H. Huth, 09.123.046; E. Scheller, 09.123.047; P. Ahnert, 09.123.048; P. Ahnert, 09.123.049; P. Ahnert, 09.121.071; P. Ahnert, 09.121.072; P. Ahnert, 09.121.073; P. Ahnert, 09.121.074; W. Zschocke, 09.123.050; A. Eichhorn, 09.123.051; E. Splittgerber, 09.123.052; D. Böhme, 09.123.053; W. Götz, 09.122.135; H. Geßner, 09.123.054; I. Meinunger, 09.123.055; W. Wenzel, 09.123.056).

008.108 Sutherland

South Africa's newest observatory.
R. P. Olowin.
Sky Telescope, Vol. 45, 347 - 350 (1973).

008.109 Swarthmore

Sproul Observatory, Swarthmore, Pennsylvania.
1 July 1971 – 30 June 1972. – Observatory report.
P. van de Kamp.
Bull. American Astron. Soc., Vol. 5, 231 (1973).

Sproul Observatory, Swarthmore, Pennsylvania, Reprints, Nos. 208 (P. van de Kamp, 07.008.137), 209 (P. van de Kamp, 07.112.004), 210 (W. D. Heintz, 07.118. 008), 211 (S. L. Lippincott, 07.118.009), 212 (J. L. Hershey, 07.118.010), 213 (S. L. Lippincott, 07.122.078), 214 (S. L. Lippincott, J. L. Hershey, 08.118.007), 215 (K. Chang, 08.118.010), 216 (P. van de Kamp, M. D. Worth, 08.117.029).

008.110 Sydney

Division of Radiophysics, C.S.I.R.O., Sydney, (Epping, New South Wales, Australia) Separate prints (M. M. Komesaroff, P. A. Hamilton, J. G. Ables, 08.141.560; J. C. Ribes, J. G. Ables, P. D. Godfrey, R. D. Brown, 09.022.044; M. W. Sinclair, N. Fourikis, J. C. Ribes, B. J. Robinson, R. D. Brown, P. D. Godfrey, 09.131.117; A. J. Shimmins, J. V. Wall, 09.141.062).

008.111 Tartu

Tartu Astronoomia Observatorium, Teated, No. 41 (H. Eelsalu, 09.155.012).

008.112 Tashkent

Chronicle.
Tsirk. Astron. Inst., *Tashkent*, No. 33 (380), p. 20 - 23 (1972). In Russian.

Tsirkulyar Astronomicheskogo Instituta, Akademiya Nauk Uzbekskoj SSR, Nos. 33 (380) (N. A. Omelina, 09. 044.017; F. G. Mustaeva, 09.075.005; G. M. Kaganovskij, 09. 032.020; M. R. Ehshmatov, Eh. Rakhmatov, 09.096.011; 09. 008.112), 34 (381) (G. M. Kaganovskij, 09.045.018; A. A. Latypov, 09.112.009), 35 (382) (N. A. Omelina, 09.044.017; F. G. Mustaeva, 09.075.005; I. M. Boroditskij, 09.041.025), 36 (383) (M. F. Bykov, 09.032.021), 37 (384) (N. A. Omelina, 09.044.017; F. G. Mustaeva, 09.075.005; A. A. Latypov, 09.112.010), 38 (385) (M. F. Bykov, 09.032.022), 39 (386) (N. A. Omelina, 09.044.017; F. G. Mustaeva, 09.075.005; A. A. Latypov, 09.112.011).

008.113 Tokyo

Time and Latitude Bulletins, Tokyo Astronomical Observatory, Vol. 46, Nos. 9 - 12 (09.044.023).

Tokyo Astronomical Bulletin, Tokyo Astronomical Observatory, Second Series, Nos. 223 (Y. Kozai, A. Tsuchiya, K. Tomita, T. Kanda, H. Sato, N. Kobayashi, Y. Torii, 09.032. 031), 224 (K. Ichimura, Y. Shimizu, E. Watanabe, T. Okada, 09.122.134), 225 (Y. Kozai, R. Manabe, 09.055.001).

Tokyo Astronomical Observatory, Reprints. Nos. 427 (S. Aoki, 09.151.001), 428 (K. Nishi, M. Makita, 09.072. 001), 429 (S. Isobe, 09.131.004), 430 (N. Kaifu, K. Akabane, M. Morimoto, 09.131.005), 431 (H. S. Hudson, K. Ohki, 07.076.022), 432 (N. Sekiguchi, 08.094.147), 433 (Y. Uchida, H. Hudson, 08.073.096), 434 (S. Isobe, 09.131.135), 435 (Y. Kozai, H. Kinoshita, 09.042.007), 438 (Y. Uchida, M. D. Altschuler, G. Newkirk, Jr., 09.074.019).

University of Tokyo, Tokyo Astronomical Observatory, Report, No. 61, Vol. 16, No. 2 (K. Saito, S. Shinozawa, 09.093.013; T. Hirayama, 09.021.004; K. Saito, A. Tojo, 09.079.001; M. Utida, 09.004.035; 09.082.047; K. Saito, S. Shinozawa, 09.097.038; R. Fukaya, 09.082.048), No. 62, Vol. 16, No. 3 (09.033.012; 09.033.013; R. Fukaya, H. Ishii, I. Morita, 09.082.049; H. Kinoshita, H. Nakai, 09.042.026; S. Nagasawa, I. Shimizu, 09.032.013; K. Nishi, K. Higashi, A. Yamaguchi, 09.022.051; H. Hara, H. Ishii, I. Kamijo, N. Miyauchi, 09.031.018; H. Tanabe, A. Takechi, A. Miyashita, Y. Mikami, 09.082.050; K. Nakajima, 09.031.019, K. Kai,

Y. Uchida, N. Shibuya, H. Hirabayashi, T. Kuwabara, 09.021. 005; K. Tomita, T. Kanda, 09.034.024).

Data Report of Hydrographic Observations. Series of Astronomy and Geodesy, Maritime Safety Agency, Tokyo, Japan, No. 7 (T. Mori, Y. Ganeko, Y. Harada, 09.096.028).

008.114 Torino

Osservatorio Astronomico di Torino, Pino Torinese. Time Service, Bulletin Nos. 3, 4 (C. Moranzino, 09.044.024).

008.115 Treviso

Osservatorio Privato Specola "Ariel", Treviso, Pubblicazione, Nos. 55 (G. Romano, 06.123.072), 56 (G. Pinto, G. Romano, 07.123.006), 57 (G. Romano, M. Perissinotto, 08.123.009), 58 (G. Romano, 07.123.017), 59 G. Romano, 07.123.019), 60 (G. Romano, 07.125.034), 61 (G. Romano, 08.141.001).

008.116 Tucson

Kitt Peak National Observatory, Tucson, Arizona and Cerro Tololo Inter-American Observatory, La Serena, Chile. – Observatory reports. L. Goldberg. Bull. American Astron. Soc., Vol. 5, 142 - 162 (1973).

National Radio Astronomy Observatory, Charlottesville, Virginia, Green Bank, West Virginia, and Tucson, Arizona. Observatory reports. D. S. Heeschen. Bull. American Astron. Soc., Vol. 5, 206 - 214 (1973). – This report covers the period July 1971 through June 1972.

008.117 Uccle

Observatoire Royal de Belgique (Koninklijke Sterrenwacht van België), Communications (Mededelingen), Série A, Nos. 20 (J. Dommanget, 08.118.002; E. L. van Dessel, 08.118.005), 21 (R. J. Dejaiffe, 09.041.002).

Observatoire Royal de Belgique (Koninklijke Sterrenwacht van België), Communications (Mededelingen), Série B, Nos. 63 (G. Evrard, C. Gonze, A. Koeckelenbergh, 08.075.004), 75 (E. van Hemelrijk, H. Debehogne, 08.084. 230).

008.118 Victoria

The Dominion Astrophysical Observatory, Victoria, B.C. J. B. Hutchings. Journ. Roy. Astron. Soc. Canada, Vol. 67, 97 - 98 (1973).

Contributions from the Dominion Astrophysical Observatory, Victoria, B.C., Nos. 163 (J. B. Hutchings, P. G. Laskarides, 07.064.002), 177 (D. Crampton, 08.112.002), 181 (J. B. Hutchings, 08.124.002), 183 (G. Hill, J. V. Barnes, 08.121.007), 184 (G. Hill, J. V. Barnes, 08.121.010), 185 (D. Crampton, J. B. Hutchings, 07.122.117), 187 (G. J. Odgers, 05.032.036), 188 (E. H. Richardson, 05.032.044; E. H. Richardson, 07.032.045), 189 (R. W. Hilditch, 08.121.

061), 190 (D. Crampton, F. D. A. Hartwick, 08.119.008), 194 (K. O. Wright, 08.121.087), 196 (D. Crampton, J. B. Hutchings, 08.142.104), 202 (E. H. Richardson, 07.034.084).

008.119 Villanova

Villanova University, Villanova, Pennsylvania. Observatory report. E. F. Jenkins. Bull. American Astron. Soc., Vol. 5, 256 - 257 (1973).

008.120 Vilnius

Vilniaus Astronomijos Observatorijos Biuletenis (Bulletin of the Vilnius Astronomical Observatory), No. 35 (V. Straižys, Z. Sviderskienė, 09.114.090).

008.121 Warsaw

Warsaw University Observatory and Astronomical Institute, Polish Academy of Sciences, Reprint Nos. 325 (B. Paczyński, R. Sienkiewicz, 08.117.026), 326 (M. Różycz-ka, 08.158.094), 327 (A. Żytkow, 08.064.031), 328 (G. Sitarski, 08.103.118).

008.122 Washington

U. S. Naval Observatory, Washington, D.C. Observatory report. K. A. Strand. Bull. American Astron. Soc., Vol. 5, 232 - 237 (1973).

Publications of the United States Naval Observatory. *Washington,* Second Series, Vol. 20, Part 5 (G. E. Kron, H. H. Guetter, B. Y. Riepe, 09.113.054).

United States Naval Observatory, *Washington, D.C.,* **Circular,** Nos. 140 (G. Kaplan, J. B. Dunham, 09.094.299), 141 (J. S. Duncombe, 09.094.300).

U.S. Naval Observatory, Washington, D.C., Time Service Publications, Series 4, Nos. 309 - 335 (09.044.040); Series 7, Nos. 262 - 288 (09.044.041).

National Aeronautics and Space Administration, Washington, D.C. − Observatory reports. N. G. Roman. Bull. American Astron. Soc., Vol. 5, 188 - 201 (1973). − Concerning the reports of Ames Research Center; Goddard Space Flight Center; Jet Propulsion Laboratory; Langley Research Center; Marshall Space Flight Center.

National Bureau of Standards, Washington, D.C. Observatory report. L. Hagan. Bull. American Astron. Soc., Vol. 5, 201 - 204 (1973).

008.123 Westford

Haystack Observatory, Northeast Radio Observatory Corporation, Westford, Massachusetts. − Observatory report. P. B. Sebring. Bull. American Astron. Soc., Vol. 5, 126 - 129 (1973). − This report covers the period from 1 July 1971 through 30 June 1972.

008.124 Wien

Zur Standortwahl und Organisations-Struktur für ein 100-cm-Ritchey-Chrétien-Teleskop. W. Weiss. SuW, Vol. 12, 79 - 82 (1973).

008.125 Williams Bay

Annual report of the University of Chicago, Yerkes Observatory, Williams Bay, Wisconsin 1971−1972. − Observatory report. W. F. van Altena. Bull. American Astron. Soc., Vol. 5, 86 - 92 (1973).

008.126 Williamstown

Hopkins Observatory, Williams College, Williamstown, Massachusetts. −Observatory report. J. M. Pasachoff. Bull. American Astron. Soc., Vol. 5, 129 - 130 (1973).

008.127 Wrocław

Wrocław Astronomical Observatory, Reprint Nos. 83 (M. Jerzykiewicz, 03.122.071), 84 (B. Rompolt, P. Rybka, A. Spodenkiewicz, B. Szczodrowska, 05.092.011), 85 (I. Garczyńska, 06.077.017), 86 (M. Jerzykiewicz, 06.122.119), 87 (J. Bem, 08.118.004), 88 (G. M. Petrov, J. Bem, 09.032.008).

008.128 Yorktown Heights, N. Y.

IBM Thomas J. Watson Research Center, Yorktown Heights, New York. − Observatory report. M. C. Gutzwiller. Bull. American Astron. Soc., Vol. 5, 130 (1973). − Research in Astronomy and Astrophysics at IBM is carried out in the General Sciences Department of the Research Center. Some of this work is done by people who have come recently from other areas of the physical sciences.

008.129 Zelenchukskaya

Chronicle. Astrofiz. Issled., Izv. Spets. Astrofiz. Obs., Vol. 4, 211 - 213 (1972). In Russian.

Astrofizicheskie Issledovaniya. Izvestiya Spetsial'-noj Astrofizicheskoj Observatorii, Vol. 4 (S. V. Rublev, 09.114.103; S. V. Rublev, 09.114.104; S. V. Rublev, 09.064.044; N. M. Chunakova, 09.064.045; R. N. Kumajgorodskaya, I. M. Kopylov, 09.114.105; K. I. Kozlova, 09.114.106; G. I. Abbasov, S. K. Zejnalov, E. L. Chentsov, 09.114.107; A. V. Kharitonov, V. G. Klochkova, 09.114.108; N. F. Vojkhanskaya, 09.122.061; Yu. P. Korovyakovsky, 09.117.022; N. V. Bystrova, 09.155.059; I. V. Gosachinsky, 09.155.060; B. P. Artamonov, L. S. Nazarova, 09.158.098; G. A. Pirog, 09.158.099; N. A. Yesepkina (*Esepkina*), 09.033.014; V. M. Spitkovsky, 09.033.015; G. B. Gelfrejkh, O. A. Golubchina, 09.033.016; S. M. Vilenchik, Ya. B. Vyatskin, A. S. Najshul, E. M. Neplokhov, 09.032.014; N. F. Nelyubin, 09.082.059; V. Ya. Golnev, V. M. Spitkovsky, 09.033.017; 09.008.129).

Soobshcheniya Spetsial'noj Astrofizicheskoj Obser-

vatorii, Akademiya Nauk SSSR, vyp. (No.) 3 (L. I. Snezhko, 09.064.075; Yu. N. Efremov, I. M. Kopylov, 09.115.026), 4 (S. V. Rublev, 09.114.159).

008.130 Zürich

Tätigkeitsbericht der Eidgenössischen Sternwarte Zürich für das Jahr 1972. M. Waldmeier. Zürich, 7 pp. (1973).

Astronomische Mitteilungen der Eidgenössischen Sternwarte Zürich, Nos. 311 (M. Waldmeier, 09.075.017), 312 (M. Waldmeier, 08.074.093), 313 (W. Stanek, 08.072. 057), 314 (M. Waldmeier, 09.074.096), 315 (M. Waldmeier, 09.073.010), 316 (M. Waldmeier, 09.073.099), 319 (M. Waldmeier, 09.075.018), 322 (M. Waldmeier, 09.079.100).

Quarterly Bulletin on Solar Activity (Zürich), Nos. 177 - 178 (M. Waldmeier, R. Howard, R. Michard, G. Olivieri, M. Bernot, 09.075.016).

009 Notes on Observatories, Planetaria, and Exhibitions

009.001 An exponential law for the establishment of observatories in the nineteenth century. D. B. Herrmann. Journ. History Astron., Vol. 4, 57 - 58 (1973). – Note.

009.002 Zur Statistik von Sternwartengründungen im 19. Jahrhundert. D. B. Herrrmann. Sterne, 49. Jahrgang, p. 48 - 52 (1973).

009.003 The new observatory in Baja California, Mexico. E. E. Mendoza V. Mercury, (Journ. Astron. Soc. Pacific), Vol. 2, No. 1, p. 9 - 10, 19 (1973).

009.004 Eine Beobachtungsnacht am Catalina Observatory. D. Lemke. SuW, Vol. 12, 70 - 74 (1973).

009.005 Meine Sternwarte. W. Isliker. Orion Schaffhausen, 31. Jahrgang, p. 58 - 62 (1973).

009.006 L'exposition mondiale Copernic à la Bibliothèque Nationale. E. Pognon. L'Astronomie, 87ᵉ année, p. 76 - 80 (1973).

009.007 Rhodes University. Department of Physics. E. E. Baart. Monthly Notes Astron. Soc. Southern Africa, Vol. 32, 2 - 3 (1973). – Report 1972.

009.008 University of Cape Town. Department of Astronomy. B. Warner. Monthly Notes Astron. Soc. Southern Africa, Vol. 32, 3 - 6 (1973). – Report 1972.

009.009 University of the Orange Free State. Boyden Observatory, Department of Astronomy. A. H. Jarrett. Monthly Notes Astron. Soc. Southern Africa, Vol. 32, 7 - 8 (1973). – Report 1972.

009.010 University of Potchefstroom. Cosmic Ray Research Unit. P. H. Stoker. Monthly Notes Astron. Soc. Southern Africa, Vol. 32, 24 (1973). – Report 1972.

009.011 University of South Africa. Department of Mathematics and Astronomy. J. Wolterbeek. Monthly Notes Astron. Soc. Southern Africa, Vol. 32, 24 (1973). – Report 1972.

009.012 Department of Natural Philosophy, University of Aberdeen. – Report for the year ending 1972 September 30. R. V. Jones. Quarterly Journ. Roy. Astron. Soc., Vol. 14, 65 - 66 (1973).

009.013 The Strasbourg Stellar Data Center. A short description. J. Jung. Stellar ages. Proc. IAU Colloquium No. 17, (see 012.015), XXX, 1 - 3 (1973).

009.014 Schul- und Volkssternwarte der Stadt Schaffhausen (Hans Rohr-Sternwarte). H. Rohr. Orion Schaffhausen, 31. Jahrgang, p. 100 (1973). – Jahres - Bericht 1972.

009.015 Volkssterrenwacht "Simon Stevin", Hoeven (N. Br.). Hemel en Dampkring, Vol. 71, 161 - 163 (1973).

009.016 Fünf Jahre Leopold Figl-Observatorium für Astrophysik auf dem Mitterschöpfl. Sternenbote, 16. Jahrgang, p. 90 - 97 (1973).

009.017 Interview sur une passion. G. Bianchi. L'Astronomie, 87ᵉ année, p. 229 - 233 (1973). Interview de G. Viscardy à l'Observatoire de Saint-Martin-de-Peille.

009.018 Stony Ridge Observatory after 10 years. D. Milon. Sky Telescope, Vol. 46, 19 - 21 (1973).

009.019 The Burke-Gaffney Observatory, Halifax, Nova Scotia. D. L. DuPuy. Journ. Roy. Astron. Soc. Canada, Vol. 67, 156 (1973).

009.020 Copernicus Astronomical Centre. J. Smak. Postępy Astron., Vol. 21, 153 - 154 (1973). In Polish.

009.021 The proposed radioastronomical centre in Toruń. S. Gorgolewski.

Postępy Astron., Vol. 21, 154 - 155 (1973). In Polish.

009.022 NESOS *(Netherlands Southern Observing Station).*
E. J. A. Meurs, G. Vleeming.
Hemel en Dampkring, Vol. 71, 212 - 215 (1973).

009.023 Evaluation of Mauna Kea, Hawaii, as an observatory site. D. Morrison, R. E. Murphy, D. P. Cruikshank, W. M. Sinton, T. Z. Martin.
Publ. Astron. Soc. Pacific, Vol. 85, 255 - 267 (1973).
Data are presented describing the qualities of Mauna Kea as an observatory site.

009.024 Ein neues Forschungszentrum für die Astronomie. H. Elsässer.
Separate print from Jahrbuch der Max-Planck-Gesellschaft zur Förderung der Wissenschaften e.V. 1972, p. 83 - 112 (1973).

009.025 People's observatories in Bulgaria.   N. S. Nikolov.

Zemlya i Vselennaya, 1973, No. 3, p. 70 - 72. In Russian.

009.026 Tätigkeitsbericht 1972 der Privatsternwarte Karlsruhe.   W. Malsch.
Separate print: Erzbergerstr. 111c, Karlsruhe (Germany). 1 p. (1973).

009.027 Aus der Arbeit der Volkssternwarten.
Sterne, 49. Jahrgang, p. 54 - 55, 122 (1973). – (1) Verbesserung zur Einstellung der Sonderprojektoren am "Zeiss-Kleinplanetarium" (*H. Weiss*); (2) Astronomiegeschichte an der Archenhold-Sternwarte im Jahre 1972 (*D. B. Herrmann*).

Astronomie auf dem Gamsberg.
SuW, Vol. 12, 10 - 11 (1973). – Concerning the possible site for an observing station of the Max-Planck-Inst. für Astronomie in Heidelberg in the southern hemisphere.

The planetarium in modern science education.
See Abstr. 014.041.

010 Societies, Associations, Organizations

010.001 American Association of Variable Star Observers (AAVSO)

No publication received.

010.002 American Astronomical Society (AAS)

Late-paper abstracts from the 139th meeting of the American Astronomical Society held in Las Cruces, New Mexico, 9–12 January 1973.
Bull. American Astron. Soc., Vol. 5, 265 - 268 (1973).

Abstracts of papers presented at the Solar Physics Division meeting held in Las Cruces, New Mexico, 8–9 January 1973.
Bull. American Astron. Soc., Vol. 5, 268 - 283 (1973).

Abstracts of papers presented at the Commission V – URSI meeting held in Socorro, New Mexico, 8–9 January 1973.
Bull. American Astron. Soc., Vol. 5, 283 - 286 (1973).

Abstracts of papers presented at the AAS 4th annual Planetary Sciences Division meeting held in Tucson, Arizona, 20–23 March 1973.
Bull. American Astron. Soc., Vol. 5, 287 - 310 (1973).

010.003 Association of Lunar and Planetary Observers (ALPO)

Announcements.

Strolling Astronomer, Vol. 23, 226 - 228, Vol. 24, 38 - 40 (1972); Vol. 24, 78 - 80, 122 - 124 (1973).

The Minor Planet Bulletin. Bulletin of the Minor Planets Section of the Association of Lunar and Planetary Observers. (To be obtained from R. G. Hodgson, Minor Planets Section, Dordt College, Sioux Center, Iowa, 51520, U.S.A.), Vol. 1, No. 1 (R. G. Hodgson, 09.098.079).

010.004 Astronomical Society of Australia (ASA)

Society business.
Proc. Astron. Soc. Australia, Vol. 2, 163 - 166 (1972).

010.005 Astronomical Society of Czechoslovakia

No publication received.

010.006 Astronomical Society of the Pacific (ASP)

Minutes of the meeting of the directors.
Mercury, (Journ. Astron. Soc. Pacific), Vol. 2, No. 2, p. 20 - 22 (1973).

New members of the Society.
Mercury, (Journ. Astron. Soc. Pacific), Vol. 2, No. 2, p. 22 - 23 (1973).

010.007   Astronomical Society of Southern Africa (ASSA)

Notices.
Monthly Notes Astron. Soc. Southern Africa, Vol. 32, 1, 21,
29 - 30, 41 (1973).

010.008   Astronomical Society of Victoria (ASV)

Society notes.
Journ. Astron. Soc. Victoria, Vol. 25, 90 - 91; Vol. 26, 31 -
32 (1972/73).

Annual report, 1972.
T. B. Tregaskis, R. J. Lawrence, D. H. Walker, A. E. Coombs,
W. G. H. Tregear, J. H. White, P. Simon, J. B. Trainor, J. D.
Patchett, G. Briggs, D. H. Whitehead, B. A. J. Clark, B. S.
Adcock, J. L. Perdrix.
Journ. Astron. Soc. Victoria, Vol. 26, 2 - 13 (1973). – Includ-
ed are reports on the activities of different sections of the
Society.

010.009   Astronomical Society of Western Australia (ASWA)

Reports of proceedings – 242nd – 245th ordinary
meetings.
Journ. Astron. Soc. Western Australia, Vol. 24, January – June
(1973).

010.010   Astronomische Gesellschaft (AG)

Wissenschaftliche Tagung der Astronomischen Gesellschaft mit 53. ordentlicher Mitgliederversammlung in Wien
vom 18.–23. September 1972.   K. Schaifers.
Mitt. Astron. Ges., No. 32, p. 7 - 8 (1973). – Bericht über die
Tagung.

Mitteilungen des Vorstandes der Astronomischen
Gesellschaft.
Mitt. Astron. Ges., No. 32, p. 281 - 283 (1973).

Frühjahrstagung der AG.
SuW, Vol. 12, 131 - 132 (1973).

010.011   Astronomisk Selskab København

No publication received.

010.012   British Astronomical Association (BAA)

Notices.
Journ. British Astron. Ass., Vol. 83, 82 - 85, 162 - 166, 242 -
244 (1973).

Meetings of the Association.
Journ. British Astron. Ass., Vol. 83, 90 - 97, 167 - 171, 245 -
253 (1973).

Lunar Section.   P. Moore.
Journ. British Astron. Ass., Vol. 83, 125 - 127, 204 - 208,
280 - 282 (1973).

Variable Star Section.   J. E. Isles.
Journ. British Astron. Ass., Vol. 83, 128 - 137, 209 - 216,
291 - 297 (1973).

Historical Section.   E. A. Beet.
Journ. British Astron. Ass., Vol. 83, 220 - 221 (1973).

New members elected.
Journ. British Astron. Ass., Vol. 83, 152 - 158, 233 - 237,
311 - 315 (1973).

The annual general meeting of the Association.
V. Barocas, N. J. Goodman, A. C. Curtis.
Journ. British Astron. Ass., Vol. 83, 86 - 89 (1973).

The programme of the Variable Star Section in
1973.   J. E. Isles.
Journ. British Astron. Ass., Vol. 83, 295 - 297 (1973).

Variable Star Section.
British Astron. Ass., Circ. No. 546 (1973).

010.013   British Interplanetary Society (BIS)

Society news.
Spaceflight, Vol. 15, 36 - 38, 75 - 77, 156, 275 - 276 (1973).

The report of the council for the year ended 31
December 1972.
Spaceflight, Vol. 15, 196 - 198 (1973).

010.014   Committee on Space Research (COSPAR)

XVᵉ session plénière COSPAR, Madrid, 15–24 mai
1972.   W. Dobaczewska.
Geodezja i Kartografia, Vol. 22, 83 - 86 (1973). In Polish.

COSPAR 1972.   B. Kołaczek, W. Zonn.
Postępy Astron., Vol. 21, 75 - 80 (1973).   In Polish. – Madrid,
1972 May.

010.015   European Space Research Organization (ESRO)

No publication received.

010.016   International Astronautical Federation (IAF)

No publication received.

010.017   International Astronomical Union (IAU)

International Astronomical Union, Information
Bulletin, Nos. 29, 30, [printed by D. Reidel, Dordrecht–
Holland], 37 + 22 pp. (1973).   C. de Jager.
Contents: General assemblies; Executive committee; Comis-
sions; IAU symposia and colloquia; Other scientific meetings;
IAU publications; Other publications; Other international
organizations; Membership.

010.018 Meteoritical Society

Abstracts of papers presented at the 35th annual meeting of the Meteoritical Society November 16–18, 1972, Chicago, Illinois.
Meteoritics, Vol. 8, 8 - 83 (1973).

The constitution of the Meteoritical Society.
Meteoritics, Vol. 8, 85 - 88 (1973).

010.019 Nederlandse Vereniging voor Weer- en Sterrenkunde

Verenigingsnieuws.
Hemel en Dampkring, Vol. 71, 118, 150 - 157, 218 (1973).

Jongerenwerkgroep.
Hemel en Dampkring, Vol. 71, 43, 156 - 157 (1973).

Werkgroepen.
Hemel en Dampkring, Vol. 71, 188 - 195, 219 - 222 (1973).

010.020 Polskie Towarzystwo Astronomiczne (PTA)

No publication received.

010.021 Polskie Towarzystwo Miłośników Astronomii (PTMA)

Chronicle of the Polish Amateur Astronomical Society.
Urania Kraków, Vol. 44, 25, 85 - 88, 119 - 122, 150 - 152, 180 - 183 (1973). In Polish.

The foundation of the Amateur Astronomers Association in 1921.   S. Mrozowski.
Urania Kraków, Vol. 44, 56 - 62 (1973). In Polish.

010.022 Royal Astronomical Society (RAS)

Meetings of the Society.
Observatory, Vol. 93, 49 - 70, 97 - 106 (1973).

Meetings of the Society.
Quarterly Journ. Roy. Astron. Soc., Vol. 14, 3 - 8 (1973).

010.023 Royal Astronomical Society of Canada (RAS Canada)

No publication received.

010.024 Royal Astronomical Society of New Zealand (RAS New Zealand)

No publication received.

010.025 Schweizerische Astronomische Gesellschaft (SAG)

Generalversammlung der SAG vom 12./13. Mai 1973 in St. Gallen.   E. Wiedemann.
Orion Schaffhausen, 31. Jahrgang, p. 87 - 88 (1973).

Generalversammlung der SAG vom 12. Mai 1973 in St. Gallen. Jahresbericht des Präsidenten.   W. Studer.
Orion Schaffhausen, 31. Jahrgang, p. 88 - 89 (1973).

Bericht des Generalsekretärs der SAG.   H. Rohr.
Orion Schaffhausen, 31. Jahrgang, p. 89 - 90 (1973).

010.026 Sociedad Astronómica de México

No publication received.

010.027 Società Astronomica Italiana (SAI)

No publication received.

010.028 Société Astronomique de France (SAF)

Les séances de la Société.   B. Clouet.
L'Astronomie, 87ᵉ année, p. 92 - 96, 130 - 132, 179 - 184, 223 - 225, 259 - 261 (1973).

La Commission du Soleil.
L'Astronomie, 87ᵉ année, p. 216 (1973).

010.029 Société Astronomique "R. Bošković"

No publication received.

010.030 Société Chronométrique de France

No publication received.

010.031 Société Belge d'Astronomie, de Météorologie et de Physique du Globe

Réunions mensuelles.
Ciel et Terre, Vol. 89, 65 (1973).

Séance mensuelle.
Ciel et Terre, Vol. 89, 152 - 153, 154 (1973).

010.032 Svenska Astronomiska Sällskapet

No publication received.

010.033 VAGO (Astronomical-Geodetical Society of the USSR)

No publication received.

010.034   **Vereniging voor Sterrenkunde, Belgie**

No pulbication received.

010.035   **Argentine Astronomical Association**

No publication received.

010.036   **Comité Belge des astronomes amateurs.**
A. Koeckelenbergh.
Ciel et Terre, Vol. 89, 151 (1973).

010.037   **Annual chronology of international astronautical events 1970 sponsored by the International Academy of Astronautics.**   R. C. Hall (Editor), with a foreword by C. S. Draper and a preface by R. C. Hall.
Astronaut. Acta, Vol. 18, 155 - 170 (1973). — Containing as appendix the 1970 international bibliography pertaining to the history of astronautics.

010.038   **Rapport d'activité de l'ADION par le Secrétaire Général.**   J.-C. Pecker.

Bull. d'Information, Ass. Développement International Obs. Nice, No. 9, p. 25 - 39 (1972). — Report 1971.

010.039   **Société Royale d'Astronomie d'Anvers, [Koninklijk Sterrenkundig Genootschap van Antwerpen].**
Cinquante-troisième rapport 1972.   J. Storms.
Imprimerie: «La Prévoyance», Antwerpen. 24 pp. (1973). In French and Flemish.

010.040   **Nachrichten der Vereinigung der Sternfreunde e.V.**
SuW, Vol. 12, 23 - 25, 56 - 57, 89 - 90, 92, 120 - 124, 152 - 154, 185 - 187 (1973).

010.041   **I. U. A. A. International Union of Amateur Astronomers. Contributions.**
Edited and directed by A. Leani.
I. U. A. A. Bull., Cremona (Italy), No. 5, 22 pp. (1973). Contents: IAU Commission 42: Bibliography and program notes on eclipsing binaries, n. 21 (2nd); Reports of the IUAA Commissions: Meteor (*K. Simmons*); Astronomical education (*K. Chilton, A. Leani*); Comets (*R. Adams*); Aurora (*J. Paton*); Occultations (*H. J. Bode*); The Scandinavian Union of Amateur Astronomers (SUAA).

011   Reports on Colloquia, Congresses, Meetings, Symposia, and Expeditions

011.001   **Conference on ancient astronomy.**   S. Mitton.
Journ. History Astron., Vol. 4, 68 - 71 (1973). London, 1972 December 7 - 8.

011.002   **The OSO-7 year of discovery.**
S. P. Maran, R. J. Thomas.
Sky Telescope, Vol. 45, 4 - 9 (1973). — Stanford, 1972 September 5 - 8.

011.003   **A symposium on solar physics.**   D. McNally.
Observatory, Vol. 93, 1 - 2 (1973).

011.004   **Gravitation conference in Armenia.**
L. P. Grishchuk.
Zemlya i Vselennaya, 1973, No. 2, p. 56 - 57. In Russian. Erevan, 1972, Oct. 11 - 14.

011.005   **European space program: It's half-speed ahead.**
D. Verguèse.
Science, Vol. 179, 984 - 985 (1973).

011.006   **Relativistische Astrophysik.**   V. Weidemann.
Naturwissenschaften, 60. Jahrgang, p. 184 - 188 (1973).
The "6. Texas Symposium on Relativistic Astrophysics"

was held in New York from Dec. 18—22, 1972. We report in some detail on quasi-stellar objects, especially the interpretation of their redshifts, apparent superlight-velocities in radio sources, newly discovered X-ray sources in binary stars and the possible existence of a black hole in Cygnus X-1.

011.007   **19. Astrometrische Konferenz des Astronomischen Rates in Moskau.**   K. -G. Steinert.
Sterne, 49. Jahrgang, p. 82 - 84 (1973). — 1972 June 27 - 30.

011.008   **Problems of lunar motions and mapping.**
J. D. Mulholland.
Sky Telescope, Vol. 45, 283 - 285 (1973).

011.009   **Variable star observers meet in Connecticut.**
J. Ashbrook.
Sky Telescope, Vol. 45, 297 (1973).

011.010   **Opening of SAAO's outstation at Sutherland.**
Monthly Notes Astron. Soc. Southern Africa, Vol. 32, 22 (1973).

011.011   **Colocviul U. A. I. ,,Asteroizi, comete, materie meteorică", Nisa, 4—6 aprilie 1972.**
C. Cristescu.

Stud. Cerc. Astron., Vol. 18, 121 - 122 (1973).

011.012 **A XIX-a conferinţă de astronomie a U.R.S.S. Moscova, 27–30 iunie 1972.** I. Rusu.
Stud. Cerc. Astron., Vol. 18, 122 - 123 (1973).

011.013 **Prima adunare astronomică europeană.** C. Popovici.
Stud. Cerc. Astron., Vol. 18, 123 - 124 (1973).

011.014 **Conferinţa naţională de astronomie, Baia Mare, 22–24 septembrie 1972.** C. Cristescu.
Stud. Cerc. Astron., Vol. 18, 125 - 126 (1973).

011.015 **Conferinţa "Cercetări ştiinţifice cu ajutorul observaţiilor sateliţilor artificiali ai pămîntului,Ulan-Bator, 26 septembrie–2 octombrie 1972.** A. Dinescu.
Stud. Cerc. Astron., Vol. 18, 126 - 127 (1973).

011.016 **Notes on the fourth Lunar Science Conference–I.** T. L. Page.
Sky Telescope, Vol. 45, 355 - 358 (1973).

011.017 **Progress of Soviet radio astronomy. Conference in Gorky.** B. A. Dubinskij.
Vestn. AN SSSR, 1972, No. 11, p. 104 - 105. In Russian.
Abstr. in Referativ. Zhurn. 51. Astron., 4.51.24 (1973).

011.018 **Origin of the solar system. Symposium in France.** V. S. Safronov.
Vestn. AN SSSR, 1972, No. 10, p. 97 - 101. In Russian.
Abstr. in Referativ. Zhurn. 51. Astron., 4.51.25 (1973).
Nizza, 1972, April 3 - 7.

011.019 **Plenum of the Astronomical Council, Sverdlovsk, 12 - 15 June 1972.** N. P. Erpylev.
Vestn. AN SSSR, 1972, No. 10, p. 122 - 123. In Russian.

011.020 **Scientific session of the Department of General Physics and Astronomy of the USSR Academy of Sciences (April 26 - 27, 1972).**
Uspekhi fiz. nauk, Vol. 108, 595 - 598 (1972). In Russian.
Abstr. in Referativ. Zhurn. 51. Astron., 4.51.38 (1973).

011.021 **Joint scientific session of the Department of General Physics and Astronomy with the Scientific Council of Physics and Chemistry of Semiconductors and the Department of Nuclear Physics of the USSR Academy of Sciences (May 24 - 25, 1972).**
Uspekhi fiz. nauk, Vol. 108, 598 - 604 (1972). In Russian.
Abstr. in Referativ. Zhurn. 51. Astron., 4.51.39 (1973).

011.022 **Materials on the All-Union conference on physics of cosmic rays, Tbilisi, 18 - 21 October 1971.**
Izv. AN SSSR. Ser. fiz., Vol. 36, 2257 - 2486 (1972). In Russian.

011.023 **Lunar dynamics and observational co-ordinate systems. A COSPAR/IAU/LSI colloquium held at** the Lunar Science Institute, Houston, on January 15 - 17, 1973. J. S. Griffith.
Journ. British Interplanet. Soc., Vol. 26, 314 - 315 (1973).
Conference report.

011.024 **Planetary science. A report on the Royal Society Copernicus quincentenary symposium.**
L. Opiela, P. L. Sowerby.
Journ. British Astron. Ass., Vol. 83, 188 - 191 (1973).

011.025 **Arbeitstagung über Astrophotographie am 14. April 1973 in Würzburg.** E. Wiedemann.
Orion Schaffhausen, 31. Jahrgang, p. 85 - 86 (1973).

011.026 **Notes on the Fourth Lunar Science Conference – II.** T. L. Page, L. B. Johnson.
Sky Telescope, Vol. 46, 14 - 17, 28 (1973).

011.027 **Nicolaus Copernicus anniversary celebration.** J. B. Trainor.
Journ. Astron. Soc. Victoria, Vol. 26, 27 - 28 (1973).

011.028 **Memorandum of the subsection of astrophysics, astronomy and space physics prepared for physics section of the 2nd Congress of Polish Science.**
B. Paczyński, S. Piotrowski.
Postępy Astron., Vol. 21, 149 - 153 (1973). In Polish.

011.029 **Summer school on observational and theoretical cosmology, Opole 1972.** P. Flin.
Postępy Astron., Vol. 21, 161 (1973). In Polish.

011.030 **4 th international conferences of the heads of planetaria.** J. Sałabun.
Urania Kraków, Vol. 44, 88 - 91 (1973). In Polish.

011.031 **10th conference of the working group "Dynamics of stellar systems".** V. A. Antonov.
Astron. Zhurn. Akad. Nauk SSSR, Vol. 50, 667 - 669 (1973). In Russian. English translation in Soviet Astron. AJ, Vol. 17, No. 3. – Alma-Ata, 1972, Oct. 23 - 26.

011.032 **Resolutions of the 1st conference of the working group on the problem 'Investigation of minor bodies of the solar system' (Riga, 10.– 11. 5. 1973).**
N. A. Belyaev.
Kometn. Tsirk., *Kiev*, No. 147 (1973). In Russian.

011.033 **Astronomical conferences and meetings.**
Astrophys. Letters, Vol. 13, 61 - 62, 123 - 124, 185 - 186, 247 - 248 (1973).

011.034 **Protokoll der 118. Sitzung der Schweiz. Geodätischen Kommission vom 10. Juni 1972 im Bernerhof in Bern mit Auszügen aus den Berichten über die Tätigkeit im Jahre 1971.**
Société Helvétique des Sciences Naturelles, (Schweiz. Naturforschende Gesellschaft). Spross+Co., Kloten. 53 pp. (1973).

012 Proceedings of Colloquia, Congresses, Meetings, and Symposia

012.001 **Abstracts of papers presented at the Third Soviet Gravitational Conference, Erevan, October 11–14, 1972.**
Sektsiya gravitatsii Ministerstva vysshego i srednego spetsial'-nogo obrazovaniya SSSR, Erevan. un-t, AN ArmSSR. Erevan. un-t, Erevan. 423 pp. Price 2 Rbl. 67 Kop. (1972). In Russian. Abstracts of these papers are published in Referativ. Zhurn. 51. Astron., No. 3 (1973). – The individual contributions within the subject scope of Astronomy and Astrophysics Abstracts are included in their corresponding categories – see abstracts 061.021, 062.015, 065.064 - 065.067, 066.023 - 066.047, 080.013, 091.024, 125.016, 125.017, 126.013, 151.027, 162.020 - 162.035.

012.002 **X- and gamma-ray astronomy.**
International Astronomical Union, Symposium No. 55, held in Madrid, Spain, 11–13 May 1972.
H. Bradt, R. Giacconi (Editors).
D. Reidel Publishing Company, Dordrecht –Holland/Boston – U.S.A. 10 + 323 pp. Price Dfl. 95.00 (1973). – The individual contributions are included in their corresponding subject categories – see abstracts 117.015, 125.018, 142.058 - 142.078, 155.043.

012.003 **Les spectres des astres dans l'infrarouge et les microondes.** Communications, présentées au dix-septième Colloque International d'Astrophysique tenu à Liège les 28, 29 et 30 juin 1971, with an introduction by N. Grevesse, A. Noels, L. Remy-Battiau.
Mém. 8° Soc. Roy. Sci. Liège, 6e Sér., Vol. 3, 629 pp. (1972). The individual contributions are included in their corresponding subject categories – see abstracts 022.035, 031.014, 064.035, 064.036, 065.068, 082.044, 113.033 - 113.035, 114.075 - 114.089, 122.050, 124.101, 131.089 - 131.109, 132.018, 132.019, 133.019, 155.044 - 155.046, 158.078 - 158.084.

012.004 **Orbital and physical parameters of double stars.**
Proceedings of Colloquium No. 18 of the International Astronomical Union held at Swarthmore College, Pennsylvania, U.S.A., April 12–15, 1972.
W. D. Heintz (Editor).
Journ. Roy. Astron. Soc. Canada, Vol. 67, 49 - 87 (1973). The individual contributions are included in their corresponding subject categories – see abstracts 117.016, 118.003 - 118.008, 119.010.

012.005 **Atoms and molecules in astrophysics.** Proceedings of the twelfth session of the Scottish Universities Summer School in Physics, 1971. A NATO Advanced Study Institute. T. R. Carson, M. J. Roberts (Editors).
Academic Press, London – New York. 14 + 367 pp. Price £ 7.50 (1972). – Review in Journ. British Interplanet. Soc., Vol. 26, 127 (1973). – The individual contributions are included in their corresponding subject categories – see abstracts 022.038 - 022.043, 022.056, 074.035, 114.092, 131.112 - 131.115, 132.020.

012.006 **Papers presented at the sixth annual general meeting of the Astronomical Society of Australia held at Monash University on 24, 25 and 26 May, 1972.**
Proc. Astron. Soc. Australia, Vol. 2, (No. 3), 122 - 162 (1972). The individual contributions are included in their corresponding subject categories – see abstracts 008.039, 022.026, 032.004 - 032.007, 033.004 - 033.006, 061.016, 065.044 - 065.046, 071.018, 072.014, 073.020, 077.014, 099.026,

114.054, 114.055, 132.011, 141.047, 143.018, 154.003, 158.053.

012.007 **Reunión 16ª de la Asociación Argentina de Astronomía (Septiembre de 1970, San Miguel).**
Compiled and prepared by Z. López-García.
Bol. As. Argentina Astron., No. 16, 67 pp. (1971). – The individual contributions are included in their corresponding subject categories – see abstracts 032.012, 042.025, 072.029, 073.029 - 073.031, 077.037, 102.010, 114.094 - 114.097, 115.013, 122.054, 131.121 - 131.123, 132.022, 133.020, 133.021, 152.005, 153.022, 155.053, 157.007, 158.088 - 158.093, 160.019, 160.020.

012.008 **The Apollo 15 lunar samples.**
J. W. Chamberlain, C. Watkins (Editors).
Reproduced and distributed by the Lunar Science Institute, Houston, Texas. 15+525 pp. Price $ 10.00 (1972). – The individual papers are included in their corresponding subject categories – see abstracts 094.169 - 094.294.

012.009 **Simpozion Copernic,** 19 februarie 1973, with a foreword by C. Popovici and an opening address by R. Răduleț.
Comitetul pentru sărbătorirea a 500 de ani de la nașterea lui Copernic.
Editura Academiei Republicii Socialiste România București. 87 pp. Price Lei 4.00 (1973). In Romanian with summaries in English and Russian. – The individual contributions are included in their corresponding subject categories – see abstracts 004.041 - 004.045, 013.005.

012.010 **Interstellar gas dynamics.**
H. J. Habing (Editor).
Translated from the English edition (see 04.012.015). Mir, Moskva. 444 pp. Price 2 Rbl. 99 Kop. (1972). In Russian. Review in Referativ. Zhurn. 51. Astron., 4.51.52 (1973).

012.011 **Papers on the XVIth Scientific Conference, November 1970. Ser. "General and Applied Physics", "Molecular and Chemical Physics".**
Moscow phys.-techn. Institute, Moskva. 144 pp. Price 70 Kop. (1972). In Russian.

012.012 **Proceedings of the Second Lunar Science Conference, Houston, Texas, January 11–14, 1971,** sponsored by the Lunar Science Institute.
A. A. Levinson (Editor).
Vol. 1: Mineralogy and petrology; Vol. 2: Chemical and isotope analyses; Vol. 3: Physical properties/Surveyor III. The MIT Press, Cambridge, Mass. – London. Geochim. Cosmochim. Acta, Suppl. 2, 35 + 2818 pp. Price £ 32.00 (1971). The individual papers are abstracted in their corresponding subject categories – see abstracts 094.303 - 094.526.

012.013 **Equatorial aeronomy.** Selected papers from the fourth international symposium, held at the Conference Centre of the University of Ibadan, Nigeria, 4–9 September 1972, with a preface by S. Matsushita, R. Cohen.
Journ. Atmosph. Terr. Phys., Vol. 35, 1025 - 1279 (1973).

012.014 **Stellar evolution.** Based on lectures given at the 3d Summer Institute for Astronomy and Astrophysics, held at the State University of New York at Stony Brook, June 18–July 16, 1969.
H.-Y. Chiu, A. Muriel (Editors).
The MIT Press, Cambridge, Mass. – London, England. 14 +

812 pp. Price $ 15.00 (1972). — The individual contributions are included in their corresponding subject categories — see abstracts 061.027, 064.056, 065.104 - 065.115, 066.067, 115.014, 117.025, 124.008, 125.023, 126.019, 141.540, 151.035.

012.015 **Stellar ages.** Proceedings of the I.A.U. Colloquium No. 17 held at the Paris–Meudon Observatory, France, September 18–22, 1972.
G. Cayrel de Strobel, A. M. Delplace (Editors).
Edited by Observatoire Paris–Meudon. (1973). — The individual contributions are included in their corresponding subject categories — see abstracts 009.013, 061.029, 061.030, 064.058 - 064.062, 065.119 - 065.137, 107.012, 112.008, 113.043, 114.132 - 114.135, 115.015 - 115.022, 116.006, 122.084, 122.085, 125.025, 126.020, 126.021, 152.008, 152.009, 155.071 - 155.077, 159.006, 162.049, 162.050.

012.016 **Cosmic plasma physics.** Proceedings of the conference on cosmic plasma physics, held at the European Space Research Institute (ESRIN), Frascati, Italy, September 20–24, 1971.
K. Schindler (Editor).
Plenum Press, New York – London. 11 + 369 pp. Price DM 97.25 (1972). — The individual contributions within the subject scope of Astronomy and Astrophysics Abstracts are included in their corresponding subject categories — see abstracts 061.033, 062.037 - 062.045, 065.141, 073.083, 073.084, 074.074 - 074.081, 080.034, 083.071 - 083.073, 084.257 - 084.259, 093.038, 099.072, 102.016, 102.017, 106.023, 106.024, 131.156, 141.547 - 141.550, 142.106, 143.045 - 143.048.

012.017 **Photo-electronic image devices.** Proceedings of the fifth symposium held at Imperial College, London, September 13–17, 1971.
J. D. McGee, D. McMullan, E. Kahan (Editors).
Advances in Electronics and Electron Physics, Vol. 33A, 33B. Academic Press, London – New York. 25 + 25 + 1189 pp. Price £ 24.00 (1972). — The individual papers within the subject scope of Astronomy and Astrophysics Abstracts are included in their corresponding subject categories — see abstracts 031.029 - 031.031, 034.057 - 034.085, 036.006, 036.007.

012.018 **Astronomy from a space platform.** Proceedings of the symposium on astronomy from a space platform December 27–28, 1971, Philadelphia, Pennsylvania.
G. W. Morgenthaler, H. D. Greyber (Editors).
Science and technology series, Vol. 28. Published by the American Astronautical Society, Tarzana, Calif. 15 + 398 pp. Price $ 17.50 (1972). — The individual contributions are included in their corresponding subject categories — see abstracts 031.039 - 031.041, 032.025 - 032.027, 033.041, 034.093, 034.094, 051.011 - 051.019, 054.019 - 054.021, 061.045, 080.044, 143.058.

012.019 **The problem of the variation of the geographical coordinates in the southern hemisphere.** Colloquium No. 1 of the IAU, held at La Plata University Observatory, Argentina, November 4–5, 1968.
O. Cáceres (Editor), with a preface and words of opening of the colloquium and of homage to Prof. F. Aguilar by J. Sahade. Published by Astron. Obs. National Univ. La Plata, Argentina, 7 + 124 pp. (1972). — The individual contributions are included in their corresponding subject categories — see abstracts 008.024, 032.032, 035.005, 041.026 - 041.028, 044.025 - 044.028, 045.020 - 045.024.

012.020 **International conference on education in and history of modern astronomy,** held at the American Museum of Natural History on August 30 and 31 and September 1, 1971.
R. Berendzen (Editor), and with a dedication to Marcel Gilles Jozef Minnaert by E. A. Müller.
Ann. New York Acad. Sci., Vol. 198, 275 pp. Price $ 24.00 (1972). — Reviews in Nature, Vol. 243, 104; 1973 (*A. J. Meadows*); Sky Telescope, Vol. 45, 114 (1973); SuW, Vol. 12, 151; 1973 (*A. Kunert*). — The individual contributions are included in their corresponding subject categories — see abstracts 004.069, 004.070, 014.019 - 014.043, 051.020, 061.046, 071.049, 104.027, 114.158, 155.091.

012.021 **Wissenschaftliche Tagung der Astronomischen Gesellschaft mit 53. ordentlicher Mitgliederversammlung in Wien, 18. - 23. September 1972.**
K. Schaifers (Editor).
Mitt. Astron. Ges., No. 32, 286 pp. Price DM 40.00 (1973). The individual contributions are included in their corresponding subject categories — see abstracts 010.010, 014.015, 022.072, 031.033 - 031.036, 032.024, 034.087 - 034.092, 041.021, 041.022, 061.038 - 061.041, 062.052, 063.039, 064.067 - 064.070, 065.147 - 065.151, 072.065, 072.066, 073.095, 074.088, 077.062, 079.102, 080.038 - 080.042, 081.021, 082.098, 082.099, 091.067, 097.091, 098.018, 100.045, 100.046, 102.022, 106.030 - 106.032, 112.012, 112.013, 113.049, 113.050, 114.149 - 114.153, 116.009, 117.031, 118.013, 121.056, 121.057, 122.108, 124.100, 124.104, 126.024, 126.025, 131.169 - 131.176, 141.101, 142.116 - 142.118, 152.010, 153.036, 153.037, 155.088, 158.120 - 158.124, 160.025.

012.022 **Proceedings of the Third Lunar Science Conference, Houston, Texas, January 10–13, 1972,** sponsored by the Lunar Science Institute.
Vol. 1: Mineralogy and petrology, E. A. King, Jr. (Editor); Vol. 2: Chemical and isotope analyses, organic chemistry, D. Heymann (Editor); Vol. 3: Physical properties, D. R. Criswell (Editor).
The MIT Press, Cambridge, Mass. – London.
Geochim. Cosmochim. Acta, Suppl. 3, 34 + 3263 + 54 + 34 (subject index) pp. Price $ 32.00 each volume, $ 90.00 the set. (1972). — Review in Sky Telescope, Vol. 45, 248 (1973). The individual papers are included in their corresponding subject categories — see abstracts 094.605 - 094.836.

012.023 **Giornate di studio dedicate al Prof. F. Zagar,** with an introduction by A. Masani.
Mem. Soc. Astron. Italiana, Nuova Ser., Vol. 43, 581 - 824 (1972/73). — The individual contributions are included in their corresponding subject categories — see abstracts 004.060, 008.000, 041.015 - 041.017, 042.035, 043.002, 044.016, 045.015 - 045.017, 061.032, 065.140, 073.082, 080.032, 080.033, 094.588, 103.110, 107.016, 114.141, 114.142, 116.008, 117.027, 122.104, 124.009, 141.097, 141.546, 142.105, 155.085.

012.024 **Physics of the moon and planets.** International symposium in Kiev, 1968 October 15 - 22.
D. Ya. Martynov, V. A. Bronshtehn (Editors).
Akademiya Nauk SSSR; Astronomicheskij Sovet; Izdatel'stvo "Nauka", Moskva. 472 pp. Price 3 Rbl. 16 Kop. (1972). In Russian. — The individual papers are included in their corresponding subject categories — see abstracts 022.082, 063. 046, 063.047, 091.072 - 091.079, 092.007 - 092.009, 093. 047 - 093.080, 094.853 - 094.895, 097.107 - 097.118, 099. 093 - 099.098, 100.053 - 100.055, 101.017, 101.018, 107. 019, 107.020.

012.025 **The use of artificial satellites for geodesy.** Third international Symposium on the use of artificial satellites for geodesy, April 15 - 17, 1971, Washington, D.C.
S. W. Henriksen, A. Mancini, B. H. Chovitz (Editors).

Geophysical Monograph 15, American Geophysical Union, Washington, D.C. 12 + 298 pp. Price DM 105.10 (1972). Contents: (1) Geometric geodesy: theory; (2) Geometric geodesy: results; (3) Physical geodesy: theory; (4) Physical geodesy: results; (5) Instrumentation and environment; (6) Extraterrestrial geodesy; (7) Data management. — The individual contributions within the subject scope of Astronomy and Astrophysics Abstracts are included in their corresponding categories — see abstracts 045.029, 045.030, 046.026 - 046.028, 052.041, 055.006 - 055.011, 081.030 - 081.037, 091.082, 091.082, 093.082, 094.911, 094.927.

012.026 **Light scattering in the earth's atmosphere.**
Transactions of the Soviet-Union conference on light scattering.
Astrophys. Inst. Academy of Sciences of the Kazakh SSR. Nauka, Alma-Ata. 316 pp. Price 2 Rbl. 24 Kop. (1972). In Russian. — Alma-Ata, 1969 November.

012.027 **UFO's — a scientific debate.** An AAAS symposium, Boston, December 1969.

C. Sagan, T. Page (Editors).
Cornell University Press, Ithaka, N. Y. — London. 22 + 310 pp. Price $·12.50 (1973). — Reviews in Journ. Roy. Astron. Soc. Canada, Vol. 67, 153 - 154; 1973 (*A. V. Douglas*); Science, Vol. 180, 593 - 595; 1973 (*P. A. Sturrock*).

012.028 **IVth international Leningrad seminar. Material of the international seminar "Uniformity of particle acceleration in different scales of the cosmos".**
(AN SSSR, Fiz.-tekhn. in-t AN SSSR, NII yader. fiz. Mosk. un-ta.) Leningrad. 296 pp. Price 1 Rbl. 80 Kop. (1972). In Russian. — Leningrad, 1972 August 16 - 18.

012.029 **Solar wind.** Proceedings of a conference, Pacific Grove, California, March 1971.
C. P. Sonett, P. J. Coleman, Jr., J. M. Wilcox (Editors).
National Aeronautics and Space Administration, Washington, D. C. NASA SP-308. (Available from the Superintendent of Documents, U. S. Government Printing Office, Washington, D. C.), 12 + 718 pp. Price $ 6.00 (1972). — Review in Sky Telescope, Vol. 45, 114 (1973).

013 Reports on Astronomy in Various Countries and Particular Fields, International Cooperation

013.001 **Astronomy in New Mexico.** W. L. Reitmeyer.
Sky Telescope, Vol. 45, 19 - 22 (1973).

013.002 **Physics and astronomy in 1972: Progress with fusion and lasers, and new discoveries with an X-ray satellite.** W. D. Metz.
Science, Vol. 179, 670 - 671 (1973).

013.003 **Stand und Fortschritt der Physik in den USA.**
Umschau, 73. Jahrgang, p. 58 - 59 (1973).

013.004 **Die Bedeutung des Copernicanismus für das physikalische und astronomische Weltbild der Gegenwart.**
H.-J. Treder.
Nicolaus Copernicus — 1473—1973, (see 003.004), p. 179 - 198 (1973).

013.005 **The significance and importance of Copernicus' astronomical work.** C. Popovici.
Simpozion Copernic, (see 012.009), p. 49 - 57 (1973). In Romanian.

013.006 **Astronomy in Mongolia.** K. Ziołkowski.
Urania Kraków, Vol. 44, 110 - 114 (1973). In Polish.

013.007 **Advances in astronomy in the year 1972.**
J. Grygar.
Říše hvězd, Vol. 54, 34 - 36, 41 - 53 (1973). In Czech.

013.008 **Infrared astronomy.** J. Lequeux.
Recherche, (*France*), Vol. 4, No. 30, p. 25 - 33 (1973). In French.
Presents a popular survey, with brief historical introduction and accounts of research in progress.

013.009 **Major contributions of scientific institutions of the Department of Sciences on the Universe and the Earth summed up in connection with the fiftieth anniversary of the USSR.** I. I. Bok.
Vestn. AN KazSSR, 1973, No. 1, p. 44 - 52. In Russian.
Abstr. in Referativ. Zhurn. 51. Astron., 7.51.63 (1973).

014 Teaching in Astronomy

014.001 **Astronomy as a school subject. An historical retrospect.** E. A. Beet.
Journ. British Astron. Ass., Vol. 83, 112 - 118 (1973).

014.002 **Astronomie und Schule.** Astronomie - Lehrpläne und Unterrichtshilfen in der DDR.
SuW, Vol. 12, 28 - 29 (1973).

014.003 **Astrophysik – ein wichtiger Teil des Physikunterrichtes.** W. Kranzer.
SuW, Vol. 12, 86 - 87 (1973).

014.004 **Das neue 63/840-mm-Schulfernrohr Telementor aus Jena.** P. Ahnert.
Jenaer Rundschau (Jena Review), 18. Jahrgang, Special Number, p. 100 - 101 (1973).

014.005 **On the principal trends of scientific research on problems of methodics in teaching astronomy.**
E. P. Levitan.
Uch. zap. Gor'kov. gos. ped. in-ta, 1972, vyp. (No.) 124, p. 74 - 77. In Russian. – Abstr. in Referativ. Zhurn. 51. Astron., 3.51.36 (1973).

014.006 **From the history of astronomical circles of Moscow.** F. Yu. Zigel'.
Zemlya i Vselennaya, 1973, No. 1, p. 73 - 76. In Russian.

014.007 **Control mechanism of a telescope.** V. M. Shuvalov.
Zemlya i Vselennaya, 1973, No. 1, p. 77 - 78. In Russian.

014.008 **School observatory in Bautzen.** H. Bernhard.
Zemlya i Vselennaya, 1973, No. 2, p. 67 - 69. In Russian. Translated from German into Russian by *E. P. Levitan.*

014.009 **Solar patrol by pupils in Omsk.** K. A. Lupoj.
Zemlya i Vselennaya, 1973, No. 2, p. 70 - 71. In Russian.

014.010 **Astronomy education – new strategies.**
F. Six, T. Wawrukiewicz, P. Campbell, R. Hackney.
Sky Telescope, Vol. 45, 286 - 290 (1973).

014.011 **Zusammenarbeit mit außerschulischen Einrichtungen bei der Durchführung von Astronomie-Kursen.** A. Kunert.
SuW, Vol. 12, 150 - 151 (1973).

014.012 **Zur unterrichtlichen Behandlung der kompakten Galaxien.** M. Schukowski.
Astron. in der Schule, 10. Jahrgang, p. 7 - 8 (1973).

014.013 **Aufgabensammlung für das Fach Astronomie.** A. Muster, H. Albert, K. Lindner.
Astron. in der Schule, 10. Jahrgang, p. 13 - 15 (1973).

014.014 **Anfertigung einer Projektionsfolie Koordinatensysteme und ihr Einsatz im Unterricht.**
H. Risse.
Astron. in der Schule, 10. Jahrgang, p. 38 - 41 (1973).

014.015 **Probleme in der Ausbildung astronomischer Beobachter.** W. Fricke.
Mitt. Astron. Ges., No. 32, p. 9 - 14 (1973). – Ansprache des Vorsitzenden der Astronomischen Gesellschaft, gehalten zur Eröffnung der Wissenschaftlichen Tagung verbunden mit der 53. ordentlichen Mitgliederversammlung der Astronomischen Gesellschaft in Wien 1972.

014.016 **Astronomische Schülerbeobachtungen unter didaktischen Gesichtspunkten betrachtet.**
H. Albert, W. Gebhardt.
Astron. in der Schule, 10. Jahrgang, p. 58 - 61 (1973).

014.017 **Ein Unterrichtsmittel zur Erarbeitung des HRD.** W. König.
Astron. in der Schule, 10. Jahrgang, p. 61 - 64 (1973).

014.018 **Über die Ausbildung in Astronomie und die Tätigkeit der Volkssternwarten und Planetarien in der Volksrepublik Bulgarien.** N. Petrow.
Astron. in der Schule, 10. Jahrgang, p. 64 - 65 (1973).

014.019 **The efforts of the American Astronomical Society in astronomy education.** G. O. Abell.
Ann. New York Acad. Sci., Vol. 198, (see 012.020), 8 - 15 (1972).

014.020 **Undergraduate astronomy education: a professor's view.** T. L. Swihart.
Ann. New York Acad. Sci., Vol. 198, (see 012.020), 16 - 18 (1972).

014.021 **Undergraduate astronomy education: a student's view.** C. J. Lada.
Ann. New York Acad. Sci., Vol. 198, (see 012.020), 19 - 29 (1972).

014.022 **Graduate astronomy education: a professor's view.** B. F. Peery, Jr.
Ann. New York Acad. Sci., Vol. 198, (see 012.020), 30 - 35 (1972).

014.023 **Graduate astronomy education: a student's view.** S. L. Shapiro.
Ann. New York Acad. Sci., Vol. 198, (see 012.020), 36 - 45 (1972).

014.024 **Manpower and employment in American astronomy.** R. Berendzen, M. T. Moslen.
Ann. New York Acad. Sci., Vol. 198, (see 012.020), 46 - 65 (1972).

014.025 **The Commission on the Teaching of Astronomy of the International Astronomical Union.**
E. A. Müller.
Ann. New York Acad. Sci., Vol. 198, (see 012.020), 66 - 76 (1972).

014.026 **Astronomical education and manpower in Canada.** R. C. Roeder.
Ann. New York Acad. Sci., Vol. 198, (see 012.020), 77 - 83 (1972).

014.027 **Astronomy education in India.** P. L. Bhatnagar.
Ann. New York Acad. Sci., Vol. 198, (see 012.020), 84 - 91 (1972).

014.028 **Astronomical education in Australia.** F. J. Kerr.
Ann. New York Acad. Sci., Vol. 198, (see 012.020), 92 - 94 (1972).

014.029 Per ardua ad astra: The education of a British astronomer. D. McNally.
Ann. New York Acad. Sci., Vol. 198, (see 012.020), 95 - 103 (1972).

014.030 The importance of astronomy in modern education. E. L. Schatzman.
Ann. New York Acad. Sci., Vol. 198, (see 012.020), 104 - 108 (1972).

014.031 University level astronomy education for nonscience concentrators. A case for astronomy.
D. G. Wentzel.
Ann. New York Acad. Sci., Vol. 198, (see 012.020), 109 - 113 (1972).

014.032 On the role of astronomy education for nonscience majors. R. Berendzen, J. Baumgardner.
Ann. New York Acad. Sci., Vol. 198, (see 012.020), 114 - 123 (1972).

014.033 Astronomical laboratory exercises. E. v. P. Smith.
Ann. New York Acad. Sci., Vol. 198, (see 012.020), 124 - 131 (1972).

014.034 The teaching of astronomy with closed-circuit television. F. K. Edmondson.
Ann. New York Acad. Sci., Vol. 198, (see 012.020), 132 - 135 (1972).

014.035 Computer graphics and animation: new modes of presenting astronomy.
M. L. Meeks, R. N. Davis, R. H. Cohen.
Ann. New York Acad. Sci., Vol. 198, (see 012.020), 136 - 145 (1972).

014.036 The case studies project on the development of modern astronomy. R. Berendzen.
Ann. New York Acad. Sci., Vol. 198, (see 012.020), 146 - 154 (1972).

014.037 University level astronomy education for nonscience concentrators. Comment. F. M. Flinsch.
Ann. New York Acad. Sci., Vol. 198, (see 012.020), 155 (1972).

014.038 Astronomy at the lower school levels.
S. P. Wyatt, Jr.

Ann. New York Acad. Sci., Vol. 198, (see 012.020), 158 - 163 (1972).

014.039 Astronomy in early high school. T. Page.
Ann. New York Acad. Sci., Vol. 198, (see 012.020), 164 - 172 (1972).

014.040 Astronomy at the upper school level.
F. G. Watson.
Ann. New York Acad. Sci., Vol. 198, (see 012.020), 173 - 177 (1972).

014.041 The planetarium in modern science education.
F. C. Jettner, J. J. Soroka.
Ann. New York Acad. Sci., Vol. 198, (see 012.020), 178 - 191 (1972).

014.042 Education in major planetariums. F. M. Branley.
Ann. New York Acad. Sci., Vol. 198, (see 012.020), 192 - 196 (1972).

014.043 Services of NASA to astronomy education.
M. H. Ahrendt.
Ann. New York Acad. Sci., Vol. 198, (see 012.020), 197 - 201 (1972).

014.044 Resource letter EMAA-1: educational materials in astronomy and astrophysics.
R. Berendzen, D. DeVorkin.
American Journ. Phys., Vol. 41, 783 - 808 (1973).

014.045 Project options in an astronomy course. D. Hoff.
American Journ. Phys., Vol. 41, 838 - 840 (1973).

014.046 Astronomy at school yesterday, today and to-morrow. E. P. Levitan.
Zemlya i Vselennaya, 1973, No. 3, p. 61 - 66. In Russian.

014.047 Young astronomers in Yaroslavl. T. L. Korovkina.
Zemlya i Vselennaya, 1973, No. 3, p. 76 - 78.
In Russian.

014.048 Astronomy and aids for teachers.
Mercury, (Journ. Astron. Soc. Pacific), Vol. 2, No. 1, p. 13 - 16; No. 2, p. 5 - 6 (1973). — All-weather observing: Student use of Palomar Sky Survey prints (*D. G. Wentzel*); A tiny revolution in astronomy education (*D. Schatz, D. Cudaback*).

015 Miscellanea

015.001 **The creation of the universe.** W. S. Krogdahl.
Sky Telescope, Vol. 45, 140 - 143 (1973).

015.002 **Cours d'astronomie de la S.A.F. 9. La Galaxie.**
M. Dumont.
L'Astronomie, 87ᵉ année, p. 18 - 31 (1973).

015.003 **Cours d'astronomie de la S.A.F. 10. Les galaxies–**
l'univers. G. Oudenot.
L'Astronomie, 87ᵉ année, p. 113 - 127 (1973).

015.004 **L'arc-en-ciel dans la littérature.** L. Dufour.
Ciel et Terre, Vol. 89, 45 - 49 (1973).

015.005 **Limitations of terrestrial life.** P. Molton.
Spaceflight, Vol. 15, 27 - 30 (1973).

015.006 **Space probe from Epsilon Boötis.**
D. A. Lunan, with an introduction by K. W. Gatland.
Spaceflight, Vol. 15, 122 - 131 (1973).
The astonishing idea that our solar system had been
visited by a space probe from another civilization was widely
reported in December. D. A. Lunan found that certain long
delayed echoes of equally spaced radio signals transmitted
from earth could be interpreted in the form of a code. It is
suggested that the signals identifying the probe's origin as the
double star Epsilon Boötis and putting its arrival here at
13,000 years in the past.

015.007 **The interpretation of signals from space.**
A. T. Lawton.
Spaceflight, Vol. 15, 132 - 137 (1973).

015.008 **Terrestrial biochemistry in perspective: Some other**
possibilities. P. M. Molton.
Spaceflight, Vol. 15, 139 - 144 (1973).

015.009 **The search for extraterrestrial intelligence.**
B. M. Oliver.
Mercury, (Journ. Astron. Soc. Pacific), Vol. 2, No. 2, p. 11 -
12 (1973).

015.010 **Constellation postcards and slides.**
Sky Telescope, Vol. 45, 368 - 369, 376 - 377
(1973).

015.011 **Die Zukunft des Raumfahrtzeitalters in sowjetischer**
Sicht – Prognosen und wissenschaftlich-kosmische
Utopien. W. Petri.
Universitas, [Wiss. Verlagsgesellschaft, Stuttgart], 27. Jahr-
gang, p. 1173 - 1184 = Veröff. Forschungsinst. Deutsch.
Museums Geschichte Naturwiss. Techn., Ser. A, No. 128
(1972).

015.012 **Is anyone out there? Evidence for the existence of**
extraterrestrial life. P. M. Molton.
Spaceflight, Vol. 15, 246 - 252 (1973).

015.013 **Reflections on CETI.** B. Belitzky.
Spaceflight, Vol. 15, 255 (1973).

Interstellar communication: Eye on the future.
Nature, Vol. 241, 363 (1973).

Applied Mathematics, Physics

021 Mathematics, Computing, Machine Programs

021.001 **Prozeßrechner-gesteuerte Sonnenbeobachtung.**
H. Wöhl.
SuW, Vol. 12, 51 - 53 (1973).

021.002 **On the selection of an optimum smoothing parameter of experimental data.**
O. B. Vasiljev, V. I. Sakharov.
Astron. Zhurn. Akad. Nauk SSSR, Vol. 50, 390 - 399 (1973).
In Russian. English translation in Soviet Astron. AJ, Vol. 17,
No. 2.
    The random (uncorrelated) errors variance may be calcul-
ated in the case of their normal distribution as a limit of the
variance of m-th differences of equidistant series when m tends
to ∞. This variance can be used for the calculations of the signal to noise ratio and also for the selection of an optimum
smoothing parameter. Such selection of an optimum
smoothing parameter of the observational data may be used
for the Whittaker and other smoothing operators. The filtering
characteristics of the frequency spectrum are examined when
smoothing of the experimental series is done using the Whittaker operator.

021.003 **About simple computational algorithms for solving
some problems of mathematical physics.**
V. A. Enalsky.
Nauchn. Informatsii, vyp. (No.) 23, p. 3 - 32 (1972). In Russian.

021.004 **An implementation of basic LISP 1.5 interpreter
for the OKITAC-5090 computer.** T. Hirayama.
Tokyo Astron. Obs., Report No. 61, Vol. 16, 386 - 392
(1973). In Japanese.

021.005 **Data processing system for 160 MHz compound
interferometer at Nobeyama Solar Radio Station.**
K. Kai, Y. Uchida, N. Shibuya, H. Hirabayashi, T. Kuwabara.
Tokyo Astron. Obs., Report No. 62, Vol. 16, 645 - 656 (1973).
In Japanese.

021.006 **The use of series expansions into eigenfunctions
in orbital theory and time series analysis.**
M. Schneider.
Inst. Astron., Phys. Geod., Techn. Univ. München, Contr. No.
99, 9 pp. (1973). — To be presented at the first international
symposium "The use of artificial satellites for geodesy and
geodynamics", Athens, May 14–21, 1973.
    A number of subroutines have been written allowing the
expansion of a great variety of functions defined over a finite
interval of the independent variable into a series of (orthonor-
mal) eigenfunctions originating from boundary value problems
These subroutines can be used for example to expand the
time dependent functions arising in orbital theory and in the
analysis of time series.

        Problemas fundamentais de fotometria fotoeléctrica.
See Abstr. 031.061.

        Reducción de observaciones fotoeléctricas (Reduc-
tion of photoelectric observations). See Abstr. 031.064.

        **A computational program for the solution of non-
LTE transfer problems by the complete linearization method.**
See Abstr. 064.082.

        **Programme de calcul des raies de l'hydrogène dans
les étoiles chaudes.** See Abstr. 064.083.

        Fourier spectroscopy in astronomy.
See Abstr. 114.052.

022 Physical Papers Related to Astronomy and Astrophysics

022.001 Elargissement et déplacement par effet de pression des raies d'absorption de quelques multiplets de l'atome neutre de manganese.
G. Pujol, P. Quercy, S. Weniger.
Journ. Quant. Spectrosc. Radiat. Transfer, Vol. 13, 9 - 20 (1973).

The profiles of many absorption lines, belonging to the violet and near ultra-violet resonance multiplets and to two intercombination multiplets of the MnI atom, have been determined in the presence of foreign gases at various pressures for several temperatures.

022.002 The electronic transition moment of the $A^1\Sigma - X^1\Sigma$ band system of BaO.
G. T. Best, H. S. Hoffman.
Journ. Quant. Spectrosc. Radiat. Transfer, Vol. 13, 69 - 78 (1973).

The relative intensities of some 50 bands of the $A^1\Sigma - X^1\Sigma$ band system of BaO have been studied, using as a source the fluorescent scattering of sunlight by a cloud of barium oxide vapor generated at an altitude of 106 km in the earth's atmosphere.

022.003 Steady-state equations of the helium atom in matrix form. E. G. Borodina.
Astron. Zhurn. Akad. Nauk SSSR, Vol. 50, 115 - 125 (1973). In Russian. English translation in Soviet Astron. AJ, Vol. 17, No. 1.

022.004 A note on the energies of some three-electron satellite spectrum lines. H. P. Summers.
Astrophys. Journ., (*Letters*), Vol. 179, L45 - L47 (1973).

Results are given of a first-order perturbation-theory calculation of the energies of some doubly excited levels in lithium-like ions with outer electron in the $n = 2, 3$, and 4 principal quantum shells. These levels give rise to observed satellite spectrum lines.

022.005 Laboratory investigation of the reaction $NO^+ + O_3 \rightarrow NO_2^+ + O_2$.
F. C. Fehsenfeld, E. E. Ferguson, C. J. Howard.
Journ. Geophys. Res., Vol. 78, 327 - 329 (1973). − Letter.

022.006 Hydrogen atom in a strong magnetic field: Bound-bound transitions.
E. R. Smith, R. J. W. Henry, G. L. Surmelian, R. F. O'Connell.
Astrophys. Journ., Vol. 179, 659 - 663, with a correction, Vol. 182, 651 (1973).

Bound-bound transition probabilities are calculated for a hydrogen atom, in magnetic fields from 10^7 to 10^8 gauss.

022.007 Shock-tube measurements of absolute gf-values for Ti I and Ti II. S. J. Wolnik, R. O. Berthel.
Astrophys. Journ., Vol. 179, 665 - 670 (1973).

Absolute gf-values for 97 Ti I and 30 Ti II lines have been determined by shock-tube emission spectroscopy. Comparisons are made with other experimental results. Implications concerning the titanium abundance in the solar photosphere are discussed.

022.008 Franck-Condon factors for the $CH^+ A^1\Pi - X^1\Sigma^+$ transition. S. Green, S. Hornstein, C. F. Bender.
Astrophys. Journ., Vol. 179, 671 - 673 (1973).

Accurate ab initio potential energy curves were used to calculate the vibrational eigenvalues and eigenfunctions for the $A^1\Pi$ and $X^1\Sigma^+$ electronic states of CH^+. The calculated

Franck-Condon factors are $q(0, 0) = 0.665$ and $q(1, 0) = 0.249$.

022.009 A shock tube determination of the CN ground state dissociation energy and the CN violet electronic transition moment. J. O. Arnold, R. W. Nicholls.
Journ. Quant. Spectrosc. Radiat. Transfer, Vol. 13, 115 - 133 (1973).

The CN ground state ($X^2\Sigma^+$) dissociation energy and the electronic transition moment of the CN violet $B^2\Sigma^+ - X^2\Sigma^+$ bands have been simultaneously determined from spectral emission measurements behind incident shock waves. The unshocked test gases were composed of various $CO_2 - CO - N_2 - Ar$ mixtures, and the temperatures behind the incident shocks ranged from 3500 to 8000°K.

022.010 Anomalous electron impact excitation of Ca^+.
J. A. Tully, D. Petrini, O. Bely.
Astron. Astrophys., Vol. 23, 15 - 17 (1973).

The Coulomb Born approximation has been used to compute collision strengths for electron impact excitation of the transitions $4s \rightarrow ns, np, nd$ and nf in Ca^+ with $5 \leq n \leq 10$. The results obtained for $4s \rightarrow np$ are anomalous in the sense that they are of a predominantly non-dipole character. This is explained by the abnormally small oscillator strengths for these allowed transitions.

022.011 Satellites to Lyman-α due to protons.
J. C. Stewart, J. M. Peek, J. Cooper.
Astrophys. Journ., Vol. 179, 983 - 986 (1973).

Line broadening of the wings of $L\alpha$ is calculated by using accurate H_2^+ wave functions. Of particular interest are the linelike satellites which appear prominently at 1233.5, 1240.5, and 1404.9 Å.

022.012 Laboratory detection of the microwave spectrum of sodium hydroxide. E. F. Pearson, M. B. Trueblood.
Astrophys. Journ., (*Letters*), Vol. 179, L145 - L146 (1973).

Rotational transitions in the ground vibrational state of NaOH are reported. These measurements yield the rotational constants $B_0 = 12,567.054$ MHz and $D_0 = 0.02872$ MHz.

022.013 Oscillator strengths for $3d^n4s \rightarrow 3d^n4p$ transitions in the iron series. C. Froese Fischer.
Journ. Quant. Spectrosc. Radiat. Transfer, Vol. 13, 201 - 207 (1973).

The observed oscillator strengths for the $3d^n4s \rightarrow 3d^n4p$ transitions in the iron series show anomalous behaviour for Cr and Mn. Theoretical Hartree-Fock gf-values are reported and a considerable discrepancy with experimental values is noted for Cr. The effect of the interaction of $3d^n4p$ with $3d^{n-1}4s4p$ is studied using a fixed core, multi-configuration Hartree-Fock approximation.

022.014 Joint diffusion of photons and particles in semi-infinite space − I. Distribution of excited atoms in semi-infinite space. A. N. Lagar'kov, N. A. Medvedeva.
Journ. Quant. Spectrosc. Radiat. Transfer, Vol. 13, 209 - 223 (1973). In Russian.

The excited atom distribution produced by the simultaneous action of two processes, namely excitation transfer by radiation in a spectral line and space movement of excited atoms, is considered. A kinetic equation describing these processes is analyzed. For steady-state conditions, an asymptotic analytical solution is obtained. This solution describes the concentration distribution of excited atoms for a plane geometry in a region which is at a distance exceeding the effective free path away from the surface limiting the volume. The influence on the general solution of either of the excitation transfer processes is found as a function of the parameters involved.

022.015 **Joint diffusion of photons and particles in semi-infinite space – II. Concentration of excited atoms ahead a shock wave front.** A. N. Lagar'kov, N. A. Medvedeva. Journ. Quant. Spectrosc. Radiat. Transfer, Vol. 13, 225 - 233 (1973). In Russian.

Steady-state distributions of excited atom concentrations are found ahead of a plane shock wave. The problem is solved for a two-level approximation. An exact analytical solution of the integro-differential equation describing the distribution of the concentration is developed by using the Wiener–Hopf method. The asymptotic behaviour in regions both far away and close to the shock front is investigated as a function of the parameters involved.

022.016 **Comments on transition probabilities of argon II.** B. van der Sijde, E. D. Tidwell. Journ. Quant. Spectrosc. Radiat. Transfer, Vol. 13, 289 - 292 (1973).

A comment is given on the method of determining transition-probability values in a previous paper of one of the authors. A revised table of values of some argon II transitions is presented.

022.017 **Quenching of $O(2^1 D_2)$ by atmospheric gases.** R. F. Heidner III, D. Husain. Nature, Phys. Sci., Vol. 241, 10 - 11 (1973).

We have recently described a sensitive technique for monitoring $O(2^1 D_2)$ directly by means of time-resolved atomic absorption spectroscopy in the vacuum ultraviolet. This work has now been extended to the determination of the absolute rate constants for the collisional quenching of $O(2^1 D_2)$ by six gases of atmospheric interest. We comment here on the applicability of these direct determinations to several processes involving $O(2^1 D_2)$ in the earth's atmosphere.

022.018 **An evaluation of molecular constants and transition probabilities for the NH free radical.** J. M. Lents. Journ. Quant. Spectrosc. Radiat. Transfer, Vol. 13, 297 - 310 (1973).

Intense NH electronic emission spectra have been obtained from the low-density plume of an argon arc-jet with ammonia introduced around the periphery. These spectra have been compared with spectra calculated by use of an IBM 360/50 computer from molecular constants, and the results of the computer comparison leading to a list of preferred molecular constants are included. Intensity measurements on emission spectra have yielded several previously unknown electronic-vibrational transition probabilities for different NH band systems. The electronic-vibrational transition probabilities available from the literature are also included.

022.019 **Line intensities of CO_2 in the 2.0 micron region.** H. D. Downing, R. H. Hunt. Journ. Quant. Spectrosc. Radiat. Transfer, Vol. 13, 311-321 (1973).

The line intensities of the three $\Sigma - \Sigma$ bands of CO_2 and their associated $\Pi - \Pi$ hot bands in the 2.0 μ region have been measured using a combination of long optical paths and low sample pressures with high resolution (0.035 cm^{-1}). The rotationless dipole moment matrix element and the Herman–Wallis coefficients are also determined for each band.

022.020 **Systematic trends in atomic transition probabilities in neutral and singly-ionized zinc, cadmium and mercury.** T. Andersen, G. Sørensen. Journ. Quant. Spectrosc. Radiat. Transfer, Vol. 13, 369 - 376

(1973).

The beam-foil technique has been used to measure mean lives of excited levels in neutral and singly ionized zinc, cadmium, and mercury. The mean lives reported were converted to f values for transitions for which reliable multiplet intensity ratios are available.

022.021 **Rotational and vibrational hydroxyl excitation in the laboratory and in the night airglow.** V. I. Krassovsky. Journ. Atmosph. Terr. Phys., Vol. 35, 705 - 711 (1973).

A comparison is made of hydroxyl rotational and vibrational excitation in the night airglow and on laboratory conditions when hydroxyl in the basic state is formed as a result of the reaction between O_3 and H. The reasons for the difference of these excitations are discussed.

022.022 **La description mathématique de certaines propriétés des systèmes de tourbillons par des réseaux statistiques. La comparaison à quelques résultats expérimentaux en mécanique des fluides, physique, météorologie, astronomie, etc.** M. Matschinski. Comptes Rendus Acad. Sci. Paris, Sér. A, Vol. 276, 947 - 950 (1973).

022.023 **Relative absorptions by the red system of the CN molecule from 4400 Å to 3 microns.** J. G. Phillips, C. M. Leung. Astrophys. Journ., Vol. 180, 607 - 615 (1973).

Absorption coefficients have been calculated for the red system of CN from 4400 Å to 3 μ at three temperatures: 1000°, 3000°, and 5000°K.

022.024 **The fundamental rotation-vibration band of CN.** J. G. Phillips. Astrophys. Journ., Vol. 180, 617 - 622 (1973).

The locations of lines of the P- and R-branches of the fundamental rotation-vibration band of the CN molecule are derived by a differencing process invoking the known structures of bands of the red system ($A\,^2\Pi - X\,^2\Sigma$).

022.025 **Accurate frequencies below 5 GHz of the lower J states of OD.** W. L. Meerts, A. Dymanus. Astrophys. Journ., (*Letters*), Vol. 180, L93 - L95 (1973).

The hyperfine Λ-doubling transitions of the low-J states of OD were measured, using the molecular beam electric-resonance technique. These measurements are used to calculate some other transitions of OD, which might also be of interest for radio astrophysics.

022.026 **Synchrotron radiation spectra.** L. J. Gleeson, K. C. Westfold. Proc. Astron. Soc. Australia, Vol. 2, 142 - 144 (1972).

022.027 **The absorption spectrum of Hg I between 370 and 900 Å.** M. W. D. Mansfield. Astrophys. Journ., Vol. 180, 1011 - 1021 (1973).

022.028 **A technique for ultralong-path absorption spectroscopy.** R. Goldstein, V. Vali, K. Fox. Astrophys. Journ., (*Letters*), Vol. 180, L129 - L131 (1973).

A method has been developed for obtaining very long optical paths for absorption spectroscopy. The method involves trapping a short laser pulse between two widely spaced (e.g., 1 km) retroreflectors.

022.029 **Determination of van der Waals broadening of Fe I emission lines induced by neutral He.** G. H. Copley, D. M. Camm. Astron. Astrophys., Vol. 24, 239 - 246 (1973).

The van der Waal's broadening coefficients for a number

of Fe I emission lines in the wavelength range of 3600–5000Å have been measured at pressures of up to 200 Torr of He in a glow discharge. The measured coefficients are approximately 50 % larger than those predicted by the formulae of Griem (1964). They are, however, in reasonable agreement with the line-broadening calculations of Fullerton and Cowley (1971).

022.030 Hydrogen Stark-broadening tables.
C. R. Vidal, J. Cooper, E. W. Smith.
Astrophys. Journ., Suppl. Ser., No. 214, Vol. 25, 37 - 135 (1973).
Tables of Stark broadening of the first four Lyman lines and the first four Balmer lines of hydrogen are presented. They are based on a recently developed "unified theory" of line broadening which generates normalized profiles covering the entire profile from the impact limit in the line center to the quasi-static limit in the line wings. The tables are presented in a convenient form for accurate numerical interpolation.

022.031 Über die Richtung der Zeit. K. Kraus.
Phys. Blätter, 29. Jahrgang, p. 9 - 19 (1973).

022.032 Pressure shift and broadening of the resonance lines of singly ionized alkaline-earth atoms and some alkali atoms in hot compressed Ar and He.
S. Y. Ch'en, P. K. Henry.
Journ. Quant. Spectrosc. Radiat. Transfer, Vol 13, 385 - 391 (1973).
With a ballistic compressor, the pressure shift and broadening of the resonance lines of singly ionized Ca. Sr and Ba were measured under various pressures of Ar and He at 2400 and 2500°K respectively. To make a comparison with the corresponding effects for alkali atoms, the first member of the principal series and that of the diffuse series of neutral Li and Na-D lines were also studied under the same experimental conditions.

022.033 Electron contribution to quasistatic Stark broadening. G. Fussmann, G. Himmel.
Journ. Quant. Spectrosc. Radiat. Transfer, Vol. 13, 393 - 399 (1973).
The contribution of quasistatic electrons to linear Stark broadening is evaluated by means of convolutions. A modified normalized microfield distribution including these electrons is derived. Hydrogen line profiles are calculated on this purely quasistatic basis. Comparisons with experimental results for Lyman-α and H$_\beta$ yield satisfactory agreement in the line wings.

022.034 A simplified Hartree-Fock method for opacity calculations. H. Kähler.
Journ. Quant. Spectrosc. Radiat. Transfer, Vol. 13, 401 - 416 (1973).
Complex atoms can be treated by using few parameters; thereby close approximation to accurate Hartree-Fock results will be obtained. This is achieved by optimization of effective potentials in the Hartree-Fock equations. Parameters are presented for ground states of neutral atoms up to $Z = 36$.

022.035 Infrared atomic spectra. Introductory report.
C. Moore-Sitterly.
17th Colloquium International Astrophys. Liège 1971, (see 012.003), p. 15 - 34 (1972).

022.036 The low terms of Cr IV. J. O. Ekberg.
Phys. Scripta, Vol. 7, 55 - 58 (1973).
The spectrum of Cr IV has been observed by using a vacuum sliding-spark discharge and a 3-m normal incidence spectrograph. 275 lines have been classified in the region from 523 to 2423 Å.

022.037 Extended analysis of Cr V. J. O. Ekberg.
Phys. Scripta, Vol. 7, 59 - 61 (1973).
The spectrum of Cr V has been observed by using a vacuum sliding spark discharge and a 3-m normal incidence spectrograph. 130 lines have been classified in the region from 433 to 1837 Å.

022.038 Atomic processes. P. G. Burke.
Atoms and molecules in astrophysics. Proc. Scottish Univ. Summer School in Physics 1971, (see 012.005), p. 1 - 63 (1972).

022.039 Highly excited atoms. I. C. Percival.
Atoms and molecules in astrophysics. Proc. Scottish Univ. Summer School in Physics 1971, (see 012.005), p. 65 - 83 (1972).

022.040 Spectral line broadening. H. van Regemorter.
Atoms and molecules in astrophysics. Proc. Scottish Univ. Summer School in Physics 1971, (see 012.005), p. 85 - 119 (1972).

022.041 Introduction to molecular spectra. H. M. Foley.
Atoms and molecules in astrophysics. Proc. Scottish Univ. Summer School in Physics 1971, (see 012.005), p. 155 - 199 (1972).

022.042 Spectral intensities from helium-like ions.
A. H. Gabriel.
Atoms and molecules in astrophysics. Proc. Scottish Univ. Summer School in Physics 1971, (see 012.005), p. 311 - 320 (1972).

022.043 Formation of molecular hydrogen on cold surfaces.
A. Schutte.
Atoms and molecules in astrophysics. Proc. Scottish Univ. Summer School in Physics 1971, (see 012.005), p. 337 - 340 (1972).

022.044 Observations of formamide at 6 cm in Sagittarius B2.
J. C. Ribes, J. G. Ables, P. D. Godfrey, R. D. Brown.
Australian Journ. Phys., Vol. 26, 79 - 84 (1973).
Six hyperfine components of the $2_{12}-2_{11}$ transition of formamide (NH_2CHO) have been measured in the laboratory. The multiplet was subsequently observed in emission in the direction of Sgr B2 with the 64 m telescope at Parkes. The relative frequencies and intensities of the three strongest components in the hyperfine structure are in excellent agreement with the laboratory measurement.

022.045 Partial cross-sections in high-energy nuclear reactions, and astrophysical applications. I. Targets with $Z \leqslant 28$. R. Silberberg, C. H. Tsao.
Astrophys. Journ., Suppl. Ser., No. 220 (I), Vol. 25, 315 - 333 (1973).
We have devised new empirical formulae for high-energy cross-sections, using measured yields of proton interactions with various target nuclei ($3 \leqslant Z_t \leqslant 28$). They are applicable at energies > 100 MeV to all products with $A \geqslant 6$ for targets ranging from Li to Ni. The new relations are useful in regions where earlier empirical formulae break down completely. Moreover, they give significantly better estimates of cross-sections that those previously available.

022.046 Partial cross-sections in high-energy nuclear reactions, and astrophysical applications. II. Targets heavier than nickel. R. Silberberg, C. H. Tsao.
Astrophys. Journ., Suppl. Ser., No. 220 (II), Vol. 25, 335 - 367 (1973).
Available experimental cross-sections for proton inter-

actions with targets ranging from copper to uranium at energies ≥ 100 MeV have been used to formulate empirical equations for calculating the production of various nuclides. In the region of spallation products, the equation is similar to Rudstam's formula. In the fission region, as well as in the region of lighter products, such expressions are presented for the first time. Equations are also given for calculating cross-sections for interactions that are largely peripheral.

022.047 Conséquences physiques possibles de l'existence d'une masse non nulle du photon sur les interactions de la lumière avec la matière et la théorie du corps noir.
M. Moles, J.-P. Vigier.
Comptes Rendus Acad. Sci. Paris, Sér. B, Vol. 276, 697 - 700 (1973).

022.048 Energy spectrum of He II in a strong magnetic field and bound-bound transition probabilities.
G. L. Surmelian, R. F. O'Connell.
Astrophys. Space Sci., Vol. 20, 85 - 91 (1973).
The ground state energy of the He II atom is determined in magnetic fields up to 10^{12} G. The 13 lowest excited states and bound-bound transition probabilities are calculated in magnetic fields from 10^7 to 10^9 G.

022.049 The effects of Stark broadening in the radio recombination line temperatures. J. P. Simpson.
Astrophys. Space Sci., Vol. 20, 187 - 203 (1973).
Observations of Orion A, M 17, and W 3 made at the National Radio Astronomy Observatory show Stark broadening in the α, β, and γ hydrogen recombination lines at 22 cm. Theoretical line profiles have been derived from model H II regions. Using Voigt profiles fitted by least squares to the observations, we found that the line intensities at 22 cm agree with the intensity ratios predicted by LTE theory.

022.050 Laws of physics and new discoveries in astronomy.
I. D. Novikov.
Zemlya i Vselennaya, 1973, No. 2, p. 7 - 10. In Russian.

022.051 Absolute intensity measurements in the vacuum ultraviolet region.
K. Nishi, K. Higashi, A. Yamaguchi.
Tokyo Astron. Obs., Report No. 62, Vol. 16, 584 - 609 (1973). In Japanese.

022.052 Secondary electrons and energy per ion-pair in a thermal gas for electron, proton and X-ray ionization. J. Bergeron, S. Collin-Souffrin.
Astron. Astrophys., Vol. 25, 1 - 8 (1973).
The total number of electron-ion pairs Φ produced by the complete absorption of an electron beam, in a pure hydrogen gas, is determined as a function of the electron energy and of the ionization degree of the gas. The average energy necessary to produce such an ion-pair is simply deduced from Φ. Ranges from $4 I_H$ (Ryd unit) to $800 I_H$ for the electron energy, and from 0.1 to 10^3 for the ratio of atomic hydrogen to electron densities are covered. The efficiency of secondary electrons in further ionizations is then given.

022.053 Radiative and dielectronic recombination coefficients for complex ions.
S. M. V. Aldrovandi, D. Péquignot.
Astron. Astrophys., Vol. 25, 137 - 140 (1973).
Radiative and dielectronic recombination coefficients are calculated for all non-hydrogenic ions of He, C, N, O, Ne, Mg, Si and S.

022.054 Stark broadening and shift of singly ionized aluminium lines. J. Heuschkel, H. J. Kusch.

Astron. Astrophys., Vol. 25, 149 - 151 (1973). – Research note.

022.055 Fluorescence of tetraphenyl-butadiene in the vacuum ultraviolet. W. M. Burton, B. A. Powell.
Applied Optics, Vol. 12, 87 - 89 (1973).
Tetraphenyl-butadiene is a useful alternative to sodium salicylate as a fluorescent wavelength converter for detectors in the vacuum uv. Measurements of the fluorescence efficiency of TPB relative to sodium salicylate over the wavelength range 735–3160 Å are presented, and the application of TPB coatings to image tube detectors is discussed.

022.056 The investigation of UV oscillator strengths in C, N and O ions. J. V. Mallow.
Atoms and molecules in astrophysics. Proc. Scottish Univ. Summer School in Physics 1971, (see 012.005), p. 347 - 352 (1972).

022.057 L'astrométrie appliquée aux nuages artificiels.
H. Debehogne, E. van Hemelrijck.
Ciel et Terre, Vol. 89, 91 - 109 (1973).

022.058 A seldom appreciated symmetry in gravity.
L. Epstein.
Observatory, Vol. 93, 70 - 74 (1973).

022.059 Determination of atomic lifetimes and absolute oscillator strengths for neutral and ionized titanium.
J. R. Roberts, T. Andersen, G. Sørensen.
Astrophys. Journ., Vol. 181, 567 - 586 (1973).
Measurements of atomic lifetimes by the beam-foil technique and branching ratios by use of a gas-flow stabilized arc have led to an experimental determination of absolute oscillator strengths of Ti II. Some lifetimes of Ti I, Ti III, and Ti IV are also presented.

022.060 Determination of atomic lifetimes and absolute oscillator strengths for neutral and ionized vanadium.
J. R. Roberts, T. Andersen, G. Sørensen.
Astrophys. Journ., Vol. 181, 587 - 604 (1973).
Measurements of atomic lifetimes by the beam-foil technique and branching ratios by use of a gas-flow stabilized arc have led to an experimental determination of absolute and relative oscillator strengths of V I and V II. Some lifetimes of V III are also presented.

022.061 Laboratory microwave spectrum of methylamine.
K. Takagi, T. Kojima.
Astrophys. Journ., *(Letters)*, Vol. 181, L91 - L93 (1973).

022.062 On the problem of excitation and ionization of neutral sodium. K. S. Tavastsherna.
Solnechnye Dannye 1973 Byull., No. 1, p. 84 - 92 (1973). In Russian.
Elementary processes for the levels $3^2 S$, $3^2 P$, $4^2 S$, $3^2 D$ and the continuum of Na I have been considered. On the basis of the solution and analysis of the stationarity equation for the $3^2 P$, $4^2 S$ and $3^2 D$ levels and the ionization equilibrium equation it is concluded on excitation and ionization of Na I. The values of relative populations of n_2/n_1 and ionization degree of n_+/n_1 at $T_e = 5000 - 15000°$, $n_e = 10^9 - 5 \times 10^{13}$ and $r_{12} = 0.02, 0.045, 0.13, 0.20, 0.50$, are given in tables.

022.063 Stark broadening of Paschen lines in a deuterium discharge. G. Himmel, F. Pinnekamp.
Journ. Quant. Spectrosc. Radiat. Transfer, Vol. 13, 555 - 566 (1973).
The Paschen lines P_6 to P_{13} emitted from a radio-frequency discharge have been measured photoelectrically. The experimental profiles are compared with refined quasistatic

calculations. These calculations include Doppler and apparatus broadening.

022.064 **Short-action repulsive forces among atoms and molecules of atmospheric gases.**
A. P. Kalinin, V. B. Leonas, A. V. Sermyagin.
Mezhplanet. sreda i fiz. magnitosfery. Nauka, Moskva, 1972, p. 196 - 201. In Russian. — Abstr. in Referativ. Zhurn. 51. Astron., 4.51.325 (1973).

022.065 **Experimental oscillator strengths of V I lines.**
K. Mie, J. Richter.
Astron. Astrophys., Vol. 25, 299 - 301 (1973).
Absolute f-values of 5 V I lines have been measured with the atomic beam absorption method. The results are in agreement with f-values which can be derived from beam foil measurements of atomic life times.

022.066 **Experiments with atomic hydrogen.** D. Kleppner.
Atomic physics and astrophysics. Brandeis Univ. Summer Inst. 1969, (see 003.007), p. 1 - 89 (1971).

022.067 **Quantum electrodynamics and the theory of the hydrogenic atom.** S. J. Brodsky.
Atomic physics and astrophysics. Brandeis Univ. Summer Inst. 1969, (see 003.007), p. 91 - 169 (1971).

022.068 **Megagauss physics.** C. M. Fowler.
Science, Vol. 180, 261 - 267 (1973).
The production, measurement, and applications of megagauss fields are surveyed.

022.069 **Free—free radiation in electron-neutral atom collisions.** S. Geltman.
Journ. Quant. Spectrosc. Radiat. Transfer, Vol. 13, 601 - 613 (1973).
Free—free absorption coefficients are calculated for the electron-neutral atom systems involving He, C, N, O, Ne, Ar, Kr and Xe. The calculations are based upon model atomic potentials which have been adjusted to fit experimental scattering cross sections or electron affinities. Some angular distributions are presented and thermal averages are evaluated in the ranges $\lambda = 0.5 - 20\,\mu m$ and $T = 500 - 20,000°K$.

022.070 **Numerical calculations of atomic structure constants. II. Radial parts, energy levels, transition probabilities for Fe XIII.** F. Bely-Dubau.
Astron. Astrophys., Vol. 25, 431 - 435 (1973).
Employing the multi-configuration Hartree-Fock method, the numerical evaluation of complex atom data is extended to the calculation of energy levels, wavelengths and parameters relative to permitted transitions in intermediate coupling with configuration interaction. Application is made to Fe XIII.

022.071 **Microwave transitions of the $O^{16}D$ molecule.**
V. K. Khersonsky.
Astron. Zhurn. Akad. Nauk SSSR, Vol. 50, 646 - 649 (1973). In Russian. English translation in Soviet Astron. AJ, Vol. 17, No. 3. — Short note.

022.072 **Ab initio Berechnung eines neuen Kandidaten für die X-ogen-Linie.** J. Barsuhn.
Mitt. Astron. Ges., No. 32, p. 119 - 120 (1973). — Abstract.

022.073 **The term analysis of atomic spectra.** B. Edlén.
Phys. Scripta, Vol. 7, 93 - 101 (1973).
After a survey of the ground configurations of atoms in different stages of ionization through the periodic system we indicate the present state of analysis for spectra of atoms and ions containing up to 28 electrons. The energy level structure of systems with up to 18 electrons is illustrated by diagrams of individual spectra and of isoelectronic sequences. A critical compilation of references to recent analyses of the spectra of elements with $Z \leqslant 28$ is appended.

022.074 **Über die Wechselwirkung der Elementarteilchen.**
H. P. Dürr.
Naturwissenschaften, 60. Jahrgang, p. 274 - 280 (1973).
A short and qualitative description of the various interactions of the elementary particles is given, with particular emphasis on features suggesting their common dynamical origin. On the basis of these findings prospects for a unified dynamical theory are discussed.

022.075 **Lifetime measurements of optical levels for 6 elements of astrophysical interest.**
J. Marek, J. Richter.
Astron. Astrophys., Vol. 26, 155 - 157 (1973).
The mean lifetimes of 25 optical levels of Mg I, Al I, Si I, Cr I, Co I, and Mn I have been determined with the phase shift method. The measured values can be used to calibrate previously determined scales of relative oscillator strengths.

022.076 **New physical laws and astronomy.**
V. L. Ginzburg.
Vopr. filosofii, 1972, No. 11, p. 14 - 20. In Russian. — Abstr. in Referativ. Zhurn. 51. Astron., 5.51.2 (1973).

022.077 **Production of ^7Be, ^9Be and ^{10}Be in the spallation of ^{13}C by protons of 150 and 600 MeV.**
G. M. Raisbeck, J. Lestringuez, F. Yiou.
Nature, Phys. Sci., Vol. 244, 28 - 30 (1973).

022.078 **Radio recombination lines from H^0 regions.**
M. Brocklehurst.
Astrophys. Letters, Vol. 14, 81 - 84 (1973).
Level populations of hydrogenic ions in conditions typical of H^0 regions are obtained. The effects of a thermal continuum, as produced by a nearby H^+ region, are included. At very low electron densities this continuum dominates electron collisions and the calculations are density independent.

022.079 **Photoionization excitation of the $CO_2^+(\widetilde{B}^2\Sigma_u^+ \rightarrow \widetilde{X}^2\Pi_g)$ 2890-Å band.**
R. W. Carlson, D. L. Judge, M. Ogawa.
Journ. Geophys. Res., Vol. 78, 3194 - 3196 (1973). — Letter.

022.080 **Semiempirical calculation of gf values: Sc II $(3d + 4s)^2 - (3d + 4s)\,4p$, a detailed example.**
R. L. Kurucz.
Smithsonian Astrophys. Obs., *Cambridge, Mass.*, Special Rep. No. 351, 7 + 57 pp. (1973).
A semiempirical procedure for calculating gf values is developed in detail. A program written by R. D. Cowan is used to produce LS transition arrays and electrostatic and spin-orbit matrices. Transition integrals are evaluated with scaled Thomas-Fermi-Dirac wavefunctions following Warner. Eigenvectors are found through a least-squares procedure that fits computed eigenvalues to observed energy levels, and then the eigenvectors are used to transform the LS transition array to the observed coupling scheme. Throughout the discussion, examples are given for Sc II $(3d + 4s)^2 - (3d + 4s)\,4p$. The final computed gf values are compared to laboratory measurements.

022.081 **Collision broadening and shift in the resonance line of calcium.** G. Smith.
Journ. Phys. B, Atomic Molecular Phys., Vol. 5, 2310 - 2319 = Dep. Astrophys. Univ. Obs. Oxford, Publ. No. 57 (1972).
The present paper describes measurements of the broadening and shift which occur when the calcium resonance line is perturbed by helium, neon, argon, krypton and xenon,

at pressures where the impact theory should apply. The results are compared with theoretical predictions for Lennard-Jones potentials and for potentials which include an additional R^{-8} dispersion term.

022.082 The equivalent widths of absorption lines of the A-band of oxygen at different pressure.
V. D. Galkin, L. N. Zhukova, L. A. Mitrofanova, V. K. Prokof'ev.
Physics of the moon and planets, (see 012.024), p. 345 - 348 (1972). In Russian.

022.083 Multiplets in astrophysics.   C. Moore-Sitterly.
Optica Pura y Aplicata, (*Spain*), Vol. 5, 147 - 158 (1973).
Astrophysical research on the composition of a star, as revealed by the lines in its spectrum, depends directly on a knowledge of the multiplet structure of the lines identified as to their chemical origin. Two aspects of this large subject are briefly reviewed: correct identifications and abundance determinations.

022.084 An astronomical view of high-pressure sodium lamps.   J. M. Fletcher, D. Crampton.
Publ. Astron. Soc. Pacific, Vol. 85, 275 - 277 = Contr. Dominion Astrophys. Obs., Victoria, No. 199 (1973).
The spectral energy distribution of a high-pressure sodium lamp is compared to that of a mercury lamp. The features in the mercury spectrum are approximately 10X stronger than those in the "sodium" spectrum. Virtually no light is emitted by these sodium lamps at the wavelengths of the D lines.

022.085 Calcul des deuxième et troisième coefficients du viriel en utilisant un puits carré de potentiel modifié.   R. E. Caligaris, J. C. Grangel.
Journ. Chim. Phys., Vol. 68, 596 - 600 = Dep. Astron. Fís., Fac. Humanidades Ciencias, Univ. Montevideo, Publ. No. 40 (1971).
In the present paper it is suggested a simple modification of the square well potential in classical theory of fluids. The second and third virial coefficients with the new potential are calculated and compared with those obtained through Lennard-Jones and the original square well potentials.

022.086 Las fuerzas radiometricas – su aplicacion al radiometro.   G. Rossel, R. Donangelo.
Dep. Astron. Fís., Fac. Humanidades Ciencias, Univ. Montevideo, Publ. No. 41, 25 pp. (1971).

An introduction to light pressure, radiometric forces, accommodation coefficients and rarefied gases is given.

022.087 Interaction lengths of energetic pions and protons in iron.   H. Crannell, C. J. Crannell, H. Whiteside, J. F. Ormes, M. J. Ryan.
Phys. Rev. D, Particles and Fields, Vol. 7, 730 - 740 (1973).
The interaction lengths of pions and protons in iron have been measured using an ionization spectrometer composed of alternating layers of iron and plastic scintillator. These measurements cover an energy range from 9.3 to 18 GeV.

022.088 The speed of light.   J. D. Mulholland.
Science, Vol. 180, 1321 - 1322 (1973). – Letter.

022.089 Photon-plasma transition $2^3S_1 - 1^3S_1$ in positronium.   S. A. Kaplan, E. B. Klejman.
Izv. vyssh. ucheb. zavedenij. Radiofizika, Vol. 16, 156 - 157 (1973). In Russian. – Abstr. in Referativ. Zhurn. 51. Astron., 7.51.208 (1973).

Mechanik, Relativität, Gravitation.
See Abstr. 003.049.

Geochemistry and cosmic chemistry of isotopes of rare gases.   See Abstr. 003.112.

Theoretical micro- and macrophysics.
See Abstr. 003.143.

On the velocity of light three centuries ago.
See Abstr. 099.023.

Errata

022.901 Erratum: "Radiative recombination coefficients for complex ions" [Astrophys. Journ., Vol. 168, 313 - 316 (1971)].   C. B. Tarter.
Astrophys. Journ., Vol. 181, 607 (1973).

022.902 Erratum: 'Ionization balance for ions of Na, Al, P, Cl, A, K, Ca, Cr, Mn, Fe and Ni' [Astron. Astrophys., Suppl. Ser., Vol. 7, 291 - 310 (1972)].   M. Landini, B. C. Fossi.
Astron. Astrophys., Suppl. Ser., Vol. 10, 281 (1973).

Instruments and Astronomical Techniques

031 Optics, Methods of Observation and Reduction

031.001 **Investigation of the influence of temperature and velocity fields on the quality of an astronomical image.** A. S. Zherbina, L. K. Zinchenko, R. L. Petrov.
Astron. Zhurn. Akad. Nauk SSSR, Vol. 50, 176 - 180 (1973).
In Russian. English translation in Soviet Astron. AJ, Vol. 17, No. 1.
The quality of an artificial star image at passing through a turbulent layer has been investigated. The distortion of the diffraction image by different temperature and velocity fields generated in the wind tunnel was studied. The dependences of star image quality on the width of the thermal layer, on its maximum temperature and on stream velocity were determined. The refractive-index structure parameter $C_p{}^2$ was calculated and compared with its value in the atmosphere for weak turbulence conditions.

031.002 **Comments on the Vasilevskis cooling curve.** K. A. Strand.
Astrophys. Journ., Vol. 179, 147 - 148 (1973).
The explanation offered by Vasilevskis that the change in the scale of plates taken with the Yerkes 40-inch (102-cm) refractor is caused by thermal conditions of the lens related to time from sunset is shown not to be valid.

031.003 **An arc lamp for optical testing.** R. E. Cox.
Sky Telescope, Vol. 45, 183 - 188 (1973).

031.004 **On image structure, and the value and challenge of very large telescopes.** R. F. Griffin.
Observatory, Vol. 93, 3 - 8 (1973).
A big telescope often gives an image having a small bright core. This paper suggests that if the telescope were bigger still, more of the light would be concentrated in the core. If the telescope were optically diffraction-limited, the core would be smaller and brighter; the resolution of most existing large telescopes is set by the optics and not by the seeing.

031.005 **Sur les dioptres asphériques en optique astronomique.** G. Lemaître.
Comptes Rendus Acad. Sci. Paris, Sér. B, Vol. 276, 145 - 148 (1973).
On se propose de rechercher les configurations de charges et d'appuis associés à un dioptre de révolution dont le profil d'épaisseur variable permet d'engendrer des dioptres déformés par élasticité; la flèche ayant pour expression une équation purement du quatrième degré. On montre ensuite la possibilité de superposer une deuxième déformation modulée par un paramètre angulaire afin d'obtenir l'anastigmatisme d'un télescope de Schmidt réflecteur hors de l'axe.

031.006 **Reconstruction of the image of a confined source.** T. W. Cole.
Astron. Astrophys., Vol. 24, 41 - 45 (1973).
Knowledge that a source is confined can be used to derive spatial frequency components of the source which are not passed by an imaging system.

031.007 **Het scheidend vermogen van een teleskoop.** M. A. M. van Venrooy.
Hemel en Dampkring, Vol. 71, 63 - 68 (1973).

031.008 **Stellarphotographie mit dem Refraktor.** H. Treutner.
Orion Schaffhausen, 31. Jahrgang, p. 22 - 26 (1973).

031.009 **Corrections for lost motion in theodolite eye-piece micrometers.** P. B. Jones.
Bull. Géod., Nouvelle Sér., Année 1973, No. 107, p. 5 - 11.

031.010 **Enhancement of spectra by digital convolution.** J. J. Lorre.
Astron. Journ., Vol. 78, 67 - 73, 159 - 161 (1973).
A method is presented for convolving digitally scanned spectra by means of computerized filtering. This method is an effective tool for improving the quality of noisy spectra and for deconvolving high-quality spectra by removing instrument signature.

031.011 **Observational optimization with modest aperture telescopes.** W. L. Sanders.
Bull. American Astron. Soc., Vol. 5, 5 (1973). – Abstr. AAS.

031.012 **Digital astronomy at Capilla Peak Observatory.** V. H. Regener.
Bull. American Astron. Soc., Vol. 5, 5 (1973). – Abstr. AAS.

031.013 **State of the art vs. practice at microwavelength.** C. L. Seeger.
Bull. American Astron. Soc., Vol. 5, 25 (1973). – Abstr. AAS.

031.014 **Recent progress in infrared and microwave techniques of astronomical interest.** Introductory report.
P. Léna.
17th Colloquium International Astrophys. Liège 1971, (see 012.003), p. 61 - 81 (1972).

031.015 **La surface polie d'un miroir de télescope vue au microscope électronique.** V. Fryder.
Orion Schaffhausen, 31. Jahrgang, p. 49 - 52 (1973).

031.016 **On the systematic pointing error in the observation of disk-shaped objects.** E. M. Nenakhova.
Astrometriya i Astrofizika, *Kiev*, vyp. (No.) 16, (see 003.006), p. 21 - 25 (1972). In Russian.
Special observations of Venus and Jupiter were made with the Wanschaff vertical circle in order to determine differences of systematic errors of pointing on the upper and lower limbs. The origin of these errors and their effect on the observed zenith distances are discussed. The author recommends a procedure which is capable of reducing and, in some cases, even eliminating the effect of setting errors.

031.017 **Holographische Prüfung optischer Spiegel.** A. F. Fercher.
Umschau, 73. Jahrgang, p. 270 - 274 (1973).

031.018 **On the reduction of the observational data with the Tokyo meridian circle.**
H. Hara, H. Ishii, I. Kamijo, N. Miyauchi.
Tokyo Astron. Obs., Report No. 62, Vol. 16, 610 - 623 (1973).
In Japanese.

031.019 Measurement of the PZT plate and identification of the stars by computer. K. Nakajima.
Tokyo Astron. Obs., Report No. 62, Vol. 16, 636 - 644 (1973). In Japanese.

031.020 Multiplicative correction of phase errors in Fourier spectroscopy. R. b. Sanderson, E. E. Bell.
Applied Optics, Vol. 12, 266 - 270 (1973).
The multiplicative correction procedure for phase errors in Fourier spectroscopy has been analyzed. It is shown that the residual error is proportional to the true spectrum measured at a resolution corresponding to the width of the cutoff through the origin of the truncation function. The importance of choosing a sufficiently broad cutoff region is demonstrated, and methods for further improving the correction are presented.

031.021 An easily constructed caustic tester.
J. F. Kielkopf.
Sky Telescope, Vol. 45, 315 - 318, with a note by R. E. Cox, 318 (1973).

031.022 Generalized histogram of the quality of an image.
S. B. Novikov, P. V. Shcheglov.
Astron. Tsirk., No. 748, p. 1 - 6 (1973). In Russian.

031.023 Les équations personnelles des observateurs à l'instrument de passage. G. Oprescu.
Stud. Cerc. Astron. Vol. 18, 33 - 36 (1973). In Romanian.

031.024 A remark on turbulence and 'production' limited telescopes. A. Greve.
Solar Physics, Vol. 29, 263 - 266 (1973).
The image degradation due to residual surface inaccuracies of the main optical element of a telescope is compared with the image degradation due to atmospheric turbulence.

031.025 On the requirements imposed upon optical characteristics of astronomical telescopes and cameras.
G. G. Slyusarev.
Trudy Gos. optich. in-ta, Vol. 41, No. 173, p. 72 - 76 (1972). In Russian. − Abstr. in Referativ. Zhurn. 51. Astron., 4.51.111 (1973).

031.026 Making of large sitall mirrors for Cassegrain and Ritchey-Chrétien telescopes. G. M. Popov.
Izv. Krymskoj Astrofiz. Obs., Vol. 46, 159 - 166 (1972). In Russian.

031.027 Contrôlons nos miroirs plans! R. Durussel.
Orion Schaffhausen, 31. Jahrgang, p. 92 - 94 (1973).

031.028 Accuracy in the reduction methods (three or five stars). High declination reduction.
H. Debehogne.
Astron. Astrophys.,Suppl. Ser., Vol. 10, 195 - 199 (1973). In French.
In this paper are given the principal results obtained by a study on the accuracy of plates reductions for the dependences method and by means of a network.

031.029 Distortion of electron images focused by almost uniform electric and magnetic fields.
J. D. H. Pilkington, K. F. Hartley.
Photo-electronic image devices, (see 012.017), p. 545 - 555 (1972).

031.030 Data reduction techniques for direct astronomical electronography. M. J. Cullum, C. L. Stephens.
Photo-electronic image devices, (see 012.017), p. 757 - 768 (1972).

031.031 The stabilization of planetary images.
W. A. Baum, D. M. Busby, T. V. Pettauer.
Photo-electronic image devices, (see 012.017), p. 781 - 788 (1972).

031.032 Ein einfaches Interferometer zur Prüfung astronomischer Optik. K.-L. Bath.
SuW, Vol. 12, 177 - 180 (1973).

031.033 Design study of a Glass Meridian Circle. E. Høg.
Mitt. Astron. Ges., No. 32, 120 - 125 (1973).

031.034 Die Eignung des Mikrodensitometers PDS 1000 zum Messen von Radialgeschwindigkeiten.
W. W. Weiss.
Mitt. Astron. Ges., No. 32, p. 130 - 138 (1973).

031.035 Zur Ermittlung des Zentrierzustandes von Zweispiegel-Systemen mittels Hartmann-Aufnahmen.
L. D. Schmadel.
Mitt. Astron. Ges., No. 32, p. 145 - 149 (1973).

031.036 Schnellphotometrie veränderlicher Sterne.
W. Strohmeier.
Mitt. Astron. Ges., No. 32, 244 - 246 (1973).

031.037 A three-lens prime focus corrector for parabolic telescope mirrors. M. Faulde, R. N. Wilson.
Astron. Astrophys., Vol. 26, 11 - 15 (1973).
The telescope parameters influencing the design of prime focus correctors are briefly reviewed to illustrate the interest of a corrector with not more than three elements of one material having good UV transmission and giving good correction for parabolic primaries or hyperbolic primaries with low excentricity. An example of such a corrector for a parabolic primary of 3.5 m aperture is given with spot-diagrams.

031.038 The use of Fourier transform techniques in the analysis of photoelectrically scanned double star images. L. A. Dicks, E. Van Rooyen.
Astrophys. Space Sci., Vol. 22, 153 - 163 (1973).
A technique for the analysis of photoelectrically scanned double star images is described. The method consists of comparing the Fourier transform of the double star profile with that of a single star profile imaged through the same telescope. If the measured profile of the double star image can be considered to be a linear superposition of two profiles, each identical in shape to the measured profile of a nearby single star, a comparison of the Fourier transforms of these profiles enables the parameters of the double star system to be determined.

031.039 Data processing for astronomy missions.
G. H. Ludwig.
Publ. American Astronaut. Soc., Sci. Techn. Ser., Vol. 28, (see 012.018), p. 85 - 92 (1972).

031.040 Automation in astronomy. E. J. Wampler.
Publ. American Astronaut. Soc., Sci. Techn. Ser., Vol. 28, (see 012.018), p. 337 - 346 (1972).

031.041 Active optics for space astronomy.
H. F. Wischnia.
Publ. American Astronaut. Soc., Sci. Techn. Ser., Vol. 28, (see 012.018), p. 347 - 361 (1972).

031.042 Optimization of the geometric factor of a telescope for the purpose of recording charged particles in cosmic space. A. A. Kolchin, V. V. Lebedev, A. S. Pyatkin, G. P. Skrebtsov.
Kosmich. Issled., Vol. 11, 465 - 469 (1973). In Russian.

031.043　Field correctors for astronomical telescopes.
　　　　C. G. Wynne.
Progress in Optics, Vol. 10, [North-Holland Publishing Company, Amsterdam], 139 - 164 (1972). – Review paper.

031.044　Determination of the value of vibration of stellar images by the photographic method.
M. R. Fedyanin, G. F. Aref'eva.
Materialy 3-j Nauch. konf. Tomsk. un-ta po mat. i mekh. Vyp. (No.) 2. Tomsk, Tomsk. un-t, 1973, p. 103. In Russian. Abstr. in Referativ. Zhurn. 51. Astron., 6.51.126 (1973).

031.045　A method for the automatic analysis of gamma ray events in astronomical spark chambers.
G. F. Bignami, G. Cioni, A. Della Ventura, P. Mussio, M. J. L. Turner, U. Volonte.
Comput. Phys. Commun., (Netherlands), Vol. 4, 299 - 314 (1972).

031.046　The brightness temperature distributions defined by a measured intensity interferogram.
P. J. Napier.
New Zealand Journ. Sci., Vol. 15, 342 - 355 (1972).
　　A previously reported technique is used to compute all the possible brightness temperature distributions that could have produced a measured intensity interferogram. Intensity interferograms, taken from the literature, of radio source P0819–30 and the stars βCru and αPsA are processed using the technique.

031.047　High resolution astronomical observations.
　　　　P. Couteau.
Atti Fondaz. G. Ronchi and Contr. Ist. Nazionale Ottica, (Italy), Vol. 27, 883 - 891 (1972). In Italian.

031.048　Comments on the note of Prof. Couteau 'High resolution astronomical observations'.　　V. Ronchi.
Atti Fondaz. G. Ronchi and Contr. Ist. Nazionale Ottica, (Italy), Vol. 27, 893 - 900 (1972). In Italian.

031.049　Intensity interferometry in the spatial domain.
　　　　R. H. Deitz, F. P. Carlson.
Journ. Optical Soc. America, Vol. 63, 274 - 280 (1973).

031.050　Diffraction-limited imaging of stellar objects using telescopes of low optical quality.　　J. C. Dainty.
Optics Commun., (Netherlands), Vol. 7, 129 - 134 (1973).
　　It is suggested that a recent method of stellar interferometry developed by Labeyrie and his colleagues (1972) may be used to obtain diffraction-limited resolution from telescopes whose aberrations are poor by conventional standards.

031.051　Improved geometry for the all reflecting Schmidt telescope.　　L. Epstein.
Applied Optics, Vol. 12, 926 (1973).
　　This is to point out an elementary but significant and often overlooked improvement upon the prototypes geometry that should be made common knowledge. The improvement allows the field of view area to be increased by approximately a factor of 2 without degrading image quality.

031.052　Comments on: Closed form solution for three-mirror telescopes, corrected for spherical aberration, coma, astigmatism, and curvature of field.　　R. Gelles.
Applied Optics, Vol. 12, 935 - 936 (1973).

031.053　New methods of processing speckle pattern star images.　K. T. Knox, B. J. Thompson.
Astrophys. Journ., (Letters), Vol. 182, L133 - L136 (1973).
　　A modification of Labeyrie's technique for processing

star images degraded by atmospheric seeing is presented. The phase information is preserved in the deconvolution process by the addition of an off-axis coherent reference wave. As a result, the final image is the actual object intensity distribution rather than the autocorrelation of the object intensity.

031.054　Solar speckle interferometry.
　　　　J. W. Harvey, J. B. Breckinridge.
Astrophys. Journ., (Letters), Vol. 182, L137 - L139 (1973).
　　The technique of speckle interferometry has been applied to solar photographs. Spatial frequencies as high as 4 cycles per arc second have been detected in sunspots.

031.055　Computations of meniscus systems.
　　　　T. S. Belorossova, D. D. Maksutov, N. V. Merman, M. A. Sosnina.
Trudy Glav. Astron. Obs. Pulkovo, Ser. 2, Vol. 77, 151 - 162 (1969). In Russian.
　　Methods of determining the constructive elements of the following meniscus systems are described: "meniscus–concave mirror", "Cassegrain I" with aluminized secondary mirror, and "Cassegrain II" with the secondary mirror ground on the 2^d surface of the meniscus. The determination is made by means of precise computations of 84 systems. The systems are calculated for identical chromatism, spherical aberration and, in case of the "meniscus–mirror" and "Cassegrain II" systems, for coma as well. The results are given in tables and on graphs.

031.056　The aureole of a star image.　　J. Piccirillo.
Publ. Astron. Soc. Pacific, Vol. 85, 278 - 280 = Publ. Goethe Link Obs., Indiana Univ., Bloomington, No. 152 (1973).
　　Telescope star-image profiles for various observing conditions are compared. Effects of mirror realuminization, secondary support diffraction spikes, and atmospheric haze are shown. A modification to the previously found inverse-square aureole is discussed.

031.057　On the determination of the azimuth of direction from observations of stars in the prime vertical.
M. A. Vajsov.
Trudy Kazan. Gorod. Astron. Obs., No. 37, p. 94 - 101 (1970). In Russian.

031.058　Determination of the azimuth of direction from observations of star groups near their elongation.
M. A. Vajsov.
Trudy Kazan. Gorod. Astron. Obs., No. 37, p. 102 - 132 (1970). In Russian.

031.059　Limitations caused by stars in the photographic detection of extended objects with low surface brightness.　　V. G. Khristich.
Trudy Astron. Obs., Leningrad, Vol. 29 (= Uchenye Zapiski Leningr. Un-ta, No. 363 = Seriya Matem. Nauk, vyp. (No.) 48), p. 98 - 105 (1973). In Russian.
　　The limitations caused by both unresolved and resolved stars are discussed for the detection of extended objects of low surface brightness using ordinary and composed photographs.

031.060　On the photogrammetric refraction.　S. V. Gromov.
Trudy Astron. Obs., Leningrad, Vol. 29 (= Uchenye Zapiski Leningr. Un-ta, No. 363 = Seriya Matem. Nauk, vyp. (No.) 48), p. 207 - 214 (1973). In Russian.

031.061　Problemas fundamentais de fotometria fotoeléctrica.
　　　　A. J. Pascoal.
Inform. Obs. Astron. Porto, No. 4, 49 pp. (1971).

031.062 **A method of determination of the coordinates of stars and planets from equal-altitude observations.**
G. S. Tyuterev.
Materialy 3-j Nauch. konf. Tomsk. un-ta po mat. i mekh. Vyp. (No.) 2. Tomsk, Tomsk. un-t, 1973, p. 99 - 100. In Russian.
Abstr. in Referativ. Zhurn. 51. Astron., 7.51.174 (1973).

031.063 **Reticule for observation of stars in the vertical of Polaris.** A. V. Butkevich.
Geod. i kartografiya, 1973, No. 1, p. 17 - 18. In Russian.
Abstr. in Referativ. Zhurn. 52. Geod. Aehros"emka, 7.52.147 (1973).

031.064 **Reducción de observaciones fotoeléctricas (Reduction of photoelectric observations).**
C. G. Morales Cabrera.
Trabajo de Grado, (Thesis), Facultad de Ingeneria, Univ. de Zulia, Maracaibo, Venezuela (1973).

The computer programs for the reduction of Strömgren uvby photometry that are used at the Universidad de Zulia are discussed and the initial results of their application presented.

031.065 **Tips für die Astropraxis.**
SuW, Vol. 12, 26 - 27, 53 - 54, 87 - 88, 116 - 118 (1973).
(1) Ein Thermostat für Frequenzwandler, (*J. Biel*), p. 26; Justierung von Spiegel- und Linsenobjektiven, (*G. Rabenschlag*), p. 26 - 27; (2) Ein Quarzgenerator mit digitalem Teil für Fernrohrantriebe, (*P. Höbel*), p. 53 - 54; Test: Astro-Stativ, Bauart Bach, (*G. D. Roth*), p. 54; (3) Der Schwarzschild-Effekt in der Amateur-Astrophotographie, (*E. Brodkorb*), p. 87 - 88; (4) Zur Praxis der Tiefkühlphotographie, (*H.-J.*

Leue, D. Tomoscheit), p. 116 - 118; Vorsicht bei der Propagierung eigener Erfahrungen! (*H. Vehrenberg*), p. 118; Verstärkung geringer Schwärzung des Negativs, (*K. Teicher*), p. 118.

Progress in optics. Vol. 10.   See Abstr. 003.126.

Automation of a transit instrument.
See Abstr. 003.145.

Speckle interferometry gives holograms of multiple star systems.   See Abstr. 034.002.

Application of an image isocon and computer to direct digitization of astronomical spectra.
See Abstr. 034.075.

Sul metodo di Nemiro per l'osservazione di coppie di stelle simmetriche rispetto allo zenit.   See Abstr. 041.015.

Interferometer baselines and poles obtained by linking radio observatories.   See Abstr. 046.018.

Optimale Ausschöpfung des Informationsgehaltes von Kometenaufnahmen durch entwicklungstechnische Kontraststeuerung.   See Abstr. 102.011.

Die Durchmesserbestimmung von Sternen mit interferometrischen Methoden.   See Abstr. 115.004.

Automatic reduction of radio astronomical maps: a map of the W43 region at 4.5 cm.   See Abstr. 141.102.

032 Astronomical Instruments

032.001 **Photographic report from Kitt Peak.**
Sky Telescope, Vol. 45, 10 - 18 (1973).

032.002 **Een teleskoop voor het infrarood.**
A. Greve.
Hemel en Dampkring, Vol. 71, 62 - 63 (1973).

032.003 **Wie steht es um das grosse ESO-Teleskop?**
E. Wiedemann.
Orion Schaffhausen, 31. Jahrgang, p. 29 (1973).

032.004 **Astronomers' facilities on the 150-inch Anglo-Australian telescope.**   P. R. Gillingham.
Proc. Astron. Soc. Australia, Vol. 2, 122 - 125 (1972).

032.005 **Pairs of spherical mirrors as prime focus correctors for the Anglo-Australian telescope.**   N. J. Rumsey.
Proc. Astron. Soc. Australia, Vol. 2, 126 - 127 (1972).

032.006 **The 100 cm optical telescope instrumentation.**
M. G. Waterworth.
Proc. Astron. Soc. Australia, Vol. 2, 127 - 132 (1972).

032.007 **Addendum: 'Progress on the 150-inch Anglo-Australian telescope' [Proc. Astron. Soc. Australia, Vol. 2, 2 - 6 (1971)].**   H. C. Minnett.
Proc. Astron. Soc. Australia, Vol. 2, 161 - 162 (1972).

032.008 **The lateral flexure of a transit circle.**
G. M. Petrov, J. Bem.
Acta Astron., Vol. 23, 49 - 52 (1973).
The investigation of the lateral flexure of the six-inch transit circle of the Wrocław Observatory is described. The method used gives the results independent of the catalog coordinates of the observed star.

032.009 **Status of the two 4-meter telescopes.**
D. L. Crawford.
Bull. American Astron. Soc., Vol. 5, 5 (1973). – Abstr. AAS.

032.010 **A study of the scale of the McCormick 26-inch refractor.**   L. W. Fredrick.
Bull. American Astron. Soc., Vol. 5, 6 (1973). – Abstr. AAS.

032.011 **Solar telescope at Winnipeg to mark quincentennial of Copernicus' birth.**   B. F. Shinn.

Journ. Roy. Astron. Soc. Canada, Vol. 67, 88 - 92 (1973).

032.012 Estudio del celostato de Lipmann en coordenadas cartesianas. T. Paneth.
Bol. As. Argentina Astron., No. 16, (see 012.007), p. 60 - 63 (1971).

032.013 On a 25-cm coudé-type coronagraph.
S. Nagasawa, I. Shimizu.
Tokyo Astron. Obs., Report No. 62, Vol. 16, 545 - 583 (1973). In Japanese.

032.014 An analysis of errors of the scale model of the large telescope on altazimuth mounting.
S. M. Vilenchik, Ya. B. Vyatskin, A. S. Najshul, E. M. Neplokhov.
Astrofiz. Issled., Izv. Spets. Astrofiz. Obs., Vol. 4, 192 - 200 (1972). In Russian.

032.015 Die 2.2-m-Teleskope des Max-Planck-Instituts für Astronomie. K. Bahner.
SuW, Vol. 12, 103 - 108 (1973).

032.016 On the measurement of the temperature distribution in the mercury basin of the floating zenith telescope (I). H. Okawa, Y. Goto.
Proc. International Latitude Obs. Mizusawa, No. 12, p. 46 - 53 (1972). In Japanese.

032.017 On the application of parallactic mountings in solar observations. G. F. Vyalshin, Yu. S. Muzalevsky.
Solnechnye Dannye 1973 Byull., No. 1, p. 100 - 101 (1973). In Russian.
The errors of parallactic mounting of solar telescopes arising due to the motion of the sun in declination are considered.

032.018 A California amateur's 12$^1/_2$-inch Newtonian-Cassegrain. H. O. Leitner.
Sky Telescope, Vol. 45, 389 - 391 (1973).

032.019 La construction d'un télescope de Schmidt.
R. Lartigau.
L'Astronomie, 87e année, p. 189 - 208 (1973).

032.020 Investigation of the screw of the eyepiece micrometer of the Bamberg transit instrument.
G. M. Kaganovskij.
Tsirk. Astron. Inst., *Tashkent*, No. 33 (380), p. 11 - 17 (1972). In Russian.

032.021 Investigation of the degree divisions of the two-minutes' scale of the Tashkent meridian circle.
M. F. Bykov.
Tsirk. Astron. Inst., *Tashkent*, No. 36 (383), 24 pp. (1972). In Russian.

032.022 Investigation of the ten-minutes' divisions of the two-minutes' scale of the Tashkent meridian circle.
M. F. Bykov.
Tsirk. Astron. Inst., *Tashkent*, No. 38 (385), 30 pp. (1973). In Russian.

032.023 First 4-meter photographs from Kitt Peak.
Sky Telescope, Vol. 46, 10 - 13 (1973).

032.024 Background paper on the one-meter Leander McCormick Observatory astrometric reflector.
L. W. Fredrick.
Mitt. Astron. Ges., No. 32, p. 129 - 130 (1973).

032.025 Solar instrumentation. G. Newkirk, Jr.
Publ. American Astronaut. Soc., Sci. Techn. Ser., Vol. 28, (see 012.018), p. 47 - 84 (1972).

032.026 High resolution imagery with the large space telescope. R. E. Danielson.
Publ. American Astronaut. Soc., Sci. Techn. Ser., Vol. 28, (see 012.018), p. 197 - 212 (1972).

032.027 Technology of space astronomical instruments.
M. Bottema.
Publ. American Astronaut. Soc., Sci. Techn. Ser., Vol. 28, (see 012.018), p. 363 - 377 (1972).

032.028 Manejo de los círculos del telescopio.
O. Betancourt.
El Universo, No. 102, Vol. 27, 32 - 33 (1973).

032.029 Simultaneous determination of the value of a revolution and errors of the screw of a position micrometer of a universal instrument. A. V. Gozhij.
Geod., kartogr. i aehrofotos"emka. Resp. mezhved. nauch.-tekhn. sb., 1972, vyp. (No.) 16, p. 21 - 25. In Russian.
Abstr. in Referativ. Zhurn. 51. Astron., 6.51.192; 52. Geod. i Aehros"emka, 6.52.136 (1973).

032.030 Analyses des erreurs de division du cercle méridien de l'Observatoire de Beograd. R. Dejaiffe.
Publ. Dep. Astron. Univ. Beograd, No. 4, p. 40 - 42 (1973).

032.031 Satellite laser ranging instruments operated at Tokyo Astronomical Observatory. Y. Kozai, A. Tsuchiya, K. Tomita, T. Kanda, H. Sato, N. Kobayashi, Y. Torii.
Tokyo Astron. Bull., Second Ser., No. 223, p. 2597 - 2605 (1973).
The instrumentation, observational procedure and data accuracy are discussed for the satellite laser ranging instrument operated at the Dodaira Station of the Tokyo Astronomical Observatory. The ranging accuracy is estimated to be ± 60 cm.

032.032 The Canadian PZT's. R. W. Tanner.
IAU Colloquium No. 1, (see 012.019), p. 21 (1972). Abstract.

032.033 120-in. Large Space Telescope (LST).
S. L. Morrison.
Journ. Spacecraft and Rockets, (*USA*), Vol. 9, 929 - 931 (1972).

032.034 A telescope for soft gamma ray astronomy.
V. Schonfelder, A. Hirner, K. Schneider.
Nuclear Instruments and Methods, (*Netherlands*), Vol. 107, 385 - 394 (1973).
A gamma ray telescope using the double Compton process is described, which measures extraterrestrial gamma ray fluxes in the energy range 1—10 MeV.

032.035 Il nuovo telescopio di 182 cm di Cima Ekar (Asiago).
C. Barbieri.
Coelum, Vol. 41, 85 - 92 (1973).

032.036 Wolter-Schwarzschild telescopes for X-ray astronomy. R. C. Chase, L. P. VanSpeybroeck.
Applied Optics, Vol. 12, 1042 - 1044 (1973).
The resolution of a Wolter-Schwarzschild telescope is intrinsically superior to the resolution of the corresponding paraboloid-hyperboloid telescope. The improvement is important for high resolution and wide field telescope designs having grazing angles larger than about 1.5°.

032.037 Astronomical theodolite with automatic telescopic

device using a stepping motor.   K. Ramsayer.
Zeitschr. Vermessungswes., Vol. 98, 114 - 118 (1973). In
German. – See Phys. Abstr., Vol. 76, No. 42265 (1973).

Isaac Newton telescope: to move or not to move.
Nature, Vol. 242, 4 - 5 (1973).

Large space telescope: Experimental definition.
Nature, Vol. 243, 488 (1973).

Das Fernrohr für Jedermann.
See Abstr. 003.106.

A new dual filter telescope.
See Abstr. 034.051.

Stellar astronomy objectives.   See Abstr. 051.012.

Future infrared space astronomical instruments.
See Abstr. 051.014.

033 Radio Telescopes and Equipment

**033.001 Polarization interferometer for 2800 MHz. Solar
noise studies with a 0.5 fan beam.**
M. B. Bell, A. E. Covington, W. A. G. Kennedy.
Solar Physics, Vol. 28, 123 - 136 (1973).
   The construction and operation of a high resolution interferometer for solar noise observations at 10.7 cm for the Al-
gonquin Radio Observatory is described. Observations of the
sun on 1972, January 6 show the range of phenomena to be
observed and a detailed study of the association of eight radio
emissive regions with all centers of solar activity is presented.

**033.002 A Brazilian radio telescope for millimeter wave-
lengths.**   P. Kaufmann, R. D'Amato.
Sky Telescope, Vol. 45, 144 - 145, 160 (1973).

**033.003 The search for signals from extraterrestrial civiliza-
tions.**   J. C. G. Walker.
Nature, Vol. 241, 379 - 381, with comments by H. L. Armstrong, Nature, Vol. 242, 355 (1973).
   Although the technology exists for exchanging radio mes-
sages with extraterrestrial civilizations, a successful search for
such civilizations among the many stars that might support
them could take more than a thousand years, even if most
habitable planets are occupied by communicative civilizations.

033.004 The new Fleurs radiotelescope.
W. N. Christiansen.
Proc. Astron. Soc. Australia, Vol. 2, 132 - 134 (1972).

033.005 Future plans for the Molonglo radio telescope.
B. Y. Mills, A. G. Little.
Proc. Astron. Soc. Australia, Vol. 2, 134 - 135 (1972).

033.006 The Llanherne low frequency radio telescope.
G. R. A. Ellis.
Proc. Astron. Soc. Australia, Vol. 2, 135 - 137 (1972).

033.007 A 65-meter telescope for millimeter wavelengths.
J. W. Findlay, S. von Hoerner.
National Radio Astronomy Observatory, Charlottesville,
Virginia. 6 + 142 pp. (1972). – The report of a design study
made by the National Radio Astronomy Observatory.

**033.008 The radio telescope at 1.37 m of the Astronomical
Institute of the Cuba Academy of Sciences.**
B. A. Dubinsky, L. I. Yurovskaya, L. Larragoiti, B. Jozkovich

(*Khoskovich*), E. Pozo.
Solnechnye Dannye 1972 Byull., No. 12, p. 64 - 71 (1973).
In Russian.
   The technical characteristics and the principle of operation of the radio telescope at 1.37 m wavelength mounted at
the Astronomical Institute of the Cuba Academy of Sciences
are given. Method and preliminary results of observations with
this telescope are described.

033.009 High-frequency noise in Schottky-barrier diodes.
T. J. Viola, Jr., R. J. Mattauch.
Proc. IEEE, Vol. 61, 393 (1973).
   A unified equivalent noise temperature equation is pre-
sented which relates the high-frequency noise in Schottky-
barrier diodes to the barrier transport mechanism.

**033.010 The influence of temporal fluctuations of the mean
integral refractive index of air on the accuracy of
radio range findings.**   D. D. Dzyaman.
Trudy Kazan. Gorod. Astron. Obs., No. 38, p. 93 - 104 (1972).
In Russian.

**033.011 Experimental investigation of the influence of radio
wave reflection from an underlying surface on the
accuracy of radio range findings.**   D. D. Dzyaman.
Trudy Kazan. Gorod. Astron. Obs., No. 38, p. 105 - 115
(1972). In Russian.

033.012 Reflector surface of the 6-meter mm-wave telescope.
Prepared by Cosmic Radio Astronomy Section.
Tokyo Astron. Obs., Report No. 62, Vol. 16, 471 - 499 (1973).
In Japanese.

033.013 Pointing accuracy of the 6-meter mm-wave telescope.
Prepared by Cosmic Radio Astronomy Section.
Tokyo Astron. Obs., Report No. 62, Vol. 16, 500 - 521 (1973).
In Japanese.

**033.014 Polarization characteristics of radio telescope an-
tennas.**   N. A. Yesepkina (*Esepkina*).
Astrofiz. Issled., Izv. Spets. Astrofiz. Obs., Vol. 4, 157 - 169
(1972). In Russian.
   Polarization characteristics of radio telescope antennas
are determined by using Müller's method with consideration
for the cross-polarization arising in an antenna for the case of
partially polarized radiation. The polarization properties of an

antenna characterized in this case by the 4 × 4 M-matrix (Müller's matrix) connecting the Stokes parameters of radiation at the input and at the output of an antenna. General expressions for the elements of this matrix are obtained. As an example the elements of the M-matrix are determined for a paraboloid antenna.

033.015 On the upper limit of the effective area of a variable profile antenna in the short-wave part of the centimeter region. V. M. Spitkovsky.
Astrofiz. Issled., Izv. Spets. Astrofiz. Obs., Vol. 4, 170 - 176 (1972). In Russian.
A procedure of calculating the effective area of a variable profile antenna (VPA) is described. Formulas for the idealized effective area of a VPA are obtained at all angles of elevation, the idealized effective area being the upper limit of the effective area of the antenna. The results of calculation of the idealized effective area of the large Pulkovo radio telescope are compared with the experimental data on testing its reflecting surface at 8-mm wavelength.

033.016 Investigation and adjustment of the large Pulkovo radio telescope by radio astronomical technique.
G. B. Gelfrejkh, O. A. Golubchina.
Astrofiz. Issled., Izv. Spets. Astrofiz. Obs., Vol. 4, 177 - 191 (1972). In Russian.
Some problems concerning the practical application of the radio astronomical technique of adjustment of the reflecting surface of the large Pulkovo radio telescope for the purpose of investigation and increasing the accuracy of its main reflector are considered. The possible and necessary time intervals for the adjustment procedure are determined, algorithms for calculation of adjustment programs are presented, the practical ways of adjustment are described.

033.017 The noise temperature of the large Pulkovo radio telescope at 6.6 cm wavelength.
V. Ya. Golnev, V. M. Spitkovsky.
Astrofiz. Issled., Izv. Spets. Astrofiz. Obs., Vol. 4, 208 - 210 (1972). In Russian. – Short note.

033.018 Automatic system of registration and data processing of observations of the radio service of the sun.
S. P. Chekalev, T. N. Aleshina, Yu. B. Vedeneev, N. M. Prytkov, V. V. Khrulev, S. A. Shmulevich.
Solnechnye Dannye 1973 Byull., No. 1, p. 107 - 110 (1973). In Russian.

033.019 A South African amateur's radio telescopes.
C. W. de Villiers.
Sky Telescope, Vol. 45, 363 (1973).

033.020 Radio astronomical method of estimating the ultrahigh frequency absorption in hydrogen under high pressure. A. G. Solov'ev.
Trudy XVI Nauch. konf. Mosk. fiz.-tekhn. in-t, 1970. Ser. "Obshch. i prikl. fiz.", "Molekulyar. i khim. fiz.", Moskva, 1972, p. 42 - 44. In Russian. – Abstr. in Referativ. Zhurn. 51. Astron., 4.51.323 (1973).

033.021 Supernova observational radio telescope.
S. A. Colgate, B. A. Blevins.
Bull. American Astron. Soc., Vol. 5, 284 (1973). – Abstr. AAS.

033.022 Radiometry at 1 mm wavelength with the NRAO 36-foot telescope.
J. D. G. Rather, E. K. Conklin, P. A. R. Ade, P. E. Clegg.
Bull. American Astron. Soc., Vol. 5, 285 - 286 (1973). Abstr. AAS.

033.023 The University of Iowa cocoa-cross.

S. D. Shawhan, W. M. Cronyn.
Bull. American Astron. Soc., Vol. 5, 286 (1973). – Abstr. AAS.

033.024 Technology and observations in radio astronomy.
W. C. Erickson, F. J. Kerr.
Adv. Electronics Electron Phys., Vol. 32, 1 - 61 (1973).
Contents: I.) Introduction; II.) Radio astronomy techniques: A.) Observational considerations; B.) Single telescopes and spectroscopy; C.) High resolution techniques; D.) Data processing; E.) Space radio astronomy; III.) Radio observations: A.) Introduction; B.) Solar radio astronomy; C.) The planets; D.) The Galaxy; E.) Extragalactic radiation; F.) Conclusion; References.

033.025 An automatic dynamic spectrum analyser for video tape recorded signals. G. R. A. Ellis.
Australian Journ. Phys., Vol. 26, 253 - 256 (1973).
A time-expansion sweep frequency spectrum analyser is described for signals in the 0–3 MHz frequency range. It is based on a standard video tape recorder used in the single frame replay mode. With a time-expansion ratio of 400 between the record time and replay time, a frequency resolution of 10 kHz together with a time resolution of 0.5 ms is realized. An example of the spectra of Jupiter radio bursts made with the analyser is presented.

033.026 Recent advances in radio astronomy.
M. P. Damon.
Spaceflight, Vol. 15, 261 - 263 (1973).

033.027 The main beam and ringlobes of an east-west rotation-synthesis array.
R. N. Bracewell, A. R. Thompson.
Astron. Journ., Vol. 182, 77 - 94 (1973).
This paper presents a study of the Fourier transform of the transfer function of a radio-telescope array which, in the spatial-frequency domain, responds on a series of equally spaced circular loci concentric about the origin. A knowledge of this response is essential to the full understanding of the brightness distributions measured in rotation synthesis with an east-west array.

033.028 A study of the amplification instability of a radiometer with semiconductors.
V. N. Brezgunov, V. A. Udal'tsov.
Trudy Fiz. in-ta. AN SSSR, Vol. 62, 68 - 76 (1972). In Russian. – Abstr. in Referativ. Zhurn. 51. Astron., 5.51.90 (1973).

033.029 Amplifier with distributed amplification for radiometry. V. N. Brezgunov, V. A. Udal'tsov.
Trudy Fiz. in-ta. AN SSSR, Vol. 62, 77 - 84 (1972). In Russian. – Abstr. in Referativ. Zhurn. 51. Astron., 5.51.91 (1973).

033.030 Parasitic modulation in radiometers.
V. M. Gudnov, V. A. Izvekova.
Trudy Fiz. in-ta. AN SSSR, Vol. 62, 85 - 97 (1972). In Russian. – Abstr. in Referativ. Zhurn. 51. Astron., 5.51.92 (1973).

033.031 Control system of the DKR-1000 FIAN antenna meridional beam.
S. N. Ivanov, Yu. P. Ilyasov, A. N. Ivanov, V. T. Solodkov, V. Ya. Shcherbinin.
Trudy Fiz. in-ta. AN SSSR, Vol. 62, 98 - 106 (1972). In Russian. – Abstr. in Referativ. Zhurn. 51. Astron., 5.51.93 (1973).

033.032 Line scanning receiver for electronic control of the DKR-1000 FIAN antenna meridional beam.

I. A. Alekseev.
Trudy Fiz. in-ta. AN SSSR, Vol. 62, 107 - 111 (1972). In Russian. − Abstr. in Referativ. Zhurn. 51. Astron., 5.51.94 (1973).

033.033 A discretely commutative phase rotator for the metre-wave range.
S. N. Ivanov, A. N. Ivanov, G. A. Pavlov, V. T. Solodkov.
Trudy Fiz. in-ta. AN SSSR, Vol. 62, 112 - 115 (1972). In Russian. − Abstr. in Referativ. Zhurn. 51. Astron., 5.51.95 (1973).

033.034 Antenna for polarization measurements together with DKR-1000.
Yu. I. Alekseev, S. M. Kutuzov, M. M. Tyaptin, V. D. Chujkov.
Trudy Fiz. in-ta. AN SSSR, Vol. 62, 121 - 127 (1972). In Russian. − Abstr. in Referativ. Zhurn. 51. Astron., 5.51.96 (1973).

033.035 A new scheme for a multiply supported radially symmetric suspension of the parabolic mirror of a radio telescope.   P. D. Kalachev.
Trudy Fiz. in-ta. AN SSSR, Vol. 62, 128 - 135 (1972). In Russian. − Abstr. in Referativ. Zhurn. 51. Astron., 5.51.97 (1973).

033.036 Transverse displacements of a parabolic antenna irradiator of a movable radio telescope.
P. D. Kalachev, I. A. Emel'yanov.
Trudy Fiz. in-ta. AN SSSR, Vol. 62, 136 - 149 (1972). In Russian. − Abstr. in Referativ. Zhurn. 51. Astron., 5.51.98 (1973).

033.037 On the alteration of the geometric parameters of the parabolic mirror of a radio telescope under radially symmetric deformations.   P. D. Kalachev.
Trudy Fiz. in-ta. AN SSSR, Vol. 62, 150 - 155 (1972). In Russian. − Abstr. in Referativ. Zhurn. 51. Astron., 5.51.99 (1973).

033.038 Longitudinal displacements of the irradiator of a parabolic antenna.   P. D. Kalachev, V. P. Nazarov.
Trudy Fiz. in-ta. AN SSSR, Vol. 62, 156 - 166 (1972). In Russian. − Abstr. in Referativ. Zhurn. 51. Astron., 5.51.100 (1973).

033.039 Simulation of radio telescope control systems.
G. G. Basistov.
Trudy Fiz. in-ta. AN SSSR, Vol. 62, 167 - 177 (1972). In Russian. − Abstr. in Referativ. Zhurn. 51. Astron., 5.51.101 (1973).

033.040 Extremal problem of radio telescope control.
V. V. Zotov, B. N. Sevryukov.
Izv. vyssh. uchebn. zavedenij. Radiofizika, Vol. 15, 1580 - 1581 (1972). In Russian. − Abstr. in Referativ. Zhurn. 51. Astron., 5.51.102 (1973).

033.041 Very long baseline interferometry: the impact on astronomy and geophysics.   I. I. Shapiro.
Publ. American Astronaut. Soc., Sci. Techn. Ser., Vol. 28, (see 012.018), p. 133 (1972). − Abstract.

033.042 Radio telescope for Britain.   I. Ridpath.
Southern Stars, Vol. 24, 134 - 136 (1973).

033.043 Comments on "Electrooptical processing for radio astronomy".   L. R. D'Addario, S. J. Wernecke, with a reply by H. Stark.
Proc. IEEE, Vol. 61, 671 - 673 (1973).

033.044 The Fleurs synthesis radiotelescope.
Australian Electronics Engineering, Vol. 5, No. 10, p. 13 - 16 (1972).
The new Fleurs synthesis telescope, operating at a wavelength of 21 cm, has an aperture more than 4000 wavelengths in diameter and is larger, in terms of wavelengths than any existing radiotelescope that is situated so as to be able to observe the Southern sky.

033.045 Cryogenic cooling of mixers for millimeter and centimeter wavelengths.
S. Weinreb, A. R. Kerr.
IEEE Journ. Solid-State Circuits, Vol. SC-8, 58 - 63 = National Radio Astron. Obs., *Green Bank,* Repr. Ser. A, No. 279 (1973).

033.046 A thermal calibrator for radiometers used in radioastronomy.   J. W. Findlay, J. Payne.
Journ. Phys. E, Sci. Instruments, Vol. 6, 152 - 154 = National Radio Astron. Obs., *Green Bank,* Repr. Ser. A, No. 283 (1973).
The construction and use of a thermal calibrator for radiometers used in radioastronomy is described. The instrument can supply noise power changes over a range of frequencies up to 7 GHz. When used under well-matched conditions, the instrument is believed to have an accuracy of 1%. It is rugged and simple to use.

033.047 The first two receivers for the radio astronomy programme on the 100 metre radiotelescope.
N. J. Keen, P. Zimmermann.
Nachrichtentechn. Zeitschr. (NTZ), Vol. 26, No. 3, p. 124 - 128 (1973). In German.

033.048 A dual frequency, dual polarized feed for radioastronomical applications.
M. E. J. Jenken, M. H. M. Knoben, K. J. Wellington.
Nachrichtentechn. Zeitschr. (NTZ), Vol. 25, 374 - 376 (1972).

033.049 Radioastronomie auf Kurzwellen.
S. Braude, A. Menj.
Bild der Wissenschaft [Deutsche Verlagsanstalt, Stuttgart], Vol. 10, 125 - 133 (1973).

033.050 New control technique in DC/DC regulators for space applications.   A. Capel.
IEEE TRANS. Aerospace Electronic Systems, Vol. AES-8, 472 - 480 (1972).

033.051 Local-oscillator-circuit optimisation for minimum distortion in double-balanced modulators.
J. G. Gardiner.
Proc. Instn. Electr. Engineers, *(GB),* Vol. 119, 1251 - 1256 (1972).

033.052 Josephson junctions as self-oscillating mixers at 34 GHz.   V. Jenkins, E. A. Parker, L. T. Little.
Electronics Letters, Vol. 8, 540 - 541 (1972).

033.053 Cylindrical antenna in a rectangular waveguide driven from a coaxial line.
A. G. Williamson, D. V. Otto.
Electronics Letters, Vol. 8, 545 - 547 (1972).

033.054 Field-effect-transistor-bridge multiplier-divider.
M. M. Abu-Zeid, H. Groendijk.
Electronics Letters, Vol. 8, 591 - 592 (1972).
A description is given of a 4-quadrant temperature-compensated multiplier-divider circuit in the form of a self-balancing F. E. T. bridge consisting of two matched F. E. T. pairs. − *JBS*

033.055 Correlation-function display and peak detection.
J. R. Jordan, M. S. Beck.
Electronics Letters, Vol. 8, 602 - 604 (1972).
A new method of correlation-function display and peak detection, based on the polarity correlator, is described.
JBS

033.056 Gain calibration of a horn antenna using pattern integration. A. Ludwig, J. Hardy, R. Norman.
California Inst. Technol., JET Propulsion Lab., Tech. Rep. 32-1572, 26 pp. (1972).
The experimental techniques are described. A spherical wave expansion method based directly on measured data yields near-field gain correction factors which are evaluated.
ACM

033.057 Precision reflectivity loss measurements of perforated-plate mesh materials by a waveguide technique.
T. Y. Otoshi.
IEEE Trans. Instrument. Measurements, Vol. IM-21, 451 - 457 (1972).
A waveguide method is described for improving the precision and accuracy of reflectivity loss measurements of perforated-plate mesh materials. *– ACM*

033.058 Harmonic synchronization of oscillators revisited.
B. N. Biswas, S. K. Ray.
IEEE Trans. Circuit Theory, Vol. CT-19, 682 - 685 (1972).
Pseudo-indirect and direct synchronization are compared with advantages seen to apply to the pseudo-indirect case. *– DJC*

033.059 Avalanche (Impatt) and transferred electron (Gunn) diode broadband microwave amplifiers.
P. W. Braddock, P. W. Manders, R. Genner.
R. R. E. Newsletter Res. Rev., No. 11, p. 4/1 - 4/9 (1972).
Description of work in progress on Impatt and Gunn diode microwave amplifiers. *– MWS*

033.060 Corrugated conical horn antennas with small flare angles. M. E. J. Jeuken.
De Ingenieur, Vol. 84 (34), 88 - 94 (1972).
A discussion of the radiation properties of corrugated conical horn antennas with small flare angles. *– DNC.*

033.061 2.2-cm variable polarization feed. W. Lavrench.
Bull. Radio Electr. Engineering Div. NRC Canada, Vol. 22, No. 1, p. 16 - 18 (1972).
A brief note on details of a 2.2-cm polarization feed using a rotating quarter-wave plate. *– MWS*

033.062 A review of microwave parametric amplifiers with particular reference to satellite communications and radio astronomy.
M. J. Adams, C. Field, M. C. McNeill, M. J. B. Scanlan.
Marconi Rev., No. 186, Vol. 35, 173 - 203 (1972).
The theory of parametric amplifiers is reviewed and the amplifiers are then classified for various applications (radar, space communication, radio astronomy) by their Idler circuits and by the method used for broadbanding. *– JWB*

033.063 Microwave low noise amplifiers for use in radar systems. H. M. Chandler.
Marconi Rev., No. 185, Vol. 35, 94 - 120 (1972).
The article discusses some of the stability factors in parametric and transistor amplifiers in low noise, small signal receiver applications. *– JWB*

033.064 Reflector profiles for the pencil-beam Cassegrain antenna. P. J. Wood.
Marconi Rev., No. 185, Vol. 35, 121 - 138 (1972).

The author suggests diffraction-optimized profiles as an alternative to ray-optics methods. Performance advantages may result in relation to gain near-in and wide-angle sidelobes and feed VSWR. Various designs are discussed. *– ACM*

033.065 Propagation and radiation characteristics of corrugated horns. P. J. B. Clarricoats, L. M. Seng.
Electronics Letters, Vol. 9, 7 - 9 (1973).
In this paper the authors compare the theoretical and experimental radiation patterns for a horn with a semi-flare angle of 110 deg. *– BMT*

033.066 Influence of horn length on radiation pattern of oblique-flare-angle corrugated horn.
P. J. B. Clarricoats, L. M. Seng.
Electronics Letters, Vol. 9, 15 - 16 (1973).
In this paper the effect of length of the corrugated section surrounding the aperture of a circular waveguide is considered. A semi-flare angle of 110 deg for the corrugated section is assumed. *– BMT*

033.067 Rectangular horn with dielectric-slab insert.
R. Ashton, R. Baldwin.
Electronics Letters, Vol. 9, 26 - 27 (1973).
A thin slab of dielectric is placed in the H-plane of a rectangular horn. This produces a reasonably constant E-plane beamwidth and low sidelobes. The H-plane is virtually unaffected. *– DNC*

033.068 Feed arrangement for axis definition of paraboloid reflector. S. Cornbleet.
Electronics Letters, Vol. 9, 66 - 67 (1973).
A method is proposed for obtaining the electrical axis of a paraboloid from the axial maxima in the field distribution. *– BMT*

033.069 Computer analysis of gradually tapered waveguide of arbitrary cross-sections.
J. B. Davies, O. J. Davies, S. S. Saad.
Electronics Letters, Vol. 9, 46 - 47 (1973).

033.070 Radiation from a paraboloid with an axially defocused feed. P. G. Ingerson, W. V. T. Rusch.
IEEE Trans. Antennas Propagation, Vol. AP-21, 104 - 106 (1973).
This communication considers the exact solution of the axial focal-region field of a paraboloid and compares the distribution with that obtained by Minnett and Thomas. In addition, the radiation patterns of a defocused paraboloid are also given. *– BMT*

033.071 Reflector antenna radiation pattern analysis by equivalent edge currents.
G. L. James, V. Kerdemelidis.
IEEE Trans. Antennas Propagation, Vol. AP-21, 19 - 24 (1973).
Equivalent edge currents are used to obtain radiation patterns of prime-focus paraboloids. Radiation patterns are shown from 0 deg to 180 deg, and compared with the measured and physical optics patterns. *– BMT*

033.072 Errors in the predicted gain of pyramidal horns.
E. V. Jull.
IEEE Trans. Antennas Propagation, Vol. AP-21, 25 - 31 (1973).
Geometrical theory of diffraction is used to derive the on-axis gain of E-plane sectoral horns. *– BMT*

033.073 The definition of cross polarization.
A. C. Ludwig.
IEEE Trans. Antennas Propagation, Vol. AP-21, 116 - 119

(1973).

An excellent communication discussing three possible methods of defining cross-polarization. — *BMT*

033.074 Eigenvalues of a class of spherical wave functions. M. S. Narasimhan.
IEEE Trans. Antennas Propagation, Vol. AP-21, 8 - 14 (1973).

Solutions for the hybrid modes in corrugated conical and quasipyramidal horns are included in the paper. — *BMT*

033.075 The 'paradisc' antenna — a novel technique to improve the axial ratio of a circularly polarized high gain antenna system. R. W. Silberberg.
IEEE Trans. Antennas Propagation, Vol. AP-21, 108 - 113 (1973).

A prime-focus circularly polarized feed is described. It consists of crossed-dipoles with a cupped backing-reflector. *BMT*

033.076 Radioastronomische Antennen (Radioastronomical antennae). A. Werner.

Jahrbuch der Schulphysik, [Aulis-Verlag, Köln], Vol. 1, 137 - 142 (1972).

A review article concerning dipoles, helix, horn and parabolic antennae, interferometer arrays, Fourier-transformation, aperture synthesis.

033.077 Empfänger für Radioteleskope (Receivers for radio telescopes). A. Werner.
Jahrbuch der Schulphysik, [Aulis-Verlag, Köln], Vol. 1, 147 - 152 (1972).

A survey of the different types of receivers and their principles, correlation and low noise preamplifiers.

Le noveau radiotélescope de Sydney.
L'Astronomie, 87ᵉ année, p. 220 - 221 (1973).

Data processing system for 160 MHz compound interferometer at Nobeyama Solar Radio Station.
See Abstr. 021.005.

High-resolution methods in radio astronomy.
See Abstr. 141.097.

034 Astronomical Accessories

034.001 Improvement of photomultiplier performance in astronomical applications. M. A. Dopita.
Astrophys. Space Sci., Vol. 18, 350 - 362 (1972).
The physics of an EMI 9558 Å photomultiplier has been investigated in some detail, and this study has led to a gain of more than thirty times in the signal-to-noise obtainable in a given observation time when used in receiver noise limited conditions. The importance of giant pulses in limiting performance is demonstrated.

034.002 Speckle interferometry gives holograms of multiple star systems.
R. H. T. Bates, P. T. Gough, P. J. Napier.
Astron. Astrophys., Vol. 22, 319 - 320 (1973). – Research note.

034.003 Using an image intensifier at a college observatory.
M. A. Seeds, A. J. Distasio.
Sky Telescope, Vol. 45, 76 - 78 (1973).

034.004 AURA's Fabry-Perot interferometer.
M. G. Smith.
Sky Telescope, Vol. 45, 208 - 211 (1973).

034.005 Semiconductor image devices for spectroscopy, autoguiding and photometry. J. V. Jelley.
Observatory, Vol. 93, 9 - 13 (1973).
The object of this paper is to describe an entirely new device, which has already been developed and used in spectroscopy, has a potential in photometry, and which is also likely to form the basis of a sensing element in autoguiding systems. Its main characteristics are a high quantum efficiency, $\epsilon \sim 0.2$, a wide dynamic range, exceedingly low noise (without cooling), unlimited integration time (electronic data storage), good linearity, and direct readout of the information in electronic digital form.

034.006 Precision spectropolarimetry of starlight: Development of a wide-band version of the Dollfus polarization modulator. J. Tinbergen.
Astron. Astrophys., Vol. 23, 25 - 48 (1973).
This paper examines the theory of the Dollfus modulator, concludes that it is intrinsically of very high quality and discusses how it can be made in an achromatic form. The results on achromatic retarders are also applicable to other types of modulator and to depolarizer theory.

034.007 The Apollo 17 far ultraviolet spectrometer experiment. W. G. Fastie.
The Moon, Vol. 7, 49 - 62 (1973). – Paper dedicated to Professor Harold C. Urey on the occasion of his 80th birthday on 29 April, 1973.
The Apollo 17 command service module in lunar orbit will carry a far ultraviolet scanning spectrometer whose prime mission will be to measure the composition of the lunar atmosphere. A detailed description of the experimental equipment which observes the spectral range 1180 to 1680 Å, the observing program and broad speculation about the possible results of the experiment, are presented.

034.008 Simultaneous determination of right ascensions and declinations with a photoelectric tracking micrometer for the meridian circle. Y. Requième.
Astron. Astrophys., Vol. 23, 453 - 460 (1973). In French.
A tracking photoelectric micrometer constructed at the Bordeaux Observatory has been mounted on the meridian circle since June 1971. It allows the automatic and simultaneous determination of the transit time and the declination of stars up to the 9.2th magnitude.

034.009 Variations of the photoresponse on small PbS-detectors. U. Fahrbach, W. Hofmann, D. Lemke.
Astron. Astrophys., Vol. 23, 461 - 462 (1973).
Surface variations of the photoresponse of 0.25 mm PbS-detectors were measured and discussed in view of use without Fabry lens.

034.010 Large window spectracon.
J. D. McGee, C. I. Coleman, E. G. Phillips.
Nature, Vol. 241, 264 (1973). – Letter.

034.011 Infrared interferometry. J. Gay, A. Journet.
Nature, Phys. Sci., Vol. 241, 32 - 33 (1973).
Infrared detectors with time constants of < 1 ns have become available recently and can be used as mixing elements in heterodyne detection techniques. In August 1972 we achieved the heterodyne detection of the sun at 10.6 μm with an infrared interferometer which includes techniques from radio and infrared astronomy. The solar diameter was resolved with an interferometer base of 1 mm.

034.012 Possibilités de dérivation et de corrélation de spectres dans les spectromètres interférentiels à grille.
G. Fortunato, A. Maréchal.
Comptes Rendus Acad. Sci. Paris, Sér. B, Vol. 276, 527 - 530 (1973).

034.013 Solobservationer med amatørinstrumenter.
P. Darnell.
Astron. Tidsskr., Årg. 6, p. 25 - 34 (1973).

034.014 A possible solar electrograph. D. Dravins.
Astrophys. Letters, Vol. 13, 243 - 245 (1973).
Some theories of solar flares predict strong electric fields during the pre-flare energy buildup, which might well be observable with existing techniques by measuring the linear polarization in spectral lines that are split due to the Stark effect.

034.015 The coudé spectrum scanner at the Lowell Observatory.
P. B. Boyce, N. M. White, R. Albrecht, A. Slettebak.
Publ. Astron. Soc. Pacific, Vol. 85, 91 - 95 (1973).
The new coudé scanner at the Lowell Observatory 42-inch telescope is described, including the system of data acquisition and reduction, and some preliminary results are presented.

034.016 Effects of polarization on the transmission of coudé-spectrometer systems. D. Clarke.
Astron. Astrophys., Vol. 24, 165 - 170 (1973).
After presenting the Mueller matrix for the effect of a reflection by a metallic mirror, various coudé designs are discussed in relation to the polarization of the light that they produce. An example is given of how a spectrometer responds to polarized light and consequently how its transmittance depends on the polarization properties of the light provided by the coudé. Comparison is made between depolarizers and half-wave plates as means of dealing with polarization to achieve the best overall transmittance of the coudé-spectrometer combination.

034.017 The polarovisor. L. V. Xanfomaliti (*Ksanfomaliti*), V. P. Dzhapiashvili.
Astron. Zhurn. Akad. Nauk SSSR, Vol. 50, 357 - 361 (1973).

In Russian. English translation in Soviet Astron. AJ, Vol. 17, No. 2.

A new device for polarimetric explorations of the moon and planets – a polarovisor is described.

034.018 Josephson junction receiver for millimeter and far infrared astronomy. B. T. Ulrich.
Bull. American Astron. Soc., Vol. 5, 25 (1973). – Abstr. AAS.

034.019 On the effective quantum efficiency of an image orthicon system applied to the detection of faint astronomical objects. J. A. Hynek, J. R. Dunlap.
Bull. American Astron. Soc., Vol. 5, 37 (1973). – Abstr. AAS.

034.020 An automatic stellar photometer with photon counter. Z. D. Mestiashvili.
Novaya tekhn. v astronomii. Vyp. (No.) 4. Nauka, Leningrad, p. 29 - 35 (1972). In Russian. – Abstr. in Referativ. Zhurn. 51. Astron., 3.51.275 (1973).

034.021 Rapid scanning photometer for the night sky.
Yu. M. Zavarzin, E. S. Andreev, D. A. Rozhkovskij, P. N. Bojko.
Novaya tekhn. v astronomii. Vyp. (No.) 4. Nauka, Leningrad, p. 35 - 38 (1972). In Russian. – Abstr. in Referativ. Zhurn. 51. Astron., 3.51.283 (1973).

034.022 Remark on magnetographic measurements.
M. Šidlichovský.
Bull. Astron. Inst. Czechoslovakia, Vol. 24, 105 - 107 (1973).

A plane-parallel plate for measuring the plasma velocity in line-forming space is usually used in a magnetograph. The error in magnetographic measurement caused by additional polarisation of light on this plate is investigated theoretically.

034.023 A two-beam multi-mode, nebular-stellar photometer.
C. Goudis, J. Meaburn.
Astrophys. Space Sci., Vol. 20, 149 - 157 (1973).

The design, construction and use of a multi-purpose photometer is discussed. It is intended to work between 9000 Å to 3700 Å on the 25-inch Turner-Newall refractor of the Athens National Observatory and eventually on the Greek 48-inch reflector now under construction.

034.024 Laser ranging instrument developed at the Tokyo Astronomical Observatory.
K. Tomita, T. Kanda.
Tokyo Astron. Obs., Report No. 62, Vol. 16, 657 - 671 (1973). In Japanese.

034.025 Optimum astronomical photoelectric photometry. Terrestrial operations in the UV-IR band up to 1μ wavelength. G. Sedmak.
Astron. Astrophys., Vol. 25, 41 - 52 (1973).

This paper deals with a theoretical analysis of the various basic photometer systems and operating modes possible in the mono-and multiband photoelectric photometry of astronomical sources by means of earth-operated photometers employing quantum detectors.

034.026 Angular dependence of the optical properties of a narrow-band interference filter.
S. Koutchmy, J. -C. Vial.
Astron. Astrophys., Vol. 25, 145 - 147 (1973). In French.

An interferential filter of 2″ diameter, centered around the Hα line, has been studied. Wavelength shift and half-height bandwidth variation of the filter are given for different incident angles.

034.027 Feedback stabilized Fabry-Perot interferometer.
G. Hernandez, O. A. Mills.

Applied Optics, Vol. 12, 126 - 130 (1973).

A method of stabilizing a Fabry-Perot interferometer with respect to a reference wavelength, while simultaneously stabilizing the parallelism of the flats, is described.

034.028 Optically contacted Fabry-Perot interferometer filter for the middle ultraviolet.
B. Bates, J. K. Conway, C. D. McKeith, H. W. Yates.
Applied Optics, Vol. 12, 140 - 142 (1973). – Letter.

034.029 Elliptic polarization in grating spectrographs.
J. M. Simon, M. C. Simon.
Applied Optics, Vol. 12, 153 - 154 (1973). – Letter.

034.030 Nebular Fabry-Perot, Pepsios, and Sisam monochromators. J. Meaburn.
Applied Optics, Vol. 12, 279 - 284 (1973).

The usefulness of the Fabry-Perot and Pepsios monochromators when studying visible nebular lines is pointed out. A method of scanning and modulating a Sisam monochromator in the visible is described. The potential performances in a variety of idealized noise conditions of Fabry-Perot, Pepsios, Sisam, and Sisam × Fabry-Perot monochromators for nebular studies are then analyzed.

034.031 Spectrometric imager. Part 2. M. Harwit.
Applied Optics, Vol. 12, 285 - 288 (1973).

This paper describes the operation of a simple spectrometric imager. The device analyzes radiation from an object into sixty-three spatial and fifteen spectral resolution elements. It can provide a fifteen-color spectrum for each point on the object, or alternately a series of fifteen pictures, each obtained at a different set of wavelengths.

034.032 Heat transfer characteristics of a linear solar collector. B. O. Seraphin.
Applied Optics, Vol. 12, 349 - 354 (1973).

The heat transfer characteristics of a linear solar energy collector are calculated as functions of dimensions, spectral quality of the selective absorber surface, optical flux concentration of the optical configuration, and thermal parameters and flow rate of the heat transfer medium. Carnot efficiency, exit temperature, and an upper limit to the amount of heat extracted are determined for systems in which liquid sodium serves as the heat transfer medium. The performance is evaluated for selective absorber surfaces representing the state of the art as well as for surfaces requiring a more mature thin-film technology.

034.033 Lens projection system for a solar simulator providing irradiance of 100 solar constants.
D. R. Buchele.
Applied Optics, Vol. 12, 355 - 358 (1973).

034.034 Use of MgF_2 and LiF photocathodes in the extreme ultraviolet. L. B. Lapson, J. G. Timothy.
Applied Optics, Vol. 12, 388 - 393 (1973).

The photoelectric yields of 2000-Å thick samples of MgF_2 and LiF have been measured at wavelengths in the range 1216–461 Å. Since the stability of response of the MgF_2 photocathodes appears to be equal to that of conventional metallic and semiconducting cathodes, it is concluded that MgF_2 would be a practical, high-efficiency photocathode for use in the extreme UV.

034.035 High resolution rocket EUV solar spectrograph.
W. E. Behring, R. J. Ugiansky, U. Feldman.
Applied Optics, Vol. 12, 528 - 532 (1973).

The design and performance of an Aerobee 150 rocket-borne solar spectrograph covering a wavelength range of 10–385 Å are discussed. In the laboratory the spectrograph has been used to record spectra of highly ionized metals with a resolution of 0.03 Å or better.

034.036 New Fourier transform all-reflection interferometer.
R. A. Kruger, L. W. Anderson, F. L. Roesler.
Applied Optics, Vol. 12, 533 - 540 (1973).

This paper describes the design and tests of an all-reflection two-beam interferometer. The interferometer consists of three reflecting diffraction gratings and two collimating mirrors. This instrument can be used as a Fourier transform spectrometer. The method of Fourier inversion is described, and examples of interferograms taken with the instrument and the spectra obtained from the interferograms are presented.

034.037 On guiding in spectrographic observations of the sun.
M. V. Kushnir.
Solnechnye Dannye 1972 Byull., No. 12, p. 98 - 101 (1973). In Russian.

A monochromatic photoelectric guider, constructed on the basis of a narrow-band monochromatic filter for the H_a line, is described.

034.038 Astronomische Übungsgeräte.
W. Becker.
SuW, Vol. 12, 145 - 149 (1973).

034.039 On the electromagnetic level for the visual zenith telescope (II). M. Ooe, S. Kuji.
Proc. International Latitude Obs. Mizusawa, No. 12, p. 29 - 38 (1972). In Japanese.

We tried to improve level measurements by constructing an electromagnetic level of pendulum with differential transformer (Sugawa, Ooe and Abe 1968). The benefits for adopting this kind of level are presented.

034.040 The automatic control unit of new PZT at Mizusawa.
K. Iwadate.
Proc. International Latitude Obs. Mizusawa, No. 12, p. 54 - 61 (1972). In Japanese.

034.041 Two types of the seeing analyzer. L. D. Parfinenko.
Solnechnye Dannye 1973 Byull., No. 1, p. 76 - 79 (1973). In Russian.

034.042 A standard for ultraviolet radiation.
G. B. Fisher, W. E. Spicer, P. C. McKernan, V. F. Pereskok, S. J. Wanner.
Applied Optics, Vol. 12, 799 - 804 (1973).

Photoemission diode standards for accurately measuring monochromatic ultraviolet light intensity (3000 Å—1100 Å) are described that are also blind to visible light ($\lambda > 3600$ Å). The standard uses an opaque photocathode of Cs_2Te. Design criteria, construction methods, and difficulties overcome in obtaining a stable, uniform, high yield photocathode responses are discussed. Cs_2Te is discussed in terms of a model for high yield photoemitters.

034.043 Solar chromatograph. J. O. Stenflo.
Applied Optics, Vol. 12, 805 - 809 (1973).

An instrument is described that produces monochromatic images differing from normal filtergrams in that the wavelength of the transmission band varies slowly and linearly across the image. The monochromatic bandwidth can be changed to almost any value, and the transmission band can easily be shifted to any part of the visible spectrum. Various applications for solar research are discussed.

034.044 Calibrations of the airglow photometers and spectrometers. B. S. Dandekar, D. J. Davis, Jr.
Applied Optics, Vol. 12, 825 - 831 (1973).

We discuss the major factors responsible for large errors such as the use of a low brightness source, bandwidths of the instruments, transmission characteristics of the optical interference filter, the effect of stray and scattered light by the instruments, as well as a step-wise procedure for minimizing the errors in a calibration of the instruments. In a test of these procedures the consideration of these factors resulted in minimizing the discrepancies in observations with photometers and spectrometers, yielding an accuracy of 20% or better in the measurements.

034.045 "Nonobjective" gratings.
I. S. Bowen, A. H. Vaughan, Jr.
Publ. Astron. Soc. Pacific, Vol. 85, 174 - 176 (1973).

The aberrations that limit the use of a transmission grating in the converging beam of a telescope can be partially compensated and the performance significantly improved if the grating is combined with a low-angle prism. A description is given of such a device which will yield spectra at dispersions of 320 Å to 1280 Å mm^{-1} of objects over a field of 150×200 mm ($1°2 \times 1°6$) for the 40-inch $f/7$ telescope of the Las Campanas Observatory.

034.046 On the determination of the absolute spectral sensitivity of photomultipliers. E. I. Terez.
Izv. Krymskoj Astrofiz. Obs., Vol. 46, 144 - 154 (1972). In Russian.

A practical procedure for calibration of a tungsten strip lamp using a standard has been worked out. A method to obtain the absolute spectral output of the system lamp—monochromator in the wavelength range from 3000 Å to 12000 Å is described.

034.047 A very simple photoelectric intensitometer for astronomical spectrographs. A. P. Kulčizky.
Izv. Krymskoj Astrofiz. Obs., Vol. 46, 155 - 158 (1972). In Russian.

A very simple photoelectric intensitometer used at the 122-cm telescope of the Crimean Astrophysical Observatory is described. The approximately logarithmic output provides a wide range of brightness available for measurements.

034.048 Ein Korrektor für Newton-Amateur-Teleskope.
E. Wiedemann.
Orion Schaffhausen, 31. Jahrgang, 96 - 97 (1973).

034.049 The Sacramento Peak Observatory universal birefringent filter. J. M. Beckers.
Bull. American Astron. Soc., Vol. 5, 269 (1973). – Abstr. AAS.

034.050 A multichannel diode array for solar observations.
R. B. Dunn, G. E. Spence.
Bull. American Astron. Soc., Vol. 5, 271 (1973). – Abstr. AAS.

034.051 A new dual filter telescope. E. N. Frazier.
Bull. American Astron. Soc., Vol. 5, 272 - 273 (1973). – Abstr. AAS.

034.052 Studies with improved digital videomagnetograph.
T. J. Janssens.
Bull. American Astron. Soc., Vol. 5, 274 (1973). – Abstr. AAS.

034.053 A multi-slit spectrograph. S. F. Martin.
Bull. American Astron. Soc., Vol. 5, 276 (1973). Abstr. AAS.

034.054 A new technique for solar magnetic field observations. R. C. Smithson.
Bull. American Astron. Soc., Vol. 5, 280 (1973). – Abstr. AAS.

034.055 Lyot filters with partial polaroids. A. M. Title.

Bull. American Astron. Soc., Vol. 5, 281 (1973). Abstr. AAS.

034.056 **Frequency-domain multiplexing multichannel photoelectric photometers.** G. Sedmak.
Astron. Astrophys., Vol. 25, 379 - 385 (1973).
This paper deals with frequency-domain multiplexing multichannel photometers. The theoretical expressions of the direct multichannel transfer function, of the inverse single channel optimum transfer function, of the precision of measurement are determined and discussed, together with a comparison between this system and the common time-domain multiplexing multichannel photoelectric photometers. A tentative two-channels FDM photometer and some numerical simulations of FDM systems are also described, which confirm the theory.

034.057 **A Lallemand electronic camera focused by a superconducting magnetic coil.**
J. P. Picat, A. Chevillot, M. Combes, P. Felenbok, B. Fort.
Photo-electronic image devices, (see 012.017), p. 1 - 6 (1972).

034.058 **Development of a new kind of Lallemand camera.**
J. Baudrand, M. Combes, P. Felenbok, B. Fort, J. P. Picat.
Photo-electronic image devices, (see 012.017), p. 7 - 12 (1972).

034.059 **Extended field Spectracon.**
J. D. McGee, H. Bacik, C. I. Coleman, B. L. Morgan.
Photo-electronic image devices, (see 012.017), p. 13 - 26 (1972).

034.060 **Sources of spurious background in the Spectracon.**
M. Oliver.
Photo-electronic image devices, (see 012.017), p. 27 - 35 (1972).

034.061 **Electronographic image tube development at the Royal Greenwich Observatory.**
D. McMullan, J. R. Powell, N. A. Curtis.
Photo-electronic image devices, (see 012.017), p. 37 - 51 (1972).

034.062 **A proximity focused ultraviolet-sensitive SEC camera tube.**
P. R. Collings, L. G. Healy, A. B. Laponsky, R. A. Shaffer.
Photo-electronic image devices, (see 012.017), p. 253 - 261 (1972).

034.063 **Pick-up storage tube having an electronic shutter, automatic exposure control, wobbling correction, and slow scanning.**
T. Hiruma, Y. Suzuki, K. Kurasawa.
Photo-electronic image devices, (see 012.017), p. 263 - 277 (1972).

034.064 **The electronic camera used in a reflection mode.**
J. P. Picat, M. Combes, P. Felenbok, B. Fort.
Photo-electronic image devices, (see 012.017), p. 557 - 562 (1972).

034.065 **Quantitative performance of single- and two-stage image tubes in spectroscopy.** K. E. Kissell.
Photo-electronic image devices, (see 012.017), p. 653 - 676 (1972).

034.066 **Evaluation of image intensifiers for astronomy.**
R. H. Cromwell, R. R. Dyvig.
Photo-electronic image devices, (see 012.017), p. 677 - 696 (1972).

034.067 **The use of electronographic-type image tubes in astronomical photometry.** M. F. Walker.
Photo-electronic image devices, (see 012.017), p. 697 - 718 (1972).

034.068 **Étude d'astres faibles en lumière totale avec la caméra electronique.**
G. Lelièvre, G. Wlérick.
Photo-electronic image devices, (see 012.017), p. 719 - 735 (1972).

034.069 **Photometry with the electronic camera.**
A. V. Hewitt, G. E. Kron, H. D. Ables.
Photo-electronic image devices, (see 012.017), p. 737 - 745 (1972).

034.070 **The analysis of direct Spectracon exposures obtained on the Isaac Newton telescope.**
H. Bacik, C. I. Coleman, M. J. Cullum, B. L. Morgan, J. Ring, C. L. Stephens.
Photo-electronic image devices, (see 012.017), p. 747 - 755 (1972).

034.071 **Problems in the use of image intensifiers in astronomical Cassegrain spectrographs.**
D. R. Palmer, A. S. Milsom.
Photo-electronic image devices, (see 012.017), p. 769 - 779 (1972).

034.072 **Improvements in the application of the image orthicon to astronomy.**
J. R. Dunlap, J. A. Hynek, W. T. Powers.
Photo-electronic image devices, (see 012.017), p. 789 - 794 (1972).

034.073 **An integrating television system for visual enhancement of faint stars.** E. W. Dennison.
Photo-electronic image devices, (see 012.017), p. 795 - 800 (1972).

034.074 **Recent developments and applications of the SEC-vidicon for astronomy.**
P. M. Zucchino, J. L. Lowrance.
Photo-electronic image devices, (see 012.017), p. 801 - 818 (1972).

034.075 **Application of an image isocon and computer to direct digitization of astronomical spectra.**
G. A. H. Walker, J. R. Auman, V. L. Buchholz, B. A. Goldberg, A. C. Gower, B. C. Isherwood, R. Knight, D. Wright.
Photo-electronic image devices, (see 012.017), p. 819 - 834 (1972).

034.076 **An image photon counting system for optical astronomy.** A. Boksenberg, D. E. Burgess.
Photo-electronic image devices, (see 012.017), p. 835 - 849 (1972).

034.077 **A photo-counting detector for stellar spectrophotometry.** J. D. McGee, B. L. Morgan, F. C. Delori, R. W. Airey, M. J. Cullum, C. L. Stephens.
Photo-electronic image devices, (see 012.017), p. 851 - 862 (1972).

034.078 **Counting image tube photoelectrons with semiconductor diodes.**
E. A. Beaver, C. E. McIlwain, J. P. Choisser, W. Wysoczanski.
Photo-electronic image devices, (see 012.017), p. 863 - 871 (1972).

034.079 **Serial read-out from image tubes incorporating sili-**

con diode arrays.
D. McMullan, G. B. Wellgate, J. Ormerod, J. Dickson.
Photo-electronic image devices, (see 012.017), p. 873 - 879 (1972).

034.080 **Further developments of magnetically focused, internal-optic image converters.** G. R. Carruthers.
Photo-electronic image devices, (see 012.017), p. 881 - 894 (1972).

034.081 **Internal-grating electronographic spectrographs for the far-ultraviolet and X-ray wavelength ranges.**
G. R. Carruthers.
Photo-electronic image devices, (see 012.017), p. 895 - 902 (1972).

034.082 **Photoelectronic image recording device optimized for high detective quantum efficiency.**
A. Choudry, G. W. Goetze, S. Nudelman, T. Y. Shen.
Photo-electronic image devices, (see 012.017), p. 903 - 910 (1972).

034.083 **Application of new ultraviolet television detectors in an astronomical satellite.**
C. C. Sturgell, J. T. Williams, W. A. Feibelman, A. Boksenberg, B. E. Anderson, G. E. MacKrell, T. J. L. Jones.
Photo-electronic image devices, (see 012.017), p. 911 - 924 (1972).

034.084 **Orbital operation and calibration of SEC-vidicons in the Celescope experiment.** W. A. Deutschman.
Photo-electronic image devices, (see 012.017), p. 925 - 935 (1972).

034.085 **The development of an intensifier-vidicon for space applications.** R. R. Beyer, H. Alsberg.
Photo-electronic image devices, (see 012.017), p. 937 - 944 (1972).

034.086 **Semi-automatic photometer for the determination of charges of heavy nuclei of cosmic rays in photo-nuclear emulsions.** V. N. Kulikov, G. A. Pyatigorskij, E. A. Yakubovskij, V. V. Varyukhin, Yu. F. Gagarin, N. S. Ivanova.
Pribory i tekhn. ehksperimenta, 1972, No. 6, p. 42 - 44.
In Russian. – Abstr. in Referativ. Zhurn. 51. Astron., 4.51.380 (1973).

034.087 **Die Einrichtung für lichtelektrische Photometrie an der Grazer Sternwarte.** H. J. Schober.
Mitt. Astron. Ges., No. 32, p. 150 - 154 (1973).

034.088 **Analyse des Regelkreises eines Seeing-Meßgerätes.**
T. V. Pettauer.
Mitt. Astron. Ges., No. 32, p. 154 - 161 (1973).

034.089 **Photographische Messung solarer Magnetfelder.**
A. Wittmann.
Mitt. Astron. Ges., No. 32, p. 163 (1973). – Abstract.

034.090 **Instrumentelle Verbesserungen im Locarno-Observatorium.** H. Wöhl.
Mitt. Astron. Ges., No. 32, p. 164 - 165 (1973).

034.091 **Protuberanzen - Spektroskopie hoher Auflösung mit Hilfe des Bildverstärkers.**
G. Stellmacher, E. Wiehr.
Mitt. Astron. Ges., No. 32, p. 166 - 167 (1973).

034.092 **Vergleichende Untersuchung von Bildverstärkern für die Aufnahme solarer Spektrogramme und Filtergramme.** P. N. Brandt, A. Wiesmeier.
Mitt. Astron. Ges., No. 32, p. 167 - 168 (1973).

034.093 **Synthetic aperture optics.** R. H. Miller.
Publ. American Astronaut. Soc., Sci. Techn. Ser., Vol. 28, (see 012.018), p. 301 - 313 (1972).

034.094 **Electronic imaging devices for astronomy from a space platform.** G. R. Carruthers.
Publ. American Astronaut. Soc., Sci. Techn. Ser., Vol. 28, (see 012.018), p. 315 - 335 (1972).

034.095 **Instrument for investigating the composition of the electron component of cosmic rays.**
E. A. Bogomolov, V. K. Karakadko, N. D. Lubyanaya, V. A. Romanov, M. G. Totubalina, M. A. Yamshchikov.
Cosmic rays No. 13, (see 003.013), p. 215 - 220 (1972).
In Russian.

034.096 **Investigation of some characteristics of the UM-92 cesium-oxygen photocathode image converters.**
K. A. Grigorian, H. V. Abrahamian (*G. V. Abramyan*), G. Lelievre, M. A. Eritsian.
Soobshch. Byurakan. Obs., vyp. (No.) 44, p. 120 - 125 (1972).
In Russian.

034.097 **On the TV-guiding of astronomical telescopes.**
V. A. Malarev, E. M. Neplokhov, I. K. Pavlov, G. A. Tambovski.
Soobshch. Byurakan. Obs., vyp. (No.) 44, p. 126 - 136 (1972).
In Russian.

034.098 **Application of the Monte Carlo method to the calibration of detectors for gamma-ray astronomy.**
R. D. Wills.
Comput. Phys. Commun., (*Netherlands*), Vol. 4, No. 1, p. 51 - 58 (1972). – See Phys. Abstr., Vol. 76, No. 13113 (1973).

034.099 **Photometric field error of the NA-MK-25 camera.**
I. M. Khaimov.
Dokl. AN TadzhSSR, Vol. 15, No. 12, p. 10 - 12 (1972).
In Romanian. – Abstr. in Referativ. Zhurn. 51. Astron., 6.51. 175 (1973).

034.100 **On the coherence length of a monochromator for use with a Michelson stellar interferometer.**
R. Q. Twiss, W. T. Welford.
Optics Commun., Vol. 7, 103 - 106 = Commun. Roy. Obs. Edinburgh, No. 144 (1973).
The drop in visibility due to path difference errors in a Michelson stellar interferometer is normally controlled by a monochromator in the system. It is shown that the effect of this monochromator on the visibility is governed by a product of two factors, one geometrical and the other a function of the resolving power of the monochromator.

034.101 **Errors of the Yale Mann measuring engine.**
D. Hoffleit.
Trans. Astron. Obs. Yale Univ., *New Haven*, Vol. 32, Part 1, 40 pp. (1973).
The Mann measuring machine at Yale is a two-screw engine designed especially for measuring cartesian coordinates on photographic plates 17 X 17 inches. Two errors are investigated: the periodic error of the X-screw and the change of the screw-length with temperature.

034.102 **A light and compact X-ray image read-out system for space applications.** J. M. Morton, W. Parkes.
Acta Electronica, (*France*), Vol. 16, 85 - 100 (1973).

034.103 **UV: eye on the moon.** L. R. Lankes.

Optical Spectra, (*USA*), Vol. 6, No. 12, p. 39 - 41 (1972).

Among the achievements of the Apollo 16 mission was man's first step in the direction of lunar-based astronomy. This article describes the Lunar Surface Ultraviolet Spectrographic Camera, which made the experiment possible.

034.104 Configuration for a multichannel Fabry-Perot spectroheliograph. L. Mertz.
Optics Commun., (*Netherlands*), Vol. 6, 282 - 283 (1972).

034.105 Star simulator for checking star guiders.
N. N. Raimov, V. S. Tataurov, G. R. Pekki, K. Ya. Kutorkina.
Soviet Journ. Optical Technology, (*USA*), Vol. 39, 271 - 272 (1972).

The optical and mechanical design, as well as specifications for a simulator designed for adjusting guiders for star-tracking systems and checking their accuracy under laboratory conditions, are described.

034.106 A simple micrometer microscope for off-axis guiding in comet photography. R. D. Lines.
Strolling Astronomer, Vol. 24, 97 - 98 (1973).

034.107 How to use a micrometer microscope to guide for comet photography. R. D. Lines.
Strolling Astronomer, Vol. 24, 98 - 99 (1973).

034.108 Photoelectric time micrometer.
J. Kakkuri, K. Kalliomäki.
Publ. Finnish Geod. Inst., *Helsinki*, No. 74, 53 pp. (1972).

Accurate observations of the satellites' positions at any time are carried out with various methods. In this paper the general principle of an electronic timing device, a time micrometer, installed in a Schmidt-Väisälä telescope of the Finnish Geodetic Institute is described.

034.109 Spectroscopy by grating synthesis.
G. A. De Biase, F. Sacchetti, D. Trevese.
Applied Optics, Vol. 11, 1163 - 1168 = Oss. Astron. Roma, Contr. Sci., Ser. III, No. 123 (1972).

Spectral information obtained by synthesis of two or more diffraction gratings is studied. Also the case of two gratings with relative motion and the case of N fixed gratings are discussed. We show that, using grating synthesis, it is possible to obtain transfer functions that increase the total obtainable information with respect to the single grating. The possibility of building a synthesis spectroscope is presented, taking into account both the gratings aberration and some construction imperfections.

034.110 Mehrfachspaltspektrograph. Zur Spektralphotometrie der Korona am 30.VI. 1973. W. Jaschek.
Sternenbote, 16. Jahrgang, p. 110 - 112 (1973).

034.111 The Apollo Alpha Spectrometer.
N. Jagoda, K. Kubierschky, R. Frank, J. Carroll.
IEEE Trans. Nuclear Sci., Vol. NS-20, No. 1, p. 90 - 98 (1973).

034.112 A position-sensitive X-ray detector for the HEAO-A satellite. D. Held, M. C. Weisskopf.
IEEE Trans. Nuclear Sci., Vol. NS-20, No. 1, p. 140 - 144 (1973).

034.113 Considerations about far infrared detectors for astronomical purposes. B. Carli, F. Melchiorri.
Infrared Phys., Vol. 13, 49 - 60 (1973).

034.114 The EUV spectrophotometer on Atmosphere Explorer.

H. E. Hinteregger, D. E. Bedo, J. E. Manson.
Radio Sci., (*USA*), Vol. 8, 349 - 359 (1973).

034.115 Use of open-structure channel electron multipliers in sounding rocket experiments. J. G. Timothy.
Rev. Sci. Instruments, Vol. 44, 207 - 211 (1973).

The performance of open-structure channel electron multipliers has been investigated in a laboratory simulation of the ascent pressure profile for a typical sounding rocket. The results of this test demonstrate that evacuation of the payload prior to launch is necessary for experiments where stability of response is an essential parameter.

034.116 Coaxial anode for background suppression in X-ray proportional counters.
A. N. Bunner, W. L. Kraushaar, D. McCammon, M. Vanderhill, F. Williamson.
Rev. Sci. Instruments, Vol. 44, 418 - 422 (1973).

034.117 Solar chromatograph. J. O. Stenflo.
Applied Optics, Vol. 12, 805 - 809 (1973).

An instrument is described that produces monochromatic images differing from normal filtergrams in that the wavelength of the transmission band varies slowly and linearly across the image. The monochromatic bandwidth can be changed to almost any value, and the transmission band can easily be shifted to any part of the visible spectrum. Various applications for solar research are discussed.

034.118 Heterodyne detection of blackbody radiation.
J. Gay, A. Journet, B. Christophe, M. Robert.
Applied Phys. Letters, Vol. 22, 448 - 449 (1973).

034.119 Absolute calibration of Apollo lunar orbital mass spectrometer.
P. R. Yeager, A. Smith, J. J. Jackson, J. H. Hoffman.
Journ. Vac. Sci. and Technol., (*USA*), Vol. 10, 348 - 354 (1973).

034.120 Spectroscopy in the Madrid astronomical observatory. M. L. Arroyo.
Optica Pura y Aplicada, (*Spain*), Vol. 5, 203 - 208 (1973). In Spanish.

034.121 Large area focusing collector for the observation of cosmic X-rays.
P. Gorenstein, A. DeCaprio, R. Chase, B. Harris.
Rev. Sci. Instruments, Vol. 44, 539 - 545 (1973).

A large area focusing X-ray collector constructed for a sounding rocket is described.

034.122 An extreme UV photometer for solar observations from Atmosphere Explorer.
D. F. Heath, J. F. Osantowski.
Radio Sci., (*USA*), Vol. 8, 361 - 367 (1973).

034.123 A multichannel spectrophotometer.
A. W. Rodgers, R. Roberts, P. T. Rudge, T. Stapinski.
Publ. Astron. Soc. Pacific, Vol. 85, 268 - 274 (1973).

A 33-channel spectrophotometer used at the Cassegrain focus of the 74-inch reflector at Mount Stromlo is described. The instrument status and data system are under remote interactive computer control. The photomultiplier cooling which is accomplished by circulating dry nitrogen at $-65°C$ through the cold boxes results in high thermal stability at the cathodes. The performance of the spectrophotometer under observing conditions is described.

034.124 Ein Infrarotmodulator für Weltraumanwendung auf der Basis eines Fabry-Perot-Interferometers.

E.-G. Lierke.
Battelle Information, (Battelle Inst., Frankfurt), No. 14, p. 7 - 12 (1972).

034.125 New active RC configuration for realising a medium-selectivity notch filter. A. M. Soliman.
Electronics Letters. Vol. 8, 522 - 524 (1972).
  A new configuration of parallel T filter and operational amplifiers to give a medium Q notch filter. — *DJC*

034.126 Contrast elements in birefringent filters.
  S. A. Schoolman.
Solar Physics, Vol. 30, 255 - 261 (1973).
  Calculations were made to determine the effects which contrast elements of various thicknesses have in Lyot filters. A filter which is significantly narrower than the line at which it is looking produces the best results. Therefore, if the filter is broad the addition of as thick an element as possible is desired. However, if the filter is already narrow, a contrast element whose thickness equals that of the second Lyot element will produce the best performance.

Corrections for lost motion in theodolite eye-piece micrometers.   See Abstr. 031.009.

A method for the automatic analysis of gamma ray events in astronomical spark chambers.
See Abstr. 031.045.

Intensity interferometry in the spatial domain.
See Abstr. 031.049.

The 100 cm optical telescope instrumentation.
See Abstr. 032.006.

Solar instrumentation.   See Abstr. 032.025.

Experimental X-ray astronomy.
See Abstr. 061.064.

Premiers résultats du levé spectrophotométrique du ciel dans l'ultraviolet à l'aide du satellite TD-1 A.
See Abstr. 114.039.

Spectrophotometric results from the Copernicus satellite. I. Instrumentation and performance.
See Abstr. 114.121.

035 Clocks and Frequency Standards

035.001 "Capture" of the mutually perturbing rates of pendulum clocks. K. Bielicka.
Acta Astron., Vol. 23, 53 - 64 (1973).
  Considering in general the problem of mutual influence of two pendulum clocks, the case has been investigated when the rates of two clocks "capture" each other i.e. they become identical. The "capture" occurs when the difference between the daily mean perturbing rates is not greater than a certain critical value. This critical value can be calculated.

035.002 Ein aperiodischer quasiperiodischer Frequenzteiler.
  G. Becker.
PTB Mitt., 83. Jahrgang, p. 13 - 16 (1973).
  An aperiodic frequency divider circuit is described which produces the pulse sequence of the desired frequency f_x by means of selected pulses of a given pulse sequence of the frequency f_0. With this procedure it is possible to obtain complicated division factors (in IC-techniques) with commercially available electronic dividers (counter circuits, flip-flops) in connection with electronic switches.

035.003 On the seasonal variation of the JJY (5 MHz) as received at Mizusawa.
T. Hara, K. Horiai, K. Iwadate.
Proc. International Latitude Obs. Mizusawa, No. 12, p. 1 - 9 (1972). In Japanese.
  The standard radio wave JJY (5 MHz) has been received for clock comparisons at a distance of 406.6 km during recent years. There are variations of travel time from transmitter to the receiving site due to fluctuations of the ionosphere.

035.004 Kodierte Zeitinformation über den Zeitmarken- und Normalfrequenzsender DCF 77.
G. Becker, P. Hetzel.
PTB Mitt., 83. Jahrgang, p. 163 - 164 (1973).

035.005 A note on coordination of remote clocks.
  H. J. M. Abraham.
IAU Colloquium No. 1, (see 012.019), p. 53 - 61 (1972).

036 Photographic Auxiliaries

036.001 **Deep-sky photography with cooled emulsions.**
O. Ernest.
Sky Telescope, Vol. 45, 189 - 195 (1973).

036.002 **On a method of testing astronomical plates in view of the possibility of detecting faint stars.**
G. V. Novikova.
Astron. Tsirk., No. 746, p. 5 - 6 (1973). In Russian.

036.003 **On the hyper-sensitization of plates by baking. I. Eastman Kodak IIaO.**
O. D. Dokuchaeva, T. A. Birulya.
Astron. Tsirk., No. 752, p. 3 - 6 (1973). In Russian.

036.004 **Color infrared photography of some astronomical objects. O. R. Norton.**
Sky Telescope, Vol. 45, 396 - 400 (1973).

036.005 **Astrofotografie. S. Marx.**

Astron. in der Schule, 10. Jahrgang, p. 30 - 32 (1973).

036.006 **Linearity and optimum working density of optical and nuclear emulsions. M. Cohen, E. Kahan.**
Photo-electronic image devices, (see 012.017), p. 53 - 65 (1972).

036.007 **Properties of commercial electron-sensitive plates for astronomical electronography.**
P. Griboval, D. Griboval, M. Marin, J. Martinez.
Photo-electronic image devices, (see 012.017), p. 67 - 82 (1972).

036.008 **An experiment with a naturally-cooled emulsion.**
J. B. Newton.
Journ. Roy. Astron. Soc. Canada, Vol. 67, 148 - 149 (1973).

Astronomical photography. A brief history.
See Abstr. 004.068.

Positional Astronomy. Celestial Mechanics

041 Positional Astronomy, Star Catalogues and Atlases

041.001 **Meridian observations made in Brorfelde (Copenhagen University Observatory) 1967 - 1969. Positions of 2246 stars brighter than 11.0 vis. mag.**
H. J. Fogh Olsen, P. Jensen, T. Knudsen.
Astron. Astrophys., Suppl. Ser., Vol. 9, 1 - 83 (1973).

This catalog presents positions for 2246 stars observed with the 7'' transit circle at Brorfelde (Copenhagen University Observatory) during the period 1967 - 1969. The positions are reduced relative to FK4, and they are selected from different lists of stars including nearby, high velocity, FK4 supp., O and B types, G and K types, and B8-B9 stars, and GC stars, mainly the weaker ones. They are located in the declination interval from $-10°$ to $+80°$ and dispersed in the magnitude interval from $3^m.0$ to $11^m.0$. The internal mean error for a single observation is $\epsilon_\alpha \cos \delta = 0^s.0152$ and $\epsilon_{\delta\,zenith} = 0''.213$.

041.002 **Frequency of star eliminations in the meridian catalogues constitutive of the Melchior-Dejaiffe ILS stars catalogue.** R. J. Dejaiffe.
Astron. Astrophys., Vol. 22, 425 - 429 (1973).

In this statistical qualitative study 106 of the 463 meridian catalogues used for the Melchior–Dejaiffe system of ILS stars declinations and proper motions are involved. The improved system for original data homogeneization and the adopted calculation procedures are described. A table presents the principal results of this study from which the indisputable quality of some catalogues is thrown into relief.

041.003 **On the systematic corrections $\Delta\alpha_\delta \cos \delta$ of the FK4 south of $-42°$ declination.** M. S. Zverev, A. A. Naumova, D. D. Polozhentsev, E. A. Stepanova.
Astron. Zhurn. Akad. Nauk SSSR, Vol. 50, 433 - 435 (1973). In Russian. English translation in Soviet Astron. AJ, Vol. 17, No. 2.

The reduction of 47000 determinations of right ascensions of SRS, BS and FK4 stars in the declination zones from $-47°$ to $-90°$ (for the FK4 from $-42°$ to $-90°$) has been completed at the Pulkovo Observatory. The corrections $\Delta\alpha_\delta \cos \delta$ to the FK4, derived from these observations, are given in a table for $2°$ zones (without smoothing).

041.004 **Effect of the interdependence of observational errors upon the estimated accuracy of astronomical observations.** V. A. Yatsenko.
Sudovozhdenie. Nauch.-tekhn. sb., 1972, vyp. (No.) 12, p. 75 - 81. In Russian. – Abstr. in Referativ. Zhurn. 51. Astron., 3.51.205 (1973).

041.005 **Method of determining the optical center and constants of a plate in photographic astrometric observations.** I. Totomanov, N. Georgiev.
Izv. Gl. upr. geod. i kartogr., 1972, No. 1, p. 12 - 20. In Bulgarian. – Abstr. in Referativ. Zhurn. 51. Astron., 3.51.212 (1973).

041.006 **Catalogue of right ascensions of 645 FKSZ stars in the FK4 system.** L. L. Vagushchenko.
Astrometriya i Astrofizika, Kiev, vyp. (No.) 16, (see 003.006), p. 3 - 13 (1972). In Russian.

041.007 **Representation of the differences of coordinates of stars in catalogues in terms of spherical harmonics.** A. N. Kurjanova.
Astrometriya i Astrofizika, Kiev, vyp. (No.) 16, (see 003.006), 13 - 21 (1972) In Russian.

A description is given of an algorithm for representing the differences of star coordinates in different catalogues in terms of spherical harmonics. A programme realizing this algorithm is described. To check the programme the systematic differences FK4 – GC have been used.

041.008 **Astrometry of the southern sky.** M. S. Zverev.
Zemlya i Vselennaya, 1973, No. 2, p. 16 - 23. In Russian.

041.009 **Sur un test de qualité appliqué à 106 catalogues méridiens.** R. Dejaiffe.
Ciel et Terre, Vol. 89, 80 - 90 (1973).

041.010 **Determination of zero points of the FK4 Catalogue from meridian observations of the moon.** V. A. Fomin.
Astron. Tsirk., No. 750, p. 3 - 4 (1973). In Russian.

041.011 **Role of the personal error in the formation of the jump in the zenith during absolute determinations of right ascensions.** O. T. Markina.
Astron. Tsirk., No. 750, p. 5 - 7 (1973). In Russian.

041.012 **The use of radio interferometers in astrometry.** N. S. Blinov, E. N. Fedoseev.
Astron. Tsirk., No. 752, p. 1 - 3 (1973). In Russian.

041.013 **The precision of the observations of Bucharest KSZ stars; a statistical study of its dependence on temperature.** M. Tudor, E. Toma.
Stud. Cerc. Astron., Vol. 18, 29 - 31 (1973). In Romanian.

041.014 **A user's guide to the Palomar Sky Survey.** J. M. Lund, R. S. Dixon.
Publ. Astron. Soc. Pacific, Vol. 85, 230 - 240 (1973).

Background information concerning the Palomar Sky Survey is presented, including some of its major uses, some interesting facts, and corrections of some errors. General equations are derived for the edges of astronomical photographs, and these equations are used in the generation of (1) specially edited versions of the SAO Star Catalog on magnetic tape, arranged by Palomar print area, and (2) maps showing the boundaries of the Palomar print areas. Both the tapes and maps are available at cost from the OSU Radio Observatory.

041.015 **Sul metodo di Nemiro per l'osservazione di coppie di stelle simmetriche rispetto allo zenit.** G. Caprioli.
Mem. Soc. Astron. Italiana, Nuova Ser., Vol. 43, 681 - 686 (1973).

Meridian-transits observations with the method of the star-pairs symmetric with respect to the zenith have been made at the Rome Astronomical Observatory. This report gives the fundaments and the characteristics of the method, and describes the computer programs for the selection of the observing list and for the reduction of the observations.

041.016   Prospects for an extension of the fundamental reference system to faint objects and radio sources.
W. Fricke.
Mem. Soc. Astron. Italiana, Nuova Ser., Vol. 43, 751 - 758 (1973).

Basic material and procedures are discussed for an improvement of the FK4 and its extension to the ninth visual magnitude resulting in a new fundamental catalogue, which I provisionally call the FK5. Arguments are given in favor of extending the proposed system FK5 to optically faint objects and radio sources and of relating the fundamental reference system to the extragalactic reference frame.

041.017   Confronto fra i cataloghi MD e GC ed analisi dei moti propri.   E. Proverbio, A. Poma.
Mem. Soc. Astron. Italiana, Nuova Ser., Vol. 43, 793 - 795 (1973).

An analysis of proper motion system of the Melchior-Dejaiffe Catalogue (1969) has been carried out. The comparison between the MD Catalogue and the General Catalogue emphasized the existence of an error in the precessional constant of the order of $-0''.003$.

041.018   On the use of radio interferometers with a large base for astrometric work.
N. S. Blinov, E. N. Fedoseev.
Astron. Zhurn. Akad. Nauk SSSR, Vol. 50, 601 - 605 (1973). In Russian. English translation in Soviet Astron. AJ, Vol. 17, No. 3.

A numerical estimate is given of the precision of the determination of coordinates of the pole, of declinations and right ascensions of point radio sources. Presented are calculations of siderial time from observations with an interferometer with a large base, the aerials of which are situated at the same latitude and scattered over a longitude of 180°.

041.019   The determination of absolute declinations of equatorial stars from micrometric measurements near the earth's equator.   E. I. Krejnin, S. A. Murri.
Astron. Zhurn. Akad. Nauk SSSR, Vol. 50, 606 - 614 (1973). In Russian. English translation in Soviet Astron. AJ, Vol. 17, No. 3.

041.020   Astrometric analysis of the field of AC +65°6955 from plates taken with the Sproul 24-inch refractor.
J. L. Hershey.
Astron. Journ., Vol. 78, 421 - 425 (1973).

Positions of 12 stars on 423 plates taken from 1937 to 1969 with the Sproul 24-inch refractor are analyzed. Small changes are found in the astrometric field which are related to the history of the objective lens. An absolute parallax of $+0''.123 \pm 0''.002$ for AC+65°6955 is determined.

041.021   Bestimmung relativer Rektaszensionen von FK4 - SUP - Sternen.   G. Polnitzky.
Mitt. Astron. Ges., No. 32, p. 125 - 126 (1973). − Abstract.

041.022   Statistik der Ausgleichsansätze zur Reduktion astrometrischer Daten.   M. G. Firneis.
Mitt. Astron. Ges., No. 32, p. 126 - 128 (1973).

041.023   Observations of Saturn with the astrolabe of the Paris Observatory during the winter 1971−72.
F. Chollet, H. Choplin, S. Débarbat, M. Feissel, S. K. Lam.
Astron. Astrophys., Vol. 26, 141 - 142 (1973). In French. Research note.

041.024   Analysis of the effect of precession, nutation and annual aberration upon the equatorial coordinates, azimuth and altitude of Polaris.
A. V. Butkevich, Tran-Zuj-Tkhoan.
Izv. vyssh. ucheb. zavedenij. Geod. i aehrofotos"emka, 1972, No. 4, p. 103 - 108. In Russian. − Abstr. in Referativ. Zhurn. 51. Astron., 5.51.216 (1973).

041.025   Right ascensions of the sun, Venus and Mars obtained from observations with the Tashkent meridian circle in 1960−1961 and 1968−1969.   I. M. Boroditskij.
Tsirk. Astron. Inst., *Tashkent*, No. 35 (382), p. 10 - 15 (1972). In Russian.

041.026   The PZT and astrolabe programs.   R. G. Hall.
IAU Colloquium No. 1, (see 012.019), p. 1 - 6 (1972).

041.027   Mount Stromlo PZT results.
H. J. M. Abraham, J. N. Boots.
IAU Colloquium No. 1, (see 012.019), p. 7 - 20 (1972).

041.028   The attachment of PZT's to FK4.   R. W. Tanner.
IAU Colloquium No. 1, (see 012.019), p. 63 - 70 (1972).

041.029   International Information Bureau on Astronomical Ephemerides.
B.I.I.E.A. (IAU − COSPAR), Paris. Information cards, Nos. 51 - 72 (1973).

041.030   Catalogues of positions of stars.
V. P. Tsesevich (Editor).
Akademiya Nauk Ukrainskoj SSR; Glavnaya Astronomicheskaya Observatoriya; Odesskaya Astronomicheskaya Observatoriya. Izdatel'stvo "Naukova Dumka", Kiev. 357 pp. Price 1 Rbl. 3 Kop. (1970). In Russian. − Contents: 7 catalogues, compiled at the Odessa Observatory, see abstracts 041.031 - 041.036, 121.075.

041.031   Catalogue of declinations of stars of the equatorial zone.   B. V. Novopashenny, G. G. Ermolayev.
Catalogues of positions of stars, *Kiev*, (see 041.030), p. 5 - 133 (1970). In Russian.

This catalogue contains declinations of 4563 stars of the Nicolayev equatorial zone (AG, $-2°$ to $+1°$) from observations made during the years 1929 to 1935 by Novopashenny with the Odessa meridian circle. The observations were reduced to the FK3 system.

041.032   Catalogue of right ascensions of 645 FKSZ stars in the FK3 system.   B. V. Novopashenny.
Catalogues of positions of stars, *Kiev*, (see 041.030), p. 135 - 155 (1970). In Russian.

The catalogue gives the right ascensions (1950.0) of 645 FKSZ stars on the FK3 system observed during 1940−1944 at the Odessa Observatory meridian circle on the differential way.

041.033   Catalogue of right ascensions of 2967 KSZ_2 stars between $-5°$ and $-25°$ declination.
B. V. Novopashenny.
Catalogues of positions of stars, *Kiev*, (see 041.030), p. 157 - 237 (1970). In Russian.

The catalogue (1950.0, FK3 system) was formed on the basis of differential observations 1954−1961 by the author with the Odessa Observatory meridian circle. The mean error of a single observation referred to the equator is $\pm 0''.027$.

041.034   Declinations of 973 KSZ stars in the $-5°$ to $-10°$ declination zone.   L. F. Cherniev.

Catalogues of positions of stars, *Kiev,* (see 041.030), p. 239 - 284 (1970). In Russian.

The observations have been carried out 1954–1961 with the Odessa Observatory meridian circle. They were reduced to the FK3 system and the equinox 1950.0. The mean error of a catalogue position is ± 0."20.

041.035 Catalogue of declinations of 192 stars of the wide-scale pairs programme in the FK4 system.
A. M. Stafeev.
Catalogues of positions of stars, *Kiev,* (see 041.030), p. 285 - 303 (1970). In Russian.

A catalogue of the new Moscow program of wide-scale pairs containing declinations of 192 stars of 6.0 mag to 8.5 mag is presented. The catalogue was observed 1966 with the Odesssa Observatory transit circle.

041.036 Catalogue of right ascensions and declinations of 645 FKSZ stars in the FK4 system. A. M. Stafeev.
Catalogues of positions of stars, *Kiev,* (see 041. 030), p. 305 - 341 (1970). In Russian.

This catalogue was observed with the Odessa Observatory Repsold transit circle 1968 October 7 – 1969 July 7. Each star was observed four times on the average.

041.037 Absolute right ascensions of 626 bright stars observed with the Freiberg-Kondratjev transit instrument at the Nikolayev Observatory during 1959–1963.
O. T. Markina, G. M. Petrov.
Trudy Glav. Astron. Obs. Pulkovo, Ser. 2, Vol. 77, 5 - 22 (1969). In Russian.

A catalogue of right ascensions of 626 bright stars for the equinox and epoch 1960.0 is given. The results of a comparison of the catalogue compiled with the FK4 catalogue are given in tables.

041.038 Catalogue of 2600 KSZ stars in the −5° to −20° declination zone compiled from observations with the meridian circle at Nikolayev. L. F. Gorel.
Trudy Glav. Astron. Obs. Pulkovo, Ser. 2, Vol. 77, 23 - 73 (1969). In Russian.

The KSZ program stars were observed with the Repsold meridian circle at the Nikolayev Observatory during 1956–1963. Positions of 2600 stars in the declination zone from −5 to −20° were determined. The observations were reduced in the FK3R system. Individual corrections for 329 FK3 stars were deduced from all the observations of reference stars obtained during the above period.

041.039 Corrections to the declinations of FK4 and PFKSZ stars. G. S. Kosin.
Trudy Glav. Astron. Obs. Pulkovo, Ser. 2, Vol. 77, 74 - 78 (1969). In Russian.

Corrections of declinations of 506 FK4 stars and 531 PFKSZ stars in the +90° to −10° declination zone are given. The corrections were derived by means of a comparison of the fundamental catalogues with the catalogues of declinations of bright stars ($Pu_{50}F$) and faint stars ($Pu_{50}Z$).

041.040 Positions of 117 additional KSZ stars in regions with extragalactic nebulae (declination zone +25° to −20°). L. F. Gorel.
Trudy Glav. Astron. Obs. Pulkovo, Ser. 2, Vol. 77, 79 - 82 (1969). In Russian.

041.041 Positions of 107 additional KSZ stars in regions with extragalactic nebulae (declination zone +25° to +90°). G. D. Baturina.
Trudy Glav. Astron. Obs. Pulkovo, Ser. 2, Vol. 77, 83 - 88 (1969). In Russian.

041.042 Catalogue of 275 stars compiled from observations during 1956–1961 at the Nikolayev Observatory.
V. N. Boyko, S. V. Shilova.
Trudy Glav. Astron. Obs. Pulkovo, Ser. 2, Vol. 77, 89 - 99 (1969). In Russian.

A catalogue of positions and proper motions of 275 stars (N) is given. As a result of the comparison of proper motions of the stars of the N catalogue and those of the Yale catalogues, it was found that systematic differences of proper motions of these catalogues stars depend on magnitude. The dependence of systematic differences of proper motions in right ascension on the spectral class of stars permits to derive approximate corrections to right ascensions of the Yale catalogues stars.

041.043 Right ascensions of the sun, Mercury and Venus observed with the transit instrument at Nikolayev during 1965. O. T. Markina, G. M. Petrov.
Trudy Glav. Astron. Obs. Pulkovo, Ser. 2, Vol. 77, 100 - 105 (1969). In Russian.

041.044 Declinations of the sun, Mercury and Venus in the FK4 system deduced from observations with the vertical circle of the Nikolayev Observatory in 1965.
I. I. Bozhko, G. K. Zimmerman.
Trudy Glav. Astron. Obs. Pulkovo, Ser. 2, Vol. 77, 106 - 108 (1969). In Russian.

041.045 Observations of the sun and planets with the Pulkovo vertical circle. G. S. Kosin.
Trudy Glav. Astron. Obs. Pulkovo, Ser. 2, Vol. 77, 109 - 112 (1969). In Russian.

041.046 Results of observations of planets with the vertical circle at the Pulkovo Observatory. G. S. Kosin.
Trudy Glav. Astron. Obs. Pulkovo, Ser. 2, Vol. 77, 113 - 114 (1969). – Mercury, Venus, Mars 1963–1964.

041.047 On the computer reduction of differential meridian observations of right ascensions.
M. S. Zverev, D. D. Polozhentsev.
Trudy Astron. Obs., *Leningrad,* Vol. 29 (= Uchenye Zapiski Leningr. Un-ta, No. 363 = Seriya Matem. Nauk, vyp. (No.) 48), p. 196 - 201 (1973). In Russian.

Bessel's formula has been transformed to an alternative form which is more suitable for computer calculations. The method is applicable for the reduction of observations in wide zones and along the whole meridian arc. The results of reductions by the classical method are compared with those based on the use of the transformed formula. It is shown that both methods are identical with respect to accidental errors.

041.048 Accuracy of outer-planet ephemerides.
R. L. Duncombe, W. J. Klepczynski, P. K. Seidelmann.
Astronaut. Aeronaut., Vol. 10, No. 8, p. 63 - 65 (1972).

041.049 Cálculo de coordenadas de estrelas.
T. Torrão, J. Osório.
Inform. Obs. Astron. Porto, No. 3, 11 pp. (1971).

041.050 Sternatlas (1975.0). S. Marx, W. Pfau.
J. A. Barth, Leipzig. 16 pp. Price MDM 30.00 (1973). – Review in Astron. in der Schule, 10. Jahrgang, p. 69 (1973).

Centre de Données Stellaires, Inform. Bull. No. 4.
See Abstr. 002.007.

On the reduction of the observational data with the Tokyo meridian circle. See Abstr. 031.018.

Measurement of the PZT plate and identification of the stars by computer.   See Abstr. 031.019.

The Canadian PZT's.   See Abstr. 032.032.

Simultaneous determination of right ascensions and declinations with a photoelectric tracking micrometer for the meridian circle.   See Abstr. 034.008.

Errors of the cosmic clock-face.
See Abstr. 044.019.

The anomalous refraction from the observations with meridian circle.   See Abstr. 082.048.

Positional photographic observations of Venus at Pulkovo in 1972.   See Abstr. 093.014.

On the problem concerning stars selection for observations on the lunar surface by the equal-altitude method. See Abstr. 094.115.

Astrographic observations of Mars at Pulkovo in 1972.   See Abstr. 097.049.

Observations of Jupiter with Danjon astrolabes in 1965, 1966, 1967.   See Abstr. 099.001.

Observational studies of Jupiter during the years 1965, 1966, 1967.   See Abstr. 099.015.

Positional photographic observations of the satellites of Jupiter at Pulkovo in 1966 - 1968. See Abstr. 099.039.

042   Celestial Mechanics

042.001   **Relative motion of near orbiting satellites.**
        J. B. Eades, Jr., J. W. Drewry.
Celestial Mechanics, Vol. 7, 3 - 30 (1973).
        The relative motion of two particles on adjacent orbits about the same primary has been investigated under the condition that both motions have the same period. The geometrical properties of the relative displacement and velocity traces, on representative planes, are studied. A complete state of the motion is given; and, the range and range-rate variations, over one or more orbits, are described.

042.002   **Stability of planar oscillations of a satellite in an elliptic orbit.**   V. A. Zlatoustov, A. P. Markeev.
Celestial Mechanics, Vol. 7, 31 - 45 (1973).
        A problem of stability of odd 2π-periodic oscillations of a satellite in the plane of an elliptic orbit of arbitrary eccentricity is considered. The motion is supposed to be only under the influence of gravitational torques. The problem of stability is solved in the non-linear mode. Terms up to the fourth order inclusive are taken into consideration in the expansion of the Hamiltonian.

042.003   **An extended canonical perturbation method.**
        J. S. Choi, B. D. Tapley.
Celestial Mechanics, Vol. 7, 77 - 90 (1973).
        A procedure is described for extending the application of canonical perturbation theories, which have been applied previously to the study of conservative systems only, to the study of non-conservative dynamical systems. As examples to illustrate the application of the method, Duffing's equation, the equation for a linear oscillator with cubic damping and the van der Pol equation are solved using the Lie-Hori perturbation algorithm.

042.004   **The perturbed Ideal Resonance Problem.**
        A. H. Jupp.
Celestial Mechanics, Vol. 7, 91 - 106 (1973).

        A perturbed form of the Ideal Resonance Problem is investigated. The perturbation manifests itself by the inclusion of the term $\mu^2 f(x, y)$ in the Hamiltonian. The new problem possesses a single degree of freedom. With a suitable choice of variables, it is shown how a formal solution to this perturbed form of the Ideal Resonance Problem can be constructed, using the method of 'parallel' perturbations. Explicit formulae for x and y are obtained, as functions of time, which include the complete first-order contributions from the perturbing function f. The solution is restricted to the region of deep resonance, but those motions in the neighbourhood of the separatrix are excluded.

042.005   **A more general system for Poisson series manipulation.**   J. R. Cherniack.
Celestial Mechanics, Vol. 7, 107 - 121 (1973).
        We describe the design of a working Poisson series processor that is more general than others in use today. We try to show that the price of generality is worth paying in active research areas in celestial mechanics.

042.006   **Studies in the application of recurrence relations to special perturbation methods. II. Comparison of the Encke and Cowell methods of integration in the restricted three-body problem.**   P. E. Moran.
Celestial Mechanics, Vol. 7, 122 - 135 (1973).
        Using the rectangular equations of motion for the restricted three-body problem a comparison is made of the Encke and Cowell methods of integration. Each set of differential equations is integrated using Taylor series expansions where the coefficients of the powers of time are determined by recurrence relationships.

042.007   **Periodic solutions of the third sort for restricted problem of three bodies and their stability.**
Y. Kozai, H. Kinoshita.
Celestial Mechanics, Vol. 7, 156 - 176 (1973).

In this paper periodic solutions of the third sort for restricted problem of three bodies in the three-dimensional space are derived numerically by starting from generating solutions obtained by one of the authors (1969) and by increasing the mass-ratio of the two primaries stepwise from zero to about 1000 for 2:1, 3:2 and 6:1 cases of commensurable mean motions. Periodic solutions both for circular and elliptic orbits of the primaries are obtained. The stability of the periodic solutions for the 2:1 circular case is discussed and it is found that none of them is linearly stable.

042.008 Out-of-plane motion about libration points: nonlinearity and eccentricity effects.
T. A. Heppenheimer.
Celestial Mechanics, Vol. 7, 177 - 194 (1973).

Out-of-plane motion about libration points is studied within the framework of the elliptic restricted three-body problem. Nonlinear motion in the circular restricted problem is given to third order in the out-of-plane amplitude A_z by Jacobi elliptic functions. Linear motion in the elliptic problem is studied using Mathieu's and Hill's equations. Additional terms needed for a complete third-order theory are found using Lindsted's method. This theory is constructed for the case of collinear libration points; for the case of triangular points, a third-order nonlinear solution is given separately in terms of Jacobi elliptic functions.

042.009 Ignorable coordinates in the Ideal Resonance Problem. B. Garfinkel.
Celestial Mechanics, Vol. 7, 205 - 224 (1973).

If a dynamical system of N degrees of freedom is reduced to the Ideal Resonance Problem, the Hamiltonian takes the form $F = B(y) + 2\mu^2 A(y)\sin^2 x_1, \mu \ll 1$ [y the momentum-vector y_k, $k = 1, 2, ..., N; x_1$ the critical argument]. A first-order global solution given in Garfinkel et al. (1971) is completed here by the construction of the functions $x_k(t)$ for $k > 1$, derivable from the new Hamiltonian $F'(y')$ and the generator $S(x, y')$ of the von Zeipel canonical transformation used in the cited paper. The solution is subject to the normality condition, derived in a previous paper for $k = 1$, and extended here to $2 \leqslant k \leqslant N$. It is shown that the condition is satisfied in the problem of the critical inclination provided it is satisfied for $k = 1$.

042.010 Exact analytical solutions basic to a class of two-body orbits. E. E. Burniston, C. E. Siewert.
Celestial Mechanics, Vol. 7, 225 - 235 (1973).

The theory of complex variables is used to establish exact analytical solutions to a class of two-body problems. In view of Lambert's theorem, two points on the conic, the chord-distance between the two points, and the time interval considered are given, and subsequently the solutions for the semi-major axis required to define the orbit are developed and expressed ultimately in terms of elementary quadratures.

042.011 Studies in the application of recurrence relations to special perturbation methods. III. Non-singular differential equations for special perturbations.
A. E. Roy, P. E. Moran.
Celestial Mechanics, Vol. 7, 236 - 255 (1973).

A set of differential equations is derived that has a number of advantages in special perturbation work. In particular, the equations remain valid for all values of the orbital eccentricity and inclination including zero. They are therefore applicable to parabolic- and hyperbolic-type orbits as well as elliptic-type; a scheme for use when the orbit is rectilinear or nearly so is provided. The equations are also much simpler in form than the Lagrange planetary equations and the transformations of the osculating elements to and from the rectangular coordinates are straight forward.

042.012 Espace des phases dans le problème plan des trois corps. B. Elmabsout.
Comptes Rendus Acad. Sci. Paris, Sér. A, Vol. 276, 495 - 498 (1973).

Généralisation d'un résultat de R. Easton dans le problème plan des troits corps.

042.013 Outcomes of tidal evolution.
C. C. Counselman III.
Astrophys. Journ., Vol. 180, 307 - 314 (1973).

A simple criterion is derived which tells the outcome of tidal evolution, and a universal diagram is constructed which shows the actual path of evolution for a system with arbitrary masses and angular velocities. The possibility that certain retrograde satellites have been lost from the solar system is reexamined by means of this diagram.

042.014 On the calculation of high-order perturbations in the motion of celestial bodies. Yu. V. Plakhov.
Izv. vyssh. ucheb. zavedenij. Geod. i aehrofotos"emka, 1972, No. 5, p. 61 - 66. In Russian. – Abstr. in Referativ. Zhurn. 62. Issled. kosmich. prostranstva, 3.62.279 (1973).

042.015 Investigation of the motion of a body with variable mass in the gravitational field of many bodies by means of a regularizing variable. Ya. G. Magnaradze.
Soobshch. AN GruzSSR, Vol. 68, No. 2, p. 325 - 328 (1972). In Russian. – Abstr. in Referativ. Zhurn. 62. Issled. kosmich. prostranstva, 3.62.283 (1973).

042.016 Motion of a satellite in the equatorial plane of a spheroid. H. W. Milnes.
Celestial Mechanics, Vol. 7, 295 - 300 (1973).

An exact, closed-form solution of the problem of the motion of a satellite in the equatorial plane of an oblate body is obtained. It is shown that the classic formula for the motion of the perihelion is a first order approximation to the exact formula.

042.017 A note on a conjecture of Wintner and its disproof by Waldvogel. J. S. Griffith.
Celestial Mechanics, Vol. 7, 315 - 320 (1973).

A conjecture of Wintner, partially disproved by Waldvogel, is shown to be completely false. The relationship between work by Kurth and Waldvogel's results is demonstrated.

042.018 Homographic motions of a Newtonian system of point masses. I: Classification. P. Havas.
Celestial Mechanics, Vol. 7, 321 - 346 (1973).

A complete classification is established of the possible types of homographic motions of a Newtonian system of point masses interacting through two-body forces which are arbitrary functions of their mutual separations. Several types of motion are included which had not been considered previously, but not all possible rigid body motions. In an appendix, the problem of homographic motions of a relativistic system of interacting point masses is outlined. For a particular form of relativistic interactions, the existence of several types of such motions is established.

042.019 On the global solution in the resonance problem of Poincaré. A. H. Jupp.
Celestial Mechanics, Vol. 7, 347 - 355 (1973).

In §§ 201 and 211 of Les méthodes nouvelles de la mécanique céleste, where Poincaré describes the passage from shallow resonance to deep resonance, he asserts an erroneous conclusion. An alternative procedure, which admits secular terms into the determining function and introduces a regularizing function, is outlined. The latter method has been successfully applied to the Ideal Resonance Problem.

042.020 **Gaussian variational equations for osculating elements of an arbitrary separable reference orbit.**
J. P. Vinti.
Celestial Mechanics, Vol. 7, 367 - 375 (1973).

If a satellite orbit is described by means of osculating Jacobi α's and β's of a separable problem, the paper shows that a perturbing force F makes them vary according to $\dot{\alpha}_k = F \cdot \partial r/\partial \beta_k$, $\dot{\beta}_k = -F \cdot \partial r/\partial \alpha_k$, $(k = 1, 2, 3)$. Here r is the position vector of the satellite and F is any perturbing force, conservative or non-conservative. Applications to the Keplerian and spheroidal reference orbits are indicated.

042.021 **Linear change of variable in normal systems.**
J. D. Mulholland.
Celestial Mechanics, Vol. 7, 384 - 387 (1973).

Linear transformations of variable in differential correction processes cannot always be accomplished in the equations of condition, yet may be desirable to reduce high correlations. The means by which this may be accomplished in the system of normal equations is derived and the process reduced to a very simple computational algorithm.

042.022 **A note on the relations between true and eccentric anomalies in the two-body problem.**
R. Broucke, P. Cefola.
Celestial Mechanics, Vol. 7, 388 - 389 (1973).

Two simple formulas are given to relate the eccentric and true anomalies in the two-body problem. The problem of the maximum difference between these two angles is also considered.

042.023 **On the translatory-rotational motion of a spheroid, asymmetric with regard to the plane perpendicular to the rotation axis, in the gravitational field of a sphere.**
A. A. Nemo.
Trudy Kazan. Gorod. Astron. Obs., No. 38, p. 74 - 92 (1972). In Russian.

042.024 **Certain particular solutions of the Clairaut equation.**
P. Lanzano.
Astrophys. Space Sci., Vol. 20, 71 - 83 (1973).

A study has been carried out to ascertain conditions on the density distribution within a fluid, rotating planet in order that the deformation of its outer shell be expressible in terms of Bessel functions of the first kind and Gaussian hypergeometric series.

042.025 **On a new form of the main part of the disturbing function in the three-body problem.** C. A. Altavista.
Bol. As. Argentina Astron., No. 16, (see 012.007), p. 40 - 41 (1971).

042.026 **On the anti-focal anomaly.**
H. Kinoshita, H. Nakai.
Tokyo Astron. Obs., Report No. 62, Vol. 16, 532 - 544 (1973). In Japanese.

042.027 **Instability of the triangular Lagrangian points of the earth—moon system.** J. Taborda.
Revista Ci. mat., Sér. A, Vol. 2, 1 - 26 (1971).

The behaviour of a small particle near the triangular Lagrangian points of the earth—moon system is studied. The necessary corrections to the corresponding restricted 3-body problem solution are introduced in order to take into account the sun's gravitational action and the actual moon's orbit eccentricity. A resonance phenomenon is found, which forces the particle to abandon the Lagrangian point neighbourhood after a time interval depending on the ratio of the two finite masses of the restricted 3-body problem. An examination of the stability near the Lagrangian point is first carried out in the limit as the ratio of the two finite masses of the restricted

3-body problem goes to zero.

042.028 **The Levi-Civita problem and the Kustaanheimo-Stiefel transformation.** G. E. O. Giacaglia.
Ist. Lombardo Accad. Sci. Lett., Rend., Sez. A, Vol. 105, 950 - 965 (1971). — Abstr. in Zentralblatt Math. Grenzgebiete, Vol. 242, No. 70009 (1973).

042.029 **Families of periodic orbits in the restricted problem of three bodies connecting families of direct and retrograde orbits.** D. S. Schmidt.
SIAM Journ. Applied Math., Vol. 22, 27 - 37 (1972).

When $\mu = 0$ the planar restricted problem of three bodies reduces to the central force problem in a rotating coordinate system. There let the mean angular velocity be a rational number $n = k/l$. Then an ordinary family of periodic orbits bifurcates from the family of direct circular orbits. The family includes a collision orbit and terminates on the family of retrograde orbits. It is shown that if $k \neq l \pm 1$ then the above statements hold also for $\mu \neq 0$ for two natural families of symmetric orbits. If $k = l \pm 1$ it is shown that the family of direct orbits of the first kind breaks up and each part connects with one of the families of the second kind.

042.030 **Expansion of the perturbation function in case of a non-central [gravitational] field.**
E. I. Timoshkova.
Vestn. Leningr. un-ta, 1972, No. 19, p. 137 - 142. In Russian. Abstr. in Referativ. Zhurn. 51. Astron., 4.51.146; 62. Issled. kosmich. prostranstva, 4.62.319 (1973).

042.031 **Stationary motions in the generalized problem of three bodies.** M. Pascal.
Astronaut. Acta, Vol. 18, 127 - 137 (1973). In French.

We study the motion of three bodies, non-punctual, free in space, attracting one another in accordance with Newton's law. Two of the three bodies are spherical, the third is a revolving body of equatorially symmetric form. We week all motions of the system by which the centres of inertia of each of the three bodies follow a uniform circular motion, the axis of the revolving solid having a fixed position in relation to defined points of the two spherical bodies. The study may be compared to that made by V. T. Kondurar and T. K. Shinkarik in the case of the limited problem.

042.032 **Horseshoe and Trojan orbits associated with Jupiter and Saturn.** E. Everhart.
Astron. Journ., Vol. 78, 316 - 328 (1973).

Numerical experiments with orbits of small inclination in the Jupiter—Saturn region find many examples of orbits associated with either Jupiter or Saturn that are horseshoe-shaped in the frame revolving with the planet-sun line. In several cases, the Saturn horseshoe patterns persisted indefinitely, one lasting at least 368000 yrs. The libration periods varied between 700 and 2200 yrs, depending on the amplitude of the excursion. Longer periods, to 20000 yrs, were accompanied by variations in eccentricities and perihelia of objects in such orbits at Saturn's distance. The Jupiter horseshoes had shorter periods and seemed to be less stable than those associated with Saturn. Trojan orbits with either planet were found about as often as the horseshoe orbits.

042.033 **Periodic solutions of the restricted three-body problem including a large number of revolutions around the smaller body.** I. V. Kurcheeva.
Mekh. upravlyaem. dvizheniya i probl. kosmich. dinamiki. Leningrad, Leningr. un-t, 1972, p. 168 - 172. In Russian. Abstr. in Referativ. Zhurn. 62. Issled. kosmich. prostranstva, 4.62.310; 51. Astron., 5.51.107 (1973).

042.034 **Periodic orbits of the second kind.**

R. B. Barrar.
Math. Journ., Indiana Univ., Vol. 22, 33 - 41 (1972).

In Volume I, Sections 42–48 of his "Les méthodes nouvelles de la mécanique céleste" (Paris 1892), H. Poincaré proves the existence of periodic orbits of the second kind for the three body problem. However, P. Stäckel and A. Wintner found fault with the proof. In this paper the author gives a modern proof of Poincaré's result.

042.035 **Invarianti adiabatici di ordine n ed un'applicazione ad un problema di meccanica celeste.** G. Aymerich.
Mem. Soc. Astron. Italiana, Nuova Ser., Vol. 43, 629 - 633 (1973).

The technique of adiabatic invariants as proposed by Coffey is applied to the classic problem of two bodies of variable mass.

042.036 **On the calculation of high-order perturbations in the motion of celestial bodies.** Yu. V. Plakhov.
Izv. vyssh. ucheb. zavedenij. Geod. i aehrofotos"emka, 1972, No. 5, p. 61 - 66. In Russian. – Abstr. in Referativ. Zhurn. 51. Astron., 5.51.109 (1973).

042.037 **Resonance phenomena in the rotational motion of natural and artificial celestial bodies.** V. V. Beletskij.
Vtoraya Chetaevsk. konf. po analit. mekh., ustojchivosti dvizheniya i optimal'n. upr., 1973. Annotatsii dokl. Kazan', 1972, p. 3 - 4. In Russian. – Abstr. in Referativ. Zhurn. 62. Issled. kosmich. prostranstva, 5.62.337 (1973).

042.038 **Note on Lie transforms and Lagrange's implicit function theorem.** J. S. Griffith.
Celestial Mechanics, Vol. 7, 395 - 397 (1973).

Some results of Feagin and Gottlieb on Lagrange's implicit function theorem are shown to be derivable from Lie transform theory.

042.039 **A nonlinear oscillator analog of rigid body dynamics.** J. L. Junkins, I. D. Jacobson, J. N. Blanton.
Celestial Mechanics, Vol. 7, 398 - 407 (1973).

A rigorous nonlinear oscillator analog of the torque-free dynamics of a general rigid body has been established. Three associated phase planes were established and their trajectories analyzed. These results are important not only as a new device for qualitative motion analyses and visualization; they demonstrate a most interesting connection between the dynamics of rigid bodies and simpler nonlinear oscillatory phenomena.

042.040 **Third and fourth order resonances in Hamiltonian systems.** K. T. Alfriend, D. L. Richardson.
Celestial Mechanics, Vol. 7, 408 - 420 (1973).

The stability of the origin of an autonomous Hamiltonian system is investigated when the system possesses a third or fourth-order resonance. The condition for instability is then given in terms of the coefficients of the higher order terms in the Hamiltonian. The transfer of energy between modes is also investigated when a near-resonant condition exists.

042.041 **Global stability and the restricted 3-body problem.** L. D. Mullins, J. H. Bartlett.
Celestial Mechanics, Vol. 7, 421 - 437 (1973).

Global stability regions are found for class i orbits of the circular restricted 3-body problem for primary masses equal and Jacobi constant $K > 15.5$. As this constant decreases, the stability region shrinks extremely rapidly.

042.042 **Proof of a conjecture of E. Strömgren.** J. Henrard.
Celestial Mechanics, Vol. 7, 449 - 457 (1973).

The conjecture of Strömgren according to which, in the restricted problem, a class of doubly asymptotic orbits are limit members of families of periodic orbits is examined in the more general framework of analytic Hamiltonian system with two degrees of freedom. Sufficient conditions for the conjecture to become a theorem are established. Theses conditions amount to a transversality condition for the doubly asymptotic orbits and are likely to be verified in the cases considered in the literature of numerical explorations of the restricted problem.

042.043 **Quasi-periodic orbits about the translunar libration point.** R. W. Farquhar, A. A. Kamel.
Celestial Mechanics, Vol. 7, 458 - 473 (1973).

Analytical solutions for quasi-periodic orbits about the translunar libration point are obtained by using the method of Lindstedt-Poincaré and computerized algebraic manipulations. The solutions include the effects of nonlinearities, lunar orbital eccentricity, and the sun's gravitational field.

042.044 **The transformational behaviour of perturbation theories.** U. Kirchgraber.
Celestial Mechanics, Vol. 7, 474 - 494 (1973).

The transformational behaviour of Hori's noncanonical perturbation theory (Hori 1971) as well as that of the theory of Krylov-Bogoliubof-Mitropolsky is studied. An integration procedure of the perturbation equations is based on the transformation properties that have been established.

042.045 **Satellite vibration-rotation motions studied via canonical transformations.** R. Pringle, Jr.
Celestial Mechanics, Vol. 7, 495 - 518 (1973).

Coupled vibration-rotation motion of a satellite is considered using a perturbation theory based on the Lie transformation method. Short-period oscillating terms are removed from the Hamiltonian function. The transformed damping forces directly affect rotational variables which were not directly influenced in the original variables. Motions and stability are more easily studied in the new variables. A dual-spin spacecraft model is used as an example.

042.046 **The anisotropic Kepler problem in two dimensions.** M. C. Gutzwiller.
Journ. Math. Phys., *New York*, Vol. 14, 139 - 152 (1973). See Phys. Abstr., Vol. 76, No. 9450 (1973).

042.047 **About a general method of celestial mechanics.** J. L. Simovljević.
Prirodno–matematichki fakultet, Beograd – Glas Srpske akademije nauka i umetnosti 283, 63 - 78 (1972). – Abstr. in Bull. Sci. Yougoslavie, Sect. A, Vol. 18, 101 (1973).

042.048 **Study of some non-stationary problems of celestial mechanics by the method of transformations.** L. M. Berkovich, B. E. Gel'fgat.
Vtoraya Chetaevsk. konf. po analit. mekh., ustojchivosti dvizheniya i optimal'n. upr., 1973. Annotatsii dokl. Kazan', 1972, p. 4. In Russian. – Abstr. in Referativ. Zhurn. 51. Astron., 6.51.134 (1973).

042.049 **An atlas of surfaces of section for the restricted problem of three bodies.** W. H. Jefferys.
Prepared by Applied Mechanics Research Laboratory, The University of Texas, Austin, Texas, 26 + 304 pp. = Publ. Dep. Astron., Univ. Texas, *Austin*, Ser. II., Vol. 3, No. 6 (1971).

Surfaces of section, plotted in configuration space, have been computed for the motion of the massless particle in the restricted problem. Nine mass ratios and a wide range of Jacobi constants are represented, and information on about half of the approximately four thousand orbits computed is given. The plots enable one to see at a glance many of the qualitative features, including stability, of large numbers of orbits.

042.050   Functional analysis, formula manipulation, and
          satellite geodesy.   J. N. Hanson.
Journ. Geophys. Res., Vol. 78, 3260 - 3270 (1973).

042.051   On the impossibility of free precession of a liquid
          mass approaching the state of relative equilibrium.
V. A. Antonov.
Trudy Astron. Obs., *Leningrad*, Vol. 29 (= Uchenye Zapiski
Leningr. Un-ta, No. 363 = Seriya Matem. Nauk, vyp. (No.)
48), p. 150 - 152 (1973). In Russian.
   It is demonstrated that a heterogeneous gravitating
liquid mass in the state of relative equilibrium behaves like
a homogeneous system.

042.052   Une théorie des perturbations en variables angles-
          actions. Application au mouvement d'un solide
autour d'un point fixe. Précession-nutation.   F. Boigey.
Journ. Mécanique, Vol. 11, 521 - 543 (1972).

042.053   Families of periodic orbits in the restricted problem
          of three bodies connecting families of direct and
retrograde orbits.   D. S. Schmidt.
S.I.A.M. Journ. Applied Math., Vol. 22, 27 - 37 (1972).

042.054   General algorithm for solving the restricted three-
          body problem by the Hill–Brown method in an
"auto-analytic" system.   L. A. Moskovkina.
Materialy 3-j Nauch. konf. Tomsk. un-ta po mat. i mekh. Vyp.
(No.) 2. Tomsk, Tomsk. un-t, 1973, p. 91 - 92. In Russian.
Abstr. in Referativ. Zhurn. 51. Astron., 7.51.136 (1973).

042.055   On the classification of states in the three-body
          problem.   T. A. Agekyan, A. I. Martynova.
Vestn. Leningr. un-ta, 1973, No. 1, p. 122 - 126. In Russian.
Abstr. in Referativ. Zhurn. 51. Astron., 7.51.137 (1973).

042.056   Taking into account perturbations in orbit im-
          provement.   K. V. Kholshevnikov.
Materialy 3-j Nauch. konf. Tomsk. un-ta po mat. i mekh. Vyp.
(No.) 2. Tomsk, Tomsk. un-t, 1973, p. 105 - 106. In Russian.
Abstr. in Referativ. Zhurn. 51. Astron., 7.51.159; 62. Issled.
kosmich. prostranstva, 7.62.280 (1973).

   Collection of problems of celestial mechanics and
cosmodynamics. Manual for college students.
See Abstr. 003.028.

Ellipsoidal figures of equilibrium.
See Abstr. 003.040.

Copernicus und die Bezugssysteme in Physik und
Himmelsmechanik.   See Abstr. 004.010.

Kepler und die Begründung der Dynamik.
See Abstr. 004.013.

Planetary elements for 10 000 000 years.
See Abstr. 091.069.

L'éphéméride analytique lunaire – ALE.
See Abstr. 094.070.

Resonances and librations of some Apollo and Amor
asteroids with the earth.   See Abstr. 098.006.

Contribution to the dynamical study of the Jovian
Galilean system. I. The intermediate solution in the non resonant case.   See Abstr. 099.032.

Mass and position limits for an hypothetical tenth
planet of the solar system.   See Abstr. 101.016.

Evolutionary processes in the solar system.
See Abstr. 107.009.

Families of isoenergetic escapes and ejections in the
problem of three bodies.   See Abstr. 117.002.

Errata

042.901   Errata: 'The perturbed Ideal Resonance Problem'
          [Celestial Mechanics, Vol. 7, 91 - 106 (1973)].
A. H. Jupp.
Celestial Mechanics, Vol. 7, 390 (1973).

042.902   Erratum: 'Improved criteria for hyperbolic-elliptic
          motion in the general three-body problem' [Publ.
Astron. Soc. Japan, Vol. 24, 391 - 408 (1972)].   J. Yoshida.
Publ. Astron. Soc. Japan, Vol. 25, 285 (1973).

043 Astronomical Constants

043.001 **Earth—moon mass ratio from Mariner 9 radio tracking data.** S. K. Wong, S. J. Reinbold.
Nature, Vol. 241, 111 - 112 (1973).

The mass ratio was determined from range and Doppler data obtained over a period of 15 weeks (June 5 to September 15, 1971). We also show the statistics from the best determination.

043.002 **Un nuovo esperimento per la determinazione della costante universale della gravitazione.** A. Marussi.
Mem. Soc. Astron. Italiana, Nuova Ser., Vol. 43, 823 - 824 (1973).

Rotation of the earth from AD 1663—1972 and the constancy of G. See Abstr. 044.003.

On numerical methods of estimating composition and accuracy of space mission tracking data when determining astronomical constants. See Abstr. 051.024.

Approaches in dG/G-determinations. See Abstr. 066.080.

Determination of the mass of Saturn from the motion of Trojans. See Abstr. 100.019.

044 Time, Rotation of the Earth

044.001 **The earth's rotation and atmospheric circulation — I. Seasonal variations.** K. Lambeck, A. Cazenave.
Geophys. Journ. Roy. Astron. Soc., Vol. 32, 79 - 93 (1973).

Analysis of the observations of the variations in the earth's speed of rotation reveals the usual semi-annual, annual and long-period terms. In addition these observations indicate the existence of a biennial term whose behaviour is intermittent. The geophysical and meteorological excitation functions have been evaluated and they are in very good agreement with the observed astronomical variations.

044.002 **Results obtained with the Paris astrolabe. Time and latitude 1970.** G. Billaud.
Astron. Astrophys., Suppl. Ser., Vol. 9, 437 - 446 (1973).
In French.

The results are given of the 1970 observations with the astrolabes OPL no. 35, OPL no. 4 and APP. The results are in the FK4 system. To express them in a system (A4) corrected for the systematic and accidental errors of the FK4 catalogue, group corrections must be added.

044.003 **Rotation of the earth from AD 1663—1972 and the constancy of G.** L. V. Morrison.
Nature, Vol. 241, 519 - 520 (1973).—Letter.

044.004 **Meeresgezeiten bremsen die Erdrotation.** P. Brosche, J. Sündermann.
Umschau, 73. Jahrgang, p. 218 - 219 (1973).

A secular deceleration of the earth's rotation turned out by astronomical observations. Many times, it has been assumed that this effect is caused by the oceanic tides, which are acting, due to bottom friction, as a permanent force on the solid earth. This paper reports on a new qualitative and quantitative treatment of this problem.

044.005 **Time-scales.** D. H. Sadler.
Journ. Navigation, *London,* Vol. 26, 235 (1973).

044.006 **The leap-second of 31 December 1972.** D. H. Sadler.
Journ. Navigation, *London,* Vol. 26, 238 - 239 (1973).

044.007 **Effects of right ascension errors on the determination of Universal Time.** G. P. Pilnik.
Astron. Zhurn. Akad. Nauk SSSR, Vol. 50, 400 - 409 (1973). In Russian. English translation in Soviet Astron. AJ, Vol. 17, No. 2.

A technique of consideration of errors of the star coordinates is proposed for determination of Universal Time and the errors $\Delta\alpha_a$ are presented.

044.008 **Discontinuous change in earth's spin rate following great solar storm of August 1972.** J. Gribbin, S. Plagemann.
Nature, Vol. 243, 26 - 27 (1973).

In August 1972 a great disturbance occurred on the sun. We have found a discontinuous change in the length of day, and a change in the rate of change of the length of day (a glitch) immediately after that event.

044.009 **Semi-diurnal tidal effects in P. Z. T. observations.** N. P. J. O'Hora.
Phys. Earth Planet. Interiors, Vol. 7, 92 - 96 (1973).

Eleven years of time and latitude observations made with the Herstmonceux Photographic Zenith Tube have been separately analysed and the two solutions reveal with great clarity the magnitudes and phases of the variations associated with tidal activity.

044.010 **Theory of precession, nutation and rotational velocity of the deformable earth (II).** S. Takagi.
Publ. International Latitude Obs. Mizusawa, Vol. 8, No. 1, p. 31 - 43 (1971).

In a previous paper (1969) we derived a system of equations to be used for studying the rotation of the earth. After having finished this work, some defects were found in the system of equations to be used in deriving the numerical expressions. We will make some revisions and will correct some misprints in the fundamental system of equations in the above mentioned paper.

044.011 **Variations of the upper atmospheric motion and the rotational motion of the earth.**   N. Kikuchi.
Proc. International Latitude Obs. Mizusawa, No. 12, p. 83 - 94 (1972). In Japanese.
   The purpose of this note is to study preliminarily the structure of yearly variations in the three-dimensional propagation of energy through the atmosphere coupling with the rotation of the earth.

044.012 **Long-term variations of the ocean and the earth's rotation (1).**   E. Onodera.
Proc. International Latitude Obs. Mizusawa, No. 12, p. 95 - 105 (1972). In Japanese.
   The aim of this paper is to study the energetics of motions in the ocean for a long-period variation. We can expect some responses in the earth's rotation due to long-period motions in the ocean, if the oceanic motion over one year period is found in a large amplitude covering the global scale. Analyses of data of the surface and the interior to 800 m depth in the ocean are performed.

044.013 **Die Rotation der Erde und unsere Uhrzeit.**
H. Müller.
Orion Schaffhausen, 31. Jahrgang, p. 79 - 84 (1973).

044.014 **The change in the earth's rotational velocity and the barometric pressure field of the northern hemisphere of the earth.**
I. V. Maximov (*Maksimov*), B. A. Sleptsov-Shevlevich.
Dokl. Akad. Nauk SSSR, Ser. Mat. Fiz., Vol. 210, 79 - 81 (1973). In Russian.

044.015 **About the influence of irregular disturbances on the variations of phase velocity of propagation of VLW and of the angular velocity of the earth's rotation.**
A. G. Fleer, L. Ya. Vorobyev, G. P. Abramova.
Geomagn. Aeronom., Vol. 13, 458 - 462 (1973). In Russian.

044.016 **Determinazione delle scale di tempo coordinato $\overline{UTC}_{Me}$ e integrato UTC_{Int}.**
F. Chlistovsky, F. Mazzoleni.
Mem. Soc. Astron. Italiana, Nuova Ser., Vol. 43, 691 - 696 (1973).
   The authors have determined for the astrometric section of Merate the physical time scales, integrated and coordinated. After the description of the performance of these time scales a statistical study was carried out regarding the precision they can assure in maintaining the physical time.

044.017 **Determination of time (TU 1).**   N. A. Omelina.
Tsirk. Astron. Inst., *Tashkent*, Nos. 33 (380), 35 (382), 37 (384), 39 (386) (1972/73). In Russian. – 1971 November - 1972 June.

044.018 **On simultaneous determination of the coordinates of the pole and irregularity of the earth's rotation.**
A. A. Korsun, E. P. Fedorov.
Astron. Zhurn. Akad. Nauk SSSR, Vol. 50, 615 - 621 (1973). In Russian. English translation in Soviet Astron. AJ, Vol. 17, No. 3.

044.019 **Errors of the cosmic clock-face.**   G. P. Pilnik.
Astron. Zhurn. Akad. Nauk SSSR, Vol. 50, 622 - 631 (1973). In Russian. English translation in Soviet Astron.

AJ, Vol. 17, No. 3.
   The errors $\Delta_{\alpha\alpha}$ of FK3, N30, FK4 are calculated from the observations of the Time Service for every tenth of the hour. These errors are analysed and compared with the fundamental catalogues' data. A method of using the star coordinates errors for the observations of Universal Time is proposed.

044.020 **On methods of combined time and latitude determinations.**   A. N. Kuznetsov.
Izv. vyssh. ucheb. zavedenij. Geod. i aehrofotos"emka, 1972, No. 4, p. 99 - 102. In Russian. – Abstr. in Referativ. Zhurn. 51. Astron., 6.51.164; 52. Geod. i Aehros"emka, 6.52.133 (1973).

044.021 **Corrections to Czechoslovak time signals.**
V. Ptáček.
Říše hvězd, Vol. 54, 22, 37, 61, 76, 101, 119, 139, 157 (1973). In Czech. – 1972 October - 1973 May.

044.022 **Time and latitude service.**
Polish Acad. Sci., Astron. Latitude Station, Borowiec, Circ. No. 124 (1972), No. 125 (1973). – 1972 October - December, 1973 January - March.

044.023 **International Time and Latitude Service at the Tokyo Astronomical Observatory during 1972.**
Tokyo Astron. Obs., Time and Latitude Bull., Vol. 46, Nos. 9 - 12, p. 55 - 81 (1972). – 1972 September – December.

044.024 **Time Service.**   C. Moranzino (Editor).
Oss. Astron. Torino (Pino Torinese), Bull. No. 3, 3 pp. (1972); No. 4, 3 pp. (1973). – Results of the time determinations 1972 September - 1973 April.

044.025 **Time and latitude in South Africa.**   J. Hers.
IAU Colloquium No. 1, (see 012.019), p. 27 - 38 (1972).

044.026 **Time and latitude results of the Danjon astrolabe program at Santiago.**   C. Anguita, F. Noël.
IAU Colloquium No. 1, (see 012.019), p. 39 - 50 (1972).

044.027 **The Brazilian program of latitude and time.**
L. Muniz Barreto.
IAU Colloquium No. 1, (see 012.019), p. 51 - 52 (1972).

044.028 **Effects of the earth tides on the motion of the earth.**
J. Mateo.
IAU Colloquium No. 1, (see 012.019), p. 103 - 122 (1972).

044.029 **Astronomische Zeit- und Breitenbestimmungen. Empfangszeiten von Zeitsignalen.**
Edited by Deutsches Hydrographisches Institut, Hamburg. 1972 October - December, 6pp. (1973).

044.030 **Der Weg zur modernen Zeitmesstechnik.**
P. Bachmann.
Sternenbote, 16. Jahrgang, p. 42 - 53 (1973).

044.031 **On the influence of an external medium on the accuracy of time determination from astronomical observations.**   M. P. Mishchenko, A. V. Shiryaev.
Trudy Astron. Obs., *Leningrad*, Vol. 29 (= Uchenye Zapiski Leningr. Un-ta, No. 363 = Seriya Matem. Nauk, vyp. (No.) 48), p. 201 - 207 (1973). In Russian.
   The influence of the external medium on astronomical time determinations with a photoelectric transit instrument is studied. The necessity of precise registration of the micrometeorological factors characterizing that medium is stressed.

044.032 **Le système amélioré de temps universel coordonné**

qui sera mis en service le premier janvier 1972.
J. M. Steele.
Rev. Hydrogr. Internationale, Monaco, Vol. 49, No. 2, p. 145 - 156 (1972).

044.033 Détermination astronomique de l'heure et heures demi-définitives de réception des signaux horaires.
L. Webrová, V. Ptáček.
Acad. Tchécoslov. Sci. Inst. Astron., Station de l'Heure, Prague, Sér. 6, Nos. 3 - 4, 12 + 13 pp. (1973). — 1972 May - August.

044.034 Astronomische Zeit- und Breitenbestimmungen. Empfangszeiten von Zeitsignalen.
Deutsche Akad. Wiss. Berlin, Zentralinst. Phys. Erde, Bereich II (Geod., Gravimetrie), Potsdam, Abt. Geod. Astron., Jahrgang 1972, Nos. 1 - 4 (1972/73). — 1972 January - August.

044.035 Time Service of the Mizusawa Observatory. Bulletins, Vol. 15, No. 1 - 12, 1970.
S. Takagi, I. Okamoto, M. Aihara, K. Yokoyama, T. Hara, G. Murakami.
Edited by the International Latitude Observatory of Mizusawa, Mizusawa-Shi, Iwate-Ken, Japan. 3 + 53 pp. (1972).

This Bulletin contains the results of time service and astronomical observations made at the Mizusawa Observatory from 1st January to 31th December, 1970.

044.036 Détermination astronomique de l'heure et de la latitude.
Obs. Neuchâtel, Bull. (B), 1972 January - December. (1973).

044.037 L'heure astronomique définitive de l'Observatoire de Neuchâtel.
Obs. Neuchâtel, Bull. (D), 1972 January - December. (1973).

044.038 Methoden zum Vergleich und zur Verbreitung von Zeitskalen.
G. Becker, B. Fischer, P. Hetzel.

Kleinheubacher Berichte, Vol. 16, 5 - 37 (1973).

At first the performances of the most important time scales are described. Then methods of time dissemination and transmission are discussed considering the special conditions in the Federal Republic of Germany. A useful combination of the method of time transmission via DCF77 and via television pulses is proposed.

044.039 Universal time and coordinates of the pole; Emission time of time signals; Coordinated universal time; Independent local atomic time scales AT (i).
Bureau International de l'Heure, (B. I. H.), Paris, Circ. D 74 - D 80 (1973). — 1972 November — 1973 May.

044.040 Daily phase values.
U.S. Naval Obs., Washington, D.C., Time Service Publ., Ser. 4, Nos. 309 - 335 (1973). — 1973 January 3 — July 4.

044.041 Preliminary times and coordinates of the pole.
U.S. Naval Obs., Washington, D.C., Time Service Publ., Ser. 7, Nos. 262 - 288 (1973). — 1973 January 4 — July 5.

044.042 National Physical Research Laboratory, C.S.I.R. Time service notice.   J. Hers.
Monthly Notes Astron. Soc. Southern Africa, Vol. 32, 28 (1973).

Les équations personnelles des observateurs à l'instrument de passage.   See Abstr. 031.023.

Contribution of the circumzenithal Vúgtk 100/1000 mm to the latitude and time services of the Geodetical Observatory Pecný.   See Abstr. 045.001.

On the "dynamical variations" of latitude and time. See Abstr. 045.002.

045 Latitude Determination, Polar Motion

045.001 **Contribution of the circumzenithal Vúgtk 100/1000 mm to the latitude and time services of the Geodetical Observatory Pecný.** J. Rambousek.
Bull. Astron. Inst. Czechoslovakia, Vol. 24, 51 - 55 (1973).

Determination of the weight of the new type of Czech astrolabe deduced from observations from 1970.2 to 1971.9 proving its ability for time and latitude services.

045.002 **On the "dynamical variations" of latitude and time.** R. d'E. Atkinson.
Astron. Journ., Vol. 78, 147 - 151 = Publ. Goethe Link Obs., Indiana Univ., *Bloomington,* No. 148 (1973).

Observations for latitude variation, reduced with the existing ephemeris, are known to require in principle a subsequent correction for the so-called "dynamical variation of latitude", and this need has been verified observationally for the leading short-period term. A corresponding correction is in fact required in time determinations also, and the appropriate period has shown up in periodogram analyses of time residuals. Application of both these corrections in full is shown to be identically the same thing as correcting the standard nutation for the difference between the earth's instantaneous pole of rotation and its pole of figure. Some secondary consequences of the new position are briefly discussed. The new values of all the nutation terms affected are tabulated.

045.003 **A correction to the excitation of the Chandler wobble by earthquakes.** F. A. Dahlen.
Geophys. Journ., Vol. 32, 203 - 217 (1973).

The purpose of this note is to point out that there is a serious and substantial error in the published numerical results of Dahlen (1971). That paper was a theoretical investigation of the hypothesis advanced by Mansinha & Smylie (1967) that the Chandler wobble of the earth is excited by earthquake activity.

045.004 **Analysis of results of observations of refractional and circumzenithal pairs of International Latitude Stations.** V. I. Sergienko, S. A. Sergienko.
Astron. Zhurn. Akad. Nauk SSSR, Vol. 50, 410 - 417 (1973). In Russian. English translation in Soviet Astron. AJ, Vol. 17, No. 2.

From observations of circumzenithal (58233) and refractional (16613) pairs of six International Latitude Stations (ILS) for the period from 1900.0–1906.0, corrections for the inclination of air layers of equal density (IALED) are obtained. The reality of the obtained corrections for IALED is shown.

045.005 **Estimate of the influence of a seasonal redistribution of air masses on the motion of the earth's poles.** N. S. Sidorenkov, A. R. Chvykov.
Astron. Zhurn. Akad. Nauk SSSR, Vol. 50, 441 - 444 (1973). In Russian. English translation in Soviet Astron. AJ, Vol. 17, No. 2. – Short note.

045.006 **The results of latitude observations obtained with the Pulkovo ZTF-135 during 1948–1954. A spectral analysis of the 1948–1967 latitude series.** V. I. Sakharov.
Trudy Glav. Astron. Obs. Pulkovo, Ser. 2, Vol. 79, p. 5 - 61, with 2 supplements, p. 81 - 187 (1972). In Russian.

045.007 **The results of latitude observations made at Blagoveshchensk during 1959–1965.** G. S. Sheptunov.
Trudy Glav. Astron. Obs. Pulkovo, Ser. 2, Vol. 79, p. 62 - 78, with five supplements, p. 188 - 217 (1972). In Russian.

045.008 **Polar motion derived from the results of time and latitude observations. Part I. Preliminary results.** S. Takagi.
Publ. International Latitude Obs. Mizusawa, Vol. 8, No. 1, p. 1 - 29 (1971).

045.009 **The observation of the vertical deflection in Japan (6).** Prepared by The Geographical Survey Institute.
Bull. Geograph. Survey Inst., *Japan,* Vol. 18, Part 1, p. 25 - 78 (1972).

045.010 **On the comparison of the polar motion obtained by the Doppler method and the ILS.** C. Sugawa, C. Kakuta.
Proc. International Latitude Obs. Mizusawa, No. 12, p. 39 - 45 (1972).

The comparison of two independent polar coordinates obtained by the Doppler measurements and the ILS are investigated, where both astronomical measurements are made by using Fourier analyses during the period from 1967.0 to 1971.5. Differences in the polar coordinates between the two methods may be considered from the viewpoint of the correction of UT2–UTC and the variation of air-drag due to nonzonal air mass distribution.

045.011 **On the comparison of latitude observations with the visual and floating zenith telescopes by time series analysis.** T. Gotō, N. Kikuchi.
Proc. International Latitude Obs. Mizusawa, No. 12, p. 62 - 82 (1972). In Japanese.

A comparison of latitude observations by time series analysis with the VZT and the FZT has been made in order to find some clues for physically understanding meteorological effects on the structures of both instruments. The results are presented.

045.012 **Some effects of the local character in the atmospheric structure on astrometry (I).** T. Gotō.
Proc. International Latitude Obs. Mizusawa, No. 12, p. 116 - 161 (1972). In Japanese.

To extend previous studies of the meteorological effects on astrometry, it is attempted to take into consideration the dynamical features of the atmosphere for the time intervals of the astronomical observation, six hours at Mizusawa, during a clear night. This paper intends to show that variations of the individual observed values of latitude at Mizusawa Latitude Observatory in the time series during the clear night may be explained by those of the meteorological elements at Mizusawa and its near stations.

045.013 **On the relation between the ILS mean closing error and the ILS z term.** C. Sugawa, H. Ishii, G. Teleki.
Proc. International Latitude Obs. Mizusawa, No. 12, p. 176 - 199 (1972).

It is concluded that the annual term in z is caused by an error in the adopted value of the semi-annual nutation as pointed out by Wako and other groups of periodic terms in z are due to meteorological effects.

045.014 **New analysis of the motion of the earth's pole.** D. H. Douglass.
Bull. American Astron. Soc., Vol. 5, 293 (1973). – Abstr. AAS.

045.015 **Su di un semplice modello meccanico per spiegare l'eccitazione e lo smorzamento del moto di Chandler.** G. Bianchini, P. Zampirollo.

Mem. Soc. Astron. Italiana, Nuova Ser., Vol. 43, 645 - 654 (1973).

Preliminary and partial study of a deterministic mechanical model of the Chandler wobble excitation and damping has been done. Here we report the first results.of an analytical and numerical analysis which lead to think that the model may be more close to the reality than we initially hoped.

045.016 **La variazione secolare della latitudine e la tettonica dei blocchi.**   E. Proverbio, V. Quesada.
Mem. Soc. Astron. Italiana, Nuova Ser., Vol. 43, 789 - 792 (1973).

Utilizing the latitude observations of the International Latitude Station, the secular polar motion and the rotation rate of the Eurasia-American and Pacific-American plates have been calculated, using Le Pichon (1968) reconstruction. From this analysis it results that the relative rate between the plates are in excellent agreement with the plate tectonics theory.

045.017 **Determinabilità astronomica azimutale di latitudine geografica con giro-teodolite.**   V. Tomelleri.
Mem. Soc. Astron. Italiana, Nuova Ser., Vol. 43, 805 - 811 (1973).

We study the possibility of the geographic latitude determination of an astro-geodetic non polar station by means of the azimuth of known stars at maximum digressions. The technique is that of the homonymous method of observation with the gyro-theodolite. Coupling the observations of the stars at east and west, nearly at the same zenith distance, we can annul or at least reduce the effects of the systematic instrumental errors. We pressume that the mean square error of such a latitude determination is about 1''.

045.018 **Results of a new reduction of observations of the Tashkent latitude series 1895–1896.**
G. M. Kaganovskij.
Tsirk. Astron. Inst., *Tashkent*, No. 34 (381), p. 1 - 20 (1972). In Russian.

045.019 **On the influence of inclined air layers of equal density on the accuracy of latitude observations at Pulkovo and Washington.**   V. G. Boltovskij, G. S. Tyuterev.
Materialy 3-j Nauch. konf. Tomsk. un-ta po mat. i mekh. Vyp. (No.) 2. Tomsk, Tomsk. un-t, 1973, p. 87. In Russian.
Abstr. in Referativ. Zhurn. 51. Astron., 5.51.217 (1973).

045.020 **The variation of the mean latitude of the Hamburg PZT.** H. Enslin.
IAU Colloquium No. 1, (see 012.019), p. 71 - 76 (1972).

045.021 **On variations in the Chandler frequency.**
H. J. M. Abraham, J. N. Boots.
IAU Colloquium No. 1, (see 012.019), p. 77 - 85 (1972).

045.022 **Secular motion of the pole and changes in geographical coordinates.**   W. Markowitz.
IAU Colloquium No. 1, (see 012.019), p. 87 - 92 (1972).

045.023 **Non-polar variation of latitude.**   S. Yumi.
IAU Colloquium No. 1, (see 012.019), p. 93 - 101 (1972).

045.024 **Uniformity of the system in latitude observation.**
S. Yumi.
IAU Colloquium No. 1, (see 012.019), p. 123 - 124 (1972).

045.025 **Reduction of astronomical latitudes and longitudes 1922–1948 into the FK4 and CIO systems.**
V. R. Ölander.
Publ. Finnish Geod. Inst., *Helsinki*, No. 73, 40 pp. (1972).

045.026 **Plate motion and the secular shift of the mean pole.**
H.-S. Liu, L. Carpenter, R. W. Agreen.
Goddard Space Flight Center, Greenbelt, Maryland, Prepr. X-592-73-162, 5 + 12 pp. (1973).
Numerical results show that the secular motion of the mean pole is $0.''0002$ year $^{-1}$ in the direction of 67°W.

045.027 **Monthly Notes of the International Polar Motion Service.**
IPMS Monthly Notes, International Latitude Obs. Mizusawa (Japan). 1972 Nos. 11–12, p. 87 - 103 (1973), 1973 Nos. 1–4, p. 1 - 32 (1973). – Announces the values of latitudes observed at the collaborating stations during 1972 November – 1973 April.

045.028 **Gravitational effects on the vertical observed by the Ottawa PZT** (*Photographic Zenith Tube*).
E. G. Woolsey.
Canadian Journ. Earth Sci., Vol. 10, 379 - 383 = Contr. Earth Phys. Branch, Ottawa, No. 433 (1973).
The deflection of the vertical due to the combined gravitational effect of the sun and moon appears in observations made with the Ottawa PZT during the years 1962–1970. The Love number $(1 + k - l)$ was determined as 1.3 ± 0.9 from latitude readings, and 0.9 ± 0.8 from longitude, where the uncertainties quoted in both cases are the 95% confidence limits. The coefficient of correlation for longitude and for latitude readings was 0.7.

045.029 **Defined Doppler satellite determinations of the earth's polar motion.**
L. K. Beuglass, R. J. Anderle.
Geophys. Monograph 15, (see 012.025), p. 181 - 186 (1972).

045.030 **Geodetic studies by laser ranging to satellites.**
D. E. Smith, R. Kolenkiewicz, P. J. Dunn.
Geophys. Monograph 15, (see 012.025), p. 187 - 196 (1972).

045.031 **Pole position for 1972 based on Doppler satellite observations.**   R. J. Anderle.
U.S. Naval Weapons Lab., Dahlgren, Virginia, NWL Techn. Rep. TR-2952, 4 + 6 + A3 + B5 + C2 + D18 + E26 + F8 pp. (1973).
The position of the earth's spin axis with respect to the crust was computed on the basis of Doppler satellite observations in 1972 for the fourth consecutive year. The standard error of the determination based on five days of observations was as low as 20 cm during periods when data from 19 stations on three satellites were used in computations. In addition, computation of satellite orbits was performed for isolated time spans dating back to 1964, and changes in latitude residuals were computed for each station for the period 1964–1972; the changes were consistent with the standard error of 10–20 cm/year.

045.032 **Coordonnées du pôle instantané rapportées à l'origine conventionnelle internationale et corrections de longitude TU1–TU0, à 0h TU.**
Bureau International de l'Heure, (B. I. H.), Paris, Circ. B/C, Nos. 201 - 207 (1973). – Valeurs interpolées et extrapolées.

Results obtained with the Paris astrolabe. Time and latitude 1970.   See Abstr. 044.002.

On methods of combined time and latitude determinations.   See Abstr. 044.020.

Time and latitude service.   See Abstr. 044.022.

International Time and Latitude Service at the

Tokyo Astronomical Observatory during 1972.
See Abstr. 044.023.

Time and latitude in South Africa.
See Abstr. 044.025.

Time and latitude results of the Danjon astrolabe
program at Santiago.   See Abstr. 044.026.

The Brazilian program of latitude and time.
See Abstr. 044.027.

Astronomische Zeit- und Breitenbestimmungen.
Empfangszeiten von Zeitsignalen.   See Abstr. 044.029.

Astronomische Zeit- und Breitenbestimmungen.
Empfangszeiten von Zeitsignalen.   See Abstr. 044.034.

Détermination astronomique de l'heure et de la
latitude.   See Abstr. 044.036.

Preliminary times and coordinates of the pole.
See Abstr. 044.041.

Residual deformation of real earth models with application to the Chandler wobble.   See Abstr. 081.009.

Quadrupolar analysis of storage and release of
elastic energy in the earth.   See Abstr. 081.018.

Determination of the free diurnal nutation of the
earth based on latitude observations in Washington for 1915–
1940.   See Abstr. 081.028.

046   Geodetic Astronomy, Navigation

046.001   **Transmission of plumb-line deflections by ground
observations in polarized light.**   M. T. Prilepin.
Bull. Géod., Nouvelle Sér., Année 1973, No. 107, p. 35 - 41.
The problem of transferring astrogeodetic plumb-line
deflections along the net by measuring angle components between the plumb-line, produced by the direction of electric
vector of a linearly polarized light wave is discussed.

046.002   **Conversion of geodetic coordinates from national
geodetic reference systems into standard earth**
system.   J. C. Bhattacharji.
Bull. Géod., Nouvelle Sér., Année 1973, No. 107, p. 65 - 72.
The method of converting geodetic coordinates from a
national geodetic reference system into the standard earth on
having known the geodetic coordinates of at least one station
in common with the considered systems, is described in detail.

046.003   **Astronomic-geodesic determinations along the base
Catania–Tromsö.**   G. Folloni, M. Unguendoli.
Boll. Geod. Sci. Affini, Anno 31, p. 501 - 529 (1972).
In Italian.

046.004   **Aquino's short-method tables.**
C. H. Cotter.
Journ. Navigation, *London,* Vol. 26, 152 - 166 (1973).

046.005   **Henry Raper's spherical traverse table.**
C. H. Cotter.
Journ. Navigation, *London,* Vol. 26, 240 - 244 (1973).

046.006   **Application of the variable increment algorithm for
solving problems of nautical astronomy.**
B. L. Bulgakov, V. D. Luginin.
Sudovozhdenie. Nauch.-tekhn. sb., 1972, vyp. (No.) 12, p. 81 -
90. In Russian. – Abstr. in Referativ. Zhurn. 51. Astron.,
3.51.206 (1973).

046.007   **Some remarks on the determination of the compass
correction by astronomical methods.**
V. G. Vasil'ev.
Sudovozhdenie. Nauch.-tekhn. sb., 1972, vyp. (No.) 12, p. 90 -
92. In Russian. – Abstr. in Referativ. Zhurn. 51. Astron.,
3.51.207 (1973).

046.008   **On the feasibility of constructing a direction finder
with a higher accuracy of the compass bearing reading. The results of testing an experimental specimen.**
M. I. Gavryuk.
Sudovozhdenie. Nauch.-tekhn. sb., 1972, vyp. (No.) 12, p. 11 -
16. In Russian. – Abstr. in Referativ. Zhurn. 51. Astron.,
3.51.244 (1973).

046.009   **Determination of the geodesic azimuth of direction
from observations of stars in the prime vertical.**
(Method of observations and their reduction).   M. A. Vajsov.
Trudy Kazan. Gorod. Astron. Obs., No. 38, p. 116 - 133
(1972). In Russian.

046.010   **Program of bright stars and auxiliary tables for
determining the azimuth from observations of star
transits in the prime vertical.**   M. A. Vajsov.
Trudy Kazan. Gorod. Astron. Obs., No. 38, p. 134 - 190
(1972). In Russian.

046.011   **L'algorithme des calculs de la longueur d'une ligne
géodésique.**   Z. Zorski.
Geodezja i Kartografia, Vol. 22, 3 - 13 (1973). In Polish.
L'auteur décrit en détail la méthode du calcul de la
longueur d'une géodésique. Cette méthode est élaborée pour
les ordinateurs électroniques. Les formules établies sont
justifiées par l'analyse numérique et des diagrammes.

046.012   **Recherches sur la configuration d'un réseau de
trilateration spatiale avec mesures laser sur satellite.**

M. Caputo, G. Folloni, L. Pieri, D. Postpischl, M. Unguendoli.
Bull. Géod., Nouvelle Sér., Année 1973, No. 108, p. 125 - 134.

We consider the method of space trilateration with only satellite laser measurements, in order to investigate the effect of the distribution of the set of points on the earth and of satellite points on the errors of the computed coordinates of the terrestrial points.

046.013 A note on computation of geodetic coordinates from geocentric (Cartesian) coordinates.
M. K. Paul.
Bull. Géod., Nouvelle Sér., Année 1973, No. 108, p. 135 - 139.

046.014 Geodetic numerical and statistical analysis of data.
D. M. J. Fubara.
Bull. Géod., Nouvelle Sér., Année 1973, No. 108, p. 157 - 186.

046.015 Four dimensional studies in earth space.
R. S. Mather.
Bull. Géod., Nouvelle Sér., Année 1973, No. 108, p. 187 - 209.

A system of reference which is directly related to observations, is proposed for four dimensional studies in earth space. Earth space is defined as the euclidian space which has the same galactic and rotational motion as the earth.

046.016 Aufgaben der theoretischen Geodäsie.
H. Moritz.
Österreich. Zeitschr. Vermessungswesen, 60. Jahrgang, p. 80 - 85 (1972).

046.017 On the application of the method of equal heights for the determination of the astronomical azimuth.
V. A. Kovalenko.
Geod., kartogr. i aehrofotos"emka. Resp. mezhved. nauch.-tekhn. sb., 1972, vyp. (No.) 16, p. 48 - 52. In Russian.
Abstr. in Referativ. Zhurn. 51. Astron., 6.51.166; 52. Geod. i Aehros"emka, 6.52.135 (1973).

046.018 Interferometer baselines and poles obtained by linking radio observatories.
T. E. Corbin, S. J. Goldstein, Jr.
Publ. Leander McCormick Obs., Univ. Virginia, *Charlottesville*, Vol. 11, (Part 27), 211 - 230 (1972).

We present here the baseline constants needed to calculate the delay and its derivatives for every pair of radio observatories listed in the American Ephemeris and Nautical Almanac for 1972. The results are presented separately for each station (observatory).

046.019 Geodetic analyses through numerical integration.
R. J. Anderle.
Separate print Naval Weapons Lab., Dahlgren, Virginia. 3 + 19 pp. (1973). – Prepared for presentation at the First International Symposium on the Use of Artificial Satellites for Geodesy and Geodynamics in Athens, Greece, May 1973.

Applying numerical integration to analysis of Doppler observations yielded geodetic coordinates to 1 m accuracy, pole position accurate to 20 cm over five days, and variations in station position over 10 years consistent with the standard error of the variation of 10 to 20 cm/year in many instances.

046.020 Geodetic and astronomical coordinates of the Cerro Tololo Inter-American Observatory.
R. S. Harrington, B. M. Blanco, V. M. Blanco.
Cerro Tololo Inter-American Obs., Contr. No. 126, 18 pp. (1973).

046.021 50 years astronomical-geodetic network of the USSR.
A. Z. Sazonov.
Geod. i kartografiya, 1972, No. 12, p. 13 - 17. In Russian.
Abstr. in Referativ. Zhurn. 52. Geod. i Aehros"emka, 6.52.67 (1973).

046.022 On the equations of condition of astronomical latitudes, longitudes and azimuths. A. E. Filippov.
Geod., kartogr. i aehrofotos"emka. Resp. mezhved. nauch.-tekhn. sb., 1972, vyp. (No.) 16, p. 105 - 108. In Russian.

046.023 Satellite tracking camera for geodetic purposes.
J. Osório, T. Torrão.
Publ. Obs. Astron. "Prof. Manuel de Barros", Faculdade Ciências do Porto, No. 27, 22 + 16 pp. (1972).

046.024 Position from gravity. R. S. Mather.
Goddard Space Flight Center, Greenbelt, Maryland, Prepr. X-592-73-164, 6 + 58 pp. (1973).

046.025 Die Orthogonaltriangulation. H. Rehse.
Vermessungstechnik, 20. Jahrgang, p. 298 - 300 = Deutsche Akad. Wiss. Berlin, Zentralinst. Phys. Erde, Potsdam, Mitt. No. 239 (1972).

046.026 Geometric accuracy obtainable from simultaneous range measurements to satellites. L. Aardoom.
Geophys. Monograph 15, (see 012.025), p. 9 - 18 (1972).

046.027 Survey improvement and calibration analysis for the Air Force Eastern Test Range with Geos C.
N. Bush.
Geophys. Monograph 15, (see 012.025), p. 83 - 91 (1972).

046.028 Determination of gravity anomalies by satellite geodesy. K. Arnold.
Geophys. Monograph 15, (see 012.025), p. 177 - 179 (1972).

046.029 On the determination of geographical coordinates of a site from simultaneous observations of the sun and Venus during the eight-year period of the planet's visibility. G. K. Prikhod'ko.
Probl. Arktiki Antarktiki. Vyp. (No.) 41. Leningrad, Gidrometeoizdat, 1973, p. 103. In Russian. – Abstr. in Referativ. Zhurn. 51. Astron., 7.51.183; 52. Geod. Aehros"emka, 7.52.146 (1973).

Functional analysis, formula manipulation, and satellite geodesy. See Abstr. 042.050.

On simultaneous determination of the coordinates of the pole and irregularity of the earth's rotation. See Abstr. 044.018.

Reduction of astronomical latitudes and longitudes 1922–1948 into the FK4 and CIO systems. See Abstr. 045.025.

Gravitational effects on the vertical observed by the Ottawa PZT. See Abstr. 045.028.

Representation of the earth potential by buried masses. See Abstr. 081.033.

Contributions to the theory of atmospheric refraction. Part II. Refraction corrections in satellite geodesy. See Abstr. 082.020.

047 Ephemerides, Almanacs, Calendars

047.001 **Events of 1973 in the graphic time table.**
Sky Telescope, Vol. 45, 33 - 35 (1973).

047.002 **Almanacco Astronomico della Rivista Coelum per l'anno 1973.**
Coelum Suppl., Vol. 40, Fasc. 11-12 [Osservatorio Astronomico Universitario, Bologna], 28 + 40 pp. Price L. 2000 (1972).

047.003 **Kalender, Datum, Uhrzeit, Wochennumerierung.**
G. Zimmermann.
SuW, Vol. 12, 16 - 17 (1973).

047.004 **The Astronomical Ephemeris for the year 1974.**
Issued by Her Majesty's Nautical Almanac Office, London; Nautical Almanac Office, United States, Naval Observatory, Washington. Her Majesty's Stationery Office, London. 8 + 562 pp. Price £ 3.50 net (1973).

047.005 **Astronomical phenomena for the year 1975.**
Issued by the Nautical Almanac Office, United States Naval Observatory.
U.S. Government Printing Office, Washington, D.C. 66 pp. Price 80 cents (1971/73).

047.006 **Astronomiskais Kalendārs 1973. Gadam.**
M. Dīriķis (Editor).
Latvijas PSR Zinātņu Akadēmija, Radioastrofizikas Observatorija, Vissavienības Astronomijas un Ģeodēzijas Biedrības Latvijas Nodaļa; Izdevniecība "Zinātne", Riga. 188 pp. Price 36 Kop. (1972). In Lettish.

047.007 **The Air Almanac 1973, September - December.**
Her Majesty's Stationery Office, London; United States Naval Observatory, Washington, 276 + A84 + F4 pp. Price £ 2.00 (1973).

047.008 **The Star Almanac for Land Surveyors for the Year 1974.**
Prepared by *H. M. Nautical Almanac Office,* published by Order of *The Science Research Council.* Her Majesty's Stationery Office, London, 16 + 76 pp. Price 45p. (1973).

047.009 **Elementi astronomici per il calendario dell'anno 1973 calcolati dall'Osservatorio Astrofisico di Arcetri–Firenze.**
Boll. Geod. Sci. Affini, Anno 32, p. 59 - 73 (1973).

047.010 **Astronomical Yearbook 1973.**
Published by the Astronomical Society of Victoria, Melbourne. 39 pp. Price 70 c (1972).

047.011 **Astronomical Yearbook, 1973.**
Academy of Sciences of the Georgian SSR. Abastumanian Astrophys. Observ., Metsniereba, Tbilisi. (1972). In Georgian.

047.012 **1973 Polaris Almanac for azimuth determination.**
Pub. No. 685. Published by the Hydrographic Office of Japan, Tokyo. 9 pp. (1972). In Japanese.

047.013 **Almanaque Nautico, 1974.**
Published by Instituto y Observatorio de Marina, San Fernando (Cádiz). Printed in Spain by Imprenta del Observatorio de Marina, San Fernando. 416 + 30 pp. (1973).

047.014 **Ephémérides Nautiques pour l'an 1974.** Ouvrage publié par le Bureau des Longitudes spécialement à l'usage des marins.
Gauthier-Villars Editeur, Paris. 479 pp. (1973).

047.015 **Kalendarzyk astronomiczny na 1973 rok.**
G. Sitarski.
Appendix to Urania Kraków, Vol. 44, No. 1, 12 pp. (1973).

047.016 **Philippine Astronomical Handbook 1973.**
Prepared under the supervision of S. V. Inciong.
Republic of the Philippines – Department of National Defense – Weather Bureau, Quezon City. 11 + 60 pp. (1972).

047.017 **Tables of Sunrise, Sunset, Twilight, Moonrise and Moonset 1973.**
Prepared under the supervision of S. V. Inciong.
Republic of the Philippines – Department of Commerce and Industry – Weather Bureau, Quezon City, 11 + 57 pp. (1972).

047.018 **Almanac for Geodetic Engineers 1973.**
Prepared under the supervision of S. V. Inciong.
Republic of the Philippines – Department of Commerce and Industry – Weather Bureau, Quezon City. 10 + 30 pp. (1972).

047.019 **Change of the almanac and the old almanac.**
M. Uchida.
Astron. Herald, *(Japan),* Vol. 66, 7 - 11 (1973). In Japanese.
The author gives a retrospect of the history of almanacs in Japan. It is 100 years since the Japanese adopted the solar almanac in place of the lunar one hitherto used. The author describes the order of the Japanese Government at that time surveying the various aspects of the movements.

047.020 **Anuário para 1973 publicado pelo Observatório Nacional Rio de Janeiro.**
Compiled and published jointly with the Observatório de São Paulo.
Ministério da Educação e Cultura, Rio de Janeiro, Ano 89, 11 + 112 + 181* + 2 pp. (1972). – This 'Anuário' is identical with the 'Anuário do Observatorio de S. Paulo'.

047.021 **Astronomical Yearbook of the USSR for the year 1976.** V. K. Abalakin (Editor).
Institut Teoreticheskoj Astronomii Akademii Nauk SSSR. Izdatel'stvo "Nauka", Leningradskoe Otdelenie, Leningrad. 718 pp. Price 7 Rbl. 25 Kop. (1973). In Russian.

047.022 **Supplement to the Astronomical Yearbook of the USSR for the year 1976.**
Akademiya Nauk SSSR, Institut Teoreticheskoj Astronomii. Izdatel'stvo "Nauka", Leningradskoe Otdelenie, Leningrad. 91 pp. (1973). In Russian.

047.023 **The American Ephemeris and Nautical Almanac for the year 1974.**
Issued by Nautical Almanac Office, United States Naval Observatory, Washington; Her Majesty's Nautical Almanac Office, Royal Greenwich Observatory, London. U.S. Government Printing Office, Washington. 8 + 562 pp. Price $ 7.30 (1972).

Errata

047.901 **Erratum: 'The sun at noon–1973' [Handbook British Astron. Ass. 1973, p. 8 - 9 (1972)].**
Journ. British Astron. Ass., Vol. 83, 199 - 200 (1973).

Space Research

051 Extraterrestrial Research, Spaceflight Related to Astronomy

051.001 **Review of the Apollo lunar exploration programme.**
B. Kołaczek.
Postępy Astron., Vol. 21, 65 - 73 (1973). In Polish.
The article presents a short review of the Apollo 11−16 missions results: lunar surface and atmosphere, observations of X and γ radiation, solar corona, zodiacal light and meteorites.

051.002 **A strategy for investigation of the outer solar system.** Outer planets, their satellites, and particles and fields at great distances from the sun.
The science advisory group: G. Münch, D. M. Hunten, A. J. Kliore, J. S. Lewis, M. B. McElroy, N. W. Spencer, P. H. Stone, G. W. Wetherill, A. G. W. Cameron, W. B. Hubbard, B. C. Murray, S. J. Peale, J. A. Van Allen, W. I. Axford, S. Gulkis, C. F. Kennel, M. D. Montgomery, E. N. Parker, C. P. Sonett, R. G. Stone, J. H. Trainor, D. G. Rea, J. E. Long, B. D. Padrick
Space Sci. Rev., Vol. 14, 347 - 362 (1973).
Following some general introductory remark a brief sketch is given of the development and current status of scientific missions to the inner planets by the U.S. and the U.S.S.R. With this perspective, the development of the U.S. program for investigation of the outer solar system is described. The mission type and sequence required to conduct a systematic exploration of the outer solar system has been developed. Technological rationales for the suggested missions are discussed in general terms.

051.003 **From the first flight of man into space to the profession "astronaut-research worker".**
E. V. Khrunov, L. S. Khachatur'yants.
Zemlya i Vselennaya, 1973, No. 2, p. 3 - 6. In Russian.

051.004 **Two possible objects in search of highly developed civilizations.** G. I. Pokrovsky.
Priroda, No. 6.73, p. 97 - 98 (1973). In Russian.

051.005 **Gravitational waves − a means of communication with highly developed civilizations.** V. B. Kudrin.
Priroda, No. 6.73, p. 98 - 99 (1973). In Russian.

051.006 **Superrelativistic interstellar flight.** J. W. Morgan.
Spaceflight, Vol. 15, 252 - 254 (1973).

051.007 **Stardrift. A navigational system for relativistic interstellar flight.** A. T. Lawton.
Spaceflight, Vol. 15, 256 - 261 (1973).

051.008 **Soviet−French cooperation in the field of space investigations.** Yu. I. Gal'perin, L. A. Vedeshin.
Vestn. AN SSSR, 1972, No. 11, p. 84 - 92. In Russian.

051.009 **Dynamics of a stratospheric observatory.**
A. M. Danilov, L. Z. Dul'kin, A. S. Zemlyakov, V. M. Matrosov, V. A. Strezhnev.
Vtoraya Chetaevsk. konf. po analit. mekh., ustojchivosti dvizheniya i optimal'n. upr., 1973. Annotatsii dokl. Kazan', 1972, p. 7. In Russian. − Abstr. in Referativ. Zhurn. 62. Issled. kosmich. prostranstva, 5.62.326 (1973).

051.010 **Astronautics in the year 1972.**

M. Grün, P. Koubský.
Říše hvězd, Vol. 54, 109 - 115 (1973). In Czech.

051.011 **Planetary astronomy objectives.** C. Sagan.
Publ. American Astronaut. Soc., Sci. Techn. Ser., Vol. 28, (see 012.018), p. 3 - 6 (1972).

051.012 **Stellar astronomy objectives.** G. W. Preston.
Publ. American Astronaut. Soc., Sci. Techn. Ser., Vol. 28, (see 012.018), p. 95 - 104 (1972).

051.013 **Extra-galactic astronomy objectives.** L. R. Doherty.
Publ. American Astronaut. Soc., Sci. Techn. Ser., Vol. 28, (see 012.018), p. 105 - 114 (1972).

051.014 **Future infrared space astronomical instruments.**
W. F. Hoffmann.
Publ. American Astronaut. Soc., Sci. Techn. Ser., Vol. 28, (see 012.018), p. 135 - 144 (1972).

051.015 **Future unmanned astronomy spacecraft.**
F. P. Simmons.
Publ. American Astronaut. Soc., Sci. Techn. Ser., Vol. 28, (see 012.018), p. 145 - 185 (1972).

051.016 **Unusual objects and high energy astronomy.**
J. P. Ostriker.
Publ. American Astronaut. Soc., Sci. Techn. Ser., Vol. 28, (see 012.018), p. 189 - 196 (1972).

051.017 **Selection of astronomy experiments for space.**
F. L. Whipple.
Publ. American Astronaut. Soc., Sci. Techn. Ser., Vol. 28, (see 012.018), p. 381 - 384 (1972).

051.018 **Manned versus unmanned space-based astronomy.**
K. G. Henize.
Publ. American Astronaut. Soc., Sci. Techn. Ser., Vol. 28, (see 012.018), p. 385 - 391 (1972).

051.019 **Strategies for space-based astronomy.**
J. E. Naugle.
Publ. American Astronaut. Soc., Sci. Techn. Ser., Vol. 28, (see 012.018), p. 393 - 395 (1972).

051.020 **Rocket astronomy.** H. Friedman.
Ann. New York Acad. Sci., Vol. 198, (see 012.020), 267 - 273 (1972).

051.021 **Organizational project of intensifying research in the field of space physics in Poland.**
S. Grzędzielski, Z. Kłos.
Postępy Astron.,Vol. 21, 156 - 159 (1973). In Polish.

051.022 **Astronautica.**
Coelum, Vol. 41, 30 - 32, 70 - 72, 110 - 112 (1973).

051.023 **Long term prospects of the development of large astronomical space stations.** W. G. Kurt.
Wiss. Zeitschr. Techn. Hochschule Karl-Marx-Stadt, No. 3, p. 325 - 328 (1972). In German.

051.024 On numerical methods of estimating composition and accuracy of space mission tracking data when determining astronomical constants.   A. A. Shiryaev.
Materialy 3-j Nauch. konf. Tomsk. un-ta po mat. i mekh. Vyp. (No.) 2. Tomsk, Tomsk. un-t, 1973, p. 106. In Russian.
Abstr. in Referativ. Zhurn. 51. Astron., 7.51.167; 62. Issled. kosmich. prostranstva, 7.62.281 (1973).

051.025 Space report.
Spaceflight, Vol. 15, 18 - 23, 30, 60 - 68, 116 - 118, 148 - 150, 187 - 191, 195, 227 - 231, 272 - 274 (1973). – Telescopes in orbit, p. 18 - 19; Future of physics, p. 19 - 20; Solar storm defined, p. 20; Joint planetary exploration, p. 20 - 21; Surface of Mars, p. 22; Moon data exchanged, p. 22; Venera 8 results, p. 23; U. K. – Canadian study of quasars, p. 30; Rescue from Skylab, p. 60 - 61; Orbital furnace, p. 61 - 63; Satellite triangulation, p. 64; Space navigation, p. 64 - 65; ESRO 4 in orbit, p. 65 - 66; Laboratory-produced 'lunar rock', p. 66; Study of Migea meteorite, p. 66 - 67; Joint flight to Mars? p. 68; Life forming molecules in space, p. 117; Pioneer 10 through asteroid belt, p. 118; Apollo mission records, p. 148; Lunokhod 2, p. 149 - 150; Mascon award, p. 188 - 189; Mapping the southern sky, p. 195; Soviet rover on Mars? p. 227; Astronomy from orbit, p. 228 - 229; Triumph of 'Copernicus', p. 229; ESRO 4 detects polar argon, p. 220 - 230; 'Tungus meteorite', p. 230; NASA seeks 'orange moon', p. 230; Orbiting computer, p. 230 - 231; Large space telescope, p. 231.

Further support for post-Apollo project.
Nature, Vol. 241, 418 (1973).

Revised HEAO programme.
Spaceflight, Vol. 15, 254 (1973).

Cosmonauts in orbit.
See Abstr. 003.056.

Annual chronology of international astronautical events 1970 sponsored by the International Academy of Astronautics.   See Abstr. 010.037. ·

Data processing for astronomy missions.
See Abstr. 031.039.

Active optics for space astronomy.
See Abstr. 031.041.

Photoelectronic image recording device optimized for high detective quantum efficiency.   See Abstr. 034.082.

Application of new ultraviolet television detectors in an astronomical satellite.   See Abstr. 034.083.

Synthetic aperture optics.   See Abstr. 034.093.

Electronic imaging devices for astronomy from a space platform.   See Abstr. 034.094.

High energy particle astronomy.
See Abstr. 143.058.

052 Astrodynamics and Navigation of Space Vehicles

052.001 Determination of a circular orbit of an artificial earth satellite from optical observations, observation time being unknown.   R. A. Zeinalov.
Astron. Zhurn. Akad. Nauk SSSR, Vol. 50, 201 - 207 (1973). In Russian. English translation in Soviet Astron. AJ, Vol. 17, No. 1.

The paper deals with a method of computation of a preliminary circular orbit of an artificial earth satellite using four non-complete observations. There are only angular coordinates (α, δ) without observation times. The observations are supposed to be situated at one turn. A numerical example illustrates the method.

052.002 Precession, nutation and the choice of reference system for close earth satellite orbits.   K. Lambeck.
Celestial Mechanics, Vol. 7, 139 - 155 (1973).

Expressions are given for the perturbations arising in the motion of close earth satellites if the orbital system introduced by Veis is used. These expressions include all terms with amplitudes greater than 10^{-8} for both long and short periods. Resonance problems can also occur under certain circumstances. Similar first order expressions obtained previously by Kozai are found to contain some errors.

052.003 On the determination of the long period tidal perturbations in the elements of artificial earth satellites.   P. Musen, T. Felsentreger.
Celestial Mechanics, Vol. 7, 256 - 279 (1973).

In the present article we develop the theory of the long period tidal effects in the motion of artificial satellites assuming the variability of elastic parameters of the earth (Love numbers) across the parallels. The dependence of Love numbers on the longitude produces perturbations of the period of one day or less and hence is neglected in the present theory. A full collection of formulas is given to facilitate the programming.

052.004 An analytical iterative algorithm for the prediction of special satellite orbit points with the Brouwer orbit theory.   R. A. Gordon.
Celestial Mechanics, Vol. 7, 280 - 290 (1973).

Employing a direct recursive algorithm in relation with analytical theories will yield a considerable saving in computer time, as opposed to simulating a point by point integration through repeated evaluations of the orbit theory. As a case in point, we shall compute the set of osculating orbiting elements corresponding to special events within the revolution of an artificial satellite.

052.005 Liapunov stability analysis and attitude response of a passively stabilized space system.
R. C. Flanagan, R. Rangarajan.
Astronaut. Acta, Vol. 18, 21 - 34 (1973).

052.006 Deceleration of an apparatus with a crew in the atmosphere after an interplanetary flight.
A. V. Klimin.
Kosmich. Issled., Vol. 11, 31 - 37 (1973). In Russian.

052.007 The existence of stable relative equilibria of an artificial satellite in a model magnetic field.
V. V. Beletsky, A. B. Novogrebelsky.
Astron. Zhurn. Akad. Nauk SSSR, Vol. 50, 327 - 335 (1973). In Russian. English translation in Soviet Astron. AJ, Vol. 17, No. 2.

The motion of a magnetized satellite in a model magnetic field is considered. With the supposition that the magnetic moment of a satellite can take arbitrary constant values, there are a great number of positions of relative equilibrium in the coordinate system connected with the vector of the magnetic intensity of the external field. Using the analogy between force functions of magnetic and gyrostatic forces, the stability of libration points is investigated.

052.008 Modification of the classical methods of Lagrange – Gauss and Laplace for the case of determination of orbits of artificial earth satellites from observations at one station. Yu. V. Surnin.
Trudy Novosib. in-ta inzh. geod., aehrofotos"emka i kartogr., Vol. 27, 49 - 55 (1972). In Russian. – Abstr. in Referativ. Zhurn. 62. Issled. kosmich. prostranstva, 3.62.289 (1973).

052.009 Stars point the way to spacecraft.
N. Ya. Kondrat'ev.
Zemlya i Vselennaya, 1973, No. 1, p. 41 - 46. In Russian.

052.010 Basic theory for PROD, a program for computing the development of satellite orbits. G. E. Cook.
Celestial Mechanics, Vol. 7, 301 - 314 (1973).

This paper presents the basic theory underlying the computer program PROD for predicting the long-term development of drag-free orbits of eccentricity up to about 0.9 under the influence of the earth's gravitational potential and the gravitational attractions of the sun and moon.

052.011 Effects of motion of the equatorial plane on the orbital elements of an earth satellite.
Y. Kozai, H. Kinoshita.
Celestial Mechanics, Vol. 7, 356 - 366 (1973).

Exact differential equations relating the perturbations to satellite orbital elements by the motion of the earth's equatorial plane are derived, and they are solved to second order in precession.

052.012 Zur Bahnberechnung von geostationären Satelliten.
W. Flury.
Celestial Mechanics, Vol. 7, 376 - 383 (1973).

A first order perturbation theory for a geostationary satellite is presented. The perturbations caused by the oblateness of the earth, the ellipticity of the earth's equator and the gravitational influence of sun and moon are considered. The resonance frequency which is slightly modified by these perturbations is determined.

052.013 Non-linear plane vibrations of a satellite in an elliptical orbit. V. G. Demin, R. B. Singkh.
Kosmich. Issled., Vol. 11, 192 - 197 (1973). In Russian.

052.014 On isochronous derivatives from some parameters of the trajectory of a space vehicle.
B. Ts. Bakhshiyan.
Kosmich. Issled., Vol. 11, 217 - 225 (1973). In Russian.

052.015 On the problem of satellite stabilization.
Eh. K. Lavrovskij.
Kosmich. Issled., Vol. 11, 329 - 330 (1973). In Russian. Brief information.

052.016 The intermediary orbit of an artificial earth satellite, constructed by the method of averaging. First-order perturbations. A. Pal.
Stud. Cerc. Astron., Vol. 18, 17 - 28 (1973). In Russian.

052.017 Systematic analysis of perturbations of orbits – The case of drag. F. McL. Mallett.
Journ. British Interplanet. Soc., Vol. 26, 291 - 304 (1973).

The perturbation equations for the case of drag may be written with either the true anomaly or the eccentric anomaly as the independent variable. Before integrations can be carried out, approximations must be made. The nature of these approximations is explored, and it is found that greater accuracy is obtained when the eccentric anomaly is the variable.

052.018 Application of the methods of mathematical statistics for investigation of the motion of an artificial earth satellite around the centre of mass. N. F. Martynova.
Mekh. upravlyaem. dvizheniya i probl. kosmich. dinamiki. Leningrad, Leningr. un-t, 1972, p. 83 - 103. In Russian. Abstr. in Referativ. Zhurn. 62. Issled. kosmich. prostranstva, 4.62.322 (1973).

052.019 Mechanics of controlled motion and problems of space dynamics.
Leningr. universitet, Leningrad. 183 pp. Price 1 Rbl. 11 Kop. (1972). In Russian. – Review in Referativ. Zhurn. 62. Issled. kosmich. prostranstva, 4.62.328 (1973).

052.020 The solar radiation pressure on the Mariner 9 Mars orbiter. R. M. Georgevic.
Astronaut. Acta, Vol. 18, 109 - 115 (1973).

The refined mathematical model of the force created by the light pressure of the sun has been used to compute the solar radiation pressure force acting on the Mariner 9 (Mariner Mars 1971) spacecraft, taking into account the reflectivity characteristics of all its components. The results have been compared with values obtained from Mariner 9 observations during the cruise phase and found to be in agreement within 0.1% of the values.

052.021 Spacecraft orbit analysis. W. Kundt.
Astrophys. Space Sci., Vol. 21, 487 - 493 (1973).

The true orbit of a spacecraft differs from its reference orbit by small deviations due to errors in planetary and initial data, radiation and impact accelerations, and correction terms to Newton's theory of gravitation. These distortions are usually small enough to have additive effects on range, at the 10 m accuracy level, throughout the mission. Closed form expressions are derived (in cylindrical coordinates) for the deviation in range due to any given perturbing acceleration.

052.022 On secular perturbations in the motion of artificial satellites caused by air drag.
E. P. Aksenov, B. N. Noskov.
Astron. Zhurn. Akad. Nauk SSSR, Vol. 50, 590 - 600 (1973). In Russian. English translation in Soviet Astron. AJ, Vol. 17, No. 3.

Perturbations of motions of artificial satellites caused by air drag are considered. Analytical formulae are obtained.

Special attention is paid to combined perturbations, i.e. to such perturbations which are conditioned by the combined influence of air drag and non-sphericity of the earth.

052.023 Problem of prediction of the rotational motion of a non-symmetrically twisted satellite.
V. S. Novoselov, L. K. Babadzhanyants, L. I. Fedorova.
Mekh. upravlyaem. dvizheniya i probl. kosmich. dinamiki. Leningrad, Leningr. un-t, 1972, p. 103 - 113. In Russian. – Abstr. in Referativ. Zhurn. 51. Astron., 5.51.145 (1973).

052.024 Application of canonical transformations for setting up methods for calculating the motion of artificial satellites. M. L. Lidov, A. I. Nejshtadt.
Vtoraya Chetaevsk. konf. po analit. mekh., ustojchivosti dvizheniya i optimal'n. upr., 1973. Annotatsii dokl. Kazan', 1972, p. 9. In Russian. – Abstr. in Referativ. Zhurn. 62. Issled. kosmich. prostranstva, 5.62.321 (1973).

052.025 On the stability of regular precessions of a satellite-gyrostat in axisymmetric gravitational fields.
M. K. Nabiullin.
Vtoraya Chetaevsk. konf. po analit. mekh., ustojchivosti dvizheniya i optimal'n. upr., 1973. Annotatsii dokl. Kazan', 1972, p. 40. In Russian. – Abstr. in Referativ. Zhurn. 62. Issled. kosmich. prostranstva, 5.62.339 (1973).

052.026 On the approximate description of the orbit of an stationary artificial earth satellite.
M. A. Vashkov'yak, M. L. Lidov.
Kosmich. Issled., Vol. 11, 347 - 359 (1973). In Russian.

052.027 The choice of the trajectory for returning to earth from the orbit of an artificial lunar satellite.
V. A. Egorov, N. I. Zolotukhina, N. A. Teslenko.
Kosmich. Issled., Vol. 11, 397 - 406 (1973). In Russian.

052.028 Spatial motion of two coupled bodies under the influence of gravitational and aerodynamical forces.
G. G. Efimenko.
Kosmich. Issled., Vol. 11, 484 - 486 (1973). In Russian.
Brief information.

052.029 Astrodynamical paradoxes. A. Drożyner.
Urania Kraków, Vol. 44, 166 - 171 (1973). In Polish.

052.030 Periodic motions of a satellite in the plane of an elliptical orbit. V. A. Sarychev, V. A. Zlatoustov.
Vtoraya Chetaevsk. konf. po analit. mekh., ustojchivosti dvizheniya i optimal'n. upr., 1973. Annotatsii dokl. Kazan', 1972, p. 13. In Russian. – Abstr in Referativ. Zhurn. 62. Issled. kosmich. prostranstva, 5.62.336 (1973).

052.031 Gravity thrust Jupiter orbiter trajectories generated by encountering the Galilean satellites.
M. A. Minovitch.
Journ. Spacecraft and Rockets, (USA), Vol. 9, 751 - 756 (1972).
A trajectory design philosophy is introduced for Jupiter orbiter missions that is based on the gravity thrust concept developed for interplanetary trajectories. This is accomplished by utilizing the moving gravitational fields of the four Galilean satellites.

052.032 A new method to compute lunisolar perturbations in satellite motions. Y. Kozai.
Smithsonian Astrophys. Obs., Cambridge, Mass., Special Rep. No. 349, 4 + 27 pp. (1973).
A new method to compute lunisolar perturbations in satellite motion is proposed. The disturbing function is expressed by the orbital elements of the satellite and the geocentric polar coordinates of the moon and the sun. The secular and long-periodic perturbations are derived by numerical integrations, and the short-periodic perturbations are derived analytically. The motion of the orbital plane for a synchronous satellite is discussed; it is concluded that the inclination cannot stay below 7°.

052.033 Sul passo d'integrazione numerica di orbite.
A. Kranjc.
Rend. Ist. Lombardo A, Vol. 105, 584 - 591 = Contr. Oss. Astron. Milano-Merate, Nuova Ser., No. 344 (1971).

052.034 Mouvement approché d'un satellite artificiel. (Méthode de Vinti). O. Godart.
Inst. d'Astron. Géophys. Georges Lemaître, Univ. Catholique Louvain, Belgique, Contr. No. 12, 27 pp. (1973).

052.035 Operational calculation of orbits [Artificial satellites]. R. Laurat.
Recherches Spatiales, (France), Vol. 12, No. 1, p. 20 - 22 (1973). In French.

052.036 Suitable orbits for various satellite missions.
K. Takahashi.
Rev. Radio Res. Lab., (Japan), Vol. 18, 345 - 353 (1972). In Japanese.
Characteristics of sun-synchronous, recurrent, near-recurrent, polar, synchronous and stationary orbits, and the relations between them are examined.

052.037 Suitable orbits for various satellite missions.
K. Takahashi.
Journ. Radio Res. Lab., (Japan), Vol. 19, 213 - 228 (1972).
Characteristics of sun-synchronous, recurrent, near-recurrent, polar, synchronous and stationary orbits, and the relations between them are examined, so that suitable orbits may be selected for various satellite missions, with special attention to sun-synchronous recurrent orbits.

052.038 Determination of perturbing moments of solar radiation pressure for a body of rotation.
E. N. Polyakhova.
Trudy Astron. Obs., Leningrad, Vol. 29 (= Uchenye Zapiski Leningr. Un-ta, No. 363 = Seriya Matem. Nauk, vyp. (No.) 48), p. 152 - 163 (1973). In Russian.
Simple approximate formulas are used for the determination of the perturbing torques due to solar radiation pressure forces for a symmetrical satellite (body of revolution). The formulas are used for the determination of the torques in the case of conical and paraboloidal satellites.

052.039 Perturbations of a nearly circular orbit.
E. I. Timoshkova, K. V. Kholshevnikov.
Trudy Astron. Obs., Leningrad, Vol. 29 (= Uchenye Zapiski Leningr. Un-ta, No. 363 = Seriya Matem. Nauk, vyp. (No.) 48), p. 163 - 176 (1973). In Russian.
The motion of a satellite in the gravitational field of a nonspherical rotating planet is considered. Expressions are given for the calculation of the secular, long-period and short-period perturbations caused by arbitrary harmonics of the planet's potential.

052.040 Four-impulse visit to a planet with return at time limitations. G. V. Ufimtsev.
Trudy Astron. Obs., Leningrad, Vol. 29 (= Uchenye Zapiski Leningr. Un-ta, No. 363 = Seriya Matem. Nauk, vyp. (No.) 48), p. 177 - 195 (1973). In Russian.

052.041 Propagation of errors in orbits computed from density layer models. F. Morrison.

Geophys. Monograph 15, (see 012.025), p. 111 - 119 (1972).

052.042 **Verwendung der Atmosphärenmodelle CIRA 65 und CIRA 72 in der Bahnbestimmung geodätischer Satelliten.** E. Nagel, C. Reigber.
Bundesministerium für Forschung und Technology, Forschungsber. (Weltraumforschung), BMFT−FBW 73-11, 70 pp. Price DM 14.70 (1973).
A detailed description of the use of CIRA 1972 model for orbit determination is given. Results of numerical integrations over a five day interval show that the perturbations due to air drag computed with the aid of CIRA 1965 and CIRA 1972 differ up to nearly 20% for the orbital elements a, e, i, Ω, M and up to 100% for the argument of perigee ω.

052.043 **Bahnbestimmung aus Richtungs- und Entfernungsmessungen künstlicher Erdsatelliten als Randwertaufgabe.** K. H. Ilk.
Bundesministerium für Forschung und Technologie, Forschungsber. (Weltraumforschung), BMFT−FBW 73-12, 62 pp. Price DM 13.05 (1973).
The orbit determination problem is formulated as a least squares adjustment of nonlinear condition equations with unknowns. To solve the problem one uses a method of successive approximation. The described method has been tested on 60 short arcs of GEOS 1 observed in the first half of 1966.

052.044 **Motion of distant AES in the luni-solar resonance region.** V. G. Sokolov.
Materialy 3-j Nauch. konf. Tomsk. un-ta po mat. i mekh. Vyp. (No.) 2. Tomsk, Tomsk. un-t, 1973, p. 96 - 97. In Russian. Abstr. in Referativ. Zhurn. 51. Astron., 7.51.140 (1973).

Problems of satellite astrometry. See Abstr. 003.140.

Resonance phenomena in the rotational motion of natural and artificial celestial bodies. See Abstr. 042.037.

Satellite vibration-rotation motions studied via canonical transformations. See Abstr. 042.045.

Stable longitudes for 12-hr eccentric orbit satellites. See Abstr. 054.023.

053 Lunar and Planetary Probes and Satellites

053.001 **Pioneer 10's progress.** R. N. Watts, Jr.
Sky Telescope, Vol. 45, 26 (1973).

053.002 **Lunokhod 2 on the moon.** R. N. Watts, Jr.
Sky Telescope, Vol. 45, 148 - 149 (1973).

053.003 **Amateurs view Apollo launch.** P. A. Valleli.
Sky Telescope, Vol. 45, 150 - 152 (1973).

053.004 **The last Apollo − 1, 2, 3.** D. Baker.
Spaceflight, Vol. 15, 42 - 47, 87 - 91, 145 - 147 (1973).

053.005 **Some problems of optimum trajectory control of a space apparatus in the Martian atmosphere.**
N. M. Ivanov, A. I. Martynov, A. A. Shilov.
Kosmich. Issled., Vol. 11, 21 - 30 (1973). In Russian.

053.006 **Bisherige Ergebnisse der Jupiter-Sonde "Pioneer 10".**
Umschau, 73. Jahrgang, p. 249 - 250 (1973).
The Pioneer 10 spacecraft, bound for Jupiter, continues to direct a stream of scientific and engineering data back to earth. The interplanetary plasma has been measured which shows some remarkable fluctuations. The asteroid belt will not constitute a dangerous area for spacecraft passing through it on future outer planet missions.

053.007 **Mission building blocks for outer solar system exploration.**
D. Herman, J. Moore, P. Tarver.
Space Sci. Rev., Vol. 14, 363 - 382 (1973).
This paper describes the technological building block nec-essary to explore the outer planets with maximum science return under strong resource constraints. It describes how missions can be planned within these constraints. Two generic spacecraft types are considered: the Mariner and the Pioneer. Following discussion of outer planet mission constraints, the evolutionary development of spacecraft, probes and propulsion building blocks is presented.

053.008 **Another Jupiter Pioneer.** R. N. Watts, Jr.
Sky Telescope, Vol. 45, 346 (1973).

053.009 **Slingshot to Saturn.**
Spaceflight, Vol. 15, 181 - 182 (1973).

053.010 **Jupiter probe—Pioneer 10.** H. Miles.
Journ. British Astron. Ass., Vol. 83, 192 - 195 (1973).

053.011 **PAET, an entry probe experiment in the earth's atmosphere.** A. Seiff, D. E. Reese, S. C. Sommer,
D. B. Kirk, E. E. Whiting, H. B. Niemann.
Icarus, Vol. 18, 525 - 563 (1973).
In this paper, our purpose is to summarize the PAET (Planetary Atmosphere Experiments Test vehicle) experiment, its objectives, implementation, analysis, and results, including probe and mission aspects, the atmosphere structure experiment, the mass spectrometer and the shock layer radiometer composition experiments, and, briefly, some of the auxiliary experiment results.

053.012 **Saturn/Uranus atmospheric entry probe: scientific objectives and instrumentation.**

H. Myers, M. L. Scheer.
Bull. American Astron. Soc., Vol. 5, 291 (1973). – Abstr. AAS.

053.013 A common Saturn/Uranus probe.
R. S. Wiltshire.
Bull. American Astron. Soc., Vol. 5, 292 (1973). – Abstr. AAS.

053.014 Off to Jupiter! R. N. Watts, Jr.
Sky Telescope, Vol. 45, 295 (1973).

053.015 Ein Einschlag-Lichtblitz-Detektor für die Jupiter–Saturn Mission.
E. Grün, H. Fechtig, G. Eichhorn.
Max-Planck-Inst. Kernphysik Heidelberg, Jahresbericht 1972, p. 237 - 239 (1973).

053.016 Engineering the space shuttle. H. Falk.
IEEE Spectrum, Vol. 10, No. 5, p. 50 - 54 (1973).
New development and design solutions by EEs will help to launch a winged space vehicle.

053.017 De Pionier 11 op weg naar Jupiter. T. de Vries.
Hemel en Dampkring, Vol. 71, 176 - 181 (1973).

053.018 Soviet Mars landers. H. Oja.
Spaceflight, Vol. 15, 242 - 245 (1973).

053.019 Objects in areocentric orbit – 1. Objects on Mars – 1. G. Falworth.
Journ. British Interplanet. Soc., Vol. 26, 433 - 435 (1973).
Tabulations of data are presented relating to space objects placed into orbit around Mars prior to 1972 Dec 31, and of space objects impacted or landed on to the surface of Mars.

053.020 Phobos/Deimos missions.
E. B. Pritchard, E. F. Harrison.

Journ. Spacecraft and Rockets, (*USA*), Vol. 9, 489 - 490 (1972).
Lander, Lander/Orbiter, and sample return missions to the satellites of Mars, Phobos and Deimos are analyzed for launch opportunities from 1977 to 1981 with both Titan-Centaur and Space Shuttle-Centaur launch systems. In addition, both single satellite and dual satellite missions are examined for the case of lander-only missions.

053.021 Scientific exploration with an out-of-ecliptic spacecraft. J. M. Wilcox.
Comments Astrophys. Space Phys., Vol. 5, 75 - 85 (1973).
An out-of-ecliptic spacecraft changes our observational view of the sun and its extended corona from the two dimensions of the ecliptic plane to the three dimensions of the spherical sun. A good perspective on this change may be obtained by considering the change in our thinking as we went from a flat earth to a spherical earth.

Mariner 10 will explore Venus and Mercury next year.
IEEE Spectrum, Vol. 10, No. 1, p. 107 - 108 (1973).

The interplanetary pioneers. See Abstr. 003.042.

Mariner 9 mission profile and project history.
See Abstr. 097.007.

Mariner 9–Image processing and products.
See Abstr. 097.008.

The exploration of Mars.
See Abstr. 097.119.

The selection of asteroids for sampling missions.
See Abstr. 098.009.

054 Artificial Earth Satellites

054.001 A Geopause satellite system concept.
J. W. Siry.
Space Sci. Rev., Vol. 14, 314 - 341 (1973).
The forthcoming 10 cm range tracking accuracy capability holds much promise in connection with a number of earth and ocean dynamics investigations. The Geopause satellite system concept offers promising approaches in connection with all of these areas. A typical Geopause satellite orbit has a 14 hour period, a mean height of about 4.6 earth radii, and is nearly circular, polar, and normal to the ecliptic.

054.002 Mean elements of GEOS 1 and GEOS 2.
B. C. Douglas, J. G. Marsh, N. E. Mullins.
Celestial Mechanics, Vol. 7, 195 - 204 (1973).
A combined analytical-numerical procedure for determining mean orbital elements is presented and applied to the orbits of GEOS 1 and GEOS 2. The precision of the mean semi-

major axes of these orbits is a few tens of centimeters when optical flash data are used to determine 2 day orbital arcs. Four day Minitrack orbits give mean semi-major axes of a few meters precision. The mean orientation parameters (i, Ω) determined from the optical data are obtained to a precision of about $0\overset{''}{.}1$.

054.003 Progress report on Skylab. R. N. Watts, Jr.
Sky Telescope, Vol. 45, 24 - 26 (1973).

054.004 The attitude measuring system of the AEROS satellite. K. Ernsberger, F. Leiss, N. Rau.
Journ. British Interplanet. Soc., Vol. 26, 28 - 45 (1973).

054.005 The effect of direct solar radiation on the attitude of the SKYNET spacecraft.
G. J. Davison, R. H. Merson.

Journ. British Interplanet. Soc., Vol. 26, 228 - 241 (1973).

054.006 **Det første bemannede rom-observatorium for sol-forskning.** O. Engvold.
Astron. Tidssk., Årg. 6, p. 3 - 5 (1973).

054.007 **Astronomy through the Skylab scientific airlocks.**
K. G. Henize, J. L. Weinberg.
Sky Telescope, Vol. 45, 272 - 276 (1973).

054.008 **Solar observers to cooperate with Skylab.**
R. N. Watts, Jr.
Sky Telescope, Vol. 45, 295 (1973).

054.009 **Sur la réduction de clichés des satellites artificiels à l'Observatoire de Bucarest.** G. Vass.
Stud. Cerc. Astron., Vol. 18, 85 - 90 (1973). In Romanian.

054.010 **Sur la rotation du satellite 1962–010 A.**
V. Mioc.
Stud. Cerc. Astron., Vol. 18, 91 - 96 (1973).

054.011 **What happened to Salyut 2?** R. N. Watts, Jr.
Sky Telescope, Vol. 45, 346 (1973).

054.012 **The problem of forecasting the rotational motion of an asymmetric twisted artificial earth satellite.**
V. S. Novoselov, L. K. Babadzhanyants, L. I. Fedorova.
Mekh. upravlyaem. dvizheniya i probl. kosmich. dinamiki. Leningrad, Leningr. un-t, 1972, p. 103 - 113. In Russian. Abstr. in Referativ. Zhurn. 62. Issled. kosmich. prostranstva, 4.62.323 (1973).

054.013 **Apollo Telescope Mount (ATM).** M. Kuperus.
Hemel en Dampkring, Vol. 71, 167 - 170 (1973).

054.014 **Analysis of the orbit of 1965-11D (Cosmos 54 rocket).** D. G. King-Hele, D. M. C. Walker.
Planet. Space Sci., Vol. 21, 1081 - 1108 (1973).
The orbit has been determined at 75 epochs during the life, using the RAE orbit determination program PROP with over 4000 observations, photographic, visual and radar. Observations from the Hewitt camera at Malvern were available for 34 of the 75 orbits and typical accuracies for these orbits are 0.0005° in inclination and 100 m in perigee height. The variations in perigee height have been analyzed to determine reliable values of density scale height, at heights between 240 and 360 km. The variations in orbital inclination have been analyzed to determine upper-atmosphere zonal winds and 15th-order harmonics in the geopotential.

054.015 **Skylab's troubled flight.** R. N. Watts, Jr.
Sky Telescope, Vol. 46, 22 - 25 (1973).

054.016 **Skylab.**
SuW, Vol. 12, 163 - 165 (1973).

054.017 **Geometric kinds of tracks of artificial satellites of the earth, moon, Mars and Jupiter.** V. L. Kalachev.
Trudy pervykh chtenij, posvyashch. razrabotke nauch. naslediya i razvitiyu idej F. A. Tsandera, 1970. Sekts. "Astrodinamika" i "Sistemy zhizneobespecheniya i astrobotan". Moskva – Riga, 1972, p. 36 - 48. In Russian. – Abstr. in Referativ. Zhurn. 62. Issled. kosmich. prostranstva, 5.62.334 (1973).

054.018 **On stationary rotations of a magnetized satellite in a magnetic field.** V. V. Beletskij, A. A. Khentov.
Vtoraya Chetaevsk. konf. po analit. mekh., ustojchivosti dvizheniya i optimal'n. upr., 1973. Annotatsii dokl. Kazan', 1972, p. 23. In Russian. – Abstr. in Referativ. Zhurn. 62. Issled. kosmich. prostranstva, 5.62.338 (1973).

054.019 **Skylab and its solar astronomy experiments.**
W. C. Schneider.
Publ. American Astronaut. Soc., Sci. Techn. Ser., Vol. 28, (see 012.018), p. 25 - 46 (1972).

054.020 **Future orbital observatory modules for stellar and galactic astronomy.** C. L. Kober.
Publ. American Astronaut. Soc., Sci. Techn. Ser., Vol. 28, (see 012.018), p. 115 - 132 (1972).

054.021 **The High Energy Astronomical Observatory.**
L. E. Peterson.
Publ. American Astronaut. Soc., Sci. Techn. Ser., Vol. 28, (see 012.018), p. 213 - 253 (1972).

054.022 **Dynamics of Alouette and ISIS satellites.**
F. R. Vigneron.
Astronaut. Acta, Vol. 18, 201 - 213 (1973). – Presented at the 22nd International Astronautical Congress of the International Astronautical Federation, Brussels, Belgium, 20–25 Sept. 1971.
This paper presents the significant flight data accumulated to date, and compares it with the existing theory of attitude and spin dynamics.

054.023 **Stable longitudes for 12-hr eccentric orbit satellites.**
C. A. Wagner.
Journ. Spacecraft and Rockets, (*USA*), Vol. 9, 757 - 763 (1972).
The accelerated longitude drift regimes of eccentric 12-hr orbits are considered, due to the resonant geopotential.

054.024 **La coopération internationale sur le satellite Américain OSO-1.** R. M. Bonnet.
Bull. d'Information, Ass. Développement International Obs. Nice, No. 9, p. 5 - 10 (1972).

054.025 **Kunstmanen.** J. Meeus.
Hemel en Dampkring, Vol. 71, 41 - 42, 185 - 187 (1973). – 1972 July - December.

054.026 **Satellite digest.**
Compiled by G. Falworth.
Spaceflight, Vol. 15, 24 - 26, 73 - 74, 80, 114 - 116, 154 - 156, 192 - 195, 234 - 237, 271 - 272 (1973). – A monthly listing of all known artificial satellites and spacecraft.

Space: Copernicus-500.
Nature, Vol. 243, 4 (1973).

Technology of space astronomical instruments.
See Abstr. 032.027.

Future unmanned astronomy spacecraft.
See Abstr. 051.015.

Zonal gravity harmonics from long satellite arcs by a seminumeric method. See Abstr. 081.024.

055 Observations of Earth Satellites, Lunar and Planetary Probes

055.001 Comparison of Pageos (66 056 01) observations made by AFU-75 and Baker-Nunn camera at Dodaira Station. Y. Kozai, R. Manabe.
Tokyo Astron. Bull., Second Ser., No. 225, p. 2613 - 2622 (1973).

Comparisons are made for Pageos satellite observations made by the AFU-75 and the Baker-Nunn cameras at the Dodaira Station of the Tokyo Astronomical Observatory in December of 1969. It is not definite that there are any systematic differences between two series of observations.

055.002 Osservazioni ottiche dei satelliti artificiali effettuate nel 1969. A. Manara.
Contr. Oss. Astron. Milano-Merate, Nuova Ser., No. 332, 19 pp. (1970).

055.003 Osservazioni ottiche dei satelliti artificiali effettuate nel 1970. A. Manara.
Contr. Oss. Astron. Milano-Merate, Nuova Ser., No. 340, 25 pp. (1971).

055.004 Visual observations of artificial earth satellites in Finland, 1972 January - 1972 December.
P. Järvi.
Published by the University of Helsinki, Finland, 9 + 107 pp. (1973).

This volume contains visual satellite observations made at the Jokioinen meteorological observatory during 1972. The reading accuracy of the observations is $0.^s1$ in time and $0.^{\circ}1$ in position.

055.005 Auswerteverfahren und Genauigkeitsmaße für Satellitenbeobachtungen mit dem SBG. K.-H. Marek.
Vermessungstechnik, 20. Jahrgang, p. 301 - 304 = Deutsche Akad. Wiss. Berlin Zentralinst. Phys. Erde, Potsdam, Mitt. No. 242 (1972).

055.006 Absolute orientation of satellite triangulation. A. A. Baldini.
Geophys. Monograph 15, (see 012.025), p. 19 - 25 (1972).

055.007 Geometrical adjustment with simultaneous laser and photographic observations on the European datum. A. Cazenave, O. Dargnies, G. Balmino, M. Lefebvre.
Geophys. Monograph 15, (see 012.025), p. 43 - 48 (1972).

055.008 Tracking-station coordinates from Geos 1 and Geos 2 optical flash data.
J. G. Marsh, B. C. Douglas, S. M. Klosko.
Geophys. Monograph 15, (see 012.025), p. 77 - 82 (1972).

055.009 Observed effects of earth-reflected radiation and hydrogen drag on the orbital accelerations of balloon satellites. E. J. Prior.
Geophys. Monograph 15, (see 012.025), p. 197 - 207 (1972).

055.010 Timing for geodetic satellites. W. Markowitz.
Geophys. Monograph 15, (see 012.025), p. 245 - 246 (1972).

055.011 Measured physical and optical properties of the Passive Geodetic Satellite (Pageos) and Echo 1.
D. S. McDougal, R. B. Lee III, D. C. Romick.
Geophys. Monograph 15, (see 012.025), p. 253 - 259 (1972).

Satellite laser ranging instruments operated at Tokyo Astronomical Observatory. See Abstr. 032.031.

Photoelectric time micrometer. See Abstr. 034.108.

Satellite tracking camera for geodetic purposes. See Abstr. 046.023.

Geometric accuracy obtainable from simultaneous range measurements to satellites. See Abstr. 046.026.

Bulletin of astronomical observations, 1969. See Abstr. 075.020.

Bulletin of astronomical observations, 1970. See Abstr. 075.021.

Simple layer model of the geopotential in satellite geodesy. See Abstr. 081.032.

Geopotential representation with sampling functions. See Abstr. 081.034.

Error model for the SAO 1969 standard earth. See Abstr. 081.037.

Theoretical Astrophysics

061 General Theoretical Problems of Astrophysics, Gravitational Instability, Neutrino Astronomy, X Ray- and Gamma Ray-Astronomy, Frequency and Origin of Elements etc.

061.001 Non-linear Compton and inverse Compton effect.
P. Stewart.
Astrophys. Space Sci., Vol. 18, 377 - 386 (1972).

A charged particle of arbitrary initial motion and position is subjected to an intense circularly polarized electromagnetic wave. The electromagnetic radiation which results is calculated exactly and the results are expressed in terms of the power observed per unit bandwidth per unit solid angle. Several special cases are considered in order to demonstrate the range of possibilities which this type of interaction may have in astrophysics and numerical examples are given for the Crab pulsar.

061.002 Gamma astronomy and cosmic rays. II.
V. L. Ginzburg.
Comments Astrophys. Space Phys., Vol. 5, 15 - 21 (1973).

061.003 Some comments about the ^{176}Lu–^{176}Hf pair.
M. Arnould.
Astron. Astrophys., Vol. 22, 311 - 315 (1973).

Some comments are presented about the possible interest of a more detailed study of the ^{176}Lu–^{176}Hf pair, which possesses the unique feature of being composed of two s-only isobars.

061.004 Viscous accretion flows and their uniqueness.
N. L. Balazs.
Monthly Notices Roy. Astron. Soc., Vol. 161, 217 - 223 (1973).

We show that the bifurcation point in a non-viscous accretion flow comes about because (a) the energy-surface of a fluid particle partaking in the flow has a saddle point for $v \neq 0$; (b) the velocity curves $v(r)$ are level curves of the energy surface; the saddle point is a bifurcation point of the level surface.

061.005 Big-bang nucleosynthesis revisited.
R. V. Wagoner.
Astrophys. Journ., Vol. 179, 343 - 360 (1973).

The results of an improved calculation of the synthesis of elements during the high-temperature phase of the expansion of big-bang universes is presented. Adoption of the viewpoint (supported by recent evidence) that most of the observed deuterium and helium is of pregalactic origin allows very general constraints to be put on any cosmological model for their production.

061.006 On a magnetized rotating sphere.
R. Ruffini, A. Treves.
Astrophys. Letters, Vol. 13, 109 - 113 (1973).

The issue of the most stable electromagnetic structure associated with a rotating sphere with a magnetic axis, aligned to the rotation axis, is here discussed. With a classical model it is shown that the configuration that minimizes the total electromagnetic energy of the system is endowed with a non-zero net charge. Some implications for the physics of collapsed objects are stressed.

061.007 Polarization in inverse Compton scattering of synchrotron radiation. S. Bonometto, A. Saggion.
Astron. Astrophys., Vol. 23, 9 - 13 (1973).

The formulae are given for calculating the linear polarization of the emission due to inverse Compton scattering of synchrotron radiation. Both the cases of optically thin and optically thick (to their own synchrotron emission) sources are studied. Numerical results are also given showing that a substantial amount of the linear synchrotron polarization is preserved by the inverse Compton effect, though losses of an order up to 75 %, 80 % can often be present. Some applications of these results to astrophysical situations are proposed. The importance of carrying out the same kind of computation for circular polarization is also stressed.

061.008 On the origin of light elements. H. Reeves, J. Audouze, W. A. Fowler, D. N. Schramm.
Astrophys. Journ., Vol. 179, 909 - 930 (1973).

A summary is given of the current beliefs regarding the origin and history of the light elements ^{2}D, ^{3}He, ^{4}He, ^{6}Li, ^{7}Li, ^{9}Be, ^{10}B, and ^{11}B in the universe. A description of the various sites of nucleosynthesis for these elements is given, and the results compared with observations.

061.009 Weakly distorted polytropes under external pressure.
G. Horedt.
Astron. Astrophys., Vol 23, 303 - 306 (1973).

The stability against radial perturbations of weakly distorted polytropic spheres, cylinders and rings under external pressure is discussed.

061.010 Consequences of a universal cosmic-ray theory for γ-ray astronomy.
A. W. Strong, A. W. Wolfendale, J. Wdowczyk.
Nature, Vol. 241, 109 - 110 (1973).

We consider the possibility of reviving a theory due to Hillas (1968) in which the observed shape of the primary spectrum up to about 3×10^{19} eV is explained in terms of an evolving sources model with constant spectral index at production, and interactions with the microwave background radiation.

061.011 Application of statistics to results in gamma ray astronomy. E. O'Mongain.
Nature, Vol. 241, 376 - 379 (1973).

Some of the published results in gamma ray astronomy are examined, and it is concluded that not all the sources which have been mentioned can be confidently considered to be present.

061.012 On the origin of deuterium.
F. Hoyle, W. A. Fowler.
Nature, Vol. 241, 384 - 386 (1973).

The origin of deuterium has always been a problem for theories of stellar nucleosynthesis. A general solution is proposed and shown to be applicable under several astrophysical circumstances in the light of new observations of the galactic abundance of deuterium.

061.013 Origin of elements.
    L. E. Snyder, D. Buhl, B. Zuckerman.
Nature, Vol. 242, 33 (1973).
    We believe the recent report in Nature, Phys. Sci., Vol. 239, 81 (1972) under this title to be misleading in the light of recent observations.

061.014 Quasi-radial pulsations of rotating relativistic polytropes.
V. V. Papoyan, D. M. Sedrakian, E. V. Chubarian.
Astrofizika, Vol. 8, 405 - 412 (1972). In Russian. English translation in Astrophysics, Vol. 8, No. 3.
    The frequences of the quasi-radial pulsations of rotating relativistic polytropes for $n = 1, 1.5, 2, 2.5, 3$ and the critical value of the relativistic parameter in the sense of dynamical instability are calculated.

061.015 Superheavy elements and the r-process.
    D. N. Schramm, E. O. Fiset.
Astrophys. Journ., Vol. 180, 551 - 570 (1973).
    Detailed, microscopic-macroscopic fission-barrier calculations have been carried out along the r-process path. It has been found that the results are qualitatively similar to the semiempirical results of Schramm and Fowler in that the r-process yield of nuclei with $A \gtrsim 290$ is strongly dependent on the surface symmetry correction to the surface energy of the nucleus. Detailed calculations following the decay of nuclei from the r-process path back to the valley of beta stability (superheavy island) have been carried out. These calculations indicate that it is possible that the predicted longest-lived nucleus $^{294}110$ might not be produced in the r-process because fission competition during the decay back might deplete the (even-N, even-Z) nucleus. The large number of assumptions and uncertainties going into superheavy elements and r-process calculations are discussed in detail.

061.016 Finite-amplitude convection under the combined effect of rotation and a magnetic field.
J. O. Murphy, R. Van der Borght.
Proc. Astron. Soc. Australia, Vol. 2, 147 - 148 (1972).

061.017 Spherical $\alpha\omega$-dynamos by a variational method.
    M. Stix.
Astron. Astrophys., Vol. 24, 275 - 281 (1973).
    Eigenvalues of the kinematic dynamo problem are obtained from a variational principle; the modes of free decay in an electrically conducting sphere are used as trial functions. An application to $\alpha\omega$-dynamos shows that a small number of trial functions often leads to quite accurate eigenvalues. The method is applied to Levy's (1972) dynamo model for the earth.

061.018 Neutrinos.   H. Massey.
    Endeavour, No. 116, Vol. 32, 86 - 92 (1973).
    Thirty years ago it seemed certain that neutrinos, even if they existed, could not be a subject of experimental study, for their essential property was one of 'non-observability'. Today, the situation is completely changed; not only is neutrino physics a flourishing branch of experimental physics, but the important role of neutrinos in astrophysics has been realized. In fact, the observation of neutrinos from extraterrestrial sources is a potentially very productive new branch of astronomy.

061.019 A new Reynolds stress tensor for astrophysical bodies.   K. L. McDonald.
Bull. American Astron. Soc., Vol. 5, 15 - 16 (1973). — Abstr. AAS.

061.020 Gamma-astronomy.   A. Gal'per, B. Luchkov.

Nauka i zhizn', 1972, No. 10, p. 120 - 125. In Russian. — Abstr. in Referativ. Zhurn. 51. Astron., 3.51.81 (1973).

061.021 Methods of a self-consistent field in astrophysics.
    D. F. Kurdgelaidze.
Third Soviet Gravitational Conference, Erevan, 1972, (see 012.001), p. 327 - 332 (1972). In Russian.

061.022 Chronometer for s-process nucleosynthesis.
    J. B. Blake, T. Lee, D. N. Schramm.
Nature, Phys. Sci., Vol. 242, 98 - 100 (1973).
    The importance of the short-lived s-process chronometer ^{205}Pb, its relation to the mechanism of s-process nucleosynthesis and its implications for cosmochronology are discussed.

061.023 Instability of a cold rotating gravitating cylinder.
    V. A. Antonov.
Dokl. Akad. Nauk SSSR, Ser. Mat. Fiz., Vol. 209, 584 - 585 (1973). In Russian.

061.024 Search for transuranic elements in the universe.
    G. B. Zhdanov.
Zemlya i Vselennaya, 1973, No. 2, p. 26 - 30. In Russian.

061.025 Self-gravitating winds.
    P. Biermann, R. Kippenhahn.
Astron. Astrophys., Vol. 25, 63 - 70 (1973).
    The general properties of stationary isothermal flows under the influence of self-gravitation are considered. We limit ourselves to cases of high symmetry: a) the planeparallel flow, b) the spherically symmetric flow and c) the rotationally symmetric flow. We find in all cases oscillatory solutions in the subsonic regime and under certain conditions smooth transitions from subsonic to supersonic flow.

061.026 γ-astronomy and cosmic rays.   V. L. Ginzburg.
    Vestn. AN SSSR, 1972, No. 10, p. 18 - 25. In Russian. — Abstr. in Referativ. Zhurn. 51. Astron., 4.51.812 (1973).

061.027 Intense magnetic fields in astrophysics.
    V. Canuto, H.-Y. Chiu.
Stellar evolution, (see 012.014), p. 735 - 806 (1972). — Reprinted from Space Sci. Rev., Vol. 12, 3 - 74 (1971). — See Abstr. 05.061.014.

061.028 Cosmic abundance of boron.
    A. G. W. Cameron, S. A. Colgate, L. Grossman.
Nature, Vol. 243, 204 - 207 (1973).
    We point out here that the cosmic abundance of boron has been greatly underestimated, and that the upward revision of its abundance has important consequences in several areas of astrophysics.

061.029 Nucleosynthesis and ages of the elements.
    H. Reeves.
Stellar ages. Proc. IAU Colloquium No. 17, (see 012.015), XXXII, 1 - 20 (1973).

061.030 Possible nucleosynthetic chronologies for the s and p-processes.   M. Arnould.
Stellar ages. Proc. IAU Colloquium No. 17, (see 012.015), XXXIII, 1 - 10 (1973).

061.031 Thermal instability due to formation of molecular or solid hydrogen.   Y. Sabano.
Sci. Rep. Tôhoku Univ., First Ser., Vol. 55, 155 - 168 (1972/73).
    The thermal stability of hydrogen gas which includes molecular or solid hydrogen is examined by linear perturbation analysis, and the resultant instability criterion is applied

to dense interstellar clouds and to proto-galaxies.

061.032 Galactic nucleosynthesis of He⁴, C¹² and N¹⁴.
A. Ferrari, R. Gallino, A. Masani.
Mem. Soc. Astron. Italiana., Nuova Ser., Vol. 43, 731 - 744 (1973).
We discuss the problem of cosmic abundances of He⁴, N¹⁴ and C¹² (O¹⁶) as produced by stellar nucleosynthesis assuming evolutionary models of the Galaxy. Using the most recent data on the compositional structure of stars at the end of their evolution, we compute a He⁴ abundance even smaller than that previously evaluated by other authors. The cosmic abundances of C¹² (O¹⁶) and N¹⁴ can be explained by the homogeneous evolution of massive stars, but also by the non-homogeneous evolution if massive stars of $M > 15\,M_\odot$ loose sufficient amount of processed material in the red giant phase.

061.033 Gyromagnetic radiation from bunched electrons.
A. Mangeney.
Cosmic plasma physics. Conference 1971, (see 012.016), p. 185 - 189 (1972).

061.034 Non-linear stabilization of synchrotron instability.
S. A. Kaplan, V. N. Tsytovich.
Izv. vyssh. ucheb. zavedenij. Radiofizika, Vol. 15, 1464 - 1468 (1972). In Russian. – Abstr. in Referativ. Zhurn. 51. Astron., 4.51.345 (1973).

061.035 ²⁴⁷Cm as a short-lived r-process chronometer.
J. B. Blake, D. N. Schramm.
Nature, Phys. Sci., Vol. 243, 138 - 140 (1973). – Letter.

061.036 Gradient-temperature instability of a gravitating medium. V. L. Poljachenko, I. G. Shukhman.
Astron. Zhurn. Akad. Nauk SSSR, Vol. 50, 649 - 651 (1973). In Russian. English translation in Soviet Astron. AJ, Vol. 17, No. 3. – Short note.
Increments of the gradient-temperature instability are calculated in a model of a homogeneous cylinder of infinite length with longitudinal temperature.

061.037 Neutrinoerne fra Solen som en mulighed for direkte observation af dens indre. P. Martinsen.
Astron. Tidssk., Årg. 6, p. 41 - 45 (1973).

061.038 Charged particle acceleration in strong dipole fields.
M. Grewing, H. Heintzmann.
Mitt. Astron. Ges., No. 32, p. 214 - 219 (1973).

061.039 Ein einfaches Hartree-Fock-Verfahren für Opazitäts-rechnungen. H. Kähler.
Mitt. Astron. Ges., No. 32, p. 221 (1973). – Abstract.

061.040 Behandlung von molekularen Reaktionsgleichge-wichten und Opazitätsrechnungen für kühle, circum-stellare Hüllen. D. Sieber.
Mitt. Astron. Ges., No. 32, p. 222 - 225 (1973).

061.041 Self-gravitating winds.
P. Biermann, R. Kippenhahn.
Mitt. Astron. Ges., No. 32, p. 236 (1973). – Abstract.

061.042 Finite-amplitude disturbances in self-gravitating media. J.-L. Tassoul, H. Dedic.
Astron. Astrophys., Vol. 26, 79 - 84 (1973).
This paper deals with the evolution of small-but finite-amplitude disturbances in a self-gravitating fluid layer of finite thickness. Contrary to the prediction of the linearized theory, the critical wavelength beyond which no small-amplitude motion exists now depends on the initial amplitude of the fluctuations. Nonlinearity slightly enlarges the range of unstable

disturbances within the fluid slab. Typical nonlinear features in the flow patterns are observed.

061.043 Comments on the paper, 'On the application of Cramér's theorem to axisymmetric incompressible turbulence', by I. Lerche. F. Krause, P. H. Roberts.
Astrophys. Space Sci., Vol. 22, 193 - 195 (1973). – Research note, with a reply by I. Lerche, p. 197 (1973).

061.044 Evaluation of astrophysical hypotheses.
P. A. Sturrock.
Astrophys. Journ., Vol. 182, 569 - 580 (1973).
The aim of this article is to set out a bookkeeping procedure for formalizing the process of assessing a hypothesis by comparison of conclusions drawn theoretically from this hypothesis with facts obtained by reduction of observational data. The formalism used is that of probability theory. The following model is used. Between observation and theory is an "interface" which comprises a number of independent items. Each item comprises a set of mutually exclusive statements. Formulae are derived which show (a) how the probability of each hypothesis should be adjusted in response to information concerning one item, and (b) how such estimates concerning more than one item may be combined. The procedure is illustrated by a "work sheet" showing how a few facts and conclusions concerning pulsars were combined to appraise the neutron-star and white-dwarf hypotheses.

061.045 X-ray astronomy – results and instruments.
H. Gursky.
Publ. American Astronaut. Soc., Sci. Techn. Ser., Vol. 28, (see 012.018), p. 255 - 288 (1972).

061.046 The rise of astrophysics. B. Strömgren.
Ann. New York Acad. Sci., Vol. 198, (see 012.020), 245 - 254 (1972).

061.047 Fermi acceleration and the energy spectra of heavy nuclei at low energies. S. Ramadurai.
Astrophys. Letters, Vol. 14, 85 - 88 (1973).
The effect of ionization loss on the energy spectra of nuclei undergoing Fermi acceleration is studied. It is found that the shapes of the energy spectra after acceleration are different for different types of nuclei, the higher the charge, the steeper the spectra at low energies. It is shown that the enhanced abundance of O, Si and Fe recently observed in solar flares may be explained by such a process.

061.048 Radiative losses in a magnetostatic and intense electromagnetic field. H. R. Standeven, P. Stewart.
Astrophys. Space Sci., Vol. 19, 181 - 187 (1972).
The average energy radiation rate is calculated exactly for the case of a charged particle injected with arbitrary momentum into an intense electromagnetic field which is propagating along a uniform magnetostatic field.

061.049 On the application of Cramér's theorem to axisymmetric, incompressible turbulence. I. Lerche.
Astrophys. Space Sci., Vol. 19, 189 - 193 (1972).
The purpose of this paper is to demonstrate that the next simplest type of velocity turbulence after isotropic turbulence, when subjected to Cramér's theorem, has built into it constraints that must be met if the turbulence is to be physically realizable. This next simplest type of velocity turbulence is homogeneous, stationary, incompressible axisymmetric turbulence.

061.050 Restrictions on the short time-constant astrophysics.
G. Cavallo, A. Ventura.
Astrophys. Space Sci., Vol. 19, 431 - 439 (1972).
The comparatively new field of short time-constant

astrophysics is investigated with the aim of checking the statement that there is a maximum amount of electromagnetic energy which can be radiated at a frequency ν_m by an object whose size is L and during a time $\Delta\tau$. Such limits are found under special assumptions.

061.051 Instability of shock waves in inhomogeneous gases.
H. Yamakazi.
Progr. Theor. Phys., (*Japan*), Vol. 48, 1860 - 1869 (1972).
The instability of an ideal-gas which propagates along the direction of density decrease, with respect to deformations of the shape of the shock front, is studied. A linearized ordinary differential equation which governs the disturbance is derived from an equation which gives the variation in shock strength through the density decrease of the gas and the divergence of the shock ray.

061.052 On the delayed neutrons at the final stage of the r-process. T. Kodama, K. Takahashi.
Phys. Letters B, (*Netherlands*), Vol. 43B, 167 - 169 (1973).
The gross theory of beta decay is applied to the r-process calculation. In particular, it is explicitly shown that the delayed-neutron emissions at the final stage of the r-process smooth out the abundance curve significantly.

061.053 Isobaric analogue states in ^{6}Li and the solar neutrino anomaly. G. R. Bishop.
Journ. Phys. A, General Phys., Vol. 6, L21 - L23 (1973).
A search is reported for the isobaric analogue state in ^{6}Li to the state in ^{6}Be proposed recently as a possible reason for the solar neutrino anomaly. Upper limits to the radiative probability of transition to such a state are obtained.

061.054 X-rays astronomy. B. Rossi.
Elettrotecnica, Vol. 59, 1229 - 1241 (1972).
In Italian.
Among the most recent progress in the astronomic field there is the possibility to study cosmic radiations, with wave length in the X-rays range, using instruments carried to very great altitudes by balloons or by launch of satellites properly realised. The results of the studies carried out till now and concerning the emission from the sun and other cosmic sources are reported in this article.

061.055 An astrophysical problem of gaseous configuration and a lemma. G. Bandyopadhyay.
Journ. Math. Phys. Sci., (*India*), Vol. 6, 439 - 451 (1972).
The equation of motion of a gravitating fluid sphere within which heat transfer is taking place through radiation and also in which there is sub-atomic energy generation, is discussed.

061.056 On the possibility of a convective vortex sheet in a fluid heated from below. R. D. Rosen.
Pure and Applied Geophys., (*Switzerland*), Vol. 101, 205 - 207 (1972). − See Phys. Abstr., Vol. 76, No. 31618 (1973).

061.057 Some astrophysical implications of second-order effects in the V-A theory of weak interactions.
A. B. Lopez-Cepero, G. Maiella.
Nuovo Cimento A, Ser. 11, Vol. 14 A, 245 - 268 (1973).

061.058 An interpretation of general anomalies of xenon and the isotopic composition of primitive xenon.
N. Takaoka.
Mass. Spectrosc., (*Japan*), Vol. 20, 287 - 302 (1972).

061.059 Extraterrestrial organic analysis. J. Oro.
Space Life Sci., (*Netherlands*), Vol. 3, 507 - 550 (1972).
Presents a review of extraterrestrial organic analysis

under the following headings: stars and sun, interstellar matter, comets, meteorites, carbonaceous compounds, terrestrial and Jovian planets and the significance for exobiology.

061.060 Maxwell theory (in media) and quantum theory in a rotating reference frame. E. Schmutzer.
Ann. Physik, Ser. 7, Vol. 29, 75 - 95 (1973).
On the basis of 3-dimensional conceptions of the electromagnetic quantities, the Maxwell theory is represented in a rotating frame of reference. For such a frame the Maxwell equations obtain additional terms depending on the angular velocity (analogous to the Coriolis term etc. in Newtonian mechanics). Using these results with the help of the Dirac equation, quantum mechanics in a rotating frame of reference is investigated.

061.061 Dynamical contraction of rotating gaseous spheroids.
T. Hara, T. Matsuda, K. Nakazawa.
Progr. Theor. Phys., (*Japan*), Vol. 49, 460 - 478 (1973).
Gravitational collapse of a non-spherical gas cloud is investigated taking into account the effects of gas pressure and rotation, but not the effect of magnetic field. It is assumed that the cloud is axisymmetric with spheroidal equidensity surfaces and contracts isothermally.

061.062 Infrared astrophysics-physical processes.
R. Johansen.
Phys. Norvegica, Vol. 6, 203 (1972). − See Phys. Abstr., Vol. 76, No. 42219 (1973).

061.063 A possible explanation for the origin of lithium, beryllium, and boron. J. Audouze, J. W. Truran.
Astrophys. Journ., Vol. 182, 839 - 846 (1973).
A simple picture is outlined which can account for the new boron abundance in the meteorites ($B/H \sim 10^{-8}$ following Cameron et al.) together with lithium and beryllium. ^{11}B, ^{10}B, and ^{7}Li are produced by large fluxes of low-energy particles (~ 10 MeV particles) accelerated by shock waves occurring in supernovae envelopes while ^{6}Li and ^{9}Be are produced by high-energy cosmic rays impinging on the interstellar gas.

061.064 Experimental X-ray astronomy. R. Novick.
Atomic physics and astrophysics. Brandeis Univ.
Summer Inst. 1969, Vol. 2, (see 003.015), 203 - 283 (1973).

061.065 Atomic processes in astrophysics.
A. Dalgarno.
Atomic physics and astrophysics. Brandeis Univ. Summer Inst. 1969, Vol. 2,(see 003.015), 285-337 (1973).

061.066 Matter in super-dense state. A. S. Kompaneets.
Zemlya i Vselennaya, 1973, No. 3, p. 13 - 17.
In Russian.

061.067 On the transformation of the wave spectrum in a medium with smooth space-time fluctuations.
V. G. Gavrilenko, N. S. Stepanov.
Izv. vyssh. ucheb. zavedenij. Radiofizika, Vol. 16, 69 - 81 (1973). In Russian. − Abstr. in Referativ. Zhurn. 51. Astron., 7.51.237 (1973).

061.068 Some simple models of particle acceleration in neutral current layers.
S. Bulanov, S. I. Syrovatskij.
IV Leningr. mezhdunar seminar "Edinoobrazie uskoreniya chastist v razlich. masshtabakh kosmosa, 1972". Leningrad, 1972, p. 101 - 108. In Russian. − Abstr. in Referativ. Zhurn. 51. Astron., 7.51.395; 62. Issled. kosmich. prostranstva, 7.62.198 (1973).

061.069 Synchrotron radiation, adiabatic invariant, gradient

drift of particles in a linearly inhomogeneous magnetic field.
S. A. Kaplan, Kh. Orazberdyev, V. Yu. Trakhtengerts.
Izv. vyssh. ucheb. zavedenij. Radiofizika, Vol. 16, 25 - 29 (1973). In Russian. – Abstr. in Referativ. Zhurn. 62. Issled. kosmich. prostranstva, 7.62.214 (1973).

061.070 Density of states for nearly-free electrons in a uniform magnetic field.
H.-P. Gail, J. Schmid-Burgk.
Phys. Letters, (Netherlands), Vol. 44A, 94 - 96 (1973).

The density of states of an electron gas in the presence of a uniform magnetic field and a weak periodic lattice-potential is calculated in the second Born-approximation.

Astrofizica centemporana.
See Abstr. 003.052.

High-energy astrophysics.   See Abstr. 003.100.

General astrophysics with elements of geophysics.
See Abstr. 003.115.

Non-radial oscillations and convective instability of a polytrope with a toroidal magnetic field.
See Abstr. 062.060.

Post-event neutron exposure of r-process material.
See Abstr. 065.018.

062 Magneto-Hydrodynamics, Plasma

062.001 A Razin-Tsytovich effect for bremsstrahlung.
D. B. Melrose.
Astrophys. Space Sci., Vol. 18, 267 - 272 (1972).

It is shown that bremsstrahlung from electrons with Lorentz factor $\gamma \gg 1$ is suppressed for $\omega < \gamma \omega_p$ in a plasma with plasma frequency ω_p compared with emission in vacuo. This suppression effect is analogous to the suppression of synchrotron radiation in a plasma (Razin-Tsytovich effect).

062.002 Stability of a gravitating fluid layer in the presence of a uniform magnetic field perpendicular to its boundary.   K. P. Das.
Publ. Astron. Soc. Japan, Vol. 25, 143 - 151 (1973).

The hydromagnetic stability of a gravitating, homogeneous, incompressible and infinitely conducting fluid layer of infinite extent but finite thickness, is considered in the presence of a uniform magnetic field perpendicular to its boundary by use of the normal mode method. A dispersion relation is obtained including the effect of viscosity but our discussion is confined to the case of an inviscid fluid. Approximate roots of the dispersion relation are obtained for a very small external magnetic field.

062.003 A self-consistent model of a simple magnetic neutral sheet system surrounded by a cold, collisionless plasma.   S. W. H. Cowley.
Cosmic Electrodynamics, Vol. 3, 448 - 501 (1973).

Following the work of Alfvén (1968), and Cowley (1971, 1972) the research presented here represents a first attempt to consider the detailed flow and field configuration of a neutral sheet system, where the current is provided by a collisionless plasma surrounding the field reversal region.

062.004 The influence of scattering by plasma ocillations on the emission spectrum from semi-opaque plasma.
V. M. Tomozov.
Astron. Zhurn. Akad. Nauk SSSR, Vol. 50, 213 - 216 (1973). In Russian. English translation in Soviet Astron. AJ, Vol. 17, No. 1.

The problem of the influence of scattering of quanta by Langmuir plasmons on the formation of the emission spectrum from semi-opaque plasma in the Rayleigh-Jeans region is considered. A method of determining the emission region parameters of the plasma and the intensity of generation of the Langmuir turbulence are developed.

062.005 Electrical and thermal conductivities of a relativistic degenerate plasma.   D. C. Kelly.
Astrophys. Journ., Vol. 179, 599 - 606 (1973).

The electrical and thermal conductivities are calculated for an electron-proton plasma. The electron distribution function is obtained by solving the Boltzmann equation for a plasma in which the electrons are relativistic and degenerate while the protons are degenerate but nonrelativistic. Such conditions are expected to prevail in neutron-star interiors.

062.006 The theory of weak screening in thermonuclear reactions.
M. D. Delano, W. D. Langer, R. A. Schwartz.
Astrophys. Letters, Vol. 13, 105 - 107 (1973).

We calculate the effects of screening of the Coulomb field on thermonuclear reaction rates in nondegenerate plasmas. We find that in the screening factor $\exp(\Delta E/k_B T)$, ΔE is 3/2 the Salpeter result.

062.007 A nonlinear, resistive boundary layer in rotating hydromagnetic flow.   D. E. Loper.
Phys. Earth Planet. Interiors, Vol. 6, 405 - 425 (1972).

A new nonlinear boundary layer in rotating hydromagnetic flows is presented. The purpose of this layer is to provide a smooth transition between the magnetic field at a rigid, electrically insulating boundary and that far from the boundary. Steady solutions are presented and their uniqueness and temporal stability are analyzed. The relevance of this layer to the core of the earth is discussed.

062.008 Solidification of a carbon-oxygen plasma.
G. L. Loumos, W. B. Hubbard.

Astrophys. Journ., Vol. 180, 199 - 206 (1973).

Numerical results are presented for the high-density liquid solid phase transition for a two-component plasma consisting of equal numbers of carbon and oxygen. Comparison with the one-component plasma of Brush, Sahlin and Teller is also made.

062.009 **On the evolution of turbulent magnetic fields in a collision dominanted plasma.**   I. Lerche.
Plasma Physics, Vol. 15, 417 - 428 (1973).

In a weakly ionized plasma, in which collisions with the neutral background control the velocity fields of the ions and electrons against pressure gradients and the Lorentz force, it is demonstrated that: (1) The energy density in the turbulent magnetic field declines as time progresses. (2) The rate of decline depends on the amount of 'helicity' present in the turbulent magnetic field. (3) The helicity must either be zero or must take on one of its two possible equipartition values, which are equal and opposite. (4) The effect of helicity is to increase the rate of decay of the long wavelength modes of the turbulent field over the decay rate obtained in his absence.

062.010 **Ionization, recombination, and population of excited levels in hydrogen plasmas.**
L. C. Johnson, E. Hinnov.
Journ. Quant. Spectrosc. Radiat. Transfer, Vol. 13, 333 - 358 (1973).

Transition rate equations for atomic hydrogen are solved to obtain coefficients for population of excited levels and for ionization and recombination. The results are based upon more accurate transition rates, cover a wider range of plasma parameters, and are obtained by a more general solution of the rate equations than previously available compilations.

062.011 **Étude par résonance paramagnétique électronique de la dissociation moléculaire dans un plasma d'oxygène.**   H. Tchen, S. Bediée.
Comptes Rendus Acad. Sci. Paris, Sér. B, Vol. 276, 519 - 522 (1973).

062.012 **Three-wave interaction in cold magnetized plasmas.**
L. Stenflo.
Planet. Space Sci., Vol. 21, 391 - 397 (1973).

The coupled equations for the resonant nonlinear interaction between three waves in cold magnetized plasmas are derived and written in a comparatively simple and symmetric form. The deficiencies of previous papers are pointed out and the possibility of explosive interactions in the solar corona is suggested.

062.013 **Resistive diffusion of force-free magnetic fields in a passive medium.**   B. C. Low.
Astrophys. Journ., Vol. 181, 209 - 226 (1973).

We consider the resistive diffusion of a force-free magnetic field in a tenuous compressible medium, which is free to move to accommodate the changing magnetic-field configuration. We show that the force-free magnetic field may evolve slowly for an extended period of time, whereupon it abruptly develops steep gradients and passes into an explosive phase. We suggest that it is this process that sets the stage for the onset of solar flares.

062.014 **On steady magnetic-field reconnection.**
E. R. Priest.
Astrophys. Journ., Vol. 181, 227 - 235 (1973).

Magnetic-field reconnection may be essential in a variety of solar and astrophysical phenomena, and there has been much debate as to the maximum allowable rate of reconnection. A more careful matching of the internal and external regions in Yeh and Axford's mechanism is here performed, with the conclusion that the highest rate of reconnection is about

$u_A/18$ (u_A = Alfvén speed).

062.015 **Propagation of small perturbations in post-Newtonian hydrodynamics.**
B. M. Berkovskij, V. I. Sadchikov.
Third Soviet Gravitational Conference, Erevan, 1972, (see 012.001), p. 291 - 293 (1972). In Russian.

062.016 **Contact discontinuities in cosmic plasma.**
K. G. Ivanov.
Geomagn. Aeronom., Vol. 13, 3 - 9 (1973). In Russian.

062.017 **On the limiting polarization of radio-waves.**
S. Grounds, D. ter Haar.
Astrophys. Space Sci., Vol. 20, 39 - 42 (1973).

We discuss the limitations of various expressions governing the determination of the limiting polarization of radiowaves and slightly extend earlier work, which was limited to the case where the wave-vector is parallel to the density gradient, to the special case where the magnetic field is parallel to the density gradient, but the wave-vector can be at an angle to both. We consider the problem especially in connection with the problem of pulsar emission.

062.018 **Plasma and fields in the environment of a rapidly moving body in the presence of an outer magnetic field.**   A. M. Moskalenko.
Geomagn. Aeronom., Vol. 13, 223 - 227 (1973). In Russian.

062.019 **About damping of hydromagnetic waves in a turbulent plasma.**   V. A. Liperovsky, S. A. Martyanov.
Geomagn. Aeronom., Vol. 13, 311 - 317 (1973). In Russian.

062.020 **One-dimensional model of a plasma layer.**
N. A. Tsyganenko.
Kosmich. Issled., Vol. 11, 332 - 335 (1973). In Russian.
Brief information.

062.021 **Screening factors for nuclear reactions. I. General theory.**
H. E. DeWitt, H. C. Graboske, M. S. Cooper.
Astrophys. Journ., Vol. 181, 439 - 456 (1973).

A generalized statistical-mechanical theory is developed to describe the effect of plasma screening on nuclear reactions. This theory recovers the previous weak- and strong-screening results of Salpeter, in the appropriate charge and screening strength limits for which they are valid. A new theoretical study of strong screening is next made for the case of equal-charge reactions, based on Monte Carlo calculations of the pair distribution function of the Coulomb fluid. This result is combined with the theory of Salpeter to construct a generalized screening function for arbitrary charge conditions. The generalized screening factor is extended to describe intermediate-screening effects.

062.022 **Screening factors for nuclear reactions. II. Intermediate screening and astrophysical applications.**
H. C. Graboske, H. E. DeWitt, A. S. Grossman, M. S. Cooper.
Astrophys. Journ., Vol. 181, 457 - 474 (1973).

The theory of intermediate screening is developed from the cluster expansion for the screening function. This systematic perturbation technique is used to extend weak-screening theory analytically to higher order in the screening parameter, and is solved numerically to yield screening-function data for intermediate and strong screening regions. The new screening theory is studied in relation to various astrophysical situations.

062.023 **Sur l'instabilité magnétogravitationnelle d'un plasma qui possède une pression anisotrope, en mouvement de rotation uniforme et sous l'influence du**

courant Hall. L'équation de dispersion (I).   M. Vasiu.
Stud. Univ. Babeş-Bolyai, Ser. Phys., Anul 18, Fasc. 1, p.
61 - 67 (1973).

062.024   **Relativistic shock hydrodynamics.**
          C. R. McKee, S. A. Colgate.
Astrophys. Journ., Vol. 181, 903 - 938 (1973).
    The special-relativistic equations for a perfect fluid are
differenced in characteristic form. Shocks are calculated
using an energy Lagrange frame where energy fractions become a coordinate. Examples are given of calculations of
shocks and expansions that agree in detail with analytical
solutions.

062.025   **Some problems of mean field electrodynamics.**
          F. Krause, P. H. Roberts.
Astrophys. Journ., Vol. 181, 977 - 992 (1973).
    The mathematical structure which describes the evolu-
tion of the ensemble average of an electromagnetic field in
an electrically conducting fluid in turbulent motion is often
called mean field electrodynamics; the theory has significant
application to cosmic magnetism. Recent studies of the
theory by Lerche have cast doubt on several earlier conclusions of Steenbeck, Krause, and Rädler, and it is the object of
this paper to resolve the major points of conflict.

062.026   **Kinematic dynamo theory. V. Comments on diverse**
          matters including historical development, isotropic
turbulence, and expansion techniques.   I. Lerche.
Astrophys. Journ., Vol. 181, 993 - 1002 (1973).
    We show that three of the four major criticisms by
Krause and Roberts (1973) of our work in turbulent kinematic dynamo theory possess fundamental errors, and are without foundation. We also correct the errors in the historical
record of the development of turbulent kinematic dynamo
theory presented in their paper.

062.027   **Hydrostatic equilibrium of hydromagnetic fields.**
          G. Yu.
Astrophys. Journ., Vol. 181, 1003 - 1008 (1973).
    The condition for a hydromagnetic field to be in hydro-
static equilibrium is discussed. It is shown that if the topology
of the wrapping pattern of lines of force of a non-force–free field
around each other changes along the field, then the configuration cannot be in hydrostatic equilibrium. A general discussion
is given, and several special cases are worked out to illustrate
the general equilibrium requirement.

062.028   **Non-linear magnetic sound in a gravitational field.**
          L. A. Ostrovskij, N. R. Rubakha.
Izv. vyssh. ucheb. zavedenij. Radiofizika, Vol. 15, 1293 -
1299 (1972). In Russian. – Abstr. in Referativ. Zhurn. 51.
Astron., 4.51.359 (1973).

062.029   **Shock waves in relativistic magnetohydrodynamics.**
          A. Lichnerowicz.
Gravitatsiya. Kiev, Nauk. dumka, 1972, p. 147 - 156, 357.
Abstr. in Referativ. Zhurn. 51. Astron., 4.51.360 (1973).

062.030   **High-frequency turbulence in cosmic plasma.**
          V. A. Gudkova, V. A. Liperovskij.
Mezhplanet. sreda i fiz. magnitosfery. Moskva, Nauka, 1972,
p. 60 - 73. In Russian. – Abstr. in Referativ. Zhurn. 62.
Issled. kosmich. prostranstva, 4.62.244 (1973).

062.031   **Energy dissipation in model experiments.**
          I. M. Podgornyj.
Mezhplanet. sreda i fiz. magnitosfery. Moskva, Nauka, 1972,
p. 44 - 49. In Russian. – Abstr. in Referativ. Zhurn. 62.
Issled. kosmich. prostranstva, 4.62.265 (1973).

062.032   **Determination of electron density in plasmas from**
          the hydrogen spectral line Hα broadened by com-
bined Stark and Zeeman effect.
Nguyen-Hoe, H. W. Drawin.
Zeitschr. Naturforschung, Vol. 28 a, 789 - 791 (1973).

062.033   **Interaction of fast particles with magneto-hydrody-**
          namical turbulence.   I. N. Toptygin.
Astrophys. Space Sci., Vol. 20, 329 - 350, 351 - 371 (1973).
In Russian and English.
    The acceleration of fast particles by Alfvén and magnetic
sound waves of small amplitude is considered. The waves exist
against the background of a strong, uniform magnetic field. We
take into account the contributions to acceleration from a
large scale random field (harmonics with $k < R^{-1}$, where R is
the Larmor radius), as well as from a small scale field ($k > R^{-1}$).
The small scale field was considered by perturbation theory,
while the large scale random field in adiabatic approximation.
The energy dependence of the diffusion coefficient in momentum space, and the time of acceleration are estimated. The
possible anisotropy of angular distribution is taken into
account. The space diffusion coefficient of particles across the
regular magnetic field is estimated. It is shown that this diffusion is due mainly to the large scale random field.

062.034   **Multi-ion plasmas in astrophysics. I: General require-**
          ments for the existence of critical points.
E. J. Weber.
Astrophys. Space Sci., Vol. 20, 391 - 400 (1973).
    The steady-state motion of a quasi-neutral n-ion plasma
is investigated using a fluid-dynamical model. The main results
obtained are that there are only two distinct ways in which
such a plasma can make a transition from a subsonic state to a
supersonic one. There is one unique possibility for which there
exists one critical point where all the ion gases have their Mach
numbers exactly to unity and where the individual ion forcing
functions (the inhomogeneous terms) are non-zero but linearly
related to each other. The other possibility which we find is
that at a critical point all inhomogeneous terms are identically
equal to zero and there exist as many critical points as there
exist simultaneous zeros of the ion forcing functions. These
results are necessary and sufficient for $n \geqslant 3$.

062.035   **Multi-ion plasmas in astrophysics. II: Motion of iso-**
          thermal plasmas in a gravitational field.
E. J. Weber.
Astrophys. Space Sci., Vol. 20, 401 - 415 (1973).
    An isothermal hydrodynamic model of the motions of a
multi-ion plasma in a gravitational field is developed and the
properties of the flow are discussed for the case of major astrophysical interest in which the gas undergoes a subsonic-super-
sonic transition. It is shown that the existence of critical points
through which the plasma has to pass will determine a large
number of the plasma parameters, especially the temperature
of the minor ions. The equation of motion of a two ion gas
(hydrogen-helium) are solved numerically and yield the interesting result that the bulk velocity of the plasma constituents are
not equal at 1 AU.

062.036   **Equations for a plasma consisting of matter and**
          antimatter.   A. H. Nelson, K. Ikuta.
Astrophys. Space Sci., Vol. 20, 439 - 458 (1973).
    A set of fluid type equations is derived to describe the
macroscopic behaviour of a plasma consisting of a mixture of
matter and antimatter. The equations are written in a form
which displays the full symmetry of the medium with respect
to particle charge and mass, a symmetry absent in normal
plasmas. This symmetry of the equations facilitates their manipulation and solution, and by way of illustration the equations
are used to analyze the propagation of electromagnetic and
acoustic waves through a matter-antimatter plasma. Some

differences from the propagation of such waves in a normal plasma are noted.

062.037 Relations between cosmic and laboratory plasma physics. H. Alfvén.
Cosmic plasma physics. Conference 1971, (see 012.016), p. 1 - 14 (1972).

062.038 Laboratory experiments on the interaction between a plasma and a neutral gas. L. Danielsson.
Cosmic plasma physics. Conference 1971, (see 012.016), p. 141 - 148 (1972).

062.039 Propagation of relativistic electromagnetic waves in a plasma. F. W. Perkins, C. E. Max.
Cosmic plasma physics. Conference 1971, (see 012.016), p. 233 - 238 (1972).

062.040 The properties of magnetic neutral sheet systems. S. W. H. Cowley.
Cosmic plasma physics. Conference 1971, (see 012.016), p. 273 - 281 (1972).

062.041 Field line motion in the presence of finite conductivity. T. J. Birmingham.
Cosmic plasma physics. Conference 1971, (see 012.016), p. 283 - 291 (1972).

062.042 Collisionless shocks. J. W. M. Paul.
Cosmic plasma physics. Conference 1971, (see 012.016), p. 293 - 303 (1972).

062.043 Non-linear evolution of firehose-unstable Alfvén waves. K. Elsässer, H. Schamel.
Cosmic plasma physics. Conference 1971, (see 012.016), p. 305 - 310 (1972).

062.044 Resonant diffusion in strongly turbulent plasmas. T. J. Birmingham, M. Bornatici.
Cosmic plasma physics. Conference 1971, (see 012.016), p. 311 - 317 (1972).

062.045 Experimental study of electron and ion heating in high-β perpendicular collisionless shock waves. M. Keilhacker, M. Kornherr, H. Niedermeyer, K.-H. Steuer.
Cosmic plasma physics. Conference 1971, (see 012.016), p. 327 - 334 (1972).

062.046 Time-dependent kinetic processes in ionized gas. S. I. Grachev.
Vestn. Leningr. un-ta, 1972, No. 19, p. 128 - 136. In Russian. Abstr. in Referativ. Zhurn. 51. Astron., 4.51.343 (1973).

062.047 On the motion of conducting gas at neutral points of a magnetic field. A. G. Khantadze.
Trudy In-t geofiz. AN GruzSSR, Vol. 28, 30 - 36 (1972). In Russian.

062.048 On the reconnection of magnetic field lines in compressible conducting fluids. T. Yeh, M. Dryer.
Astrophys. Journ., Vol. 182, 301 - 315 (1973).
A previous study by Yeh and Axford (1970) of the reconnection of magnetic field lines in an incompressible fluid is extended to a compressible fluid. The authors discuss only the convective flows surrounding the diffusion region, leaving out the responsive flows inside the diffusion region to future studies. They compute the relationship between the outgoing flow and the incoming flow in Sonnerup's configuration for an isothermal, compressible fluid.

062.049 Re-connexion of magnetic lines of force: evolution in incompressible MHD fluids.
S. Fukao, T. Tsuda.
Planet. Space Sci., Vol. 21, 1151 - 1178 (1973).
Time-dependent incompressible MHD (magnetohydrodynamic) solutions in two dimensions are obtained numerically to study the evolutionary process involving a re-connexion of magnetic lines of force. Given an initial antiparallel magnetic field, or a current sheet, to which there is an injection of fluid in a transverse direction, we seek to see how the process of re-connexion builds up.

062.050 Normal Doppler shifted cyclotron radiation from a cold plasma. R. P. Singh.
Planet. Space Sci., Vol. 21, 1268 - 1269 (1973). – Research note.

062.051 Coherent gyromagnetic emission as a radiation mechanism. D. B. Melrose.
Australian Journ. Phys., Vol. 26, 229 - 247 (1973).
The properties of coherent gyromagnetic emission of radiation which can escape from a source of astrophysical interest by a bi-Maxwellian distribution of nonrelativistic electrons are considered.

062.052 Kritik an Syrovatskiis Mechanismus der dynamischen Dissipation von Magnetfeldern.
U. Anzer.
Mitt. Astron. Ges., No. 32, p. 171 - 172 (1973).

062.053 Effect of finite resistivity on the dynamic stability of a composite plasma.
P. K. Bhatia, P. N. Gupta.
Phys. Scripta, Vol. 7, 179 - 182 (1973).
The dynamic stability of a composite hydromagnetic plasma has been investigated to include simultaneously the effects of collisions with neutrals, Hall currents and finite resistivity. The prevalent magnetic field is assumed to be uniform and horizontal. It is found that the influence of the simultaneous inclusion of the effects of finite resistivity, Hall currents and frictional effects with neutrals is destabilizing.

062.054 Ionization and heating of a low density plasma by energetic particles.
S. M. V. Aldrovandi, D. Péquignot.
Astron. Astrophys., Vol. 26, 33 - 43 (1973).
The ionization and thermal equilibria of a gas with cosmic abundances heated by energetic particles are calculated in the range of temperature $10^2 - 10^8$ °K and in the range of number density $1 - 10^8$ cm^{-3}. The elements considered are H, He, C, N, O, Ne, Mg, Si and S. The cooling rates and line intensities of astrophysical interest are given as functions of temperature and total density. The reabsorption of the photons emitted by hydrogen and helium influences strongly the line intensities.

062.055 Electrodynamics of nonequilibrium plasma.
A. G. Sitenko, H. Wilhelmsson.
Phys. Scripta, Vol. 7, 189 (1973).

062.056 Transition probability approach to the theory of plasmas. S. Ichimaru, I. P. Yakimenko.
Phys. Scripta, Vol. 7, 198 - 208 (1973).

062.057 Five wave interaction–a possibility for enhancement of optical or microwave radiation by nonlinear coupling to explosively unstable plasma waves.
H. Wilhelmsson, V. P. Pavlenko.
Phys. Scripta, Vol. 7, 213 - 216 (1973).
A rigorous solution is obtained to the problem of five wave interaction in a plasma medium for the case where two of the waves are transverse electromagnetic waves, whereas the

remaining three waves are longitudinal plasma or beam waves. The effects of nonlinear instability are studied in a coherent wave description. As a result enhancement of radiative waves may occur in media of interest to high power laser and present day astrophysical research.

062.058 **On the stabilization of explosive instabilities by nonlinear frequency shifts.** V. N. Oraevskii, H. Wilhelmsson, E. Ya. Kogan, V. P. Pavlenko. Phys. Scripta, Vol. 7, 217 - 221 (1973).

062.059 **Einführende Betrachtungen zur neueren Entwicklung der Turbulenzphysik des Plasmas.** J. Wilhelm. Festschrift des wissenschaftlichen Kolloquiums zum 65. Geburtstag von Robert Rompe, [Akademie-Verlag, Berlin], p. 107 - 117 (1973).

062.060 **Non-radial oscillations and convective instability of a polytrope with a toroidal magnetic field.** N. K. Sood, S. K. Trehan. Astrophys. Space Sci., Vol. 19, 441 - 467 (1972). The authors examine the non-radial modes of oscillation, belonging to spherical harmonics of orders $l = 1$ and $l = 3$, of a gaseous polytrope with a toroidal magnetic field.

062.061 **On the stability of the spheroids extracted along the axis of rotation by a toroidal magnetic field.** R. S. Hovanesian (*Oganesyan*), M. G. Abramian. Astrofizika, Vol. 8, 599 - 608 (1972). In Russian. — English translation in Astrophysics, Vol. 8, No. 4. The problem of the stability of liquid spheroids extracted along the axis of rotation by a toroidal magnetic field is considered. The stability concerning small surface perturbations of the type $n = m = 2$ is studied. It is established that the figures extracted by a magnetic field are stable configurations.

062.062 **Plasma oscillations in magnetic and gravitational fields.** A. Z. Dolginov, M. A. Zelikman. Astrofizika, Vol. 9, 99 - 106 (1973). In Russian. — English translation in Astrophysics, Vol. 9, No. 1. The dispersion relations are obtained for low-frequency plasma oscillations in magnetic and gravitational fields. This kind of oscillations can exist in solar and stellar plasmas.

062.063 **Temperature determination from the relative line-to-continuum intensity of the He II 4686 and 3203 lines.** A. Delcroix, S. Volonte. Journ. Phys. B, Atomic Molecular Phys., Vol. 6, L4 - L7 = Commun. Dép. Astrophys. Fac. Sci. Mons, Mons Astrophys. Papers No. 32 (1973). Calculations of the temperature sensitive line-to-continuum intensity ratio of the He II 4686 and 3203 lines are presented for the temperature range below 80000 K where a density dependence sets in. When electron density is such that partial LTE holds, the method provides very accurate temperature measurements. The effect of small hydrogen admixtures is also outlined.

062.064 **Stability of a vortex sheet between collisionless plasmas [solar wind-magnetosphere boundary].** R. Rajaram, G. L. Kalra, J. N. Tandon. Journ. Plasma Phys., Vol. 9, 249 - 260 (1973). The problem of stability of a vortex sheet between two collisionless plasma media described by Chew et al. (1956) equations is examined. The relevance of these investigations to the internal structure of the solar wind and to the stability of the magnetosphere-solar wind boundary is discussed.

062.065 **Aligned rotating magnetospheres. I. General analysis.** E. T. Scharlemann, R. V. Wagoner. Astrophys. Journ., Vol. 182, 951 - 960 (1973). The authors consider the problem of finding self-consistent steady-state axisymmetric solutions for the electron and ion densities and velocities, as well as the electromagnetic field, outside a uniformly rotating, perfectly conducting star. The special case of a toroidal magnetic field which is a linear function of the poloidal magnetic potential A is treated in detail.

062.066 **A note on plasma acceleration caused by plasma turbulence.** H. Motz, V. N. Tsytovich. Plasma Phys., Vol. 14, 583 - 589 (1972).

062.067 **Wave propagation and amplification properties of the molecular electronic plasma.** O. E. H. Rydbeck, Å. Hjalmarson. Ionosph. Res. Lab., Pennsylvania State Univ., University Park, Penn., Sci. Rep. PSU-IRL-SCI-403, 3 + 43 pp. (1972).

062.068 **Propagation and damping of magnetohydrodynamic waves in interplanetary space.** I. N. Toptygin. IV Leningr. mezhdunar. seminar "Edinoobrazie uskoreniya chastits v razlich. masshtabakh kosmosa, 1972". Leningrad, 1972, p. 267 - 292. In Russian. — Abstr. in Referativ. Zhurn. 51. Astron., 7.51.400; 62. Issled. kosmich. prostranstva, 7.62.202 (1973).

Elementare Plasmaphysik.   See Abstr. 003.025.

Principles of plasma physics.   See Abstr. 003.075.

An introduction to the theory of plasma turbulence. See Abstr. 003.117.

Hydrogen Stark-broadening tables. See Abstr. 022.030.

Note on variational principles governing small departures from equilibrium in gaseous spheres and in horizontal fluids.   See Abstr. 065.167.

Wave-trains in the solar wind. I: General theory and its application to an ideal, isotropic, one-fluid plasma. See Abstr. 074.053.

Comments on geophysical plasma instabilities and their importance.   See Abstr. 084.283.

Errata

062.901 **Erratum: 'Quantum theory of line formation in a magnetic field'** [Solar Physics, Vol. 27, 319 - 329 (1972)].   E. Landi Degl'Innocenti, M. Landi Degl'Innocenti. Solar Physics, Vol. 29, 528 (1973).

063 Radiative Transfer

063.001 **Combined-operations method for diffuse reflection by an isotropic, non-coherent scattering homogeneous sphere.** T. H. Kho, K. K. Sen.
Astrophys. Space Sci., Vol. 18, 363 - 376 (1972).

Combined-operations method has been utilised to solve the problem of diffuse reflection by a homogeneous, isotropic, non-coherent scattering spherical medium. The source function is considered to be frequency independent. The auxiliary equation has been formulated, the scattering function defined, and the integro-differential equation for this function deduced. A method for obtaining the emergent intensity and the internal source function for non-zero internal source distribution has been suggested for a given line profile.

063.002 **A direct solution of the radiative transfer equation: Application to Rayleigh and Mie atmospheres.**
J. Canosa, H. R. Penafiel.
Journ. Quant. Spectrosc. Radiat. Transfer, Vol. 13, 21 - 39 (1973).

A direct method is given for the solution of the spherical harmonics approximation to the equation of radiative transfer in plane-parallel atmospheres. Test computations performed for Rayleigh and Mie scattering phase functions show that the direct method is unconditionally stable and solves efficiently problems both for optically thin and very thick atmospheres. Timing comparisons with the method of Chandrasekhar for Rayleigh atmospheres and with an integral-equation iterative method for Mie atmospheres are quite favorable to the proposed method.

063.003 **The applicability of an approximate expression for radiative heating.** R. D. Cess, V. Ramanathan.
Journ. Quant. Spectrosc. Radiat. Transfer, Vol. 13, 79 - 81 (1973).

The purpose of the present note is to discuss an approximate expression for radiative heating within a gas, which employs a radiative response time, and to appraise its applicability through comparison with a more detailed approach. Specific attention is directed to infrared radiation.

063.004 **On the resolvent kernel for the transfer problem in an homogeneous sphere.**
S. J. Wilson, K. K. Sen.
Journ. Quant. Spectrosc. Radiat. Transfer, Vol. 13, 83 - 85 (1973).

A scheme for obtaining the resolvent kernel of the integral equation for radiative transfer problems in an isotropically scattering sphere is proposed. Unlike the earlier attempts, the spherical problem is not transformed to its plane-parallel analogue.

063.005 **Multiple scattering of polarized light in a semi-infinite atmosphere with small true absorption.**
H. Domke.
Astron. Zhurn. Akad. Nauk SSSR, Vol. 50, 126 - 136 (1973).
In Russian. English translation in Soviet Astron. AJ, Vol. 17, No. 1.

The molecular scattering of light is assumed to be accompanied with small true absorption i.e. the particle albedo λ is close to unity. The linear terms of expansions in $(1-\lambda)^{1/2}$ of functions characterizing the emergent polarized light from a semi-infinite atmosphere are derived in terms of functions for conservative scattering. In the special case of Rayleigh scattering the functions are expressed in terms of the familiar scalar functions $H_e(\mu)$ and $H_r(\mu)$. The errors of the formulae are estimated.

063.006 **Non-grey radiative heat transfer in the picket-fence approximation.** S. K. Loyalka, A. Abadir.
Journ. Quant. Spectrosc. Radiat. Transfer, Vol. 13, 135 - 144 (1973).

The radiative heat transfer between parallel plates separated by an absorbing gas with a "picket-fence" absorption coefficient is treated by a variational technique. For the heat transfer, a simple rational algebraic expression is obtained and the variational results are found in excellent agreement with the numerically "exact" results reported recently by Reith et al (1971). An expression for the extrapolation distance is also deduced and the results are compared with the values reported earlier by Bond and Siewert (1970).

063.007 **Theorems on symmetries and flux conservation in radiative transfer using the matrix operator theory.**
G. W. Kattawar.
Journ. Quant. Spectrosc. Radiat. Transfer, Vol. 13, 145 - 153 (1973).

The matrix operator approach to radiative transfer is shown to be a very powerful technique in establishing symmetry relations for multiple scattering in inhomogeneous atmospheres. Symmetries are derived for the reflection and transmission operators using only the symmetry of the phase function. These results will mean large savings in computer time and storage for performing calculations for realistic planetary atmospheres using this method. The results have also been extended to establish a condition on the reflection matrix of a boundary in order to preserve reciprocity. Finally energy conservation is rigorously proven for conservative scattering in inhomogeneous atmospheres.

063.008 **One-dimensional line radiative transfer.**
K. G. Harstad.
Journ. Quant. Spectrosc. Radiat. Transfer, Vol. 13, 155 - 165 (1973).

Integrations over solid angle and frequency are performed in the expressions for the radiant heat flux and local energy loss of a line in a region of strong variations of the source function in one direction. Approximations are given for coefficients and kernels in the resulting forms which involve integrals over the physical coordinate.

063.009 **Probabilistic model for the resolvent kernel in diffusion problems in spherical-shell media.**
S. J. Wilson, K. K. Sen.
Journ. Quant. Spectrosc. Radiat. Transfer, Vol. 13, 255 - 266 (1973).

A probabilistic model of the resolvent kernel for the integral equation arising in diffusion problems in nonhomogeneous, isotropically scattering, spherical-shell media is proposed. An analogue of the X and Y functions of Chandrasekhar is given for the spherical-shell geometry.

063.010 **Equivalence relationships between diffuse radiation fields for finite slabs bounded by a perfect specular reflector and a perfect absorber.**
J. Casti, H. Kagiwada, R. Kalaba.
Journ. Quant. Spectrosc. Radiat. Transfer, Vol. 13, 267 - 272 (1973).

As part of a study of computational methods in radiative transfer problems, we present some simple algebraic relationships between source functions, as well as transmitted, reflected, and internal intensities, for a finite slab bounded by a completely absorbing layer and the same slab bounded by a perfect specular reflector. The paper also provides numerical checks on these equivalence relationships.

063.011   **Light scattering in a homogeneous sphere.**
V. V. Sobolev.
Astrofizika, Vol. 8, 197 - 212 (1972). In Russian. English translation in Astrophysics, Vol. 8, No. 2.

Formulae are obtained for the mean number of photon scatterings in a sphere and its luminosity, the distribution of energy sources being arbitrary. Three particular forms of the source distribution are considered in more detail: 1) uniform distribution, 2) external illumination of the sphere by parallel beams and 3) point source located at an arbitrary distance from the center of the sphere.

063.012   **Noncoherent scattering. III.**
N. B. Yengibarian (*Engibaryan*), A. G. Nikogosian.
Astrofizika, Vol. 8, 213 - 225 (1972). In Russian. English translation in Astrophysics, Vol. 8, No. 2.

The problem of noncoherent radiation transfer in a spectral line through an isothermic medium is considered. The systems of functional equations for the auxiliary functions, which are a generalization of Ambartsumian's well-known functions φ and ψ have been derived. A knowledge of these functions enables us to determine the reflection and transmission coefficients, as well as the radiation field in the medium.

063.013   **Simultaneous radiative and conductive heat transfer in non-gray media.**
D. G. Doornink, R. G. Hering.
Journ. Quant. Spectrosc. Radiat. Transfer, Vol. 13, 323-332 (1973).

Steady-state energy transfer through non-gray radiating and conducting media enclosed by black walls of unequal temperature is studied. Temperature distributions and total heat transfer results are presented for materials which absorb radiation (a) of low frequency, (b) of high frequency, (c) within a finite band width, and (d) of all frequencies (gray). The influence of optical thickness (τ_0) and conduction to a radiation interaction parameter (N) are examined and the results for non-gray materials are compared with those for a gray analysis. Exact results are compared with those determined by using the optically-thin and the optically-thick approximations as well as with those evaluated for purely conductive and purely radiative transfer.

063.014   **Radiative transfer in a nongray spherical layer: Simplified rectangular model.**
A. L. Crosbie, H. K. Khalil.
Journ. Quant. Spectrosc. Radiat. Transfer, Vol. 13, 359 - 367 (1973).

The problem of radiative transfer in a nongray, absorbing –emitting spherical layer is investigated. The nongray radiative transfer problem is reduced to a gray solution without the use of any approximation (such as the Planck or Rosseland means) for an isothermal layer and for radiative equilibrium.

063.015   **Scattering of resonance radiation in a sphere.**
D. I. Nagirner.
Astrofizika, Vol. 8, 353 - 368 (1972). In Russian. English translation in Astrophysics, Vol. 8, No. 3.

The scattering of resonance line radiation in a homogeneous sphere with spherically symmetric radiation sources is considered. The resolvent of the integral equation of the problem and the number of scatterings of photons are investigated. For the case of conservative scattering and optically thick sphere asymptotic forms of the resolvent, the mean number and the dispersion of the number of scatterings have been found. An asymptotic formula for the emergent intensity is also derived under the assumption of uniform source distribution. The results can be used in the theoretical study of planetary nebulae.

063.016   **Radiation absorption as a result of photoionization in a magnetic field.**   G. G. Pavlov.
Astron. Zhurn. Akad. Nauk SSSR, Vol. 50, 320 - 326 (1973). In Russian. English translation in Soviet Astron. AJ, Vol. 17, No. 2.

The coefficient of photoionization absorption in a magnetic field is calculated. It is shown that in the frequency dependence of the photoeffect cross section there arise peaks situated at equal distance (the Larmor frequency).

063.017   **Radiative transfer through carbon ablation layers.**
H. F. Nelson.
Journ. Quant. Spectrosc. Radiat. Transfer, Vol. 13, 427 - 445 (1973).

The composition, emittance, and transmittance of carbon plasmas have been calculated for temperatures of 3000 and 5000°K, thicknesses from 0.01 to 1.0 cm, and pressures from 0.1 to 10.0 atmospheres. The carbon plasmas are assumed to be isothermal and in local thermodynamic equilibrium. The emittance and transmittance results are applied to a simple model of the ablation layer in the stagnation shock layer for two cases: (1) Jupiter entry and (2) earth entry.

063.018   **Formation des raies spectrales et étude des courbes de croissance dans une atmosphère diffusante semi-infinie.**   Y. Fouquart, J. Lenoble.
Journ. Quant. Spectrosc. Radiat. Transfer, Vol. 13, 447 - 459 (1973).

The equivalent width of an absorption line formed by reflection from a semi-infinite planetary atmosphere with an arbitrary phase function is studied. An analytic expression, for a Lorentz line, is derived from the distribution of photon optical path. An approximate expression for the distribution of photon optical path, which is useful for computations of synthetic spectra and equivalent widths, is proposed and some examples of the "curve of growth" are presented for different phase functions.

063.019   **The superposition of layers in radiative transfer.**
H. G. Horak, R. F. Beebe, H. A. Beebe.
Bull. American Astron. Soc., Vol. 5, 26 (1973). – Abstr. AAS.

063.020   **Two dimensional implicit radiation hydrodynamics.**
M. T. Sandford II.
Bull. American Astron. Soc., Vol. 5, 26 (1973). – Abstr. AAS.

063.021   **Tables of light scattering by a polydisperse system of spherical particles.**
É. G. Yanovitskij, Z. O. Dumanskij.
Naukova dumka, Kiev. 124 pp. Price 30 Kop. (1972). In Russian. – Review in Referativ. Zhurn. 51. Astron., 3.51.256 (1973).

063.022   **Extinction and scattering by several types of silicate sphere of radius $0.05-1.0$ μm, for the wavelength range $0.21-50$ μm.**
G. E. Bromage, K. Nandy, B. N. Khare.
Astrophys. Space Sci., Vol. 20, 213 - 224 (1973).

The exact calculation of scattering and absorption by various sub-micron sized silicate spheres is presented, using accurately determined optical constants in the wavelength range from 50 μm to 0.21 μm. The extinction features near 10 μm and 20 μm for various samples are discussed.

063.023   **The scattering of resonance-line radiation in the limit of large optical depth.**   J. P. Harrington.
Monthly Notices Roy. Astron. Soc., Vol. 162, 43 - 52 (1973).

It is shown that the essential features of the transfer of resonance-line radiation in very optically thick media of low density are described by the Poisson equation. The solution of this equation is presented in the form of an eigen-function expansion. Simple closed expressions are given for the mean number of scatterings for escape and for the line profile at both the centre and surface of the medium.

063.024 Matrix operator theory of radiative transfer. 1: Rayleigh scattering.
G. N. Plass, G. W. Kattawar, F. E. Catchings.
Applied Optics, Vol. 12, 314 - 329 (1973).

An entirely rigorous method for the solution of the equations for radiative transfer based on the matrix operator theory is reviewed. Both the general theory and its history together with the method of calculation are discussed. As a first example of the method numerous curves are given for both the reflected and transmitted radiance for Rayleigh scattering from a homogeneous layer for a range of optical thickness from 0.0019 to 4096, surface albedo $A = 0, 0.2$ and 1, and cosine of solar zenith angle $\mu = 1, 0.5397$, and 0.1882. It is shown that the matrix operator approach contains the doubling method as a special case.

063.025 Induced acceleration of relativistic electrons by Compton scattering.
Yu. P. Ochelkov, V. M. Charugin.
Astron. Tsirk., No. 759, p. 1 - 2 (1973). In Russian.

063.026 An improved separability approximation for line radiative transport in nonhomogeneous media.
R. E. Boughner.
Journ. Quant. Spectrosc. Radiat. Transfer, Vol. 13, 499 - 508 (1973).

A simple modification to the constant half-width approximation of Wilson and Greif is introduced which permits a more accurate evaluation of line radiative transport in non-homogeneous gases. To demonstrate the method's accuracy, comparisons are made with Wilson and Greif and numerical frequency integrated results for the line equivalent width and radiative flux in a planar slab with prescribed Planck function and line half-width spatial variations.

063.027 Measurement of Mie scattering intensities from monodispersed spherical particles as a function of wavelength.
A. Cohen, V. E. Derr, G. T. McNice, R. E. Cupp.
Applied Optics, Vol. 12, 779 - 782 (1973).

Mie scattering intensities as a function of the size parameter are measured by use of a tunable dye laser and mono-dispersed spherical particles. The experimental results are compared with the Mie single scattering theory; a discrepancy in the exact position of the maxima and minima was detected. Agreement between experiment and theory was improved by applying a correction to the manufacturer's index of refraction function for the particles.

063.028 Photon bubbles.
K. H. Prendergast, E. A. Spiegel.
Comments Astrophys. Space Phys., Vol. 5, 43 - 50 (1973).

Can we expect that bubbles of photons form and move up through a scattering atmosphere just as gas bubbles do in a fluidized bed? If such a process is possible in astrophysical contexts it would help to explain some puzzling observations in objects having hot, extended atmospheres. But before discussing such matters we have to ask whether the notion of photon bubbles is reasonable.

063.029 Integral kernels and exact solutions to some radiative problems in spherical geometries.
J. Schmid-Burgk.
Astrophys. Journ., Vol. 181, 865 - 874 (1973).

Problems of radiative transfer in spherical geometries are investigated in the gray approximation by using a two-para-

meter family of absorption coefficients as functions of radial depth. The parameters describe the extent to which the atmosphere is actually dominated by geometric effects. With these absorption coefficients, kernels for the Fredholm equation for the mean intensity are calculated in analytic form. Exact solutions to the resulting Wiener-Hopf equation are presented for the case that the "degree of sphericity" remains constant throughout the atmosphere. Milne's problem is used for demonstration. The extent to which the radiation field remains anisotropic down to arbitrarily large optical depths (thus invalidating the unmodified Eddington approximation) is shown as well as the difference in surface intensities between plane-parallel and extended atmospheres. An approximate expression for the mean intensity is derived from the exact solution.

063.030 On the relation of the coherence theory to the radiative transfer equation.
G. I. Ovchinnikov, V. I. Tatarskij.
Izv. vyssh. ucheb. zavedenij. Radiofizika, Vol. 15, 1419 - 1421 (1972). In Russian. – Abstr. in Referativ. Zhurn. 51. Astron., 4.51.329 (1973).

063.031 Angular quadrature perturbations in radiative transfer theory.   C. J. Cannon.
Journ. Quant. Spectrosc. Radiat. Transfer, Vol. 13, 627 - 633 (1973).

A new numerical method is presented for solving the general equation of radiative transfer. The approximation, which replaces the integral term over angle in the transfer equation by a quadrature sum, is studied; an estimate of the error involved is obtained and this error can then be used to evaluate a correction to the radiation field originally determined. This process may then be continued as a perturbation series. Examples are given in the context of spectral line formation in slab geometry.

063.032 Probabilistic radiative transfer: mean number of scatterings.   G. D. Finn.
Journ. Quant. Spectrosc. Radiat. Transfer, Vol. 13, 683 - 697 (1973).

Integral relations are formulated for the mean numbers of scatterings for escaping and for non-escaping photons. Numerical solutions are obtained and discussed for cases of transfer through finite and semi-infinite atmospheres. The association between these functions and the source function is also discussed.

063.033 Scattering and transmission matrices of partially polarized radiation in a Rayleigh atmosphere bounded by a specular reflector.   S. Ueno.
Bull. American Astron. Soc., Vol. 5, 304 (1973). – Abstr. AAS.

063.034 Non-coherent scattering in transfer problems in spherical shell media. I. Frequency-independent source function.   T. H. Kho, K. K. Sen.
Astrophys. Space Sci., Vol. 21, 39 - 57 (1973).

The Combined Operations Method is utilised to solve diffuse reflection and transmission problems in inhomogeneous, isotropically and non-coherently scattering, spherical shell media. The source function is assumed to be frequency independent. The N-solution of an auxiliary equation is sought, and tractable equations for the scattering and transmission functions are established. The solution of the problem for a scattering and emitting medium have been considered for a perfectly absorbing core.

063.035 Probabilistic model for radiative transfer problems in cylindrical shell media with complete redistribution in frequency.   T. K. Leong, K. K. Sen.

Astrophys. Space Sci., Vol. 21, 59 - 72 (1973).

A probabilistic model for solving transfer problems in non-homogeneous, isotropic, and non-coherent scattering cylindrical shell media has been proposed. The source function is considered to be frequency independent. The scattering and transmission functions have been defined for the case of complete redistribution in frequency. A tractable integro-differential equation for the scattering function has been derived.

063.036 **Non-coherent scattering in transfer problems in spherical shell media. II. Frequency-dependent source function.** T. H. Kho, K. K. Sen.
Astrophys. Space Sci., Vol. 21, 237 - 255 (1973).

In the present paper, the Combined Operations Method has been utilised to solve the diffuse radiation problems in inhomogeneous, non-coherently scattering, spherical shell media. A redistribution function is introduced so as to account for the frequency dependent source function. The relevant auxiliary equation has been formulated and its N-solution sought. A scattering and emitting medium with a partially transparent core has been considered.

063.037 **Phase velocity effects in non-linear Compton scattering.** P. Stewart.
Astron. Astrophys., Vol. 25, 457 - 460 (1973).

An intense circularly polarized electromagnetic wave whose phase velocity is greater than $c = 3 \times 10^{10}$ cm/s can cause a charged particle to radiate energy at a much greater rate than if the phase velocity is equal to c. The power radiated can in certain cases depend on the fourth power of electric field.

063.038 **Redistribution of resonance radiation. II. The effect of magnetic fields.**
A. Omont, E. W. Smith, J. Cooper.
Astrophys. Journ., Vol. 182, 283 - 300 (1973).

Previously obtained results for scattering of radiation in the presence of collisions are restated in a density matrix formalism which employs an irreducible-tensor description of the radiation field. The redistribution is then extended to include the effect of a weak magnetic field. By averaging over a finite bandwidth which is on the order of the Doppler width, simplified expressions of physical significance for the scattering in the Doppler core and the Lorentz wings are obtained. Expressions are also obtained for the corresponding source function of radiative transfer theory.

063.039 **Streufunktionen nichtkugelförmiger Teilchen.**
R. Zerull.
Mitt. Astron. Ges., No. 32, p. 162 (1973). – Abstract.

063.040 **On the dependence of the two-level source function on its own radiation field.**
R. Steinitz, R. A. Shine.
Monthly Notices Roy. Astron. Soc., Vol. 162, 197 - 206 (1973).

The consequences of the universally made assumption that the stimulated emission profile is identical to the absorption profile are quantitatively investigated for a two-level atom with Doppler redistribution.

063.041 **Circular polarization by single scattering of unpolarized light from loss-less, non-spherical particles.**
L. W. Bandermann, J. C. Kemp.
Monthly Notices Roy. Astron. Soc., Vol. 162, 367 - 377 (1973).

Elliptically polarized light can be obtained by scattering of unpolarized light from loss-less, non-spherical particles whose dimensions are not necessarily small compared with a wavelength. The authors represent such a particle by a finite or infinite number of Rayleigh spheres and thereby derive approx-imate expressions for the total scattered intensity and for the circularly polarized component.

063.042 **Probabilistic model for time-dependent transfer problems in finite cylindrical shell medium.**
T. K. Leong.
Astrophys. Space Sci., Vol. 19, 369 - 385 (1972).

A time dependent probabilistic model for the description of photon diffusion in infinite cylindrical shell media has been proposed. Four basic probability functions are defined and an integro-differential equation for the scattering function is derived. The expressions for the emergent intensities expressed in terms of the scattering, transmission, back scattering, and back transmission functions are obtained for two special cases.

063.043 **Note on the modified two-stream approximation of Sagan and Pollack.** D. R. Lyzenga.
Icarus, Vol. 19, 240 - 243 (1973).

The modified two-stream approximation is derived from the radiative transfer equation using a two-point gaussian quadrature and neglecting terms of order $l \geqslant 3$ in the Legendre expansion of the phase function. The results are compared with the Eddington approximation and with exact results for the special cases of perfect absorption, perfect scattering, and a semi-infinite layer with isotropic scattering.

063.044 **Absorption and emission of radiation by nonspherical particles.** J. M. Greenberg.
Journ. Colloid Interface Sci., Vol. 39, 513 - 519 = Dudley Obs., *Albany, N. Y.*, Repr. No. C 45 (1972).

The absorption and emission cross sections of finely divided particles are shown to depend strongly on particle shapes not only when the particles are comparable to the wavelength but also when the particles are so small that the Rayleigh approximation is applicable. Some detailed comparisons are shown for some infrared absorption spectra of enstatite and ice.

063.045 **Non-coherent scattering. IV. An infinite medium.**
N. B. Yengibarian (*Engibaryan*), A. G. Nicoghossian (*Nikogosyan*).
Astrofizika, Vol. 9, 79 - 94 (1973). In Russian. – English translation in Astrophysics, Vol. 9, No. 1.

The problem of resonance radiation transfer in an infinite medium is solved. The problem is considered for a sufficient general assumption on the form of the redistribution function, when the latter is independent of the angle of scattering. The degree of excitation on different distances from the source, as well as the distribution of the length of the photon path depending on the original frequency are determined. Results of numerical calculations are given.

063.046 **The possibility of photon escape from a semi-infinite atmosphere with total redistribution according to frequencies.** E. W. Elst.
Physics of the moon and planets, (see 012.024), p. 204 (1972). In Russian. – Abstract.

063.047 **Calculation of diffuse reflexion and transmission of light through a semi-infinite atmosphere.**
A. K. Kolesov.
Physics of the moon and planets, (see 012.024), p. 204 - 206 (1972). In Russian.

063.048 **Infrared extinction cross sections of silicate grains.**
R. F. Knacke, R. K. Thomson.
Publ. Astron. Soc. Pacific, Vol. 85, 341 - 347 (1973).

Extinction cross sections for plagioclase and pyroxene silicate particles have been calculated for frequencies between 10 and 1400 cm^{-1}. Results are compared to infrared observations of grains.

063.049 Reflection and transmission of light by a semi-infinite atmosphere with anisotropic scattering.
A. K. Kolesov.
Trudy Astron. Obs., *Leningrad,* Vol. 29 (= Uchenye Zapiski Leningr. Un-ta, No. 363 = Seriya Matem. Nauk, vyp. (No.) 48), p. 6 - 16 (1973). In Russian.

The coefficients of reflection and transmission of light by a semi-infinite atmosphere are calculated for several special forms of the scattering indicatrix. A method, due to V. V. Sobolev, based on the expression of reflection and transmission coefficients in terms of Chandrasekhar's H-function and polynomials, is used in the calculations.

063.050 Functions describing the scattering of resonance radiation in an infinite medium.
D. I. Nagirner, A. B. Schneeweiss (*Shnejvajs*).
Trudy Astron. Obs., *Leningrad,* Vol. 29 (= Uchenye Zapiski Leningr. Un-ta, No. 363 = Seriya Matem. Nauk, vyp. (No.) 48), p. 16 - 32 (1973). In Russian.

The Green functions (resolvents) are tabulated of the integral equations of resonance radiation transfer in a plane infinite medium and in a homogeneous infinite medium. The radiation is scattered by two-level atoms with complete frequency redistribution. The cases of Lorentz and Doppler profiles are considered. Exact and asymptotic expressions for the functions under consideration are listed. The range of validity and the accuracy of the asymptotics are estimated.

063.051 Diffuse reflection of light from a semi-infinite

atmosphere with a three-term scattering indicatrix.
V. M. Loskutov.
Trudy Astron. Obs., *Leningrad,* Vol. 29 (=Uchenye Zapiski Leningr. Un-ta, No. 363 = Seriya Matem. Nauk, vyp. (No.) 48), p. 33 - 39 (1973). In Russian.

Expressions are given for the determination of the azimuth-dependent components of the intensity of radiation in the problem of diffuse reflection by a semi-infinite atmosphere with the scattering indicatrix given by Kolesov and Sobolev (1969). These expressions are used for computations assuming two special forms of the indicatrix. The results and the brightness distribution along the disk of a planet are given in tables.

Emission, absorption and transfer of radiation in related atmospheres.   See Abstr. 003.024.

Light scattering functions for small particles with applications in astronomy.   See Abstr. 003.124.

Radiation field in the deep layers of planetary atmospheres.   See Abstr. 091.001.

Radiative transfer in atmospheres of Algol-type binaries. Departures from the state of LTE in late-type components.   See Abstr. 121.053.

Radiative transport in interstellar masers.
See Abstr. 131.203.

064 Stellar Atmospheres, Stellar Envelopes

064.001 Grain escape velocities from cool stars.
R. C. Gilman.
Monthly Notices Roy. Astron. Soc., Vol. 161, 3 P - 4 P (1973).
Wickramasinghe obtained grain escape velocities ~ 1000 km s^{-1} assuming that the drag of the circumstellar gas was negligible. It is shown that the drag of the circumstellar gas is not negligible, and when it is included, the escape velocities are more likely < 100 km s^{-1}.

064.002 An experiment on using an electrospectrometer for determining electron concentration in stellar atmospheres.
V. G. Karetnikov, T. N. Korotkikh, Ju. A. Medvedev.
Astron. Zhurn. Akad. Nauk SSSR, Vol. 50, 225 - 227 (1973).
In Russian. English translation in Soviet Astron. AJ, Vol. 17, No. 1. – Short note.

064.003 Optically thin stellar winds in early-type stars.
J. P. Cassinelli, J. I. Castor.
Astrophys. Journ., Vol. 179, 189 - 207 (1973).
The role of the radiation field in the heating, cooling, and transfer of momentum to expanding atmospheres of hot stars is examined. The general equations for radial, steady-state flow are presented, and the conditions for the existence of a transonic flow are discussed. Numerical solutions are carried out for a case in which the run of mean intensity is assumed to be that of freely streaming radiation. Limits are deduced for the ratio of the velocity of escape to the thermal speed at the sonic point. True absorptive opacity is found to be essential for the transition to supersonic flow. A two-point boundary-value technique is presented which ensures that the atmosphere approaches the usual static behavior at the base and becomes transonic at a larger radius.

064.004 Analyses of light-ion spectra in stellar atmospheres. II. The calcium II K-line in B stars. D. Mihalas.
Astrophys. Journ., Vol. 179, 209 - 220 (1973).
A calculation of the Ca II K-line strength in the middle B stars has been performed by means of a self-consistent solution of the equations of transfer and steady-state statistical equilibrium. The multilevel model ion, with many explicitly computed bound-bound transitions, is treated as part of a multi-ion system (Ca II, Ca III, Ca IV). Results are given for two abundances and microturbulent velocities, and are used to assess systematic errors in LTE diagnostics of these parameters. A discussion of the K-line strengths in ι Her and γ Peg indicates that they are compatible with a Ca abundance in these stars equal to the solar abundance.

064.005 Electric conductivity in the atmosphere of early-type stars. M. Kopecký, P. Kotrč.
Bull. Astron. Inst. Czechoslovakia, Vol. 24, 39 - 49 (1973).
Using the Kuklin and Nagasawa methods the values of the electric conductivity in Mihala's models of atmospheres of early-type stars have been determined. The difference between the Kuklin and Nagasawa methods in computing the electric conductivity is discussed. The electric conductivity as a function of optical depth, acceleration of gravity at the surface of the star and of the effective temperature of the stars is also discussed, in models of Mihalas's stellar atmospheres, as well as in models of stellar atmospheres of de Jager and Neven.

064.006 On the atmospheric abundances of seven Am SB2 systems. D. J. Stickland.
Monthly Notices Roy. Astron. Soc., Vol. 161, 193 - 211 (1973).
The observed characteristics of the metallic-line A type stars are described and the theories to account for their apparent abundance anomalies are reviewed. The mounting of an observing programme for the determination of the compositions of the component stars of a number of double-lined (SB2) systems and the reduction of the spectroscopic and spectrophotometric data are discussed. An account is given of a method for the determination of the atmospheric parameters and the component star luminosity ratios in SB2 systems and also of the abundance analysis procedure through the differential curve of growth.

064.007 The metallic-line star 15 UMa and the F 5 V star 5 And. M. A. Falipou.
Astron. Astrophys., Vol. 22, 445 - 451 (1973).
A detailed model atmosphere analysis is presented for the Am star 15 UMa and the F 5 V star 5 And, assuming LTE. The hydrogen line profiles and the stellar ionization equilibrium indicate that the atmospheres can be represented by a model with: $\theta_{eff} = 0.71$ and $\log g = 4.2$ for 15 UMa and $\theta_{eff} = 0.81$, $\log g = 3.8$ for 5 And. In order to obtain the microturbulent parameter and the abundances, a differential curve of growth analysis was applied, the sun first, and then 5 And being taken as comparison stars. It is shown that 15 UMa whose line profiles are rotationally broadened, is a classical Am star.

064.008 On ultraviolet absorption by molecular hydrogen in stellar atmospheres. F. Praderie, T. P. Stecher.
Astron. Astrophys., Vol. 23, 49 - 50 (1973).
Quasimolecular absorption by molecular hydrogen in the singlet $X^1 \Sigma_g^+ - B^1 \Sigma_u^+$ is shown to be a major contributor to opacity in main sequence A stars at wavelengths shorter than 1600 Å. The importance of this absorber increases for higher surface gravity. This absorption partially masks the Si I photoionization discontinuity at 1520 Å, rendering it less apparent, in accordance with observation.

064.009 Chromospheric heating of very hot stars by radiation driven sound waves. II. A. G. Hearn.
Astron. Astrophys., Vol. 23, 97 - 103 (1973).
An estimate is made of the flux of radiation driven sound waves available for chromospheric heating of very hot stars. The equilibrium energy of radiation driven sound waves propagating in a uniform density atmosphere with viscous damping is calculated and this solution is used to estimate the flux in a star with an effective temperature of 50000°K (type O5). It is estimated that 8.6×10^7 erg cm^{-2} s^{-1} is available.

064.010 The line dispersion function for enhanced damping.
A. Wittmann.
Astron. Astrophys. Suppl. Ser., Vol. 9, 209 - 212 (1973).
In analogy to the Taylor series expansion of the line dispersion function $F(a, v)$ for small damping parameter a (similar to that of the related Voigt function), an approximation of $F(a, v)$ for large a is provided in order to facilitate frequent computations of that function.

064.011 The production of discrete, quantized outflow velocities by radiation pressure in stars, Seyfert nuclei, and quasi-stellar objects. J. D. Scargle.
Astrophys. Journ., Vol. 179, 705 - 719 = Contr. Lick Obs., No. 372 (1973).
A number of stars, Seyfert galaxies, and quasi-stellar objects are ejecting gas at discrete velocities which, in some cases, form arithmetic sequences. This mass loss is probably driven by radiation pressure; the discreteness and ordered velocity structure can be understood as the result of the pres-

ence of opacity in closely spaced pairs of lines. The value of the "quantum" of velocity in a number of specific cases predicted by the theory is in good agreement with the observations.

064.012 **Analyses of light-ion spectra in stellar atmospheres. III. Nitrogen III in the O stars.**
D. Mihalas, D. G. Hummer.
Astrophys. Journ., Vol. 179, 827 - 845 (1973).
An analysis of the N III emission lines in O stars has been carried out on the basis of a detailed solution of the coupled statistical-equilibrium and transfer equations for a multiline, multilevel, multi-ion ensemble. Our calculations, using static, plane-parallel models reproduce successfully the observed emission at $\lambda\lambda 4634, 4640, 4641$ ($3p\,^2P^0 - 3d\,^2D$) and absorption at $\lambda\lambda 4097, 4103$ ($3s\,^2S - 3p\,^2P^0$).

064.013 **Monte Carlo radiative-transfer solutions for cool stellar photospheres.**
M. T. Sandford II, T. A. Pauls.
Astrophys. Journ., Vol. 179, 875 - 883 (1973).
Numerical evaluations of the exact solution for the uniformly emitting, Rayleigh scattering, conservative, plane-parallel, finite atmosphere are presented for optical thicknesses $\tau_1 = 0.20$, 1.00 and 2.00, and are used to evaluate the precision of solutions made with the adjoint Monte Carlo method. This method is employed to calculate the limb darkening and polarization produced by Rayleigh scattering in an M-star model atmosphere, at four wavelengths.

064.014 **Comparison of Celescope magnitudes with model-atmosphere predictions for A, F, and G supergiants.**
S. B. Parsons, E. Peytremann.
Astrophys. Journ., Vol. 180, 71 - 79 (1973).
Ultraviolet data from OAO-2 are analyzed with respect to three different sets of model atmospheres. We find that (1) A0−A2 supergiants are in good agreement with model atmospheres; (2) late A and early F supergiants are fainter than predicted; and (3) the two G0 supergiants are considerably brighter than expected.

064.015 **Cooling effect of CO in stellar atmospheres.**
H. R. Johnson.
Astrophys. Journ., Vol. 180, 81 - 89 = Publ. Goethe Link Obs., *Bloomington*, No. 146 (1973).
We investigate the effects of the CO infrared opacity ($1.5-5.0\,\mu$) in stellar atmospheres. For model atmospheres with effective temperatures between 3000° and 5000°K, CO lowers the boundary temperature by amounts up to ~1000°K and usually back-warms the photosphere slightly. We speculate on the possible application of the results to the temperature minimum of the sun and to grain formation in cooler stars.

064.016 **On the possibility of acceleration of matter in hot stars by absorption in spectral lines.**
I. F. Malov.
Astrofizika, Vol. 8, 227 - 233 (1972). In Russian. English translation in Astrophysics, Vol. 8, No. 2.
The maximum values of momentum transferred by radiation to the matter by absorption in spectral lines in Wolf-Rayet stars and OB-supergiants are obtained. The results are shown in figures and listed in a table. It is shown that the acceleration mechanism connected with an absorption in spectral lines does not play the main role in the observed outflow of matter from Wolf-Rayet stars and OB-supergiants.

064.017 **The atmosphere of non-rotating baryon stars.**
G. S. Saakian, D. M. Sedrakian.
Astrofizika, Vol. 8, 283 - 293 (1972). In Russian. English translation in Astrophysics, Vol. 8, No. 2.
A model of a hot baryon star atmosphere, composed of electron-proton gas and radiation, is considered. When the luminosity is of the order of $L = 1.3 \times 10^{38}$ erg/sec ($M = M_\odot$, $R = 10$ km and the surface temperature $T_R = 2 \times 10^7\,°$K) we get a sufficiently extended atmosphere. The density of particles on the surface is $n = 7 \times 10^{18}$ cm^{-3}, and on a distance of the order of 10^3 km is $n = 7 \times 10^{12}$ cm^{-3}. These characteristics of the atmosphere coincide with those of pulsars.

064.018 **On the relation between optical scale height and density scale height in a stellar atmosphere.**
H. G. van Bueren.
Astron. Astrophys., Vol. 23, 247 - 252 (1973).
In the present paper the optical scale height is expressed in terms of the density scale height H. Whereas this relation depends on the variation of the absorption coefficient with density and temperature, in reasonable approximation a simple proportionality between the two scale heights can be shown to exist, be it in a limited region of optical depth ($10^{-4} < \tau < 0.5$).

064.019 **A numerical method for inverting a single absorption line profile.** A. G. Hearn, J. N. Holt.
Astron. Astrophys., Vol. 23, 347 - 355 (1973).
A numerical method of inverting a single absorption line profile is described. It determines the largest set of significant parameters which describe the atmosphere forming the line. The method is general in application, but it has been developed using a sodium D_2 line profile measured at the centre of the disc of the sun at high spatial resolution.

064.020 **Molecular abundances in stellar atmospheres. II.**
T. Tsuji.
Astron. Astrophys., Vol. 23, 411 - 431 (1973).
Chemical equilibria of 36 elements are solved for the physical conditions of cool stellar atmospheres. It is found that the molecular species formed (monoxide, dioxide, halide etc.) and the degree of molecular association (i.e. the fraction of atoms locked in molecules) are well correlated with the position of each element (both group and atomic weight) in the periodic table.

064.021 **The reality of microturbulence.** G. Worrall.
Nature, Phys. Sci., Vol. 241, 7 (1973).
A few comments to the controversy stimulated by recent questioning of the reality of microturbulence in stellar atmospheres and of the assumption of local thermodynamic equilibrium.

064.022 **Conditions for the production of promethium in stellar atmospheres.**
H. R. E. Tjin a Djie, R. J. Takens, E. P. J. van den Heuvel.
Astrophys. Letters, Vol. 13, 215 - 220 (1973).
Physical conditions for the production of promethium in rare-earth-rich stellar atmospheres have been examined. Production by the spontaneous fission of superheavy elements sets very strong requirements on the time elapsed since the superheavy elements were formed. It appears that reactions of flare-accelerated particles with neodymium offer a more likely mechanism for the production of observable quantities of Pm.

064.023 **A non-LTE study of silicon line formation in early-type main-sequence atmospheres.** L. W. Kamp.
Astrophys. Journ., Vol. 180, 447 - 468 (1973).
We have computed populations of 16 levels of Si III−V and radiation fields in all connecting transitions; in particular the first six Si III triplet levels, including the $\lambda 4553$ line, and the first six Si IV levels including $\lambda 4089$. The computations were done for four non-LTE H - He model atmospheres, provided by Auer and Mihalas, with $T_{eff} = 25{,}000°$, $30{,}000°$, $35{,}000°$, and $40{,}000°$K, all with $\log g = 4$. The results have

been compared with LTE calculations for the same models and with observations.

064.024 Application of the complete-linearization method to the problem of non-LTE line formation.
L. Auer.
Astrophys. Journ., Vol. 180, 469 - 472 (1973).

It is shown that the equations of statistical equilibrium may be incorporated directly into the transfer equation, thus reducing the basic matrix size for the complete-linearization method. Treatment of weak transitions and development of starting solutions are also discussed.

064.025 Stellar model chromospheres. I. On the temperature minima of F, G, and K stars.
J. L. Linsky, T. Ayres.
Astrophys. Journ., Vol. 180, 473 - 481 (1973).

Brightness temperatures are deduced for the H_{1V} and K_{1V} features of the Ca II resonance lines in Procyon (F5 IV–V), Arcturus (K2 IIIp), and the sun (G2 V).

064.026 Wolf-Rayet stars. V. The temperature stratification.
L. V. Kuhi.
Astrophys. Journ., Vol. 180, 783 - 789 (1973).

The behavior of C III λ5696 with changing spectral subclass is interpreted in terms of the expanding-envelope model to indicate that in the WC stars the temperature must decrease with radius over the bulk of the envelope.

064.027 On the ultraviolet line blanketing in early type atmospheres. D. Eberlein, M. Scholz, G. Traving.
Astron. Astrophys., Vol. 24, 295 - 298 (1973).

Ultraviolet line blanketing in early type atmospheres is discussed. Taking into account realistic damping constants and approximative non-LTE effects, heating of the atmosphere turns out to be smaller than found in the Princeton computations.

064.028 A search for density inversions in stellar envelopes.
J. G. Eoll.
Bull. American Astron. Soc., Vol. 5, 1 (1973). – Abstr. AAS.

064.029 Discovery of upper photospheric temperature inversions or chromospheres in early A stars.
J. L. Linsky, R. A. Shine, T. R. Ayres, F. Praderie.
Bull. American Astron. Soc., Vol. 5, 3 - 4, with a correction, p. 310 (1973). – Abstr. AAS.

064.030 On the Balmer discontinuity in the O-stars.
C. M. Anderson, F. H. Schiffer.
Bull. American Astron. Soc., Vol. 5, 9 - 10 (1973). – Abstr. AAS.

064.031 A picket fence model for stars with extended atmospheres. J. P. Cassinelli.
Bull. American Astron. Soc., Vol. 5, 10 (1973). – Abstr. AAS.

064.032 Formation of spectral lines in spherical stellar atmospheres. P. Kunasz, D. G. Hummer.
Bull. American Astron. Soc., Vol. 5, 11 (1973). – Abstr. AAS.

064.033 The envelopes of Be stars. C. R. Kitchin.
Monthly Notices Roy. Astron. Soc., Vol. 161, 381 - 388 (1973).

The method for determining the envelope radii for Be stars presented in an earlier paper has been extended. It now gives comparatively unambiguous models for the envelopes and provides some measure of the size of the region of the envelope which is emitting or absorbing a line.

064.034 Some measurements of Be star envelopes.

C. R. Kitchin.
Monthly Notices Roy. Astron. Soc., Vol. 161, 389 - 392 (1973).

Measurements on 48 lines from the envelopes of 11 Be stars have been made. The envelope model and the radius of the emitting/absorbing region has been determined by the method outlined in earlier papers. The results may indicate that a change in the nature of Be star envelopes occurs at a distance of about 30 stellar radii.

064.035 Opacity probability distribution functions for application to non-grey late-type stars model atmospheres. M. Querci, F. Querci, T. Tsuji.
17th Colloquium International Astrophys. Liège 1971, (see 012.003), p. 179 - 186 (1972).

064.036 A kinematic study of spectral features in cool star envelopes. N. J. Woolf.
17th Colloquium International Astrophys. Liège 1971, (see 012.003), p. 209 - 216 (1972).

064.037 Non-local convection model for stellar envelopes.
E. Ergma.
Nauchn. Informatsii, vyp. (No.) 23, p. 33 - 46 (1972). In Russian.

A method for calculating non-local convection is presented. The model of convection used is a variation of that given by Ulrich (1970). Envelope models for a main-sequence star with M = $1.25 M_\odot$, L = $2.14 L_\odot$, lg T_{ef} = 3.836 and for a giant with M = $3 M_\odot$, L = $100 L_\odot$, lg T_{ef} = 3.700 have been calculated.

064.038 Thermal convective instability in a magnetized isothermal atmosphere. M. Singla, S. P. Talwar.
Astron. Astrophys., Vol. 24, 441 - 446 (1973).

The problem of thermal instability in a uniformly rotating magnetized atmosphere is investigated. The configuration is assumed isothermal and characterized by a constant Alfvén speed. The heat-loss mechanism is taken to depend on temperature only.

064.039 Construction of the photospheric model of an A5 V star according to the modified method of Mustel.
E. E. Belyaeva, L. L. Shishkina.
Trudy Kazan. Gorod. Astron. Obs., No. 38, p. 26 - 36 (1972). In Russian.

064.040 On the numerical solution of an equation in the theory of stellar photospheres.
E. E. Belyaeva, L. L. Shishkina.
Trudy Kazan. Gorod. Astron. Obs., No. 38, p. 37 - 45 (1972). In Russian.

064.041 X-ray and radio emission from stellar coronae.
M. Landini, B. C. Fossi.
Astron. Astrophys., Vol. 25, 9 - 16 (1973).

Recent models of solar type coronae have been analyzed in order to give expected X-ray and radio waves fluxes. Hard X-ray (1 – 8 Å), soft X-ray (44 – 60 Å) and metric fluxes are computed for a selected number of stars. Soft X-ray emission from stellar coronae within 200 pc may account for a large fraction of the soft X-ray background.

064.042 Possible production of the soft X-ray background by the coronae of red giants. J. G. Hills.
Astrophys. Letters, Vol. 14, 69 - 71 (1973).

We find that if stellar mass loss occurs predominantly by means of stellar winds there is a direct proportionality between the optical luminosity of most stellar systems and the integrated soft X-ray luminosity of the coronae of stars evolving off the main sequence in these systems.

064.043 About the influence of a magnetic field on the model atmosphere of a magnetic star.
J. Staude.
Astron. Nachr., Vol. 294, 113 - 121 (1973).

A quasi-static magnetic field similar to that of a large single sunspot is assumed to be embedded in a photospheric model of an Ao star. The differences of the model atmosphere within the magnetic spot relative to the photosphere are calculated by means of the equations of hydromagnetics. The differences are small for $\tau \geq 1$ but increase towards smaller τ. An outline is given how to extend and generalize the present model calculations.

064.044 On the "limiting decrement" method for determination of the electron temperatures in the atmospheres of Wolf−Rayet stars (Study of Wolf−Rayet stars. V).
S. V. Rublev.
Astrofiz. Issled., Izv. Spets. Astrofiz. Obs., Vol. 4, 32 - 41 (1972). In Russian.

For application to the Pickering He II series a general «limiting decrement» method is described suitable in principle for evaluating electron temperatures of various stars from the relative strengths of emission lines in their spectra. Evaluating for 5 bright Wolf−Rayet stars (HD 191765, 192163, 193077, 192103, and 192641) has given values of T_e (He III) in the range of 19−31 thousand degrees. The difficulties are pointed out of practical applications of the method in the case of high-temperature objects. A comparison is made with the results of other evaluations.

064.045 The atmosphere of γ Peg. I. Physical parameters.
N. M. Chunakova.
Astrofiz. Issled., Izv. Spets. Astrofiz. Obs., Vol. 4, 42 - 49 (1972). In Russian.

Physical parameters of the atmosphere of γ Peg are derived using three $1.1−1.4$ Å mm^{-1} spectrograms.

064.046 Influence of the blanketing effect in the atmospheres of M-stars on V and R magnitudes.
A. V. Dragunova.
Astron. Tsirk., No. 750, p. 1 - 3 (1973). In Russian.

064.047 Line effects on the radiative acceleration in supergiant stars. E. Böhm-Vitense.
Astrophys. Journ., Vol. 181, 379 - 385 (1973).

The radiative acceleration g_r is studied for supergiants with $T_{eff} \leq 8000°$K. Nongray effects, including line absorption, raise g_r by a factor of at least 6 − in comparison with the gray case − at the surface of the stars. The luminosity limit for stable stars set by g_r will probably be lowered by this factor in comparison with earlier estimates based on electron scattering only. No instability is found for red supergiants.

064.048 Radiative opacity in stellar atmospheres. IV. Remarks on the accuracy of metal absorption coefficients. S. Matsushima, L. D. Travis.
Astrophys. Journ., Vol. 181, 387 - 391 (1973).

The random errors intrinsic to the determination of metal absorption coefficients based on the quantum-defect theory are estimated by comparing independently calculated values. The errors averaged over the entire spectral region appear to be about 30 percent or less for sodium, aluminum, and magnesium, but increase to a factor of 2 for carbon, nitrogen, oxygen, and silicon. However, for stars of normal abundances such errors seem to remain within 20 percent for the spectroscopically important wavelengths between 1250 and 3500 Å.

064.049 The last Balmer line and Hγ in model B stars.
D. Fischel, D. A. Klinglesmith.
Astrophys. Journ., Vol. 181, 841 - 850 (1973).

Profiles of the last few Balmer lines (H12−H27) and the full profiles of the Hγ line were computed for a grid of hydrogen-line−blanketed model atmospheres. The use of the last hydrogen line and the equivalent width of Hγ provide a quick means of estimating the effective temperature and surface gravity of B stars. Regression curves relating the quantum number of the last hydrogen line to the electron density in the model atmosphere are given. Furthermore, it is shown that the difference in the gravity determination deduced from Hγ equivalent widths using the Edmonds, Schlüter, and Wells formalism and the Griem wing formula is, at worst, 0.18 in the common logarithm.

064.050 Model atmospheres for Betelgeuse.
T. D. Faÿ, H. R. Johnson.
Astrophys. Journ., Vol. 181, 851 - 864 = Publ. Goethe Link Obs., Indiana Univ., *Bloomington*, No. 149 (1973).

We present a series of stellar atmospheric models at effective temperatures of 3800° and 3500°K and compare these in detail with scanner observations of Betelgeuse (α Ori, M 2 Iab). The atmospheres are hydrostatic, flux-constant, LTE atmospheres which include the opacity of H_2O, CO, CN, and atomic line blanketing. Comparison of our predicted strengths of observed CO and CN features with observations and of our predicted column densities of CO, OH, NH, and H_2O with published column densities suggests that C/H may be less than its solar value by about a factor of 10 and C/O may be less than 0.6 in Betelgeuse.

064.051 Conditions for carbon monoxide vibration-rotation LTE in late stars. R. I. Thompson.
Astrophys. Journ., Vol. 181, 1039 - 1054 (1973).

Collisional vibration and rotation transition rates are found for carbon monoxide due to H, H_2, He, and electrons. These rates are compared to the radiative rates of transition in carbon monoxide to establish temperature-pressure regions in which vibration-rotation LTE is a good approximation and those regions in which possible non-LTE effects may occur. Comparison of these data with current late-star model atmospheres indicates that possible vibrational non-LTE effects may occur in supergiants, but in all late-star computed atmospheres rotational LTE should be a good approximation. The further question of total molecular LTE is examined and rate equations set up. Rough order-of-magnitude estimates are made on the conditions for total LTE in CO.

064.052 On the thermal instability and energetic relations in convective envelopes of slowly rotating stars.
I. M. Yarovskaya.
Mezhplanet. sreda i fiz. magnitosfery. Moskva, Nauka, 1972, p. 149 - 163. In Russian. − Abstr. in Referativ. Zhurn. 51. Astron., 4.51.672 (1973).

064.053 Abundance of lithium in the atmospheres of variable M stars of type SR. M. E. Boyarchuk.
Izv. Krymskoj Astrofiz. Obs., Vol. 46, 47 - 53 (1972). In Russian.

Twenty spectrograms of seven variable M stars with dispersion 12 Å/mm were obtained with the coudé spectrograph of the 2.6-m telescope of the Crimean Observatory in 1968 - 1969. A spectroscopic survey has been made by the curves-of-growth method. The line Li λ6707.9 was identified. The equivalent widths of 10 absorption lines of Ca I and of 20 absorption lines of Fe I were measured. The parameters $v_{D Li}$, T_B and lg (N_{Li}/N_{Ca}) were determined.

064.054 Abundance of lithium in the atmospheres of two dM5 stars. M. E. Boyarchuk.
Izv. Krymskoj Astrofiz. Obs., Vol. 46, 54 - 58 (1972). In Russian.

Nine spectrograms of the flare star AD Leo (M4.5 V) and of Barnard's star BD +4° 3561 (M5 VI) have been investigated by the curves-of-growth method. The line Li λ6707.9 was identified. The equivalent widths of 10 absorption lines of Ca I and of 20 absorption lines of Fe I were measured. The following physical parameters of the atmospheres have been obtained: parameter of Doppler velocity v_{DLi}, excitation temperature $\Theta = 5040/T_B$, and relative abundance of Li, lg N_{Li}/N_{Ca}.

064.055 On the influence of radiation losses on the spectrum of turbulence in a stellar atmosphere.
O. P. Hollandskij.
Izv. Krymskoj Astrofiz. Obs., Vol. 46, 83 - 89 (1972).
In Russian.
   A solution of the spectral equation of isotropic turbulence with the non–linear term in Kovásznay's form (1948) modified by Panchev (1969) and with a term accounting for the energy dissipation by radiation is analysed.

064.056 Population I helium abundances.   S. E. Strom.
   Stellar evolution, (see 012.014), p. 419 - 425 (1972).

064.057 The ultraviolet flux envelopes of main-sequence B stars.   A. B. Underhill.
Astron. Astrophys., Vol. 25, 175 - 185 (1973).
   Flux envelopes on an absolute energy scale from 1100–6000 Å prepared from OAO-II scans and from published ground-based material are presented for λ Leporis, B 0.5 V, η Ursae Majoris, B 3 V, γ Ursae Majoris, A 0 V, and α Canis Majoris A 1 V from rocket scans. These, with flux envelopes for ζ Draconis, B 6 III, and α Leonis, B 7 V, (Underhill, 1973) are intercompared and compared with reference flux envelopes predicated by LTE theory from lightly line-blanketed model atmospheres.

064.058 Abundance changes in the envelopes of moderate mass stars at late stages of evolution.   U. Uus.
Stellar ages. Proc. IAU Colloquium No. 17, (see 012.015), XIII, 1 - 4 (1973).

064.059 Stellar chromospheres and circumstellar envelopes.
L. V. Kuhi.
Stellar ages. Proc. IAU Colloquium No. 17, (see 012.015), XLIII, 1 - 29 (1973).

064.060 Dynamics of supergiant atmospheres and ages.
M. Hack.
Stellar ages. Proc. IAU Colloquium No. 17, (see 012.015), XLV, 1 - 2 (1973).

064.061 Chemical abundances of field population II A-type stars.   K. Kodaira.
Stellar ages. Proc. IAU Colloquium No. 17, (see 012.015), LI, 1 - 5 (1973).

064.062 Metallicity and microturbulence in G and K giant stars.   S. Andersen, B. Gustafsson, P. Kjaergaard.
Stellar ages. Proc. IAU Colloquium No. 17, (see 012.015), LIV, 1 - 10 (1973).

064.063 The theoretical relation between UBV colors and T_{eff} and luminosity for extreme pop. II giants.
E. Böhm-Vitense.
Bull. American Astron. Soc., Vol. 5, 266 (1973). – Abstr. AAS.

064.064 Curvature effects in extended stellar atmospheres – pure absorption.   A. Peraiah.
Astrophys. Space Sci., Vol. 21, 223 - 235 (1973).
   The effects of curvature in an atmosphere with pure absorption are investigated. Numerical solution of the transfer equation has been obtained in the framework of the Discrete Space Theory of Radiative Transfer. Two cases have been con-

sidered: (a) the atmosphere is irradiated at the bottom and there is no incident radiation at the top of the atmosphere; and (b) no radiation is incident on either side of the atmosphere.

064.065 Microturbulence in A stars as derived from line profiles.   M. A. Smith.
Astrophys. Journ., Vol. 182, 159 - 175 = Contr. Lick Obs., No. 331 (1973).
   This study represents an attempt to fix a scale of the "microturbulent velocity" in main-sequence A stars by examining the profiles of weak and saturated lines of six sharp-lined Am stars. Although the study is photographic, so that it is difficult to determine accurate core and wing parameters, studies of the loci of half-intensity widths and core shapes provide useful information.

064.066 The role of convection in stellar atmospheres. II. Cool main-sequence stars and metal-deficient subdwarfs.   L. D. Travis, S. Matsushima.
Astrophys. Journ., Vol. 182, 189 - 207 (1973).
   The authors extend their model atmosphere study based on Spiegel's improved theory of convection to relatively cool metal-deficient subdwarfs. Their primary objective is to investigate the effects of convection and metal deficiency on the structure of these models and subsequent changes in the observable features. They compare with observations various color indices predicted by models based on different theories.

064.067 Linienbreiten von Ca II K_2 und $H\alpha$ und die Chromosphären später Sterne.   D. Reimers.
Mitt. Astron. Ges., No. 32, p. 194 - 196 (1973).

064.068 The atmosphere of the hydrogen-poor stars HD 144941 and CPD–69° 2698.
K. Hunger, J. P. Kaufmann.
Mitt. Astron. Ges., No. 32, p. 241 (1973).

064.069 Die Atmosphäre des wasserstoffarmen Sterns HD 135485.   D. Schönberner.
Mitt. Astron. Ges., No. 32, p. 242 (1973). – Abstract.

064.070 The atmosphere of the hydrogen-deficient star HD 96446.   R. Wolf.
Mitt. Astron. Ges., No. 32, p. 242 (1973). – Abstract.

064.071 On the stationary mass outflow from stars. II. The results for a 30 $M_\odot$ star.   A. Żytkow.
Acta Astron., Vol. 23, 121 - 134 (1973).
   Models of spherically symmetric, stationary outflowing stellar envelopes with radiation pressure in continuum as the main driving force were constructed for a 30 $M_\odot$ star by means of integrations of the equations of conservation of mass, momentum and energy together with the radiative transport equation. In addition results of model calculations for a 1 $M_\odot$ star with luminosity at the surface $L = 4 \times 10^4 L_\odot > L_{crit} = 3.88 \times 10^4 L_\odot$ are also briefly presented.

064.072 Microturbulence in atmospheres of F, G, K type stars. II. Changes with temperature, luminosity and metal abundance.   R. Głębocki.
Acta Astron., Vol. 23, 135 - 157 (1973).
   A catalogue of stars of spectral types F and later with turbulent velocities determined by use of the curve of growth method is completed. It contains 375 stars of different luminosity classes. Temperatures and luminosities of all stars from the catalogue are estimated and changes in microturbulence in the H-R diagram are discussed. A possible correlation between metal deficiency and the velocity of microturbulence is discussed.

064.073 The atmosphere of the hydrogen-deficient star HD 96446. R. E. A. Wolf.
Astron. Astrophys., Vol. 26, 127 - 136 (1973).

A quantitative fine analysis of the hydrogen-deficient star HD 96446 has been performed by using a grid of flux-constant model atmospheres. 6 coudé-spectrograms (dispersion 12.3 Å/mm) have been taken with the ESO 152 cm telescope at La Silla. Opacity sources considered are H I, H⁻, He I, He II, He⁻, electron scattering, and Lyman- and Balmer-lines. The following quantities are computed: the profiles of Hγ and Hδ, the equivalent widths and profiles of five He I-lines, the ionization equilibria of He I/He II and Si II/Si III, and the turbulent velocity. Mass, radius and luminosity are determined to be $M/M_\odot = 7.6$, $R/R_\odot = 3.6$, $\log L/L_\odot = 3.70$.

064.074 A model-atmosphere abundance analysis of the B9 V star Nu Capricorni. S. J. Adelman.
Astrophys. Journ., Vol. 182, 531 - 538 (1973).

A complete model-atmosphere abundance analysis of the moderately sharp-lined star ν Cap has been performed. The values of its derived abundances are nearly identical with those of the sun and other normal main-sequence stars.

064.075 On methods of calculating absorption line profiles. L. I. Snezhko.
Soobshch. Spets. Astrofiz. Obs. AN SSSR, *Zelenchukskaya*, vyp. (No.) 3, p. 3 - 16 (1971). In Russian.

064.076 The atmosphere of Epsilon Leonis. P. M. Williams. Monthly Notices Roy. Astron. Soc., Vol. 162, 235 - 242 (1973).

Using model atmospheres and high dispersion (2.2 Å mm⁻¹) spectrograms observed with the Isaac Newton Telescope, the G0 II type giant ϵ Leonis is analysed relative to the sun and ϵ Vir. It is found to have a solar-type composition, but a significantly higher microturbulence (2.6 km s⁻¹) than either ϵ Vir (1.8 km s⁻¹) or the sun (assumed to have zero microturbulence).

064.077 The relative importance of Rayleigh scattering by atoms and negative ion absorption in the atmospheres of late-type stars. S. P. Tarafdar, M. S. Vardya.
Monthly Notices Roy. Astron. Soc., Vol. 162, 299 - 305 (1973).

The Rayleigh scattering by He, C and N has been compared with the absorption by negative ions and with electron scattering and has been found to be important at temperatures and pressures appropriate to the atmospheres of late-type hydrogen deficient giant and supergiant stars.

064.078 Spectral line formation in extended atmospheres — II. A. Peraiah.
Monthly Notices Roy. Astron. Soc., Vol. 162, 321 - 327 (1973).

Line profiles have been computed for a model of two-level atom both for plane parallel and spherically symmetric geometry assuming both LTE and non-LTE.

064.079 The thermal radiation spectra of supermassive stars and X-ray sources.
A. F. Illarionov, R. A. Sunyaev.
Astrophys. Space Sci., Vol. 19, 47 - 60, 61 - 74 (1972). In Russian and English.

In the external layers of supermassive stars and thermal sources of X-ray radiation electron scattering contributes more to the opacity than free-free processes. In this paper a case is considered, in which both density and temperature vary in a power-law manner with the depth of the layer.

064.080 On the possibility of radiative acceleration of gas in stellar atmospheres. Mechanism of radiative conduc- tivity. M. V. Konyukov, I. F. Malov.
Trudy fiz. in-ta. AN SSSR, Vol. 62, 188 - 193 (1972). In Russian. – Abstr. in Referativ. Zhurn. 51. Astron., 6.51.565 (1973).

064.081 Molecular dissociation equilibria in SC stars. A. E. Greene.
Contr. Perkins Obs., Ohio State Univ.– Ohio Wesleyan Univ., Ser. II, No. 31, 129 pp. (1972).

The spectra of SC stars show strong atomic features due to s-process elements, molecular bands of CN, remarkably strong Na–D lines, and a marked depression of light in the violet spectral region reminiscent of N-type carbon stars. However, the bands of most of the molecules that one normally observes in cool star spectra are extremely weak. It has been pointed out that the weakness of these bands in SC stars can be accounted for by an atmospheric oxygen to carbon ratio that is near unity. It is the purpose of this investigation to calculate the molecular dissociation equilibria in gas mixtures with O/C near unity and conditions of temperature and pressure that are, hopefully, similar to those present in cool giant stars. The results of the calculations based on 21 cases are presented.

064.082 A computational program for the solution of non-LTE transfer problems by the complete linearization method. L. H. Auer, J. N. Heasley, R. W. Milkey.
Kitt Peak National Obs., Contr. No. 555, 3 + 183 pp. (1972).

We present a computer program for the simultaneous solution of the equations of radiative transfer and statistical equilibrium by the complete linearization technique. This code will treat bound-bound and bound-free radiative transitions for atoms of a single species in several stages of ionization.

064.083 Programme de calcul des raies de l'hydrogène dans les étoiles chaudes. A. Delcroix.
Dep. Astrophys. Fac. Sci. Mons, Commun. No. 34, 84 pp. (1973).

L'analyse de l'atmosphère des étoiles de type jeune se fait par l'analyse des profils des raies de l'hydrogène et spécialement par les raies de la série de Balmer puisque ces raies se situent dans le visible. Le programme décrit dans ce rapport calcule les profils des raies de l'hydrogène en unités de flux pour les étoiles de type jeune.

064.084 Plasma perturbations as a possible mechanism of mass loss from hot stars.
B. Basu, R. Bandyopadhaya.
Indian Journ. Phys., Vol. 46, 513 - 520 (1972).

The rate of mass loss from hot stars has been calculated with the linearized macroscopic plasma perturbation theory. Numerical estimates of the loss rates have been made for Be stars, hot supergiants and Wolf-Rayet stars.

064.085 On the solution of the radiative transfer equation in extended hot stellar atmospheres. (Special case).
V. I. Stebnev.
Trudy Kazan. Gorod. Astron. Obs., No. 37, p. 73 - 75 (1970). In Russian.

064.086 On the absorption coefficients by ionized helium in the atmospheres of Wolf-Rayet stars.
V. I. Stebnev.
Trudy Kazan. Gorod. Astron. Obs., No. 37, p. 76 - 85 (1970). In Russian.

064.087 On the temperature distribution in extended stellar atmospheres. V. ?. Merezhin, V. I. Stebnev.
Trudy Kazan. Gorod. Astron. Obs., No. 37, p. 86 - 93 (1970). In Russian.

064.088 The continuous absorption coefficient of Ne^-.
T. L. John, A. R. Williams.
Phys. Letters A,(*Netherlands*), Vol. 43A, 32 - 38 (1973).

The continuous absorption coefficient of Ne^- is shown to be of the same order of magnitude as He^- and thus is unlikely to be an important source of continuous absorption in stellar atmospheres.

064.089 On the possibility of radiative acceleration of gas in stellar atmospheres. I. F. Malov.
Trudy fiz. in-ta. AN SSSR, Vol. 62, 33 - 41 (1972). In Russian. – Abstr. in Referativ. Zhurn. 51. Astron., 7.51.497 (1973).

Stellar chromospheres. See Abstr. 003.069.

Introduction to stellar atmospheres & interiors. See Abstr. 003.098.

Stellar atmospheres and interplanetary plasma. Technical details of radio astronomical reception. See Abstr. 003.138.

Hydrogen Stark-broadening tables. See Abstr. 022.030.

Semiempirical calculation of gf values: Sc II (3d + 4s)^2 – (3d + 4s) 4p, a detailed example. See Abstr. 022.080.

Behandlung von molekularen Reaktionsgleichgewichten und Opazitätsrechnungen für kühle, circumstellare Hüllen. See Abstr. 061.040.

Integral kernels and exact solutions to some radiative problems in spherical geometries. See Abstr. 063.029.

Self-similar isothermal flow in generalized Roche model. See Abstr. 065.020.

Formation of dust in low-temperature stars. See Abstr. 065.056.

Abundance changes in the envelope of a 5 $M_\odot$ red supergiant. See Abstr. 065.070.

The *UBVr* colors of extreme Pop II giants. See Abstr. 065.073.

The role of convection in stellar atmospheres. I. Observable effects of convection in the solar atmosphere. See Abstr. 080.008.

Speckle interferometry: Color-dependent limb darkening evidenced on Alpha and Omicron Ceti. See Abstr. 113.024.

Spectroscopic studies of O-type stars. II. Comparison with non-LTE models. See Abstr. 114.017.

Spectroscopic studies of O-type stars. III. The effective-temperature scale. See Abstr. 114.018.

On ultraviolet stellar fluxes. IV. Importance of bound-free absorption of S I in B to K stars. See Abstr. 114.020.

Study of the abundances of heavy elements in F–G type stars. I. Differential analysis of two metal deficient stars: HR 646 and HR 860. See Abstr. 114.035.

A study of Ca II K$_2$ and H$_a$ line widths in late type stars. See Abstr. 114.048.

Contribution to the study of supermetallicity in late-type giants. See Abstr. 114.060.

A model atmosphere analysis of the super-supergiant HR 5171. See Abstr. 114.074.

The $^{12}C/^{13}C$ ratio in the atmosphere of Arcturus. See Abstr. 114.082.

Infrared observations and atmospheres of cool stars. I. Support mechnisms for circumstellar shell. See Abstr. 114.083.

Metallicism in border regions of the Am domain. II. Analysis of the Fm stars. See Abstr. 114.093.

Nature of the light variation of the peculiar A-star HD 221568. See Abstr. 114.100.

On the relative abundance of helium in the envelopes of Wolf–Rayet stars (Study of Wolf–Rayet stars. III). See Abstr. 114.103.

The atmosphere of the supergiant 6 Cas. I. Spectral material and its photometric processing with the aid of an electronic digital computer. See Abstr. 114.107.

C$_2$ in Eta Aquilae spectrum. See Abstr. 114.111.

On the observability of CO and CO$^+$ in Eta Aquilae. See Abstr. 114.112.

Some magnetic null lines of astrophysical interest. See Abstr. 114.118.

A study of B 6 stars. See Abstr. 114.129.

Spectralphotometry and quantitative analysis of the hydrogen-deficient stars HD 144941 and CPD–69°2698. See Abstr. 114.130.

Ricerche sulle stelle con regioni attive. See Abstr. 114.141.

Radiative transfer in atmospheres of Algol-type binaries. Departures from the state of LTE in late-type components. See Abstr. 121.053.

An expanding circumstellar cloud of Zeta Aurigae. See Abstr. 121.062.

Velocity gradients and microturbulence in cepheids. See Abstr. 122.032.

Atmosphärenmodelle kühler Weisser Zwerge. See Abstr. 126.024.

Condensation nuclei for interstellar grains and the IR emission of late type stars. See Abstr. 131.090.

Observations of optical nebulae at 2695 MHz. See Abstr. 132.004.

On ionization instability as the cause of planetary nebulae. See Abstr. 133.008.

Galactic magnetic fields: cellular or filamentary structure? See Abstr. 158.028.

065 Stellar Structure, Stellar Evolution, Stellar Nucleosynthesis

065.001 **Charged particle thermonuclear reactions in nucleosynthesis.** J. W. Truran.
Astrophys. Space Sci., Vol. 18, 306 - 323 (1972).

Thermonuclear reaction rates are calculated at temperatures consistent with nucleosynthesis conditions in stars and supernovae ($10^9 \lesssim T \lesssim 10^{10}$ K). They are compared with sums over experimentally determined resonance strengths for a number of charged-particle reactions involving medium mass nuclei ($24 \lesssim A \lesssim 40$).

065.002 **Post-Newtonian equilibrium configurations of rotating polytropes.** M. J. Miketinac, R. J. Barton.
Astrophys. Space Sci., Vol. 18, 437 - 448 (1972).

Post-Newtonian equations are solved numerically for stellar models with a polytropic pressure-density relation for the case of uniform rotation, no meridional currents, and axial symmetry. The solution is obtained by following Stoeckly's numerical technique. Parameters characterizing the critical configuration are determined and compared with the values obtained recently by Fahlman and Anand, who followed Chandrasekhar's series expansion method.

065.003 **Formation of neutron star spots and its connection with pulsars. II. Close similarities between radiation from the sun and pulsars.** M. Fujimoto, T. Murai.
Publ. Astron. Soc. Japan, Vol. 25, 75 - 90 (1973).

Models of neutron star spots for pulsars are discussed on the basis of observational data on the mean pulse profile, polarization, and the time-structure of individual radio pulses. The circular polarization and time-structure of radiation from the sun and of radiation from pulsars are compared, and it is suggested that pulsar radiation may be due to "noise storms" or consecutive bursts at neutron star spots with a local bipolar magnetic field. The slowing-down of pulsars is interpreted in terms of the ejection of diffuse matter from the active spot regions.

065.004 **Formation of stars in a rotating cloud with magnetic field.** T. Nakano.
Publ. Astron. Soc. Japan, Vol. 25, 91 - 100 (1973).

The decrease of the magnetic flux caused by ambipolar diffusion in a contracted cloud is investigated in connection with star formation. Star formation by fragmentation of a rotating disk is discussed and the smallest mass of stars formed in such a cloud is found to be several times $0.01 \, M_\odot$. Formation of a galactic cluster and an association is also discussed.

065.005 **Helium red giants.** V. Trimble, B. Paczyński.
Astron. Astrophys., Vol. 22, 9 - 12 (1973).

Models of helium red giants have been obtained for stars in the mass range $0.7-10 \, M_\odot$. The models have degenerate carbon-oxygen cores, helium burning shell sources, and helium envelopes. Their absolute bolometric magnitudes are -3 to -7, and their effective temperatures range from $\log T_e = 3.0$ to 3.8. The relation of the models to observed helium-rich red giants (R Corona Borealis variables) is discussed.

065.006 **Evolution from the main sequence to the helium flash for population II stars.**
P. Demarque, J. G. Mengel.
Astron. Astrophys., Vol. 22, 121 - 128 (1973).

Evolutionary turnoffs from the main sequence for stars of low mass with the composition parameters $0.50 \gtrless Y \gtrless 0$ and $Z = 10^{-3}$ and 10^{-4} are presented and compared with previous results. The evolution of red giants to the helium flash is also investigated in an attempt to determine more reliably: (a) the effect of chemical composition on the position and slope of the giant branch; (b) the dependence on the original value of

Y of the core mass and maximum luminosity at the helium flash; (c) the dependence of red-giant lifetimes on the total mass of the star. The position and extent of the luminosity dip discovered by Thomas (1967) is also briefly discussed.

065.007 **The hydromagnetic oscillations and stability of self-gravitating masses. III. Magnetic polytropes.**
S. D. Grover, S. Singh, J. N. Tandon.
Astron. Astrophys., Vol. 22, 133 - 137 (1973).

A variational principle is developed to study the oscillations and stability of magnetic polytropes. It is then applied to study the special case of the $n = 1$ polytrope using five parameter trial functions. We obtain in addition to the non-trivial Kelvin mode, two pulsational modes and two convectively unstable modes. The effect of magnetic field on these modes is discussed in detail for $l = 1$, 2 and 3.

065.008 **About stability and radial pulsations of rotating neutron stars.**
Yu. L. Vartanian, A. V. Ovsepian, G. S. Ajian (*Adzhyan*).
Astron. Zhurn. Akad. Nauk SSSR, Vol. 50, 48 - 59 (1973).
In Russian. English translation in Soviet Astron. AJ, Vol. 17, No. 1.

Radial pulsations of rotating cold neutron stars that are near the state of stability loss are studied by the energetic method. The integral parameters and frequency of radial pulsations for different equilibrium configurations are calculated. It is shown that change of the period of pulsation of stable neutron stars due to rotation is very small. The gravitational defect of mass for rotating configurations is calculated. Configurations which consist of real neutron gas are investigated also.

065.009 **Condensation of stars and formation of a magnetic field in protogalaxies.**
G. S. Bisnovatyj-Kogan, A. A. Ruzmaikin, R. A. Sunyaev.
Astron. Zhurn. Akad. Nauk SSSR, Vol. 50, 210 - 213 (1973).
In Russian. English translation in Soviet Astron. AJ, Vol. 17, No. 1.

A scheme of star formation and generation of a magnetic field in a protogalaxy is considered. The magnetic field generated by dynamo action in contracting protostars is found to be sufficient for the transmission of the superfluous angular momentum to the surrounding medium. Therefore star formation in a nonmagnetic protogalaxy is possible.

065.010 **Star formation and evolution in spiral galaxies.**
W. J. Quirk, B. M. Tinsley.
Astrophys. Journ., Vol. 179, 69 - 83 (1973).

Evolutionary models for regions of M31 and M33 and the solar neighborhood are based on a stellar birthrate suggested by the dynamics of spiral structure: we assume that stars are formed very efficiently until the gas content reaches equilibrium at its present value, which takes about 10^9 years; thereafter, the birthrate just equals the rate at which gas enters the system from stellar mass-loss or infall of intergalactic matter. Each star is followed in the H-R diagram from the main sequence to death as an invisible remnant. Integrated magnitudes, colors, mass-to-light ratio, gas content, helium and metal abundance, are computed in steps of 10^9 years.

065.011 **A linear nonadiabatic analysis of radial oscillations in models for Beta Cephei stars.** W. R. Davey.
Astrophys. Journ., Vol. 179, 235 - 240 = Contr. Univ. Waterloo Obs., No. 17 (1973).

A linearized, fully nonadiabatic analysis was used to examine the pulsational stability toward radial oscillations of stellar evolutionary models passing through that portion of the H–R diagram occupied by β Cephei stars. Equilibrium models in three distinct phases of evolution were constructed: post–main-sequence models of 10 and 15 $M_\odot$, pre–main-sequence models of 15 $M_\odot$, and highly evolved models of 2.5$M_\odot$.

065.012 Advanced evolution of massive stars. III. Hydrostatic carbon-burning nucleosynthesis and energy generation. W. D. Arnett.
Astrophys. Journ., Vol. 179, 249 - 256 (1973).

Carbon-burning nucleosynthesis and energy generation in hydrostatic stars is considered, using a realistic reaction network. Useful approximations are presented and compared with the numerical results. It is found that the hydrostatic nucleosynthesis resembles that at a suitably chosen, constant temperature, and that the "no-sodium approximation" works well for massive stars ($M \gtrsim 15\ M_\odot$).

065.013 Determination of properties of cold stars in general relativity by a variational method.
M. Nauenberg, G. Chapline, Jr.
Astrophys. Journ., Vol. 179, 277 - 287 (1973).

Approximate analytic formulae are obtained for the mass, the number of baryons, and the radius of a cold star by applying the energy variational principle in general relativity to a uniform-density sphere of baryons. Conditions for equilibrium and for stability are expressed by transcendental equations which generalize familiar results of Newtonian theory. These formulae allow one to understand readily how various assumptions concerning the equation of state affect the properties of a neutron star. By making use of the condition that the speed of sound is less than the speed of light, and of properties of neutron matter near nuclear densities, we derive an upper limit to the maximum mass of a stable neutron star.

065.014 A post-Newtonian study of differentially rotating polytropes. F. H. Seguin.
Astrophys. Journ., Vol. 179, 289 - 308 (1973).

Chandrasekhar's post-Newtonian equations of hydrodynamics are applied to a self-gravitating polytropic fluid in rapid, axisymmetric, differential rotation, and used to construct an iterative method of the "self-consistent field method" type which can be used on the computer to find equilibrium configurations. The geometrical characteristics of the rotation are investigated.

065.015 Models for rapidly rotating pre-main sequence stars.
D. L. Moss.
Monthly Notices Roy. Astron. Soc., Vol. 161, 225 - 237 (1973).

Approximate models of rapidly rotating largely convective low mass stars contracting homologously to the main sequence are constructed. The effects of rotation on times for contraction to the main sequence are discussed, and it is concluded that discrepancies in the estimates of ages of young clusters may be increased if a coeval origin of the cluster members is retained.

065.016 The history of star formation and the colors of late-type galaxies.
L. Searle, W. L. W. Sargent, W. G. Bagnuolo.
Astrophys. Journ., Vol. 179, 427 - 438 (1973).

We have calculated the U, B, V colors of simple model galaxies which are characterized by three parameters: α, which specifies the mass distribution of stars at birth, β, the reciprocal decay time of an assumed exponential decay in the rate of star formation, and their age. The computations are based on theoretical evolutionary tracks for individual stars. We propose that the range in colors of the late-type galaxies reflects the range in decay times and point out that, if so, the frequency of type II supernovae will depend on both the mass and the color of the parent galaxy. Finally we discuss the question whether the bluest known galaxies, all of which are dwarfs, are young. We tentatively conclude that, instead, they are galaxies which undergo intermittent and unusually intense bursts of star formation.

065.017 Stellar evolution at high mass based on the Ledoux criterion for convection.
R. Stothers, C.-W. Chin.
Astrophys. Journ., Vol. 179, 555 - 568 (1973).

Theoretical evolutionary sequences of models for stars of 15 and 30 $M_\odot$ have been computed from the zero-age main sequence to the end of core helium burning, with the use of the Ledoux criterion for convective instability. The sensitivity of the models to the most important physical input parameters has been tested.

065.018 Post-event neutron exposure of r-process material.
J. B. Blake, D. N. Schramm.
Astrophys. Journ., Vol. 179, 569 - 583 (1973).

The effects of neutron fluxes on r-process material following the r-process event are considered. The calculations examine the case of neutron exposure occurring after the r-process material has decayed back to or near the valley of β-stability. The effects of these neutron fluxes on the abundances of lead, bismuth, and the nucleochronologically important actinides are examined.

065.019 Some comments on secular stability criteria and applications. J. Demaret, P. Ledoux.
Astron. Astrophys., Vol. 23, 111 - 116 (1973).

The general and formally exact criterion for secular stability is discussed to point out the interrelation between secular and dynamical stability. Two approximate criteria relating essentially to the stability of stellar nuclear burning are recovered easily from the general criterion. A general relativistic extension of the rigorous criterion shows that supermassive stars cannot be secularly unstable unless dynamical instability has set in previously. In that case, the secular instability is no more than the consequence of the correlation stressed above and cannot be very significant since the dynamical time-scale becomes soon much shorter than the secular time-scale.

065.020 Self-similar isothermal flow in generalized Roche model. M. P. Ranga Rao, S. C. Purohit.
Astron. Astrophys., Vol. 23, 155 - 157 (1973).

We investigate the self-similar isothermal flow of a gas moving under the gravitational attraction of a central body of fixed mass (generalized Roche model) behind a spherical shock wave and propagating in a non-uniform stationary atmosphere. The numerical solutions are compared with the corresponding solutions for adiabatic flow.

065.021 Spectra of rapidly rotating objects.
S.-L. Su, R. M. Spector.
Astrophys. Journ., Vol. 180, 143 - 157 (1973).

The general problem of the altered spectra to be expected from luminous, rapidly rotating objects is considered. For the cases of Briet-Wigner and blackbody spectra, Doppler effects are considered as well as limb darkening. Detailed results are given for typical physical parameters in these cases. The motivation for these investigations provided by pulsars is indicated.

065.022 Rapidly rotating stars. VIII. Zero-viscosity polytropic sequences. P. Bodenheimer, J. P. Ostriker.
Astrophys. Journ., Vol. 180, 159 - 169 = Contr. Lick Obs., No. 381 (1973).

The sequence of uniformly rotating (Maclaurin) spheroids of uniform density is generalized to the case of sequences of axisymmetric centrally condensed configurations, with fixed distributions of angular momentum. Numerical results are presented for such sequences of differentially rotating polytropes with indices 0, 3/2, and 3, and differing assumptions regarding the distribution of angular momentum.

065.023 **On the oscillations and stability of rapidly rotating stellar models. III. Zero-viscosity polytropic sequences.** J. P. Ostriker, P. Bodenheimer.
Astrophys. Journ., Vol. 180, 171 - 180 = Contr. Lick Obs., No. 382 (1973).

The frequencies of oscillation of the lowest "radial" and nonradial modes are calculated for differentially rotating, axisymmetric, stellar models along the generalized polytropic sequences in the preceding paper. The stability of the models to nonaxisymmetric perturbations is discussed and compared with the properties of the classical Maclaurin spheroids.

065.024 **Evolution of low-mass stars. V. Minimum mass for the deuterium main sequence.**
A. S. Grossman, H. C. Graboske.
Astrophys. Journ., Vol. 180, 195 - 198 (1973).

An evolutionary sequence of stellar models has been calculated to determine the minimum mass and maximum time duration of stars on the low-mass end of the deuterium main sequence. The chemical composition chosen is ($X = 0.68$, $Y = 0.29$), with the terrestrial value $X_D = 1.9 \times 10^{-4}$, and $l/Hp = 1$. Evolutionary tracks have been calculated for 0.008, 0.01, 0.012, 0.015, and 0.02 $M_\odot$.

065.025 **Rotating magnetosphere: A simple relativistic model.**
F. C. Michel.
Astrophys. Journ., Vol. 180, 207 - 225 (1973).

We solve exactly for the electromagnetic field configuration about an aligned rotating magnetic moment ("star") for the case that there is sufficient plasma about the object to "freeze in" the magnetic fields, but in the limit that the inertia and pressure of the plasma particles can be neglected. These solutions are expected to be valid except near the light cylinder, although there is no mathematical difficulty in solving the resultant equations at or even beyond the light cylinder.

065.026 **On diffusion of radiation in a stellar shell expanding with constant velocity.** V. V. Vityazev.
Astrofizika, Vol. 8, 235 - 245 (1972). In Russian. English translation in Astrophysics, Vol. 8, No. 2.

In one-dimensional approximation the diffusion of radiation in two layers moving relative to one another with constant velocity is considered. Complete redistribution in frequency is assumed. In the case of finite layers the source function, emergent radiation intensities and radiation pressure are found numerically. In the case of semi-infinite layers the equation for the emergent radiation intensity and an explicit expression for the source function are derived.

065.027 **Secular stability. V. The perturbation of chemical abundances.** M. L. Aizenman, J. Perdang.
Astron. Astrophys., Vol. 23, 209 - 214 (1973).

Calculations of the secular stability of stars near the main sequence have been modified to include perturbations of chemical abundances. It is found that these perturbations give rise to an additional spectrum which appears to be continuous. The usual secular spectrum is practically unchanged from the one obtained when abundance perturbations are neglected.

065.028 **Low mass B stars with low surface gravity.**
V. Trimble
Astron. Astrophys., Vol. 23, 281 - 283 (1973).

A series of static models of stars with inert, isothermal helium cores ($X = 0.0; Z = 0.03$) and thin hydrogen-burning shells has been calculated. The mass range covered is 0.26–0.5 $M_\odot$. All core masses are more than 98% of the total stellar mass. The envelope composition is $X = 0.7$, $Z = 0.03$.

065.029 **An approximate equation of state for stellar material.**
P. P. Eggleton, J. Faulkner, B. P. Flannery.
Astron. Astrophys., Vol. 23, 325 - 330 = Lick Obs. Bull., No. 632 (1973).

A simple type of formula is described which approximates the Fermi-Dirac integrals for electron density, pressure, and internal energy to about 0.1 % or better over the entire range of their arguments: the accuracy depends on the chosen level of approximation. The method is recommended for the rapid calculation of stellar interiors. A further simple formula is given for a very crude approximation to the effect of pressure ionization on the equation of state in stellar interiors.

065.030 **Secular instabilities of pure helium stars.**
Y. Osaki, C. J. Hansen.
Astron. Astrophys., Vol. 23, 475 - 478 (1973).

Models of pure helium stars have been reanalyzed for secular stability. A simple explanation for the behavior of eigenvalues along the helium sequence is given using a two-zone model.

065.031 **Mixing in stellar models.**
R. K. Ulrich, R. T. Rood.
Nature, Phys. Sci., Vol. 241, 111 - 112 (1973).

Dilke and Gough (1972) propose that the absence of a detectable flux of neutrinos from the sun and geological ice ages is a consequence of occasional mixing in the solar core. Here we present stellar model calculations which raise serious doubts whether their mechanism can solve the solar neutrino problem; our results also have important implications for semi-convection theory.

065.032 **Nuclear reactions in carbon stars.**
A. Gélinas, M. Tassoul, G. Beaudet.
Astron. Astrophys., Vol. 24, 111 - 120 (1973).

The aim of the present work is to determine the effect on stellar evolution of the network of reactions that can take place during carbon burning. Initial models of low mass stars ($0.8\,M_0 < M < 1.25\,M_0$) with a homogeneous composition ($C^{12} : O^{16} = 1$ by mass) are considered.

065.033 **Wie stirbt ein Stern?** D. Wyler.
Orion Schaffhausen, 31. Jahrgang, p. 12 - 13 (1973).
Compiled from a lecture by G. A. Tammann, Aarau, 1972 December 5.

065.034 **Sternentstehung und Protosterne.**
I. Appenzeller, W. Tscharnuter.
SuW, Vol. 12, 12 - 15 (1973).

065.035 **Hydrogen flash in stars.**
F. Perri, A. G. W. Cameron.
Nature, Vol. 242, 395 - 396 (1973). – Letter.

065.036 **Late stages of stellar evolution.** B. Paczyński.
Postępy Astron., Vol. 21, 9 - 24 (1973). In Polish.

Evolutionary phases following the exhaustion of helium in the stellar cores are discussed. The models of red giants with hydrogen and helium shell sources and their relevance to the problem of formation of planetary nebulae is considered. The problem of carbon ignition or detonation in degenerate carbon cores and its relevance to the supernova explosions and the origin of pulsars is discussed.

065.037 **The stability of rotating barion stars.**
G. G. Harutyunian (*Arutyunyan*), D. M. Sedrakian.
Astrofizika, Vol. 8, 419 - 423 (1972). In Russian. English translation in Astrophysics, Vol. 8, No. 3.

Quasi-radial pulsations of rotating barion stars are considered. It is shown that the central density for rotating and static stars coincide where stellar configurations lose stability.

065.038 Analytical expressions of the parameters of rotating stars.
O. H. Huseynov (*O. Kh. Gusejnov*), F. K. Kasumov.
Astrofizika, Vol. 8, 425 - 432 (1972). In Russian. English translation in Astrophysics, Vol. 8, No. 3.

Analytical expressions of the parameters of rotating stars in the particular case of the pseudo-polytrope — introduced as "stepenar" — have been obtained. A connection between polytrope and "stepenar" along the most essential energetical parameters has been established.

065.039 On the stability of a rotating inhomogeneous star.
Yu. V. Vandakurov.
Astrofizika, Vol. 8, 433 - 439 (1972). In Russian. English translation in Astrophysics, Vol. 8, No. 3.

The stability of a rigidly rotating chemically inhomogeneous star with respect to small-scale perturbations is considered. It is shown that to stabilize non-axisymmetric oscillations in the radiative zone of the star, coincidence of isothermal and isobaric surfaces is necessary. The instability can also be suppressed in the presence of a comparatively weak toroidal magnetic field.

065.040 Energy in rotating neutron stars. R. M. Avakian,
G. G. Harutyunian (*Arutyunyan*), G. S. Saakian.
Astrofizika, Vol. 8, 476 - 479 (1972). In Russian. English translation in Astrophysics, Vol. 8, No. 3.

Integral parameters of rotating neutron stars and addition of mass and energy due to the deformation of rotation are calculated. It is shown that the addition of this energy is enough to provide the observed luminosity of pulsars.

065.041 Core-helium-burning stars in extremely young clusters. J. W. Robertson.
Astrophys. Journ., Vol. 180, 425 - 433 (1973).

Recent theoretical core-helium-burning evolutionary sequences for stars in the mass range $8-30\,M_\odot$ are compared with H-R diagrams of extremely young clusters in the Galaxy and the Magellanic Clouds.

065.042 Models of population I clump giants.
D. J. Faulkner, R. D. Cannon.
Astrophys. Journ., Vol. 180, 435 - 446 (1973).

The main-sequence, giant-branch, and core-helium-burning evolution is investigated for low-mass stars with original composition $(X, Y, Z) = (0.68, 0.30, 0.02)$. Stars of 1.5 and $2.0\,M_\odot$ are evolved to the helium flash with a careful treatment of the pause on the giant branch caused by the hydrogen-burning shell passing through the abundance discontinuity left behind by surface convection. Double-energy-source models are obtained at four combinations of total and helium core mass, and one of these is evolved through the core-helium-burning phase. Isochrones for 5×10^8, 10^9, and 2×10^9 years are constructed and compared with the observed C-M diagram of NGC 7789. It is found that the core-helium-burning models reproduce all the features of the clump in a satisfactory way.

065.043 Exponential growth rates of convective motion in an advanced red giant. W. M. Fawley.
Astrophys. Journ., Vol. 180, 483 - 485 (1973).

In this paper, *e*-folding times are determined for the fundamental adiabatic convective mode of $1\,M_\odot$ red-giant models.

065.044 Supersonic convection and the structure of T Tauri stars. A. J. R. Prentice.
Proc. Astron. Soc. Australia, Vol. 2, 152 - 153 (1972).

065.045 The evolution of stars to the white dwarf stage.
G. Wegner.
Proc. Astron. Soc. Australia, Vol. 2, 153 - 154 (1972).

065.046 Dust shell models for infrared sources.
A. R. Hyland, R. A. Gingold.
Proc. Astron. Soc. Australia, Vol. 2, 155 - 157 (1972).

065.047 Evolution of stars with suppressed core convection.
R. Stothers, C.-w. Chin.
Astrophys. Journ., Vol. 180, 901 - 905 (1973).

Stellar evolution on the upper main sequence has been computed for models of stars with cores assumed to be in radiative equilibrium, up to the point of central helium ignition. Observational data are used to rule out the hypothesis of evolution with radiative cores (in upper main-sequence stars) and, by implication, of magnetic fields that are sufficiently strong to have suppressed the core convection.

065.048 Neutrino-pair emission in dense matter: A many-body approach. E. Flowers.
Astrophys. Journ., Vol. 180, 911 - 935 (1973).

We formulate the description of neutrino-pair emission in dense matter in terms of the correlation functions of the background matter, and consider in detail, at densities and temperatures expected for neutron stars, the neutrino-pair bremsstrahlung emission from electrons in a static lattice and the one-phonon corrections to this process.

065.049 Some effects of Urca reactions upon degenerate carbon burning. R. G. Couch, W. D. Arnett.
Astrophys. Journ., (*Letters*), Vol. 180, L101 - L105 (1973).

The ignition of ^{12}C in degenerate cores is studied using a realistic reaction network. Convection is taken into account, and Urca neutrino losses are calculated explicitly.

065.050 URCA process and the evolution of carbon stellar core. B. Paczyński.
Acta Astron., Vol. 23, 1 - 21 (1973).

The influence of URCA neutrino cooling on the evolution of degenerate carbon-oxygen cores is studied.

065.051 Thermal pulses in helium shell-burning stars. II.
P. R. Wood, D. J. Faulkner.
Astrophys. Journ., Vol. 181, 147 - 156 (1973).

The shell-burning evolution of pure-helium stars in the mass range $0.5-1.0\,M_\odot$ is studied, both including and neglecting neutrino energy losses due to the universal Fermi interaction. The significance of the results for the central stars of planetary nebulae is discussed, and a possible explanation is given for carbon features in the spectra of the nuclei of advanced planetary nebulae, and of some white dwarfs. A possible relationship with the hydrogen-deficient carbon stars is also discussed.

065.052 Nonradial pulsation of general-relativistic stellar models. VI. Corrections.
J. R. Ipser, K. S. Thorne.
Astrophys. Journ., Vol. 181, 181 - 182 (1973).

The errors corrected here do not invalidate any of the final results obtained in previous papers of this series.

065.053 Secular stability of an $8\,M_\odot$ star during central helium burning. A. Noels, M. Gabriel.
Astron. Astrophys., Vol. 24, 201 - 208 (1973).

We study the secular stability of an $8\,M_\odot$ population I star during the phase of central He burning. An instability appears soon after the onset of the He burning. The stability is however rapidly restored and is maintained throughout the remaining of the phase. The reason of the instability is discussed.

065.054 Pulsational instability of a star of $0.5\,M_\odot$ during core hydrogen burning. A. Boury, A. Noels.
Astron. Astrophys., Vol. 24, 255 - 258 (1973).

The stability of a 0.5 $M_\odot$ star towards radial oscillations is investigated. The evolutionary stages starting at the ignition of hydrogen at the center up to its near exhaustion in the core have been covered. All models are found unstable, not only because of nuclear reactions but also because of the effects of convection. In consequence, the use of the available theory of convection is open to question.

065.055 **Evolution of stars on the stage of growth of a carbon-oxygen core.** U. Uus.
Astron. Zhurn. Akad. Nauk SSSR, Vol. 50, 297 - 304 (1973). In Russian. English translation in Soviet Astron. AJ, Vol. 17, No. 2.

A simple method for the calculation of stellar models with dense isothermal cores and thin nuclear burning shells is proposed. The results of evolutionary calculations for stars with masses 1.5, 2, 3 and 5 $M_\odot$ and a chemical composition of the envelope $X = 0.70$, $Z = 0.04$ at the stage of growth of the carbon-oxygen core, starting from the mass of the core $M_c \approx$ 0.6 $M_\odot$ and terminating at $M_c \approx 1.4 M_\odot$, are presented.

065.056 **Formation of dust in low-temperature stars.** V. S. Bychkova.
Astron. Zhurn. Akad. Nauk SSSR, Vol. 50, 427 - 430 (1973). In Russian. English translation in Soviet Astron. AJ, Vol. 17, No. 2.

Some aspects of formation of graphite grain mixtures in long-period variables during the minimum of light are considered. The amount of grains which can grow up to the standard size $r = 5 \times 10^{-6}$ cm is evaluated.

065.057 **Convection in luminosity class III red giants.** R. K. Ulrich.
Bull. American Astron. Soc., Vol. 5, 1 (1973). – Abstr. AAS.

065.058 **Evolution of N-type carbon stars: Some implications from photometry of the intermediate-age cluster NGC 2660.** F. D. A. Hartwick, J. E. Hesser.
Bull. American Astron. Soc., Vol. 5, 14 (1973). – Abstr. AAS.

065.059 **On the accuracy of the Thomas-Fermi atom for opacities.** L. D. Cloutman.
Bull. American Astron. Soc., Vol. 5, 15 (1973). – Abstr. AAS.

065.060 **Dynamical r-process calculations.** D. N. Schramm.
Bull. American Astron. Soc., Vol. 5, 27 (1973). Abstr. AAS.

065.061 **A new strong screening theory with effects on thermonuclear detonation in evolved cores.** H. C. Graboske.
Bull. American Astron. Soc., Vol. 5, 27 (1973). – Abstr. AAS.

065.062 **The adiabatic stability of stars containing magnetic fields – I. Toroidal fields.** R. J. Tayler.
Monthly Notices Roy. Astron. Soc., Vol. 161, 365 - 380 (1973).

Conditions for the stability against adiabatic perturbations of a star containing a purely toroidal magnetic field are established. Because the criteria are complicated, it is difficult to draw general conclusions from them. It is however possible to show that instability must occur close to the axis of symmetry of the star, if there is a non-zero electric current density on the axis. This proves that a large class of configurations are unstable. It is not easy to predict the ultimate effect of such instabilities. They have very short growth times compared to characteristic times of stellar evolution and, even if the instability is initially contained as an oscillation of finite amplitude, it may lead to a considerably enhanced decay of the magnetic field.

065.063 **Influence of interactions on stability and pulsations of rotating neutron stars.** A. V. Ovsepyan.
Dokl. AN ArmSSR, Vol. 55, No. 1, p. 36 - 41 (1972). In Russian. – Abstr. in Referativ. Zhurn. 51. Astron., 3.51.554 (1973).

065.064 **Hot rotating neutron stars with sources of internal energy.**
R. M. Avakyan, G. G. Arutyunyan, G. S. Saakyan, D. M. Sedrakyan.
Third Soviet Gravitational Conference, Erevan, 1972, (see 012.001), p. 284 - 287 (1972). In Russian.

065.065 **Characteristics of rotating neutron stars depending upon the number of barions.** G. G. Arutyunyan.
Third Soviet Gravitational Conference, Erevan, 1972, (see 012.001), p. 287 - 290 (1972). In Russian.

065.066 **Supermassive oblique rotator: electrodynamics and evolution.** L. M. Ozernoj, V. V. Usov.
Third Soviet Gravitational Conference, Erevan, 1972, (see 012.001), p. 340 - 341 (1972). In Russian.

065.067 **Dynamical stability of rotating stellar models.** V. V. Papoyan, D. M. Sedrakyan, É. V. Chubaryan.
Third Soviet Gravitational Conference, Erevan, 1972, (see 012.001), p. 342 - 343 (1972). In Russian.

065.068 **A new cosmological model: Formation of organic molecules, planets, and comets.** F. M. Johnson.
17th Colloquium International Astrophys. Liège 1971, (see 012.003), p. 609 - 627 (1972).

065.069 **16 $M_\odot$, 32 $M_\odot$, 64 $M_\odot$ stars evolution. I. Evolution from the main sequence up to the onset of core helium burning.** V. I. Varshavsky, A. V. Tutukov.
Nauchn. Informatsii, vyp. (No.) 23, p. 47 - 73 (1972).

Evolutionary sequences of 16, 32 and 64 $M_\odot$ stellar models with initial chemical composition $X = 0.602$, $Y = 0.354$, $Z = 0.044$ ($X_{12} = 0.00619$, $X_{16} = 0.01847$) were calculated using three different conditions for convective neutrality in the semiconvective zone.

065.070 **Abundance changes in the envelope of a 5 $M_\odot$ red supergiant.** U. Uus.
Nauchn. Informatsii, vyp. (No.) 23, p. 85 - 96 (1972). In Russian.

The evolution of a 5 $M_\odot$ star ($X = 0.70$, $Z = 0.04$) on the carbon-oxygen core growing stage has been computed. Abundance changes in the envelope have the nature of successive establishment of equilibrium ratios between concentrations of C, N, O isotopes.

065.071 **The interaction between core and envelope in stars with central helium burning.** D. Lauterborn.
Astron. Astrophys., Vol. 24, 421 - 427 (1973).

The interaction between core and envelope in stars with central helium burning is investigated. If core and envelope are treated as independent systems, feedback terms arise. All feedback terms are written down and discussed in detail. The approximative treatment of feedback terms in the earlier papers of Lauterborn, Refsdal, and Weigert is found to be fully justified. The problem of secular instabilities in models with central helium burning is rediscussed.

065.072 **Possible thermal instability of hydrogen burning shells.** A. J. C. Bolton, P. P. Eggleton.
Astron. Astrophys., Vol. 24, 429 - 434 (1973).

A simplified model of a nuclear burning shell is investigated for its stability to radial perturbations on a thermal (Kelvin-Helmholtz) time scale.

065.073 **The $UBVr$ colors of extreme Pop II giants.**
E. Böhm-Vitense.
Astron. Astrophys., Vol. 24, 447 - 458 (1973).

The $UBVr$ colors of Pop II giants with metal abundances
1/100 and 1/1000 times the solar metal abundances are computed. The mean intensity J, the source function for the Rayleigh scattering, has been determined, considering not only the
continuous absorption of hydrogen and the metals but also the
line absorption. The $UBVr$ colors are given in the range 4200°
$\lesssim T_{eff} \lesssim 7700°$ and $4.6 \lessgtr L/L_\odot < 1800$. From this the relation
between $B - V$ and T_{eff} is determined for different luminosities
and also the ultraviolet excess.

065.074 **Initial asymptotic branch evolution of population II**
stars. A. V. Sweigart.
Astron. Astrophys., Vol. 24, 459 - 464 (1973).

The asymptotic branch evolution for population II stars
of 0.6 and 0.8 $M_\odot$ has been followed from helium exhaustion
in the core through the initial relaxation cycles with a detailed
treatment of the envelope physics.

065.075 **Temperature dependent nuclear mass and its appli-**
cation to astrophysical problems. T. Ohnishi.
Astrophys. Space Sci., Vol. 20, 225 - 239 (1973).

Temperature dependent nuclear binding energies are
proposed. The binding energies for some 170 nuclides in
$40 \lessgtr A \lessgtr 66$ are computed at several temperatures according to this model. The abundances of the nuclides in the
above atomic mass region are then calculated for example by
using these temperature dependent binding energies.

065.076 ^{187} **Re, recycling r-process elements through stars,**
and the age of the Galaxy. R. J. Talbot, Jr.
Astrophys. Space Sci., Vol. 20, 241 - 249 (1973).

The enhanced β-decay rate of ionized ^{187}Re in stars has
been studied within the context of a detailed numerical
model of the production of r-process elements and their recycling through stars during the course of galactic evolution.
It is concluded that the enhanced decay rate does not significantly reduce the Re-Os chronometer age for the Galaxy.

065.077 **Evolution of massive population II stars.**
V. Trimble, B. Paczyński, B. A. Zimmerman.
Astron. Astrophys., Vol. 25, 35 - 40 (1973).

Evolutionary tracks are presented for 2, 3, 10, 12, and
30 $M_\odot$ stars having $X = 0.7$ and $Z = 0.001$.

065.078 **Secular stability IV. The effect of shell sources on**
the secular spectrum. M. L. Aizenman, J. Perdang.
Astron. Astrophys., Vol. 25, 53 - 57 (1973).

The method of successive approximations used to discuss
the onset of complex roots in the secular spectrum is extended
to the case where shell sources are present in the star. Parameters describing the intensity and position of such sources are
defined, and their influence on the secular spectrum is discussed. Secular eigenvalues are determined for different
choices of these parameters.

065.079 **Central gravitational field of stars and evolution to**
red giants. W. Höppner, A. Weigert.
Astron. Astrophys., Vol. 25, 99 - 103 (1973).

In order to determine the effect mainly responsible for
the evolution into the red giant region after central hydrogen
burning, models are considered in which the physical properties are altered artificially. In this way, the effects of a strong
gravitational field and of inhomogeneities in the molecular
weight and the energy generation rate are separated as far as
possible. It is shown that only a strong central gravitational
field yields models with properties similar to those of red giants.

065.080 **Exploding massive objects.**
I. Appenzeller, W. Tscharnuter.
Astron. Astrophys., Vol. 25, 125 - 128 (1973).

Hydrodynamic model calculations of the evolution of
non-rotating massive objects with population I chemical composition for different mass values have been carried out. The
numerical results indicate that all such objects in the mass interval $4.1 \lessgtr 10^{-5} M/M_\odot \lessgtr 7$ terminate their evolution by violent thermonuclear explosions.

065.081 **Deviations from Einstein's principle of equivalence**
and neutron stars. K. Fritze.
Astron. Nachr., Vol. 294, 91 - 93 (1973).

The equations of stellar structure are derived for relativistic stellar models assuming deviations from the weak principle of equivalence. It is estimated that observable deviations
from normal neutron star models will occur if the deviation
constant ϵ exceeds 10^{-3}.

065.082 **The effect of interstellar medium parameters on the**
accretion by neutron stars.
P. R. Amnuel, O. H. Guseinov (*O. Kh. Gusejnov*).
Astron. Nachr., Vol. 294, 139 - 146 (1973).

The effect of the magnetic field trapped into the interstellar gas on the accretion by neutron stars is considered. The
account of the magnetic field decreases the accretion by 2
orders. As a result of accretion the luminosity of the neutron
stars is found to be $\sim 10^{28} - 10^{29}$ erg/s and so their observation
as X-ray sources is hardly probable.

065.083 **Evolución estelar.** A. Paluzíe Borrell.
Urania Barcelona, Año 57, No. 275, p. 123 - 143
(1972).

065.084 **Non-linear pulsations of upper main sequence**
stars–I. A perturbation approach.
J. C. B. Papaloizou.
Monthly Notices Roy. Astron. Soc., Vol. 162, 143 - 168
(1973).

The non-linear pulsations of vibrationally unstable main
sequence stars of between 70 and 170 $M_\odot$ are investigated
using a perturbation technique. More general as well as periodic solutions are considered. Consideration of the behaviour
close to resonance is given and the relationship of the periodic
solutions to more general solutions discussed.

065.085 **On the law of escape of a strong thermal wave on**
the surface of a star. I. A. Klimishin.
Astron. Tsirk., No. 754, p. 3 - 4 (1973). In Russian.

065.086 **Non-linear pulsations of upper main sequence**
stars–II. Direct numerical integrations.
J. C. B. Papaloizou.
Monthly Notices Roy. Astron. Soc., Vol. 162, 169 - 187
(1973).
The non-linear pulsations of vibrationally unstable main
sequence stars of between 70 and 210 $M_\odot$ are studied using
numerical methods directly. Non-linear solutions are investigated for various amplitudes for short periods of time. The
results are found to be in qualitative agreement with those
obtained in a previous paper using a perturbation technique.
The results are found to be different from those given by
other authors.

065.087 **Models for initially homogeneous carbon-rich stars.**
A. H. Boozer, P. C. Joss, E. E. Salpeter.
Astrophys. Journ., Vol. 181, 393 - 407 (1973).

Stellar models between 0.75 and 1.45 $M_\odot$, initially consisting of 50 percent ^{12}C and 50 percent ^{16}O (by mass), have
been constructed. The evolutionary computations included

the most recent nuclear reaction rates and neutrino loss rates (excluding bremsstrahlung neutrinos), screening corrections to the reaction rates, and Coulomb corrections to the thermodynamic properties of the nuclei.

065.088 **Models for carbon-rich stars with helium envelopes.**
P. C. Joss, J. I. Katz, R. C. Malone, E. E. Salpeter.
Astrophys. Journ., Vol. 181, 409 - 428 (1973).
    Stellar models with carbon-rich interiors and helium envelopes of varying fractional mass have been constructed. The models have masses of 0.75, 1.02, and 1.37 $M_\odot$. A few $1.02\,M_\odot$ models with hydrogen-rich outer envelopes of small fractional mass have also been constructed. The evolution of these models is compared with the observed evolution of the nuclei of planetary nebulae.

065.089 **On the "critical luminosity" in stellar interiors and stellar surface boundary conditions.**
P. C. Joss, E. E. Salpeter, J. P. Ostriker.
Astrophys. Journ., Vol. 181, 429 - 438 (1973).
    General relations are derived among the so-called critical luminosity, and the actual radiative and convective luminosities within a star in hydrostatic and local thermodynamic equilibrium. A general boundary condition for an outer layer of small fractional mass is derived which expresses the surface luminosity in terms of temperature, density, and opacity at the base of the layer and a parameter. Explicit expressions are obtained for a perfect gas plus blackbody radiation and Kramers opacity plus electron scattering.

065.090 **On the stability of axisymmetric systems to axisymmetric perturbations in general relativity. IV. Allowance for gravitational radiation in an odd-parity mode.**
S. Chandrasekhar, J. L. Friedman.
Astrophys. Journ., Vol. 181, 481 - 495 (1973).
    The variational principle derived in a previous paper (1972) is clarified; and it is shown how it may be used to treat the damping of the axisymmetric oscillations of a uniformly rotating star, by the emission of gravitational radiation in an odd-parity mode.

065.091 **The effect of gravitational radiation on the secular stability of uniformly rotating fluid masses.**
B. D. Miller.
Astrophys. Journ., Vol. 181, 497 - 512 (1973).
    The secular effect of gravitational radiation-reaction on the oscillations of uniformly rotating polytropes and the Maclaurin spheroids is examined. There are examples of oscillations which become unstable, which are damped, and which are unaffected.

065.092 **On the evolution of the secularly unstable, viscous Maclaurin spheroids.**
W. H. Press, S. A. Teukolsky.
Astrophys. Journ., Vol. 181, 513 - 517 (1973).
    Previous investigations, which are superficially contradictory, are here reconciled. With new numerical results, a consistent picture emerges: A secularly unstable, viscous Maclaurin spheroid slowly and monotonically deforms itself into a stable, Jacobi ellipsoid. The intermediate configurations are Riemann S-type ellipsoids.

065.093 **The oscillations and the stability of rotating masses with magnetic fields. III.**
R. K. Kochhar, S. K. Trehan.
Astrophys. Journ., Vol. 181, 519 - 521 (1973).
    It is shown that when a magnetic field is present along the axis of rotation, the point of bifurcation, where the Jacobi ellipsoids branch off from the Maclaurin spheroids, occurs at a value of eccentricity which is higher than the value ($e = 0.81267$) that obtains in the absence of a magnetic field. This

is in contrast to the effect of a toroidal magnetic field which, as has been shown earlier, leaves the point of bifurcation unaffected.

065.094 **The early evolution of stars – I, II.**
S. E. Strom, K. M. Strom.
Sky Telescope, Vol. 45, 279 - 282, 359 - 361 (1973).

065.095 **Stability of super-massive stars.**
M.-K. Fujimoto, W. Unno.
Publ. Astron. Soc. Japan, Vol. 25, 243 - 252 (1973).
    Stabilities of super-massive stars for a non-adiabatic perturbation are discussed in the framework of the post-Newtonian approximation. Special attention is paid to the case where the thermal time-scale is shorter than the dynamical time-scale.

065.096 **On approximate evaluation of the Fermi-Dirac functions.**   Y. Ogawa, M. Nishida.
Publ. Astron. Soc. Japan, Vol. 25, 281 - 283 (1973).
    It is shown that the rational Chebyshev approximations for the Fermi-Dirac functions of orders $-1/2$, $1/2$, and $3/2$ are preferable in the computation of stellar configurations involving partially degenerate matter.

065.097 **Positron-annihilation radiation from neutron stars.**
R. Ramaty, G. Borner, J. M. Cohen.
Astrophys. Journ., Vol. 181, 891 - 894 (1973).
    We propose that the recently observed 473 ± 30 keV spectral feature from the galactic center is gravitationally redshifted positron-annihilation radiation produced at the surfaces of neutron stars.

065.098 **Equation of state of matter at supernuclear densities.**   Y. C. Leung, C. G. Wang.
Astrophys. Journ., Vol. 181, 895 - 902 (1973).
    The qualitative features of the equation of state of matter at supernuclear densities are deduced through a careful examination of the nature of particle interactions at short distances and by the introduction of an "effective baryon mass spectrum". It is found that the equation of state begins to take on a particularly simple form (the "asymptotic form") at a relatively low matter density of 10^{17} g cm^{-3}.

065.099 **Relativistic stellar stability: an empirical approach.**
W.-T. Ni.
Astrophys. Journ., Vol. 181, 939 - 956 (1973).
    The "PPN formalism" – which encompasses the post-Newtonian limit of nearly every metric theory of gravity – is used to analyze stellar stability. This analysis enables one to infer, for any given gravitation theory, the extent to which post-Newtonian effects induce instabilities in white dwarfs, in neutron stars, and in supermassive stars. It also reveals the extent to which our current empirical knowledge of post-Newtonian gravity (based on solar-system experiments) actually guarantees that relativistic instabilities exist.

065.100 **A variational principle for magnetoelastic, rotating stars in general relativity.**   M. W. Munn.
Astrophys. Journ., Vol. 181, 957 - 976 (1973).
    We show here that an extremal energy variational principle can be formulated which, as conditions for the extremum, reproduces the gravitational field equations, the equation of magnetoelastic equilibrium, and an expression for the angular velocity of the star. The proof is carried out for elastic, rotating stars which contain magnetic fields and is done in the Newtonian theory and in general relativity.

065.101 **Sources of internal energy of baryon stars.**
R. M. Avakyan, G. S. Saakyan, D. M. Sedrakyan.
Uch. zap. Erevan. un-t. Estestv. n., 1972, No. 2 (120), p. 16 -

22. In Russian. – Abstr. in Referativ. Zhurn. 51. Astron., 4.51.670 (1973).

065.102 Neutrino thermal conductivity in collapsing stars.
V. S. Imshennik, D. K. Nadëzhin.
Zhurn. ehksperim. i teor. fiz., Vol. 63, 1548 - 1561 (1972).
In Russian. – Abstr. in Referativ. Zhurn. 51. Astron., 4.51.671 (1973).

065.103 Interaction of proto-stars in a collapsing cluster.
T. Arny, P. Weissman.
Astron. Journ., Vol. 78, 309 - 315 (1973).
The collapse and early evolution of a cluster of proto-stars is studied with an n-body integration. Each proto-star is assigned an initial radius which decreases with time. As the cluster collapses, random transverse velocities rapidly grow and orbital mixing occurs which prevent the fragments from reaching the center simultaneously. Roughly 50 % of all fragments suffer very close encounters or physical collisions during the initial collapse of the cloud, unless the fragment radii shrink four times as fast as the cluster itself. This suggests that the usual picture of isolated collapse from interstellar cloud to pre-main-sequence star is in need of more critical analysis.

065.104 Normal stellar evolution. I. Iben, Jr.
Stellar evolution, (see 012.014), p. 1 - 106 (1972).

065.105 Evolution near the main sequence. P. Demarque.
Stellar evolution, (see 012.014), p. 107 - 128 (1972).

065.106 Stellar evolution from main sequence to white dwarf or carbon ignition. B. Paczyński.
Stellar evolution, (see 012.014), p. 129 - 140 (1972). – Reprinted from Acta Astron., Vol. 20, 47 - 58 (1970). – See Abstr. 03.065.079.

065.107 Structure of massive main-sequence stars.
R. Stothers.
Stellar evolution, (see 012.014), p. 141 - 153 (1972).

065.108 Stellar stability and stellar pulsation. N. Baker.
Stellar evolution, (see 012.014), p. 155 - 172 (1972).

065.109 Variable stars – realistic star models.
R. F. Christy.
Stellar evolution, (see 012.014), p. 173 - 210 (1972).

065.110 Neutron stars. A. G. W. Cameron.
Stellar evolution, (see 012.014), p. 329 - 350 (1972).

065.111 Stellar opacity. T. R. Carson.
Stellar evolution, (see 012.014), p. 427 - 491 (1972).

065.112 Transport mechanisms in stars. E. A. Spiegel.
Stellar evolution, (see 012.014). p. 493 - 519 (1972).

065.113 Thermonuclear reactions and nucleosynthesis.
J. W. Truran.
Stellar evolution, (see 012.014), p. 521 - 568 (1972).

065.114 Instability problem in nuclear burning shells.
W. K. Rose.
Stellar evolution, (see 012.014), p. 569 - 592 (1972).

065.115 Stellar magnetism and rotation. L. Mestel.
Stellar evolution, (see 012.014), p. 643 - 734 (1972).

065.116 Solid core in neutron stars.
V. Canuto, S. M. Chitre.
Nature, Phys. Sci., Vol. 243, 63 - 65 (1973).
We present calculations designed to show that under conditions of pressure and density that typically prevail in the interior of a neutron star, a system of strongly interacting baryons minimize the energy by arranging the constituents in a lattice structure.

065.117 The beaming of radiation from an accreting magnetic neutron star and the X-ray pulsars.
Yu. N. Gnedin, R. A. Sunyaev.
Astron. Astrophys., Vol. 25, 233 - 239 (1973).
As a result of accretion, a neutron star with a strong magnetic field $H \sim 10^{10} - 10^{12}$ Gauss may exhibit characteristics of an X-ray pulsar with a knife radiation pattern. The infalling gas is channeled to the magnetic pole regions where its kinetic energy is released as it strikes the star surface. The cyclotron radiation at high gyrofrequency harmonics (lying in the X-ray range) determines the rate of energy removal from the hot plasma. Radiation at high harmonics is characterized by a strong beaming perpendicular to the magnetic field lines of force.

065.118 Secular stability of static models during central helium burning. M. Gabriel, A. Noels.
Astron. Astrophys., Vol. 25, 313 - 317 (1973).
A sequence of static helium burning models has been computed. The slope of the hydrogen profile, dX/dm, above the hydrogen burning shell has been chosen as the free parameter. We have also computed sequences of envelope models. All models are secularly stable. The relation between secular stability and linear series concept is discussed.

065.119 Stellar ages – a brief introduction. I. Iben, Jr.
Stellar ages. Proc. IAU Colloquium No. 17, (see 012.015), I, 1 - 11 (1973).

065.120 Time table of star formation in the Large Magellanic Cloud. C. Payne-Gaposhkin.
Stellar ages. Proc. IAU Colloquium No. 17, (see 012.015), III, 1 - 12 (1973).

065.121 Stellar ages – a rough review. I. Iben, Jr.
Stellar ages. Proc. IAU Colloquium No. 17, (see 012.015), XI, 1 - 33 (1973).

065.122 The influence of the chemical composition, rotation, magnetic fields and semiconvection on the calculated lifetime of main sequence stars. A. Tutukov.
Stellar ages. Proc. IAU Colloquium No. 17, (see 012.015), XIV, 1 - 3 (1973).

065.123 Age and initial chemical composition of supergiant stars as function of distance from the galactic center.
G. Barbaro, C. Chiosi.
Stellar ages. Proc. IAU Colloquium No. 17, (see 012.015), XV, 1 - 15 (1973).

065.124 On the ages and masses of carbon stars.
G. Barbaro, N. Dallaporta.
Stellar ages. Proc. IAU Colloquium No. 17, (see 012.015), XVI, 1 - 15 (1973).

065.125 Age calibrations for population I compositions.
P. M. Hejlesen, H. E. Jørgensen, J. O. Petersen, L. Rømcke.
Stellar ages. Proc. IAU Colloquium No. 17, (see 012.015), XVII, 1 - 36 (1973).

065.126 · **Age determination and theoretical luminosity functions for clusters of an intermediate population II.**
P. M. Hejlesen.
Stellar ages. Proc. IAU Colloquium No. 17, (see 012.015), XVIII, 1 - 23 (1973).

065.127 **The period-age relationship for δ Cephei stars.**
E. Meyer-Hofmeister.
Stellar ages. Proc. IAU Colloquium No. 17, (see 012.015), XIX, 1 - 3 (1973).

065.128 **Remarks on stellar abundances.** P. Demarque.
Stellar ages. Proc. IAU Colloquium No. 17, (see 012.015), XXI, 1 - 9 (1973).

065.129 **Some very tentative evidence for recent star formation in elliptical galaxies.** S. van den Bergh.
Stellar ages. Proc. IAU Colloquium No. 17, (see 012.015), XXXIV, 1 - 7 (1973).

065.130 **Abundance perturbations, stellar evolution, and secular stability.** M. L. Aizenman, J. Perdang.
Stellar ages. Proc. IAU Colloquium No. 17, (see 012.015), XXXV, 1 - 11 (1973).

065.131 **An observational check on evolutionary time scales for massive hot stars.** M. L. Burnichon.
Stellar ages. Proc. IAU Colloquium No. 17, (see 012.015), XLII, 1 - 3 (1973).

065.132 **Ages and chemical compositions of stars.**
B. E. J. Pagel.
Stellar ages. Proc. IAU Colloquium No. 17, (see 012.015), XLVII, 1 - 19 (1973).

065.133 **Metallicism and evolution of the Am stars.**
B. Hauck.
Stellar ages. Proc. IAU Colloquium No. 17, (see 012.015), XLIX, 1 - 3 (1973).

065.134 **Star formation and the chemical history of galaxies.**
L. Searle.
Stellar ages. Proc. IAU Colloquium No. 17, (see 012.017), LII, 1 - 15 (1973).

065.135 **On metal rich and SMR (*super-metal-rich*) stars.**
M. Grenon.
Stellar ages. Proc. IAU Colloquium No. 17, (see 012.015), LV, 1 - 6 (1973).

065.136 **Age from spectroscopic properties. Concluding remarks.** R. Cayrel.
Stellar ages. Proc. IAU Colloquium No. 17, (see 012.015), LIX, 1 - 7 (1973). In French.

065.137 **The problem of the ages of the stars and the connections with other time dependent processes (a summary and conclusion of the I.A.U. Colloquium No. 17 on the ages of the stars).** E. Schatzman.
Stellar ages. Proc. IAU Colloquium No. 17, (see 012.015), LX, 1 - 10 (1973).

065.138 **Effect of computational techniques on the main sequence evolution of a star of 5 $M_\odot$.**
K. Mimura.
Sci. Rep. Tôhoku Univ.,First Ser.,Vol. 55, 124 - 138 (1972/73).
In preparing a program to compute the evolutionary stellar structure, some technical points associated mainly with the initial value problem and how to specify the outer boundary of the convective core have been investigated. Six evolutionary sequences have been constructed for a star of 5 $M_\odot$

with an initial composition given by $X = 0.602$, $Y = 0.354$ and $Z = 0.044$.

065.139 **Toroidal and poloidal oscillations of magnetic rotating stars in a decay field.** K. Maezawa.
Sci. Rep. Tôhoku Univ.,First Ser., Vol. 55, 139 - 154 (1972/73).
This paper deals with the magnetohydrodynamic effect of small oscillations of a magnetic rotating star. It is assumed that the star is a homogeneous incompressible sphere rotating uniformly about its axis and initially has a decaying dipole magnetic field with longest decay time. Magnetohydrodynamic oscillations about the initial field are studied by using Elsasser's field functions when small toroidal and poloidal perturbations are applied. Characteristic frequencies of these oscillations are given numerically.

065.140 **On the angular momentum of stars and the main sequence stellar models.** N. Virgopia.
Mem. Soc. Astron. Italiana, Nuova Ser., Vol. 43, 813 - 821 (1973).
Starting from the relation between effective temperatures and spectral types for main sequence stars, the problem of stellar rotation is discussed by considering two theoretical ZAMS population I stars together with the observed main sequence stars. It seems that under the hypothesis of rigid rotation, all stars initially satisfied the same angular momentum-mass relationship.

065.141 **Stellar magnetohydrodynamics.** L. Mestel.
Cosmic plasma physics. Conference 1971, (see 012.016), p. 203 - 213 (1972).

065.142 **Crystallization density of cold dense neutron matter.**
D. Schiff.
Nature, Phys. Sci., Vol. 243, 130 - 133 (1973). − Letter.

065.143 **Evolution of supermassive stars with a strong magnetic field.** G. S. Bisnovatyj-Kogan, S. I. Blinnikov.
Astron. Zhurn. Akad. Nauk SSSR, Vol. 50, 475 - 480 (1973). In Russian. English translation in Soviet Astron. AJ, Vol. 17, No. 3.
The evolution of supermassive stars with a magnetic field is considered in the presence of pulsar-like radiation and loss of rotational momentum.

065.144 **BD−10°4662 interpreted as a post−T Tauri star.**
G. H. Herbig.
Astrophys. Journ., Vol. 182, 129 - 138 = Contr. Lick Obs., No. 360 (1973).
BD−10°4662, known to have flared on at least four occasions, has been found to be a close (1.'3) double of types about K5p V and K7p V. Both stars have Hα in emission as well as very strong absorption Li I λ6707. It is argued that the components of −10°4662 are stars of about 2 $M_\odot$, approximately $1-2 \times 10^5$ years old, that have now evolved to a position near the bottom of the vertical branch of their Hayashi tracks, and hence have already passed through their T Tauri stages.

065.145 **Thermal instability of the helium-burning shell in massive stars.** R. Stothers, C.-w. Chin.
Astrophys. Journ., Vol. 182, 209 - 214 (1973).
Nonlinear numerical calculations of stellar evolution at high mass show that thermal instability develops temporarily in the helium-burning shell, shortly after the ignition of shell helium.

065.146 **A simplified model for oscillatory secular modes.**
R. J. Defouw.
Astrophys. Journ., Vol. 182, 215 - 224 (1973).
A star is considered to consist of two zones, one of which

contains a nuclear source. The interaction between the two zones can lead to slow oscillations during which the star remains near hydrostatic equilibrium. Applications are made to stars on the main sequence and to stars with helium-burning shells.

065.147 **Stability of periodic oscillations of stars.**
K. v. Sengbusch.
Mitt. Astron. Ges., No. 32, p. 228 - 232 (1973).

065.148 **Was bleibt vom Vogt-Russell-Theorem?**
H. Kähler.
Mitt. Astron. Ges., No. 32, p. 240 (1973). — Abstract.

065.149 **Beispiele für Mehrfachlösungen bei Sternmodellen.**
M. L. Roth, A. Weigert.
Mitt. Astron. Ges., No. 32, p. 240 (1973). — Abstract.

065.150 **Zentrales Schwerefeld und Entwicklung ins Riesenstadium.** W. Höppner, A. Weigert.
Mitt. Astron. Ges., No. 32, p. 240 (1973). — Abstract.

065.151 **Protosterne von 1 $M_\odot$ und 60 $M_\odot$.**
I. Appenzeller, W. Tscharnuter.
Mitt. Astron. Ges., No. 32, p. 272 - 273 (1973).

065.152 **Linear series of stellar models. III. Hydrogen-helium star of 10 $M_\odot$.** M. Kozłowski, B. Paczyński.
Acta Astron., Vol. 23, 65 - 77 (1973).
Two linear series of models for a star of 10 $M_\odot$ are constructed. The models consist of hydrogen-rich envelopes and helium-rich cores. The two series have a somewhat different profile of hydrogen distribution.

065.153 **On thermal waves in stars, II.** I. A. Klimishin.
Astrophys. Space Sci., Vol. 22, 3 - 12 (1973).
The regularities of strong thermal waves moving in nonuniform media are considered in some detail. The application of thermal wave theory to nova outburst interpretation is discussed.

065.154 **Relativistic astrophysics. IV. Relativistic rotating stars.** M. A. Abramowicz.
Postępy Astron., Vol. 21, 87 - 114 (1973). In Polish.
Papers describing the principal known properties of relativistic rotating stars are reviewed. It was intended to do this in a general way without reference to any particular coordinate system. Some results are presented for the first time.

065.155 **Neutron star moments of inertia: Theoretical implications of the observational data.**
B. Carter, H. Quintana.
Astrophys. Letters, Vol. 14, 105 - 109 (1973).
The moment of inertia of a neutron star is shown to depend very sensitively on the form of the equation of state for neutron star matter — much more so than the mass. Consequently it is possible to use existing observational data, derived mainly from the Crab nebula pulsar, to rule out some of the 'softer' theoretical equations of state that have been proposed for neutron star matter. This provides indirect evidence for the form of the short-range repulsive forces between nucleons.

065.156 **Magnetic fields in rapidly rotating stars.**
M. Maheswaran, H. A. B. M. de Silva.
Monthly Notices Roy. Astron. Soc., Vol. 162, 289 - 293 (1973).
The authors discuss the question as to why a magnetic field should not be observed in a rapidly rotating pole-on star.

065.157 **Pinch instabilities in magnetic stars.**
G. A. E. Wright.

Monthly Notices Roy. Astron. Soc., Vol. 162, 339 - 358 (1973).
It is suggested that hydromagnetic 'pinch' instabilities are capable of efficient magnetic flux destruction in the early stages of a star's history. The stability of an axisymmetric magnetic field in the convectively stable region of a star is investigated.

065.158 **The equilibrium, stability and evolution of a rotating magnetized gaseous disk.**
G. S. Bisnovatyi-Kogan, S. I. Blinnikov.
Astrophys. Space Sci., Vol. 19, 93 - 118, 119 - 144 (1972).
In Russian and English.
Exact solutions for the equilibrium of a rotating gaseous disk with poloidal magnetic field are obtained. The stability of the disk with respect to uniform expansion and contraction is investigated by means of a variational principle.

065.159 **Irregular nebular variables and neutrino emission.**
P. Raychaudhuri.
Astrophys. Space Sci., Vol. 19, 297 - 301 (1972).
The suggestion that a less massive star will reach the main sequence stage earlier than a massive star due to the effect of neutrino emission according to the photon-neutrino coupling theory is supported by the observed behaviour of H-R diagram of irregular nebular variables.

065.160 **Meridian circulation in differentially rotating stars.**
B. E. McDonald.
Astrophys. Space Sci., Vol. 19, 309 - 349 (1972).
We formulate the first order theory of meridian circulation in radiative zones of approximate chemical homogeneity so as to allow calculation of circulation velocities in a star subject to an arbitrary axially symmetric angular velocity $\omega(r, \theta)$. We find two types of steady state configurations: (1) rotation according to a special class of distributions $\omega(r, \theta)$ which drive no currents, and (2) rotation according to arbitrary $\omega(r, \theta)$, but with a correction to the molecular weight μ such that the net circulation velocity is zero. We find that a small μ gradient can quench the circulation in many cases. In particular, we conclude that a large differential rotation in the sun might have escaped disruption by meridian currents, for a μ stratification of a few parts in 10^3.

065.161 **Electric field around an accreting star.**
A. M. Anile, A. Treves.
Astrophys. Space Sci., Vol. 19, 411 - 415 (1972).
The accretion of matter on a neutron star is considered. The electric field is studied in the case of a null magnetic field and of a dipole field. The relevance of the results for models of X-ray sources is examined.

065.162 **The role of the Coriolis force on the stability of rotating magnetic stars and the origin of convective motions.** K. Sakurai.
Astrophys. Space Sci., Vol. 19, 417 - 421 (1972).
Based on the method of the energy principle, the effect of the Coriolis force on the stability of rotating magnetic stars is examined and the conditions for instability are derived. It is shown that, in these stars, the effect of this force is to inhibit the onset of convective motion. Discussion is given on the possibility of hydromagnetic dynamo processes in respect to the convective motion inside these stars.

065.163 **Stellar evolution toward pre-supernova state. II. Carbon and oxygen stars of 1.5 $M_\odot$ and 2.6 $M_\odot$.**
S. Ikeuchi, K. Nakazawa, T. Murai, R. Hoshi, C. Hayashi.
Progr. Theor. Phys., (Japan), Vol. 48, 1870 - 1884 (1972).
The evolution of carbon-oxygen stars of 1.5 $M_\odot$ and 2.6 $M_\odot$ has been computed from a pre-carbon-burning stage to a pre-supernova stage for the two alternative cases, with

and without neutrino loss.

065.164 Neutron stars in the scalar-tensor theory of gravitation. K. Yokoi.
Progr. Theor. Phys., (*Japan*), Vol. 48, 1760 - 1761 (1972).
The author applies the scalar-tensor theory of Brans and Dicke to neutron star studies. He takes into account the nuclear forces using the Kodama—Yamada mass formula for compressible nuclei. He assumes, for simplicity, that the neutron star is static and spherically symmetric and composed of neutrons only, He concludes that the difference between scalar tensor theory and general relativity is rather small in comparison with uncertainties caused by various equations of state.

065.165 Gravitational collapse with charge and small asymmetries. I. Scalar perturbations. J. Bicak.
General Relativity and Gravitation, Vol. 3, 331 - 349 (1972).

065.166 Magnetic field in the plasmasphere of a compact star. R. H. Cohen. B. Coppi, A. Treves.
Nuovo Cimento B, Ser. 11, Vol. 13 B, 59 - 81 (1973).
The main characteristics of the plasmasphere which is likely to surround a compact star,with a strong magnetic field and a relatively large velocity of rotation, are pointed out.

065.167 Note on variational principles governing small departures from equilibrium in gaseous spheres and in horizontal fluids. P. Smeyers.
Ric. Astron., Specola Vaticana, *Castel Gandolfo*, Vol. 8, (No. 17), 359 - 366 (1973).
This note deals with the analogy between a variational principle of Chandrasekhar and Lebovitz governing adiabatic radial and non-radial oscillations of a gaseous sphere, and a variational principle of Tolstoy governing waves in stratified horizontal fluids.

065.168 Models of main-sequence stars. T. Angelov.
Publ. Dep. Astron. Univ. Beograd, No. 4, p. 10 - 39 (1973).
This paper presents results of the computation of models for main-sequence stars having homogeneous chemical composition, in the phase of hydrogen reactions. Fourteen models having convective core and radiative envelope are considered, with masses 4, 5, 8, 10 and 16 $M_\odot$ and chemical structure $0.650 \leqslant X \leqslant 0.900, 0.01 \leqslant Z \leqslant 0.045$. These models are compared with models of other authors and with observational results.

065.169 Convection in the presence of magnetic fields (in stars). R. S. Peckover.
Comput. Phys. Commun., (*Netherlands*), Vol. 4, 339 - 344 (1972).
Convection driven by horizontal temperature gradients in an electrically conducting, viscous, Boussinesq fluid has been studied on a computer. Results of numerical experiments with a Grashof number of 10^4 are presented and compared with linear theory. The astrophysical relevance of these results is discussed.

065.170 Neutron radiative-capture cross-sections for 's-process' calculations in stars.
V. Benzi, R. D'Orazi, G. Reffo.
Nuovo Cimento B, Ser. 11, Vol. 13B, 226 - 248 (1973).
An estimate of the effective neutron cross-sections for a number of nuclei involved in the s-process in stars is given. The effective cross-sections were obtained as weighted averages of neutron radiative-capture cross-sections over Maxwellian energy distributions at various temperatures.

065.171 Possibility of a phase transition to a pion conden-

sate in neutron stars.
S. Barshay, G. Vagradov, G. E. Brown.
Phys. Letters B, (*Netherlands*), Vol. 43B, 359 - 361 (1973).

065.172 Elastic perturbation theory in general relativity and a variation principle for a rotating solid star.
B. Carter.
Commun. Math. Phys., Vol. 30, 261 - 286 (1973). – See Phys. Abstr., Vol. 76, No. 37876 (1973).

065.173 Solidification of neutron matter.
V. Canuto, S. M. Chitre.
Phys. Rev. Letters, Vol. 30, 999 - 1002 (1973).
A t-matrix calculation of the ground-state energy of cold neutron matter has been performed and it is shown that neutron matter solidifies at a density of the order of 1.6×10^{15} g/cm^3, thus implying the existence of a solid core inside heavy neutron stars.

065.174 Radial pulsations of pre-white-dwarf-stars. II. Pulsational stability of ^{12}C shell-burning stars.
M. P. Marshall, H. M. Van Horn.
Astrophys. Journ., Vol. 182, 901 - 913 (1973).
Radial pulsation periods, eigenfunctions, and stability integrals have been computed numerically for a number of initially pure ^{12}C shell-burning star models selected from the evolutionary sequences of Kutter and Savedoff. All of the models were found to be pulsationally stable, even during times of maximum nuclear burning. The conditions under which nuclear shell-burning can excite pulsations are discussed, and the relevance of these results for the nuclei of planetary nebulae and for the mechanism of nebular ejection are briefly considered.

065.175 Secular stability with departures from ^{3}He equilibrium in the proton-proton chain.
R. J. Defouw.
Astrophys. Journ., Vol. 182, 983 - 988 (1973).
When departures from ^{3}He equilibrium in the proton-proton chain are taken into account, the classical secular modes are modified and new modes are introduced. The corresponding modes of a two-zone model consisting of a core and an envelope are examined and found to be stable.

065.176 Neutrino energy loss in neutron star matter.
N. Itoh, T. Tsuneto.
Progr. Theor. Phys., (*Japan*), Vol. 48, 1849 - 1859 (1972).
The theory of the energy loss due to neutrino processes in neutron stars is formulated in such a form that one can include the effect of strong interaction systematically.

065.177 Gravitational contraction of protostars. I. Volume energy losses. I. G. Kolesnik.
Astrometriya i Astrofizika, *Kiev,* Vyp. (No.) 18, (see 003. 016), p. 45 - 58 (1973). In Russian.
This article deals with the cooling mechanisms of protostars by CI, CII, OI, FeII, SiII, molecular hydrogen and dust grains.

065.178 Inversion of the density gradient in the nuclei of stars. S. I. Petrusevich.
Trudy Kazan. Gorod. Astron. Obs., No. 37, p. 66 - 72 (1970). In Russian.

065.179 Stabilité vibrationnelle d'étoiles supermassives stabilisées dynamiquement par une rotation uniforme ou différentielle. J. Demaret.
Acad. Roy. Belgique, Bull. Cl. Sci., Vol. 58, 68 - 85 (1972).

065.180 Rote Riesen mit rasch rotierenden, dreiachsigen Kernen. J. U. C. Parra.

Thesis, Univ. Göttingen, 102 pp. (1972).

065.181   Caracteres evolutivas de las estrellas de la población II (Evolutionary characteristics of population II stars).   F. J. Fuenmayor Suarez.
Trabajo de Grado, (Thesis), Facultad de Ciencias, Univ. de Los Andes, Mérida, Venezuela (1973).

A review of the present state of understanding of population II stellar evolution is given. The evolution of a typical star from the initial main sequence to the asymptotic branch phase is described and how the evolutionary track depends on the initial mass, the initial composition, the allowance for relativistic effects, the inclusion of neutrino losses, and the treatment of semi-convection is discussed.

065.182   Analytische Methoden zum besseren Verständnis numerisch simulierter Sternentwicklungsphasen.
P. R. Kolbeck.
Thesis, Math.-Naturwiss. Fakultät, Univ. Göttingen. 156 pp. (1972).

Introduction to the physics of stellar interiors.
See Abstr. 003.074.

Nuclear reactions in cosmic bodies.
See Abstr. 003.076.

Physics of stellar interiors.   See Abstr. 003.113.

Megagauss physics.   See Abstr. 022.068.

On the origin of light elements.
See Abstr. 061.008.

Cosmic abundance of boron.
See Abstr. 061.028.

Possible nucleosynthetic chronologies for the s and p-processes.   See Abstr. 061.030.

Galactic nucleosynthesis of He^4, C^{12} and N^{14}.
See Abstr. 061.032.

Evaluation of astrophysical hypotheses.
See Abstr. 061.044.

Instability of shock waves in inhomogeneous gases.
See Abstr. 061.051.

Dynamical contraction of rotating gaseous spheroids.
See Abstr. 061.061.

Electrical and thermal conductivities of a relativistic degenerate plasma.   See Abstr. 062.005.

Screening factors for nuclear reactions. I. General theory.   See Abstr. 062.021.

Screening factors for nuclear reactions. II. Intermediate screening and astrophysical applications.
See Abstr. 062.022.

On the thermal instability and energetic relations in convective envelopes of slowly rotating stars.
See Abstr. 064.052.

The theoretical relation between UBV colors and T_{eff} and luminosity for extreme pop. II giants.
See Abstr. 064.063.

The role of convection in stellar atmospheres. II. Cool main-sequence stars and metal-deficient subdwarfs.
See Abstr. 064.066.

The T-ρ-abundance history of material undergoing gravitational collapse.   See Abstr. 066.020.

Elastic general relativistic systems.
See Abstr. 066.091.

The Hertzsprung-Russell diagram and stellar ages.
See Abstr. 115.015.

Age from location in the H-R diagram. I. Concluding remarks.   See Abstr. 115.016.

Origin of stellar magnetic fields.
See Abstr. 116.001.

Photometric investigations of magnetic stars.
See Abstr. 116.002.

Stellar rotation and age determinations.
See Abstr. 116.006.

Evolution of contact binaries and star models for cataclysmic binaries.   See Abstr. 117.004.

Period changes in β Cephei stars: Comparison of observation with theory.   See Abstr. 122.049.

V1057 Cygni and pre-main-sequence evolution.
See Abstr. 122.137.

Nonlinear cepheid pulsation calculations and comparison with linear theory.   See Abstr. 122.138.

Pulsational stability of stars in thermal imbalance.
See Abstr. 122.139.

The magnetohydrodynamic stability of white dwarfs and neutron stars.   See Abstr. 126.014.

Mass deficiency of white dwarfs and unstable neutron stars.   See Abstr. 126.018.

A simple probabilistic theory of fragmentation.
See Abstr. 131.019.

Ultraviolet stars and the interstellar gas.
See Abstr. 131.057.

On the gravitational collapse of interstellar magnetic clouds.   See Abstr. 131.134.

An atlas of Stokes parameters based on a synchrotron radiation model.   See Abstr. 141.531.

Radio emission from pulsars and surface temperature of neutron stars.   See Abstr. 141.532.

Rotating neutron stars: a model for pulsars.
See Abstr. 141.533.

Density shock waves driven by star formation — a mechanism.   See Abstr. 151.009.

Chemical composition of globular cluster stars and morphological properties of their horizontal branches.
See Abstr. 154.001.

The effects of a variation in the CNO abundances on the position of initial horizontal-branch models. See Abstr. 154.018.

The abundances and ages of F and G stars in the solar neighborhood.   See Abstr. 155.030.

Galactic three-dimensional shock waves and its effect on the formation of stars.   See Abstr. 155.066.

Time variation of metal abundance in galaxies. Super-metal-rich stage.   See Abstr. 158.129.

066 Relativistic Astrophysics (without Cosmology), Background Radiation, Gravitation Theory

066.001 On relativistic vortex motion.
S. I. Vainshtein, A. A. Ruzmaikin.
Astron. Zhurn. Akad. Nauk SSSR, Vol. 50, 12 - 18 (1973).
In Russian. English translation in Soviet Astron. AJ, Vol. 17, No. 1.

The relativistic generalization of Helmholtz' equation and the equivalent integral theorem on circulation of velocity are deduced for an isentropic fluid. It is shown that potential motions in such a fluid cannot generate a vortex. The relativistic vortex 'dynamo' is analysed.

066.002 The internal composition of a rotating baryon configuration.
G. G. Arutunian, D. M. Sedrakian, E. V. Chubarian.
Astron. Zhurn. Akad. Nauk SSSR, Vol. 50, 60 - 64 (1973).
In Russian. English translation in Soviet Astron. AJ, Vol. 17, No. 1.

The basic parameters and internal characteristics of rotating configurations in the frame of Einstein's theory have been calculated. The binding energy in the rotating case has also been calculated.

066.003 A measurement of the gravitational deflection of radio waves by the sun during 1972 October.
J. M. Riley.
Monthly Notices Roy. Astron. Soc., Vol. 161, 11P - 14P (1973).

The deflection of radio waves in the sun's gravitational field has been measured by observing the radio source 3C 279 before and after its occultation by the sun in 1972. The observed deflection was 1.04 ± 0.08 times that predicted by general relativity.

066.004 Origin of the microwave background.
D. Layzer, R. Hively.
Astrophys. Journ., Vol. 179, 361 - 369 (1973).

This communication explores an alternative to the conventional interpretation of the cosmic microwave background as the remnant of a primordial fireball. We postulate that the radiation was generated by ordinary astronomical processes (thermonuclear reactions in stars or gravitational collapse of objects of galactic mass) and subsequently thermalized by interaction with dust grains.

066.005 Relativistic shocks: The Taub adiabat.
K. S. Thorne.
Astrophys. Journ., Vol. 179, 897 - 907 (1973).

This paper presents a detailed analysis of the adiabat for a relativistic shock wave in a fluid. This analysis extends and supplements previous analyses by Lichnerowicz.

066.006 Remarques concernant le déplacement vers le rouge.
É. Argence.
Comptes Rendus Acad. Sci. Paris, Sér. A, Vol. 276, 77 - 80 (1973).

Établissement de relations de validité concernant le déplacement vers le rouge, dans le cadre d'une théorie à connexion affine.

066.007 Relativistic behaviour of circumnavigating clocks.
R. K. Pathria.
Nature, Vol. 241, 263 (1973).

The author presents a calculation, made directly in the rotating frame of reference, which not only vindicates Hafele's result (1970, 1972) but also generalizes it to more practical situations.

066.008 Black holes in an expanding universe.
M. Demiański, J. P. Lasota.
Nature, Phys. Sci., Vol. 241, 53 - 55 (1973).

The question of gravitational collapse to a black hole is treated in the context of an expanding universe.

066.009 Comment on "Non-velocity redshifts and photon-photon interactions.
R. Aldrovandi, S. Caser, R. Omnés.
Nature, Vol. 241, 340 - 341 (1973).

It has been claimed recently that some redshifts can be explained by considering a new type of process in which the incident photon scatters inelastically on the 3K radiation to give four less energetic photons. Here we describe the detailed calculation of this process.

066.010 Gravitational waves and relativistic disks.
J. C. Jackson.
Nature, Vol. 241, 513 - 515 (1973).

Relativistic disks enhance the intensity of radiation in directions near their plane. The gravitational pulses which Weber claims to have detected may arise in such a disk situated in the galactic nucleus, with its plane aligned with that of the Galaxy. A typical enhancement factor is 30.

066.011 Ejection of matter and gravitational radiation from orbiting bodies. A. A. Jackson IV, R. A. Matzner.
Nature, Phys. Sci., Vol. 241, 139 - 140 (1973).

If a particle of mass m is in a circular orbit of radius r around a central body of mass $M \gg m$ and if M suddenly undergoes a decrease of mass, carried to infinity by massless particles, say, the orbit of m will be perturbed as the shell of emitted particles passes its radius. The central acceleration of m changes during the period when $\dot{M} \neq 0$. The usual formula for quadrupole gravitational radiation indicates that a large radiation rate can be produced.

066.012 Flying clocks and the Sagnac effect.
R. Schlegel.
Nature, Vol. 242, 180 (1973).

In their "flying clock" experiment (1972) Hafele and Keating observed an on-earth directional dependence of the relativistic time dilation. I have argued (1971) that such a dependence is contrary to special relativity theory, but my neglect of an effect invalidated my argument. A seeming inconsistency then still arises in considering the clocks from the earth reference frame. But if the Sagnac effect is taken into account in the synchronization of clocks in the earth frame the contradiction disappears; one finds, rather, a further exemplar of consistency in the theory of relativity.

066.013 Another test of space-time curvature and relativity.
H. W. Grayson.
Nature, Vol. 242, 317 (1973).

By looking far enough out into space, astronomers should be able to detect a redshift in starlight. This redshift when it is discovered will provide additional confirmation of Einsteinian space-time curvature and relativity.

066.014 Spin and torsion may avert gravitational singularities. A. Trautman.

Nature, Phys. Sci., Vol. 242, 7 - 8 (1973). — Letter.

066.015 **On a Lagrangian proposed by Pecker, Roberts and Vigier.** H. Chew.
Nature, Phys. Sci., Vol. 242, 8 - 9 (1973). — Letter.

066.016 **Relativistic effects for moving terrestrial clocks.** E. G. C. Burt.
Nature, Phys. Sci., Vol. 242, 94 - 95 (1973). — Letter.

066.017 **Relativistic astrophysics. Part III. Gravitational collapse, singularities and black holes.** J. P. Lasota.
Postępy Astron., Vol. 21, 25 - 35 (1973). In Polish.
Basis of the black holes theory and problems related to the existence of singularities are discussed.

066.018 **Observational aspects of black holes.** B. Krygier, J. Krempeć.
Postępy Astron., Vol. 21, 37 - 52 (1973). In Polish.
The possibility of the existence of black holes in elliptic and spiral galaxies, in globular clusters and binary stellar systems is described. Individual cases are discussed in detail with emphasis on the possibility of their detection.

066.019 **Spectrum of a radiation source moving along a stable circular orbit near a rotating "black hole".**
A. G. Polnarev.
Astrofizika, Vol. 8, 461 - 471 (1972). In Russian. English translation in Astrophysics, Vol. 8, No. 3.
The source of electromagnetic radiation following some stable circular orbit which lies in the equatorial plane of a rotating "black hole" has been treated. To a distant observer a spectrum averaged over the period of orbital motion was obtained, if the radiation is nearly monochromatic in the rest frame of the source.

066.020 **The T-ρ-abundance history of material undergoing gravitational collapse.** S. W. Bruenn.
Bull. American Astron. Soc., Vol. 5, 27 (1973). — Abstr. AAS.

066.021 **The cosmic background radiation at 1.32 mm.** D. J. Hegyi, W. A. Traub, N. P. Carleton.
Bull. American Astron. Soc., Vol. 5, 34 (1973). — Abstr. AAS.

066.022 **Gravitation, cosmology, cosmogony (general relativity in the physical description of the world).**
I. É. Gurevich, A. D. Chernin.
Fiz. nashikh dnej. Moskva, Znanie, 1972, p. 5 - 80. In Russian. — Abstr. in Referativ. Zhurn. 51. Astron., 3.51.775 (1973).

066.023 **Interaction of a rotating particle with a gravitational wave.** V. I. Golikov.
Third Soviet Gravitational Conference, Erevan, 1972, (see 012.001), p. 38 - 41 (1972). In Russian.

066.024 **Emission of gravitational waves by a system of mass points.** I. I. Gutman.
Third Soviet Gravitational Conference, Erevan, 1972, (see 012.001), p. 47 - 50 (1972). In Russian.

066.025 **A new catalogue of gravitational theories.** D. D. Ivanenko.
Third Soviet Gravitational Conference, Erevan, 1972, (see 012.001), p. 68 - 69 (1972). In Russian.

066.026 **On a generalization of Kerr's metric.** A. A. Koppel.
Third Soviet Gravitational Conference, Erevan, 1972, (see 012.001), p. 76 - 78 (1972). In Russian.

066.027 **Non-static spherically symmetric solutions of gener-**
al relativity equations. M. P. Korkina.
Third Soviet Gravitational Conference, Erevan, 1972, (see 012.001), p. 82 - 84 (1972). In Russian.

066.028 **On the canonical quantization of Friedmann's metric matched with Kruskal's metric.**
Yu. N. Barabanenko.
Third Soviet Gravitational Conference, Erevan, 1972, (see 012.001), p. 207 - 209 (1972). In Russian.

066.029 **Microgeon with Kerr metric.** A. Ya. Burinskij.
Third Soviet Gravitational Conference, Erevan, 1972, (see 012.001), p. 217 - 220 (1972). In Russian.

066.030 **On the problem of particle creation from a vacuum by a gravitational field.**
A. A. Grib, B. A. Levitskij, V. M. Mostepanenko.
Third Soviet Gravitational Conference, Erevan, 1972, (see 012.001), p. 225 - 228 (1972). In Russian.

066.031 **On gravitational collapse and on black and white holes.** M. E. Gertsenshtejn, K. P. Stanyukovich.
Third Soviet Gravitational Conference, Erevan, 1972, (see 012.001), p. 301 - 303 (1972). In Russian.

066.032 **Generalization of Taub's solution.** Ya. Gorskij.
Third Soviet Gravitational Conference, Erevan, 1972, (see 012.001), p. 303 - 304 (1972). In Russian.

066.033 **Long gravitational waves in a closed universe.**
L. P. Grishchuk, A. G. Doroshkevich, V. M. Yudin.
Third Soviet Gravitational Conference, Erevan, 1972, (see 012.001), p. 304 - 306 (1972). In Russian.

066.034 **Birth of particles and polarization of vacuum in an anisotropic gravitational field.**
Ya. B. Zel'dovich, A. A. Starobinskij.
Third Soviet Gravitational Conference, Erevan, 1972, (see 012.001), p. 314 - 315 (1972). In Russian.

066.035 **The gravitational field of a rotating radiating source.**
D. Kramer.
Third Soviet Gravitational Conference, Erevan, 1972, (see 012.001), p. 321 - 323 (1972). In Russian.

066.036 **Energy-momentum tensor for an elastic medium and the equilibrium condition of internal stresses in general relativity.** T. M. Kuchina.
Third Soviet Gravitational Conference, Erevan, 1972, (see 012.001), p. 332 - 334 (1972). In Russian.

066.037 **The propagation of waves in a weak gravitational field.** I. R. Pijr, V. V. Myurk.
Third Soviet Gravitational Conference, Erevan, 1972, (see 012.001), p. 343 - 345 (1972). In Russian.

066.038 **Relativistic libration of planets.**
Ya. I. Pugachev, V. T. Rykov.
Third Soviet Gravitational Conference, Erevan, 1972, (see 012.001), p. 346 - 347 (1972). In Russian.

066.039 **On the instability of the relativistic stellar collapse in the presence of an external source of gravitational waves.** N. R. Sibgatullin.
Third Soviet Gravitational Conference, Erevan, 1972, (see 012.001), p. 358 - 361 (1972). In Russian.

066.040 **Correlation of "gravitational" signals in Weber's experiments with solar and terrestrial magnetic activity.** R. A. Adamyants, A. D. Alekseev, N. I. Kolosnitsyn.
Third Soviet Gravitational Conference, Erevan, 1972, (see

012.001), p. 375 (1972). In Russian.

066.041 **Gravitons: possible sources and the flux to be expected in the earth's vicinity.**
L. F. Vladimirova, Yu. S. Vladimirov.
Third Soviet Gravitational Conference, Erevan, 1972, (see 012.001), p. 379 (1972). In Russian.

066.042 **Spin-orbital effects in gravitational fields.**
E. N. Epikhin, I. Pulido, N. V. Mitskévić.
Third Soviet Gravitational Conference, Erevan, 1972, (see 012.001), p. 380 - 383 (1972). In Russian.

066.043 **Scalar-tensor theory of gravitation and observable phenomena.**
N. A. Zajtsev, S. M. Kolesnikov, A. G. Radynov.
Third Soviet Gravitational Conference, Erevan, 1972, (see 012.001), p. 384 - 387 (1972). In Russian.

066.044 **Detection of gravitational waves by the method of scattering of radiation on elastic vibrations.**
S. A. Zel'dovich, U. Kh. Kopvillem, V. R. Nagibarov, V. V. Samartsev.
Third Soviet Gravitational Conference, Erevan, 1972, (see 012.001), p. 387 - 390 (1972). In Russian.

066.045 **Gravitational and inertial effects in solid bodies.**
U. Kh. Kopvillem.
Third Soviet Gravitational Conference, Erevan, 1972, (see 012.001), p. 396 - 398 (1972). In Russian.

066.046 **Conclusions drawn from the data on classical effects.**
N. M. Polievktov-Nikoladze.
Third Soviet Gravitational Conference, Erevan, 1972, (see 012.001), p. 404 - 407 (1972). In Russian.

066.047 **The reality of the gravitational waves.**
R. I. Khrapko.
Third Soviet Gravitational Conference, Erevan, 1972, (see 012.001), p. 408 - 410 (1972). In Russian.

066.048 **Mach's principle – a critical review.**
M. Reinhardt.
Zeitschr. Naturforschung, Vol. 28a, 529 - 537 (1973).
After a short historical introduction it is discussed how far Mach's principle is incorporated into general relativity.

066.049 **Black holes in binary systems. Observational appearance.** N. I. Shakura, R. A. Sunyaev.
Astron. Astrophys., Vol. 24, 337 - 355 (1973).
The attention of the reader is drawn to the case where the outflow of matter from the surface of the visible component of a binary and its accretion by the black hole should lead to an appreciable observational effect.

066.050 **The problem of anisotropy of inertia.** Z. Horák.
Bull. Astron. Inst. Czechoslovakia, Vol. 24, 143 - 149 (1973).

066.051 **Fine-scale anisotropy of the microwave background: an upper limit at λ = 3.5 millimeters.**
P. E. Boynton, R. B. Partridge.
Astrophys. Journ., Vol. 181, 243 - 253 (1973).
Using the 36-foot NRAO telescope at λ = 3.5 mm, we have set an upper limit of $0.0043°$K, with 90 percent confidence, on the fluctuations in the cosmic microwave background. The angular scale of the measurement was $\sim 80''$. If discrete sources produce all of the microwave background, their number must exceed ~ 0.35 Mpc^{-3}.

066.052 **The significance of Painlevé's coordinates.**

I.-M. Ganea.
Stud. Cerc. Astron., Vol. 18, 97 - 102 (1973). In Romanian.
The author analyses the sense in which it can be considered that Painlevé's coordinates are more "objective" than others.

066.053 **The possibility of measuring gravitational redshift by means of earth satellites.** A. Korchak.
Comments Astrophys. Space Phys., Vol. 5, 37 - 42 (1973).
Reliable recording of gravitational frequency shift by means of satellites can be made either with the use of a receiving station outside the earth's ionosphere and magnetosphere or with a radical improvement of the phase radio methods.

066.054 **Trägheitsfreie Mechanik und Webersches Potential.**
D.-E. Liebscher.
Gerlands Beiträge Geophys., Vol. 82, 3 - 12 (1973).
Mechanics invariant with respect to the full kinematic group of euclidean space are considered. Special points are the induction of mass and inertial forces, the Kepler problem embedded in the cosmic background and the pure two-body problem.

066.055 **Trägheitsrelativität und Trägheitsinduktion.**
H.-J. Treder.
Gerlands Beiträge Geophys., Vol. 82, 13 - 24 (1973).

066.056 **On singularities in general relativity and cosmology.**
V. L. Ginzburg.
Gravitatsiya, Kiev, Nauk. dumka, 1972, p. 40 - 45. In Russian.
Abstr. in Referativ. Zhurn. 51. Astron., 4.51.937 (1973).

066.057 **On the generalized Einstein theory of gravitation.**
G. S. Saakyan.
Gravitatsiya, Kiev, Nauk. dumka, 1972, p. 216 - 231. In Russian. – Abstr. in Referativ. Zhurn. 51. Astron., 4.51.953 (1973).

066.058 **Experimental test of the symmetry of gravitational radiation.** J. Weber.
Gravitatsiya, Kiev, Nauk. dumka, 1972, p. 333 - 337, 359.
Abstr. in Referativ. Zhurn. 51. Astron., 4.51.961 (1973).

066.059 **On the detection of gravitational radiation from some extraterrestrial sources by a heterodyne receiver.** V. B. Braginskij, V. S. Nazarenko.
Gravitatsiya, Kiev, Nauk. dumka, 1972, p. 9 - 16. In Russian.
Abstr. in Referativ. Zhurn. 51. Astron., 4.51.962 (1973).

066.060 **On the possibility of detecting gravitational effects by super-radiation.** U. Kh. Kopvillem.
Gravitatsiya, Kiev, Nauk. dumka, 1972, p. 100 - 112. In Russian. – Abstr. in Referativ. Zhurn. 51. Astron., 4.51.963 (1973).

066.061 **Synchrotron gravitational radiation.**
A. G. Doroshkevich, I. D. Novikov, A. G. Polnarev.
Zhurn. ehksperim. i teor. fiz., Vol. 63, 1533 - 1547 (1972).
In Russian. – Abstr. in Referativ. Zhurn. 51. Astron., 4.51.965 (1973).

066.062 **Gravitational radiation of the system sun—earth—moon.** M. M. Abdil'din, M. S. Omarov.
Prikl. i teor. fizika. Vyp. (No.) 3. Alma-Ata, 1972, p. 15– 18. In Russian. – Abstr. in Referativ. Zhurn. 51. Astron., 4.51.967 (1973).

066.063 **On variational principles in general relativity.**
A. H. Taub.
Gravitatsiya, Kiev, Nauk. dumka, 1972, p. 276 - 287, 358.

Abstr. in Referativ. Zhurn. 51. Astron., 4.51.971 (1973).

066.064 Aberration of light and relativity.
E. Yu. Stepanova, R. V. Kunitskij.
Uch. zap. Gor'kov. gos. ped. in-t, 1972, vyp. (No.) 124, p. 64 - 69. In Russian. – Abstr. in Referativ. Zhurn. 51. Astron. 4.51.974 (1973).

066.065 On the difference between the physical structure of relativistic corrections in Einstein's and Birkhoff's gravitation theories. O. S. Ivanitskaya.
Izv. AN BSSR. Ser. fiz.-mat. n., 1972, No. 6, p. 65 - 72. In Russian. – Abstr. in Referativ. Zhurn. 51. Astron., 4.51.978 (1973).

066.066 Necessity of new experimental tests of general relativity. B. M. Chikhachev.
Gravitatsiya, Kiev, Nauk. dumka, 1972, p. 312 - 320. In Russian. – Abstr. in Referativ. Zhurn. 51. Astron., 4.51.979 (1973).

066.067 Relativistic stars and gravitational waves – an account for non-relativists. K. S. Thorne.
Stellar evolution, (see 012.014), p. 593 - 641 (1972).

066.068 Lorentz-covariant reference-tetrad theories of gravitation. H.-J. Treder.
Ann. Physik, Ser. 7, Vol. 28, 238 - 244 (1972). In German.

066.069 Die Einstein-Gruppe und die Symmetrie-Gruppen der Gravitationstheorie. H.-J. Treder.
Ann. Physik, Ser. 7, Vol. 28, 333 - 340 (1973).

066.070 Lorentz-kovariante Bezugstetradentheorien. II.
H. -J. Treder.
Ann. Physik, Ser. 7, Vol. 28, 366 - 368 (1973). – Short communication.

066.071 Linearity and parametrisation of gravitational effects. S. Deser, B. Laurent.
Astron. Astrophys., Vol. 25, 327 - 328 (1973).
It is argued on two grounds that the conventional parametrised post-Newtonian formalism is unsuited for the discussion of gravitational non-linearities.

066.072 Zur Frage der Zeitabhängigkeit der Naturgesetze.
H.-J. Treder.
Veröff. Forschungsbereich Kosm. Phys., Akad. Wiss. DDR, No. 1, p. 61 - 74 (1973).

066.073 On the effects of gravitational and electromagnetic wave packet scattering in the gravitational field of a "black hole". N. R. Sibgatullin.
Dokl. Akad. Nauk SSSR, Ser. Mat. Fiz., Vol. 209, 815 - 818 (1973). In Russian.

066.074 Kinemetric invariants and their relation to the chronometric invariants of Einstein's theory of gravitation. A. L. Zelmanov.
Dokl. Akad. Nauk SSSR, Ser. Mat. Fiz., Vol. 209, 822 - 825 (1973). In Russian.

066.075 The virial theorem in general relativity.
S. Bonazzola.
Astrophys. Journ., Vol. 182, 335 - 340 (1973).
The formulation of the virial theorem in general relativity is given for spherical and stationary axisymmetric spacetimes. The Newtonian virial theorem is found in the first approximation. In the nonstationary case, the Newtonian virial theorem is modified by adding a gravitational pressure term.

066.076 Search for small-scale anisotropy in the 2.7° K cosmic background radiation at a wavelength of 3.56 centimeters.
R. L. Carpenter, S. Gulkis, T. Sato.
Astrophys. Journ., (*Letters*), Vol. 182, L61 - L64 (1973).
Drift scans were made of a selected track on the sky to search for small-scale anisotropy in the microwave background radiation. The upper bound on the small anisotropy along our track is $\Delta T/T < 7.15 \times 10^{-4}$ (90 percent confidence). Some implications of this result are discussed.

066.077 Singularities in general relativity and cosmology.
Z. Klimek.
Postępy Astron., Vol. 21, 115 - 125 (1973). In Polish.
This article contains a review of problems connected with the definition of singularities in general relativity and with the conditions of their existence.

066.078 Matter and antimatter from general relativity.
M. Sachs.
Nuovo Cimento Lettere, Ser. 2, Vol. 5, 947 - 950 (1972).
A factorization of the usual tensor formulation of general-relativity theory into a quaternion field representation was found to lead to an explicit field relationship between the inertial mass of an elementary particle and the geometrical variables representing the remaining matter of an assumed closed physical system.

066.079 Gauge invariance and the quantization of mass (of gravitational charge). L. Motz.
Nuovo Cimento B, Ser. 11, Vol. 12B, 239 - 255 (1972).
From a quantization condition the author deduces that the fundamental particle in nature (the uniton) has an inertial mass equal to about 10^{-5} g. Cosmological implications of the uniton are also discussed and it is suggested that unitons can clear up the solar-neutrino discrepancy, the energy output of QSO's and the 'missing mass' in the universe.

066.080 Approaches in dG/G-determinations.
E. Groten, S. Thyssen-Bornemisza.
Pure and Applied Geophys., (*Switzerland*), Vol. 99, No. 7, p. 5 - 11 (1972).

066.081 Maxwell equations in a spherically symmetric black-hole background and radiation by a radially moving charge. J. Tiomno.
Nuovo Cimento Lettere, Ser. 2, Vol. 5, 851 - 855 (1972).
The amount of radiation emitted by a particle falling into a Schwarzschild black hole from rest at infinity is calculated numerically.

066.082 Radial motion of a spinning test body in the field of a black hole. R. F. O'Connell.
Phys. Rev. D, Particles and Fields, Vol. 6, 3035 - 3036 (1972).
The gravitational radiation emitted by a test particle falling radially into a Schwarzschild black hole has been analyzed by Zerilli. Here the author investigates the effect of assigning spin to both the test particle and the source. He finds that more (less) gravitational radiation is emitted if the spins are antiparallel (parallel).

066.083 On the nonsymmetric gravitational collapse of an infinite, dust-filled cylinder.
M. J. Miketinac, R. J. Barton.
International Journ. Theor. Phys., (*GB*), Vol. 6, 437 - 442 (1972).
Nonsymmetric terms or angle-dependent terms in the power series expansion of Einstein's equations are considered. It is shown, first, that they do not influence the time evolution of the symmetric terms and, second, that they do not remain bounded as the cylinder collapses.

066.084 Mass formula for Kerr black holes. L. Smarr.
Phys. Rev. Letters, Vol. 30, 71 - 73 (1973).
A new mass formula for Kerr black holes is deduced,
and is contrasted to the mass formula which is obtained by
integrating term by term the mass differential and which
consists of three terms interpreted, respectively, as the surface
energy, rotational energy, and electromagnetic energy of the
charged rotating black hole.

066.085 Black holes and spinning test bodies.
S. N. Rasband.
Phys. Rev. Letters, Vol. 30, 111 - 114 (1973).

**066.086 An example of creation of matter in a gravitational
field.** R. I. Khrapko.
Soviet Phys. – JETP, (USA), Vol. 35, 441 - 442 (1972).
An example is constructed for the creation of matter in
the form of a compact body, and of the subsequent destruction of the matter in an external nonstationary gravitational
field, within the framework of classical general relativity without using ideas of quantum theory.

**066.087 Amplification of waves reflected from a rotating
'black hole''.** A. A. Starobinskij.
Zhurn. ehksperim. i teor. fiz., Vol. 64, 48 - 57 (1973). In Russian. – Abstr. in Referativ. Zhurn. 51. Astron., 6.51.799
(1973).

**066.088 Gravitational collapse with a physical singularity on
an isotropic hypersurface.**
V. K. Pinus, A. L. Frenkel'.
Zhurn. ehksperim. i teor. fiz., Vol. 64, 43 - 47 (1973). In Russian. – Abstr. in Referativ. Zhurn. 51. Astron., 6.51.801
(1973).

066.089 The collapse in the scalar-tensor gravitational theory.
G. E. Gorelik.
Izv. vyssh. ucheb. zavedenij. Fizika, 1973, No. 1, p. 56 - 60.
In Russian. – Abstr. in Referativ. Zhurn. 51. Astron., 6.51.802
(1973).

066.090 Post-Newtonian hydrodynamics in Jordan's theory.
A. A. Baranov.
Astrofizika, Vol. 9, 95 - 98 (1973). In Russian. – English
translation in Astrophysics, Vol. 9, No.1.
Post-Newtonian hydrodynamics is derived in Jordan's
theory of gravitation. The propagation of small perturbations
is considered. It is pointed out that a test of Jordan's theory
is possible by means of precise acoustical measurements.

066.091 Elastic general relativistic systems.
E. N. Glass, J. Winicour.
Journ. Math. Phys., New York, Vol. 13, 1934 - 1940 (1972).
A theory of elastic deformations of general relativistic
systems is presented. The theory is quite comprehensive in
scope and applicable to fully relativistic situations such as the
elastic behavior of neutron stars.

**066.092 On the general metric for a static sphere of a perfect
fluid in canonical coordinates.** B. Kuchowicz.
Phys. Letters A, (Netherlands), Vol. 42A, 485 - 486 (1973).
General expressions for a metric corresponding to a static
and spherically symmetric distribution of a noncharged, perfect fluid are given in standard canonical coordinates.

**066.093 Position dependent Robertson-Walker solutions of
the Brans-Dicke field equations.** C. B. G. McIntosh.
Phys. Letters A, (Netherlands), Vol. 43A, 33 - 34 (1973).
Cosmological solutions of the Brans-Dicke gravitational
field equations are given in which the metric is the flat Robertson-Walker one and in which the scalar field, energy density

and pressure are functions of position as well as time.

066.094 Generalization of the Einstein theory (of gravitation).
P. Rastall.
Phys. Rev. D, Particles and Fields, Vol. 6, 3357 - 3359 (1972).

066.095 Collapsed Schwarzschild fields and thermodynamics.
L. Basano, A. Morro.
Nuovo Cimento Lettere, Ser. 2, Vol. 6, 193 - 196 (1973).
The problem of the transcendence of the second law of
thermodynamics in connection with black-hole physics has
been analysed. The authors shall briefly summarize here both
the problem and the proposed solution.

066.096 Rotating incoherent matter in general relativity.
J. Pachner.
Canadian Journ. Phys., Vol. 51, 477 - 490 (1973).
Under the assumption of rotational symmetry a method
is developed for the numerical integration of exact Einstein
equations describing the time evolution of a rotating incoherent matter.

**066.097 Use of the Schwarzschild metric in the Klein-Gor-
don equation.** P. D. P. Smith.
Phys. Letters A, (Netherlands), Vol. 43A, 144 (1973).
It is shown that substitution of the Schwarzschild metric
tensor into the Klein-Gordon equation predicts the usual
perihelion advance of classical general relativity.

**066.098 Gravitational collapse and higher-order gravita-
tional Lagrangians.** F. C. Michel.
Ann. Phys., (USA), Vol. 76, 281 - 298 (1973).
A general form of higher-order contributions to the
Einstein field equations is displayed. The additional terms
may either stabilize or destabilize self-gravitating objects in
gravitational collapse depending on the sign of the coefficient
introducing the quadratic term.

066.099 Entropy and black-hole dynamics. W. Israel.
Nuovo Cimento Lettere, Ser. 2, Vol. 6, 267 - 269
(1973).

066.100 Surface geometry of charged rotating black holes.
L. Smarr.
Phys. Rev. D, Particles and Fields, Vol. 7, 289 - 295 (1973).
Invariant measures of the surface geometry of a charged
rotating (Kerr-Newman) black hole are examined.

**066.101 Baryon-antibaryon phase transition at high tempera-
ture.** A. Cisneros.
Phys. Rev. D, Particles and Fields, Vol. 7, 362 - 367 (1973).
Present experimental data on nucleon-antinucleon scattering allow a study of the possibility of a phase transition in
a nucleon-antinucleon gas at high temperature.

**066.102 Balloon measurements of the far-infrared back-
ground radiation.** D. Muehlner, R. Weiss.
Phys. Rev. D, Particles and Fields, Vol. 7, 326 - 344 (1973).

**066.103 Group structure and field equations in Einstein uni-
verse.** D. Kramer.
Acta Phys. Polonica B, Vol. B4, No. 1, p. 11 - 20 (1973).
In German.
The isometric group in the Einstein universe has two
Casimir operators whose eigenvalue equations and physical
field equations are compared. This leads to a discrete spec-
trum for the eigenvalues of the energy.

**066.104 The averaged Lagrangian and high-frequency gravi-
tational waves.** M. A. H. MacCallum, A. H. Taub.
Commun. Math. Phys., Vol. 30, 153 - 169 (1973).

066.105 **The gravitational field of a uniformly rotating sphere in third approximation.** J. McCrea.
Proc. Roy. Irish Acad., Ser. A, Vol. 73, 25 - 45 (1973).
A model universe is constructed which represents a sphere rotating with constant angular velocity. The metric satisfies Einstein's equations with an error of order k_4^* and k may be identified with the mass/radius ratio of the sphere ($k = 10^{-6}$ for a body like the sun).

066.106 **Local structure of space-time singularity and gravitational collapse.** E. P. T. Liang.
Nuovo Cimento Lettere, Ser. 2, Vol. 6, 459 - 463 (1973).
Some findings concerning the singularities of static space-times are reported and a pragmatic, though rather ad hoc recipe of constructing the singularities of general space-times is proposed. The bearing of the local singularity structure on the problem of gravitational collapse is briefly discussed.

066.107 **Artificial satellites to test general relativity theory.** G. Maugin.
Aeronaut. and Astronaut., (France), No. 40, p. 19 - 23 (1973). In French.
A possible application of space research to test general-relativity theory is presented. A description is given of the experiment proposed by Schiff (1971) and presently achieved at Stanford. The latter permits the experimental study of the relativistic precession of a gyroscope which equips an artificial satellite.

066.108 **Gyromagnetic ratio of a massive body.** J. M. Cohen, J. Tiomno, R. M. Wald.
Phys. Rev. D, Particles and Fields, Vol. 7, 998 - 1001 (1973).

066.109 **Gravitational radiation from a mass projected into a Schwarzschild black hole.** R. Ruffini.
Phys. Rev. D, Particles and Fields, Vol. 7, 972 - 976 (1973).
Gravitational radiation emitted by a particle projected with non zero kinetic energy from infinite distance into a Schwarzschild black hole is examined.

066.110 **Primordial 2.7° radiation as evidence against secular variation of Planck's constant.** P. D. Noerdlinger.
Phys. Rev. Letters, Vol. 30, 761 - 762 (1973).
The blackbody form of the 2.7° cosmic microwave radiation appears to be inconsistent with variation of Planck's constant by as much as 10% since the epoch $z \sim 1000$.

066.111 **Extraction of energy and charge from a black hole.** J. D. Bekenstein.
Phys. Rev. D, Particles and Fields, Vol. 7, 949 - 953 (1973).
Misner has shown that in the scattering of massless wave fields by a Kerr black hole, certain modes are amplified at the expense of the rotational energy of the hole. The author shows here that the existence of this effect can be deduced from simple considerations based on Hawking's theorem that the area of a black hole can never decrease.

066.112 **Vector and tensor radiation from Schwarzschild relativistic circular geodesics.** R. A. Breuer, R. Ruffini, J. Tiomno, C. V. Vishveshwara.
Phys. Rev. D, Particles and Fields, Vol. 7, 1002 - 1007 (1973).

066.113 **Polarization of synchrotron radiation from relativistic Schwarzschild circular geodesics.** R. A. Breuer, C. V. Vishveshwara.
Phys. Rev. D, Particles and Fields, Vol. 7, 1008 - 1017 (1973).

066.114 **Standing pion waves in superdense matter.**
R. F. Sawyer, A. C. Yao.
Phys. Rev. D, Particles and Fields, Vol. 7, 1579 - 1586 (1973).

066.115 **A four-dimensional Green's function approach to the calculation of gravitational radiation from a particle falling into a black hole.** K. P. Chung.
Nuovo Cimento B, Ser. 11, Vol. 14B, 293 - 308 (1973).
A four-dimensional Green's function method to calculate the gravitational radiation from a particle moving in the field of a black hole is described in detail.

066.116 **Gravitational-wave observations as a tool for testing relativistic gravity.** D. M. Eardley, D. L. Lee, A. P. Lightman, R. V. Wagoner, C. M. Will.
Phys. Rev. Letters, Vol. 30, 884 - 886 (1973).

066.117 **Scalar-tensor theories and conformal invariance.** J. O'hanlon, B. O. J. Tupper.
Nuovo Cimento B, Ser. 11, Vol. 14B, 190 - 202 (1973).

066.118 **Do black holes exist?** A. G. W. Cameron.
Recherche, (France), No. 32, Vol. 4, 215 - 222 (1973). In French.

066.119 **Minimal and nonminimal gravitational interactions, and the asymmetric energy momentum tensor.** K. Hayashi.
General Relativity and Gravitation, (GB), Vol. 4, 1 - 11 (1973).

066.120 **A method for calculating the space-time metric inside a collapsing or expanding sphere.** S. Banerji.
General Relativity and Gravitation, (GB), Vol. 4, 13 - 21 (1973).

066.121 **Black holes in static vacuum space-times.** H. Muller Zum Hagen, D. C. Robinson, H. J. Seifert.
General Relativity and Gravitation, (GB), Vol. 4, 53 - 78 (1973).

066.122 **A new interpretation of the Kerr-Schild metric.** M. Misra.
Nuovo Cimento Lettere, Ser. 2, Vol. 6, 715 - 716 (1973).

066.123 **Correlation of reported gravitational radiation events with terrestrial phenomena.** J. A. Tyson, C. G. Maclennan, L. J. Lanzerotti.
Phys. Rev. Letters, Vol. 30, 1006 - 1009 (1973).
Reports results of a statistical cross-correlation study between 262 of Weber's gravitational radiation events and various geophysical, meteorological, and other phenomena.

066.124 **Equivalence of massive Brans-Dicke and Einstein theories of gravitation.** R. Acharya, P. A. Hogan.
Nuovo Cimento Lettere, Ser. 2, Vol. 6, 668 - 672 (1973).

066.125 **Gravitational radiation from relativistic phase transitions.** J. Winicour.
Astrophys. Journ., Vol. 182, 919 - 934 (1973).
A semi-Newtonian treatment of the generation and detection of gravitational waves is presented. This is used to estimate the gravitational radiation from relativistic increases in pressure. The results are applied to various astrophysical processes, and the corresponding signal-to-noise ratios for a Weber antenna are obtained. It is concluded that observable events at the galactic center must have yields in excess of $1 M_\odot$.

066.126 **On the temperature of the microwave background radiation at a large redshift.** J. N. Bahcall, P. C. Joss, R. Lynds.

Astrophys. Journ., (*Letters*), Vol. 182, L95 - L98 (1973).

It is shown that the temperature of the microwave background radiation at a redshift of the order of 2.5 is certainly less than 200°K and probably less than 45°K. Further detailed studies of the absorption spectra of large-redshift quasars can improve these limits.

066.127 Some remarks concerning the response of macroscopic systems to gravitational radiation.
M. Weinstein.
Phys. Rev. D, Particles and Fields, Vol. 6, 3383 - 3389 (1972).

Since Weber first reported the apparent detection of gravitational radiation there has been considerable discussion of how gravitational detectors work. In this paper the author presents an alternative approach to the problem which emphasizes the important point that the design and calibration of such detectors can be done without reference to a theoretical understanding of the complicated systems involved.

066.128 Tensorielle Massen und Masseninduktion.
H.-J. Treder.
Gerlands Beiträge Geophys., Vol. 82, 92 - 106 (1973).

In mechanics without inertia the effective inert masses of the particles are homogeneous functions of the interaction potential. In such mechanics Poincaré's principle of the relativity of accelerations is fulfilled, and the dynamical equations are only dependent on relative quantities. In relative mechanics the total momentum of a closed system vanishes identically. In consequence of this behaviour of the total momentum the virial and the total angular momentum of a closed system are independent on the point of reference. The virial and the angular momentum are proportional to the longitudinal or transversal parts, respectively, of the inert masses.

066.129 Simultaneity, time and space in the theory of relativity. L. Ya. Arifov.
Dokl. Akad. Nauk SSSR, Ser. Mat. Fiz., Vol. 210, 1320 - 1322 (1973). In Russian.

066.130 Gravitation. W. Thirring.
Essays in Phys., (*GB*), Vol. 4, 125 - 163 (1972).

066.131 Singularities and collisions of Newtonian gravitational systems. D. G. Saari.
Arch. Ration. Mech. Anal., (*Germany*), Vol. 49, 310 - 320 (1973).

066.132 Gravity waves: correlation with geomagnetic storms
W. D. Metz.
Science, Vol. 180, 1161 - 1162 (1973).

066.133 Anisotropy and inhomogeneity in the cosmic background radiation. E. K. Conklin.
Diss. Dep. Electrical Engineering, Stanford Univ. 1969. [Available from Univ. Microfilms Inc., Ann Arbor, Mich.], 128 pp. (1972).

066.134 Two-dimensional spaces in general relativity.
R. F. Polishchuk.
Vestn. Mosk. un-ta. Fiz., astron., 1973, No. 1, p. 3 - 7. In Russian. – Abstr. in Referativ. Zhurn. 51. Astron., 7.51.818 (1973).

066.135 Neue Bestimmung der Gravitations-Rotverschiebung des Sirius-Begleiters (New determination of the gravitational redshift for the companion of Sirius).
A. Werner.
Der math. u. naturwiss. Unterricht, [Verlag Dümmler, Bonn], 26. Jahrgang, p. 243 - 244 (1973).

Comparison of the older and new values (see 06.126.

016) and observational aspects concerning the gravitational redshift for Sirius B.

Gravitation waves in Einstein's theory of gravitation. See Abstr. 003.132.

On a magnetized rotating sphere.
See Abstr. 061.006.

Relativistic shock hydrodynamics.
See Abstr. 062.024.

The stability of rotating barion stars.
See Abstr. 065.037.

On the stability of axisymmetric systems to axisymmetric perturbations in general relativity. IV. Allowance for gravitational radiation in an odd-parity mode.
See Abstr. 065.090.

The effect of gravitational radiation on the secular stability of uniformly rotating fluid masses.
See Abstr. 065.091.

Relavistic stellar stability: an empirical approach.
See Abstr. 065.099.

Neutrino thermal conductivity in collapsing stars.
See Abstr. 065.102.

Relativistic astrophysics. IV. Relativistic rotating stars. See Abstr. 065.154.

Gravitational collapse with charge and small asymmetries. I. Scalar perturbations. See Abstr. 065.165.

Elastic perturbation theory in general relativity and a variation principle for a rotating solid star.
See Abstr. 065.172.

Is the solar system gravitationally closed?
See Abstr. 091.039.

Black holes in binary systems: observational appearances. See Abstr. 117.015.

The case for a black hole in BM Orionis.
See Abstr. 121.069.

Accretion onto black holes: The emergent radiation spectrum. See Abstr. 131.044.

Black holes and absorption redshifts in quasi-stellar objects. See Abstr. 141.076.

X-ray astronomy (III): Searching for a black hole.
See Abstr. 142.037.

2U 0900-40 a black hole? See Abstr. 142.139.

Ultrahigh energy photons, electrons, and neutrinos, the microwave background, and the universal cosmic-ray hypothesis. See Abstr. 143.030.

Errata

066.901 Erratum: 'Note concerning gravitation and electromagnetism' [Astrophys. Space Sci., Vol. 17, 368 - 377 (1972)]. C. C. Leiby, Jr.
Astrophys. Space Sci., Vol. 21, 510 (1973).

Sun

071 Solar Photosphere, Spectrum

071.001 Fraunhofer lines with large Zeeman splitting.
J. W. Harvey.
Solar Physics, Vol. 28, 9 - 13 (1973).
A list of solar spectral lines having simple Zeeman triplet splitting with Landé g-factors equal to or greater than 2.5 is presented.

071.002 The absorption spectrum of atmospheric water vapor in the vicinity of the He 10830 Å triplet.
J. B. Breckinridge, D. N. B. Hall.
Solar Physics, Vol. 28, 15 - 21 (1973).
Wavelengths of clean atmospheric water lines, and some solar lines, in the wavelength interval 10750 Å to 10900 Å have been measured to an accuracy approaching ± 1 mÅ. Strengths and wavelengths have been measured for all atmospheric water lines with absorption coefficients $> 5 \times 10^{-4}$ cm^{-1} gm^{-1} cm^{-2} at ~280 K, that lie within 15 Å of the He I 10830 Å feature.

071.003 New identifications of disk emission lines in the Ca II H and K line wings.
O. Engvold, H. D. Halvorsen.
Solar Physics, Vol. 28, 23 - 25 (1973). – Research note.

071.004 Magnesium II doublet profiles of chromospheric inhomogeneities at the center of the solar disk.
P. Lemaire, A. Skumanich.
Astron. Astrophys., Vol. 22, 61 - 68 (1973).
An analysis of a balloon spectrum of the sun obtained on June 24, 1970, with 7″ angular resolution and 25 mÅ spectral resolution respectively is presented. Average cell, network and plages profiles near the center of the solar disk are identified and compared with profiles computed on the basis of recent chromospheric models.

071.005 The temperature of the sun from CN.
G. A. Porfirjeva.
Astron. Zhurn. Akad. Nauk SSSR, Vol. 50, 221 - 222 (1973).
In Russian. English translation in Soviet Astron. AJ, Vol. 17, No. 1.
For a preliminary interpretation of the result obtained earlier on the independence of the turbulence velocity, determined from half-width of line contours of the CN molecule, and from the value sin ϑ, rotational and vibrational temperatures in the centre of the solar disk and for sin $\vartheta = 0.95$ are calculated. They turned out to be equal to each other.

071.006 Observations of the variation of temperature with latitude in the upper solar photosphere. II. Magnetic-field comparison, implications for solar-oblateness measurements, and harmonic analysis. R. C. Canfield.
Astrophys. Journ., Vol. 179, 643 - 650 (1973).
It is shown that there is a close relationship between the latitudinal variation of upper-photospheric temperature and that of photospheric magnetic field. This correlation, when used with 1966 magnetic-field data, implies that very little of Dicke and Goldenberg's solar-oblateness signal was due to pole-equator temperature differences at small optical depths. In the Appendix, Legendre-polynomial representations of the temperature-difference data are given.

071.007 Saturation effects in Fraunhofer lines of neutral iron. C. R. Cowley, J. Toney.
Astron. Astrophys., Vol. 22, 441 - 443 (1973).
The phenomena of saturation has been studied by comparing the ratios of theoretical intensities to equivalent width ratios. The study is confined to pairs of lines within the same multiplet. Empirical damping constants are obtained which are larger than those predicted by traditional theoretical treatments by about one order of magnitude.

071.008 Theoretical study of the Fraunhofer lines polarization: The case of Ca I 4227.
S. Dumont, A. Omont, J.-C. Pecker.
Solar Physics, Vol. 28, 271 - 288 (1973).
The measurements by Brückner (1963) of the Ca I 4227 polarization at the sun's limb provides us with a test for the theory of line polarization. Computations are developed taking into account: (a) the transfer polarization, due to the anisotropy of radiation field; (b) the depolarizing collisions acting in the wings. The magnetic field is not taken into account and the theory is not valid in the Doppler core. In the wings a very good fit is obtained, using appropriate source-functions fitting the observed profiles at the center of the disk, and from center to limb.

071.009 Contribution to the observation of the photospheric oscillations. E. Fossat, G. Ricort.
Solar Physics, Vol. 28, 311 - 317 (1973).
Observations of the 300 s photospheric oscillation on large solar surfaces (up to 5′20″ in diameter) using a sodium optical resonance cell seem to show that the power at long horizontal wavelengths is larger than previous results would indicate. In order to get more information about the spatial distribution of the energy, a new observational method has been perfected, which will allow us to obtain the spatio-temporal power spectrum. In some of our observations, a long-period oscillation (about 40 min) appears, with an amplitude comparable to that of the 300-s oscillation, and which seems to be correlated with the occurrence of chromospheric flares.

071.010 Brightness fluctuations in the K-line wings.
M. Y. Cha, F. Q. Orrall.
Solar Physics, Vol. 28, 333 - 341 (1973).
A power-spectrum and cross-spectrum analysis has been made of measurements of temporal fluctuations of intensity observed in the K-line wing (2.07 Å from line center) and of simultaneous measurements of temporal fluctuations of Doppler displacement of the cores of λ 3931.122 Fe I and λ 3933 Ca II (K₃). The measurements were made in a quiet region near the center of the sun's disk.

071.011 The solar temperature distribution with latitude.
R. J. Rutten.
Solar Physics, Vol. 28, 347 - 349 (1973). – Research note.

071.012 Reduction of images of the solar photosphere by the method of filtration of spatial frequencies and discrimination according to brightness. L. D. Parfinenko.
Solnechnye Dannye 1972 Byull., No. 11, p. 89 - 91 (1972/73).
In Russian.

Preliminary results of image processing by the TV method are reported.

071.013 **On the equivalent widths of the spectral lines of solar granulation elements.** V. M. Sobolev.
Solnechnye Dannye 1972 Byull., No. 11, p. 95 - 103 (1972/73). In Russian.

Some results of the reduction of 6543 spectrograms taken during the third flight of the solar stratospheric station on July 30, 1970 are given. Equivalent widths of the spectral lines of several chemical elements were measured for five granules and five neighbouring intergranular intervals.

071.014 **Solar oscillations at 9.6 mm.**
W. L. H. Shuter, W. H. McCutcheon.
Nature, Phys. Sci., Vol. 241, 140 - 142 (1973).

Observations were made in an attempt to study the solar oscillatory phenomenon in the mid-range of other reported results ($\lambda = 9.6$ mm). The power spectrum shows a peak (0.0033 Hz) of 1 K amplitude at 300 s. We take the value 1 K at 300 s as being an upper limit to any periodic signal from the sun.

071.015 **Das Licht als Informationsträger aus dem Weltall.**
G. Schmidtke, W. Schweizer.
Umschau, 73. Jahrgang, p. 185 - 186 (1973).

071.016 **Helium abundance of the sun.** J. Hirshberg.
Rev. Geophys. Space Phys., Vol. 11, 115 - 131 (1973).

The solar abundance of helium (more specifically the ratio of solar helium to hydrogen) is a basic quantity in understanding many astrophysical and space physical problems. We here review critically the four methods that have been used to estimate the ratio of helium to hydrogen; the solar neutrino flux, spectral intensity of helium lines in prominences and the chromosphere, elemental abundance of solar cosmic rays, and variations of solar wind He/H.

071.017 **Intermediate-coupling line strengths in the iron spectrum and the solar abundance of iron.** J. E. Ross.
Astrophys. Journ., Vol. 180, 599 - 606 (1973).

Line strengths for the $3d^6 4s^2 - 3d^6 4s(^6D)4p$ resonance transitions in Fe I are computed in intermediate coupling. For lines permitted in LS coupling, gf-values computed in intermediate coupling are in excellent agreement with recent experimental data and a considerable improvement over gf-values computed in LS coupling is obtained. The solar abundance of iron deduced from the weak intercombination resonance lines is found to be log $N(\text{Fe})/N(\text{H}) + 12 = 7.4$, in accord with recent determinations.

071.018 **A numerical method for the inversion of a single line profile.** J. N. Holt.
Proc. Astron. Soc. Australia, Vol. 2, 150 - 151 (1972).

071.019 **Evaluation of damping constants and turbulent velocity in the solar photosphere by Voigt's method.**
O. N. Mitropolskaya, G. F. Sitnik.
Astron. Zhurn. Akad. Nauk SSSR, Vol. 50, 343 - 347 (1973). In Russian. English translation in Soviet Astron. AJ, Vol. 17, No. 2.

Profiles of 9 lines of atoms and ions were obtained photoelectrically at different distances from the center of the solar disc. Damping constants and turbulent velocities were calculated from these profiles by Voigt's method.

071.020 **The dynamics of solar granulation.** S. Musman.
Bull. American Astron. Soc., Vol. 5, 2 (1973).
Abstr. AAS.

071.021 **Balmer lines and the Harvard-Smithsonian Reference Atmosphere.** G. Elste, M. Hartoog.
Bull. American Astron. Soc., Vol. 5, 20 - 21 (1973). – Abstr. AAS.

071.022 **Structural changes and regularities in the distribution of calcium flocculi on the solar surface in the course of cycle 19.** P. Ambrož.
Bull. Astron. Inst. Czechoslovakia, Vol. 24, 80 - 88, 112a - h (1973).

The paper presents an analysis of the longitudinal distribution of Ca II flocculi in the equatorial zone of ±20 heliographic degrees.

071.023 **On the empirical determination of the source function.** E. A. Gurtovenko, G. L. Fedorchenko.
Astrometriya i Astrofizika, *Kiev*, vyp. (No.) 16, (see 003.006), p. 61 - 68 (1972). In Russian.

A method of empirical determination of the source function for the upper photosphere based on center-to-limb observations of central line intensities and known gf-values for the multiplet lines is developed. The method is tested using observational data by de Jager and Neven.

071.024 **Centre-to-limb change of faint Fraunhofer line profiles. II. Asymmetry and width of lines.**
E. A. Gurtovenko.
Astrometriya i Astrofizika, *Kiev*, vyp. (No.) 16, (see 003.006), p. 77 - 92 (1972). In Russian.

The profiles of fifty very faint Fraunhofer lines are studied. The accuracy of the observations and data processing are analyzed. Asymmetry of the lines is highly diverse. The predominance of the asymmetry with the sloped red wing appears to be the only regularity. This effect almost disappears near the limb. Turbulence velocities as deduced from the lines of various elements differ essentially. The anomalously large width for most lines can be explained only by the hyperfine and isotopic structure.

071.025 **TV registration of the solar spectrum. III.**
L. D. Parfinenko.
Solnechnye Dannye 1972 Byull., No. 12, p. 72 - 78 (1973). In Russian.

The apparatus used for simultaneous television registration of the solar spectrum and image is described. Photographs of the observational results are presented. The statistical reduction of radial velocity fields and brightness fluctuations is also given.

071.026 **On the location of dark surges ejections.**
V. G. Banin, S. A. Afanasiev.
Solnechnye Dannye 1972 Byull., No. 12, p. 79 - 81 (1973). In Russian.

The location of dark surges ejection is discussed briefly on the base of known results and a sample of a surge. The dark surges are confirmed to be ejected from the photosphere, not from sunspots.

071.027 **Time-frequency spectrum of atmospheric oscillations of the solar limb.**
O. B. Vasilyev, U. I. Iljasov.
Solnechnye Dannye 1972 Byull., No. 12, p. 82 - 90 (1973). In Russian.

Time-frequency spectra of atmospheric oscillations of the solar limb at wavelengths from some units to some thousands of Hertz are investigated. The oscillations may be easily interpreted with the classification of Krat. The results of the analysis of the spectra of the solar limb oscillations are compared with the conclusions of the statistical theory of propagation of electromagnetic radiation in a turbulent atmosphere.

071.028 Absence of the Phillips bands in the solar photo-spheric spectrum.   K. Sinha.
Bull. Astron. Inst. Czechoslovakia, Vol. 24, 136 - 138 (1973).

The absence of Phillips bands in the solar photospheric spectrum and the presence of Swan bands seems to point to a non-LTE path for the formation and disappearance of C_2 molecules in the solar photosphere.

071.029 The photometric effect in the solar photosphere as observed on August 7, 1972.   A. A. Kalinyak.
Astron. Tsirk., No. 759, p. 3 - 4 (1973). In Russian.

071.030 On the distribution of the azimuth of a magnetic field in the solar photosphere under filaments.
B. A. Ioshpa.
Solnechnye Dannye 1973 Byull., No. 1, p. 79 - 84 (1973). In Russian.

Two azimuth maps of the magnetic field have been analysed in order to derive the dependence between the mean direction of the quiescent H_α-filaments. It is shown that the angle between the mean direction of the transversal component of the magnetic field and that of the filament has a minimum near the filament.

071.031 A proposed correction to the solar abundances of carbon and oxygen utilizing new and accurate theoretical forbidden transition probabilities.
C. A. Nicolaides, O. Sinanoğlu.
Solar Physics, Vol. 29, 17 - 22 (1973).

Previously published solar abundances of oxygen and carbon can be corrected to be $\log N(O) = 8.93$ and $\log N(C) = 8.60$ on the hydrogen log-scale when new accurate forbidden electric quadrupole transition probabilities A_Q (s^{-1}) are used. Such A_Q's, based on the new atomic structure and electron correlation theory, developed recently by Sinanoğlu and co-workers, are reported for the $(^1S_0-^1D_2)$ lines of [C I], [N II], [O I] and [O III] and the $(^2P-^2D)$ lines of [N I] and [O II]. The available experimental values are also given for comparison.

071.032 An early observation of λ8542 of the Ca II infrared triplet.   J. A. Eddy.
Solar Physics, Vol. 29, 23 - 24 (1973). – Research note.

071.033 Observation and interpretation of phase lags in the five-minute oscillation.
R. C. Canfield, S. Musman.
Bull. American Astron. Soc., Vol. 5, 269 (1973). – Abstr. AAS.

071.034 Cinematography of solar granulation.
R. B. Dunn, G. R. Mann, G. W. Simon.
Bull. American Astron. Soc., Vol. 5, 271 (1973). – Abstr. AAS.

071.035 Non-LTE profiles of the aluminum I autoionization lines.   G. D. Finn.
Bull. American Astron. Soc., Vol. 5, 272 (1973). – Abstr. AAS.

071.036 The solar brightness temperature at 350 and 450 μ.
D. Y. Gezari, R. R. Joyce, M. Simon.
Bull. American Astron. Soc., Vol. 5, 273 (1973). – Abstr. AAS.

071.037 Solar speckle interferometry.
J. W. Harvey, J. B. Breckinridge.
Bull. American Astron. Soc., Vol. 5, 273 (1973). – Abstr. AAS.

071.038 High resolution rocket observations of solar line profiles between 166.0 and 154.7 nm.
J. L. Kohl, W. H. Parkinson, E. M. Reeves.
Bull. American Astron. Soc., Vol. 5, 274 (1973). – Abstr. AAS.

071.039 Inversion of the limb darkening equation in the presence of noise.

C. V. Kunasz, J. T. Jefferies, O. R. White.
Bull. American Astron. Soc., Vol. 5, 274 (1973). – Abstr. AAS.

071.040 One- and multi-component models of the upper photosphere based on the 3883 Å band head of CN.
G. H. Mount, J. L. Linsky.
Bull. American Astron. Soc., Vol. 5, 277 (1973). – Abstr. AAS.

071.041 A search for the roots of photospheric magnetic fields.   G. W. Simon, J. B. Zirker.
Bull. American Astron. Soc., Vol. 5, 280 (1973). – Abstr. AAS.

071.042 The formation of Mg I 4571 Å in the solar atmosphere. II: The effect of one-dimensional macroscopic velocity fields.   R. C. Altrock, C. J. Cannon.
Solar Physics, Vol. 29, 275 - 286 (1973).

An analysis of the 4571 Å line of neutral magnesium is presented in which one-dimensional macroscopic velocity fields are included. It is shown that gradients over restricted heights in the vertical and horizontal components of the velocity field of order -0.005 s^{-1} and -0.004 s^{-1}, respectively, result in asymmetries in the computed line profile similar to those observed. The results indicate that for the Mg I 4571 Å line model calculations that do not include one-dimensional flow velocities may safely be compared with frequency-averaged observations.

071.043 A search for continuous ultraviolet opacity sources in the sun's photosphere.
E. Landi Degl'Innocenti, G. Noci.
Solar Physics, Vol. 29, 287 - 297 (1973).

Experimental results on limb darkening and specific intensities imply more ultraviolet continuous opacity than that predicted by theoretical calculations. Some atomic and molecular processes, not yet studied from this standpoint are investigated as to their importance on the continuous absorption coefficient. The negative results obtained suggest some arguments about the importance of iron as photo-absorber.

071.044 Positions of filament feet in relation to the super-granular calcium network.
S. Płocieniak, B. Rompolt.
Solar Physics, Vol. 29, 399 - 401 (1973). – Research Note.

071.045 The extreme-ultraviolet spectrum of a solar active region.   A. K. Dupree, M. C. E. Huber, R. W. Noyes, W. H. Parkinson, E. M. Reeves, G. L. Withbroe.
Astrophys. Journ., Vol. 182, 321 - 333 (1973).

Extreme-ultraviolet spectra (280–1370 Å) of the brightest point in McMath region 10266 and of the quiet solar atmosphere are presented as measured by the Harvard scanning spectrometer on OSO-6. Line identifications and physical parameters of the active region are discussed.

071.046 Center-to-limb polarization measurements on the quiet sun's disk.   D. L. Mickey, F. Q. Orrall.
Bull. American Astron. Soc., Vol. 5, 277 (1973). – Abstr. AAS.

071.047 On methods of study of solar granulation fields in the presence of atmospheric disturbances.
M. B. Kerimbekov.
Izv. AN AzSSR. Ser. fiz.-tekhn. i mat. n., 1972, No. 2, p. 82 - 88. In Russian. – Abstr. in Referativ. Zhurn. 51. Astron., 5.51.479 (1973).

071.048 Solar absorption in the CO fundamental region.   A. Goldman, D. G. Murcray, F. H. Murcray, W. J. Williams.
Astrophys. Journ., Vol. 182, 581 - 584 (1973).

Infrared solar spectra have been obtained with spectral resolution of 0.3 cm^{-1}, in the 4.7-μ region, from a balloon-borne grating spectrometer. The spectra obtained from altitudes above 20 km are interpreted in terms of solar CO $\Delta\nu$ = 1 vibration-rotation lines at 4500°K.

071.049 The history of astronomical spectroscopy II. Quantitative chemical analysis and the structure of the solar atmosphere.   D. H. Menzel.
Ann. New York Acad. Sci., Vol. 198, (see 012.020), 235 - 244 (1972).

071.050 Studies of the fine structure of the solar atmosphere at the Skalnaté Pleso Observatory.   J. Sýkora.
Kozmos, Vol. 4, 14 - 18 (1973). In Slovak.

071.051 Observations of the infrared sunspot spectrum between 11340 Å and 24778 Å.   D. N. B. Hall.
Kitt Peak National Obs., Contr. No. 556, 9 + 116 pp. (1970). Thesis presented to the Department of Astronomy, Harvard University, Cambridge, Mass.

071.052 Evidence for polarized radiation from the sun in the far infrared.   G. Dall'Oglio, E. Gandolfi, B. Melchiorri, F. Melchiorri, V. Natale.
Infrared Physics, Vol. 13, 1 - 6 (1973).
    During a balloon borne experiment intended to study the state of polarization of the far infrared radiation in the 100—2000 micron range, a signal was detected from the sun corresponding to a linear polarization of at least 6.4 % assuming a constant polarization over the whole bandwidth observed (wavelengths longer than 500 microns cannot contribute significantly to the observed signal even if they are completely polarized). The vibrational plane is parallel to the sun equator within 5°.

071.053 On resolution enhancement of line spectra by deconvolution.   A. Goldman, P. Alon.
Applied Spectrosc., (*USA*), Vol. 27, 50 - 51 (1973).
    The authors have studied the resolution enhancement of digitized infrared solar spectra obtained with a balloon-borne grating spectrometer.

071.054 Isotopes of rubidium in the sun.   O. Hauge.
Phys. Norvegica, Vol. 6, 202 (1972). – See Phys. Abstr., Vol. 76, No. 42134 (1973).

071.055 Detection of the ^{13}C, ^{17}O, and ^{18}O isotope bands of CO in the infrared solar spectrum.   D. N. B. Hall.
Astrophys. Journ., Vol. 182, 977 - 982 (1973).
    Preliminary analysis of a particularly clean spectral interval (2140–2147 cm^{-1}) supports the result that the solar ^{13}C/^{12}C abundance ratio is terrestrial to within ±15 percent and indicates that the solar ^{18}O/^{16}O and ^{17}O/^{16}O abundance ratios are terrestrial to within 35 percent and a factor of 2.5, respectively.

071.056 Effect of a traveling sound wave on the profiles of spectral lines. I. Central intensity oscillations of the Fraunhofer lines.   R. I. Kostik.
Astrometriya i Astrofizika, *Kiev*, Vyp. (No.) 18, (see 003. 016), p. 94 - 99 (1973). In Russian.
    An expression is found for the profile of a spectral line in a plane homogeneous medium with a traveling sound wave. Central intensities of λ5304.185 CrI, λ5305.866 CrII, λ6098.250 FeI, λ6238.390 FeII lines are computed for different moments of the period.

071.057 First observations of the granulation at 1.65 μ, center to limb variation of the contrast.
P. J. Turon, P. Léna.
Solar Physics, Vol. 30, 3 - 14 (1973).
    Brightness fluctuations at 1.65 μ have been recorded by

means of a 64-element array. Photographs and quantitative analysis show the existence of a strong contrast variation from the center to the limb. Seeing and instrumental effects are discussed. A model M.T.F. is utilized to compute a fore-shortening correction. We find a definite variation of the observed rms which goes from 1.48% ± 0.15, at the center, to 1.05% ± 0.15 at μ = 0.7 (after foreshortening correction).

071.058 Statistical analysis of a solar granulation plate.   C. Aime.
Solar Physics, Vol. 30, 15 - 18 (1973).
    Two-dimensional autocorrelation function and power spectrum per unit area are given for a solar granulation plate taken at the Pic-du-Midi Observatory. A comparison is made between our result and the power per unit wave number taken from the Schwarzschild stratoscope data.

071.059 Studies of granular velocities. III. The influence of finite spectral and spatial resolution upon the measurement of granular Doppler shifts.
J. P. Mehltretter.
Solar Physics, Vol. 30, 19 - 28 (1973).
    It is the purpose of this paper to analyze the method of measuring Doppler shifts in the presence of finite spectral and spatial resolution, and to formulate conditions under which the measured velocities are free from systematic errors, and can be successfully corrected for finite spatial resolution using the same spread function valid also for brightness fluctuations.

071.060 The formation of MgI 4571 Å in the solar atmosphere. III: The Holweger solar model.
R. C. Altrock, C. J. Cannon.
Solar Physics, Vol. 30, 31 - 33 (1973). – Research note.

071.061 Microturbulence and the effect of departures from LTE on photospheric iron lines.   H. Holweger.
Solar Physics, Vol. 30, 35 - 37 (1973).
    It is shown that depth-dependent departures from LTE such as obtained by Athay and Lites (1972) will not notably affect the solar curve-of-growth of FeI. This implies that both abundance and microturbulence may be determined from this curve-of-growth assuming LTE, and excludes that microturbulence is an artefact produced by non-LTE effects.

071.062 Is there horizontal phase propagation of 5-min oscillations at high velocities?
F.-L. Deubner, N. Hayashi.
Solar Physics, Vol. 30, 39 - 46 = Mitt. Fraunhofer Inst., *Freiburg*, No. 119 (1973).
    New observations of the photospheric 5-min oscillations are presented which prove that the physical reality of the very high horizontal phase propagation velocities observed in connection with the oscillations cannot be maintained. Instead, a statistical model is proposed to explain the observed phase relations.

071.063 The five-minute period oscillation in magnetically active regions.   A. G. Michalitsanos.
Solar Physics, Vol. 30, 47 - 61 (1973).
    The magnetohydrodynamic frequency-wavelength relation, derived by McLellan and Winterberg (1968), has been evaluated for an isothermal atmosphere. It is shown that the frequency band in which vertical wave propagation is impossible in the non-magnetic photosphere, becomes smaller when an inclined uniform magnetic field is introduced, and that low frequency magnetically coupled internal-gravity waves do not propagate vertically if the horizontal wavelengths associated with this mode are greater than a critical wavelength which decreases with field strength. It is also

demonstrated that an inclined magnetic field will inhibit the resonance that occurs at the critical frequency ω_g in the non-magnetic atmosphere which is a result consistent with recent observations of the 'wiggly line structure' in active regions.

Energy distribution in the solar spectrum and the solar constant.   See Abstr. 003.079.

Shock-tube measurements of absolute gf-values for Ti I and Ti II.   See Abstr. 022.007.

On the problem of excitation and ionization of neutral sodium.   See Abstr. 022.062.

Solar speckle interferometry.
See Abstr. 031.054.

A numerical method for inverting a single absorption line profile.   See Abstr. 064.019.

Photospheric convective network as a determining factor in sunspot and group development and stabilization. See Abstr. 072.006.

Determination of the temperature distribution in a photospheric facula by solving an integral equation by a gradient-random search method.   See Abstr. 072.062.

An analysis of the solar extreme-ultraviolet spectrum between 50 and 300 Å.   See Abstr. 076.015.

Extreme ultraviolet line intensities from the sun. See Abstr. 076.035.

Equator—pole temperature difference and the solar oblateness.   See Abstr. 080.006.

Response of solar atmosphere to a granular excitation.   See Abstr. 080.010.

A mechanism for the production of light and dark contrasts in radiatively controlled lines.   See Abstr. 080.018.

Eine statistische Deutung der horizontalen Phasengeschwindigkeit der 5 - Min. - Oszillationen.
See Abstr. 080.040.

072 Sunspots, Faculae, Solar Activity

072.001 Measurements of the magnetic field vector of a sunspot. K. Nishi, M. Makita.
Publ. Astron. Soc. Japan, Vol. 25, 51 - 63 (1973).

A new polarimetric observation of the Fraunhofer line λ 6302.5 Å was made along the line profile considering the reduction of the instrumental polarization. The existence of a nearly horizontal magnetic field in the penumbra was confirmed. It was also found that the field direction was along the filamentary structure of the penumbra.

072.002 Polarization of red system CN lines in sunspots. J. W. Harvey.
Solar Physics, Vol. 28, 43 - 47 (1973).

The relative intensities of the Zeeman components of molecular spectral lines are not necessarily symmetric in a strong magnetic field. This leads to non-zero net polarization for molecular lines formed in sunspots. The effect is particularly striking for lines of the (0, 0) band of the red system of CN.

072.003 Relative umbral intensity of a large sunspot. N. Mykland.
Solar Physics, Vol. 28, 49 - 60 (1973).

Simultaneous observations of relative umbral intensities in four wavelength regions are presented. In the visual wavelength region the umbral intensities show lower values than given by most authors. By observing the same spot during different seeing conditions the method of correction for stray light is found to be consistent within the accuracy of the method. In addition, a new simple correction method is suggested.

072.004 Observations of moving magnetic features near sunspots. K. Harvey, J. Harvey.
Solar Physics, Vol. 28, 61 - 71 (1973).

The properties of small (< 2″) moving magnetic features near certain sunspots are studied with several time series of longitudinal magnetograms and Hα filtergrams. A model to help understand the observations is proposed.

072.005 The magnetic properties of solar surges. J.-R. Roy.
Solar Physics, Vol. 28, 95 - 114 (1973).

High resolution on- and off-band Hα filtergrams of disk solar surges obtained with the Vacuum Tower Telescope of the Sacramento Peak Observatory have been compared to magnetic data.

072.006 Photospheric convective network as a determining factor in sunspot and group development and stabilization. V. Bumba, P. Ranzinger, J. Suda.
Bull. Astron. Inst. Czechoslovakia, Vol. 24, 22 - 38, 56a, b (1973).

Using the high-resolution photographs of sunspot groups obtained at the Ondřejov Observatory, the regularities in spot distribution and forms are investigated in their relation to the supergranular network. The instrument and method of observation used, as well as the obtained observational material is described. Three main and two secondary characteristics ("quantized") values of spot distances and diameters, which may be mutualy composed in different formations, are found: 19000 km, 27000 km, 35000 km, 46000 km and 54000 km.

072.007 A large and very complex sunspot group. D. Capper.
Sky Telescope, Vol. 45, 61 (1973).

072.008 Sunspot observations by means of a vidicon camera (I). K. Matsumaru.
Solar Physics, Vol. 28, 351 - 360 (1973).

In our electronic method, the circular sunspots are recorded on a magnetic video tape recorder, and then the recorded images of them are reproduced on a picture monitor and the video signals are simultaneously displayed on an oscilloscope. By means of a line selector, the waveform of a single scanning line of the total raster may be photographed immediately. As typical examples we selected five circular spots of medium scale from many recorded data and explained their characteristics.

072.009 On some characteristics of umbral fine structure. F. Kneer.
Solar Physics, Vol. 28, 361 - 367 = Mitt. Fraunhofer Inst., *Freiburg*, No. 115 (1973).

Photographic spectra of the umbra of a sunspot (1971, August 24, Rome No. 6205) around 6150 Å show fine bright threads which were identified as the spectra of a lightbridge, of the bright end of a penumbral filament and of umbral dots, respectively. An attempt is made to measure the magnetic field in an umbral dot.

072.010 Width of emission cores of the line K Ca II in sunspots. R. B. Teplitskaja, S. A. Efendieva.
Solar Physics, Vol. 28, 369 - 375 (1973).

Emission core widths of K Ca II line in the umbra and penumbra of 9 sunspots and in their vicinity are measured. All sunspots are located near the solar disc center. Data on variation of widths along the 'mean' sunspot radius are obtained.

072.011 Are penumbral filaments convection rolls? D. J. Mullan.
Astron. Astrophys., Vol. 24, 103 - 105 (1973).

The occurrence of strong magnetic fields in dark filaments in sunspot penumbrae is shown to be only marginally consistent with the hypothesis of penumbral convection rolls.

072.012 The large sunspot of 1972, August. V. Barocas.
Journ. British Astron. Ass., Vol. 83, 119 - 121 (1973).

072.013 Sonnenaufnahmen in der Großstadt. W. Brückner.
SuW, Vol. 12, 21 - 22 (1973).

072.014 Giant cells and the solar cycle. P. R. Wilson.
Proc. Astron. Soc. Australia, Vol. 2, 144 - 146 (1972).

072.015 A high dispersion spectrum 6610 Å to 6770 Å of a large sunspot. O. Engvold.
Astron. Astrophys., Suppl. Ser., Vol. 10, 11 - 45 (1973).

A high dispersion spectrum (8.5 mm/Å) in the range 6610–6770 Å of a large sunspot (Roma No 5367) has been recorded photographically at Oslo Solar Observatory. Nearly simultaneous broad band observations of umbral/photospheric contrast and drift curves across the solar limb were made in the same spectral region using a pinhole photometer. The intensity profile of the umbral line spectrum is presented (corrected for parasitic light). A total of 1649 umbral lines are detected within the observed spectral range. Of these we find 1256 lines from the TiO molecule, 2 lines of CaH and 14 atomic lines. The wavelengths and central intensities of umbral lines are tabulated.

072.016 Generation of umbral flashes and running penumbral waves. R. L. Moore.
Bull. American Astron. Soc., Vol. 5, 1 (1973). – Abstr. AAS.

072.017 Seven color photometry of umbral cores with the Bartol coudé telescope.
D. J. Mullan, A. A. Wyller.
Bull. American Astron. Soc., Vol. 5, 20 (1973). – Abstr. AAS.

072.018 Solar high-resolution radio measurements of active regions at a wavelength of 2.8 cm.
C. J. Grebenkemper, D. M. Rust.
Bull. American Astron. Soc., Vol. 5, 21 (1973). – Abstr. AAS.

072.019 Determination of precise coordinates of sunspots from photographic plates of partial eclipse phases.
K. V. Kuimov, M. A. Livshits, S. B. Men'shikova.
Radio astronomy observations of the solar eclipse on 20 May, 1966, (see 003.002), p. 23 - 33 (1972). In Russian.

072.020 The structure of discrete sources of the sunspot group No. 57 from observation of the solar eclipse on 20 May, 1966.
G. P. Apushkinskij, V. G. Nagnibeda.
Radio astronomy observations of the solar eclipse on 20 May, 1966, (see 003.002), p. 33 - 38 (1972). In Russian.

072.021 Stellar analogies with solar activity. F. Yu. Zigel'.
Solntse, ehlektrichestvo, zhizn'. Moskva, Mosk. un-t, 1972, p. 24 - 25. In Russian. – Abstr. in Referativ. Zhurn. 51. Astron., 3.51.516 (1973).

072.022 On the problem of solar activity recurrence.
K. P. Butusov.
Solntse, ehlektrichestvo, zhizn'. Moskva, Mosk. un-t, 1972, p. 33 - 35. In Russian. – Abstr. in Referativ. Zhurn. 51. Astron., 3.51.517 (1973).

072.023 On the structure of the sunspot penumbra.
G. F. Vjalshin.
Solnechnye Dannye 1972 Byull., No. 11, p. 57 - 61 (1972/73). In Russian.
The structure of a large sunspot in the group N 359 (numeration from "Solar Data") was investigated using the photographs taken during the third flight of the Soviet stratospheric solar station on July 30, 1970.

072.024 On the structure of the sunspot umbra.
R. N. Ikhsanov.
Solnechnye Dannye 1972 Byull., No. 11, p. 62 - 71 (1972/73). In Russian.
The sunspot umbra was studied from the photographs taken during the flight of the Soviet stratospheric station on July 30, 1970. Two types of bright points were found to exist with diameters of about 150–180 km and 300 km. The structure of the umbra was found to represent an hierarchical network, the size of cells being $0.''4$, $1.''2$ and $3.''7$.

072.025 The fine structure of a sunspot penumbra and its variation with time. R. N. Ikhsanov.
Solnechnye Dannye 1972 Byull., No. 11, p. 72 - 80 (1972/73). In Russian.
The penumbra of a large sunspot has been investigated using the photographs taken during the third flight of the Soviet stratospheric station on July 30, 1970. The great variety of the bright filament forms is characteristic for the sunspot penumbra. Six types of bright filaments have been detected.

072.026 On the hierarchic structure of the rope system of the magnetic field of a sunspot. R. N. Ikhsanov.

Solnechnye Dannye 1972 Byull., No. 11, p. 81 - 88 (1972/73). In Russian.
Charts of isophotes for five photographs of a large sunspot have been plotted by the equidensity method. The photographs were taken during the third flight of the Soviet stratospheric station on July 30, 1970.

072.027 On detecting rapid variations in the fine structure of a spot penumbra.
V. L. Lentsman, L. D. Parfinenko.
Solnechnye Dannye 1972 Byull., No. 11, p. 92 - 94 (1972/73). In Russian.
Rapid variations of the fine structure of a spot penumbra are detected by the TV method.

072.028 Relation of the large-scale distribution of activity on the solar surface and of the fluctuations of some activity indices in the course of cycle 19. P. Ambrož.
Bull. Astron. Inst. Czechoslovakia, Vol. 24, 88 - 95 (1973).
The paper presents a comparison of the pattern of the average surface density of Ca II flocculi, described by the index S, and of the fluctuations of some indices of activity (area of sunspots, relative number and area of prominences).

072.029 Algunas características de los perfiles de la línea K del Ca II en fulguraciones sobre manchas solares.
J. R. Seibold.
Bol. As. Argentina Astron., No. 16, (see 012.007), p. 47 - 51 (1971).

072.030 Annual and three-monthly variations in solar activity and cosmic ray intensity.
E. V. Kolomeets, Yu. A. Shakhova.
Geomagn. Aeronom., Vol. 13, 219 - 222 (1973). In Russian.

072.031 Puentes de luz. M. Vázquez.
Urania Barcelona, Año 57, No. 275, p. 42 - 53 (1972).

072.032 Einige Bemerkungen zur Sonnenfleckenstatistik.
P. Ahnert.
Sterne, 49. Jahrgang, p. 78 - 82 (1973).

072.033 On the contrast of solar faculae near the limb.
I. F. Nikulin.
Astron. Tsirk., No. 745, p. 5 - 7 (1973). In Russian.

072.034 The periodicity of large sunspot groups.
M. Kopecký.
Bull. Astron. Inst. Czechoslovakia, Vol. 24, 113 - 118 (1973).
The periodicity of sunspot groups with an average area larger than 500 millionths of the solar hemisphere surface and sunspot groups with the maximum area larger than 1500 millionths is studied. Their 11-year period, the double maximum of the 11-year period according to Gnevyshev, the "butterfly" diagrams and the long period of the sunspots are studied.

072.035 On the physical relation between the magnetic field and the brightness in the sunspot umbrae.
H. I. Abdussamatov.
Bull. Astron. Inst. Czechoslovakia, Vol. 24, 118 - 120 (1973).

072.036 A comment on the seasonal variations of solar activity. P. Ambrož.
Bull. Astron. Inst. Czechoslovakia, Vol. 24, 130 - 132 (1973).
The paper presents a statistical study of the fluctuation of monthly average values of the relative number. It was found that fluctuations with half-periods in intervals of 120–140 and 180–200 days occur most frequently.

072.037 **Molecular abundances in sunspots.**
V. P. Gaur, M. C. Pande, B. M. Tripathi.
Bull. Astron. Inst. Czechoslovakia, Vol. 24, 138 - 143 (1973).
Results of dissociation equilibrium calculations for
Zwaan's sunspot model are given. It appears that many di- and
tri-atomic molecules form in sufficient abundances in the spots.

072.038 **Short periodicities in solar activity.**
K. R. Rao.
Solar Physics, Vol. 29, 47 - 53 (1973).
Several indices of solar activity are subjected to a high
pass filter and power spectral analysis to verify the existence
of shorter periodicities in solar activity. Though all these indices show the presence of short periodicities, above 95% confidence level, the common indices like sunspot number, fail
to show these periodicities. The basic parameters given by
Kopecký (1967), however, reveal the presence of 5.6- and
3.5-yr periodicities.

072.039 **Étude morphologique et cinématique des structures
fines d'une tache solaire.** R. Muller.
Solar Physics, Vol. 29, 55 - 73 (1973).
A sequence of 34 photographs of the main spot of the
group H 26 (*Daily Maps of the Sun*, Freiburg 1970, Rome
number 5847) has been obtained with the 38 cm refractor of
the Pic-du-Midi Observatory, showing throughout a resolution
very close or equal to $0''.3$. An interval of 3 hr is covered. The
pictures taken at intervals of 6 min approximately permit to
study the fine structure of the penumbra and associated phenomena.

072.040 **Spectral analysis of sunspot flares.**
M. E. Machado, J. R. Seibold.
Solar Physics, Vol. 29, 75 - 92 (1973).
We have qualitatively analyzed, in the H and K lines spectral region, 31 flares covering part of umbrae or penumbrae
of sunspots. A strong narrowing of the emission lines has been
observed over the umbrae, and the lines are, in general, much
weaker than in common flares suggesting that the optical
thickness is quite low in these parts. We have calculated the
Stark broadening of the Hϵ line from the general theory, and
it has been applied to obtain the electron density in 9 flare
spectra. In all cases it has been found that $n_e > 10^{13}$ cm^{-3}.
Goldberg's method has been applied to find the kinetic temperature from the H and K lines of Ca II, and from the ratio
between the central intensities of the lines we have calculated
the optical thickness in the K line.

072.041 **Spatial distribution of emerging flux regions.**
D. L. Glackin.
Publ. Astron. Soc. Pacific, Vol. 85, 241 - 248 (1973).
The absence of preferential longitudes for emerging flux
regions (EFRs) on the sun indicates that sunspot preferential
longitudes, if they exist, are due to favorable longitudes for
EFRs to become spot groups rather than to the distribution
of points of emergence of EFRs.

072.042 **On the annual variation of solar faculae.**
F. Loewe.
Gerlands Beiträge Geophys., Vol. 82, 25 - 26 (1973).
The sizes of solar faculae are biggest in July in an average
of 80 years and in three subdivisions. The summer values exceed those of the winter. The same applies to 100 years of relative sunspot numbers.

072.043 **A study on the solar activity and its influence on
cosmic ray variations.** T. S. Razmadze.
Trudy In-t geofiz. AN GruzSSR, Vol. 28, 95 - 103 (1972).
In Russian. — Abstr. in Referativ. Zhurn. 51. Astron.,
4.51.643 (1973).

072.044 **The influence of the choice of the sunspot model on
the determination of the turbulent velocity.**
E. A. Baranovsky, N. N. Stepanyan.
Izv. Krymskoj Astrofiz. Obs., Vol. 46, 106 - 114 (1972).
In Russian.
Turbulent velocities are determined for seven sunspot
models. Two methods are used for the determination of the
turbulent velocities. A rather strong dependence of the derived
value of the turbulent velocity on the model properties is
obtained.

072.045 **On the interpretation of total magnetic vector
measurements in sunspots.** V. A. Kotov.
Izv. Krymskoj Astrofiz. Obs., Vol. 46, 115 - 127 (1972).
In Russian.
A large change of the magnetic field with depth in a spot
was found on the basis of total vector H measurements at two
levels corresponding to the depths of the λ 5250 Fe I and
λ 6103 Ca I line formation. A considerable discrepancy
between the height gradients of the vertical field inside the
spot was obtained (1) from the measured vertical field at two
levels ($\sim \pm 3$ Gauss/km) and (2) from the use of the equation
div $\mathbf{H} = 0$ ($\sim \pm 0.5$ Gauss/km).

072.046 **Introductory review of solar activity.**
J. W. Evans.
Progr. Astronaut. Aeronaut., Vol. 30, (see 003.009), p. 3 - 17
(1972). — Presented at the AIAA observation and prediction
of solar activity conference, Huntsville, Ala., Nov. 16—18,
1970.

072.047 **Large-scale organization of solar activity in time and
space.** H. W. Dodson, E. R. Hedeman.
Progr. Astronaut. Aeronaut., Vol. 30, (see 003.009), p. 19 - 31
(1972). — Presented at the AIAA observation and prediction
of solar activity conference, Huntsville, Ala., Nov. 16—18,
1970.

072.048 **The solar wind cycle, the sunspot cycle, and the
corona.** J. Hirshberg.
Astrophys. Space Sci., Vol. 20, 473 - 481 (1973).
The author notes that the shape of the corona typical of
a 'maximum' eclipse occurs 1.5 yr before sunspot maximum,
compared with 2 yr as might be expected from Leighton's
'standard' model. Further, he argues that the phase of the solar
wind cycle can be determined from geomagnetic observations.
Using this phase, a solar cycle variation of 100 km s^{-1} in the
solar wind velocity and 1 γ in the magnetic field intensity
becomes apparent. In general, the solar wind cycle lags the
coronal-eclipse-form cycle by 3 yr, compared with the 2 yr
that might be expected from model calculations.

072.049 **Fine structure in the sunspot spectrum — 2 to 70
years.** R. G. Currie.
Astrophys. Space Sci., Vol. 20, 509 - 518 (1973).
Application of a new data adaptive approach to power
spectrum estimation has yielded evidence for a double solar
cycle line in the Zürich sunspot time history. There is significant power from 8 to 15 yr in the spectrum with the primary
line at 11.1 yr and three attendant multiplets that may be
significant. The first four harmonics of the solar cycle are detected too. Quite marginal evidence for a peak at ~ 65 yr in
the spectrum is presented. These results closely correspond to
those recently found in the geomagnetic spectrum.

072.050 **Magnetic fields and proton flares — 7 July and
2 September 1966.**
A. M. Zvereva, A. B. Severny.
Air Force Cambridge Res. Lab., Hanscom Field, Bedford, Mass.,
AFCRL-71-0605, Translations, No. 95, 7 + 69 pp. (1971).

Translated from Izv. Krymskoj Astrofiz. Obs., Vol. 41–42, 97 - 157 (1970) – see 05.072.007.

072.051 **Solar activity.** E. Tandberg-Hanssen.
Rev. Geophys. Space Phys., Vol. 11, 469 - 504 (1973).
The complex of solar activity can be divided into two parts: (1) the problem of the generation of magnetic fields and of the solar cycle (the 11-year period) and (2) the surface manifestations of solar activity. The first problem concerns the solar dynamo, and the author discusses in some detail dynamo maintenance. When the magnetic fields are transported to the sun's surface, they interact with the motions of the atmospheric plasma to produce the manifestation of solar activity: plages, sunspots, flares, prominences, and the active corona. Models of plages and several types of prominences are discussed, and the first steps toward theories for solar flares are indicated.

072.052 **Magnetic outflow – a stage in the development of an active region.**
R. Allen, S. Edberg, B. Labonte, N. R. Sheeley.
Bull. American Astron. Soc., Vol. 5, 268 (1973). – Abstr. AAS.

072.053 **Recent observations of the sun with a 3840 Å filter.**
G. A. Chapman.
Bull. American Astron. Soc., Vol. 5, 270 (1973). – Abstr. AAS.

072.054 **Test for planetary influence on solar activity.**
L. A. Dingle, G. Van Hoven, P. A. Sturrock.
Bull. American Astron. Soc., Vol. 5, 271 (1973). – Abstr. AAS.

072.055 **Multi-channel observations of sunspot oscillations.**
G. L. Epstein, R. W. Hobbs.
Bull. American Astron. Soc., Vol. 5, 272 (1973). – Abstr. AAS.

072.056 **H-alpha bright points.** B. Labonte.
Bull. American Astron. Soc., Vol. 5, 274 - 275 (1973). – Abstr. AAS.

072.057 **Solar magnetic field at sunspot minima.**
L. Svalgaard.
Bull. American Astron. Soc., Vol. 5, 280 - 281 (1973). Abstr. AAS.

072.058 **Analysis of EFRs: Hα filtergrams vs. magnetograms.**
J. Vorpahl.
Bull. American Astron. Soc., Vol. 5, 281 (1973). – Abstr. AAS.

072.059 **High resolution observations of a solar active region at 3.71 and 11.1 cm wavelength.**
R. W. Hobbs, S. D. Jordan, S. P. Maran, W. J. Webster, Jr.
Bull. American Astron. Soc., Vol. 5, 284 - 285 (1973). Abstr. AAS.

072.060 **Solar energy cycle and its relation to geomagnetic activity.** J. Kangas, P. Raychaudhuri.
Astrophys. Space Sci., Vol. 21, 3 - 5 (1973).
It is suggested that recurrent and nonrecurrent geomagnetic disturbances which are related to the release of solar magnetic energy in the form of unipolar and bipolar magnetic regions, respectively, are connected with the variations in the solar energy source.

072.061 **Videomagnetograph studies of solar magnetic fields. I: Magnetic field diffusion in weak plage regions.**
R. C. Smithson.
Solar Physics, Vol. 29, 365 - 382 (1973).
Observations of magnetic field diffusion in weak plage regions have been made using the analog videomagnetograph at the California Institute of Technology. Points of magnetic flux were found to have a mean lifetime of three to four days, and to disperse primarily by means of two mechanisms: a random walk with a step time short compared to 24 h, and a sudden transport of magnetic flux over distances of 5000 to 20000 km during a time span of one to three hours. The second mechanism is probably the predominant one. Similar observations have been made using K_3 spectroheliograms.

072.062 **Determination of the temperature distribution in a photospheric facula by solving an integral equation by a gradient-random search method.**
O. G. Badalyan, A. G. Prudkovsky.
Astron. Zhurn. Akad. Nauk SSSR, Vol. 50, 558 - 563 (1973). In Russian. English translation in Soviet Astron. AJ, Vol. 17, No. 3.
It was found that the maximal difference between the temperatures of the facula and undisturbed photosphere is $+400°K$ at $\tau_{5000} = 0.6$, and the minimal one is $-240°K$ at $\tau_{5000} \approx 2.5-3.0$. In the upper layers ($\tau \approx 0.01$) the difference of the facula and the photospheric temperatures does not exceed $50°K$.

072.063 **The latitudinal zonation on the sun.**
N. I. Kozhevnikov.
Astron. Zhurn. Akad. Nauk SSSR, Vol. 50, 564 - 567 (1973). In Russian. English translation in Soviet Astron. AJ, Vol. 17, No. 3.
The distribution with respect to heliographic latitude of the maximum numbers of flares and of the duration of sunspot groups is investigated. It is found that the distribution curves show a clear periodicity with solar latitude.

072.064 **Courbes de l'activité solaire en 1972.**
M. Waitz, W. Groubé.
L'Astronomie, 87e année, p. 263 - 266 (1973).

072.065 **On the decay of sunspots.**
F. Meyer, H. U. Schmidt.
Mitt. Astron. Ges., No. 32, p. 173 - 175 (1973).

072.066 **Magnetfeld - Entwicklung eines Aktivitätsgebietes vor und nach Fleckenbildung.** M. Roßbach.
Mitt. Astron. Ges., No. 32, p. 175 - 177 (1973).

072.067 **On the magnetic classification of sunspot groups.**
G. R. Greatrix, G. H. Curtis.
Observatory, Vol. 93, 114 - 116 (1973).
Results published in an earlier paper suggested that the relationship between solar flares and sunspots is not symmetrical about the equator of the sun. A proposed new system of classification of sunspot groups avoids this apparent asymmetry and is therefore physically more satisfactory.

072.068 **Relation between the main elements of the 22-year cycle of solar activity.** A. Bonov.
Izv. Sekts. astron. Blg. AN, Vol. 5, 33 - 40 (1972). In Bulgarian. – Abstr. in Referativ. Zhurn. 51. Astron., 5.51.537 (1973).

072.069 **Correlations between the elements of the 22-year cycle of solar activity.** A. Bonov.
Izv. Sekts. astron. Blg. AN, Vol. 5, 41 - 46 (1972). In Bulgarian. – Abstr. in Referativ. Zhurn. 51. Astron., 5.51.538 (1973).

072.070 **Two maxima of the 20th cycle of solar activity.**
W. Szymański.
Postępy Astron., Vol. 21, 145 - 147 (1973). In Polish.

072.071 **August 1972 solar activity.**
L. Křivský, J. Olmr, J. Klimeš.
Říše hvězd, Vol. 54, 1 - 6 (1973). In Czech.

072.072 **Magnetic polarities and maximum field strengths of selected sunspots groups with time distances of about half an hour during the periods 1971 July 19 – July 23, 1971 August 19 – August 26.**   H. Künzel.
Zentralinstitut für Solar-Terrestrische Physik (Heinrich-Hertz-Institut), Deutsche Akad. Wiss. Berlin, HHI Suppl. Ser. Solar Data, Vol. 3, 69 - 94 (1972).

In continuation to the first tables of magnetic data of selected spotgroups within the years 1968/69 (see abstr. 05. 072.039) the following tables contain the results of measurements of maximum magnetic field strengths of several main spots in two selected spotgroups with time distances of about half an hour during the periods 1971 July 19 – July 23 and 1971 August 19 – August 26. In addition to the magnetic data of the main spots the distribution of the magnetic polarities of the remaining spots in the two groups derived by visual measurements are given.

072.073 **On the possibility of constructing a radiative sunspot model in magnetohydrostatic equilibrium.** D. J. Mullan.
Solar Physics, Vol. 30, 75 - 81 (1973).

The author wishes to point out that by using a different theory of convection, due to Öpik (1950), it is possible to compute a radiative sunspot model in which the field becomes no greater than 9000 G. By applying two boundary conditions, (I) depth of spot equals depth of convection zone, (II) magnetic field has zero gradient at the base of the spot, he shows that a radiative spot has a unique effective temperature for a given Wilson depression, Δ. For $\Delta = 650$ km, he finds $T_e = 3800$ K; for $\Delta = 150$ km, $T_e = 3950$ K. According to his model, spots having T_e cooler than these values should not exist.

072.074 **Can oscillations grow in a sunspot umbra?** D. J. Mullan, H. S. Yun.
Solar Physics, Vol. 30, 83 - 91 (1973).

The authors have extended Moore's analysis to examine the depth-dependence of overstable oscillations in a recently computed umbral model. Electrical conductivity is evaluated taking full account of partial ionization and magnetic fields. The umbral model used is based on Öpik's cellular convection model. The interaction between the vertical magnetic field and convection is included by varying the diameter of the cell, and not its height.

072.075 **The east-west asymmetry in the number of spot-groups in relation to their classification.** R. Bartsch.
Solar Physics, Vol. 30, 93 - 102 (1973).

It is shown that the east-west asymmetry is quite different for spot-groups of different classes of evolution and that there exists a conspicuous difference between the ascending and descending branches with regard to the east-west asymmetry and the asymmetry between corresponding lunes. Some results point to the existence of systematic errors in the classification of the spot-groups.

072.076 **Periodicities in solar activity.** T. W. Cole.
Solar Physics, Vol. 30, 103 - 110 = Division Radiophys., CSIRO, Sydney, Radiophys. Publ. RPP 1647 (1973).

The techniques of power spectral analysis are used to determine significant periodicities in the annual mean relative sunspot numbers. The main conclusion is that a period of 10.45 yr is very basic and can be associated with an excitation of new solar cycles. When combined with a period of 11.8 yr, associated here with the free-running length of a solar cycle, the mean cycle length of 11.06 yr and a phase variation of 190 yr are explained. Similarly the amplitude variations with periods 88 and 59 yr (previously described as the 80-yr cycle) are due to an amplitude modulation of the solar cycle by a period of 11.9 ± 0.3 yr.

072.077 **Test of an analysis of the form of the relief of some physical parameters of active regions of the sun.** V. V. Kasinskij, L. A. Plyusnina.
Issled. po geomagnetizmu, aehron. i fiz. Solntsa. Vyp. (No.) 26. Moskva, Nauka, 1973, p. 54 - 66. In Russian. – Abstr. in Referativ. Zhurn. 51. Astron., 7.51.426 (1973).

072.078 **Ratio of central residual intensities of the H and K Ca II lines in an active region.**   S. A. Efendieva.
Issled. po geomagnetizmu, aehron. i fiz. Solntsa. Vyp. (No.) 26. Moskva, Nauka, 1973, p. 67 - 77. In Russian. – Abstr. in Referativ. Zhurn. 51. Astron., 7.51.427 (1973).

072.079 **The different run of maximum magnetic field strengths of opposed polarity in sunspot groups before and after flares.**   G. V. Kuklin, I. A. Nikiforova.
Issled. po geomagnetizmu, aehron. i fiz. Solntsa. Vyp. (No.) 26. Moskva, Nauka, 1973, p. 90 - 104. In Russian. – Abstr. in Referativ. Zhurn. 51. Astron., 7.51.443 (1973).

Solar speckle interferometry.
See Abstr. 031.054.

Observations of the infrared sunspot spectrum between 11340 Å and 24778 Å.   See Abstr. 071.051.

First observations of the granulation at 1.65 μ, center to limb variation of the contrast.
See Abstr. 071.057.

A correlation analysis among the Ca II resonance and subordinate lines based on high resolution spectrograms.   See Abstr. 073.053.

Chromospheric force-free magnetic fields associated with bi-polar sunspot groups.   See Abstr. 073.067.

On the role of sunspot-group-satellites as proton flare predecessors.   See Abstr. 073.111.

Die Sonnenaktivität im Jahre 1971.
See Abstr. 075.017.

Investigation of the polarization of solar radio emission at 9.0 cm with development of an active region.   See Abstr. 077.038.

Solar active regions at 9 and 3.5 mm wavelengths under disturbed conditions.   See Abstr. 077.059.

Solar rotation and solar activity.
See Abstr. 080.026.

Changes in solar rotation due to the solar energy generation cycle of 11 years.   See Abstr. 080.043.

A search for periodic variations in geomagnetic acti-vitiy and their solar cycle dependence.
See Abstr. 084.266.

Planetary resonances, bi-stable oscillation modes and solar activity cycles.   See Abstr. 091.065.

Planets, sunspots and earthquakes.
See Abstr. 091.068.

Intensity of Jupiter's atmospheric belts and solar activity.   See Abstr. 099.031.

Solar acitivity and cosmic rays in 1963—1965.
See Abstr. 143.062.

Frequency spectrum of intensity variations of cosmic rays and solar activity.   See Abstr. 143.075.

On the relation between the 27-day cosmic-ray variations and various indices of solar activity during 1957 - 1970. See Abstr. 143.077.

Forbush decreases and their relation with solar activity and the parameters of interplanetary matter.
See Abstr. 143.078.

Errata

072.901   Erratum: 'The cooling of a sunspot. II.' [Solar Physics, Vol. 27, 363 - 372 (1972)].
P. R. Wilson.
Solar Physics, Vol. 30, 280 (1973).

073 Solar Chromosphere, Flares, Prominences

073.001 **The He⁺ λ 4686 line in the low chromosphere.**
S. P. Worden, J. M. Beckers, T. Hirayama.
Solar Physics, Vol. 28, 27 - 34 (1973).
We report an unsuccessful search for the He⁺ λ 4686 line in the low chromosphere. However, at the location of this line we detect a number of other chromospheric emission lines. This leads us to the conclusion that the He⁺ λ 4686 identification made in the past, as well as other identifications, are probably in error. Additionally the region of the neutral helium λ 4713 line is also studied.

073.002 **On the random nature of the eruption of magnetic flux at the solar surface.**
R. Howard, S. J. Edberg.
Solar Physics, Vol. 28, 73 - 75 (1973). – Research note.

073.003 **Flares associated with EFR's (Emerging Flux Regions).** J. A. Vorpahl.
Solar Physics, Vol. 28, 115 - 122 (1973).
The author examined a moderately active sunspot group, McMath 9735, and found that 15 of 16 flares observed in 1968, October 20–21 occurred near, and were preceded by, at least one of several EFR's (Emerging Flux Regions) in the area.

073.004 **Two-component temperature analysis of OSO-5 X-ray flare data.** J. R. H. Herring, I. J. D. Craig.
Solar Physics, Vol. 28, 169 - 174 (1973).
Using data covering the 2.6–10 Å wavelength range from the OSO-5 satellite a four-parameter model of the emitting region in a flare process is derived. The thermal emission spectrum of Landini and Fossi is used to calculate the plasma parameters, electron temperature and emission measure. The X-ray flare data is explained by a model which treats the source volume as two time-varying temperature regions.

073.005 **Directivity of high-energy X-ray emission during flares.** K. J. H. Phillips.
Observatory, Vol. 93, 17 - 18 (1973).

073.006 **The reconnection rate of magnetic fields.**
E. N. Parker.
Astrophys. Journ., Vol. 180, 247 - 252 (1973).
Solar flares, and the absence of intense small-scale magnetic fields in the turbulent solar photosphere, suggest that the reconnection rate, or merging speed, of two oppositely directed magnetic fields is of the general order of the Alfvén speed. Hence reconnection is sufficiently rapid, with or without microinstabilities, that it plays a major role in solar flares and in the reduction of small-scale turbulent fields.

073.007 **The extreme-ultraviolet spectrum of Fe XV in a solar flare.** R. D. Cowan, K. G. Widing.
Astrophys. Journ., Vol. 180, 285 - 292 (1973).
Approximately 13 lines of Fe XV are identified between 200 and 500 Å in the slitless spectrum of a solar flare. The population of the metastable levels of $3s3p\ ^3P^0$ is calculated as a function of electron density, and $n_e = 4 \times 10^{10}$ in the flare is derived from the intensity ratio of the lines at 234 and 244 Å. Several new singlet lines of Fe XV are identified in nonflare solar slit spectra below 100 Å. The problem of the identification of the strong enhancing line at 69.66 Å is discussed.

073.008 **Spectral investigation of the chromosphere. II. The nature of the mottles and a model of the overall structure.** U. Grossmann-Doerth, M. von Uexküll.
Solar Physics, Vol. 28, 319 - 332 = Mitt. Fraunhofer Inst., *Freiburg*, No. 116 (1973).
Highly resolved Hα spectra and filtergrams obtained at the Fraunhofer Observatory on Capri were analyzed by a method whose principles have been described before. As a result the tentative conclusion of our previous work has been confirmed: The mottles of the chromospheric fine structure are clouds superimposed on the low chromosphere. Furthermore, it is proposed that the latter is identical with the interior of the supergranular cells whose spatial averages lend themselves to an interpretation in terms of a spherically symmetric model.

073.009 **Filter observations of prominences in the D₃ and Hα lines.** I. S. Kim, G. M. Nikolsky.
Solar Physics, Vol. 28, 377 - 388 (1973).
D₃ and Hα pictures of prominences were obtained with a 21-in. Lyot coronagraph and a Fabry-Perot etalon used as a narrow band filter. The monochromatic images of quiescent, 'quasi-quiescent' and loop-prominences were studied.

073.010 **A secondary polar zone of solar prominences.**
M. Waldmeier.
Solar Physics, Vol. 28, 389 - 398 = Astron. Mitt. Sternw. Zürich, No. 315 (1973).
The polar prominences are concentrated in a zone, which in the period between sunspot minimum and maximum is shifted from about 45° heliographic latitude towards the pole. Cycle No. 20 has shown an anomaly never observed before, as on the northern hemisphere two zones of polar prominences were formed, the second zone following the first one at an interval of 2.5 yrs. We investigated whether the anomalous appearance of a second polar zone might be related to a corresponding anomaly in the main zone.

073.011 **Dynamics and localization of surges in the chromosphere.** Yu. V. Platov.
Solar Physics, Vol. 28, 477 - 485 (1973).
The dynamics of small surges has been studied using filter and spectral observations in the Hα line. The surge evolution allows the division into two stages which follows from the observational results. The first stage includes an abrupt Hα line broadening and the upward acceleration of the surge material. The second one is the inertial motion of the surge plasma along the magnetic force lines. Surges are probably produced by the plasma raking-up which is connected with the growth of the local magnetic field.

073.012 **Note on heliumlike silicon and sulfur lines observed in the X-ray spectra of solar flares.**
G. A. Doschek, J. F. Meekins.
Solar Physics, Vol. 28, 517 - 518 (1973). – Research note – erratum.

073.013 **Characteristics of electron and high-energy proton flares.** E. T. Sarris, S. D. Shawhan.
Solar Physics, Vol. 28, 519 - 532 (1973).
High-energy proton ($E_p > 55$ MeV) and electron ($E_e > 50$ keV) events were observed by University of Iowa experiments on the satellites Explorer 33 and 35. The solar X-ray (2–12 Å) flares associated with the energetic proton events were found to have in general higher peak fluxes, considerably longer decay times and smaller rise to decay time ratios than the X-ray flares associated with the electron events.

073.014 **Some notes on flare patrol.**
R. Falciani, M. Rigutti.

Solar Physics, Vol. 28, 539 - 542 (1973).

Using data from ESSA Research Laboratory bulletins an estimate of the uncertainty for the times of beginning, maximum and ending, and maximum areas of Hα flarès reported during 1967 is deduced and discussed. The need for an improvement of the reliability of the solar patrol service is also indicated.

073.015 **Simultaneous determination of the electron temperature and density in the chromospheric-coronal transition region of the sun.**
M. Malinovsky, L. Heroux, S. Sahal-Bréchot.
Astron. Astrophys., Vol. 23, 291 - 294 (1973).

The electron temperature and density in the solar atmosphere in the region of emission of O V have been obtained from rocket measurements of EUV spectral line intensity ratios.

073.016 **Determination of the temperature in a solar X-ray flare.** G. C. Rumi.
Journ. Atmosph. Terr. Phys., Vol. 35, 639 - 646 (1973).

The equivalent temperature in a solar X-ray flare for $\lambda \approx 2$ Å is deduced from the measurement of the X-ray flux by Explorer 37 and the simultaneous measurement of the ionospheric electron density by VLF and LF records on the Rugby-Turin link. These data yield the value of the effective photoionization cross section in the band 1 Å $< \lambda < 8$ Å.

073.017 **Protuberanser.** B. M. Rustad.
Astron. Tidssk., Årg. 6, p. 6 - 17 (1973).

073.018 **Energy flux into the solar chromosphere from the five-minute oscillation.** Yu. D. Zhugzhda.
Astrophys. Letters, Vol. 13, 173 - 176 (1973).

The formula for the energy flux of non-adiabatic oscillations in an inhomogeneous atmosphere is deduced. The energy flux of 300-sec oscillations at the 450 km level is estimated. In the case of a horizontal wavelength of 3×10^3 km the flux is sufficient to heat the solar atmosphere.

073.019 **Microwave pulsations from solar flares.**
A. Maxwell, J. FitzWilliam.
Astrophys. Letters, Vol. 13, 237 - 242 (1973).

Fourier analysis of the pulsations that occur in type IV microwave bursts reveals components with periods in the range 1 - 200 sec, some of which have surprisingly large amplitudes. The characteristics of the pulsating bursts may possibly be accounted for by theories of solar radio pulsations of the type proposed by Chiu.

073.020 **Shock waves and the interpretation of the chromospheric calcium K-line.** L. E. Cram.
Proc. Astron. Soc. Australia, Vol. 2, 146 - 147 (1972).

073.021 **Observation of an instability in a "quiescent" prominence.** G. Stellmacher, E. Wiehr.
Astron. Astrophys., Vol. 24, 321 - 324 (1973).

A cavity-like instability has been observed in a "quiescent" prominence. H_α-slit-yaw pictures are shown together with Ca$^+$ λ 8542 spectra taken with an image intensifier. The instability propagates almost with the phase velocity of MHD compressive waves. A violation of the lateral stability criterion in the Kippenhahn-Schlüter model is indicated, producing a material outflow along the lines of force.

073.022 **Metal lines in the solar flare on July 12, 1961 and properties of the region of their emission.**
N. S. Shilova.
Astron. Zhurn. Akad. Nauk SSSR, Vol. 50, 336 - 342 (1973).
In Russian. English translation in Soviet Astron. AJ, Vol. 17, No. 2.

The intensities of lines of neutral and singly ionized metals in the spectrum of the solar flare of importance 3$^+$ on July 12, 1961 have been studied. These lines characterize the emission of the coolest flare regions.

073.023 **Multi-wavelength studies of the evolution of active regions.** R. W. Hobbs, G. L. Epstein, S. B. Modali.
Bull. American Astron. Soc., Vol. 5, 20 (1973). – Abstr. AAS.

073.024 **Kα line emission during solar flares.**
K. J. H. Phillips, W. M. Neupert.
Bull. American Astron. Soc., Vol. 5, 20 (1973). – Abstr. AAS.

073.025 **Powerful chromospheric flares and rotating discontinuities in the solar wind.**
V. I. Afanasieva, K. G. Ivanov.
Geomagn. Aeronom., Vol. 13, 10 - 13 (1973). In Russian.

073.026 **Trends of development of the proton active region of 24 January 1971.** L. Křivský.
Bull. Astron. Inst. Czechoslovakia, Vol. 24, 96 - 100 (1973).

The development of the activity in the region where a proton flare occurred on 24 Jan. 1971, is presented using the method of summation curves.

073.027 **Remark on the equilibrium of helical magnetic structures in prominences.** M. Šidlichovský.
Bull. Astron. Inst. Czechoslovakia, Vol. 24, 102 - 105 (1973).

A simple model for the formation of prominences through a pinch effect is proposed.

073.028 **Role of electron impact in the excitation of prominence luminescence in Ca$^+$ lines.**
P. N. Polupan.
Astrometriya i Astrofizika, *Kiev*, vyp. (No.) 16, (see 003.006), p. 68 - 77 (1972). In Russian.

The mechanism of electron impact is of secondary significance in the prominence luminescence excitation in Ca$^+$ lines as compared with resonance scattering of photospheric radiation. Its contribution changes considerably with the 1000° temperature change in prominences.

073.029 **Fulguración cromosférica del 27 de febrero 1969.**
H. Grossi Gallegos.
Bol. As. Argentina Astron., No. 16, (see 012.007), p. 42 - 46 (1971).

073.030 **Desaparición brusca de un filamento asociado a la fulguración del 8 de marzo de 1970.**
M. Rovira, M. Machado.
Bol. As. Argentina Astron., No. 16, (see 012.007), p. 47 (1971).

073.031 **Sobre el problema de oscilación de filamentos.**
H. Grossi Gallegos, M. Machado, M. Peralta.
Bol. As. Argentina Astron., No. 16, (see 012.007), p. 51 - 52, 55 - 59 (1971).

073.032 **Taking into account the instrumental profile in investigations of the spectra of solar prominences.**
G. A. Zakharova.
Solnechnye Dannye 1972 Byull., No. 12, p. 91 - 95 (1973).
In Russian.

Several methods of taking into account the instrumental profiles are considered. The problem of necessity of corresponding reductions for the hydrogen and D$_3$ helium lines is discussed.

073.033 **The Stark effect on plasma oscillations as a possible cause of broadening of hydrogen lines in solar flares.**
V. M. Tomozov.

Astron. Tsirk., No. 749, p. 1 - 4 (1973). In Russian.

073.034 On the populations of quantum levels of hydrogen atoms in chromospheric flares.
E. V. Kurochka, L. N. Kurochka.
Bull. Astron. Inst. Czechoslovakia, Vol. 24, 132 - 136 (1973).
It is shown that the relative population of the hydrogen atom levels cannot be used to determine the electron temperature in the flares. The problem of the overexcitation of the levels of hydrogen atoms in the flares, which as is shown in this paper occurs only at low values of optical thickness, has also been solved.

073.035 The structure of the lower chromosphere in the Na and He (D_1, D_2, D_3) lines from observations with the 53-cm coronograph. V. I. Makarov, N. S. Shilova.
Solnechnye Dannye 1973 Byull., No. 1, p. 69 - 76 (1973). In Russian.

073.036 Determination of the characteristics of optical inhomogeneities by cinematographing the sun.
U. I. Iljasov.
Solnechnye Dannye 1973 Byull., No. 1, p. 92 - 96 (1973). In Russian.
The results of determining the characteristics (height, dimension, velocity) of optical inhomogeneities from cinematographing the sun are given.

073.037 Studies of the solar chromosphere from millimetre and sub-millimetre observations. I: Isophotometric mapping. J. E. Beckman, C. D. Clark.
Solar Physics, Vol. 29, 25 - 39 (1973).
Spectroheliograms were obtained in bands centred at 1.2 mm, 0.8 mm and 0.4 mm wavelength during 1969 and 1971. In order to obtain photometrically valid data, a specialized set of reduction techniques was employed. Comparison of our maps with those of other observers at 3 mm and 8.6 mm wavelength suggests that the chromospheric brightness temperature increments above active regions show a monotonic increase with increasing height above the photosphere over the normal increase in brightness temperature in the quiet chromosphere. Within the limit of angular resolution of 3' available, no evidence was recorded of normal limb brightening in our three passbands but the presence of isophotes at 50% of the central disc temperature consistently circumscribing the optical limb implies a narrow spike of sub-millimetre brightening close to the limb.

073.038 Non-thermal ionization and recombination processes during solar flares.
M. Landini, B. C. Fossi, R. Pallavicini.
Solar Physics, Vol. 29, 93 - 105 (1973).
Ionization and recombination processes are studied for a plasma of which the electrons follow a power-law energy distribution. The rates for collisional ionization, radiative and dielectronic recombination and for autoionization are evaluated. Numerical computations are performed for H-like, He-like and Li-like ions from neon to nickel as a function of the spectral index of the electron distribution. The ionization equilibrium is evaluated as well as the ratios of fluxes emitted in two lines pertaining to two successive ionization stages of the same element. A comparison with a few experimental data is made and the possibility of a non-thermal interpretation of X-ray line emission during solar flares is discussed.

073.039 The extreme ultraviolet emissions of solar flares: a comparison between OSO-6 spectroheliograph observations and SFDs.
R. F. Donnelly, A. T. Wood, Jr., R. W. Noyes.
Solar Physics, Vol. 29, 107 - 123 (1973).
The time structure and intensity of OSO-6 observations

of EUV bursts were studied in relation to the corresponding 10-1030Å enhancements deduced from SFD data. Impulsive EUV emissions from lines normally emitted from either the chromosphere or from the chromosphere-corona transition region rise simultaneously with the 10-1030 Å flash, to within the time resolution of the OSO-6 observations. An interpretation of EUV flare emissions is given.

073.040 Spectra of solar flares from 8.5 Å to 16 Å.
G. A. Doschek, J. F. Meekins, R. D. Cowan.
Solar Physics, Vol. 29, 125 - 141 (1973).
X-ray spectra of solar flares in the spectral range from 8.5 Å to ~ 16 Å have been obtained from a Naval Research Laboratory crystal spectrometer flown on the sixth Orbiting Solar Observatory (OSO-6). A list of emission features is presented and tentative identifications of some of the features are suggested. The time-behavior of the emission lines during flares is discussed, and the possibility of determining electron densities in flare plasmas using density sensitive lines of highly ionized iron is considered. Approximate calculations are performed for a density sensitive line of Fe XXII.

073.041 On some transient Hα features associated with metric type III bursts.
F. Axisa, M. J. Martres, M. Pick, I. Soru-Escaut.
Solar Physics, Vol. 29, 163 - 182 (1973).
The purpose of this article is to reconsider the question of the chromospheric type III association by using a selection of cases for which the chromospheric situation was simple enough to avoid any ambiguous association. It has been found that the type III bursts are essentially as frequently associated with an Hα flare brightening as with a particular kind of Hα absorbing feature. Furthermore the dual character of this association leads to a possible causal link between three distinct phenomena, i.e. the type III emission, the flare, and a class of Hα absorbing feature.

073.042 The brightnesses at different levels in solar active regions. S. I. Gopasyuk, T. T. Tsap.
Izv. Krymskoj Astrofiz. Obs., Vol. 46, 90 - 100 (1972). In Russian.
Relations between the brightness of different lines and between brightnesses and magnetic fields at different levels in active regions are studied.

073.043 On the heating of solar active regions.
S. I. Gopasyuk, T. T. Tsap.
Izv. Krymskoj Astrofiz. Obs., Vol. 46, 101 - 105 (1972). In Russian.
Heating of solar active regions by shock waves, hydromagnetic waves and Joule losses in partially ionized gas is considered. It is shown that Joule heating plays the most important role in the heating of active regions.

073.044 Extreme ultraviolet observations of solar flares.
A. T. Wood, Jr., R. W. Noyes, E. M. Reeves.
Progr. Astronaut. Aeronaut., Vol. 30, (see 003.009), p. 117 - 125 (1972). — Presented at the AIAA observation and prediction of solar activity conference, Huntsville, Ala., Nov. 16–18, 1970.

073.045 Magnetic models of solar flares. P. A. Sturrock.
Progr. Astronaut. Aeronaut., Vol. 30, (see 003.009), p. 163 - 176 (1972). — Presented at the AIAA observation and prediction of solar activity conference, Huntsville, Ala., Nov. 16–18, 1970.

073.046 Early recognition of major solar flares in Hα.
S. F. Martin, H. E. Ramsey.
Progr. Astronaut. Aeronaut., Vol. 30, (see 003.009), p. 371 - 387 (1972). — Presented at the AIAA observation and predic-

tion of solar activity conference, Huntsville, Ala., Nov. 16–18, 1970.

073.047 **Prediction of proton flares and Forbush effects.**
L. Křivský.
Progr. Astronaut. Aeronaut., Vol. 30, (see 003.009), p. 389 - 409 (1972).

073.048 **Quantitative short-term prediction of proton and nonproton flares.** E. I. Mogilevsky.
Progr. Astronaut. Aeronaut., Vol. 30, (see 003.009), p. 411 - 419 (1972).

073.049 **Forecasting flares from inferred magnetic fields.**
J. J. Lemmon.
Progr. Astronaut. Aeronaut., Vol. 30, (see 003.009), p. 421 - 428 (1972). – Presented at the AIAA observation and prediction of solar activity conference, Huntsville, Ala., Nov. 16–18, 1970.

073.050 **Predicting activity levels for specific locations within solar active regions.** J. B. Smith, Jr.
Progr. Astronaut. Aeronaut., Vol. 30, (see 003.009), p. 429 - 442 (1972). – Presented at the AIAA observation and prediction of solar activity conference, Huntsville, Ala., Nov. 16–18, 1970.

073.051 **Spatial structure of a two-phase solar flare observed in the EUV by OSO 7.**
R. J. Thomas, W. M. Neupert, R. D. Chapman.
Bull. American Astron. Soc., Vol. 5, 266 (1973). – Abstr. AAS.

073.052 **Magnetic fields of the loop prominences of August 2, 7, and 11, 1972.** V. Bar, D. M. Rust.
Bull. American Astron. Soc., Vol. 5, 269 (1973). – Abstr. AAS.

073.053 **A correlation analysis among the Ca II resonance and subordinate lines based on high resolution spectrograms.** D. R. Brown.
Bull. American Astron. Soc., Vol. 5, 269 (1973). – Abstr. AAS.

073.054 **Some new observations of the solar UV chromosphere.** E. C. Bruner, Jr.
Bull. American Astron. Soc., Vol. 5, 269 (1973). – Abstr. AAS.

073.055 **Radiative damping of internal gravity waves.**
P. A. Clark, A. Clark, Jr.
Bull. American Astron. Soc., Vol. 5, 270 (1973). – Abstr. AAS.

073.056 **Evidence for thin-target X-ray emission in a small solar flare on 26 February 1972.**
D. Datlowe, R. P. Lin.
Bull. American Astron. Soc., Vol. 5, 270 (1973). – Abstr. AAS.

073.057 **Development of solar active regions.**
A. K. Dupree, D. J. Bechis.
Bull. American Astron. Soc., Vol. 5, 271 (1973). – Abstr. AAS.

073.058 **On the energetics and momentum balance of pole-equator temperature differences in the sun.**
B. R. Durney.
Bull. American Astron. Soc., Vol. 5, 271 (1973). – Abstr. AAS.

073.059 **Measurements of the solar brightness temperature in the far infrared.**
J. A. Eddy, R. H. Lee, R. M. MacQueen, W. G. Mankin.
Bull. American Astron. Soc., Vol. 5, 271 (1973). – Abstr. AAS.

073.060 **Observations of the solar flare of 7 August 1972.**
G. L. Epstein, R. W. Hobbs, S. P. Maran, H. P. Jones.
Bull. American Astron. Soc., Vol. 5, 272 (1973). – Abstr. AAS.

073.061 **Far infrared observations of solar flares.**
H. S. Hudson.
Bull. American Astron. Soc., Vol. 5, 274 (1973). – Abstr. AAS.

073.062 **Magnesium II doublet profiles of chromospheric inhomogeneities at the center of the solar disk.**
P. Lemaire, A. Skumanich.
Bull. American Astron. Soc., Vol. 5, 275 (1973). – Abstr. AAS.

073.063 **Magnetic observations related to the 3B flare of 4 August.** W. Livingston.
Bull. American Astron. Soc., Vol. 5, 276 (1973). – Abstr. AAS.

073.064 **The San Manuel effect – a progress report.**
W. Livingston, L. Ramsey.
Bull. American Astron. Soc., Vol. 5, 276 (1973). – Abstr. AAS.

073.065 **Magnetic fields in active prominences.**
J. M. Malville, E. Tandberg-Hanssen.
Bull. American Astron. Soc., Vol. 5, 276 (1973). – Abstr. AAS.

073.066 **Spatial relationship between λ 5303 and Hβ components of a loop prominence system.**
M. K. McCabe.
Bull. American Astron. Soc., Vol. 5, 277 (1973). – Abstr. AAS.

073.067 **Chromospheric force-free magnetic fields associated with bi-polar sunspot groups.**
R. X. Meyer, E. B. Mayfield.
Bull. American Astron. Soc., Vol. 5, 277 (1973). – Abstr. AAS.

073.068 **The excitation of chromospheric helium.**
R. W. Milkey, J. N. Heasley, H. A. Beebe.
Bull. American Astron. Soc., Vol. 5, 277 (1973). – Abstr. AAS.

073.069 **Trapped oscillations in the chromosphere in the presence of a magnetic field.** Y. Nakagawa.
Bull. American Astron. Soc., Vol. 5, 277 (1973). – Abstr. AAS.

073.070 **The prominence-corona interface.**
F. Q. Orrall, R. J. Speer.
Bull. American Astron. Soc., Vol. 5, 278 (1973). – Abstr. AAS.

073.071 **Hα chromospheric oscillations above sunspot umbrae.** G. L. Phillis, H. E. Ramsey, A. M. Title.
Bull. American Astron. Soc., Vol. 5, 278 - 279 (1973).
Abstr. AAS.

073.072 **Hα profiles from the 2 August events.**
S. A. Schoolman.
Bull. American Astron. Soc., Vol. 5, 279 (1973). – Abstr. AAS.

073.073 **The effect of small and large scale sine waves upon chromospheric line profiles.**
R. Shine, L. Oster.
Bull. American Astron. Soc., Vol. 5, 279 - 280 (1973).
Abstr. AAS.

073.074 **Flare rates and coronal density evolution of active regions.** S. P. Smith, G. L. Withbroe.
Bull. American Astron. Soc., Vol. 5, 280 (1973). – Abstr. AAS.

073.075 **On the visibility of Hα fibrils.**
R. Steinitz, K. B. Gebbie.
Bull. American Astron. Soc., Vol. 5, 280 (1973). – Abstr. AAS.

073.076 **Mass flow in solar flares.** P. A. Sturrock.
Bull. American Astron. Soc., Vol. 5, 280 (1973).
Abstr. AAS.

073.077 **Fine structure, evolution and oscillation of Hα**

mottles.   K. Tanaka.
Bull. American Astron. Soc., Vol. 5, 281 (1973). – Abstr. AAS.

073.078 Motions and heating of the August flares.
   K. Tanaka, H. Zirin.
Bull. American Astron. Soc., Vol. 5, 281 (1973). – Abstr. AAS.

073.079 The promise of self-consistent flare data.
   F. W. Ward, R. F. Carnevale.
Bull. American Astron. Soc., Vol. 5, 281 (1973). – Abstr. AAS.

**073.080 A physical mechanism for the production of solar
   prominences.**   M. L. White.
Bull. American Astron. Soc., Vol. 5, 282 (1973). – Abstr. AAS.

073.081 The flares of August.   H. Zirin, K. Tanaka.
   Bull. American Astron. Soc., Vol. 5, 282 (1973).
Abstr. AAS.

073.082 Ricerche sulle protuberanze solari.
   G. Godoli, S. Motta, V. Sciuto, R. A. Zappalà.
Mem. Soc. Astron. Italiana, Nuova Ser., Vol. 43, 773 - 778
(1973).
   The model proposed by Kippenhahn-Schlüter applied to
the actual conditions of the solar atmosphere allows the interpretation of some characteristics of solar quiescent prominences as the values and the signs of the inclination of the prominence planes and the prominence sudden disappearances.

073.083 Soft X-ray spectral studies of solar flare plasmas.
   G. A. Doschek, J. F. Meekins, R. W. Kreplin, T. A.
Chubb, H. Friedman.
Cosmic plasma physics. Conference 1971, (see 012.016),
p. 165 - 174 (1972).

**073.084 Similarities between solar flares and laboratory hot
   plasma phenomena.**
W. H. Bostick, V. Nardi, W. Prior.
Cosmic plasma physics. Conference 1971, (see 012.016),
p. 175 - 184 (1972).

**073.085 High-resolution photography of the solar chromo-
   sphere. X: Physical parameters of Hα mottles.**
R. J. Bray.
Solar Physics, Vol. 29, 317 - 325 (1973).
   High-resolution filtergrams of the quiet chromosphere,
taken at seven wavelengths in Hα with the aid of a computer-
controlled 1/8 Å filter, have been used to derive the contrast
of ten bright and dark mottles as functions of wavelength. The
contrast profiles of bright and dark mottles are strikingly different. Comparison between observation and theory yields
values for the source function, optical thickness, line broadening parameter, and line-of-sight velocity for both bright and
dark mottles.

**073.086 High-resolution photography of the solar chromo-
   sphere. XI: Hα contrast profiles of mottles near the
limb.**   R. E. Loughhead.
Solar Physics, Vol. 29, 327 - 332 (1973).
   Curves showing the contrast of individual bright and dark
mottles near the limb at various wavelengths in the Hα line
have been plotted from observations made through a 1/8 Å
filter. The results cannot be explained in terms of the 'cloud'
model of mottles proposed by Beckers (1964) and employed
with some success by Grossmann-Doerth and von Uexküll
(1971) and Bray (1973) to interpret the profiles of bright and
dark mottles near the centre of the disk.

073.087 Comments on the Zirin-Frazier controversy.
   C.-C. Cheng, K. J. H. Phillips, A. M. Wilson.
Solar Physics, Vol. 29, 383 - 384 (1973). – Research Note.

**073.088 Observations of an active limb prominence in the
   Hβ line.**   J. Burns.
Solar Physics, Vol. 29, 403 - 408 (1973).
   Measurements of the Hβ line intensity in a large active
prominence indicate that α, the fraction of length in the line
of sight which contains emitting material, is less or equal to
0.1.

073.089 A comment on the flare activity in August 1972.
   L. Fritzová-Svestková, Z. Švestka.
Solar Physics, Vol. 29, 417 - 419 = Mitt. Fraunhofer Inst.,
Freiburg, No. 120 (1973). – Research Note.

073.090 On the ionisation of hydrogen in optical flares.
   J. C. Brown.
Solar Physics, Vol. 29, 421 - 427 (1973).
   Non-steady state and non-LTE effects on the ionisation
equilibrium of hydrogen in optical flares are considered in
terms of a two-level hydrogen atom. It is shown that, just as
in the quiet low chromosphere, the ionisation equation is controlled by spontaneous recombination to the second level
and by photoionisation from this level by photospheric radiation, and is independent of the nature of the flare energy input mechanism.

**073.091 A possible mechanism for acceleration of eruptive
   prominences.**   Lyu Van Lyong.
Astron. Zhurn. Akad. Nauk SSSR, Vol. 50, 549 - 557 (1973).
In Russian. English translation in Soviet Astron. AJ, Vol. 17,
No. 3.
   The motion of eruptive prominences is explained by
stretching out of strong magnetic lines of a coronal field which
supported the prominence when it was a quiescent one.

**073.092 La zone de transition chromosphère-couronne, une
   région de l'atmosphère solaire encore bien embarrassante.**   J.-P. Rozelot.
L'Astronomie, 87ᵉ année, p. 234 - 240 (1973).

073.093 Radio emissions from solar flares.   A. Maxwell.
   Sky Telescope, Vol. 46, 4 - 9 (1973).

**073.094 New observations of the solar ultraviolet chromo-
   sphere.**
E. C. Bruner, Jr., R. W. Parker, E. Chipman, R. Stevens.
Astrophys. Journ., (*Letters*), Vol. 182, L33 - L34 (1973).
   The authors present some of the results of a rocket flight
which obtained a stigmatic spectrum of the sun in the region
1190 to 1320 Å. Lines which are formed in the chromosphere
and transition zone show strong fluctuations with position on
the disk. The correspondence between the H Lα profile and
chromospheric details seen in the Ca K-line is demonstrated.

073.095 Die Struktur der solaren Chromosphäre.
   U. Grossmann-Doerth, M. von Uexküll.
Mitt. Astron. Ges., No. 32, p. 187 - 194 (1973).

**073.096 Distribution of marked prominences according to
   heliographic latitude on both hemispheres of the
sun during the 11-year solar activity cycles No. 18 and No. 19.**
A. Bonov, V. Dermendzhiev.
Izv. Sekts. astron. Blg. AN, Vol. 5, 21 - 31 (1972). In Bulgarian. – Abstr. in Referativ. Zhurn. 51. Astron., 5.51.539 (1973).

**073.097 Some studies on solar optical flares reported under
   new classification.**
R. K. Mitra, S. K. Sarkar, M. K. Das Gupta.
Indian Journ. Radio and Space Phys., Vol. 1, 170 - 174
(1972).
   Solar optical flares, numbering 8230, reported during the
period September 1967 to November 1969, which covers the

peak phase of the current solar cycle were examined in relation to (1) percentage occurrences under different classes, (2) position on the solar disk, (3) rise-time and duration, and (4) associated visual indications, numbering 26.

073.098 Enrichment of heavy nuclei in the 17 April 1972 solar flare. R. L. Fleischer, H. R. Hart, Jr.
Phys. Rev. Letters, Vol. 30, 31 - 34 (1973).

Polycarbonate and glass detectors exposed on Apollo 16 to the 17 April 1972 solar flare were used to measure the spectrum of iron-group cosmic-ray nuclei down to ~0.02 MeV/ nucleon. The enrichment of iron relative to lighter nuclei previously seen at higher energies increases markedly in this new, very-low-energy region.

073.099 Solar prominences 1964–1971. M. Waldmeier.
Astron. Mitt. Sternw. Zürich, No. 316, 12 pp. (1972).

The latitude distribution of prominence areas is given month by month for the years 1964 through 1971. The development of the polar zone and their poleward migration is shown clearly. As a consequence of a phase shift between the activities of the two hemispheres, the southern polar zone is, relative to the northern one, delayed by about one year. The most remarkable fact is a multiple structure of the polar zone.

073.100 Charge states and energy-dependent composition of solar-flare particles.
D. Braddy, J. Chan, P. B. Price.
Phys. Rev. Letters, Vol. 30, 669 - 671 (1973).

During a weak solar flare energy spectra of He, O, and Fe from 0.2 to ~ 30 MeV per nucleon were measured with glass and plastic detectors exposed on the Apollo 16 spacecraft. The spectra were very steep and the abundance ratios Fe/O and O/He decreased rapidly with energy.

073.101 Physics of solar flares. E. Tandberg-Hanssen.
Earth Extraterr. Sci., Vol. 2, 89 - 98 (1973).

Discusses the flare phenomenon in the framework of the preexisting magnetic fields and distinguish three flare components, the 'plage flare', the 'prominence flare' and the 'sunspot flare'.

073.102 A morphological study of solar spicules.
D. K. Lynch, J. M. Beckers, R. B. Dunn.
Solar Physics, Vol. 30, 63 - 70 (1973).

From improved spicule filtergrams obtained with the Sacramento Peak vacuum telescope we measured some spicule properties. The spicule diameter of 950 km was well resolved. A small decrease of diameter with height was observed confirming older observations. The expansion of the spicule was found to be at least an order of magnitude less than reported by Mouradian. Spicule counts are very sensitive to the threshold intensity of the observations. Counts, and their dependence on threshold intensity, height and wavelength are reported.

073.103 Spectroscopic investigation of the chromosphere. III: Hα line profile from the interior of supergranular cells. U. Grossmann-Doerth, M. von Uexküll.
Solar Physics, Vol. 30, 71 - 74 = Mitt. Fraunhofer Inst., *Freiburg*, No. 117 (1973).

The line profile of Hα as emanating from the interior of supergranular cells was measured at $\sin \theta = 0$, 0.6, 0.8 and 0.9. The measurements are described and the results presented.

073.104 A kinematic model of a solar flare.
Y. Nakagawa, S. T. Wu, S. M. Han.
Solar Physics, Vol. 30, 111 - 120 (1973).

The possibility of identifying the optical flare with the response of the chromosphere to a supersonic disturbance, i.e., a shock, propagating downward is examined. The undisturbed chromosphere is represented by the Harvard-Smithsonian Reference Atmosphere and the evolution of the shock is evaluated with the use of the CCW (Chisnell, Chester, Whitham) approximation based on the theory of characteristics. It is shown that the chromosphere is heated by the shock and that radiation is enhanced, and that the enhanced radiation terminates the shock around the height of the temperature minimum. Numerical results obtained and possible future improvements of this type of study are discussed.

073.105 Multidirectional scanning of active regions with a slit-jaw spectrograph and a solar chromatograph. I: Description of the method and some preliminary results for the flare event of August 4th 1972.
U. Kusoffsky, G. Pålsgård.
Solar Physics, Vol. 30, 121 - 127 (1973).

At the Swedish Solar Observatory in Anacapri we have simultaneously used the following combination of instruments in our investigation of active regions: (1) A spectrograph with an image rotator placed in front of the slit; (2) A subtractive double dispersive spectrograph (solar chromatograph); (3) A Hα ± 0.5 Å patrol instrument. Scans over the 3b flare of August 4th 1972 are used to illustrate the method. The illustrations clearly show downflowing matter connected with bright knots and filaments in the emitting area, possibly in accordance with Hyder's infall-impact mechanism.

073.106 The limb flare of August 11, 1972.
M. Waldmeier.
Solar Physics, Vol. 30, 129 - 137 = Astron. Mitt. Sternw. Zürich, No. 318 (1973).

A limb flare is described that occurred above a complex and very active sunspot. Four stages can be distinguished: the flash-phase, the spray-phase, the surge-phase and the loop-phase. Each of them had a duration that was longer than that of the preceding one. In the spray ascending speeds up to 745 km s^{-1} and accelerations up to 1.3 km s^{-2} were recorded. The loop-phase has been observed in the coronal lines 5303 and 5694 Å. The yellow line, being very weak before the flare, became extremely strong in the loop and surpassed five times the intensity of the green line. X-ray bursts and ionospheric disturbances of long duration demonstrate that not only the flare itself but also the loop was a source of X-rays. Most of the radio-bursts can be ascribed to specific features in the Hα-records of the event.

073.107 Possible mechanism of surge formation in the solar atmosphere. Yu. V. Platov, B. V. Somov, S. I. Syrovatskii.
Solar Physics, Vol. 30, 139 - 147 (1973).

The possibility of surge formation as a result of plasma 'raking-up', the latter being associated with the growth of the local magnetic field in the solar atmosphere, is considered in this paper. The question is treated numerically in the MHD approximation for the case of a dipolar magnetic field. It is shown that the field growth results in the appearance of relatively dense condensations stretched along the axis of the dipole. Simultaneously the plasma acquires the upward velocity along force lines which leads to the formation of a surge. Some properties of this surge model are discussed.

073.108 On long gravitational waves in the chromospheric temperature minimum. V. V. Kasinskij.
Issled. po geomagnetizmu, aehron. i fiz. Solntsa. Vyp. (No.) 26. Moskva, Nauka, 1973, p. 38 - 53. In Russian. – Abstr. in Referativ. Zhurn. 51. Astron., 7.51.371 (1973).

073.109 Non-thermal electron population in solar flares.
H. S. Hudson.

IV Leningr. mezhdunar. seminar "Edinoobrazie uskoreniya chastits v razlich. masshtabakh kosmosa, 1972". Leningrad, 1972, p. 127 - 143. In Russian. – Abstr. in Referativ. Zhurn. 51. Astron., 7.51.423 (1973).

073.110 Temperature and X-ray radiation of plasma during the injection of electrons accelerated in solar flares. Yu. N. Starbunov, Yu. E. Charikov.
IV Leningr. mezhdunar. seminar "Edinoobrazie uskoreniya chastits v razlich. masshtabakh kosmosa, 1972". Leningrad, 1972, p. 145 - 165. In Russian. – Abstr. in Referativ. Zhurn. 51. Astron., 7.51.424 (1973).

073.111 On the role of sunspot-group satellites as proton flare predecessors. V. V. Kasinskij.
Issled. po geomagnetizmu, aehron. i fiz. Solntsa. Vyp. (No.) 26. Moskva, Nauka, 1973, p. 118 - 128. In Russian. – Abstr. in Referativ. Zhurn. 51. Astron., 7.51.428 (1973).

073.112 Some regularities of time sequences of flares. V. A. Belyaev, G. V. Kuklin.
Issled. po geomagnetizmu, aehron. i fiz. Solntsa. Vyp. (No.) 26. Moskva, Nauka, 1973, p. 105 - 117. In Russian. – Abstr. in Referativ. Zhurn. 51. Astron., 7.51.444 (1973).

General bibliography of solar prominence research 1880 - 1970. See Abstr. 003.151.

A possible solar electrograph. See Abstr. 034.014.

Resistive diffusion of force-free magnetic fields in a passive medium. See Abstr. 062.013.

On steady magnetic-field reconnection. See Abstr. 062.014.

Kritik an Syrovatskiis Mechanismus der dynamischen Dissipation von Magnetfeldern. See Abstr. 062.052.

Stellar model chromospheres. I. On the temperature minima of F, G, and K stars. See Abstr. 064.025.

The absorption spectrum of atmospheric water vapor in the vicinity of the He 10830 Å triplet. See Abstr. 071.002.

Solar oscillations at 9.6 mm. See Abstr. 071.014.

Structural changes and regularities in the distribution of calcium flocculi on the solar surface in the course of cycle 19. See Abstr. 071.022.

Observation and interpretation of phase lags in the five-minute oscillation. See Abstr. 071.033.

Spectral analysis of sunspot flares. See Abstr. 072.040.

Introductory review of solar activity. See Abstr. 072.046.

Magnetic fields and proton flares – 7 July and 2 September 1966. See Abstr. 072.050.

The latitudinal zonation on the sun. See Abstr. 072.063.

Thermal instability of coronal neutral sheets and the formation of quiescent prominences. See Abstr. 074.002.

Some comments on the low intensity Hα emission observed by J.-L. Leroy in the solar corona. See Abstr. 074.013.

Thick target X-ray bremsstrahlung from partially ionized targets in solar flares. See Abstr. 076.003.

Superthermal plasma nodules and their relation to solar flares. See Abstr. 076.005.

Solar gamma ray lines observed during the solar activity of August 2 to August 11, 1972. See Abstr. 076.008.

The solar albedo of hard X-ray flares. See Abstr. 076.014.

X-ray line emission associated with solar flares. See Abstr. 076.016.

X-ray observations from the August 2, 1972 flare. See Abstr. 076.017.

Impulsive EUV spectra of solar flares. See Abstr. 076.018.

On the determination of non-thermal electron spectra in solar flares from the observed hard X-ray emission. See Abstr. 076.019.

Spatial distribution of soft X-ray and EUV emission associated with a flare of importance 1B on August 2, 1972. See Abstr. 076.022.

Extreme ultraviolet emission from chromospheric inhomogeneities. An analysis of the extreme ultraviolet flash spectrum of the sun. See Abstr. 076.028.

Continuous energy injection at numerous bright points during soft X-ray flare enhancement. See Abstr. 076.030.

19 – 20 May 1969, an example of type III emission during the impulsive phase of flares. See Abstr. 076.032.

Development of moving type IV solar radio bursts and relation to expanding magnetic bottles from flare regions. See Abstr. 077.042.

A correlation between 10.7 cm (2800 MHz) solar radio emission and chromospheric flares. See Abstr. 077.045.

Energetic solar particles and their relation to optical flares. See Abstr. 078.001.

Estimation of an upper limit for the solar neutron emission during large flares. See Abstr. 078.002.

Detection of relativistic solar particles before the Hα maximum of a solar flare. See Abstr. 078.007.

The differential energy spectra of solar-flare ^{1}H, ^{3}He, and ^{4}He. See Abstr. 078.013.

Solar flare particle propagation: Comparison of a new analytic solution with spacecraft measurements. See Abstr. 078.017.

Solar-flare generated cosmic ray emission of 24 January 1971. See Abstr. 078.022.

The continuous emission of low energy cosmic rays during solar flares.   See Abstr. 078.027.

Several solar aspects of flare-associated particle events.   See Abstr. 078.029.

Propagation of particles from solar flares in a medium with a sharply changing diffusion coefficient. See Abstr. 078.037.

A comparison of theoretical and experimental estimates of the solar proton diffusion coefficient during three flare events.   See Abstr. 078.038.

Generation of electrons and protons during solar flares.   See Abstr. 078.040.

Corpuscular radiation and active prominences. See Abstr. 078.041.

Solar flare cosmic rays at and beyond the modulation boundary.   See Abstr. 078.046.

Acceleration of relativistic particles by shock waves in interplanetary space.   See Abstr. 078.066.

Study of solar flares, cosmic dust and lunar erosion with vesicular basalts.   See Abstr. 094.276.

Exposure ages of Apollo 15 samples by means of microcrater statistics and solar flare particle tracks. See Abstr. 094.277.

Stable rare gas isotopes produced by solar flares in single particles of Apollo 11 and Apollo 12 fines. See Abstr. 094.429.

Nuclear track studies of dynamic surface processes on the moon and the constancy of solar activity. See Abstr. 094.504.

Solar flares, the lunar surface, and gas-rich meteorites.   See Abstr. 094.518.

On the configuration of interplanetary shock waves produced by powerful chromospheric flares (based on space probes).   See Abstr. 106.001.

Errata

073.901   Erratum: 'Current limitation in solar flares'
          [Astrophys. Journ., Vol. 176, 487 - 496 (1972)].
D. F. Smith, E. R. Priest.
Astrophys. Journ., Vol. 180, 667 (1973).

074   Solar Corona, Solar Wind

074.001· **On the kinetic temperature of He^{++} in the solar wind.**   A. Barnes, R. J. Hung.
Cosmic Electrodynamics, Vol. 3, 416 - 436 (1973).

We discuss several processes which may influence the temperature of He^{++} in the solar wind, and consider whether they can account for the observations that $T_{He} \sim 4T_p$. Preferential heating of helium by nonhydromagnetic modes may be required to raise T_{He} above T_p enough that hydromagnetic-wave dissipation would preferentially raise T_{He}. Another process that might tend to make $T_{He} \sim 4T_p$ is pitch-angle scattering of He and protons in colliding streams.

074.002 **Thermal instability of coronal neutral sheets and the formation of quiescent prominences.**
M. A. Raadu, M. Kuperus.
Solar Physics, Vol. 28, 77 - 94 (1973).

By developing the theory of the structure and evolution of current sheets under coronal conditions we can attempt to gain a comprehensive understanding of the structure, evolution, and mass and energy balance of quiescent prominences. A stability analysis for coronal material permeated by a vertical magnetic field rooted in the photosphere, indicates that a condensation will take the form of a thin vertical wedge of cool matter. The development of a finite condensation is followed and it is shown that photospheric line tying is only important in the initial stages. A perturbation analysis of vertical motions at the neutral sheet shows that thermal instability can lead to overstable oscillations. Cooling of coronal material can lead to both upward and downward mass motions, and gravitational energy release is important to the thermal balance of prominences. Relevant optical and radio observations are discussed.

074.003 **The solar wind and the temperature-density structure of the solar corona.**   G. W. Pneuman.
Solar Physics, Vol. 28, 247 - 262 (1973).

The influence of the solar wind on large-scale temperature and density distributions in the lower corona is studied. This influence is most profoundly felt through its effect upon the geometry of coronal magnetic fields since the presence of expansion divides the corona into magnetically 'open' and 'closed' regions. An approximate method for calculating the temperature and density distribution in a known magnetic field geometry is outlined and numerical estimates are carried out for representative coronal conditions.

074.004 **The heating of the solar corona. I. Observation of ion energies in the transition zone.**
B. C. Boland, S. F. T. Engstrom, B. B. Jones, R. Wilson.
Astron. Astrophys., Vol. 22, 161 - 169 (1973).

A high resolution solar spectrum has been obtained over the range 1200−2200 Å using an echelle spectrograph launched in a sun-pointing Skylark rocket. Profiles of emission lines were recorded due to the ions Si II, C II, Si III and C IV, embracing the temperature range $10^4–10^5$ °K. The measured Doppler widths correspond to "temperatures" very much in excess of the electron temperatures deduced from ionization balance theory. Since the ion-electron relaxation times are very short, this clearly demonstrates the presence of a nonthermal energy component in the particular region of the transition zone.

074.005 **On the relation between solar wind structure and solar wind rotational and tangential discontinuities.**
J. M. Turner.
Journ. Geophys. Res., Vol. 78, 59 - 70 (1973).

Solar wind plasma and magnetic field data were used to identify rotational and tangential discontinuities during the first 40 days of the flight of Mariner 5. Of the 40 rotational discontinuities found, 36 were clustered in three distinct 3- to 6-day intervals. These three intervals were characterized by high solar wind bulk velocities, high magnetic field magnitudes, low densities, high correlation between velocity and magnetic field changes, and the presence of smooth Alfvén waves. An examination of the qualitative features of fast and slow plasma streams revealed differences in the behavior of the magnetic field and velocity and their fluctuations, the field magnitude, and the density. It is suggested that these differences are correlated with the occurrence of rotational discontinuities.

074.006 **Interaction of singly charged interstellar helium ions with the solar wind.**
C. S. Wu, R. E. Hartle, K. W. Ogilvie.
Journ. Geophys. Res., Vol. 78, 306 - 309 (1973). – Letter.

074.007 **Turbulent heating of colliding streams in the solar wind.**   M. L. Goldstein, A. Eviatar.
Astrophys. Journ., Vol. 179, 627 - 636 (1973).

Turbulent heating of colliding plasma streams has previously been observed in the solar wind. The original data were interpreted in terms of a fluid model. We argue that a plasma-kinetic description is the more appropriate theoretical approach and is necessary in order to better understand the microscopic physical phenomena that underlie all fluid models. We used microscopic solar-wind parameters characteristic of conditions during the observations, together with the quasi-linear plasma-kinetic theory, to compute the expected magnetic field and temperature enhancements in the interaction region between two counterstreaming plasma beams.

074.008 **Heat conductivity, plasma instabilities, and radio star scintillations in the solar wind.**   F. Perkins.
Astrophys. Journ., Vol. 179, 637 - 642 (1973).

The motion of a typical solar-wind electron is periodic because of the combination of magnetic mirror and electrostatic reflections. The periodic motion reduces the electron heat conductivity from the Spitzer formula and causes magnetoacoustic waves with an inward-directed phase velocity to become unstable. A nonlinear theory of the instability quantitatively accounts for radio-star scintillation data and an increased rate of energy exchange between electrons and ions above the Spitzer value.

074.009 **Electrostatic turbulence at colliding plasma streams as the source of ion heating in the solar wind.**
K. Papadopoulos.
Astrophys. Journ., Vol. 179, 931 - 938 (1973).

We propose a mechanism for the nonthermal ion heating observed in colliding solar-wind streams at 1 a.u. based on an electrostatic, short-wavelength instability between the ions in the observed colliding plasma streams. The predictions are consistent with observations.

074.010 **Nonthermal turbulent heating in the solar envelope.**   K. Papadopoulos.
Astrophys. Journ., Vol. 179, 939 - 947 (1973).

It is the purpose of the present note to point out that compressive MHD pulses, in the form of fast magnetosonic waves or solitons, can produce a strong electron-ion coupling capable of maintaining $T_e \sim T_p$, and provide a nonthermal heat source in various astrophysical plasmas. Nonthermal heating in the solar envelope is used as a typical example.

074.011 Dissipation of hydromagnetic waves with application to the outer solar corona. I. Collisionless protons and collisional electrons. R. J. Hung, A. Barnes.
Astrophys. Journ., Vol. 180, 253 - 269 (1973).

We discuss the theory of hydromagnetic-wave dissipation for conditions likely to occur in most of the outer solar corona ($2\,R_\odot \lesssim r \lesssim 4\,R_\odot$). Magnetoacoustic waves of period ~5 min are dissipated by ion Landau damping, just as in a perfectly collisionless plasma, even though ion-electron collisions are not neglected. For certain propagation directions, dissipation by (collisional) electron thermal conduction may occur. The implications of these effects for heating the outer corona are discussed.

074.012 Dissipation of hydromagnetic waves with application to the outer solar corona. II. Transition from collisional to collisionless electrons. R. J. Hung, A. Barnes.
Astrophys. Journ., Vol. 180, 271 - 284 (1973).

We discuss the theory of hydromagnetic wave dissipation in a region of transition between collisional and collisionless electron behavior. For waves of period 5 min, the period of maximum photospheric noise, this transition occurs in the solar corona at roughly $4\,R_\odot$ heliocentric distance. Electron viscous dissipation on the collisional side of the transition passes continuously into electron Landau damping as the electron collision frequency passes through the wave frequency.

074.013 Some comments on the low intensity Hα emission observed by J.-L. Leroy in the solar corona. Y. Öhman.
Solar Physics, Vol. 28, 399 - 402 (1973).

Some comments are presented on the important observations of faint prominences made recently by Leroy at the Pic du Midi Observatory. The writer draws attention to the very probable connection with faintly luminous Hα obscuring prominences which appear sometimes as dark lanes and markings in ordinary prominences.

074.014 Energy budget in coronal holes. G. Noci.
Solar Physics, Vol. 28, 403 - 407 (1973).

It is shown that the constancy of the ratio between conductive flux and pressure squared as one goes from quiet regions to 'holes' (regions of exceptionally low density and temperature) in the solar corona, observed in the case of the first well-studied coronal hole, implies that a strong solar wind is likely to originate in coronal holes.

074.015 Improved three-dimensional mapping of the electron density distribution of the solar corona. R. M. Perry, M. D. Altschuler.
Solar Physics, Vol. 28, 435 - 456 (1973).

Three-dimensional maps of the distribution of coronal electron density can now be computed with two radial functions in the series expansion for the density. With the improved maps we can determine the topological variation of the electron density with radial distance, and thus can (1) distinguish coronal condensations from coronal streamers, (2) trace the structure of a streamer as a function of height, and (3) determine the non-radial orientation of a streamer. We summarize the previous work in concise mathematical notation, show examples of the improved maps derived from two radial functions, and discuss in detail the expectations and limitations of the method.

074.016 On the Aller's admixture radiation effect during the compression process in the solar corona and generation of coronal formations. R. E. Guseinov.
Solar Physics, Vol. 28, 457 - 463 (1973).

The cooling due to Aller's admixture radiation in the energy balance equation describing the coronal gas condensation is taken into account. It is shown that the compression

mechanism does not apply for the explanation of the quiescent prominences formation but is quite suitable for definite kinds of active prominences and flares of coronal origin.

074.017 On an anomalous polarization of the corona. M. M. Molodensky.
Solar Physics, Vol. 28, 465 - 475 (1973).

At present, data exist showing that in some regions of the corona the polarization degree has been found to be higher than the maximum possible value determined by Thomson scattering. Besides this, there exist regions where the direction of the prevailing vibration of the E-vector does not coincide with the tangential one. This may be caused by the velocity of scattered electrons. The theory of polarization taking the velocity into account is given, and the above-mentioned data are discussed.

074.018 An investigation of the structure of coronal active regions. J. H. Parkinson.
Solar Physics, Vol. 28, 487 - 493 (1973).

The temperature and density structure of a 'typical' coronal active region is deduced from X-ray observations of several active regions. An emission measure-temperature distribution is deduced from high resolution X-ray spectra obtained with a rocket observation of two similar regions. These observations are combined to give a model of a 'typical' active region, the temperature varying from 2 to 6×10^6 K with corresponding densities between 2×10^9 and 10^{10} cms^{-3}.

074.019 Flare-produced coronal MHD-fast-mode wavefronts and Moreton's wave phenomenon. Y. Uchida, M. D. Altschuler, G. Newkirk, Jr.
Solar Physics, Vol. 28, 495 - 516 (1973).

The propagation characteristics of MHD fast-mode disturbances, which can emanate from flare regions, are computed for realistic conditions of the solar corona at the times of particular flares. We use the coronal (electron) density distribution calculated from daily K-coronameter data, and the coronal magnetic field calculated under the current-free approximation from magnetograph measurements of the photospheric magnetic field. We compare the path and time-development of an MHD fast-mode wavefront emitted from the flare region (as calculated from a realistic model corona for the day of the observed Moreton wave event) with actual observations of the Moreton wave event.

074.020 Le vent solaire: Un bref aperçu. J.-P. Rozelot.
L'Astronomie, 87ᵉ année, p. 85 - 89 (1973).

074.021 On the variation of the coronal λ5303 intensity relative to the interplanetary and solar magnetic sector structure, and to geomagnetic activity. A. Gulbrandsen.
Planet. Space Sci., Vol. 21, 703 - 707 (1973).

An analysis on the variation of coronal λ5303 intensity relative to the solar magnetic sector boundaries is presented. The location of the boundaries has been extrapolated from the observed interplanetary sector structure. The results indicate that in the years 1962−1964 the solar activity is in general high to the west and low to the east of a solar sector boundary.

074.022 Direct measurements of solar-wind fluctuations between 0.0048 and 13.3 Hz. T. W. J. Unti, M. Neugebauer, B. E. Goldstein.
Astrophys. Journ., Vol. 180, 591 - 598 (1973).

Power spectra to a frequency of 13.3 Hz have been calculated from the OGO-5 Faraday cup measurements of the ion-charge flux of the solar wind. Twenty-five out of 32 spectra are fairly regular and are presented in linearized form. Their power levels and slopes are consistent with previously reported density spectra at much lower frequencies.

074.023 Dissipation of hydromagnetic waves with application to the outer solar corona. III. Transition from collisional to collisionless protons. R. J. Hung, A. Barnes.
Astrophys. Journ., Vol. 181, 183 - 208 (1973).

We develop the theory of hydromagnetic wave dissipation in a region of transition between collisional and collisionless proton behavior. Dissipation by proton-proton collisions and electron thermal conduction on the collisional side of the transition passes continuously into proton Landau damping on the collisionless side of the transition. For waves of period 5 min in the solar corona, this transition occurs within 2 $R_\odot$ heliocentric distance.

074.024 Observations of the F corona and inner zodiacal light during the 1972 July 10 total solar eclipse.
M. T. Sandford II, J. K. Theobald, H. G. Horak.
Astrophys. Journ., (*Letters*), Vol. 181, L15 - L17 (1973).

A rocket-borne, four-color, six-channel polarimeter launched from Poker Flat, Alaska, traversed the moon's umbra during the 1972 July 10 solar eclipse. The brightness and linear polarization of the F corona and inner zodiacal light were measured from altitudes ranging between 310 and 330 km at solar elongations to 30° east and west of the sun.

074.025 Polarization measurements in the green coronal line. D. L. Mickey.
Astrophys. Journ., (*Letters*), Vol. 181, L19 - L21 (1973).

Linear polarization up to 15 percent in the Fe XIV $\lambda 5303$ line has been measured photoelectrically in the solar corona. The polarization deviates only slightly from the radial direction and increases with increasing height.

074.026 Excitation of the Fe XIII spectrum in the solar corona. D. R. Flower, G. Pineau des Forêts.
Astron. Astrophys., Vol. 24, 181 - 192 (1973).

The statistical equilibrium equations of the $^3P_{0,1,2}$, $^1D_2$ and $^1S_0$ levels of the ground $3s^2\ 3p^2$ configuration of Fe^{+12} have been solved for the ranges of electron temperature, electron density and geometrical dilution factor believed to be relevant to Fe XIII emission from the solar corona. Particular attention is paid to assessing the remaining uncertainties in the atomic parameters.

074.027 Fe XVII emission from the solar corona.
M. Loulergue, H. Nussbaumer.
Astron. Astrophys., Vol. 24, 209 - 213 (1973).

Observations of Fe XVII in the 13–17 Å region from several rocket flights are combined and compared against theoretical intensities in the individual lines. The emission observed at 17.06 Å and attributed to an electric dipole intercombination transition is shown to be a blend of two lines at 17.05 Å and 17.10 Å; and the emission at 13.8 Å is a blend of two lines at 13.82 Å and 13.88 Å. The intensity ratios among the most commonly observed lines are insensitive to variations in T_e or N_e, the absolute emissivities are given as a function of these parameters.

074.028 Photometric analysis of monochromatic photographs of the solar corona taken in the green line (5303 Å) and the red line (6374 Å).
J. P. Picat, B. Fort, M. Dantel, J. L. Leroy.
Astron. Astrophys., Vol. 24, 259 - 265 (1973).

We describe a photometric analysis of monochromatic photographs of the solar corona taken in the green line (5303 Å) and the red line (6374 Å) with a coronograph at the Pic-du-Midi Observatory. We first describe the morphology of the structures observed in the red line and the limits to the observations in the red and green lines, and then present intensity maps (in units of the mean photospheric radiation intensity) of the corona. We can thus present a series of numerical results which are independent of all theoretical assumptions (except that only one structure is observed along the line of sight) and provide the data for a study of the physical conditions in the coronal plasma. We also give the intensity gradients observed in the red and green lines corresponding to both the general coronal emission and to the emission of the individual structures.

074.029 Coronal densities and temperatures derived from monochromatic images in the red and green lines.
B. Fort, J. P. Picat, M. Dantel, J. L. Leroy.
Astron. Astrophys., Vol. 24, 267 - 273 (1973).

In a preceding article (Picat et al., 1973), the authors presented the results of a detailed photometric analysis of monochromatic photographs in the red and green coronal lines taken with the coronograph of the Pic-du-Midi Observatory. Starting from these results, we calculate the electron densities and temperatures of the observed structures (arches and fans) using the most recent solutions of the ionisation and statistical equilibrium equations for Fe X and Fe XIV in coronal conditions.

074.030 Airborne white light polarimetry of the outer corona during July 1972 solar eclipse.
C. F. Keller.
Bull. American Astron. Soc., Vol. 5, 19 - 20 (1973). – Abstr. AAS.

074.031 The solar wind in the outer solar system.
M. D. Montgomery.
Space Sci. Rev., Vol. 14, 559 - 575 (1973).

The properties of the solar wind including magnetic fields, plasma, and plasma waves are briefly reviewed with emphasis on conditions near and beyond the orbit of Jupiter. An extrapolation of the steady-state wind to large distances, evolution of disturbances and structure, modulation of cosmic rays, interactions with planetary bodies (bow shocks and magnetosheaths), and interactions with interstellar neutral helium and hydrogen are briefly discussed. Some comments on instrumentation requirements to observationally define the above phenomena are also included.

074.032 The effect of instationarity in the chemical composition of plasma streams from the sun.
I. S. Veselovsky.
Geomagn. Aeronom., Vol. 13, 166 - 168 (1973). In Russian. Brief information.

074.033 Detection of solar wind at synchronous orbit.
S. E. DeForest.
Journ. Geophys. Res., Vol. 78, 1195 - 1197 (1973). – Letter.

074.034 On the March 7–8, 1970, event. V. Formisano.
Journ. Geophys. Res., Vol. 78, 1198 - 1202 (1973). Letter.

074.035 Abundances in the solar corona. H. Olthof.
Atoms and molecules in astrophysics. Proc. Scottish Univ. Summer School in Physics 1971, (see 012.005), p. 321 - 325 (1972).

074.036 Test for detection of fine structure of the solar wind velocity. N. A. Lotova, I. V. Chashey.
Astrophys. Space Sci., Vol. 20, 251 - 262 (1973).

A new method of search and analysis of the fine structure in the velocity of interplanetary plasma irregularities is developed.

074.037 Observations of the "white-light" corona during partial phases of the solar eclipse on July 10, 1972.
G. M. Nikolsky.
Solnechnye Dannye 1972 Byull., No. 12, p. 96 - 97 (1973).

In Russian.

The "white-light" solar corona was visible up to a height of $0.5-1\ R_\odot$ a few minutes before and after the totality in observations with a small Lyot coronograph. The "double-eclipse" method is found to be useful for solar corona investigations.

074.038 Effect of the general magnetic field of the sun on the geoefficiency of the solar wind.
M. I. Pudovkin, E. V. Pudovkina.
Solnechnye Dannye 1972 Byull., No. 12, p. 106 - 110 (1973). In Russian.

Periodic variations of the geoefficiency of the solar wind are considered. A distinct cyclic modulation of the efficiency of the solar wind is shown to exist. This modulation may be explained as result of the existence of the altering quasi-dipolar magnetic field of the sun.

074.039 The height gradient of the coronal emission line 5303 Å.   V. Rušin.
Bull. Astron. Inst. Czechoslovakia, Vol. 24, 121 - 129 (1973).

The paper treats the data of surface photometry, obtained from the spectrogrammes of the coronal station at Lomnický Štít, over the period 1966 – 1971. If the gradient is determined from the relation $G = \Delta (\log I)/\Delta h$, its magnitude fluctuates in relatively wide limits and differs considerably from the values obtained by other authors. It was found that this wide scale of gradient values is not random, but that it corresponds to the different degrees of solar activity and their manifestations in the solar corona.

074.040 Alfvén waves in a two-fluid model of the solar wind.
J. V. Hollweg.
Astrophys. Journ., Vol. 181, 547 - 566 (1973).

The author presents a two-fluid model for the solar wind which includes the presence of Alfvén waves which originate at the sun. The effective pressure of the Alfvén waves is included, as well as a model representation for proton heating via nonlinear damping of the Alfvén waves. The effects of rotation in the solar equatorial plane are self-consistently included.

074.041 Method of precise determination of the position of the polarization plane in the corona.
M. M. Molodensky.
Solnechnye Dannye 1973 Byull., No. 1, p. 97 - 99 (1973). In Russian.

074.042 Observations of the inner F and K coronas below λ 2220.   F. Q. Orrall, R. J. Speer.
Solar Physics, Vol. 29, 41 - 46 (1973).

The inner coronal continuum has been observed and measured below λ 2220 on the slitless spectra obtained by Speer *et al.* (1970) at the 7 March 1970 eclipse. These observations set some constraints on the brightness of the inner F corona and hence on the scattering efficiency of the inner interplanetary dust cloud particles in the far ultraviolet. They neither confirm nor reject the possibility that the inner dust cloud has the same sharp upturn in scattering efficiency below λ 2000 observed in the zodiacal light and in the interstellar medium.

074.043 Solar wind density model from km-wave type III bursts.   H. Alvarez, F. T. Haddock.
Solar Physics, Vol. 29, 197 - 209 (1973).

The analysis of type III bursts observed from the OGO-5 satellite between 3.5 MHz and 50 kHz (λ 6 km) gives an empirical expression for the frequency drift rate as a function of frequency that is valid from 75 kHz to 550 MHz. Using this expression and some simplifying assumptions we obtain indirectly an empirical formula for the electron density distribution of the solar wind to 1 AU which is consistent with pub-

lished values of electron density and with observed type III burst drift rates.

074.044 Restrictions on radial magnetic field and flow solutions for the solar wind.
M. S. Gussenhoven, R. L. Carovillano.
Solar Physics, Vol. 29, 233 - 241 (1973).

In Parker's original model, the solar wind is represented as a spherically symmetric hydrodynamic flow. The velocity is radially directed and decoupled from the magnetic field. The simple extension of this model to include a dependence on the polar angle, θ, is shown to be invalid for radial flow and radial magnetic field. This work demonstrates how *ad hoc* symmetry conditions imposed to simplify a non-linear problem can be incompatible with the basic hydromagnetic equations.

074.045 Stability of solar wind against electromagnetic streaming instability.   B. Buti.
Astrophys. Journ., Vol. 181, 1055 - 1063 (1973).

The stability of the solar wind in the presence of small concentrations of helium against high-frequency electromagnetic streaming instability is explored. For relative abundances of helium to hydrogen up to 20 percent for the expected magnetic fields, densities, and temperatures of the solar wind, this instability can be triggered only if the relative streaming of protons with respect to α-particles exceeds the electron thermal velocity. The solar wind cannot satisfy this requirement and hence is stable against this instability.

074.046 Nonlinear model of high-speed solar wind streams.
A. J. Hundhausen.
Journ. Geophys. Res., Vol. 78, 1528 - 1542 (1973).

A hydrodynamic model describing the generation and propagation of high-speed plasma streams in the solar wind is presented. The model is based upon numerical integrations of the conservation equations for a time-dependent, spherically symmetric, radial flow of interplanetary plasma.

074.047 Power spectrum of density irregularities in the solar wind plasma.   B. J. Rickett.
Journ. Geophys. Res., Vol. 78, 1543 - 1552 (1973).

Observations of interplanetary scintillations are compared with predictions based on a spatial electron density spectrum extrapolated from the power law shape at wave numbers $10^{-6}-10^{-5}$ rad/km measured from spacecraft. The comparison suggests a model in which the high wave number density spectrum lies below the extrapolated power law but has a relatively flat part before the cutoff.

074.048 On the continuum fluid approach to the solar wind–moon interaction problem.   H. Pérez-de-Tejada.
Journ. Geophys. Res., Vol. 78, 1711 - 1714 (1973). – Letter.

074.049 Prediction of coronal and interplanetary magnetic fields.   K. H. Schatten.
Progr. Astronaut. Aeronaut., Vol. 30, (see 003.009), p. 179 - 196 (1972). – Presented at the AIAA observation and prediction of solar activity conference, Huntsville, Ala., Nov. 16–18, 1970.

074.050 The solar wind: a review.
M. Dryer, S. Cuperman.
Progr. Astronaut. Aeronaut., Vol. 30, (see 003.009), p. 197 - 229 (1972).

074.051 Predicting the solar wind speed.   J. T. Gosling.
Progr. Astronaut. Aeronaut., Vol. 30, (see 003.009), p. 231 - 245 (1972). – Presented at the AIAA observation and prediction of solar activity conference, Huntsville, Ala., Nov. 16–18, 1970.

074.052 **The propagation of Alfvén waves and their directional anisotropy in the solar wind.**
H. J. Völk, W. Aplers.
Astrophys. Space Sci., Vol. 20, 267 - 285 (1973).
The propagation of solar Alfvén waves in interplanetary space is studied in the approximation of geometrical optics. Ray paths and the change of wave vectors and amplitudes along the rays are determined assuming an Archimedean-spiral interplanetary magnetic field. In particular, the Alfvénic fluctuations in the 2 directions perpendicular to the magnetic field direction are calculated under the assumption that the Alfvén waves are produced at the sun and emitted with an isotropic directional distribution from a reference level close to the sun. Our results are compared with spacecraft observations made by Belcher and Davis (1971) that show an anisotropy of a similar character. Finally, arguments are presented to explain the discrepancy between the calculated high anisotropy and the measured low anisotropy in terms of finite amplitude effects and wave-scattering.

074.053 **Wave-trains in the solar wind. I: General theory and its application to an ideal, isotropic, one-fluid plasma.**
D. J. Olbers, A. K. Richter.
Astrophys. Space Sci., Vol. 20, 373 - 389 (1973).
The main results of Whitham's averaged Lagrangian method for the treatment of linear wave-trains in a weakly inhomogeneous, moving medium are presented briefly. This method is then applied to an ideal, isotropic, one-fluid plasma which can be taken for the lowest order approximation for the interplanetary solar wind expansion.

074.054 **Note on the solar wind heating (two fluid models).**
S. Cuperman.
Astrophys. Space Sci., Vol. 20, 519 - 520 (1973). – Research note.

074.055 **Kinetic models of the solar and polar winds.**
J. Lemaire, M. Scherer.
Rev. Geophys. Space Phys., Vol. 11, 427 - 468 (1973).
In this paper the application of the kinetic theory to the collisionless regions of the polar and solar winds is discussed. A brief historical review is given to illustrate the evolution of the theoretical models proposed to explain the main phenomenon and observations. The parallelism between the development of the solar wind models and the evolution of the polar wind theory is stressed especially. A kinetic method, based on the quasi neutrality and the zero current condition in a stationary plasma with open magnetic field lines, is described. The applicability of this approach on the solar and polar winds is illustrated by comparison of the predicted results with the observations. The kinetic models are also compared with hydrodynamic ones.

074.056 **Close connexion between flare-generated coronal and interplanetary shock waves.** S. Pintér.
Nature, Phys. Sci., Vol. 243, 96 - 97 (1973).
The author shows that there is a very close relation between the flare-generated coronal and interplanetary shock waves. He has studied fourteen interplanetary shock waves, whose fronts intersected detectors located on heliocentric space probes. The studied interplanetary shock waves were always connected with flares, accompanied by type II radio bursts. Seven typical cases are listed in a table.

074.057 **Evolution of a magnetic disturbance in the solar corona with a general Ohm's law.**
M. D. Altschuler, D. F. Smith, P. Swarztrauber, E. R. Priest.
Bull. American Astron. Soc., Vol. 5, 268 - 269 (1973). Abstr. AAS.

074.058 **Are coronal holes M regions?**

B. Bell, G. Noci.
Bull. American Astron. Soc., Vol. 5, 269 (1973). – Abstr. AAS.

074.059 **Fast transient events observed in the green coronal emission line.**
H. L. Demastus, W. J. Wagner, R. D. Robinson.
Bull. American Astron. Soc., Vol. 5, 270 - 271 (1973). Abstr. AAS.

074.060 **The 11 August loops, a coronal continuum event.**
R. Fisher.
Bull. American Astron. Soc., Vol. 5, 272 (1973). – Abstr. AAS.

074.061 **The solar cycle variation of the solar wind.**
J. Hirshberg.
Bull. American Astron. Soc., Vol. 5, 273 (1973). – Abstr. AAS.

074.062 **Coronal emission line profile analysis from airborne eclipse observations of 30 May 1965.**
D. H. Liebenberg, R. Bessey, B. Watson.
Bull. American Astron. Soc., Vol. 5, 275 (1973). – Abstr. AAS.

074.063 **Airborne video recorded coronal emission line profiles of λ 5303 at the 10 July 1972 total solar eclipse.**
D. H. Liebenberg, M. Hoffman, W. M. Sanders, J. M. Beckers.
Bull. American Astron. Soc., Vol. 5, 275 (1973). – Abstr. AAS.

074.064 **Coronal magnetic fields and energetic particles.**
G. Newkirk, Jr.
Bull. American Astron. Soc., Vol. 5, 278 (1973). – Abstr. AAS.

074.065 **Coronal magnetic fields and the variation of geomagnetic activity.** L. Oster, M. D. Altschuler.
Bull. American Astron. Soc., Vol. 5, 278 (1973). – Abstr. AAS.

074.066 **The solar wind and the temperature-density structure of the solar corona.** G. W. Pneuman.
Bull. American Astron. Soc., Vol. 5, 279 (1973). – Abstr. AAS.

074.067 **The abundance of iron in the corona.**
A. B. C. Walker, Jr., H. R. Rugge, K. Weiss.
Bull. American Astron. Soc., Vol. 5, 281 - 282 (1973). Abstr. AAS.

074.068 **On solar wind measurements of the Pioneer–Venus mission.** R. R. Lewis.
Bull. American Astron. Soc., Vol. 5, 308 (1973). – Abstr. AAS.

074.069 **Anomalously low proton temperatures in the solar wind following interplanetary shock waves – evidence for magnetic bottles?**
J. T. Gosling, V. Pizzo, S. J. Bame.
Journ. Geophys. Res., Vol. 78, 2001 - 2009 (1973).
Occasionally, anomalously low values of the solar wind proton temperature T_p are observed when the solar wind velocity v is high. A large fraction of such measurements by the Vela 3 satellites follow the passage of interplanetary shocks by some 20–60 hours. Of 24 post-shock events in which v exceeded 400 km sec^{-1} and for which Vela 3 measurements are available, 12 exhibited plasma states of anomalously low T_p, high v. The anomalously low proton temperatures result from the adiabatic cooling of the plasma within the magnetic bottle.

074.070 **Double ion streams in the solar wind.** W. C. Feldman, J. R. Asbridge, S. J. Bame, M. D. Montgomery.
Journ. Geophys. Res., Vol. 78, 2017 - 2027 (1973).
It is the purpose of this paper to report the first observation of interpenetrating ion streams in the solar wind. The plasma measurements were made from March 18, 1971,

through June 12, 1971, using the IMP 6 electrostatic analyzer of the Los Alamos Scientific Laboratory.

074.071 Solar wind temperature and speed.
L. F. Burlaga, K. W. Ogilvie.
Journ. Geophys. Res., Vol. 78, 2028 - 2034 (1973).

074.072 Evolution of large-scale solar wind structures beyond 1 AU. A. J. Hundhausen.
Journ. Geophys. Res., Vol. 78, 2035 - 2042 (1973).

The evolution of solar wind speed fluctuations beyond 1 AU is considered under the assumption that turbulent dissipation is negligible (except at shock fronts). Both a dimensional argument, giving the distance scale on which a shell of high-speed solar wind is 'smoothed' into the ambient, and numerical solutions of the nonlinear fluid equations indicate that the solar wind structures observed near 1 AU persist to well beyond 5–10 AU.

074.073 Plasma radiation from collisionless MHD shock waves and the high-frequency waves in the upstream solar wind. D. F. Smith.
Journ. Geophys. Res., Vol. 78, 2302 - 2307 (1973).
Brief report.

074.074 Thermal energy transport in the solar wind.
M. D. Montgomery.
Cosmic plasma physics. Conference 1971, (see 012.016), p. 61 - 72 (1972).

074.075 The solar wind near the sun: the solar envelope.
L. F. Burlaga.
Cosmic plasma physics. Conference 1971, (see 012.016), p. 73 - 79 (1972).

074.076 Influence of neutral interstellar matter on the expansion of the solar wind. H. J. Fahr.
Cosmic plasma physics. Conference 1971, (see 012.016), p. 81 - 91 (1972).

074.077 Hydrogen – helium expansion from the sun.
E. J. Weber.
Cosmic plasma physics. Conference 1971, (see 012.016), p. 93 - 100 (1972).

074.078 Heating of the solar wind ions.
N. D'Angelo, V. O. Jensen.
Cosmic plasma physics. Conference 1971, (see 012.016), p. 101 (1972). – Abstract.

074.079 On the generation of shock pairs in the solar wind.
V. Formisano, J. K. Chao.
Cosmic plasma physics. Conference 1971, (see 012.016), p. 103 - 104 (1972).

074.080 Spectral anisotropy of Alfvén-waves in the solar wind. H. J. Völk, W. Alpers.
Cosmic plasma physics. Conference 1971, (see 012.016), p. 105 - 111 (1972).

074.081 Observations of coronal magnetic field strengths and flux tubes and their stability.
H. Rosenberg.
Cosmic plasma physics. Conference 1971, (see 012.016), p. 191 - 193 (1972).

074.082 Coronal density and temperature gradients.
R. G. Athay.
Solar Physics, Vol. 29, 357 - 364 (1973).

Mean density models of the solar corona show evidence for two distinctive density regimes characterized by different density gradients. High density gradients are identified with regions of predominantly open magnetic lines of force and low density gradients are identified with regions of predominantly closed magnetic lines of force.

074.083 A coronal hole and its identification as the source of a high velocity solar wind stream.
A. S. Krieger, A. F. Timothy, E. C. Roelof.
Solar Physics, Vol. 29, 505 - 525 (1973).

X-ray images of the solar corona, taken on November 24, 1970, showed a magnetically open structure in the low corona which extended from N20 W20 to the south pole. Analysis of the measured X-ray intensities shows the density scale height within the structure to be typically a factor of two less than that in the surrounding large scale magnetically closed regions. The structure is identified as a coronal hole. Since there have been several predictions that such a region should be the source of a high velocity stream in the solar wind, wind measurements for the appropriate period were traced back to the sun by the method of instantaneous ideal spirals. A striking agreement was found between the Carrington longitude of the solar source of a recurrent high velocity solar wind stream and the position of the hole.

074.084 Temperature variations of coronal regions near a prominence.
Ts. Tchultem (*Chultem*).
Astron. vestn., Vol. 6, 242 - 248 (1972). In Russian.

The paper presents the results of the treatment of the data obtained in observations of the coronal line λ 5303 Å before the appearance, during the development and after the disappearance of an active prominence at the edge of the disk.

074.085 Polarization of the integral radiation of the corona during the solar eclipse on March 7, 1970.
D. L. Astavin-Razumin.
Astron. vestn., Vol. 6, 255 - 256 (1972). In Russian.

074.086 Results of an observation of a coronal condensation in photographic range during the eclipse of September 22, 1968. V. I. Bystritskij, V. P. Vasiljev.
Astron. vestn., Vol. 7, 32 - 37 (1973). In Russian.

A coronal condensation, observed on the eastern edge of limb during the total solar eclipse on September 22, 1968 has been studied by photographic photometry. Photographs of the corona in 2 spectral ranges (3950–4900 Å and 5600–7600 Å) obtained with the help of the 5-m coronograph during the total phase have been discussed.

074.087 Some early attempts at photographing the daylight corona. J. Muirden.
Journ. British Astron. Ass., Vol. 83, 275 - 279 (1973).

074.088 Polarimetrische Untersuchungen der Sonnenkorona bei der totalen Finsternis vom 7. März 1970 in Mexiko. H. Haupt, R. Weinberger.
Mitt. Astron. Ges., No. 32, p. 197 - 198 (1973). – Abstract.

074.089 Influence of strong magnetic fields on the structure of the solar corona. B. V. Somov.
Trudy XVI Nauch. konf. Mosk. fiz.-tekhn. in-t, 1970. Ser. "Obshch. i prikl. fiz.", "Molekulyar. i khim. fiz.". Moskva, 1972, p. 68 - 74. In Russian. – Abstr. in Referativ. Zhurn. 51. Astron., 5.51.491 (1973).

074.090 Thermalization of solar wind.
S. Krishan, D. Rankin.
Astrophys. Space Sci., Vol. 19, 207 - 217 (1972).

Three kinetic equations describing the linear and nonlinear wave-particle interaction for an anisotropic solar wind plasma have been developed. These equations have been solved numerically to find the variation in $T_\perp/T_\parallel$ with respect to time where $T_\perp$ and $T_\parallel$ are the perpendicular and parallel temperatures with respect to the ambient magnetic field of the solar wind.

074.091 Single-fluid model of the distant solar wind.
M. K. Wallis.
Journ. Geophys. Res., Vol. 78, 3155 - 3158 (1973). – Letter.

074.092 Traveling regions of high solar wind density observed in early August 1972. T. A. Croft.
Journ. Geophys. Res., Vol. 78, 3159 - 3166 (1973). – Letter.

074.093 On the role of fluctuations in the interplantary magnetic field on heat conduction in the solar wind.
S. Cuperman, N. Metzler.
Journ. Geophys. Res., Vol. 78, 3167 - 3168 (1973). – Letter.

074.094 Comments on paper by E. Leer and T. E. Holzer, 'Collisionless solar wind protons: A comparison of kinetic and hydrodynamic descriptions.'
J. C. Brandt, C. L. Wolff.
Journ. Geophys. Res., Vol. 78, 3197 - 3198, with a reply by E. Holzer, E. Leer, p. 3199 - 3201 (1973).

074.095 Empirical closure relation for the Vlasov moment equations (solar wind).
W. C. Feldman, J. R. Asbridge, S. J. Bame, H. R. Lewis.
Phys. Rev. Letters, Vol. 30, 271 - 274 (1973).

Empirical statistical relationships are found between each of the third moments and the two second moments of solar-wind proton velocity distributions. These can be used to close the Vlasov moment equations when applied to the behavior of the solar-wind plasma over a sufficiently large temporal or spatial scale.

074.096 The spatial distribution of the source function of the white-light corona. M. Waldmeier.
Astron. Mitt. Sternw. Zürich, No. 314, 9 pp. (1972).

The intensity at any point of the white-light corona is the integral over the source function along the line of sight. From a spherical model of the corona this source function has been calculated for distances up to r = 5.0. The results can be used to calculate the deviation of the electron-density in localized structures like condensations, streamers, holes, etc. from that in the model coroña, provided that their extension along the line of sight is known.

074.097 Quiet and collisionless solar wind I.C.G.L. approximation and the evaporative type method.
M. Fridman.
Bull. Soc. Roy. Sci. Liège, (*Belgium*), Vol. 41, 511 - 520 (1972).

074.098 Transition probabilities for some forbidden lines.
R. H. Garstang.
Optica Pura y Aplicada, (*Spain*), Vol. 5, 192 - 194 (1973).

Transition probabilities are given for forbidden lines of solar coronal interest in the $2p^3$ configurations of Si VIII and S X, and the $3p^3$ configurations of Fe XII and Ni XIV. Transition probabilities are also given for three electric quadrupole transitions ($3p^6$-$3p^5 4p$) in Ar I.

074.099 On the thermal conductivity of the quiet solar-wind plasma and consequences.
S. Cuperman, N. Metzler.
Astrophys. Journ., Vol. 182, 961 - 975 (1973).

The structure, space dependence and consequences of an anomalously low (electron) thermal conductivity in the quiet solar wind are investigated. The solar wind is described by fluid equations but the thermal conductivity includes effects coming from both the inhibition of heat transport across the spiraling interplanetary magnetic field and electromagnetic turbulence (noncollisional processes). The results of the investigation are presented.

074.100 Non-thermal solar wind heating by supra-thermal ions. H. J. Fahr.
Solar Physics, Vol. 30, 193 - 206 = Forschungsber. Astron. Inst. Bonn, 72-10 (1973).

The effect of a new energy source due to energies transferred from supra-thermal secondary ions on the temperature profile of the solar wind has been considered. For this purpose a solution of a tri-fluid model of the solar wind including solar electrons, protons, and α-particles, and starting with the boundary conditions of Hartle and Barnes at 0.5 AU is given.

074.101 Variations of α-particle abundance in the solar wind. G. Moreno, F. Palmiotto.
Solar Physics, Vol. 30, 207 - 210 (1973). – Research note.

074.102 Solar-wind properties at the earth as predicted by the two-fluid model. B. R. Durney.
Solar Physics, Vol. 30, 223 - 234 (1973).

The two-fluid equations for the solar wind are written down in a simplified form, similar to that suggested by Roberts (1971) for the one-fluid model. For a variety of values of the density and temperature at the base of the corona the solutions of the two-fluid solar wind model are computed and the predicted and observed solar wind parameters at the earth are compared.

074.103 Method and results of an observation of the 3C 144 radio source occultation by the far-off solar corona with the DKR-1000 cross radio telescope of the Physical Institute of the Academy of Sciences.
S. N. Ivanov, Yu. P. Ilyasov.
Trudy fiz. in-ta. AN SSSR, Vol. 62, 116 - 120 (1972). In Russian. – Abstr. in Referativ. Zhurn. 51. Astron., 7.51.382 (1973).

074.104 On the nature of the solar wind.
M. V. Konyukov.
Trudy fiz. in-ta. AN SSSR, Vol. 62, 20 - 32 (1972). In Russian. – Abstr. in Referativ. Zhurn. 51. Astron., 7.51.393 (1973).

074.105 Estimate of the value of the solar coronal magnetic field from measurements of the linear polarization of solar radio bursts. Yu. I. Alekseev.
Trudy fiz. in-ta. AN SSSR, Vol. 62, 61 - 67 (1972). In Russian. – Abstr. in Referativ. Zhurn. 51. Astron., 7.51.417 (1973).

Introduction to the solar wind.
See Abstr. 003.037.

The OSO-7 year of discovery.
See Abstr. 011.002.

Mehrfachspaltspektrograph. Zur Spektralphotometrie der Korona am 30.VI.1973. See Abstr. 034.110.

Multi-ion plasmas in astrophysics. I: General requirements for the existence of critical points.
See Abstr. 062.034.

Non-linear evolution of firehose-unstable Alfvén waves. See Abstr. 062.043.

Stability of a vortex sheet between collisionless plasmas [solar wind-magnetosphere boundary].

See Abstr. 062.064.

Introductory review of solar activity.
See Abstr. 072.046.

The solar wind cycle, the sunspot cycle, and the corona.   See Abstr. 072.048.

Simultaneous determination of the electron temperature and density in the chromospheric-coronal transition region of the sun.   See Abstr. 073.015.

Powerful chromospheric flares and rotating discontinuities in the solar wind.   See Abstr. 073.025.

The prominence-corona interface.
See Abstr. 073.070.

Flare rates and coronal density evolution of active regions.   See Abstr. 073.074.

La zone de transition chromosphère-couronne, une région de l'atmosphère solaire encore bien embarrassante.
See Abstr. 073.092.

On the interpretation of solar radio-burst positions in a scattering corona.   See Abstr. 077.014.

Radial variation of magnetic fluctuations and the cosmic-ray diffusion tensor in the solar wind.
See Abstr. 078.043.

Cosmic-ray current and features in the solar wind.
See Abstr. 078.051.

The effect of the earth's bow shock and magnetosheath on the interaction of a discontinuity in the solar wind with the magnetosphere.   See Abstr. 084.221.

About the interaction between the earth's magnetic field and the antiparallel field of the solar wind.
See Abstr. 084.230.

On solar wind interaction with the earth's magnetosphere.   See Abstr. 084.262.

Ionospheric currents induced by solar wind interaction with planetary atmospheres.   See Abstr. 091.014.

The effect of a uniform external pressure on the ionospheric boundary of a non-magnetic planet in a steady solar wind.   See Abstr. 091.015.

Comet-like interaction of Venus with the solar wind.   See Abstr. 093.038.

Photoelectrons and solar wind/lunar limb interaction.   See Abstr. 094.052.

Solar plasma interaction with Mars: Preliminary results.   See Abstr. 097.006.

Comets in the solar wind.   See Abstr. 102.016.

Comets and the structure of the solar wind.
See Abstr. 102.025.

Solar wind interaction with comet Bennett (1969i).
See Abstr. 103.102.

Tail peculiarities in comet Bennett caused by solar wind disturbances.   See Abstr. 103.102.

On the analysis of the observations of interplanetary scintillations obtained with three spaced receivers.
See Abstr. 106.011.

The Pioneer 9 electric field experiment. III. Radial gradients and storm observations.   See Abstr. 106.016.

On the relation between the pattern and wind velocities in interplanetary scintillations.
See Abstr. 106.029.

Some recent aspects of spectroscopy at UV and X-ray wavelengths.   See Abstr. 114.092.

Interaction of the interstellar medium with the solar wind.   See Abstr. 131.088.

Interaction between the interstellar medium and solar wind plasma.   See Abstr. 131.119.

Modulation of galactic cosmic rays by the solar wind, asymmetrically with regard to the helio-latitude.
See Abstr. 143.025.

Peculiarities of the influence of the solar wind on cosmic-ray intensity in 1969.   See Abstr. 143.069.

075 Solar Patrol

075.001 The forecasting importance of the radio patrol of the sun. N. P. Tsimakhovich.
Solntse, ehlektrichestvo, zhizn'. Moskva, Mosk. un-t, 1972, p. 25 - 27. In Russian. — Abstr. in Referativ. Zhurn. 51. Astron., 3.51.532 (1973).

075.002 Survey of current solar forecast centers. P. Simon, P. S. McIntosh.
Progr. Astronaut. Aeronaut., Vol. 30, (see 003.009), p. 343 - 357 (1972).

075.003 Identification and adjustment of psychological factors to improve solar patrol observing. R. M. Pickett.
Progr. Astronaut. Aeronaut., Vol. 30, (see 003.009), p. 359 - 370 (1972). — Presented at the AIAA observation and prediction of solar activity conference, Huntsville, Ala., Nov. 16–18, 1970.

075.004 Definitive Sonnenflecken-Relativzahlen für 1972. R. A. Naef.
Orion Schaffhausen, 31. Jahrgang, p. 98 (1973).

075.005 Solar activity. F. G. Mustaeva.
Tsirk. Astron. Inst., *Tashkent*, Nos. 33 (380), 35 (382), 37 (384), 39 (386) (1972/73). In Russian. — 1971 November - 1972 June.

075.006 Solar activity in 1972. J. Mergentaler.
Urania Kraków, Vol. 44, 116 - 118 (1973). In Polish.

075.007 Visual observation of the sun in the year 1972 in Czechoslovakia. L. Schmied.
Říše hvězd, Vol. 54, 134 - 135 (1973). In Czech.

075.008 Map of the sun.
Edited by Fraunhofer Institut, Freiburg. 1973 January 1 - June 30.

075.009 Solar observations made at Catania Astrophysical Observatory during 1972. G. Godoli, V. Sciuto, R. A. Zappalà, E. Catinoto, G. Domina, G. Celeani, G. Sapienza, S. Sciuto.
Oss. Astrofis. Catania, Pubbl. No. 150, 119 pp. (1973).
This bulletin includes all the data deduced from solar observations made during 1972 at Catania Astrophysical Observatory: Sunspots; Hα and K faculae; Hα flares; Hα quiescent prominences; K quiescent prominences; Hα active prominences on disc and at limb; Hα disc and limb patrol hours.

075.010 Solar photospheric observations. F. Bruin, H. Hourani, N. G. Bustati.
Lee Obs., American Univ. Beirut, Monthly Bull. Astron. Section, 1972 October - 1973 April (1972/73).
Sunspot relative numbers; Heliographic mean position and classification of the sunspot groups; Number of facular zones.

075.011 Solare Beobachtungsergebnisse (Solar Data). C.-U. Wagner, A. Böhme, F. Fürstenberg, D. Scholz, S. Böhm.
Zentralinst. für Solar-Terrestrische Physik (Heinrich-Hertz-Inst.), Deutsche Akad. Wiss. Berlin, HHI Solar Data, Vol. 23, September – December (1972); Vol. 24, January – March (1973). – Solar radio emission.

075.012 Actividad solar julio - diciembre 1972. R. Barbarroja, V. Capolongo, J. A. Gutierrez, O. F. Liesche, L. A. Mansilla.
Obs. Astron. Municipal, Rosario — Argentina, Dep. Fis. Solar, Contr. Ser. 1, No. 6, 31 pp. (1972).

075.013 Fenomeni solari. S. Delli Santi.
Coelum, Vol. 41, 37 - 38, 79, 116 (1973). – 1972 September - 1973 February.

075.014 Osservatorio Magnetico de l'Aquila. Bollettino magnetico.
Coelum, Vol. 41, 39, 80, 118 (1973). – 1972 August - 1973 January.

075.015 Centro Universitario Fenomeni fluttuanti – Firenze. Test P.
Coelum, Vol. 41, 40 (1973). – 1972 September - October.

075.016 Sunspots (sunspot relative numbers and sunspot-areas); Synoptic charts of solar magnetic fields (Mount Wilson Observatory); **Eruptions chromosphériques brillantes; Intensité de la couronne solaire; Solar radio emission.**
M. Waldmeier, R. Howard, R. Michard, G. Olivieri, M. Bernot.
Quarterly Bull. Solar Activity (published by Eidgen. Sternw. Zürich), Nos. 177 - 178, p. 149 - 240 (1972/73). – Observations of the co-operating observatories for 1972 January - June are given.

075.017 Die Sonnenaktivität im Jahre 1971. M. Waldmeier.
Vierteljahrsschrift Naturforsch. Ges. Zürich, Jahrgang 117, p. 153 - 168 = Astron. Mitt. Sternw. Zürich, No. 311 (1972).
The present paper gives the frequency numbers of sunspots, photospheric faculae and prominences as well as the intensity of the coronal line 5303 Å and of the solar radio emission at the wavelength of 10.7 cm, all characterizing the solar activity in the year 1971.

075.018 Sunspot relative numbers for 1972. M. Waldmeier.
Astron. Mitt. Sternw. Zürich, No. 319, 10 pp. (1973)
This paper gives the daily values of the sunspot relative numbers for the whole disc, for the central zone and the daily number of sunspot groups.

075.019 Summary of daily observational results of solar phenomena, cosmic ray, geomagnetic variation, ionosphere, radio wave propagation and airglow during October 1969 through December 1971.
Rep. Ionosph. and Space Res., (*Japan*), Vol. 26, 175 - 206 (1972).
The diagrams in this paper illustrate the summary of daily observational results of solar phenomena, cosmic ray, geomagnetic variation, ionosphere, radio wave propagation and airglow observed in Japan.

075.020 Bulletin of astronomical observations, 1969.
Republic of the Philippines — Department of Commerce and Industry — Weather Bureau, Quezon City. 3 + 99 pp. (1970).
Relative sunspot (Wolf) numbers; Position and classification of sunspot groups; Solar rotations and evolution of sunspot groups; Artificial satellites; Lunar occultations.

075.021 Bulletin of astronomical observations, 1970.
Republic of the Philippines — Department of Com-

merce and Industry – Weather Bureau, Quezon City. 3 + 102 pp. (1971).

Relative sunspot (Wolf) numbers; Position and classification of sunspot groups; Solar rotations and evolution of sunspot groups; Artificial satellites; Lunar occultations.

075.022 **Solar phenomena.** M. Cimino, M. Torelli, V. Croce, R. Flamini, F. Casamassima.
Oss. Astron. Roma, Monthly Bull. Nos. 176, 177, 179 (1973). 1972 December, 1973 January, March: Daily total areas of sunspot-groups; Heliographic position, classification and area of sunspot-groups; Longitudinal sunspot magnetic fields; Hours of K-line cinematographic patrol; Hours of Hα cinematographic patrol; S.C.N.A. and S.E.A.; Explanation.

075.023 **Daily Hα chromosphere pictures, daily K_{232} chromosphere pictures, daily white light photosphere pictures.** M. Cimino (Editor).
Photographic Journ. of the Sun, Oss. Astron. Roma, Nos. 57, 58, 66, 68 (1972/73). – 1972 February 24 - April 18; 1972 October 26 - 1972 November 22; 1972 December 20 - 1973 January 16. – Rotations 1585, 1586, 1594, 1596.

075.024 **Solar activity in 1972.** M. C. Papathanassiou.
Published by Eugenides Foundation, Athens. 13 pp. (1973).

075.025 **Curve of the solar radiation from observations of the Observatory of the Department of Astronomy at the Kiev University in Lesnikakh.**
Kometn. Tsirk., *Kiev*, Nos. 141, 144, 147 (1973). In Russian.

075.026 **Definitieve zonnevlekkengetallen 1972.**
Hemel en Dampkring, Vol. 71, 163 (1973).

075.027 **Indices of geomagnetic activity.**
Journ. Atmosph. Terr. Phys., Vol. 35, 191 - 192, 384, 589, 1023 - 1024, 1427 - 1428 (1973). – 1972 August - 1973 March.

075.028 **Solar and solar system activity.**
R. J. J. Langton, J. R. Smith, K. F. Tapping.
Journ. British Astron. Ass., Vol. 83, 138 - 141, 217 - 219, 298 - 300 (1973). – Report of Radio Astron. Section of the British Astron. Ass. – 1972 September - 1973 February.

075.029 **Geomagnetic and solar data.**
J. V. Lincoln (Editor).
Journ. Geophys. Res., Vol. 78, 337, 780, 1243 - 1244, 1739, 2375 - 2378, 3202 (1973). – 1972 September - 1973 February.

075.030 **L'activité solaire.** M.-J. Martres.
L'Astronomie, 87ᵉ année, p. 52 - 54, 98 - 99, 132 - 133, 185 - 186, 225 - 227, 262 (1973). – Rotations 1589 - 1595.

075.031 **Sunspot numbers.**
Sky Telescope, Vol. 45, 61, 131, 195, 259, 328, 400; Vol. 46, 61 (1973). – 1972 November – 1973 May.

075.032 **Daily maps of the sun and geophysical graphs.**
Solnechnye Dannye 1972 Byull., No. 11, p. 1 - 56; No. 12, p. 1 - 55; 1973, No. 1, p. 1 - 68. In Russian.

075.033 **Magnetic fields of sunspots.**
Prilozhenie k Byulletenyu "Solnechnye Dannye", 1972, Nos. 11 - 12; 1973, No. 1. In Russian.

076 Solar UV, X Rays, Gamma Radiation

076.001 High spatial resolution photographs of the sun in Lα radiation. D. K. Prinz.
Solar Physics, Vol. 28, 35 - 42 (1973).

Photographs of the sun in predominantly Lα radiation (centered at 1215.67 Å) with 3″ spatial resolution were taken from an Aerobee rocket shortly after fourth contact by the moon on the eclipse day of 1972 July 10. This preliminary reporting of the results describes the instrument and shows two of the photographs taken. Densitometer traces across the disk are presented giving the flux incident on the earth from active regions, cell boundaries, and filaments.

076.002 Time variations in the X-ray emission of solar active regions. J. H. Parkinson.
Solar Physics, Vol. 28, 137 - 150 (1973).

The X-ray emission of individual solar active regions is found from OSO-5 data to vary on three main timescales. Flare associated events typically last for times of minutes to an hour. Events lasting several hours, with several peaks in the X-ray emission, are accompanied by a simplification in the magnetic field, and often mark a turning point in the life of the region. Smooth changes over periods of several days are associated with the general development of a region. Examples of these last two variations are presented.

076.003 Thick target X-ray bremsstrahlung from partially ionized targets in solar flares. J. C. Brown.
Solar Physics, Vol. 28, 151 - 158 (1973).

The effect of partial ionization of a thick target bremsstrahlung source on the emitted X-ray intensity is analyzed. It is shown that a totally ionized target produces an X-ray burst only about one third as intense as that from an unionized target. In the case of a solar flare plasma target, the ionization decreases with increasing depth in the flare.

076.004 Spectral development of a solar X-ray burst observed on OSO-7.
D. L. McKenzie, D. W. Datlowe, L. E. Peterson.
Solar Physics, Vol. 28, 175 - 182 (1973).

The UCSD solar X-ray instrument on the OSO-7 satellite observes X-ray bursts in the 2–300 keV range with 10.24 s time resolution. Spectra obtained from the proportional counter and scintillation counter are analyzed for the event of November 16, 1971, at 0519 UT in terms of thermal (exponential spectrum) and non-thermal (power law) components.

076.005 Superthermal plasma nodules and their relation to solar flares. L. D. de Feiter, C. de Jager.
Solar Physics, Vol. 28, 183 - 186 (1973).

We define superthermal plasma nodules as bright points (diameter ≲ 20″), visible on high resolution X-ray heliograms. Flares appear to show a strong tendency to occur at the places of these nodules. There are indications that (part of) the hot plasma produced by consecutive flares is accumulated and confined in the superthermal plasma nodules, and that with increasing energy content of a nodule the probability for a drastic change of its magnetic structure increases, thus reducing the possibility for more flares to occur.

076.006 Solar activity and XUV emission.
C. W. Allen, S. Yousef.
Monthly Notices Roy. Astron. Soc., Vol. 161, 181 - 191 (1973).

An attempt has been made to segregate the solar activity variation of solar XUV emission into the components: (Q) due to solar cycle changes in the Quiet Sun radiation; and (AR) due to radiation from the Active Regions themselves. The segregation has been achieved by making correlations with sunspot number of various observations of XUV fluxes and intensities from rockets and satellites.

076.007 High-resolution ultraviolet solar spectra in the region 2765–2822 Å.
A. Greve, C. D. McKeith, N. E. McKeith.
Solar Physics, Vol. 28, 289 - 309 (1973).

Wavelengths and identifications of the near ultraviolet solar spectrum are presented. The data were obtained during the rocket flight of an interferometer spectrograph with a spectral resolution of 0.03 Å.

076.008 Solar gamma ray lines observed during the solar activity of August 2 to August 11, 1972.
E. L. Chupp, D. J. Forrest, P. R. Higbie, A. N. Suri, C. Tsai, P. P. Dunphy.
Nature, Vol. 241, 333 - 335 (1973).

We report preliminary observations from OSO-7 on the emission of gamma ray lines associated with the solar flares on August 4 and 7, 1972. This is the first observation of such solar gamma rays.

076.009 New observations of Fe XVII in the solar X-ray spectrum. J. H. Parkinson.
Astron. Astrophys., Vol. 24, 215 - 218 (1973).

New observations of the Fe XVII lines between 13 and 18 Å in the solar X-ray spectrum are presented. The relative intensities of the Fe XVII lines are in good agreement with calculations of Loulergue and Nussbaumer (1973) which include cascade processes.

076.010 X-ray flares on the sun. S. L. Mandel'shtam.
Vestn. AN SSSR, 1972, No. 9, p. 26 - 36. In Russian. – Abstr. in Referativ. Zhurn. 51. Astron., 3.51.483; 62. Issled. kosmich. prostranstva, 3.62.130 (1973).

076.011 Solar Lyman alpha changes and related hydrogen density distribution at the earth's exobase (1969–1970). A. Vidal-Madjar, J. E. Blamont, B. Phissamay.
Journ. Geophys. Res., Vol. 78, 1115 - 1144 (1973).

The University of Paris experiment, operating on board the Oso 5 spacecraft since January 22, 1969, observes the solar Lyman α line integrated over the whole solar disk. The hydrogen and deuterium resonance cells yield information about the solar flux at the center of the line and on the blue wing. The center of the solar line as observed from the spacecraft is strongly absorbed by the earth's geocorona. Therefore the hydrogen resonance data contain solar, as well as geocoronal, information. Comparison of the data with sophisticated exospheric models has been made, and the results are presented.

076.012 Broad band solar EUV absorption in the earth's upper atmosphere.
K. H. Allen, W. A. Rense.
Journ. Geophys. Res., Vol. 78, 1219 - 1224 (1973). – Letter.

076.013 Hard X-ray solar bursts observed from the OSO-6 satellite.
D. Brini, F. Evangelisti, M. T. Fuligni Di Grande, G. Pizzichini, A. Spizzichino, G. R. Vespignani.
Astron. Astrophys., Vol. 25, 17 - 27 (1973).

A selection of hard X-ray bursts detected in the range 20–200 keV from aboard the satellite OSO-6 is presented. A description is given of the treatment of data and the information that can be extracted from them, as afforded by the instrumental characteristics.

076.014 **The solar albedo of hard X-ray flares.**
N. Santangelo, H. Horstman, E. Horstman-Moretti.
Solar Physics, Vol. 29, 143 - 148 (1973).

The calculations of Compton backscattering from the solar surface of flare X-rays performed by Tomblin (1972) are extended to higher energies. It is shown that the effect is even more pronounced in the 40 keV region and that it can lead to substantial corrections to the observed X-ray spectra.

076.015 **An analysis of the solar extreme-ultraviolet spectrum between 50 and 300 Å.**
M. Malinovsky, L. Heroux.
Astrophys. Journ., Vol. 181, 1009 - 1030 (1973).

Photoelectric data on the solar spectrum between 50 and 300 Å obtained with a grazing-incidence rocket spectrometer flown on 1969 April 4 are given. The spectrometer viewed the entire solar disk. The intensities of the spectral lines were analyzed in detail to verify previous identifications of solar lines and to suggest new identifications. The method of Pottasch was also applied to determine the relative abundances of the elements O, Ne, Mg, Si, S, Fe, and Ni present in the sun.

076.016 **X-ray line emission associated with solar flares.**
W. M. Neupert.
Progr. Astronaut. Aeronaut., Vol. 30, (see 003.009), p. 127 - 140 (1972). — Presented at the AIAA observation and prediction of solar activity conference, Huntsville, Ala., Nov. 16–18, 1970.

076.017 **X-ray observations from the August 2, 1972 flare.**
D. W. Datlowe, L. E. Peterson, K. Tanaka, H. Zirin.
Bull. American Astron. Soc., Vol. 5, 270 (1973). — Abstr. AAS.

076.018 **Impulsive EUV spectra of solar flares.**
R. F. Donnelly, L. A. Hall.
Bull. American Astron. Soc., Vol. 5, 271 (1973). — Abstr. AAS.

076.019 **On the determination of non-thermal electron spectra in solar flares from the observed hard X-ray emission.**   S. R. Kane.
Bull. American Astron. Soc., Vol. 5, 274 (1973). — Abstr. AAS.

076.020 **Analysis of Si II solar UV emission lines.**
H. C. McAllister, J. T. Jefferies.
Bull. American Astron. Soc., Vol. 5, 276 (1973). — Abstr. AAS.

076.021 **Ultraviolet solar spectrum recorded by echelle spectrograph (1970–1800 Å).**
H. C. McAllister, P. Smith.
Bull. American Astron. Soc., Vol. 5, 277 (1973). — Abstr. AAS.

076.022 **Spatial distribution of soft X-ray and EUV emission associated with a flare of importance IB on August 2, 1972.**   W. M. Neupert.
Bull. American Astron. Soc., Vol. 5, 277 - 278 (1973). Abstr. AAS.

076.023 **A model for impulsive solar X-ray bursts.**
V. Petrosian, P. A. Sturrock.
Bull. American Astron. Soc., Vol. 5, 278 (1973). — Abstr. AAS.

076.024 **X-ray spectra of multi-temperature plasmas.**
W. T. Zaumen, L. W. Acton, R. C. Catura.
Bull. American Astron. Soc., Vol. 5, 282 (1973). — Abstr. AAS.

076.025 **Spectrometer for investigating bursts of solar gamma radiation in the energy region 0.03–0.3 MeV.**
M. I. Kudryavtsev, O. B. Likin, A. S. Melioransky, I. A. Savenko, V. V. Smirnov, V. M. Shamolin.
Geomagn. Aeronom., Vol. 13, 406 - 410 (1973). In Russian.

076.026 **25 jaar UV-waarnemingen van de zon.**   A. Greve.
Hemel en Dampkring, Vol. 71, 170 - 172 (1973).

076.027 **The near ultra-violet flux of the Harvard Smithsonian Reference Atmosphere.**   R. A. Bell.
Solar Physics, Vol. 29, 299 - 300 (1973). — Research Note.

076.028 **Extreme ultraviolet emission from chromospheric inhomogeneities. An analysis of the extreme ultraviolet flash spectrum of the sun.**
G. E. Brueckner, K. R. Nicolas.
Solar Physics, Vol. 29, 301 - 315 (1973).

An Aerobee 170 rocket carried five slitless extreme ultraviolet (XUV) spectroheliographs into the March 7, 1970, solar eclipse. The analysis is based on a comparison of the extent of flash spectrum crescents from emission lines formed in the chromosphere-corona transition zone with two simple but fundamentally different models describing this region. The observations can be satisfactorily described by an inhomogeneous model where cool spicules are surrounded by a transition zone which has the same temperature and density structure as the chromospheric coronal transition zone customarily used in spherically symmetric models of the quiet sun.

076.029 **X-rays spectroheliograms in lines of Mg XI and Mg XII.**
C. Bonnelle, C. Senemaud, G. Senemaud, G. Chambe, M. Guionnet, J. C. Henoux, R. Michard.
Solar Physics, Vol. 29, 341 - 355 (1973).

Spectroheliograms in the $L\alpha$ Mg XII line and in the Mg XI resonance (R) $1s^2\ ^1S_0 - 1s2p\ ^1P_1$ line, intercombination (I) $1s^2\ ^1S_0 - 1s2p\ ^3P_{1,2}$ line, and the forbidden (F) $1s^2\ ^1S_0 - 1s2s\ ^3S_1$ line, have been obtained. The observed intensities of the Mg XI R line, Mg XII $L\alpha$ line and Mg X $1s^2 2s\ ^2S_{1/2} - 1s2p\ ^1P$ $2s \times ^2P_{1/2,\ 3/2}\ S$ line are not well explained by an isothermal model. Good agreement between computed and observed intensities is obtained using the non-isothermal model proposed here.

076.030 **Continuous energy injection at numerous bright points during soft X-ray flare enhancement.**
W. M. Glencross.
Solar Physics, Vol. 29, 429 - 439 (1973).

The common assumption that a single volume of plasma produces X-ray emission during solar flares is difficult to reconcile with the very complex structure observed in $H\alpha$ spectroheliograms. Data presented in this paper show that a number of secondary peaks in intensity are usually observed throughout the soft X-ray emission. These can be explained by a model in which the X-ray emission comes from many relatively short-lived volumes of hot plasma.

076.031 **Solar X-ray spectra observed from the 'Intercosmos - 4' satellite and the 'Vertical - 2' rocket.**
Yu. I. Grineva, V. I. Karev, V. V. Korneev, V. V. Krutov, S. L. Mandelstam, L. A. Vainstein, B. N. Vasilyev, I. A. Zhitnik.
Solar Physics, Vol. 29, 441 - 446 (1973).

Results are given of the detailed analysis of fourteen Fe XXV-XXIII lines ($\lambda = 1.850 - 1.870$ Å) in the spectra of a solar flare on 16 Nov. 1970. The spectra were obtained with a resolution of about 4×10^{-4} Å, which revealed lines not previously observed and allowed the measurement of line profiles. The measured values of the wavelengths and emission fluxes are presented and compared with theoretical calculations.

076.032 **19 – 20 May 1969, an example of type III emission during the impulsive phase of flares.**   J. Vorpahl.
Solar Physics, Vol. 29, 447 - 460 (1973).

The author examined two multiple impulsive events in May 1969 and compared their impulsive $H\alpha$, hard X-ray and microwave components to the observed type III emission.

076.033 Space observations of the variability of solar irradiance in the near and far ultraviolet. D. F. Heath.
Journ. Geophys. Res., Vol. 78, 2779 - 2792 (1973).
    Satellite observations of UV solar irradiance in selected
wavelength bands between 1200 and 3000 A have been made
continuously by photometers consisting of broad band sensors
operated on Nimbus 3 and 4. The change in irradiance with
solar rotation was found to increase with decreasing wavelengths. Different types of observed variations in UV solar irradiance can be classified in accordance with characteristic
times; in order of increasing periods, they are (1) flare-associated enhancements, (2) 27-day variations due to solar rotation, (3) a possible biennial effect, and (4) long-term variations associated with the 11-year solar cycle.

076.034 Measurements of the intensity of solar Lyman-al-
pha-emission by non-optical methods using the
Vertical-1 rocket.   L. Martini, N. M. Shutte, K. I. Gringauz,
B. Stark.
Cosmic Res. (U.S.A.), Vol. 10, 227 - 231 (1972). – See Phys.
Abstr., Vol. 76, No. 25394 (1973).

076.035 Extreme ultraviolet line intensities from the sun.
B. E. Woodgate, D. E. Knight, R. Uribe, P. Sheather,
J. Bowles, R. Nettleship.
Proc. Roy. Soc. London, Ser. A, Vol. 332, 291 - 309 (1973).
    An extreme ultraviolet spectrometer experiment by
University College London for observation of whole-sun spec-
trum-line fluxes, and flown on Orbiting Solar Observatory-6,
is described. Results for the four lines with absolute calibration are shown for the first 6 months of flight, from August
1969 to February 1970. Relations between the fluxes and the
10.7 cm radio flux and Zürich sunspot number are presented.

076.036 Revision of data on short-wave solar emission.
G. S. Ivanov-Kholodnyj, B. N. Velichanskij.
Issled. po geomagnetizmu, aehron. i fiz. Solntsa. Vyp. (No.)
26. Moskva, Nauka, 1973, p. 14 - 25. In Russian. – Abstr. in
Referativ. Zhurn. 51. Astron., 7.51.363 (1973).

076.037 Peculiarities of the solar X-ray burst detected on
December 10, 1970 by the spectrometric apparatus
"Rifma".   G. E. Kocharov, Yu. E. Charikov.
IV Leningr. mezhdunar. seminar "Edinoobrazie uskoreniya
chastits v razlich. masshtabakh kosmosa, 1972". Leningrad,
1972, p. 167 - 182. In Russian. – Abstr. in Referativ. Zhurn.
51. Astron., 7.51.425; 62. Issled. kosmich. prostranstva,
7.62.136 (1973).

On the dependence of the two-level source function
on its own radiation field.   See Abstr. 063.040.

Das Licht als Informationsträger aus dem Weltall.
See Abstr. 071.015.

The extreme-ultraviolet spectrum of a solar active
region.   See Abstr. 071.045.

Two-component temperature analysis of OSO-5
X-ray flare data.   See Abstr. 073.004.

Directivity of high-energy X-ray emission during
flares.   See Abstr. 073.005.

The extreme-ultraviolet spectrum of Fe XV in a solar flare.   See Abstr. 073.007.

Note on heliumlike silicon and sulfur lines observed
in the X-ray spectra of solar flares.   See Abstr. 073.012.

Determination of the temperature in a solar X-ray
flare.   See Abstr. 073.016.

The extreme ultraviolet emissions of solar flares: a
comparison between OSO-6 spectroheliograph observations
and SFDs.   See Abstr. 073.039.

Spectra of solar flares from 8.5 Å to 16 Å.
See Abstr. 073.040.

Extreme ultraviolet observations of solar flares.
See Abstr. 073.044.

Spatial structure of a two-phase solar flare observed
in the EUV by OSO 7.   See Abstr. 073.051.

Some new observations of the solar UV chromosphere.   See Abstr. 073.054.

Development of solar active regions.
See Abstr. 073.057.

New observations of the solar ultraviolet chromosphere.   See Abstr. 073.094.

An investigation of the strucutre of coronal active
regions.   See Abstr. 074.018.

D-region recombination coefficients and the short
wavelength X-ray flux during a solar flare.
See Abstr. 083.028.

Some recent aspects of spectroscopy at UV and
X-ray wavelengths.   See Abstr. 114.092.

077 Solar Radio Radiation

077.001 Spatial distributions of intensity and polarization over the source of microwave impulsive bursts.
Y. Naito, T. Takakura.
Publ. Astron. Soc. Japan, Vol. 25, 65 - 73 (1973).

The spatial distributions of the intensity and the polarization degree on the half of the radio source of microwave impulsive bursts are computed by using the dipole model as the magnetic field in the source proposed by Takakura and Scalise (1970). This model suggests double peaks, if the source is near the center of the solar disk, and suggests a more complex structure if it is away from the center.

077.002 The intensity decrease of microwave bursts.
E. Fürst.
Solar Physics, Vol. 28, 159 - 168 (1973).

A typical microwave burst on 1968 January 11, 1700 UT is used to demonstrate that the radiation spectrum at maximum phase can be described by gyromagnetic absorption. A model for the source is derived from the observed spectrum. With the aid of this model, we try to explain the decreasing phase of the burst intensity.

077.003 Detailed correlation of type III radio bursts with Hα activity. I. Active region of 22 May 1970.
T. B. H. Kuiper, J. M. Pasachoff.
Solar Physics, Vol. 28, 187 - 196 (1973).

We compare observations of type III impulsive radio bursts made at the Clark Lake Radio Observatory with high-spatial-resolution cinematographic observations taken at the Big Bear Solar Observatory. Use of the log-periodic radio interferometer allows us to localize the radio emission uniquely. This study concentrates on the particularly active region close to the limb on 22 May 1970. Sixteen of the 17 groups were associated with some Hα activity, 11 of them with the start of such activity.

077.004 Observations on the time and frequency structure of solar decameter radio bursts. C. V. Sastry.
Solar Physics, Vol. 28, 197 - 209 (1973).

Solar radio bursts were observed with a 4-channel radiometer and polarization analyzer at wavelengths around 12 m. The time and frequency resolutions were 10 ms and 100 kHz respectively. Observations on the duration, time profile and frequency splitting are described.

077.005 Landau damping of type III solar radiobursts.
C. C. Harvey, M. G. Aubier.
Astron. Astrophys., Vol. 22, 1 - 8 (1973).

The propagation of a plasma wave in an inhomogeneous convecting solar wind plasma is studied by means of ray theory. Assuming the plasma hypothesis for the generation of type III solar radiobursts, the observable effects of the damping are investigated. Using electron temperature and spatial density distributions thought to be appropriate for the solar wind, together with a maxwellian velocity distribution, it is found that the effect of Landau damping is to cause a relatively abrupt cut-off of the type III emission. The possible modification of the high energy 'wing' of the velocity distribution is investigated, and a model for the decay of type III radiobursts is proposed.

077.006 The radio diameter of the sun from interferometer measurements at 9 mm wavelength.
P. S. Nicholson, E. A. Parker.
Observatory, Vol. 93, 13 - 16 (1973).

We have attempted to measure the diameter of the sun using two small parabolic reflectors arranged as an interferometer.

077.007 A U-like radio burst observed with high space-time resolution.
C. Caroubalos, P. Couturier, T. Prokakis.
Astron. Astrophys., Vol. 23, 131 - 138 (1973).

On July 19th, 1971, a U-like burst has been recorded on 169 MHz by the Nançay radioheliograph. High space-time resolution records and spectral data are available on that event of which only the upper part has been observed. Using Fokker's (1970) method and the available coronal magnetic field map, the velocity and trajectory of the exciter are derived. The widening and the reduction in brightness of the descending branch as well as details of the dynamic spectra are interpreted consistently assuming the presence of a helmet streamer associated with a neutral sheet (Pneuman, 1972) located at the upper part of the magnetic arch which guides the exciter. This model can be used to interpret most kinds of U-bursts.

077.008 The height of 9.1 cm solar emission from latitude shift. W. Graf, R. N. Bracewell.
Solar Physics, Vol. 28, 425 - 433 (1973).

Active regions that are sources of microwave solar emission should appear displaced in latitude relative to the associated optical feature and from this displacement it should be possible to deduce the height. It was found possible, in an investigation based on 1264 cases in 1970, to obtain a height with a precision as good as or better than has generally been achieved in previous studies based on the rate of motion in longitude.

077.009 The solar outburst on August 7, 1972 at 17 GHz and 35 GHz.
E. Fürst, O. Hachenberg, W. Hirth.
Solar Physics, Vol. 28, 533 - 537 (1973).

The radio-burst on August 7, 1972, is discussed. On 17 GHz the peak flux was about 25000 SFU. Considering the decreasing phase of the burst, it was found that an exponential decrease as well as a power-law decrease can be used. The degree of circular polarization shows an increase to about 25% during the ascending phase of the burst, while in the phase of maximal radiation and during the decrease the polarization degree was small.

077.010 Harmonic structure in a solar type V burst.
A. O. Benz.
Nature, Phys. Sci., Vol. 242, 38 - 39 (1973).

On October 25, 1972, at 1136 UT a sudden explosive flare (1b) occurred, as reported by the Zürich Hα partol. The author concludes that this type V burst shows at its beginning a harmonic frequency structure of the relation 2 : 3. This first observation of definite harmonic structure in a type V burst gives strong support to the theory explaining this type as plasma wave emission.

077.011 High resolution space-time structure and centre-limb distribution of solar type I sources observed at 169 MHz. J. L. Bougeret.
Astron. Astrophys., Vol. 24, 53 - 58 (1973).

The one-dimensional brightness distribution over individual type I solar bursts is studied with high time-resolution from photographic recordings and isophote plots (Nançay radioheliograph). Three main space-time shapes are distinguished according to the time-evolution of the brightness distribution peak: stationary (65%), drifting (15%) and splitted (0.4%). An interpretation in which the space-time characteristics are attributed to propagation effects – excitation source real size much smaller than the observed size — could be relevant of the results.

077.012 Solar energetic particles and wide-band continuum storms from metric to hectometric frequencies.
K. Sakurai.
Planet. Space Sci., Vol. 21, 17 - 22 (1973).
   In association with solar flares accompanying type IV radio bursts of U-shaped spectrum, solar cosmic rays (MeV) and energetic electrons (keV) were generated. After acceleration, they were first stored in or near the flare regions and then gradually emitted into outer space. It seems that the streams of keV electrons generated the continuum radio emissions from metric to hectometric frequencies while passing through the outer coronal regions.

077.013 Solar microwave bursts and polar cap absorption.
D. L. Croom.
Planet. Space Sci., Vol. 21, 707 - 709 (1973). — Research note.

077.014 On the interpretation of solar radio-burst positions in a scattering corona. A. C. Riddle.
Proc. Astron. Soc. Australia, Vol. 2, 148 - 150 (1972).

077.015 Solar radio emission at 9.1 cm wavelength.
W. Graf, R. N. Bracewell.
Bull. American Astron. Soc., Vol. 5, 21 (1973). — Abstr. AAS.

077.016 Radio astronomy observations of the solar eclipse on 20 May, 1966. Introduction.
G. B. Gel'frejkh, N. G. Peterova.
Radio astronomy observations of the solar eclipse on 20 May, 1966, (see 003.002), p. 5 - 15 (1972). In Russian.

077.017 Observations of the solar eclipse on May 20, 1966 on the large Pulkovo radio telescope.
Sh. B. Akhmedov, V. N. Borovik, N. V. Vinogradova, V. Ya. Gol'nev, V. N. Ikhsanova, V. G. Nagnibeda, N. G. Peterova.
Radio astronomy observations of the solar eclipse on 20 May, 1966, (see 003.002), p. 16 - 23 (1972). In Russian.

077.018 The structure of radio emission sources associated with bipolar sunspot groups.
G. B. Gel'frejkh, A. F. Dravskikh, A. A. Starshinov.
Radio astronomy observations of the solar eclipse on 20 May, 1966, (see 003.002), p. 38 - 41 (1972). In Russian.

077.019 Observations of the solar eclipse on 20 May, 1966 at $\lambda = 4.5$ cm. N. G. Peterova.
Radio astronomy observations of the solar eclipse on 20 May, 1966, (see 003.002), p. 42 - 49 (1972). In Russian.

077.020 Polarization measurements of the solar eclipse on 20 May, 1966 at $\lambda\lambda = 2$ and 4.9 cm.
G. B. Gel'frejkh, A. N. Korzhavin, G. F. Shemyakin.
Radio astronomy observations of the solar eclipse on 20 May, 1966, (see 003.002), p. 50 - 67 (1972). In Russian.

077.021 Determination of intensity, position and dimensions of discrete radio emission sources at $\lambda\lambda = 10.2$, 30 cm and 1.37 m from observation of the solar eclipse on 20 May, 1966. Yu. F. Yurovskij.
Radio astronomy observations of the solar eclipse on 20 May, 1966, (see 003.002), p. 67 - 79 (1972). In Russian.

077.022 Determination of the dimensions of a discrete source by a diffraction image recorded at 1.37 m during the solar eclipse on May 20, 1966.
Yu. F. Yurovskij, L. I. Yurovskaya.
Radio astronomy observations of the solar eclipse on 20 May,

1966, (see 003.002), p. 80 - 85 (1972). In Russian.

077.023 Observation of the solar eclipse at frequencies 204 and 3000 MHz.
S. A. Amiantov, A. A. Gnezdilov, A. E. Manchenko.
Radio astronomy observations of the solar eclipse on 20 May, 1966, (see 003.002), p. 85 - 90 (1972). In Russian.

077.024 Some observational data on the solar radio eclipse on 20 May, 1966 in Gor'kij.
M. M. Kobrin, T. B. Pyatunina, V. M. Fridman.
Radio astronomy observations of the solar eclipse on 20 May, 1966, (see 003.002), p. 90 - 96 (1972). In Russian.

077.025 Observations of the solar eclipse on 20 May, 1966 at $\lambda\lambda = 3.0–3.3$, 5.3 and 7.8 cm.
I. F. Belov, Yu. A. Grachev, D. A. Dmitrenko, G. A. Lavrinov, E. I. Lebedev, A. M. Starodubtsev, K. M. Strezhneva, B. V. Timofeev, V. M. Fridman, O. I. Yudin.
Radio astronomy observations of the solar eclipse on 20 May, 1966, (see 003.002), p. 97 - 101 (1972). In Russian.

077.026 Some data on observations of the solar eclipse on 20 May, 1966 at $\lambda\lambda = 10.2$, 46 and 163 cm.
Yu. B. Vedeneev, N. M. Prytkov, V. Ya. Yashkov.
Radio astronomy observations of the solar eclipse on 20 May, 1966, (see 003.002), p. 101 - 104 (1972). In Russian.

077.027 Observation of the solar eclipse on 20 May, 1966 at 1.24 cm. L. I. Fedoseev.
Radio astronomy observations of the solar eclipse on 20 May, 1966, (see 003.002), p. 104 - 105 (1972). In Russian.

077.028 Observation of the solar eclipse on 20 May, 1966 at $\lambda = 34$ cm. O. G. Gontarev.
Radio astronomy observations of the solar eclipse on 20 May, 1966, (see 003.002), p. 106 - 109 (1972). In Russian.

077.029 Observations of the solar eclipse on 20 May, 1966 at $\lambda = 0.5$ and 1.5 cm.
A. T. Nesmyanovich, V. V. Chmil', Yu. A. Chesnok.
Radio astronomy observations of the solar eclipse on 20 May, 1966, (see 003.002), p. 109 - 113 (1972). In Russian.

077.030 Observations of the solar eclipse on 20 May, 1966 at $\lambda = 3.04$ cm. A. F. Dravskikh, Z. V. Dravskikh.
Radio astronomy observations of the solar eclipse on 20 May, 1966, (see 003.002), p. 114 (1972). In Russian.

077.031 Observations of the solar eclipse on 20 May, 1966 in Pulkovo at $\lambda = 3.2$ cm. V. N. Borovik.
Radio astronomy observations of the solar eclipse on 20 May, 1966, (see 003.002), p. 114 - 122 (1972). In Russian.

077.032 Observations of the occultation of an active region by the moon during the solar eclipse on 20 May, 1966. V. P. Nefed'ev.
Radio astronomy observations of the solar eclipse on 20 May, 1966, (see 003.002), p. 122 - 126 (1972). In Russian.

077.033 Observations of the solar eclipse on 20 May, 1966 at $\lambda = 4$ cm. G. B. Gel'frejkh, A. N. Korzhavin.
Radio astronomy observations of the solar eclipse on 20 May, 1966, (see 003.002), p. 127 - 129 (1972). In Russian.

077.034 Spectra of type IV radio bursts and magnetic fields of active regions on the sun for some proton and nonproton flares.
S. T. Akin'yan, L. B. Demkina, O. S. Korolev, E. I. Mogilevskij.
Radio astronomy observations of the solar eclipse on 20 May, 1966, (see 003.002), p. 129 - 136 (1972). In Russian.

077.035 **A recalibration of the quiet sun millimeter spectrum based on the moon as an absolute radiometric standard.** J. L. Linsky.
Solar Physics, Vol. 28, 409 - 418 (1973).

The solar millimeter continuum between 1 and 20 mm is recalibrated using observations of the average lunar brightness temperature at the center of lunar disk and new moon brightness temperatures. A least-squares parabolic regression curve is proposed for the solar millimeter continuum.

077.036 **Further evidence for a complex limb structure in the solar radial brightness distribution at mm wavelengths.** P. N. Swanson, F. L. Wefer, W. J. Decker, J. P. Hagen.
Solar Physics, Vol. 28, 419 - 424 (1973).

A computer program to convolve numerically any azimuthally symmetric, solar radial brightness distribution with standard antenna patterns of small half power beamwidths has been used to find a solar brightness distribution which is a good fit to the eclipse curve obtained during the 7 March, 1970 partial solar eclipse with the NRAO 36-ft antenna at 3.5 mm. This brightness distribution is compared with the brightness distribution at 3.2 mm determined by the Pennsylvania State University Radio Astronomy Observatory group during the same eclipse.

077.037 **Fenómenos homólogos en 408 MHz.** M. Ferrari, V. Esterkin.
Bol. As. Argentina Astron., No. 16, (see 012.007), p. 53 - 55 (1971).

077.038 **Investigation of the polarization of solar radio emission at 9.0 cm with development of an active region.** Sh. B. Akhmedov, N. V. Dedelova.
Solnechnye Dannye 1972 Byull., No. 12, p. 56 - 63 (1973). In Russian.

The peculiarities in time variations of the flux density and the rate of polarization of local radio sources at 9 cm have been studied. The polarized radio emission of a local source is shown to appear and disappear simultaneously with the corresponding sunspot group.

077.039 **Scattering effects on the relative positions and intensities of fundamental and harmonic emission of solar radio bursts.** Y. Leblanc.
Astrophys. Letters, Vol. 14, 41 - 45 (1973).

The effect of coronal inhomogeneities of the fundamental (F) and harmonic (H) radiation of type-II and type-III radio bursts has been computed at 30 MHz (30 F) and 60 MHz (60 H), in the case when the two emissions originate from the same region, and at 60 MHz (60 F and 60 H) for the two emissions originating from two different levels. The results are compared with observations.

077.040 **About the possible nature of the fine structure of the sporadic radio radiation of the sun and other cosmic objects with high density of electromagnetic radiation.** L. I. Dorman, M. E. Katz, A. K. Yukhimuk.
Geomagn. Aeronom., Vol. 13, 201 - 207 (1973). In Russian.

077.041 **On quasi-periodical fluctuations of solar radio emission at 60 cm wavelength.** V. V. Pakhomov, S. D. Snegirev.
Astron. Tsirk., No. 753, p. 6 - 8 (1973). In Russian.

077.042 **Development of moving type IV solar radio bursts and relation to expanding magnetic bottles from flare regions.** K. Sakurai.
Nature, Phys. Sci., Vol. 243, 46 - 48 (1973).

The author has analysed observations of type IV bursts associated with proton flares to obtain the time difference as a function of the longitude positions of associated flares on the solar disk. Further, he has analysed a relation between peak flux intensity of moving type IVm bursts (~200 MHz) and the solar longitude value of associated flares. In general, it seems that in the initial stage of their development type IV radio bursts consist of two distinctive parts: one is stationary and confined in high frequency range (500~10,000 MHz or more), and the other is classified as a moving type IVm burst in metric and decametric frequency range.

077.043 **High resolution radio observations of the sun at 3.71 and 11.1 cm.** R. W. Hobbs, S. D. Jordan, W. J. Webster, Jr.
Nature, Phys. Sci., Vol. 243, 48 - 50 (1973). – Letter.

077.044 **The 350 MHz radio sun during 1972 November. Two interesting events.** A. N. Kelly.
Monthly Notes Astron. Soc. Southern Africa, Vol. 32, 9 - 10 (1973).

077.045 **A correlation between 10.7 cm (2800 MHz) solar radio emission and chromospheric flares.** G. Mariş.
Stud. Cerc. Astron., Vol. 18, 73 - 77 (1973). In Romanian.

The author has calculated coefficients of correlation between the 10.7 cm solar radio flux and a daily flare index for each solar rotation in the interval 1960–1971.

077.046 **Non-existence of linear polarization in type III solar bursts at 80 MHz.** R. J.-M. Grognard, D. J. McLean.
Solar Physics, Vol. 29, 149 - 161 = Radiophys. Publ., CSIRO, Sydney, RPP 1642 (1973).

A search for linear polarization showing the effect of Faraday rotation has been made at 80 MHz in type III solar radio bursts. A novel autocorrelation technique was employed. The results were entirely negative, contrary to what was expected on the ground of earlier, less sophisticated experiments. However, there are convincing theoretical reasons why no linear polarization should be expected.

077.047 **The prevalence of second harmonic radiation in type III bursts observed at kilometric wavelengths.** F. T. Haddock, H. Alvarez.
Solar Physics, Vol. 29, 183 - 196 (1973).

We present the analysis of 64 type III solar bursts that drifted from 3.5 MHz down to the range 350–50 kHz between March 1968 and February 1970. Bursts arrival times were predicted by a simple model and then compared with observations. The results show that, as the bursts drift, the fundamental often disappears below a certain frequency range while second harmonic remains. Below about 1 MHz the second harmonic occurrence predominates.

077.048 **Absolute calibration of solar radio flux density in the microwave region.** H. Tanaka, J. P. Castelli, A. E. Covington, A. Krüger, T. L. Landecker, A. Tlamicha.
Solar Physics, Vol. 29, 243 - 262 (1973).

The absolute calibration of solar radio flux density in the microwave region, which showed considerable discrepancies until 1966, has become completely uniform through international cooperative work. A complete history is described to avoid confusion, and correction factors are derived to convert the published values into absolute values for long series of routine observations. It is also shown that the most reliable calibration can be made by using a large pyramidal horn and by using sky and room temperature as calibration standards

077.049 **Distribution of the radio brightness near the solar limb from observations of the solar eclipses on the 22nd of September 1968 and the 7th of March 1970 at 10 cm**

wavelength. N. Ya. Nikolaev, Yu. F. Yurovsky.
Izv. Krymskoj Astrofiz. Obs., Vol. 46, 128 - 135 (1972).
In Russian.

077.050 Some characteristics of local radio sources from observations of the solar eclipse on the 7th of March 1970 in Cuba. Yu. F. Yurovsky, L. I. Yurovskaya.
Izv. Krymskoj Astrofiz. Obs., Vol. 46, 136 - 143 (1972).
In Russian.

The analysis of the observational data obtained during the solar eclipse on the 7th of March 1970 has shown that the sources of the S-component at 10 cm wavelength correspond to all sunspots. The maximum radio emission above a unipolar sunspot does not show an apparent displacement relative to the radius passing through the centre of the spot. In complex groups the brightest part of the radio source is displaced from the preceding sunspot to the following smaller ones, which are of opposite polarity.

077.051 Recent advances in solar radio astronomy. N. R. Labrum.
Progr. Astronaut. Aeronaut., Vol. 30, (see 003.009), p. 93 - 116 (1972).

077.052 Radio type III bursts accompanied by energetic solar electrons. L. G. Evans, J. Fainberg.
Bull. American Astron. Soc., Vol. 5, 272 (1973). – Abstr. AAS.

077.053 Coronal densities determined from hectometric radio observations of solar storm activity.
J. Fainberg, R. Fitzenreiter.
Bull. American Astron. Soc., Vol. 5, 272 (1973). – Abstr. AAS.

077.054 Interferometry of solar radio sources at 15,000 wavelengths. K. R. Lang, H. Zirin.
Bull. American Astron. Soc., Vol. 5, 275 (1973). – Abstr. AAS.

077.055 Type II emission at hectometric frequencies. H. H. Malitson, R. G. Stone.
Bull. American Astron. Soc., Vol. 5, 276 (1973). – Abstr. AAS.

077.056 The coherent amplification of radio emission in type III bursts. B. Prasad.
Bull. American Astron. Soc., Vol. 5, 279 (1973). – Abstr. AAS.

077.057 Solar longitude dependence of observed low frequency type III radio bursts. K. Sakurai.
Bull. American Astron. Soc., Vol. 5, 279 (1973). – Abstr. AAS.

077.058 Search for circular polarized emission from solar hemispheres at microwaves.
P. Kaufmann, E. Scalise, Jr., R. E. Schaal, J. R. D. Lépine, D. Basu, A. L. Ibañez.
Solar Physics, Vol. 29, 393 - 397 (1973).

Solar mappings with moderate angular resolution at 7 GHz seem to support the hemispherical dependence of the sense of circular polarization being right-handed for the southern hemisphere, and left-handed for the northern hemisphere. One explanation of the effect can be found by taking into account the 'missing' fields from active centers, emerging from one hemisphere and immersing into the other.

077.059 Solar active regions at 9 and 3.5 mm wavelengths under disturbed conditions.
M. R. Kundu, S.-Y. Liu.
Solar Physics, Vol. 29, 409 - 415 (1973).

Some properties of solar active regions at 9 and 3.5 mm wavelengths under disturbed conditions are discussed. New regions develop or weak regions intensify at millimeter wavelengths as a result of flares at distant sites. The spectra of the peak flux density of moderately strong bursts observed at

9 mm show a sharp drop toward the shorter millimeter wavelengths. The weak bursts at 3.5 mm manifest mainly as heating phenomena.

077.060 Interferometer observation of pulsating sources associated with a type IV solar radio burst.
K. Kai, A. Takayanagi.
Solar Physics, Vol. 29, 461 - 475 (1973).

An extremely complex outburst, part of which showed unusually rapid intensity fluctuations of a few second interval, was observed on 1970 November 5 with the 160 MHz interferometer of the Nobeyama Solar Radio Station. Two alternative mechanisms responsible for the pulsating phenomenon are suggested: (1) gyroresonance absorption of continuum radiation by a fast particle beam injected in a quasi-periodic manner into a large region of weak magnetic field, or (2) magnetohydrodynamic oscillation of the continuum source itself, which is intrinsically much smaller than observed.

077.061 Submillimetre-wavelength solar observations at sea level using a Fourier spectrometer.
A. S. Vardanyan, A. N. Vystavkin, I. A. Iskhakov, Yu. I. Kolesov, V. N. Listvin, A. Ya. Smirnov, A. V. Sokolov, E. V. Sukhonin.
Astron. Zhurn. Akad. Nauk SSSR, Vol. 50, 657 - 659 (1973).
In Russian. English translation in Soviet Astron. AJ, Vol. 17, No. 3. – Short note.

077.062 Circular polarized emission from solar active regions at millimeter wavelengths.
S. Edelson, F. I. Shimabukuro, E. B. Mayfield.
Mitt. Astron. Ges., No. 32, p. 178 - 182 (1973).

077.063 The 810 MHz solar radio emission in the years 1957–1967. S. Zięba.
Acta Astron., Vol. 23, 159 - 167 (1973).

Observations of solar radio emission at 810 MHz have been made at the Cracov Observatory since October 1957. From the records obtained the daily values of solar flux density were compiled for the period of 1957–1967. Special care was taken to correct the daily values for observational effects and to reduce them to the absolute scale.

077.064 Observation of a type II solar radio burst to 37 $R_\odot$.
H. H. Malitson, J. Fainberg, R. G. Stone.
Astrophys. Letters, Vol. 14, 111 - 114 (1973).

For the first time, type II radio emission has been detected above a height of 5 solar radii from the center of the sun. A type II burst with clearly defined fundamental and second harmonic components was observed with the IMP–6 satellite on June 30, 1971, from 0542 to 0934 UT at frequencies from 1850 to 292 kHz. For a reasonable density distribution in the outer corona and assuming a plasma wave origin for the burst, this frequency range corresponds to heights of 14 to 37 solar radii. The derived average velocity of the burst is 635 km/sec.

077.065 Summary of daily solar microwave fluxes measured at the Heinrich-Hertz-Institut (1954–1971).
F. Fürstenberg, A. Krüger.
Zentralinstitut für Solar-Terrestrische Physik (Heinrich-Hertz-Institut), Deutsche Akad. Wiss. Berlin, HHI Suppl. Ser. Solar Data, Vol. 3, 1 - 67 (1972).

After achieving an international unification of the solar microwave spectrum by the URSI-Working Group on the absolute calibration of solar flux density the corrected series of the daily flux observations by the HHI are presented.

077.066 Correlation and spectral analysis of daily solar radio flux. M. El-Raey, P. Scherrer.
Solar Physics, Vol. 30, 149 - 158 (1973).

Correlation and spectral analysis of solar radio flux density and sunspot number near the maximum of the sunspot cycle has indicated the existence of (a) long period amplitude modulation of the slowly varying component (SVC) of radio emission; (b) coronal storage over a period of the order of three solar rotations; (c) fast decay (one solar rotation period or less) of gyromagnetic emissions from radio sources; (d) shift in location of chromospheric sources compared to those of either the upper corona or the photosphere.

077.067 **Polarization inversions in the radio emission at 237 MHz of McMath zone 11482.** P. Santin.
Solar Physics, Vol. 30, 159 - 161 (1973). — Research note.

077.068 **On the source of the slowly varying component at centimeter and millimeter wavelengths.**
F. I. Shimabukuro, G. A. Chapman, E. B. Mayfield, S. Edelson.
Solar Physics, Vol. 30, 163 - 173 (1973).
The general features of the slowly varying component at centimeter and millimeter wavelengths are explained by magneto-ionic thermal emission. A model of an active region is constructed in which the electron temperature and density profile is based on recent EUV measurements, and the current-free magnetic field configuration is derived from a longitudinal magnetogram and scalar potential theory. In the model, the contributions of the reflected component of the inward extraordinary wave is important in determining the characteristic features of the radio flux and polarization. Emission by the mechanism of resonance absorption does not appear to be a significant factor in this model.

077.069 **Decay time of type III solar bursts observed at kilometric wavelengths.**
H. Alvarez, F. T. Haddock.
Solar Physics, Vol. 30, 175 - 182 (1973).
Type III bursts were observed between 3.5 MHz and 50 kHz by the University of Michigan radio astronomy experiment aboard the OGO-5 satellite. Decay times were measured and then combined with published data ranging up to about 200 MHz. The observed decay times increase with decreasing frequency but at a rate considerably slower than that expected from electron-proton Coulomb collisions. At 50 kHz values differ by about a factor of 100. Using Hartle and Sturrock's solar wind model, Coulomb collisional frequencies were computed and compared with the apparent collisional frequencies deduced from the observations.

Polarization interferometer for 2800 MHz. Solar noise studies with a 0'.5 fan beam.
See Abstr. 033.001.

Solar high-resolution radio measurements of active regions at a wavelength of 2.8 cm. See Abstr. 072.018.

Microwave pulsations from solar flares.
See Abstr. 073.019.

On some transient Hα features associated with metric type III bursts. See Abstr. 073.041.

Radio emissions from solar flares.
See Abstr. 073.093.

Solar wind density model from km-wave type III bursts. See Abstr. 074.043.

Estimate of the value of the solar coronal magnetic field from measurements of the linear polarization of solar radio bursts. See Abstr. 074.105.

19 — 20 May 1969, an example of type III emission during the impulsive phase of flares. See Abstr. 076.032.

On the generation of high-energy particles in solar flares. See Abstr. 078.024.

Energetic solar electrons accompanying type III bursts observed at 1 A.U. See Abstr. 078.034.

The radio radiation of the moon and sun at 2.25 mm wavelength and of Jupiter at 2.1 mm wavelength. See Abstr. 094.872.

078 Solar Cosmic Radiation

078.001 Energetic solar particles and their relation to optical flares. S. Biswas, B. Radhakrishnan.
Solar Physics, Vol. 28, 211 - 231 (1973).

It has been recently suggested by several investigators that the accelerated charged particles provide the energy of the optical flare by the ionization loss process. We have examined this mechanism assuming different forms of the spectrum of the accelerated protons at lower chromosphere. The flux and the energy spectrum of protons of energy 0.1–100 MeV have been calculated at successive heights, from 10^3 to 40×10^3 km from the solar surface taking into account the ionization loss, pitch angle distribution and density distribution of the neutral and ionized hydrogen in the chromosphere and lower corona. Hence the energy spectrum of the protons escaping from the sun and the amount of energy dissipated in the solar chromosphere are computed. The calculated results are compared with the observational data on the solar event of September 28, 1961.

078.002 Estimation of an upper limit for the solar neutron emission during large flares. E. Kirsch.
Solar Physics, Vol. 28, 233 - 246 (1973).

Solar neutron emission during large flares is investigated by using neutron monitor data from the mountain stations Chacaltaya (Bolivia), Mina Aguilar (Argentine), Pic-du-Midi (France) and Jungfraujoch (Switzerland). Registrations from such days on which large flares appeared around the local noon time of the monitor station are superimposed with the time of the optical flare as reference point. No positive evidence for a solar neutron emission was found with this method. However, by using an extrapolation of the neutron transport functions given by Alsmiller and Boughner a rough estimation of mean upper limits for the solar neutron flux is possible. The flux limits are compared with Lingenfelter's model calculations.

078.003 Propagation anisotropies of solar flare protons and electrons at low energies in interplanetary space. K. R. Pyle.
Journ. Geophys. Res., Vol. 78, 12 - 28 (1973).

Flux anisotropies in interplanetary space were investigated for protons with $E > 0.66$ Mev and electrons with $E > 400$ kev. Data were taken from the University of Chicago charged-particle telescope aboard the deep-space probe Pioneer 7 and from the Goddard Space Flight Center magnetometer aboard the same spacecraft. In the first part the study shows that the proton and electron anisotropies are almost always directed from east of the magnetic field direction. This paper investigates the origin of the eastward anisotropy at low energies, during periods both early and late in the history of a solar particle event.

078.004 Pitch angle distribution of solar flare particles in interplanetary space.
R. H. Maurer, S. P. Duggal, M. A. Pomerantz.
Journ. Geophys. Res., Vol. 78, 29 - 36 (1973).

During the initial anisotropic phase of solar particle events, the flux in the radial direction is considerably less than that from the garden hose field direction. An analysis based on the theory of charged particle transport in random magnetic fields has been conducted to determine the pitch angle distribution of particles with respect to the garden hose field lines. The theoretical result is verified by the analysis of relativistic solar particle observations i.e., those of May 4, 1960; November 12, 1960; November 18, 1968; and February 25, 1969.

078.005 Strong pitch angle diffusion and magnetospheric solar protons. D. J. Williams, F. T. Heuring.
Journ. Geophys. Res., Vol. 78, 37 - 50 (1973).

Observations of 1.2- to 2.2- and 2.2- to 8.2-Mev solar protons at low altitudes are studied with emphasis on the interaction zone between the stable trapping regions and the high-latitude polar cap region.

078.006 Record-breaking cosmic ray storm stemming from solar activity in August 1972.
M. A. Pomerantz, S. P. Duggal.
Nature, Vol. 241, 331 - 333 (1973).

An unusual centre of solar activity was the seat of the most severe disturbances of the current solar cycle early in August 1972. Of the great variety of physical phenomena associated with this remarkable region (sunspot group 331, Mc-Math plage region 11976, centred at Carrington, longitude 009 degrees), we will concentrate on a description of the gross features of the greatest cosmic ray storm ever observed that commenced on August 4, and of which sunspot group 331 was the ultimate source.

078.007 Detection of relativistic solar particles before the Hα maximum of a solar flare.
T. Mathews, L. J. Lanzerotti.
Nature, Vol. 241, 335 - 338 (1973).

We report here the first direct detection of relativistic solar protons before the Hα maximum phase of a solar flare and in near coincidence with the maximum phase of the white-light flare. These observations have important implications for theories of particle acceleration during flares and for the appropriate particle injection times to be used in the modelling of solar particle events by diffusion theories.

078.008 Additional evidence for the existence of a very high energy solar particle component.
S. M. Schindler, P. D. Kearney.
Nature, Phys. Sci., Vol. 242, 56 - 57 (1973).

We present here initial results of a statistical analysis of ~ 6,300 h of data obtained from the CSU underground solar cosmic ray experiment. The data correspond to the interval March 1971–June 1972. We draw attention to an apparently significant trend existing in the results. The purpose of this work is to establish further the existence of a very high energy flare-generated particle flux, beyond the past limits set by the vertical geomagnetic cutoff (~ 15 GeV). The net result will be an obvious extension of the flare-particle spectrum into this region.

078.009 Numerical studies of the transport of solar protons in interplanetary space. S. Webb, J. J. Quenby.
Planet. Space Sci., Vol. 21, 23 - 42 (1973).

Numerical solutions of the Fokker-Planck equation governing the transport of solar protons are obtained using the Crank-Nicholson technique with the diffusion coefficient represented by $K_r = K_0 r^b$ where r is radial distance from the sun and b can take on positive or negative values.

078.010 On the concentration of heavy nuclei in solar cosmic radiation. S. S. Konyakhina, L. V. Kurnosova, V. I. Logachev, L. A. Razorenov, M. I. Fradkin.
Kosmich. Issled., Vol. 11, 162 - 163 (1973). In Russian. Brief information.

078.011 Satellite measurements of the charge composition of solar cosmic rays in the $6 \leq Z \leq 26$ interval.
B. J. Teegarden, T. T. von Rosenvinge, F. B. McDonald.
Astrophys. Journ., Vol. 180, 571 - 581 (1973).

We report measurements of the charge composition of solar cosmic rays during two flares occurring in 1971 April and September. The results were derived from a solid-state dE/dx versus E telescope which was part of the Goddard cosmic-ray experiment on the IMP-VI spacecraft. We compare our results with the spectroscopically determined coronal and photospheric values.

078.012 Measurements of the iron-group abundance in energetic solar particles.
D. L. Bertsch, C. E. Fichtel, C. J. Pellerin, D. V. Reames.
Astrophys. Journ., Vol. 180, 583 - 589 (1973).

The abundance of iron-group nuclei in the energetic solar particles was measured twice in the 1971 January 24 event and once in the 1971 September 2 event. Including earlier results from the 1966 September 2 event, the experimental series being discussed in this article has found the iron-group abundance to be in the range from 3–6 percent of the oxygen nuclei in the energy interval from 21 to 50 MeV per nucleon, in those events where the iron-group abundance could be measured. The abundance for the iron-group nuclei is consistent with the present solar spectroscopic abundance estimates.

078.013 The differential energy spectra of solar-flare ^{1}H, ^{3}He, and ^{4}He. W. F. Dietrich.
Astrophys. Journ., Vol. 180, 955 - 973 (1973).

The differential energy spectra of ^{3}He from the 1969 November 2 and the 1971 January 25 flares have been measured by the University of Chicago charged-particle telescope on board the IMP 5 satellite in the energy range of ~9 to 90 MeV nucleon^{-1}. Assuming a power-law spectrum in energy per nucleon, the spectral indices for the two flares for protons, ^{4}He, and ^{3}He are determined as well as the ^{3}He/^{4}He abundance ratio in the energy interval 10–50 MeV.

078.014 On the propagation of solar cosmic rays during maximum and minimum solar activity.
A. G. Zusmanovich, E. V. Kolomeets.
Prikl. i teor. fizika. Vyp. (No.) 3. Alma-Ata, 1972, p. 119 - 124. In Russian. – Abstr. in Referativ. Zhurn. 51. Astron., 3.51.478 (1973).

078.015 Choice of the model of propagation for calculating the injection spectrum of solar protons.
L. I. Miroshnichenko.
Geomagn. Aeronom., Vol. 13, 26 - 30 (1973). In Russian.

078.016 Registration of solar cosmic rays simultaneously near Venus and in the magnetosphere of the earth.
S. N. Vernov, T. A. Ivanova, S. N. Kuznetsov, Yu. I. Logachev, G. B. Lopatina, E. N. Sosnovets.
Geomagn. Aeronom., Vol. 13, 164 - 166 (1973). In Russian. Brief information.

078.017 Solar flare particle propagation: Comparison of a new analytic solution with spacecraft measurements.
J. E. Lupton, E. C. Stone.
Journ. Geophys. Res., Vol. 78, 1007 - 1018 (1973).

A new radial solution has been obtained to the Fokker-Planck equation for solar flare particle propagation that includes the effects of convection, energy change, and anisotropic diffusion with κ_r = constant. With an outer boundary at ~ 2.7 AU, a solar wind velocity of ~400 km/sec, and $\kappa_r \approx 2$ to 8×10^{20} cm^2/sec, the complete solution gives reasonable fits to the time profiles of 1- to 10-MeV protons from 'classical' flare-associated events observed with the Caltech solar and galactic cosmic ray experiment aboard Ogo 6.

078.018 Anisotropies in the interplanetary intensity of solar protons $E_p > 0.3$ MeV.
W. G. Innanen, J. A. Van Allen.

Journ. Geophys. Res., Vol. 78, 1019 - 1035 (1973).

By using Explorer 35 interplanetary observations of solar protons of $E_p > 0.3$ MeV during ten selected solar events (1967–1970) the time dependence of intensity and of the angular distribution of intensity has been studied for the first time in the sub-MeV range of energy. The respective contributions of diffusive and convective transport are resolved.

078.019 Access of solar protons to the earth's polar caps.
J. F. Fennell.
Journ. Geophys. Res., Vol. 78, 1036 - 1046 (1973).

Energetic solar proton observations ($E_p \gtrsim 300$ keV) in the interplanetary medium by Explorer 33 and Explorer 35 and over the polar caps by Injun 5 during the period September 1968 through March 1970 have been examined in detail. The solar proton intensities observed over the polar regions were compared with the interplanetary intensities on an absolute basis.

078.020 Comments on a paper by R. B. McKibben, 'The azimuthal propagation of low-energy solar-flare protons'. G. M. Simnett.
Journ. Geophys. Res., Vol. 78, 1235 - 1238, with a reply by R. B. McKibben, p. 1239 - 1242 (1973). – Letters.

078.021 Alpha particles in solar cosmic rays over the last 80,000 years.
L. J. Lanzerotti, R. C. Reedy, J. R. Arnold.
Science, Vol. 179, 1232 - 1234 (1973).

Present-day (1967 to 1969) fluxes of alpha particles from solar cosmic rays, determined from satellite measurements, were used to calculate the production rates of cobalt-57, cobalt-58, and nickel-59 in lunar surface samples. Comparison with the activities of nickel-59 (half-life, 8×10^4 years) measured in lunar samples indicate that the long-term and present-day fluxes of solar alpha particles are comparable within a factor of approximately 4.

078.022 Solar-flare generated cosmic ray emission of 24 January 1971. J. Ilenčík, L. Křivský.
Bull. Astron. Inst. Czechoslovakia, Vol. 24, 100 - 102 (1973).

An analysis has been made of the solar-flare generated cosmic ray emission of 24. 1. 1971 which caused the radioactivity of rocks on the moon.

078.023 Estimate of the probability of observing streams of solar cosmic ray protons in the earth's orbit.
I. V. Getzelev, V. I. Tkachenko.
Geomagn. Aeronom., Vol. 13, 208 - 211 (1973). In Russian.

078.024 On the generation of high-energy particles in solar flares. K. Sakurai.
Planet. Space Sci., Vol. 21, 793 - 798 (1973).

This paper discusses the relationship between some characteristics of microwave type IV radio bursts and solar cosmic ray protons of MeV energy. Brief discussion is given on the propagation of solar cosmic rays in the solar envelope after ejection from the flare regions.

078.025 Intensity increase of solar protons in July 1970.
N. V. Pereslegina, G. P. Lyubimov.
Kosmich. Issled., Vol. 11, 236 - 244 (1973). In Russian.

078.026 The determination of a probable interval for the mean transit time from the sun to the earth of the solar SSC particles. I. Niţă.
Stud. Cerc. Astron., Vol. 18, 79 - 84 (1973). In Romanian.

A probable interval has been obtained for the mean transit time of solar SSC particles by processing the observational data for the period 1966–1971 by the correlation method.

078.027 **The continuous emission of low energy cosmic rays during solar flares.** J. Feit.
Solar Physics, Vol. 29, 211 - 231 (1973).

A new type of diffusion equation is presented in which the shape of the emission curve and the time during which the emission of energetic flare particles from the solar surface occurs can be prescribed. An analysis of 13 solar events is given.

078.028 **On the acceleration of heavy nuclei on the sun.**
S. S. Konyakhina, L. V. Kurnosova, V. I. Logachëv, L. A. Rasorënov, M. I. Fradkin.
Kratkie soobshch. po fiz., 1972, No. 7, p. 73 - 79. In Russian.
Abstr. in Referativ. Zhurn. 51. Astron., 4.51.627 (1973).

078.029 **Several solar aspects of flare-associated particle events.** Z. Švestka.
Progr. Astronaut. Aeronaut., Vol. 30, (see 003.009), p. 141 - 162 (1972).

078.030 **Propagation of solar cosmic rays in the solar wind.** J. R. Jokipii.
Progr. Astronaut. Aeronaut., Vol. 30, (see 003.009), p. 247 - 261 (1972). – Presented at the AIAA observation and prediction of solar activity conference, Huntsville, Ala., Nov. 16–18, 1970.

078.031 **Measurements of solar protons, helium and heavy nuclei in the Aug. 4, 1972 solar event.**
S. Biswas, D. L. Bertsch, C. E. Fichtel, C. J. Pellerin, D. V. Reames.
Bull. American Astron. Soc., Vol. 5, 269 (1973). – Abstr. AAS.

078.032 **Composition of cosmic rays in the January 24 and September 2, 1971 solar events.**
C. E. Fichtel, D. L. Bertsch, S. Biswas, C. J. Pellerin, D. V. Reames.
Bull. American Astron. Soc., Vol. 5, 272 (1973). – Abstr. AAS.

078.033 **The solar particle events of August, 1972.**
M. A. I. van Hollebeke, F. B. McDonald.
Bull. American Astron. Soc., Vol. 5, 273 (1973). – Abstr. AAS.

078.034 **Energetic solar electrons accompanying type III bursts observed at 1 A.U.**
R. P. Lin, K. A. Anderson.
Bull. American Astron. Soc., Vol. 5, 275 - 276 (1973). Abstr. AAS.

078.035 **The variability of the charge composition of solar cosmic rays.**
T. T. von Rosenvinge, B. J. Teegarden, F. B. McDonald.
Bull. American Astron. Soc., Vol. 5, 279 (1973). – Abstr. AAS.

078.036 **Directional diffusion coefficients of solar protons inside and outside the bow shock.**
P. Verzariu, S. M. Krimigis.
Planet. Space Sci., Vol. 21, 971 - 982 (1973).

The directional diffusion coefficients of low-energy ($\geqslant 0.3$ MeV) solar protons inside and outside the bow shock are examined during the solar flare event of 24 January 1969.

078.037 **Propagation of particles from solar flares in a medium with a sharply changing diffusion coefficient.**
I. N. Toptygin.
Geomagn. Aeronom., Vol. 13, 393 - 398 (1973). In Russian.

078.038 **A comparison of theoretical and experimental estimates of the solar proton diffusion coefficient during three flare events.**
S. Webb, A. Balogh, J. J. Quenby, J. F. Sear.

Solar Physics, Vol. 29, 477 - 503 (1973).

The propagation time for solar protons observed during the events of January 24, February 25 and March 17, 1969 are compared with those estimated from numerical solutions of the Fokker-Planck transport equation, using values of the diffusion coefficient of the form $K_r = K_0 r^b$ where r is radial distance from the sun, K_0 is obtained from the plasma-field parameters near the earth and b varies from -3 to $+1$.

078.039 **Low-energy protons of solar origin and investigation of the interplanetary medium.**
V. N. Lutsenko, N. F. Pisarenko.
Mezhplanet. sreda i fiz. magnitosfery. Moskva, Nauka, 1972, p. 127 - 148. In Russian. – Abstr. in Referativ. Zhurn. 51. Astron., 5.51.507 (1973).

078.040 **Generation of electrons and protons during solar flares.** L. E. Gajnova, E. V. Kolomeets.
Prikl. i teor. fizika. Vyp. (No.) 3. Alma-Ata, 1972, p. 130 - 134. In Russian. – Abstr. in Referativ. Zhurn. 51. Astron., 5.51.524 (1973).

078.041 **Corpuscular radiation and active prominences.**
I. Klechek.
Izv. AN SSSR. Ser. fiz., Vol. 36, 2278 - 2280 (1972). In Russian. – Abstr. in Referativ. Zhurn. 51. Astron., 5.51.557 (1973).

078.042 **Low-energy protons of solar origin and investigation of the interplanetary medium.**
V. N. Lutsenko, N. F. Pisarenko.
Mezhplanet. sreda i fiz. magnitosfery. Nauka, Moskva, 1972, p. 127 - 148. In Russian. – Abstr. in Referativ. Zhurn. 62. Issled. kosmich. prostranstva, 5.62.279 (1973).

078.043 **Radial variation of magnetic fluctuations and the cosmic-ray diffusion tensor in the solar wind.**
J. R. Jokipii.
Astrophys. Journ., Vol. 182, 585 - 600 (1973).

The radial evolution of the power spectra of fluctuations in the interplanetary magnetic field is studied under the assumptions that the fluctuations are frozen-in magnetic fluctuations or that they are Alfvén waves with wave vector parallel to the average field. These are used to obtain the radial variation of the cosmic-ray diffusion tensor, under the assumption that the tensor is determined principally by $P_\perp(k_\parallel)$.

078.044 **Relation between the chemical composition of solar cosmic radiation and the spectrum of duration of a burst.** N. P. Tsimakhovich.
Cosmic rays No. 13, (see 003.013), p. 69 - 72 (1972). In Russian.

078.045 **Cosmic-ray burst on January 28, 1967.**
N. P. Chirkov, G. F. Krymsky, I. S. Samsonov, V. I. Ipat'ev.
Cosmic rays No. 13, (see 003.013), p. 77 - 80 (1972). In Russian.

078.046 **Solar flare cosmic rays at and beyond the modulation boundary.** J. R. Jokipii, E. C. Stone.
Journ. Geophys. Res., Vol. 78, 3150 - 3154 (1973). – Letter.

078.047 **On the isotope composition of helium in solar corpuscular streams.** B. S. Boltenkov, V. N. Gartmanov, G. E. Kocharov, V. O. Najdenov, Yu. N. Starbunov.
Izv. AN SSSR. Ser. fiz., Vol. 36, 2319 - 2323 (1972). In Russian. – Abstr. in Referativ. Zhurn. 51. Astron., 6.51.460 (1973).

078.048 Study of cosmic radiation in the moon's vicinity with the lunar satellites Luna 10, 11, 12.
N. L. Grigorov, V. G. Kurt, V. N. Lutsenko, V. L. Maduev, N. F. Pisarenko, I. A. Savenko.
Mezhplanet. sreda i fiz. magnitosfery. Moskva, Nauka, 1972, p. 109 - 126. In Russian. – Abstr. in Referativ. Zhurn. 51. Astron., 6.51.464 (1973).

078.049 Scattering of particles in the interplanetary space and properties of solar corpuscular streams.
I. N. Toptygin.
Izv. AN SSSR. Ser. fiz., Vol. 36, 2258 - 2264 (1972). In Russian. – Abstr. in Referativ. Zhurn. 51. Astron., 6.51.466; 62. Issled. kosmich. prostranstva, 6.62.226 (1973).

078.050 On the theory of cosmic-ray transfer with anisotropic scattering of particles and with convection.
V. F. Zakharchenko.
Izv. AN SSSR. Ser. fiz., Vol. 36, 2265 - 2270 (1972). In Russian. – Abstr. in Referativ. Zhurn. 51. Astron., 6.51.467; 62. Issled. kosmich. prostranstva, 6.62.227 (1973).

078.051 Cosmic-ray current and features in the solar wind.
A. M. Altukhov, G. F. Krymskij, A. I. Kuz'min, G. V. Skripin, I. A. Transkij.
Izv. AN SSSR. Ser. fiz., Vol. 36, 2285 - 2291 (1972). In Russian. – Abstr. in Referativ. Zhurn. 51. Astron., 6.51.468; 62. Issled. kosmich. prostranstva, 6.62.225 (1973).

078.052 On solar cosmic-ray propagation in interplanetary medium.
G. P. Lyubimov, N. V. Pereslegina, N. N. Kontor.
Izv. AN SSSR. Ser. fiz., Vol. 36, 2297 - 2305 (1972). In Russian. – Abstr. in Referativ. Zhurn. 51. Astron., 6.51.469; 62. Issled. kosmich. prostranstva, 6.62.229 (1973).

078.053 On a possibility of studying cosmic-ray variations in the past. V. A. Dergachev, G. E. Kocharov.
Izv. AN SSSR. Ser. fiz., Vol. 36, 2312 - 2318 (1972). In Russian. – Abstr. in Referativ. Zhurn. 51. Astron., 6.51.475 (1973).

078.054 Galactic cosmic-ray modulation with non-spherically symmetrical solar wind taking into account anisotropic conditions near the sun.
L. I. Dorman, Z. Kobilinski.
Izv. AN SSSR, Ser. fiz., Vol. 36, 2332 - 2345 (1972). In Russian. – Abstr. in Referativ. Zhurn. 51. Astron., 6.51.476; 62. Issled. kosmich. prostranstva, 6.62.228 (1973).

078.055 Cosmic-ray events in 1970 - 1971 from observations in the stratosphere.
A. N. Charakhch'yan, G. A. Basilevskaya, E. V. Vashenyuk, G. L. Petrova, Yu. I. Stozhkov, V. D. Khor'kov, T. N. Charakhch'yan.
Izv. AN SSSR. Ser. fiz., Vol. 36, 2363 - 2368 (1972). In Russian. – Abstr. in Referativ. Zhurn. 51. Astron., 6.51.478 (1973).

078.056 Energy spectra of solar protons from observations in the stratosphere. G. A. Bazilevskaya, Yu. I. Stozhkov, A. N. Charakhch'yan, T. N. Charakhch'yan.
Izv. AN SSSR. Ser. fiz., Vol. 36, 2369 - 2375 (1972). In Russian. – Abstr. in Referativ. Zhurn. 51. Astron., 6.51.479 (1973).

078.057 Cosmic rays from the far side of the sun.
L. Křivský.
Vesmír, Vol. 52, 149 - 150 (1973). In Czech.

078.058 Comparative characteristics of the soft component of solar and galactic cosmic radiation according to rocket and stratospheric measurements on the island with the coordinates 71°2 N and 155°0 E and in Apatity.
Eh. V. Vashenyuk, L. L. Lazutin, V. F. Tulinov, V. V. Tulyakov.
Izv. AN SSSR. Ser. fiz., Vol. 36, 2387 - 2390 (1972). In Russian. – Abstr. in Referativ. Zhurn. 62. Issled. kosmich. prostranstva, 6.62.223 (1973).

078.059 Upper limit to the 1−20 MeV solar neutron flux.
J. A. Lockwood, S. O. Ifedili, R. W. Jenkins.
Solar Physics, Vol. 30, 183 - 191 (1973).
The upper limit on the quiet time solar neutron flux from 1−20 MeV has been measured to be less than $2 \times 10^{-3} n$ $cm^{-2} s^{-1}$ at the 95% confidence level. This result is deduced from the OGO-6 neutron detector measurements of the 'day-night' effect near the equator at low altitudes for the period from June 7, 1969, to December 23, 1969. The OGO-6 detector had very low (<4%) counting rate contributions from locally produced neutrons in the detecting system and the spacecraft and from charged-particle interactions in the neutron sensor.

078.060 Energy losses of solar cosmic rays in interplanetary space. I. D. Palmer.
Solar Physics, Vol. 30, 235 - 242 (1973).
A simple model of solar cosmic ray propagation which includes diffusion, convection, and energy loss by adiabatic deceleration is studied. A Monte Carlo technique is employed to investigate the variation of mean particle energy in the interplanetary medium after the impulsive release of mono-energetic particles at the sun. Results are compared with an observation by Murray et al. (1971) of a 'knee' in the energy spectrum of solar protons.

078.061 Evidence for confinement of low-energy cosmic rays ahead of interplanetary shock waves.
R. A. R. Palmeira, F. R. Allum.
Solar Physics, Vol. 30, 243 - 253 (1973).
Short-lived (~ 15 min), low-energy proton increases associated with the passage of interplanetary shock waves have been previously reported. In the present paper the concurrent particle and magnetic field data, taken by detectors on Explorer 34, for four of these events are examined in a fine time scale (~ 1 min). Our results further support the view that these impulsive events are due to confinement of the solar cosmic-ray particles in the region just ahead (~ 10^6 km) of the advancing shock front. Data from the Pioneer 7 spacecraft for a similar event are shown to be consistent with this interpretation.

078.062 On the nature of the inhomogeneous structure of the circumsolar plasma.
I. S. Bajkov, N. A. Lotova.
Trudy fiz. in-ta. AN SSSR, Vol. 62, 53 - 60 (1972). In Russian. – Abstr. in Referativ. Zhurn. 51. Astron., 7.51.394 (1973).

078.063 The influence of ionization losses on particle acceleration on the sun and in space. P. Velinov.
IV Leningr. mezhdunar. seminar "Edinoobrazie uskoreniya chastits v razlich. masshtabakh kosmosa, 1972". Leningrad, 1972, p. 109 - 120. In Russian. – Abstr. in Referativ. Zhurn. 51. Astron., 7.51.396; 62. Issled. kosmich. prostranstva, 7.62.204 (1973).

078.064 On the enrichment of solar cosmic rays by heavy nuclei. S. S. Konyakhina, L. V. Kurnosova, V. I. Logachev, L. A. Razorenov, M. I. Fradkin.
IV Leningr. mezhdunar. seminar "Edinoobrazie uskoreniya chastits v razlich. masshtabakh kosmosa, 1972". Leningrad,

1972, p. 121 - 126. In Russian. − Abstr. in Referativ. Zhurn. 51. Astron., 7.51.397; 62. Issled. kosmich. prostranstva, 7.62.205 (1973).

078.065 Diffusion, convection and adiabatic cooling of solar cosmic rays.
V. N. Vasil'ev, I. N. Toptygin, L. G. Fridgant.
IV Leningr. mezhdunar. seminar "Edinoobrazie uskoreniya chastits v razlich. masshtabakh kosmosa, 1972". Leningrad, 1972, p. 213 - 235. In Russian. − Abstr. in Referativ. Zhurn. 51. Astron., 7.51.398; 62. Issled. kosmich. prostranstva, 7.62.206 (1973).

078.066 Acceleration of relativistic particles by shock waves in interplanetary space.
L. I. Dorman, N. S. Kaminer, A. E. Kuz'micheva.
IV Leningr. mezhdunar. seminar "Edinoobrazie uskoreniya chastits v razlich. masshtabakh kosmosa, 1972". Leningrad, 1972, p. 251 - 264. In Russian. − Abstr. in Referativ. Zhurn. 51. Astron., 7.51.399; 62. Issled. kosmich. prostranstva, 7.62.201 (1973).

Fermi acceleration and the energy spectra of heavy nuclei at low energies. See Abstr. 061.047.

Characteristics of electron and high-energy proton flares. See Abstr. 073.013.

Enrichment of heavy nuclei in the 17 April 1972 solar flare. See Abstr. 073.098.

Coronal magnetic fields and energetic particles. See Abstr. 074.064.

Solar energetic particles and wide-band continuum storms from metric to hectometric frequencies. See Abstr. 077.012.

Solar electrical discharges. See Abstr. 082.013.

Neutron measurements in space. See Abstr. 082.096.

Energy spectra of ancient solar flare particles and the origin of gas rich meteorites. See Abstr. 105.110.

079 Solar Eclipses

079.001 How often can we meet total solar eclipses at a fixed station on the earth?
K. Saito, A. Tojo.
Tokyo Astron. Obs., Report No. 61, Vol. 16, 391 - 415 (1973). In Japanese.

079.002 Some hints for photographers of total solar eclipses.
R. Mack, L. Weinstein, G. East.
Sky Telescope, Vol. 45, 322 - 326 (1973).

079.003 Eclipses totales de sol y aficionados mexicanos.
F. Diego Q.
El Universo, No. 102, Vol. 27, 3 - 4 (1973). − Concerning 1963 July 20, 1970 March 7, 1972 July 10.

079.004 Problematik von Sonnenfinsternisbeobachtungen.
H. Haupt.
Sternenbote, 16. Jahrgang, p. 66 - 71 (1973).

079.005 Finsternistheorie − Sonnenfinsternis 1973 06 30.
M. Firneis.
Sternenbote, 16. Jahrgang, p. 71 - 77 (1973).

079.006 On shadow bands accompanying total solar eclipses.
A. L. Stanford, Jr.
American Journ. Phys., Vol. 41, 731 - 733 (1973).
    A simple model is presented for production of shadow bands that is consistent with observed data, and which is also relatively easy to subject to experimental verification.

Lunar limb profiles for solar eclipses. See Abstr. 094.300.

079.100 Solar eclipse 1972 July 10

The solar eclipse of July 10, 1972. M. Waldmeier.
Astron. Mitt. Sternw. Zürich, No. 322, 15 pp. (1973).

Observations of the solar eclipse on July 10, 1972 by the expedition of the Kiev Department of the Astronomical and Geodetical Society.
D. V. Pyaskovsky, A. T. Nesmyanovich, L. N. Kurochka.
Astron. Tsirk., No. 756, p. 7 - 8 (1973). In Russian.

Limb darkening at the extreme solar limb from observations during the eclipse of 10 July 1972.
H. L. Poss, W. Rosen.
Bull. American Astron. Soc., Vol. 5, 20 (1973). − Abstr. AAS.

July 10, 1972 eclipse report from Charlottetown.
L. Lindsay.
Mercury, (Journ. Astron. Soc. Pacific), Vol. 2, No. 1, p. 8 (1973).

Pursuing the lunar shadow. A. N. Cox.
Sky Telescope, Vol. 45, 88 - 89 (1973).

Observations of the "white-light" corona during partial phases of the solar eclipse on July 10, 1972. See Abstr. 074.037.

Airborne video recorded coronal emission line profiles of λ 5303 at the 10 July 1972 total solar eclipse. See Abstr. 074.063.

079.101   Solar eclipse 1973 June 30

Conditions de l'éclipse totale de soleil du 30 juin 1973, au voisinage d'un point situé sur la côte est de la partie méridionale du Lac Rodolphe (Kenya).   H. Debehogne. Ciel et Terre, Vol. 89, 123 - 124 (1973).

El eclipse total de sol del 30 de junio de 1973. C. H. Smiley, translated from Sky Telescope, Vol. 44, 282 - 283 (1972) by A. D. Lara. El Universo, No. 102, Vol. 27, 5 - 7 (1973).

De totale zoneclips van 30 juni 1973 in Suriname. W. H. C. Carton. Hemel en Dampkring, Vol. 71, 160 - 161 (1973).

Solar eclipse of 1973 June 30 and its northern limit in England.   J. Meeus. Journ. British Astron. Ass., Vol. 83, 110 - 111 (1973).

Photographie et cinématographie astronomiques. Luminations et focales à adopter lors des éclipses totales de soleil.   G. Bianchi. L'Astronomie, 87e année, p. 46 - 50 (1973).

L'éclipse totale de soleil du 30 juin 1973. J. Rösch. L'Astronomie, 87e année, p. 81 - 84 (1973).

A propos de l'éclipse du 30 juin 1973. B. Morando. L'Astronomie, 87e année, p. 213 - 215 (1973).

Solar eclipse: June 30, 1973. Sky Telescope, Vol. 45, 161 - 163 (1973).

Swift as the moon's shadow. D. H. Liebenberg, M. M. Hoffman. Sky Telescope, Vol. 45, 351 - 353 (1973).

079.102   Solar eclipse 1970 March 7

Astrometrische Ergebnisse der Expedition zur totalen Sonnenfinsternis vom 7. März 1970 in Mexiko. M. G. Firneis, H. Haupt. Mitt. Astron. Ges., No. 32, p. 196 - 197 (1973). – Abstract.

Sky color and darkness at the total solar eclipse of March 7, 1970.   W. H. Glenn. Strolling Astronomer, Vol. 24, 92 - 97 (1973).

Polarization of the integral radiation of the corona during the solar eclipse on March 7, 1970. See Abstr. 074.085.

Polarimetrische Untersuchungen der Sonnenkorona bei der totalen Finsternis vom 7. März 1970 in Mexiko. See Abstr. 074.088.

Further evidence for a complex limb structure in the solar radial brightness distribution at mm wavelengths. See Abstr. 077.036.

Distribution of the radio brightness near the solar limb from observations of the solar eclipses on the 22nd of September 1968 and the 7th of March 1970 at 10 cm wavelength.   See Abstr. 077.049.

Some characteristics of local radio sources from observations of the solar eclipse on the 7th of March 1970 in Cuba.   See Abstr. 077.050.

079.103   Solar eclipse 1968 September 22

Run of direct and scattered radiation during the solar eclipse from observations at the station Rudnyj (Kustanaj Region).   V. I. Gubanova, P. M. Timofeeva. Fizika. Vyp. (No.) 5. Alma-Ata, 1971, p. 114 - 116. In Russian. – Abstr. in Referativ. Zhurn. 51. Astron., 3.51.473 (1973).

Actinometric observations during the total solar eclipse.   K. N. Kopylets, M. A. Yugaj. Fizika, Vyp. (No.) 5. Alma-Ata, 1971, p. 165 - 169. In Russian. – Abstr. in Referativ. Zhurn. 51. Astron., 3.51.474 (1973).

Distribution of the radio brightness near the solar limb from observations of the solar eclipses on the 22nd of September 1968 and the 7th of March 1970 at 10 cm wavelength.   See Abstr. 077.049.

079.104   Solar eclipse 1966 May 20

Radio astronomy observations of the solar eclipse on 20 May, 1966.   See Abstr. 003.002.

079.105   Solar eclipse 1955 June 20

Solar eclipse effect on sporadic E ionization, 2. See Abstr. 083.002.

079.106   Solar eclipse 1878 July 29

The great eclipse of 1878.   J. A. Eddy. Sky Telescope, Vol. 45, 340 - 346 (1973).

079.107   Solar eclipse 1965 May 30

Coronal emission line profile analysis from airborne eclipse observations of 30 May 1965.   See Abstr. 074.062.

080   Solar Figure, Internal Constitution, Rotation, Miscellanea

080.001   **Solar neutrinos and the influence of opacity, thermal instability, additional neutrino sources, and a central black hole on solar models.**   R. Stothers, D. Ezer.
Astrophys. Letters, Vol. 13, 45 - 48 (1973).

Adoption of the 'best' available opacities, of hypothetical additional neutrino sources, or of a central black hole in theoretical models for the sun only increases the discrepancy with the null results of Davis's ^{37}Cl experiment to detect solar neutrinos. No thermal instabilities in any of the solar models have yet been found.

080.002   **The solar neutrino problem – A progress (?) report.**   V. Trimble, F. Reines.
Rev. Modern Phys., Vol. 45, 1 - 5 (1973).

The conflict between observation and theoretical prediction of the flux of electron neutrinos from the sun has advanced in the past year from being merely difficult to understand to being impossible to live with. We review here attempts to explore the nature of the conflict, to seek possible ways out of it, and to inquire into additional experiments that have the capability either of resolving the conflict or at least of deciding which branch of physics or astrophysics is responsible for it.

080.003   **The effect of mechanical waves on empirical solar models.**   P. Ulmschneider, W. Kalkofen.
Solar Physics, Vol. 28, 3 - 7 (1973).

Empirical solar models contain the effect of heating due to radiative energy loss from acoustic waves. We estimate here the temperature difference between the radiative equilibrium model and the empirical model. The temperature difference between the equator and the poles caused by a hypothetical difference in the heating is estimated.

080.004   **The energy spectrum of small-scale solar magnetic fields.**   Y. Nakagawa, E. R. Priest.
Astrophys. Journ., Vol. 179, 949 - 963 (1973).

On the basis of observations that magnetic flux is transported by fluid motions at the photospheric level of the sun, the possible interpretation of the energy spectra of small-scale solar magnetic fields in terms of a passive response of the (longitudinal) magnetic field to turbulent fluid motions is examined. In consideration of the prevailing physical conditions, a theory is developed which accounts for the two-dimensional passive response of longitudinal magnetic fields to a three-dimensionally isotropic turbulent convection at the level of vanishing vertical velocity. The theoretical results are compared with the energy spectra of small-scale magnetic fields obtained from longitudinal magnetograms for typical active, quiet, and mixed regions of the sun.

080.005   **Solar oblateness and equatorial brightening.**   R. H. Dicke.
Astrophys. Journ., Vol. 180, 293 - 305 (1973).

Equatorial brightening near the limb associated with an elevated temperature in the upper photosphere is discussed. It is not possible to obtain satisfactory agreement with the observations simultaneously at both of the color bands employed for the observations. The energy requirements of such hypothetical excess photospheric temperatures are severe, and the stresses needed for a force balance do not seem to be present.

080.006   **Equator–pole temperature difference and the solar oblateness.**
R. W. Noyes, T. R. Ayres, D. N. B. Hall.
Solar Physics, Vol. 28, 343 - 345 (1973). – Research note.

080.007   **Solar neutrino problem: No low energy ^{3}He+ ^{3}He resonance.**
P. D. Parker, D. J. Pisano, M. E. Cobern, G. H. Marks.
Nature, Phys. Sci., Vol. 241, 106 - 108 (1973).

From a study of the ^{6}Li(^{3}He, t)^{6}Be reaction, experimental evidence is presented against the existence of the ^{3}He+^{3}He resonance corresponding to a level at 11.5 MeV in ^{6}Be which has been suggested as an ad hoc solution to the solar neutrino problem. The ^{6}Li(^{3}He, t) ^{6}Be data establish a limit on the spectroscopic factor S(^{6}Be*$\rightarrow^3$He+^{3}He) of less than 6×10^{-3} for any state in this region of ^{6}Be.

080.008   **The role of convection in stellar atmospheres. I. Observable effects of convection in the solar atmosphere.**   L. D. Travis, S. Matsushima.
Astrophys. Journ., Vol. 180, 975 - 985 (1973).

A numerical method has been developed to apply Spiegel's theory of convection based on the generalization of the mixing-length formalism for model-atmosphere calculations. The method is used to construct solar model atmospheres under various physical conditions in order to investigate the effect of convective flux on various observable quantities.

080.009   **Thermal plasma fluctuations and the bound-state ^{7}Be $(e^-, \nu)^7$Li rate in the sun.**
W. D. Watson, E. E. Salpeter.
Astrophys. Journ., Vol. 181, 237 - 240 (1973).

The influence of thermal fluctuations in the screening on the capture rate of bound, K-shell electrons by ^{7}Be is investigated for conditions in the sun where most of the predicted neutrino flux in the Davis experiment is produced. A two-parameter variational wave-function and a Monte Carlo technique are employed in the calculations.

080.010   **Response of solar atmosphere to a granular excitation.**   N. C. J. Chen.
Bull. American Astron. Soc., Vol. 5, 2 (1973). – Abstr. AAS.

080.011   **Four sources of error in solar model calculations.**   C. A. Rouse.
Bull. American Astron. Soc., Vol. 5, 2 (1973). – Abstr. AAS.

080.012   **A theory for the 5-minute oscillations of large horizontal scale.**   C. L. Wolff.
Bull. American Astron. Soc., Vol. 5, 20 (1973). – Abstr. AAS.

080.013   **The sun as a gravitational lens.**   N. V. Mitskevich.
Third Soviet Gravitational Conference, Erevan, 1972, (see 012.001), p. 401 - 404 (1972). In Russian.

080.014   **Solar magnetic sector structure: relation to circulation of the earth's atmosphere.**   J. M. Wilcox, P. H. Scherrer, L. Svalgaard, W. O. Roberts, R. H. Olson.
Science, Vol. 180, 185 - 186 (1973).

080.015   **On the cycle of solar magnetic activity.**   A. I. Ol.
Solnechnye Dannye 1972 Byull., No. 12, p. 102 - 105 (1973). In Russian.

The cycle of solar magnetic activity is shown to consist in the development of UM-regions at the end of an odd cycle, development of BM- and UM-regions in an even cycle and development of BM-regions in the next odd cycle. The mean duration of this cycle is approximately equal to 26 years.

080.016   **Solar neutrinos and a central magnetic field in the**

sun.
S. M. Chitre, D. Ezer, R. Stothers.
Astrophys. Letters, Vol. 14, 37 - 40 (1973).

A strong, centrally concentrated magnetic field in the sun could explain, at least in part, the low solar neutrino flux observed by Davis.

080.017 The fine structure of the latitude zone of the sun.
N. I. Kozhevnikov.
Astron. Tsirk., No. 756, p. 5 - 7 (1973). In Russian.

080.018 A mechanism for the production of light and dark contrasts in radiatively controlled lines.
K. B. Gebbie, R. Steinitz.
Solar Physics, Vol. 29, 3 - 15 (1973).

It is argued that visible contrasts can arise even in a line that is controlled wholly by an external radiation field. Lateral differences in the local shapes of the line absorption profile are shown to account for such contrasts. Two cases are treated explicitly: (a) a profile locally broadened by mass flow, and (b) a profile locally narrower due to the suppression of turbulent velocities, as might result from the presence of magnetic fields.

080.019 Magnetic fields in solar active regions.   D. M. Rust.
Progr. Astronaut. Aeronaut., Vol. 30, (see 003.009), p. 33 - 49 (1972). – Presented at the AIAA observation and prediction of solar activity conference, Huntsville, Ala., Nov. 16–18, 1970.

080.020 Recent solar magnetograph results.   J. W. Harvey.
Progr. Astronaut. Aeronaut., Vol. 30, (see 003.009), p. 51 - 63 (1972). – Presented at the AIAA observation and prediction of solar activity conference, Huntsville, Ala., Nov. 16–18, 1970.

080.021 Inference of solar magnetic polarities from H-alpha observations.   P. S. McIntosh.
Progr. Astronaut. Aeronaut., Vol. 30, (see 003.009), p. 65 - 92 (1972). – Presented at the AIAA observation and prediction of solar activity conference, Huntsville, Ala., Nov. 16–18, 1970.

080.022 Major variations in solar luminosity?
A. G. W. Cameron.
Rev. Geophys. Space Phys., Vol. 11, 505 - 510 (1973).

One possible explanation for the failure to detect neutrinos emitted from the solar interior is that the core of the sun may be in a temporarily expanded state. If so, the solar luminosity is currently much less than its normal value, which may account for the fact that the earth is now in the middle of a major glacial period. The astrophysical background of this situation is described, and the terrestrial consequences are discussed.

080.023 A one-dimensional approximation to the macroturbulent velocity field in the solar atmosphere.
R. C. Altrock, C. J. Cannon.
Bull. American Astron. Soc., Vol. 5, 268 (1973). – Abstr. AAS.

080.024 Internal time scales in stratified spin-down.
A. Clark, Jr.
Bull. American Astron. Soc., Vol. 5, 270 (1973). – Abstr. AAS.

080.025 Solar rotation determined from OSO-6 EUV spectroheliograms.   W. Henze, A. K. Dupree.
Bull. American Astron. Soc., Vol. 5, 273 (1973). – Abstr. AAS.

080.026 Solar rotation and solar activity.   R. Howard.
Bull. American Astron. Soc., Vol. 5, 273 - 274 (1973). – Abstr. AAS.

080.027 Line formation in multidimensional atmospheres with rapid depth variation of absorption coefficient.
H. P. Jones.
Bull. American Astron. Soc., Vol. 5, 274 (1973). – Abstr. AAS.

080.028 Photodisintegration of ^8B in the interior of the sun.
R. Mitalas.
Observatory, Vol. 93, 107 - 110 (1973).

Neither the high energy photons of the Planck distribution at $T_6 = 15$, nor the gamma rays produced by nuclear reactions in the sun are able to cause photodisintegration of ^8B in the sun to any extent. Hence the solar neutrino flux is not affected.

080.029 The mean solar magnetic field observed at the Mt. Wilson Solar Observatory.
P. H. Scherrer, J. M. Wilcox, R. F. Howard.
Bull. American Astron. Soc., Vol. 5, 279 (1973). – Abstr. AAS.

080.030 Differential rotation, magnetic fields and the solar neutrino flux.   D. Bartenwerfer.
Astron. Astrophys., Vol., 25, 455 - 456 (1973).

It is shown that differential rotation and magnetic fields yield smaller solar neutrino fluxes than previously calculated if one chooses appropriate angular velocity distributions or appropriate magnetic fields.

080.031 The solar dynamo and estimates of the magnetic diffusivity and the α-effect.   H. Köhler.
Astron. Astrophys., Vol. 25, 467 - 476 (1973).

The dynamo equations including the α-effect are applied to axisymmetric magnetic fields in the solar convection zone. The unknown functions as magnetic diffusivity, α-effect and the law of rotation are discussed by adapting the theoretical butterfly diagrams to the observed ones. The estimates give an appreciable lower value for the magnetic diffusivity than that used in the literature. Computations with different laws of rotation show that the angular velocity must increase inside the sun. The magnitude of the α-effect is estimated.

080.032 Ricerche sulla rotazione solare.
G. Belvedere, G. Godoli, S. Motta, L. Paternò.
Mem. Soc. Astron. Italiana, Nuova Ser., Vol. 43, 637 - 644 (1973).

The possibility is pointed out of interpreting in the same way: 1) the discrepancy between the solar rotation velocity deduced by the tracer method and that deduced by the spectroscopic method; 2) the dual behaviour of photospheric magnetic fields as far as their rotational velocity is concerned; 3) the preferential longitudes.

080.033 Solar semi-diameter measurements made in 1971 on the basis of a photographic method.   S. Leone.
Mem. Soc. Astron. Italiana, Nuova Ser., Vol. 43, 779 - 787 (1973).

The author presents a procedure for calculating the value ω_o (instantaneous) and the value $\omega_o{}^*$ (corresponding to the radius vector $r = 1$) of the geocentrical solar semi-diameter, on the basis of particular photo-measurements corrected for the differential effects of refraction and aberration.

080.034 Divers solar rotations.   J. M. Wilcox.
Cosmic plasma physics. Conference 1971, (see 012.016), p. 157 - 164 (1972).

080.035 The solar diameter at 5000 Å and Hα from photoelectric drift scans.   A. Wittmann.
Solar Physics, Vol. 29, 333 - 340 (1973).

An improved method is described for the measurement of both the solar radius and the height of the chromosphere in any desired wavelength. Possible sources of uncertainty are discussed and a comparison with other methods is made.

080.036 Short-periodic oscillations of the magnetic field of the sun as a star.
B. A. Ioshpa, V. N. Obridko, B. D. Shelting.
Solar Physics, Vol. 29, 385 - 392 (1973).
Correlation analysis applied to recordings of the magnetic field and velocity of the sun as a star reveals oscillations close to 300 s. The power spectrum of these oscillations is discussed.

080.037 Solar neutrinos, Martian rivers, and Praesepe.
C. Sagan, A. T. Young.
Nature, Vol. 243, 459 - 460 (1973). − Letter.

080.038 Solare Magnetfeldmessungen. E. H. Schröter.
Mitt. Astron. Ges., No. 32, p. 55 - 64 (1973). − Review paper presented at the assembly of the Astron. Ges., Wien, 1972 Sept.

080.039 Die Deutung des solaren Wechselfeldes mit dem α - Effekt und eine Abschätzung der magnetischen Diffusivität. H. Köhler.
Mitt. Astron. Ges., No. 32, p. 168 - 171 (1973).

080.040 Eine statistische Deutung der horizontalen Phasengeschwindigkeit der 5-Min.-Oszillationen.
F.-L. Deubner.
Mitt. Astron. Ges., No. 32, p. 182 (1973). − Abstract.

080.041 On the nature and origin of the solar five-minute oscillations. H. U. Schmidt, M. Stix.
Mitt. Astron. Ges., No. 32, p. 182 - 186 (1973),

080.042 Ausbreitung mechanischer Wellen in der Sonnenatmosphäre. R. Wolf.
Mitt. Astron. Ges., No.32, p. 219 - 221 (1973).

080.043 Changes in solar rotation due to the solar energy generation cycle of 11 years.
J. Kangas, P. Raychaudhuri.
Astrophys. Space Sci., Vol. 22, 123 - 126 (1973).
It is suggested that the observed differences in the periods of variation of some solar phenomena (solar brightness, appearance of sunspot maximum and interplanetary sector structure) occurring close to 27 days are due to differences in the rotation periods of the solar regions in which these phenomena are originated. Changes in periods during the solar cycle can be attributed to changes in the solar energy generation.

080.044 Solar astronomy objectives. F. Q. Orrall.
Publ. American Astronaut. Soc., Sci. Techn. Ser., Vol. 28, (see 012.018), p. 7 - 23 (1972).

080.045 What is the penetrating ability of the neutrino?
L. A. Mikaelyan.
JETP Letters, (USA), Vol. 16, 221 - 222 (1972).
The question whether the negative results of Davis' experiments might be attributed to the fact that the neutrino in solar matter loses an appreciable fraction of its energy is discussed.

080.046 A simple integration through the solar atmosphere.
R. J. Doyle.
American Journ. Phys., Vol. 41, 412 - 414 (1973).
The author describes an elementary astrophysical project which can be performed using a desk calculator or small computer. A simple integration is performed through the solar atmosphere to find the temperature-pressure relationship.

080.047 Search for solar-neutrino related M1 transitions in ⁶Li using 180° electron scattering.
L. W. Fagg, W. L. Bendel, N. Ensslin, E. C. Jones, Jr.
Phys. Letters B, (Netherlands), Vol. 44B, 163 - 164 (1973).

080.048 Neutrino spectrum and the solar-neutrino experiment. S. Pakvasa, K. Tennakone.
Nuovo Cimento Lettere, Ser. 2, Vol. 6, 675 - 676 (1973).

080.049 Magnetic fields of the sun and stars.
A. B. Severnyj.
Zemlya i Vselennaya, 1973, No. 3, p. 2 - 11. In Russian.

080.050 Temperature difference between pole and equator of the sun. A. Peraiah.
Solar Physics, Vol. 30, 29 - 30 (1973). − Research note.

Energy distribution in the solar spectrum and the solar constant. See Abstr. 003.079.

La description mathématique de certaines propriétés des systèmes de tourbillons par des réseaux statistiques. See Abstr. 022.022.

Scientific exploration with an out-of-ecliptic spacecraft. See Abstr. 053.021.

Neutrinoerne fra Solen som en mulighed for direkte observation af dens indre. See Abstr. 061.037.

The theory of weak screening in thermonuclear reactions. See Abstr. 062.006.

Mixing in stellar models. See Abstr. 065.031.

Meridian circulation in differentially rotating stars. See Abstr. 065.160.

Secular stability with departures from ³He equilibrium in the proton-proton chain. See Abstr. 065.175.

Observations of the variation of temperature with latitude in the upper solar photosphere. II. Magnetic-field comparison, implications for solar-oblateness measurements, and harmonic analysis. See Abstr. 071.006.

The formation of Mg I 4571 Å in the solar atmosphere. III: The Holweger solar model. See Abstr. 071.060.

On the possibility of constructing a radiative sunspot model in magnetohydrostatic equilibrium. See Abstr. 072.073.

Effect of the general magnetic field of the sun on the geoefficiency of the solar wind. See Abstr. 074.038.

Errata

080.901 Corrigendum: 'What cooks with solar neutrinos? '
[Nature, Vol. 238, 24 - 26 (1972)]. W. A. Fowler.
Nature, Vol. 242, 424 (1973).

Earth

081 Figure, Composition, and Gravity of the Earth

081.001 On the normal gravity field of the earth and the moon. D. V. Zagrebin.
Astron. Zhurn. Akad. Nauk SSSR, Vol. 50, 181 - 185 (1973). In Russian. English translation in Soviet Astron. AJ, Vol. 17, No. 1.
The earth's and moon's normal gravity fields are expressed in terms of Lamé functions.

081.002 The rapid calculation of potential anomalies. R. L. Parker.
Geophys. Journ. Roy. Astron. Soc., Vol. 31, 447 - 455 (1973).
It is shown how a series of Fourier transforms can be used to calculate the magnetic or gravitational anomaly caused by an uneven, non-uniform layer of material. Modern methods for finding Fourier transforms numerically are very fast and make this approach attractive in situations where large quantities of observations are available.

081.003 Review of Precambrian paleomagnetic data for Europe. H. Spall.
Earth Planet. Sci. Letters, Vol. 18, 1 - 8 (1973).
Some 36 paleomagnetic poles are available from Precambrian rock units from Europe, west of the Urals. They allow us to amplify Neuvonen's suggestions, and speculate on the pole path for the interval 1200–2000 my. In order to link younger Precambrian poles with Phanerozoic data, one interpretation is that a closed loop is required during the interval 500–1400 my.

081.004 Spherical harmonic analyses of the palaeomagnetic field.
N. P. Benkova, A. N. Khramov, T. N. Cherevko, N. V. Adam.
Earth Planet. Sci. Letters, Vol. 18, 141 - 147 (1973).
Analytical models of the palaeomagnetic field have been constructed for a number of geological periods (Quaternary, Neogene, Jurassic, Triassic, Permian and Permo-Carboniferous) by spherical harmonic analysis using the present-day world map as a basis and (for the earlier periods) using also a palaeogeographic reconstruction. The use of the palaeogeographic chart for the earlier periods simplifies the models, and its use appears to be valid.

081.005 Temperature gradients at the core-mantle interface. G. C. Kennedy, G. H. Higgins.
The Moon, Vol. 7, 14 - 21 (1973). – Paper dedicated to Professor Harold C. Urey on the occasion of his 80th birthday on 29 April, 1973.
Heat flowing out of the core must flow into the mantle. If the earth's magnetic field is owing to adiabatic magnetohydrodynamic circulation of the outer core, whole mantle convection or melting at the core mantle boundary is required to keep the inner core from becoming isothermal, thereby preventing adiabatic circulation.

081.006 The geophysical consequences of Professor Lyttleton. K. Runcorn.
Nature Vol. 241, 521 - 523 (1973). – Letter.

081.007 Mantle plumes, palaeomagnetism and polar wandering. M. W. McElhinny.
Nature, Vol. 241, 523 - 524 (1973). – Letter.

081.008 The need for a standard model of the earth's structure. R. O. Vicente.
Bull. Géod., Nouvelle Sér., Année 1973, No. 107, p. 105 - 106.

081.009 Residual deformation of real earth models with application to the Chandler wobble.
M. Israel, A. Ben-Menahem, S. J. Singh.
Geophys. Journ., Vol. 32, 219 - 247 (1973).

081.010 The implications for geophysics of modern cosmologies in which G is variable. P. S. Wesson.
Quarterly Journ. Roy. Astron. Soc., Vol. 14, 9 - 64 (1973).

081.011 On the estimate of accuracy of the determination of planetary characteristics of the earth's gravitational field. V. V. Buzuk, I. G. Vovk.
Trudy Novosib. in-ta inzh. geod., aehrofotos"emka i kartogr., Vol. 27, 3 - 14 (1972). In Russian. – Abstr. in Referativ. Zhurn. 52. Geod. Aehros"emka, 3.52.70 (1973).

081.012 Gravity measurements in the Sea of Japan. P. A. Stroev, Ju. A. Pavlov, V. L. Panteleev, V. O. Bagramjants.
Trudy Gos. Astron. Inst. Shternberga, Vol. 43, vyp. (No.) 1, 116 pp. (1972). In Russian.

081.013 Physical background of the geoidal figure. G. Barta.
Nature, Vol. 243, 156 - 158 (1973).
The author has approximated the equatorial section of the geoid by zonal spherical harmonics. It emerges that the equatorial section of the geoid can be approximated by the sum of two rigorously symmetrical geometrical forms. By varying the direction of the axes of the two symmetrical geometric figures, it is possible to choose a best fit solution. The geoid figure thus computed is similar to and even identical with the measured figure as regards both the position and the size of anomalies. The geoid is the result of two effects originating at great depth and in all probability is not linked with the distribution of oceans and continents on the surface of the globe. The starting point of this study is the coincidence of the direction of eccentricity of the magnetic dipole with the great positive geoid anomaly over Australia.

081.014 Fluid tidal effects on satellite orbit and other temporal variations in the geopotential.
K. Lambeck, A. Cazenave.
Groupe Recherches Géod. Spatiale, Bull. No. 7, 2 + 42 pp. (1973).
Perturbations in the motion of close earth satellites arise from the atmospheric and ocean tides. The principal perturbations are due to the semi-diurnal ocean tides; if they are neglected in the orbital analysis for solid earth tides, the resultant Love numbers tend to be too small. Variations in the geopotential arising from seasonal variations in the earth's inertia tensor have also been estimated.

081.015 Photometry of the earth with the Zond space stations. N. P. Lavrova, A. B. Sandomirskij.
Izv. vyssh. ucheb. zavedenij. Geod. i aehrofotos"emka, 1972, No. 4, p. 109 - 114. In Russian. – Abstr. in Referativ. Zhurn.

62. Issled. kosmich. prostranstva, 4.62.283 (1973).

081.016 Longman tidal formulas: resolution of horizontal components. H. N. Pollack.
Journ. Geophys. Res., Vol. 78, 2598 - 2600 (1973).
Formulas are presented to compute, without reference to tables, the north-south and east-west components of the tidal accelerations due to the moon and sun at any point on the earth's surface at any given time.

081.017 Physical state of the earth's core. J. A. Jacobs.
Nature, Phys. Sci., Vol. 243, 113 - 114 (1973).

081.018 Quadrupolar analysis of storage and release of elastic energy in the earth.
D. Pines, J. Shaham.
Nature, Phys. Sci., Vol. 243, 122 - 127 (1973).
The reservoir of elastic energy in the crust and mantle of the earth has been assumed to be too small to be important for most geophysical processes. But here we report that we have calculated the global quadrupolar mechanical energy stored in the earth to be ~ 10^{32} erg and that it may play an important role in various processes in the crust and mantle, including polar wandering and seismic activity.

081.019 Formation of the earth's core. H. G. Tolland.
Nature, Phys. Sci., Vol. 243, 141 - 142 (1973).
The author considers the possibility of core formation after accretion from an initial homogeneous mixture of about 82% by volume of silicate and 18% of iron plus sulphide.

081.020 On the possible direction of the earth's evolutionary process. I. V. Kirillov.
Astron. vestn., Vol. 7, 113 - 117 (1973). In Russian.
The paper considers the background of the idea about the earth's expansion put forward by the author more than two decades ago on the basis of the analysis of global geomorphological features of earth's surface. A survey of arguments in favour of the possibility of the expansion of the earth and the increase of its mass during its geological history is presented.

081.021 Potentialentwicklung nach Laméschen Funktionen. H. G. Walter.
Mitt. Astron. Ges., No. 32, p. 226 - 228 (1973).

081.022 On the relation between earthquakes and gravitational waves. T. Mitani.
Mem. Japan Astron. Study Ass., No. 19, Vol. 5, 235 - 266 (1972). In Japanese.

081.023 The probabilistic background of some statistical methods in physical geodesy. S. L. Lauritzen.
Geod. Inst., København, Danmark, Meddelelse No. 48, 96 pp. (1973).

081.024 Zonal gravity harmonics from long satellite arcs by a seminumeric method. C. A. Wagner.
Journ. Geophys. Res., Vol. 78, 3271 - 3280 (1973).

081.025 A solution of the geodetic boundary value problem to order e^3. R. S. Mather.
Goddard Space Flight Center, Greenbelt, Maryland, Prepr. X-592-73-11, 6 + 99 + A 28 pp. (1973).

081.026 Hydrostatic figure of the earth: Theory and results. M. A. Khan.
Goddard Space Flight Center, Greenbelt, Maryland, Prepr. X-592-73-105, 1 + 33 pp. (1973).

081.027 11th order resonance terms in the geopotential from the orbit of Vanguard 3. C. A. Wagner.

Goddard Space Flight Center, Greenbelt, Maryland, Prepr. X-592-73-130, 6 + 27 pp. (1973).

081.028 Determination of the free diurnal nutation of the earth based on latitude observations in Washington for 1915–1940. A. I. Emets, Ya. S. Yatskiv.
Astrometriya i Astrofizika, *Kiev,* Vyp. (No.) 18, (see 003. 016), p. 3 - 8 (1973). In Russian.
The paper deals with the power spectrum of the latitude variations in Washington within the frequency range close to the frequency of free diurnal nutation. For comparison of the power spectrum an analysis of artificial random series is given.

081.029 Das Geoid aus Beobachtungen der Satellitenaltimetrie. K. Arnold.
Deutsche Akad. Wiss. Berlin, Zentralinst. Phys. Erde, Potsdam, Veröff. No. 7, 46 pp. (1972).
A theory for the perturbations of satellite orbits by the earth's gravity potential has been developed; geoid undulations prove to be the parameters of the potential. The error equations of satellite altimetry are established.

081.030 Analysis of methods for computing an earth gravitational model from a combination of terrestrial and satellite data. J. Hopkins.
Geophys. Monograph 15, (see 012.025), p. 93 - 98 (1972).

081.031 Nature of the satellite-determined gravity anomalies. M. A. Khan.
Geophys. Monograph 15, (see 012.025), p. 99 - 106 (1972).

081.032 Simple layer model of the geopotential in satellite geodesy. K.-R. Koch.
Geophys. Monograph 15, (see 012.025), p. 107 - 109 (1972).

081.033 Representation of the earth potential by buried masses. G. Balmino.
Geophys. Monograph 15, (see 012.025), p. 121 - 124 (1972).

081.034 Geopotential representation with sampling functions. C. A. Lundquist, G. E. O. Giacaglia.
Geophys. Monograph 15, (see 012.025), p. 125 - 131 (1972).

081.035 Satellite-satellite tracking for estimating geopotential coefficients.
C. F. Martin, T. V. Martin, D. E. Smith.
Geophys. Monograph 15, (see 012.025), p. 139 - 144 (1972).

081.036 Improvement of zonal harmonics by the use of observations of low-inclination satellites Dial, SAS, and Peole. A. Cazenave, F. Forestier, F. Nouel, J. L. Pieplu.
Geophys. Monograph 15, (see 012.025), p. 145 - 150 (1972).

081.037 Error model for the SAO 1969 standard earth. C. F. Martin, N. A. Roy.
Geophys. Monograph 15, (see 012.025), p. 161 - 167 (1972).

Doppelplanet Erde–Mond.
See Abstr. 003.057.

Physique et dynamique planétaires. Géodynamique.
Vol. 4. See Abstr. 003.085.

Geomorphology. Abstracts of papers.
See Abstr. 003.136.

Theory of precession, nutation and rotational velocity of the deformable earth (II).
See Abstr. 044.010.

A nonlinear, resistive boundary layer in rotating

hydromagnetic flow.   See Abstr. 062.007.

A new cosmological model: Formation of organic molecules, planets, and comets.   See Abstr. 065.068.

On planetary cores.   See Abstr. 091.060.

Cometary collisions and geological periods. See Abstr. 102.004.

Errata

081.901   **Errata: 'Equations for 15th-order geopotential coefficients from the orbit of Transit 1B'** [Planet. Space Sci., Vol. 20, 1213 - 1228 (1972)]. H. Hiller, D. G. King-Hele. Planet. Space Sci., Vol. 21, 535 (1973).

082 The Earth's Atmosphere including Refraction, Scintillation, Extinction, Airglow, Site Testing

082.001 Stellar component of the night glow.
A. S. Sharov, N. A. Lipaeva.
Astron. Zhurn. Akad. Nauk SSSR, Vol. 50, 107 - 114 (1973).
In Russian. English translation in Soviet Astron. AJ, Vol. 17, No. 1.

The value of the stellar component of the night glow in the B system based on star counts is determined. Corrections to the photometric scales of catalogues used for star counts lead to a considerable (about 1.5 times) increase of the estimate of night sky brightness compared with the results by Roach and Megill.

082.002 Observations of the He II 304-Å radiation in the night sky.
F. Paresce, S. Bowyer, S. Kumar.
Journ. Geophys. Res., Vol. 78, 71 - 79 (1973).

The intensity and spatial variations of the He II 304-Å radiation in the night sky were measured to an altitude of 264 km from a sounding rocket launched from Thumba, India, on March 10, 1970. The data obtained are presented in the form of an all-sky map and are compared with theoretical predictions. The data can be fit with a constant density plasmasphere model bounded at the magnetic shell $L = 4$ in combination with a tenuous gas of helium ions in the plasma sheet.

082.003 Interpretation of OGO 5 Lyman alpha measurements in the upper geocorona.
J. L. Bertaux, J. E. Blamont.
Journ. Geophys. Res., Vol. 78, 80 - 91 (1973).

Lyman α intensity measurements (1216 Å) were obtained by a photometer on board the OGO 5 spacecraft outside the geocorona. The Lyman α emission originating from hydrogen atoms in the upper part of the geocorona was derived from the total measured intensity after subtraction of the extraterrestrial emission. The hydrogen density distribution between 5 and 16 R_E is consistent with the evaporative atmospheric model of J. W. Chamberlain on the night side.

082.004 Is there enough solar extreme ultraviolet radiation to maintain the global mean thermospheric temperature? R. G. Roble, R. E. Dickinson.
Journ. Geophys. Res., Vol. 78, 249 - 257 (1973).

The global mean temperature profile of the neutral thermosphere above 120 km is calculated using the solar EUV flux tabulation of H. E. Hinteregger (1970) for wavelengths less than 1310 Å and the data of K. G. Widing et al. (1970) for wavelengths greater than 1310 Å. The electron temperature in the ionosphere is also calculated using the solar EUV flux and is compared with the data of J. V. Evans (1971).

082.005 Far ultraviolet spectra and altitude profiles of the dawn airglow.
G. J. Rottman, P. D. Feldman, H. W. Moos.
Journ. Geophys. Res., Vol. 78, 258 - 264 (1973).

082.006 Comparison of the correlation of incoherent scatter and ionosonde measurements of temperature with calcium plage and 2800-Megahertz intensities.
J. L. Rohrbaugh, W. E. Swartz, J. S. Nisbet.
Journ. Geophys. Res., Vol. 78, 281 - 287 (1973). – Brief report.

082.007 Atomic hydrogen concentrations in the mesosphere and the hydroxyl emissions.
W. F. J. Evans, E. J. Llewellyn.

Journ. Geophys. Res., Vol. 78, 323 - 326 (1973). – Letter.

082.008 Low-intensity Balmer emissions from the interstellar medium and geocorona.
R. J. Reynolds, F. L. Roesler, F. Scherb.
Astrophys. Journ., Vol. 179, 651 - 657 (1973).

Galactic and nongalactic components of the diffuse Hα and Hβ nightsky emissions have been resolved with a Fabry-Perot spectrometer. The nongalactic component of both lines accounts for most of the emission at galactic latitudes greater than 30°. The intensities of the galactic component yield values for the average ionization rate per hydrogen atom that are between 10^{-15} and 10^{-14} s^{-1} assuming steady-state ionization.

082.009 Site testing. P. Fellgett.
Observatory, Vol. 93, 34 - 37 (1973). – Letter.

082.010 Mode d'épaississement d'une basse couche convective matinale en ciel clair. R. Rosset,
P. Mascart, H. Isaka, R.-G. Soulage.
Comptes Rendus Acad. Sci. Paris, Sér. B, Vol. 276, 223 - 226 (1973).

Cette note illustre sur un exemple l'aspect discontinu de l'épaississement de la couche convective matinale en ciel clair. Le phénomène est lié à la stabilité statique et à la turbulence de la couche stable sus-jacente.

082.011 Mise en évidence d'une diminution temporaire de l'ozone de la haute atmosphère au moment du lever du soleil. P. Rigaud.
Comptes Rendus Acad. Sci. Paris, Sér. B, Vol. 276, 445 - 447 (1973).

Des mesures photométriques stellaires montrent une importante diminution de l'ozone de la haute atmosphère au moment du lever du soleil. On explique cette diminution par le fait que le rayonnement solaire n'éclaire pas l'atmosphère au même moment pour toutes les longueurs d'onde.

082.012 The evaluation of night time seeing from polar star trails. E. Moroder, A. Righini.
Astron. Astrophys., Vol. 23, 307 - 310 (1973).

In this paper, using the theory of propagation of electromagnetic waves in turbulent media and the results of Fourier optics, the resolving power of large telescopes is evaluated from image motion as measured on the polar trails obtained with small size site testing telescopes.

082.013 Solar electrical discharges. E. W. Crew.
Nature, Vol. 241, 39 (1973). – Letter.

082.014 Response of a general circulation model of the atmosphere to removal of the Arctic ice-cap.
R. L. Newson.
Nature, Vol. 241, 39 - 40 (1973). – Letter.

082.015 Night sky background measurement at 6 to 0.3 mm. K. D. Williamson, A. G. Blair, L. L. Catlin, R. D. Hiebert, E. G. Loyd, H. V. Romero.
Nature, Phys. Sci., Vol. 241, 79 - 80 (1973).

On May 17, 1972, at 0112 HST a far infrared radiometer cooled by superfluid helium was launched on a rocket from the Kauai Test Range Facility, Kauai, Hawaii. Photometric measurements of the night sky background were successfully made in three spectral regions: 6 to 0.8 mm, 6 to 0.6 mm, and 6 to 0.3 mm. Here we describe the results of this experiment, which yielded a background flux consistent with a 2.7 K blackbody source.

082.016 High resolution spectra of the stratosphere between 30 and 200 cm^{-1}. J. P. Baluteau, E. Bussoletti.
Nature, Phys. Sci., Vol. 241, 113 - 114 (1973). − Letter.

082.017 Supersonic generation of atmospheric waves. T. Beer.
Nature, Vol. 242, 34 (1973).

The author draws an analogy between the supersonic motion of the moon's shadow and the supersonic motion of the earth's terminator. The terminator is supersonic between ±45° latitudes at all altitudes below 100 km and may therefore generate gravity waves in this region.

082.018 Infrared photography of OH airglow structures. A. W. Peterson, L. M. Kieffaber.
Nature, Vol. 242, 321 - 322 (1973).

Infrared airglow emission is bright, patchy and erratic. The primary source of this emission was identified by Meinel as rotation-vibration bands of the OH molecule. We have carried out a study at 2.2 μm and a similar study at 1.65 μm. At both wavelengths many small bright patches were resolved which had lifetimes of tens of minutes. Surprising results were obtained from a series of 60 consecutive 15-min photographs, taken during the moonless portions of the nights of December 1 and 2 at Capilla Peak Observatory. The photographs all show bright cloud-like structures which moved on the sky and varied in brightness.

082.019 Effect of dimerization on the transmission of water vapor in the near-infrared. S. S. Penner.
Journ. Quant. Spectrosc. Radiat. Transfer, Vol. 13, 383 - 384 (1973).

We emphasize the possible importance of dimerization of H_2O on the optical properties of the atmosphere. A representative estimate for increased absorption is given for the 2.7 μ region.

082.020 Contributions to the theory of atmospheric refraction. Part II. Refraction corrections in satellite geodesy. J. Saastamoinen.
Bull. Géod., Nouvelle Sér., Année 1973, No. 107, p. 13 - 34.

A detailed description of refraction in satellite photogrammetry with vertical photography of the earth's surface taken from an orbiting satellite, or with photography of an orbiting satellite taken from the surface of the earth against the stellar background is reported.

082.021 On the seasonal variation of upper atmospheric sodium. G. Fiocco, G. Visconti.
Journ. Atmosph. Terr. Phys., Vol. 35, 165 - 171 (1973).

The origin of free Na atoms in the upper atmosphere has been attributed to sublimation from dust. In this paper we consider the energetic budget for the dust particles and estimate their temperature T_p and the related Na evaporation rate.

082.022 6300 Å night airglow enhancements in low latitudes. V. R. Rao, P. V. Kulkarni.
Journ. Atmosph. Terr. Phys., Vol. 35, 193 - 206 (1973).

At low latitudes, in the nocturnal variation of 6300 Å night airglow, quite often increases in intensity followed by decreases are observed. From a large amount of data obtained at Mt. Abu, India, with three photoelectric photometers, these enhancements are studied and the dynamics of the ionosphere in low latitudes during those hours is investigated.

082.023 Sunrise changes in concentrations of minor neutral constituents in the mesosphere.

M. R. Bowman, L. Thomas.
Journ. Atmosph. Terr. Phys., Vol. 35, 347 - 352 (1973).
Short paper.

082.024 Increase of Na twilight emission after the earth's crossing of the orbital planes of comets Halley and Encke. G. Visconti, G. Fiocco.
Journ. Atmosph. Terr. Phys., Vol. 35, 353 - 356 (1973).

Measurements of Na twilight emission indicate an increase after the crossing by the earth of the orbital planes of comets Halley and Encke. Data support the hypotheses that dust particles of cometary origin enter the atmosphere and fragment into small grains, and that Na is produced by sublimation from the grains.

082.025 Neutral thermosphere temperatures from density scale height measurements.
G. P. Newton, D. T. Pelz.
Journ. Geophys. Res., Vol. 78, 725 - 732 (1973).

During each 4-min satellite interrogation period the Explorer 32 density gages measured the atmospheric density approximately every 2 sec. At equatorial latitudes ($-27°$ to $+27°$) the temperatures have a diurnal maximum value of approximately 900°K, in agreement with the Jacchia 1965 model, and a diurnal minimum value of approximately 650°K, below the model values.

082.026 Low latitude density variations in the earth's neutral atmosphere between 200 and 400 km, from August 1969 to May 1970.
D. E. Knight, R. Uribe, B. E. Woodgate.
Planet. Space Sci., Vol. 21, 253 - 271 (1973).

082.027 Observation of mesospheric ozone at low latitudes. P. B. Hays, R. G. Roble.
Planet. Space Sci., Vol. 21, 273 - 279 (1973).

Stellar ultraviolet light near 2500 Å is attenuated in the earth's upper atmosphere due to strong absorption in the Hartley continuum of ozone. The intensity of stars in the Hartley continuum region has been monitored by the University of Wisconsin stellar photometers aboard the OAO-2 satellite during occultation of the star by the earth's atmosphere. The results of approximately 12 stellar occultations, obtained in low latitudes, are presented, giving the nighttime vertical number density profile of ozone in the 60- to 100-km region.

082.028 Spatial and temporal variations of the Lyman-alpha airglow and related atomic hydrogen distributions. R. R. Meier, P. Mange.
Planet. Space Sci., Vol. 21, 309 - 327 (1973).

Spatial variations of the atomic hydrogen Lyman-alpha airglow were observed at 550 km from the OSO-4 spacecraft. We present an analysis of the data from the Lyman-alpha detectors. The fundamental model that we adopt is one in which geocoronal atomic hydrogen resonantly scatters solar radiation. The variation of the airglow with direction of view and solar zenith angle is computed from models and compared with the observations in order to arrive at the distribution of atomic hydrogen.

082.029 Stellar occultation measurements of molecular oxygen in the lower thermosphere. P. B. Hays, R. G. Roble.
Planet. Space Sci., Vol. 21, 339 - 348 (1973).

Stellar ultraviolet light near 1500 Å is attenuated in the earth's upper atmosphere due to strong absorption in the Schumann-Runge continuum of molecular oxygen. The intensity of stars in the Schumann-Runge continuum region has been monitored by the University of Wisconsin stellar photometers aboard the OAO-2 satellite during occultation of the star by the earth's atmosphere. The results of 14 stellar occultations obtained in low and middle latitudes are presented giving the night-time vertical number density profile of molecular oxygen in the 140-200 km region.

082.030 On empirical models of the upper atmosphere in the polar regions. P. W. Blum, I. Harris.
Planet. Space Sci., Vol. 21, 377 - 381 (1973).

082.031 Photochemistry of minor constituents in the troposphere. H. Levy II.
Planet. Space Sci., Vol. 21, 575 - 591 (1973).

082.032 The diurnal variation of atomic hydrogen.
B. A. Tinsley.
Planet. Space Sci., Vol. 21, 686 - 691 (1973). – Research note.

082.033 Molecular photoionisation at 584 Å and 304 Å.
J. Fryar, R. Browning.
Planet. Space Sci., Vol. 21, 709 - 711 (1973). – Research note.

082.034 Aspherical model of the density of the upper atmosphere. M. I. Vojskovskij, I. I. Volkov,
N. I. Gryazev, B. V. Kugaenko, V. M. Sinitsyn, P. E. Ehl'yasberg.
Kosmich. Issled., Vol. 11, 70 - 79 (1973). In Russian.

082.035 Sky illumination at Blue Mesa Observatory.
J. Cuffey.
Bull. American Astron. Soc., Vol. 5, 25 (1973). – Abstr. AAS.

082.036 The mean nocturnal height profile of the atmospheric emission at λ 6300 Å.
M. M. Gogoshev, K. B. Serafimov.
Geomagn. Aeronom., Vol. 13, 95 - 99 (1973). In Russian.

082.037 About the radiation of vibration-excited hydroxyl in the upper atmosphere. T. G. Komkova.
Geomagn. Aeronom., Vol. 13, 100 - 103 (1973). In Russian.

082.038 Temperature variations in the upper atmosphere in the period of minimum solar activity from data of outer sounding of the ionosphere.
A. Sh. Sire, V. S. Agalakov, S. I. Torbin.
Geomagn. Aeronom., Vol. 13, 104 - 109 (1973). In Russian.

082.039 About seasonal variations of the upper atmosphere's radiation in the Hα line of atomic hydrogen.
N. M. Martsvaladze, L. M. Fishkova.
Geomagn. Aeronom., Vol. 13, 192 - 193 (1973). In Russian.
Brief information.

082.040 Observations of the helium II 304-A and helium I 584-A atmospheric dayglow radiation.
S. Kumar, S. Bowyer, M. Lampton.
Journ. Geophys. Res., Vol. 78, 1107 - 1114 (1973).
Two photometers with bandpasses of 170–500 and 170–800 A were employed to observe dayglow emissions in the extreme ultraviolet over an altitude range of 90–186 km. We have identified the emissions observed with these photometers as resonantly scattered He I 584-A and He II 304-A radiations.

082.041 Twilight airglow. 1. Photoelectrons and [O I] 5577-Angstrom radiation. P. B. Hays, W. E. Sharp.
Journ. Geophys. Res., Vol. 78, 1153 - 1166 (1973).
A payload consisting of a number of experiments to study the earth's atmosphere was launched from White Sands on February 8, 1971, at dawn. The differential photoelectron flux spectrum was measured as a function of altitude. The ion and electron density distributions were measured simultaneously. An optical measurement of [O I] 5577-A radiation was made.

082.042 Excitation of oxygen permitted line emissions in the tropical nightglow.
B. A. Tinsley, A. B. Christensen, J. Bittencourt, H. Gouveia, P. D. Angreji, H. Takahashi.
Journ. Geophys. Res., Vol. 78, 1174 - 1186 (1973).
The ultraviolet oxygen emissions at 1304 and 1356 A in the tropical nightglow seen from Ogo 4 by Hicks and Chubb and Barth and Schaffner are accompanied by emissions at 7774 and 4368 A. Simultaneous [O I] 6300-A measurements were also made. A theoretical value for the partial rate coefficient for 7774 emission by radiative recombination has been obtained, and from the ionospheric data and a model atmosphere the expected rates of radiative recombination and ion recombination were calculated.

082.043 Effect of atomic oxygen on the N_2 vibrational temperature in the lower thermosphere.
E. L. Breig, M. E. Brennan, R. J. McNeal.
Journ. Geophys. Res., Vol. 78, 1225 - 1228 (1973). – Letter.

082.044 A search for spectral features in the submillimeter background radiation.
J. C. Mather, M. W. Werner, P. L. Richards.
17th Colloquium International Astrophys. Liège 1971, (see 012.003), p. 607 - 608 (1972).

082.045 Light pollution. Outdoor lighting is a growing threat to astronomy. K. W. Riegel.
Science, Vol. 179, 1285 - 1291 (1973).
It is the author's purpose in this article to delineate astronomical dark sky requirements for scientifically useful observing, to examine what conditions will probably prevail for the next generation or two of observing astronomers, and to suggest changes in public policy that would alleviate some of the actual and projected damage to the astronomical observing environment.

082.046 Model computations on seeing.
C. D. Andriesse, T. Wierstra.
Astron. Astrophys., Vol. 24, 465 - 469 (1973).
Seeing disks, considered as blurred interference patterns of corrugated wavefronts, are computed from a wavefront model characterized by strength and scale of the atmospheric turbulence. Star images are computed assuming that the amplitude on the wavefront varies only weakly.

082.047 A study of conditions of astronomical observation at Kiso District in Nagano Prefecture.
Prepared by the Galactic Astronomy Section and the Photometry Section.
Tokyo Astron. Obs., Report No. 61, Vol. 16, 424 - 433 (1973). In Japanese.

082.048 The anomalous refraction from the observations with meridian circle. R. Fukaya.
Tokyo Astron. Obs., Report No. 61, Vol. 16, 464 - 469 (1973). In Japanese.

082.049 On the air temperatures of some heights with a steel tower. R. Fukaya, H. Ishii, I. Morita.
Tokyo Astron. Obs., Report No. 62, Vol. 16, 522 - 531 (1973). In Japanese.

082.050 Airglow observations at the Ogasawara Islands.
H. Tanabe, A. Takechi, A. Miyashita, Y. Mikami.
Tokyo Astron. Obs., Report No. 62, Vol. 16, 624 - 635 (1973). In Japanese.

082.051 **Blue moon: Is this a property of background aerosol?** W. M. Porch, D. S. Ensor, R. J. Charlson, J. Heintzenberg.
Applied Optics, Vol. 12, 34 - 36 (1973).

Stellar extinction measurements made at three astronomical observatories showed that on ~50% of the nights the extinction due to aerosol light scattering increased rather than decreased with increasing wavelength (anomalous extinction) for wavelengths close to 500 nm. This extinction behavior is analyzed in this paper and limits are established for the aerosol characteristics necessary for this phenomenon to exist.

082.052 **Spectral radiance in the S20-range and luminance of the clear and overcast night sky.**
D. H. Höhn, W. Büchtemann.
Applied Optics, Vol. 12, 52 - 61 (1973).

The spectral radiance of the night sky was investigated in the wavelength range from 0.40 μm to 0.80 μm. A series of measurements was taken in the Austrian Alps in 1968. Statistical reductions were made for the total data assembly as well as for date groups with different types of night sky cover, for moonless nights, and for nights with moon. The mean spectral radiance and the standard deviation in the above mentioned wavelength range are presented and discussed together with the distribution functions of the spectral radiance at 0.40 μm, 0.45 μm, 0.5577 (OI) μm, and 0.80 (OH) μm, of the photopic and scotopic luminances and of the S20 radiance. Finally the correlation coefficients regarding each pair of photopic, scotopic, and S20 responses were calculated together with the corresponding regression coefficients.

082.053 **Criterion for the choice of exposure time in atmospheric turbulence investigation with an optical wave.** A. Arrigucci, L. Stefanutti.
Applied Optics, Vol. 12, 136 - 138 (1973). – Letter.

082.054 **Stratospheric aerosol measurements with implications for global climate.**
L. Elterman, R. B. Toolin, J. D. Essex.
Applied Optics, Vol. 12, 330 - 337 (1973).

We present measurement results obtained in New Mexico with bistatic optical probing of the atmosphere using a searchlight beam. The data yield vertical profiles of the aerosol attenuation coefficient. Because they approximate proportionality to aerosol concentration, these profiles provide information concerning the aerosol layer structure and its parameters. During a 9-day period in October and November 1970, a series of forty-one such profiles was obtained which includes altitudes 12–25 km, selected for study because of the relatively high aerosol content of this stratospheric region and its relation to global climate.

082.055 **Balloon-borne spectroscopic observation of the infrared hydroxyl airglow.**
R. P. Lowe, E. A. Lytle.
Applied Optics, Vol. 12, 579 - 583 (1973).

A balloon-borne grating spectrometer has been used to study the spectrum of the airglow between 1.8 μm and 3.6 μm and its diurnal variation. The principal features identified are the bands of the $\Delta v = 2$ and $\Delta v = 1$ sequences of the vibration-rotation spectrum of OH.

082.056 **Rocket observations of the extreme ultraviolet dayglow.** R. W. Carlson, D. L. Judge.
Planet. Space Sci., Vol. 21, 879 - 880 (1973).

The ultraviolet dayglow in the wavelength region 750–1050 Å was investigated over the altitude range 100–800 km using a thin film filter photometer. From the airglow spectrum obtained by Carruthers and Page, one of the dominant features in this wavelength range is OII 834 Å. It is pointed out that the major excitation mechanism for this transition is photoionization excitation of atomic oxygen.

082.057 **On the scattering of Lα-quanta in the earth's atmosphere.**
V. A. Teodoronskij, L. G. Titarchuk, S. D. Chuvakhin.
Kosmich. Issled., Vol. 11, 306 - 314 (1973). In Russian.

082.058 **On the distribution of neutral hydrogen in the upper atmosphere of the earth.** S. D. Chuvakhin.
Kosmich. Issled., Vol. 11, 338 - 340 (1973). In Russian.
Brief information.

082.059 **On the nature of the low-frequency component in stellar image vibration.** N. F. Nelyubin.
Astrofiz. Issled., Izv. Spets. Astrofiz. Obs., Vol. 4, 201 - 207 (1972). In Russian.

A problem of the low-frequency component in the stellar image scintillation is investigated. On the basis that the main contribution is made by the ground layer of the atmosphere some physical phenomena in the layer are considered resulting in low-frequency fluctuations of the apparent positions of stars. Quantitative estimates are obtained of the effect of density fluctuations, of an inclination angle of layers of equal density, and of wind velocity in the ground layer on the amount of the image scintillation.

082.060 **Determinación de la densidad de la alta atmosfera mediante satélites artificiales.** S. Alonso.
Urania Barcelona, Año 57, No. 275, p. 54 - 77 (1972).

Density values in the upper atmosphere are derived through the rate of change of Explorer I's semi-latus rectum. Results thus obtained are compared with those derived by evaluating the rate of change of the orbital period, a more commonly applied method.

082.061 **Atmosphärisch-optische Erscheinungen.**
B. Albers.
SuW, Vol. 12, 137 - 142 (1973).

082.062 **On the global increase of turbidity of the earth's atmosphere in October 1969.**
N. B. Divari, Yu. D. Mateshvili.
Astron. Tsirk., No. 744, p. 5 - 7 (1973). In Russian.

082.063 **On results of double-beam site testing at the Crimean Astrophysical Observatory of the USSR Academy of Sciences.** S. B. Novikov, V. V. Rodionov, S. P. Yatsenko.
Astron. Tsirk., No. 748, p. 6 - 7 (1973). In Russian.

082.064 **On the coefficient of atmospheric extinction from spectrophotometric observations on the Crimea (Crimean Station of the Sternberg Astronomical Institute).**
I. G. Glushneva, V. T. Doroshenko.
Astron. Tsirk., No. 754, p. 1 - 3 (1973). In Russian.

082.065 **Atmospheric oscillations of a solar image at various heights over the earth's surface.**
Sh. P. Darchia, P. G. Kovadlo, V. I. Ivanov.
Solnechnye Dannye 1973 Byull., No. 1, p. 102 - 106 (1973). In Russian.

Atmospheric oscillations of the solar limb were investigated by rising the telescope to various heights over the earth's surface up to 310 m.

082.066 **Optical absorption by atmospheric aerosols.**
A. P. Waggoner, M. B. Baker, R. J. Charlson.
Applied Optics, Vol. 12, 896 (1973). – Letter.

082.067 **Transparenz und Trübung der Atmosphäre.**
P. W. Hodge, N. Laulainen.
Umschau, 73. Jahrgang, p. 325 - 331 (1973).

The world-wide pollution of the atmosphere is increasing rapidly. Concerning to the project ASTRA a group of scientists at the University of Washington was sampling and analyzing unpublished data by sending out specific requests to all observatories in the world for determining the international atmospheric transparency.

082.068 Density and composition of the lower thermosphere.
K. Moe.
Journ. Geophys. Res., Vol. 78, 1633 - 1644 (1973).
This paper presents a new median model of the lower thermosphere. It is based on a critical evaluation of all the measurements of density and composition in this region and carries forward the process of re-examination begun by von Zahn in his 1970 paper. In the resulting model, the O/O_2 ratio is 1.6 at 120 km. The O/N_2 ratios are 0.69 at 150 km and 1.76 at 200 km, in the midst of recent measurements. The median density at 150 km is 1.83 $\mu g/m^3$.

082.069 Night airglow zenith intensity variations at El Leoncito Observatory, Argentina.
E. Ciner, L. L. Smith.
Journ. Geophys. Res., Vol. 78, 1654 - 1662 (1973).

082.070 Photodissociation continuums of N_2 and O_2.
G. R. Cook, M. Ogawa, R. W. Carlson.
Journ. Geophys. Res., Vol. 78, 1663 - 1667 (1973).

082.071 Magnetic control of the near equatorial neutral thermosphere.   A. E. Hedin, H. G. Mayr.
Journ. Geophys. Res., Vol. 78, 1688 - 1691 (1973). – Brief report.

082.072 Response of the neutral upper atmosphere to variations in solar activity.   G. M. Keating, J. S. Levine.
Progr. Astronaut. Aeronaut., Vol. 30, (see 003.009), p. 313 - 340 (1972). – Presented at the AIAA observation and prediction of solar activity conference, Huntsville, Ala., Nov. 16–18, 1970.

082.073 Anomalous absorption in the atmosphere for 2.7 mm radiation.
C. J. Gibbins, A. C. Gordon-Smith, H. A. Gebbie.
Nature, Vol. 243, 397 - 398 (1973).
Measured values of atmospheric attenuation of electromagnetic radiation in the "windows" at 2 and 3 mm are greater by factors of between two and three than is calculated from monomeric water vapour and oxygen line strengths. To investigate these anomalies, the authors have made measurements at 110 and 113 GHz (2.73 and 2.65 mm) of the variation of zenith sky noise temperatures with atmospheric conditions, using Dicke radiometers.

082.074 Discovery of rapidly-varying airglow structures in the photographic infrared.
A. W. Peterson, L. M. Kieffaber.
Bull. American Astron. Soc., Vol. 5, 265 (1973). – Abstr. AAS.

082.075 Observations of atmospheric ozone at 110836 MHz.
W. J. Wilson, F. I. Shimabukuro.
Bull. American Astron. Soc., Vol. 5, 286 (1973). – Abstr. AAS.

082.076 The intensity and polarization of the twilight sky.
W. Blattner, H. Horak.
Bull. American Astron. Soc., Vol. 5, 303 - 304 (1973). Abstr. AAS.

082.077 Daytime ion chemistry of N_2^+.   P. D. Feldman.
Journ. Geophys. Res., Vol. 78, 2010 - 2016 (1973).
Rocket measurements of the emission of the (0,0) first negative band of N_2^+ at 3914 Å in the day airglow have been

made between 120 and 300 km. The observations are described and the results are discussed.

082.078 Magnetic storm characteristics of the thermosphere.
H. G. Mayr, H. Volland.
Journ. Geophys. Res., Vol. 78, 2251 - 2264 (1973).

082.079 A thermosphere composition measurement using a quadrupole mass spectrometer with a side energy focusing quasi-open ion source.
H. B. Niemann, N. W. Spencer, G. A. Schmitt.
Journ. Geophys. Res., Vol. 78, 2265 - 2277 (1973).
The atomic oxygen concentration in the altitude range 130–240 km has been determined through the use of a quadrupole mass spectrometer with a strongly focusing ion source. The result obtained confirms the thesis that consideration of these surface effects is significant in quantifying spectrometer measurements of reactive gases.

082.080 Conjugate photoelectron excitation of O I 4368 airglow emission.
A. B. Christensen, N. R. Teixeira, P. D. Angreji.
Journ. Geophys. Res., Vol. 78, 2315 - 2323 (1973).
Brief report.

082.081 The aeronomic dissociation of nitric oxide.
S. Cieslik, M. Nicolet.
Planet. Space Sci., Vol. 21, 925 - 938 (1973).
The effects of the dissociation of nitric oxide is considered in the mesosphere and stratosphere since there is an important source of NO in the stratosphere (Nicolet, 1970) which is due to the reaction of nitrous oxide with atomic oxygen in its first excited level, namely $N_2O + O\,(^1D) \rightarrow 2\,NO$.

082.082 Chemospheric processes of nitric oxide in the mesosphere and stratosphere.
G. Brasseur, M. Nicolet.
Planet. Space Sci., Vol. 21, 939 - 961 (1973).

082.083 Measurement of the ozone concentration from 55 to 95 km at sunset.   D. E. Miller, P. Ryder.
Planet. Space Sci., Vol. 21, 963 - 970 (1973).
The purpose of this paper is to report the results of three attempts at measuring the concentration of ozone in the mesosphere and lower thermosphere.

082.084 Atomic oxygen transport in the thermosphere.
F. S. Johnson, B. Gottlieb.
Planet. Space Sci., Vol. 21, 1001 - 1009 (1973).

082.085 Diurnal, annual and solar cycle variations of hydroxyl and sodium nightglow intensities in the Europe-Africa sector.   R. H. Wiens, G. Weill.
Planet. Space Sci., Vol. 21, 1011 - 1027 (1973).

082.086 Remarks concerning the diffusion of ion clouds in the earth's upper atmosphere.   W. M. Pickering.
Planet. Space Sci., Vol. 21, 1073 - 1075 (1973). – Research note.

082.087 Some problems in the theory of determining geodetic refraction by dispersion method.
M. T. Prilepin.
Bull. Géod., Nouvelle Sér., Année 1973, No. 108, p. 115 - 123.
The effect of the second refraction index gradient and that of humidity on the accuracy of determining refraction by dispersion method is analyzed. The feasibility of obtaining those parameters instrumentally is suggested.

082.088 Atmospheric composition changes and the F2-layer

seasonal anomaly.　R. Rüster, J. W. King.
Journ. Atmosph. Terr. Phys., Vol. 35, 1317 - 1322 (1973).

082.089　Enhancement of upper atmospheric sodium from sporadic dust influxes.　G. Visconti.
Journ. Atmosph. Terr. Phys., Vol. 35, 1331 - 1340 (1973).

A model is proposed to explain the dependence of upper atmospheric sodium from extraterrestrial dust influxes. With this model we consider the diffusion of the dust in the upper atmosphere, the Na production and the evolution of the Na layer. The results are used to explain the formation of the Na layer in the upper atmosphere normally observed and the enhancement in density attributed to dust influxes due to the crossing of cometary orbital planes.

082.090　Streams of soft electrons in the upper atmosphere of the polar region.
V. F. Tulinov, Yu. M. Dzhuchenko (*Zhuchenko*), V. A. Lipovetsky, G. F. Tulinov, V. M. Feigin.
Geomagn. Aeronom., Vol. 13, 513 - 514 (1973). In Russian. Brief information.

082.091　Die Evolution der Erdatmosphäre.
M. Schidlowski, C. Junge.
Phys. Blätter, 29. Jahrgang, p. 203 - 212 (1973).

082.092　Results of air temperature, density and pressure measurements obtained with the aid of foil cloud sensors in the height region between 80 and 95 km.
G. Rose, H. U. Widdel.
Planet. Space Sci., Vol. 21, 1131 - 1140 (1973).

Results of neutral air density, temperature and pressure measurements obtained between November 1968 and June 1972 in the height range between 80 and 95 km over Arenosillo, Spain are given. A simple foil cloud payload, primarily designed for wind finding purposes, was used. The results obtained are compared with CIRA Standard Atmosphere data and ground based radio wave absorption measurements. An estimate of the accuracy of the method is given.

082.093　Results of twilight probing in the epochs of appearance of noctilucent clouds in June - July 1969.
R. A. Kurbanov, A. N. Lukin, V. A. Romejko, B. N. Jurtchenko (*Yurchenko*).
Astron. vestn., Vol. 5, 253 - 256 (1971). In Russian.

Twilight probing of the upper atmosphere in the epochs of appearance of noctilucent clouds has been performed with the help of electrophotometers on the basis of selenium-cadmium photoresistors. Anomalous twilight variation in the night of the appearance of noctilucent clouds has been revealed.

082.094　Observations of noctilucent clouds by the Tomsk branch of the Astronomical-Geodetical Society of the USSR (VAGO) in 1969 - 1970.
N. P. Fast.
Astron. vestn., Vol. 6, 131 - 133 (1972). In Russian.

082.095　Rate constants for the reactions of hydroxyl and hydroperoxyl radicals with ozone.　W. B. DeMore.
Science, Vol. 180, 735 - 737 (1973).

082.096　Neutron measurements in space.　J. A. Lockwood.
Space Sci. Rev., Vol. 14, 663 - 719 (1973).

The experimental measurements of the neutron flux and energy spectrum in space since 1964 are reviewed and related to the theoretical predictions. A discussion of the neutron sources is presented. The difficulties associated with neutron measurements of both the atmospheric neutron leakage flux and solar neutrons are included. Particular emphasis is placed upon the neutron leakage flux and energy measurements at

energies greater than about 1 MeV. The possibilities of CRAND as a source for the energetic trapped protons are discussed in light of recent measurements of the 10–100 MeV neutron flux. The current status of the solar neutron flux observations is also presented.

082.097　Results of observations of seeing conditions on Mt. Majdanak.　V. S. Shevshenko.
Astron. Zhurn. Akad. Nauk SSSR, Vol. 50, 632 - 644 (1973). In Russian. English translation in Soviet Astron. AJ, Vol. 17, No. 3.

082.098　Seeing-Vergleich zwischen Gamsberg/Südwestafrika und La Silla/Chile.
K. Birkle, H. Elsässer, T. Neckel, G. Schnur, B. Schwarze.
Mitt. Astron. Ges., No. 32, p. 143 - 145 (1973).

082.099　Lichtelektrische Photometrie der Nachthimmelshelligkeit mit dem Ballonteleskop THISBE.
A. Frey, W. Hofmann, D. Lemke.
Mitt. Astron. Ges., No. 32, p. 161 (1973). – Abstract.

082.100　Atmospheric water vapour at Mt. Kobau and Calgary and its relevance to infrared astronomical measurements.　T. A. Clark, G. Irwin.
Journ. Roy. Astron. Soc. Canada, Vol. 67, 142 - 147 = Rothney Astrophys. Obs. Publ. No. 3 (1973).

082.101　A simple propagation law for artificial night-sky illumination.　P. J. Treanor.
Observatory, Vol. 93, 117 - 120 (1973).

082.102　Detection of nitric oxide in the lower atmosphere.
R. A. Toth, C. B. Farmer, R. A. Schindler, O. F. Raper, P. W. Schaper.
Nature, Phys. Sci., Vol. 244, 7 - 8 (1973).

The authors describe the results of spectroscopic measurements made in the lower stratosphere, in which absorptions due to NO have been identified.

082.103　Investigation of the astroclimate in Central Asia.
V. S. Shevchenko.
Young stellar groups. Astroclimate, (see 003.012), p. 84 - 115 (1972). In Russian.

082.104　On the formation of optical inhomogeneities in the inversion layer of the atmosphere.
V. G. Khetselius.
Young stellar groups. Astroclimate, (see 003.012), p. 116 - 136 (1972). In Russian.

082.105　Astroclimate observations during the Majdanak Alpine expedition.　V. G. Khetselius.
Young stellar groups. Astroclimate, (see 003.012), p. 137 - 143 (1972). In Russian.

082.106　Connection and calibration of astroclimatic systems.
V. E. Slutskij.
Young stellar groups. Astroclimate, (see 003.012), p. 144 - 163 (1972). In Russian.

082.107　Calculation of different meteorological data and division into regions of the USSR territory according to the coefficient T.　E. A. Kusaev.
Young stellar groups. Astroclimate, (see 003.012), p. 164 - 175 (1972). In Russian.

082.108　Photographic parallax heights of infrared airglow structures.　A. W. Peterson, L. M. Kieffaber.
Nature, Vol. 244, 92 - 93 (1973).

(89 ± 3) km measured at Albuquerque during the night of May 2–3, 1973.

082.109 On the density of radiation into space of the system earth–atmosphere.
A. V. Pavlov, V. D. Permyakov, Yu. F. Sitnikov.
Kosmich. Issled., Vol. 11, 447 - 450 (1973). In Russian.

082.110 Measurements of protons with an energy up to 50 MeV in the atmosphere.   V. E. Dudkin,
A. A. Levkovskij, V. I. Ostroumov, V. M. Petrov.
Kosmich. Issled., Vol. 11, 492 - 494 (1973). In Russian.
Brief information.

082.111 Calculations of neutron flux spectra induced in the earth's atmosphere by galactic cosmic rays.
T. W. Armstrong, K. C. Chandler, J. Barish.
Journ. Geophys. Res., Vol. 78, 2715 - 2726 (1973).
   Calculations have been carried out to determine the neutron flux induced in the earth's atmosphere by galactic protons and alpha particles at solar minimum for a geomagnetic latitude of 42° N. Neutron flux spectra in the energy range $\sim 10^{-8}$ to $\sim 10^5$ Mev at various depths in the atmosphere were calculated by using Monte Carlo and discrete ordinates methods, and various comparisons with experimental data are presented. The magnitude and shape of the calculated neutron leakage spectrum at the particular latitude considered support the theory that the cosmic ray albedo neutron decay (Crand) mechanism is the source of the protons trapped in the inner radiation belt.

082.112 Time dependent worldwide distribution of atmospheric neutrons and of their products. 1. Fast neutron observations.   M. Merker, E. S. Light, H. J. Verschell, R. B. Mendell, S. A. Korff.
Journ. Geophys. Res., Vol. 78, 2727 - 2740 (1973).

082.113 Time dependent worldwide distribution of atmospheric neutrons and of their products. 2. Calculation.   E. S. Light, M. Merker, H. J. Verschell, R. B. Mendell, S. A. Korff.
Journ. Geophys. Res., Vol. 78, 2741 - 2762 (1973).

082.114 Time dependent worldwide distribution of atmospheric neutrons and of their products. 3. Neutrons from solar protons.   R. B. Mendell, H. J. Verschell, M. Merker, E. S. Light, S. A. Korff.
Journ. Geophys. Res., Vol. 78, 2763 - 2778 (1973).

082.115 Thermospheric wind effects on the distribution of helium and argon in the earth's upper atmosphere.
C. A. Reber, P. B. Hays.
Journ. Geophys. Res., Vol. 78, 2977 - 2991 (1973).

082.116 Millisecond time scale atmospheric light pulses associated with solar and magnetospheric activity.
H. Ögelman.
Journ. Geophys. Res., Vol. 78, 3033 - 3039 (1973).
Brief report.

082.117 Tropical UV arcs: Comparison of brightness with $f_0 F_2$.   R. R. Meier, C. B. Opal.
Journ. Geophys. Res., Vol. 78, 3189 - 3193 (1973). – Letter.

082.118 The refraction of light rays in an atmosphere with arbitrary parameters.   I. F. Kushtin.
Geod., kartogr. i aehrofotos"emka. Resp. mezhved. nauch.-tekhn. sb., 1972, vyp. (No.) 16, p. 59 - 65. In Russian.
Abstr. in Referativ. Zhurn. 51. Astron., 6.51.162; 52. Geod. i Aehros"emka, 6.52.138 (1973).

082.119 The radiation balance of the earth's surface and inclinations of isodioptric surfaces.   V. V. Kirichuk.
Geod., kartogr. i aehrofotos"emka. Resp. mezhved. nauch.-tekhn. sb., 1972, vyp. (No.) 16, p. 42 - 47. In Russian.
Abstr. in Referativ. Zhurn. 51. Astron., 6.51.167 (1973).

082.120 A supplemental catalog of atmospheric densities from satellite-drag analysis.
L. G. Jacchia, J. W. Slowey.
Smithsonian Astrophys. Obs., *Cambridge, Mass.*, Special Rep. No. 348, 5 + 323 pp. (1972).
   The present catalog of densities derived from satellite-drag analysis extends and supplements similar earlier publications by the authors. The densities were computed from nine artificial satellites for effective heights ranging from 300 to 1130 km, and are presented together with pertinent data that permit the location in time and space of the point to which they refer. The intervals covered vary between 1 and 6 years for the different satellites and average about 3.5 years for the nine satellites. The cutoff date for the data included is, in most cases, January 14, 1970.

082.121 The artificial night-sky illumination in Italy.
F. C. Bertiau, E. de Graeve, P. J. Treanor.
Vatican Obs. Publ., Vol. 1, (No. 4), 159 - 179 (1973).
   The selection of sites for astronomical observatories requires a knowledge of artificial night-sky illumination over large regions. This can be derived by means of a propagation law of town-light in the night sky. Using this law, we have mapped the zenith artificial intensity for Italy and its islands.

082.122 Astronomical refraction investigations today.
G. Teleki.
Publ. Dep. Astron. Univ. Beograd, No. 4, p. 5 - 9 (1973).

082.123 The atmospheres of the earth and the terrestrial planets: their origin and evolution. A. J. Meadows.
Phys. Rep. Phys. Letters C, (*Netherlands*), Vol. 5C, 197 - 236 (1972).
   The atmospheres of the earth, Venus and Mars are compared, and it is shown that a consistent model of their evolution can be obtained on the basis of the degassing from the solid planet.

082.124 Contribution to the study of atmospheric refraction index and photogrammetric refraction in the area of Greece by means of meteorological data.   E. N. Patmios.
Meteorologika, Publ. Meteorol. Inst. Univ. Thessaloniki No. 20, 2 + 51 pp. (1972). – Thesis Fac. Phys. Math, Thessaloniki, Greece.

082.125 Una determinazione della densita' dell'alta atmosfera usando il programma "Perlo".
A. Manara, I. Almàr, E. I. Almàr.
Rend. Ist. Lombardo A, Vol. 105, 105 - 121 = Contr. Oss. Astron. Milano-Merate, Nuova Ser., No. 335 (1971).
   Exposition of a method to determine the density of the upper-atmosphere using observations of artificial satellites near the celestial equator. Numerical values of air density at 603 km, from the satellite 1965-53-F, and at 284 km, from 1965-11-D, are calculated for June 1966.

082.126 Sondages automatiques de la basse atmosphère.
O. Godart, G. Schayes, J-P. Peters.
Inst. d'Astron. Géophys. Georges Lemaître, Univ. Catholique Louvain, Belgique, Contr. No. 11, 23 pp. (1972).

082.127 Some results of an investigation on the astroclimate in the Mongolian People's Republic.
D. Khaltar.
Astrometriya i Astrofizika, *Kiev*, Vyp. (No.) 18, (see 003.

016), p. 20 - 24 (1973). In Russian.

082.128 On the naked-eye threshold observation of stars
at different background brightness.
I. A. Zabelina.
Optiko-mekh. prom-st', 1973, No. 1, p. 16 - 19. In Russian.
Abstr. in Referativ. Zhurn. 51. Astron., 7.51.126 (1973).

082.129 On the problem of application of a high-level meteorological tower in astronomy.
Sh. P. Darchiya, P. G. Kovadlo, V. I. Ivanov.
Issled. po geomagnetizmu, aehron. i fiz. Solntsa. Vyp. (No.)
26. Moskva, Nauka, 1973, p. 273 - 278. In Russian. – Abstr.
in Referativ. Zhurn. 51. Astron., 7.51.135 (1973).

Evaluation of Mauna Kea, Hawaii, as an observatory
site. See Abstr. 009.023.

Laboratory investigation of the reaction $NO^+ + O_3 \rightarrow$
$NO_2^+ + O_2$. See Abstr. 022.005.

Quenching of $O(2^1D_2)$ by atmospheric gases.
See Abstr. 022.017.

Rotational and vibrational hydroxyl excitation in
the laboratory and in the night airglow.
See Abstr. 022.021.

An astronomical view of high-pressure sodium
lamps. See Abstr. 022.084.

The stabilization of planetary images.
See Abstr. 031.031.

On the photogrammetric refraction.
See Abstr. 031.060.

Variations of the upper atmospheric motion and the
rotational motion of the earth. See Abstr. 044.011.

Some effects of the local character in the atmospheric structure on astrometry (I). See Abstr. 045.012.

PAET, an entry probe experiment in the earth's atmosphere. See Abstr. 053.011.

Analysis of the orbit of 1965-11 D (Cosmos 54
rocket). See Abstr. 054.014.

Solar Lyman alpha changes and related hydrogen
density distribution at the earth's exobase (1969–1970).
See Abstr. 076.011.

Broad band solar EUV absorption in the earth's
upper atmosphere. See Abstr. 076.012.

An observation of polar aurora and airglow from
the ISIS-II spacecraft. See Abstr. 084.031.

Physical characteristics of the "blue clouds" of the
planets Mars, Earth and Venus. See Abstr. 097.114.

Atmospheric extinction – effects of variations.
See Abstr. 113.026.

Errata

082.901 Errata: 'On the possibility of a simultaneous meas-
urement of wind speed, wind direction, air density
and air temperatures at heights which correspond to the upper
D-region (max. 95 km) with chaff cloud sensors' [Planet.
Space Sci., Vol. 20, 877 - 889 (1972)].
G. Rose, H. U. Widdel.
Planet. Space Sci., Vol. 21, 535 (1973).

082.902 Erratum: 'Molecular photoionisation at 584 Å and
304 Å' [Planet. Space Sci., Vol. 21, 709 - 711
(1973)]. J. Fryar, R. Browning.
Planet. Space Sci., Vol. 21, 1080 (1973).

083 Ionosphere

083.001 On Cerenkov radiation by electron beam injected into the ionosphere.
Ju. K. Alekhin, V. I. Karpman.
Cosmic Electrodynamics, Vol. 3, 395 - 405, 406 - 415 (1973).
In Russian and English.

The spontaneous Cerenkov emission of longitudinal waves produced by a single particle and cylindrical electron beam moving inside the magnetosphere along the magnetic field is considered. The electric field of the radiation is calculated and compared with existing experimental results on artificial electron beams injected into space; the comparison shows a qualitative agreement between the theoretical results and experimental data.

083.002 Solar eclipse effect on sporadic E ionization, 2.
R. N. Datta.
Journ. Geophys. Res., Vol. 78, 320 - 322 (1973). – Letter.

083.003 Multistatic incoherent scatter measurements of ionospheric drift velocity.
G. N. Taylor, H. Rishbeth, P. J. S. Williams.
Nature, Vol. 242, 109 - 111 (1973).

We present here multi-component incoherent scatter drift measurements which are coincident in space and time; these were achieved by receiving the scattered signals at more than one site.

083.004 On what ionospheric workers should know about the plasmapause-plasmasphere.
D. L. Carpenter, C. G. Park.
Rev. Geophys. Space Phys., Vol. 11, 133 - 154 (1973).

083.005 A satellite study of the mid-latitude trough in electron density and VLF radio emissions during the magnetic storm of 25–27 May 1967.
Y. (Kabasakal) Tulunay, A. R. W. Hughes.
Journ. Atmosph. Terr. Phys., Vol. 35, 153 - 163 (1973).

083.006 The ionospheric effects of geomagnetic sudden commencements as measured with an HF Doppler sounder at Hawaii.
Y.-N. Huang, K. Najita, P. Yuen.
Journ. Atmosph. Terr. Phys., Vol. 35, 173 - 181 (1973).

083.007 The seasonal variation of the height and thickness of the E-region. J. D. Whitehead.
Journ. Atmosph. Terr. Phys., Vol. 35, 183 - 185 (1973).
Short paper.

083.008 Calculated distributions of hydrogen and helium ions in the low-latitude ionosphere.
R. J. Moffett, W. B. Hanson.
Journ. Atmosph. Terr. Phys., Vol. 35, 207 - 222 (1973).

The simultaneous time-dependent continuity equations for O^+, H^+ and He^+ in the low latitude F-region are solved. The calculated profiles of O^+ and H^+ concentrations at 1630 LT are in fair agreement with the observations of Hanson and his co-workers, and with satellite results.

083.009 Global electron density distributions from the Ariel 3 satellite at mid-latitudes during quiet magnetic periods. Y. (Kabasakal) Tulunay.
Journ. Atmosph. Terr. Phys., Vol. 35, 233 - 254 (1973).

083.010 The negative-ion composition of the daytime D-region.
L. Thomas, P. M. Gondhalekar, M. R. Bowman.

Journ. Atmosph. Terr. Phys., Vol. 35, 397 - 404 (1973).

083.011 Monte Carlo simulation of a model ionosphere – II. Energy flow and energy dissipation.
K. Schlegel.
Journ. Atmosph. Terr. Phys., Vol. 35, 415 - 424 (1973).

083.012 Type variation of solar sudden field anomaly (SFA) on 164 kHz as an indicator of seasonal structure changes in the D-region.
V. Letfus, E. M. Apostolov, G. Nestorov.
Journ. Atmosph. Terr. Phys., Vol. 35, 571 - 576 (1973).
Short paper.

083.013 The causes of storm-time increases of the F-layer at mid-latitudes. J. V. Evans.
Journ. Atmosph. Terr. Phys., Vol. 35, 593 - 616 (1973).

083.014 ISIS-1 satellite observations of the ionosphere at high southern latitudes.
D. Eccles, J. W. King, A. J. Slater.
Journ. Atmosph. Terr. Phys., Vol. 35, 625 - 632 (1973).

083.015 Ariel 3 satellite observations of the ionosphere at high southern latitudes. H. D. Hopkins.
Journ. Atmosph. Terr. Phys., Vol. 35, 633 - 637 (1973).

Electron density probe data from the Ariel 3 satellite are presented in the form of polar maps which show the spatial variation of the topside ionosphere at high southern latitudes for particular universal times in the winter and summer of 1967/1968.

083.016 The topside ionosphere at mid-latitudes during local sunrise. H. Soicher.
Journ. Atmosph. Terr. Phys., Vol. 35, 657 - 668 (1973).

Electron-density and scale height distributions obtained from more than 1000 Alouette I ionograms have been used to study the structure and variations of the mid-latitude topside ionosphere during the sunrise period from 21 May to 4 July 1963. The electron densities at fixed topside altitudes decreased with increasing illumination at all altitudes and latitudes until about ground sunrise.

083.017 A theoretical study of lunar variations in $f_0 F2$ at low latitude.
D. N. Anderson, S. Matsushita, J. D. Tarpley.
Journ. Atmosph. Terr. Phys., Vol. 35, 753 - 759 (1973).

083.018 A diffusion model for the electron density distribution along the earth's magnetic field in an F-region plasma cloud. P. B. Rao.
Journ. Atmosph. Terr. Phys., Vol. 35, 795 - 803 (1973).

083.019 On the solar Lyman-α control of the ionospheric absorption at 2775 kHz. J. Laštovička.
Journ. Atmosph. Terr. Phys., Vol. 35, 815 - 820 (1973).
Short paper.

083.020 The nature of seasonal changes in the effects of magnetic storms on mid-latitude F-layer electron concentration. P. H. Spurling, K. L. Jones.
Journ. Atmosph. Terr. Phys., Vol. 35, 921 - 927 (1973).

083.021 The electron content of the southern mid-latitude ionosphere, 1965–1971. J. E. Titheridge.
Journ. Atmosph. Terr. Phys., Vol. 35, 981 - 1001 (1973).

083.022 **Total electron content measurements during visible auroras.** B. J. Watkins, E. A. Essex.
Journ. Atmosph. Terr. Phys., Vol. 35, 1009 - 1013 (1973).
Short paper.

083.023 **Enhancements of ionospheric total electron content in the southern auroral zone associated with magnetospheric substorms.** E. A. Essex, B. J. Watkins.
Journ. Atmosph. Terr. Phys., Vol. 35, 1015 - 1018 (1973).
Short paper.

083.024 **Metallic ions in the equatorial ionosphere.** A. C. Aikin, R. A. Goldberg.
Journ. Geophys. Res., Vol. 78, 734 - 745 (1973).
A metallic ion layer centered at 92 km is found to contain Mg^+, Fe^+, Ca^+, K^+, Al^+, Na^+, and possibly Si^+ ions. The layer is explained in terms of a similarly shaped altitude distribution of neutral atoms that are photoionized and charge exchanged with NO^+ and O_2^+.

083.025 **Coupling between the F-region and protonosphere: Numerical solution of the time-dependent equations.** R. J. Moffett, J. A. Murphy.
Planet. Space Sci., Vol. 21, 43 - 52 (1973).
This paper describes a new method of solution of the time-dependent continuity and momentum equations for H^+ and O^+ in mid-latitude magnetic field tubes from the F-region to the equator. As an example of application of the method a suggestion by Park (1971) about observed night-time enhancements of $N_m F2$ is examined.

083.026 **Lower ionosphere electron densities from rocket measurements employing LF radio propagation and DC probe techniques.** J. E. Hall.
Planet. Space Sci., Vol. 21, 119 - 131 (1973).

083.027 **Plasma sheet particle precipitation: A kinetic model.** J. Lemaire, M. Scherer.
Planet. Space Sci., Vol. 21, 281 - 289 (1973).
Ionospheric and plasma sheet particle densities, fluxes and bulk velocities along an auroral magnetic field line have been calculated for an ion-exosphere model.

083.028 **D-region recombination coefficients and the short wavelength X-ray flux during a solar flare.**
S. Ananthakrishnan, M. A. Abdu, L. R. Piazza.
Planet. Space Sci., Vol. 21, 367 - 375 (1973).
This paper presents the results of a calculation to obtain information on the nature of the X-ray emission during a solar flare observed on 22nd January 1972 based on phase and amplitude effects observed at six different frequencies.

083.029 **The day-sector polar F-layer during a magnetospheric substorm.** A. L. Snyder, S.-I. Akasofu, C. P. Pike.
Planet. Space Sci., Vol. 21, 399 - 407 (1973).

083.030 **A theoretical study of the ionospheric F region equatorial anomaly–I. Theory.** D. N. Anderson.
Planet. Space Sci., Vol. 21, 409 - 419 (1973).

083.031 **A theoretical study of the ionospheric F region equatorial anomaly–II. Results in the American and Asian sectors.** D. N. Anderson.
Planet. Space Sci., Vol. 21, 421 - 442 (1973).

083.032 **Direct measurements of plasma convection in the upper ionosphere.**
Yu. I. Gal'perin, V. N. Ponomarev.
Kosmich. Issled., Vol. 11, 88 - 94 (1973). In Russian.

083.033 **Investigations of the ionosphere made with Inter-**
cosmos 2. I. K. I. Gringauz, K. B. Serafimov, K. G. Shmelovskij, Ya. Shmilauehr.
Kosmich. Issled., Vol. 11, 95 - 100 (1973). In Russian.

083.034 **Coefficients of reflection of hydromagnetic waves from charged ionospheres.**
L. L. Van'yan, A. A. Kozhevnikov, Yu. G. Turbin.
Kosmich. Issled., Vol. 11, 152 - 154 (1973). In Russian.
Brief information.

083.035 **Amorçage d'une décharge en champs croisés: application à l'ionosphère en zone aurorale.**
F. Akoum, D. Quemada.
Comptes Rendus Acad. Sci. Paris, Sér. B, Vol. 276, 677 - 680 (1973).

083.036 **Three-dimensional analytical model of distribution of electron concentration in the quiet ionosphere.**
A. V. Gurevich, D. I. Fishchuk, E. E. Tsedilina.
Geomagn. Aeronom., Vol. 13, 31 - 40 (1973). In Russian.

083.037 **Dynamical model of interaction of the ionospheric F-region with the plasmosphere.** M. A. Kutimskaya, V. M. Polyakov, N. N. Klimov, G. M. Kuznetsova, G. I. Gershengorn.
Geomagn. Aeronom., Vol. 13, 41 - 46 (1973). In Russian.

083.038 **The influence of the changing ionospheric-protonospheric plasma flux on the night F-region of the ionosphere.** G. S. Ivanov-Kholodny, A. V. Mikhailov.
Geomagn. Aeronom., Vol. 13, 47 - 51 (1973). In Russian.

083.039 **Analytical model of the nocturnal instationary F2-region of the ionosphere at mean latitudes.**
M. N. Fatkullin, M. G. Deminov.
Geomagn. Aeronom., Vol. 13, 52 - 58 (1973). In Russian.

083.040 **The position of the ionospheric boundary in the period from 1963 to 1968.**
N. M. Rudneva, Ya. I. Feldshtein.
Geomagn. Aeronom., Vol. 13, 116 - 121 (1973). In Russian.

083.041 **About the ionization of the night E-region of the ionosphere.**
L. N. Rubtzev (*Rubtsov*), B. G. Solovei.
Geomagn. Aeronom., Vol. 13, 178 - 180 (1973). In Russian.
Brief information.

083.042 **The changes of the parameters of the outer ionosphere caused by the magnetic storm of March 8 - 10, 1970.** G. K. Solodovnikov, V. M. Migunov, G. N. Tkachev, Yu. G. Ivanov, V. M. Lazarenko, M. N. Chalaya, A. R. Yagovkin.
Geomagn. Aeronom., Vol. 13, 183 - 185 (1973). In Russian.
Brief information.

083.043 **About the connection of large ionospheric irregularities with solar activity.**
M. A. Ovsyankin, E. M. Smetanina.
Geomagn. Aeronom., Vol. 13, 185 - 186 (1973). In Russian.
Brief information.

083.044 **Distribution of NO_2^+ in the lower ionosphere.** A. C. Aikin, R. A. Goldberg.
Journ. Geophys. Res., Vol. 78, 1229 - 1231 (1973). – Letter.

083.045 **The low-latitude and equatorial outer ionosphere during the magnetic storm of January 2–4, 1964.**
M. N. Fatkullin, E. S. Zayarnaya, L. F. Mamonova.
Geomagn. Aeronom., Vol. 13, 228 - 232 (1973). In Russian.

083.046 About the diffusion in the F layer of the ionosphere.
E. E. Tsedilina.
Geomagn. Aeronom., Vol. 13, 233 - 241 (1973). In Russian.

083.047 Calculations of the function of ion formation for
mean solar activity.
L. A. Akatova, B. N. Velichansky, G. S. Ivanov-Kholodny,
M. K. Ivelskaya, N. N. Klimov.
Geomagn. Aeronom., Vol. 13, 249 - 255 (1973). In Russian.

083.048 The calculation of $N(z)$-profiles of the ionosphere
using data on radio wave absorption.
N. P. Danilkin, O. A. Maltzeva.
Geomagn. Aeronom., Vol. 13, 256 - 260 (1973). In Russian.

083.049 The sporadic E layer as an accidental process in
time.   O. Ovezgeldiev.
Geomagn. Aeronom., Vol. 13, 267 - 271 (1973). In Russian.

083.050 Definition of chemical reaction constants in the
ionospheric D-region.
L. V. Zelenkova, M. I. Pudovkin, A. S. Sokolov.
Geomagn. Aeronom., Vol. 13, 276 - 282 (1973). In Russian.

083.051 Investigation of moving ionospheric disturbances.
V. I. Drobdzev (Drobzhev), V. A. Rybin.
Geomagn. Aeronom., Vol. 13, 369 - 371 (1973). In Russian.
Brief information.

083.052 Total electron content of the equatorial ionosphere.
R. G. Rastogi, R. P. Sharma, V. Shodhan.
Planet. Space Sci., Vol. 21, 713 - 720 (1973).
Total electron content (N_t) variations in the ionosphere
above the magnetic equator (Thumba dip 0.6° S) obtained by
the Faraday rotation measurements of beacon signals from
S 66 satellites are described for the period December 1965 –
August 1968. Combining the data of Thumba and Ahmedabad
the diurnal development of the equatorial anomaly in N_t is
described.

083.053 Investigations of the ionosphere made with Intercosmos 2. II. Investigations of the equatorial anomaly of the F-region and the outer ionosphere by means of
spherical traps.
G. L. Gdalevich, B. N. Goroshankin, I. S. Kutiev, D. T. Samardzhiev, K. B. Serafimov.
Kosmich. Issled., Vol. 11, 245 - 253 (1973). In Russian.

083.054 Investigations of the ionosphere made with Inter-
cosmos 2. III. Measurements of the electron tem-
perature in the ionosphere by high-frequency sounding.
V. V. Afonin, G. L. Gdalevich, K. I. Gringauz, Ya. Kajnarova,
Ya. Shmilauehr.
Kosmich. Issled., Vol. 11, 254 - 266 (1973). In Russian.

083.055 Investigations of the ionosphere made with Inter-
cosmos 2. IV. Measurements with the help of cylin-
drical Langmuir probes.
K. Bishoff, G. L. Gdalevich, V. F. Gubskij, I. D. Dmitrieva, G.
Tsimmerman.
Kosmich. Issled., Vol. 11, 267 - 272 (1973).   In Russian.

083.056 Direct measurements of drift velocity of ions in the
upper ionosphere during a magnetic storm. I. Me-
thodical questions and some results of measurements during a
magnetically quiet period of time.
Yu. I. Gal'perin, V. N. Ponomarev, A. G. Zosimova.
Kosmich. Issled., Vol. 11, 273 - 283 (1973). In Russian.

083.057 Direct measurements of drift velocity of ions in the
upper ionosphere during a magnetic storm. II. Results of measurements during the magnetic storm on November 3, 1967.
Yu. I. Gal'perin, V. N. Ponomarev, A. G. Zosimova.
Kosmich. Issled., Vol. 11, 284 - 296 (1973). In Russian.

083.058 Rocket measurements of the ion composition.
A. D. Zhlud'ko.
Kosmich. Issled., Vol. 11, 297 - 305 (1973). In Russian.

083.059 Some results of measurements of the parameters of
inhomogeneities of the ionospheric electron density
with Cosmos 381.   G. G. Getmantsev, G. P. Komrakov,
V. P. Ivanov, I. V. Popkov, V. N. Tyukin.
Kosmich. Issled., Vol. 11, 335 - 338 (1973). In Russian.
Brief information.

083.060 Solar and lunar tidal effects on the low-latitude
ionosphere – a review.   S. Matsushita.
Journ. Atmosph. Terr. Phys., Vol. 35, 1027 - 1034 (1973).

083.061 Ionospheric effects of solar activity.   G. C. Reid.
Progr. Astronaut. Aeronaut., Vol. 30, (see 003.009),
p. 293 - 312 (1972). – Presented at the AIAA observation and
prediction of solar activity conference, Huntsville, Ala., Nov.
16–18, 1970.

083.062 The ionospheric electric field during substorms –
an interpretation based on non-uniform reconnection in the geomagnetic tail.   S. W. H. Cowley.
Astrophys. Space Sci., Vol. 20, 491 - 497 (1973).
It is suggested that changes in the electric field in the
night-side auroral zone and polar cap observed during the expansion phase of a substorm are related to a change in the
magnetospheric flow pattern. During the substorm growth
phase the flow appears to be fairly uniform across the width
of the magnetosphere (uniform electric field across the tail),
while at expansion the observations are consistent with the
magnetospheric potential drop in the tail falling across a
narrow region near the dusk magnetopause. Such non-uniform
electric fields in the tail have been predicted by recent theoretical work. A rather speculative interpretation of events during
a magnetospheric substorm is presented.

083.063 Seasonal and sunspot cycle variations of F region
electron temperatures and protonospheric heat
fluxes.   J. V. Evans.
Journ. Geophys. Res., Vol. 78, 2344 - 2349 (1973). – Letter.

083.064 The behaviour of the topside ionosphere during
magnetically disturbed conditions.
P. M. Gondhalekar.
Journ. Atmosph. Terr. Phys., Vol. 35, 1293 - 1298 (1973).

083.065 Impedance model of the day-time ionosphere.
A. B. Orlov, I. M. Sternina.
Geomagn. Aeronom., Vol. 13, 416 - 421 (1973). In Russian.

083.066 Reflection of variations of neutral composition at
levels of the turbopause in the n_e (h) profiles of the
mid-latitude F2 region.   M. N. Fatkullin, A. Muradov.
Geomagn. Aeronom., Vol. 13, 422 - 426 (1973). In Russian.

083.067 Seasonal and daily variations of altitude and electron concentration of the maximum of the F2 layer
in planetary scale.   N. M. Boenkova.
Geomagn. Aeronom., Vol. 13, 427 - 432 (1973). In Russian.

083.068 Morphology of the subauroral F2$_s$.
A. P. Mamrukov, E. K. Zikrach.
Geomagn. Aeronom., Vol. 13, 433 - 436 (1973). In Russian.

083.069 **Fast latitude variation of the aeronomical state in the near-equatorial ionosphere at very high activity of the sun.** L. A. Tshepkin (*Shchepkin*).
Geomagn. Aeronom., Vol. 13, 517 - 518 (1973). In Russian. Brief information.

083.070 **Disturbance of the density of the F layer caused by a small-scale dynamo effect.** Yu. S. Vardanyan.
Geomagn. Aeronom., Vol. 13, 518 - 520 (1973). In Russian. Brief information.

083.071 **Parametric instabilities generated in the ionosphere by intense radio waves.**
C. Oberman, F. Perkins, E. Valeo.
Cosmic plasma physics. Conference 1971, (see 012.016), p. 25 - 26 (1972).

083.072 **Deformation and striation of barium clouds in the ionosphere.**
F. W. Perkins, N. J. Zabusky, J. H. Doles.
Cosmic plasma physics. Conference 1971, (see 012.016), p. 55 - 60 (1972).

083.073 **Nonlinear theory of cross-field and two-stream instabilities in the equatorial electrojet.**
A. Rogister.
Cosmic plasma physics. Conference 1971, (see 012.016), p. 335 - 337 (1972).

083.074 **Lunar tidal oscillations in the horizontal ionospheric drift at the equator.** R. K. Misra.
Planet. Space Sci., Vol. 21, 1109 - 1114 (1973).
The effect of lunar tides on the apparent ionospheric drift velocity for an equatorial station Thumba is computed by using nearly six years of data at fixed solar hours. Significant tides are observed in the E-region drifts, particularly around 12.00 hr and in the F-region drifts around 15.00 hr.

083.075 **Thermospheric observations combining chemical seeding and ground-based techniques — II. Ionospheric drifts and the Sq current system.**
D. Rees, G. Haerendel, D. G. Felgate, K. H. Lloyd, C. H. Low.
Planet. Space Sci., Vol. 21, 1237 - 1249 (1973).
Two Skylark sounding rockets carrying chemical seeding payloads were launched from Woomera, South Australia in October 1969. In conjunction with these firings, the F-region drifts were determined with the Buckland Park aerial array and the results compared with the observed motion of the barium ion clouds. The local ionospheric Sq current system was calculated both from the observed ionospheric parameters and from ground-based magnetograms and the differences between the two results are discussed.

083.076 **Orientational dependence of certain RF impedance probes in the ionosphere.** R. H. Bishop.
Planet. Space Sci., Vol. 21, 1262 - 1267 (1973). — Research note.

083.077 **Wave-like disturbances in the ionosphere.** W. J. Raitt, D. H. Clark.
Nature, Vol. 243, 508 - 509 (1973).
Useful data were retrieved from 2,200 of the first 2,800 orbits of ESRO-1A, and these data were searched for the occurrence of wave-like disturbances. Evidence of such disturbances was found on 120 orbits. All 120 examples occurred after the satellite entered the dawn sector and were usually restricted to the latitude range ±60°.

083.078 **Measurements of electric fields in the ionosphere.** Z. Kłos.
Postępy Astron., Vol. 21, 127 - 139 (1973). In Polish.
This paper deals with the methods of measurement of steady electric fields in the terrestrial upper atmosphere. First, different methods of the measurement are reviewed, then their applicability to practical purposes is considered. Measurements with a dual Langmuir probe are discussed in greater detail.

083.079 **Simultaneous measurements of the ion density and of the intensity of corpuscular streams at heights between 10 and 70 km.** Yu. A. Bragin, V. F. Tulinov, L. N. Smirnykh, S. G. Yakovlev.
Kosmich. Issled., Vol. 11, 488 - 489 (1973). In Russian. Brief information.

083.080 **Errors in ion and electron temperature measurements due to grid plane potential nonuniformities in retarding potential analyzers.**
P. D. Goldan, E. J. Yadlowsky, E. C. Whipple, Jr.
Journ. Geophys. Res., Vol. 78, 2907 - 2916 (1973).

Solar microwave bursts and polar cap absorption. See Abstr. 077.013.

Is there enough solar extreme ultraviolet radiation to maintain the global mean thermospheric temperature? See Abstr. 082.004.

Comparison of the correlation of incoherent scatter and ionosonde measurements of temperature with calcium plage and 2800-Megahertz intensities. See Abstr. 082.006.

6300 Å night airglow enhancements in low latitudes. See Abstr. 082.022.

Sunrise changes in concentrations of minor neutral constituents in the mesosphere. See Abstr. 082.023.

Neutral thermosphere temperatures from density scale height measurements. See Abstr. 082.025.

Atmospheric composition changes and the $F2$-layer seasonal anomaly. See Abstr. 082.088.

Tropical UV arcs: Comparison of brightness with f_0F_2. See Abstr. 082.117.

A study of the relationship between geomagnetic storms and ionospheric disturbances at mid-latitudes. See Abstr. 084.219.

Can the ionosphere regulate magnetospheric convection? See Abstr. 084.271.

Possible low ionosphere response to very hard X-rays from Cygnus X-3 bursts in September 1972. See Abstr. 142.122.

Errata

083.901 **Erratum: 'Potential double layers in the ionosphere'** [Cosmic Electrodynamics, Vol. 3, 349 - 376 (1972)].
L. P. Block.
Astrophys. Space Sci., Vol. 20, 521 (1973).

084 Aurorae, Geomagnetic Field, Radiation Belts

Aurorae

084.001 **Observed relationship between electric fields and auroral particle precipitation.**
D. A. Gurnett, L. A. Frank.
Journ. Geophys. Res., Vol. 78, 145 - 170 (1973).

084.002 **Photometric observations of auroras during August, 1972.** E. G. Mullen, D. J. Davis, Jr.
Sky Telescope, Vol. 45, 130 - 131 (1973).

084.003 **Auroral activity during 1971.** J. Paton.
Observatory, Vol. 93, 47 - 48 (1973). – Note.

084.004 **The auroral oval. – A reevaluation.**
R. H. Eather.
Rev. Geophys. Space Phys., Vol. 11, 155 - 167 (1973). – Topical review.

084.005 **Auroral helium precipitation.** D. L. Reasoner.
Rev. Geophys. Space Phys., Vol. 11, 169 - 180 (1973). – Topical review.

084.006 **Balloon observations of auroral-zone X-rays in conjugate regions.** J. R. Barcus, R. R. Brown, R. H. Karas, K. Brønstad, H. Trefall, M. Kodama, T. J. Rosenberg.
Journ. Atmosph. Terr. Phys., Vol. 35, 497 - 511 (1973).

084.007 **Simultaneous observations of low energy electron fluxes and the polar red emission at 6300 Å.**
M. K. Andrews, J. R. Strömman.
Journ. Atmosph. Terr. Phys., Vol. 35, 537 - 543 (1973).

084.008 **On the morphology of auroral-zone X-ray events – II. Events during the early morning hours.**
G. Kremser, K. Wilhelm, W. Riedler, K. Brønstad, H. Trefall, S. L. Ullaland, J. P. Legrand, J. Kangas, P. Tanskanen.
Journ. Atmosph. Terr. Phys., Vol. 35, 713 - 733 (1973).
Auroral-zone electron precipitation during early morning hours (0200-0600 hr magnetic local time) has been analysed with the aid of X-ray measurements from northern Scandinavia together with recordings of geomagnetic variations and cosmic noise absorption.

084.009 **On the morphology of auroral-zone X-ray events – III. Large-scale observations in the midnight-to-morning sector.** G. Maral, K. Brønstad, H. Trefall, G. Kremser, H. Specht, P. Tanskanen, J. Kangas, W. Riedler, J. P. Legrand.
Journ. Atmosph. Terr. Phys., Vol. 35, 735 - 751 (1973).
A model has been constructed for auroral electron precipitation events, describing in general terms the morphology of the acceleration and precipitation processes for energetic electrons. We use simultaneous balloon recordings of X-rays over Iceland and northern Scandinavia to test the validity of this model, when applied to large-scale observations of electron precipitation events.

084.010 **Some effects of magnetospheric acceleration mechanisms on variations in ultraviolet intensity height profiles, and on consequent rocket spectrograph sensitivities.**
J. R. Catchpoole.
Journ. Atmosph. Terr. Phys., Vol. 35, 861 - 870 (1973).
An estimate is made of typical height profiles of auroral ultraviolet radiation; and such distributions for the emission of radiation of wavelength 1304 Å are presented. These profiles are compared with a previously used model based on visible measurements, in order to produce factors which can be used to modify earlier estimates of the response of a spectrograph used in rocket experiments.

084.011 **Correlation between pulsations in auroral luminosity variations and X-rays.** T. Sørensen, J. Bjordal, H. Trefall, G. J. Kvifte, H. Pettersen.
Journ. Atmosph. Terr. Phys., Vol. 35, 961 - 969 (1973).
Simultaneous balloon observations of > 25 keV bremsstrahlung X-rays and auroral N_2^+ emissions at 4278 Å show that quasi-periodic 5–15 sec pulsations in the two phenomena are correlated. The correlation with light seems to be limited to X-rays < 50 keV.

084.012 **Measurement of auroral Birkeland currents and energetic particle fluxes.** P. A. Cloutier, B. R. Sandel, H. R. Anderson, P. M. Pazich, R. J. Spiger.
Journ. Geophys. Res., Vol. 78, 640 - 647 (1973).

084.013 **Aurora and the poleward edge of the main ionospheric trough.**
H. F. Bates, A. E. Belon, R. D. Hunsucker.
Journ. Geophys. Res., Vol. 78, 648 - 658 (1973).

084.014 **Differences in auroral intensity at conjugate points.** H. C. Stenbaek-Nielsen, E. M. Wescott, T. N. Davis, R. W. Peterson.
Journ. Geophys. Res., Vol. 78, 659 - 671 (1973).
Conjugate auroral all-sky camera data obtained in 18 flights through the auroral zone near the College, Alaska, magnetic meridian show hemispherical differences in auroral frequency and intensity. The hemispherical difference is attributed to the 800-γ difference in magnetic field strength between the conjugate areas. The evidence for hemispheric differences in auroral particle precipitation adds substantially to earlier indications of variation with longitude of auroral precipitation patterns.

084.015 **Electron intensities over two auroral arcs.**
D. A. Bryant, G. M. Courtier, G. Bennett.
Planet. Space Sci., Vol. 21, 165 - 177 (1973).
The paper discusses electron intensities observed on two rocket flights over auroral arcs. It is concluded that the electrons producing the auroral arcs were accelerated at the boundary between two source plasmas in the magnetosphere. The possible identity of the source plasmas is discussed.

084.016 **Time dependent studies of the aurora – I. Ion density and composition.** R. A. Jones, M. H. Rees.
Planet. Space Sci., Vol. 21, 537 - 557 (1973).
Ion densities and composition are investigated in a time varying model aurora. The auroral ion density and composition depend on the energy distribution as well as the magnitude of the bombarding flux of electrons, parameters that vary with time. Our model computations show that a steady state is generally not attained at all altitudes and for all species during a typical auroral event.

084.017 **Auroral heating and the composition of the neutral atmosphere.**
P. B. Hays, R. A. Jones, M. H. Rees.
Planet. Space Sci., Vol. 21, 559 - 573 (1973).

084.018 **Upper atmospheric temperatures from Doppler line widths – V. Auroral electron energy spectra and**

fluxes deduced from the 5577 and 6300 Å atomic oxygen emissions.   H. H. Zwick, G. G. Shepherd.
Planet. Space Sci., Vol. 21, 605 - 621 (1973).

084.019   **Observation of stable auroral red arcs from Southern Africa.**
E. H. Carman, M. P. Heeran, R. W. H. Stevenson.
Planet. Space Sci., Vol. 21, 683 - 686 (1973). – Research note.

084.020   **The behavior of midday auroras during substorms.**
S.-I. Akasofu, D. S. Kimball.
Planet. Space Sci., Vol. 21, 696 - 698 (1973). – Research note.

084.021   **A discussion of the energy transfer from $N_2(A^3\Sigma_u^+)$ to $O(^1S)$ in pulsating aurora.**   A. Brekke.
Planet. Space Sci., Vol. 21, 698 - 702 (1973). – Research note

084.022   **The aurorae of the day side of the oval in the period of substorms.**
G. V. Starkov, Ya. I. Feldshtein, N. F. Shevnina.
Geomagn. Aeronom., Vol. 13, 86 - 90 (1973). In Russian.

084.023   **About the daily variation of the length of auroral rays.**   N. I. Dzyubenko.
Geomagn. Aeronom., Vol. 13, 91 - 94 (1973). In Russian.

084.024   **Relationship of southward-drifting auroral arcs to the magnetospheric electric field and substorm activity.**   S. Subbarao, G. Rostoker.
Journ. Geophys. Res., Vol. 78, 1100 - 1106 (1973).
The magnitude of the westward-directed auroral zone electric field is inferred from the measurement of the southward drift of moderately bright auroral arcs, using a sample of 43 events recorded over a period of 15 months at Fort Smith.

084.025   **Simultaneous occurrence of hydrogen arcs and mid-latitude stable auroral red arcs.**
E. W. Kleckner, R. J. Hoch.
Journ. Geophys. Res., Vol. 78, 1187 - 1193 (1973).
Observational evidence of the simultaneous occurrences of hydrogen and mid-latitude stable auroral red arcs is presented for seven occasions during the period September 1967 to May 1971.

084.026   **Auroral oval.**   Y. I. Feldstein.
Journ. Geophys. Res., Vol. 78, 1210 - 1213 (1973). Letter.

084.027   **The occurrence of aurorae in the course of the 11-year solar cycle.**   Z. Pokorný.
Bull. Astron. Inst. Czechoslovakia, Vol. 24, 107 - 108 (1973).
The method of superposition was used to investigate the occurrence of aurorae in the course of the 11-year solar cycle. By analysing data from the years 1897–1951, obtained from the Yerkes Observatory, it was found that aurorae occur most frequently during year +2 following the maximum of the relative sunspot number.

084.028   **About the nature of an auroral electrojet.**
L. L. Vanyan, A. S. Debabov, I. L. Osipova.
Geomagn. Aeronom., Vol. 13, 325 - 329 (1973). In Russian.

084.029   **Motion of aurorae during a substorm.**
A. M. Lyatskaya, V. B. Lyatzky.
Geomagn. Aeronom., Vol. 13, 375 - 377 (1973). In Russian. Brief information.

084.030   **A uniform belt of diffuse auroral emission seen by the ISIS-2 scanning photometer.**
A. T. Y. Lui, C. D. Anger.
Planet. Space Sci., Vol. 21, 799 - 809 (1973).

The scanning auroral photometer on board the ISIS-2 satellite, which measures the auroral emissions at 5577 and 3914 Å, can survey a broad region during a single pass over the auroral zone. In this paper we consider one of the most obvious and most persistent features of the high latitude regions as seen by this instrument – a continuous belt of diffuse auroral emission.

084.031   **An observation of polar aurora and airglow from the ISIS-2 spacecraft.**
G. G. Shepherd, C. D. Anger, L. H. Brace, J. R. Burrows, W. J. Heikkila, J. Hoffman, E. J. Maier, J. H. Whitteker.
Planet. Space Sci., Vol. 21, 819 - 829 (1973).
This is a preliminary but comprehensive report on coordinated data obtained with the ISIS-2 spacecraft, launched 1 April 1971, into a near circular 1400 km orbit. Data obtained during a 30-min pass over the south pole depict the nightside oval and polar cap, as well as mid-latitude airglow effects; these data are described and discussed.

084.032   **The diffuse aurora.**
A. T. Y. Lui, P. Perreault, S.-I. Akasofu, C. D. Anger.
Planet. Space Sci., Vol. 21, 857 - 861 (1973).
The purpose of this paper is : (1) to identify the diffuse aurora observed by the ISIS-2 scanning auroral photometer in the simultaneous ground all-sky photographs, and (2) to describe some aspects of its temporal behavior from ground-based observations.

084.033   **The excitation of atomic oxygen to the $O(^1S)$ level by energy transfer from $N_2(A^3\Sigma_u^+)$ molecules in aurora.**   K. Henriksen.
Planet. Space Sci., Vol. 21, 863 - 871 (1973).
The part that the energy transfer reaction $N_2(A^3\Sigma_u^+)$ + $O(^3P) \rightarrow N_2(X^1\Sigma_g^+) + O(^1S)$ plays in the excitation of the auroral green line has been investigated.

084.034   **A global view at the polar region on 18 December 1971.**   C. D. Anger, A. T. Y. Lui.
Planet. Space Sci., Vol. 21, 873 - 878 (1973).
This paper reports some unique characteristics of auroral emissions in the polar region observed by ISIS-2 during the recovery phase of the intense magnetic storm which occurred in December 1971.

084.035   **Pitch angle diffusion of low-energy auroral electrons.**
B. A. Whalen, I. B. McDiarmid.
Journ. Geophys. Res., Vol. 78, 1608 - 1614 (1973).

084.036   **Distributions and characteristics of high-latitude field-aligned electron precipitation.**   F. W. Berko.
Journ. Geophys. Res., Vol. 78, 1615 - 1626 (1973).

084.037   **VHF Doppler spectra of radar echoes associated with a visual auroral form: Observations and implications.**   B. B. Balsley, W. L. Ecklund, R. A. Greenwald.
Journ. Geophys. Res., Vol. 78, 1681 - 1687 (1973).
Detailed analysis of a relatively uncomplicated premidnight auroral event shows that there can be a close correspondence between visual auroral and radar auroral forms.

084.038   **Twin payload observations of incident and backscattered auroral electrons.**
D. L. Reasoner, C. R. Chappell.
Journ. Geophys. Res., Vol. 78, 2176 - 2186 (1973).

084.039   **Auroral electrons of energy less than 1 Kev observed at rocket altitudes.**   R. L. Arnoldy, L. W. Choy.
Journ. Geophys. Res., Vol. 78, 2187 - 2200 (1973).

084.040   **Spatial separation of 3914- and 3160-Å emissions**

of nitrogen in an aurora. M. Henrist.
Journ. Geophys. Res., Vol. 78, 2350 - 2352 (1973). – Letter.

084.041 Comment on paper by D. C. Cartwright, S. Trajmar, and W. Williams, 'Vibrational population of the $A^3\Sigma_u^+$ and $B^3\Pi_g$ states of N_2 in normal auroras'.
D. E. Shemansky, A. L. Broadfoot.
Journ. Geophys. Res., Vol. 78, 2357 - 2364, with a reply by D. C. Cartwright, S. Trajmar, W. Williams, p. 2365 - 2373 (1973).

084.042 Photometric investigation of the 4278 Å and 5577 Å emissions in aurora. K. Henriksen.
Journ. Atmosph. Terr. Phys., Vol. 35, 1341 - 1350 (1973).
The excitation of the green line is found consistent with major contribution from energy transfer from N_2 ($A^3\Sigma_u^+$) molecules and dissociative recombination of O_2^+ ions. The green/blue intensity ratio of the auroral pulsation amplitudes is found to be of the same magnitudes as in bright aurora, and in these displays the intensity ratio is increasing with intensity.

084.043 Measurements of auroral ionization at high latitudes on Cosmos 348.
N. V. Jorgio (*Dzhordzhio*), L. G. Oldekop, N. I. Fedorova.
Geomagn. Aeronom., Vol. 13, 437 - 441 (1973). In Russian.

084.044 Cosmic radio noise absorption and hydrogen emission in the auroral substorm.
B. P. Kilfoyle, F. Jacka.
Australian Journ. Phys., Vol. 26, 225 - 228 (1973).
From a study of records from Mawson, Kiruna, and Murmansk it is shown that slowly varying ionospheric absorption and Hβ emission are characteristic of the region behind the midnight poleward bulge of the auroral substorm, in accord with the model of Akasofu (1968).

084.045 Auroral ion velocity distributions using a relaxation model. J.-P. St-Maurice, R. W. Schunk.
Planet. Space Sci., Vol. 21, 1115 - 1130 (1973).
For application to the auroral ionosphere we have calculated ion velocity distributions for a weakly-ionized plasma subjected to crossed electric and magnetic fields. By replacing the Boltzmann collision integral with a simple relaxation model, we have been able to obtain an exact solution to Boltzmann's equation.

084.046 Time dependent studies of the aurora – II. Spec-
troscopic morphology. M. H. Rees, R. A. Jones.
Planet. Space Sci., Vol. 21, 1213 - 1235 (1973).
The temporal morphology of auroral spectral emission features is investigated. For a given energy distribution of bombarding electrons but a time varying flux magnitude, the emission rates of various auroral radiations exhibit a nonlinear time response due to the variety of reactions that contribute to excitation.

084.047 Balmer-line emission from auroral protons.
D. Ellis, R. Ptak, R. Stoner.
Astrophys. Journ., Vol. 182, 637 - 647 (1973).
The authors present results of calculations of cross-sections for proton-on-hydrogen-atom collisions involving excitation, ionization, and charge transfer, in the energy range from 1 keV to 150 keV. Excitation events are treated in a first-order quantum-impact parameter approach, while values for all types of events are found with a semiclassical model. These cross-sections are used to calculate the Balmer-line emission expected from auroral protons.

084.048 Observations of the auroral oval and a westward traveling surge from the Isis 2 satellite and the Alaskan meridian all-sky cameras.
C. D. Anger, A. T. Y. Lui, S.-I. Akasofu.
Journ. Geophys. Res., Vol. 78, 3020 - 3026 (1973).
Brief report.

084.049 Red auroras in the morning sector.
S.-I. Akasofu, F. Yasuhara.
Journ. Geophys. Res., Vol. 78, 3027 - 3032 (1973).
Brief report.

084.050 Measurements of the composition of energetic auroral ions. W. I. Axford, F. Buhler, H. J. A. Chivers, P. Eberhardt, J. Geiss.
IV Leningr. mezhdunar. seminar "Edinoobrazie uskoreniya chastits v razlich. masshtabakh kosmosa, 1972". Leningrad, 1972, p. 183 - 196. In Russian. – Abstr. in Referativ. Zhurn. 62. Issled. kosmich. prostranstva, 7.62.240 (1973).

084.051 Bewegungsvorgänge in Radio-Polarlichtern.
P. Czechowsky.
Thesis, Math.-Naturwiss. Fakultät, Univ. Göttingen. 67 pp. (1972).

Total electron content measurements during visible auroras. See Abstr. 083.022.

Geomagnetic Field

084.201 **Semiannual variation of geomagnetic activity.**
C. T. Russell, R. L. McPherron.
Journ. Geophys. Res., Vol. 78, 92 - 108 (1973).

084.202 **Substorm variations of the magnetotail plasma sheet from $X_{SM} \approx -6\,R_E$ to $X_{SM} \approx -60\,R_E$.**
E. W. Hones, Jr., J. R. Asbridge, S. J. Bame, S. Singer.
Journ. Geophys. Res., Vol. 78, 109 - 132 (1973).

084.203 **Drift shell splitting by internal geomagnetic multipoles.**
J. G. Roederer, H. H. Hilton, M. Schulz.
Journ. Geophys. Res., Vol. 78, 133 - 144 (1973).

084.204 **Analyses of techniques for measuring DC and AC electric fields in the magnetosphere.** F. S. Mozer.
Space Sci. Rev., Vol. 14, 272 - 313 (1973).
Methods for measuring the amplitudes and directions of DC electric fields and the directions, power spectra, and dispersion relations of AC electric fields in the magnetosphere are discussed with emphasis on their applicability in various regimes of the magnetospheric plasma.

084.205 **Observation de l'entrée des protons solaires dans la magnétosphère aux très hautes latitudes.**
J.-J. Berthelier, M. Pirre.
Comptes Rendus Acad. Sci. Paris, Sér. B, Vol. 276, 279 - 282 (1973).
Pendant la phase initiale de l'événement d'absorption consécutif à l'éruption solaire du 28 janvier 1967, nous avons pu calculer, à partir de mesures faites au sol ou à basse altitude et dans le milieu interplanétaire, à quelle distance de la terre les protons précipités dans les calottes polaires ont pénétré dans la queue de la magnétosphère.

084.206 **A method of determination of the magnetosphere's longitudinal conductivity by means of pulsations of the magnetic field of Pi-2 type.** V. M. Davydov.
Dokl. Akad. Nauk SSSR, Ser. Mat. Fiz., Vol. 208, 1071 - 1073 (1973). In Russian.

084.207 **The mechanism of geomagnetic field reversal.**
T. Rikitake.
Phys. Earth Planet. Interiors, Vol. 6, 340 - 345 (1972).
That the frequency distribution of geomagnetic polarity intervals is described by a Poisson law has raised the supposition that the occurrence of geomagnetic field reversal is a stochastic process. It is shown, however, that even deterministic dynamo models lead to such a distribution. It seems likely that geomagnetic polarity reversals occur as a result of a complicated energy exchange between units consisting of dynamo models and also of conversion between kinetic and magnetic energies.

084.208 **Magnetospheric substorms: Some problems and controversies.**
V. M. Vasyliunas, R. A. Wolf.
Rev. Geophys. Space Phys., Vol. 11, 181 - 189 (1973). — Topical review.

084.209 **Magnetic pulsations within the magnetosphere: A review.** D. Orr.
Journ. Atmosph. Terr. Phys., Vol. 35, 1 - 50 (1973).

084.210 **Investigation into the origin of an ELF discrete signal.** R. A. Fleming.
Journ. Atmosph. Terr. Phys., Vol. 35, 187 - 189 (1973).
A high intensity geomagnetic disturbance at extremely low frequency received simultaneously at two spaced stations supports the theory that extra-terrestrial particles, in this case solar protons, contribute to ELF signal generation.

084.211 **Lunar and solar geomagnetic tides in declination at Alibag.** K. S. R. Rao, S. J. Reddy, K. N. Rao.
Journ. Atmosph. Terr. Phys., Vol. 35, 255 - 262 (1973).
The solar and lunar geomagnetic tides in declination at Alibag have been determined by spectral analysis of discrete Fourier transforms following the method of Black and by the well known Chapman-Miller method.

084.212 **Annual and semi-annual variations in the electron density of the inner magnetosphere deduced from whistler dispersion.** M. Hayakawa, J. Ohtsu.
Journ. Atmosph. Terr. Phys., Vol. 35, 339 - 345 (1973).

084.213 **Evening/forenoon asymmetry in the 27-day oscillation of the low-latitude magnetic field.**
B. N. Bhargava.
Journ. Atmosph. Terr. Phys., Vol. 35, 567 - 570 (1973).
Power spectra of low-latitude horizontal intensity, computed from series derived from observations restricted to different local times, reveal a large evening/forenoon asymmetry in the magnitude of the 27-day oscillation of the field.

084.214 **Nonadiabatic particle motion in the magnetosphere.**
G. Morfill.
Journ. Geophys. Res., Vol. 78, 588 - 596 (1973).
This paper investigates the nonadiabatic motion of solar protons in the magnetosphere and the effects on the solar proton cutoff and the polar cap intensity structure. Calculations are made in a model field and provide a basis to which wave-particle interactions and other scattering effects may be added.

084.215 **Proton scattering in the region near the earth's bow shock.** S. L. Ossakow, G. W. Sharp.
Journ. Geophys. Res., Vol. 78, 607 - 616 (1973).
We present Lockheed spectrometer observations of bow shock proton scattering lengths and UCLA magnetometer data for the first 20 orbits of OGO 5, covering the period from March 6, 1968, to April 24, 1968.

084.216 **The behaviour of ULF waves and particles in the magnetosphere.** D. J. Southwood.
Planet. Space Sci., Vol. 21, 53 - 65 (1973).

084.217 **Geomagnetic variations with the period of a sidereal day.** S. R. C. Malin.
Planet. Space Sci., Vol. 21, 145 - 146 (1973).
· Recent determinations of sidereal-day variations in the geomagnetic field have been attributed to galactic sources of ionisation. Alternative mechanisms that could equally well account for the observed variations are presented, based on solar and lunar tides.

084.218 **Field-aligned currents between 400 and 3000 km in auroral and polar latitudes.**
B. Theile, H. M. Praetorius.
Planet. Space Sci., Vol. 21, 179 - 187 (1973).
We present some properties of magnetic field variations transverse to the geomagnetic field measured aboard the German research satellite AZUR in auroral and polar latitudes. These data are interpreted in terms of time dependent field-aligned currents.

084.219 **A study of the relationship between geomagnetic storms and ionospheric disturbances at mid-latitudes.**
M. Mendillo.
Planet. Space Sci., Vol. 21, 349 - 358 (1973).

084.220 Precise calculation of the magnetosphere surface for a tilted dipole.
J. Y. Choe, D. B. Beard, E. C. Sullivan.
Planet. Space Sci., Vol. 21, 485 - 498 (1973).

084.221 The effect of the earth's bow shock and magneto-sheath on the interaction of a discontinuity in the solar wind with the magnetosphere.
B. J. Rigby, J. S. Mainstone.
Planet. Space Sci., Vol. 21, 499 - 506 (1973).
A theoretical model is proposed for the interaction of a plane discontinuity in the solar wind with the magnetosphere.

084.222 The magnetic field in the earth's tail.
Y. H. Huang.
Planet. Space Sci., Vol. 21, 528 - 532 (1973). – Research note.

084.223 Influence of a pressure anisotropy on the oscillations of the magnetosphere's tail.
A. I. Ershkovich, S. A. Mart'yanov.
Kosmich. Issled., Vol. 11, 80 - 87 (1973). In Russian.

084.224 About oscillations of the magnetic tail of the earth in quasi-hydrodynamical approximation.
A. I. Ershkovich, S. A. Martyanov.
Geomagn. Aeronom., Vol. 13, 110 - 115 (1973). In Russian.

084.225 About the mean free path in the transition region beyond the boundary of the magnetosphere.
M. S. Kovner.
Geomagn. Aeronom., Vol. 13, 168 - 171 (1973). In Russian.
Brief information.

084.226 The variability of quiet-sun daily variations from day to day and the direction of the interplanetary magnetic field. V. I. Afanasieva.
Geomagn. Aeronom., Vol. 13, 193 - 195 (1973). In Russian.
Brief information.

084.227 Effects of interplanetary magnetic sector structure on auroral zone and polar cap magnetic activity.
J. L. Burch.
Journ. Geophys. Res., Vol. 78, 1047 - 1057 (1973).
For quiet ($C9 \leq 2$) and quiet to moderately disturbed ($C9 \leq 4$) conditions in the period 1964−1968, the sector structure dependence of auroral zone positive and negative bay activity, as measured by the AE components AU and AL, is noted.

084.228 Electron pitch angle distributions throughout the magnetosphere as observed on Ogo 5.
H. I. West, Jr., R. M. Buck, J. R. Walton.
Journ. Geophys. Res., Vol. 78, 1064 - 1081 (1973).
A survey of the equatorial pitch angle distributions of energetic electrons is provided for all local times out to radial distances of 20 R_E on the night side of the earth and to the magnetopause on the day side of the earth.

084.229 Variations diurnes de la composante verticale du champ magnétique dans les régions de haute latitude en fonction de la composante Est-Ouest du champ magnétique interplanétaire. A. Berthelier.
Comptes Rendus Acad. Sci. Paris, Sér. B, Vol. 276, 693 - 696 (1973).

084.230 About the interaction between the earth's magnetic field and the antiparallel field of the solar wind.
M. S. Kovner.
Geomagn. Aeronom., Vol. 13, 302 - 305 (1973). In Russian.

084.231 About radiation at the boundary of the magneto-

sphere. M. S. Kovner.
Geomagn. Aeronom., Vol. 13, 306 - 310 (1973). In Russian.

084.232 About the dynamics of longitudinal currents in the magnetosphere.
V. A. Gudkova, L. M. Zeleny, V. A. Liperovsky.
Geomagn. Aeronom., Vol. 13, 318 - 324 (1973). In Russian.

084.233 About the division of a variable geomagnetic field into a normal and an abnormal part.
M. N. Berdichevsky, M. S. Zhdanov.
Geomagn. Aeronom., Vol. 13, 339 - 342 (1973). In Russian.

084.234 Absorption of hydromagnetic waves in a plasma layer of the magnetosphere's tail. A. S. Potapov.
Geomagn. Aeronom., Vol. 13, 377 - 379 (1973). In Russian.
Brief information.

084.235 The electric field in the magnetosphere on quiet conditions according to ground observations of whistling atmospherics. Ya. P. Sobolev.
Geomagn. Aeronom., Vol. 13, 379 - 382 (1973). In Russian.
Brief information.

084.236 About the connection of the period of geomagnetic pulsations Pc3, 4 with the parameters of the inter-planetary medium in the earth's orbit.
A. V. Gulielmi, T. A. Plyasova-Bakunina, R. V. Tshepetnov (*Shchepetnov*).
Geomagn. Aeronom., Vol. 13, 382 - 384 (1973). In Russian.
Brief information.

084.237 Linear relationships in geomagnetic variation studies. F. E. M. Lilley, D. J. Bennett.
Phys. Earth Planet. Interiors, Vol. 7, 9 - 14 (1973).
In studies of geomagnetic variations, it is common to seek empirical relationships between different components which are observed to be obeyed independently of time. This paper explores the theoretical significance of such linear relationships.

084.238 Average high latitude magnetic field: Variation with interplanetary sector and with season – I.
Disturbed conditions. R. Langel.
Planet. Space Sci., Vol. 21, 839 - 855 (1973).
High latitude magnetic field data from 16 northern observatories are averaged during periods of magnetic disturbance level Kp = 2⁻ to 3⁺. Within this disturbance level, variations between interplanetary magnetic field sector and geomagnetic season are delineated.

084.239 Magnetic field signatures of substorms on high-latitude field lines in the nighttime magnetosphere.
D. H. Fairfield.
Journ. Geophys. Res., Vol. 78, 1553 - 1562 (1973).

084.240 Magnetosheath observations at high northern latitudes by Heos 2. P. C. Hedgecock,
P. Cerulli, A. Coletti, A. Egidi, R. Marconero, V. Domingo,
D. Köhn, D. E. Page, B. G. Taylor, K.-P. Wenzel.
Journ. Geophys. Res., Vol. 78, 1715 - 1718 (1973). – Letter.

084.241 Some problems of the physics of the magneto-sphere.
A. I. Ershkovich, G. A. Skuridin, V. P. Shalimov.
Mezhplanet. sreda i fiz. magnitosfery. Moskva, Nauka, 1972,
p. 3 - 25. In Russian. – Abstr. in Referativ. Zhurn. 62.
Issled. kosmich. prostranstva, 4.62.252 (1973).

084.242 On the calculation of the form of the magneto-sphere's boundary. S. A. Mart'yanov.

Mezhplanet. sreda i fiz. magnitosfery. Moskva, Nauka, 1972, p. 164 - 178. In Russian. – Abstr. in Referativ. Zhurn. 62. Issled. kosmich. prostranstva, 4.62.256 (1973).

084.243 **Oscillations of the earth's magnetic tail.**
A. I. Ershkovich, A. A. Nusinov.
Mezhplanet. sreda i fiz. magnitosfery. Moskva, Nauka, 1972, p. 191 - 195. In Russian. – Abstr. in Referativ. Zhurn. 62. Issled. kosmich. prostranstva, 4.62.260 (1973).

084.244 **Magnetic measurements in model experiments.**
Eh. M. Dubinin, G. G. Managadze, I. M. Podgornyj.
Mezhplanet. sreda i fiz. magnitosfery. Moskva, Nauka, 1972, p. 33 - 43. In Russian. – Abstr. in Referativ. Zhurn. 62. Issled. kosmich. prostranstva, 4.62.266 (1973).

084.245 **On the transport of charged particles in the earth's magnetosphere.** V. P. Shalimov.
Mezhplanet. sreda i fiz. magnitosfery. Moskva, Nauka, 1972, p. 179 - 190. In Russian. – Abstr. in Referativ. Zhurn. 62. Issled. kosmich. prostranstva, 4.62.270 (1973).

084.246 **Hydromagnetic waves directed by the geomagnetic field.** L. L. Van'yan, L. A. Abramov, M. B. Gokhberg, V. L. Yudovich.
Mezhplanet. sreda i fiz. magnitosfery. Moskva, Nauka, 1972, p. 26 - 32. In Russian. – Abstr. in Referativ. Zhurn. 62. Issled. kosmich. prostranstva, 4.62.275 (1973).

084.247 **Effects of solar particle events on the energetic particle population in the earth's magnetosphere.**
G. A. Paulikas.
Progr. Astronaut. Aeronaut., Vol. 30, (see 003.009), p. 265 - 283 (1972). – Presented at the AIAA observation and prediction of solar activity conference, Huntsville, Ala., Nov. 16–18, 1970.

084.248 **Geomagnetic response to solar activity.**
G. D. Mead.
Progr. Astronaut. Aeronaut., Vol. 30, (see 003.009), p. 285 - 292 (1972). – Presented at the AIAA observation and prediction of solar activity conference, Huntsville, Ala., Nov. 16–18, 1970.

084.249 **A self-consistent two-dimensional approach to magnetospheric structures.** M. Soop, K. Schindler.
Astrophys. Space Sci., Vol. 20, 287 - 305 (1973).
A simple self-consistent description of the geomagnetic tail is given and its consequences are explored. The model is two-dimensional, ignoring the spatial variation along the direction of the tail current. A discussion of equilibria is based on the assumption of an isotropic pressure. One of the conclusions is that the transition between the dipolar and the tail region will take place over a fairly short distance. The analytical results are supplemented by a numerical example. Properties of configurations containing field line loops are investigated. A stability discussion includes the effect of the net magnetic flux through the neutral sheet.

084.250 **Three dimensional dynamo theory in the magnetosphere.** R. Pratap, V. Sarabhai, K. N. Nair.
Astrophys. Space Sci., Vol. 20, 307 - 327 (1973).
We have solved in this paper the three dimensional dynamo equation consistent with the conditions in the magnetosphere. The conductivity we have adopted here is that for a fully ionised but highly rarefied gas in a magnetic field. The velocity field is based on the measurements of the convection patterns made by different satellites. The solution obtained of the dynamo equation is presented here in the most general form so that it can be used when the various parameters are known to a higher degree of accuracy in future. We have then

computed the components of the current as well as the iso-intensity curves in the midday-midnight meridional plane as well as on the dawn-dusk meridional plane. These theoretical results have then been matched with observations.

084.251 **Energy and momentum theorems in magnetospheric processes.** R. L. Carovillano, G. L. Siscoe.
Rev. Geophys. Space Phys., Vol. 11, 289 - 353 (1973).
This review deals with the several energy and momentum theorems that relate to magnetospheric processes that have been developed. The region of primary consideration in this paper is the magnetospheric domain that extends between the ionosphere and the interplanetary medium. Both energy theorems and momentum theorems with their applications are presented.

084.252 **Polar cap magnetic variations and their relationship with the interplanetary magnetic sector structure.**
L. Svalgaard.
Journ. Geophys. Res., Vol. 78, 2064 - 2078 (1973).

084.253 **Correspondence of solar field sector direction and polar cap geomagnetic field changes for 1965.**
W. H. Campbell, S. Matsushita.
Journ. Geophys. Res., Vol. 78, 2079 - 2087 (1973).

084.254 **Magneto-ionospheric effect of a substorm in the Tbilisi region.** G. N. Pushkova, L. A. Yudovich, V. I. Petviashvili, Ya. I. Feldshtein.
Geomagn. Aeronom., Vol. 13, 478 - 481 (1973). In Russian.

084.255 **The geomagnetic tail in experiments with three-dimensional terrellas.**
E. M. Dubinin, I. M. Podgorny.
Geomagn. Aeronom., Vol. 13, 532 - 535 (1973). In Russian. Brief information.

084.256 **Seasonal changes of variations of the Z-component in the near-polar region in connection with the sign of the Y_{SE}-component of the interplanetary magnetic field.** P. V. Sumaruk, Ya. I. Feldshtein.
Geomagn. Aeronom., Vol. 13, 545 - 546 (1973). In Russian. Brief information.

084.257 **Magnetospheric substorms.**
F. V. Coroniti, C. F. Kennel.
Cosmic plasma physics. Conference 1971, (see 012.016), p. 15 - 23 (1972).

084.258 **Universal instability associated with the plasmapause and its role in geomagnetic micropulsations.**
H. Kikuchi.
Cosmic plasma physics. Conference 1971, (see 012.016), p. 45 - 54 (1972).

084.259 **The structure of the earth's bow shock.**
B. Bertotti, D. Parkinson, K. Schindler, P. Goldberg.
Cosmic plasma physics. Conference 1971, (see 012.016), p. 319 - 325 (1972).

084.260 **Geomagnetic line spectra – 2 to 70 years.**
R. G. Currie.
Astrophys. Space Sci., Vol. 21, 425 - 438 (1973).
Application of a new data adaptive approach to power spectrum estimation has yielded greatly improved knowledge of the geomagnetic spectrum in the range 2 to 70 yr. The first successful line spectrum detection of the solar and double solar cycle variations in absolute geomagnetic element data are presented; also detected are the first four harmonics of the solar cycle and, excepting one, the first nine harmonics of the double solar cycle. Finally, evidence is found for a ~ 60 yr line.

The implications of these results for a variety of problems in planetary and space physics are discussed.

084.261 Geomagnetic tail oscillations in quasihydrodynamics.
A. I. Ershkovich, S. A. Martjanov.
Astrophys. Space Sci., Vol. 21, 439 - 450, 451 - 460 (1973).
In Russian and English.

The effect of pressure anisotropy on geomagnetic tail oscillations is studied using the Chew-Goldberger-Low equations. It is shown that anisotropy of the solar wind plasma pressure may result in generation of waves, assuming no waves in the case of isotropy. Anisotropy of the plasma pressure in the magnetospheric tail may also cause a basic distortion of the oscillation spectra.

084.262 On solar wind interaction with the earth's magnetosphere. M. S. Kovner, Y. I. Feldstein.
Planet. Space Sci., Vol. 21, 1191 - 1211 (1973).

Hydrodynamic and electrodynamic problems of solar wind interaction with the earth's magnetosphere on the dayside are investigated.

084.263 Semi-annual modulation of earth's magnetic field in the equatorial electrojet region.
B. N. Bhargava, D. R. K. Rao, B. R. Arora.
Planet. Space Sci., Vol. 21, 1251 - 1255 (1973). — Research note.

084.264 Magnetospheric dayside cusp: A topside view of its 6300-Ångstrom atomic oxygen emission.
G. G. Shepherd, F. W. Thirkettle.
Science, Vol. 180, 737 - 739 (1973).

An interference filter photometer on the ISIS-II spacecraft generates global maps of the atomic oxygen emission at 6300 Å from the ionosphere. The most prominent feature observed is a band of permanent red aurora on the dayside of the earth, centered on magnetic noon at about 78 degrees magnetic (invariant) latitude, brighter than the quiet-time nightside aurora.

084.265 A theoretical prediction of ion plasma oscillations in a neutral sheet. E. C. Bowers.
Astrophys. Space Sci., Vol. 21, 399 - 423 (1973).

Starting from the Vlasov equation the steady state and stability properties of the electron sheet in the Cowley neutral sheet model of the geomagnetic tail are considered. Electrostatic ion plasma oscillations propagating from dusk to dawn are found to be unstable provided the thermal spread normal to the current is sufficiently large. It is shown how a localisation of the cross tail electric field could lead to the instability first appearing around midnight.

084.266 A search for periodic variations in geomagnetic activity and their solar cycle dependence.
H. Hauska, S. Abdel-Wahab, E. Dyring.
Phys. Scripta, Vol. 7, 135 - 140 (1973).

Periodic variations in geomagnetic activity have been investigated, in the range from 4 to 40 days, by means of power spectrum analysis during the period 1932 to 1969. Main periodicities are found to exist at periods of 28.7, 13.3, 9.1 and 6.9 days. The variation of the power density at different frequencies—particularly the 27-, 13-, 9- and 6-day variation—with the solar cycle has been investigated. An average value of the period of the 27-day variation has been determined by means of a modified power spectrum technique. A high correlation with the solar cycle has been found and possibilities of a multiple cycle have also been discussed.

084.267 Periodic variations in geomagnetic activity and sector structure of the interplanetary magnetic field.
H. Hauska.

Phys. Scripta, Vol. 7, 183 - 188 (1973).

Periodic variations, in the period range from 4–40 days, in geomagnetic activity are explained as due to the effects of the solar and interplanetary magnetic field structure. The dependence of the geomagnetic activity of the different components of the interplanetary magnetic field is studied. Linear relations for this dependence are given. An attempt is made to predict the interplanetary sector structure, during 1964, by using the polarity of solar magnetic fields.

084.268 The earth's magnetic tail in experiments with a three dimensional terrella.
E. M. Dubinin, I. M. Podgorny.
IV Leningr. mezhdunar. seminar "Edinoobrazie uskoreniya chastits v razlich. masshtabakh kosmosa", 1972. Leningrad, p. 294. — Abstr. in Referativ. Zhurn. 62. Issled. kosmich. prostranstva, 5.62.287 (1973).

084.269 Equatorial paleopoles and behavior of the dipole field during polarity transitions.
P. Steinhauser, S. A. Vincenz.
Earth Planet. Sci., Letters, Vol. 19, 113 - 119 (1973).

Data on 23 field reversals of Recent, Tertiary and Upper Mesozoic age are examined with regard to the longitudinal and latitudinal distribution of paleomagnetic poles during a polarity change.

084.270 Intensity variations of low-energy protons and electrons in the outer magnetosphere at the sudden beginning of a magnetic storm.
M. I. Panasyuk, Eh. N. Sosnovets, L. V. Tverskaya.
Kosmich. Issled., Vol. 11, 431 - 435 (1973). In Russian.

084.271 Can the ionosphere regulate magnetospheric convection? F. V. Coroniti, C. F. Kennel.
Journ. Geophys. Res., Vol. 78, 2837 - 2851 (1973).

084.272 Self-consistent calculation of the motion of a sheet of ions in the magnetosphere.
R. K. Jaggi, R. A. Wolf.
Journ. Geophys. Res., Vol. 78, 2852 - 2866 (1973).

084.273 Satellite studies of magnetospheric substorms on August 15, 1968. 1. State of the magnetosphere.
R. L. McPherron.
Journ. Geophys. Res., Vol. 78, 3044 - 3053 (1973).

084.274 Satellite studies of magnetospheric substorms on August 15, 1968. 2. Solar wind and outer magnetosphere. R. L. McPherron, G. K. Parks, D. S. Colburn, M. D. Montgomery.
Journ. Geophys. Res., Vol. 78, 3054 - 3061 (1973).

084.275 Satellite studies of magnetospheric substorms on August 15, 1968. 3. Some features of magnetospheric convection. D. L. Carpenter, C. R. Chappell.
Journ. Geophys. Res., Vol. 78, 3062 - 3067 (1973).

084.276 Satellite studies of magnetospheric substorms on August 15, 1968. 4. Ogo 5 magnetic field observations. R. L. McPherron, M. P. Aubry, C. T. Russell, P. J. Coleman, Jr.
Journ. Geophys. Res., Vol. 78, 3068 - 3078 (1973).

084.277 Satellite studies of magnetospheric substorms on August 15, 1968. 5. Energetic electrons, spatial boundaries, and wave-particle interactions at Ogo 5.
M. G. Kivelson, T. A. Farley, M. P. Aubry.
Journ. Geophys. Res., Vol. 78, 3079 - 3092 (1973).

084.278 Satellite studies of magnetospheric substorms on

August 15, 1968. 6. Ogo 5 energetic electron observations—pitch angle distributions in the nighttime magnetosphere. H. I. West, Jr., R. M. Buck, J. R. Walton.
Journ. Geophys. Res., Vol. 78, 3093 - 3102 (1973).

084.279 Satellite studies of magnetospheric substorms on August 15, 1968. 7. Ogo 5 energetic proton observations—spatial boundaries.
R. M. Buck, H. I. West, Jr., R. G. D'Arcy, Jr.
Journ. Geophys. Res., Vol. 78, 3103 - 3118 (1973).

084.280 Satellite studies of magnetospheric substorms on August 15, 1968. 8. Ogo 5 plasma wave observations. F. L. Scarf, R. W. Fredricks, C. F. Kennel, F. V. Coroniti.
Journ. Geophys. Res., Vol. 78, 3119 - 3130 (1973).

084.281 Satellite studies of magnetospheric substorms on August 15, 1968. 9. Phenomenological model for substorms. R. L. McPherron, C. T. Russell, M. P. Aubry.
Journ. Geophys. Res., Vol. 78, 3131 - 3149 (1973).

084.282 Quiet time magnetospheric field depression at $2.3-3.6 R_E$. M. Sugiura.
Journ. Geophys. Res., Vol. 78, 3182 - 3185 (1973). — Letter.

084.283 Comments on geophysical plasma instabilities and their importance. D. J. Williams.
Comments Astrophys. Space Phys., Vol. 5, 57 - 65 (1973).
Direct in situ observations and studies of naturally occurring plasmas in the near-earth space environment will provide a major impetus towards the understanding of solar active regions (including flares), other planetary environments (such as the Jovian magnetosphere) and more distant phenomena such as pulsars, quasars, and magnetized plasmas throughout the cosmos.

084.284 Magnetic results 1969, Eskdalemuir, Hartland and Lerwick Observatories.
Natural Environment Research Council, Inst. of Geol. Sci., Geomagnetic Bull. No. 3, [London: Her Majesty's Stationery Office], 4 + 154 pp. Price £ 4.00 (1973).

084.285 Magnetic results 1970, Eskdalemuir, Hartland and Lerwick Observatories.
Natural Environment Research Council, Inst. of Geol. Sci., Geomagnetic Bull. No. 4, [London: Her Majesty's Stationery Office], 9 + 154 pp. Price £ 4.00 (1973).

Interplanetary medium and physics of the magnetosphere. See Abstr. 003.135.

The interplanetary medium and physics of the magnetosphere. See Abstr. 003.139.

Gravity waves: correlation with geomagnetic storms. See Abstr. 066.132.

Solar energy cycle and its relation to geomagnetic activity. See Abstr. 072.060.

Strong pitch angle diffusion and magnetospheric solar protons. See Abstr. 078.005.

Directional diffusion coefficients of solar protons inside and outside the bow shock. See Abstr. 078.036.

Magnetic storm characteristics of the thermosphere. See Abstr. 082.078.

The causes of storm-time increases of the F-layer at mid-latitudes. See Abstr. 083.013.

A diffusion model for the electron density distribution along the earth's magnetic field in an F-region plasma cloud. See Abstr. 083.018.

Enhancements of ionospheric total electron content in the southern auroral zone associated with magnetospheric substorms. See Abstr. 083.023.

Coupling between the F-region and protonosphere: Numerical solution of the time-dependent equations. See Abstr. 083.025.

The ionospheric electric field during substorms — an interpretation based on non-uniform reconnection in the geomagnetic tail. See Abstr. 083.062.

Influence du champ magnétique interplanétaire sur les perturbations magnétiques des régions de haute latitude: mise en évidence d'une asymétrie par rapport à la direction terre—soleil. See Abstr. 106.006.

Interaction between the interplanetary medium and the geomagnetosphere. See Abstr. 106.013.

Interplanetary magnetic field and geomagnetic Dst variations. See Abstr. 106.014.

Magnetotail response to sudden changes in the interplanetary magnetic field. See Abstr. 106.015.

Alfvén wave refraction by interplanetary inhomogeneities. See Abstr. 106.017.

The southern component of the interplanetary magnetic field and magnetospheric substorms. See Abstr. 106.021.

Radiation Belts

084.401 **Low-energy solar protons in the pseudo-trapping region of the magnetosphere.**
A. Bewick, G. P. Haskell, R. J. Hynds, G. Morfill.
Journ. Geophys. Res., Vol. 78, 597 - 606 (1973).

084.402 **Distribution of radiation portions in the radiation belts of the earth in years of maximum solar activity.**
O. I. Savun, I. N. Senchuro, P. I. Shavrin, V. I. Shumshurov.
Kosmich. Issled., Vol. 11, 119 - 123 (1973). In Russian.

084.403 **About the damping velocity of the ring current and polar disturbances.** A. D. Shevnin.
Geomagn. Aeronom., Vol. 13, 122 - 127 (1973). In Russian.

084.404 **Super-Alfvén transfer of hydromagnetic impulses in the radiation belt of the earth.** A. V. Gulielmi.
Geomagn. Aeronom., Vol. 13, 128 - 131 (1973). In Russian.

084.405 **Observations of ring current protons at low altitudes.**
P. F. Mizera, J. B. Blake.
Journ. Geophys. Res., Vol. 78, 1058 - 1062 (1973).
Variable intensities of geomagnetically trapped protons with energies between 12.4 and 500 keV were observed during times encompassing the magnetic storms on March 20 and 24, 1969. These proton fluxes were measured for $1.0 < L < 1.1$ near the magnetic equator at local midnight with solid state detectors and an electrostatic analyzer on the low-altitude satellite OV1-17.

084.406 **Electrostatic turbulent loss of ring current protons.**
M. Nambu.
Journ. Geophys. Res., Vol. 78, 1203 - 1205 (1973). − Letter.

084.407 **Exponential distribution of protons of the outer radiation belt from adiabatic invariants.**
Yu. I. Gubar, V. P. Shabansky.
Geomagn. Aeronom., Vol. 13, 355 - 357 (1973). In Russian.
Brief information.

084.408 **Observations of noise bands associated with the**
upper hybrid resonance by the IMP 6 radio astronomy experiment.** S. R. Mosier, M. L. Kaiser, L. W. Brown.
Journ. Geophys. Res., Vol. 78, 1673 - 1679 (1973). − Brief report.

084.409 **Measurements of streams of charged particles at heights from 200 to 300 km with an orbital station.**
S. S. Konyakhina, L. V. Kurnosova, V. I. Logachev, L. A. Razorenov, V. G. Sinitsina, M. I. Fradkin.
Izv. AN SSSR. Ser. fiz., Vol. 36, 2441 - 2446 (1972). In Russian. − Abstr. in Referativ. Zhurn. 62. Issled. kosmich. prostranstva, 4.62.268 (1973).

084.410 **Equilibrium structure of radiation belt electrons.**
L. R. Lyons, R. M. Thorne.
Journ. Geophys. Res., Vol. 78, 2142 - 2149 (1973).

084.411 **Observation of a current-driven plasma instability at the outer zone−plasma sheet boundary.**
F. L. Scarf, R. W. Fredricks, C. T. Russell, M. Kivelson, M. Neugebauer, C. R. Chappell.
Journ. Geophys. Res., Vol. 78, 2150 - 2165 (1973).

084.412 **Existence of geomagnetically trapped electrons at altitudes below the inner radiation belt.**
S. Hayakawa, T. Kato, T. Kohno, T. Murakami, F. Nagase, K. Nishimura, Y. Tanaka.
Journ. Geophys. Res., Vol. 78, 2341 - 2343 (1973). − Letter.

084.413 **Spatial distribution of trapped particles in the outer magnetosphere.**
U. Kasimov, V. P. Shabansky.
Geomagn. Aeronom., Vol. 13, 511 - 513 (1973). In Russian. Brief information.

084.414 **Differential energy spectrum of low-energy protons in the inner regions of the radiation belt.**
M. I. Panasyuk, Eh. N. Sosnovets.
Kosmich. Issled., Vol. 11, 436 - 440 (1973). In Russian.

On what ionospheric workers should know about the plasmapause-plasmasphere. See Abstr. 083.004.

085 Solar-Terrestrial Relations

085.001 **Influx of stratospheric air masses into the lower troposphere after solar flares.** R. Reiter.
Naturwissenschaften, 60. Jahrgang, p. 152 - 153 (1973).
Short communication.

085.002 **On the influence of solar activity on the earth's seismicity.** A. D. Sytinskii.
Dokl. Akad. Nauk SSSR, Ser. Mat. Fiz., Vol. 208, 1078 - 1081 (1973). In Russian.

085.003 **Sudden enhancement of atmospherics.** V. Barocas.
Journ. British Astron. Ass., Vol. 83, 98 - 109 (1973). – Presidential address given during the annual general meeting of the British Astron. Ass.

085.004 **Some studies on the association of solar optical flares and microwave bursts with sudden ionospheric disturbances.**
M. K. Das Gupta, R. K. Mitra, S. K. Sarkar.
Journ. Atmosph. Terr. Phys., Vol. 35, 805 - 813 (1973).
The association of solar optical flares (numbering 8230) reported under the dual-classification and also of different types of microwave bursts (numbering 1670) with the sudden ionospheric disturbances (SID's) during the peak phase of current solar cycle (cycle 20) have been examined. Important results are obtained.

085.005 **Variation of atmospheric radio noise level with sunspot number.** P. J. Joglekar, R. A. Agarwala.
Proc. IEEE, Vol. 61, 252 - 253 (1973).
Analysis of atmospheric radio noise data collected during 20–24-h periods at Delhi, India, shows that the noise levels decrease linearly with sunspot number.

085.006 **Les variations du rayonnement solaire, les glaciations associées et leurs conséquences géophysiques, géographiques et biogéographiques.** A. Dauvillier.
Ciel et Terre, Vol. 89, 110 - 122 (1973).

085.007 **Solar corpuscular radiation and the atmospheric circulation.** N. S. Sidorenkov.
Astron. Tsirk., No. 752, p. 7 - 8 (1973). In Russian.

085.008 **Relationship between solar activity and the number of mist days.** A. V. Zavriev.
Vestn. Belorus. un-ta, 1972, ser. 2, No. 3, p. 71 - 73. In Russian. – Abstr. in Referativ. Zhurn. 51. Astron., 4.51.649 (1973).

085.009 **Some problems of solar terrestrial relations and the physics of the atmosphere.**
Trudy In-t geofiz. AN GruzSSR, Vol. 28, Tbilisi, Metsniereba. 246 pp. Price 1 Rbl. 55 Kop. (1972). In Russian. – Review in Referativ. Zhurn. 51. Astron., 4.51.650 (1973).

085.010 **Längstwellenausbreitung während der Sonnenaktivität im August 1972.** G. Becker.
PTB Mitt., 83. Jahrgang, p. 147 - 150 (1973).

085.011 **Solar activity and its relation with some parameters of the phenomena of the geophysical complex.**
T. S. Razmadze, A. M. Chkhetiya.
Cosmic rays No. 13, (see 003.013), p. 80 - 86 (1972).

In Russian.
Results of a detailed analysis of solar activity in 1957–1965 are compared with the data on solar activity with variations of cosmic ray activity, and geomagnetic, ionospheric and terrestrial currents. Cosmic ray variations are investigated with reference to the asymmetry of solar hemispheres. All storms in the IGY period are classified and a symbolic diagram of time sequence is presented.

085.012 **Inter-relation between processes on the sun and their ramifications on earth.** N. W. Puschkov.
Wiss. Zeitschr. Techn. Hochschule Karl - Marx - Stadt, No. 3, p. 337 - 339 (1972). In German.

085.013 **International Geophysical Year, World Data Center A: Catalogue of data on solar-terrestrial physics in World Data Center A subcenters. Solar and interplanetary phenomena.**
World Data Center A Upper Atmosph. Geophys. Rep. UAG-20, 271 pp. (1972).
Ionospheric phenomena, flare-associated events, geomagnetic phenomena, aurora, cosmic rays, airglow, prepared by World Data Center A for solar-terrestrial physics. – *DMCL*

085.014 **Preliminary compilation of data for retrospective world interval July 26–August 14, 1972.**
J. V. Lincoln, H. I. Leighton.
World Data Center A Solar-Terrestrial Phys. Rep. UAG-21, 128 pp. (1972).

085.015 **On a relation of the atmospheric electric field with solar flares and geomagnetic phenomena.**
Yu. A. Bragin, A. F. Konenko, I. I. Nesterova, A. Kh. Filippov, G. I. Éndikov, A. N. Fedorov, V. L. Yanchukovskij.
Vopr. issled. nizhn. ionosfery. Novosibirsk, 1972, p. 135 - 139. In Russian. – Abstr. in Referativ. Zhurn. 51. Astron., 7.51.468 (1973).

085.016 **Application of natural orthogonal functions to the analysis of the solar activity effect on the temperature field of the northern hemisphere.**
G. I. Sukhomazova, V. F. Loginov.
Issled. po geomagnetizmu, aehron. i fiz. Solntsa. Vyp. (No.) 21. Irkutsk, 1972, p. 126 - 140. In Russian. – Abstr. in Referativ. Zhurn. 51. Astron., 7.51.469 (1973).

085.017 **On the helio-geophysical nature of wave processes.**
B. F. Osins'ka.
Geol. uzberezhzhya i dna Chorn. ta Azov. moriv u mezhakh URSR. Mizhvid. resp. nauk zb., 1972, vip. (No.) 6, p. 161 - 165, 190. In Ukrainian. – Abstr. in Referativ. Zhurn. 51. Astron., 7.51.470 (1973).

085.018 **Data on solar-geophysical activity associated with the major ground level cosmic ray events of 24 January and 1 September 1971.** H. E. Coffey, J. V. Lincoln.
World Data Center A Solar Terrestrial Phys. Rep. UAG-24, Parts 1–2, 462 pp. (1972).

Observation de l'entrée des protons solaires dans la magnétosphère aux très hautes latitudes.
See Abstr. 084.205.

Planetary System

091 Physics of the Planetary System (Planetary Atmospheres, Figure, Interior, Magnetic Fields, Rotation, etc.)

091.001 **Radiation field in the deep layers of planetary atmospheres.** A. S. Anikonov.
Astron. Zhurn. Akad. Nauk SSSR, Vol. 50, 137 - 145 (1973). In Russian. English translation in Soviet Astron. AJ, Vol. 17, No. 1.
The fundamental functions $\Phi^m(\tau)$ of radiative transfer theory in a semi-infinite atmosphere with anisotropic scattering are investigated.

091.002 **The integration of equations of the theory of the figure of planets.**
V. N. Zharkov, A. B. Makalkin, V. P. Trubitsyn.
Astron. Zhurn. Akad. Nauk SSSR, Vol. 50, 150 - 162 (1973). In Russian. English translation in Soviet Astron. AJ, Vol. 17, No. 1.
A method is proposed for the numerical integration of equations of the theory of the figure of a rotating planet to the third order of the oblateness for any density distribution within the planet. The accuracy of the method is analysed. The parameters of the figure e, k, h and the gravitational moments J_2, J_4, J_6 are calculated for available models of Jupiter and Saturn. Two following moments J_8 and J_{10} and corrections to the moments J_2, J_4 and J_6 for the changes of the mean radius and angular velocity of rotation of these planets are estimated. The dependence of the gravitational moments on the redistribution of the density within a planet is investigated for the five-layer model.

091.003 **Potential atmospheric composition of smaller bodies in the solar system and some aspects of planetary evolution.** W. E. McGovern.
Journ. Geophys. Res., Vol. 78, 274 - 280 (1973).
A simple relationship indicating the maximum potential atmospheric surface density for a planetary body with a given radius, density, temperature structure, and atmospheric composition is developed. This result in conjunction with calculated exospheric temperatures for Mercury, Pluto, Triton, Titan, and the Galilean satellites of Jupiter permits an assessment of their ability to retain an appreciable atmosphere dominated by a particular constituent. These techniques are extended to the area of planetary evolution to (1) determine the stability of Jupiter at various locations in the solar system and (2) determine the minimum Jovian planetesimal radius capable of retaining a H_2-dominated atmosphere under various solar conditions.

091.004 **On the determination of the polarization transfer function by spectrophotometry of natural formations on a planetary surface from space.**
K. J. Kondratiev, O. I. Smoktii.
Dokl. Akad. Nauk SSSR, Ser. Mat. Fiz., Vol. 208, 77 - 80 (1973). In Russian.

091.005 **A survey of the outer planets Jupiter, Saturn, Uranus, Neptune, Pluto, and their satellites.**
R. L. Newburn, Jr., S. Gulkis.
Space Sci. Rev., Vol. 14, 179 - 271 (1973).
A survey of current knowledge about Jupiter, Saturn, Uranus, Neptune, Pluto, and their satellites is presented. The best available numerical values are given for physical parameters, including orbital and body properties, atmospheric composition and structure, and photometric parameters. The more acceptable current theories of these bodies are outlined with thorough referencing offering access to the details. The survey attempts to cover the literature through May 1, 1972.

091.006 **On the He$-$H$_2$ thermal opacity in planetary atmospheres.** L. Trafton.
Astrophys. Journ., Vol. 179, 971 - 976 (1973).
The pressure-induced absorption coefficient for He$-$H$_2$ mixtures is poorly known for the range of physical conditions in the atmospheres of the major planets, largely because of the uncertainty in the overlap parameter of the induced dipole moment. We have reduced this uncertainty by examining the extent to which the published measurements and the existing theory specify this parameter.

091.007 **Measurements on the infrared lines of planetary gases at low temperatures. I. ν_3-fundamental of methane.** P. Varanasi, S. Sarangi, L. Pugh.
Astrophys. Journ., Vol. 179, 977 - 982 (1973).
Intensities and half-widths of the lines $R(0)$, $R(1)$, and $R(2)$ in the ν_3-fundamental of $^{12}CH_4$ have been measured at various temperatures between 82° and 295°K. Lorentz half-widths are presented for broadening by hydrogen and by helium and in self-broadening. The dependence of line intensities and widths upon temperature is determined.

091.008 **Evidence for convection in planetary interiors from first-order topography.**
R. E. Lingenfelter, G. Schubert.
The Moon, Vol. 7, 172 - 180 (1973). – Paper dedicated to Professor Harold C. Urey on the occasion of his 80th birthday on 29 April, 1973.
Center of mass-center of figure offsets are known for the earth, moon, Mars and Venus. Such an offset requires a density distribution asymmetric about the center of mass. Observational evidence indicates that the terrestrial, lunar and Martian offsets result from crusts of variable thickness rather than lateral density inhomogeneities and that the thickness variations are more likely caused by internal convection than impact.

091.009 **Bode's law.** M. Lecar.
Nature, Vol. 242, 318 - 319 (1973).
M. W. Ovenden has outlined a theory intended to provide a dynamical explanation for Bode's law. The author suggests instead that the approximately constant spacing ratio expressed in Bode's mnemonic can be generated by a sequence of random numbers subject to the constraint that adjacent planets cannot be "too close to each other".

091.010 **Where are the satellites of the inner planets?**

J. A. Burns.
Nature, Phys. Sci., Vol. 242, 23 - 25 (1973).

Simple calculations indicate that both Mercury and Venus, perhaps as well as Pluto, could have had substantial satellites earlier in the solar system's life, these satellites having been subsequently eliminated by tidal friction.

091.011 **Simple model for scanning-angle distribution of planetary albedo gamma-rays.** F. W. Stecker.
Nature, Phys. Sci., Vol. 242, 59 - 60 (1973).

The author discusses a simple model for calculating the planetary albedo scanning-angle distribution for γ-radiation above 100 MeV and compares the results with present observations.

091.012 **Strukturschema der Planeten im Sonnensystem.** H. Kündig.
Orion Schaffhausen, 31. Jahrgang, p. 30 (1973).

091.013 **Die Atmosphären der Planeten.** H. Volland.
SuW, Vol. 12, 43 - 47 (1973).

091.014 **Ionospheric currents induced by solar wind interaction with planetary atmospheres.**
P. A. Cloutier, R. E. Daniell, Jr.
Planet. Space Sci., Vol. 21, 463 - 474 (1973).

In a steady-state model for the interaction of the solar wind with the atmosphere of a non-magnetic planet, the magnetized solar wind acts as a dynamo over the dayside of the planet and induces Ohmic currents in the planet's ionosphere. A model for the dynamo mechanism and for the induced current configuration is developed.

091.015 **The effect of a uniform external pressure on the ionospheric boundary of a non-magnetic planet in a steady solar wind.** P. K. Mukherjee, S. P. Talwar.
Planet. Space Sci., Vol. 21, 507 - 511 (1973).

The interaction of the solar wind with non-magnetic planets has been investigated theoretically taking account of the dilution of a planet's gravity field with distance and a uniform static external pressure of the incident solar wind. Numerical results are obtained for the shape of the boundary for different values of the scale height of the ionosphere, and for ratios of static to dynamic pressure of the incident solar wind.

091.016 **Radiative transfer in a spherical multi-layer atmosphere of a planet.** L. G. Titarchuk.
Kosmich. Issled., Vol. 11, 130 - 139 (1973). In Russian.

091.017 **On the stability of the planetary system.** J. Birn.
Astron. Astrophys., Vol. 24, 283 - 293 (1973).

Numerical experiments of the n-body type have been carried out to test the stability of the planetary system. The orbits of the planets have been followed by means of numerical integration, starting with various initial configurations. It has been found that stable orbits occur only within certain stability zones, in the central part of which the present orbits lie. The stability limits coincide with commensurabilities of the type $(n+1)/n$. The results are confirmed by analytical discussion by means of the Jacobian integral. The stability limits of orbits of planetoids are also discussed.

091.018 **Precessional constants of the giant planets.**
V. N. Zharkov, V. P. Trubitsyn, A. B. Makalkin.
Astron. Zhurn. Akad. Nauk SSSR, Vol. 50, 430 - 433 (1973). In Russian. English translation in Soviet Astron. AJ, Vol. 17, No. 2.

The precessional constants (dynamical ellipticities) are calculated for available models of Jupiter, Saturn, Uranus and Neptune.

091.019 **Formation of spectral lines in planetary atmospheres – V. Collision narrowed profiles of quadrupole lines in hydrogen atmospheres.** G. E. Hunt, J. S. Margolis.
Journ. Quant. Spectrosc. Radiat. Transfer, Vol. 13, 417 - 426 (1973).

We present comparisons of the line profile, equivalent width and center-to-limb variation of the equivalent width of quadrupole lines for a characteristic scattering and nonscattering hydrogen atmosphere, where the line profile has been computed by the Dicke and Galatry shapes. Our computations show that the results produced by the Dicke profile may be very different from the features predicted by the Galatry profile, which has been shown to give results in good agreement with experiment.

091.020 **Calculation of transport coefficients for planetary atmospheres formed by mixtures of CO_2–N_2.**
M. D. Zdunkevich, V. B. Leonas.
Teplofiz. vysokikh temperatur, Vol. 10, 1110 - 1112 (1972). In Russian. – Abstr. in Referativ. Zhurn. 51. Astron., 3.51.257 (1973).

091.021 **Structure of the inner planets and cosmochemistry of the planetary system.** B. Yu. Levin.
Ocherki sovrem. geokhimii i analit. khimii. Nauka, Moskva, p. 8 - 16 (1972). In Russian. – Abstr. in Referativ. Zhurn. 51. Astron., 3.51.291 (1973).

091.022 **A new variant of calculation of intensity of intrinsic thermal radiation of the system planet – atmosphere (the spherical case).**
M. V. Anolik, A. M. Bunakova.
Probl. fiz. atmosf., No. 10. Leningrad, Leningr. un-t, 1972, p. 79 - 86. In Russian. – Abstr. in Referativ. Zhurn. 51. Astron., 3.51.300 (1973).

091.023 **Corrections for overlapping of absorption lines in the Q-branches of the 15 μ band of CO_2 in calculating the radiative energy influx into the upper planetary atmospheres.** A. F. Nerushev, G. M. Shved.
Probl. fiz. atmosf., No. 10. Leningrad, Leningr. un-t, 1972, p. 123 - 125. In Russian. – Abstr. in Referativ. Zhurn. 51. Astron., 3.51.301 (1973).

091.024 **On the interaction of a magnetic field with the proper rotation of a body.** A. P. Ryabushko.
Third Soviet Gravitational Conference, Erevan, 1972, (see 012.001), p. 352 - 353 (1972). In Russian.

091.025 **Mars 3 investigates plasma.** O. L. Vajsberg.
Zemlya i Vselennaya, 1973, No. 1, p. 28 - 31. In Russian.

091.026 **Elemental and isotopic abundances of the volatile elements in the outer planets.** A. G. W. Cameron.
Space Sci. Rev., Vol. 14, 392 - 400 (1973).

Certain fundamental scientific problems of a cosmological as well as cosmogonic character, may be solved by the insertion of entry probes into the atmospheres of the outer planets. It is recommended that attempts be made to determine the elemental and isotopic abundances of H, D, He³, He⁴, C, N, O, S, and the rare gas elements.

091.027 **Chemistry of the outer solar system.** J. S. Lewis.
Space Sci. Rev., Vol. 14, 401 - 411 (1973).

Data on the composition of the satellites of the outer planets and the composition and structure of planetary atmospheres are briefly reviewed in light of simple models for the origin of the solar system and the planets. Some crucial tests of present theories are suggested.

091.028 The gravitational fields of the major planets.
S. J. Peale.
Space Sci. Rev., Vol. 14, 412 - 423 (1973).

The constraints placed on models of the interiors of the major planets by the non-spherical components of their gravitational fields are explained, and several methods of determining these non-spherical components are described and evaluated.

091.029 The significance of atmospheric measurements for interior models of the major planets.
W. B. Hubbard.
Space Sci. Rev., Vol. 14, 424 - 432 (1973).

We present a discussion of proposed models for interior processes in Jupiter and Saturn, and discuss how these models can be tested by atmospheric measurements by space vehicles. The importance of measurements at Uranus and Neptune is also discussed.

091.030 The dynamics of the atmospheres of the major planets. P. H. Stone.
Space Sci. Rev., Vol. 14, 444 - 459 (1973).

The literature on the dynamics of Jupiter's atmosphere is reviewed and used as a basis for suggesting what observations would yield useful information about Jovian dynamics. The atmospheres of Saturn, Uranus and Neptune are discussed from the same point of view.

091.031 The ionospheres of the major planets.
M. B. McElroy.
Space Sci. Rev., Vol. 14, 460 - 473 (1973).

Physical and chemical processes which affect the equilibrium distribution of ionization in the atmospheres of Jupiter, Saturn, Uranus and Neptune are reviewed. Attention is directed to the probable importance of dissociative ionization of H_2 as a source of H^+. A number of potentially important loss mechanisms for H^+ are discussed including a possible reaction of H^+ with vibrationally excited H_2.

091.032 Imaging of the outer planets and satellites.
B. C. Murray.
Space Sci. Rev., Vol. 14, 474 - 496 (1973).

The purpose of this paper is twofold: First, to illustrate how imaging can be used as a part of scientific experimentation, and especially to illustrate proposed imaging experiments for the outer planets and their satellites. Second, to show that imaging yields highly communicable results having not only scientific significance but broad intellectual and cultural meaning as well.

091.033 Thermal radio emission from the major planets.
S. Gulkis.
Space Sci. Rev., Vol. 14, 497 - 510 (1973).

The use of long-wavelength radio measurements of brightness temperature to remotely measure the thermal structure of the atmospheres of the major planets at great depths (> 10 atm.) is discussed. Data are presented which show that the gross features of Jupiter's and Saturn's microwave spectra, as determined from ground based observations, can be explained in terms of thermal emission from ammonia in deep convective atmospheres of He and H_2.

091.034 Magnetospheres of the planets. C. F. Kennel.
Space Sci. Rev., Vol. 14, 511 - 533 (1973).

Scaling laws for possible outer planet magnetospheres are derived. These suggest that convection and its associated auroral effects will play a relatively smaller role than at earth, and that there is a possibility that the outer planets could have significant radiation belts of energetic trapped particles.

091.035 Radio physics of the outer solar system.
R. G. Stone.
Space Sci. Rev., Vol. 14, 534 - 551 (1973).

The remote sensing of low frequency nonthermal radio emission is the astronomy of field and particle phenomena. Observations conducted from space lead to information about the composition and dynamic processes occurring in planetary magnetospheres as well as within the interplanetary and interstellar medium. The potential of this technique is demonstrated by considering observations obtained from earth orbit missions.

091.036 Planetary magnetism in the outer solar system.
C. P. Sonett.
Space Sci. Rev., Vol. 14, 552 - 558 (1973).

A brief review of the salient considerations which apply to the existence of magnetic fields in connection with planetary and subplanetary objects in the outer solar system is given. Consideration is given to internal dynamo fields, fields which might originate from interaction with the solar wind or magnetospheres (externally driven dynamos) and lastly fossil magnetic fields such as have been discovered on the moon.

091.037 On the distribution of satellite bodies according to the mean distances in the systems sun, Jupiter, Saturn and Uranus. Yu. K. Gulak.
Astrometriya i Astrofizika, *Kiev*, vyp. (No.) 16, (see 003.006), p. 92 - 99 (1972). In Russian.

091.038 Cores of the terrestrial planets. K. E. Bullen.
Nature, Vol. 243, 68 - 70 (1973).

The compositions of the cores of the terrestrial planets have been re-examined following a calculation by O. G. Soroktin that the iron oxide Fe_2O is stable at pressures reached in the earth's core and his suggestion that the outer core may consist of Fe_2O. It is shown that the idea of an Fe_2O outer core in the earth can be fairly well reconciled with a common overall composition for the planets earth, Venus and Mars.

091.039 Is the solar system gravitationally closed?
B. Bertotti.
Astrophys. Letters, Vol. 14, 51 - 53 (1973).

A cosmological spectrum of gravitational waves produces a random walk in the orbital elements of a planet or a satellite. Upper limits to the spectral density at the revolution frequencies can be established in this way. We stress the opportunity of taking this effect into account in analyzing future experiments.

091.040 New techniques for determining sizes of satellites and asteroids. D. Morrison.
Comments Astrophys. Space Phys., Vol. 5, 51 - 56 (1973).

In this comment attention is drawn to three new techniques that are yielding radii for dozens of small objects and that, by virtue of their independence of angular size, are capable of extension to even smaller and fainter objects.

091.041 Statistical mechanics of light elements at high pressure. III. Molecular hydrogen.
W. L. Slattery, W. B. Hubbard.
Astrophys. Journ., Vol. 181, 1031 - 1038 (1973).

The high-pressure thermodynamic properties of a molecular hydrogen fluid are calculated in the classical domain using a Monte Carlo approach. A recent theoretical intermolecular potential of the $\exp -6$ form which gives good agreement with low-pressure experimental data, and fair agreement with high-pressure data, is used. Melting curves and adiabats with relevance to planetary models are calculated. We note a possible application of our results to the theory of pressure-induced opacity in molecular hydrogen.

091.042 Evolution of satellite resonances by tidal dissipation.

R. Greenberg.
Astron. Journ., Vol. 78, 338 - 346 (1973).

Analysis of a realistic model shows how satellites' gravitational interaction can halt their differential tidal evolution when resonant commensurabilities of their orbital periods are reached. The success of this study lends support to the hypothesis that orbit—orbit resonances among satellites in the solar system, including the Titan—Hyperion case, did evolve as a result of tidal energy dissipation.

091.043 **New infrared spectra of the Jovian planets from 12000 to 4000 cm⁻¹ by Fourier transform spectroscopy. I. Study of Jupiter in the $3\nu_3$ CH$_4$ band.**
J. P. Maillard, M. Combes, T. Encrenaz, J. Lecacheux.
Astron. Astrophys., Vol. 25, 219 - 232 (1973).

New spectra of Jupiter from $12000-4000$ cm⁻¹ were obtained in May 1972, using a Fourier transform Michelson interferometer and the Haute Provence Observatory's 193 cm telescope. Laboratory spectra of CH$_4$ at very high resolution have been recorded in the range $7500-12000$ cm⁻¹. New wave numbers of the $3\nu_3$ J manifolds are given. We fitted the observational data to synthetic profiles based upon the reflecting layer model, and convoluted by the instrumental profile and by the rotational broadening effect. Using the method of Margolis and Fox (1969), we estimated half-widths, the rotational temperature and the CH$_4$ abundance.

091.044 **Analysis of spikes in occultation curves: a critique of Brinkmann's method.**
L. Wasserman, J. Veverka.
Icarus, Vol. 18, 599 - 604 (1973).

Brinkmann's method of deriving the composition of a planetary atmosphere from the timing of occultation curve spikes is discussed in detail. Contrary to the statement made in Brinkmann's paper, it is shown that not only must the spikes be timed, but the intensity of the background occultation curve must be determined at the points at which the spikes occur.

091.045 **Topography on satellite surfaces and the shape of asteroids.** T. V. Johnson, T. R. McGetchin.
Icarus, Vol. 18, 612 - 620 = Contr. Planet. Astron. Lab., Dep. Earth Planet. Sci., Mass. Inst. Technology, No. 41 (1973).

Calculations of the topography and shape of planetary bodies are presented for two sets of models. One set of models deals with the effects of static loading on bodies, taking into account strengths of materials, density, and size. The other set considers the effects of creep deformation on model bodies of differing composition, size and temperature. The results of application of these models to asteroids and satellites of the major planets are given.

091.046 **On the reduction of occultation light curves: applications to the outer planets.**
J. Veverka, L. Wasserman.
Bull. American Astron. Soc., Vol. 5, 289 (1973). – Abstr. AAS.

091.047 **An exact expression for the temperature structure of a simple planetary atmosphere.**
B. R. Barkstrom.
Bull. American Astron. Soc., Vol. 5, 302 - 303 (1973). Abstr. AAS.

091.048 **Loss of hydrogen from primitive atmospheres.**
D. M. Hunten.
Bull. American Astron. Soc., Vol. 5, 303 (1973). – Abstr. AAS.

091.049 **Some new laboratory measurements of the hydrogen quadrupole absorption lines.** J. S. Margolis.
Bull. American Astron. Soc., Vol. 5, 303 (1973). – Abstr. AAS.

091.050 **A new laboratory technique for high resolution absorption spectroscopy over ultra long paths.**
K. Fox, R. Goldstein, V. Vali.
Bull. American Astron. Soc., Vol. 5, 303 (1973). – Abstr. AAS.

091.051 **On the negative polarization branch.**
W. M. Sinton.
Bull. American Astron. Soc., Vol. 5, 303 (1973). – Abstr. AAS.

091.052 **Path length distributions for photons diffusely reflected from a non-conservative planetary atmosphere.** J. F. Appleby, W. M. Irvine.
Bull. American Astron. Soc., Vol. 5, 303 (1973). – Abstr. AAS.

091.053 **Satellites and asteroids: a review of recent work.**
D. Morrison.
Bull. American Astron. Soc., Vol. 5, 304 (1973). – Abstr. AAS.

091.054 **Effective temperatures and infrared continua of the planets and satellites.** F. J. Low, K. R. Armstrong.
Bull. American Astron. Soc., Vol. 5, 306 (1973). – Abstr. AAS.

091.055 **Spectral reflectivity of frosts.**
H. H. Kieffer, W. D. Smythe.
Bull. American Astron. Soc., Vol. 5, 307 (1973). – Abstr. AAS.

091.056 **Spectral reflectivities of ices.** L. A. Lebofsky.
Bull. American Astron. Soc., Vol. 5, 307 - 308 (1973). – Abstr. AAS.

091.057 **Determination of radii of satellites and asteroids from radiometry and photometry.** D. Morrison.
Icarus, Vol. 19, 1 - 14 (1973).

Equations and graphical solutions for radius and albedo of small solar system objects are presented for cases where the object is at opposition, in equilibrium with the insolation, and has unit values for phase integral and infrared emissivities. Each of these assumptions is then discussed, and expressions are given for the dependence of the derived parameters on the assumptions. Applications are then discussed to Saturn's satellites Iapetus and Rhea and to asteroids (1) Ceres, (4) Vesta, and (324) Bamberga.

091.058 **A numerical method for determining the temperature structure of planetary atmospheres.**
J. B. Pollack, G. Ohring.
Icarus, Vol. 19, 34 - 42 (1973).

A numerical method for calculating the time-average, vertical temperature structure of planetary atmospheres is presented. It is assumed that the atmospheres are in radiative—convective equilibrium, which is a good first approximation to many situations. Numerical tests of the rate of convergence and accuracy of the answer are presented. As an application of this procedure, we have calculated some model atmospheres of Jupiter.

091.059 **Planetary brightness temperature measurements at 8.6 mm and 3.1 mm wavelengths.**
B. L. Ulich, J. R. Cogdell, J. H. Davis.
Icarus, Vol. 19, 59 - 82 (1973).

New measurements of the sun, moon, Mercury, Venus, Mars, Jupiter, and Saturn at 3.1 and 8.6 mm wavelengths are given. The temperatures reported for the planets at 3.1 mm wavelength are higher than previous measurements in this wavelength range and change the interpretation of some planetary spectra. For Jupiter, the need to recalculate the spectrum with recent models is demonstrated. The flux density scale proposed by Dent (1972) has been revised according to a more accurate determination of the millimeter brightness temperature of Jupiter.

091.060 **On planetary cores.** K. E. Bullen.
The Moon, Vol. 7, 384 - 395 (1973). – Paper dedicated to Professor Harold C. Urey on the occasion of his 80th birthday on 29 April, 1973.
This article outlines a variety of recent calculations which bear on the structures of the cores of the terrestrial planets. Brief comments are made on the moon.

091.061 **High-resolution infrared spectroscopy of planetary atmospheres.** K. Fox.
Molecular Spectroscopy: Modern research, [Academic Press Inc., London], p. 79 - 114 (1972). – Review paper.

091.062 **The outer solar system.** A. G. W. Cameron.
Science, Vol. 180, 701 - 708 (1973).

091.063 **The profiles and curves of growth for the absorption lines formed in a scattering medium.**
V. G. Teifel, L. A. Usoltseva.
Astron. Zhurn. Akad. Nauk SSSR, Vol. 50, 568 - 575 (1973). In Russian. English translation in Soviet Astron. AJ, Vol. 17, No. 3.
Numerical calculations of the profiles of absorption lines formed in a homogeneous, plane, semi-infinite scattering aerosol layer and in the overcloud layer of pure gas in a planetary atmosphere were carried out.

091.064 **On circular polarization of light scattered from planets.** S. A. Khejfets.
Astron. Zhurn. Akad. Nauk SSSR, Vol. 50, 660 - 661 (1973). In Russian. English translation in Soviet Astron. AJ, Vol. 17, No. 3. – Short note.

091.065 **Planetary resonances, bi-stable oscillation modes and solar activity cycles.** H. P. Sleeper, Jr.
Contractor Report CR-2035, U. S. National Aeronautics and Space Administration, Greenbelt, Maryland. [Available from National Technical Information Service, Springfield, Virginia, USA]. Price $ 3.00 (1972). – Comments on this report by B. A. J. Clark in Journ. Astron. Soc. Victoria, Vol. 25, 88 - 90 (1972).

091.066 **Observational constraint on the structure of hydrogen planets.** W. B. Hubbard.
Astrophys. Journ., (*Letters*), Vol. 182, L35 - L38 (1973).
A measurement of temperature, pressure, and chemical composition of the convective portion of the atmosphere of a hydrogen-helium planet can be coupled to a theoretical many-body calculation in the metallic interior to derive a constraint on the internal temperature distribution.

091.067 **Stabilitätsuntersuchungen an Planetensystemen.** J. Birn.
Mitt. Astron. Ges., No. 32, p. 212 (1973). – Abstract.

091.068 **Planets, sunspots and earthquakes.** J. Gribbin.
Observatory, Vol. 93, 121 (1973). – Letter.

091.069 **Planetary elements for 10 000 000 years.**
C. J. Cohen, E. C. Hubbard, C. Oesterwinter.
Celestial Mechanics, Vol. 7, 438 - 448 (1973).
In 1950 Brouwer and van Woerkom published a secular theory of the variations of the planetary elements in analytical form. In the present paper we provide a graphical representation of this theory in the form of element plots for a time span of ten million years.

091.070 **Bode's law and the resonant structure of the solar system.** S. F. Dermott.
Nature, Phys. Sci., Vol. 244, 18 - 21 (1973).

In this discussion of Bode's law the author considers whether the law could have arisen by chance, and how the near resonances between a number of orbits in the solar system may have developed.

091.071 **Radioastronomy of the planets.** A. Boischot.
Industries Atomiques et Spatiales, (*France*), Vol. 16, No. 5, p. 65 - 74 (1972). In French.

091.072 **Scattering of light in the atmospheres of planets.** V. V. Sobolev.
Physics of the moon and planets, (see 012.024), p. 199 - 204 (1972). In Russian.

091.073 **New determination of the diameters of planets and satellites.** A. Dollfus.
Physics of the moon and planets, (see 012.024), p. 207 - 209 (1972). In Russian.

091.074 **The thermal history of the terrestrial planets.** S. V. Maeva.
Physics of the moon and planets, (see 012.024), p. 223 - 228 (1972). In Russian.

091.075 **Estimates of some characteristics of the general circulation in the atmospheres of terrestrial planets.**
G. S. Golitsyn.
Physics of the moon and planets, (see 012.024), p. 393 (1972). In Russian. – Abstract.

091.076 **Photometric and spectral observations of planets in the region of 8 - 14 μ.** V. I. Moroz.
Physics of the moon and planets, (see 012.024), p. 408 (1972) In Russian. – Abstract.

091.077 **On the quality of a planetary image.**
V. N. Dudinov.
Physics of the moon and planets, (see 012.024), p. 411 - 413 (1972). In Russian.

091.078 **The present-day status and problems of investigation of the giant planets.** V. G. Tejfel'.
Physics of the moon and planets, (see 012.024), p. 426 - 430 (1972). In Russian.

091.079 **Figures and inner structure of hydrogen-helium planets.**
V. N. Zharkov, V. P. Trubitsyn, A. A. Kalachnikov.
Physics of the moon and planets, (see 012.024), p. 430 - 433 (1972). In Russian.

091.080 **Numerical solution for the composition of a thermosphere in the presence of a steady subsolar-to-antisolar circulation with application to Venus.**
R. E. Dickinson, E. C. Ridley.
Journ. Atmosph. Sci., Vol. 29, 1557 - 1570 (1972).
Considers the thermosphere of a nonrotating planet with a large-scale circulation from dayside to nightside driven by differential solar heating. A numerical model is developed and a method of solution derived for the distribution of N components in the presence of sources and sinks due to photodissociation. The model is integrated for parameters appropriate to the Venusian upper atmosphere, assuming only CO_2 is carried upward on the dayside through the bottom boundary at 0.1 mb.

091.081 **Remote sensing of the turbulence characteristics of a planetary atmosphere by radio occultation of a space probe.** R. Woo, A. Ishimaru.
Radio Sci., (*USA*), Vol. 8, 103 - 108 (1973).
Analyzes the effects of small-scale turbulence on radio

waves propagating through a planetary atmosphere. The analysis provides a technique for inferring the turbulence characteristics of a planetary atmosphere from the radio signals received from a spacecraft as it is occulted by the planet. The planetary turbulence is assumed to be localized and smoothly varying with the structure constant varying exponentially with altitude.

091.082 Prospects for planetary geodesy. W. M. Kaula.
Geophys. Monograph 15, (see 012.025), p. 279 - 281 (1972).

The solar system. See Abstr. 003.073.

Physique et dynamique planétaires. Géodynamique.
Vol. 4. See Abstr. 003.085.

Results of observations of planets with the vertical circle at the Pulkovo Observatory. See Abstr. 041.046.

Certain particular solutions of the Clairaut equation.
See Abstr. 042.024.

Radiative transfer through carbon ablation layers.
See Abstr. 063.017.

Formation des raies spectrales et étude des courbes de croissance dans une atmosphère diffusante semi-infinie.
See Abstr. 063.018.

The atmospheres of the earth and the terrestrial planets: their origin and evolution. See Abstr. 082.123.

Test of a photographic equidensitometry of the moon and planets. See Abstr. 094.895.

Water vapor from a lunar breccia: implications for evolving planetary atmospheres. See Abstr. 094.933.

Formation of spectral lines in planetary atmosphere IV. Theoretical evidence for structure of the Jovian clouds from spectroscopic observations of methane and hydrogen quadrupole lines. See Abstr. 099.045.

Eftersøgningen af fremmede planetsystemer.
See Abstr. 117.029.

Kosmische Radioquellen (Cosmic radio sources).
See Abstr. 141.138.

092 Mercury

092.001 Ultra-violet argon dayglow lines in the atmosphere of Mercury. M. Zeilik, A. Dalgarno.
Planet. Space Sci., Vol. 21, 383 - 389 (1973).
We investigate the response of an atmosphere of argon to solar ultra-violet radiation. With the assumption that Mercury has an argon atmosphere that is optically thick to ionizing radiation the intensities of the ultra-violet dayglow lines resulting from photoelectron impact are calculated. For most of the model atmospheres, the predicted intensities are above the detection threshold of the 1973 Venus-Mercury ultra-violet spectrometer of Broadfoot, McElroy and Belton.

092.002 A new upper limit for an atmosphere of CO_2 on Mercury. R. F. Poppen, U. Fink, H. P. Larson.
Bull. American Astron. Soc., Vol. 5, 302 (1973). – Abstr. AAS.

092.003 Mercury and Venus: high-resolution radar topographic profiles.
G. H. Pettengill, D. B. Campbell, R. B. Dyce, R. P. Ingalls.
Bull. American Astron. Soc., Vol. 5, 302 (1973). – Abstr. AAS.

092.004 Optical properties of Mercury's surface layer.
A. V. Morozhenko, E. G. Yanovitsky.
Astron. vestn., Vol. 7, 57 - 64 (1973). In Russian.
A parabolic approximation of phase observations of the integral brightness of Mercury has been performed by the methods of statistic analysis. Spectral values of geometrical and spherical albedo and phase function are derived from observational data.

092.005 Radio interferometry of moving sources in the presence of confusion. An application to Mercury at 21-centimeter wavelength. F. H. Briggs, F. D. Drake.
Astrophys. Journ., Vol. 182, 601 - 607 (1973).
The effects of background confusion sources on radio interferometer observations of moving objects are examined quantitatively. A brightness temperature of 363°± 35°K was measured for Mercury on 1971 August 14—15.

092.006 De Mercuriusovergang van 10 november 1973.
J. Meeus.
Hemel en Dampkring, Vol. 71, 215 - 217 (1973).

092.007 On the inner structure and chemical composition of Mercury. S. V. Kozlovskaya.
Physics of the moon and planets, (see 012.024), p. 228 - 231 (1972). In Russian.

092.008 The upper boundary of the night temperature on Mercury's surface. B. Murray.
Physics of the moon and planets, (see 012.024), p. 231 (1972). In Russian. – Abstract.

092.009 Photographic measurements of the rotation of Mercury. B. A. Smith.
Physics of the moon and planets, (see 012.024), p. 231 - 232 (1972). In Russian.

093 Venus

093.001 Comment on"The composition of the Venus cloud tops in light of recent spectroscopic data".
L. D. G. Young, A. T. Young.
Astrophys. Journ., (*Letters*), Vol. 179, L39 - L43 (1973).

Laboratory spectra of aqueous HCl solutions are compared with spectra of Venus. The suggestion that the clouds of Venus are composed of aqueous HCl solutions is not confirmed.

093.002 Natural radioactive element content in Venusian rock. Results of a Venera 8 space probe experiment.
A. P. Vinogradov, Iu. A. Surkov, F. F. Kirnozov, V. N. Glazov.
Dokl. Akad. Nauk SSSR, Ser. Mat. Fiz., Vol. 208, 576 - 579 (1973). In Russian.

093.003 Venus: Radar determination of gravity potential.
I. I. Shapiro, G. H. Pettengill, G. N. Sherman, A. E. E. Rogers, R. P. Ingalls.
Science, Vol. 179, 473 - 476 (1973).

We describe a method for the determination of the gravity potential of Venus from multiple-frequency radar measurements. The method is based on the strong frequency dependence of the absorption of radio waves in Venus' atmosphere. The absorption-sensitive Haystack Observatory data have been analyzed under the assumption of uniform surface reflectivity to yield a gravity equipotential contour for the equatorial region and a tentative upper bound of 6×10^{-4} on the fractional difference of Venus' principal equatorial moments of inertia.

093.004 Venus: Microwave opacity of the minor atmospheric constituents. C. de Bergh.
Astron. Astrophys., Vol. 23, 467 - 470 (1973).

Some extra microwave opacity in the troposphere of Venus relative to the opacity that would be produced by a pure CO_2 atmosphere has been observed and it is suggested here that, after detailed interpretations of radio, radar and Mariner 5 occultation experiments, water vapor could be considered as the only microwave absorber responsible for the excess in opacity.

093.005 Venus: How much do we know? M. Marov.
Spaceflight, Vol. 15, 48 - 50 (1973).

093.006 The planet Venus: a new periodic spectrum variable.
L. G. Young, A. T. Young, J. W. Young, J. T. Bergstralh.
Astrophys. Journ., (*Letters*), Vol. 181, L5 - L8 (1973).

The apparent strength of CO_2 absorption in the spectrum of Venus varies by 20 percent, in a period of 4 days. The variations are synchronous over the disk, and thus represent a fundamental dynamical mode of the atmosphere.

093.007 The lower Venus atmosphere.
G. E. Hunt, J. T. Bartlett.
Endeavour, No. 115, Vol. 32, 39 - 43 (1973).

Although the planet Venus is very similar to the earth in many respects, recent American and Soviet spacecraft missions and earth-based observations have conclusively demonstrated that its atmosphere is very different from our own. The atmosphere is primarily composed of carbon dioxide but the composition and structure of the clouds which obscure the planet's surface are still unknown.

093.008 Venus: New microwave measurements show no atmospheric water vapor.
M. A. Janssen, R. E. Hills, D. D. Thornton, W. J. Welch.

Science, Vol. 179, 994 - 997 (1973).

Two sets of passive radio observations of Venus—measurements of the spectrum of the disk temperature near the 1-centimeter wavelength, and interferometric measurements of the planetary limb darkening at the 1.35-centimeter water vapor resonance—show no evidence of water vapor in the lower atmosphere of Venus.

093.009 Does spectroscopic evidence require two scattering layers in the Venus atmosphere?
J. L. Regas, R. W. Boese, L. P. Giver, J. H. Miller.
Journ. Quant. Spectrosc. Radiat. Transfer, Vol. 13, 461 - 463 (1973).

The phase variation of lines in the 7820 and 7883 Å CO_2 bands has been interpreted by Hunt using an inhomogeneous, anisotropic scattering model of the Venus atmosphere. He concluded that the Venus atmosphere contains two scattering layers. We show that the observed phase variation may be due to the strong backward lobe in the Venus cloud phase function and that two cloud layers are not neccessarily required.

093.010 There is evidence for two scattering layers in the Venus atmosphere. G. E. Hunt.
Journ. Quant. Spectrosc. Radiat. Transfer, Vol. 13, 465 - 466 (1973). – Note.

093.011 Model of the Venus atmosphere.
M. Ya. Marov, O. L. Ryabov.
In-t prikl. mat. AN SSSR. Preprint No. 39. Moskva. 47 pp. (1972). In Russian. – Review in Referativ. Zhurn. 51.. Astron., 3.51.305 (1973).

093.012 On the cloud layer of Venus.
Yu. A. Surkov, B. M. Andrejchikov, O. M. Kalinkina, I. M. Grechishcheva.
Ocherki sovrem. geokhimii i analit. khimii. Moskva, Nauka, 1972, p. 17 - 21. In Russian. – Abstr. in Referativ. Zhurn. 62. Issled. kosmich. prostranstva, 3.62.191 (1973).

093.013 On the observations of the transit of Venus over the sun with particular emphasis on the December 9, 1874 event observed in Japan – Part II. (Collective review).
K. Saito, S. Shinozawa.
Tokyo Astron. Obs., Report No. 61, Vol. 16, 259 - 385 (1973). In Japanese.

093.014 Positional photographic observations of Venus at Pulkovo in 1972. T. P. Kiseleva.
Astron. Tsirk., No. 758, p. 1 - 2 (1973). In Russian.

093.015 Spectral observations of Venus at inferior conjunction in June 1972. O. G. Taranova.
Astron. Tsirk., No. 758, p. 7 - 8 (1973). In Russian.

093.016 About the cloud layer of Venus.
Yu. A. Surkov, B. M. Andrejchikov, O. M. Kalinkina, I. M. Grechishcheva.
Ocherki sovrem. geokhimii i analit. khimii. Nauka, Moskva, 1972, p. 17 - 21. In Russian. – Abstr. in Referativ. Zhurn. 51 Astron., 4.51.399 (1973).

093.017 Are the clouds of Venus sulfuric acid?
A. T. Young.
Icarus, Vol. 18, 564 - 582 (1973).

Sulfuric acid precipitation may explain some peculiarities in Venera and Mariner data. Because sulfuric acid solutions are in good agreement with the Venus data, and because no other material that has been proposed is even consistent with the polarimetric and spectroscopic data, H_2SO_4 must be considered the most probable constituent of the Venus clouds.

093.018 Atmospheric ion losses by Venus and Mars to the solar wind.
P. A. Cloutier, R. E. Daniell, Jr., D. M. Butler, F. C. Michel.
Bull. American Astron. Soc., Vol. 5, 299 (1973). – Abstr. AAS.

093.019 Are the clouds of Venus sulfuric acid?
A. T. Young, L. G. Young.
Bull. American Astron. Soc., Vol. 5, 299 (1973). – Abstr. AAS.

093.020 Sulfuric acid in the clouds of Venus. G. T. Sill.
Bull. American Astron. Soc., Vol. 5, 299 (1973).
Abstr. AAS.

093.021 Aircraft observations of the near infrared spectrum of Venus: implications for cloud composition.
J. B. Pollack, E. Erickson, C. Chackerian, Jr., F. Witteborn, A. Summers, B. Baldwin.
Bull. American Astron. Soc., Vol. 5, 299 (1973). – Abstr. AAS.

093.022 A search for H_2O and O_2 on Venus.
W. A. Traub, N. P. Carleton.
Bull. American Astron. Soc., Vol. 5, 299 - 300 (1973).
Abstr. AAS.

093.023 Observations of the Venus water vapor line at 8197 Å over the disk of Venus. E. S. Barker.
Bull. American Astron. Soc., Vol. 5, 300 (1973). – Abstr. AAS.

093.024 A photochemical haze model for the clouds of Venus. R. G. Prinn.
Bull. American Astron. Soc., Vol. 5, 300 (1973). – Abstr. AAS.

093.025 Spectra of Venus with "Connes' interferometer" analyzed for best temperature and pressure fits.
K. E. Dierenfeldt, U. Fink, H. P. Larson.
Bull. American Astron. Soc., Vol. 5, 300 (1973). – Abstr. AAS.

093.026 A comparison of Venus cloud models determined by spectroscopic investigations.
J. L. Regas, L. P. Giver, R. W. Boese, J. H. Miller.
Bull. American Astron. Soc., Vol. 5, 300 (1973). – Abstr. AAS.

093.027 Weather on Venus? R. A. Schorn, E. S. Barker.
Bull. American Astron. Soc., Vol. 5, 300 - 301 (1973). – Abstr. AAS.

093.028 Venus CO_2 observations at inferior conjunction: Variations with time and position on the disk.
E. S. Barker, R. A. Schorn.
Bull. American Astron. Soc., Vol. 5, 301 (1973). – Abstr. AAS.

093.029 Further observations of weather on Venus.
L. G. Young.
Bull. American Astron. Soc., Vol. 5, 301 (1973). – Abstr. AAS.

093.030 Venus: ultraviolet polarization variations.
D. L. Coffeen, A. L. Baker.
Bull. American Astron. Soc., Vol. 5, 301 (1973). – Abstr. AAS.

093.031 Evaluation of the circulation patterns of the upper cloud deck of Venus.
R. Beebe, H. Reitsema, E. Reese, A. Scott.
Bull. American Astron. Soc., Vol. 5, 301 (1973). – Abstr. AAS.

093.032 Relative spectrophotometry of ultraviolet clouds on Venus. J. H. Woodman, E. S. Barker.
Bull. American Astron. Soc., Vol. 5, 301 (1973). – Abstr. AAS.

093.033 Some properties of oblique angle radar returns from Venus at 70 cm wavelength.
D. B. Campbell, T. Hagfors.
Bull. American Astron. Soc., Vol. 5, 302 (1973). – Abstr. AAS.

093.034 Results from the radar imaging of Venus done during the 1972 inferior conjunction at 70 cm wavelength. D. B. Campbell.
Bull. American Astron. Soc., Vol. 5, 302 (1973). – Abstr. AAS.

093.035 Radio interferometric observations of Venus near 1.35 cm wavelength – implications for the middle atmosphere. M. A. Janssen.
Bull. American Astron. Soc., Vol. 5, 302 (1973). – Abstr. AAS.

093.036 General atmospheric circulation driven by polar and diurnal surface temperature variations.
I. O. Bohachevsky.
Icarus, Vol. 19, 118 - 125 (1973).
    Described is a global circulation model for the Venus atmosphere that includes the effects of both polar cooling and diurnal temperature variation. It is based on a linearized Boussinesq approximation and boundary conditions derived from theoretical and empirical considerations. The time-dependent, three-dimensional flow field is deduced without any a priori assumptions about its configuration.

093.037 Measurements of temperature, pressure, and velocity of wind in the Venus atmosphere on the automatic space probe Venera 8. M. Ya. Marov, V. S. Avduevskij, V. V. Kerzhanovich, M. K. Rozhdestvenskij, N. F. Borodin, O. L. Ryabov.
Dokl. Akad. Nauk SSSR, Ser. Mat. Fiz., Vol. 210, 559 - 562 (1973). In Russian.

093.038 Comet-like interaction of Venus with the solar wind. M. K. Wallis.
Cosmic plasma physics. Conference 1971, (see 012.016), p. 137 - 140 (1972).

093.039 Dissipation of the Venus atmosphere.
G. M. Nedyalkova, I. E. Turchinovich.
Astron. Zhurn. Akad. Nauk SSSR, Vol. 50, 661 - 663 (1973). In Russian. English translation in Soviet Astron. AJ, Vol. 17, No. 3. – Short note.

093.040 Venusdichotomie im August 72.
G. Glitscher, G. Klingelhöfer.
SuW, Vol. 12, 183 (1973).

093.041 Mathematical model of the climate on Venus.
S. Zilitinkevich, A. Monin.
Nauka i zhizn', 1972, No. 12, p. 55 - 57. In Russian. – Abstr. in Referativ. Zhurn. 51. Astron., 5.51.71 (1973).

093.042 Gas dissipation from the neutral Venus atmosphere.
G. M. Nedyalkova, I. E. Turchinovich.
Teor. mikro- i makrofizika. Leningrad, 1972, p. 75 - 84. In Russian. – Abstr. in Referativ. Zhurn. 51. Astron., 5.51.319 (1973).

093.043 The Venus ionosphere and the problem of its dissipation. I. E. Turchinovich.
Teor. mikro- i makrofizika. Leningrad, 1972, p. 85 - 90. In Russian. – Abstr. in Referativ. Zhurn. 51. Astron., 5.51.320 (1973).

093.044 The O I 1304- and 1356-Å emissions from the atmosphere of Venus. D. J. Strickland.
Journ. Geophys. Res., Vol. 78, 2827 - 2836 (1973).
    In this paper a theoretical prediction is made for the

expected emissions of atomic oxygen at 1304 Å and 1356 Å from the upper atmosphere of Venus. The work is an extension of an analysis by Strickland et al. (1972) of Mariner 6 and 7 1304-Å data and a current analysis by the author of Mariner 9 1304-Å data.

093.045 Electromagnetic wave propagation in the Venusian ionosphere. D. C. Agarwal.
Indian Journ. Phys., Vol. 46, 183 - 188 (1972).

The author deals with some aspects of electromagnetic wave propagation in the Venusian ionosphere. Using recent data, the effective collision frequency and high frequency conductivity have been calculated. The real part of the complex refractive index is then calculated for $\nu_e = 0$. The results are discussed in some detail.

093.046 Infrared radiative heating and cooling in the Venusian mesosphere. I. Global mean radiative equilibrium. R. E. Dickinson.
Journ. Atmosph. Sci., Vol. 29, 1531 - 1556 (1972).

Global mean sources and sinks of radiative energy are calculated for the upper atmosphere of Venus. Especially considered is the region between 90 and 130 km, where the equilibrium temperature is largely controlled through infrared absorption and emission by vibrational-rotational bands of CO_2. Source functions for bands deviating from thermodynamic equilibrium are determined as part of the calculation. Radiative transfer in the region of non-overlapping lines is calculated by summing the contribution of individual Voigt lines.

093.047 Planet Venus. A. D. Kuz'min.
Physics of the moon and planets, (see 012.024), p. 236 - 243 (1972). In Russian.

093.048 Investigation of the chemical composition of the Venus atmosphere with the automatic station Venera 4. A. P. Vinogradov, Yu. A. Surkov, K. P. Florenskij.
Physics of the moon and planets, (see 012.024), p. 244 - 250 (1972). In Russian.

093.049 Thermodynamic characteristics of the lower atmosphere of Venus from results of an experiment with Venera 4. V. V. Mikhnevich, V. A. Sokolov.
Physics of the moon and planets, (see 012.024), p. 251 - 254 (1972). In Russian.

093.050 Model of the Venus atmosphere from data of direct measurements. V. S. Avduevskij, M. Ya. Marov, M. K. Rozhdestvenskij.
Physics of the moon and planets, (see 012.024), p. 254 - 261 (1972). In Russian.

093.051 Atmosphere and ionosphere of Venus from data obtained by Mariner 5 in the S-region during a radio eclipse. A. Kliore, G. S. Levy, D. L. Cain, G. Fjeldbo, S. J. Rasool.
Physics of the moon and planets, (see 012.024), p. 262 - 268 (1972). In Russian.

093.052 Radio eclipse measurements of the Venus atmosphere made with Mariner 5 in the 10 cm wavelength region. A. Kliore.
Physics of the moon and planets, (see 012.024), p. 269 - 272 (1972). In Russian.

093.053 Investigation of the ultraviolet radiation with Venera 4. V. G. Kurt, E. K. Sheffer, S. B. Dostovalov.
Physics of the moon and planets, (see 012.024), p. 275 - 278

(1972). In Russian.

093.054 Plasma near Venus. Comparison of results obtained with the help of Venera 4 and Mariner 5.
T. K. Breus, K. I. Gringauz.
Physics of the moon and planets, (see 012.024), p. 279 - 283 (1972). In Russian.

093.055 The magnetic field in the vicinity of Venus.
Sh. Sh. Dolginov, E. G. Eroshenko, L. N. Zhuzgov.
Physics of the moon and planets, (see 012.024), p. 283 - 288 (1972). In Russian.

093.056 Radar observations of Venus at 3.8 cm wavelength.
J. V. Evans, T. Hagfors, R. P. Ingalls, D. Karp, W. E. Morrow, G. H. Pettengill, A. E. E. Rogers, I. I. Shapiro, W. B. Smith, F. S. Weinstein.
Physics of the moon and planets, (see 012.024), p. 289 - 308 (1972). In Russian.

093.057 Analysis of reflected signals obtained by radar investigations of Venus using a computer.
Yu. N. Aleksandrov.
Physics of the moon and planets, (see 012.024), p. 309 - 314 (1972). In Russian.

093.058 Determination of the rotational elements of Venus and of the coordinates of the surface regions with increased reflectivity in the radio region. V. K. Golovkov.
Physics of the moon and planets, (see 012.024), p. 314 - 319 (1972). In Russian.

093.059 Comparison of the determinations of the rotational velocity of Venus by the methods of radar measurements, the optical Doppler effect and spot changes.
J. Rösch.
Physics of the moon and planets, (see 012.024), p. 319 - 323 (1972). In Russian.

093.060 On the retrograde rotation of Venus. N. Bonev.
Physics of the moon and planets, (see 012.024), p. 323 - 324 (1972). In Russian.

093.061 Rapid motions of ultraviolet clouds on Venus.
B. A. Smith.
Physics of the moon and planets, (see 012.024), p. 324 - 325 (1972). In Russian.

093.062 On the modern level of volcanic activity on Venus.
D. P. Cruikshank.
Physics of the moon and planets, (see 012.024), p. 335 - 338 (1972). In Russian.

093.063 On a spectrophotometry of individual regions of Venus. O. M. Starodubtseva.
Physics of the moon and planets, (see 012.024), p. 338 - 345 (1972). In Russian.

093.064 Distribution of the radio brightness across the Venus disk at 8 mm wavelength. B. Ya. Gol'nev, Yu. N. Parijskij, P. A. Fridman, O. N. Shivris.
Physics of the moon and planets, (see 012.024), p. 348 - 352 (1972). In Russian.

093.065 The radio radiation of Venus and Jupiter at 2 and 8 mm wavelengths. V. A. Efanov, A. G. Kislyakov, I. G. Moiseev, A. I. Naumov.
Physics of the moon and planets, (see 012.024), p. 352 - 355 (1972). In Russian.

093.066 Interferometric observations of Venus with high

resolution at 3.1 cm wavelength.   G. L. Berge.
Physics of the moon and planets, (see 012.024), p. 355 - 358
(1972). In Russian.

093.067 Some results of a combined reduction of measure-
         ments with Venera 4 and terrestrial radio astro-
nomical and radar measurements.
Yu. N. Vetukhnovskaya, A. D. Kuz'min.
Physics of the moon and planets, (see 012.024), p. 359 - 365
(1972). In Russian.

093.068 On the interpretation of radar measurements of
         Venus in the microwave radio region.
N. N. Krupenio, A. P. Naumov.
Physics of the moon and planets, (see 012.024), p. 365 - 367
(1972). In Russian.

093.069 Absorption of radio waves of the cm region in the
         Venus atmosphere.   O. N. Rzhiga.
Physics of the moon and planets, (see 012.024), p. 367 - 371
(1972). In Russian.

093.070 Analysis of radio wave propagation in the Venus
         atmosphere.   O. I. Yakovlev.
Physics of the moon and planets, (see 012.024), p. 372 - 374
(1972). In Russian.

093.071 Estimates of the water content in the Venus atmo-
         sphere from data of radio astronomical measure-
ments and space probes.   A. E. Basharinov, B. G. Kutuza.
Physics of the moon and planets, (see 012.024), p. 375 - 377
(1972). In Russian.

093.072 On the absorption of radio waves in the Venus
         ionosphere. A. N. Kazantsev, V. A. Danilin.
Physics of the moon and planets, (see 012.024), p. 379
(1972). In Russian. – Abstract.

093.073 The influence of the horizontal inhomogeneity of
         the Venus atmosphere on the accuracy of measure-
ment of its parameters by the radio eclipse method.
A. N. Kazantsev, D. S. Lukin, V. A. Shkol'nikov.
Physics of the moon and planets, (see 012.024), p. 379 - 382
(1972). In Russian.

093.074 On the refraction of radio waves and the field
         strength in the atmosphere of Venus.
A. N. Kazantsev, D. S. Lukin, Yu. G. Spiridonov, V. A.
Shkol'nikov.
Physics of the moon and planets, (see 012.024), p. 382
(1972). In Russian. – Abstract.

093.075 Thermal conditions and convective motions in the
         lower layers of the Venus atmosphere.
V. S. Avduevskij, F. S. Zavelevich, M. Ya. Marov, A. I.
Nojkina, V. I. Polezhaev.
Physics of the moon and planets, (see 012.024), p. 383 - 388
(1972). In Russian.

093.076 The greenhouse effect in the convective atmosphere
         of Venus in the light of Venera 4 data.
V. I. Aleshin, I. G. Zarnitsyna, T. N. Fedoseeva.
Physics of the moon and planets, (see 012.024), p. 389 - 390
(1972). In Russian.

093.077 The greenhouse effect in the Venus atmosphere.
         G. M. Strelkov, N. F. Kukharskaya.
Physics of the moon and planets, (see 012.024), p. 390
(1972). In Russian. – Abstract.

093.078 Some optical properties of the Venus atmosphere

and the radiative equilibrium.
A. S. Ginzburg, E. M. Fejgel'son.
Physics of the moon and planets, (see 012.024), p. 391
(1972). In Russian. – Abstract.

093.079 Optical properties of the Venus atmosphere.
         V. V. Sobolev.
Physics of the moon and planets, (see 012.024), p. 391 - 393
(1972). In Russian.

093.080 Measurement of cosmic rays and search for radia-
         tion belts near Venus.   S. N. Vernov, A. E.
Chudakov, P. V. Vakulov, E. V. Gorchakov, P. P. Ignat'ev,
N. N. Kontor, Yu. I. Logachev, G. P. Lyubimov, A. G.
Nikolaev, N. V. Pereslegina.
Physics of the moon and planets, (see 012.024), p. 397 - 402
(1972). In Russian.

093.081 Propagation of radio waves through the lower
         atmosphere of Venus.   K. R. Richter.
Report NASA-TM-X-66046, National Aeronautics and Space
Administration, Greenbelt, Maryland. [Available from NTIS,
Springfield, Va.], 31 pp. (1972).

093.082 Estimate of the gravity anomalies of Venus.
         P. Baldi, E. Boschi, M. Caputo.
Geophys. Monograph 15, (see 012.025), p. 275 - 278 (1972).

093.083 Results of direct measurements of illumination in
         the atmosphere and on the surface of the planet
Venus during the flight of the automatic space probe Venera 8.
V. S. Avduevskij, M. Ya. Marov, B. E. Moshkin, A. P. Ekonomov.
Dokl. Akad. Nauk SSSR, Ser. Mat. Fiz., Vol. 210, 799 - 802
(1973). In Russian.

093.084 Estimates of optical characteristics of the Venus
         atmosphere important for photographing its
clouds and surface.
Yu. L. Biryukov, A. S. Panfilov, L. G. Titarchuk.
Kosmich. ikonika. Moskva, Nauka, 1973, p. 106 - 117. In
Russian. – Abstr. in Referativ. Zhurn. 51. Astron., 7.51.269
(1973).

         Numerical solution for the composition of a ther-
mosphere in the presence of a steady subsolar-to-antisolar
circulation with application to Venus.   See Abstr. 091.080.

         Mercury and Venus: high-resolution radar topo-
graphic profiles.   See Abstr. 092.003.

         The monochromatic and radiometric albedo of
Mars and Venus.   See Abstr. 097.107.

         Spectral investigations of the atmospheres of Mars
and Venus.   See Abstr. 097.108.

         Estimates of the turbulence intensity in the atmo-
spheres of Mars and Venus.   See Abstr. 097.109.

         Physical characteristics of the "blue clouds" of the
planets Mars, Earth and Venus.   See Abstr. 097.114.

         The optical properties of Venus and the Jovian
planets I. The atmosphere of Jupiter according to polarimetric
observations.   See Abstr. 099.043.

093.901 Erratum: 'Venus: a perspective at the beginning of
         planetary exploration' [Icarus, Vol. 16, 415 - 461
(1972)].   M. Ya. Marov.
Icarus, Vol. 18, 669 (1973).

094 Moon

094.001 **Rb—Sr ages and initial strontium in basalts from Apollo 15.**
D. A. Papanastassiou, G. J. Wasserburg.
Earth Planet. Sci. Letters, Vol. 17, 324 - 337 (1973).

We report on the Rb—Sr age and the initial $^{87}Sr/^{86}Sr$ of six basaltic rocks from the Apollo 15 mission. These data represent the most extensive Rb—Sr ages presently available from the Hadley Rille landing site. We have carried out a series of tracer and standard calibrations in order to further establish the validity of the analytical procedures used by us. All the results to date show that subsequent to the formation of the moon the lunar basalts evolved in a magma reservoir with Rb/Sr $\sim 10^{-2}$ and cannot be derived from a mantle which has a chondritic Rb/Sr.

094.002 **The kinetics of ulvöspinel reduction: Synthetic study and applications to lunar rocks.**
R. H. McCallister, L. A. Taylor.
Earth Planet. Sci. Letters, Vol. 17, 357 - 364 (1973).

The kinetics of Fe_2TiO_4 reduction to $FeTiO_3 + Fe$ were studied using $CO—CO_2$ gas mixtures with fO_2 measured by a solid ceramic (calcia-zirconia) oxygen electrolyte cell. Comparison of the synthetic textures with the variety of textures found in reduced lunar ulvöspinels makes possible qualitative estimates of the rate of reduction.

094.003 **The inverse problem of the moon's electrical conductivity.** B. A. Hobbs.
Earth Planet. Sci. Letters, Vol. 17, 380 - 384 (1973).

Data obtained from magnetometers on the moon's surface and on lunar orbiting satellites enable the electromagnetic response of the moon to some magnetic fluctuations in the solar wind to be determined. It has been shown that the calculated response from various, quite widely differing, conductivity profiles agrees with this observed response to some measure. Using the first order theory in the transverse electric mode of induction, the inversion methods of Backus and Gilbert are applied to the lunar magnetometer data. By assuming the data to be error free, a limit is placed on the fineness of detail that can be resolved in the conductivity profile, and it is thereby shown that only a two-layer moon is discernible from the data.

094.004 **Constrained least-squares analysis of petrologic problems with an application to lunar sample 12040.**
M. J. Reid, A. J. Gancarz, A. L. Albee.
Earth Planet. Sci. Letters, Vol. 17, 433 - 445 (1973).

This paper presents a systematic treatment of the application of least-squares analysis to petrologic problems including the direct utilization of physical constraints and weighting factors in the problem, and the assessment of uncertainties in the solution. As an example, least-squares analysis is used to examine, in detail, the mass balance equations for lunar rock 12040 and to determine the consistency of the available analytical data.

094.005 **Extinct lunar radioactivities: Xenon from ^{244}Pu and ^{129}I in Apollo 14 breccias.**
C. J. Behrmann, R. J. Drozd, C. M. Hohenberg.
Earth Planet. Sci. Letters, Vol. 17, 446 - 455 (1973).

Two Apollo 14 breccias have been found to contain xenon from the spontaneous fission of 82 my ^{244}Pu. A third contains 60 times as much fission xenon as local uranium can account for and is probably of similar character. One of the breccias shows a ^{129}Xe excess most likely due to the decay of 17 my ^{129}I.

094.006 **Lunar gravity derived from long-period satellite motion — a proposed method.** A. J. Ferrari.
Celestial Mechanics, Vol. 7, 46 - 76 (1973).

A new method has been devised to determine the spherical harmonic coefficients of the lunar gravity field. This method consists of a two-step data reduction and estimation process. In the first step, a weighted least-squares empirical orbit determination scheme is applied to Doppler tracking data from lunar orbits to estimate long-period Kepler elements and rates. In the second step, the Kepler element rates are used as input to a second least-squares processor that estimates lunar gravity coefficients using the long-period Lagrange perturbation equations. Pseudo Doppler data have been generated simulating two different lunar orbits. This analysis included the perturbing effects of the L1 lunar gravity field, the earth, the sun, and solar radiation pressure. Orbit determinations were performed on these data and long-period orbital elements obtained. The Kepler element rates from these solutions were used to recover L1 lunar gravity coefficients.

094.007 **Identification, distribution and significance of lunar volcanic domes.** E. I. Smith.
The Moon, Vol. 6, 3 - 31 (1973).

Over 300 previously unrecognized volcanic domes were identified on Lunar Orbiter photographs using the following criteria: (1) the recognition of land forms on the moon similar in morphology to terrestrial volcanic domes, (2) structural control, (3) geomorphic discordance, and (4) the recognition of land forms modified by dome-like swellings.

094.008 **Displaced mass, depth, diameter, and effects of oblique trajectories for impact craters formed in dense crystalline rocks.** D. E. Gault.
The Moon, Vol. 6, 32 - 44 (1973).

Empirical formulae are presented for calculating the displaced mass, depth, diameter, and effects of oblique trajectories for impact craters formed in dense crystalline rocks. The formulae are applicable to craters with diameters from approximately $10^{-3} - 10^3$ cm that require, respectively, impact kinetic energies of approximately 10 to 10^{16} ergs for their formation. The experimental results are in poor agreement with Öpik's theoretical calculations and raise questions on the validity of his theoretical model.

094.009 **Simultaneous impact and lunar craters.**
V. R. Oberbeck.
The Moon, Vol. 6, 83 - 92 (1973).

The existence of large terrestrial impact crater doublets and Martian crater doublets that have been inferred to be impact craters demonstrates that simultaneous impact of two or more bodies occurs at nearly the same point on planetary surfaces. The purpose of this paper is to present some preliminary results of a series of simultaneous impact cratering experiments and to show that the craters produced are similar in structure to many of the large lunar craters.

094.010 **Thermal radiation properties of Apollo 14 fines.**
R. C. Birkebak, J. P. Dawson.
The Moon, Vol. 6, 93 - 99 (1973).

The thermal radiation properties as a function of bulk density, angle of illumination and wavelength are presented for lunar fines from the Apollo 14 mission. The density range covered is from 1095 kg/m^3 to 1590 kg/m^3 and a wavelength range of 0.36—14.5 μm. The solar albedo and total emittance were calculated from spectral values and are compared to Apollo 11 and 12 values.

094.011 **Viscosity of the moon. I: After mare formation.**
J. Arkani-Hamed.
The Moon, Vol. 6, 100 - 111 = Lunar Sci. Inst., *Houston, Texas,* Contr. No. 99 (1973).

Using data from the present gravitational potential and surface topography of the moon, it is possible to determine a lower limit of about 5 b.y. for the relaxation time of the mascons. Assuming that the moon has behaved as a Maxwellian viscoelastic body since the formation of the mascons, this relaxation time indicates a value of about 10^{27} poise for the viscosity of the lunar interior. Such a high viscosity implies that there has been no convection current inside the upper 800 km of the moon since the formation of the mascons.

094.012 **Viscosity of the moon. II: During mare formation.**
J. Arkani-Hamed.
The Moon, Vol. 6, 112 - 124 (1973).

The Apollo 15 mission provided reliable data on the depths of maria Serenitatis and Smythii. Using the present depth values and the excess masses of the associated mascons of these maria together with the ages of their final fillings the average viscosity of the upper part of the lunar interior is determined for the mare formation period (from 3.8 to 3.3 b.y. ago) and the period after mare formation (since 3.3 b.y. ago). It is found that the lower limit of the average viscosity within the first period is about 10^{25} poise and within the second period is about 8×10^{26} poise.

094.013 **On a possible relation between lunar transient phenomena and the earth-shine.** F. Link.
The Moon, Vol. 6, 125 - 126 (1973).

The brightness of the earth-shine and the frequency of some categories of lunar transient phenomena show an enhancement near the full moon, when the latter enters or exits the bow-shock front of the magnetosphere.

094.014 **Physical librations due to the third and fourth degree harmonics of the lunar gravity potential.**
D. H. Eckhardt.
The Moon, Vol. 6, 127 - 134 (1973).

The existence of third and fourth harmonics of the lunar gravity potential gives rise to sizable lunar physical librations. Using one recent set of potential estimates, the following effects are noted: the mean sub-earth point is displaced from the earthward principal moment of inertia axis by $168''$; the inclination of the lunar equator to the ecliptic is decreased by $14.''5$; and a six year period libration in longitude, with amplitude $13.''1$, is induced.

094.015 **Stress differences in the moon as an evidence for a cold moon.** J. Arkani-Hamed.
The Moon, Vol. 6, 135 - 163 = Lunar Sci. Inst., *Houston, Texas,* Contr. No. 97 (1973).

Assuming that the lateral variations of density in the lunar crust, the crustal density anomalies, are responsible for the lateral undulations of the lunar gravitational potential, we compute these anomalies for four different lunar models, which include an entirely solid moon and three different solid lunar models with partially molten layers located within 600 km depth. The stress differences created by the density anomalies are determined for these models.

094.016 **Boulder tracks and nature of lunar soil.**
H. J. Hovland, J. K. Mitchell.
The Moon, Vol. 6, 164 - 175 (1973).

Boulder tracks from 19 different locations on the moon, observable in Lunar Orbiter photographs, have been examined. Measurements of the track width indicate that some of the boulders sank considerably deeper than others. It is suggested that lunar surface materials vary from place to place; the state of compaction (density of lunar soil) is probably one of the

significant variables. Using bearing capacity theory, modified to be applicable to the rolling boulder problem by theoretical studies and extensive testing, the friction angle of the lunar soil was estimated.

094.017 **On the model of the accumulation of the moon compatible with the data on the composition and the age of lunar rocks.** E. L. Ruskol.
The Moon, Vol. 6, 176 - 189, 190 - 201 (1973). — In Russian and English.

It is suggested that the overall early melting of the lunar surface is not necessary for the explanation of facts and that the structure of highlands is more complicated than a solidified anorthositic 'plot'. The early heating of the interior of the moon up to 1000K is really needed for the subsequent thermal history with the maximum melting 3.5×10^9 yr ago, to give the observed ages for mare basalts.This may be considered as an indication that the moon during the accumulation retained a portion of its gravitational energy converted into heat, which may occur only at rapid processes.

094.018 **Precision of selenodetic frames of reference.**
I. V. Gavrilov, V. S. Kisliuk.
The Moon, Vol. 6, 202 - 211 (1973).

The number of reference points for the fixing of a selenodetic reference frame in the moon's body is estimated. It is shown that, for this purpose, from 40 to 100 reference points are sufficient. Precision of the selenodetic coordinate transformations from one system to another is also analyzed.

094.019 **Chemical composition of some Apollo 14 lunar samples.**
H. B. Wiik, J. A. Maxwell, J.-L. Bouvier.
Earth Planet. Sci. Letters, Vol. 17, 365 - 368 (1973).

Major, minor and trace element data, hitherto unpublished, are given for Apollo 14 samples 14163, 14259, 14303, 14305 and 14306, with brief details of the analytical procedures used. As found previously for the Apollo 11 and 12 samples, volatile constituents such as water, fluorine and carbon dioxide are present in very low concentrations, but the Apollo 14 samples are characterised by much higher contents of sodium, potassium and phosphorus. The values for TiO_2 continue the downward trend previously noted.

094.020 **Determination of lunar libration by laser moon-ranging.** O. Calame.
Astron. Astrophys., Vol. 22, 75 - 80(1973). In French.

The theoretical work is intended to show the advantage of laser measurements of distances from an observer on earth to a lunar reflector in order to determine the moon's motion around its center of mass (physical libration). It consits of two steps: the first is intended to determine the selenocentric coordinates of the lunar reflector used and the relative configurations of the observatory and the moon which should be selected to obtain the best accuracy; the second deals with the libration itself from measurements supposed performed on several reflectors (three at least), during one year or more.

094.021 **Statistical distribution of the albedo over the lunar disk.** N. N. Evsjukov.
Astron. Zhurn. Akad. Nauk SSSR, Vol. 50, 172 - 175 (1973). In Russian. English translation in Soviet Astron. AJ, Vol. 17, No. 1.

A histogram of the albedo distribution over the lunar disk is given; its peculiarities are discussed.

094.022 **Dynamical figure of the moon and density distribution of the lunar interior.** G. A. Meshcheryakov.
Astron. Zhurn. Akad. Nauk SSSR, Vol. 50, 186 - 200 (1973). In Russian. English translation in Soviet Astron. AJ, Vol. 17, No. 1.

The moments of inertia of the moon and the equalized values of its dynamical flattenings, of the principal parameter of the physical libration of the moon 'f', of Stokes' constants c_{20} and c_{22} are obtained by means of the least squares method equalizing the low harmonics of the lunar gravitational field and of the dynamical flattenings of the moon. The ellipsoid of inertia of the moon is completely constructed. A three-axial ellipsoid, which is an approach to the selenoid, is determined. A discussion of the results obtained is executed.

094.023   Lunar crater Copernicus: Search for debris of impacting body at Apollo 12 site.
J. W. Morgan, R. Ganapathy, J. C. Laul, E. Anders.
Geochim. Cosmochim. Acta, Vol. 37, 141 - 154 (1973).

In an attempt to characterize meteoritic material at the Apollo 12 site, 4 KREEP (moon soil components rich in K, rare earth elements (REE), and P) concentrates from soil 12033 have been analyzed by neutron activation analysis. These contain a meteoritic component in which siderophile Ir, Re and Sb are depleted by about a factor of 2, while volatile Se, Zn, Ag and Bi are depleted by a factor of more than 5 relative to Au.

094.024   A lunar core of Fe−Ni−S.   R. Brett.
Geochim. Cosmochim. Acta, Vol. 37, 165 - 170 (1973).

Crystalline rocks from all lunar landing sites contain metallic Fe; conditions in the portions of the lunar interior from which the magmas were derived were therefore such that the silicates were at, or close to, equilibrium with a metal phase. Three possibilities exist: (1) the metal is disseminated through the lunar interior, (2) has segregated sufficiently to form pockets or layer(s), or (3) the metal segregated early in lunar history to form a core. This paper considers the implications of the last two possibilities.

094.025   Major element chemistry of glasses in Apollo 14 soil 14156.
A. M. Reid, W. I. Ridley, R. S. Harmon, P. Jakeš.
Geochim. Cosmochim. Acta, Vol. 37, 695 - 699 (1973).

Glasses in a soil sample (14156) from the middle layer of the trench at the Fra Mauro landing site show a wide range of compositions clustered around certain preferred compositions. The results are compared with those of analyzed glasses sampled during other Apollo missions.

094.026   Revision of lunar Rb−Sr ages.
J. R. de Laeter, M. J. Vernon, W. Compston.
Geochim. Cosmochim. Acta, Vol. 37, 700 - 702 (1973). Note.

094.027   Thermophysical properties of Apollo 12 fines.
C. J. Cremers.
Icarus, Vol. 18, 294 - 303 (1973).

The vacuum thermal conductivity of the Apollo 12 fines is found to vary from about 10^{-3} W/m-°K at 100°K to about 3×10^{-3} W/m-°K at 400°K. The conductivity of the fines is found to be close to that of terrestrial basalt both under vacuum and at higher pressures.

094.028   Niobian rutile in an Apollo 14 KREEP fragment.
P. F. Hlava, M. Prinz, K. Keil.
Meteoritics, Vol. 7, 479 - 485 (1972).

Niobian rutile was found in a KREEP lithic fragment of basaltic texture. Rare earth elements were not detected, in contrast with lunar niobian rutile of Marvin (1971).

094.029   Thermal diffusivity of lunar rocks under atmospheric and vacuum conditions.
N. Fujii, M. Osako.
Earth Planet. Sci. Letters, Vol. 18, 65 - 71 (1973).

Thermal diffusivity of three lunar rocks (10049 and 10069; type A, Apollo 11 and 14311; Apollo 14) and a terrestrial basalt (alkaline olivine basalt, Oki-dôgo, Japan) was measured under one atmosphere and in vacuum conditions ($10^{-3} \sim 10^{-5}$ mmHg) in the temperature range from 85 to 850°K. One of the purposes of this paper is to clarify the effect of porosity and temperature variation of thermal diffusivity of lunar crystalline rocks.

094.030   Crater frequency age determinations for the proposed Apollo 17 site at Taurus-Littrow.
R. Greeley, D. E. Gault.
Earth Planet. Sci. Letters, Vol. 18, 102 - 108 (1973).

Crater frequency distributions determined for surfaces in the Taurus-Littrow region of the moon and compared with crater counts and radiometric age dates for Apollo 11, 12 and 14 landing sites indicate that the surface of the proposed Apollo 17 landing site was formed between 2.5 and 2.8 by ago.

094.031   Temperatures in the lunar interior and some implications.   A. Duba, A. E. Ringwood.
Earth Planet. Sci. Letters, Vol. 18, 158 - 162 (1973).

Data on the electrical conductivity of olivine and pyroxene obtained under redox conditions similar to those that exist in the moon indicate that the moon is at temperatures near the melting point at depths of 600-900 km. This temperature profile, combined with information on the distribution of radioactive elements and evidence of extensive differentiation of the moon, lead to the conclusion that the moon accreted at temperatures between 600−1000°C.

094.032   Visit to Taurus-Littrow.   R. W. Sinnott.
Sky Telescope, Vol. 45, 79 - 84 (1973).

094.033   Orange soil and other Apollo 17 results.
R. N. Watts, Jr.
Sky Telescope, Vol. 45, 146 - 148 (1973).

094.034   Morphology of the structure of a metallic fragment of lunar matter.   R. I. Mints, T. M. Petukhova.
Dokl. Akad. Nauk SSSR, Ser. Mat. Fiz., Vol. 208, 1315 - 1317 (1973). In Russian.

094.035   The tsunami model of the origin of ring structures concentric with large lunar craters.
R. B. Baldwin.
Phys. Earth Planet. Interiors, Vol. 6, 327 - 339 (1972).

The largest lunar craters are normally surrounded by one or more ring anticlines which are accompanied on their inner edges by ring synclines. Four arguments are presented which tend to make probable the tsunami theory of the origin of these structures. The presence of smaller raised rings within some of these great craters is consistent with the tsunami model.

094.036   Some conparative aspects of lunar origin.
A. E. Ringwood.
Phys. Earth Planet. Interiors, Vol. 6, 366 - 376 (1972).

A widely held assumption that the distribution of mass and angular momentum in the earth−moon system is anomalous when compared with the corresponding distributions in other planet−satellite systems is critically examined. In order to make valid comparisons, the light gases which were originally associated with the satellites of Jupiter, Saturn, Uranus and Neptune, and which have since escaped, should be taken into account.

094.037   Metallic particles of high cobalt content in Apollo 15 soil samples   H. J. Axon, J. I. Goldstein.
Earth Planet. Sci. Letters. Vol. 18, 173 - 180 (1973).

Single phase α-kamacite containing >3.2 wt% Co and γ-taenite containing from 30–60 wt% Ni from the Apollo 15 soils – 15031, 15071, 15081, 15261 and 15271 – have been examined by metallographic and electron microprobe techniques. In addition two phase $a + \gamma$ particles from soils 14003, 15071, 15261 and 15271 with Ni and Co contents well outside the meteoritic range have also been examined.

094.038   Chemical compositions and petrogenetic relationships in Apollo 15 mare basalts.
B. W. Chappell, D. H. Green.
Earth Planet. Sci. Letters, Vol. 18, 237 - 246 (1973).

We report the compositions, in terms of major elements and some trace elements, of thirteen basalts, including examples of the four petrographic types. These basalts are compared with Apollo 11 and 12 mare basalts. Knowledge of the crystallization behaviour of the basalts at both high and low pressures is used to discuss possible genetic relationships among the Apollo 15 basalts themselves and the more general problems of petrogenesis and source region for mare basalt.

094.039   The composition and origin of the moon.
D. L. Anderson.
Earth Planet. Sci. Letters, Vol. 18, 301 - 316 = Contr. Div. Geol. Planet. Sci., California Inst. Technology, *Pasadena*, No. 2193 (1973).

Many of the properties of the moon, including its "enrichment" in Ca, Al, Ti, U, Th, Ba, Sr and the REE and "depletion" in Fe, Rb, K, Na and other volatiles can be explained by early condensation processes in the solar nebula.

094.040   Particle track record of Apollo 15 green soil and rock.   R. L. Fleischer, H. R. Hart, Jr.
Earth Planet. Sci. Letters, Vol. 18, 357 - 364 (1973).

Track densities, track stability, and uranium contents have been measured in lunar samples containing abundant green glass spherules from stations 6a and 7 of the Apollo 15 mission.

094.041   Harold Urey and the moon.   H. E. Newell.
The Moon, Vol. 7, 1 - 5 (1973). – Paper dedicated to Professor Harold C. Urey on the occasion of his 80th birthday on 29 April, 1973.

094.042   The use of the Saros in lunar dynamical studies.
A. E. Roy.
The Moon, Vol. 7, 6 - 13 (1973). – Paper dedicated to Professor Harold C. Urey on the occasion of his 80th birthday on 29 April, 1973.

It is shown that the near-periodicity in the earth-moon-sun system demonstrated by the possibility of using the Saros to predict eclipses, suggests that the Saros can also be used in a fast and accurate method of special perturbations which can be applied for long term study of the evolution of the moon's orbit.

094.043   Why is the moon grey?   W. F. Libby.
The Moon, Vol. 7, 46 - 48 (1973). – Paper dedicated to Professor Harold C. Urey on the occasion of his 80th birthday on 29 April, 1973.

094.044   The velocity structure of the lunar crust.
R. L. Kovach, J. S. Watkins.
The Moon, Vol. 7, 63 - 75 (1973). – Paper dedicated to Professor Harold C. Urey on the occasion of his 80th birthday on 29 April, 1973.

Seismic refraction data, obtained at the Apollo 14 and 16 sites, when combined with other lunar seismic data, allow a compressional wave velocity profile of the lunar near-surface and crust to be derived.

094.045   Particle track record of the Luna missions.
G. M. Comstock, R. L. Fleischer, H. R. Hart, Jr.
The Moon, Vol. 7, 76 - 83 (1973). – Paper dedicated to Professor Harold C. Urey on the occasion of his 80th birthday on 29 April, 1973.

Measurements are reported of particle-track densities in 100-200 μ crystalline grains taken from one level of the soil column returned from the lunar highlands between Mare Fecunditatis and Mare Crisium by Luna 20 and from two levels in that from Mare Fecundatis by Luna 16.

094.046   Density and stress distribution in the moon.
J. Arkani-Hamed.
The Moon, Vol. 7, 84 - 126 (1973). – Paper dedicated to Professor Harold C. Urey on the occasion of his 80th birthday on 29 April, 1973.

A model is presented for the lateral variations of density within the moon. The model gives rise to a gravitational potential which is equal to the observed potential at the lunar surface, moreover, it minimizes the total shear-strain energy of the moon.

094.047   Magnetism of the moon.   R. Smoluchowski.
The Moon, Vol. 7, 127 - 131 (1973). – Paper dedicated to Professor Harold C. Urey on the occasion of his 80th birthday on 29 April, 1973.

The Malkus theory of a precessionally driven magneto-turbulence in a liquid core is applied to the moon. It is shown that a lunar magnetic field requires the presence of a non-metallic core at least 2500K or of an iron core at at least 2000K. A new mechanism is proposed which is based on tidal effects in the outer solid and liquid shells whose existence is suggested by measurements of lunar radioactivity. This mechanism could account for the generation of local rather than poloidal fields at low latitudes in agreement with observation.

094.048   Green spherules from Apollo 15: Inferences about their origin from inert gas measurements.
S. Lakatos, D. Heymann, A. Yaniv.
The Moon, Vol. 7, 132 - 148 (1973). – Paper dedicated to Professor Harold C. Urey on the occasion of his 80th birthday on 29 April, 1973.

Green spherules from the 'clod' 15426 and from fines 15421 contain about 100 times less trapped inert gases than normal bulk fines from Apollo 15. These spherules have apparently never been directly exposed to the solar wind. Spherules from other fines contain about 10 times more trapped gas than those from the 'clod'. The trapped gases can be of solar - wind origin, but this origin requires a two-stage model for the spherules from the clods. The trapped gases may also be assumed to represent primordial lunar gas. The composition of this gas is then similar to the 'solar' or 'unfractionated' component of gas-rich meteorites, but unlike that in most of the carbonaceous chondrites. The gas content of the spherules from fines suggests strongly that all spherules were at one time in 'clod' -like material.

094.049   Lunar evolution: How well do we know it now?
V. R. Murthy, S. K. Banerjee.
The Moon, Vol. 7, 149 - 171 (1973). – Paper dedicated to Professor Harold C. Urey on the occasion of his 80th birthday on 29 April, 1973.

The currently known astronomical, chemical and magnetic data are not uniquely indicative of an extensively and globally molten moon. We argue here for an accretional layering in the moon, but at temperatures below solidus. We conclude that the characteristics exhibited of the lunar materials are reconcilable with a 'cold' moon such as discussed by Urey over the past two decades.

094.050 **Geochemistry of the lunar highlands.**
S. R. Taylor.
The Moon, Vol. 7, 181 - 195 (1973). – Paper dedicated to Professor Harold C. Urey on the occasion of his 80th birthday on 29 April, 1973.

The aim of this study is to attempt to unravel the apparently complex chemical compositions of the breccias, in order to shed light on the following questions: (1) Are the highlands simple or complex chemically; (2) was the observed chemically fractionated crust formed after accretion of the moon; or (3) was it due to a late accretion of a chemically distinct Al-rich layer; or (4) is the present lunar crust the result of both heterogeneous accretion and later melting and chemical fractionation.

094.051 **A determination of the intensity of the ancient lunar magnetic field.**
W. A. Gose, D. W. Strangway, G. W. Pearce.
The Moon, Vol. 7, 196 - 201 (1973). – Paper dedicated to Professor Harold C. Urey on the occasion of his 80th birthday on 29 April, 1973.

Thermal demagnetization of lunar breccia 15498,36 shows that the natural remanent magnetization is a simple thermoremanence carried by metallic iron. Using the classical Thellier-Thellier method the strength of the magnetizing field at the time of sample formation was found to be 2100 ± 80 gammas.

094.052 **Photoelectrons and solar wind/lunar limb interaction.** D. R. Criswell.
The Moon, Vol. 7, 202 - 238 (1973). – Paper dedicated to Professor Harold C. Urey on the occasion of his 80th birthday on 29 April, 1973.

It is suggested that boundary conditions for solar wind/lunar limb interactions are active. The 'whole-moon' limb does not evoke a shock cone because warm ($\simeq 13$ eV/electron) solar wind electrons are replaced by cool ($\lesssim 2$ eV/electron) photoelectrons that are ejected from the generally smooth areas of the lunar terminator illuminated at glazing angles by the sun. Conversely, directly illuminated highland areas exchange hot photoelectrons (> 20 eV/electron) for warm solar wind electrons. The hot electrons generate a localized pressure increase in the adjacent solar wind flow which evokes a shock streamer in the solar wind.

094.053 **The filling of the lunar mare basins.**
L. B. Ronca.
The Moon, Vol. 7, 239 - 248 (1973). – Paper dedicated to Professor Harold C. Urey on the occasion of his 80th birthday on 29 April, 1973.

The surface of each mare is not a homogeneous geomorphological unit, but displays a variety of geomorphologies. The interpretation of this phenomenon depends on the assumptions one is willing to accept.

094.054 **The Apollo 16 lunar samples: Petrographic and chemical description.**
Apollo 16 Preliminary Examination Team: P. W. Gast, W. C. Phinney, M. B. Duke, E. K. Jackson, N. J. Hubbard, P. Butler, R. B. Laughon, S. O. Agrell, M. N. Bass, R. Brett, W. D. Carrier, U. S. Clanton, A. L. Eaton, J. Head, G. H. Heiken, F. Horz, G. E. Lofgren, D. S. McKay, D. A. Morrison, W. R. Muehlberger, J. S. Nagle, A. M. Reid, W. I. Ridley, C. Simonds, D. Stuart-Alexander, J. L. Warner, R. J. Williams, H. Wilshire, B. M. Bansal, J. A. Brannon, A. M. Landry, J. M. Rhodes, K. V. Rodgers, J. E. Wainwright, L. Bennett, R. S. Clark, J. E. Keith, G. D. O'Kelley, R. W. Perkins, L. A. Rancitelli, W. R. Portenier, M. K. Robbins, E. Schonfeld, E. K. Gibson, C. F. Lewis, C. B. Moore, D. R. Moore.
Science, Vol. 179, 23 - 34 (1973).

In this article we summarize the chemical and petrographic characteristics of a representative suite of the Apollo 16 rock and soil specimens. At the present time no clear-cut correlation of any of the observed characteristics with position in the site has been observed. This generalization is based on a detailed examination of only a portion of the returned samples.

094.055 **Apollo 16 exploration of Descartes: A geological summary.**
Apollo Field Geology Investigation Team: G. E. Ulrich, W. R. Muehlberger, G. A. Swann, R. L. Sutton, M. H. Hait, H. G. Wilshire, R. M. Batson, E. L. Boudette, R. E. Eggleton, D. P. Elston, V. L. Freeman. T. A. Hall, H. E. Holt, J. A. Jordan, K. B. Larson, V. S. Reed, G. G. Schaber, J. P. Schafer, R. L. Tyner, E. W. Wolfe, C. A. Hodges, E. D. Jackson, D. J. Milton, D. Stuart-Alexander, C. M. Duke, J. W. Young, A. W. England, J. W. Head, J. J. Rennilson, L. T. Silver.
Science, Vol. 179, 62 - 69 (1973).

The Apollo 16 results have demonstrated again that the moon is far more complex than predicted on the basis of early studies. The remarkable suite of feldspathic crystalline rocks and breccias from the largest lateral and vertical range sampled to date helps to clarify the origin and history of a significant part of the lunar highlands and make possible more precise statements of new questions.

094.056 **Volatile-rich lunar soil: Evidence of possible cometary impact.** E. K. Gibson, Jr., G. W. Moore.
Science, Vol. 179, 69 - 71 (1973).

A subsurface Apollo 16 soil, 61221, is much richer in volatile compounds than soils from any other locations or sites as shown by thermal analysis-gas release measurements. These volatile components may have been brought to this site by a comet, which may have formed North Ray crater.

094.057 **Breccias from the lunar highlands: Preliminary petrographic report on Apollo 16 samples 60017 and 63335.** S. J. Kridelbaugh, G. A. McKay, D. F. Weill.
Science, Vol. 179, 71 - 74 (1973).

Lunar samples 60017,4 and 63335,14 are composed of microbreccias and devitrified glass. These components are predominantly anorthositic, with the exception of a cryptocrystalline clast found in the microbreccia portion of 63335,14 which contains 2.7 percent potassium oxide and 66.7 percent silicon dioxide.

094.058 **Spinel troctolite and anorthosite in Apollo 16 samples.**
M. Prinz, E. Dowty, K. Keil, T. E. Bunch.
Science. Vol. 179, 74 - 76 (1973).

A spinel troctolite and an anorthosite from the Apollo 16 landing site represent contrasting types of "primitive" lunar cumulates. The two rock types probably formed from the same parent magma type, a high-alumina magnesian basalt, with the troctolite forming earlier by crystal settling, and the anorthosite later, possibly by flotation.

094.059 **Lunar shape via the Apollo laser altimeter.**
W. L. Sjogren, W. R. Wollenhaupt.
Science, Vol. 179, 275 - 278 (1973).

Data from the Apollo 15 and 16 laser altimeters reveal the first accurate elevation differences between distant features on both sides of the moon. The large far-side depression observed in the Apollo 15 data is not present in the Apollo 16 data. The offset of the center of gravity from the optical center is about 2 kilometers toward the earth and 1 kilometer eastward.

094.060 **Detection of radon emanation from the crater Aristarchus by the Apollo 15 alpha particle spectrometer.** P. Gorenstein, P. Bjorkholm.

Science, Vol. 179, 792 - 794 (1973).

A significant increase in radon-222 activity was detected from a region containing the crater Aristarchus. The result is interpreted as probably indicating internal activity at the site. By analogy with terrestrial processes, increased radon emanation may be associated with the emission of other volatiles.

094.061 **Lunar surface radioactivity: Preliminary results of the Apollo 15 and Apollo 16 gamma-ray spectrometer experiments.** A. E. Metzger, J. I. Trombka, L. E. Peterson, R. C. Reedy, J. R. Arnold.
Science, Vol. 179, 800 - 803 (1973).

Gamma-ray spectrometers on the Apollo 15 and Apollo 16 missions have been used to map the moon's radioactivity over 20 percent of its surface.

094.062 **Influence of ephemerides errors on various determinations using laser lunar ranging.** A. G. Orszag.
Astron. Astrophys., Vol. 23, 441 - 451 (1973). In French.

The purpose of this work has been to evaluate the possible use of laser ranging at a lunar reflector for determination of various parameters relating to ranging station and moon center of mass.

094.063 **Endogenic cratering distribution on the moon.** E. B. Grudewicz.
Nature, Vol. 241, 186 - 187 (1973).

The author has made crater counts and measurements for the Hyginus Rille region from a medium resolution Lunar Orbiter V photograph. Incremental and cumulative analyses of the crater-size frequency distribution show that the mare regions north and south of the rille are similar in age.

094.064 **On-surface and laboratory size measurements of fine lunar particles.** L. D. Jaffe, J. N. Strand.
Nature, Phys. Sci., Vol. 241, 57 - 59 (1973).

Before the return of lunar material to earth, some determinations of lunar soil particle size were made from Surveyor television pictures. The resolution of the pictures permitted direct measurement and counting only of particles larger than 1 mm in diameter. Various techniques, summarized below, were used to evaluate sizes of particles smaller than the resolution limit.

094.065 **Distribution of methane and carbide in Apollo 11 fines.**
P. H. Cadogan, G. Eglinton, J. R. Maxwell, C. T. Pillinger.
Nature, Phys. Sci., Vol 241, 81 - 83 (1973).

We have examined the distribution of both CH_4 and carbide by fractionating a sample of Apollo 11 fines (10086D) according to the size of the particles, and their magnetic susceptibility and density. Concentrations of CH_4 and carbide (as CD_4) in the resulting fractions were determined by gas chromatographic analysis of the gases released by DCl dissolution.

094.066 **Volatilization studies on a terrestrial basalt and their applicability to volatilization from the lunar surface.**
W. C. Storey.
Nature, Phys. Sci., Vol. 241, 154 - 157 (1973).

The author shows that mass loss by volatilization of Na and K may reach 10^{-6} and 10^{-7} of $cm^{-2}s^{-1}$, respectively.

094.067 **Plutonic or metamorphic equilibration in Apollo 16 lunar pyroxenes.** A. Peckett, G. M. Brown.
Nature, Vol. 242, 252 - 255 (1973). – Letter.

094.068 **The origin of the moon.** J. H. Fremlin.
Nature, Vol. 242, 317 - 318 (1973).

The author has some quantitative doubts concerning Anderson's proposals that the moon condensed from material off the median plane of the initial solar nebula. To him, it seems easier to believe that the material of the planets was collected by the sun in passing through a cold dust cloud.

094.069 **Carbon compounds in pyrolysates and amino acids in extracts of Apollo 14 lunar samples.**
V. E. Modzeleski, J. E. Modzeleski, M. A. J. Mohammed, L. A. Nagy, B. Nagy, W. S. McEwan, H. C. Urey, P. B. Hamilton.
Nature, Phys. Sci., Vol. 242, 50 - 52 (1973).

The analysis of seven samples brought back by the Apollo 14 astronauts is described. Carbon was present in the gases evolved to between 76 and 161 p.p.m. Glycine, aspartic acid, glutamic acid and serine were among the most abundant amino acids found.

094.070 **L'éphéméride analytique lunaire – ALE.**
J. Henrard.
Ciel et Terre, Vol. 89, 1 - 27 (1973).

094.071 **Preliminary data on lunar soil collected by the Luna 20 unmanned spacecraft.** A. P. Vinogradov.
Geochim.Cosmochim. Acta, Vol. 37, 721 - 729 (1973). Presented at a meeting of the Presidium of the U.S.S.R. Academy of Sciences, 11 May, 1972. – Translated from Geokhimiya, 1972, p. 763 - 774.

094.072 **Visible and near-infra-red transmission and reflectance measurements of the Luna 20 soil.**
J. B. Adams, P. M. Bell, J. E. Conel, H. K. Mao, T. B. McCord, D. B. Nash.
Geochim. Cosmochim. Acta, Vol. 37, 731 - 743 (1973).

094.073 **$^{207}Pb/^{206}Pb$ ages of individual mineral phases in Luna 20 material by ion microprobe mass analysis.**
C. A. Andersen, J. R. Hinthorne.
Geochim. Cosmochim. Acta, Vol. 37, 745 - 754 (1973).

094.074 **Optical and chemical analysis of iron in Luna 20 plagioclase.** P. M. Bell, H. K. Mao.
Geochim. Cosmochim. Acta, Vol. 37, 755 - 759 (1973).

094.075 **Oxide minerals in lithic fragments from Luna 20 fines.**
R. Brett, R. C. Gooley, E. Dowty, M. Prinz, K. Keil.
Geochim. Cosmochim. Acta, Vol. 37, 761 - 773 (1973).

094.076 **Petrology of fine-grained rock fragments and petrologic implications of single crystals from the Luna 20 soil.** K. L. Cameron, J. J. Papike, A. E. Bence, S. Sueno.
Geochim. Cosmochim. Acta, Vol. 37, 775 - 793 (1973).

094.077 **Chemistry and surface morphology of soil particles from Luna 20 LRL sample 22003.** J. L. Carter.
Geochim. Cosmochim. Acta, Vol. 37, 795 - 803 (1973).

094.078 **An unusual basalt fragment in Luna 20 sample L2010.** M. C. Michel-Lévy, Z. Johan.
Geochim. Cosmochim. Acta, Vol. 37, 805 - 809 (1973).

094.079 **Oxygen isotopic composition of the Luna 20 soil.** R. N. Clayton.
Geochim. Cosmochim. Acta, Vol. 37, 811 - 813 (1973).

094.080 **Petrology of Luna 20 regolith from the lunar highlands.** M. L. Crawford, P. W. Weigand.
Geochim. Cosmochim. Acta, Vol. 37, 815 - 823 (1973).

094.081 **Fossil track and thermoluminescence studies of Luna 20 material.** G. Crozaz, R. Walker, D. Zimmerman.
Geochim. Cosmochim. Acta, Vol. 37, 825 - 830 (1973).

094.082 Luna 20 pyroxenes: exsolution and phase transformation as indicators of petrologic history.
S. Ghose, I. S. McCallum, E. Tidy.
Geochim. Cosmochim. Acta, Vol. 37, 831 - 839 (1973).

094.083 Major element compositions of Luna 20 glass particles. B. P. Glass.
Geochim. Cosmochim. Acta, Vol. 37, 841 - 846 (1973).

094.084 Chemistry and thermal history of metal particles in Luna 20 soils. J. I. Goldstein, P. J. Blau.
Geochim. Cosmochim. Acta, Vol. 37, 847 - 855 (1973).

094.085 Luna 20: mineral chemistry of spinel, pleonaste, chromite, ulvöspinel, ilmenite and rutile.
S. E. Haggerty.
Geochim. Cosmochim. Acta, Vol. 37, 857 - 867 (1973).

094.086 Rare earths, other trace elements and iron in Luna 20 samples.
P. A. Helmke, D. P. Blanchard, J. W. Jacobs, L. A. Haskin.
Geochim. Cosmochim. Acta, Vol. 37, 869 - 874 (1973).

094.087 Inert gases in a terra sample: measurements in six grain-size fractions and two single particles from Luna 20. D. Heymann, S. Lakatos, J. R. Walton.
Geochim. Cosmochim. Acta, Vol. 37, 875 - 885 (1973).

094.088 The age and petrology of two Luna 20 fragments and inferences for widespread lunar metamorphism.
F. A. Podosek, J. C. Huneke, A. J. Gancarz, G. J. Wasserburg.
Geochim. Cosmochim. Acta, Vol. 37, 887 - 904 = Contr. Div. Geol. Planet. Sci., California Inst. Technology, *Pasadena*, No. 2315 (1973).

094.089 Oxygen and bulk element abundances in Luna 20 fines.
M. Janghorbani, D. E. Gillum, W. D. Ehmann.
Geochim. Cosmochim. Acta, Vol. 37, 905 - 908 (1973).

094.090 Chemical composition of Luna 20 soil and rock fragments. D. Y. Jérome, J.-C. Philippot.
Geochim. Cosmochim. Acta, Vol. 37, 909 - 914 (1973).

094.091 The mineralogy and petrology of the Luna 20 soil sample. S. J. Kridelbaugh, D. F. Weill.
Geochim. Cosmochim. Acta, Vol. 37, 915 - 926 (1973).

094.092 Chemical composition of Luna 20 rocks and soil and Apollo 16 soils. J. C. Laul, R. A. Schmitt.
Geochim. Cosmochim. Acta, Vol. 37, 927 - 942 (1973).

094.093 Luna 20: mineralogy and petrology of fragments less than 125 μm size. H. O. A. Meyer.
Geochim. Cosmochim. Acta, Vol. 37, 943 - 952 (1973).

094.094 Luna 20 soil: abundance of 17 trace elements.
J. W. Morgan, U. Krähenbühl, R. Ganapathy, E. Anders.
Geochim. Cosmochim. Acta, Vol. 37, 953 - 961 (1973).

094.095 A lunar differentiation model in light of new chemical data on Luna 20 and Apollo 16 soils.
D. F. Nava, J. A. Philpotts.
Geochim. Cosmochim. Acta, Vol. 37, 963 - 973 (1973).

094.096 Radiation damage in Luna 20 soil.
P. P. Phakey, P. B. Price.
Geochim. Cosmochim. Acta, Vol. 37, 975 - 977 (1973).

094.097 Mineralogy, petrology and chemistry of lithic frag-

ments from Luna 20 fines: origin of the cumulate ANT suite and its relationship to high-alumina and mare basalts. M. Prinz, E. Dowty, K. Keil, T. E. Bunch.
Geochim. Cosmochim. Acta, Vol. 37, 979 - 1006 (1973).

094.098 The halogens in Luna 16 and Luna 20 soils.
G. W. Reed, Jr., S. Jovanovic.
Geochim. Cosmochim. Acta, Vol. 37, 1007 - 1009 (1973).

094.099 Luna 20 soil: abundance and composition of phases in the 45-125 micron fraction.
A. M. Reid, J. L. Warner, W. I. Ridley, R. W. Brown.
Geochim. Cosmochim. Acta, Vol. 37, 1011 - 1030 (1973).

094.100 Petrology of some lithic fragments from Luna 20.
E. Roedder, P. W. Weiblen.
Geochim. Cosmochim. Acta, Vol. 37, 1031 - 1052 (1973).

094.101 Comparison of the magnetic properties of glass from Luna 20 with similar properties of glass from the Apollo missions.
F. E. Senftle, A. N. Thorpe, C. C. Alexander, C. L. Briggs.
Geochim. Cosmochim. Acta, Vol. 37, 1053 - 1062 (1973).

094.102 Carbon chemistry of Luna 16 and Luna 20 samples.
B. R. Simoneit, P. C. Wszolek, P. Christiansen, R. F. Jackson, A. L. Burlingame.
Geochim. Cosmochim. Acta, Vol. 37, 1063 - 1074 (1973).

094.103 Compositional and X-ray data for Luna 20 feldspar.
I. M. Steele, J. V. Smith.
Geochim. Cosmochim. Acta, Vol. 37, 1075 - 1077 (1973).

094.104 U-Th-Pb measurements of Luna 20 soil.
M. Tatsumoto.
Geochim. Cosmochim. Acta, Vol. 37, 1079 - 1086 (1973).

094.105 The Luna 20 lithic fragments, and the composition and origin of the lunar highlands.
G. J. Taylor, M. J. Drake, J. A. Wood, U. B. Marvin.
Geochim. Cosmochim. Acta, Vol. 37, 1087 - 1106 (1973).

094.106 Oxygen and silicon isotope ratios of the Luna 20 soil. H. P. Taylor, Jr., S. Epstein.
Geochim. Cosmochim. Acta, Vol. 37, 1107 - 1109 = Contr. Div. Geol. Planet. Sci., California Inst. Technology, *Pasadena*, No. 2268 (1973).

094.107 Ergebnisse der Apollo- und Luna-Mondflüge.
O. Müller.
SuW, Vol. 12, 4 - 9 (1973).

094.108 Apollo 17 und Mariner 9 – zwei erfolgreich beendete Raumfahrtunternehmen der USA. H. W. Köhler.
SuW, Vol. 12, 67 - 69 (1973).

094.109 Survival of micro-organisms on the moon.
P. M. Molton.
Spaceflight, Vol. 15, 51 (1973).

094.110 Lunar permafrost: Dielectric identification.
R. Alvarez.
Science, Vol. 179, 1122 - 1123 (1973).
A simulator of lunar permafrost at 100° K exhibits a dielectric relaxation centered at approximately 300 Hz. If permafrost exists in the moon between 100° and 213° K it should present a relaxation peak at approximately 300 Hz. For temperatures up to 263° K it may go up to 20 kHz.

094.111 Optical properties of Apollo 12 moon samples.

B. O'Leary, F. Briggs.
Journ. Geophys. Res., Vol. 78, 792 - 797 (1973).

We present the photometric phase function, color, normal albedo, polarimetric phase function, and spectrophotometry of the Apollo 12 soil. With a few minor exceptions, the optical properties of the Apollo 12 soil are very similar to those of the Apollo 11 soil and of lunar mare surfaces.

094.112   A photometric investigation of the packing state of Apollo 11 lunar regolith samples.   L. Wilson.
Planet. Space Sci., Vol. 21, 113 - 118 (1973).

The angular light scattering properties of an Apollo 11 lunar regolith 'fines' sample have been determined experimentally for both flat and undulating sample surface preparations. The light scattering curves, whose shapes are known to be a function of the porosity and slope distribution of the measured surface, have been compared with corresponding earth-based lunar measurements. The comparison method involves the numerical fitting of theoretical photometric functions to both the astronomical and laboratory data.

094.113   Zinc, lead, chlorine and FeOOH-bearing assemblages in the Apollo 16 sample 66095: Origin by impact of a comet or a carbonaceous chondrite?
A. El Goresy, P. Ramdohr, M. Pavićević, O. Medenbach, O. Müller, W. Gentner.
Earth Planet. Sci. Letters, Vol. 18, 411 - 419 (1973).

Sample 66095,89 collected from station 6 from the lunar highlands in the Descartes site shows evidence of mild to severe shock. These shock features are accompanied by an unusual enrichment in the volatile elements Cl, Zn and Pb and by the presence of FeOOH. The formation of this unique assemblage and the introduction of the material rich in volatile elements is very probably genetically connected with an impact of a carbonaceous chondrite or a comet.

094.114   Particle track record in Apollo 15 deep core from 54 to 80 cm depths.   R. L. Fleischer, H. R. Hart, Jr.
Earth Planet. Sci. Letters, Vol. 18, 420 - 426 (1973).

Particle track measurements have been made in nearly 500 individual grains from 13 levels in the 54—80 cm depth range of the Apollo 15 deep core. They reveal a wide range of track densities at all depths and some systematic variations within layers, indicating that both predepositional mixing and subsequent layering are present and that separate sub-layers exist within larger regions where no sub-layers are visible.

094.115   On the problem concerning stars selection for observations on the lunar surface by the equal-altitude method.   A. N. Sanovich.
Astron. Zhurn. Akad. Nauk SSSR, Vol. 50, 418 - 421 (1973). In Russian. English translation in Soviet Astron. AJ, Vol. 17, No. 2.

094.116   Translational-precessional motion of the moon under the action of gravitation of the earth and sun.
G. F. Osipov.
Astron. Zhurn. Akad. Nauk SSSR, Vol. 50, 435 - 441 (1973). In Russian. English translation in Soviet Astron. AJ, Vol. 17, No. 2.

The periodical solution of the restricted three-body problem in the generalized Hill's variant in the first approximation by Lyapunov's method is obtained.

094.117   Erste Ergebnisse der Apollo-16-Mondproben.
Umschau, 73. Jahrgang, p. 238 - 239 (1973).

The chemical characteristics of the Apollo 16 rocks are simple and straighforward and show high abundances of Al and Ca. Some rock samples with low contents of Al are similar to those of the KREEP basalts found at the Apollo 12, 14, 15 sites.

094.118   The Corralitos Lunar Transient Phenomena (LTP) surveillance program (1966–1972).
J. R. Dunlap, J. A. Hynek.
Bull. American Astron. Soc., Vol. 5, 37 (1973). – Abstr. AAS.

094.119   Parameters characterizing the level ellipsoid of the moon and the selenoidal undulations.
V. V. Buzuk.
Geod. i kartografiya, 1972, No. 9, p. 13 - 19. In Russian. Abstr. in Referativ. Zhurn. 51. Astron., 3.51.375 (1973).

094.120   Radiological interpretation of anomalous values of the ages of terrestrial and lunar rocks.
É. K. Gerling, I. M. Morozova, Yu. V. Nikitin, G. V. Ovchinnikova, V. D. Sprintsson.
Ocherki sovrem. geokhimii i analit. khimii. Moskva, Nauka, 1972, p. 429 - 440. In Russian. – Abstr. in Referativ. Zhurn. 51. Astron., 3.51.376 (1973).

094.121   Rare earth abundances in lunar soil and rocks from the Oceanus Procellarum.
P. W. Gast, N. J. Hubbard.
Ocherki sovrem. geokhimii i analit. khimii. Moskva, Nauka, 1972, p. 43 - 48. In Russian. – Abstr. in Referativ. Zhurn. 51. Astron., 3.51.377; 62. Issled. kosmich. prostranstva, 3.62.173 (1973).

094.122   Preliminary data on lunar soil returned by Luna 20.
A. P. Vinogradov.
Vestn. AN SSSR, 1972, No. 10, p. 26 - 40. In Russian. Abstr. in Referativ. Zhurn. 62. Issled. kosmich. prostranstva, 3.62.136 (1973).

094.123   Apollo 17 age determinations.
G. Turner, P. H. Cadogan, C. J. Yonge.
Nature, Vol. 242, 513 - 515 (1973).

The authors determined cosmic-ray exposure ages for the samples 75055 and 76055 on the basis of the ratio of cosmogenic ^{38}Ar to ^{37}Ar produced artificially from Ca. The exposure ages are: 98 ± 5 m.y. (75055) and 136 ± 7 m.y. (76055).

094.124   Orange soil from the moon.
G. M. Brown, J. G. Holland, A. Peckett.
Nature, Vol. 242, 515 - 516 (1973).

The discovery of an orange-coloured soil on the moon by the Apollo 17 astronauts last December led to much excitement and speculation. The authors show how it was possible to reach a precise conclusion from chemical analysis, and to clarify the situation regarding the orange soil. They conclude that the material was not produced by volcanic processes.

094.125   Lunar tides and magnetism.   R. Smoluchowski.
Nature, Vol. 242, 516 - 517 (1973).

An admittedly very simplified model leads to the conclusion that the outer layer of the moon could have been magnetized by local, more or less randomly oriented, magnetic fields produced by tidal currents, that there would be no overall poloidal field and that the magnetization would be observable primarily near the lunar equator where the tide producing force has its maximum.

094.126   Some scientific results of the Apollo 16 flight.
D. Yu. Gol'dovskij.
Zemlya i Vselennaya, 1973, No. 1, p. 37 - 39. In Russian.

094.127   Magnetic phases in lunar fines: metallic Fe or ferric oxides?
F.-D. Tsay, S. L. Manatt, S. I. Chan.
Geochim. Cosmochim. Acta, Vol. 37, 1201 - 1211 (1973).

The authors show that the electron spin resonance (ESR) results they have obtained from lineshape analyses, intensity measurements, and temperature dependence studies all indicate unambiguously that the intense ESR signals observed for the lunar fines (Apollo 11, 12, 14 and 15) originate from metallic Fe particles and not from hematite, magnetite or any other Fe^{3+} oxides.

094.128   Natural exoelectronic emission of anorthosite rocks returned by the Luna 20 probe.
R. I. Mints, I. I. Milman, V. I. Kriuk, L. S. Tarasov.
Dokl. Akad. Nauk SSSR, Ser. Mat. Fiz., Vol. 209, 586 - 588 (1973). In Russian.

094.129   Lunar science: Analyzing the Apollo legacy.
A. L. Hammond.
Science, Vol. 179, 1313 - 1315 (1973).– Research news.

094.130   Lunar cinder cones.
T. R. McGetchin, J. W. Head.
Science, Vol. 180, 68 - 71 (1973).

Data on terrestrial eruptions of pyroclastic material and ballistic considerations suggest that in the lunar environment (vacuum and reduced gravity) low-rimmed pyroclastic rings are formed rather than the high-rimmed cinder cones so abundant on the earth. Dark blanketing deposits in the Taurus-Littrow region are interpreted as being at least partly composed of lunar counterparts of terrestrial cinder cones.

094.131   On systematic errors of the corrections for limb profile irregularities in the charts of the marginal zone of the moon.   D. P. Duma, L. N. Kisjun.
Astrometriya i Astrofizika, *Kiev*, vyp. (No.) 16, (see 003.006), p. 25 - 30 (1972). In Russian.

The corrections to the inclination of the moon's orbit with respect to the ecliptic and to the eccentricity have been analysed for systematic errors in the charts of the marginal zone. A part of libration effect is found to be due to the charts used in the reduction of observations, but not to the figure of the moon itself. Corrections to the moon's semi-diameter depending upon the optical libration in longitude and latitude have been revealed for the north and south as well as for the east and west limbs of the moon. The corrections for profile irregularities taken from the charts of Hayn, Weimer, Nefedjev and Watts are not free from some systematic errors.

094.132   Deformation of the selenodetic reference network due to errors of the rotation parameters of the moon.   V. S. Kisliuk.
Astrometriya i Astrofizika, *Kiev*, vyp. (No.) 16, (see 003.006), p. 30 - 40 (1972). In Russian.

When compiling a catalogue of the coordinates of details on the moon's surface, one should adopt certain values of the inclination of the moon's equator to the ecliptic and the function of the principal moments of inertia f. The paper deals with the effect of errors of these values on the accuracy of the coordinates of lunar details.

094.133   Systematic differences of selenodetic catalogues depending on the method used for reduction of measurements.   V. S. Kisliuk.
Astrometriya i Astrofizika, *Kiev*, vyp. (No.) 16, (see 003.006), p. 40 - 46 (1972). In Russian.

To compare the ACIC and DOD-66 catalogues the author made use of 27 common points whose coordinates had been derived in both catalogues from the same measurements. Angles of rotation, displacements of the origins relative to one another and deformation of the systems of these catalogues have been determined. The systematic differences of the selenodetic networks obtained by the methods of the ACIC and the AMS have proved to be substantial.

094.134   Approximation of the geometrical figure of the moon by means of spherical harmonics.
I. V. Gavrilov, G. T. Yanovitskaya.
Astrometriya i Astrofizika, *Kiev*, vyp. (No.) 16, (see 003.006), p. 46 - 52 (1972). In Russian.

To represent the figure of the moon by means of spherical harmonics the lunar surface was divided into 398 equal areas. Absolute heights of the visible lunar surface were picked up from the hypsometrical chart of the moon prepared at the Kiev Observatory. Absolute heights of the far side and the marginal zone of the moon were estimated using the relations between heights of the lunar and equipotential surfaces derived from data of Luna 10 and Orbiters 1 - 4. Values of 36 harmonic coefficients are calculated in five variants.

094.135   Ortho and para-armalcolite samples in Apollo 17.
S. E. Haggerty.
Nature, Phys. Sci., Vol. 242, 123 - 125 (1973).

Optical and selected electron microprobe analyses of polished thin sections, of one coarse grained ilmenite basalt (70035) and of 147 predominantly basalt particles in four soil samples (74242, 74243, 75082 and 75083) that range from coarse grained to vitrophyric have been carried out on the material returned. Armalcolite in these samples differs in several respects.

094.136   The moon as a proposed radiometric standard for microwave and infrared observations of extended sources.   J. L. Linsky.
Astrophys. Journ., Suppl. Ser., No. 216, Vol. 25, 163 - 203 (1973).

Measured values of the average midnight and morning terminator infrared brightness temperatures of the central portion of the lunar disk can quite accurately determine the mean surface temperature despite likely horizontal and vertical inhomogeneities of the thermal properties of the lunar soil. The accuracy with which the moon can be used as an absolute radiometric standard for extended sources is estimated based on the likely range of lunar thermal and electromagnetic properties.

094.137   Meteoritisches Material auf dem Mond.
J. W. Morgan.
Umschau, 73. Jahrgang, p. 277 - 278 (1973).

Meteoritic material on the lunar surface results both from large crater-forming impacts, and from a steady influx of micrometeorites, probably of cometary origin. Precise analysis of lunar material allows important deductions to be made concerning the influx rate and composition of meteorites, and the early history of the earth–moon system.

094.138   Sinus Iridum.   S. V. Landau.
Zemlya i Vselennaya, 1973, No. 2, p. 45. In Russian.

094.139   The extralunar component in the Apollo-16 regolith.   P. A. Baedecker, C.-L. Chou, L. L. Sundberg, R. Bild, E. Grudewicz, J. T. Wasson.
Meteoritics, Vol. 8, 13 (1973). – Abstract.

094.140   Noble gas concentrations in Apollo 15 and 16 deep drill cores.   D. D. Bogard, L. E. Nyquist.
Meteoritics, Vol. 8, 16 (1973). – Abstract.

094.141   Information from lunar microcraters pertaining to interplanetary grains.
D. E. Brownlee, P. W. Hodge, F. Hörz.
Meteoritics, Vol. 8, 18 (1973). – Abstract.

094.142   Composition and origin of glasses and chondrules in Apollo 15 rake samples from Spur Crater.

T. E. Bunch, M. Prinz, K. Keil, E. Dowty.
Meteoritics, Vol. 8, 21 - 22 (1973). – Abstract.

094.143 ^{236}U and the lunar neutron flux.
     H. Diamond, P. R. Fields, D. N. Metta, D. J. Rokop.
Meteoritics, Vol. 8, 27 (1973). – Abstract.

094.144 Oxygen abundances and the oxygen-silicon relation-
     ship in lunar samples and meteorites.
W. D. Ehmann, D. E. Gillum, M. Janghorbani.
Meteoritics, Vol. 8, 30 (1973). – Abstract.

094.145 Phase B, zirkelite, zirconolite: are they the same
     mineral? A. El Goresy.
Meteoritics, Vol. 8, 31 (1973). – Abstract.

094.146 Apollo 15 opaque minerals: Geochemistry, mineral-
     ogy, and subsolidus reduction.
A. El Goresy, L. A. Taylor, P. Ramdohr.
Meteoritics, Vol. 8, 32 (1973). – Abstract.

094.147 Abrasion and catastrophic rupture of lunar rocks:
     Implications to the flux of micrometeoroids and
energetic particles at 1 AU.
D. E. Gault, F. Hörz, J. B. Hartung.
Meteoritics, Vol. 8, 37 (1973). – Abstract.

094.148 Inorganic gases from lunar samples.
     E. K. Gibson, Jr., G. W. Moore.
Meteoritics, Vol. 8, 38 - 39 (1973). – Abstract.

094.149 Metallic particles of high cobalt content in Apollo
     15 soil samples. J. I. Goldstein, H. J. Axon.
Meteoritics, Vol. 8, 40 (1973). – Abstract.

094.150 Apollo 15 rake samples: Breccias.
     L. Grossman, I. M. Steele, J. V. Smith.
Meteoritics, Vol. 8, 42 - 43 (1973). – Abstract.

094.151 Initial lunar temperature profiles and the accumula-
     tion time of the proto-moon. M. Hallam.
Meteoritics, Vol. 8, 43 (1973). – Abstract.

094.152 Extinct lunar radioactivities: ^{244}Pu and ^{129}I xenon
     in Apollo 14 breccias. C. M. Hohenberg.
Meteoritics, Vol. 8, 44 - 45 (1973). – Abstract.

094.153 Ion microprobe analysis of plagioclase in lunar anor-
     thosite fragments. C. Meyer, Jr., D. H. Anderson.
Meteoritics, Vol. 8, 56 - 57 (1973). – Abstract.

094.154 Ancient meteoritic components in the lunar rego-
     lith.
J. W. Morgan, R. Ganapathy, U. Krähenbühl, E. Anders.
Meteoritics, Vol. 8, 58 (1973). – Abstract.

094.155 Simulated lunar spray structures. G. Mueller.
Meteoritics, Vol. 8, 59 (1973). – Abstract.

094.156 Lunar fluorine and fluorapatite.
     G. W. Reed, Jr., S. Jovanovic.
Meteoritics, Vol. 8, 65 - 66 (1973). – Abstract.

094.157 Trace element abundances in Apollo 15, Apollo 16
     and Luna 20 soils.
C. C. Schnetzler, J. A. Philpotts.
Meteoritics, Vol. 8, 68 (1973). – Abstract.

094.158 Apollo 15 rake samples: Basalts.
     J. V. Smith, I. M. Steele, L. Grossman.
Meteoritics, Vol. 8, 71 (1973). – Abstract.

094.159 Ultrabasic lunar samples. I. M. Steele, J. V. Smith.
Meteoritics, Vol. 8, 72 (1973). – Abstract.

094.160 Igneous clasts in Apollo 15 breccia rake samples.
     I. M. Steele, L. Grossman, J. V. Smith.
Meteoritics, Vol. 8, 73 (1973). – Abstract.

094.161 Track studies in glasses and minerals from Apollo-15
     soils.
D. Storzer, D. Heymann, P. Horn, T. Kirsten, G. Poupeau.
Meteoritics, Vol. 8, 74 - 75 (1973). – Abstract.

094.162 Luna 20 lithic fragments.
     G. J. Taylor, M. J. Drake.
Meteoritics, Vol. 8, 75 - 76 (1973). – Abstract.

094.163 The significance of Zr partitioning in Apollo 15 il-
     menite and ulvöspinel and the subsolidus reduction
of ulvöspinel. L. A. Taylor, R. H. McCallister, R. J. Williams.
Meteoritics, Vol. 8, 76 - 77 (1973). – Abstract.

094.164 Multielement analyses of lunar samples and the de-
     gree of oxydation of lunar and meteoritic matter.
H. Wänke, H. Palme, B. Spettel, F. Teschke.
Meteoritics, Vol. 8, 78 - 79 (1973). – Abstract.

094.165 Geographic, geophysical, and chemical asymmetry
     of the moon: Why? J. A. Wood.
Meteoritics, Vol. 8, 82 - 83 (1973). – Abstract.

094.166 Analyse des premiers échos laser obtenus sur le ré-
     flecteur de Luna 21. V. K. Abalakin, O. Calame,
Y. L. Kokurin, J. D. Mulholland, A. Orszag, E. C. Silverberg.
Comptes Rendus Acad. Sci. Paris, Sér. B, Vol. 276, 673 - 676
(1973).

094.167 Comments on lunar origin. E. J. Öpik.
     Irish Astron. Journ., Vol. 10, 190 - 238 (1972).
Review paper concerning the following topics: 1. Introduction
and general outline; 2. Factors of orbital evolution; 3. The
auxiliary "Model Zero" (MZ); 4. Orbital evolution along the
outgoing branch; 5. Capture models; 6.Break-up; 7. Capture
probability; 8. Capture with break-up; 9. Tidal evolution of
circular fragmented rings; 10. Tidal evolution of elliptical fragmented rings; 11. Precession and collisions in a ring of fragments; 12. Concluding note.

094.168 Internal constitution and evolution of the moon.
     S. C. Solomon, M. N. Toksöz.
Phys. Earth Planet. Interiors, Vol. 7, 15 - 38 (1973).
     We begin with a critical summary of the available con-
straints on lunar constitution and evolution. We next reconsider in detail possible models for the thermal evolution in
the moon. Using several estimates of present-day temperature
and the confirmed thickness of a low-density crust, we examine density models for the lunar interior that match the
moon's mean density and moment of inertia. Finally, we briefly consider a few of the consequences of an iron-rich core in
the moon.

094.169 Petrology of the 2–4 mm sized soil fragments from
     Apollo 15.
K. L. Cameron, J. W. Delano, A. E. Bence, J. J. Papike.
The Apollo 15 lunar samples, (see 012.008), p. 1 - 4 (1972).

094.170 The source area of Apollo 15 "green glasses".
     A. Carusi, G. Cavarretta, F. Cinotti, G. Civitelli, A.
Coradini, M. Fulchignoni, R. Funiciello, A. Taddeucci, R.
Trigila.
The Apollo 15 lunar samples, (see 012.008), p. 5 - 9 (1972).

094.171 **Optical evidence for average pyroxene composition of Apollo 15 samples.** J. B. Adams, T. B. McCord.
The Apollo 15 lunar samples, (see 012.008), p. 10 - 13 (1972).

094.172 **Partitioning of Ti and Al between pyroxenes, garnets, oxides, and liquid.** J. Akella, F. R. Boyd.
The Apollo 15 lunar samples, (see 012.008), p. 14 - 19 (1972).

094.173 **Petrology of Apollo 15 sample 15486.**
A. L. Albee, A. A. Chodos, A. J. Gancarz.
The Apollo 15 lunar samples, (see 012.008), p. 20 - 25 (1972).

094.174 **Zoned olivine crystals in an Apollo 15 lunar rock.**
P. M. Bell, H. K. Mao.
The Apollo 15 lunar samples, (see 012.008), p. 26 - 28 (1972).

094.175 **Crystallography of lunar feldspars and pyroxenes from 15076,55.**
B. Berking, H. Jagodzinski, M. Korekawa, R. Schmid.
The Apollo 15 lunar samples, (see 012.008), p. 29 - 33 (1972).

094.176 **Apollo 15 glasses of impact origin.**
J. B. Best, J. A. Minkin.
The Apollo 15 lunar samples, (see 012.008), p. 34 - 39 (1972).

094.177 **Petrology, mineralogy and classification of Apollo 15 mare basalts.**
G. M. Brown, C. H. Emeleus, J. G. Holland, A. Peckett, R. Phillips.
The Apollo 15 lunar samples, (see 012.008), p. 40 - 44 (1972).

094.178 **Size frequency distributions and petrographic observations of Apollo 15 samples.**
J. C. Butler, E. A. King, Jr., M. F. Carman.
The Apollo 15 lunar samples, (see 012.008), p. 45 - 47 (1972).

094.179 **Chemical and petrographic characteristics of the regolith at the Apollo 15 landing site.**
M. H. Carr, C. E. Meyer.
The Apollo 15 lunar samples, (see 012.008), p. 48 - 50 (1972).

094.180 **Morphology and chemistry of glass surface of breccia 15015,36.** J. L. Carter.
The Apollo 15 lunar samples, (see 012.008), p. 51 - 53 (1972).

094.181 **Relationship of exposure age to size distribution and particle types in the Apollo 15 drill core.**
U. S. Clanton, D. S. McKay, R. M. Taylor, G. H. Heiken.
The Apollo 15 lunar samples, (see 012.008), p. 54 - 56 (1972).

094.182 **Viscous flow of lunar compositions.**
M. Cukierman, D. R. Uhlmann.
The Apollo 15 lunar samples, (see 012.008), p. 57 - 59 (1972).

094.183 **Petrologic examination of breccia 15465 and its implications as to the nature of the Apennine Front.**
J. W. Delano.
The Apollo 15 lunar samples, (see 012.008), p. 60 - 61 (1972).

094.184 **Anorthosite in the Apollo 15 rake sample from Spur Crater.** E. Dowty, K. Keil, M. Prinz.
The Apollo 15 lunar samples, (see 012.008), p. 62 - 66 (1972).

094.185 **Mineralogical and chemical studies of breccia 15086.** J. C. Drake, C. Klein, Jr.
The Apollo 15 lunar samples, (see 012.008), p. 67 - 69 (1972).

094.186 **Mineralogy and petrology of two Apollo 15 mare basalts.** P. Gay, I. D. Muir, G. G. Price.
The Apollo 15 lunar samples, (see 012.008), p. 70 - 72 (1972).

094.187 **Major element composition of Apollo 15 glasses.**
B. P. Glass.
The Apollo 15 lunar samples, (see 012.008), p. 73 - 77 (1972).

094.188 **Metallic particles from 3 Apollo 15 soils.**
J. I. Goldstein, H. J. Axon.
The Apollo 15 lunar samples, (see 012.008), p. 78 - 81 (1972).

094.189 **Significance of Apollo 15 mare basalts and 'primitive' green glasses in lunar petrogenesis.**
D. H. Green, A. E. Ringwood.
The Apollo 15 lunar samples, (see 012.008), p. 82 - 84 (1972).

094.190 **An enstatite chondrite from Hadley Rille.**
S. E. Haggerty.
The Apollo 15 lunar samples, (see 012.008), p. 85 - 87 (1972).

094.191 **The mineral chemistry of some decomposition and reaction assemblages associated with Cr-Zr, Ca-Zr and Fe-Mg-Zr titanates.** S. E. Haggerty.
The Apollo 15 lunar samples, (see 012.008), p. 88 - 91 (1972).

094.192 **Chemical characteristics of spinels in some Apollo 15 basalts.** S. E. Haggerty.
The Apollo 15 lunar samples, (see 012.008), p. 92 - 97 (1972).

094.193 **High voltage (HVEM) electron petrographic study of Apollo 15 rocks.**
A. H. Heuer, G. L. Nord, Jr., S. V. Radcliffe, R. M. Fisher, J. S. Lally, J. M. Christie, D. T. Griggs.
The Apollo 15 lunar samples, (see 012.008), p. 98 - 102 (1972).

094.194 **Phase equilibria and origin of Apollo 15 basalts etc.**
D. J. Humphries, G. M. Biggar, M. J. O'Hara.
The Apollo 15 lunar samples, (see 012.008), p. 103 - 107 (1972).

094.195 **Mineralogical notes on Apollo 15 samples.**
J. Jedwab.
The Apollo 15 lunar samples, (see 012.008), p. 108 - 109 (1972).

094.196 **Petrology and chemistry of some Apollo 15 crystalline rocks.**
V. C. Juan, J. C. Chen, C. K. Huang, P. Y. Chen, C. M. Wang Lee.
The Apollo 15 lunar samples, (see 012.008), p. 110 - 115 (1972).

094.197 **Petrology and chemistry of some Apollo 15 regoliths.**
V. C. Juan, J. C. Chen, C. K. Huang, P. Y. Chen, C. M. Wang Lee.
The Apollo 15 lunar samples, (see 012.008), p. 116 - 122 (1972).

094.198 **Glass compositions in breccias 15028 and 15059.**
S. J. Kridelbaugh, R. A. F. Grieve, D. F. Weill.
The Apollo 15 lunar samples, (see 012.008), p. 123 - 127 (1972).

094.199 **Petrology of some Apollo 15 mare basalts.**
I. Kushiro.
The Apollo 15 lunar samples, (see 012.008), p. 128 - 130 (1972).

094.200 **Petrology of mare/rille basalts 15555 and 15065.**
J. Longhi, D. Walker, E. N. Stolper, T. L. Grove, J. F. Hays.
The Apollo 15 lunar samples, (see 012.008), p. 131 - 134 (1972).

094.201 **Mineralogy and petrology of lunar samples 15264,19, 15274,12, and 15314,59.** B. Mason.
The Apollo 15 lunar samples, (see 012.008), p. 135 - 136 (1972).

094.202 **Mineralogy and petrology of polymict breccia 15498.** B. Mason.
The Apollo 15 lunar samples, (see 012.008), p. 137 - 139 (1972).

094.203 **Crystal chemistry of zoned clinopyroxenes from lunar rock 15058.**
A. Morawski, D. J. Vaughan, R. G. Burns.
The Apollo 15 lunar samples, (see 012.008), p. 140 - 143 (1972).

094.204 **Subsolidus relations of pyroxenes from Apollo 15 basalts.** J. J. Papike, A. E. Bence, M. A. Ward.
The Apollo 15 lunar samples, (see 012.008), p. 144 - 148 (1972).

094.205 **Classification and distribution of rock types at Spur Crater.**
W. C. Phinney, J. L. Warner, C. H. Simonds, G. E. Lofgren.
The Apollo 15 lunar samples, (see 012.008), p. 149 - 153 (1972).

094.206 **Olivine-rich, true spinel-bearing anorthosites from Apollo 15 & Luna 20 soils – possible fragments of the earliest formed lunar crust.** J. B. Reid, Jr.
The Apollo 15 lunar samples, (see 012.008), p. 154 - 157 (1972).

094.207 **Mineralogy and petrology of Apollo 15 rake samples: I. Basalts.**
I. M. Steele, J. V. Smith, L. Grossman.
The Apollo 15 lunar samples, (see 012.008), p. 158 - 160 (1972).

094.208 **Mineralogy and petrology of Apollo 15 rake samples. II. Breccias.** I. M. Steele, J. V. Smith, L. Grossman.
The Apollo 15 lunar samples, (see 012.008), p. 161 - 164 (1972).

094.209 **Anorthositic lithic fragments in Apollo 15 soils and fractional crystallization in the early lunar crust.**
G. J. Taylor.
The Apollo 15 lunar samples, (see 012.008), p. 165 - 168 (1972).

094.210 **Opaque mineralogy of Apollo 15 rocks: Experimental investigations of elemental partitionings and subsolidus reduction.** L. A. Taylor, R. H. McCallister.
The Apollo 15 lunar samples, (see 012.008), p. 169 - 173 (1972).

094.211 **Apollo 15 regolith and breccias.**
W. von Engelhardt, J. Arndt, H. Schneider.
The Apollo 15 lunar samples, (see 012.008), p. 174 - 178 (1972).

094.212 **Apollo 15 glasses and the distribution of non-mare crustal rock types.**
J. Warner, W. I. Ridley, A. M. Reid, R. W. Brown.
The Apollo 15 lunar samples, (see 012.008), p. 179 - 181 (1972).

094.213 **Ferromagnetic and paramagnetic resonance of magnetic phases and Fe^{3+} in Apollo 15 samples: a comparison.** R. A. Weeks.

The Apollo 15 lunar samples, (see 012.008), p. 182 - 186 (1972).

094.214 **Petrology of pyroxene vitrophyre 15597.**
P. W. Weigand.
The Apollo 15 lunar samples, (see 012.008), p. 187 - 188 (1972).

094.215 **On bytownite 15085,36.**
E. Wenk, A. Glauser, H. Schwander.
The Apollo 15 lunar samples, (see 012.008), p. 189 - 190 (1972).

094.216 **Secondary ion analysis of pyroxenes from two porphyritic lunar basalts.** A. E. Bence, B. Autier.
The Apollo 15 lunar samples, (see 012.008), p. 191 - 194 (1972).

094.217 **Elemental composition of Apollo 15 samples.**
A. O. Brunfelt, K. S. Heier, B. Nilssen, E. Steinnes, B. Sundvoll.
The Apollo 15 lunar samples, (see 012.008), p. 195 - 197 (1972).

094.218 **Elemental analyses of lunar soil samples from Apollo 15 mission.**
M. K. Carron, C. S. Annell, R. P. Christian, F. Cuttitta, E. J. Dwornik, D. T. Ligon, Jr., H. J. Rose, Jr.
The Apollo 15 lunar samples, (see 012.008), p. 198 - 201 (1972).

094.219 **Geochemistry of green glass spheres from Apollo 15 samples.** G. Cavarretta, R. Funiciello, H. Giles, G. D. Nicholls, A. Taddeucci, J. Zussman.
The Apollo 15 lunar samples, (see 012.008), p. 202 - 205 (1972).

094.220 **Chemical composition of some Apollo 15 igneous rocks.** R. P. Christian, C. S. Annell, M. K. Carron, F. Cuttitta, E. J. Dwornik, D. T. Ligon, Jr., H. J. Rose, Jr.
The Apollo 15 lunar samples, (see 012.008), p. 206 - 209 (1972).

094.221 **The distribution of K, Ti, Zr, U and Hf in Apollo 14 and 15 materials.**
S. E. Church, B. M. Bansal, H. Wiesmann.
The Apollo 15 lunar samples, (see 012.008), p. 210 - 213 (1972).

094.222 **Elemental abundance studies of Apollo 15 and some Fra Mauro formation lunar samples.**
W. D. Ehmann, M. Janghorbani, D. E. Gillum.
The Apollo 15 lunar samples, (see 012.008), p. 214 - 216 (1972).

094.223 **Rare earths and other trace elements in Apollo 15 samples.** P. A. Helmke, L. A. Haskin.
The Apollo 15 lunar samples, (see 012.008), p. 217 - 220 (1972).

094.224 **Bulk and REE abundances in anorthosites and noritic fragments.** J. C. Laul, H. Wakita, R. A. Schmitt.
The Apollo 15 lunar samples, (see 012.008), p. 221 - 224 (1972).

094.225 **Bulk and REE abundances in three Apollo 15 igneous rocks and six basaltic rake samples.**
J. C. Laul, R. A. Schmitt.
The Apollo 15 lunar samples, (see 012.008), p. 225 - 228 (1972).

094.226 **Elemental abundances of Apollo 15 four soils, a clod and five breccia rocks and two soils of Apollo 16.** J. C. Laul, D. L. Showalter, R. A. Schmitt.
The Apollo 15 lunar samples, (see 012.008), p. 229 - 232 (1972).

094.227 **Chemical composition of some Apollo 15 lunar samples.** J. A. Maxwell, J.-L. Bouvier, H. B. Wiik.
The Apollo 15 lunar samples, (see 012.008), p. 233 - 238 (1972).

094.228 **Trace elements in Apollo 15 samples: Implications for meteorite influx and volatile depletion on the moon.**
J. W. Morgan, U. Krähenbühl, R. Ganapathy, E. Anders.
The Apollo 15 lunar samples, (see 012.008), p. 239 (1972). Abstract.

094.229 **Alkali and alkaline earth elements, La and U in Apollo 14 and Apollo 15 samples.** O. Müller.
The Apollo 15 lunar samples, (see 012.008), p. 240 - 243 (1972).

094.230 **Abundances of the primordial radioelements K, Th, and U in Apollo 15 samples, as determined by non-destructive gamma-ray spectrometry.**
G. D. O'Kelley, J. S. Eldridge, K. J. Northcutt.
The Apollo 15 lunar samples, (see 012.008), p. 244 - 246 (1972).

094.231 **Trace element comparisons between mare and Apennine-Front nonmare samples.**
G. W. Reed, Jr., S. Jovanovic.
The Apollo 15 lunar samples, (see 012.008), p. 247 - 249 (1972).

094.232 **Major element chemistry of Apollo 15 mare basalts.**
J. M. Rhodes.
The Apollo 15 lunar samples, (see 012.008), p. 250 - 252 (1972).

094.233 **K, U, and Th concentrations in rake sample 15382 by non-destructive gamma-ray spectroscopy.**
E. Schonfeld, G. D. O'Kelley, J. S. Eldridge, K. J. Northcutt.
The Apollo 15 lunar samples, (see 012.008), p. 253 - 254 (1972).

094.234 **Chemical analysis of lunar samples 15101,65 and 15211,6.** J. H. Scoon.
The Apollo 15 lunar samples, (see 012.008), p. 255 - 256 (1972).

094.235 **Analysis of lunar samples 15065, 15301 and 15556, with isotopic data for $^7Li/^6Li$.**
A. Strasheim, J. H. J. Coetzee, P. F. S. Jackson, F. W. E. Strelow, F. T. Wybenga, A. J. Gricius, M. L. Kokot.
The Apollo 15 lunar samples, (see 012.008), p. 257 - 259 (1972).

094.236 **Chemistry of pyroxenes from Apollo soil 15501,53.**
H. C. J. Taylor, J. L. Carter.
The Apollo 15 lunar samples, (see 012.008), p. 260 - 261 (1972).

094.237 **Composition of the lunar highlands II. The Apennine Front.** S. R. Taylor, M. Gorton, P. Muir, W. Nance, R. Rudowski, N. Ware.
The Apollo 15 lunar samples, (see 012.008), p. 262 - 264 (1972).

094.238 **Multielement analyses and a comparison of the degree of oxydation of lunar and meteoritic matter.**
H. Wänke, H. Palme, B. Spettel, F. Teschke.
The Apollo 15 lunar samples, (see 012.008), p. 265 - 267 (1972).

094.239 **Geochemical features of Apollo 15 materials.**
J. P. Willis, A. J. Erlank, J. J. Gurney, L. H. Ahrens.
The Apollo 15 lunar samples, (see 012.008), p. 268 - 271 (1972).

094.240 **The absorption of atomic hydrogen on 15101,68.**
D. A. Cadenhead, B. R. Jones.
The Apollo 15 lunar samples, (see 012.008), p. 272 - 274 (1972).

094.241 **Analysis of organogenic compounds in Apollo 15 samples.**
D. A. Flory, J. Oro', S. Wikstrom, D. Beaman, D. Nooner.
The Apollo 15 lunar samples, (see 012.008), p. 275 - 279 (1972).

094.242 **Inert gases in fines from the Hadley-Apennine region.** J. L. Jordan, S. Lakatos, D. Heymann.
The Apollo 15 lunar samples, (see 012.008), p. 280 - 281 (1972).

094.243 **Total nitrogen abundances in five Apollo-15 samples (Hadley-Apennine region) by neutron activation analysis.** B. K. Kothari, P. S. Goel.
The Apollo 15 lunar samples, (see 012.008), p. 282 - 283 (1972).

094.244 **Inert gases in green glass from Apollo 15.**
S. Lakatos, D. Heymann.
The Apollo 15 lunar samples, (see 012.008), p. 284 - 285 (1972).

094.245 **Apollo 15 lunar samples: LM exhaust products in the SESC 15013.**
B. R. Simoneit, P. C. Wszolek, A. L. Burlingame.
The Apollo 15 lunar samples, (see 012.008), p. 286 - 290 (1972).

094.246 **Carbon, nitrogen and sulfur released during pyrolysis of bulk Apollo 15 fines.**
S. Chang, J. Smith, H. Sakai, C. Petrowski, K. A. Kvenvolden, I. R. Kaplan.
The Apollo 15 lunar samples, (see 012.008), p. 291 - 293 (1972).

094.247 **Pyrolysis study of carbon in lunar fines and rocks.** D. J. DesMarais, J. M. Hayes, W. G. Meinschein.
The Apollo 15 lunar samples, (see 012.008), p. 294 - 298 (1972).

094.248 **Analysis for amino acid precursors of a sample of lunar soil subjected to rocket exhaust on Apollo 15.**
S. W. Fox, K. Harada, P. E. Hare.
The Apollo 15 lunar samples, (see 012.008), p. 299 - 301 (1972).

094.249 **Isotopic composition of carbon and hydrogen in some Apollo 14 and 15 samples.**
I. Friedman, K. G. Hardcastle, J. D. Gleason.
The Apollo 15 lunar samples, (see 012.008), p. 302 - 306 (1972).

094.250 **Thermal analysis—inorganic gas release studies on Apollo 14, 15, and 16 lunar samples.**

E. K. Gibson, Jr., G. W. Moore.
The Apollo 15 lunar samples, (see 012.008), p. 307 - 310 (1972).

094.251   Carbon compounds in Apollo 15 lunar samples.
J. E. Modzeleski, V. E. Modzeleski, L. A. Nagy, B. Nagy, P. B. Hamilton, W. S. McEwan, H. C. Urey.
The Apollo 15 lunar samples, (see 012.008), p. 311 - 315 (1972).

094.252   Carbon and nitrogen in Apollo 15 lunar samples.
C. B. Moore, C. F. Lewis, E. K. Gibson, Jr.
The Apollo 15 lunar samples, (see 012.008), p. 316 - 318 (1972).

094.253   Distribution of carbon and sulfur in hydrolyzed Apollo 15 lunar fines.
H. Sakai, S. Chang, C. Petrowski, J. Smith, I. R. Kaplan.
The Apollo 15 lunar samples, (see 012.008), p. 319 - 323 (1972).

094.254   Carbon chemistry of the Apollo 15 deep drill stem and a glass-rich sample related to the uniformity of the regolith and lunar surface processes.
P. C. Wszolek, R. J. Jackson, A. L. Burlingame.
The Apollo 15 lunar samples, (see 012.008), p. 324 - 328 (1972).

094.255   Rare gas and particle track studies of Apollo 15 samples: Hadley Rille and special soils.
C. Behrmann, G. Crozaz, R. Drozd, C. M. Hohenberg, C. Ralston, R. M. Walker, D. Yuhas.
The Apollo 15 lunar samples, (see 012.008), p. 329 - 332 (1972).

094.256   Irradiation studies of lunar soils: 15100, Luna 20, and compacted soil from breccia 14307.
J. L. Berdot, G. C. Chétrit, J. C. Lorin, P. Pellas, G. Poupeau.
The Apollo 15 lunar samples, (see 012.008), p. 333 - 335 (1972).

094.257   Apollo 15 regolith: a predominantly accretion or mixing model?
N. Bhandari, J. N. Goswami, D. Lal.
The Apollo 15 lunar samples, (see 012.008), p. 336 - 341 (1972).

094.258   Noble gases in the Apollo 15 drill cores.
D. D. Bogard, L. E. Nyquist.
The Apollo 15 lunar samples, (see 012.008), p. 342 - 346 (1972).

094.259   Strontium isotope geochemistry of Apollo 15 basalts.   W. Compston, J. R. de Laeter, M. J. Vernon
The Apollo 15 lunar samples, (see 012.008), p. 347 - 351 (1972).

094.260   Charged-particle track parameters of Apollo 15 lunar glasses.   S. A. Durrani, H. A. Khan.
The Apollo 15 lunar samples, (see 012.008), p. 352 - 356 (1972).

094.261   Concentrations of cosmogenic radionuclides in Apollo 15 rocks and soil.
J. S. Eldridge, G. D. O'Kelley, K. J. Northcutt.
The Apollo 15 lunar samples, (see 012.008), p. 357 - 359 (1972).

094.262   Lunar actinides: ^{236}U, ^{237}Np, ^{244}Pu, ^{239}Pu and ^{238}Pu.
P. R. Fields, H. Diamond, D. N. Metta, D. J. Rokop.

The Apollo 15 lunar samples, (see 012.008), p. 360 - 363 (1972).

094.263   Depth variation of Ar^{37} and Ar^{39} in lunar material.
E. L. Fireman.
The Apollo 15 lunar samples, (see 012.008), p. 364 - 367 (1972).

094.264   Particle track record of Apollo 15 green soil and rock.   R. L. Fleischer, H. R. Hart, Jr.
The Apollo 15 lunar samples, (see 012.008), p. 368 - 370 (1972).

094.265   Particle track record in Apollo 15 deep core from 54 to 80 cm depths.   R. L. Fleischer, H. R. Hart, Jr.
The Apollo 15 lunar samples, (see 012.008), p. 371 - 373 (1972).

094.266   The $^{40}Ar-^{39}Ar$ and cosmic ray exposure ages of Apollo 15 crystalline rocks, breccias and glasses.
L. Husain.
The Apollo 15 lunar samples, (see 012.008), p. 374 - 377 (1972).

094.267   In situ $^{40}Ar/^{39}Ar$ ages of breccia 14301, and concentration gradients of helium, neon, and argon isotopes in Apollo 15 samples.   G. H. Megrue.
The Apollo 15 lunar samples, (see 012.008), p. 378 - 379 (1972).

094.268   Rb-Sr systematics for chemically defined Apollo 15 materials.
L. E. Nyquist, P. W. Gast, S. E. Church, H. Wiesmann, B. Bansal.
The Apollo 15 lunar samples, (see 012.008), p. 380 - 384 (1972).

094.269   Track analysis of rocks 15058, 15555, 15641 and 14307.   G. Poupeau, P. Pellas, J. C. Lorin, G. C. Chétrit, J. L. Berdot.
The Apollo 15 lunar samples, (see 012.008), p. 385 - 387 (1972).

094.270   Uranium-thorium-lead isotopes and the nature of the mare surface debris at Hadley-Apennine.
L. T. Silver.
The Apollo 15 lunar samples, (see 012.008), p. 388 - 390 (1972).

094.271   U-Th-Pb, Rb-Sr, and K measurements on some Apollo 15 and Apollo 16 samples.   M. Tatsumoto, C. E. Hedge, R. J. Knight, D. M. Unruh, B. R. Doe.
The Apollo 15 lunar samples, (see 012.008), p. 391 - 395 (1972).

094.272   Distribution of Pb-U-Th in lunar anorthosite 15415 and inferences about its age.
F. Tera, L. A. Ray, G. J. Wasserburg.
The Apollo 15 lunar samples, (see 012.008), p. 396 - 401 (1972).

094.273   Sulphur concentrations and isotope ratios in Apollo 14 and 15 samples.   H. G. Thode, C. E. Rees.
The Apollo 15 lunar samples, (see 012.008), p. 402 - 403 (1972).

094.274   Cosmonuclides in lunar soil from Apollo 15.
Y. Yokoyama, J. L. Reyss, F. Guichard, J. Sato.
The Apollo 15 lunar samples, (see 012.008), p. 404 - 406 (1972).

094.275   Micrometeoroid craters smaller than 100 microns.

D. E. Brownlee, F. Hörz, J. B. Hartung, D. E. Gault.
The Apollo 15 lunar samples, (see 012.008), p. 407 - 411 (1972).

094.276 **Study of solar flares, cosmic dust and lunar erosion with vesicular basalts.**
I. D. Hutcheon, D. Braddy, P. P. Phakey, P. B. Price.
The Apollo 15 lunar samples, (see 012.008), p. 412 - 414 (1972).

094.277 **Exposure ages of Apollo 15 samples by means of microcrater statistics and solar flare particle tracks.**
E. Schneider, D. Storzer, H. Fechtig.
The Apollo 15 lunar samples, (see 012.008), p. 415 - 419 (1972).

094.278 **Difficulties in separating the stable component of natural remanent magnetization in lunar rocks.**
S. K. Banerjee, K. A. Hoffman, J. P. Mellema.
The Apollo 15 lunar samples, (see 012.008), p. 420 - 424 (1972).

094.279 **Magnetic properties of Apollo 15 rocks and fines.**
D. W. Collinson, S. K. Runcorn, A. Stephenson.
The Apollo 15 lunar samples, (see 012.008), p. 425 - 429 (1972).

094.280 **Magnetism of Apollo 15 samples.**
W. A. Gose, G. W. Pearce, D. W. Strangway, J. Carnes.
The Apollo 15 lunar samples, (see 012.008), p. 430 - 434 (1972).

094.281 **Ferromagnetic resonance of small multidomain iron particles in an 0.5-cm fragment of lunar glass,**
15434,62. D. L. Griscom, C. L. Marquardt.
The Apollo 15 lunar samples, (see 012.008), p. 435 - 437 (1972).

094.282 **Remanent magnetism in four Apollo 15 igneous rock fragments.** R. B. Hargraves, N. Dorety.
The Apollo 15 lunar samples, (see 012.008), p. 438 - 439 (1972).

094.283 **Mössbauer analyses of Apollo 15 samples.**
G. P. Huffman, F. C. Schwerer, R. M. Fisher.
The Apollo 15 lunar samples, (see 012.008), p. 440 - 441 (1972).

094.284 **Summary of rock magnetism of Apollo 15 lunar materials.** T. Nagata, R. M. Fisher, F. C. Schwerer, M. D. Fuller, J. R. Dunn.
The Apollo 15 lunar samples, (see 012.008), p. 442 - 445 (1972).

094.285 **Magnetic hysteresis classification for the lunar surface.** P. Wasilewski.
The Apollo 15 lunar samples, (see 012.008), p. 446 - 452 (1972).

094.286 **Thermoluminescence of Apollo 15 lunar samples: (I) −196 to +250°C.** I. M. Blair, R. A. Jahn, J. A. Edgington, S. A. Durrani, M. Phillips, A. Kacperek.
The Apollo 15 lunar samples, (see 012.008), p. 453 - 456 (1972).

094.287 **Thermoluminescence of Apollo 15 lunar samples: (II) 20 to 550°C.**
S. A. Durrani, W. Prachyabrued, J. A. Edgington, I. M. Blair.
The Apollo 15 lunar samples, (see 012.008), p. 457 - 461 (1972).

094.288 **Rayleigh wave studies of two Apollo 15 rocks.**
B. R. Tittmann, R. M. Housley, E. H. Cirlin, M. Abdel-Gawad.
The Apollo 15 lunar samples, (see 012.008), p. 462 - 465 (1972).

094.289 **Electrostatic interparticle adhesion in Apollo 15 fines.** S. K. Asunmaa, G. Arrhenius.
The Apollo 15 lunar samples, (see 012.008), p. 466 - 469 (1972).

094.290 **Infrared structural characterization of single grains from two Apollo 15 dusts.**
P. A. Estep, J. J. Kovach, C. Karr, Jr.
The Apollo 15 lunar samples, (see 012.008), p. 470 - 474 (1972).

094.291 **Infrared emission spectra of Apollo 15 soils.**
L. M. Logan, G. R. Hunt, J. W. Salisbury.
The Apollo 15 lunar samples, (see 012.008), p. 475 - 476 (1972).

094.292 **Frequency and temperature dependence of the electrical properties of a soil sample from Apollo 15.**
G. R. Olhoeft, A. L. Frisillo, D. W. Strangway.
The Apollo 15 lunar samples, (see 012.008), p. 477 - 481 (1972).

094.293 **Infrared studies of Apollo 15 fines.**
C. H. Perry, R. P. Lowndes.
The Apollo 15 lunar samples, (see 012.008), p. 482 - 485 (1972).

094.294 **Secondary electron emission and Auger electron spectroscopy from Apollo 15 lunar samples.**
R. F. Willis, M. Anderegg, B. Feuerbacher, B. Fitton.
The Apollo 15 lunar samples, (see 012.008), p. 486 - 489 (1972).

094.295 **Some results of a morphometric investigation of the near side of the moon.**
G. A. Lejkin, M. P. Popova, Zh. F. Rodionova.
Astron. Tsirk., No. 755, p. 3 - 6 (1973). In Russian.

094.296 **Far ultraviolet reflectivity of lunar dust samples: Apollo 11, 12, and 14.**
R. L. Lucke, R. C. Henry, W. G. Fastie.
Astron. Journ., Vol. 78, 263 - 267, 283 (1973).
    The reflectivity of three samples of lunar fines has been measured over the spectral range 1200−1600 Å and 2500−5800 Å. Values of the lunar albedo are inferred from these data, and comparisons are made with direct observation of the moon in the 1050−3600 Å range from various spacecraft and also with ground-based observations in the 3000−5800 Å range. Conclusions relating to the value of the far ultraviolet solar flux are drawn.

094.297 **Monochromatic phase curves and albedos for the lunar disk.** A. P. Lane, W. M. Irvine.
Astron. Journ., Vol. 78, 267 - 277 = Contr. Five College Obs., Univ. Mass., *Amherst, Mass.,* No. 159 (1973).
    Photoelectric observations of the entire lunar disk were made in 1964−1965 over phase angles $6° \leq i \leq 120°$ in nine narrow bands $(0.35−1.0\mu)$ and in *UBV*. Phase curves are presented as a function of wavelength. Geometric albedos, phase coefficients, phase integrals, and Bond albedos are presented at all wavelengths, excluding any opposition effect.

094.298 **Fluorine in lunar samples: implications concerning lunar fluorapatite.**
G. W. Reed, Jr., S. Jovanovic.

Geochim. Cosmochim. Acta, Vol. 37, 1457 - 1462 (1973).

In this note we discuss the source and magnitude of error in our previously (1971) reported F concentrations. We also report our F data on Apollo 14 and 15 samples, and make some observations on the F content in relation to the 'moonwide' correlation between Cl and P_2O_5 concentrations already reported (Reed et al., 1972).

094.299 **Rectangular coordinates of the moon 1971–1980.**
G. Kaplan, J. B. Dunham.
United States Naval Obs., *Washington, D. C.*, Circ. No. 140, 123 pp. (1973).

094.300 **Lunar limb profiles for solar eclipses.**
J. S. Duncombe.
United States Naval Obs., *Washington, D. C.*, Circ. No. 141, 33 pp. (1973).

This circular contains selected lunar limb profiles designed for use in solar eclipse predictions. These profiles allow the eclipse observer to predict the effect of lunar limb irregularities on the time of second and third contacts for any site along the eclipse path. The arguments of these profiles range in topocentric longitude from +6° to −6°; topocentric latitude is 0°.

094.301 **Lunar composition from Apollo orbital measurements.**   I. Adler, J. I. Trombka, L. I. Yin, P. Gorenstein, P. Bjorkholm, J. Gerard.
Naturwissenschaften, 60. Jahrgang, p. 231 - 242 (1973).

Several spectrometers carried in the Service Module of the Apollo 15 and Apollo 16 spacecraft were employed for the compositional mapping of the lunar surface. A large scale compositional map of over 20 percent of the lunar surface was obtained for the first time. It was possible to demonstrate interesting chemical differences between the maria and the highlands, to find specific areas of high radioactivity and to learn something about the composition of the moon's hidden side. Further the same devices were used to obtain useful astronomical data during the return to earth.

094.302 **Schröter's Valley and Herodotus.**   P. Moore.
Journ. British Astron. Ass., Vol. 83, 185 (1973).

094.303 **Preliminary data on lunar ground brought to earth by automatic probe "Luna–16".**
A. P. Vinogradov.
Proc. Second Lunar Sci. Conference, (see 012.012), Vol. 1, 1 - 16 (1971).

094.304 **Lunar locations and orientations of rock samples from Apollo missions 11 and 12.**
R. L. Sutton, G. G. Schaber.
Proc. Second Lunar Sci. Conference, (see 012.012), Vol. 1, 17 - 26 (1971).

094.305 **Surface lineaments at the Apollo 11 and Apollo 12 landing sites.**   G. G. Schaber, G. A. Swann.
Proc. Second Lunar Sci. Conference, (see 012.012), Vol. 1, 27 - 38 (1971).

094.306 **Tranquillityite: a new silicate mineral from Apollo 11 and Apollo 12 basaltic rocks.**
J. F. Lovering, D. A. Wark, A. F. Reid, N. G. Ware, K. Keil, M. Prinz, T. E. Bunch, A. El Goresy, P. Ramdohr, G. M. Brown, A. Peckett, R. Phillips, E. N. Cameron, J. A. V. Douglas, A. G. Plant.
Proc. Second Lunar Sci. Conference, (see 012.012), Vol. 1, 39 - 45 (1971).

094.307 **The crystal structure of pyroxferroite from Mare Tranquillitatis.**   C. W. Burnham.

Proc. Second Lunar Sci. Conference, (see 012.012), Vol. 1, 47 - 57 (1971).

094.308 **Apollo 12 clinopyroxenes: high temperature X-ray diffraction studies.**
C. T. Prewitt, G. E. Brown, J. J. Papike.
Proc. Second Lunar Sci. Conference, (see 012.012), Vol. 1, 59 - 68 (1971).

094.309 **Comparative electron petrography of Apollo 11, Apollo 12, and terrestrial rocks.**
J. M. Christie, J. S. Lally, A. H. Heuer, R. M. Fisher, D. T. Griggs, S. V. Radcliffe.
Proc. Second Lunar Sci. Conference, (see 012.012), Vol. 1, 69 - 89 (1971).

094.310 **Cation distributions and cooling history of clinopyroxenes from Oceanus Procellarum.**
S. S. Hafner, D. Virgo, D. Warburton.
Proc. Second Lunar Sci. Conference, (see 012.012), Vol. 1, 91 - 108 (1971).

094.311 **Correlated electron microscopy and diffraction of lunar clinopyroxenes from Apollo 12 samples.**
H. Fernández-Morán, M. Ohtsuki, A. Hibino.
Proc. Second Lunar Sci. Conference, (see 012.012), Vol. 1, 109 - 116 (1971).

094.312 **Studies of lunar plagioclases, tridymite, and cristobalite.**   D. E. Appleman, H.-U. Nissen, D. B. Stewart, J. R. Clark, E. Dowty, J. S. Huebner.
Proc. Second Lunar Sci. Conference, (see 012.012), Vol. 1, 117 - 133 (1971).

094.313 **Lunar bytownite from sample 12032,44.**
H.-R. Wenk, G. L. Nord.
Proc. Second Lunar Sci. Conference, (see 012.012), Vol. 1, 135 - 140 (1971).

094.314 **Note on tridymite in rock 12021.**
W. A. Dollase, R. A. Cliff, G. W. Wetherill.
Proc. Second Lunar Sci. Conference, (see 012.012), Vol. 1, 141 - 142 (1971).

094.315 **Minor elements in Apollo 11 and Apollo 12 olivine and plagioclase.**   J. V. Smith.
Proc. Second Lunar Sci. Conference, (see 012.012), Vol. 1, 143 - 150 (1971).

094.316 **Uranium-enriched phases in Apollo 11 and Apollo 12 basaltic rocks.**   J. F. Lovering, D. A. Wark.
Proc. Second Lunar Sci. Conference, (see 012.012), Vol. 1, 151 - 158 (1971).

094.317 **Distribution of uranium in Apollo 11 rock 10017.**
C. M. Rice, S. H. U. Bowie.
Proc. Second Lunar Sci. Conference, (see 012.012), Vol. 1, 159 - 166 (1971).

094.318 **Radioactive halos and the lunar environment.**
R. V. Gentry.
Proc. Second Lunar Sci. Conference, (see 012.012), Vol. 1, 167 - 168 (1971).

094.319 **Zirconium fractionation in Apollo 11 and Apollo 12 rocks.**
G. Arrhenius, J. E. Everson, R. W. Fitzgerald, H. Fujita.
Proc. Second Lunar Sci. Conference, (see 012.012), Vol. 1, 169 - 176 (1971).

094.320 **Metallic inclusions and metal particles in the Apollo 12 lunar soil.** J. I. Goldstein, H. Yakowitz.
Proc. Second Lunar Sci. Conference, (see 012.012), Vol. 1, 177 - 191 (1971).

094.321 **Opaque minerals in certain lunar rocks from Apollo 12.** E. N. Cameron.
Proc. Second Lunar Sci. Conference, (see 012.012), Vol. 1, 193 - 206 (1971).

094.322 **Opaque phases in Apollo 12 samples.** P. R. Simpson, S. H. U. Bowie.
Proc. Second Lunar Sci. Conference, (see 012.012), Vol. 1, 207 - 218 (1971).

094.323 **The opaque minerals in the lunar rocks from Oceanus Procellarum.** A. El Goresy, P. Ramdohr, L. A. Taylor.
Proc. Second Lunar Sci. Conference, (see 012.012), Vol. 1, 219 - 236 (1971).

094.324 **Single crystal X-ray investigation of deformation in terrestrial and lunar ilmenite.** J. A. Minkin, E. C. T. Chao.
Proc. Second Lunar Sci. Conference, (see 012.012), Vol. 1, 237 - 246 (1971).

094.325 **Luminescence petrography of the Apollo 12 rocks and comparative features in terrestrial rocks and meteorites.** R. F. Sippel.
Proc. Second Lunar Sci. Conference, (see 012.012), Vol. 1, 247 - 263 (1971).

094.326 **Mineralogical, petrological, and chemical features of four Apollo 12 lunar microgabbros.** C. Klein, Jr., J. C. Drake, C. Frondel.
Proc. Second Lunar Sci. Conference, (see 012.012), Vol. 1, 265 - 284 (1971).

094.327 **Mineralogy and petrology of some Apollo 12 samples.** M. R. Dence, J. A. V. Douglas, A. G. Plant, R. J. Traill.
Proc. Second Lunar Sci. Conference, (see 012.012), Vol. 1, 285 - 299 (1971).

094.328 **Apollo 12 igneous rocks 12004, 12008, 12009, and 12022: A mineralogical and petrological study.** R. Brett, P. Butler, Jr., C. Meyer, Jr., A. M. Reid, H. Takeda, R. Williams.
Proc. Second Lunar Sci. Conference, (see 012.012), Vol. 1, 301 - 317 (1971).

094.329 **Mineralogy, petrology, and chemistry of some Apollo 12 samples.** K. Keil, M. Prinz, T. E. Bunch.
Proc. Second Lunar Sci. Conference, (see 012.012), Vol. 1, 319 - 341 (1971).

094.330 **Mineralogical studies of Apollo 12 samples.** L. S. Walter, B. M. French, K. F. J. Heinrich, P. D. Lowman, Jr., A. S. Doan, I. Adler.
Proc. Second Lunar Sci. Conference, (see 012.012), Vol. 1, 343 - 358 (1971).

094.331 **Mineralogy and petrology of some Apollo 12 lunar samples.** P. E. Champness, A. C. Dunham, F. G. F. Gibb, H. N. Giles, W. S. MacKenzie, E. F. Stumpfl, J. Zussman.
Proc. Second Lunar Sci. Conference, (see 012.012), Vol. 1, 359 - 376 (1971).

094.332 **Mineralogical and petrographic investigation of some Apollo 12 samples.** P. Gay, M. G. Bown, I. D. Muir, G. M. Bancroft, P. G. L. Williams.
Proc. Second Lunar Sci. Conference, (see 012.012.), Vol. 1, 377 - 392 (1971).

094.333 **Mineralogy, chemistry, and origin of the KREEP component in soil samples from the Ocean of Storms.** C. Meyer, Jr., R. Brett, N. J. Hubbard, D. A. Morrison, D. S. McKay, F. K. Aitken, H. Takeda, E. Schonfeld.
Proc. Second Lunar Sci. Conference, (see 012.012), Vol. 1, 393 - 411 (1971).

094.334 **Mineralogy-petrology of lunar samples. Microprobe studies of samples 12021 and 12022; viscosity of melts of selected lunar compositions.** D. F. Weill, R. A. Grieve, I. S. McCallum, Y. Bottinga.
Proc. Second Lunar Sci. Conference, (see 012.012), Vol. 1, 413 - 430 (1971).

094.335 **Nature, occurrence, and exotic origin of "gray mottled" (luny rock) basalts in Apollo 12 soils and breccias.** A. T. Anderson, Jr., J. V. Smith.
Proc. Second Lunar Sci. Conference, (see 012.012), Vol. 1, 431 - 438 (1971).

094.336 **Composition of five Apollo 11 and Apollo 12 rocks and one Apollo 11 soil and some petrogenic considerations.** A. E. J. Engel, C. G. Engel, A. L. Sutton, A. T. Myers.
Proc. Second Lunar Sci. Conference, (see 012.012), Vol. 1, 439 - 448 (1971).

094.337 **Vaporization from heated lunar samples and the investigation of lunar erosion by volatilized alkalis.** J. J. Naughton, J. V. Derby, V. A. Lewis.
Proc. Second Lunar Sci. Conference, (see 012.012), Vol. 1, 449 - 457 (1971).

094.338 **Lunar "basalts": some comparisons with terrestrial and meteoritic analogs, and a proposed classification and nomenclature.** W. G. Melson, B. Mason.
Proc. Second Lunar Sci. Conference, (see 012.012), Vol. 1, 459 - 467 (1971).

094.339 **Lunar crystalline rocks: petrology and geology.** J. L. Warner.
Proc. Second Lunar Sci. Conference: (see 012.012), Vol. 1, 469 - 480 (1971).

094.340 **Petrology of some Apollo 12 crystalline rocks.** I. Kushiro, Y. Nakamura, K. Kitayama, S. Akimoto.
Proc. Second Lunar Sci. Conference, (see 012.012), Vol. 1, 481 - 495 (1971).

094.341 **Equilibrium relations among phases occurring in lunar rocks.** A. Muan, J. Hauck, E. F. Osborn, J. F. Schairer.
Proc. Second Lunar Sci. Conference, (see 012.012), Vol. 1, 497 - 505 (1971).

094.342 **Petrology of silicate melt inclusions, Apollo 11 and Apollo 12 and terrestrial equivalents.** E. Roedder, P. W. Weiblen.
Proc. Second Lunar Sci. Conference, (see 012.012), Vol. 1, 507 - 528 (1971).

094.343 **Petrogenetic significance of pyroxenes in two Apollo 12 samples.** L. S. Hollister, W. E. Trzcienski, Jr., R. B. Hargraves, C. G. Kulick.

Proc. Second Lunar Sci. Conference, (see 012.012), Vol. 1, 529 - 557 (1971).

094.344 **Crystallization histories of clinopyroxenes in two porphyritic rocks from Oceanus Procellarum.**
A. E. Bence, J. J. Papike, D. H. Lindsley.
Proc. Second Lunar Sci. Conference, (see 012.012), Vol. 1, 559 - 574 (1971).

094.345 **Accumulation of olivine in rock 12040 and other basaltic fragments in the light of analysis and syntheses.** R. C. Newton, A. T. Anderson, J. V. Smith.
Proc. Second Lunar Sci. Conference, (see 012.012), Vol. 1, 575 - 582 (1971).

094.346 **Picrite basalts, ferrobasalts, feldspathic norites, and rhyolites in a strongly fractionated lunar crust.**
G. M. Brown, C. H. Emeleus, J. G. Holland, A. Peckett, R. Phillips.
Proc. Second Lunar Sci. Conference, (see 012.012), Vol. 1, 583 - 600 (1971).

094.347 **Experimental petrology and petrogenesis of Apollo 12 basalts.** D. H. Green, A. E. Ringwood, N. G. Ware, W. O. Hibberson, A. Major, E. Kiss.
Proc. Second Lunar Sci. Conference, (see 012.012), Vol. 1, 601 - 615 (1971).

094.348 **Lunar lavas and the achondrites: petrogenesis of protohypersthene basalts in the maria lava lakes.**
G. M. Biggar, M. J. O'Hara, A. Peckett, D. J. Humphries.
Proc. Second Lunar Sci. Conference, (see 012.012), Vol. 1, 617 - 643 (1971).

094.349 **H₂O in lunar processes: The stability of hydrous phases in lunar samples 10058 and 12013.**
R. W. Charles, D. A. Hewitt, D. R. Wones.
Proc. Second Lunar Sci. Conference, (see 012.012), Vol. 1, 645 - 664 (1971).

094.350 **Composition and grain-size characteristics of fines from the Apollo 12 double-core tube.**
G. A. Sellers, C. C. Woo, M. L. Bird, M. B. Duke.
Proc. Second Lunar Sci. Conference, (see 012.012), Vol. 1, 665 - 678 (1971).

094.351 **Relative proportions and probable sources of rock fragments in the Apollo 12 soil samples.**
U. B. Marvin, J. A. Wood, G. J. Taylor, J. B. Reid, Jr., B. N. Powell, J. S. Dickey, Jr., J. F. Bower.
Proc. Second Lunar Sci. Conference, (see 012.012), Vol. 1, 679 - 699 (1971).

094.352 **Investigations of the natural history of the regolith at the Apollo 12 site.**
W. Quaide, V. Oberbeck, T. Bunch, G. Polkowski.
Proc. Second Lunar Sci. Conference, (see 012.012), Vol. 1, 701 - 718 (1971).

094.353 **Mineralogical and chemical data on Apollo 12 lunar fines.** C. Frondel, C. Klein, Jr., J. Ito.
Proc. Second Lunar Sci. Conference, (see 012.012), Vol. 1, 719 - 726 (1971).

094.354 **Glasses and sialic components in Mare Procellarum soil.** K. Fredriksson, J. Nelen, A. Noonan, C. A. Andersen, J. R. Hinthorne.
Proc. Second Lunar Sci. Conference, (see 012.012), Vol. 1, 727 - 735 (1971).

094.355 **The lunar regolith as sampled by Apollo 11 and Apollo 12: grain size analyses, modal analyses, and origins of particles.**
E. A. King, Jr., J. C. Butler, M. F. Carman, Jr.
Proc. Second Lunar Sci. Conference, (see 012.012), Vol. 1, 737 - 746 (1971).

094.356 **Mineralogical and chemical studies of lunar fines 10084,148 and 12070,98.**
Y. K. Kim, S. M. Lee, J. H. Yang, J. H. Kim, C. K. Kim.
Proc. Second Lunar Sci. Conference, (see 012.012), Vol. 1, 747 - 753 (1971).

094.357 **Apollo 12 soil and breccia.**
D. S. McKay, D. A. Morrison, U. S. Clanton, G. H. Ladle, J. F. Lindsay.
Proc. Second Lunar Sci. Conference, (see 012.012), Vol. 1, 755 - 773 (1971).

094.358 **Pyroxenes and olivines in crystalline rocks from the Ocean of Storms.**
N. L. Carter, L. A. Fernandez, H. G. Avé Lallemant, I. S. Leung.
Proc. Second Lunar Sci. Conference, (see 012.012), Vol. 1, 775 - 795 (1971).

094.359 **The petrology of unshocked and shocked Apollo 11 and Apollo 12 microbreccias.**
E. C. T. Chao, J. A. Boreman, G. A. Desborough.
Proc. Second Lunar Sci. Conference, (see 012.012), Vol. 1, 797 - 816 (1971).

094.360 **Shock-induced features of Apollo 12 microbreccias.**
C. B. Sclar.
Proc. Second Lunar Sci. Conference, (see 012.012), Vol. 1, 817 - 832 (1971).

094.361 **Shock metamorphism and origin of regolith and breccias at the Apollo 11 and Apollo 12 landing sites.**
W. von Engelhardt, J. Arndt, W. F. Müller, D. Stöffler.
Proc. Second Lunar Sci. Conference, (see 012.012), Vol. 1, 833 - 854 (1971).

094.362 **Opaque mineralogy and textural features of Apollo 12 samples and a comparison with Apollo 11 rocks.**
L. A. Taylor, G. Kullerud, W. B. Bryan.
Proc. Second Lunar Sci. Conference, (see 012.012), Vol. 1, 855 - 871 (1971).

094.363 **Chemistry and surface morphology of fragments from Apollo 12 soil.** J. L. Carter.
Proc. Second Lunar Sci. Conference, (see 012.012), Vol. 1, 873 - 892 (1971).

094.364 **Matrix characteristics and origin of lunar breccia samples 12034 and 12073.**
A. C. Waters, R. V. Fisher, R. E. Garrison, D. Wax.
Proc. Second Lunar Sci. Conference, (see 012.012), Vol. 1, 893 - 907 (1971).

094.365 **Surface micrography of lunar fines compared with tektites and terrestrial volcanic analogs.**
S. V. Margolis, V. Barnes, P. Cloud, R. V. Fisher.
Proc. Second Lunar Sci. Conference, (see 012.012), Vol. 1, 909 - 921 (1971).

094.366 **Surface morphology of free-growing ilmenites and chromites from vuggy rocks 10072,31 and 12036,2.** J. Jedwab.

Proc. Second Lunar Sci. Conference, (see 012.012), Vol. 1, 923 - 935 (1971).

094.367 Glassy spheroids in lunar fines from Apollo 12 samples 12070,37; 12001,73; and 12057,60.
M. Fulchignoni, R. Funiciello, A. Taddeucci, R. Trigila.
Proc. Second Lunar Sci. Conference, (see 012.012), Vol. 1, 937 - 948 (1971).

094.368 Devitrified glass fragments from Apollo 11 and Apollo 12 lunar samples. G. Lofgren.
Proc. Second Lunar Sci. Conference, (see 012.012), Vol. 1, 949 - 955 (1971).

094.369 Chromatographic and mineralogical study of lunar fines and glass. C. R. Masson, J. Götz, W. D. Jamieson, J. L. McLachlan, A. Volborth.
Proc. Second Lunar Sci. Conference, (see 012.012), Vol. 1, 957 - 971 (1971).

094.370 Response of Apollo 12 lunar dust to reagents simulative of those in the weathering environment of earth. W. D. Keller, W. H. Huang.
Proc. Second Lunar Sci. Conference, (see 012.012), Vol. 1, 973 - 981 (1971).

094.371 Apollo 12 lunar sample inventory.
Proc. Second Lunar Sci. Conference, (see 012.012), Vol. 1, 983 (1971).

094.372 Model history of the lunar surface.
H. C. Urey, K. Marti, J. W. Hawkins, M. K. Liu.
Proc. Second Lunar Sci. Conference, (see 012.012), Vol. 2, 987 - 998 (1971).

094.373 Chemical composition and origin of nonmare lunar basalts. N. J. Hubbard, P. W. Gast.
Proc. Second Lunar Sci. Conference, (see 012.012), Vol. 2, 999 - 1020 (1971).

094.374 Volatile and siderophile elements in lunar rocks: comparison with terrestrial and meteoritic basalts.
E. Anders, R. Ganapathy, R. R. Keays, J. C. Laul, J. W. Morgan.
Proc. Second Lunar Sci. Conference, (see 012.012), Vol. 2, 1021 - 1036 (1971).

094.375 Trace element studies of rocks and soils from Oceanus Procellarum and Mare Tranquillitatis.
P. A. Baedecker, R. Schaudy, J. L. Elzie, J. Kimberlin, J. T. Wasson.
Proc. Second Lunar Sci. Conference, (see 012.012), Vol. 2, 1037 - 1061 (1971).

094.376 Analyses of Apollo 12 specimens: compositional variations, differentiation processes, and lunar soil mixing models.
G. G. Goles, A. R. Duncan, D. J. Lindstrom, M. R. Martin, R. L. Beyer, M. Osawa, K. Randle, L. T. Meek, T. L. Steinborn, S. M. McKay.
Proc. Second Lunar Sci. Conference, (see 012.012), Vol. 2, 1063 - 1081 (1971).

094.377 Trace element chemistry of lunar samples from the Ocean of Storms. S. R. Taylor, R. Rudowski, P. Muir, A. Graham, M. Kaye.
Proc. Second Lunar Sci. Conference, (see 012.012), Vol. 2, 1083 - 1099 (1971).

094.378 Alkali, alkaline earth, and rare-earth element concentrations in some Apollo 12 soils, rocks, and separated phases. C. C. Schnetzler, J. A. Philpotts.
Proc. Second Lunar Sci. Conference, (see 012.012), Vol. 2, 1101 - 1122 (1971).

094.379 Some interelement relationships between lunar rocks and fines, and stony meteorites.
J. P. Willis, L. H. Ahrens, R. V. Danchin, A. J. Erlank, J. J. Gurney, P. K. Hofmeyr, T. S. McCarthy, M. J. Orren.
Proc. Second Lunar Sci. Conference, (see 012.012), Vol. 2, 1123 - 1138 (1971).

094.380 Meteoritic material in lunar samples: characterization from trace elements.
J. C. Laul, J. W. Morgan, R. Ganapathy, E. Anders.
Proc. Second Lunar Sci. Conference, (see 012.012), Vol. 2, 1139 - 1158 (1971).

094.381 Abundances of the primordial radionuclides K, Th, and U in Apollo 12 lunar samples by nondestructive gamma-ray spectrometry: Implications for origin of lunar soils.
G. D. O'Kelley, J. S. Eldridge, E. Schonfeld, P. R. Bell.
Proc. Second Lunar Sci. Conference, (see 012.012), Vol. 2, 1159 - 1168 (1971).

094.382 Elemental abundances of lunar soil and rocks from Apollo 12. G. H. Morrison, J. T. Gerard, N. M. Potter, E. V. Gangadharam, A. M. Rothenberg, R. A. Burdo.
Proc. Second Lunar Sci. Conference, (see 012.012), Vol. 2, 1169 - 1185 (1971).

094.383 Apollo 12 samples: chemical composition and its relation to sample locations and exposure ages, the two component origin of the various soil samples and studies on lunar metallic particles.
H. Wänke, F. Wlotzka, H. Baddenhausen, A. Balacescu, B. Spettel, F. Teschke, E. Jagoutz, H. Kruse, M. Quijano-Rico, R. Rieder.
Proc. Second Lunar Sci. Conference, (see 012.012), Vol. 2, 1187 - 1208 (1971).

094.384 Comparison of the analytical results from the Surveyor, Apollo, and Luna missions.
A. L. Turkevich.
Proc. Second Lunar Sci. Conference, (see 012.012), Vol. 2, 1209 - 1215 (1971).

094.385 Elemental composition of some Apollo 12 lunar rocks and soils.
F. Cuttitta, H. J. Rose, Jr., C. S. Annell, M. K. Carron, R. P. Christian, E. J. Dwornik, L. P. Greenland, A. W. Helz, D. T. Ligon, Jr.
Proc. Second Lunar Sci. Conference, (see 012.012), Vol. 2, 1217 - 1229 (1971).

094.386 Bulk elemental composition of Apollo 12 samples: five igneous and one breccia rocks and four soils.
H. Wakita, R. A. Schmitt.
Proc. Second Lunar Sci. Conference, (see 012.012), Vol. 2, 1231 - 1236 (1971).

094.387 Major element abundances in Apollo 12 rocks and fines by 14 MeV neutron activation.
W. D. Ehmann, J. W. Morgan.
Proc. Second Lunar Sci. Conference, (see 012.012), Vol. 2, 1237 - 1245 (1971).

094.388 Spark mass spectrometric analysis of major and minor elements in six lunar samples.
M. Bouchet, G. Kaplan, A. Voudon, M.-J. Bertoletti.
Proc. Second Lunar Sci. Conference, (see 012.012), Vol. 2, 1247 - 1252 (1971).

094.389 **Elemental composition of lunar surface material (part 2).** A. A. Smales, D. Mapper, M. S. W. Webb, R. K. Webster, J. D. Wilson, J. S. Hislop.
Proc. Second Lunar Sci. Conference, (see 012.012), Vol. 2, 1253 - 1258 (1971).

094.390 **Chemical analyses of lunar samples 12040 and 12064.** J. H. Scoon.
Proc. Second Lunar Sci. Conference, (see 012.012), Vol. 2, 1259 - 1260 (1971).

094.391 **The halogens and other trace elements in Apollo 12 samples and the implications of halides, platinum metals, and mercury on surfaces.** G. W. Reed, S. Jovanovic.
Proc. Second Lunar Sci. Conference, (see 012.012), Vol. 2, 1261 - 1276 (1971).

094.392 **Multielement neutron activation analysis of trace elements in lunar fines.**
A. Travesí, J. Palomares, J. Adrada.
Proc. Second Lunar Sci. Conference, (see 012.012), Vol. 2, 1277 - 1280 (1971).

094.393 **Determination of 40 elements in Apollo 12 materials by neutron activation analysis.**
A. O. Brunfelt, K. S. Heier, E. Steinnes.
Proc. Second Lunar Sci. Conference, (see 012.012), Vol. 2, 1281 - 1290 (1971).

094.394 **Radioanalytical determination of elemental compositions of lunar samples.**
M. Vobecký, J. Frána, J. Bauer, Z. Řanda, J. Benada, J. Kunclí.
Proc. Second Lunar Sci. Conference, (see 012.012), Vol. 2, 1291 - 1300 (1971).

094.395 **Analyses of Apollo 11 and Apollo 12 rocks and soils by neutron activation.**
D. P. Kharkar, K. K. Turekian.
Proc. Second Lunar Sci. Conference, (see 012.012), Vol. 2, 1301 - 1305 (1971).

094.396 **Rare-earth elements in Apollo 12 lunar materials.** L. A. Haskin, P. A. Helmke, R. O. Allen, M. R. Anderson, R. L. Korotev, K. A. Zweifel.
Proc. Second Lunar Sci. Conference, (see 012.012), Vol. 2, 1307 - 1317 (1971).

094.397 **Abundances of the 14 rare-earth elements and 12 other trace elements in Apollo 12 samples: five igneous and one breccia rocks and four soils.**
H. Wakita, P. Rey, R. A. Schmitt.
Proc. Second Lunar Sci. Conference, (see 012.012), Vol. 2, 1319 - 1329 (1971).

094.398 **Rhenium and osmium abundance determinations and meteoritic contamination levels in Apollo 11 and Apollo 12 lunar samples.** J. F. Lovering, T. C. Hughes.
Proc. Second Lunar Sci. Conference, (see 012.012), Vol. 2, 1331 - 1335 (1971).

094.399 **Search for rhenium isotopic anomalies in lunar surface material by neutron bombardment.**
W. Herr, U. Herpers, R. Michel, A. A. Abdel Rassoul, R. Woelfle.
Proc. Second Lunar Sci. Conference, (see 012.012), Vol. 2, 1337 - 1341 (1971).

094.400 **Total carbon and nitrogen abundances in Apollo 12 lunar samples.**
C. B. Moore, C. F. Lewis, J. W. Larimer, F. M. Delles, R. C. Gooley, W. Nichiporuk, E. K. Gibson, Jr.
Proc. Second Lunar Sci. Conference, (see 012.012), Vol. 2, 1343 - 1350 (1971).

094.401 **Thermal analysis-inorganic gas release studies of lunar samples.** E. K. Gibson, Jr., S. M. Johnson.
Proc. Second Lunar Sci. Conference, (see 012.012), Vol. 2, 1351 - 1366 (1971).

094.402 **Mass spectrometric investigation of the vaporization process of Apollo 12 lunar samples.**
G. De Maria, G. Balducci, M. Guido, V. Piacente.
Proc. Second Lunar Sci. Conference, (see 012.012), Vol. 2, 1367 - 1380 (1971).

094.403 **Active and inert gases in Apollo 12 and Apollo 11 samples released by crushing at room temperature and by heating at low temperatures.**
J. Funkhouser, E. Jessberger, O. Müller, J. Zähringer.
Proc. Second Lunar Sci. Conference, (see 012.012), Vol. 2, 1381 - 1396 (1971).

094.404 **Carbon and sulfur isotope studies on Apollo 12 lunar samples.** I. R. Kaplan, C. Petrowski.
Proc. Second Lunar Sci. Conference, (see 012.012), Vol. 2, 1397 - 1406 (1971).

094.405 **The carbon and hydrogen content and isotopic composition of some Apollo 12 materials.**
I. Friedman, J. R. O'Neil, J. D. Gleason, K. Hardcastle.
Proc. Second Lunar Sci. Conference, (see 012.012), Vol. 2, 1407 - 1415 (1971).

094.406 **Oxygen isotope fractionation in Apollo 12 rocks and soils.** R. N. Clayton, N. Onuma, T. K. Mayeda.
Proc. Second Lunar Sci. Conference, (see 012.012), Vol. 2, 1417 - 1420 (1971).

094.407 **O^{18}/O^{16}, Si^{30}/Si^{28}, D/H, and C^{13}/C^{12} ratios in lunar samples.** S. Epstein, H. P. Taylor, Jr.
Proc. Second Lunar Sci. Conference, (see 012.012), Vol. 2, 1421 - 1441 = Contr. Div. Geol. Planet. Sci., California Inst. Technology, *Pasadena*, No. 1985 (1971).

094.408 **Vanadium isotopic composition and contents in lunar rocks and dust from the Ocean of Storms.**
M. E. Lipschutz, H. Balsiger, I. Z. Pelly.
Proc. Second Lunar Sci. Conference, (see 012.012), Vol. 2, 1443 - 1450 (1971).

094.409 **Variations in beryllium and chromium contents in lunar fines compared with crystalline rocks.**
R. E. Sievers, K. J. Eisentraut, D. J. Griest, M. F. Richardson, W. R. Wolf, W. D. Ross, N. M. Frew, T. L. Isenhour.
Proc. Second Lunar Sci. Conference, (see 012.012), Vol. 2, 1451 - 1459 (1971).

094.410 **Li, B, Mg, and Ti isotopic abundances and search for trapped solar wind Li in Apollo 11 and Apollo 12 material.** O. Eugster, R. Bernas.
Proc. Second Lunar Sci. Conference, (see 012.012), Vol. 2, 1461 - 1469 (1971).

094.411 **Rubidium—strontium chronology and chemistry of lunar material from the Ocean of Storms.**
W. Compston, H. Berry, M. J. Vernon, B. W. Chappell, M. J. Kaye.
Proc. Second Lunar Sci. Conference, (see 012.012), Vol. 2, 1471 - 1485 (1971).

094.412 **Sr isotopic measurements in Apollo 12 samples.** M. L. Bottino, P. D. Fullagar, C. C. Schnetzler,

J. A. Philpotts.
Proc. Second Lunar Sci. Conference, (see 012.012), Vol. 2,
1487 - 1491 (1971).

094.413 Rb–Sr and U, Th–Pb measurements on Apollo 12 material.
R. A. Cliff, C. Lee-Hu, G. W. Wetherill.
Proc. Second Lunar Sci. Conference, (see 012.012), Vol. 2,
1493 - 1502 (1971).

094.414 Lunar astrology – U–Th distributions and fission-track dating of lunar samples.
D. Burnett, M. Monnin, M. Seitz, R. Walker, D. Yuhas.
Proc. Second Lunar Sci. Conference, (see 012.012), Vol. 2,
1503 - 1519 (1971).

094.415 U–Th–Pb systematics of Apollo 12 lunar samples.
M. Tatsumoto, R. J. Knight, B. R. Doe.
Proc. Second Lunar Sci. Conference, (see 012.012), Vol. 2,
1521 - 1546 (1971).

094.416 Lead isotopes and volatile transfer in the lunar soil.
J. M. Huey, H. Ihochi, L. P. Black, R. G. Ostic,
T. P. Kohman.
Proc. Second Lunar Sci. Conference, (see 012.012), Vol. 2,
1547 - 1564 (1971).

094.417 Activation analysis determination of uranium and ^{204}Pb in Apollo 11 lunar fines.
A. Turkevich, G. W. Reed, Jr., H. R. Heydegger, J. Collister.
Proc. Second Lunar Sci. Conference, (see 012.012), Vol. 2,
1565 - 1570 (1971).

094.418 Isotopic abundances of actinide elements in Apollo 12 samples.
P. R. Fields, H. Diamond, D. N. Metta, C. M. Stevens, D. J. Rokop.
Proc. Second Lunar Sci. Conference, (see 012.012), Vol. 2,
1571 - 1576 (1971).

094.419 Isotopic composition of thorium and uranium in Apollo 12 samples.
J. N. Rosholt, M. Tatsumoto.
Proc. Second Lunar Sci. Conference, (see 012.012), Vol. 2,
1577 - 1584 (1971).

094.420 A search for superheavy elements in lunar material.
J. J. Wesolowski, W. John, D. Nease.
Proc. Second Lunar Sci. Conference, (see 012.012), Vol. 2,
1585 - 1589 (1971).

094.421 Kr^{81}–Kr and K–A^{40} ages, cosmic-ray spallation products, and neutron effects in lunar samples from Oceanus Procellarum. K. Marti, G. W. Lugmair.
Proc. Second Lunar Sci. Conference, (see 012.012), Vol. 2,
1591 - 1605 (1971).

094.422 Concentrations and isotopic abundances of the rare gases in lunar matter.
H. Hintenberger, H. W. Weber, N. Takaoka.
Proc. Second Lunar Sci. Conference, (see 012.012), Vol. 2,
1607 - 1625 (1971).

094.423 Rare gas measurements in three mineral separates of rock 12013,10,31. W. A. Kaiser.
Proc. Second Lunar Sci. Conference, (see 012.012), Vol. 2,
1627 - 1641 (1971).

094.424 Spallogenic Ne, Kr, and Xe from a depth study of 12002. E. C. Alexander, Jr.

Proc. Second Lunar Sci. Conference, (see 012.012), Vol. 2,
1643 - 1650 (1971).

094.425 Location and variation of trapped rare gases in Apollo 12 lunar samples.
T. Kirsten, F. Steinbrunn, J. Zähringer.
Proc. Second Lunar Sci. Conference, (see 012.012), Vol. 2,
1651 - 1669 (1971).

094.426 The irradiation history of lunar samples.
D. S. Burnett, J. C. Huneke, F. A. Podosek, G. Price
Russ III, G. J. Wasserburg.
Proc. Second Lunar Sci. Conference, (see 012.012), Vol. 2,
1671 - 1679 (1971).

094.427 Breccia 10065: release of inert gases by vacuum crushing at room temperature.
D. Heymann, A. Yaniv.
Proc. Second Lunar Sci. Conference, (see 012.012), Vol. 2,
1681 - 1692 (1971).

094.428 Stepwise heating analyses of rare gases from pile-irradiated rocks 10044 and 10057.
P. K. Davis, R. S. Lewis, J. H. Reynolds.
Proc. Second Lunar Sci. Conference, (see 012.012), Vol. 2,
1693 - 1703 (1971).

094.429 Stable rare gas isotopes produced by solar flares in single particles of Apollo 11 and Apollo 12 fines.
A. Yaniv, G. J. Taylor, S. Allen, D. Heymann.
Proc. Second Lunar Sci. Conference, (see 012.012), Vol. 2,
1705 - 1715 (1971).

094.430 Lunar atmosphere as a source of lunar surface elements. R. H. Manka, F. C. Michel.
Proc. Second Lunar Sci. Conference, (see 012.012), Vol. 2,
1717 - 1728 (1971).

094.431 Calculation of cosmogenic radionuclides in the moon and comparison with Apollo measurements.
T. W. Armstrong, R. G. Alsmiller, Jr.
Proc. Second Lunar Sci. Conference, (see 012.012), Vol. 2,
1729 - 1745 (1971).

094.432 Cosmogenic radionuclide concentrations and exposure ages of lunar samples from Apollo 12.
G. D. O'Kelley, J. S. Eldridge, E. Schonfeld, P. R. Bell.
Proc. Second Lunar Sci. Conference, (see 012.012), Vol. 2,
1747 - 1755 (1971).

094.433 Erosion and mixing of the lunar surface from cosmogenic and primordial radionuclide measurements in Apollo 12 lunar samples.
L. A. Rancitelli, R. W. Perkins, W. D. Felix, N. A. Wogman.
Proc. Second Lunar Sci. Conference, (see 012.012), Vol. 2,
1757 - 1772 (1971).

094.434 Depth variation of cosmogenic nuclides in a lunar surface rock and lunar soil.
R. C. Finkel, J. R. Arnold, M. Imamura, R. C. Reedy, J. S. Fruchter, H. H. Loosli, J. C. Evans, A. C. Delany, J. P. Shedlovsky.
Proc. Second Lunar Sci. Conference, (see 012.012), Vol. 2,
1773 - 1789 (1971).

094.435 Some cosmogenic and primordial radionuclides in Apollo 12 lunar surface materials. R. C. Wrigley.
Proc. Second Lunar Sci. Conference, (see 012.012), Vol. 2,
1791 - 1796 (1971).

094.436 Spallogenic ^{53}Mn ($T \sim 2 \times 10^6 \, y$) in lunar surface

material by neutron activation.
W. Herr, U. Herpers, R. Woelfle.
Proc. Second Lunar Sci. Conference, (see 012.012), Vol. 2,
1797 - 1802 (1971).

094.437 Tritium in lunar material.
P. Bochsler, P. Eberhardt, J. Geiss, H. Loosli, H.
Oeschger, M. Wahlen.
Proc. Second Lunar Sci. Conference, (see 012.012), Vol. 2,
1803 - 1812 (1971).

094.438 Radioactive rare gases and tritium in lunar rocks and
in the sample return container.
R. W. Stoenner, W. Lyman, R. Davis, Jr.
Proc. Second Lunar Sci. Conference, (see 012.012), Vol. 2,
1813 - 1823 (1971).

094.439 Tritium and argon radioactivities and their depth
variations in Apollo 12 samples.
J. D'Amico, J. DeFelice, E. L. Fireman, C. Jones, G. Spannagel.
Proc. Second Lunar Sci. Conference, (see 012.012), Vol. 2,
1825 - 1839 (1971).

094.440 Survey of lunar carbon compounds. I. The presence
of indigenous gases and hydrolysable carbon compounds in Apollo 11 and Apollo 12 samples.
P. I. Abell, P. H. Cadogan, G. Eglinton, J. R. Maxwell, C. T.
Pillinger.
Proc. Second Lunar Sci. Conference, (see 012.012), Vol. 2,
1843 - 1863 (1971).

094.441 Lunar pigments: porphyrin-like compounds from an
Apollo 12 sample.
G. W. Hodgson, E. Bunnenberg, B. Halpern, E. Peterson, K. A.
Kvenvolden, C. Ponnamperuma.
Proc. Second Lunar Sci. Conference, (see 012.012), Vol. 2,
1865 - 1874 (1971).

094.442 Absence of porphyrins in an Apollo 12 lunar surface
sample.
J. H. Rho, A. J. Bauman, T. F. Yen, J. Bonner.
Proc. Second Lunar Sci. Conference, (see 012.012), Vol. 2,
1875 - 1877 (1971).

094.443 The search for organic compounds in various Apollo
12 samples by mass spectrometry.
G. Preti, R. C. Murphy, K. Biemann.
Proc. Second Lunar Sci. Conference, (see 012.012), Vol. 2,
1879 - 1889 (1971).

094.444 Preliminary organic analysis of the Apollo 12 cores.
A. L. Burlingame, J. S. Hauser, B. R. Simoneit, D. H.
Smith, K. Biemann, N. Mancuso, R. Murphy, D. A. Flory,
M. A. Reynolds.
Proc. Second Lunar Sci. Conference, (see 012.012), Vol. 2,
1891 - 1899 (1971).

094.445 Study of carbon compounds in Apollo 11 and Apol-
lo 12 returned lunar samples.
W. Henderson, W. C. Kray, W. A. Newman, W. E. Reed, B. R.
Simoneit, M. Calvin.
Proc. Second Lunar Sci. Conference, (see 012.012), Vol. 2,
1901 - 1912 (1971).

094.446 Abundances and distribution of organogenic elements and compounds in Apollo 12 lunar samples.
J. Oró, D. A. Flory, J. M. Gibert, J. McReynolds, H. A. Lichtenstein, S. Wikstrom.
Proc. Second Lunar Sci. Conference, (see 012.012), Vol. 2,
1913 - 1925 (1971).

094.447 Search for alkanes containing 15 to 30 carbon atoms
per molecule in Apollo 12 lunar fines.
J. M. Mitchell, T. J. Jackson, R. P. Newlin, W. G. Meinschein,
E. Cordes, V. J. Shiner, Jr.
Proc. Second Lunar Sci. Conference, (see 012.012), Vol. 2,
1927 - 1928 (1971).

094.448 A search for biogenic structures in the Apollo 12
lunar samples.   J. W. Schopf.
Proc. Second Lunar Sci. Conference, (see 012.012), Vol. 2,
1929 - 1930 (1971).

094.449 Search for viable organisms in lunar samples: further
biological studies on Apollo 11 core, Apollo 12
bulk, and Apollo 12 core samples.
V. I. Oyama, E. L. Merek, M. P. Silverman, C. W. Boylen.
Proc. Second Lunar Sci. Conference, (see 012.012), Vol. 2,
1931 - 1937 (1971).

094.450 Microbial assay of lunar samples.
G. R. Taylor, W. Ellis, P. H. Johnson, K. Kropp,
T. Groves.
Proc. Second Lunar Sci. Conference, (see 012.012), Vol. 2,
1939 - 1948 (1971).

094.451 Apollo 12 lunar sample inventory.
Proc. Second Lunar Sci. Conference, (see 012.012),
Vol. 2, 1949; Vol. 3, 2797 (1971).

094.452 Lunar core tube sampling.
W. N. Houston, J. K. Mitchell.
Proc. Second Lunar Sci. Conference, (see 012.012), Vol. 3,
1953 - 1958 (1971).

094.453 Disturbance in samples recovered with the Apollo
core tubes.
W. D. Carrier III, S. W. Johnson, R. A. Werner, R. Schmidt.
Proc. Second Lunar Sci. Conference, (see 012.012), Vol. 3,
1959 - 1972 (1971).

094.454 Cone penetration resistance test – an approach to
evaluating in-place strength and packing of lunar
soils.   N. C. Costes, G. T. Cohron, D. C. Moss.
Proc. Second Lunar Sci. Conference, (see 012.012), Vol. 3,
1973 - 1987 (1971).

094.455 Particle size and shape distribution for lunar fines
sample 12057,72.   H. Heywood.
Proc. Second Lunar Sci. Conference, (see 012.012), Vol. 3,
1989 - 2001 (1971).

094.456 The formation of spherical glass particles on the
lunar surface.   J. O. Isard.
Proc. Second Lunar Sci. Conference, (see 012.012), Vol. 3,
2003 - 2008 (1971).

094.457 Interaction of gases with lunar materials: preliminary
results.
E. L. Fuller, Jr., H. F. Holmes, R. B. Gammage, K. Becker.
Proc. Second Lunar Sci. Conference, (see 012.012), Vol. 3,
2009 - 2019 (1971).

094.458 Particle size and shape distributions of lunar fines
by CESEMI.
H. Görz, E. W. White, R. Roy, G. G. Johnson, Jr.
Proc. Second Lunar Sci. Conference, (see 012.012), Vol. 3,
2021 - 2025 (1971).

094.459 Ultramicroscopic features in micron-sized lunar dust
grains and cosmophysics.

J. Borg, M. Maurette, L. Durrieu, C. Jouret.
Proc. Second Lunar Sci. Conference, (see 012.012), Vol. 3,
2027 - 2040 (1971).

094.460 **Morphology and petrostatistics of regular particles in Apollo 11 and Apollo 12 fines.** G. Mueller.
Proc. Second Lunar Sci. Conference, (see 012.012), Vol. 3,
2041 - 2047 (1971).

094.461 **Compositions, homogeneity, densities, and thermal history of lunar glass particles.**
C. H. Greene, L. D. Pye, H. J. Stevens, D. E. Rase, H. F. Kay.
Proc. Second Lunar Sci. Conference, (see 012.012), Vol. 3,
2049 - 2055 (1971).

094.462 **Exoelectron emission and surface characteristics of lunar materials.** R. B. Gammage, K. Becker.
Proc. Second Lunar Sci. Conference, (see 012.012), Vol. 3,
2057 - 2067 (1971).

094.463 **Lunar glass I: densification and relaxation studies.** R. Roy, D. M. Roy, S. Kurtossy, S. P. Faile.
Proc. Second Lunar Sci. Conference, (see 012.012), Vol. 3,
2069 - 2078 (1971).

094.464 **Neutron diffraction study of lunar materials.** S. J. Pickart, H. Alperin.
Proc. Second Lunar Sci. Conference, (see 012.012), Vol. 3,
2079 - 2082 (1971).

094.465 **Auger electron spectroscopy of lunar materials.** G. L. Connell, R. F. Schneidmiller, P. Kraatz, Y. P. Gupta.
Proc. Second Lunar Sci. Conference, (see 012.012), Vol. 3,
2083 - 2092 (1971).

094.466 **Some results from the Apollo 12 Suprathermal Ion Detector.**
J. W. Freeman, Jr., H. K. Hills, M. A. Fenner.
Proc. Second Lunar Sci. Conference, (see 012.012), Vol. 3,
2093 - 2102 (1971).

094.467 **Mössbauer instrumental analysis of Apollo 12 lunar rock and soil samples.**
C. L. Herzenberg, R. B. Moler, D. L. Riley.
Proc. Second Lunar Sci. Conference, (see 012.012), Vol. 3,
2103 - 2123 (1971).

094.468 **Mössbauer studies of Apollo 12 samples.** R. M. Housley, R. W. Grant, A. H. Muir, Jr., M. Blander, M. Abdel-Gawad.
Proc. Second Lunar Sci. Conference, (see 012.012), Vol. 3,
2125 - 2136 (1971).

094.469 **Infrared vibrational spectroscopic studies of minerals from Apollo 11 and Apollo 12 lunar samples.**
P. A. Estep, J. J. Kovach, C. Karr, Jr.
Proc. Second Lunar Sci. Conference, (see 012.012), Vol. 3,
2137 - 2151 (1971).

094.470 **Microchemical, microphysical, and adhesive properties of lunar material, II.**
J. J. Grossman, J. A. Ryan, N. R. Mukherjee, M. W. Wegner.
Proc. Second Lunar Sci. Conference, (see 012.012), Vol. 3,
2153 - 2164 (1971).

094.471 **Pressure-volume properties of two Apollo 12 basalts.** D. R. Stephens, E. M. Lilley.
Proc. Second Lunar Sci. Conference, (see 012.012), Vol. 3,
2165 - 2172 (1971).

094.472 **Some physical properties of Apollo 12 lunar samples.** T. Gold, B. T. O'Leary, M. Campbell.
Proc. Second Lunar Sci. Conference, (see 012.012), Vol. 3,
2173 - 2181 (1971).

094.473 **Optical properties of mineral separates, glass, and anorthositic fragments from Apollo mare samples.**
J. B. Adams, T. B. McCord.
Proc. Second Lunar Sci. Conference, (see 012.012), Vol. 3,
2183 - 2195 (1971).

094.474 **Spectral directional reflectance of lunar fines as a function of bulk density.**
R. C. Birkebak, C. J. Cremers, J. P. Dawson.
Proc. Second Lunar Sci. Conference, (see 012.012), Vol. 3,
2197 - 2202 (1971).

094.475 **Far infrared properties of lunar rock.** P. A. Ade, J. A. Bastin, A. C. Marston, S. J. Pandya, E. Puplett.
Proc. Second Lunar Sci. Conference, (see 012.012), Vol. 3,
2203 - 2211 (1971).

094.476 **Physical characterization of lunar glasses and fines.** W. B. White, E. W. White, H. Görz, H. K. Henisch, G. W. Fabel, R. Roy, J. N. Weber.
Proc. Second Lunar Sci. Conference, (see 012.012), Vol. 3,
2213 - 2221 (1971).

094.477 **Luminescence of Apollo 11 and Apollo 12 lunar samples.** N. N. Greenman, H. G. Gross.
Proc. Second Lunar Sci. Conference, (see 012.012), Vol. 3,
2223 - 2233 (1971).

094.478 **Luminescence and reflectance of Apollo 12 samples.** D. B. Nash, J. E. Conel.
Proc. Second Lunar Sci. Conference, (see 012.012), Vol. 3,
2235 - 2244 (1971).

094.479 **Radiation dose rates and thermal gradients in the lunar regolith: thermoluminescence and DTA of Apollo 12 samples.** H. P. Hoyt, Jr., M. Miyajima, R. M. Walker, D. W. Zimmerman, J. Zimmerman, D. Britton, J. L. Kardos.
Proc. Second Lunar Sci. Conference, (see 012.012), Vol. 3,
2245 - 2263 (1971).

094.480 **Luminescence of Apollo lunar samples.** J. E. Geake, G. Walker, A. A. Mills, G. F. J. Garlick.
Proc. Second Lunar Sci. Conference, (see 012.012), Vol. 3,
2265 - 2275 (1971). – Paper 1 of three collaborative papers.

094.481 **Thermoluminescence of lunar samples and terrestrial plagioclases.**
G. F. J. Garlick. W. E. Lamb, G. A. Steigmann, J. E. Geake.
Proc. Second Lunar Sci. Conference, (see 012.012), Vol. 3,
2277 - 2283 (1971). – Paper 2 of three collaborative papers.

094.482 **Polarimetric properties of the lunar surface and its interpretation. Part 3: Apollo 11 and Apollo 12 lunar samples.**
A. Dollfus, J. E. Geake, C. Titulaer.
Proc. Second Lunar Sci. Conference, (see 012.012), Vol. 3,
2285 - 2300 (1971). – Paper 3 of three collaborative papers.

094.483 **Apollo 12 multispectral photography experiment.** A. F. H. Goetz, F. C. Billingsley, J. W. Head, T. B. McCord, E. Yost.
Proc. Second Lunar Sci. Conference, (see 012.012), Vol. 3,
2301 - 2310 (1971).

094.484 Thermal conductivity of fines from Apollo 12.
C. J. Cremers, R. C. Birkebak.
Proc. Second Lunar Sci. Conference, (see 012.012), Vol. 3,
2311 - 2315 (1971).

094.485 Thermal expansion of lunar rocks.
W. S. Baldridge, G. Simmons.
Proc. Second Lunar Sci. Conference, (see 012.012), Vol. 3,
2317 - 2321 (1971).

094.486 Elastic wave velocities of Apollo 12 rocks at high
pressures.
H. Kanamori, H. Mizutani, Y. Hamano.
Proc. Second Lunar Sci. Conference, (see 012.012), Vol. 3,
2323 - 2326 (1971).

094.487 Elastic properties of Apollo 12 rocks.
H. Wang, T. Todd, D. Weidner, G. Simmons.
Proc. Second Lunar Sci. Conference, (see 012.012), Vol. 3,
2327 - 2336 (1971).

094.488 Surface elastic wave propagation studies in lunar
rocks.   B. R. Tittmann, R. M. Housley.
Proc. Second Lunar Sci. Conference, (see 012.012), Vol. 3,
2337 - 2343 (1971).

094.489 Elastic and thermal properties of Apollo 11 and
Apollo 12 rocks.
N. Warren, E. Schreiber, C. Scholz, J. A. Morrison, P. R.
Norton, M. Kumazawa, O. L. Anderson.
Proc. Second Lunar Sci. Conference, (see 012.012), Vol. 3,
2345 - 2360 (1971).

094.490 Specific heats of the lunar breccia (10021) and
olivine dolerite (12018) between 90° and 350°
Kelvin.   R. A. Robie, B. S. Hemingway.
Proc. Second Lunar Sci. Conference, (see 012.012), Vol. 3,
2361 - 2365 (1971).

094.491 Electrical properties of Apollo 11 and Apollo 12
lunar samples.   T. J. Katsube, L. S. Collett.
Proc. Second Lunar Sci. Conference, (see 012.012), Vol. 3,
2367 - 2379 (1971).

094.492 Dielectric behavior of lunar samples: electromagnetic probing of the lunar interior.
D. H. Chung, W. B. Westphal, G. Simmons.
Proc. Second Lunar Sci. Conference, (see 012.012), Vol. 3,
2381 - 2390 (1971).

094.493 The Apollo 12 magnetometer experiment: internal
lunar properties from transient and steady magnetic
field measurements.   P. Dyal, C. W. Parkin.
Proc. Second Lunar Sci. Conference, (see 012.012), Vol. 3,
2391 - 2413 (1971).

094.494 Lunar electrical conductivity from Apollo 12 magnetometer measurements: compositional and thermal inferences.   C. P. Sonett, G. Schubert, B. F. Smith, K.
Schwartz, D. S. Colburn.
Proc. Second Lunar Sci. Conference, (see 012.012), Vol. 3,
2415 - 2431 (1971).

094.495 Magnetic properties of individual glass spherules,
Apollo 11 and Apollo 12 lunar samples.
S. Sullivan, A. N. Thorpe, C. C. Alexander, F. E. Senftle,
E. Dwornik.
Proc. Second Lunar Sci. Conference, (see 012.012), Vol. 3,
2433 - 2449 (1971).

094.496 Magnetism of two Apollo 12 igneous rocks.

G. W. Pearce, D. W. Strangway, E. E. Larson.
Proc. Second Lunar Sci. Conference, (see 012.012), Vol. 3,
2451 - 2460 (1971).

094.497 Magnetic properties and remanent magnetization of
Apollo 12 lunar materials and Apollo 11 lunar mi-
crobreccia.
T. Nagata, R. M. Fisher, F. C. Schwerer, M. D. Fuller, J. R. Dunn.
Proc. Second Lunar Sci. Conference, (see 012.012), Vol. 3,
2461 - 2476 (1971).

094.498 Magnetic properties of some lunar crystalline rocks
returned by Apollo 11 and Apollo 12.
R. B. Hargraves, N. Dorety.
Proc. Second Lunar Sci. Conference, (see 012.012), Vol. 3,
2477 - 2483 (1971).

094.499 Evidence for an ancient lunar magnetic field.
C. E. Helsley.
Proc. Second Lunar Sci. Conference, (see 012.012), Vol. 3,
2485 - 2490 (1971).

094.500 Magnetic properties of Apollo 12 lunar samples
12052 and 12065.   C. S. Grommé, R. R. Doell.
Proc. Second Lunar Sci. Conference, (see 012.012), Vol. 3,
2491 - 2499 (1971).

094.501 Magnetic resonance properties of lunar samples:
mostly Apollo 12.
J. L. Kolopus, D. Kline, A. Chatelain, R. A. Weeks.
Proc. Second Lunar Sci. Conference, (see 012.012), Vol. 3,
2501 - 2514 (1971).

094.502 Magnetic resonance studies of Apollo 11 and Apollo
12 samples.
F.-D. Tsay, S. I. Chan, S. L. Manatt.
Proc. Second Lunar Sci. Conference, (see 012.012), Vol. 3,
2515 - 2528 (1971).

094.503 Clean lunar rock surfaces; unpaired electron density
and adsorptive capacity for oxygen.
D. Haneman, D. J. Miller.
Proc. Second Lunar Sci. Conference, (see 012.012), Vol. 3,
2529 - 2541 (1971).

094.504 Nuclear track studies of dynamic surface processes
on the moon and the constancy of solar activity.
G. Grozaz, R. Walker, D. Woolum.
Proc. Second Lunar Sci. Conference, (see 012.012), Vol. 3,
2543 - 2558 (1971).

094.505 The particle track record of the Ocean of Storms.
R. L. Fleischer, H. R. Hart, Jr., G. M. Comstock,
A. O. Evwaraye.
Proc. Second Lunar Sci. Conference, (see 012.012), Vol. 3,
2559 - 2568 (1971).

094.506 The particle track record of lunar soil.
G. M. Comstock, A. O. Evwaraye, R. L. Fleischer,
H. R. Hart, Jr.
Proc. Second Lunar Sci. Conference, (see 012.012), Vol. 3,
2569 - 2582 (1971).

094.507 The exposure history of the Apollo 12 regolith.
G. Arrhenius, S. Liang, D. Macdougall, L. Wilkening,
N. Bhandari, S. Bhat, D. Lal, G. Rajagopalan, A. S. Tamhane,
V. S. Venkatavaradan.
Proc. Second Lunar Sci. Conference, (see 012.012), Vol. 3,
2583 - 2598 (1971).

094.508 Spontaneous fission record of uranium and extinct

transuranic elements in Apollo samples.
N. Bhandari, S. Bhat, D. Lal, G. Rajagopalan, A. S. Tamhane, V. S. Venkatavaradan.
Proc. Second Lunar Sci. Conference, (see 012.012), Vol. 3, 2599 - 2609 (1971).

094.509 High resolution time averaged (millions of years) energy spectrum and chemical composition of iron-group cosmic ray nuclei at 1 A.U. based on fossil tracks in Apollo samples. N. Bhandari, S. Bhat, D. Lal, G. Rajagopalan, A. S. Tamhane, V. S. Venkatavaradan.
Proc. Second Lunar Sci. Conference, (see 012.012), Vol. 3, 2611 - 2619 (1971).

094.510 Ultra-heavy cosmic rays in the moon.
P. B. Price, R. S. Rajan, E. K. Shirk.
Proc. Second Lunar Sci. Conference, (see 012.012), Vol. 3, 2621 - 2627 (1971).

094.511 The lunar-surface orientation of some Apollo 12 rocks. F. Hörz, J. B. Hartung.
Proc. Second Lunar Sci. Conference, (see 012.012), Vol. 3, 2629 - 2638 = Lunar Sci. Inst., Houston, Texas, Contr. No. 14 (1971).

094.512 Meteorite impact craters, crater simulations, and the meteoroid flux in the early solar system.
M. R. Bloch, H. Fechtig, W. Gentner, G. Neukum, E. Schneider.
Proc. Second Lunar Sci. Conference, (see 012.012), Vol. 3, 2639 - 2652 (1971).

094.513 Influence of target temperature on crater morphology and implications on the origin of craters on lunar glass spheres. J. L. Carter, D. S. McKay.
Proc. Second Lunar Sci. Conference, (see 012.012), Vol. 3, 2653 - 2670 (1971).

094.514 Search for stable, fractionally charged particles (quarks) in lunar material.
C. M. Stevens, J. P. Schiffer, W. A. Chupka.
Proc. Second Lunar Sci. Conference, (see 012.012), Vol. 3, 2671 - 2674 (1971).

094.515 Evolution of mare surface. T. Gold.
Proc. Second Lunar Sci. Conference, (see 012.012), Vol. 3, 2675 - 2680 (1971).

094.516 Surveyor III material analysis program.
N. L. Nickle.
Proc. Second Lunar Sci. Conference, (see 012.012), Vol. 3, 2683 - 2697 (1971).

094.517 Examination of returned Surveyor III camera visor for alpha radioactivity.
T. E. Economou, A. L. Turkevich.
Proc. Second Lunar Sci. Conference, (see 012.012), Vol. 3, 2699 - 2703 (1971).

094.518 Solar flares, the lunar surface, and gas-rich meteorites. D. J. Barber, R. Cowsik, I. D. Hutcheon, P. B. Price, R. S. Rajan.
Proc. Second Lunar Sci. Conference, (see 012.012), Vol. 3, 2705 - 2714 (1971).

094.519 Microbiological sampling of returned Surveyor III electrical cabling.
M. D. Knittel, M. S. Favero, R. H. Green.
Proc. Second Lunar Sci. Conference, (see 012.012), Vol. 3, 2715 - 2719 (1971).

094.520 Surveyor III: bacterium isolated from lunar-retrieved TV camera. F. J. Mitchell, W. L. Ellis.
Proc. Second Lunar Sci. Conference, (see 012.012), Vol. 3, 2721 - 2733 (1971).

094.521 Discoloration and lunar dust contamination of Surveyor III surfaces. W. F. Carroll, P. M. Blair.
Proc. Second Lunar Sci. Conference, (see 012.012), Vol. 3, 2735 - 2742 (1971).

094.522 Examination of returned Surveyor III surface sampler. R. F. Scott, K. A. Zuckerman.
Proc. Second Lunar Sci. Conference, (see 012.012), Vol. 3, 2743 - 2751 (1971).

094.523 X-ray probe, SEM, and optical property analysis of the surface features of Surveyor III materials.
D. L. Anderson, B. E. Cunningham, R. G. Dahms, R. G. Morgan.
Proc. Second Lunar Sci. Conference, (see 012.012), Vol. 3, 2753 - 2765 (1971).

094.524 Results of the Surveyor III sample impact examination conducted at the Manned Spacecraft Center.
B. G. Cour-Palais, R. E. Flaherty, R. W. High, D. J. Kessler, D. S. McKay, H. A. Zook.
Proc. Second Lunar Sci. Conference, (see 012.012), Vol. 3, 2767 - 2780 (1971).

094.525 Micrometeoroid flux from Surveyor glass surfaces.
D. Brownlee, W. Bucher, P. Hodge.
Proc. Second Lunar Sci. Conference, (see 012.012), Vol. 3, 2781 - 2789 (1971).

094.526 Replication electron microscopy on Surveyor III unpainted aluminum tubing. E. A. Buvinger.
Proc. Second Lunar Sci. Conference, (see 012.012), Vol. 3, 2791 - 2795 (1971).

094.527 ^{39}Ar–^{40}Ar-Alter von Apollo 15 und Apollo 16 Mondgesteinen. P. Horn, T. Kirsten.
Max-Planck-Inst. Kernphysik Heidelberg, Jahresbericht 1972, p. 209 - 210 (1973).

094.528 Solare und kosmogene Edelgase in Apollo 16–Mondstaubproben. T. Kirsten.
Max-Planck-Inst. Kernphysik Heidelberg, Jahresbericht 1972, p. 211 - 212 (1973).

094.529 Edelgase im Apollo 15 Bohrkern des Mondregolithen. W. Hübner, D. Heymann, T. Kirsten.
Max-Planck-Inst. Kernphysik Heidelberg, Jahresbericht 1972, p. 212 - 213 (1973). – Abstract.

094.530 Diffusionsmessungen an Mondstaub-Einzelkörnern und künstlich implantierten Gläsern.
H. Ducati, S. Kalbitzer, J. Kiko, T. Kirsten, H. W. Müller.
Max-Planck-Inst. Kernphysik Heidelberg, Jahresbericht 1972, p. 213 - 214 (1973).

094.531 Messung der individuellen Variationen des He-Gehaltes in Apollo 14 - Mondproben mit der Helium-Mikrosonde. J. Deubner.
Max-Planck-Inst. Kernphysik Heidelberg, Jahresbericht 1972, p. 215 - 217 (1973).

094.532 Aufbau einer Edelgas-Ionensonde zur Messung von Konzentrationsprofilen in lunaren Proben.
H. W. Müller, G. Frey.
Max-Planck-Inst. Kernphysik Heidelberg, Jahresbericht 1972, p. 217 - 218 (1973).

094.533 Chemisch gebundene Stickstoffgehalte von Apollo 16- und Apollo 15-Mondstaubproben. O. Müller.
Max-Planck-Inst. Kernphysik Heidelberg, Jahresbericht 1972, p. 218 - 219 (1973).

094.534 Geochemie und Mineralogie der kristallinen Gesteine aus dem Mt. Hadley–Apenninen–Landegebiet.
A. El Goresy, L. A. Taylor, P. Ramdohr.
Max-Planck-Inst. Kernphysik Heidelberg, Jahresbericht 1972, p. 221 - 224 (1973).

094.535 Krateruntersuchungen an Mondproben und Experimente zur Kratersimulation.
E. Schneider, D. Storzer, J. B. Hartung, A. Mehl, K. Nagel, H. Fechtig, W. Gentner.
Max-Planck-Inst. Kernphysik Heidelberg, Jahresbericht 1972, p. 226 - 228 (1973).

094.536 Mikrometeoritischer Fluß, Entwicklung von Mikrokrater-Populationen und Erosionsraten von Mondgestein, sowie Bestimmung von Exponierungsalter von Apollo 16-Steinen aus Kraterstatistiken. G. Neukum.
Max-Planck-Inst. Kernphysik Heidelberg, Jahresbericht 1972, p. 228 - 231 (1973).

094.537 Partikel- und Spaltspuren in lunaren Gläsern.
D. Storzer.
Max-Planck-Inst. Kernphysik Heidelberg, Jahresbericht 1972, p. 231 - 232 (1973). – Abstract.

094.538 The infrared moon: a review. R. W. Shorthill.
Progr. Astronaut. Aeronaut., Vol. 28, (see 003.008), 3 - 49 (1972).

094.539 Microwave emission from the moon.
D. O. Muhleman.
Progr. Astronaut. Aeronaut., Vol. 28, (see 003.008), 51 - 81 (1972).

094.540 Radar mapping of lunar surface roughness.
T. W. Thompson, S. H. Zisk.
Progr. Astronaut. Aeronaut., Vol. 28, (see 003.008), 83 - 117 (1972).

094.541 Lunar thermal aspects from Surveyor data.
L. D. Stimpson, J. W. Lucas.
Progr. Astronaut. Aeronaut., Vol. 28, (see 003.008), 121 - 150 (1972).

094.542 Lunar surface temperatures from Apollo 11 data.
P. J. Hickson.
Progr. Astronaut. Aeronaut., Vol. 28, (see 003.008), 151 - 167 (1972).

094.543 Development of an in situ thermal conductivity measurement for the lunar heat flow experiment.
M. G. Langseth, Jr., E. M. Drake, D. Nathanson, J. A. Fountain.
Progr. Astronaut. Aeronaut., Vol. 28, (see 003.008), 169 - 204 (1972).

094.544 The Apollo 15 lunar heat flow measurement.
M. G. Langseth, Jr., S. P. Clark, Jr., J. Chute, Jr., S. Keihm.
Progr. Astronaut. Aeronaut., Vol. 28, (see 003.008), 205 - 212 (1972).

094.545 Thermal properties of granulated materials.
A. E. Wechsler, P. E. Glaser, J. A. Fountain.
Progr. Astronaut. Aeronaut., Vol. 28, (see 003.008), 215 - 241 (1972).

094.546 Thermal property measurements on lunar material returned by Apollo 11 and 12 missions.
K.-i. Horai, G. Simmons.
Progr. Astronaut. Aeronaut., Vol. 28, (see 003.008), 243 - 267 (1972).

094.547 Thermal characteristics of lunar surface roughness.
D. F. Winter, J. A. Bastin, D. A. Allen.
Progr. Astronaut. Aeronaut., Vol. 28, (see 003.008), 269 - 299 (1972).

094.548 Thermal history of the moon.
R. T. Reynolds, P. E. Fricker, A. L. Summers.
Progr. Astronaut. Aeronaut., Vol. 28, (see 003.008), 303 - 337 (1972).

094.549 Production of lunar fragmental material by meteoroid impact. A. H. Marcus.
Icarus, Vol. 18, 621 - 633 (1973).
It is the purpose of this note to extend the analyses by Shoemaker et al. (1969) and Shoemaker (1970) to a more complete and realistic analytical model of the rate of production of regolith material by size.

094.550 Ancient lunar mega-regolith and subsurface structure. W. K. Hartmann.
Icarus, Vol. 18, 634 - 636 (1973).
Effects of intense pre-mare cratering on subsurface structure and seismic properties are considered. A mega-regolith of fragmental material (possibly bonded at depth) exists not only in the terrae but possibly in subsurface layers under some maria.

094.551 Significance of a primitive lunar basaltic composition present in Apollo 15 soils and breccias.
D. H. Green, A. E. Ringwood.
Earth Planet. Sci. Letters, Vol. 19, 1 - 8 (1973).
Distinctive spherules and fragments of 'Green Glass' previously described from Apollo 15 soils, have olivine-rich magnesian picritic compositions. Experimental studies of high pressure melting relations in a compositionally similar Apollo 12 basalt (12040) lead to prediction of the nature of liquidus phases of Apollo 15 Green Glass at various pressures. It is argued that the Apollo 15 Green Glass unit was a product of very high degrees of partial melting of pyroxenite source rock with magma segregation occurring at 15 kb, $T = 1450°C$.

094.552 Petrology of the 2–4 mm soil fraction from the Hadley-Apennine region of the moon.
K. L. Cameron, J. W. Delano, A. E. Bence, J. J. Papike.
Earth Planet. Sci. Letters, Vol. 19, 9 - 21 (1973).
Regolith fragments in the size range 2–4 mm have been examined from four Apollo 15 sample stations, two on the Apennine Front (stations 2 and 6) and two on the mare surface (stations 4 and 9A). A co-operative study has been made of these fragments including both their mineralogy-petrology, reported herein, and their $^{40}Ar-^{39}Ar$ and cosmic-ray exposure ages, reported by Husain (1972; in preparation).

094.553 Basaltic vitrophyre 15597: an undifferentiated melt sample. P. W. Weigand, L. S. Hollister.
Earth Planet. Sci. Letters, Vol. 19, 61 - 74 (1973).
Apollo 15 sample 15597, from the rim of Hadley Rille, is a pyroxene vitrophyre consisting primarily of acicular pyroxene phenocrysts and glass matrix. Textural evidence, including the vitrophyric nature of the sample itself, the unusual compositions of the pigeonite centers (high Mg and low Ca) and chromites (high Cr), and the extreme chemical zonation of the pyroxenes all give strong evidence that this rock arrived at the lunar surface in an essentially entirely liquid state, and that eruption was followed by metastable pigeonite nucleation,

rapid metastable growth and continued metastable pyroxene nucleation, and final solidification. It thus may represent one of the best examples of a mare basalt completely unaffected by local differentiation.

094.554 **Alteration of an Apollo 12 sample by adsorption of water vapor.**
H. F. Holmes, E. L. Fuller, Jr., R. B. Gammage.
Earth Planet. Sci. Letters, Vol. 19, 90 - 96 (1973).

This article is concerned with the specific surface area of an Apollo 12 sample as determined by the inert gases nitrogen and argon, the reactivity of the sample toward water vapor, and the subsequent determination (by adsorption of nitrogen and argon) of any changes induced in the sample by its reaction with water.

094.555 **ALPO selected areas program: Alphonsus.**
C. Vaucher.
Strolling Astronomer, Vol. 24, 60 - 68 (1973).

094.556 **Apollo 16 rocks: Petrology and classification.**
H. G. Wilshire, D. E. Stuart-Alexander, E. D. Jackson.
Journ. Geophys. Res., Vol. 78, 2379 - 2392 (1973).

The Apollo 16 rocks are classified in three broad intergradational groups: (1) crystalline rocks, subdivided into igneous rocks and metaclastic rocks, (2) glass, and (3) breccias, which are subdivided into five groups on the basis of clast and matrix colors. Most of the rocks were derived by impact brecciation of an anorthosite-norite suite but may represent ejecta from more than one major basin.

094.557 **Paramagnetic resonance spectra of Ti^{3+}, Fe^{3+}, and Mn^{2+} in lunar plagioclases.** R. A. Weeks.
Journ. Geophys. Res., Vol. 78, 2393 - 2401 (1973).

At least three paramagnetic and one ferromagnetic components have been resolved in the electron magnetic resonance spectra of plagioclase fractions of two lunar samples, 14053-47 and 14321-166. One of the paramagnetic components is attributed to Fe^{3+} in crystal sites of low symmetries. A second paramagnetic component is due to the $M_s = 1/2 \leftrightarrow M_s = -1/2$ transition of Mn^{2+}. The third paramagnetic component is attributed to Ti^{3+} in crystal sites for which symmetry is lower than axial.

094.558 **A new theory of lunar magnetism.**
S. K. Runcorn, H. C. Urey.
Science, Vol. 180, 636 - 638 (1973).

In the hypothesis advanced here it is supposed that the field, in which rocks at the lunar surface acquired the remanent magnetization found through the Apollo project, arose from permanent magnetization of the deep interior of the moon.

094.559 **Far ultraviolet reflectivity of lunar dust samples: Apollo 11, 12, and 14.**
R. L. Lucke, R. C. Henry, W. G. Fastie.
Bull. American Astron. Soc., Vol. 5, 266 (1973). — Abstr. AAS.

094.560 **Preliminary results of the infrared mapping experiment on Apollo 17.** F. J. Low, W. W. Mendell.
Bull. American Astron. Soc., Vol. 5, 266 (1973). — Abstr. AAS.

094.561 **On the growth of the earth–moon system.**
F. L. Whipple.
Bull. American Astron. Soc., Vol. 5, 292 (1973). — Abstr. AAS.

094.562 **Eddy current heating of the lunar crust during early solar system evolution.**
C. P. Sonett, D. S. Colburn, K. Schwartz.
Bull. American Astron. Soc., Vol. 5, 292 (1973). — Abstr. AAS.

094.563 **Lunar electromagnetic sounding with the Apollos 12 and 15 lunar surface magnetometers.**
B. F. Smith, C. P. Sonett, D. S. Colburn, G. Schubert, K. Schwartz.
Bull. American Astron. Soc., Vol. 5, 292 (1973). — Abstr. AAS.

094.564 **Lunar crater statistics: crater origins and modification.** C. A. Wood.
Bull. American Astron. Soc., Vol. 5, 292 - 293 (1973). Abstr. AAS.

094.565 **Monochromatic phase curves and albedos for the lunar disk.** A. P. Lane, W. M. Irvine.
Bull. American Astron. Soc., Vol. 5, 293 (1973). — Abstr. AAS.

094.566 **Mare Serenitatis: Lunar surface types defined by remote observations.**
T. Thompson, E. Whitaker, R. Shorthill.
Bull. American Astron. Soc., Vol. 5, 293 (1973). — Abstr. AAS.

094.567 **Prediction of Apollo 17 and Luna 21 soil composition from telescopic observations.**
C. Pieters, T. B. McCord, J. B. Adams.
Bull. American Astron. Soc., Vol. 5, 293 (1973). — Abstr. AAS.

094.568 **Preliminary results from the infrared scanning radiometer on Apollo 17.** F. J. Low, W. W. Mendell.
Bull. American Astron. Soc., Vol. 5, 293 (1973). — Abstr. AAS.

094.569 **Preliminary results of the Apollo Lunar Sounder experiment.**
T. Thompson, R. Phillips, W. Brown, S. Ward, W. Peeples, G. Schaber, R. Eggleton, G. Adams, P. Jackson.
Bull. American Astron. Soc., Vol. 5, 308 - 309 (1973). Abstr. AAS.

094.570 **Laser transit-time measurements between the earth and the moon with a transportable system.**
C. G. Lehr, S. J. Criswell, J. P. Ouellette, P. W. Sozanski, J. D. Mulholland, P. J. Shelus.
Science, Vol. 180, 954 - 955 (1973).

A high-radiance, pulsed laser system with a transportable transmitting unit was used at Agassiz Station, Harvard College Observatory, Harvard, Massachusetts, to measure the transit times of 25-nanosecond, 10-joule, 530-nanometer pulses from the earth to the Apollo 15 retroreflector on the moon and back.

094.571 **Detection of a nonuniform distribution of polonium-210 on the moon with the Apollo 16 alpha particle spectrometer.**
P. Bjorkholm, L. Golub, P. Gorenstein.
Science, Vol. 180, 957- 959 (1973).

The polonium-210 activity of the lunar surface is significantly larger than the activity of its progenitor radon-222. This result establishes unequivocally that radon emanation from the present-day moon varies considerably within the 21-year half-life of lead-210, the parent nuclide of polonium-210. There are large variations and well-localized enhancements in polonium-210 activity over much of the moon's surface.

094.572 **Monte Carlo calculations of lunar regolith thickness distributions.**
V. R. Oberbeck, W. L. Quaide, M. Mahan, J. Paulson.
Icarus, Vol. 19, 87 - 107 (1973).

The purpose of this paper is to present results of a Monte Carlo computer simulation of regolith evolution that was designed to consider the full effect of the buffering regolith through calculation of the amount of debris produced by any given crater as a function of the amount of debris present at the site of the crater at the time of crater formation.

094.573 Induced magnetosphere of the moon, 1. Theory.
G. Schubert, C. P. Sonett, K. Schwartz, H. J. Lee.
Journ. Geophys. Res., Vol. 78, 2094 - 2110 (1973).

An analytic solution for the magnetic field in the space defined by a spherical moon and its downstream cylindrical cavity formed by the solar wind is derived for interplanetary magnetic fields both parallel and perpendicular to the cavity axis. By superposition, the solution is obtained for arbitrary orientations of the interplanetary field.

094.574 Role of pressure transients in the detection and identification of lunar surface gas sources.
F. G. Hall.
Journ. Geophys. Res., Vol. 78, 2111 - 2132 (1973).

The dynamic behavior of neutral gases emitted from lunar surface sources is investigated. Pressure transients to be expected from several types of possible lunar surface sources are constructed.

094.575 Thermal history and evolution of the moon.
M. N. Toksöz, S. C. Solomon.
The Moon, Vol. 7, 251 - 278 (1973). — Paper dedicated to Professor Harold C. Urey on the occasion of his 80th birthday on 29 April,1973.

Theoretical lunar temperature models are computed taking into account different initial conditions to represent possible accretion models and various abundances of heat sources to correspond to different compositions. Differentiation and convection are simulated in the numerical computational scheme.

094.576 Topology of induced lunar magnetic fields.
K. Schwartz, G. Schubert.
The Moon, Vol. 7, 279 - 292 (1973). — Paper dedicated to Professor Harold C. Urey on the occasion of his 80th birthday on 29 April, 1973.

Using the asymmetric theory of lunar induction derived by Schubert et al. (1973), we have obtained the total and induced magnetic field line structure within the moon and the diamagnetic cavity. Total field distributions are shown for orientations of the oscillating interplanetary field parallel, perpendicular and at 45° to the cavity axis. Induced field lines are shown only for the orientations of the interplanetary field parallel and orthogonal to the cavity axis.

094.577 Conjectures about the evolution of the moon.
T. Gold.
The Moon, Vol. 7, 293 - 306 (1973). — Paper dedicated to Professor Harold C. Urey on the occasion of his 80th birthday on 29 April, 1973.

The principal questions about the derivation of the lunar surface have not yet been settled: is it the surface left over from the process of accumulation of the moon, or is it a surface generated by magmatic processes on the moon and subsequently altered by further infall from outside? The evidence derived from many sources now favors the former.

094.578 Anisotropy of absorption bands in some lunar, meteoritic, and terrestrial pyroxenes. A. J. Cohen.
The Moon, Vol. 7, 307 - 321 (1973). — Paper dedicated to Professor Harold C. Urey on the occasion of his 80th birthday on 29 April, 1973.

The optical anisotropy of individual spin-forbidden transitions of Fe^{3+} and Fe^{2+} ions and spin-allowed transitions of Ti^{3+} and Fe^{2+} ions are observed in complex clinopyroxene crystals in lunar rock in both the pigeonite core and the augite overgrowth. Spectral bands are compared to similar ones in terrestrial augite from Maui, Hawaii, meteoritic augite in the Angra dos Reis achondrite, meteoritic shocked hypersthene in the Tatahauine achondrite and meteoritic diopside in the Nakhla achondrite. The maxima of Fe^{3+} ion-ligand charge-transfer bands in the ultraviolet are compared in these minerals.

094.579 Orbital mapping of the lunar magnetic field.
L. R. Sharp, P. J. Coleman, Jr., B. R. Lichtenstein, C. T. Russell, G. Schubert.
The Moon, Vol. 7, 322 - 341 (1973). — Paper dedicated to Professor Harold C. Urey on the occasion of his 80th birthday on 29 April, 1973.

Magnetometer data obtained during the first four lunations after the deployment of the Apollo 15 subsatellite have been used to construct contour maps of the lunar magnetic field referred to 100 km altitude. These contour maps cover a relatively small band on the lunar surface. Within the region covered there is a marked near side-far side asymmetry.

094.580 Darkening of silicate rock powders by solar wind sputtering. B. Hapke.
The Moon, Vol. 7, 342 - 355 (1973). — Paper dedicated to Professor Harold C. Urey on the occasion of his 80th birthday on 29 April, 1973.

Darkening of lunar igneous rock powders by the formation of solar wind-sputtered glass films is a real process which occurs on the moon. The time scale for darkening of undisturbed lunar soil is of the order of 50000–100000 yr. Comparison of the rates of the formation of glasses on the lunar surface by solar wind sputter-deposition, meteorite impact melting and impact vaporization-deposition indicates that these processes are of comparable importance under the present flux of meteorites. Thus the formation of glass by sputter-deposition must be regarded as a major process on the lunar surface.

094.581 Electrical conductivity, internal temperatures and thermal evolution of the moon.
A. Duba, A. E. Ringwood.
The Moon, Vol. 7, 356 - 376 (1973). — Paper dedicated to Professor Harold C. Urey on the occasion of his 80th birthday on 29 April, 1973.

The most important conclusion resulting from the present study is that the temperatures in the lunar interior at depths of 500–900 km are much higher than previously believed, and close to the solidus temperatures which were attained during the generation of maria basalts. Taking estimates of the radioactivity of the moon's deep interior obtained by 3 different methods, we conclude that the moon originally accreted at rather high temperatures – probably between 800°C and 1000°C. This inference has profound implications for early lunar thermal history and for problems of lunar origin. We restrict attention to certain cosmochemical implications.

094.582 Properties of the solar nebula and the origin of the moon. A. G. W. Cameron.
The Moon, Vol. 7, 377 - 383 (1973). — Paper dedicated to Professor Harold C. Urey on the occasion of his 80th birthday on 29 April, 1973.

The basic geochemical model of the structure of the moon proposed by Anderson, in which the moon is formed by differentiation of the calcium, aluminium, titanium-rich inclusions in the Allende meteorite, is accepted, and the conditions for formation of this moon within the solar nebula models of Cameron and Pine are discussed.

094.583 Lunar structure and dynamics – results from the Apollo passive seismic experiment.
G. Latham, M. Ewing, J. Dorman, Y. Nakamura, F. Press, N. Toksöz, G. Sutton, F. Duennebier, D. Lammlein.
The Moon, Vol. 7, 396 - 421 (1973). — Paper dedicated to Professor Harold C. Urey on the occasion of his 80th birthday on 29 April, 1973.

Analysis of seismic signals from man-made impacts, moonquakes, and meteoroid impacts has established the presence of a lunar crust, approximately 60 km thick in the region of the Apollo seismic network; an underlying zone of nearly constant seismic velocity extending to a depth of about 1000 km, referred to as the mantle; and a lunar core, beginning at a depth of about 1000 km, in which shear waves are highly attenuated suggesting the presence of appreciable melting. Seismic velocities in the crust reach 7 km s^{-1} beneath the lower-velocity surface zone.

094.584 Moon: 'Ghost' craters formed during mare filling.
D. P. Cruikshank, W. K. Hartmann, C. A. Wood.
The Moon, Vol. 7, 440 - 452 (1973). – Paper dedicated to Professor Harold C. Urey on the occasion of his 80th birthday on 29 April, 1973.

This paper discusses formation of 'pathological' cases of crater morphology due to interaction of craters with molten lavas. Terrestrial observations of such a process are discussed. Some specific lunar examples are discussed, including unusual shallow rings resembling experimental craters deformed by isostatic filling.

094.585 Progress in remote optical analysis of lunar surface composition. T. B. McCord, J. B. Adams.
The Moon, Vol. 7, 453 - 474 = Contr. MIT Planet. Astron. Lab., *Cambridge, Mass.*, No. 69 (1973). – Paper dedicated to Professor Harold C. Urey on the occasion of his 80th birthday on 29 April, 1973.

This article covers three general subjects. First, a brief discussion of the optical properties of ferro-silicate minerals; second, a more extensive description of telescope observations of the moon; and third, an interpretation of optical measurement of Apollo samples.

094.586 Lunar and terrestrial impact crater spherules.
P. W. Hodge.
The Moon, Vol. 7, 483 - 486 (1973). – Paper dedicated to Professor Harold C. Urey on the occasion of his 80th birthday on 29 April, 1973.

Comparisons between the chemistry of impact debris, especially spherical droplets of impact-formed material, are made for lunar samples and for soil samples taken from terrestrial impact sites. Differences are assigned to differences in the chemistry of impacting bodies and in the surface rocks, and to the influence of the atmosphere in the terrestrial cases.

094.587 Apollo 15 and 16 results of the integrated geochemical experiment.
I. Adler, J. I. Trombka, P. Lowman, R. Schmadebeck, H. Blodget, E. Eller, L. Yin, R. Lamothe, G. Osswald, J. Gerard, P. Gorenstein, P. Bjorkholm, H. Gursky, B. Harris, J. Arnold, A. Metzger, R. Reedy.
The Moon, Vol. 7, 487 - 504 (1973). – Paper dedicated to Professor Harold C. Urey on the occasion of his 80th birthday on 29 April, 1973.

A number of experiments carried in orbit on the Apollo 15 and 16 spacecraft were used in the compositional mapping of the lunar surface. The observations involved measurements of secondary (fluorescent) X-rays, gamma rays and alpha particle emissions. A large scale compositional map of over 20 % of the lunar surface was obtained for the first time. It was possible to demonstrate significant chemical differences between the mare and the highlands, to find specific areas of high radioactivity and to learn something about the composition of the moon's hidden side.

094.588 Application of the virial tensor to the determination of stress differences in the moon. F. Bocchio.
Mem. Soc. Astron. Italiana, Nuova Ser., Vol. 43, 669 - 674 (1973).

The volume average of the stress differences in the moon, due to the deviation from a hydrostatic J_2, is deduced by application of the second order virial tensor theorem. The paper takes into account a recently proposed density model and some recent data on the gravitational field of the moon deduced from satellite orbit perturbations.

094.589 Alkali- und Erdalkalielemente, La und U in Apollo 14-, 15- und 16- Mondmaterial. O. Müller.
Max-Planck-Inst. Kernphysik Heidelberg, Jahresbericht 1972, p. 220 (1973).

094.590 Distribution of craters on the moon and Mars in connection with their origin. N. Bonev.
Izv. Sekts. astron. Blg. AN, Vol. 5, 11 - 14 (1972). In Bulgarian. Abstr. in Referativ. Zhurn. 51. Astron., 4.51.412 (1973).

094.591 Gravitational field, relief and some questions concerning the internal structure of the moon.
A. I. Frolov.
Astron. vestn., Vol. 5, 201 - 216 (1971). In Russian.

All major features of the gravitational field of the moon find simple explanation in changes of the thickness of crust of the constructed models. The obtained results support the conclusions made by Soviet geologists about differentiated age of continental and mare areas of the moon: continents belong to oldest formations, while mare are younger formations now being in the stage of active development.

094.592 Observations of instationary phenomena on the moon based on materials of Russian observers.
P. V. Florenskij, V. M. Chernov.
Astron. vestn., Vol. 7, 38 - 44 (1973). In Russian.

094.593 On the physical nature of the albedo of the lunar surface. N. N. Evsyukov.
Astron. vestn., Vol. 7, 65 - 72 (1973). In Russian.

An attempt of physical interpretation of the distribution of albedo on the moon is made on the basis of the map of albedo of the visible hemisphere of the moon obtained by the author.

094.594 Volatile elements in Apollo 16 samples: Possible evidence for outgassing of the moon.
U. Krahenbuhl, R. Ganapathy, J. W. Morgan, E. Anders.
Science, Vol. 180, 858 - 861 (1973).

Several Apollo 16 breccias, including one containing goethite, are strikingly enriched in volatile elements such as bromine, cadmium, germanium, antimony, thallium, and zinc. Similar but smaller enrichments are found in all highland soils. It appears that volcanic processes took place in the lunar highlands, involving the release of volatiles including water.

094.595 The geology of the moon. A. J. W. Gleadow.
Journ. Astron. Soc. Victoria, Vol. 25, 81 - 87 (1972).

094.596 Simultaneous thermal and X-ray analysis of lunar samples and of minerals.
G. Bayer, H. G. Wiedemann.
Naturwissenschaften, 60. Jahrgang, p. 299 - 300 (1973). Short communication.

094.597 Observational equations of the libration problem in a lunar horizontal system. J. Mietelski.
Acta Astron., Vol. 23, 179 - 188 (1973).

A general matrix form of observation equations of $O-C$ type for the libration problem has been obtained on the basis of the principal Cracovian transformation of the geocentric equatorial system into a selenocentric horizontal one. Determinations of azimuth and altitude of a star with a lunar theodolite have been assumed. All the formulae have been tested numerically.

094.598 Transient lunar events and their connection with solar activity and tidal effect. P. V. Florenskij.
Solntse, ehlektrichestvo, zhizn', Moskva. Mosk. un-t, 1972, p. 36. In Russian.

094.599 Preliminary data on the lunar ground returned by the automatic station Luna 20. A. P. Vinogradov.
Vestn. AN SSSR, 1972, No. 10, p. 26 - 40. In Russian. – Abstr. in Referativ. Zhurn. 51. Astron., 5.51.364 (1973).

094.600 Apparent loss of angular momentum in the earth-moon system. D. H. Weinstein, J. Keeney.
Nature, Vol. 244, 83 - 84 (1973).
Pannella and others have presented palaeontological counts of the number of days per solar year and per synodic month during geologic time. The observations extend back to 1.7×10^9 yr with some incomplete data to 2.8×10^9 yr. It appears that there is a continual loss of momentum from the oldest point of 1.7×10^9 yr (a Precambrian stromatolite) to the present.

094.601 A response to a comment on U–Pb systematics in lunar basalts. F. Tera, G. J. Wasserburg.
Earth Planet. Sci. Letters, Vol. 19, 213 - 217 (1973).

094.602 Apollo 16 neutron stratigraphy. G. P. Russ III.
Earth Planet. Sci. Letters, Vol. 19, 275 - 289 (1973).
Models for the development of the regolith at the Apollo 16 site are fit to the isotopic data for Gd and Sm from the drill stem, and the layering implied by these models is compared with geological descriptions of the site and the layers observed in the cores by X-ray radiography.

094.603 Carrying out the experiment of selecting a sample from the lunar surface with the automatic lunar station Luna 20. V. P. Bulekov, L. Eh. Graf, D. D. Dryuchenko, B. V. Zakhar'ev, Eh. A. Motovilov, M. I. Smorodinov, Yu. N. Strelov, V. V. Shvarev.
Kosmich. Issled., Vol. 11, 460 - 464 (1973). In Russian.

094.604 Electrostatic charging of the lunar surface and possible consequences. K. Knott.
Journ. Geophys. Res., Vol. 78, 3172 - 3175 (1973). – Letter.

094.605 Geology of Hadley Rille. K. A. Howard, J. W. Head, G. A. Swann.
Proc. Third Lunar Sci. Conference, (see 012.022), Vol. 1, 1 - 14 (1972).

094.606 Lineaments of the Apennine Front – Apollo 15 landing site. E. W. Wolfe, N. G. Bailey.
Proc. Third Lunar Sci. Conference, (see 012.022), Vol. 1, 15 - 25 (1972).

094.607 Geology of the Apollo 14 landing site. R. L. Sutton, M. H. Hait, G. A. Swann.
Proc. Third Lunar Sci. Conference, (see 012.022), Vol. 1, 27 - 38 (1972).

094.608 New geological findings in Apollo 15 lunar orbital photography. F. El-Baz.
Proc. Third Lunar Sci. Conference, (see 012.022), Vol. 1, 39 - 61 (1972).

094.609 Significant results from Apollo 14 lunar orbital photography. F. El-Baz, S. A. Roosa.
Proc. Third Lunar Sci. Conference, (see 012.022), Vol. 1, 63 - 83 (1972).

094.610 Astronaut observations from lunar orbit and their geologic significance. F. El-Baz, A. M. Worden, V. D. Brand.
Proc. Third Lunar Sci. Conference, (see 012.022), Vol. 1, 85 - 104 (1972).

094.611 Mössbauer spectroscopy of lunar regolith returned by the automatic station Luna 16. T. V. Malysheva.
Proc. Third Lunar Sci. Conference, (see 012.022), Vol. 1, 105 - 114 (1972).

094.612 Petrology of Apollo 14 high-alumina basalt. I. Kushiro, Y. Ikeda, Y. Nakamura.
Proc. Third Lunar Sci. Conference, (see 012.022), Vol. 1, 115 - 129 (1972).

094.613 Petrography and crystallization history of basalts 14310 and 14072. J. Longhi, D. Walker, J. F. Hays.
Proc. Third Lunar Sci. Conference, (see 012.022), Vol. 1, 131 - 139 (1972).

094.614 Mineral-chemical variations in Apollo 14 and Apollo 15 basalts and granitic fractions. G. M. Brown, C. H. Emeleus, J. G. Holland, A. Peckett, R. Phillips.
Proc. Third Lunar Sci. Conference, (see 012.022), Vol. 1, 141 - 157 (1972).

094.615 Petrology of Fra Mauro basalt 14310. W. I. Ridley, R. Brett, R. J. Williams, H. Takeda, R. W. Brown.
Proc. Third Lunar Sci. Conference, (see 012.022), Vol. 1, 159 - 170 (1972).

094.616 Some textures in Apollo 12 lunar igneous rocks and in terrestrial analogs. H. I. Drever, R. Johnston, P. Butler, Jr., F. G. F. Gibb.
Proc. Third Lunar Sci. Conference, (see 012.022), Vol. 1, 171 - 184 (1972).

094.617 Equilibrium studies with a bearing on lunar rocks. A. Muan, J. Hauck, T. Löfall.
Proc. Third Lunar Sci. Conference, (see 012.022), Vol. 1, 185 - 196 (1972).

094.618 Experimental petrology and petrogenesis of Apollo 14 basalts. D. H. Green, A. E. Ringwood, N. G. Ware, W. O. Hibberson.
Proc. Third Lunar Sci. Conference, (see 012.022), Vol. 1, 197 - 206 (1972).

094.619 Role of water in the evolution of the lunar crust; an experimental study of sample 14310; an indication of lunar calc-alkaline volcanism. C. E. Ford, G. M. Biggar, D. J. Humphries, G. Wilson, D. Dixon, M. J. O'Hara.
Proc. Third Lunar Sci. Conference, (see 012.022), Vol. 1, 207 - 229 (1972).

094.620 Electron petrography of Apollo sample 14310. D. K. Smith, P. A. Thrower, W. P. Hoffman.
Proc. Third Lunar Sci. Conference, (see 012.022), Vol. 1, 231 - 241 (1972).

094.621 Mineralogical evidence for subsolidus vapor-phase transport of alkalis in lunar basalts. B. J. Skinner, H. Winchell.
Proc. Third Lunar Sci. Conference, (see 012.022), Vol. 1, 243 - 249 (1972).

094.622 Petrographic features and petrologic significance of melt inclusions in Apollo 14 and 15 rocks.
E. Roedder, P. W. Weiblen.
Proc. Third Lunar Sci. Conference, (see 012.022), Vol. 1, 251 - 279 (1972).

094.623 Uranium and potassium fractionation in pre-Imbrian lunar crustal rocks.
J. F. Lovering, D. A. Wark, A. J. W. Gleadow, D. K. B. Sewell.
Proc. Third Lunar Sci. Conference, (see 012.022), Vol. 1, 281 - 294 (1972).

094.624 Electron microprobe investigations of the oxidation states of Fe and Ti in ilmenite in Apollo 11, Apollo 12, and Apollo 14 crystalline rocks.
M. Pavićević, P. Ramdohr, A. El Goresy.
Proc. Third Lunar Sci. Conference, (see 012.022), Vol. 1, 295 - 303 (1972).

094.625 Apollo 14: Subsolidus reduction and compositional variations of spinels. S. E. Haggerty.
Proc. Third Lunar Sci. Conference, (see 012.022), Vol. 1, 305 - 332 (1972).

094.626 Fra Mauro crystalline rocks: Mineralogy, geochemistry and subsolidus reduction of the opaque minerals.
A. El Goresy, L. A. Taylor, P. Ramdohr.
Proc. Third Lunar Sci. Conference, (see 012.022), Vol. 1, 333 - 349 (1972).

094. 627 Mineralogical and petrographic features of two Apollo 14 rocks. P. Gay, M. G. Bown, I. D. Muir.
Proc. Third Lunar Sci. Conference, (see 012.022), Vol. 1, 351 - 362 (1972).

094.628 The major element compositions of lunar rocks as inferred from glass compositions in the lunar soils.
A. M. Reid, J. Warner, W. I. Ridley, D. A. Johnston, R. S. Harmon, P. Jakeš, R. W. Brown.
Proc. Third Lunar Sci. Conference, (see 012.022), Vol. 1, 363 - 378 (1972).

094.629 Analysis of Fra Mauro samples and the origin of the Imbrium Basin. M. R. Dence, A. G. Plant.
Proc. Third Lunar Sci. Conference, (see 012.022), Vol. 1, 379 - 399 (1972).

094.630 Electron petrography of Apollo 14 and 15 rocks.
J. S. Lally, R. M. Fisher, J. M. Christie, D. T. Griggs, A. H. Heuer, G. L. Nord, Jr., S. V. Radcliffe.
Proc. Third Lunar Sci. Conference, (see 012.022), Vol. 1, 401 - 422 (1972).

094.631 Crystallography and chemical trends of orthopyroxene-pigeonite from rock 14310 and coarse fine 12033. H. Takeda, W. I. Ridley.
Proc. Third Lunar Sci. Conference, (see 012.022), Vol. 1, 423 - 430 (1972).

094.632 Pyroxenes as recorders of lunar basalt petrogenesis: Chemical trends due to crystal-liquid interaction.
A. E. Bence, J. J. Papike.
Proc. Third Lunar Sci. Conference, (see 012.022), Vol. 1, 431 - 469 (1972).

094.633 Pyroxenes from breccia 14303.
P. W. Weigand, L. S. Hollister.
Proc. Third Lunar Sci. Conference, (see 012.022), Vol. 1, 471 - 480 (1972).

094.634 Fe^{2+}-Mg site distribution in Apollo 12021 clinopyroxenes: Evidence for bias in Mössbauer measurements, and relation of ordering to exsolution.
E. Dowty, M. Ross, F. Cuttitta.
Proc. Third Lunar Sci. Conference, (see 012.022), Vol. 1, 481 - 492 (1972).

094.635 Distinct subsolidus cooling histories of Apollo 14 basalts. K. Schürmann, S. S. Hafner.
Proc. Third Lunar Sci. Conference, (see 012.022), Vol. 1, 493 - 506 (1972).

094.636 Clinopyroxenes from Apollo 12 and 14: Exsolution, domain structure, and cation order.
S. Ghose, G. Ng, L. S. Walter.
Proc. Third Lunar Sci. Conference, (see 012.022), Vol. 1, 507 - 531 (1972).

094.637 Crystal field spectra of lunar pyroxenes.
R. G. Burns, R. M. Abu-Eid, F. E. Huggins.
Proc. Third Lunar Sci. Conference, (see 012.022), Vol. 1, 533 - 543 (1972).

094.638 Crystal-field effects of iron and titanium in selected grains of Apollo 12, 14, and 15 rocks, glasses, and fine fractions. P. M. Bell, H. K. Mao.
Proc. Third Lunar Sci. Conference, (see 012.022), Vol. 1, 545 - 553 (1972).

094.639 X-ray investigations of lunar plagioclases and pyroxenes. H. Jagodzinski, M. Korekawa.
Proc. Third Lunar Sci. Conference, (see 012.022), Vol. 1, 555 - 568 (1972).

094.640 Lunar plagioclase: A mineralogical study.
H.-R. Wenk, M. Ulbrich, W. F. Müller.
Proc. Third Lunar Sci. Conference, (see 012.022), Vol. 1, 569 - 579 (1972).

094.641 Twin laws, optic orientation, and composition of plagioclases from rocks 12051, 14053, and 14310.
E. Wenk, A. Glauser, H. Schwander, V. Trommsdorff.
Proc. Third Lunar Sci. Conference, (see 012.022), Vol. 1, 581 - 589 (1972).

094.642 Plagioclase and Ba—K phases from Apollo samples 12063 and 14310.
W. E. Trzcienski, Jr., C. G. Kulick.
Proc. Third Lunar Sci. Conference, (see 012.022), Vol. 1, 591 - 602 (1972).

094.643 Crystallographic studies of lunar plagioclases from samples 14053, 14163, 14301, and 14310.
M. Czank, K. Girgis, A. B. Harnik, F. Laves, R. Schmid, H. Schulz, L. Weber.
Proc. Third Lunar Sci. Conference, (see 012.022), Vol. 1, 603 - 613 (1972).

094.644 On the amount of ferric iron in plagioclases from lunar igneous rocks. K. Schürmann, S. S. Hafner.
Proc. Third Lunar Sci. Conference, (see 012.022), Vol. 1, 615 - 621 (1972).

094.645 Metamorphism of Apollo 14 breccias.
J. L. Warner.
Proc. Third Lunar Sci. Conference, (see 012.022), Vol. 1, 623 - 643 (1972).

094.646 Apollo 14 breccias: General characteristics and classification. E. C. T. Chao, J. A. Minkin, J. B. Best.

Proc. Third Lunar Sci. Conference, (see 012.022), Vol. 1, 645 - 659 (1972).

094.647 Apollo 14 breccia 14313: A mineralogic and petrologic report.
R. J. Floran, K. L. Cameron, A. E. Bence, J. J. Papike.
Proc. Third Lunar Sci. Conference, (see 012.022), Vol. 1, 661 - 671 (1972).

094.648 Chondrules in Apollo 14 samples and size analyses of Apollo 14 and 15 fines.
E. A. King, Jr., J. C. Butler, M. F. Carman.
Proc. Third Lunar Sci. Conference, (see 012.022), Vol. 1, 673 - 686 (1972).

094.649 Petrology and chemistry of some Apollo 14 lunar samples.
V. C. Juan, J. C. Chen, C. K. Huang, P. Y. Chen, C. M. Wang Lee.
Proc. Third Lunar Sci. Conference, (see 012.022), Vol. 1, 687 - 705 (1972).

094.650 Chondrules of lunar origin.
G. Kurat, K. Keil, M. Prinz, C. E. Nehru.
Proc. Third Lunar Sci. Conference, (see 012.022), Vol. 1, 707 - 721 (1972).

094.651 Lunar glasses, breccias, and chondrules.
J. Nelen, A. Noonan, K. Fredriksson.
Proc. Third Lunar Sci. Conference, (see 012.022), Vol. 1, 723 - 737 (1972).

094.652 Vapor phase crystallization in Apollo 14 breccia.
D. S. McKay, U. S. Clanton, D. A. Morrison, G. H. Ladle.
Proc. Third Lunar Sci. Conference, (see 012.022), Vol. 1, 739 - 752 (1972).

094.653 Apollo 14 regolith and fragmental rocks, their compositions and origin by impacts.
W. von Engelhardt, J. Arndt, D. Stöffler, H. Schneider.
Proc. Third Lunar Sci. Conference, (see 012.022), Vol. 1, 753 - 770 (1972).

094.654 Mineralogy and origin of Fra Mauro fines and breccias. W. Quaide, R. Wrigley.
Proc. Third Lunar Sci. Conference, (see 012.022), Vol. 1, 771 - 784 (1972).

094.655 Mineralogy, petrology, and chemical composition of lunar samples 15085, 15256, 15271, 15471, 15475, 15476, 15535, 15555, and 15556.
B. Mason, E. Jarosewich, W. G. Melson, G. Thompson.
Proc. Third Lunar Sci. Conference, (see 012.022), Vol. 1, 785 - 796 (1972).

094.656 Experimental petrology and origin of Fra Mauro rocks and soil.
D. Walker, J. Longhi, J. F. Hays.
Proc. Third Lunar Sci. Conference, (see 012.022), Vol. 1, 797 - 817 (1972).

094.657 Thermal and mechanical history of breccias 14306, 14063, 14270, and 14321.
A. T. Anderson, T. F. Braziunas, J. Jacoby, J. V. Smith.
Proc. Third Lunar Sci. Conference, (see 012.022), Vol. 1, 819 - 835 (1972).

094.658 Petrology and origin of lithic fragments in the Apollo 14 regolith. B. N. Powell, P. W. Weiblen.
Proc. Third Lunar Sci. Conference, (see 012.022), Vol. 1,

837 - 852 (1972).

094.659 Inclusions and interface relationships between glass and breccia in lunar sample 14306,50.
J. F. Wosinski, J. P. Williams, E. J. Korda, W. T. Kane, G. B. Carrier, J. W. H. Schreurs.
Proc. Third Lunar Sci. Conference, (see 012.022), Vol. 1, 853 - 864 (1972).

094.660 Rock 14068: An unusual lunar breccia.
R. T. Helz.
Proc. Third Lunar Sci. Conference, (see 012.022), Vol. 1, 865 - 886 (1972).

094.661 The magnesian spinel-bearing rocks from the Fra Mauro formation.
M. C.-M.-Levy, C. Levy, R. Caye, R. Pierrot.
Proc. Third Lunar Sci. Conference, (see 012.022), Vol. 1, 887 - 894 (1972).

094.662 Deformation of silicates in some Fra Mauro breccias.
H. G. Avé Lallemant, N. L. Carter.
Proc. Third Lunar Sci. Conference, (see 012.022), Vol. 1, 895 - 906 (1972).

094.663 Apollo 14 glasses of impact origin and their parent rock types. E. C. T. Chao, J. B. Best, J. A. Minkin.
Proc. Third Lunar Sci. Conference, (see 012.022), Vol. 1, 907 - 925 (1972).

094.664 Chemistry and particle track studies of Apollo 14 glasses. B. P. Glass, D. Storzer, G. A. Wagner.
Proc. Third Lunar Sci. Conference, (see 012.022), Vol. 1, 927 - 937 (1972).

094.665 Structure of lunar glasses by Raman and soft X-ray spectroscopy.
G. W. Fabel, W. B. White, E. W. White, R. Roy.
Proc. Third Lunar Sci. Conference, (see 012.022), Vol. 1, 939 - 951 (1972).

094.666 Metallic mounds produced by reduction of material of simulated lunar composition and implications on the origin of metallic mounds on lunar glasses.
J. L. Carter, D. S. McKay.
Proc. Third Lunar Sci. Conference, (see 012.022), Vol. 1, 953 - 970 (1972).

094.667 Compositions and mineralogy of lithic fragments in 1–2 mm soil samples 14002,7 and 14258,33.
I. M. Steele, J. V. Smith.
Proc. Third Lunar Sci. Conference, (see 012.022), Vol. 1, 971 - 981 (1972).

094.668 Apollo 14 soils: Size distribution and particle types. D. S. McKay, G. H. Heiken, R. M. Taylor, U. S. Clanton, D. A. Morrison, G. H. Ladle.
Proc. Third Lunar Sci. Conference, (see 012.022), Vol. 1, 983 - 994 (1972).

094.669 Noritic fragments in the Apollo 14 and 12 soils and the origin of Oceanus Procellarum.
G. J. Taylor, U. B. Marvin, J. B. Reid, Jr., J. A. Wood.
Proc. Third Lunar Sci. Conference, (see 012.022), Vol. 1, 995 - 1014 (1972).

094.670 Chemical and petrographic characterization of Fra Mauro soils. M. H. Carr, C. E. Meyer.
Proc. Third Lunar Sci. Conference, (see 012.022), Vol. 1, 1015 - 1027 (1972).

094.671 **Chromatographic and mineralogical study of Apollo 14 fines.** C. R. Masson, I. B. Smith, W. D. Jamieson, J. L. McLachlan, A. Volborth.
Proc. Third Lunar Sci. Conference, (see 012.022), Vol. 1, 1029 - 1036 (1972).

094.672 **Metallic particles in the Apollo 14 lunar soil.**
J. I. Goldstein, H. J. Axon, C. F. Yen.
Proc. Third Lunar Sci. Conference, (see 012.022), Vol. 1, 1037 - 1064 (1972).

094.673 **Study of excess Fe metal in the lunar fines by magnetic separation, Mössbauer spectroscopy, and microscopic examination.**
R. M. Housley, R. W. Grant, M. Abdel-Gawad.
Proc. Third Lunar Sci. Conference, (see 012.022), Vol. 1, 1065 - 1076 (1972).

094.674 **On lunar metallic particles and their contribution to the trace element content of Apollo 14 and 15 soils.**
F. Wlotzka, E. Jagoutz, B. Spettel, H. Baddenhausen, A. Balacescu, H. Wänke.
Proc. Third Lunar Sci. Conference, (see 012.022), Vol. 1, 1077 - 1084 (1972).

094.675 **Glassy particles in Apollo 14 soil 14163,88: Peculiarities and genetic considerations.**
G. Cavarretta, A. Coradini, R. Funiciello, M. Fulchignoni, A. Taddeucci, R. Trigila.
Proc. Third Lunar Sci. Conference, (see 012.022), Vol. 1, 1085 - 1094 (1972).

094.676 **Mineralogy, petrology, and surface features of some fragmental material from the Fra Mauro site.**
C. Klein, Jr., J. C. Drake.
Proc. Third Lunar Sci. Conference, (see 012.022), Vol. 1, 1095 - 1113 (1972).

094.677 **A new titanium and zirconium oxide from the Apollo 14 samples.**
C. Levy, M. C.-M.-Levy, P. Picot, R. Caye.
Proc. Third Lunar Sci. Conference, (see 012.022), Vol. 1, 1115 - 1120 (1972).

094.678 **Electron microscopy of some experimentally shocked counterparts of lunar minerals.**
C. B. Sclar, S. P. Morzenti.
Proc. Third Lunar Sci. Conference, (see 012.022), Vol. 1, 1121 - 1132 (1972).

094.679 **Distribution of elements between different phases of Apollo 14 rocks and soils.**
A. O. Brunfelt, K. S. Heier, B. Nilssen, B. Sundvoll, E. Steinnes.
Proc. Third Lunar Sci. Conference, (see 012.022), Vol. 2, 1133 - 1147 (1972).

094.680 **Oxygen and bulk element composition studies of Apollo 14 and other lunar rocks and soils.**
W. D. Ehmann, D. E. Gillum, J. W. Morgan.
Proc. Third Lunar Sci. Conference, (see 012.022), Vol. 2, 1149 - 1160 (1972).

094.681 **Nonmare basalts: Part II.**
N. J. Hubbard, P. W. Gast, J. M. Rhodes, B. M. Bansal, H. Wiesmann, S. E. Church.
Proc. Third Lunar Sci. Conference, (see 012.022), Vol. 2, 1161 - 1179 (1972).

094.682 **Bulk, rare earth, and other trace elements in Apollo 14 and 15 and Luna 16 samples.**
J. C. Laul, H. Wakita, D. L. Showalter, W. V. Boynton, R. A. Schmitt.
Proc. Third Lunar Sci. Conference, (see 012.022), Vol. 2, 1181 - 1200 (1972).

094.683 **Compositional characteristics of some Apollo 14 clastic materials.** M. M. Lindstrom, A. R. Duncan, J. S. Fruchter, S. M. McKay, J. W. Stoeser, G. G. Goles, D. J. Lindstrom.
Proc. Third Lunar Sci. Conference, (see 012.022), Vol. 2, 1201 - 1214 (1972).

094.684 **Compositional data for twenty-one Fra Mauro lunar materials.** H. J. Rose, Jr., F. Cuttitta, C. S. Annell, M. K. Carron, R. P. Christian, E. J. Dwornik, L. P. Greenland, D. T. Ligon, Jr.
Proc. Third Lunar Sci. Conference, (see 012.022), Vol. 2, 1215 - 1229 (1972).

094.685 **Composition of the lunar uplands: Chemistry of Apollo 14 samples from Fra Mauro.**
S. R. Taylor, M. Kaye, P. Muir, W. Nance, R. Rudowski, N. Ware.
Proc. Third Lunar Sci. Conference, (see 012.022), Vol. 2, 1231 - 1249 (1972).

094.686 **Multielement analyses of lunar samples and some implications of the results.**
H. Wänke, H. Baddenhausen, A. Balacescu, F. Teschke, B. Spettel, G. Dreibus, H. Palme, M. Quijano-Rico, H. Kruse, F. Wlotzka, F. Begemann.
Proc. Third Lunar Sci. Conference, (see 012.022), Vol. 2, 1251 - 1268 (1972).

094.687 **Major, minor, and trace element data for some Apollo 11, 12, 14, and 15 samples.**
J. P. Willis, A. J. Erlank, J. J. Gurney, R. H. Theil, L. H. Ahrens.
Proc. Third Lunar Sci. Conference, (see 012.022), Vol. 2, 1269 - 1273 (1972).

094.688 **Rare earths and other trace elements in Apollo 14 samples.**
P. A. Helmke, L. A. Haskin, R. L. Korotev, K. E. Ziege.
Proc. Third Lunar Sci. Conference, (see 012.022), Vol. 2, 1275 - 1292 (1972).

094.689 **Apollo 14: Some geochemical aspects.** J. A. Philpotts, C. C. Schnetzler, D. F. Nava, M. L. Bottino, P. D. Fullagar, H. H. Thomas, S. Schuhmann, C. W. Kouns.
Proc. Third Lunar Sci. Conference, (see 012.022), Vol. 2, 1293 - 1305 (1972).

094.690 **Precise determination of rare-earth elements in the Apollo 14 and 15 samples.**
A. Masuda, N. Nakamura, H. Kurasawa, T. Tanaka.
Proc. Third Lunar Sci. Conference, (see 012.022), Vol. 2, 1307 - 1313 (1972).

094.691 **Provenance of Apollo 12 KREEP.** J. T. Wasson, P. A. Baedecker.
Proc. Third Lunar Sci. Conference, (see 012.022), Vol. 2, 1315 - 1326 (1972).

094.692 **Beryllium and chromium abundances in Fra Mauro and Hadley-Apennine lunar samples.**
K. J. Eisentraut, M. S. Black, F. D. Hileman, R. E. Sievers, W. D. Ross.
Proc. Third Lunar Sci. Conference, (see 012.022), Vol. 2, 1327 - 1333 (1972).

094.693 Chemical analyses of lunar samples 14003, 14311, and 14321.   J. H. Scoon.
Proc. Third Lunar Sci. Conference, (see 012.022), Vol. 2, 1335 - 1336 (1972).

094.694 Analysis of lunar samples 14163, 14259, and 14321 with isotopic data for ^{7}Li/^{6}Li.
A. Strasheim, P. F. S. Jackson, J. H. J. Coetzee, F. W. E. Strelow, F. T. Wybenga, A. J. Gricius, M. L. Kokot, R. H. Scott.
Proc. Third Lunar Sci. Conference, (see 012.022), Vol. 2, 1337 - 1342 (1972).

094.695 The extralunar component in lunar soils and breccias.
P. A. Baedecker, C.-L. Chou, J. T. Wasson.
Proc. Third Lunar Sci. Conference, (see 012.022), Vol. 2, 1343 - 1359 (1972).

094.696 Trace elements in Apollo 15 samples: Implications for meteorite influx and volatile depletion on the moon.
J. W. Morgan, U. Krähenbühl, R. Ganapathy, E. Anders.
Proc. Third Lunar Sci. Conference, (see 012.022), Vol. 2, 1361 - 1376 (1972).

094.697 Major impacts on the moon: Characterization from trace elements in Apollo 12 and 14 samples.
J. W. Morgan, J. C. Laul, U. Krähenbühl, R. Ganapathy, E. Anders.
Proc. Third Lunar Sci. Conference, (see 012.022), Vol. 2, 1377 - 1395 (1972).

094.698 The abundances of components of the lunar soils by a least-squares mixing model and the formation age of KREEP.   E. Schonfeld, C. Meyer, Jr.
Proc. Third Lunar Sci. Conference, (see 012.022), Vol. 2, 1397 - 1420 (1972).

094.699 ESCA-investigation of lunar regolith from the Seas of Fertility and Tranquility.
A. P. Vinogradov, V. I. Nefedov, V. S. Urusov, N. M. Zhavoronkov.
Proc. Third Lunar Sci. Conference, (see 012.022), Vol. 2, 1421 - 1427 (1972).

094.700 O^{18}/O^{16}, Si^{30}/Si^{28}, C^{13}/C^{12}, and D/H studies of Apollo 14 and 15 samples.
S. Epstein, H. P. Taylor, Jr.
Proc. Third Lunar Sci. Conference, (see 012.022), Vol. 2, 1429 - 1454 (1972).

094.701 Oxygen isotopic compositions and oxygen concentrations of Apollo 14 and Apollo 15 rocks and soils.
R. N. Clayton, J. M. Hurd, T. K. Mayeda.
Proc. Third Lunar Sci. Conference, (see 012.022), Vol. 2, 1455 - 1463 (1972).

094.702 Isotopic abundance ratios and concentrations of selected elements in Apollo 14 samples.
I. L. Barnes, B. S. Carpenter, E. L. Garner, J. W. Gramlich, E. C. Kuehner, L. A. Machlan, E. J. Maienthal, J. R. Moody, L. J. Moore, T. J. Murphy, P. J. Paulsen, K. M. Sappenfield, W. R. Shields.
Proc. Third Lunar Sci. Conference, (see 012.022), Vol. 2, 1465 - 1472 (1972).

094.703 Deuterium content of lunar material.
L. Merlivat, G. Nief, E. Roth.
Proc. Third Lunar Sci. Conference, (see 012.022), Vol. 2, 1473 - 1477 (1972).

094.704 Sulphur concentrations and isotope ratios in lunar samples.   C. E. Rees, H. G. Thode.
Proc. Third Lunar Sci. Conference, (see 012.022), Vol. 2, 1479 - 1485 (1972).

094.705 Apollo 14 mineral ages and the thermal history of the Fra Mauro formation.
W. Compston, M. J. Vernon, H. Berry, R. Rudowski, C. M. Gray, N. Ware, B. W. Chappell, M. Kaye.
Proc. Third Lunar Sci. Conference, (see 012.022), Vol. 2, 1487 - 1501 (1972).

094.706 Apollo 14 and 15 samples: Rb-Sr ages, trace elements, and lunar evolution.
V. R. Murthy, N. M. Evensen, Bor-Ming Jahn, M. R. Coscio, Jr.
Proc. Third Lunar Sci. Conference, (see 012.022), Vol. 2, 1503 - 1514 (1972).

094.707 Rb-Sr systematics for chemically defined Apollo 14 breccias.   L. E. Nyquist, N. J. Hubbard, P. W. Gast, S. E. Church, B. M. Bansal, H. Wiesmann.
Proc. Third Lunar Sci. Conference, (see 012.022), Vol. 2, 1515 - 1530 (1972).

094.708 U—Th—Pb and Rb—Sr measurements on some Apollo 14 lunar samples.
M. Tatsumoto, C. E. Hedge, B. R. Doe, D. M. Unruh.
Proc. Third Lunar Sci. Conference, (see 012.022), Vol. 2, 1531 - 1555 (1972).

094.709 The ages of lunar material from Fra Mauro, Hadley Rille, and Spur Crater.
L. Husain, O. A. Schaeffer, J. Funkhouser, J. Sutter.
Proc. Third Lunar Sci. Conference, (see 012.022), Vol. 2, 1557 - 1567 (1972).

094.710 K—Ar dating of lunar fines: Apollo 12, Apollo 14, and Luna 16.
R. O. Pepin, J. G. Bradley, J. C. Dragon, L. E. Nyquist.
Proc. Third Lunar Sci. Conference, (see 012.022), Vol. 2, 1569 - 1588 (1972).

094.711 Ar^{40}-Ar^{39} systematics in rocks and separated minerals from Apollo 14.
G. Turner, J. C. Huneke, F. A. Podosek, G. J. Wasserburg.
Proc. Third Lunar Sci. Conference, (see 012.022), Vol. 2, 1589 - 1612 (1972).

094.712 ^{40}Ar—^{39}Ar ages of Apollo 14 and 15 samples.
D. York, W. J. Kenyon, R. J. Doyle.
Proc. Third Lunar Sci. Conference, (see 012.022), Vol. 2, 1613 - 1622 (1972).

094.713 Uranium and extinct Pu^{244} effects in Apollo 14 materials.   G. Crozaz, R. Drozd, H. Graf, C. M. Hohenberg, M. Monnin, D. Ragan, C. Ralston, M. Seitz, J. Shirck, R. M. Walker, J. Zimmerman.
Proc. Third Lunar Sci. Conference, (see 012.022), Vol. 2, 1623 - 1636 (1972).

094.714 ^{237}Np, ^{236}U, and other actinides on the moon.
P. R. Fields, H. Diamond, D. N. Metta, D. J. Rokop, C. M. Stevens.
Proc. Third Lunar Sci. Conference, (see 012.022), Vol. 2, 1637 - 1644 (1972).

094.715 ^{204}Pb in Apollo 14 samples and inferences regarding primordial Pb lunar geochemistry.
R. O. Allen, Jr., S. Jovanovic, G. W. Reed, Jr.
Proc. Third Lunar Sci. Conference, (see 012.022), Vol. 2, 1645 - 1650 (1972).

094.716 **Abundances of primordial and cosmogenic radio-nuclides in Apollo 14 rocks and fines.**
J. S. Eldridge, G. D. O'Kelley, K. J. Northcutt.
Proc. Third Lunar Sci. Conference, (see 012.022), Vol. 2, 1651 - 1658 (1972).

094.717 **Primordial radioelements and cosmogenic radio-nuclides in lunar samples from Apollo 15.**
G. D. O'Kelley, J. S. Eldridge, K. J. Northcutt, E. Schonfeld.
Proc. Third Lunar Sci. Conference, (see 012.022), Vol. 2, 1659 - 1670 (1972).

094.718 **Gamma-ray measurments of Apollo 12, 14, and 15 lunar samples.**
J. E. Keith, R. S. Clark, K. A. Richardson.
Proc. Third Lunar Sci. Conference, (see 012.022), Vol. 2, 1671 - 1680 (1972).

094.719 **Lunar surface processes and cosmic ray characterization from Apollo 12—15 lunar sample analyses.**
L. A. Rancitelli, R. W. Perkins, W. D. Felix, N. A. Wogman.
Proc. Third Lunar Sci. Conference, (see 012.022), Vol. 2, 1681 - 1691 (1972).

094.720 **Cosmic-ray produced radioisotopes in Apollo 12 and Apollo 14 samples.** F. Begemann, W. Born, H. Palme, E. Vilcsek, H. Wänke.
Proc. Third Lunar Sci. Conference, (see 012.022), Vol. 2, 1693 - 1702 (1972).

094.721 **Argon, radon, and tritium radioactivities in the sample return container and the lunar surface.**
R. W. Stoenner, R. M. Lindstrom, W. Lyman, R. Davis, Jr.
Proc. Third Lunar Sci. Conference, (see 012.022), Vol. 2, 1703 - 1717 (1972).

094.722 **Cosmogenic nuclides in football-sized rocks.**
M. Wahlen, M. Honda, M. Imamura, J. S. Fruchter, R. C. Finkel, C. P. Kohl, J. R. Arnold, R. C. Reedy.
Proc. Third Lunar Sci. Conference, (see 012.022), Vol. 2, 1719 - 1732 (1972).

094.723 **Cosmonuclides in lunar rocks.**
Y. Yokoyama, R. Auger, R. Bibron, R. Chesselet, F. Guichard, C. Leger, H. Mabuchi, J. L. Reyss, J. Sato.
Proc. Third Lunar Sci. Conference, (see 012.022), Vol. 2, 1733 - 1746 (1972).

094.724 **Radioactivities in returned lunar materials.**
E. L. Fireman, J. D'Amico, J. DeFelice, G. Spannagel.
Proc. Third Lunar Sci. Conference, (see 012.022), Vol. 2, 1747 - 1761 (1972).

094.725 **Study on the cosmic ray produced long-lived Mn-53 in Apollo 14 samples.**
W. Herr, U. Herpers, R. Woelfle.
Proc. Third Lunar Sci. Conference, (see 012.022), Vol. 2, 1763 - 1769 (1972).

094.726 **Alpha spectrometry of a surface exposed lunar rock.**
G. Lambert, T. Grjebine, J. C. Le Roulley, P. Bristeau.
Proc. Third Lunar Sci. Conference, (see 012.022), Vol. 2, 1771 - 1777 (1972).

094.727 **Vanadium isotopic composition and the concentrations of it and ferromagnesian elements in lunar material.** P. Rey, H. Balsiger, M. E. Lipschutz.
Proc. Third Lunar Sci. Conference, (see 012.022), Vol. 2, 1779 - 1786 (1972).

094.728 **Rare-gas analyses on neutron irradiatied Apollo 12 samples.**
E. C. Alexander, Jr., P. K. Davis, J. H. Reynolds.
Proc. Third Lunar Sci. Conference, (see 012.022), Vol. 2, 1787 - 1795 (1972).

094.729 **Noble gas studies on regolith materials from Apollo 14 and 15.** D. D. Bogard, L. E. Nyquist.
Proc. Third Lunar Sci. Conference, (see 012.022), Vol. 2, 1797 - 1819 (1972).

094.730 **Trapped solar wind noble gases in Apollo 12 lunar fines 12001 and Apollo 11 breccia 10046.**
P. Eberhardt, J. Geiss, H. Graf, N. Grögler, M. D. Mendia, M. Mörgeli, H. Schwaller, A. Stettler, U. Krähenbühl, H. R. von Gunten.
Proc. Third Lunar Sci. Conference, (see 012.022), Vol. 2, 1821 - 1856 (1972).

094.731 **Inert gases from Apollo 12, 14, and 15 fines.**
D. Heymann, A. Yaniv, S. Lakatos.
Proc. Third Lunar Sci. Conference, (see 012.022), Vol. 2, 1857 - 1863 (1972).

094.732 **The rare gas record of Apollo 14 and 15 samples.**
T. Kirsten, J. Deubner, P. Horn, I. Kaneoka, J. Kiko, O. A. Schaeffer, S. K. Thio.
Proc. Third Lunar Sci. Conference, (see 012.022), Vol. 2, 1865 - 1889 (1972).

094.733 **Exposure ages and neutron capture record in lunar samples from Fra Mauro.**
G. W. Lugmair, K. Marti.
Proc. Third Lunar Sci. Conference, (see 012.022), Vol. 2, 1891 - 1897 (1972).

094.734 **Classification and source of lunar soils; clastic rocks; and individual mineral, rock, and glass fragments from Apollo 12 and 14 samples as determined by the concentration gradients of the helium, neon, and argon isotopes.**
G. H. Megrue, F. Steinbrunn.
Proc. Third Lunar Sci. Conference, (see 012.022), Vol. 2, 1899 - 1916 (1972).

094.735 **Isotopic anomalies in lunar rhenium.**
R. Michel, U. Herpers, H. Kulus, W. Herr.
Proc. Third Lunar Sci. Conference, (see 012.022), Vol. 2, 1917 - 1925 (1972).

094.736 **A comparison of noble gases released from lunar fines (# 15601.64) with noble gases in meteorites and in the earth.**
B. Srinivasan, E. W. Hennecke, D. E. Sinclair, O. K. Manuel.
Proc. Third Lunar Sci. Conference, (see 012.022), Vol. 2, 1927 - 1945 (1972).

094.737 **Thermal release of helium, neon, and argon from lunar fines and minerals.**
H. Baur, U. Frick, H. Funk, L. Schultz, P. Signer.
Proc. Third Lunar Sci. Conference, (see 012.022), Vol. 2, 1947 - 1966 (1972).

094.738 **Atmospheric Ar^{40} in lunar fines.**
A. Yaniv, D. Heymann.
Proc. Third Lunar Sci. Conference, (see 012.022), Vol. 2, 1967 - 1980 (1972).

094.739 **Volatilized lead from Apollo 12 and 14 soils.**
B. R. Doe, M. Tatsumoto.
Proc. Third Lunar Sci. Conference, (see 012.022), Vol. 2, 1981 - 1988 (1972).

094.740 Trace element relations between Apollo 14 and 15
and other lunar samples, and the implications of a
moon-wide Cl-KREEP coherence and Pt-metal noncoherence.
G. W. Reed, Jr., S. Jovanovic, L. Fuchs.
Proc. Third Lunar Sci. Conference, (see 012.022), Vol. 2,
1989 - 2001 (1972).

094.741 Thermal volatilization studies on lunar samples.
E. K. Gibson Jr., N. J. Hubbard.
Proc. Third Lunar Sci. Conference, (see 012.022), Vol. 2,
2003 - 2014 (1972).

094.742 The nature and effect of the volatile cloud produced
by volcanic and impact events on the moon as derived from a terrestrial volcanic model.
J. J. Naughton, D. A. Hammond, S. V. Margolis, D. W. Muenow.
Proc. Third Lunar Sci. Conference, (see 012.022), Vol. 2,
2015 - 2024 (1972).

094.743 Analysis of single particles of lunar dust for dissolved
gases.   F. M. Ernsberger.
Proc. Third Lunar Sci. Conference, (see 012.022), Vol. 2,
2025 - 2027 (1972).

094.744 Inorganic gas release and thermal analysis study of
Apollo 14 and 15 soils.
E. K. Gibson, Jr., G. W. Moore.
Proc. Third Lunar Sci. Conference, (see 012.022), Vol. 2,
2029 - 2040 (1972).

094.745 Total nitrogen contents of some Apollo 14 lunar
samples by neutron activation analysis.
P. S. Goel, B. K. Kothari.
Proc. Third Lunar Sci. Conference, (see 012.022), Vol. 2,
2041 - 2050 (1972).

094.746 Total carbon, nitrogen, and sulfur in Apollo 14 lunar
samples.   C. B. Moore, C. F. Lewis, J. Cripe,
F. M. Delles, W. R. Kelly, E. K. Gibson, Jr.
Proc. Third Lunar Sci. Conference, (see 012.022), Vol. 2,
2051 - 2058 (1972).

094.747 Chemically bound nitrogen abundances in lunar
samples, and active gases released by heating at
lower temperatures (250 to 500°C).   O. Müller.
Proc. Third Lunar Sci. Conference, (see 012.022), Vol. 2,
2059 - 2068 (1972).

094.748 Survey of lunar carbon compounds: II. The carbon
chemistry of Apollo 11, 12, 14, and 15 samples.
P. H. Cadogan, G. Eglinton, J. N. M. Firth, J. R. Maxwell,
B. J. Mays, C. T. Pillinger.
Proc. Third Lunar Sci. Conference, (see 012.022), Vol. 2,
2069 - 2090 (1972).

094.749 Analysis of organogenic compounds in Apollo 11,
12, and 14 lunar samples.
D. A. Flory, S. Wikstrom, S. Gupta, J. M. Gibert, J. Oró.
Proc. Third Lunar Sci. Conference, (see 012.022), Vol. 2,
2091 - 2108 (1972).

094.750 Amino acid precursors in lunar fines from Apollo 14
and earlier missions.
S. W. Fox, K. Harada, P. E. Hare.
Proc. Third Lunar Sci. Conference, (see 012.022), Vol. 2,
2109 - 2118 (1972).

094.751 Amino acid analyses of Apollo 14 samples.
C. W. Gehrke, R. W. Zumwalt, K. Kuo, W. A. Aue,
D. L. Stalling, K. A. Kvenvolden, C. Ponnamperuma.
Proc. Third Lunar Sci. Conference, (see 012.022), Vol. 2,
2119 - 2129 (1972).

094.752 Compounds of carbon and other volatile elements in
Apollo 14 and 15 samples.
P. T. Holland, B. R. Simoneit, P. C. Wszolek, A. L. Burlingame.
Proc. Third Lunar Sci. Conference, (see 012.022), Vol. 2,
2131 - 2147 (1972).

094.753 Spectrofluorometric search for porphyrins in Apollo
14 surface fines.
J. H. Rho, E. A. Cohen, A. J. Bauman.
Proc. Third Lunar Sci. Conference, (see 012.022), Vol. 2,
2149 - 2155 (1972).

094.754 A first look at the lunar orbital gamma-ray data.
A. E. Metzger, J. I. Trombka, L. E. Peterson, R. C.
Reedy, J. R. Arnold.
Proc. Third Lunar Sci. Conference, (see 012.022), Vol. 3,
Frontispiece, 4 pp. (1972).

094.755 The Apollo 15 X-ray fluorescence experiment.
I. Adler, J. Gerard, J. Trombka, R. Schmadebeck,
P. Lowman, H. Blodget, L. Yin, E. Eller, R. Lamothe, P.
Gorenstein, P. Bjorkholm, B. Harris, H. Gursky.
Proc. Third Lunar Sci. Conference, (see 012.022), Vol. 3,
2157 - 2178 (1972).

094.756 Observation of lunar radon emanation with the
Apollo 15 alpha particle spectrometer.
P. Gorenstein, P. Bjorkholm.
Proc. Third Lunar Sci. Conference, (see 012.022), Vol. 3,
2179 - 2187 (1972).

094.757 Analysis and interpretation of lunar laser altimetry.
W. M. Kaula, G. Schubert, R. E. Lingenfelter, W. L.
Sjogren, W. R. Wollenhaupt.
Proc. Third Lunar Sci. Conference, (see 012.022), Vol. 3,
2189 - 2204 (1972).

094.758 Lunar orbital mass spectrometer experiment.
J. H. Hoffman, R. R. Hodges, Jr., D. E. Evans.
Proc. Third Lunar Sci. Conference, (see 012.022), Vol. 3,
2205 - 2216 (1972).

094.759 Water vapor, whence comest thou?
J. W. Freeman, Jr., H. K. Hills, R. R. Vondrak.
Proc. Third Lunar Sci. Conference, (see 012.022), Vol. 3,
2217 - 2230 (1972).

094.760 Lunar atmosphere measurements.
F. S. Johnson, J. M. Carroll, D. E. Evans.
Proc. Third Lunar Sci. Conference, (see 012.022), Vol. 3,
2231 - 2242 (1972).

094.761 Some surface characteristics and gas interactions of
Apollo 14 fines and rock fragments.
D. A. Cadenhead, N. J. Wagner, B. R. Jones, J. R. Stetter.
Proc. Third Lunar Sci. Conference, (see 012.022), Vol. 3,
2243 - 2257 (1972).

094.762 Microphysical, microchemical, and adhesive properties of lunar material III: Gas interaction with lunar
material.   J. J. Grossman, N. R. Mukherjee, J. A. Ryan.
Proc. Third Lunar Sci. Conference, (see 012.022), Vol. 3,
2259 - 2269 (1972).

094.763 Magnetic fields near the moon.
P. J. Coleman, Jr., B. R. Lichtenstein, C. T. Russell,
L. R. Sharp, G. Schubert.
Proc. Third Lunar Sci. Conference, (see 012.022), Vol. 3,
2271 - 2286 (1972).

094.764 Surface magnetometer experiments: Internal lunar properties and lunar field interactions with the solar plasma. P. Dyal, C. W. Parkin, P. Cassen.
Proc. Third Lunar Sci. Conference, (see 012.022), Vol. 3, 2287 - 2307 (1972).

094.765 The induced magnetic field of the moon: Conductivity profiles and inferred temperature.
C. P. Sonett, B. F. Smith, D. S. Colburn, G. Schubert, K. Schwartz.
Proc. Third Lunar Sci. Conference, (see 012.022), Vol. 3, 2309 - 2336 (1972).

094.766 Iron–titanium–chromite, a possible new carrier of remanent magnetization in lunar rocks.
S. K. Banerjee.
Proc. Third Lunar Sci. Conference, (see 012.022), Vol. 3, 2337 - 2342 (1972).

094.767 Magnetic properties of Apollo 14 rocks and fines. D. W. Collinson, S. K. Runcorn, A. Stephenson, A. J. Manson.
Proc. Third Lunar Sci. Conference, (see 012.022), Vol. 3, 2343 - 2361 (1972).

094.768 On the remanent magnetism of lunar samples with special reference to 10048,55 and 14053,48.
J. R. Dunn, M. Fuller.
Proc. Third Lunar Sci. Conference, (see 012.022), Vol. 3, 2363 - 2386 (1972).

094.769 Magnetic properties of Apollo 14 breccias and their correlation with metamorphism.
W. A. Gose, G. W. Pearce, D. W. Strangway, E. E. Larson.
Proc. Third Lunar Sci. Conference, (see 012.022), Vol. 3, 2387 - 2395 (1972).

094.770 Evidence of lunar surface oxidation processes: Electron spin resonance spectra of lunar materials and simulated lunar materials. D. L. Griscom, C. L. Marquardt.
Proc. Third Lunar Sci. Conference, (see 012.022), Vol. 3, 2397 - 2415 (1972).

094.771 Natural remanent magnetization in lunar breccia 14321. R. B. Hargraves, N. Dorety.
Proc. Third Lunar Sci. Conference, (see 012.022), Vol. 3, 2417 - 2421 (1972).

094.772 Rock magnetism of Apollo 14 and 15 materials. T. Nagata, R. M. Fisher, F. C. Schwerer, M. D. Fuller, J. R. Dunn.
Proc. Third Lunar Sci. Conference, (see 012.022), Vol. 3, 2423 - 2447 (1972).

094.773 Remanent magnetization of the lunar surface. G. W. Pearce, D. W. Strangway, W. A. Gose.
Proc. Third Lunar Sci. Conference, (see 012.022), Vol. 3, 2449 - 2464 (1972).

094.774 Temperature-dependent magnetic properties of individual glass spherules, Apollo 11, 12, and 14 lunar samples. A. N. Thorpe, S. Sullivan, C. C. Alexander, F. E. Senftle, E. J. Dwornik.
Proc. Third Lunar Sci. Conference, (see 012.022), Vol. 3, 2465 - 2478 (1972).

094.775 Mössbauer studies of Apollo 14 lunar samples. T. C. Gibb, R. Greatrex, N. N. Greenwood, M. H. Battey.
Proc. Third Lunar Sci. Conference, (see 012.022), Vol. 3, 2479 - 2493 (1972).

094.776 Nuclear magnetic resonance properties of lunar samples. D. Kline, R. A. Weeks.
Proc. Third Lunar Sci. Conference, (see 012.022), Vol. 3, 2495 - 2501 (1972).

094.777 Magnetic phases in lunar material and their electron magnetic resonance spectra: Apollo 14.
R. A. Weeks.
Proc. Third Lunar Sci. Conference, (see 012.022), Vol. 3, 2503 - 2517 (1972).

094.778 Moonquakes and lunar tectonism results from the Apollo passive seismic experiment.
G. Latham, M. Ewing, J. Dorman, D. Lammlein, F. Press, N. Toksöz, G. Sutton, F. Duennebier, Y. Nakamura.
Proc. Third Lunar Sci. Conference, (see 012.022), Vol. 3, 2519 - 2526 (1972).

094.779 Structure, composition, and properties of lunar crust. M. N. Toksöz, F. Press, A. Dainty, K. Anderson, G. Latham, M. Ewing, J. Dorman, D. Lammlein, G. Sutton, F. Duennebier.
Proc. Third Lunar Sci. Conference, (see 012.022), Vol. 3, 2527 - 2544 (1972).

094.780 Measurements of the acoustical parameters of rock powders and the Gold–Soter lunar model.
B. W. Jones.
Proc. Third Lunar Sci. Conference, (see 012.022), Vol. 3, 2545 - 2555 (1972).

094.781 Elastic wave velocities and thermal diffusivities of Apollo 14 rocks.
H. Mizutani, N. Fujii, Y. Hamano, M. Osako.
Proc. Third Lunar Sci. Conference, (see 012.022), Vol. 3, 2557 - 2564 (1972).

094.782 Elastic velocity and Q factor measurements on Apollo 12, 14, and 15 rocks.
B. R. Tittmann, M. Abdel-Gawad, R. M. Housley.
Proc. Third Lunar Sci. Conference, (see 012.022), Vol. 3, 2565 - 2575 (1972).

094.783 Elastic properties of Apollo 14 and 15 rocks. T. Todd, H. Wang, W. S. Baldridge, G. Simmons.
Proc. Third Lunar Sci. Conference, (see 012.022), Vol. 3, 2577 - 2586 (1972).

094.784 Applications to lunar geophysical models of the velocity-density properties of lunar rocks, glasses, and artificial lunar glasses.
N. Warren, O. L. Anderson, N. Soga.
Proc. Third Lunar Sci. Conference, (see 012.022), Vol. 3, 2587 - 2598 (1972).

094.785 Thermal expansion of Apollo lunar samples and Fairfax diabase.
W. S. Baldridge, F. Miller, H. Wang, G. Simmons.
Proc. Third Lunar Sci. Conference, (see 012.022), Vol. 3, 2599 - 2609 (1972).

094.786 Thermal conductivity of Apollo 14 fines. C. J. Cremers.
Proc. Third Lunar Sci. Conference, (see 012.022), Vol. 3, 2611 - 2617 (1972).

094.787 Viscous flow behavior of lunar compositions 14259 and 14310.
M. Cukierman, P. M. Tutts, D. R. Uhlmann.
Proc. Third Lunar Sci. Conference, (see 012.022), Vol. 3, 2619 - 2625 (1972).

094.788 **Crystallization behavior and glass formation of selected lunar compositions.**
G. Scherer, R. W. Hopper, D. R. Uhlmann.
Proc. Third Lunar Sci. Conference, (see 012.022), Vol. 3, 2627 - 2637 (1972).

094.789 **Direct observation of the lunar photoelectron layer.**
D. L. Reasoner, W. J. Burke.
Proc. Third Lunar Sci. Conference, (see 012.022), Vol. 3, 2639 - 2654 (1972).

094.790 **Photoemission from lunar surface fines and the lunar photoelectron sheath.**
B. Feuerbacher, M. Anderegg, B. Fitton, L. D. Laude, R. F. Willis, R. J. L. Grard.
Proc. Third Lunar Sci. Conference, (see 012.022), Vol. 3, 2655 - 2663 (1972).

094.791 **Secondary electron emission characteristics of lunar surface fines.** M. Anderegg, B. Feuerbacher,
B. Fitton, L. D. Laude, R. F. Willis.
Proc. Third Lunar Sci. Conference, (see 012.022), Vol. 3, 2665 - 2669 (1972).

094.792 **Lunar dust motion.** D. R. Criswell.
Proc. Third Lunar Sci. Conference, (see 012.022), Vol. 3, 2671 - 2680 (1972).

094.793 **An explanation of transient lunar phenomena from studies of static and fluidized lunar dust layers.**
G. F. J. Garlick, G. A. Steigmann, W. E. Lamb, J. E. Geake.
Proc. Third Lunar Sci. Conference, (see 012.022), Vol. 3, 2681 - 2687 (1972).

094.794 **Lunar ash flow with heat transfer.**
S. I. Pai, T. Hsieh, J. A. O'Keefe.
Proc. Third Lunar Sci. Conference, (see 012.022), Vol. 3, 2689 - 2711 (1972).

094.795 **Effects of microcratering on the lunar surface.**
D. E. Gault, F. Hörz, J. B. Hartung.
Proc. Third Lunar Sci. Conference, (see 012.022), Vol. 3, 2713 - 2734 (1972).

094.796 **Lunar microcraters and interplanetary dust.**
J. B. Hartung, F. Hörz, D. E. Gault.
Proc. Third Lunar Sci. Conference, (see 012.022), Vol. 3, 2735 - 2753 (1972).

094.797 **Simulated microscale erosion on the lunar surface by hypervelocity impact, solar wind sputtering, and thermal cycling.**
J. A. M. McDonnell, D. G. Ashworth, R. P. Flavill, R. C. Jennison
Proc. Third Lunar Sci. Conference, (see 012.022), Vol. 3, 2755 - 2765 (1972).

094.798 **Microcraters on lunar rocks.** D. A. Morrison,
D. S. McKay, G. H. Heiken, H. J. Moore.
Proc. Third Lunar Sci. Conference, (see 012.022), Vol. 3, 2767 - 2791 (1972).

094.799 **Lunar craters and exposure ages derived from crater statistics and solar flare tracks.**
G. Neukum, E. Schneider, A. Mehl, D. Storzer, G. A. Wagner, H. Fechtig, M. R. Bloch.
Proc. Third Lunar Sci. Conference, (see 012.022), Vol. 3, 2793 - 2810 (1972).

094.800 **Collision controlled radiation history of the lunar regolith.** N. Bhandari, J. N. Goswami, S. K. Gupta, D. Lal, A. S. Tamhane, V. S. Venkatavaradan.
Proc. Third Lunar Sci. Conference, (see 012.022), Vol. 3, 2811 - 2829 (1972).

094.801 **The particle track record of Fra Mauro.**
H. R. Hart, Jr., G. M. Comstock, R. L. Fleischer.
Proc. Third Lunar Sci. Conference, (see 012.022), Vol. 3, 2831 - 2844 (1972).

094.802 **Studies bearing on the history of lunar breccias.**
I. D. Hutcheon, P. P. Phakey, P. B. Price.
Proc. Third Lunar Sci. Conference, (see 012.022), Vol. 3, 2845 - 2865 (1972).

094.803 **Track studies of Apollo 14 rocks, and Apollo 14, Apollo 15, and Luna 16 soils.** J. L. Berdot, G. C. Chetrit, J. C. Lorin, P. Pellas, G. Poupeau.
Proc. Third Lunar Sci. Conference, (see 012.022), Vol. 3, 2867 - 2881 (1972).

094.804 **Track metamorphism in extraterrestrial breccias.**
J. C. Dran, J. P. Duraud, M. Maurette, L. Durrieu, C. Jouret, C. Legressus.
Proc. Third Lunar Sci. Conference, (see 012.022), Vol. 3, 2883 - 2903 (1972).

094.805 **Radiation effects in soils from five lunar missions.**
P. P. Phakey, I. D. Hutcheon, R. S. Rajan, P. B. Price.
Proc. Third Lunar Sci. Conference, (see 012.022), Vol. 3, 2905 - 2915 (1972).

094.806 **Solar flare and galactic cosmic ray studies of Apollo 14 and 15 samples.** G. Crozaz, R. Drozd, C. M. Hohenberg, H. P. Hoyt, Jr., D. Ragan, R. M. Walker, D. Yuhas.
Proc. Third Lunar Sci. Conference, (see 012.022), Vol. 3, 2917 - 2931 (1972).

094.807 **Charge assignment to cosmic ray heavy ion tracks in lunar pyroxenes.**
T. Plieninger, W. Krätschmer, W. Gentner.
Proc. Third Lunar Sci. Conference, (see 012.022), Vol. 3, 2933 - 2939 (1972).

094.808 **Track consortium report on rock 14310.**
D. E. Yuhas, R. M. Walker, H. Reeves, G. Poupeau, P. Pellas, J. C. Lorin, G. C. Chetrit, J. L. Berdot, P. B. Price, I. D. Hutcheon, H. R. Hart, Jr., R. L. Fleischer, G. M. Comstock, D. Lal, J. N. Goswami, N. Bhandari.
Proc. Third Lunar Sci. Conference, (see 012.022), Vol. 3, 2941 - 2947 (1972).

094.809 **Thermoluminescence of Apollo 14 lunar samples following irradiation at −196°C.**
I. M. Blair, J. A. Edgington, R. Chen, R. A. Jahn.
Proc. Third Lunar Sci. Conference, (see 012.022), Vol. 3, 2949 - 2953 (1972).

094.810 **Thermoluminescence of Apollo 12 samples: Implications for lunar temperature and radiation histories.**
S. A. Durrani, W. Prachyabrued, C. Christodoulides, J. H. Fremlin, J. A. Edgington, R. Chen, I. M. Blair.
Proc. Third Lunar Sci. Conference, (see 012.022), Vol. 3, 2955 - 2970 (1972).

094.811 **Luminescence of lunar material excited by electrons.**
J. E. Geake, G. Walker, A. A. Mills, G. F. J. Garlick.
Proc. Third Lunar Sci. Conference, (see 012.022), Vol. 3, 2971 - 2979 (1972).

094.812 **Luminescence of Apollo 14 and Apollo 15 lunar samples.** N. N. Greenman, H. G. Gross.
Proc. Third Lunar Sci. Conference, (see 012.022), Vol. 3,

2981 - 2995 (1972).

094.813 Thermoluminescence of individual grains and bulk
         samples of lunar fines.
H. P. Hoyt, Jr., R. M. Walker, D. W. Zimmerman, J. Zimmerman.
Proc. Third Lunar Sci. Conference, (see 012.022), Vol. 3,
2997 - 3007 (1972).

094.814 Spectral emission of natural and artificially induced
         thermoluminescence in Apollo 14 lunar sample
14163,147.   C. Lalou, G. Valladas, U. Brito, A. Henni,
T. Ceva, R. Visocekas.
Proc. Third Lunar Sci. Conference, (see 012.022), Vol. 3,
3009 - 3020 (1972).

094.815 Electronic spectra of pyroxenes and interpretation
         of telescopic spectral reflectivity curves of the
moon.   J. B. Adams, T. B. McCord.
Proc. Third Lunar Sci. Conference, (see 012.022), Vol. 3,
3021 - 3034 (1972).

094.816 Far infrared properties of lunar rock.
         P. E. Clegg, S. J. Pandya, S. A. Foster, J. A. Bastin.
Proc. Third Lunar Sci. Conference, (see 012.022), Vol. 3,
3035 - 3045 (1972).

094.817 Infrared and Raman spectroscopic studies of struc-
         tural variations in minerals from Apollo 11, 12, 14,
and 15 samples.
P. A. Estep, J. J. Kovach, P. Waldstein, C. Karr, Jr.
Proc. Third Lunar Sci. Conference, (see 012.022), Vol. 3,
3047 - 3067 (1972).

094.818 Midinfrared emission spectra of Apollo 14 and 15
         soils and remote compositional mapping of the
moon.
L. M. Logan, G. R. Hunt, S. R. Balsamo, J. W. Salisbury.
Proc. Third Lunar Sci. Conference, (see 012.022), Vol. 3,
3069 - 3076 (1972).

094.819 Far infrared and Raman spectroscopic investiga-
         tions of lunar materials from Apollo 11, 12, 14,
and 15.   C. H. Perry, D. K. Agrawal, E. Anastassakis,
R. P. Lowndes, N. E. Tornberg.
Proc. Third Lunar Sci. Conference, (see 012.022), Vol. 3,
3077 - 3095 (1972).

094.820 Reflectance and absorption spectra of Apollo 11
         and Apollo 12 samples.
I. I. Antipova-Karataeva, Yu. I. Stacheev, L. S. Tarasov.
Proc. Third Lunar Sci. Conference, (see 012.022), Vol. 3,
3097 - 3101 (1972).

094.821 Polarimetric properties of the lunar surface and its
         interpretation. Part 5: Apollo 14 and Luna 16 lunar
samples.   E. Bowell, A. Dollfus, J. E. Geake.
Proc. Third Lunar Sci. Conference, (see 012.022), Vol. 3,
3103 - 3126 (1972).

094.822 Lunar surface properties as determined from earth-
         shine and near-terminator photography.
D. D. Lloyd, J. W. Head.
Proc. Third Lunar Sci. Conference, (see 012.022), Vol. 3,
3127 - 3142 (1972).

094.823 Optical properties of lunar glass spherules from
         Apollo 14 fines.   K. J. Rao, A. R. Cooper.
Proc. Third Lunar Sci. Conference, (see 012.022), Vol. 3,
3143 - 3155 (1972).

094.824 Dielectric properties of Apollo 14 lunar samples at

microwave and millimeter wavelengths.
H. L. Bassett, R. G. Shackelford.
Proc. Third Lunar Sci. Conference, (see 012.022), Vol. 3,
3157 - 3160 (1972).

094.825 Dielectric properties of Apollo 14 lunar samples.
         D. H. Chung, W. B. Westphal, G. R. Olhoeft.
Proc. Third Lunar Sci. Conference, (see 012.022), Vol. 3,
3161 - 3172 (1972).

094.826 Electrical conductivity and Mössbauer study of
         Apollo lunar samples.
F. C. Schwerer, G. P. Huffman, R. M. Fisher, T. Nagata.
Proc. Third Lunar Sci. Conference, (see 012.022), Vol. 3,
3173 - 3185 (1972).

094.827 Grain size analysis, optical reflectivity measure-
         ments, and determination of high-frequency electri-
cal properties for Apollo 14 lunar samples.
T. Gold, E. Bilson, M. Yerbury.
Proc. Third Lunar Sci. Conference, (see 012.022), Vol. 3,
3187 - 3193 (1972).

094.828 CESEMI studies of Apollo 14 and 15 fines.
         H. Görz, E. W. White, G. G. Johnson, Jr., M. W.
Pearson.
Proc. Third Lunar Sci. Conference, (see 012.022), Vol. 3,
3195 - 3200 (1972).

094.829 Scanning electron microscope and energy dispersive
         X-ray analysis of the surface features of Surveyor 3
television mirror.   J. C. Mandeville, H. Y. Lem.
Proc. Third Lunar Sci. Conference, (see 012.022), Vol. 3,
3201 - 3212 (1972).

094.830 Core sample depth relationships: Apollo 14 and 15.
         W. D. Carrier III, S. W. Johnson, L. H. Carrasco,
R. Schmidt.
Proc. Third Lunar Sci. Conference, (see 012.022), Vol. 3,
3213 - 3221 (1972).

094.831 Strength and compressibility of returned lunar soil.
         W. D. Carrier III, L. G. Bromwell, R. T. Martin.
Proc. Third Lunar Sci. Conference, (see 012.022), Vol. 3,
3223 - 3234 (1972).

094.832 Mechanical properties of lunar soil: Density, porosi-
         ty, cohesion, and angle of internal friction.
J. K. Mitchell, W. N. Houston, R. F. Scott, N. C. Costes, W. D.
Carrier III, L. G. Bromwell.
Proc. Third Lunar Sci. Conference, (see 012.022), Vol. 3,
3235 - 3253 (1972).

094.833 Lunar soil porosity and its variation as estimated
         from footprints and boulder tracks.
W. N. Houston, H. J. Hovland, J. K. Mitchell, L. I. Namiq.
Proc. Third Lunar Sci. Conference, (see 012.022), Vol. 3,
3255 - 3263 (1972).

094.834 Lunar sample inventory for Apollo 11, 12, 14, and
         15.
Proc. Third Lunar Sci. Conference, (see 012.022), Vol. 1,
Appendix, 20 pp. (1972).

094.835 Lunar sample cross reference, (1970 and 1971
         Proceedings).
Proc. Third Lunar Sci. Conference, (see 012.022), Vol. 2,
Appendix, 11 pp. (1972).

094.836 Lunar sample cross reference, (1972 Proceedings).
         Proc. Third Lunar Sci. Conference, (see 012.022),

Vol. 3, Appendix, 15 pp. (1972).

094.837 Some inter-area chemical characteristics of the lunar surface. L. H. Ahrens.
Comments Earth Sci. Geophys., Vol. 2, 179 - 186 (1972).

094.838 Spectral emittance of Apollo-12 lunar fines. R. C. Birkebak.
Trans. ASME, (*USA*), Ser. C, Vol. 93, 323 - 324 (1972).
See Phys. Abstr., Vol. 76, No. 12997 (1973).

094.839 The geology of the moon. G. Fielder.
Contemporary Phys., (*GB*), Vol. 14, 39 - 54 (1973).
As a result of tremendous strides forward, the views of a cold, rigid moon built predominantly from stony meteorites has had to be revised. The interior of the moon is hot; the distant past has seen not only major impacts but also extensive volcanism altering the lunar surface, and major motions within the moon. Even today, weak moonquakes remain to remind one of the past upheavals that accompanied the geological processes now being unravelled through detailed studies of the lunar rocks.

094.840 Mare Humorum: an integrated study of spectral reflectivity.
T. V. Johnson, C. Pieters, T. B. McCord.
Icarus, Vol. 19, 224 - 229 = Publ. Planet. Astron. Lab., MIT, Cambridge, Mass., No. 50 (1973).
A detailed study was made of the spectral reflectivity $(0.3-1.1 \mu m)$ of 31 areas (10–20 km in diam) in the Humorum basin region.

094.841 Lunar surface. Z. Kopal.
Vesmír, Vol. 52, 35 - 37 (1973). In Czech.

094.842 Thermal history of the moon. V. Čermák.
Vesmír, Vol. 52, 101 - 104 (1973). In Czech.

094.843 Geologic history of the moon and Mars.
K. Beneš.
Vesmír, Vol. 52, 99 - 100 (1973). In Czech.

094.844 The lunar surface layer. J. A. Bastin.
Rep. Progr. Phys., Vol. 36, 289 - 346 (1973).
This article describes the application of the concepts and laws of physics to the study of the formation, structure and properties of the surface layers of the moon. Both impact and internal mechanisms for the origin of the lunar surface features are considered. The effect of solar electromagnetic radiation on the moon's surface is described, from which information about the chemical composition, microstructure and thermal properties of the surface may be deduced.

094.845 Role of convection in the moon.
P. Cassen, R. T. Reynolds.
Journ. Geophys. Res., Vol. 78, 3203 - 3215 (1973).
The purpose of this paper is to examine the constraints that the possibility of solid convection places on thermal history models and what may be inferred about the role of convection by existing and future observations. The models treated in detail cannot be claimed to be definitive in view of the large number of parameters and initial conditions that must be specified for a thermal history calculation. They do, however, demonstrate the factors that must be considered in order to assess the role of solid convection.

094.846 Spatial distribution of $^{40}Ar/^{39}Ar$ ages in lunar breccia 14301. G. H. Megrue.
Journ. Geophys. Res., Vol. 78, 3216 - 3221 (1973).

094.847 Removal of a constraint on the composition of the lunar interior. D. L. Anderson.
Journ. Geophys. Res., Vol. 78, 3222 - 3225 (1973).
The purpose of this note is to point out that the now generally accepted constraint on the CaO and Al_2O_3 content of the lunar interior is not valid.

094.848 Interference patterns of a horizontal electric dipole over layered dielectric media.
L. Tsang, J. A. Kong, G. Simmons.
Journ. Geophys. Res., Vol. 78, 3287 - 3300 (1973).
The surface electrical properties experiment will be used on Apollo 17 to detect subsurface layering and to measure both dielectric constant and loss tangent of the lunar subsurface.

094.849 Experimental results on combined ultraviolet-proton excitation of moon rock luminescence.
D. B. Nash.
Journ. Geophys. Res., Vol. 78, 3512 - 3514 (1973). – Letter.

094.850 Unusual aspect of Messier and Pickering.
R. C. Parish.
Strolling Astronomer, Vol. 24, 102 - 103 (1973).

094.851 A modified fission process for the formation of the moon. C. C. Mason.
Strolling Astronomer, Vol. 24, 107 - 114 (1973).

094.852 Lunar notes. H. D. Jamieson, C. Vaucher.
Strolling Astronomer, Vol. 24, 114 - 119, 121 (1973).
A summary of findings by the bright and banded craters program; The selected areas program: Endymion, Gassendi, Piton, and Aristillus.

094.853 Physical properties of the lunar surface. A. Dollfus.
Physics of the moon and planets, (see 012.024), p. 11 - 24 (1972). In Russian.

094.854 Investigation of the lunar surface with the Soviet automatic stations Luna 9 and Luna 13.
A. P. Vinogradov, Yu. A. Surkov, K. P. Florenskij, I. I. Cherkasov, V. V. Shvarev.
Physics of the moon and planets, (see 012.024), p. 25 - 31 (1972). In Russian.

094.855 Results of lunar landings of the Surveyor series. L. D. Jaffe.
Physics of the moon and planets, (see 012.024), p. 31 - 47 (1972). In Russian.

094.856 Composition, structure and history of the lunar ground. B. Hapke.
Physics of the moon and planets, (see 012.024), p. 48 (1972). In Russian. – Abstract.

094.857 Determination of the density of the moon's upper layer from given surface temperatures during an eclipse and the lunar night. V. S. Troitskij, O. B. Shchuko, V. N. Gol'dberg, L. V. Drobova.
Physics of the moon and planets, (see 012.024), p. 48 - 52 (1972). In Russian.

094.858 On the microrelief of the moon.
N. P. Barabashov, L. A. Akimov.
Physics of the moon and planets, (see 012.024), p. 52 - 56 (1972). In Russian.

094.859 Transfer processes on the lunar surface. T. Gold.
Physics of the moon and planets, (see 012.024), p. 57 (1972). In Russian. – Abstract.

094.860 **Variation of optical properties and pulverization of minerals under the influence of ion bombardment.** Sh. S. Radzhabov, U. A. Arifov, R. A. Ashmyanskij, M. Yu. Borukhov, D. D. Gruich, V. P. Peshekhonov, R. R. Rakhimov.
Physics of the moon and planets, (see 012.024), p. 57 - 61 (1972). In Russian.

094.861 **Surface diffusion and migration on dispersed silicon and basalt at slight heating in a vacuum.** M. Yu. Borukhov, G. V. Krasheninnikova.
Physics of the moon and planets, (see 012.024), p. 61 - 63 (1972). In Russian.

094.862 **Investigation of the polarization of moonlight and the nature of the lunar surface.** A. Dollfus, E. Bowell.
Physics of the moon and planets, (see 012.024), p. 63 - 67 (1972). In Russian.

094.863 **Investigation of the lunar surface by means of a polarovisor-discriminator.** V. P. Dzhapiashvili, A. N. Korol', L. V. Ksanfomaliti, V. K. Lokhov.
Physics of the moon and planets, (see 012.024), p. 68 - 71 (1972). In Russian.

094.864 **Aerological investigation of the volcanic caps of Kamchatka by polarization and spectral methods.** Yu. N. Lipskij, M. M. Pospergelis, G. S. Shtejnberg, V. V. Novikov, L. V. Gromova, S. V. Landau, A. N. Sanovich, M. F. Shabanov.
Physics of the moon and planets, (see 012.024), p. 71 - 75 (1972). In Russian.

094.865 **On the degree of continuity of reflectivity of lunar surface details.** A. V. Markov.
Physics of the moon and planets, (see 012.024), p. 76 - 79 (1972). In Russian.

094.866 **Some results of an investigation of the lunar surface luminescence at the Kharkov Astronomical Observatory.** V. S. Tsvetkova, L. A. Akimov.
Physics of the moon and planets, (see 012.024), p. 79 - 82 (1972). In Russian.

094.867 **Infrared investigations of the composition of the lunar surface.** D. P. Cruikshank.
Physics of the moon and planets, (see 012.024), p. 83 - 91 (1972). In Russian.

094.868 **On the infrared radiation of the moon in the region of $3.5-3.9\,\mu$.** G. A. Lejkin, T. E. Shvidkovskaya.
Physics of the moon and planets, (see 012.024), p. 91 - 95 (1972). In Russian.

094.869 **On ultraviolet measurements of the moon in the region of 1950−2750 Å.** V. A. Krasnopol'skij, G. A. Lejkin, M. U. Aganina, T. E. Shvidkovskaya.
Physics of the moon and planets, (see 012.024), p. 96 - 99 (1972). In Russian.

094.870 **On the photometric relief of the lunar continental shield.** Yu. N. Lipskij, V. V. Shevchenko.
Physics of the moon and planets, (see 012.024), p. 99 - 104 (1972). In Russian.

094.871 **The radio radiation of the moon in the region of 1.25−2.5 cm.** V. M. Plechkov.
Physics of the moon and planets, (see 012.024), p. 104 - 106 (1972). In Russian.

094.872 **The radio radiation of the moon and sun at 2.25 mm wavelength and of Jupiter at 2.1 mm wavelength.** A. I. Naumov, A. G. Kislyakov, V. N. Voronov.
Physics of the moon and planets, (see 012.024), p. 106 - 110 (1972). In Russian.

094.873 **The radio radiation of the moon in the regions of millimeter and submillimeter waves.** L. I. Fedoseev, L. V. Lubyako, L. M. Kukin.
Physics of the moon and planets, (see 012.024), p. 111 - 113 (1972). In Russian.

094.874 **The spectrum of the reflection coefficient of radio waves on the lunar surface with material properties changing with depth.** T. V. Tikhonova, V. S. Troitskij.
Physics of the moon and planets, (see 012.024), p. 114 (1972). In Russian. − Abstract.

094.875 **Some results of a theoretical investigation of the polarization of radio radiation of the rough moon.** A. V. Alekseev, T. N. Aleshina, V. L. Krotikov, L. A. Mol'-kova.
Physics of the moon and planets, (see 012.024), p. 114 - 118 (1972). In Russian.

094.876 **Investigation of the dielectric properties of terrestrial rocks at super-high frequencies in order to define exactly the composition of lunar matter.** R. Belov, L. N. Bondar', K. A. Goronina, L. V. Lubyako, S. A. Shmulevich.
Physics of the moon and planets, (see 012.024), p. 118 - 123 (1972). In Russian.

094.877 **Determination of the lunar albedo at 6 m wavelength.** T. Hagfors.
Physics of the moon and planets, (see 012.024), p. 123 - 126 (1972). In Russian.

094.878 **Investigation of the lunar surface by the method of scattering of radio waves of lunar satellites.** O. I. Yakovlev.
Physics of the moon and planets, (see 012.024), p. 127 - 129 (1972). In Russian.

094.879 **Cartography of the moon.** Yu. N. Lipskij, Yu. P. Pskovskij.
Physics of the moon and planets, (see 012.024), p. 130 - 134 (1972). In Russian.

094.880 **The libration effect of the radius of the moon.** K. Kozieł.
Physics of the moon and planets, (see 012.024), p. 135 - 139 (1972). In Russian.

094.881 **Problems of morphometric investigations of the lunar surface.** Zh. F. Rodionova, L. I. Volchkova.
Physics of the moon and planets, (see 012.024), p. 139 - 144 (1972). In Russian.

094.882 **Hypsographic curve of the moon.** I. V. Gavrilov.
Physics of the moon and planets, (see 012.024), p. 144 - 148 (1972). In Russian.

094.883 **Distribution and origin of rocks in the landing area of Luna 13.** I. Ya. Koval'skaya, G. A. Lejkin.
Physics of the moon and planets, (see 012.024), p. 149 - 150 (1972). In Russian.

094.884 **Some conclusions on the morphometry of lunar**

segments photographed by Luna 12.
K. P. Florenskij, I. M. Taborko.
Physics of the moon and planets, (see 012.024), p. 150 - 153 (1972). In Russian.

094.885 **Investigation of the topography of the landing sites of Luna 9 and Luna 13.** B. N. Rodionov.
Physics of the moon and planets, (see 012.024), p. 153 - 157 (1972). In Russian.

094.886 **Cosmogony of the moon.** E. L. Ruskol.
Physics of the moon and planets, (see 012.024), p. 160 - 167 (1972). In Russian.

094.887 **New results of investigations aboard Explorer 35.** N. Ness.
Physics of the moon and planets, (see 012.024), p. 167 - 178 (1972). In Russian.

094.888 **Determination of the thermal flux from the lunar interior with inhomogeneous structure of the lunar surface layer.** T. V. Tikhonov, V. S. Troitskij.
Physics of the moon and planets, (see 012.024), p. 178 - 182 (1972). In Russian.

094.889 **Morphological features of the moon and convection in its mantle.** S. Miyamoto.
Physics of the moon and planets, (see 012.024), p. 183 - 184 (1972). In Russian.

094.890 **On general peculiarities of the lunar volcanism.** I. V. Melekestsev, G. S. Shtejnberg, E. N. Ehrlich.
Physics of the moon and planets, (see 012.024), p. 184 - 186 (1972). In Russian.

094.891 **On the origin of large lunar craters and circular maria.** G. S. Shtejnberg.
Physics of the moon and planets, (see 012.024), p. 187 - 188 (1972). In Russian.

094.892 **On some peculiarities of the interaction between the atmosphere and litosphere on the moon.**
A. M. Gutkin, M. S. Markov, Ts. M. Rajtburd, M. V. Slonimskaya.
Physics of the moon and planets, (see 012.024), p. 189 - 191 (1972). In Russian.

094.893 **On discrete sizes of circular maria and thalassoids of the moon.** V. G. Trifonov, P. V. Florenskij.
Physics of the moon and planets, (see 012.024), p. 191 - 194 (1972). In Russian.

094.894 **Some questions concerning the inner structure of the moon. (Pilot lunar models).**
V. N. Zharkov, V. L. Pan'kov.
Physics of the moon and planets, (see 012.024), p. 195 - 197 (1972). In Russian.

094.895 **Test of a photographic equidensitometry of the moon and planets.**
V. F. Kartashov, V. G. Tejfel', L. A. Usol'tseva.
Physics of the moon and planets, (see 012.024), p. 459 - 467 (1972). In Russian.

094.896 **Lunar composition from Apollo orbital measurements.** I. Adler, J. I. Trombka.
IEEE Trans. Nuclear Sci., Vol. NS-20, No. 1, p. 24 - 32 (1973).
Several spectrometers carried in the Service Module of the Apollo 15 and Apollo 16 spacecraft were employed for the compositional mapping of the lunar surface. It was possible to demonstrate interesting chemical differences between the mare and the highlands, to find specific areas of high radioactivity and to learn something about the composition of the moon's hidden side.

094.897 **Extralunar sources for carbon on the moon.** J. M. Hayes.
Space Life Sci., Vol. 3, 474 - 483 (1972).

094.898 **The search for indigenous lunar organic matter.** C. Sagan.
Space Life Sci., Vol. 3, 484 - 489 (1972).

094.899 **Lunar organic analysis implications for chemical evolution.** C. Ponnamperuma.
Space Life Sci., Vol. 3, 493 - 496 (1972).

094.900 **Lunar carbon chemistry: relations to and implications for terrestrial organic geochemistry.**
G. Eglinton, J. R. Maxwell, C. T. Pillinger.
Space Life Sci., Vol. 3, 497 - 506 (1972).

094.901 **Study of carbon compounds in Apollo 12 and 14 lunar samples.**
P. T. Holland, B. R. Simoneit, P. C. Wszolek, A. L. Burlingame.
Space Life Sci., Vol. 3, 551 - 561 (1972).

094.902 **Distribution and isotopic abundance of biogenic elements in lunar samples.** I. R. Kaplan.
Space Life Sci., Vol. 3, 383 - 403 (1972).

094.903 **Compounds of the organogenic elements in Apollo 11 and 12 lunar samples: a review.**
E. K. Gibson, Jr., C. B. Moore.
Space Life Sci., Vol. 3, 404 - 414 (1972).

094.904 **Analyses of the returned lunar surface fines for porphyrins.**
J. H. Rho, A. J. Bauman, E. A. Cohen, T. F. Yen, J. Bonner.
Space Life Sci., Vol. 3, 415 - 418 (1972).

094.905 **A quest for porphyrins in lunar soil: samples from Apollo 11, 12 and 14.**
G. W. Hodgson, K. Kvenvolden, E. Peterson, C. Ponnamperuma.
Space Life Sci., Vol. 3, 419 - 424 (1972).

094.906 **Amino acid precursors in lunar samples.** S. W. Fox, K. Harada, P. E. Hare.
Space Life Sci., Vol. 3, 425 - 431 (1972).

094.907 **Problems in the search for amino acids in lunar fines.** P. B. Hamilton, B. Nagy.
Space Life Sci., Vol. 3, 432 - 438 (1972).

094.908 **Results and problems of moon research.** K. P. Florenskij.
Wiss. Zeitschr. Techn. Hochschule Karl-Marx-Stadt, No. 3, p. 317 - 319 (1972). In German.

094.909 **Chemistry of the moon.** J. S. Fruchter, J. R. Arnold.
Annual Rev. Phys. Chemistry, [Annual Reviews Inc., Palo Alto Calif.], Vol. 23, 485 - 508 (1972). – Review article.

094.910 **The composition and origin of the moon.** D. L. Anderson.
Report NASA-CR-128194, California Inst. Techn., Pasadena. [Available from NTIS, Springfield, Va.], 58 pp. (1972).
A model is presented of the moon as a high temperature

condensate from the solar nebula. The Ca, Al, and Ti rich compounds condense first in a cooling nebula.

094.911 Lunar gravity model obtained by using spherical harmonics with mascon terms.
M. H. Kaplan, B. G. Kunciw.
Geophys. Monograph 15, (see 012.025), p. 265 - 273 (1972).

094.912 Bistatic-radar estimation of surface-slope probability distributions with applications to the moon.
M. N. Parker, G. L. Tyler.
Radio Sci., (*USA*), Vol. 8, 177 - 184 (1973).
A method for extracting surface-slope frequency distributions from bistatic-radar data has been developed and applied to the lunar surface.

094.913 Microprobe analyses of glasses in lunar soils.
R. W. Brown, W. I. Ridley, J. L. Warner, A. M. Reid.
Seventh National Conference on 'Electron probe analysis'.
Summaries, San Francisco 1972, 50A, 2 pp. (1972).

094.914 Intracrystalline variations of major and minor elements in lunar pyroxenes (electron probe analysis).
A. E. Bence.
Seventh National Conference on 'Electron probe analysis'.
Summaries, San Francisco 1972, 51A, 3 pp. (1972).

094.915 Vapor-phase crystallization in lunar breccias (microprobe analysis).
D. S. McKay, U. S. Clanton, G. H. Ladle.
Seventh National Conference on 'Electron probe analysis'.
Summaries, San Francisco 1972, 52A, 5 pp. (1972).

094.916 Lunar range measurements with a high-radiance frequency-doubled neodymium-glass laser system.
C. G. Lehr, J. P. Quellette, P. W. Sozanski, J. T. Williams, S. J. Criswell, M. Mattei.
Applied Optics, Vol. 12, 946 - 947 (1973).

094.917 Velocity and internal friction in lunar rocks.
B. R. Tittmann.
Journ. Phys., Vol. 33, Nos. 11 - 12 Suppl., p. C6/271 - 275 (1972).
Reviews some recent experimental progress towards explaining the anomalous elastic and anelastic properties of lunar rocks.

094.918 The surface reduction of lunar fines, 14163,111.
D. A. Cadenhead, B. R. Jones.
Journ. Colloid and Interface Sci., (*USA*), Vol. 42, 650 - 653 (1973). – See Phys. Abstr., Vol. 76, No. 42179 (1973).

094.919 Systematic errors of Watt's charts and their effect on the determination of the orbital elements of the moon. L. N. Kizyun.
Astrometriya i Astrofizika, *Kiev*, Vyp. (No.), 18, (see 003. 016), p. 9 - 20 (1973). In Russian.
Corrections to the moon's orbital elements are calculated from meridian observations carried out at the United States Naval Observatory during 1925–1968.5 and processed before and after the application of corrections for limb profile irregularities from Watt's charts. The results of a statistical analysis substantiated the data of a previous study on the systematic errors with a very complicated structure in the charts.

094.920 Preliminary estimates of the porosity of the upper soil layer for selected details of the lunar surface.
V. V. Botvinova, L. R. Lisina.
Astrometriya i Astrofizika, *Kiev*, Vyp. (No.) 18, (see 003. 016), p. 25 - 32 (1973). In Russian.

Using the opposition effect, the "compaction parameter" (g) is determined for 13 details of the moon's surface observed by Gehrels et al. (1964). The opposition effect was reproduced by the theoretical function derived by Morozhenko and Yanovitsky (1971). The measured values of g are within the limits of 0.1–0.5. An attempt was made to investigate the relation between the "compaction parameter" and the age of the selected details.

094.921 The lunar ellipsoid according to measurements of absolute heights.
Sh. T. Khabibullin, Yu. A. Chikanov.
Trudy Kazan. Gorod. Astron. Obs., No. 37, p. 23 - 39 (1970). In Russian.

094.922 The physical libration of the moon for a model with a viscous nucleus and absolutely solid envelope.
S. S. Peruanskij.
Trudy Kazan. Gorod. Astron. Obs., No. 37, p. 40 - 54 (1970). In Russian.

094.923 Lunar materials: their mineralogy, petrology and chemistry. C. Klein, Jr.
Earth Sci. Rev., (*Netherlands*), Vol. 8, 169 - 204 (1972).

094.924 Teneurs en ^{27}Rb-^{27}Sr, terres rares et K, Rb, Sr, Ba dans le sol lunaire ramené par la sonde soviétique Luna 20. M. Loubet, J.-L. Birk, C. J. Allègre.
Comptes Rendus Acad. Sci. Paris, Sér. D, Vol. 275, 1095 - 1097 (1972).

094.925 Apollo 11 and 12 mare basalts and gabbros: Classification, compositional variations, and possible petrogenetic relations. O. B. James, T. L. Wright.
Bull. Geol. Soc. America, Vol. 83, 2357 - 2382 (1972).

094.926 Possible europium-normal rare-earth abundances estimated from the lunar samples, and the terrestrial analogue. A. Masuda, T. Tanaka.
Beiträge Mineral. Petrol.,(*Germany*), Vol. 34, 336 - 342 (1972).

094.927 Geodesy results obtainable with lunar retroreflectors. J. E. Faller, P. L. Bender, C. O. Alley, D. G. Currie, R. H. Dicke, W. M. Kaula, G. J. F. MacDonald, J. D. Mulholland, H. H. Plotkin, E. C. Silverberg, D. T. Wilkinson.
Geophys. Monograph 15, (see 012.025), p. 261 - 264 (1972).

094.928 Apollo mission 16 lunar photography index maps.
Prepared and published by the Defense Mapping Agency Aerospace Center, St. Louis AFS, Missouri 63118, for the National Aeronautics and Space Administration, Greenbelt, Maryland. 6 sheets (1972).

094.929 Data users note: Apollo 16 lunar photography.
W. S. Cameron, F. J. Doyle, M. A. Niksch, K. Hug, L. Levenson, K. Michlovitz.
Published by National Space Science Data Center, Goddard Space Flight Center National Aeronautics and Space Administration, Greenbelt, Maryland. NSSDC 73-01, 47 +A7 +B6 pp. (1973).
The purposes of this Data Users Note are to announce the availability of Apollo 16 pictorial data and to aid an investigator in the selection of Apollo 16 photographs for study. As background information, the note includes a brief description of the Apollo 16 mission and mission objectives.

094.930 Lunokhod 2 investigates the moon.
R. O. Kuz'min.
Zemlya i Vselennaya, 1973, No. 3, p. 34 - 39. In Russian.

094.931 **Avalanche mode of motion: implications from lunar examples.** K. A. Howard.
Science, Vol. 180, 1052 - 1055 (1973).

A large avalanche (21 square kilometers) at the Apollo 17 landing site moved out several kilometers over flat ground beyond its source slope. If not triggered by impacts, then it was as "efficient" as terrestrial avalanches attributed to air-cushion sliding.

094.932 **Apollo 17 seismic profiling: Probing the lunar crust.** R. L. Kovach, J. S. Watkins.
Science, Vol. 180, 1063 - 1064 (1973).

Apollo 17 seismic data are interpreted to determine the structure of the lunar crust to a depth of several kilometers.

094.933 **Water vapor from a lunar breccia: implications for evolving planetary atmospheres.**
D. A. Cadenhead, W. G. Buergel.
Science, Vol. 180, 1166 - 1168 (1973).

The exposure of a typical complex lunar breccia to hydrogen after a thorough outgassing produces a fully reduced surface state. Subsequent outgassing over a wide temperature range results in the production of water vapor formed from the chemisorbed hydrogen and oxygen from the lunar sample; the proposed mechanism has been confirmed in terms of the chemisorption of deuterium and the release of heavy water. It is also proposed that such a process could play an important role in the early history of many planets where an oxygen-rich soil is exposed to a reducing atmosphere.

094.934 **Al-Khwarizmi: a new-found basin on the lunar far side.** F. El-Baz.
Science, Vol. 180, 1173 - 1176 (1973).

Apollo 16 and Apollo 17 photographs of the far side of the moon reveal a double-ringed basin 500 kilometers in diameter centered at 1°N, 112°E. The structure is very old and subdued; it is probably Pre-Nectarian in age and appears to have been filled and modified by younger events.

094.935 **Lunar volcanism: Age of the glass in the Apollo 17 orange soil.**
L. Husain, O. A. Schaeffer.
Science, Vol. 180, 1358 - 1360 (1973).

The formation age of the glass in the orange soil brought back by the Apollo 17 astronauts from the Taurus-Littrow valley has been measured by the ^{40}Ar-^{39}Ar stepwise heating technique to be $3710 \pm 60 \times 10^6$ years. The orange glass is thus much older than expected.

094.936 **Phenocryst fabric in lunar basalt sample 12052 from the Ocean of Storms.**
W. R. Greenwood, D. A. Morrison, A. L. Clark.
Bull. Geol. Soc. America, Vol. 83, 2809 - 2816 (1972).

094.937 **Moon rocks. First color portfolio of lunar thin sections.** J. Sinkankas, J. Kath.
Mineral. Digest, (USA), Vol. 2, 6 - 20 (1972).

094.938 **Mössbauer spectrum of ilmenite ($FeTiO_3$) below the Néel temperature.**
B. N. Warner, C. Terry, J. A. Morel, P. N. Shive.
Journ. Geomagn. Geoelectr., (Japan), Vol. 23, 399 - 400 (1971).

094.939 **Structure of Sierra Madera, Texas, as a guide to central peaks of lunar craters.**
K. A. Howard, T. W. Offield, H. G. Wilshire.
Bull. Geol. Soc. America, Vol. 83, 2795 - 2808 (1972).

094.940 **Apollo 17, exploration at Taurus-Littrow.**
E. W. Wolfe, V. L. Freeman, W. R. Muehlberger, J. W. Head, H. H. Schmidt, J. R. Sevier.
Geotimes, (USA), Vol. 17, 14 - 18 (1972).

094.941 **Sedimentology of clastic rocks returned from the moon by Apollo 15.** J. F. Lindsay.
Bull. Geol. Soc. America, Vol. 83, 2957 - 2970 (1972).

094.942 **Nature of the density reversal beneath the lunar maria.** C. C. Mason.
Bull. Geol. Soc. America, Vol. 83, 3725 - 3730 (1972).

094.943 **Mineralogy and petrography of some Apollo 12 samples.** B. Mason, W. G. Melson, E. P. Henderson, E. Jarosewich, J. Nelen.
Smithsonian Contr. Earth Sci., No. 9, p. 1 - 4 (1972).

094.944 **Microhardness of a lunar iron particle and high-purity iron samples.**
E. P. Henderson, R. D. Buchheit, J. L. McCall.
Smithsonian Contr. Earth Sci., No. 9, p. 5 - 11 (1972).

094.945 **Geologische Betrachtungen an Mondgesteinsproben.** K. Beneš, G. N. Katterfeld, S. S. Schulz.
Geologie, (Germany), Vol. 21, 247 - 269 (1972).

094.946 **The organic analysis and carbon chemistry of lunar samples: their significance for exobiology.**
S. Chang, R. S. Young.
Space Life Sci., (Netherlands), Vol. 3, 315 - 319 (1972).

094.947 **An evaluation of pyrolytic techniques with regard to the Apollo 11, 12 and 14 lunar samples analyses.**
B. Nagy, M. A. Jabbar Mohammed, V. E. Modzeleski.
Space Life Sci., (Netherlands), Vol. 3, 323 - 329 (1972).

094.948 **Review of methods used in lunar organic analysis: extraction and hydrolysis techniques.**
K. A. Kvenvolden.
Space Life Sci., (Netherlands), Vol. 3, 330 - 341 (1972).

094.949 **Gas-liquid chromatography in lunar organic analysis.** C. W. Gehrke.
Space Life Sci., (Netherlands), Vol. 3, 342 - 353 (1972).

094.950 **Ion-exchange chromatography in lunar organic analysis.** P. E. Hare.
Space Life Sci., (Netherlands), Vol. 3, 354 - 359 (1972).

094.951 **Search for biogenic structures and viable organisms in lunar samples: a review.** V. I. Oyama.
Space Life Sci., (Netherlands), Vol. 3, 377 - 382 (1972).

094.952 **Research for amino acids in lunar samples.**
C. W. Gehrke, R. W. Zumwalt, K. Kuo, J. J. Rash, W. A. Aue, D. L. Stalling, K. A. Kvenvolden, C. Ponnamperuma.
Space Life Sci., (Netherlands), Vol. 3, 439 - 449 (1972).

094.953 **Aromatic and heteroatom-containing organic compounds in the lunar samples.** R. C. Murthy.
Space Life Sci., (Netherlands), Vol. 3, 450 - 454 (1972).

094.954 **Terrestrial contamination in Apollo lunar samples.** D. A. Flory, B. R. Simoneit.
Space Life Sci., (Netherlands), Vol. 3, 457 - 468 (1972).

094.955 **In situ synthesis during organic analysis of lunar samples.** K. Biemann.
Space Life Sci., (Netherlands), Vol. 3, 469 - 473 (1972).

094.956 **Determination of selenographic coordinates of**

points on the lunar surface from single photographs made with Zond 6.
Ya. L. Ziman, V. F. Baratova, I. V. Isavnina.
Kosmich. ikonika. Moskva, Nauka, 1973, p. 224 - 233. In Russian. — Abstr. in Referativ. Zhurn. 51. Astron., 7.51.302; 62. Issled. kosmich. prostranstva, 7.62.141 (1973).

094.957 Some present-day aspects of cosmic mineralogy. II.
E. K. Lazarenko, A. A. Yasinskaya.
Mineral. sb. L'vov un-ta, 1972, No. 26, vyp. (No.) 1, p. 14 - 34. In Russian. — Abstr. in Referativ. Zhurn. 51. Astron., 7.51.317 (1973).

094.958 Questions of the physical cartography of the moon and investigation of the Martian surface at the symposium on physics of the moon and planets.
Yu. S. Tyuflin.
Geod. i kartografiya, 1973, No. 2, p. 74 - 75. In Russian. Abstr. in Referativ. Zhurn. 62. Issled. kosmich. prostranstva, 7.62.11 (1973).

094.959 On the determination of some astronomical, selenodetic and gravitational parameters of the moon.
E. P. Aleksashin, Ya. L. Ziman, I. V. Isavnina, V. A. Krasikov, B. V. Nepoklonov, B. N. Rodionov, A. P. Tishchenko.
Kosmich. ikonika. Moskva, Nauka, 1973, p. 179 - 190. In Russian. — Abstr. in Referativ. Zhurn. 62. Issled. kosmich. prostranstva, 7.62.140 (1973).

New and full moons: 1001 B. C. to A. D. 1651. See Abstr. 003.054.

The lunar rocks.   See Abstr. 003.081.

Geology of the moon. See Abstr. 003.094.

Some peculiarities of the Fourier spectra of basalts returned by Luna 16 and Apollo 11 according to results of investigations in the USSR and materials of the conference in Houston (USA).   See Abstr. 003.099.

Problems of lunar geology.   See Abstr. 003.103.

Problems of lunar motions and mapping. See Abstr. 011.008.

Notes on the fourth Lunar Science Conference—I. See Abstr. 011.016.

Planetary science. A report on the Royal Society

Copernicus quincentenary symposium. See Abstr. 011.024.

Quasi-periodic orbits about the translunar libration point.   See Abstr. 042.043.

On the continuum fluid approach to the solar wind—moon interaction problem.   See Abstr. 074.048.

A recalibration of the quiet sun millimeter spectrum based on the moon as an absolute radiometric standard. See Abstr. 077.035.

On the normal gravity field of the earth and the moon.   See Abstr. 081.001.

Evidence for convection in planetary interiors from first-order topography.   See Abstr. 091.008.

Composition of metal in type III carbonaceous chondrites and its relevance to the source-assignment of lunar metal.   See Abstr. 105.042.

The Ca—Al relationship in stony meteorites and some lunar materials.   See Abstr. 105.047.

Microbreccias, impact glasses and spherules from Lonar crater, India; lunar analogs.   See Abstr. 105.077.

Composition of metal in type III carbonaceous chondrites and its relevance to the source-assignment of lunar metal.   See Abstr. 105.079.

The chemical compositions of meteorites, Apollo 14, 15, 16, and Luna 20 lunar samples as seen through the optics of a flaming atomic absorption spectrophotometer. See Abstr. 105.102.

Meteoritic, lunar and Lonar impact chondrules. See Abstr. 105.137.

Some modern aspects of cosmic mineralogy. II. See Abstr. 105.150.

The nature and significance of terrestrial impact structures.   See Abstr. 105.159.

Europium anomaly in plagioclase feldspar: experimental results and semiquantitative model. See Abstr. 105.160.

095 Lunar Eclipses

095.001 **Infrared atlas charts of the eclipsed moon.**
R. W. Shorthill.
The Moon, Vol. 7, 22 - 45 (1973). – Paper dedicated to Professor Harold C. Urey on the occasion of his 80th birthday on 29 April, 1973.

The objective of this atlas is to present the thermal response of the lunar surface observed during an eclipse of the moon with accurate position data. The observations were made at a wavelength of 11 μm with an angular resolution of 10″, equivalent to 17 km at the disk center. Over a thousand thermally anomalous regions (hot spots) were detected. Forty four charts make up this atlas which are identical in coverage and projection to the Lunar Atlas Charts (LAC) series. The appendix is a list of the published infrared atlantes. These include isothermal contour maps and images of the day-time, eclipsed and night-time moon, and catalogues of thermal anomalies of the eclipsed and night-time moon.

095.002 **Photoelectric observations of the lunar eclipse of**

1972 January 30. B. P. Carroll.
Journ. British Astron. Ass., Vol. 83, 183 - 184 (1973).

095.003 **Brightness of future lunar eclipses.**
A. M. Bakharev, V. M. Chernov.
Astron. vestn., Vol. 6, 249 - 251 (1972). In Russian.

095.004 **Total lunar eclipse of August 6, 1971.**
V. M. Chernov.
Astron. vestn., Vol. 6, 257 - 258 (1972). In Russian.

095.005 **Total lunar eclipse of August 6, 1971 observed in Tadzhikistan.** A. M. Bakharev, U. Yusupov.
Astron. vestn., Vol. 7, 45 - 46 (1973). In Russian.

095.006 **Photoelectric observations of the total lunar eclipse on January 30, 1972.**
B. O. Galperin, V. A. Golubev, A. I. Martynjuk.
Astron. vestn., Vol. 7, 124 - 125 (1973). In Russian.

096 Lunar Occultations

096.001 **Observations of the moon in the Pleiades.**
Sky Telescope, Vol. 45, 256 - 259 (1973).

096.002 **A May occultation of Vesta by the moon.**
D. W. Dunham.
Sky Telescope, Vol. 45, 261 (1973).

096.003 **Sterbedekkingen zichtbaar te Utrecht, juli - december 1973.** J. Meeus.
Hemel en Dampkring, Vol. 71, 76 - 77 (1973).

096.004 **Rakende sterbedekkingen, juli - december 1973.**
J. Meeus.
Hemel en Dampkring, Vol. 71, 116 - 117 (1973).

096.005 **Sur les premiers résultats d'observations des occultations d'étoiles par la lune selon le principe de la double image.** C. Meyer.
Comptes Rendus Acad. Sci. Paris, Sér. B, Vol. 276, 627 - 629 (1973).

096.006 **Observation de l'occultation rasante de 98 Tau du 1ᵉʳ septembre 1972.** J. Bourgeois.
Ciel et Terre, Vol. 89, 125 - 128 (1973).

096.007 **Vorausberechnete Sternbedeckungen durch den Mond 1973.**
Astron. Nachr., Vol. 294, 131 - 136 (1973).

096.008 **Sternbedeckungen 1971, beobachtet auf der Sternwarte Sonneberg.** P. Ahnert.
Astron. Nachr., Vol. 294, 137 - 138 (1973).

096.009 **Grazing occultation of ZC 3453, 1972 May 9.**

M. D. Overbeek.
Monthly Notes Astron. Soc. Southern Africa, Vol. 32, 25 - 26 (1973).

096.010 **Quelques occultations rasantes visibles en France, août–décembre 1973.** J. Meeus.
L'Astronomie, 87ᵉ année, p. 217 - 219 (1973).

096.011 **Observations of occultations of stars by the moon in 1971 in Tashkent.**
M. R. Ehshmatov, Eh. Rakhmatov.
Tsirk. Astron. Inst., *Tashkent*, No. 33 (380), p. 18 - 20 (1972). In Russian.

096.012 **Results of observations of occultations of stars by the moon in Gorky in 1968 - 1969.**
E. G. Demidovich, A. P. Poroshin.
Astron. vestn., Vol. 6, 69 - 71 (1972). In Russian.

096.013 **Occultation of the Pleiades by the moon observed on August 6, 1972.** A. M. Bakharev.
Astron. vestn., Vol. 7, 126 (1973). In Russian.

096.014 **Den norske seksjon for okkultasjoner 1972.**
H. Brubak.
Astron. Tidssk., Årg. 6, p. 80 - 81 (1973).

096.015 **Grazing occultation of 139 Tauri on 1972 March 21.**
L. V. Morrison.
Journ. British Astron. Ass., Vol. 83, 272 - 274 (1973).

096.016 **Les occultations de Saturne de 1973–1974.**
J. Meeus.
L'Astronomie, 87ᵉ année, p. 251 - 258 (1973).

096.017 **Grazing occultation of SAO 76530, 1972 August 4.**
J. Hers.
Monthly Notes Astron. Soc. Southern Africa, Vol. 32, 31 - 32 (1973).

096.018 **Grazing occultation of ZC 912, 1972 September 2.**
J. Hers.
Monthly Notes Astron. Soc. Southern Africa, Vol. 32, 33 - 34 (1973).

096.019 **Slow fading during lunar occultations.**
M. D. Overbeek.
Monthly Notes Astron. Soc. Southern Africa, Vol. 32, 41 - 42 (1973). – Letter.

096.020 **Ocultaciones. Cuarta parte: Ocultaciones rasantes.**
F. Diego Q.
El Universo, No. 102, Vol. 27, 24 - 30 (1973).

096.021 **Report of occultation observations in 1970.**
S. Kimura.
Mem. Japan Astron. Study Ass., No. 19, Vol. 5, 232 - 234 (1972). In Japanese.

096.022 **Occultation of Vesta by the moon.** D. W. Dunham.
IAU Circ., No. 2497 (1973).

096.023 **Occultations of Ve 2-45 by the moon.**
L. V. Morrison.
IAU Circ., No. 2521 (1973).

096.024 **GX 5-1.**
J. A. Hoffman, P. J. N. Davison, L. V. Morrison.
IAU Circ., No. 2538 (1973).

096.025 **Occultation of Ceres by the moon.**
D. W. Dunham.
IAU Circ., No. 2540 (1973).

096.026 **36 Sextantis.** H. Povenmire.
IAU Circ., No. 2545 (1973).

096.027 **Eclipses of stars by the moon.** J. Witkowski.
Urania Kraków, Vol. 44, 98 - 104, 130 - 138 (1973).
In Polish.

096.028 **Occultation observations in 1971.**
T. Mori, Y. Ganeko, Y. Harada.
Data Rep. Hydrographic Observations, Ser. Astron. Geod., *Tokyo,* (Pub. No. 691), No. 7, p. 1 - 22 (1972).

096.029 **Occultation of stars by the moon.** J. Osório.
Anais Faculdade Ciências do Porto, Vol. 54, Fasc. 3, 26 pp. = Publ. Obs. Astron. Porto, No. 25 (1971).
This paper presents the results of occultation observations during the years 1969–70 at the Astronomical Observatory of the University of Porto.

096.030 **Occultation of stars by the moon.**
J. Osório, N. Rego.
Publ. Obs. Astron. "Prof. Manuel de Barros", Faculdade Ciências do Porto, No. 28, 13 pp. (1972).
The paper gives the results of occultations observed in 1971 at the Astronomical Observatory of the University of Porto.

Bulletin of astronomical observations, 1969.
See Abstr. 075.020.

Bulletin of astronomical observations, 1970.
See Abstr. 075.021.

Lunar occultations from Cerro Tololo I. The carbon star, TX Piscium. See Abstr. 115.008.

The angular diameter of Upsilon Capricorni and an occultation of SAO 118655. See Abstr. 115.009.

Occultation observations of six radio sources with flat spectra. See Abstr. 141.068.

Lunar occultations of the galactic center region in H I, OH and H$_2$CO lines. I. Observations and contour maps. See Abstr. 155.016.

097 Mars

097.001 Observations of Mars from earth between 1965 and 1969. C. C. Counselman III.
Icarus, Vol. 18, 1 - 7 (1973).

Since the 1965 Symposium on Planetary Atmospheres and Surfaces, ground-based observations of Mars have added significantly to our knowledge of that planet. The results of these investigations are reviewed.

097.002 Martian topography and surface properties as seen by radar: The 1971 opposition. G. S. Downs, R. M. Goldstein, R. R. Green, G. A. Morris, P. E. Reichley.
Icarus, Vol. 18, 8 - 21 (1973).

Taking advantage of the favorable opposition of 1971, the Goldstone radar system, operating at 2388 MHz, was used to scan the Martian surface. Measurements of altitude and reflected power were taken approximately every 3 days. Each measurement represents an area 8 km E-W × 80 km N-S, the highest resolution attained to date. Altitude measurements obtained on different observing days were combined to produce altitude profiles for three complete rotations, each at different latitudes. Large-scale variations in altitudes cover a range of 14 km. Altitude changes of 5 in 30 km of longitude were observed.

097.003 Topography and radar scattering properties of Mars. G. H. Pettengill, I. I. Shapiro, A. E. E. Rogers.
Icarus, Vol. 18, 22 - 28 (1973).

High-resolution radar observations of Mars at a wavelength of 3.8 cm have been carried out at the Haystack Observatory for a period of about 6 months surrounding the 1971 opposition. The relative surface height variation with longitude over a band of Martian latitudes between about $-14°$ and $-22°$ has been derived from these observations with an error of about 75 m in the most favorable cases. The mean equatorial radius of Mars as determined from the combined radar data of 1967, 1969 and 1971 was found to be 3394 ± 2 km.

097.004 CO_2 distribution on Mars. T. D. Parkinson, D. M. Hunten.
Icarus, Vol. 18, 29 - 53 (1973).

Ground-based observations of the CO_2 distribution on Mars were made this past opposition from Cerro Tololo Interamerican Observatory. Almost complete coverage of the Martian surface from 40°N to 60°S was obtained. Agreement with previous Kitt Peak observations is good, and confirmation of a pressure anomaly in the Tharsis region has been obtained. The ridge whose eastern slope is Syrtis Major stops at about 15°S, in agreement with the 1971 radar data.

097.005 Preliminary results on plasma electrons from Mars-2 and Mars-3. K. I. Gringauz, V. V. Bezrukikh, G. I. Volkov, T. K. Breus, I. S. Musatov, L. P. Havkin, G. P. Sloutchonkov.
Icarus, Vol. 18, 54 - 58, with a correction in Vol. 18, 669 (1973).

A preliminary analysis is presented of the first results on plasma electrons near Mars obtained by retarding potential analyzers onboard the Mars-2 and Mars-3 orbiters. Two zones of significantly increased electron density or temperature were uncovered; one near Mars, which is interpreted as the solar wind shock front in interaction with Mars, and the other $> 10^5$ km from Mars.

097.006 Solar plasma interaction with Mars: Preliminary results. O. L. Vaisberg, A. V. Bogdanov, N. F. Borodin, A. A. Zertzalov, B. V. Polenov, S. Romanov.
Icarus, Vol. 18, 59 - 63 (1973).

A region with a high intensity of ions with energies $E < 150$ eV has been observed near Mars. The shape of the outer boundary of the region, and the sharp increase of the ion velocity temperature at the boundary suggests the presence of a collisionless shock wave on the sunward side of the planet. The ion flux appears to be disturbed even ahead of the shock wave, and this may be the result of an interaction with the extended outer atmosphere of Mars.

097.007 Mariner 9 mission profile and project history. R. H. Steinbacher, N. R. Haynes.
Icarus, Vol. 18, 64 - 74 (1973).

A general background of the Mariner Mars 1971 project and the significant events of the Mariner 9 mission are presented.

097.008 Mariner 9 – image processing and products. E. C. Levinthal, W. B. Green, J. A. Cutts, E. D. Jahelka, R. A. Johansen, M. J. Sander, J. B. Seidman, A. T. Young, L. A. Soderblom.
Icarus, Vol. 18, 75 - 101 (1973).

The purpose of this paper is to describe the system for the display, processing, and production of image-data products created to support the Mariner 9 Television Experiment. This paper describes the systems that carried out the processes and delivered the products necessary for real-time and near-real-time analyses. References are made to the computer algorithms used for the different levels of decalibration and analysis.

097.009 Mariner 9 ultraviolet spectrometer experiment: Observations of ozone on Mars. A. L. Lane, C. A. Barth, C. W. Hord, A. I. Stewart.
Icarus, Vol. 18, 102 - 108 (1973).

The Mariner 9 ultraviolet spectrometer has observed the 2550 Å ozone spectral absorption feature on Mars. Mariner 9 did not observe ozone at any time in the equatorial region, nor at the south polar cap during its summer season. However, ozone was found in the north polar region beginning at a latitude of 45°N and extending northward. Ozone later appeared in the southern hemisphere southward of 50°S as the Mars autumnal equinox approached. The presence of ozone on Mars seems to be coupled to the water vapor content of its atmosphere.

097.010 On the implications of the shape of Mars. S. K. Runcorn.
Icarus, Vol. 18, 109 - 112 (1973).

Determinations of the ellipticity of the surface of Mars reveal a large discrepancy with the dynamical value obtained from the precession of the orbits of the satellites. The discrepancy might be explained by solid state convection in the deep interior of Mars. Observations from occultations of the Mariner spacecraft have confirmed the existence of the discrepancy between the ellipticities; thus, the role of convection in the Martian interior may be a key to its evolution.

097.011 On the Martian dust storms. G. S. Golitsyn.
Icarus, Vol. 18, 113 - 119 (1973).

A short review is given of the observational data on the great Martian dust storms. It is noted that these storms are observed at the time of the great oppositions, when Mars is at its perihelion. Those problems of Martian meteorology and micrometeorology are discussed which have to be studied for a better understanding of the generation, development and decay of a dust storm.

097.012 Exchange of water vapor between the atmosphere and surface of Mars. C. B. Leovy.
Icarus, Vol. 18, 120 - 125 (1973).

A model for exchange of water from the atmosphere to condensing CO_2 caps is developed. The rate of water condensation in the caps is assumed to be proportional to the meridional heat flux. It follows that the amount of water condensed in the caps varies inversely with the amount of CO_2 condensed.

097.013 The vertical thermal structure of the Martian atmosphere: Modification by motions.
S. L. Blumsack, P. J. Gierasch.
Icarus, Vol. 18, 126 - 133 (1973).

The discrepancy between observations of the Martian atmospheric thermal structure and the radiative-convective equilibrium theory is considered and three processes that could account for this difference are analyzed.

097.014 Remote sensing photometric studies of Mars in 1971. P. Boyce.
Icarus, Vol. 18, 134 - 141 (1973).

Results of several studies are presented which indicate the following: 1. The Minnaert function is generally valid for the bright areas on Mars. 2. Some dust was apparently present in the Martian atmosphere before the onset of the dust storm. 3. The south polar cap varied in brightness as a function of LCM indicating a strong limb-darkening effect. 4. The onset of the dust storm did not significantly change the brightness of Mars.

097.015 New optical measurements of planetary diameters. IV. Size of the north polar cap of Mars.
A. Dollfus.
Icarus, Vol. 18, 142 - 155 (1973).

The telescopic determination of the size, shape, and variation of the north polar cap of Mars was significantly improved by the use of the double-image micrometer and high quality photographic plates. This report presents the results of a survey of micrometric measurements on planets and satellites by the double-image technique.

097.016 Slope angle and frost formation on Mars.
S. R. Balsamo, J. W. Salisbury.
Icarus, Vol. 18, 156 - 163 (1973).

We present here a model for the development of persistent frost layers near the equator, which depends upon the importance of slope angle in controlling surface temperatures.

097.017 A new look at the Martian "violet haze" problem. II. "Blue clearing" in 1969. D. T. Thompson.
Icarus, Vol. 18, 164 - 170 (1973).

The 1969 apparition of Mars has been extensively observed by the International Planetary Patrol Program. Using this material, the distribution of "blue clearing" in phase angle has been examined for four Martian bright area-dark area pairs.

097.018 Gravity field of Mars from Mariner 9 tracking data.
J. Lorell, G. H. Born, E. J. Christensen, P. B. Esposito, J. F. Jordan, P. A. Laing, W. L. Sjogren, S. K. Wong, R. D. Reasenberg, I. I. Shapiro, G. L. Slater.
Icarus, Vol. 18, 304 - 316 (1973).

Further reduction of Doppler tracking data from Mariner 9 confirms our earlier conclusion that the gravity field of Mars is considerably rougher than the fields of either the earth or the moon. The largest positive gravity anomaly uncovered is in the Tharsis region. The value obtained for the inverse mass of Mars (3 098 720 $\pm$ 70 $M_\odot^{-1}$) is in good agreement with prior determinations from Mariner flyby trajectories. The direction found for the rotational pole of Mars, referred to the mean

equinox and equator of 1950.0 is characterized by $\alpha = 317°.3 \pm 0°.2$, $\delta = 52°.7 \pm 0°.2$. This result is in excellent agreement with Sinclair's recent value, determined from earth-based observations of Mars' satellites.

097.019 On radiation conditions of the Martian surface and dusty atmosphere. A. S. Ginzburg.
Dokl. Akad. Nauk SSSR, Ser. Mat. Fiz., Vol. 208, 295 - 298 (1973). In Russian.

097.020 Preliminary results of measurements of the infrared temperature of the Martian surface from the automatic interplanetary station Mars 3.
V. I. Moroz, L. V. Ksanfomaliti, A. M. Kasatkin, G. N. Krasovskii, N. A. Parfentiev, V. D. Davydov, G. F. Filippov.
Dokl. Akad. Nauk SSSR, Ser. Mat. Fiz., Vol. 208, 299 - 302 (1973). In Russian.

097.021 Mars from Mariner 9. B. C. Murray.
Sci. American, Vol. 228, No. 1, p. 48 - 63, 66 - 69 (1973).

The first spacecraft to go into orbit around another planet provides evidence that Mars is just beginning to heat up internally. Systems of channels and gullies suggest erosion by water or some other agent.

097.022 Preliminary results of measurements of H_2O content in the Martian atmosphere from the automatic interplanetary station Mars 3.
V. I. Moroz, A. E. Nadzhip, A. B. Gilvarg, F. A. Korolev, V. S. Zhegulev.
Dokl. Akad. Nauk SSSR, Ser. Mat. Fiz., Vol. 208, 797 - 800 (1973). In Russian.

097.023 Preliminary results of determination of elevations on Mars according to the λ 2 μ bands of CO_2 from the automatic interplanetary station Mars 3.
V. I. Moroz, L. V. Ksanfomaliti, A. M. Kasatkin, B. S. Kunashev, K. A. Tsoi.
Dokl. Akad. Nauk SSSR, Ser. Mat. Fiz., Vol. 208, 1048 - 1051 (1973). In Russian.

097.024 The new Mars: Volcanism, water, and a debate over its history. A. L. Hammond.
Science, Vol. 179, 463 - 465 (1973).

097.025 Mariner 9 ultraviolet spectrometer experiment: Seasonal variation of ozone on Mars.
C. A. Barth, C. W. Hord, A. I. Stewart, A. L. Lane, M. L. Dick, G. P. Anderson.
Science, Vol. 179, 795 - 796 (1973).

Ozone is observed to be present in the polar regions of Mars and to have a seasonal variation. Ozone is not observed in the equatorial region during any season.

097.026 Large scale surface structure of Mars.
W. Schlosser, W. Haupt.
Astron. Astrophys., Vol. 23, 471 - 473 (1973).

Cloud observations of Mars are compared with radar profile scans and gravity field measurements of the planet. A large scale surface structure is indicated, which influences the meteorology of the planet.

097.027 Stability of CO_2 in the Martian atmosphere and under radiolysis. D. A. Parkes.
Nature, Vol. 241, 110 - 111 (1973).

CO_2 does show an apparent stability under reactor radiolysis; isotope-exchange experiments have confirmed that both dissociation and recombination occur. Similar behavior should be expected on Mars and it is pertinent to look for a common explanation.

097.028 Martian centre of mass – centre of figure offset.
G. Schubert, R. E. Lingenfelter.
Nature, Vol. 242, 251 - 252 (1973).
A differentiated crust and possible internal convection are suggested by this first report of a Martian centre of mass—centre of figure offset.

097.029 De marsopnamen van de Mariner 9 zijn beëindigd.
T. de Vries.
Hemel en Dampkring, Vol. 71, 3 - 10 (1973).

097.030 Die neue Marskarte der NASA.   E. Wiedemann.
Orion Schaffhausen, 31. Jahrgang, p. 16 - 18 (1973).

097.031 Mars 1971 (Opposition 10 août 1971).
S. Cortesi.
Orion Schaffhausen, 31. Jahrgang, p. 18 - 20 (1973). – Rapport No. 23 du «Groupement planétaire SAS».

097.032 Een onderzoek van de Mariner 9-foto's van Mars.
T. de Vries.
Hemel en Dampkring, Vol. 71, 99 -103 (1973).

097.033 An alpha particle experiment for chemical analysis of the Martian surface and atmosphere.
T. E. Economou, A. L. Turkevich, J. H. Patterson.
Journ. Geophys. Res., Vol. 78, 781 - 791 (1973).
The paper describes the capabilities of an α particle experiment specially designed to provide important analytical information about the Martian surface and atmosphere. The development instrument, although only in preprototype stage, could easily be brought to flight configuration and could be used on missions to Mars or other extraterrestrial bodies.

097.034 Polar wandering on Mars?
B. C. Murray, M. C. Malin.
Science, Vol. 179, 997 - 1000 (1973).
Polar wandering during the past 10^8 years may be recorded by unique quasi-circular structures in the polar regions of Mars. Polar wandering on Mars is likely if deep convection is involved in the origin of the very large constructional volcanic features located near the equator.

097.035 Mars as seen by Mariner 9.   B. A. Smith.
Bull. American Astron. Soc., Vol. 5, 35 - 36 (1973).
Abstr. AAS.

097.036 Mars 3: four months of work.
V. I. Moroz, L. V. Ksanfomaliti.
Vestn. AN SSSR, 1972, No. 9, p. 10 - 25. In Russian.
Abstr. in Referativ. Zhurn. 51. Astron., 3.51.327 (1973).

097.037 Tentative estimate of the field of the emerging thermal emission from Mars.
A. M. Bunakova, K. Ya. Kondrat'ev.
Probl. fiz. atmosf., No. 10. Leningrad, Leningr. un-t, 1972, p. 73 - 78. In Russian. – Abstr. in Referativ. Zhurn. 51. Astron., 3.51.328 (1973).

097.038 On the near approach of Mars in 1877 and the so-called "Saigo" star. K. Saito, S. Shinozawa.
Tokyo Astron. Obs., Report No. 61, Vol. 16, 434 - 463 (1973). In Japanese.

097.039 Absolute measurements and computed values for Martian irradiance between 10.5 and 12.5 μm.
L. M. Logan, S. R. Balsamo, G. R. Hunt.
Icarus, Vol. 18, 451 - 458 (1973).

The combination of seasonal and orbital changes in Martian insolation result in complex latitude dependent surface temperature variations that affect the total radiance of the planet as seen from the earth. These surface temperature variations have been calculated, based upon a computer simulation of the thermal environment of the planet. The temperature variations are then integrated to yield the total radiance of the planet as seen from the earth as a function of time. The absolute radiance of Mars was measured on April 4, 1971, with a balloon-borne radiometer system operating in the wavelength range between 10.5 and 12.5 μm.

097.040 Mars: Components of infrared spectra and the composition of the dust cloud.
G. R. Hunt, L. M. Logan, J. W. Salisbury.
Icarus, Vol. 18, 459 - 469 (1973).
Inrared spectra of Mars are made up of three separate components, each of which may dominate the spectrum under different Martian meteorological and observational conditions. By means of laboratory examples we show that both the shape and spectral contrast of the spectral curves change dramatically, depending on which component is dominant. Each experimental condition has been experienced during either the Mariner 69 or 71 observations.

097.041 High altitude infrared spectroscopic evidence for bound water on Mars.   J. R. Houck, J. B. Pollack, C. Sagan, D. Schaack, J. A. Decker, Jr.
Icarus, Vol. 18, 470 - 480 (1973).
The reflectivity of the Martian surface has been measured between 2.0 and 4.0 μm. A broad absorption band is observed which has a minimum at 2.85 μm. The position and shape of the band is compared to features in numerous terrestrial minerals. Bound water in the Martian surface material is the most likely cause of the observed band.

097.042 Mariner 9 ultraviolet spectrometer experiment: 1971 Mars' dust storm.   K. Pang, C. W. Hord.
Icarus, Vol. 18, 481 - 488 (1973).
This paper reports on the spectral and temporal variations of the optical thickness of the Martian atmosphere over the south polar region during the subsiding phase of the storm.

097.043 Ultraviolet observations of Mars made by the Orbiting Astronomical Observatory.
J. Caldwell.
Icarus, Vol. 18, 489 - 496 (1973).
Ultraviolet albedos of Mars in the region $\lambda\lambda 2000-3600$ Å are discussed.

097.044 Water and the Martian W cloud.   S. J. Peale.
Icarus, Vol. 18, 497 - 501 (1973).
The diurnal brightening of the W cloud region of Mars during the flyby of Mariner 6 and 7 is likely due to the formation of water ice clouds.

097.045 Mars and Jupiter: Radio emission at 1.35 cm.
M. A. Janssen, W. J. Welch.
Icarus, Vol. 18, 502 - 504 (1973).
We report observations of the radio disk temperatures of Mars and Jupiter made during October 1971, at a wavelength of 1.35 cm. The mean disk temperature of Jupiter is $136 \pm 5°$K that of Mars is $181 \pm 11°$K.

097.046 On the effect of electron–neutral particle collisions upon the refraction of high-frequency radio waves by the lower atmosphere and ionosphere of Mars.
P. T. McCormick, R. C. Whitten.
Planet. Space Sci., Vol. 21, 881 - 883 (1973). – Research note.

097.047 **A fuming atmosphere for Mars?** V. A. Firsoff.
Observatory, Vol. 93, 85 - 88 (1973). _ Letter.

097.048 **Investigation of the scattered ultraviolet radiation in the upper atmosphere of Mars with Mars 3.**
V. G. Kurt, A. S. Smirnov, S. D. Chuvakhin.
Kosmich. Issled., Vol. 11, 315 - 320 (1973). In Russian.

097.049 **Astrographic observations of Mars at Pulkovo in 1972.** N. A. Shakht, T. P. Kiseleva.
Astron. Tsirk., No. 758, p. 5 - 7 (1973). In Russian.

097.050 **Periodic insolation variations on Mars.**
B. C. Murray, W. R. Ward, S. C. Yeung.
Science, Vol. 180, 638 - 640 (1973).
Previously unrecognized insolation variations on Mars are a consequence of periodic variations in eccentricity, first established by the theory of Brouwer and Van Woerkom (1950). Such annual insolation variations, characterized by both 95,000-year and 2,000,000-year periodicities, may actually be recorded in newly discovered layered deposits in the polar regions of Mars. An additional north-south variation in seasonal insolation, but not average annual insolation, exists with 51,000-year and 2,000,000-year periodicities.

097.051 **The Martian yellow cloud of July 1971.**
C. F. Capen, L. J. Martin.
Bull. American Astron. Soc., Vol. 5, 266 (1973). − Abstr. AAS.

097.052 **The Mariner 9 control net of Mars.** M. E. Davies.
Bull. American Astron. Soc., Vol. 5, 293 - 294 (1973). − Abstr. AAS.

097.053 **Martian gravity analysis..... some recent results.**
R. D. Reasenberg, I. I. Shapiro, J. Lorell, P. Laing.
Bull. American Astron. Soc., Vol. 5, 294 (1973). − Abstr. AAS.

097.054 **Mars: geologic interpretation of Tharsis Bouguer gravity anomalies.**
R. S. Saunders, J. E. Conel, R. J. Phillips.
Bull. American Astron. Soc., Vol. 5, 294 (1973). − Abstr. AAS.

097.055 **Mariner 9 ultraviolet topographic measurements of Mars.** C. W. Hord.
Bull. American Astron. Soc., Vol. 5, 294 (1973). − Abstr. AAS.

097.056 **The traditional features of Mars compared with the geologic map of the planet.** P. M. Millman.
Bull. American Astron. Soc., Vol. 5, 294 (1973). − Abstr. AAS.

097.057 **Climatic change on Mars.**
C. Sagan, O. B. Toon, P. Gierasch.
Bull. American Astron. Soc., Vol. 5, 294 (1973). − Abstr. AAS.

097.058 **The furrowed terrain of Mars.** R. S. Saunders.
Bull. American Astron. Soc., Vol. 5, 295 (1973). Abstr. AAS.

097.059 **Significance of Martian dune features.**
J. A. Cutts, R. S. U. Smith.
Bull. American Astron. Soc., Vol. 5, 295 (1973). − Abstr. AAS.

097.060 **Salt weathering on Mars?** M. C. Malin.
Bull. American Astron. Soc., Vol. 5, 295 (1973). Abstr. AAS.

097.061 **The oxidation of the Martian surface.**
R. Huguenin, T. B. McCord, J. B. Adams.
Bull. American Astron. Soc., Vol. 5, 295 (1973). − Abstr. AAS.

097.062 **Variable features on Mars: Mariner 9 global results.**
C. Sagan, J. Veverka, P. Fox, R. French, R. Dubisch, P. Gierasch, L. Quam, J. Lederberg, E. Levinthal, R. Tucker, B. Eross, J. Pollack.
Bull. American Astron. Soc., Vol. 5, 295 (1973). − Abstr. AAS.

097.063 **Variable features on Mars: Mariner 9 observations of Promethei Sinus.**
C. Sagan, J. Veverka, P. Fox, R. French, R. Dubisch, P. Gierasch, L. Quam, J. Lederberg, E. Levinthal, R. Tucker, B. Eross, J. Pollack.
Bull. American Astron. Soc., Vol. 5, 295 - 296 (1973). Abstr. AAS.

097.064 **Variable features on Mars: comparison of Mariner 6 and 7 with Mariner 9 results.**
J. Veverka, C. Sagan, L. Quam, R. Tucker, B. Eross.
Bull. American Astron. Soc., Vol. 5, 296 (1973). − Abstr. AAS.

097.065 **Interpretation of diurnal contrast changes on Mars.**
D. T. Thompson.
Bull. American Astron. Soc., Vol. 5, 296 (1973). − Abstr. AAS.

097.066 **Martian paleowind distributions from Mariner 9 photographs.**
R. E. Arvidson, T. A. Mutch, K. L. Jones.
Bull. American Astron. Soc., Vol. 5, 296 (1973). − Abstr. AAS.

097.067 **Behavior of the Martian polar caps since 1905.**
W. A. Baum, L. J. Martin.
Bull. American Astron. Soc., Vol. 5, 296 (1973). − Abstr. AAS.

097.068 **Periodic variations in Martian dust storm activity and the layered deposits of the polar regions.**
G. A. Briggs.
Bull. American Astron. Soc., Vol. 5, 296 (1973). − Abstr. AAS.

097.069 **Nature and origin of Martian polar layered deposits.**
J. A. Cutts.
Bull. American Astron. Soc., Vol. 5, 296 - 297 (1973). Abstr. AAS.

097.070 **The long-term stability of solid carbon dioxide on Mars.** B. C. Murray.
Bull. American Astron. Soc., Vol. 5, 297 (1973). − Abstr. AAS.

097.071 **Infrared spectra of Mars from the NASA CV-990 aircraft.** H. P. Larson, U. Fink, G. Michel.
Bull. American Astron. Soc., Vol. 5, 297 (1973). − Abstr. AAS.

097.072 **Mariner 9 observations of the north polar hood and underlying terrain.** G. A. Briggs.
Bull. American Astron. Soc., Vol. 5, 297 (1973). − Abstr. AAS.

097.073 **Ozone and the polar hood of Mars.** C. A. Barth.
Bull. American Astron. Soc., Vol. 5, 297 (1973). Abstr. AAS.

097.074 **Water vapor variations in the atmosphere of Mars from Mariner 9 IRIS.** V. G. Kunde.
Bull. American Astron. Soc., Vol. 5, 297 (1973). − Abstr. AAS.

097.075 **Mariner 9 IRIS observations of Martian ice clouds.**
R. J. Curran, B. J. Conrath, V. G. Kunde, R. A. Hanel.
Bull. American Astron. Soc., Vol. 5, 297 (1973). − Abstr. AAS.

097.076 **Martian atmospheric temperature fields from the Mariner 9 infrared spectroscopy experiment.**
B. J. Conrath.
Bull. American Astron. Soc., Vol. 5, 297 - 298 (1973). Abstr. AAS.

097.077 Some recent results of Mariner 9 occultation measurements of Mars. A. J. Kliore, D. L. Cain, G. Fjeldbo, B. L. Seidel, M. J. Sykes, P. M. Woiceshyn.
Bull. American Astron. Soc., Vol. 5, 298 (1973). – Abstr. AAS.

097.078 The distribution of clouds on Mars in 1969 and 1971. L. J. Martin, W. A. Baum, P. C. Crump.
Bull. American Astron. Soc., Vol. 5, 298 (1973). – Abstr. AAS.

097.079 Martian thermally driven polar symmetric winds with an Ekman layer and surface drag.
J. A. Pirraglia.
Bull. American Astron. Soc., Vol. 5, 298 (1973). – Abstr. AAS.

097.080 Photometry of the 1971 Martian dust storm.
P. B. Boyce.
Bull. American Astron. Soc., Vol. 5, 298 (1973). – Abstr. AAS.

097.081 Multicolor polarimetric observations of the great 1971 Martian duststorm.
J. Veverka, J. Goguen, W. Liller.
Bull. American Astron. Soc., Vol. 5, 298 (1973). – Abstr. AAS.

097.082 Mariner 9 television observations of Phobos and Deimos. J. B. Pollack, J. Veverka, M. Noland, C. Sagan, T. C. Duxbury, C. H. Acton, Jr., G. H. Born, W. K. Hartmann, B. A. Smith.
Bull. American Astron. Soc., Vol. 5, 306 - 307 (1973).
Abstr. AAS.

097.083 Mapping Phobos. T. C. Duxbury.
Bull. American Astron. Soc., Vol. 5, 307 (1973).
Abstr. AAS.

097.084 Infrared surface absorptions in the solar system: Martian polar caps, Galilean satellites, rings of Saturn, and laboratory ices. U. Fink, H. P. Larson.
Bull. American Astron. Soc., Vol. 5, 307 (1973). – Abstr. AAS.

097.085 Towards a more habitable Mars – or – the coming Martian spring. J. A. Burns, M. Harwit.
Icarus, Vol. 19, 126 - 130 (1973).
Two schemes are presented in which matter is moved closer to Mars so as to increase the equinoctial precession period of Mars. In such a way the Martian springs, during which Mars is habitable in the long winter model of Sagan (1971), could be extended and, in fact, prolonged indefinitely. The schemes involve using solar energy to move the satellite Phobos and/or material from the asteroid belt.

097.086 Theoretical calculation of Martian airmass.
J. H. Woodman, E. S. Barker.
Icarus, Vol. 19, 131 - 136 (1973).
Analysis of spectroscopic observations of Mars requires values of the effective Martian airmass to obtain true abundances. We have computed detailed values corresponding to specified slits superimposed on the disk of the planet, giving useful output in the form of curves presenting the average airmass for different regions of the planet and various conditions of planet diameter, seeing and phase angle.

097.087 Contrast measurement data of some Martian maria in July - August 1971. A. R. Gajduk.
Astron. vestn., Vol. 6, 212 - 217 (1972). In Russian.
Contrast of continent–mare measurement data from photographs obtained at the time of the great opposition of Mars in 1971 are presented. The dependence of contrast on the incidence angle of light is given.

097.088 Mars in 1971. E. H. Collinson.
Journ. British Astron. Ass., Vol. 83, 283 - 290

(1973). – Report of the Mars Section of the British Astron. Ass.

097.089 The planet Mars in 1973. C. F. Capen.
Sky Telescope, Vol. 46, 53 - 60 (1973).

097.090 Electrical breakdown caused by dust motion in low-pressure atmospheres: Considerations for Mars.
H. F. Eden, B. Vonnegut.
Science, Vol. 180, 962 - 963 (1973).
Electrification of agitated dust can cause visible breakdown in a carbon dioxide atmosphere at low pressure in a laboratory experiment. Dust storms on earth become electrified, with accompanying breakdown phenomena. Martian dust storms may reduce the atmospheric conductivity by capturing fast ions on particles, and, by electrifying, may cause discharges in the relatively low pressure atmosphere.

097.091 Photoelektrisches Area Scanning auf Mars während der Opposition 1971.
P. B. Boyce, R. Albrecht.
Mitt. Astron. Ges., No. 32, p. 208 - 209 (1973).

097.092 The traditional features of Mars compared with the geologic map of the planet. P. M. Millman.
Journ. Roy. Astron. Soc. Canada, Vol. 67, 115 - 122 (1973). Presented at the 4th Annual Meeting, Division for Planetary Sciences, A. A. S., March 20 – 23, 1973, Tucson, Arizona.
Twelve traditional albedo features, consistently plotted on the maps of Mars over the last century, are selected for comparison with the geologic maps of Mars based on the Mariner program. The boundary agreement between the geologic terrains and the traditional features is generally poor or non-existent, in contrast to what is observed on the surface of the moon. A correlation between strong slopes and strong albedo features is suggested.

097.093 Alpine glacial features of Mars.
J. Kane, J. Kasold, M. Suda, P. Metcalf, S. Caccamo.
Nature, Vol. 244, 20 - 21 (1973). – Letter.

097.094 Landing on Mars. C. Sagan.
Nature, Vol. 244, 61 (1973). – Letter.

097.095 Mariner 9, Tercera y última parte.
R. N. Watts, Jr., translated from Sky Telescope, Vol. 43, 208 - 213 (1972) by A. D. Lara.
El Universo, No. 102, Vol. 27, 11 - 13 (1973).

097.096 On the extent of the Martian ionosphere.
S. J. Bauer, R. E. Hartle.
Journ. Geophys. Res., Vol. 78, 3169 - 3171 (1973). – Letter.

097.097 Landingsplaatsen van de Vikingschepen op Mars.
T. de Vries.
Hemel en Dampkring, Vol. 71, 208 - 212 (1973).

097.098 A rotating chart of Mars. S. R. Brzostkiewicz.
Urania Kraków, Vol. 44, 172 - 175 (1973). In Polish.

097.099 Investigations of Mars from the Soviet automatic stations Mars 2 and 3. M. Ya. Marov, G. I. Petrov.
Icarus, Vol. 19, 163 - 179 (1973).
Preliminary results from the Mars 2 and 3 orbiters and landers are presented, including descriptions of the spacecraft and their operation; orbital elements; landing sequencing; and scientific results on the distribution of temperature, water vapor, dielectric constant, elevation, and photometric properties over the surface of Mars. Further results are given on ultraviolet emission, interplanetary and circum-Martian plas-

mas, high altitude ultraviolet clouds, ionospheric electron densities, and a zone of thermalized ions. A surface equatorial magnetic field strength of about 60 γ is deduced. Orbital photography implies a value of the optical oblateness close to the dynamical value. Interesting twilight phenomena have been uncovered.

097.100 Mars: Evidence for dynamic processes from Mariners 6 and 7. W. E. Elston, E. I. Smith.
Icarus, Vol. 19, 180 - 194 (1973).
The paper deals with preliminary analysis of some surface features of the equatorial region of Mars photographed by Mariners 6 and 7. The photographs document a three-stage evolution of that part of the Martian surface.

097.101 The ground surface of the planet Mars. A. Dollfus.
Aeronaut. Astronaut. (*France*), No. 38, p. 5 - 22 (1972). In French.
The telescopic study of the planet Mars has shown that the ground surface is dotted with dark spots whose configuration changes seasonally; the ground is powdery; the polar regions are covered with a volatile deposit, sublimating in summer.

097.102 Mars, the planet with shield volcanoes of the Hawaiian type. K. Beneš.
Říše hvězd, Vol. 54, 54 - 56 (1973). In Czech.

097.103 Spectroscopy and aeronomy of O_2 on Mars. T. D. Parkinson, D. M. Hunten.
Journ. Atmosph. Sci., Vol. 29, 1380 - 1390 (1972).
High-dispersion photoelectric spectra have been taken of four oxygen A band lines. Synthetic spectra have been calculated, and ratios of the Martian spectra with the synthetic spectra have been taken to enhance any possible Martian lines.

097.104 The north polar hood of Mars in 1969. B. Salmon.
Strolling Astronomer, Vol. 24, 90 - 93 (1973).

097.105 A 1969 photovisual chart of Mars — ALPO report III. C. F. Capen.
Strolling Astronomer, Vol. 24, 99 - 102 (1973).

097.106 The red planet Mars chart.
Available from the National Geographic Society, Washington, D.C. Price: heavy paper $ 2.00, plastic $ 3.00 (1973). — Review in Strolling Astronomer, Vol. 24, 103 - 104; 1973 (*C. F. Capen*).

097.107 The monochromatic and radiometric albedo of Mars and Venus. W. M. Irvine.
Physics of the moon and planets, (see 012.024), p. 326 - 330 (1972). In Russian.

097.108 Spectral investigations of the atmospheres of Mars and Venus. T. Owen, H. P. Mason.
Physics of the moon and planets, (see 012.024), p. 330 - 334 (1972). In Russian.

097.109 Estimates of the turbulence intensity in the atmospheres of Mars and Venus.
G. S. Golitsyn, V. I. Tatarskij.
Physics of the moon and planets, (see 012.024), p. 394 - 397 (1972). In Russian.

097.110 Physics of the planet Mars. I. K. Koval'.
Physics of the moon and planets, (see 012.024), p. 404 - 408 (1972). In Russian.

097.111 The diurnal temperature variation in the aerosol-gaseous atmosphere and in the Martian ground.
V. I. Aleshin, T. N. Fedoseeva.
Physics of the moon and planets, (see 012.024), p. 409 (1972). In Russian. — Abstract.

097.112 Optical properties of the Martian atmosphere from polarization observations. A. V. Morozhenko.
Physics of the moon and planets, (see 012.024), p. 409 (1972) In Russian. — Abstract.

097.113 The role of water vapour in the meteorology of Mars. S. Miyamoto.
Physics of the moon and planets, (see 012.024), p. 409 - 410 (1972). In Russian.

097.114 Physical characteristics of the "blue clouds" of the planets Mars, earth and Venus.
Ch. I. Villmann, N. I. Grishin.
Physics of the moon and planets, (see 012.024), p. 413 - 418 (1972). In Russian.

097.115 Identification of some details on photographs of Mars obtained with Mariner 4 and from the earth.
B. A. Smith.
Physics of the moon and planets, (see 012.024), p. 418 - 421 (1972). In Russian.

097.116 On craters on photographs of the Martian surface obtained in 1965 with Mariner 4. N. Bonev.
Physics of the moon and planets, (see 012.024), p. 421 - 422 (1972). In Russian.

097.117 The question of liquid water on Mars. B. Murray.
Physics of the moon and planets, (see 012.024), p. 422 (1972). In Russian. — Abstract.

097.118 Criteria for the existence of H_2O crystals on Mars. V. D. Davydov.
Physics of the moon and planets, (see 012.024), p. 423 - 424 (1972). In Russian.

097.119 The exploration of Mars. J. Buj.
Rev. Esp. Electron. (*Spain*), No. 219, Vol. 20, 30 - 33 (1973). In Spanish.

097.120 Contrasts of equatorial Martian maria obtained from observations with the 2-m telescope at the Shemakha Astrophysical Observatory in 1971. A. R. Gajduk.
Astrometriya i Astrofizika, *Kiev*, Vyp. (No.) 18, (see 003. 016), p. 32 - 38 (1973). In Russian.
Spectral contrast values for 26 regions of the Martian maria in the equatorial zone of the southern hemisphere are given.

097.121 Numerical experiment of radiative-convective equilibrium of the Martian atmosphere.
S. Moriyama.
Journ. Meteorol. Soc. Japan, Vol. 50, 181 - 193 (1972).

097.122 La découverte de Mars. A. Ducrocq.
Sci. et Avenir, (*France*), No. 307, p. 718 - 770 (1972).

097.123 Carbonic acid and life on Mars. V. K. Tsyskovskij.
Zemlya i Vselennaya, 1973, No. 3, p. 54. In Russian.

097.124 Aeolian processes on Mars: erosive velocities, settling velocities, and yellow clouds. R. E. Arvidson.
Bull. Geol. Soc. America, Vol. 83, 1503 - 1508 (1972).

097.125 Nature of the Martian surface as inferred from the particle-size distribution of lunar-surface material: discussion and reply.
A. T. Buller, J. McManus, C. C. Mason.
Bull. Geol. Soc. America, Vol. 83, 3833 - 3838 (1972).

097.126 Observation of a shock wave on Mars.
O. L. Vajsberg, A. V. Bogdanov, A. A. Zertsalov, S. A. Romanov.
IV Leningr. mezhdunar. seminar "Edinoobrazie uskoreniya chastits v razlich. masshtabakh kosmosa, 1972". Leningrad, 1972, p. 295, 296. − Abstr. in Referativ. Zhurn. 62. Issled. kosmich. prostranstva 7.62.177 (1973).

Mariner 9 silenced; transmitted Mars data.
IEEE Spectrum, Vol. 10, No. 1, p. 108 (1973).

Targets for Viking.
Nature, Vol. 243, 53 (1973).

Mars unveiled.
Spaceflight, Vol. 15, 151 - 153 (1973).

Quenching of $O(2^1 D_2)$ by atmospheric gases.
See Abstr. 022.017.

Solar neutrinos, Martian rivers, and Praesepe.
See Abstr. 080.037.

Atmospheric ion losses by Venus and Mars to the solar wind. See Abstr. 093.018.

Simultaneous impact and lunar craters.
See Abstr. 094.009.

Apollo 17 and Mariner 9 − zwei erfolgreich beendete Raumfahrtunternehmen der USA. See Abstr. 094.108.

Distribution of craters on the moon and Mars in connection with their origin. See Abstr. 094.590.

Geologic history of the moon and Mars.
See Abstr. 094.843.

Questions of the physical cartography of the moon and investigation of the Martian surface at the symposium on physics of the moon and planets. See Abstr. 094.958.

098   Minor Planets

098.001 Further evidence for collisions among asteroids.
D. C. McAdoo, J. A. Burns.
Icarus, Vol. 18, 285 - 293 (1973).
The 64 asteroids with reliably known rotational properties [rotation period P, magnitude $B(1.0)$ and maximum change of magnitude Δm] are studied. A plot of $B(1.0)$ vs P illustrates that smaller asteroids tend to rotate faster than larger asteroids. The mean P for all 64 asteroids is 8.8hr. The results are interpreted in terms of a model in which collisions break asteroids into irregular fragments.

098.002 A study of commensurable motion in the asteroid belt. R. Giffen.
Astron. Astrophys., Vol. 23, 387 - 403 (1973).
Schubart's averaging method for studying the secular effects of commensurable motion is applied to the 2/1 Hecuba gap and the 3/2 Hilda group in the asteroid belt. The development of this method, which is based on the planar, elliptic restricted three-body problem, and the assumptions used to arrive at this model are discussed. The existence of a non-classical integral for the averaged elliptical problem permits the use of invariant curves in the study of these two resonances, analogous to Hénon's study of the circular problem.

098.003 Planetoïde nr. 15, Eunomia.
V. Cladder.
Hemel en Dampkring, Vol. 71, 119 - 120 (1973).

098.004 Asteroid spectral reflectivities. C. R. Chapman, T. B. McCord, T. V. Johnson.
Astron. Journ., Vol. 78, 126 - 140 = Contr. M.I.T. Planet. Astron. Lab., *Cambridge, Mass.*, No. 55 (1973).
We measured spectral reflectivities (0.3−1.1 μm) for 32 asteroids. There are at least 14 different curve types. Several asteroids show probable color variations with rotation, especially 6 Hebe. A sample of 102 asteroids with reliably known colors is derived from the reflectivities and from earlier colorimetry. Several correlations of colors and spectral curve types with orbital and physical parameters are examined.

098.005 Positions of minor planets Baumeia 813, Triberga 619, and Jessonda 459.
H. K. Herglotz, J. W. McCarter, G. Pruckmayr.
Astron. Journ., Vol. 78, 141 (1973).
Three minor planets were photographed near their opposition in 1970. Positions were determined relative to identified stars from the SAO catalog, with a precision of 1 sec of arc or better.

098.006 Resonances and librations of some Apollo and Amor asteroids with the earth. W.-H. Ip, R. Mehra.
Astron. Journ., Vol. 78, 142 - 147 (1973).
The orbital evolution of the Apollo asteroids 1620 Geographos and 1685 Toro and Amor asteroids 433 Eros, 1221 Amor, and 1627 Ivar are investigated by numerical integration. All these asteroids, with the exception of Geographos, exhibit systematic orbital couplings with the earth during the time interval studied (1600−2350 A.D.).

098.007 Precise astrographic positions of minor planets ob-

tained at the Poznań University Observatory.
B. Morkowska.
Acta Astron., Vol. 23, 43 - 48 (1973).

The paper contains results of precise observations of the minor planets (1) Ceres, (3) Juno, and (7) Iris taken in the year 1967.

098.008 **Les noms des astéroides.** M.-A. Combes.
L'Astronomie, 87ᵉ année, p. 164 - 178 (1973).

098.009 **The selection of asteroids for sampling missions.**
J. T. Wasson.
Meteoritics, Vol. 8, 81 (1973). – Abstract.

098.010 **Determination of the parameters of motion and rotation of an asteroid from distance measurements of a space station on its surface.** Yu. V. Batrakov, L. F. Bragar'.
Kosmich. Issled., Vol. 11, 226 - 235 (1973). In Russian.

098.011 **Minor planet observations in 1963 at the Bucharest Observatory.**
G. Bocşa, C. Cristescu, I. Gheţu, B. Milet.
Stud. Cerc. Astron., Vol. 18, 105 - 114 (1973).

098.012 **Comets and minor planets observed during 1972 at Bucharest and Nice Observatory.** G. Bocşa,
C. Cristescu, I. Gheţu, V. I. Vlasceanu, B. Milet.
Stud. Cerc. Astron., Vol. 18, 115 - 119 (1973). – Concerning the minor planets 433 (Eros), 1036 (Ganymed), 1685 (Toro).

098.013 **The establishment of a set of standardized spectral reflectivity curves for the meteorite classes with applications to the interpretation of asteroid spectra.**
M. J. Gaffey, T. B. McCord.
Bull. American Astron. Soc., Vol. 5, 307 (1973). – Abstr. AAS.

098.014 **Comparisons of meteorite and asteroid spectral reflectivities.** J. W. Salisbury, C. R. Chapman.
Bull. American Astron. Soc., Vol. 5, 308 (1973). – Abstr. AAS.

098.015 **Spectrophotometry of (43) Ariadne: a possible chondritic composition.**
T. V. Johnson, D. L. Matson.
Bull. American Astron. Soc., Vol. 5, 308 (1973). – Abstr. AAS.

098.016 **Radii and albedos of nine asteroids.**
D. P. Cruikshank, D. Morrison.
Bull. American Astron. Soc., Vol. 5, 308 (1973). – Abstr. AAS.

098.017 **Polarimetric observations of 9 Metis, 15 Eunomia, 89 Julia, and other asteroids.** J. Veverka.
Icarus, Vol. 19, 114 - 117 (1973).
White-light polarization curves for a number of asteroids are presented. All have well-developed negative branches, suggestive of relatively dark, texturally complex surfaces—possibly regoliths. Geometric albedos, estimated from the positive branch of the polarization curves, show that a significant dispersion in surface albedo exists within the asteroid belt.

098.018 **A study of commensurable motion in the asteroid belt.** R. Giffen.
Mitt. Astron. Ges., No. 32, p. 202 - 206 (1973).

098.019 **Ephemerides of minor planets for 1974.**
Editor: Institut Teoreticheskoj Astronomii Akademii Nauk SSSR, under the editorship of G. A. Chebotarev.
Izdatel'stvo "Nauka", Leningradskoe Otdelenie, Leningrad.
191 pp. Price 2 Rbl. 44 Kop. (1973). In Russian and English.
Contents: Introduction, p. 3 - 8; Information on new elements, p. 9; Elements, p. 10 - 42; Opposition dates, p. 43 - 53; Ephemerides, p. 54 - 172; Ephemerides of bright planets, p.

173 - 184; Ephemerides of some unusual planets, p. 185 - 188, Critical list, p. 189.

098.020 **Positions of minor planets.**
G. Soulié, R. Dumont.
Astron. Astrophys., Suppl. Ser., Vol. 11, 67 - 76 (1973).
In French.

098.021 **Photographic observations of comets and minor planets.** B. G. Jørgensen, B. Reipurth.
Astron. Astrophys., Suppl. Ser., Vol. 11, 107 - 117 (1973).
The observed minor planets are: 1 Ceres, 2 Pallas, 3 Juno, 4 Vesta, 40 Harmonia.

098.022 **1972 XA.** P. Wild, B. Milet, M. Schürer.
IAU Circ., No. 2476 (1973).

098.023 **1972 RA.** A. R. Klemola, J. Gibson, U. Gibson.
IAU Circ., No. 2478 (1973).

098.024 **Occultation of BD +2°2913 by Pallas on 1973 February 6.** G. E. Taylor.
IAU Circ., No. 2480 (1973).

098.025 **1972 XA.**
T. Seki, H. L. Giclas, M. L. Kantz, P. Wild, B. G. Marsden.
IAU Circ., No. 2480 (1973).

098.026 **Occultation by Pallas.**
V. V. Kallarakal, F. J. Josties, J. A. Howell.
IAU Circ., No. 2483 (1973).

098.027 **1972 RA.**
J. Gibson, U. Gibson, A. R. Klemola.
IAU Circ., No. 2487 (1973).

098.028 **1972 XA.** D. Ruhnow, S. Petermann, L. Kohoutek, M. Dieckvoss, P. Wild, B. Milet.
IAU Circ., No. 2487 (1973).

098.029 **Occultation of BD +2°2913 by Pallas.**
C. F. Lillie, E. Chipman.
IAU Circ., No. 2488 (1973).

098.030 **1971 UA.** L. Kohoutek, M. Dieckvoss.
IAU Circ., No. 2488 (1973).

098.031 **Occultation of BD +2°2913 by Pallas.**
K. A. Janes, H. Reitsema.
IAU Circ., No. 2494 (1973).

098.032 **1953 EA.**
IAU Circ., No. 2495 (1973).

098.033 **1932 HA (Apollo).** B. G. Marsden.
IAU Circ., Nos. 2499, 2503 (1973).

098.034 **1968 AA.** B. G. Marsden.
IAU Circ., No. 2503 (1973).

098.035 **Occultation of BD +2°2913 by Pallas.**
J. A. Clark, E. F. Milone, D. J. I. Fry, F. J. Howell, G. Emerson.
IAU Circ., No. 2506 (1973).

098.036 **Fast-moving object Kowal.**
C. T. Kowal, E. Helin.
IAU Circ., No. 2507 (1973).

098.037 **Fast-moving object Wild (= 1973 EB).**

P. Wild.
IAU Circ., Nos. 2508, 2517 (1973).

098.038 **Object Kowal.** C. T. Kowal, E. Helin.
IAU Circ., No. 2509 (1973).

098.039 **Fast-moving object Gibson (= 1973 EC).**
J. Gibson.
IAU Circ., Nos. 2510, 2517 (1973).

098.040 **Object Wild.** P. Wild.
IAU Circ., No. 2511 (1973).

098.041 **Object Gibson.** J. Gibson, U. Gibson,
B. G. Marsden, K. Aksnes.
IAU Circ., No. 2513 (1973).

098.042 **Object Kowal.** P. Wild, B. G. Marsden.
IAU Circ., No. 2515 (1973).

098.043 **Object Wild.** P. Wild.
IAU Circ., No. 2515 (1973).

098.044 **1932 HA (Apollo).** R. E. McCrosky, C. Y. Shao.
IAU Circ., No. 2516 (1973).

098.045 **1932 HA (Apollo).**
R. E. McCrosky, C. Y. Shao, B. G. Marsden.
IAU Circ., No. 2517 (1973).

098.046 **1973 EA (object Kowal).** P. Wild.
IAU Circ., No. 2517 (1973).

098.047 **1971 UA.** B. G. Marsden.
IAU Circ., No. 2519 (1973).

098.048 **1953 EA.** E. Roemer, L. M. Vaughn.
IAU Circ., No. 2520 (1973).

098.049 **1932 HA (Apollo).** E. Roemer.
IAU Circ., No. 2520 (1973).

098.050 **Object Antal.** A. Mrkos.
Acta Univ. Carolinae Math. Phys., Vol. 13, No. 1, p. 91 - 95 (1973).

098.051 **1973 EC (object Gibson).**
IAU Circ., No. 2523 (1973).

098.052 **1973 EA.** P. Wild.
IAU Circ., No. 2524 (1973).

098.053 **1973 EC.** J. Gibson, U. Gibson.
IAU Circ., No. 2525 (1973).

098.054 **1932 HA (Apollo).** B. G. Marsden.
IAU Circ., No. 2526 (1973).

098.055 **1973 EA.** C. T. Kowal.
IAU Circ., No. 2532 (1973).

098.056 **1973 EC.** J. Gibson, U. Gibson, B. G. Marsden.
IAU Circ., No. 2533 (1973).

098.057 **1973 EA.** B. G. Marsden.
IAU Circ., No. 2535 (1973).

098.058 **1973 EC.** J. Gibson, U. Gibson, J. A. Bruwer.
IAU Circ., No. 2536 (1973).

098.059 **1973 EA.** C. Y. Shao, R. E. McCrosky.

IAU Circ., No. 2536 (1973).

098.060 **1932 HA (Apollo).** C. Y. Shao, B. G. Marsden.
IAU Circ., No. 2549 (1973).

098.061 **1973 EC.** J. Gibson, U. Gibson.
IAU Circ., No. 2551 (1973).

098.062 **Asteroid reflectivities from polarization curves: Calibration of the "slope-albedo" relationship.**
J. Veverka, M. Noland.
Icarus, Vol. 19, 230 - 239 (1973).
   The relationship between the slope of the positive branch of a polarization curve and the normal reflectance of the surface has been calibrated using a wide range of published data. It is determined that asteroid albedos can be inferred meaningfully by this method without prior knowledge of the detailed mineralogical composition of asteroid surfaces.

098.063 **Forces acting upon an asteroid or planet moving across a meteorite stream.** N. A. Barricelli.
Phys. Norvegica, Vol. 6, 199 (1972). − See Phys. Abstr., Vol. 76, No. 42186 (1973).

098.064 **Minor Planet Circulars, (MPC), Nos. 3407 - 3534** (1973).
Edited by Cincinnati Observatory, under the supervision of P. Herget.
   A repository of nearly all new data for numbered and unnumbered minor planets: Observations, elements and ephemerides, identifications, newly assigned numbers and names, occultations.

098.065 **Observations of asteroids in the year 1971.**
M. Procházková.
Contr. Observations People's Obs. Prague, Vol. 8, Ser. 5, No. 1. 5 pp. (1973).
   A list of accurate positions obtained from the photographs taken at the Kleť Observatory during July and August 1971 is given.

098.066 **Estimates of accuracy of the main methods for improving asteroid orbits.** L. E. Sukhoplyueva.
Materialy 3-j Nauch. konf. Tomsk. un-ta po mat. i mekh. Vyp. (No.) 2. Tomsk, Tomsk. un-t, 1973, p. 97 - 98. In Russian.
Abstr. in Referativ. Zhurn. 51. Astron., 7.51.152 (1973).

098.067 **Improving asteroid orbits with the filtration method.** L. E. Sukhoplyueva.
Materialy 3-j Nauch. konf. Tomsk. un-ta po mat. i mekh. Vyp. (No.) 2. Tomsk, Tomsk. un-t, 1973, p. 98. In Russian.
Abstr. in Referativ. Zhurn. 51. Astron., 7.51.153 (1973).

098.068 **On determining the rotational and orbital motion of an asteroid using measured distances to the point on its surface.** L. F. Bragar',
Materialy 3-j Nauch. konf. Tomsk. un-ta po mat. i mekh. Vyp. (No.) 2. Tomsk, Tomsk. un-t, 1973, p. 86 - 87. In Russian.
Abstr. in Referativ. Zhurn. 51. Astron., 7.51.154 (1973).

098.069 **Fast moving object Wild, 1972 XA.**
Kometn. Tsirk., *Kiev*, No. 141 (1973). In Russian.

098.070 **Object Koval, 1973.**
Kometn. Tsirk., *Kiev*, No. 143 (1973). In Russian.

098.071 **Asteroid Apollo, 1932 HA.**
Kometn. Tsirk., *Kiev*, No. 145 (1973). In Russian.

098.072 **New asteroid of group Apollo, 1973 EA.**
Kometn. Tsirk., *Kiev*, No. 149 (1973). In Russian.

098.073 **1972 XA (object Wild).**
    Japan Astron. Study Ass. Circ. 286 - 288 (1973).
In Japanese.

098.074 **1973 EA (object Kowal).**
    Japan Astron. Study Ass. Circ. 286 - 288 (1973).
In Japanese.

098.075 **1973 EB (object Wild).**
    Japan Astron. Study Ass. Circ. 286 - 288 (1973).
In Japanese.

098.076 **1973 EC (object Gibson).**
    Japan Astron. Study Ass. Circ. 286 - 288 (1973).
In Japanese.

098.077 **1932 HA (Apollo).**
    Japan Astron. Study Ass. Circ. 286 - 288 (1973).
In Japanese.

098.078 **1953 EA.**
    Japan Astron. Study Ass. Circ. 286 - 288 (1973).
In Japanese.

098.079 **Useful work on minor planets.**   R. G. Hodgson.
    Minor Planet Bull., Vol. 1, No. 1, p. 1 - 4 (1973).

The faintest asteroids.
Southern Stars, Vol. 24, 136 (1973).

Asteroizi şi comete.   See Abstr. 003.095.

Horseshoe and Trojan orbits associated with Jupiter and Saturn.   See Abstr. 042.032.

New techniques for determining sizes of satellites and asteroids.   See Abstr. 091.040.

Topography on satellite surfaces and the shape of asteroids.   See Abstr. 091.045.

Determination of radii of satellites and asteroids from radiometry and photometry.   See Abstr. 091.057.

Determination of the mass of Saturn from the motion of Trojans.   See Abstr. 100.019.

Determination of the mass of Saturn from the motion of Trojans.   See Abstr. 100.045.

The missing planet.   See Abstr. 101.004.

099 Jupiter

099.001 **Observations of Jupiter with Danjon astrolabes in 1965, 1966, 1967.** S. Débarbat, P. Grudler.
Astron. Astrophys., Vol. 22, 81 - 84 (1973). In French.

This paper contains the results for Jupiter observed with the Danjon astrolabe (Paris and Besançon Observatories) through the years 1965, 1966, 1967. The observations have been corrected for defective illumination ($\Delta\alpha_1$, $\Delta\delta_1$) and ($\Delta\alpha_2$, $\Delta\delta_2$) according to Débarbat and Guinot (1970). The results are analysed in order to study the effectiveness of observational techniques.

099.002 **The β Scorpii occultation by Jupiter. I. The Jovian diameter.** J. Lecacheux, M. Combes, L. Vapillon.
Astron. Astrophys., Vol. 22, 289 - 292 (1973).

The occultation of β Scorpii by Jupiter on May 13, 1971 is used to derive a value for the equatorial radius of the planet. At the altitude of the occulting layer, the equatorial radius is 71802 ± 55 km. At cloud level, the radius is estimated to be 71350 ± 170 km.

099.003 **Polarimetric observations of the Jovian planets. III. Jupiter.** A. V. Morozhenko.
Astron. Zhurn. Akad. Nauk SSSR, Vol. 50, 163 - 166 (1973). In Russian. English translation in Soviet Astron. AJ, Vol. 17, No. 1.

The polarization of the whole disk of Jupiter and of its center in 7 fractions of the spectrum within the interval of 0.37 to 0.80 μ for the phase angles from 0°. 4 to 10°. 3 was measured.

099.004 **The abundance of CH_3D and the D/H ratio in Jupiter.** R. Beer, F. W. Taylor.
Astrophys. Journ., Vol. 179, 309 - 327 (1973).

From observations of the ν_2 parallel band of mono-deuterated methane (CH_3D) in the Jovian atmosphere, we have deduced a CH_3D abundance and mixing ratio and a value for the D/H ratio in this planet, with due regard to the problems of Jovian atmospheric structure and deuterium fractionation. We find the D/H ratio to be significantly less than the terrestrial value and discuss some of the implications to the early history of the solar system.

099.005 **Interpretation of hydrogen quadrupole and methane observations of Jupiter and the radiative properties of the visible clouds.** G. E. Hunt.
Monthly Notices Roy. Astron. Soc., Vol. 161, 347 - 363 (1973).

We describe the most detailed and physically realistic analysis yet of the published observations of the hydrogen quadrupole lines and the R-branch of the $3\nu_3$ CH_4 band within the framework of our present understanding of the physical processes involved. The results give a self-consistent model for the structure composition of the Jovian atmosphere and the radiative properties of the visible clouds at the centre of the disc.

099.006 **Observations of Jupiter at 13-cm wavelength during 1969 and 1971.**
S. Gulkis, B. Gary, M. Klein, C. Stelzried.
Icarus, Vol. 18, 181 - 191 (1973).

We observed Jupiter at 13-cm wavelength in 1969 and 1971 in order to investigate the magnetosphere of the planet, and to search for interactions between the magnetosphere and the solar wind. The results are reported in detail.

099.007 **Theory of decametric radio emissions from Jupiter.**
C. S. Wu, R. A. Smith, J. S. Zmuidzinas.

Icarus, Vol. 18, 192 - 205 (1973).

In this paper we present a new theory of Jovian decametric radio emission phenomena introducing concepts distinctly different from those previously suggested in the literature. Our aim is to develop a framework for more extensive future theoretical investigations.

099.008 **Jupiter's radiation belts.**
N. Brice, T. R. McDonough.
Icarus, Vol. 18, 206 - 219 (1973).

A model for the production and loss of energetic electrons in Jupiter's radiation belt is presented. It is postulated that the electrons originate in the solar wind and are diffused in toward the planet by perturbations which violate the particles' third adiabatic invariant. The winds required to diffuse the energetic particles across the orbit of the satellite Io in a time equal to their drift period are also estimated. If Io is nonconducting, modest winds are required, but if Io is conducting, only small winds are needed. It is concluded that both protons and electrons are diffused in from the solar wind to small distances without serious losses occurring due to the particles being swept up by the satellites.

099.009 **Réception des ondes décamétriques de Jupiter.**
M. Morell.
Icarus, Vol. 18, 220 - 221 (1973).

L'on commente les enregistrements de la radioémission de Jupiter, qui ont été obtenus pour la première fois à Barcelone. Ces enregistrements sont, en principe, cohérents avec la période du Système III (1967.0), et on trouve satisfaisante, aussi l'oscillation que l'on observe, des radio sources A et B, autour de leurs longitudes moyennes respectives.

099.010 **On the radio emission of Callisto.**
A. D. Kuzmin, B. Ya. Losovsky.
Icarus, Vol. 18, 222 - 223 (1973).

A model of an icy surface and interior for Callisto gives a predicted thermal radio emission in good agreement with experimental radio astronomical data. The radio brightness temperature of an icy surface will not depend on wavelength. This may be a method to test icy surface hypotheses. The brightness temperatures of other satellites with icy surfaces will be equal to 200 - 220°K and will not depend on wavelength.

099.011 **Thermal properties of the Galilean satellites.**
D. Morrison, D. P. Cruikshank.
Icarus, Vol. 18, 224 - 236 (1973).

Radiometry in the 20-μm band of eclipses of each of the four Galilean satellites of Jupiter provides information about the thermal properties of the uppermost surface layers of these bodies. Their thermal inertias are all smaller than those of the moon or of Mercury. The results of all our eclipse observations are discussed in detail.

099.012 **Ten-micron eclipse observations of Io, Europa and Ganymede.** O. L. Hansen.
Icarus, Vol. 18, 237 - 246 (1973).

Eclipse observations of Jupiter's satellites Io, Europa, and Ganymede have been obtained in an 8 to 14-μm band pass during 1971. The simplest thermal model able to explain the data for each satellite is a two-layer surface structure with an upper layer, only a few millimeters thick, having low thermal conductivity consistent with fine rock powder or frost, and a subsurface having high thermal conductivity consistent with solid rock or dense ice.

099.013 **Energetic particles in Jupiter's magnetosphere.**

C. K. Goertz.
Astrophys. Letters, Vol. 13, 95 - 96 (1973).

It is shown that electron streams of sufficient energy to explain the intense Jovian decametric radiation are subject to a Pierce-Buneman type instability. It is unlikely that the Io-controlled radiation can be generated by these streams.

099.014 Mutual phenomena of Jupiter's satellites in 1973–74. R. T. Brinkmann, R. L. Millis.
Sky Telescope, Vol. 45, 93 - 95 (1973).

099.015 Observational studies of Jupiter during the years 1965, 1966, 1967. S. Débarbat.
Astron. Astrophys., Vol. 22, 329 - 336 (1973). In French.

This paper contains an analysis of Jupiter astrolabe observations, details of which have been published elsewhere (Débarbat and Grudler, 1972). The differences (astrolabe − American Ephemeris) have been investigated in three ways: corrections for defective illumination, instrumental comparisons (astrolabe, meridian circle, astrograph), systematic errors. Numerical results are plotted for five observatories.

099.016 Observations of 7.9-micron limb brightening on Jupiter. F. C. Gillett, J. A. Westphal.
Astrophys. Journ. (*Letters*), Vol. 179, L153 - L154 (1973).

Initial observations of limb brightening on Jupiter in the 7.7-μ CH_4 band are presented.

099.017 Infrared spectra of the Galilean satellites of Jupiter. U. Fink, N. H. Dekkers, H. P. Larson.
Astrophys. Journ., (*Letters*), Vol. 179, L155 - L159 (1973).

Spectra of the four Galilean satellites from 1 to 4 μ were obtained with a Michelson interferometer. The spectra show that the albedo of Io is quite flat and shows no absorptions in the region observed. Europa and Ganymede have large amounts of water ice on their surface. Callisto shows some faint ice absorptions. Upper limits of 0.5 cm-atm (STP) corresponding to 6×10^{-8} atm partial surface pressure were set for CH_4 and NH_3 on all four satellites.

099.018 Fine structure of the Jupiter radio bursts. G. R. A. Ellis.
Nature, Vol. 241, 387 - 389 (1973).

An improvement in each of the parameters time resolution and frequency resolution of at least an order of magnitude seemed to be essential to reveal the fine structure of the Jupiter bursts. In the observations reported here a time resolution of 0.5 ms combined with a frequency resolution of 10 kHz has been attained at radio frequencies using the technique of time-expansion spectrographic analysis.

099.019 L'interdépendance des ranimations de la S.E.B. et des sources radio de Jupiter. C. Botton.
L'Astronomie, 87e année, p. 35 - 45 (1973).

099.020 Onderlinge verschijnselen der Jupitermanen. J. Meeus.
Hemel en Dampkring, Vol. 71, 61 - 62 (1973).

099.021 Jupitermonde und ihre Schatten. R. Sopper.
SuW, Vol. 12, 20 - 21 (1973).

099.022 Position and velocity components for Jupiter VIII–XII. K. Aksnes.
Astron. Journ., Vol. 78, 121 (1973).

New initial conditions for Jupiter VIII–XII for the epoch 10.0 October 1972 are derived from those published by Herget for five different epochs between 1914 and 1951.

099.023 On the velocity of light three centuries ago.
S. J. Goldstein, Jr., J. D. Trasco, T. J. Ogburn III.

Astron. Journ., Vol. 78, 122 - 125 (1973).

Observations of the times of eclipses of Jupiter's satellite Io by Picard and Roemer were reduced by the principle of least squares with modern orbits for earth and Jupiter. The best-fitting value for the light travel time across one astronomical unit does not differ from the currently accepted value by one part in 200. The rms deviation between the observations and our model is 118 sec.

099.024 A new high-energy component of Jupiter's decametric radio emission. H. R. Miller, A. G. Smith.
Astrophys. Letters, Vol. 13, 177 - 180 (1973).

A pulse-height analysis of the decameter-wavelength observations of Jupiter has revealed that L bursts are composed of two components. The low-energy component is well described by an exponential while the high-energy component, which exhibits a striking departure from the exponential, is found to be clearly associated with the Jovian sources A, B and C.

099.025 A review of Jovian atmospheric dynamics. T. Maxworthy.
Planet. Space Sci., Vol. 21, 623 - 641 (1973).

A brief review is presented of available knowledge of the fluid motions within Jupiter's atmosphere. Evidence is presented to support the contention that the observed cloud masses are probably not simply convected by the main zonal flows. It is likely that an understanding of wave motions within the atmosphere will be of great importance in interpreting data gathered both from the ground and from spacecraft.

099.026 Dynamic spectra of Jupiter radio bursts. G. R. A. Ellis.
Proc. Astron. Soc. Australia, Vol. 2, 157 - 158 (1972).

099.027 Observations of the satellites Jupiter VI and VII. A. R. Klemola.
Astron. Journ., Vol. 78, 226 = Lick Obs. Bull., No. 633 (1973).

Observations of positions are reported for Jupiter VI and VII based on measurements of photographs taken with the blue lens of the 51-cm double astrograph on six nights during the period 1967 to 1969. Positions are given also for six asteroids of magnitude 14−17 which appear in the field of Jupiter.

099.028 An analysis of high-resolution spectral recordings of Jovian decametric radiation.
D. S. Krausche, G. R. Lebo, R. S. Flagg, J. R. Kennedy.
Bull. American Astron. Soc., Vol. 5, 36 - 37 (1973). − Abstr. AAS.

099.029 Jovian satellite–satellite eclipses and occultations in 1973-74. R. T. Brinkmann, R. L. Millis.
Bull. American Astron. Soc., Vol. 5, 37 (1973). − Abstr. AAS.

099.030 The helium abundance on Jupiter.
D. M. Hunten, G. Münch.
Space Sci. Rev., Vol. 14, 433 - 443 (1973).

Methods of determining helium on Jupiter (and the Jovian planets) are critically surveyed. Current information is consistent with solar abundance, $He/H_2 = 0.11$ by number. Methods usable from spacecraft flying by are discussed.

099.031 Intensity of Jupiter's atmospheric belts and solar activity. Z. Pokorný.
Bull. Astron. Inst. Czechoslovakia, Vol. 24, 109 - 110 (1973).

Using Chree's superposition method the relationship between Jupiter's photometric coefficient of activity, which gives the intensity of the atmospheric belts of the planet, and solar activity is investigated.

099.032 Contribution to the dynamical study of the Jovian

Galilean system. I. The intermediate solution in the non resonant case. J.-L. Sagnier.
Astron. Astrophys., Vol. 25, 113 - 124 (1973). In French.

We intend to study the motion of the Galilean satellites of Jupiter on the basis of an approach due to Ferraz Mello (1966), in the light of recent results obtained by Brumberg (1969). Both have suggested that the $(N + 1)$-body planetary problem should be subdivided into the determination of two particular families of solutions, respectively depending on $2N$ (the "intermediate solution") and $4N$ arbitrary constants (leading to the general solution). Our present aim is the determination of this intermediate solution, limiting our study to the second order terms and to non-resonant planetary effects.

099.033 Periodicities in the Jovian decametric emission.
M. L. Kaiser, J. K. Alexander.
Astrophys. Letters, Vol. 14, 55 - 58 (1973).

We have examined the power spectrum from nearly 17 yr of 22-MHz observations of Jupiter's decametric emissions in order to investigate the effect of modulation by the planet's five innermost satellites and to determine the periodicities of long-term phenomena possibly associated with solar activity or earth—Jupiter geometry effects. Whereas Io apparently stimulates a radio storm every 13 hr upon successive sweeps of a given Jovian longitude, we can detect no comparable effects due to other satellites.

099.034 Structure and time variations of the Jovian ionosphere. T. Tanaka, K. Hirao.
Planet. Space Sci., Vol. 21, 751 - 762 (1973).

The problem of the ionospheric formation in the Jovian upper atmosphere is examined. By adopting two plausible atmospheric models, we solve coupled time-dependent continuity equations for ions H_2^+, H_e^+, H^+, H_3^+ and H_eH^+ simultaneously. It is shown that both radiative and three body association of H^+ to H_2 are important for the determination of the structure of the Jovian ionosphere. It is also shown that diurnal variation with large-amplitude can exist in the Jovian ionosphere.

099.035 On long-period variations of the Jovian atmospherical activity. L. P. Sorokina.
Astron. Tsirk., No. 747, p. 3 - 4 (1973). In Russian.

099.036 Maximal contrasts on the Jovian disk in 1962—1969. L. P. Sorokina.
Astron. Tsirk., No. 749, p. 4 - 7 (1973). In Russian.

099.037 Possible explanation of the differential rotation of the atmospheres of Jupiter and Saturn. I.
R. S. Iroshnikov.
Astron. Tsirk., No. 751, p. 1 - 4 (1973). In Russian.

099.038 Possible explanation of the differential rotation of the atmospheres of Jupiter and Saturn. II.
R. S. Iroshnikov.
Astron. Tsirk., No. 751, p. 4 - 5 (1973). In Russian.

099.039 Positional photographic observations of the satellites of Jupiter at Pulkovo in 1966 - 1968.
T. P. Kiseleva.
Astron. Tsirk., No. 758, p. 2 - 5 (1973). In Russian.

099.040 On convection and gravitational layering in Jupiter and in stars of low mass. E. E. Salpeter.
Astrophys. Journ., (*Letters*), Vol. 181, L83 - L86 (1973).

Neutral helium is probably insoluble in metallic hydrogen below some critical temperature. When Jupiter has cooled sufficiently, helium droplets can form, grow, and fall under gravity, leading to chemical separation and gravitational layering in spite of rapid convection. Applications to cool stars are discussed.

099.041 Photometric observation of the occultation of β_1 Scorpii by Jupiter on 1971 May 13. S. J. Elwin.
Journ. British Astron. Ass., Vol. 83, 172 - 178 (1973).

099.042 Metallic hydrogen: simulating Jupiter in the laboratory. W. D. Metz.
Science, Vol. 180, 398 - 399 (1973).

099.043 The optical properties of Venus and the Jovian planets I. The atmosphere of Jupiter according to polarimetric observations. A. V. Morozhenko, E. G. Yanovitskii.
Icarus, Vol. 18, 583 - 592 (1973).

Results are given for polarization measurements of both the entire Jupiter disk and its centre for seven wavelength regions in the $0.373-0.800\ \mu m$ range. Interpretation of these observations is based on two model atmospheres. An approximate method is used for the determination of parameters of the Jovian atmosphere. This method was tested by evaluation of the parameters for the Venus cloud layer.

099.044 On the level of H_2 quadrupole absorption in the Jovian atmosphere. J. S. Margolis, G. E. Hunt.
Icarus, Vol. 18, 593 - 598 (1973).

We show in this note that the application of the Curtis—Godson approximation to the case of collision narrowed lines is valid, at least for the case of an atmosphere based on a Jovian model. We apply our methods to the determination of the amount of hydrogen in the line of sight for Jupiter. We derive a mixing ratio (by volume) of approx 7×10^{-4} for methane to hydrogen in agreement with earlier results.

099.045 Formation of spectral lines in planetary atmosphere IV. Theoretical evidence for structure of the Jovian clouds from spectroscopic observations of methane and hydrogen quadrupole lines. G. E. Hunt.
Icarus, Vol. 18, 637 - 648 (1973).

In this study we have compared, under the same continuum conditions, pressure-broadened and collision-narrowed absorption lines. The conclusions provide insight and understanding of these physical processes, and together with our previous studies (Hunt, 1972), they form a comprehensive theory of spectral line formation in a cloudy planetary atmosphere.

099.046 Wave propagation in the magnetosphere of Jupiter.
H. B. Liemohn.
Astrophys. Space Sci., Vol. 20, 417 - 429 (1973).

The purpose of this research is to develop a systematic procedure for identifying the spatial regimes of various modes of propagation that may be encountered by flyby missions to Jupiter. In order to systematically study the propagation properties of the magnetoplasma, the well known Clemmow-Mullaly-Allis diagram of plasma physics has been utilized. This diagram divides the complex modes of propagation into various regions or ponds in which a characteristic type of propagation is readily identified and analyzed. For specified propagation frequencies and selected magnetoplasma models, similar propagation ponds can be identified in the configuration space around the planet. Loci of propagation cutoffs and resonances are clearly identified and assist in determining the distribution of radio noise. These properties provide a useful basis for speculation about the distribution of local radio noise and its relevant source mechanisms.

099.047 On limits to Jupiter's magnetospheric diffusion rates. J. D. Mihalov.
Astrophys. Space Sci., Vol. 20, 483 - 490 (1973).

X-ray fluxes at earth estimated from hypothetical fluxes and spectra of energetic particles trapped in Jupiter's magnetic field are found to be 1/170000 times the upper limit X-ray flux from Jupiter based on published results from a rocket experiment. Detection of the calculated X-ray flux from Jupiter does not necessarily provide information on an energetic trapped proton component because the X-ray flux due to the hypothetical trapped energetic proton fluxes alone is comparable in magnitude to that due alone to trapped energetic electron fluxes at Jupiter.

099.048 The 1971 apparition of Jupiter.
P. K. Mackal.
Strolling Astronomer, Vol. 24, 41 - 57 (1973).

099.049 Mutual phenomena of Jupiter's satellites, June 6–October 30, 1973.   P. W. Budine.
Strolling Astronomer, Vol. 24, 71 - 75 (1973).

099.050 Periodicities in the Jovian decametric emission.
J. K. Alexander, M. L. Kaiser.
Bull. American Astron. Soc., Vol. 5, 287 (1973). – Abstr. AAS.

099.051 A search for narrow-band ammonia lines in the Jovian microwave spectrum.
S. Gulkis, M. J. Klein, R. L. Poynter, R. B. Read.
Bull. American Astron. Soc., Vol. 5, 287 (1973). – Abstr. AAS.

099.052 Broad-band measurements of the Jovian spectrum from 20 to 24 GHz.   M. J. Klein.
Bull. American Astron. Soc., Vol. 5, 287 (1973). – Abstr. AAS.

099.053 Structure of the upper atmosphere of Jupiter from multichannel observations of the β Sco occultation.
L. Wasserman, J. Veverka, J. Elliot, C. Sagan, W. Liller.
Bull. American Astron. Soc., Vol. 5, 287 (1973). – Abstr. AAS.

099.054 The far ultraviolet spectrum of Jupiter.
H. W. Moos, J. Giles, W. McKinney, C. Freer.
Bull. American Astron. Soc., Vol. 5, 287 (1973). – Abstr. AAS.

099.055 Ammonia abundance estimates and the UV albedo of Jupiter.   M. G. Tomasko.
Bull. American Astron. Soc., Vol. 5, 287 - 288 (1973). Abstr. AAS.

099.056 Photochemistry of NH_3 in the Jovian atmosphere.
D. F. Strobel.
Bull. American Astron. Soc., Vol. 5, 288 (1973). – Abstr. AAS.

099.057 Measurement of the D/H ratio in the atmosphere of Jupiter with a PEPSIOS spectrometer.
J. T. Trauger, F. L. Roesler, N. P. Carleton, W. A. Traub.
Bull. American Astron. Soc., Vol. 5, 288 (1973). – Abstr. AAS.

099.058 Variability of the Jovian clouds from observations of the H_2 quadrupole lines.
J. T. Bergstralh, G. E. Hunt.
Bull. American Astron. Soc., Vol. 5, 288 (1973). – Abstr. AAS.

099.059 Spatial variations in band strength as a probe of the Jovian atmosphere.
R. A. Stokes, R. W. Avery, J. J. Michalsky, Jr.
Bull. American Astron. Soc., Vol. 5, 288 (1973). – Abstr. AAS.

099.060 Scattering model interpretation of the limb-darkening behavior of Jupiter's south tropical zone and Great Red Spot as observed in the 8880 Å absorption band of methane.   A. E. Clements.
Bull. American Astron. Soc., Vol. 5, 288 (1973). – Abstr. AAS.

099.061 IR spectrophotometry of Jupiter and Saturn.
A. B. Binder, D. W. McCarthy, Jr.
Bull. American Astron. Soc., Vol. 5, 288 - 289 (1973). Abstr. AAS.

099.062 Five-micron maps of Jupiter.
C. S. L. Keay, G. H. Rieke, F. J. Low.
Bull. American Astron. Soc., Vol. 5, 289 (1973). – Abstr. AAS.

099.063 Photopolarimetry and imaging of Jupiter in connection with the Pioneer missions.
T. Gehrels, D. L. Coffeen, C. E. Kenknight, W. Swindell, M. G. Tomasko.
Bull. American Astron. Soc., Vol. 5, 289 (1973). – Abstr. AAS.

099.064 H_2 pressure-induced lines in the spectrum of the major planets.   M. J. S. Belton, H. Spinrad.
Bull. American Astron. Soc., Vol. 5, 291 (1973). – Abstr. AAS.

099.065 Galilean satellites: identification of water frost.
C. B. Pilcher, S. T. Ridgway, T. B. McCord.
Bull. American Astron. Soc., Vol. 5, 306 (1973). – Abstr. AAS.

099.066 The post-eclipse brightening of Io.
D. P. Cruikshank, R. E. Murphy.
Bull. American Astron. Soc., Vol. 5, 306 (1973). – Abstr. AAS.

099.067 Limb darkening of Io and Ganymede from eclipse light curves and occultation diameters.
T. F. Greene, R. W. Shorthill, D. W. Smith.
Bull. American Astron. Soc., Vol. 5, 306 (1973). – Abstr. AAS.

099.068 The occultation of SAO 186800 by Ganymede (JIII) on June 6, 1972.   B. O'Leary.
Bull. American Astron. Soc., Vol. 5, 306 (1973). – Abstr. AAS.

099.069 Jovian satellite – satellite eclipses and occultations.
R. T. Brinkmann.
Icarus, Vol. 19, 15 - 29 (1973).
    Toward the end of 1973 and in the first part of 1974 Jovian satellite–satellite eclipses and occultations will occur. From observation of a few of these events the ephemerides of the satellites can be improved, radii and limb darkening curves determined, and crude information about the degree and extent of albedo fluctuations deduced.

099.070 Surface color variations of the Galilean satellites.
F. N. Owen, F. J. Lazor.
Icarus, Vol. 19, 30 - 33 (1973).
    Color variations of the four Galilean satellites have been monitored during the summer of 1971 with the McDonald Observatory area-scanning photometer. All were found to vary with orbital phase, with the exception of Europa in B–V.

099.071 Location of the Jovian magnetic dipole.
P. M. McCulloch, M. M. Komesaroff.
Icarus, Vol. 19, 83 - 86 (1973).
    Measurements of the position of Jupiter's radio centroid at 11 cm were made using the Parkes telescope. They indicate that any systematic displacement of the Jovian magnetic dipole along the rotation axis is to the north of the center of the planet, contrary to a model proposed by Warwick. Any offset of the dipole normal to the rotation axis does not exceed 0.1 of a planetary radius.

099.072 Study of a Jovian plasmasphere and the occurrence of Jupiter radio bursts.
L. Conseil, Y. Leblanc, G. Antonini, D. Quemada.
Cosmic plasma physics. Conference 1971, (see 012.016), p. 27 - 35 (1972).

099.073 One some results obtained by spectrophotometry of the methane absorption band (7250 Å) on Jupiter's disk.
V. V. Avramchuk, N. B. Ibragimov, E. F. Slivenko.
Astron. vestn., Vol. 6, 218 - 222 (1972). In Russian.
The observed absorption process in the methane band along Jupiter's disk is satisfactorily explained in the frame of a single-layer semi-infinite atmosphere. Optical parameters of the cloud region of the planet for the near-infrared of the spectrum have been evaluated.

099.074 Investigation of molecular absorption in the atmospheres of major planets.
L. A. Bugaenko, L. S. Galkin, A. V. Morozhenko.
Astron. vestn., Vol. 6, 223 - 227 (1972). In Russian.
The methane and ammonia absorption bands for individual details of Jupiter's and Saturn's disks have been measured at visual and near-infrared regions.

099.075 Photometric research of the Jovian atmospheric activity in the period of 1962−1969.
N. N. Petrova, L. P. Sorokina.
Astron. vestn., Vol. 7, 9 - 15 (1973). In Russian.
Photometric results of the investigation of the atmospheric activity of Jupiter during the period of 1962−1969 in seven wavelengths are briefly cited.

099.076 Occultation of β Sco by Jupiter on May 13, 1971.
N. A. Nesterko, V. E. Solovjev, A. L. Chikarenko.
Astron. vestn., Vol. 7, 47 (1973). In Russian.

099.077 On a possible relation of the rotation period of the central zone of Jupiter with the variation of its equatorial diameter. M. A. Klyakotko.
Astron. vestn., Vol. 7, 52 - 53 (1973). In Russian.

099.078 Some new experimental data on the existence of unknown satellites of Jupiter. S. S. Gamburg.
Astron. vestn., Vol. 7, 53 (1973). In Russian.

099.079 Structure of Jupiter and Saturn.
W. B. Hubbard, R. Smoluchowski.
Space Sci. Rev., Vol. 14, 599 - 662 (1973).
Understanding of the planetary interiors depends upon our knowledge of the equations of state and of the transport properties of matter at high pressures and temperatures. The present status of this knowledge in relation to hydrogen and helium is discussed in detail including electrical and thermal conductivity, viscosity, diffusivity, etc. On this basis the various possible models of the internal structure of Jupiter and of Saturn are presented and their agreement with observational constraints such as the multipole gravitational coefficients analyzed. Relevance of planetary magnetic fields, basic atmospheric information and the Great Red Spot of Jupiter to the models of the interiors are discussed.

099.080 Initial development of the 1971 June South Equatorial Belt disturbance on Jupiter. R. B. Minton.
Journ. British Astron. Ass., Vol. 83, 263 - 271 (1973).

099.081 Photochemical reactions in the Jovian atmosphere.
P. Molton, J. C. Gilbert.
Journ. British Interplanet. Soc., Vol. 26, 385 - 407 (1973).
Chemical kinetic data relevant to reactions in the Jovian atmosphere are summarized.

099.082 An iterative method to infer the Jovian atmospheric structure from infrared measurements.
T. Encrenaz, D. Gautier.
Astron. Astrophys., Vol. 26, 143 - 147 (1973).
An iterative method of solution of the radiative transfer equation is applied to a synthetic spectrum of the infrared flux of Jupiter. It is shown that both thermal profile and hydrogen to helium mixing ratio of the Jovian atmosphere could be inferred from spectral measurements of the outgoing flux between 18 and 50 microns.

099.083 Analytical theory of the motion of the fifth satellite of Jupiter. G. T. Arazov.
Izv. AN AzSSR. Ser. Fiz.-tekhn. i mat. n., 1972, No. 2, p. 75 - 81. In Russian. − Abstr. in Referativ. Zhurn. 51. Astron., 5.51.126 (1973).

099.084 Ammonia absorption relevant to the albedo of Jupiter. I. Experimental results.
K. A. Dick, A. O. Ziko.
Astrophys. Journ., Vol. 182, 609 - 613 (1973).
The absorption coefficient of ammonia between 2100 and 2250 Å has been studied at temperatures of 295° and 195°K. No evidence was found for high-resolution transmission windows in this spectral region. The existence of such windows can therefore not be invoked to explain the observed albedo of Jupiter.

099.085 Mutual phenomena of Jupiter's satellites.
K. Aksnes.
IAU Circ., No. 2509 (1973).

099.086 Jupiter's radiation belts and the sweeping effect of its satellites. G. D. Mead, W. N. Hess.
Journ. Geophys. Res., Vol. 78, 2793 - 2811 (1973).
In this study we first examine the known characteristics of Jupiter's magnetosphere and summarize some of the expected characteristics of trapped protons and electrons. We then examine the role of Jupiter's satellites in the inner magnetosphere and calculate characteristic absorption lifetimes for trapped particles. Next we calculate a radial diffusion coefficient, assuming that particle diffusion is caused by violation of the third invariant due to magnetic fluctuations associated with fluctuations in the solar wind. Finally, we evaluate these results in light of our present knowledge of the radiation belts of both earth and Jupiter.

099.087 Photoelectron excitation of the Jupiter dayglow.
J. J. Olivero, J. N. Bass, A. E. S. Green.
Journ. Geophys. Res., Vol. 78, 2812 - 2826 (1973).
The photoelectron impact contribution to the Jupiter dayglow is estimated for an atmospheric model with and without helium. Primary photoelectron production rates are calculated at specific altitudes for zero solar zenith angle and a 10.7-cm flux of 150. Extensive excitation, dissociation, and ionization cross-section data for H_2, He, and H are used to model the energy deposition of photoelectrons produced locally.

099.088 A comment on Jovian greenhouse models.
L. Trafton.
Icarus, Vol. 19, 244 - 246 (1973).
Radiative greenhouse models of Jupiter's atmosphere seriously overestimate the temperature of the lower cloud level because they neglect the convective transport of heat.

099.089 On the determination of Jupiter's mass from observations of its VIth satellite. L. E. Bykova.
Materialy 3-j Nauch. konf. Tomsk. un-ta po mat. i mekh. Vyp. (No.) 2. Tomsk, Tomsk. un-t, 1973, p. 88 - 89. In Russian. Abstr. in Referativ. Zhurn. 51. Astron., 6.51.294 (1973).

099.090 Estimate of medium sizes of particles of the cloud layer in Jupiter's atmosphere. O. R. Bolkvadze.
Soobshch. AN GruzSSR, Vol. 69, No. 1, p. 53 - 55 (1973). In Russian. − Abstr. in Referativ. Zhurn. 51. Astron.,

6.51.300 (1973).

099.091 Jupiter in 1967–68: Rotation periods.
P. W. Budine.
Strolling Astronomer, Vol. 24, 81 - 89 (1973).

099.092 A request for observations of spots in Jupiter's equatorial zone. E. J. Reese, with a discussion by P. K. Mackal.
Strolling Astronomer, Vol. 24, 119 - 120 (1973).

099.093 Investigation of the peculiarities of molecular absorption in the spectrum of Jupiter.
A. N. Aksenov, Z. N. Grigor'eva, V. G. Tejfel', G. A. Kharitonova.
Physics of the moon and planets, (see 012.024), p. 433 - 438 (1972). In Russian.

099.094 Results of three-year observations of the absorption bands of methane (6190 Å) and ammonia (6441 and 6478 Å) on the Jovian disk. V. V. Avramchuk.
Physics of the moon and planets, (see 012.024), p. 439 - 443 (1972). In Russian.

099.095 Photometric investigations of Jupiter's atmospheric activity. L. P. Sorokina, N. V. Priboeva.
Physics of the moon and planets, (see 012.024), p. 443 - 444 (1972). In Russian.

099.096 Continued activity of Jupiter and comparison of eruptions 1871–1880 and 1961–1965.
S. K. Vsekhsvyatskij.
Physics of the moon and planets, (see 012.024), p. 444 - 447 (1972). In Russian.

099.097 Complex motion in Jupiter's atmosphere.
B. A. Smith.
Physics of the moon and planets, (see 012.024), p. 447 - 448 (1972). In Russian.

099.098 Results of observations of Jupiter in the cm-wavelength region.
N. S. Soboleva, Yu. N. Parijskij.
Physics of the moon and planets, (see 012.024), p. 449 - 451 (1972). In Russian.

099.099 The Lyman-alpha albedo of Jupiter.
L. Wallace, D. M. Hunten.
Astrophys. Journ., Vol. 182, 1013 - 1031 (1973).
    Simplified, but fairly accurate, models are derived for the distributions of H and CH_4 in Jupiter's upper atmosphere. A method is developed and tested for replacing this inhomogeneous scattering atmosphere by a single equivalent layer. The scattering from this layer is then computed by means of Chandrasekhar's X and Y functions. The average albedo of the planet for Lyman α is given as a function of the eddy diffusion coefficient, K, that characterizes the model atmospheres. Comparison with available data suggests that K is less than $10^6 cm^2 s^{-1}$. Line profiles and limb-darkening curves are also given.

099.100 The equilibration of deuterium in the Jovian atmosphere. R. Beer, F. W. Taylor.
Astrophys. Journ., (*Letters*), Vol. 182, L131 - L132 (1973).
    Recent observations of deuterium in the Jovian atmosphere in the CH_3D and HD phases permit a reexamination of the mechanism of deuterium fractionation in Jupiter. It is shown that catalysis must play an important role in the reaction paths and that, most probably, the catalytic agents are embedded in the lower cloud deck.

099.101 An analysis of the intensity fluctuations produced by the interaction of the decametric radio emission from Jupiter with the interplanetary medium.
D. L. Thompson.
Diss. Dep. Phys. Florida State Univ., College of Arts and Sciences 1969. [Available from Univ. Microfilms Inc., Ann Arbor, Mich.], 156 pp. (1972).

099.102 Theory of motion of the VII satellite of Jupiter. T. V. Bordovitsyna.
Materialy 3-j Nauch. konf. Tomsk. un-ta po mat. i mekh. Vyp. (No.) 2. Tomsk, Tomsk. un-t, 1973, p. 87 - 88. In Russian. Abstr. in Referativ. Zhurn. 51. Astron., 7.51.149 (1973).

    An automatic dynamic spectrum analyser for video tape recorded signals. See Abstr. 033.025.

    The integration of equations of the theory of the figure of planets. See Abstr. 091.002.

    New infrared spectra of the Jovian planets from 12000 to 4000 cm^{-1} by Fourier transform spectroscopy. I. Study of Jupiter in the 3 ν_3 CH_4 band. See Abstr. 091.043.

    A numerical method for determining the temperature structure of planetary atmospheres. See Abstr. 091.058.

    The radio radiation of Venus and Jupiter at 2 and 8 mm wavelengths. See Abstr. 093.065.

    The radio radiation of the moon and sun at 2.25 mm wavelength and of Jupiter at 2.1 mm wavelength. See Abstr. 094.872.

    Mars and Jupiter: Radio emission at 1.35 cm. See Abstr. 097.045.

    Infrared surface absorptions in the solar system: Martian polar caps, Galilean satellites, rings of Saturn, and laboratory ices. See Abstr. 097.084.

    Evolutionary aspects of the atmospheres of Titan and the Galilean satellites. See Abstr. 100.043.

    Preliminary results of observations of discrete sources and Jupiter at 2 cm in Pulkovo. See Abstr. 141.006.

100 Saturn

100.001 Some comments on the magnetosphere and plasma environment of Saturn. F. L. Scarf.
Cosmic Electrodynamics, Vol. 3, 437 - 447 (1973).

Some properties of a model magnetosphere for Saturn are studied in order to determine the bounds that can be set on the surface field strength and the trapped particle population. The primary observational constraint is that non-thermal radiation similar to the Jovian radio emissions must be undetectable from earth. It is argued that for a Saturn surface field of approximately one gauss, those particles that are energized as they diffuse in from the magnetopause with conservation of magnetic moment will produce synchrotron radiation levels that are undetectable at a range of 9.5 AU. The plasma instabilities that heat the oncoming wind particles at the bow shock and others that can limit the stably-trapped flux levels are also discussed briefly.

100.002 Titan revisited. J.S. Lewis, R. G. Prinn.
Comments Astrophys. Space Phys., Vol. 5, 1 - 7 = Contr. M. I. T. Planet. Astron. Lab. No. 70 (1973).

100.003 On the analytical theory of Saturn's satellites. Enceladus-Dione. M. Rapaport.
Astron. Astrophys., Vol. 22, 179 - 186 (1973). In French.

We consider Saturn's satellites with near commensurability relations between their mean motions, and specially the system Enceladus-Dione, in the potential of the oblate planet Saturn. We use Hori's modification of von Zeipel's method for eliminating the short period terms. A non-negligible term appears in the results and permits to obtain a better coherence between the values of elements that Struve obtained from the theory and from the utilisation of observations.

100.004 Optical properties and structure of Saturn's atmosphere. II. The latitudinal variations of absorption at band CH$_4$ 0.62 μ and the peculiarities of the planet in the near ultraviolet.
V. G. Teifel, L. A. Usoltseva, G. A. Kharitonova.
Astron. Zhurn. Akad. Nauk SSSR, Vol. 50, 167 - 171 (1973). In Russian. English translation in Soviet Astron. AJ, Vol. 17, No. 1.

100.005 Photometric properties of Saturn's rings. W. M. Irvine, A. P. Lane.
Icarus, Vol. 18, 171 - 176 = Contr. Five College Obs., *Amherst, Mass.*, No. 105 (1973).

The spectral reflectivity of Saturn's rings between 0.36 and 1.06 μm is derived from observations of the combined light of the Saturn system and the previously determined spectrum of the disk of Saturn. The rings are red relative to the sun for wavelengths $\lambda \lesssim 0.7 \, \mu$m; at longer wavelengths, the spectral reflectivity declines. The amplitude of the opposition effect shows a maximum at both ends of our spectral range.

100.006 The brightness temperature of Saturn at decimeter wavelengths.
M. J. Yerbury, J. J. Condon, D. L. Jauncey.
Icarus, Vol. 18, 177 - 180 (1973).

Observations of the planet Saturn at wavelengths of 49.5 and 94.3 cm are reported. The equivalent disk brightness temperatures were found to be 400 ± 65°K and 540 ± 110° K, respectively. It is suggested that the enhanced portion of the spectrum of the disk brightness temperature favours the idea that the observed long wavelength radiation comes from the planet's atmosphere.

100.007 UBV photometry of Iapetus. R. L. Millis.

Icarus, Vol. 18, 247 - 252 (1973).

UBV observations of Iapetus on 34 nights have revealed a 0.30-mag difference in the depths of two successive minima in the light curve of this satellite. The difference can be explained by a two-hemisphere model with the leading darker hemisphere having a much stronger variation in brightness with solar phase angle than the trailing brighter hemisphere.

100.008 Saturn's rings – A survey.
A. F. Cook, F. A. Franklin, F. D. Palluconi.
Icarus, Vol. 18, 317 - 337 (1973).

In this review paper we first discuss the dimensions of major ring features and of the disk of the planet. We then summarize the observed photometric parameters, and because frozen H$_2$O appears to be a major ring constituent, we compare the appropriate photometric properties of various forms of snow with those of the ring. We examine several ring models, noting certain characteristics that any model should supply.

100.009 Maximum d'ouverture de l'anneau de Saturne.
J. Meeus.
L'Astronomie, 87e année, p. 90 - 91 (1973).

100.010 Saturne: Présentation 1971/72 (Opposition 26 novembre 1971). F. Jetzer, A. Materni.
Orion Schaffhausen, 31. Jahrgang, p. 21 (1973). – Rapport No. 24 du «Groupement planétaire SAS».

100.011 Optical scattering properties of Saturn's ring.
M. J. Price.
Astron. Journ., Vol. 78, 113 - 120 (1973).

Reliable data defining the photometric function of the Saturn ring system at visual (V) wavelengths are interpreted in terms of a simple scattering model. To facilitate the analysis, new photographic photometry of the ring has been carried out utilizing the Lowell Observatory plate collection. Homogeneous measurements of the mean surface brightness (rings A and B together), covering almost the complete range in planetocentric solar declination angle, are presented.

100.012 Observations of Saturn at wavelengths of 6.2, 11.1 and 21.2 cm. E. Gérard, I. Kazès.
Astrophys. Letters, Vol. 13, 181 - 184 (1973).

Observations of the planet Saturn have been made at 6.2, 11.1 and 21.2 cm in 1970 and 1971 when the rings were wide open (inclination > 21°). The disk brightness temperature increases slowly with wavelength according to a power law; there is no evidence for non-thermal emission in the decimeter spectrum at the time of the observations.

100.013 An upper limit to the 11.2 m-λ flux of Saturn using VLBI.
S. D. Shawhan, T. A. Clark, J. P. Basart, W. M. Cronyn.
Bull. American Astron. Soc., Vol. 5, 36 (1973). – Abstr. AAS.

100.014 New kind of ring around Saturn?
T. R. McDonough, N. M. Brice.
Nature, Vol. 242, 513 (1973).

Atoms and molecules lost by Titan are forced by the planet's gravitational field to orbit Saturn until ionized or until they are recaptured by Titan, forming a gaseous torus encompassing Titan's orbit.

100.015 The dark-side illumination of Saturn's rings.
K. A. Hämeen-Anttila.
Astrophys. Space Sci., Vol. 20, 159 - 164 (1973).

Using Focas and Dollfus' (1969) measurements, the effective optical thickness of Saturn's rings along the cross-section studied is evaluated from intensity of radiation transmitted through the rings.

100.016 The spectral reflectivity of Saturn's rings and disk at 0.4 – 0.8 microns.
G. A. Kharitonova, V. G. Teifel.
Astron. Tsirk., No. 747, p. 1 - 3 (1973). In Russian.

100.017 Temperatures of Saturn's rings.
R. E. Murphy.
Astrophys. Journ., (Letters), Vol. 181, L87 - L90 (1973).

The brightness temperatures at 20 μ of the three components of Saturn's rings are $89° \pm 3°$ K, $94° \pm 2°$ K, and $89° \pm 4°$ K for the A, B, and C rings, respectively. The 20-μ brightness temperature of the center of Saturn's disk is $97° \pm 2°$ K.

100.018 Upper limit to the 11.4 m flux of Saturn using VLBI.
S. D. Shawhan, T. A. Clark, W. M. Cronyn, J. P. Basart.
Nature, Phys. Sci., Vol. 243, 65 - 66 (1973).

We report on a series of interferometric observations of Saturn using large phased dipole arrays at 11.4 m wavelength (26.3 MHz). We interpret our results as negative for both decametric continuum and noise storm emission from source regions much less than the planetary disk size. This leads to an upper limit value of approximately 14 f.u. from a source less than 1 arc s in diameter located in a region ± 40 min in RA and $3.5°$ in declination about Saturn's optical position.

100.019 Determination of the mass of Saturn from the motion of Trojans. H. Scholl.
Astron. Astrophys., Vol. 25, 203 - 209 (1973).

An analysis of the observations of three Trojans yields the following values for the reciprocal mass of Saturn: (588) Achilles 3500.5 ± 1.7, (624) Hektor 3498.6 ± 3.0, (659) Nestor 3499.2 ± 2.9.

100.020 Methane absorption in the atmosphere of Saturn: rotational temperature and abundance from the $3\nu_3$ band. J. T. Bergstralh.
Icarus, Vol. 18, 605 - 611 (1973).

Three high-dispersion spectra of Saturn, in the methane $3\nu_3$ band at 1.1 μm, were obtained during September and October, 1970. Tracings of these spectra have been measured, and reduced by a curve of growth technique which assumes a reflecting-layer model and Lorentzian line profiles. The reductions yield a range of rotational temperatures from 122 to $142°$ K, and methane line-of-sight abundances, ηN, from 86 ± 14 to 51 ± 11 m amagat, depending on the value of the Lorentz halfwidth, α, used in computation of the curves of growth.

100.021 The greenhouse of Titan. C. Sagan.
Icarus, Vol. 18, 649 - 656 (1973).

Both non-gray radiative equilibrium and gray convective equilibrium calculations for Titan indicate that the discrepancy between the equilibrium temperature of an atmosphereless Titan and the observed infrared temperatures can be explained by a massive molecular hydrogen greenhouse effect. It is considered that the present atmosphere is in equilibrium between outgassing and blow-off on the one hand and accretion from protons trapped in a hypothetical Saturnian magnetic field on the other; or exhibits uncompensated blow-off of outgassing products.

100.022 Titan: polarimetric evidence for an optically thick atmosphere? J. Veverka.
Icarus, Vol. 18, 657 - 660 (1973).

The disk-integrated polarization of Titan has been meas-

ured at phase angles ranging from $0°.4$ to $6°.1$. The observed polarization is positive throughout this interval. This fact, when combined with published photometric data, suggests a model in which an optically thin Rayleigh atmosphere overlies an opaque cloud deck.

100.023 The polarization of Titan. B. Zellner.
Icarus, Vol. 18, 661 - 664 (1973).

New polarization observations of Titan in three spectral regions are presented. The results are not consistent with scattering from either an ordinary planetary surface or a pure molecular atmosphere. Apparently an opaque cloud layer with a strongly uv-absorbing constituent is needed.

100.024 Saturn central meridian ephemeris: 1973.
J. E. Westfall.
Strolling Astronomer, Vol. 24, 57 - 60 (1973).

100.025 Planetenphotographie mit kleinen Fernrohren.
P. Hückel.
Orion Schaffhausen, 31. Jahrgang, p. 91 (1973).

100.026 Polarization observations of Saturn made in ultraviolet and visual light since 1968.
J. S. Hall, L. A. Riley.
Bull. American Astron. Soc., Vol. 5, 289 (1973). – Abstr. AAS.

100.027 Saturn: variation of circular polarization with phase angle.
J. B. Swedlund, R. W. Avery, J. J. Michalsky, Jr., R. A. Stokes.
Bull. American Astron. Soc., Vol. 5, 289 (1973). – Abstr. AAS.

100.028 Constraints on Saturn's Q from satellite orbit theory. R. J. Greenberg.
Bull. American Astron. Soc., Vol. 5, 289 (1973). – Abstr. AAS.

100.029 Multiple scattering in Saturn's rings. Y. Kawata.
Bull. American Astron. Soc., Vol. 5, 289 - 290 (1973). – Abstr. AAS.

100.030 New estimates of Saturn ring particle sizes based on eclipse cooling curves.
H. H. Aumann, H. H. Kieffer.
Bull. American Astron. Soc., Vol. 5, 290 (1973). – Abstr. AAS.

100.031 Saturn's rings: Properties from observations of Saturn's λ 8900 CH₄ band viewed through the rings.
L. Trafton.
Bull. American Astron. Soc., Vol. 5, 290 (1973). – Abstr. AAS.

100.032 Optical thickness of Saturn's rings. I. Ferrin.
Bull. American Astron. Soc., Vol. 5, 290 (1973). Abstr. AAS.

100.033 Linear polarization and transmittance of Saturn's rings. J. C. Kemp, R. E. Murphy.
Bull. American Astron. Soc., Vol. 5, 290 (1973). – Abstr. AAS.

100.034 Temperatures of Saturn's rings. R. E. Murphy.
Bull. American Astron. Soc., Vol. 5, 290 (1973). Abstr. AAS.

100.035 Radar observations of the rings of Saturn.
R. Goldstein.
Bull. American Astron. Soc., Vol. 5, 290 (1973). – Abstr. AAS.

100.036 Albedos and densities of the inner satellites of Saturn. D. Morrison.
Bull. American Astron. Soc., Vol. 5, 304 (1973). – Abstr. AAS.

100.037 UBV photometry of Enceladus, Tethys, and Dione.

O. G. Franz, R. L. Millis.
Bull. American Astron. Soc., Vol. 5, 304 (1973). – Abstr. AAS.

100.038 Greenhouse models of the atmosphere of Titan.
J. B. Pollack.
Bull. American Astron. Soc., Vol. 5, 304 - 305 (1973).
Abstr. AAS.

100.039 Interpretation of Titan's infrared spectrum in terms of a high-altitude haze layer. L. Trafton.
Bull. American Astron. Soc., Vol. 5, 305 (1973). – Abstr. AAS.

100.040 The continuum albedo of Titan.
J. Caldwell, D. R. Larach, R. E. Danielson.
Bull. American Astron. Soc., Vol. 5, 305 (1973). – Abstr. AAS.

100.041 On the spectrum of Titan. G. Münch.
Bull. American Astron. Soc., Vol. 5, 305 (1973).
Abstr. AAS.

100.042 The reflectivity of Titan from 3000–4350 Å.
E. S. Barker, L. M. Trafton.
Bull. American Astron. Soc., Vol. 5, 305 (1973). – Abstr. AAS.

100.043 Evolutionary aspects of the atmospheres of Titan and the Galilean satellites. S. H. Gross.
Bull. American Astron. Soc., Vol. 5, 305 - 306 (1973).
Abstr. AAS.

100.044 Greenhouse models of the atmosphere of Titan.
J. B. Pollack.
Icarus, Vol. 19, 43 - 58 (1973).

The greenhouse effect is calculated for a series of model atmospheres of Titan containing varying proportions of methane, hydrogen, helium, and ammonia. The pressure induced transitions of hydrogen and methane are the major sources of infrared opacity. For each model atmosphere we first computed its temperature structure with a radiative–convective equilibrium computer program and then generated its brightness temperature spectrum to compare with observed values.

100.045 Determination of the mass of Saturn from the motion of Trojans. H. Scholl.
Mitt. Astron. Ges., No. 32, p. 206 - 207 (1973). – Abstract.

100.046 Die Bedeckung des Sternes SAO 93826 durch den Saturnring. R. Albrecht.
Mitt. Astron. Ges., No. 32, p. 209 - 211 (1973).

100.047 Saturn: a study of the $3\nu_3$ methane band.
L. Trafton.
Astrophys. Journ., Vol. 182, 615 - 636 (1973).

The author presents photometric spectra of various manifolds belonging to the R-branch of Saturn's $3\nu_3$ band. The measurements were acquired primarily during the winter of 1970 and were made along Saturn's central meridian, but they excluded the superposed rings and their immediate neighborhood. Analysis of the manifold shapes in terms of a reflecting-layer model proves largely self-consistent. A method for detecting and evaluating the influence of scattering on these manifolds is suggested.

100.048 A first look at atmospheric dynamics and temperature variations on Titan.
C. B. Leovy, J. B. Pollack.
Icarus, Vol. 19, 195 - 201 (1973).

Pollack (1973) has used a radiative equilibrium model to match radiometric data for Titan and infers the atmospheric mass, composition, opacity, and gross vertical thermal structure. These results are used to estimate the atmospheric temperature variations by means of scaling analysis, taking into account dynamics both for a baroclinic wave regime and for an axially symmetric circulation regime.

100.049 Les anneaux de Saturne en 1969. Étude morphologique et photométrique. I. Obtention et dépouillement des photographies. P. Guérin.
Icarus, Vol. 19, 202 - 211 (1973).

Some excellent photographs of Saturn in yellow and violet light have been taken in 1969 with the 105-cm reflecting telescope of the Pic du Midi Observatory. They revealed the existence, inside the C-ring, of an extremely faint, fourth ring. The reduction of microphotometric tracings along the major axis of the rings have given the photometric curves of the rings in the two colours above.

100.050 Les anneaux de Saturne en 1969. Étude morphologique et photométrique. II. Déconvolution des courbes photométriques brutes. G. Coupinot.
Icarus, Vol. 19, 212 - 223 (1973).

The objective of the present work is to improve the resolution of P. Guérin's photographs of the rings of Saturn by deconvoluting the raw photometric curves.

100.051 Report on Saturn in 1969–70: Some corrections and additions.
Strolling Astronomer, Vol. 24, 89 - 90 (1973).

100.052 Planetological fragments – 9. Iapetus, an unusual satellite of Saturn.
Strolling Astronomer, Vol. 24, 120 - 122 (1973).

100.053 The thickness of Saturn's rings from observational data in 1966 on the Pic du Midi Observatory.
A. Dollfus, J. Focas.
Physics of the moon and planets, (see 012.024), p. 451 - 452 (1972). In Russian.

100.054 Observations of the rings of Saturn during the transit of the earth across the plane of the rings in 1966. R. I. Kiladze.
Physics of the moon and planets, (see 012.024), p. 453 - 455 (1972). In Russian.

100.055 Some preliminary conclusions from available observational results of the international Saturn patrol in 1966. M. S. Bobrov.
Physics of the moon and planets, (see 012.024), p. 455 - 457 (1972). In Russian.

100.056 Results of photographic observations of Saturn and their interpretation.
V. V. Avramchuk, V. D. Krugov.
Astrometriya i Astrofizika, *Kiev*, Vyp. (No.) 18, (see 003. 016), p. 39 - 45 (1973). In Russian.

Results are presented of photographic observations of Saturn in 1971. The photographs were made in 5 narrow bands between 3620 and 6250 Å. The optical parameters of the cloud layer and overcloud atmosphere of the planet are determined.

Observations of Saturn with the astrolabe of the Paris Observatory during the winter 1971–72.
See Abstr. 041.023.

The integration of equations of the theory of the figure of planets. See Abstr. 091.002.

Spectral reflectivities of ices.
See Abstr. 091.056.

Determination of radii of satellites and asteroids

from radiometry and photometry.   See Abstr. 091.057.

Infrared surface absorptions in the solar system: Martian polar caps, Galilean satellites, rings of Saturn, and laboratory ices.   See Abstr. 097.084.

Possible explanation of the differential rotation of the atmospheres of Jupiter and Saturn. I.
See Abstr. 099.037.

Possible explanation of the differential rotation of the atmospheres of Jupiter and Saturn. II.
See Abstr. 099.038.

IR spectrophotometry of Jupiter and Saturn.
See Abstr. 099.061.

H_2 pressure-induced lines in the spectrum of the major planets.   See Abstr. 099.064.

Investigation of molecular absorption in the atmospheres of major planets.   See Abstr. 099.074.

Structure of Jupiter and Saturn.
See Abstr. 099.079.

Observations of Uranus and Saturn by a new method of radio interferometry of faint moving sources.
See Abstr. 101.020.

Errata

100.901   Erratum: 'The measurement of Saturn's radio emission at 8.2 mm and evaluation of the optical thickness of its rings' [Astron. vestn., Vol. 5, 78 - 81 (1971)].
A. D. Kuz'min, B. Ya. Losovskij.
Astron. vestn., Vol. 6, 71 (1972). In Russian.

101   Uranus, Neptune, Pluto, Transplutonian Planet

101.001   **Temperatures of Uranus and Neptune at 24 microns.**
D. Morrison, D. P. Cruikshank.
Astrophys. Journ., Vol. 179, 329 - 331 (1973).
The temperatures of Uranus and Neptune, as measured in the 17- to 28-μ spectral band, are $54.7° \pm 1.8°$K and $57.2° \pm 1.6°$K, respectively. The temperature difference is significant at the 1.5 σ level.

101.002   **Uranus atmosphere: Structure and composition.**
R. G. Prinn, J. S. Lewis.
Astrophys. Journ., Vol. 179, 333 - 342 = Contr. MIT Planet. Astron. Lab., No. 54 (1973).
The best available data on the geometric albedo of Uranus are shown to be compatible with a model in which particulate matter is present in the atmosphere. Cloud structures compatible with simple models for the accretion of Uranus and with thermal-balance studies of its atmosphere are discussed. A methane haze layer is suggested by the observations, and a deep, dense ammonia cloud layer far below the methane clouds is considered likely.

101.003   **Limb brightening on Uranus: A prediction.**
M. J. S. Belton, M. J. Price.
Astrophys. Journ., Vol. 179, 965 - 970 (1973).
A test is proposed of the hypothesis that clouds are absent from the visible atmosphere of Uranus. It consists of measuring the wavelengths at which the variation of intensity across the disk changes from limb darkening to limb brightening and vice versa.

101.004   **The missing planet.**
W. McD. Napier, R. J. Dodd.
Nature, Vol. 242, 250 - 251 (1973).
We eliminate mechanisms which are incapable of providing the energy needed both to break up the hypothetical planet and to remove most of its mass beyond the solar system. The mechanisms for disrupting a planet might be chemical, gravitational or nuclear in nature.

101.005   **The tenth planet – A search and a problem.**
T. Clarke.
Journ. Roy. Astron. Soc. Canada, Vol. 67, L1 - L2 (1973).

101.006   **En planet utenfor Pluto?**   B. R. Pettersen.
Astron. Tidssk., Årg. 6, p. 1 - 2 (1973).

101.007   **Multicolor photoelectric photometry of Neptune.**
J. F. Appleby.
Astron. Journ., Vol. 78, 110 - 112 = Contr. Five College Obs., Univ. Mass., *Amherst, Mass.*, No. 156 (1973).
Narrow-band and *UBV* photoelectric measurements of the magnitude at unit distance and geometric albedo of Neptune are presented for wavelengths $0.314 \leq \lambda \leq 0.627\,\mu$.

101.008   **The orientation of the rotational axis of Pluto.**
L. Andersson.
Bull. American Astron. Soc., Vol. 5, 36 (1973). – Abstr. AAS.

101.009   **Observation of the 3–0 band of molecular hydrogen in the spectrum of Uranus.**   B. L. Lutz.
Bull. American Astron. Soc., Vol. 5, 36 (1973). – Abstr. AAS.

101.010   **Observations of Pluto.**   C. Cristescu, B. Milet.
Stud. Cerc. Astron., Vol. 18, 103 (1973).

101.011   **Scanner observations of the quadrupole H_2 lines in**

the spectrum of Uranus.   L. Trafton.
Bull. American Astron. Soc., Vol. 5, 290 - 291 (1973).
Abstr. AAS.

**101.012   The scattering mean free path in the Uranian atmo-
sphere.   M. J. Price.**
Bull. American Astron. Soc., Vol. 5, 291 (1973). — Abstr. AAS.

**101.013   Further analysis of the limb darkening curves of
Uranus.   E. S. Light, R. E. Danielson.**
Bull. American Astron. Soc., Vol. 5, 291 (1973). — Abstr. AAS.

**101.014   Inhomogeneous models of the atmosphere of Ura-
nus.   R. E. Danielson, P. G. Wannier.**
Bull. American Astron. Soc., Vol. 5, 291 (1973). — Abstr. AAS.

101.015   Photometry of Neptune.   J. F. Appleby.
Bull. American Astron. Soc., Vol. 5, 291 (1973).
Abstr. AAS.

**101.016   Mass and position limits for an hypothetical tenth
planet of the solar system.**
D. Rawlins, M. Hammerton.
Monthly Notices Roy. Astron. Soc., Vol. 162, 261 - 270
(1973).
   This paper describes an analysis of the residuals of Neptune, conducted in order to delimit the possible range of mass
and position for an hypothetical tenth planet of the solar system. Results are presented which offer a restricted range of
possible values; and their possible significance is discussed.

**101.017   Rotational temperature and pressure in the atmo-
sphere of Uranus.**
V. G. Tejfel', G. A. Kharitonova.
Physics of the moon and planets, (see 012.024), p. 458 (1972).
In Russian. — Abstract.

101.018   On the oblateness of Neptune.   N. Bonev.
Physics of the moon and planets, (see 012.024),
p. 458 - 459 (1972). In Russian.

**101.019   Molecular hydrogen on Uranus. Observation of the
3—0 quadrupole band.   B. L. Lutz.**
Astrophys. Journ., Vol. 182, 989 - 998 (1973).
   The author reports the observation of the $S(0)$ and $S(1)$
lines of the 3—0 quadrupole band of H_2 in the spectrum of
Uranus. The measured equivalent widths are interpreted in
terms of a reflecting-layer model using Fink and Belton's
(1969) collision-narrowed curves of growth and combined
with a similar analysis of Giver and Spinrad's data to form a
self-consistent rotational temperature and line-of-sight abundance of H_2.

**101.020   Observations of Uranus and Saturn by a new method
of radio interferometry of faint moving sources.**
F. H. Briggs.
Astrophys. Journ., Vol. 182, 999 - 1011 (1973).
   An efficient radio interferometric technique for sub-
tracting the effects of background confusion from observations of slowly moving sources is developed. High-resolution
observations of Saturn at 21-cm wavelength by this method
are consistent with the planet being uniformly bright with a
brightness temperature of $230° \pm 15°$ K and show no indications of radiation belts. The equivalent disk brightness temperature of Uranus at 21 cm is $280° \pm 60°$ K.

**H_2 pressure-induced lines in the spectrum of the
major planets.   See Abstr. 099.064.**

The orbit of Halley's comet.
See Abstr. 103.103.

102   Comets

**102.001   Contributions to the kinematics of type I tails of
comets.   K. Wurm, A. Mammano.**
Astrophys. Space Sci., Vol. 18, 273 - 286 (1972).
   Observed irregular 'oscillations' of the ion tail axis in
comets have by some authors been brought in connection
with changes in the flow conditions in the solar wind. We are
defending in this paper — by arguments resting on well known
observations — the conception that these 'oscillations' are
caused by slight variations in the emission conditions for the
ions at their source which has always its place close to the
cometary nucleus.

**102.002   The motion of dust and gas in the heads of comets
with type II tails.   K. Wurm, A. Mammano.**
Astrophys. Space Sci., Vol. 18, 491 - 503 (1972).
   Photographs of comet Bennett 1969i taken in the dust-
scattered continuum reveal that the dust particles, leading to
the formation of the type II tail, leave the vicinity of the nuc-

leus only within a certain cone with the aperture in the direction to the sun. Three parabolic envelopes embracing the nucleus are formed by the dust (vertex always about on the radius vector) reaching distances from the nucleus of 30000,
60000 and 100000 km. The cone of expulsion of the dust is
identical with the cone of expulsion for the ions leading to
the formation of the type I tail. Dust - and ion envelopes have,
however, different kinematical properties. Comet Bennett is
compared with comet Halley 1910; they are related in many
respects although comet Halley had a lower dust production
than the comet Bennett. We ascribe to the dust particles of
the tail II from the beginning of the expulsion an electrical
charge.

**102.003   Lyman-α radiation in the hydrogen atmospheres of
comets. A model with multiple scattering.**
H. U. Keller.
Astron. Astrophys., Vol. 23, 269 - 280 (1973).

On the basis of excess energies of 1 eV and more for the hydrogen atoms after dissociation the high central intensities of Code's cometary Ly-α isophote measurements can be explained. A short description of a model for the outer optically thin parts of the hydrogen atmospheres is given. By comparing this to measurements of comet Bennett (Bertaux and Blamont) the production rate, the outflow velocity and the lifetime of the hydrogen atoms are found. In the second part of the paper a model for the optically thick part of the hydrogen atmosphere is evaluated. Using the Monte-Carlo method the brightness distribution of a gas-cloud which is illuminated from one side and fluorescing in the light of the resonance line wavelength is computed taking multiple scattering into account.

102.004 Cometary collisions and geological periods.
H. C. Urey.
Nature, Vol. 242, 32 - 33 (1973).

A table gives some estimates for the effect of a cometary collision with the earth. Were the ages of Tertiary times determined by the fall of comets which produced the tektite fields? A table lists the ages of these recent geologic periods and the ages of tektites. Rough agreement exists.

102.005 The origin of Jupiter's family of comets.
S. Vaghi.
Astron. Astrophys., Vol. 24, 107 - 110 (1973).

Some indications about the original orbits of the comets of Jupiter's family before their capture by the planet are derived by considering their Tisserand constants. The results are discussed in connection with the hypotheses on the origin of the comets of the family.

102.006 Amateurastronomen ontdekten kometen.
W. Schmidt.
Hemel en Dampkring, Vol. 71, 68 - 74 (1973).

102.007 Non-linear waves in type-1 comet tails.
A. I. Ershkovich, A. A. Chernikov.
Planet. Space Sci., Vol. 21, 663 - 670 (1973).

We show that helical forms and cloud structures in type-1 comet tails may be interpreted as a result of non-linear evolution of the Kelvin-Helmholtz instability. Though Biermann (1951, 1953) came to the idea of the solar wind through observations of those clouds, the mechanism of the cloud formation has not been known until now.

102.008 Comets and nongravitational forces. V.
B. G. Marsden, Z. Sekanina, D. K. Yeomans.
Astron. Journ., Vol. 78, 211 - 225 (1973).

The problem of the variation of the nongravitational forces with heliocentric distance is considered. Calculations are presented for nine short-period and five long-period comets, the variation of the forces with distance being determined from a law based on the vaporization rate of water snow. Results obtained earlier are modified to conform to the new law, and the relative values of the forces on different comets are interpreted. The effect of emissivity of the cometary nucleus on the vaporization rate is also discussed. Particular attention is paid to the matter of deriving "original" and "future" orbits of long-period comets when nongravitational forces are taken into account.

102.009 The comet − meteor stream complex.
D. A. Mendis.
Astrophys. Space Sci., Vol. 20, 165 - 176 (1973).

The genetic relationship between short-period comets and meteor streams is investigated. It is shown that mechanisms exist for the radial and the longitudinal focussing of particles in meteor streams with characteristic time scales of agglomeration significantly smaller than those of any of the known dispersive processes. Consequently, it is claimed that meteor streams may not merely form a sink for short-period comets but may also form a source.

102.010 Sobre la determinación de fuerzas no gravitatorias que actúan sobre los cometas. P. E. Zadunaisky.
Bol. As. Argentina Astron., No. 16, (see 012.007), p. 37 - 40 (1971).

102.011 Optimale Ausschöpfung des Informationsgehaltes von Kometenaufnahmen durch entwicklungstechnische Kontraststeuerung.
W. Högner, H. Löchel, N. Richter.
Sterne, 49. Jahrgang, p. 72 - 77 (1973).

102.012 Examination of several ideas of comet origins.
E. Everhart.
Astron. Journ., Vol. 78, 329 - 337 (1973).

The results of numerical experiments on the evolution of orbits are used to examine several ideas that comets originate within the solar system. It is shown that comets whose entire orbit lies beyond Jupiter can cross the Jupiter barrier with the help of Saturn's perturbations. Crossing a barrier depends on the current value of the Jacobi quantity. The numerical experiments show that there must be many comets in Trojan orbits, horseshoe orbits, and mid-range orbits between Jupiter and Saturn.

102.013 On the origin of short-period comets.
P. C. Joss.
Astron. Astrophys., Vol. 25, 271 - 273 (1973).

On the basis of Oort's theory of comet origin and numerical calculations of a capture mechanism proposed by Everhart, the observed number of short-period comets cannot be accounted for.

102.014 A new model for cometary nuclei. C. R. O'Dell.
Icarus, Vol. 19, 137 - 146 (1973).

A new model for the nucleus of comets is presented, hypothesizing formation at large heliocentric distances from many independent solid bodies. It is shown that such a configuration would collapse to a single assemblage if it is to survive into the inner solar system.

102.015 Orientation-dependent effects in Oort's theory of comet origin. II. Anisotropies in the distribution of long-period comet orbits. P. C. Joss.
Icarus, Vol. 19, 147 - 153 (1973).

The statistical significance of anisotropies in the distribution of orbital orientations among the long-period and nearly parabolic comets is evaluated. A numerical model for the distribution of orbital orientations is constructed, based on Oort's theory of comet origin and the assumption that the observed anisotropies are caused by multiple planetary perturbations over the course of many perihelion passages. The model, which is restricted to comets with perihelion distances less than 0.3 AU, does not predict any significant anisotropies.

102.016 Comets in the solar wind. L. Biermann.
Cosmic plasma physics. Conference 1971, (see 012.016), p. 123 - 135 (1972).

102.017 Wave motion in type I comet tails.
M. Dobrowolny, N. D'Angelo.
Cosmic plasma physics. Conference 1971, (see 012.016), p. 149 - 156 (1972).

102.018 Does the Oort cloud exist? Yu. P. Pskovsky.
Priroda, No. 5.73, p. 128 (1973). In Russian.
Letter.

102.019 **New data for the benefit of the interstellar origin of comets.** V. V. Radzievskij, V. P. Tomanov.
Astron. vestn., Vol. 7, 73 - 82 (1973). In Russian.

There exists a statistical dependence of the number of comets and their mean brightness on the angular distance of the perihelion from the sun's apex. The predicted effects are completely supported by observational data. The properties of interstellar bodies, the capture of which on the Nölke axis is probable enough, are studied and the required conditions are clarified.

102.020 **New statistical regularities in the system of long-period comets.** V. P. Tomanov.
Astron. vestn., Vol. 7, 83 - 87 (1973). In Russian.

The observed distribution of perihelions of nearly parabolic comets with due regard to the selection effects has been studied, and true existence of well expressed maximum of perihelia in the area of the solar apex has been demonstrated. Statistical dependence of the mean perihelion distance of comets and the mean value of their magnitude on angular distance of perihelia from the solar apex has been obtained.

102.021 **The effect of multiple encounters on short-period comet orbits.** B. E. Lowrey.
Astron. Journ., Vol. 78, 428 - 437 (1973).

The observed orbital elements of short-period comets are found to be consistent with the hypothesis of derivation from long-period comets as long as two assumptions are made. First, the distribution of short-period comets has been randomized by multiple encounters with Jupiter and second, the short-period comets have low velocities of encounter with Jupiter. Some 16% of the observed short-period comets have lower encounter velocities than is allowed mathematically using Laplace's method. This may be due to double-encounter processes with Jupiter and Saturn.

102.022 **Probleme bei Bahnberechnungen von periodischen Kometen.** G. Schrutka-Rechtenstamm.
Mitt. Astron. Ges., No. 32, p. 207 - 208 (1973). – Abstract.

102.023 **Reduction of observations of comets during the 19th century.** E. D. Kondrat'eva.
Trudy Kazan. Gorod. Astron. Obs., No. 37, p. 141 - 145 (1970). In Russian.

102.024 **Beobachtungen der Kometen und ihre Interpretation.** J. Rahe.
Diss. Techn. Univ. Berlin, 180 pp. (1971).

102.025 **Comets and the structure of the solar wind.** V. P. Tarashchuk.
Kometn. Tsirk., *Kiev*, No. 140 (1973). In Russian.

Comets and meteors. See Abstr. 003.026.

Asteroizi și comete. See Abstr. 003.095.

The incentive of a bold hypothesis: Hyperbolic meteors and comets. See Abstr. 104.027.

On the process of accretion in the formation of the planets and comets. See Abstr. 107.008.

Comets and the formation of planets. See Abstr. 107.017.

Neue Gesichtspunkte zur Entstehung des Planetensystems. See Abstr. 107.018.

103   Comets: Listed Objects

**103.001   Comets and minor planets observed during 1972
at Bucharest and Nice Observatory.**   G. Bocşa,
C. Cristescu, I. Gheţu, V. I. Vlăsceanu, B. Milet.
Stud. Cerc. Astron., Vol. 18, 115 - 119 (1973). – Concerning
1972h, 1972d, 1972j, 1971c.

**103.002   Definitive Bezeichnungen der Kometen des Jahres
1971.**   R. A. Naef.
Orion Schaffhausen, 31. Jahrgang, p. 98 (1973).

103.003   Kometer 1972.   H. Q. Rasmusen.
Astron. Tidssk., Årg. 6, p. 82 - 83 (1973).

103.004   Comet notes.   J. B. Trainor.
Journ. Astron. Soc. Victoria, Vol. 26, 13 - 15
(1973).

**103.005   Photographic observations of comets and minor
planets.**   B. G. Jørgensen, B. Reipurth.
Astron. Astrophys., Suppl. Ser., Vol. 11, 107 - 117 (1973).
The observed comets are: Ikeya-Seki 1967n, Honda
1968c, Tago-Sato-Kosaka 1969g, Bennett 1969i, Abe 1970g.

103.006   Roman numeral designations of comets in 1971.
IAU Circ., No. 2496 (1973).

103.007   Comets in the year 1972.   J. Bouška.
Vesmír, Vol. 52, 156 (1973).   In Czech.

103.008   Periodic comets in the year 1974.   J. Bouška.
Říše hvězd, Vol. 54, 81 - 82 (1973). In Czech.

**103.009   Observations of comets and asteroids at the Kleť
Observatory in the year 1971.**   A. Mrkos.
Acta Univ. Carolinae Math. Phys., Vol. 13, No. 1, p. 91 - 95
(1973).
Precise positions of comets P/Whipple 1969c, Abe 1970g,
P/Wolf-Harrington 1970o, Kojima 1970r, Toba 1971a and
object Antal.

103.010   Observations of comets.
R. L. Waterfield, R. H. South, G. H. Rutter, N.
Wood, D. Griffiths, I. M. Purcell.
British Astron. Ass., Circ. No. 547 (1973). – Concerning
P/Kearns-Kwee 1971c, Sandage 1972h, Kojima 1972j, Heck-
Sause 1973a, Kohoutek 1973e, Kohoutek 1973f.

103.011   Possible comet Torres.
Kometn. Tsirk., Kiev, No. 141 (1973). In Russian.

103.012   Observations at the Skalnaté Pleso Observatory.
M. Antal.
Kometn.Tsirk., Kiev, No. 142 (1973). In Russian. – Concerning comet Giacobini-Zinner, 1972d; comet Sandage, 1972h;
comet Kojima, 1972j.

103.013   Observations at the Kleť Observatory.
A. Mrkos.
Kometn. Tsirk., Kiev, No. 142 (1973). In Russian. – Concerning comet Kojima, 1972j; periodic comet Kearns–Kwee,
1971c; periodic comet Reinmuth 1, 1972i.

**103.014   Observations of comets at the Skalnaté Pleso Ob-
servatory.**   M. Antal.
Kometn. Tsirk., Kiev, No. 143 (1973). In Russian. – Concerning comet Kojima, 1972j; comet Heck–Sause, 1973a.

103.015   Observations at the Kleť Observatory.
Kometn. Tsirk., Kiev, No. 143 (1973). In Russian.
Concerning comet Reinmuth 1, 1972i; comet Kearns–Kwee,
1971c; comet Kojima, 1972j; comet Heck–Sause, 1973a.

103.016   Observations of comets in Alma–Ata.
D. A. Rozhkovskij, D. E. Gorodetskij, F. K. Rspaev.
Kometn. Tsirk., Kiev, No. 144 (1973). In Russian. – Concern-
ing comet Tempel 1, 1972a; comet Sandage, 1972h; comet
Giacobini–Zinner, 1972d.

103.017   Comet near the sun?
Kometn. Tsirk., Kiev, No. 149 (1973). In Russian.

103.018   Comet notes.   E. Roemer.
Mercury, (Journ. Astron. Soc. Pacific), Vol. 2, No. 2,
p. 17 - 19 (1973).

**103.019   Definitive designation of comets passed through
the perihelion in 1971.**
Kometn. Tsirk., Kiev, No. 143 (1973). In Russian.

103.100   Comet 1973a Heck-Sause

New comet Heck-Sause.
A. Heck, G. Sause, B. G. Marsden, S. W. Milbourn.
British Astron. Ass., Circ. No. 546 (1973).

Comet Heck-Sause 1973a.
R. L. Waterfield, N. Wood, I. M. Purcell, B. G. Marsden.
British Astron. Ass., Circ. No. 547 (1973).

Comet Heck-Sause.   A. Heck, G. Sause.
IAU Circ., No. 2479 (1973).

Comet Heck-Sause (1973a).   M. Koishikawa,
R. E. McCrosky, C. Y. Shao, B. G. Marsden, T. Seki.
IAU Circ., No. 2481 (1973).

Comet Heck-Sause (1973a).   T. Urata.
IAU Circ., No. 2482 (1973).

Comet Heck-Sause (1973a).   T. Urata,
M. Koishikawa, T. Seki, Y. Andrillat.
IAU Circ., No. 2483 (1973).

Comet Heck-Sause (1973a).   A. Heck, G. Sause,
F. Dossin, R. E. McCrosky, C. Y. Shao, M. Sugano, T. Urata,
N. Kojima, T. Seki, B. Milet.
IAU Circ., No. 2485 (1973).

Comet Heck-Sause (1973a).
N. Kojima, T. Seki, B. G. Marsden.
IAU Circ., No. 2489 (1973).

Comet Heck-Sause (1973a).   M. Koishikawa,
A. Watanabe, K. Aizawa, N. Kojima, T. Seki, J. Bortle.
IAU Circ., No. 2494 (1973).

Comet Heck-Sause (1973a).   K. Suzuki,
T. Urata, B. Milet, K. Tsuchiya, A. Mrkos, R. Petrovičová,
R. L. Waterfield, N. Wood.
IAU Circ., No. 2495 (1973).

Comet Heck-Sause (1973a).   W. L. W. Sargent,

C. T. Kowal, N. Kojima, T. Seki.
IAU Circ., No. 2501 (1973).

Comet Heck-Sause (1973a).   W. Ferreri, S. Vaghi,
R. L. Waterfield, I. M. Purcell, H. L. Giclas, M. L. Kantz,
T. Seki, H. Debehogne, G. Roland.
IAU Circ., No. 2511 (1973).

Comet Heck-Sause (1973a).   L. Briqueu, B. Milet,
S. Vaghi, W. Ferreri, A. Mrkos, R. Petrovičová, L. Petrik.
IAU Circ., No. 2519 (1973).

Comet Heck-Sause (1973a).   M. Antal, A. Mrkos,
R. Petrovičová.
IAU Circ., No. 2529 (1973).

Comet Heck-Sause (1973a).   H. Debehogne,
G. Roland, N. S. Chernykh, L. I. Chernykh.
IAU Circ., No. 2543 (1973).

Comet Heck-Sause (1973a).   A. Heck, G. Sause,
F. Dossin, J.-M. Vreux, K. Suzuki, T. Urata.
IAU Circ., No. 2545 (1973).

Comet Heck-Sause (1973a).
Japan Astron. Study Ass. Circ. 286 - 288 (1973). In Japanese.

New comet Heck–Sause, 1973a.
Kometn. Tsirk., *Kiev*, No. 141 (1973). In Russian.

Comet Heck-Sause, 1973a.
Kometn. Tsirk., *Kiev*, No. 142 (1973). In Russian.

Comet Heck–Sause, 1973a.
Kometn.Tsirk., *Kiev*, No. 143 (1973). In Russian.

Comet Heck–Sause, 1973a.
V. M. Kovalenko, A. Mrkos.
Kometn. Tsirk., *Kiev*, No. 144 (1973). In Russian.

Comet Heck–Sause, 1973a.
M. Antal, V. M. Kovalenko, O. N. Kovalenko, N. S. Chernykh,
G. R. Kastel'.
Kometn. Tsirk., *Kiev*, No. 145 (1973). In Russian.

Comet Heck–Sause, 1973a.
V. M. Kovalenko, O. N. Kovalenko, A. Mrkos.
Kometn. Tsirk., *Kiev*, No. 146 (1973). In Russian.

Comet Heck–Sause, 1973a.
E. I. Shchukin.
Kometn. Tsirk., *Kiev*, No. 148 (1973). In Russian.

Comet Heck–Sause, 1973a.
Kometn. Tsirk., *Kiev*, No. 149 (1973). In Russian.

Neuer Komet Heck-Sause (1973a).
E. Wiedemann.
Orion Schaffhausen, 31. Jahrgang, p. 14 (1973).

Erstaufnahme des Kometen Heck-Sause (1973a).
Orion Schaffhausen, 31. Jahrgang, p. 53 (1973).

Comet Heck-Sause (1973a).
Sky Telescope, Vol. 45, 143 (1973).

The first new comet of 1973.
F. Dossin, A. Heck.
Sky Telescope, Vol. 45, 291 (1973).

103.101   Comet 1972b Grigg-Skjellerup

Observations of meteors associated with comet
Grigg-Skjellerup.   See Abstr. 104.007.

103.102   Comet 1970 II Bennett

Observations of comets.   S. Świerkowska.
Acta Astron., Vol. 23, 177 - 178 (1973).
Observed positions of comets are presented in this paper:
Bennett 1969 i (observations from April 14 to May 18, 1970)
and Abe 1970g (observations from September 12 to October
7, 1970).

Preliminary interpretation of spectropolarimetric
observations of comet Bennett.   L. A. Bugaenko,
O. I. Bugaenko, L. S. Galkin, V. P. Konopleva, A. V.
Morozhenko.
Astrometriya i Astrofizika, *Kiev*, Vyp. (No.) 18, (see 003.
016), p. 78 - 82 (1973). In Russian.
The results of polarization measurements in monochromatic light of the head of the comet Bennett (1969i) are discussed. The observations were carried out in ten spectral regions
($\lambda\lambda 360-750\,m\mu$) from March 31 till April 10, 1970.

Photometry of the comet Bennett (1969i) by the
equidensity method.   N. M. Bronnikova.
Astrometriya i Astrofizika, *Kiev*, Vyp. (No.) 18, (see 003.
016), p. 84 - 94 (1973). In Russian.

Interpretation of hydrogen Lyman-alpha observations of comets Bennett and Encke.
J. L. Bertaux, J. E. Blamont, M. Festou.
Astron. Astrophys., Vol. 25, 415 - 430 (1973).
The Lyman-alpha emission of the hydrogen cloud surrounding comet Bennett was mapped from the OGO–5 satellite observations during the whole month of April 1970. The
measured distribution of intensity was compared to a theoretical model of hydrogen cloud in which atoms leave radially the
cometary head with a Maxwellian velocity distribution. The
ejection rate of H atoms in comet Encke was estimated to be
$\simeq 5 \times 10^{26}$ atom·sterad^{-1} s^{-1}.

Tail peculiarities in comet Bennett caused by solar
wind disturbances.   K. Jockers, Rh. Lüst.
Astron. Astrophys., Vol. 26, 113 - 121 (1973).
Peculiarities observed in the plasma tail of comet Bennett
between March 30 and April 7, 1970, were shown to be related
to solar wind events. In particular two series of exposures taken
from stations at different geographic longitudes are discussed
which show the development of a secondary tail. One of these
is clearly connected with an interplanetary shock wave. There
seems to be a correlation between kinks in the tail axis and
changes of the solar wind direction.

The hydrogen production rates of comet Bennett
(1969i) in the first half of April 1970.   H. U. Keller.
Bull. American Astron. Soc., Vol. 5, 266 (1973). – Abstr. AAS.

Comet Bennett 1970 II.
Z. Sekanina, F. D. Miller.
Science, Vol. 179, 565 - 567 (1973).
The model for dust comets, formulated by Finson and
Probstein, which had previously been tested only on comet
Arend-Roland 1957 III, has been successfully applied to three
calibrated photographic plates of comet Bennett. The size
distribution, emission rate, and initial velocities of dust parti-
cles emitted from the comet's nucleus are given.

Solar wind interaction with comet Bennett (1969i).
L. F. Burlaga, J. Rahe, B. Donn, M. Neugebauer.
Solar Physics, Vol. 30, 211 - 222 (1973).

This paper examines the relations between the solar-wind and comet Bennett during the period March 23 to April 5, 1970. A large kink was observed in the ion tail of the comet on April 4, but no solar-wind stream was observed in the ecliptic plane which could have caused the kink. Thus, either there was no correlation between the solar wind at the earth and that at comet Bennett (which was 40° above the ecliptic) or the kink was caused by something other than a high-speed stream. The fine structure visible in photographs of the kink favors the second of these alternatives.

103.103 Comet 1910 II Halley

Les variations orbitales de la comète de Halley.
M.-A. Combes.
L'Astronomie, 87ᵉ année, p. 103 - 112 (1973).

The cause of the residuals in the motion of Halley's comet. T. Kiang.
Monthly Notices Roy. Astron. Soc., Vol. 162, 271 - 287 (1973).

A remarkable periodicity in the residuals in perihelion time of Halley's comet was recently attributed by Brady to the action of a trans-plutonian planet. The main idea of this paper is that, on the contrary, this periodicity can be shown to be an inherent property of the 3-body configuration Sun—Jupiter—Halley.

The orbit of Halley's comet.
T. Kiang, P. A. Wayman.
Nature, Vol. 241, 520 - 521 (1973).

We turned our attention to the question of how a long periodicity could arise in the departures from a theoretical orbit of Halley's comet through the effect, say, of a tangential impulse occurring once per revolution of the comet in its orbit, that is, every 76 yr. We considered a simplified numerical model consisting of the sun, Jupiter and Halley's comet. Numerical solutions of the equation have been found to show a long-period variation (600—700 yr) for a range of initial conditions.

Halley's comet in 1986.   A. C. Gilmore.
Southern Stars, Vol. 24, 139 - 141 (1973).

103.104 Comet 1969 IV Churyumov-Gerasimenko

Peculiarities in the future evolution of the orbit of comet Churyumov-Gerasimenko.   K. I. Churyumov.
Astrometriya i Astrofizika, Kiev, vyp. (No.) 16,(see 003.006), p. 52 - 61 (1972).   In Russian.

The equations of motion of comet Churyumov-Gerasimenko have been integrated numerically by Cowell's method for the interval 1969—2192. The evolution of its orbit is discussed in detail.

103.105 Comet 1972d Giacobini-Zinner

Periodic comet Giacobini-Zinner (1972d).
T. Seki, N. Kojima.
IAU Circ., No. 2479 (1973).

Periodic comet Giacobini-Zinner (1972d).

M. Antal.
IAU Circ., No. 2503 (1973).

Periodic comet Giacobini—Zinner, 1972d.
L. Markova, V. Kovalenko, O. Kovalenko.
Kometn. Tsirk., Kiev, No. 140 (1973). In Russian.

Elements of the orbit of comet Giacobini-Zinner according to observations from 1959 - 1965 and its motion from 1965 to 1972.   Yu. V. Evdokimov.
Trudy Kazan. Gorod. Astron. Obs., No. 37, p. 133 - 140 (1970). In·Russian.

Motion of comet Giacobini-Zinner before 1900 and later than 1965.   Yu. V. Evdokimov.
Trudy Kazan. Gorod. Astron. Obs., No. 38, p. 64 - 67 (1972). In Russian.

103.106 Comet 1968 I Ikeya-Seki

The light curve of comet Ikeya-Seki 1968 I.
C. S. Morris, J. E. Bortle.
Publ. Astron. Soc. Pacific, Vol. 85, 249 - 252 (1973).

The results of a reduction of 243 visual and photoelectric magnitude estimates of comet Ikeya-Seki 1968 I are presented. These results indicate that comet Ikeya-Seki was an intrinsically bright comet. A detailed study of the light curve of this comet shows that before and after perihelion there were fairly large variations in the brightness of the comet. Around perihelion, however, the brightness of the comet was almost constant.

103.107 Comet 1971a Toba

Comet magnitude analysis: Toba 1971a.
C. S. Morris.
Strolling Astronomer, Vol. 24, 68 - 71 (1973).

103.108 Comet 1969d Fujikawa

Comet Fujikawa, 1969d.
J. E. Bortle.
Strolling Astronomer, Vol. 24, 75 - 76 (1973).

103.109 Comet 1971 II Encke

Periodic comet Encke.   B. G. Marsden.
IAU Circ., No. 2547 (1973).

Elements and ephemeris of comet Encke.
N. A. Bokhan.
Kometn. Tsirk., Kiev, No. 145 (1973). In Russian.

Interpretation of hydrogen Lyman-alpha observations of comets Bennett and Encke.   See Abstr. 103.102.

Comet Encke: Meteor metallic ion identification by mass spectrometer.   See Abstr. 104.015.

103.110 Comet 1970r Kojima

Studio dell'orbita della cometa 1970r.

L. Buffoni, A. Manara.
Mem. Soc. Astron. Italiana, Nuova Ser., Vol. 43, 675 - 680 (1973).

From a previous analysis comet 1970r has appeared to undergo a heavy perturbation from the planet Jupiter. The energy and orbital elements of the comet are studied as long as it is approaching the planet.

103.111 Comet 1953 VI Harrington 2

Search ephemerides of the periodic comet Harrington (1953 VI) for its returns in 1973, 1980, and 1987—88. G. Sitarski.
Acta Astron., Vol. 23, 169 - 173 (1973).

Two apparitions of the comet were linked using 30 observations made in 1953 and 1960; all the perturbations (also in differential coefficients) were taken into account. The process of improvement yielded very exact orbital elements. The ephemerides of the comet for its next three returns were computed including the perturbations from Mercury to Pluto. The conditions of visibility are presented in graphs.

103.112 Comet 1970 XV Abe

Photographic observations of comet Abe (1970g) and comet Tago-Sato-Kosaka (1969g).
J. Bem, T. Jastrzębski.
Acta Astron., Vol. 23, 175 - 176 (1973).

Intensity variations of the radiation of comet Abe, 1970 XV in the ultraviolet and visible regions.
K. I. Churyumov, F. I. Kravtsov.
Kometn. Tsirk., *Kiev*, No. 146 (1973). In Russian.

Observations of comets.   See Abstr. 103.102.

103.113 Comet 1969 IX Tago-Sato-Kosaka

Photometry of the comet Tago-Sato-Kosaka (1969g).   N. M. Bronnikova.
Astrometriya i Astrofizika, *Kiev*, Vyp. (No.) 18, (see 003. 016), p. 83 - 84 (1973). In Russian.

Photographic observations of comet Abe (1970g) and comet Tago-Sato-Kosaka (1969g).   See Abstr. 103.112.

103.114 Comet 1972j Kojima

Comet Kojima (1972j).   T. Urata, H. Hatanaka, A. Mrkos, R. Petrovičová, J. Gibson, U. Gibson.
IAU Circ., No. 2476 (1973).

Comet Kojima (1972j).   T. Seki, H. L. Giclas, M. L. Kantz, Miyamoto, B. G. Marsden.
IAU Circ., No. 2478 (1973).

Comet Kojima (1972j).   A. Mrkos, R. Petrovičová, T. Seki, B. Milet.
IAU Circ., No. 2484 (1973).

Comet Kojima (1972j).   K. Tsuchiya, T. Urata, B. Milet, J. Gibson, J. Bortle.
IAU Circ., No. 2485 (1973).

Comet Kojima (1972j).   T. Seki.
IAU Circ., No. 2489 (1973).

Comet Kojima (1972j).   M. Antal, K. Suzuki, T. Urata, H. Hatanaka, V. Zappalà, W. Ferreri, R. L. Waterfield, N. Wood, B. Milet, A. Mrkos.
IAU Circ., No. 2497 (1973).

Comet Kojima (1972j).   N. Kojima, T. Seki, W. Ferreri, H. L. Giclas, M. L. Kantz.
IAU Circ., No. 2513 (1973).

Comet Kojima (1972j).   M. Antal, B. Milet, W. Ferreri, A. Mrkos.
IAU Circ., No. 2527 (1973).

Comet Kojima (1972j).   B. G. Marsden.
IAU Circ., No. 2530 (1973).

Comet Kojima (1972j).   R. L. Waterfield, R. H. S. South, G. H. Rutter, W. Ferreri.
IAU Circ., No. 2534 (1973).

Comet Kojima (1972j).
Japan Astron. Study Ass. Circ. 286 - 288 (1973). In Japanese.

Comet Kojima, 1972j.
Kometn. Tsirk., *Kiev*, No. 140 (1973). In Russian.

Komet Kojima, 1972j.
V. M. Kovalenko, O. N. Kovalenko.
Kometn. Tsirk., *Kiev*, No. 141 (1973). In Russian.

Observations of the Alpine expedition of the Sternberg—Institute (Alma—Ata). Comet Kojima, 1972j.
V. M. Kovalenko, O. N. Kovalenko.
Kometn. Tsirk., *Kiev*, No. 142 (1973). In Russian.

Comet Kojima, 1972j.
Kometn. Tsirk., *Kiev*, No. 144 (1973). In Russian.

Comet Kojima, 1972j.
V. M. Kovalenko, O. N. Kovalenko, M. Antal.
Kometn. Tsirk., *Kiev*, No. 145 (1973). In Russian.

Comet Kojima, 1972j.
Kometn. Tsirk., *Kiev*, No. 146 (1973). In Russian.

Ephemeris of comet Kojima, 1972j.
Kometn. Tsirk., *Kiev*, No. 148 (1973). In Russian.

103.115 Comet 1972h Sandage

Comet Sandage (1972h).
R. L. Waterfield, G. H. Rutter.
IAU Circ., No. 2476 (1973).

Comet Sandage (1972h).   T. Seki, T. Urata.
IAU Circ., No. 2487 (1973).

Comet Sandage (1972h).   M. Antal, T. Seki.
IAU Circ., No. 2495 (1973).

Comet Sandage (1972h).
R. L. Waterfield, N. Wood.
IAU Circ., No. 2510 (1973).

Comet Sandage (1972h).   A. Mrkos.
IAU Circ., No. 2521 (1973).

Comet Sandage (1972h).   R. Petrovičová,
A. Mrkos, M. Antal, M. J. Hendrie, R. L. Waterfield.
IAU Circ., No. 2534 (1973).

Comet Sandage, 1972h.
Kometn. Tsirk., *Kiev*, No. 141 (1973). In Russian.

Comet Sandage, 1972h.
M. Antal.
Kometn. Tsirk., *Kiev*, No. 146 (1973). In Russian.

103.116   Comet 1971c Kearns-Kwee

Periodic comet Kearns-Kwee (1971c).
N. S. Chernykh, L. V. Zhuravleva, B. Milet.
IAU Circ., No. 2476 (1973).

Periodic comet Kearns-Kwee (1971c).
T. Seki, N. Kojima.
IAU Circ., No. 2479 (1973).

Periodic comet Kearns-Kwee (1971c).
B. Milet, V. Zappalà.
IAU Circ., No. 2482 (1973).

Periodic comet Kearns-Kwee (1971c).
R. L. Waterfield, N. Woods, G. H. Rutter, A. Mrkos, B. Milet,
J. Bortle.
IAU Circ., No. 2485 (1973).

Periodic comet Kearns-Kwee (1971c).
N. S. Chernykh, L. I. Chernykh.
IAU Circ., No. 2489 (1973).

Periodic comet Kearns-Kwee (1971c).   N. Kojima,
T. Seki, V. Zappalà, K. Suzuki, B. Milet, A. Mrkos.
IAU Circ., No. 2501 (1973).

Periodic comet Kearns-Kwee (1971c).
V. Zappalà.
IAU Circ., No. 2513 (1973).

Periodic comet Kearns—Kwee, 1971c.
Kometn. Tsirk., *Kiev*, No. 140 (1973). In Russian.

Comet Kearns—Kwee, 1971c.
Kometn. Tsirk., *Kiev*, No. 141 (1973). In Russian.

Observations at the Crimean Astrophysical Observatory. Comet Kearns—Kwee, 1971c.
Kometn. Tsirk., *Kiev*, No. 142 (1973). In Russian.

103.117   Comet 1972*l* Araya

Comet Araya 1972*l*.
G. Araya, J. Gibson, Z. Sekanina.
British Astron. Ass., Circ. No. 546 (1973).

Comet Araya (1972*l*).   C. U. Cesco, J. Gibson,
Z. Sekanina.
IAU Circ., No. 2477 (1973).

Comet Araya (1972*l*).   J. Gibson.
IAU Circ., No. 2486 (1973).

Comet Araya (1972*l*).   J. Gibson, U. Gibson.
IAU Circ., No. 2489 (1973).

Comet Araya (1972*l*).   C. Torres.
IAU Circ., No. 2497 (1973).

Comet Araya (1972*l*).   J. Gibson, U. Gibson.
IAU Circ., No. 2511 (1973).

Comet Araya (1972*l*).   J. Gibson, U. Gibson.
IAU Circ., No. 2514 (1973).

Comet Araya (1972 *l*).   G. Araya, C. Bolelli,
B. M. Blanco.
IAU Circ., No. 2529 (1973).

Comet Araya (1972*l*).   J. Gibson, U. Gibson,
B. G. Marsden.
IAU Circ., No. 2538 (1973).

Comet Araya (1972*l*).
Japan Astron. Study Ass. Circ. 286 - 288 (1973). In Japanese.

Comet Araya, 1972*l*.
Kometn. Tsirk., *Kiev*, No. 141 (1973). In Russian.

103.118   Comet 1972k Gehrels

Periodic comet Gehrels 1972k.
T. Gehrels, B. G. Marsden.
British Astron. Ass., Circ. No. 546 (1973).

Periodic comet Gehrels (1972k).
H. L. Giclas, M. L. Kantz.
IAU Circ., No. 2478 (1973).

Periodic comet Gehrels (1972k).
C. Y. Shao, R. E. McCrosky.
IAU Circ., No. 2513 (1973).

Periodic comet Gehrels (1972k).   B. G. Marsden.
IAU Circ., No. 2545 (1973).

P/comet Gehrels (1972k).
Japan Astron. Study Ass. Circ. 286 - 288 (1973). In Japanese.

Elements and ephemeris of comet Gehrels, 1972k.
Kometn. Tsirk., *Kiev*, No. 148 (1973). In Russian.

103.119   Comet 1973b Tuttle-Giacobini-Kresák

Comet P/Tuttle-Giacobini-Kresák 1973b.
E. Roemer, J. Q. Latta.
British Astron. Ass., Circ. No. 547 (1973).

Periodic comet Tuttle-Giacobini-Kresák (1973b).
E. Roemer, J. Q. Latta.
IAU Circ., No. 2486 (1973).

Periodic comet Tuttle-Giacobini-Kresák (1973b).
J. Bortle.
IAU Circ., No. 2541 (1973).

Periodic comet Tuttle-Giacobini-Kresák (1973b).
F. Seiler, T. Kleine, L. Kohoutek.
IAU Circ., No. 2542 (1973).

Periodic comet Tuttle-Giacobini-Kresák (1973b).
F. Seiler, T. Kleine, C. Y. Shao.
IAU Circ., No. 2543 (1973).

P/comet Tuttle-Giacobini-Kresák (1973b).
Japan Astron. Study Ass. Circ. 286 - 288 (1973). In Japanese.

Rediscovery of comet Tuttle-Giacobini-Kresák, 1973b.
Kometn. Tsirk., *Kiev*, No. 142 (1973). In Russian.

Outburst of comet Tuttle–Giacobini–Kresák, 1973b.
Kometn. Tsirk., *Kiev*, No. 147 (1973). In Russian.

Outburst of comet Tuttle–Giacobini–Kresák, 1973b.
Kometn. Tsirk., *Kiev*, No. 148 (1973). In Russian.

Brightness outburst of comet Tuttle–Giacobini– Kresák, 1973b.
Kometn. Tsirk., *Kiev*, No. 149 (1973). In Russian.

103.120   Comet 1973c Wild

Comet P/Wild 1973c.   E. Roemer, J. Q. Latta.
British Astron. Ass., Circ. No. 547 (1973).

Periodic comet Wild (1973c).
E. Roemer, J. Q. Latta.
IAU Circ., No. 2490 (1973).

P/comet Wild (1973c).
Japan Astron. Study Ass. Circ. 286 - 288 (1973). In Japanese.

Rediscovery of the periodic comet Wild, 1973c.
Kometn. Tsirk., *Kiev*, No. 143 (1973). In Russian.

103.121   Comet 1973d Swift-Gehrels

Comet P/Swift-Gehrels 1973d.
T. Gehrels, R. Adams, B. G. Marsden.
British Astron. Ass., Circ. No. 547 (1973).

Comet Gehrels (1973d).   T. Gehrels, R. Adams,
R. E. McCrosky, C. Y. Shao.
IAU Circ., No. 2491 (1973).

Comet Gehrels (1973d).   T. Gehrels, C. D. Vesely,
R. Sather, R. C. Capen, R. E. McCrosky, C. Y. Shao.
IAU Circ., No. 2492 (1973).

Comet Gehrels (1973d).
R. E. McCrosky, C. Y. Shao, B. G. Marsden.
IAU Circ., No. 2500 (1973).

Comet Gehrels (1973d).
R. E. McCrosky, C. Y. Shao, B. G. Marsden.
IAU Circ., No. 2503 (1973).

Periodic comet Swift-Gehrels (1973d).
E. Roemer, B. G. Marsden, J. Mikolas.
IAU Circ., No. 2517 (1973).

P/comet Swift-Gehrels (1889 VI = 1973d).
Japan Astron. Study Ass. Circ. 286 - 288 (1973). In Japanese.

New comet Gehrels, 1973d.
Kometn. Tsirk., *Kiev*, No. 142 (1973). In Russian.

Comet Gehrels, 1973d.

Kometn. Tsirk., *Kiev*, No. 143 (1973). In Russian.

Periodic comet Swift–Gehrels, 1973d.
Kometn. Tsirk., *Kiev*, No. 145 (1973). In Russian.

103.122   Comet 1973g Reinmuth 2

Periodic comet Reinmuth 2.
E. Rabe, B. G. Marsden.
IAU Circ., No. 2493 (1973).

Periodic comet Reinmuth 2 (1973g).
E. Roemer, G. Reskin.
IAU Circ., No. 2532 (1973).

Rediscovery of comet Reinmuth 2, 1973g.
Kometn. Tsirk., *Kiev*, No. 146 (1973). In Russian.

103.123   Comet 1972i Reinmuth 1

Periodic comet Reinmuth 1 (1972i).
A. Mrkos, N. Kojima, T. Seki.
IAU Circ., No. 2487 (1973).

Periodic comet Reinmuth 1 (1972i).   A. Mrkos.
IAU Circ., No. 2498 (1973).

P/comet Reinmuth 1 (1972i).
Japan Astron. Study Ass. Circ. 286 - 288 (1973). In Japanese.

103.124   Comet 1973e Kohoutek

New comet Kohoutek 1973e.   L. Kohoutek,
N. Wood, R. L. Waterfield, B. G. Marsden.
British Astron. Ass., Circ. No. 547 (1973).

Comet Kohoutek (1973e).   L. Kohoutek.
IAU Circ., No. 2501 (1973).

Comet Kohoutek (1973e).   L. Kohoutek,
T. Seki, M. Antal, B. G. Marsden.
IAU Circ., No. 2504 (1973).

Comet Kohoutek (1973e).
N. Kojima, T. Seki, H. Kosai.
IAU Circ., No. 2506 (1973).

Comet Kohoutek (1973e).   N. Kojima, T. Seki,
H. L. Giclas, N. Wood, R. L. Waterfield, K. Ike.
IAU Circ., No. 2510 (1973).

Comet Kohoutek (1973e).
N. Kojima, H. Kosai, L. Kohoutek.
IAU Circ., No. 2511 (1973).

Comet Kohoutek (1973e).   N. Kojima, T. Seki,
B. Milet, H. L. Giclas.
IAU Circ., No. 2521 (1973).

Comet Kohoutek (1973e).   H. L. Giclas,
M. L. Kantz.
IAU Circ., No. 2523 (1973).

Comet Kohoutek (1973e).   R. L. Waterfield,
I. M. Purcell, N. Wood, G. H. Rutter, R. Petrovičová, A. Mrkos.

IAU Circ., No. 2531 (1973).

Comet Kohoutek (1973e). A. Mrkos.
IAU Circ., No. 2538 (1973).

Comet Kohoutek (1973e). J. Bortle,
B. G. Marsden.
IAU Circ., No. 2540 (1973).

Comet Kohoutek (1973e).
Japan Astron. Study Ass. Circ. 286 - 288 (1973). In Japanese.

New comet Kohoutek, 1973e.
Kometn. Tsirk., *Kiev*, No. 143 (1973). In Russian.

Comet Kohoutek, 1973e.
Kometn. Tsirk., *Kiev*, No. 144 (1973). In Russian.

Comet Kohoutek, 1973e.
Kometn. Tsirk., *Kiev*, No. 145 (1973). In Russian.

Comet Kohoutek, 1973e.
A. Mrkos.
Kometn. Tsirk., *Kiev*, No. 146 (1973). In Russian.

Ephemeris of comet Kohoutek, 1973e.
Kometn. Tsirk., *Kiev*, No. 147 (1973). In Russian.

103.125 Comet 1925 II Schwassmann-Wachmann 1

Periodic comet Schwassmann-Wachmann 1.
P. Herget.
IAU Circ., No. 2501 (1973).

Ephemeris of comet Schwassmann–Wachmann 1.
Kometn. Tsirk., *Kiev*, No. 144 (1973). In Russian.

103.126 Comet 1973j Brooks 2

Periodic comet Brooks 2.
B. G. Marsden, Z. Sekanina.
IAU Circ., No. 2505 (1973).

Periodic comet Brooks 2.
Y. V. Evdokimov, I. Y. Evdokimov, I. N. Platonov.
IAU Circ., No. 2518 (1973).

Elements and ephemeris of comet Brooks 2.
Yu. V. Evdokimov, I. Yu. Evdokimov, I. N. Platonov.
Kometn. Tsirk., *Kiev*, No. 144 (1973). In Russian.

Rediscovery of comet Brooks 2, 1973j.
Kometn. Tsirk., *Kiev*, No. 149 (1973). In Russian.

103.127 Comet 1973f Kohoutek

New comet Kohoutek 1973f. L. Kohoutek,
R. L. Waterfield, D. Griffiths, G. H. Rutter, B. G. Marsden,
S. W. Milbourn.
British Astron. Ass., Circ. No. 547 (1973).

Comet Kohoutek (1973f). L. Kohoutek.
IAU Circ., No. 2511 (1973).

Comet Kohoutek (1973f).

L. Kohoutek, B. G. Marsden.
IAU Circ., Nos. 2514, 2515 (1973).

Comet Kohoutek (1973f). C. Y. Shao.
IAU Circ., No. 2516 (1973).

Comet Kohoutek (1973f).
H. Kosai, K. Hurukawa, T. Hirayama.
IAU Circ., No. 2517 (1973).

Comet Kohoutek (1973f).
R. L. Waterfield, D. Griffiths, G. H. Rutter.
IAU Circ., No. 2519 (1973).

Comet Kohoutek (1973f). L. Kohoutek.
IAU Circ., No. 2522 (1973).

Comet Kohoutek (1973f).
IAU Circ., No. 2527 (1973).

Comet Kohoutek (1973f). A. Mrkos,
M. L. Kantz, H. L. Giclas.
IAU Circ., No. 2532 (1973).

Comet Kohoutek (1973f).
T. Seki, N. Kojima.
IAU Circ., No. 2534 (1973).

Comet Kohoutek (1973f).
T. Seki, N. Kojima.
IAU Circ., No. 2537 (1973).

Comet Kohoutek (1973f).
B. G. Marsden, D. K. Yeomans.
IAU Circ., No. 2541 (1973).

Comet Kohoutek (1973f). N. S. Chernykh,
L. I. Chernykh, S. I. Gerasimenko, N. P. Marchenko.
IAU Circ., No. 2543 (1973).

Comet Kohoutek (1973f). V. Vanýsek, J. Rahe.
IAU Circ., No. 2549 (1973).

Comet Kohoutek (1973f).
Japan Astron. Study Ass. Circ. 286 - 288 (1973). In Japanese

New comet Kohoutek, 1973f.
Kometn. Tsirk., *Kiev*, No. 144 (1973). In Russian.

Comet Kohoutek, 1973f.
Kometn. Tsirk., *Kiev*, No. 145 (1973). In Russian.

Comet Kohoutek, 1973f.
Kometn. Tsirk., *Kiev*, No. 146 (1973). In Russian.

Comet Kohoutek, 1973f.
Kometn. Tsirk., *Kiev*, No. 147 (1973). In Russian.

103.128 Comet 1971 IX Shajn-Schaldach

Observations of comets. C. Torres.
IAU Circ., No. 2520 (1973).

103.129 Comet 1971 VI Wolf-Harrington

Observations of comets. C. Torres.
IAU Circ., No. 2520 (1973).

103.130 Comet 1929 I Schwassmann-Wachmann 2

Periodic comet Schwassmann-Wachmann 2.
B. G. Marsden.
IAU Circ., No. 2527 (1973).

103.131 Comet 1905 II Borrelly

Periodic comet Borrelly.
D. K. Yeomans, L. M. Belous.
IAU Circ., No. 2531 (1973).

103.132 Comet 1973h Huchra

Comet Huchra (1973h). J. P. Huchra, R. Green.
IAU Circ., No. 2533 (1973).

Comet Huchra (1973h). J. P. Huchra.
IAU Circ., No. 2534 (1973).

Comet Huchra (1973h).
T. Seki, R. E. McCrosky, C. Y. Shao, N. Kojima.
IAU Circ., No. 2536 (1973).

Comet Huchra (1973h).
C. Y. Shao, R. E. McCrosky.
IAU Circ., No. 2539 (1973).

Comet Huchra (1973h).
K. Suzuki, T. Urata, R. E. McCrosky, C. Y. Shao.
IAU Circ., No. 2540 (1973).

Comet Huchra (1973h). T. Seki, N. Kojima.
IAU Circ., No. 2542 (1973).

Comet Huchra (1973h). R. Green, J. P. Huchra,
H. L. Giclas, M. L. Kantz, C. Y. Shao, R. E. McCrosky.
IAU Circ., No. 2546 (1973).

Comet Huchra (1973h). T. Seki, M. Antal,
N. Wood, R. L. Waterfield.
IAU Circ., No. 2551 (1973).

New comet Huchra, 1973h.
Kometn. Tsirk., *Kiev*, No. 146 (1973). In Russian.

Comet Huchra, 1973h.
Kometn. Tsirk., *Kiev*, No. 147 (1973). In Russian.

Comet Huchra, 1973h.
Kometn. Tsirk., *Kiev*, No. 148 (1973). In Russian.

103.133 Comet 1973i Clark

Comet Clark. A. C. Gilmore, M. Clark.
IAU Circ., No. 2544 (1973).

Comet Clark (1973i). M. Clark, A. C. Gilmore,
P. M. Kilmartin, R. E. Millington.
IAU Circ., No. 2545 (1973).

Comet Clark (1973i). A. C. Gilmore, R. E. Millington, P. M. Kilmartin, B. G. Marsden.
IAU Circ., No. 2548 (1973).

Periodic comet Clark (1973i). M. Clark, P. M.
Kilmartin, A. C. Gilmore, R. E. Millington, B. G. Marsden.
IAU Circ., No. 2550 (1973).

New comet Clark, 1973i.
Kometn. Tsirk., *Kiev*, No. 147 (1973). In Russian.

Periodic comet Clark, 1973i.
Kometn. Tsirk., *Kiev*, No. 149 (1973). In Russian.

103.134 Comet 1972c Tempel 2

Periodic comet Tempel 2 (1972c). B. G. Marsden.
IAU Circ., No. 2544 (1973).

Ephemeris of the periodic comet Tempel 2, 1972c.
Kometn. Tsirk., *Kiev*, No. 149 (1973). In Russian.

103.135 Comet 1970 III Kohoutek

**Possible moment of disruption of the nucleus of
comet Kohoutek, 1970 III.**
V. A. Golubev.
Kometn. Tsirk., *Kiev*, No. 147 (1973). In Russian.

103.136 Comet 1930 VI Schwassmann-Wachmann 3

**Former and future motion of comet Schwassmann-
Wachmann 3.**
N. A. Belyaev, S. D. Shaporev.
Kometn. Tsirk., *Kiev*, No. 148 (1973). In Russian.

103.137 Comet 1973k Sandage

New comet Sandage, 1973k.
Kometn. Tsirk., *Kiev*, No. 149 (1973). In Russian.

103.138 Comet 1969 II Gunn

Ephemeris of the periodic comet Gunn, 1969 II.
Kometn. Tsirk., *Kiev*, No. 149 (1973). In Russian.

104 Meteors, Meteor Streams

104.001 The telescopic radiant areas of the Perseids and the Orionids. V. Porubčan.
Bull. Astron. Inst. Czechoslovakia, Vol. 24, 1 - 8 (1973).

The form and size of the radiant areas of the Perseid and Orionid meteor streams is studied on the basis of the single-station telescopic observations of the radiant published by Bakharev. Guth's method of computing individual radiant positions from the relation between angular lengths of the meteors and their radiant elongations is applied and discussed. The results indicate that the radiants of faint meteors are more dispersed than those obtained from photographic observations of brighter meteors.

104.002 The structure of the Eta Aquarid meteor stream. A. Hajduk.
Bull. Astron. Inst. Czechoslovakia, Vol. 24, 9 - 13 (1973).

The structural features of the Eta Aquarid meteor stream have been studied on the basis of the variation in the density distribution along and across the stream. The variations in meteor rates at consecutive returns are found to be mainly due to a varying contribution of smaller particles. The observed displacement of the maxima in solar longitude, the size distribution of particles and other characteristics appear to be in agreement with the concept of a common origin of the Eta Aquarid and Orionid streams in comet Halley.

104.003 Data on three significant fireballs photographed within the European Network in 1971.
Z. Ceplecha, M. Ježková, J. Boček, T. Kirsten, J. Kiko.
Bull. Astron. Inst. Czechoslovakia, Vol. 24, 13 - 22 (1973).

Detailed data on three significant fireballs photographed within the European Network of all-sky cameras in 1971 are given, including heights, distances, velocities, geographical coordinates, light curves, radiants, orbits and impact coordinates. Operational procedures of the German part of the European Network are briefly described. Some minor changes of the positional reduction of all-sky-camera films are mentioned.

104.004 Statistical model of meteor streams. III. Stream search among 19303 radio meteors. Z. Sekanina.
Icarus, Vol. 18, 253 - 284 (1973).

Using a computerized technique of stream search, based on the statistical model of meteor streams, we have detected 72 additional streams in a sample of 19303 radio meteor orbits. The streams are found to have a tendency to cluster, partly along the ecliptic and partly in high-inclination orbits. Identification of the detected radio streams with previously known streams is presented, and plans for future work are briefly outlined.

104.005 Radio-echo measurements of the flux of the Quadrantid, Perseid and Geminid meteor streams.
D. W. Hughes.
Monthly Notices Roy. Astron. Soc., Vol. 161, 113 - 125 (1973).

This paper deals with the results of 17 MHz backscatter radio-echo measurements made at the University of Sheffield yielding the cumulative influx (particles $m^{-2} s^{-1}$) and the mass distribution index of the meteoroids in the Quadrantid, Perseid and Geminid meteor streams and also in the sporadic meteor background. These data are compared with other information from visual, radio-echo and satellite observations.

104.006 Perseid observations by Dutch and Belgian amateurs. B. C. J. Apeldoorn.
Sky Telescope, Vol. 45, 228 - 229 (1973).

104.007 Observations of meteors associated with comet Grigg-Skjellerup. W. J. Baggaley.
Observatory, Vol. 93, 23 - 26 (1973).

The predicted close approach of the earth to the orbit of periodic comet Grigg-Skjellerup (1902 II − 1922 I) indicated the possibility of the terrestrial accretion of meteoric material. An observed increase in the rate of radio-meteor echoes over the normal sporadic activity on the four days 1972 April 21, 22, 23 and 24, between 16^h and 20^h N.Z.S.T., indicated an influx of shower meteoroids corresponding to radio magnitudes brighter than +8.5.

104.008 De Perseïden van 1972, een enorm succes. B. Apeldoorn.
Hemel en Dampkring, Vol. 71, 23 - 29 (1973).

104.009 Airborne observations of the 1972 Giacobinids. P. M. Millman.
Journ. Roy. Astron. Soc. Canada, Vol. 67, 35 - 38 (1973).

104.010 Some considerations on the origin of the Leonid meteor shower (1833−1966).
Yu. V. Evdokimov, E. D. Kondrat'eva.
Trudy Kazan. Gorod. Astron. Obs., No. 38, p. 68 - 73 (1972).
In Russian.

104.011 Preliminary results of artificial meteor ablation of an olivine mineral sample.
M. B. Blanchard, G. G. Cunningham.
Meteoritics, Vol. 8, 15 (1973). − Abstract.

104.012 Some airborne observations of the 1972 Giacobinid meteor shower. I. Halliday, P. M. Millman.
Meteoritics, Vol. 8, 44 (1973). − Abstract.

104.013 Variations in the distribution and physical properties of Perseid meteors. J. A. Russell.
Meteoritics, Vol. 8, 66 - 67 (1973). − Abstract.

104.014 Observations of the Quadrantid meteors in 1973.
Sky Telescope, Vol. 45, 327, 328 (1973).

104.015 Comet Encke: Meteor metallic ion identification by mass spectrometer.
R. A. Goldberg, A. C. Aikin.
Science, Vol. 180, 294 - 296 (1973).

Metal ions have been detected in the upper atmosphere during the period of the Beta Taurids meteor shower. The abundances of these ions relative to Si^+ show agreement in most instances with abundances in chondrites. A notable exception is 45^+, which, if it is Sc^+, is 100 times more abundant than neutral scandium found in chondrites.

104.016 The effect of Yarkovsky-Radzievsky and the evolution of meteoric streams.
L. A. Katasev, N. V. Kulikova.
Astron. vestn., Vol. 6, 237 - 241 (1972). In Russian.

Values for the Yarkovsky−Radzievsky effect for various angles of the inclination of meteoric particle rotation equator to the orbital plane have been obtained. The influence of this effect on the evolution of the motion of meteoric particles belonging to the Draconids stream has been evaluated. The time of fall on the sun of particles of 0.1 and 1 cm in radius for direct and reverse axial rotation has been calculated with due regard for the Yarkovsky−Radzievsky effect.

104.017 Meteor streams of the ε Lyrids, α Coronids and

φ Draconids in 1969.
V. V. Martynenko.
Astron. vestn., Vol. 6, 252 - 254 (1972). In Russian.

104.018 **Radar studies of small condensations within a meteor stream.**
V. N. Lebedinets, V. N. Korpusov, A. K. Sosnova.
Astron. vestn., Vol. 7, 16 - 20 (1973). In Russian.
A yearly cycle of corresponding measurements of velocities and radiants of meteors was accomplished, and a catalogue of the orbits of 20000 meteoric bodies was obtained. The analysis of 16800 orbits permitted to reveal 715 meteor streams and associations, at least 366 of which are evidently real streams.

104.019 **About the mineralogical density of sporadic meteoric bodies.**   V. V. Benukh (*Benyukh*).
Astron. vestn., Vol. 7, 21 - 29 (1973). In Russian.
Mineralogical density of 2267 sporadic meteoric bodies has been studied by photographic observations of meteors, and a comparison of these data with the composition of meteorites and the earth's envelope has been made.

104.020 **The structure of the Perseids radiants in 1970.**
V. V. Martynenko.
Astron. vestn., Vol. 7, 48 - 51 (1973). In Russian.

104.021 **On methods to reveal meteor streams and associations.**   A. K. Sosnova, V. N. Lebedinets.
Astron. vestn., Vol. 7, 88 - 94 (1973). In Russian.
Special methods for the selection of orbits of meteor streams and associations with the help of electronic computers have been developed to study the general catalogue of the orbits of meteor bodies obtained from the results of radar measurements of radiants and the velocities of individual meteors.

104.022 **Once more on meteor satellites of the earth.**
M. P. Ananieva, K. V. Kostylev.
Astron. vestn., Vol. 7, 95 - 98 (1973). In Russian.
The system of differential equations for the ballistic motion of small meteoric particles in the earth's atmosphere has been numerically solved. The change in their temperature and mass as well as radiational losses are considered. The formation of relatively stable meteor satellites of the earth is possible with specific geocentric orbits available.

104.023 **Duration of meteors and the position of the point of their maximum brightness.**   I. S. Shestaka.
Astron. vestn., Vol. 7, 99 - 104 (1973). In Russian.
Dependencies of the durations of meteors and the relative position of their maximum brightness points on parameters characterising the meteoric body and its flight in the atmosphere have been studied on the basis of the analysis of meteor photographs.

104.024 **Registration of the angular velocity of meteors by a photographic scanning method.**
A. K. Stanyukovich, E. L. Stanyukovich.
Astron. vestn., Vol. 7, 105 - 106 (1973). In Russian.

104.025 **The structure of the Perseids' radiants in 1971.**
V. V. Martynenko, N. V. Smirnov.
Astron. vestn., Vol. 7, 118 - 123 (1973). In Russian.

104.026 **Evolution of orbits and radiants of the meteor stream of Jupiter's family.**
E. I. Kazimirchak-Polonskaya, A. K. Terent'eva.
Astron. Zhurn. Akad. Nauk SSSR, Vol. 50, 576 - 589 (1973). In Russian. English translation in Soviet Astron. AJ, Vol. 17, No. 3.

The evolution of orbits of three groups and 188 radiants of the α-Virginids stream has been investigated over the interval of 1860–2060, the perturbations from Venus–Neptune being taken into account. The semimajor axis, the revolution period and the aphelion distance are shown to be stable. Large changes are found in the elements of the spatial orientation of the meteor stream orbit having essential influence on the evolution of the radiants and the encounter conditions of the stream and the earth. It turns out that Jupiter is able to decrease the inclination of the orbit almost up to zero in the depths of its sphere of action. It is shown that the magnitude and the character of the perturbations from Jupiter are the principal factors which determine the sizes of the radiation areas of the shower and affect the duration of its visibility.

104.027 **The incentive of a bold hypothesis: Hyperbolic meteors and comets.**   F. L. Whipple.
Ann. New York Acad. Sci., Vol. 198, (see 012.020), 219 - 224 (1972).

104.028 **On the meteoric variation of cosmic-ray intensity.**
R. G. Lazarev, N. T. Svetashkova.
Materialy 3-j Nauch. konf. Tomsk. un-ta po mat. i mekh. Vyp. (No.) 2. Tomsk, Tomsk. un-t, 1973, p. 88 - 89. In Russian. Abstr. in Referativ. Zhurn. 51. Astron., 6.51.371 (1973).

104.029 **Generalization of Staude's formula for the rate of meteor bodies.**   M. K. Nazarenko.
Materialy 3-j Nauch. konf. Tomsk. un-ta po mat. i mekh. Vyp. (No.) 2. Tomsk, Tomsk. un-t, 1973, p. 92 - 93. In Russian. Abstr. in Referativ. Zhurn. 51. Astron., 6.51.372 (1973).

104.030 **Rate indices for meteor bodies and meteors.**
M. K. Nazarenko.
Materialy 3-j Nauch. konf. Tomsk. un-ta po mat. i mekh. Vyp. (No.) 2. Tomsk, Tomsk. un-t, 1973, p. 94 - 95. In Russian. Abstr. in Referativ. Zhurn. 51. Astron., 6.51.373 (1973).

104.031 **On the spatial distribution of meteoric matter in the vicinity of the earth's orbit.**   M. K. Nazarenko.
Materialy 3-j Nauch. konf. Tomsk. un-ta po mat. i mekh. Vyp. (No.) 2. Tomsk, Tomsk. un-t, 1973, p. 95 - 96. In Russian. Abstr. in Referativ. Zhurn. 51. Astron., 6.51.374 (1973).

104.032 **Observations of the 1972 Draconid meteor shower at Tomsk.**   N. P. Fast, E. P. Litvinova.
Materialy 3-j Nauch. konf. Tomsk. un-ta po mat. i mekh. Vyp. (No.) 2. Tomsk, Tomsk. un-t, 1973, p. 100 - 101. In Russian. Abstr. in Referativ. Zhurn. 51. Astron., 6.51.375 (1973).

104.033 **A great fireball of 12 August 1972.**   J. Mach.
Říše hvězd, Vol. 54, 98 (1973).   In Czech.

104.034 **Bolid Praha.**   Z. Ceplecha.
Říše hvězd, Vol. 54, 96 (1973). In Czech.
A fireball of -12^m was observed on 2 January 1973 in Central Bohemia. The position of the radiant, geocentric and heliocentric orbit and spectrum are given.

104.035 **Observations of the Draconids near Saratov.**
M. B. Bogdanov, Yu. V. Mikhajlov.
Kometn. Tsirk., *Kiev,* No. 140 (1973). In Russian.

104.036 **On the possible mineralogical composition of the meteor bodies of the Draconids.**   V. V. Benyukh.
Kometn. Tsirk., *Kiev,* No. 140 (1973). In Russian.

Osservazione fotografica di un bolide.
Coelum, Vol. 41, 76 - 77 (1973).

Comets and meteors.   See Abstr. 003.026.

105 Meteorites, Meteorite Craters

105.001 Debye-Scherrer investigations of experimentally shocked silicates. F. Hörz, W. L. Quaide.
The Moon, Vol. 6, 45 - 82 = Lunar Sci. Inst., *Houston, Texas*, Contr. No. 2 (1973).
The Moon, Vol. 6, 45 - 82 (1973).

Small ballistic ranges were used to perform controlled laboratory shock experiments on 12 selected silicates [quartz (30–310 kb), oligoclase (30–340 kb), andesine (40–100 kb), olivine (80–500 kb), forsterite (50–150 kb), enstatite (60–150 kb), biotite (10–90 kb), hornblende (50–150 kb), garnet (40–160 kb), kunzite (60–150 kb), beryl (60–140 kb), topaz (60–150 kb)]. At least 4 pressure points per mineral are available. Debye-Scherrer investigations of shocked materials revealed a gradual lattice breakdown of crystalline matter under shock. Individual mineral species behave selectively.

105.002 Abundance patterns of thirteen trace elements in primitive carbonaceous and unequilibrated ordinary chondrites. D. R. Case, J. C. Laul, I. Z. Pelly, M. A. Wechter, F. Schmidt-Bleek, M. E. Lipschutz.
Geochim. Cosmochim. Acta, Vol. 37, 19 - 33 (1973).

A neutron activation analysis technique was used to determine Au, Re, Co, Mo, As, Sb, Ga, Se, Te, Hg, Zn, Bi and Tl in 11 carbonaceous chondrites, 12 unequilibrated ordinary chondrites (UOC), and 4 equilibrated ordinary chondrites. The first 6 elements are 'undepleted', the next 3 'normally-depleted' and the last 4 'strongly-depleted'. The results are reported in tables and are discussed in detail.

105.003 Geochemical evidence for the origin of moldavites. V. Bouška, J. Benada, Z. Řanda, J. Kunčíř.
Geochim. Cosmochim. Acta, Vol. 37, 121 - 131 (1973).

Forty-eight moldavites and samples of rocks from the impact crater of Ries were analyzed using non-destructive neutron activation analysis. The following elements have been determined: La, Ce, Sm, Eu, Lu, Sc, Co, Cs, Hf and Th; and Rb and Cr in two moldavites.

105.004 Inter-element relationships between trace elements in primitive carbonaceous and unequilibrated ordinary chondrites.
R. K. Kurimoto, I. Z. Pelly, J. C. Laul, M. E. Lipschutz.
Geochim. Cosmochim. Acta, Vol. 37, 209 - 224 (1973).

The general pattern of inter-element relationships among 13 trace elements in carbonaceous and unequilibrated ordinary chondrites indicate pronounced differences in the formation conditions of these two sorts of primitive chondrites. The inter-element relationships among 26 trace elements in unequilibrated ordinary chondrites are consistent with the hypothesis that their formation proceeded via a two-stage process: a metal–silicate fractionation which took place in the solar nebula followed by a thermal fractionation which could have occurred either during condensation and accretion of solid material from the nebula or, after accretion, in the chondritic parent bodies.

105.005 Fine structures of mutually normalized rare-earth patterns of chondrites.
A. Masuda, N. Nakamura, T. Tanaka.
Geochim. Cosmochim. Acta, Vol. 37, 239 - 248 (1973).

Rare earth element (REE) abundances in ten chondrites (nine falls and one find) were determined very accurately by mass-spectrometric stable isotope dilution techniques. All of the chondrites have different relative and absolute REE patterns. Except for Eu and, rarely, for Ce, the REE abundances in chondrites are smoothly fractionated from sample to sample.

105.006 Gallium and germanium in the metal and silicates of L- and LL-chondrites. C.-L. Chou, A. J. Cohen.
Geochim. Cosmochim. Acta, Vol. 37, 315 - 327 (1973).

105.007 Chemical fractionations in meteorites – VI. Accretion temperatures of H-, LL-, and E-chondrites, from abundance of volatile trace elements.
J. C. Laul, R. Ganapathy, E. Anders, J. W. Morgan.
Geochim. Cosmochim. Acta, Vol. 37, 329 - 357 (1973).

Extending our earlier work on 11 L-chondrites (Geochim. Cosmochim. Acta, Vol. 35, 337 - 363, 1971), we have measured 9 volatile elements (Ag, Bi, Cs, In, Rb, Tl, Se, Cd, Zn) by neutron activation analysis in 11 LL- and 10 E-chondrites; the first 6 elements also in 22 H-chondrites.

105.008 Noble gases in eleven H-chondrites.
R. Ganapathy, E. Anders.
Geochim. Cosmochim. Acta, Vol. 37, 359 - 362 (1973).

He, Ne, Ar and Xe were measured in aliquots of 11 H-chondrites, to complement trace element studies on the same meteorites (Laul et al., 1972). Bielokrynitschie, Charsonville, Pultusk and Supuhee have lost radiogenic gases before cosmic-ray exposure, and Doroninsk, during exposure.

105.009 Heterocyclic compounds recovered from carbonaceous chondrites.
C. E. Folsome, J. G. Lawless, M. Romiez, C. Ponnamperuma.
Geochim. Cosmochim. Acta, Vol. 37, 455 - 465 (1973).

N-heterocyclic compounds were recovered from the Murchison, C2, Murray, C2, and Orgueil C1 carbonaceous chondrites. Combined gas chromatography–low resolution mass spectrometry and gas chromatography–high resolution mass spectrography of trimethylsilyl derivatives provided data permitting identification of some of the extracted material. No biological N-heterocyclics nor triazines were recovered. The relevance of these findings to chemical evolution is discussed.

105.010 The organic analysis of the Murchison meteorite.
R. L. Levy, M. A. Grayson, C. J. Wolf.
Geochim. Cosmochim. Acta, Vol. 37, 467 - 483 (1973).

The organic compounds released from the Murchison carbonaceous chondrite following vaporization–pyrolysis at 150, 300 and 430°C were investigated. The total organic yield was 272 ppm and consisted of n-alkanes, alkenes, aromatic hydrocarbons and thioaromatics. The composition and yield at all three temperatures are compared with those obtained by an identical analysis on another carbonaceous chondrite, Allende, and two terrestrial rocks.

105.011 Spallation production of ^{3}He, ^{21}Ne, and ^{38}Ar from target elements in the Bruderheim chondrite.
D. D. Bogard, P. J. Cressy, Jr.
Geochim. Cosmochim. Acta, Vol. 37, 527 - 546 (1973).

The concentrations of noble gas isotopes of He, Ne and Ar have been measured in eight mineral separates of the Bruderheim chondrite. The cosmic-ray-produced nuclides ^{21}Ne and ^{38}Ar were correlated by a computer least-squares fitting program with the elemental composition in each separate of potential targets for nuclear production. One of the results is the exposure age of the Bruderheim specimen of 18.5×10^6 years.

105.012 Noble gas and carbon abundances of the Haverö, Dingo Pup Donga, and North Haig ureilites.
D. D. Bogard, E. K. Gibson, Jr., D. R. Moore, N. L. Turner, R. B. Wilkin.

Geochim. Cosmochim. Acta, Vol. 37, 547 - 557 (1973).

Ureilites are an uncommon class of feldspar-poor achondrites represented by six known specimens, three falls and three finds. Total carbon determinations on the Haverö, Dingo Pup Donga, and North Haig ureilites yield values of 2.07, 3.17, and 5.58 wt.%, respectively. The first two meteorites also contain dominant cosmic-ray-produced He and Ne, and show ^{3}He exposure ages of ~23 m.y. and ~7 m.y., respectively.

105.013 High-temperature condensates in chondrites and the environment in which they formed.
L. Grossman, S. P. Clark, Jr.
Geochim. Cosmochim. Acta, Vol. 37, 635 - 649 (1973).

Chemical compositions of melilites and titaniferous pyroxenes in calcium- and aluminum-rich inclusions in carbonaceous chondrites are consistent with their origin as high-temperature condensates from a gas of solar composition. The high-temperature inclusions constitute evidence that accretion of grains to cm-sized objects occurred at a very early stage in the evolution of the solar nebula.

105.014 Argon 40—argon 39 chronology of four calcium-rich achondrites.   F. A. Podosek, J. C. Huneke.
Geochim. Cosmochim. Acta, Vol. 37, 667 - 684 (1973).

Results are presented from thermal-release argon 40—argon 39 dating experiments on four calcium-rich achondrites, Pasamonte, St. Severin, Guareña, and Petersburg. The results indicate that as a group the calcium-rich achondrites have experienced a more diverse and extensive thermal history than have the chondritic meteorites.

105.015 A preliminary sorting-out of "Ness County" meteorites.   W. F. Read.
Meteoritics, Vol. 7, 417 - 428 (1972).

With the exception of a distinctive olivine-bronzite chondrite recognized by H. H. Nininger: "Ness County (1938)", most of the numerous stony meteorites found in Ness County, Kansas, have been lumped together under the designation "Ness County (1894)". It is the purpose of this paper to show that at least three separate falls are covered by this name.

105.016 Mineralogy and petrology of the Yilmia enstatite chondrite.   P. R. Buseck, E. F. Holdsworth.
Meteoritics, Vol. 7, 429 - 447 (1972).

Yilmia, a new enstatite chondrite contains moderately well defined radiating and granular chondrules. Major phases include enstatite, plagioclase ($Ab_{80}An_{16}Or_4$), silica, silicon-rich kamacite and titanian troilite. Minor phases are many and varied: sinoite, silicon-rich taenite, schreibersite, graphite, osbornite, oldhamite, "normal" and zincian daubreelite, ferroan alabandite and a new FeZnMn monosulfide.

105.017 The Washougal meteorite.
D. Y. Jérome, M. C. Michel-Lévy.
Meteoritics, Vol. 7, 449 - 461 (1972).

The Washougal, Washington, U.S.A., howardite fell in 1939. We studied its mineralogy optically and determined the ranges of composition of plagioclase and pyroxenes from measurements of densities and indices of refraction.

105.018 The Seminole, Gaines Co., Texas, meteorite.
G. I Huss, P. R. Buseck, C. B. Moore.
Meteoritics, Vol. 7, 463 - 468 (1972).

The Seminole meteorite was found near Seminole, Texas in 1961 and recognized as a meteorite in 1963. Analyses showed the two fragments to be a single black olivine-bronzite, or H4-group chondrite, showing brecciation and a moderate number of individual chondrules.

105.019 The Gosnells iron — a fragment of the Mount Dooling octahedrite.

J. R. De Laeter, G. J. H. McCall, S. J. B. Reed.
Meteoritics, Vol. 7, 469 - 477 (1972).

A 1.5 kg iron found in 1960 at Gosnells, near Perth, Western Australia, belongs to Wasson's chemical group I-An3 and is structurally unusual, being best described as a heat-altered granular coarse octahedrite. It is chemically and structurally very similar to the Mount Dooling iron, found in 1909 about 400 km away.

105.020 Potentiostatic study of iron meteorite corrosion.
S. L. Tackett, A. J. Goudy.
Meteoritics, Vol. 7, 487 - 494 (1972).

A potentiostat was used to study the electrolytic corrosion of iron meteorites in a neutral solution. Iron, nickel, and cobalt ions, the products of corrosion, were soluble in the electrolyte solution, and were determined after each electrolysis by atomic absorption spectrophotometry. None of the six meteorites observed started to dissolve at a lower potential than pure iron, nor at a higher potential than pure nickel.

105.021 The Oro Grande, New Mexico, chondrite and its lithic inclusion.
R. V. Fodor, K. Keil, E. Jarosewich.
Meteoritics, Vol. 7, 495 - 507 (1972).

The Oro Grande, New Mexico, U.S.A., chondrite was found in 1971. A mineralogical and chemical analysis was made. On the basis of composition and texture, the Oro Grande meteorite is classified as an H5 chondrite.

105.022 The Ransom meteorites.   W. F. Read.
Meteoritics, Vol. 7, 509 - 513 (1972).

A number of stony meteorite specimens, presumably from a single fall, have been recovered from an area about 5 - 6 miles north of Ransom, Kansas. This paper presents a map of discovery locations, so far as known.

105.023 The Haverö ureilite.   K. J. Neuvonen, B. Ohlson, H. Papunen, T. A. Häkli, P. Ramdohr.
Meteoritics, Vol. 7, 515 - 531 (1972).

The Haverö ureilite fell on August 2, 1971 on the Island of Haverö, Finland, lat 60°14'44"N., long 22°03'43"E. The petrology, textural features and origin are discussed.

105.024 Cosmogenic radionuclides in the Haverö meteorite.
P. J. Cressy, Jr.
Meteoritics, Vol. 7, 533 - 536 (1972).

An 87-gram sample of the Haverö ureilite has been analyzed by non-destructive gamma-ray spectrometry. The results of the measurements of cosmogenic radionuclides are reported and discussed.

105.025 Morphologies of iron crystals from the Haverö meteorite.   J. Jedwab.
Meteoritics, Vol. 7, 537 - 546 (1972).

Studies of unpolished chips of the Haverö meteorite using the scanning electron microscope (SEM) and the electron microprobe (EMP), show two types of metallic iron particles: A: discrete convex globules of 5 to 50 microns and B: flattened contorted crystals, less than one micron.

105.026 Forms of carbon in the new Haverö ureilite of Finland.   G. P. Vdovykin.
Meteoritics, Vol. 7, 547 - 552 (1972).

Based on its mineral composition and structure the Haverö meteorite is a ureilite, the sixth meteorite in this rare group. Elementary carbon in this meteorite is represented by diamond, lonsdaleite, graphite and chaoite microcrystals in the intergrowths.

105.027 The chemical composition of the Haverö meteorite and the genesis of the ureilites.   H. B. Wiik.

Meteoritics, Vol. 7, 553 - 557 (1972).

The chemical composition of Haverö is presented and compared with the composition of the other five ureilites.

105.028 Argon-37, argon-39, and tritium radioactivities in the Haverö meteorite.
E. L. Fireman, G. Spannagel.
Meteoritics, Vol. 7, 559 - 564 (1972).

The ^{37}Ar and ^{39}Ar radioactivities were measured in a dissolved and in a melted sample of Haverö. The ^{3}H activity combined with the ^{3}He content gives a ^{3}He/2^{3}H exposure age for Haverö of $(29.5 \pm 2.5) \times 10^6$ years.

105.029 The highly reflecting and opaque components in the mineral content of the Haverö meteorite.
P. Ramdohr.
Meteoritics, Vol. 7, 565 - 571 (1972).

The Haverö meteorite is similar to the ureilite, Novo Urei. The minerals found are: iron, troilite, diamond, graphite, oldhamite, niningerite, traces of chromite and, perhaps, amorphous carbon.

105.030 Elemental abundances in the Haverö meteorite.
D. E. Gillum, M. Janghorbani, M. D. Miller, L. L. Chyi, W. D. Ehmann.
Meteoritics, Vol. 7, 573 - 578 (1972).

Abundances of 15 major, minor and trace elements have been determined in powders and interior chips derived from the Haverö ureilite. The values are in close agreement with mean values for other ureilites.

105.031 The chemistry of Haverö ureilite.
H. Wänke, H. Baddenhausen, B. Spettel, F. Teschke, M. Quijano-Rico, G. Dreibus, H. Palme.
Meteoritics, Vol. 7, 579 - 590 (1972).

The concentrations of 41 major, minor and trace elements were determined in Haverö ureilite. The results are reported in 4 tables.

105.032 Haverö ureilite: Evidence for recrystallization and partial reduction. F. Wlotzka.
Meteoritics, Vol. 7, 591 - 600 (1972).

Haverö is a recrystallized silicate rock with inclusions of a foreign material introduced late in its history; these inclusions contain the planetary primordial gases.

105.033 The Haverö ureilite: Petrographic notes.
U. B. Marvin, J. A. Wood.
Meteoritics, Vol. 7, 601 - 610 (1972).

105.034 Production de ^{26}Al dans Fe et Si par protons de 0.6 et 24 GeV.
S. Regnier, M. Lagarde, G. N. Simonoff, Y. Yokoyama.
Earth Planet. Sci. Letters, Vol. 18, 9 - 12 (1973).

Cross sections for ^{26}Al formation in iron and silicon targets bombarded with 0.6 and 24 GeV protons have been measured by using highly selective chemical separation and γ-γ spectrometry. These values are used in order to fit the excitation function Fe (p,X)^{26}Al by analogy with known equivalent nuclear reactions.

105.035 Étude de la thermoluminescence de la météorite Saint Séverin. G. Valladas, C. Lalou.
Earth Planet. Sci. Letters, Vol. 18, 168 - 171 (1973).

To attempt to obtain by thermoluminescence measurements the irradiation age of the Saint-Séverin meteorite, a study of its natural and artificial thermoluminescence has been done.

105.036 La radioactivité de la météorite Tillaberi (chute du 25 avril 1970) mesurée par spectrométrie γ.

J. Tobailem, D. Nordemann, C. Lalou.
Comptes Rendus Acad. Sci. Paris, Sér. B, Vol. 276, 59 - 61 (1973).

Les activités des radionuclides ^{54}Mn, ^{22}Na, ^{60}Co, ^{26}Al et ^{40}K ont été mesurées par spectrométrie γ à faible mouvement propre, dans un fragment de 0,146 kg de la météorite de pierre Tillaberi, tombée le 25 avril 1970. Les résultats obtenus sont discutés.

105.037 Allende meteorite carbonaceous phase: Intractable nature and scanning electron morphology.
A. J. Bauman, J. R. Devaney, E. M. Bollin.
Nature, Vol. 241, 264 - 267 (1973).

We have examined the ultrastructure of the Allende matrix by scanning electron microscopy and isolated the carbonaceous phase for study by mass spectrometry and oxidative thermal analysis. Energy-dispersive X-ray spectroscopy was used to determine the composition of μm-size areas for elements heavier than sodium. The Allende chondrite used in this study was specimen 818 from the Center for Meteorite Studies, Arizona State University at Tempe.

105.038 Temperature gradients and atmospheric ablation rates for the Barwell meteorite.
D. W. Sears, A. A. Mills.
Nature, Phys. Sci., Vol. 242, 25 - 26 (1973).

We have been studying the fusion crust of the Barwell meteorite in order to quantify the way meteorites behave in the atmosphere. We have found that differences occur in the temperature gradients associated with specimens derived from various faces of the original stone. We have also determined the corresponding ablation rates (which seem to be the first to be reported for any stony meteorite), the mass loss and the effective heating time for this meteorite.

105.039 Shock history of mesosiderites.
A. V. Jain, M. E. Lipshutz.
Nature, Phys. Sci., Vol. 242, 26 - 28 (1973).

We obtained samples of 18 meteorites and each specimen was polished, etched and studied metallographically. We then determined the crystallographic character of kamacite in each specimen by back-reflexion X-ray diffraction. Eight specimens were also examined by reflected light microscopy. The kamacite in 14 of the samples shows no evidence for unusual deformation or exposure to high shock pressures.

105.040 Source of extraterrestrial spheroids.
J. Rosinski, C. T. Nagamoto, T. C. Kerrigan.
Journ. Atmosph. Terr. Phys., Vol. 35, 95 - 100 (1973).

Masses of magnetic spheroids were determined by means of an oscillating fiber microbalance. The wide range of densities found $(1.33-6.45 \text{ g/cm}^3)$ is caused mainly by the presence of internal cavities which were detected by X-ray photography. True densities were used to calculate the source of the measured flux of magnetic spheroids during the first part of October 1969. It was shown that different meteor showers contributed to the spheroid flux.

105.041 Investigation of the Canyon Diablo metallic spheroids and their relationship to the breakup of the Canyon Diablo meteorite.
P. J. Blau, H. J. Axon, J. I. Goldstein.
Journ. Geophys. Res., Vol. 78, 363 - 374 (1973).

The purpose of this paper is to describe the structure and the cooling history of the Canyon Diablo spheroids and to determine the processes by which they, and, presumably, the lunar spheroids also, were formed. To do this, some of the processes that accompanied the breakup of the iron meteorite as it penetrated first the atmosphere and then the surface of the earth must be considered.

105.042 **Composition of metal in type III carbonaceous chondrites and its relevance to the source-assignment of lunar metal.** L. H. Fuchs, E. Olsen.
Earth Planet. Sci. Letters, Vol. 18, 379 - 384 (1973).

The metal in seven type III carbonaceous chondrites has been measured for concentrations of Ni, Co and Cr. Ni and Co contents of kamacite from several of the type III's studied fall outside of the range for these elements in bulk meteoritic metal and are relevant to the assignment of a meteoritic vs a non-meteoritic origin for lunar metal particles in the fines and breccias.

105.043 **Cordierite glass formed by shock in a cordierite-garnet-gneiss from the Ries crater, Germany.**
V. Stähle.
Earth Planet. Sci. Letters, Vol. 18, 385 - 390 (1973).

A glass of cordierite composition was found in a highly shocked garnet gneiss collected from Otting quarry in the Ries crater. Despite the variation in color no compositional variation from grain to grain was found. In the sample investigated, all cordierite grains were transformed to isotropic cordierite glass.

105.044 **On the origin of eucrites and diogenites.**
T. S. McCarthy, A. J. Erlank, J. P. Willis.
Earth Planet. Sci. Letters, Vol. 18, 433 - 442 (1973).

Eleven eucrites have been analyzed for major, minor and some trace (K, Sr, Zr, Y, Ba and Ni) constituents. These data are interpreted in terms of an igneous fractionation model according to which the observed enrichment trends of various elements in eucrite liquids are considered to be indicative of the simultaneous fractionation of plagioclase and pyroxene. The achondrite Binda, a monomict breccia of howarditic composition, is interpreted as a possible precursor to the eucrite liquids.

105.045 **Identification of some Ukrainian meteorites.**
R. Ganapathy, E. Anders.
Ocherki sovrem. geokhimii i analit. khimii. Moskva, Nauka, 1972, p. 72 - 74. In Russian. – Abstr. in Referativ. Zhurn. 51. Astron., 3.51.411 (1973).

105.046 **Cosmogenic isotopes of helium, neon, and argon in 33 iron meteorites.** R. Nord, J. Zähringer.
Ocherki sovrem. geokhimii i analit. khimii. Moskva, Nauka, 1972, p. 59 - 71. In Russian. – Abstr. in Referativ. Zhurn. 51. Astron., 3.51.416 (1973).

105.047 **The Ca–Al relationship in stony meteorites and some lunar materials.** L. H. Ahrens.
Ocherki sovrem. geokhimii i analit. khimii. Moskva, Nauka, 1972, p. 49 - 52. In Russian. – Abstr. in Referativ. Zhurn. 51. Astron., 3.51.428 (1973).

105.048 **Water in meteorites.** G. P. Vdovykin.
Ocherki sovrem. geokhimii i analit. khimii.
Moskva, Nauka, 1972, p. 53 - 58. In Russian. – Abstr. in Referativ. Zhurn. 51. Astron., 3.51.429 (1973).

105.049 **In a meteorite crater – search for diamonds.**
V. L. Masajtis.
Zemlya i Vselennaya, 1973, No. 1, p. 32 - 36. In Russian.

105.050 **Refractory trace elements in Ca-Al-rich inclusions in the Allende meteorite.** L. Grossman.
Geochim. Cosmochim. Acta, Vol. 37, 1119 - 1140 (1973).

The condensation temperatures are calculated for a number of refractory trace metals from a gas of solar composition at 10^{-3} and 10^{-4} atm. total pressure. Instrumental neutron activation analysis of Ca-Al-rich inclusions in the Allende carbonaceous chondrite reveals enrichments of 22.8 ± 2.2 in the concentrations of Ir, Sc and the rare earths relative to Cl chondrites. Such enrichments cannot be due to magmatic differentiation processes.

105.051 **The question of Eaton: terrestrial versus meteoritic copper.**
P. R. Buseck, E. Holdsworth, G. R. Scott.
Geochim. Cosmochim. Acta, Vol. 37, 1249 - 1254 (1973).

The major phases and inclusions of Eaton closely resemble those in commercial yellow brass. Eaton contains α and β Cu-Zn, small Pb inclusions around the Cu-Zn crystals and larger Ca aluminosilicate inclusions similar to those from sand casting molds. Based on these data Eaton does not appear to be a meteorite.

105.052 **The isotopic composition of 'graphitic' carbon from iron meteorites and some remarks on the troilitic sulfur of iron meteorites.** P. Deines, F. E. Wickman.
Geochim. Cosmochim. Acta, Vol. 37, 1295 - 1319 (1973).

The carbon isotopic composition of graphite inclusions and finely dispersed carbon from twelve iron meteorites has been studied. The analyses are given as parts permil deviations from the PDB standard. The large variability in the C^{13} concentration encountered within individual irons contrasts with the isotopic constancy of sulfur observed in meteorites.

105.053 **Abundance of 17 trace elements in carbonaceous chondrites.**
U. Krähenbühl, J. W. Morgan, R. Ganapathy, E. Anders.
Geochim. Cosmochim. Acta, Vol. 37, 1353 - 1370 (1973).

Seventeen trace elements (Ag, Au, Bi, Br, Cd, Cs, Ge, In, Ir, Rb, Re, Sb, Se, Te, Tl, U and Zn) were measured by neutron activation analysis in 8 Cl samples (1 Alais, 3 Ivuna, 4 Orgueil) and in 3 C2 samples (one each of Mighei, Murchison, Murray). The results show far less scatter than earlier literature data. The new data suggest significant revisions in cosmic abundance for the following elements (old values in parentheses): Zn 1250 (1500), Cd 1.51 (2.12), Ir 0.72 (0.43) atoms/10^6 Si atoms.

105.054 **Solubilities of noble gases in magnetite: implications for planetary gases in meteorites.**
M. S. Lancet, E. Anders.
Geochim. Cosmochim. Acta, Vol. 37, 1371 - 1388 (1973).

Solubilities of noble gases in magnetite were determined by growing magnetite in a noble-gas atmosphere between 450 and 700°K. It seems that equilibrium solubility may be able to account for four features of planetary gas: elemental ratios, amounts, correlations with other volatiles, and retentive siting. It cannot account for the isotopic fractionation of planetary gas, however.

105.055 **Spatial distribution of elements in tektites and comparable materials by charged particle activation analysis.** N. A. Askouri, S. A. Durrani, J. H. Fremlin.
Journ. Geophys. Res., Vol. 78, 1245 - 1252 (1973).

The spatial distribution of elements in tektites from Indochina and Czechoslovakia and in certain terrestrial materials has been studied by charged particle activation analysis and β autoradiography in an attempt to throw light on the mode of formation of tektites.

105.056 **Depth variation of cosmogenic noble gases in the ~ 120-kg Keyes chondrite.**
R. J. Wright, L. A. Simms, M. A. Reynolds, D. D. Bogard.
Journ. Geophys. Res., Vol. 78, 1308 - 1318 (1973).

Cosmic ray spallation produced ^{3}He, ^{21}Ne, ^{22}Ne, and ^{38}Ar have been measured as a function of depth in three mutually perpendicular cores (56, 51, and 21 cm) taken from the ~120-kg L chondrite found near Keyes, Oklahoma. The data obtained represent the first experimental determination of the depth variation in the production rates of these cosmogenic nuclides in a single stone meteorite.

105.057 Hammond Downs, a new chondrite from the Tenham area, Queensland, Australia. B. Mason.
Meteoritics, Vol. 8, 1 - 7 (1973).

Among a collection of meteorites from the area of the Tenham shower was a 27 kg stone which proved to be different from the other Tenham stones. It is a bronzite, H4, chondrite, the principal minerals being olivine, clinobronzite, nickel-iron, and troilite. It has a highly chondritic structure, with devitrified glass within the chondrules, and without visible plagioclase.

105.058 Allende C-3 chondrite carbonaceous phase: Scanning electron morphology, differential thermal analysis, solvent properties, and spark source and electron impact mass spectrometry. A. J. Bauman, J. R. Devaney.
Meteoritics, Vol. 8, 13 - 14 (1973). — Abstract.

105.059 Mesosiderites: A search for fractionation trends. R. Bild, R. Schaudy, C.-L. Chou, P. A. Baedecker, J. T. Wasson.
Meteoritics, Vol. 8, 14 (1973). — Abstract.

105.060 The origin of chondrules: Experimental investigation of metastable liquid silicates.
M. Blander, K. Keil, H. N. Planner, L. S. Nelson.
Meteoritics, Vol. 8, 15 (1973). — Abstract.

105.061 Rare earth and other abundances in calcium-poor achondrites. W. V. Boynton, R. A. Schmitt.
Meteoritics, Vol. 8, 16 - 17 (1973). — Abstract.

105.062 The paleomagnetic record in carbonaceous meteorites. A. Brecher.
Meteoritics, Vol. 8, 17 - 18 (1973). — Abstract.

105.063 Microhardness of iron meteorites.
V. F. Buchwald.
Meteoritics, Vol. 8, 19 (1973). — Abstract.

105.064 Zerhamra, a new iron meteorite find from Sahara.
V. F. Buchwald.
Meteoritics, Vol. 8, 20 (1973). — Abstract.

105.065 Restudy of Fe-Mg equilibration temperature for ordinary chondrite pyroxenes.
T. E. Bunch, E. Olsen.
Meteoritics, Vol. 8, 22 - 23 (1973). — Abstract.

105.066 A new enstatite chondrite from Yilmia, Western Australia. P. R. Buseck, E. F. Holdsworth.
Meteoritics, Vol. 8, 23 (1973). — Abstract.

105.067 Fractionation of siderophilic and volatile elements in H-group chondrites.
C.-L. Chou, P. A. Baedecker, J. T. Wasson.
Meteoritics, Vol. 8, 24 (1973). — Abstract.

105.068 Scopas surveys at Lake Wanapitei, Ontario, Canada. J. F. Clark.
Meteoritics, Vol. 8, 24 - 25 (1973). — Abstract.

105.069 Rare earth patterns in pallasitic olivines.
T. D. Cooper, R. A. Schmitt.
Meteoritics, Vol. 8, 25 - 26 (1973). — Abstract.

105.070 Chemistry-corrected ^{21}Ne and ^{38}Ar exposure ages of stone meteorites. P. J. Cressy, D. D. Bogard.
Meteoritics, Vol. 8, 26 (1973). — Abstract.

105.071 Laguna Guatavita (Colombia): not meteoritic, probable salt collapse crater.
R. S. Dietz, J. F. McHone.
Meteoritics, Vol. 8, 27 - 28 (1973). — Abstract.

105.072 Hudson Bay arc as an astrobleme: A negative search.
R. S. Dietz, J. P. Barringer.
Meteoritics, Vol. 8, 28 - 29 (1973). — Abstract.

105.073 Fluorine in meteorites.
G. Dreibus, H. Palme, H. Wänke.
Meteoritics, Vol. 8, 29 - 30 (1973). — Abstract.

105.074 The Yilmia type II enstatite chondrite.
A. El Goresy, J. F. Lovering.
Meteoritics, Vol. 8, 31 (1973). — Abstract.

105.075 Cosmogenic rare gas production rates in chondritic meteorites. D. E. Fisher.
Meteoritics, Vol. 8, 33 (1973). — Abstract.

105.076 Carbonaceous and non-carbonaceous lithic inclusions in the Plainview, Texas, chondrite.
R. V. Fodor, K. Keil.
Meteoritics, Vol. 8, 33 - 34 (1973). — Abstract.

105.077 Microbreccias, impact glasses and spherules from Lonar crater, India; lunar analogs.
K. Fredriksson, A. Dube, D. Milton.
Meteoritics, Vol. 8, 34 (1973). — Abstract.

105.078 The Lonar meteorite crater, India.
K. Fredriksson, D. Milton, A. Dube, M. S. Balasundaram.
Meteoritics, Vol. 8, 35 (1973). — Abstract.

105.079 Composition of metal in type III carbonaceous chondrites and its relevance to the source-assign- ment of lunar metal. L. H. Fuchs, E. Olsen.
Meteoritics, Vol. 8, 35 - 36 (1973). — Abstract.

105.080 The Pretoria Salt-Pan: astrobleme or crystovolcano?
R. F. Fudali, D. P. Gold.
Meteoritics, Vol. 8, 36 (1973). — Abstract.

105.081 Diameter, depth, displaced mass, and effects of oblique trajectories for impact craters formed in crys- talline rocks. D. E. Gault, J. A. Wedekind.
Meteoritics, Vol. 8, 37 (1973). — Abstract.

105.082 K-Ar and fission track dating of "Darwin crater" glass.
W. Gentner, T. Kirsten, D. Storzer, G. A. Wagner.
Meteoritics, Vol. 8, 37 - 38 (1973). — Abstract.

105.083 North American (?) microtektites.
B. P. Glass, R. N. Baker, J. Barone.
Meteoritics, Vol. 8, 39 - 40 (1973). — Abstract.

105.084 The metal in diogenites.
R. C. Gooley, C. B. Moore.
Meteoritics, Vol. 8, 41 (1973). — Abstract.

105.085 Composition of coexisting pentlandite and awaruite in the Allende meteorite.
E. Holdsworth, W. Nichiporuk, C. B. Moore.
Meteoritics, Vol. 8, 45 (1973). — Abstract.

105.086 A census of the meteorites of Roosevelt County, N.M. G. I Huss, I. E. Wilson.
Meteoritics, Vol. 8, 45 - 46 (1973). — Abstract.

105.087 **Angra Dos Reis (stone) mineralogy and its bearing on the genesis of chondrites.** R. Hutchison.
Meteoritics, Vol. 8, 46 (1973). − Abstract.

105.088 **More data on the eucrite series.**
R. Hutchison, R. F. Symes.
Meteoritics, Vol. 8, 46 (1973). − Abstract.

105.089 **The Washougal meteorite.**
D. Y. Jérome, M. C. Michel-Lévy.
Meteoritics, Vol. 8, 47 (1973). − Abstract.

105.090 **Comparative study of the distribution of trace elements in meteoritic matter.**
S. Jovanovic, G. W. Reed, Jr.
Meteoritics, Vol. 8, 47 - 48 (1973). − Abstract.

105.091 **$(Fe, Cr)_{1+x}(Ti, Fe)_2 S_4$, a new mineral in the Bustee enstatite achondrite.** K. Keil, R. Brett.
Meteoritics, Vol. 8, 48 - 49 (1973). − Abstract.

105.092 **The determination of zinc in iron meteorites.**
W. R. Kelly, C. B. Moore,
Meteoritics, Vol. 8, 49 - 50 (1973). − Abstract.

105.093 **Abundance of 17 trace elements in carbonaceous chondrites.**
U. Krähenbühl, J. W. Morgan, R. Ganapathy, E. Anders.
Meteoritics, Vol. 8, 50 (1973). − Abstract.

105.094 **The Lancè chondrite: Further evidence for the complex development of chondrites.** G. Kurat.
Meteoritics, Vol. 8, 51 - 52 (1973). − Abstract.

105.095 **Elemental abundances and inter-element relationships of trace elements in primitive meteorites.**
R. K. Kurimoto, I. Z. Pelly, C. M. Binz, M. E. Lipschutz.
Meteoritics, Vol. 8, 52 - 53 (1973). − Abstract.

105.096 **Criteria for abiogenicity: Clues from carbonaceous chondrites.** K. A. Kvenvolden.
Meteoritics, Vol. 8, 53 (1973). − Abstract.

105.097 **Sulfides of the LL chondrites.**
D. E. Lange, C. B. Moore.
Meteoritics, Vol. 8, 54 (1973). − Abstract.

105.098 **Xenon isotopes in carbonaceous chondrites.**
O. K. Manuel, E. W. Hennecke, D. D. Sabu.
Meteoritics, Vol. 8, 54 - 55 (1973). − Abstract.

105.099 **Ages of the Allende chondrules and inclusions: Do they set a time scale for the formation of the solar system?** K. Marti.
Meteoritics, Vol. 8, 55 - 56 (1973). − Abstract.

105.100 **Chemical composition of 36 individual chondrules from the Allende meteorite.**
C. Menninga, L. A. Rancitelli, R. H. Beauchamp.
Meteoritics, Vol. 8, 56 (1973). − Abstract.

105.101 **The Nedagolla meteorite.**
G. Miyake, J. I. Goldstein.
Meteoritics, Vol. 8, 57 - 58 (1973). − Abstract.

105.102 **The chemical compositions of meteorites, Apollo 14, 15, 16, and Luna 20 lunar samples as seen through the optics of a flaming atomic absorption spectrophotometer.** D. F. Nava.
Meteoritics, Vol. 8, 59 - 60 (1973). − Abstract.

105.103 **Distribution of manganese between the troilite and metal phases of iron meteorites.** W. Nichiporuk.
Meteoritics, Vol. 8, 60 (1973). − Abstract.

105.104 **Fossil meteorites.** H. H. Nininger.
Meteoritics, Vol. 8, 61 (1973). − Abstract.

105.105 **The St. Mary's County, Maryland, LL3 chondrite.**
A. F. Noonan, E. Jarosewich, R. S. Clarke, Jr.
Meteoritics, Vol. 8, 61 - 62 (1973). − Abstract.

105.106 **Chromium and phosphorus enrichment in the metal of type II carbonaceous chondrites.**
E. Olsen, L. H. Fuchs, W. C. Forbes.
Meteoritics, Vol. 8, 62 (1973). − Abstract.

105.107 **Oxygen isotope abundances in C2 and C3 meteorites.**
N. Onuma, R. N. Clayton, T. K. Mayeda.
Meteoritics, Vol. 8, 63 (1973). − Abstract.

105.108 **Role of ^{53}Mn in meteoritics.**
P. P. Parekh, M. Heimann, W. Herr.
Meteoritics, Vol. 8, 63 - 64 (1973). − Abstract.

105.109 **Noble gas chronology of meteorites.**
F. A. Podosek, J. C. Huneke.
Meteoritics, Vol. 8, 64 (1973). − Abstract.

105.110 **Energy spectra of ancient solar flare particles and the origin of gas meteorites.**
R. S. Rajan, D. Macdougall, P. P. Phakey.
Meteoritics, Vol. 8, 64 - 65 (1973). − Abstract.

105.111 **Coexisting bronzite and clinobronzite in the Steinbach meteorite.**
A. M. Reid, R. J. Williams, H. Takeda.
Meteoritics, Vol. 8, 66 (1973). − Abstract.

105.112 **The Kramer Creek, Colorado meteorite.**
T. E. Schmidt.
Meteoritics, Vol. 8, 67 (1973). − Abstract.

105.113 **Microstructure of an experimentally shocked brittle-ductile interface and the origin of "brecciated" silicate inclusions in iron meteorites.** C. B. Sclar.
Meteoritics, Vol. 8, 68 - 69 (1973). − Abstract.

105.114 **The nature of dark-etching rim zones in meteoritic taenite.** E. R. D. Scott.
Meteoritics, Vol. 8, 69 - 70 (1973). − Abstract.

105.115 **Two more chemical groups of iron meteorites, IIIE and IIIF.** E. R. D. Scott, J. T. Wasson.
Meteoritics, Vol. 8, 70 - 71 (1973). − Abstract.

105.116 **Experimental hypervelocity craters in quartz and glass sand: Distribution and shock metamorphism of ejected mass.**
D. Stöffler, J. A. Wedekind, G. Polkowski, D. E. Gault.
Meteoritics, Vol. 8, 74 (1973). − Abstract.

105.117 **Magnetic properties of chondrite meteorite fusion crust.** P. J. Wasilewski.
Meteoritics, Vol. 8, 79 - 80 (1973). − Abstract.

105.118 **Magnetite in carbonaceous chondrites.**
P. J. Wasilewski.
Meteoritics, Vol. 8, 80 - 81 (1973). − Abstract.

105.119 **Aluminum-26 in ureilites.**
L. L. Wilkening, G. F. Herman, E. Anders.
Meteoritics, Vol. 8, 82 (1973). − Abstract.

105.120 **Iodine-129/xenon-129 age of magnetite from the Orgueil meteorite.** G. F. Herzog, E. Anders,
E. C. Alexander, Jr., P. K. Davis, R. S. Lewis.
Science, Vol. 180, 489 - 491 (1973).

Magnetite from the Orgueil C1 chondrite is only 2.0 ± 2.4 million years older by the iodine-xenon method than the next oldest meteorite, the Karoonda C4 chondrite. This age ties the primitive C1 chondrites to the extensive iodine-xenon chronology of normal chondrites.

105.121 **Minor and trace elements in some meteoritic minerals.** R. O. Allen, Jr., B. Mason.
Geochim. Cosmochim. Acta, Vol. 37, 1435 - 1456 (1973).

The mineral phases including olivine, orthopyroxene, clinopyroxene, troilite, nickel-iron, plagioclase, chromite and the phosphates were separated from several meteorites. These were a hypersthene chondrite (Modoc), a bronzite chondrite (Guareña), an enstatite chondrite (Khairpur), and two eucrites (Haraiya and Moore County); diopside was separated from the Nakhla achondrite. The purified minerals were analyzed for trace and minor elements by spark source mass spectrometry and instrumental neutron activation analysis.

105.122 **Chemical fractionations in meteorites—VII. Cosmothermometry and cosmobarometry.**
J. W. Larimer.
Geochim. Cosmochim. Acta, Vol. 37, 1603 - 1623 (1973).

The fractional condensation of Bi, Cd, In, Pb and Tl from a cooling gas of cosmic composition is calculated. Predicted absolute and relative abundances of the elements are in good to excellent agreement with the analytical data. This strongly suggests that the presently observed abundances were established at the time of accretion. The elements may therefore be used as cosmothermometers to predict accretion temperatures. The condensation curves of all these elements are pressure-dependent but are confined to fall in the temperature interval 400 to 600°K owing to the absence of Fe_3O_4 and the presence of FeS in ordinary chondrites. Absolute upper and lower limits on the total pressure can thus be deduced.

105.123 **Meteoritenortung über Süddeutschland.**
J. Kiko, T. Kirsten, Z. Ceplecha.
Max-Planck-Inst. Kernphysik Heidelberg, Jahresbericht 1972, p. 193 (1973). − Abstract.

105.124 **Ein Meteoritenkrater in Süditalien?**
P. Horn, P. Sighinolfi, B. Kleinmann, F. Fogia.
Max-Planck-Inst. Kernphysik Heidelberg, Jahresbericht 1972, p. 193 - 194 (1973). − Abstract.

105.125 **Yilmia: Ein neuer Enstatit-Chondrit aus Westaustralien.** J. F. Lovering, A. El Goresy.
Max-Planck-Inst. Kernphysik Heidelberg, Jahresbericht 1972, p. 194 (1973). − Abstract.

105.126 **Messung von ^{26}Al in Eisenmeteoriten.** H. Kammer.
Max-Planck-Inst. Kernphysik Heidelberg, Jahresbericht 1972, p. 194 - 195 (1973). − Abstract.

105.127 **Zur Frage der Eisenmeteoritenentstehung − Einige Simulationsexperimente.**
M. R. Bloch, O. Müller.
Max-Planck-Inst. Kernphysik Heidelberg, Jahresbericht 1972, p. 195 - 196 (1973). − Abstract.

105.128 **Fortgang der Sägearbeiten am Meteoriten Mundrabilla II.** O. Medenbach.

Max-Planck-Inst. Kernphysik Heidelberg, Jahresbericht 1972, p. 196 - 197 (1973). − Abstract.

105.129 **Spaltspurenalter von Mikrotektiten aus Tiefseesedimenten aus dem Karibischen Meer.**
G. A. Wagner, D. Storzer.
Max-Planck-Inst. Kernphysik Heidelberg, Jahresbericht 1972, p. 205 (1973). − Abstract.

105.130 **Spaltspurenalter und stratigraphisches Vorkommen von Georgia-Tektiten.** D. Storzer, G. A. Wagner.
Max-Planck-Inst. Kernphysik Heidelberg, Jahresbericht 1972,

105.131 **Das Alter des Darwin-Kraters in Tasmanien.**
W. Gentner, T. Kirsten, D. Storzer, G. A. Wagner.
Max-Planck-Inst. Kernphysik Heidelberg, Jahresbericht 1972, p. 206 (1973). − Abstract.

105.132 **Cl-, Br-, Na- und K-Gehalte in Tektiten, Muong Nong Typ-Gläsern und anderen natürlichen Gläsern mittels Neutronenaktivierung.** O. Müller.
Max-Planck-Inst. Kernphysik Heidelberg, Jahresbericht 1972, p. 207 (1973). − Abstract.

105.133 **Evidence of the extinct nuclide ^{146}Sm in "Juvinas" achondrite.** K. Notsu, H. Mabuchi, O. Yoshioka, J. Matsuda, M. Ozima.
Earth Planet. Sci. Letters, Vol. 19, 29 - 36 (1973).

The search for the isotopic anomaly of Nd due to the disintegration of ^{147}Sm and ^{146}Sm was carried out on separated mineral phases of Juvinas achondrite. An internal isochron of 4.3 ± 2.5 billion years was obtained for ^{147}Sm−^{143}Nd pair. This provides the possibility of dating the planetary materials. Anomalies were observed for ^{142}Nd, giving the isotopic ratio ^{146}Sm/^{144}Sm at the time of solidification to be 0.39 ± 0.14. This is the first positive evidence of extinct ^{146}Sm which may be used as a p-process chronometer.

105.134 **Calculation of meteoroid impacts on moon and earth.**
L. W. Bandermann, S. F. Singer.
Icarus, Vol. 19, 108 - 113 (1973).

Concise derivations are given for the expected flux of meteoroids to the surfaces of the earth and the moon. Contrary to other published results, we find an accretion rate which is lower for the near side of the moon than for the far side and which is lower for the moon than for the earth, for all earth−moon distances.

105.135 **Raketenflüge: Staubexperimente in Höhen zwischen 70 und 110 km.** H.-J. Hoffmann, H. Fechtig.
Max-Planck-Inst. Kernphysik Heidelberg, Jahresbericht 1972, p. 234 (1973). − Abstract.

105.136 **Mikrometeoritenexperiment in der Sonnensonde HELIOS.**
H. Dietzel, H. Fechtig, E. Grün, P. Gammelin, J. Kissel.
Max-Planck-Inst. Kernphysik Heidelberg, Jahresbericht 1972, p. 236 - 237 (1973).

105.137 **Meteoritic, lunar and Lonar impact chondrules.**
K. Fredriksson, A. Noonan, J. Nelen.
The Moon, Vol. 7, 475 - 482 (1973). − Paper dedicated to Professor Harold C. Urey on the occasion of his 80th birthday on 29 April, 1973.

Impact-generated silicate spherules from the Lonar Crater, India, and from all Apollo sites are analogous to meteoritic chondrules (and some microtektites). Thus, the impact origin of chondrules, first proposed by Urey (1952), is a mechanism strongly supported by physical evidence from both the moon and the earth. Chondrites appear to be essentially impact breccias similar to lunar and Lonar microbreccias. The implications of this with regard to size and composition of the meteorite parent bodies are reviewed as well as the possible variations of element fractionation by volatilization-condensation.

105.138 Reflection of meteorite-generated shock waves from the earth's surface.
L. G. Gvozdeva, A. K. Stanyukovich.
Astron. vestn., Vol. 6, 228 - 236 (1972). In Russian.

Different regimes of the reflection of large meteorite-generated shock waves are studied. It is shown that analysing local destruction one cannot restrict the studies to the analysis of the incident shock waves only and neglect reflection processes. Four possible regimes of reflection are considered: normal reflection arising in the epicentre of destruction, glancing incidence, regular and irregular, i.e. Mach reflection. A detailed analysis of the most difficult case of Mach reflection is given.

105.139 Investigation of supposed meteoritic craters.
V. I. Koval.
Astron. vestn., Vol. 7, 30 - 31 (1973). In Russian.

105.140 Some features of surface scattering of the Sikhote-Alin meteorite shower. V. I. Tsvetkov.
Astron. vestn., Vol. 7, 107 - 110 (1973). In Russian.

105.141 Lonar Lake, India: An impact crater in basalt.
K. Fredriksson, A. Dube, D. J. Milton, M. S. Balasundaram.
Science, Vol. 180, 862 - 864 (1973).

Discovery of shock-metamorphosed material establishes the impact origin of Lonar Crater. As the only known terrestrial impact crater in basalt, Lonar Crater provides unique opportunities for comparison with lunar craters. In particular, microbreccias and glass spherules from Lonar Crater have close analogs among the Apollo specimens.

105.142 He, Ne and Ar in chondritic Ni–Fe as irradiation hardness sensors.
L. Nyquist, H. Funk, L. Schultz, P. Signer.
Geochim. Cosmochim. Acta, Vol. 37, 1655 - 1685 (1973).

The goal of this investigation of L, LL, and H chondrites was the correlation of the spallogenic noble gas abundance patterns in the metallic Ni–Fe to those in iron meteorites and to the production model of Signer and Nier (1960, 1962). The primary result of the present investigation is the unambiguous establishment that the $^3He/^{21}Ne$ and $^{22}Ne/^{21}Ne$ ratios are irradiation hardness sensors in bulk chondrites. Correlation to the previously established relationships for iron meteorites allows lower limits to be placed on the preatmospheric masses of the chondritic meteoroids.

105.143 Uranium and thorium in achondrites.
J. W. Morgan, J. F. Lovering.
Geochim. Cosmochim. Acta, Vol. 37, 1697 - 1707 (1973).

The abundances of U and Th in 19 achondrites and two pallasite olivines have been measured by radiochemical neutron activation analysis. The present communication summarizes the data and discusses the results in detail.

105.144 Aluminum-26 in meteorites – VII. Ureilites, their unique radiation history.
L. L. Wilkening, G. F. Herman, E. Anders.
Geochim. Cosmochim. Acta, Vol. 37, 1803 - 1810 (1973).

Cosmogenic ^{26}Al activities have been measured by $\gamma - \gamma$ coincidence counting in the three ureilites which had not previously been studied. The values in dpm/kg are: Dingo Pup Donga, 38.4 ± 2.4; North Haig, 39.3 ± 4.8; Dyalpur, 55.8 ± 4.8.

105.145 Spatial variations of cosmic rays derived from the radio activity of meteorites with known orbits.

A. K. Lavrukhina, V. D. Gorin, G. K. Ustinova.
Izv. AN SSSR. Ser. fiz., Vol. 36, 2306 - 2311 (1972). In Russian. – Abstr. in Referativ. Zhurn. 51. Astron.,5.51.423 (1973).

105.146 Thermal models of inhomogeneously accreted meteorite parent bodies.
J. M. Herndon, M. W. Rowe.
Nature, Phys. Sci., Vol. 244, 40 - 41 (1973).

Recent studies suggest that iron was fractionated early in the planetary formation process and that the terrestrial planets were inhomogeneously accreted (Lewis, 1972). Considering the differences in thermal properties and distribution of radioactive heat sources that result from initial core formation followed by silicate accretion, the authors present thermal calculations on a set of models of meteorite parent bodies in which initial formation of a Ni-Fe core is assumed.

105.147 Thermal history of the nakhlites by the $^{40}Ar–^{39}Ar$ method. F. A. Podosek.
Earth Planet. Sci. Letters, Vol. 19, 135 - 144 (1973).

This paper reports the results of thermal-release argon analyses of neutron-irradiated samples of the two nakhlite meteorites, Lafayette and Nakhla.

105.148 North American microtektites from the Caribbean Sea and their fission track age.
B. P. Glass, R. N. Baker, D. Storzer, G. A. Wagner.
Earth Planet. Sci. Letters, Vol. 19, 184 - 192 (1973).

Over 6000 microscopic glass spherules between 125 μm and 1 mm in diameter were found in a sediment core (RC9-58) from the Caribbean Sea. The authors believe that the microscopic glass objects are microtektites belonging to the North American strewnfield, based on their geographical location, appearance, physical properties, stratigraphic age (middle Upper Eocene), fission track age ($\sim$ 34.6 my) and major element compositions.

105.149 Chemical composition and rare gas content of four new detected Antarctic meteorites.
Mak. Shima, Mas. Shima, H. Hintenberger.
Earth Planet. Sci. Letters, Vol. 19, 246 - 249 (1973).

Within a region of 5 km $\times$ 10 km on a downhill slope of the Yamato Mounties, in 1969 the Japanese Expedition Team collected many stones. 9 of them were recognized as meteorites. On 4 of these findings the authors determined the chemical composition and the rare gas content. Exposure ages are 1.7, 31, 25 and 4.3 my respectively.

105.150 Some modern aspects of cosmic mineralogy. II.
E. K. Lazarenko, A. A. Yasinskaya.
Mineral. sb. L'vov. un-ta, 1972, No. 26, vyp. (No.) 1, p. 14 - 34. In Russian. – Abstr. in Referativ. Zhurn. 62. Issled. kosmich. prostranstva, 6.62.166 (1973).

105.151 $^{207}Pb–^{206}Pb$ isochron and age of chondrites.
J. M. Huey, T. P. Kohman.
Journ. Geophys. Res., Vol. 78, 3227 - 3244 (1973).

The isotopic composition of lead was measured in 16 chondrites and 1 achondrite, and the lead contents were determined by isotope dilution. Model ages for individual meteorites are calculated. We compare the chondrite data with measurements on the single achondrite and with literature data on iron meteorites to seek information on the relative ages of different major classes of meteorites.

105.152 Solar wind xenon in some carbonaceous chondrites.
D. D. Sabu.
Journ. Geophys. Res., Vol. 78, 3245 - 3248 (1973).

Isotopic composition of xenon in several carbonaceous chondrites can be understood in terms of a mixture of solar wind and average carbonaceous chondrite xenon.

105.153 Ages of eight recently fallen meteorites.
S. Smith, E. L. Fireman.
Journ. Geophys. Res., Vol. 78, 3249 - 3259 (1973).
Radioactivities and He, Ne, and Ar stable isotopes were measured in eight recently fallen meteorites, and cosmic-ray exposure ages from several pairs of isotopes were obtained.

105.154 Tektite ablation: Some confirming calculations.
J. A. O'Keefe III, E. W. Adams, J. D. Warmbrod, A. D. Silver, W. S. Cameron.
Journ. Geophys. Res., Vol. 78, 3491 - 3496 (1973). – Brief report.

105.155 Chemical and microprobe investigations of the Allende-meteorite.
H. Malissa, Jr., F. Hermann, F. Kluger, W. Kiesl.
Mikrochim. Acta, No. 3, p. 434 - 450 (1972).

105.156 Relativkonduktometrische, simultane Kohlenstoff- und Schwefelbestimmung in Meteoriten.
H. Malissa, M. Grasserbauer, E. Waldmann.
Mikrochim. Acta, No. 3, p. 455 - 458 (1972).

105.157 Stratigraphical evidence for the terrestrial age of australites. J. F. Lovering, B. Mason, G. E. Williams, D. H. McColl.
Journ. Geol. Soc. Australia, Vol. 18, 409 - 418 (1972).

105.158 The Popigay meteorite crater.
V. L. Masajtis, M. V. Mikhajlov, T. V. Selivanovskaya.
International Geol. Rev., (USA), Vol. 14, 327 - 331 (1972).

105.159 The nature and significance of terrestrial impact structures. M. R. Dence.
Contr. Earth Phys. Branch, Ottawa, No. 393, 13 pp. (1972).
Topographic, structural and petrographic characteristics, geophysical properties and styles of deformation distinguish terrestrial impact structures from volcanic and tectonic phenomena. Quaternary craters with associated meteorites are classed as certain impact sites, older structures with petrographic evidence of shock deformation as probable, and less thoroughly studied structures as possible or suspect. Comparisons of data from terrestrial and also lunar impact events with those obtained from cratering and equation-of-state experiments are leading to refinements of the classification and theory of cratering phenomena, and to a better understanding of the response of crustal materials to large-scale shock events.

105.160 Europium anomaly in plagioclase feldspar: experimental results and semiquantitative model.
D. F. Weill, M. J. Drake.
Science, Vol. 180, 1059 - 1060 (1973).
The partition of europium between plagioclase feldspar and magmatic liquid is considered in terms of the distribution coefficients for divalent and trivalent europium. A model equation is derived giving the europium anomaly in plagioclase as a function of temperature and oxygen fugacity.

105.161 Time differences in the formation of meteorites as determined from the ratio of lead-207 to lead-206.
M. Tatsumoto, R. J. Knight, C. J. Allegre.
Science, Vol. 180, 1279 - 1283 (1973).
Measurements of the lead isotopic composition and the uranium, thorium, and lead concentrations in meteorites were made in order to obtain more precise radiometric ages of these members of the solar system.

105.162 Geophysical investigation of the Versailles, Kentucky, astrobleme. C. R. Seeger.

Bull. Geol. Soc. America, Vol. 83, 3515 - 3518 (1972).

105.163 Forest Vale meteorite.
A. F. Noonan, K. A. Fredriksson, J. Nelen.
Smithsonian Contr. Earth Sci., No. 9, p. 57 - 64 (1972).

105.164 The Estherville meteorite.
J. Nelen, B. Mason.
Smithsonian Contr. Earth Sci., No. 9, p. 55 - 56 (1972).

105.165 Iron meteorite compositions.
R. S. Clarke, Jr., E. Jarosewich.
Smithsonian Contr. Earth Sci., No. 9, p. 65 - 66 (1972).

105.166 The Nejo, Ethiopia, meteorite.
R. S. Clarke, Jr., E. Jarosewich, J. Nelen.
Smithsonian Contr. Earth Sci., No. 9, p. 67 - 68 (1972).

105.167 The Nakhon Pathom meteorite.
J. A. Nelen, K. Fredriksson.
Smithsonian Contr. Earth Sci., No. 9, p. 69 - 74 (1972).

105.168 Evaluation de la perte de masse dans l'atmosphère de la pallasite de Krasnojarsk (Sibérie) (trouvée en 1749). Y. Cantelaube.
Comptes Rendus Acad. Sci. Paris, Sér. D, Vol. 276, 1093 - 1094 (1973).

105.169 Modelling of nuclear reactions in isotropically irradiated thick targets.
A. K. Lavrukhina, G. K. Ustinova, V. V. Malyshev, L. M. Satarova.
Atom. ehnergiya, Vol. 34, No. 1, p. 23 - 28, 58 (1973). In Russian. – Abstr. in Referativ. Zhurn. 51. Astron., 7.51.352 (1973).

Meteorites and their origins. See Abstr. 003.083.

The opaque minerals in stony meteorites. See Abstr. 003.104.

Some comments about the $^{176}Lu-^{176}Hf$ pair. See Abstr. 061.003.

^{247}CM as a short-lived r-process chronometer. See Abstr. 061.035.

Oxygen abundances and the oxygen-silicon relationship in lunar samples and meteorites. See Abstr. 094.144.

Multielement analyses of lunar samples and the degree of oxydation of lunar and meteoritic matter. See Abstr. 094.164.

Multielement analyses and a comparison of the degree of oxydation of lunar and meteoritic matter. See Abstr. 094.238.

Micrometeoroid craters smaller than 100 microns. See Abstr. 094.275.

Luminescence petrography of the Apollo 12 rocks and comparative features in terrestrial rocks and meteorites. See Abstr. 094.325.

Lunar "basalts": some comparisons with terrestrial and meteoritic analogs, and a proposed classification and nomenclature. See Abstr. 094.338.

Volatile and siderophile elements in lunar rocks:

comparison with terrestrial and meteoritic basalts.
See Abstr. 094.374.

Some interelement relationships between lunar
rocks and fines, and stony meteorites.   See Abstr. 094.379.

Meteoritic material in lunar samples: characterization from trace elements.   See Abstr. 094.380.

Rhenium and osmium abundance determinations
and meteoritic contamination levels in Apollo 11 and Apollo
12 lunar samples.   See Abstr. 094.398.

Ultramicroscopic features in micron-sized lunar dust
grains and cosmophysics.   See Abstr. 094.459.

Meteorite impact craters, crater simulations, and the
meteoroid flux in the early solar system.
See Abstr. 094.512.

Influence of target temperature on crater morphology and implications on the origin of craters on lunar glass
spheres.   See Abstr. 094.513.

Solar flares, the lunar surface, and gas-rich meteorites.   See Abstr. 094.518.

Micrometeoroid flux from Surveyor glass surfaces.
See Abstr. 094.525.

Anisotropy of absorption bands in some lunar,
meteoritic, and terrestrial pyroxenes.   See Abstr. 094.578.

Lunar and terrestrial impact crater spherules.
See Abstr. 094.586.

Chondrules in Apollo 14 samples and size analyses
of Apollo 14 and 15 fines.   See Abstr. 094.648.

Chondrules of lunar origin.   See Abstr. 094.650.

Lunar glasses, breccias, and chondrules.
See Abstr. 094.651.

A comparison of noble gases released from lunar
fines (# 15601.64) with noble gases in meteorites and in the
earth.   See Abstr. 094.736.

Track metamorphism in extraterrestrial breccias.
See Abstr. 094.804.

Scanning electron microscope and energy dispersive
X-ray analysis of the surface features of Surveyor 3 television
mirror.   See Abstr. 094.829.

Some present-day aspects of cosmic mineralogy. II.
See Abstr. 094.957.

The establishment of a set of standardized spectral
reflectivity curves for the meteorite classes with applications
to the interpretation of asteroid spectra.
See Abstr. 098.013.

Comparisons of meteorite and asteroid spectral
reflectivities.   See Abstr. 098.014.

Errata

105.901   Corrigendum: 'Evidence for association between Ir
and Al in L chondrites' [Nature, Phys. Sci., Vol.
239, 10 - 11 (1972)].   T. W. Osborn.
Nature, Phys. Sci., Vol. 242, 111 (1973).

106 Interplanetary Matter, Interplanetary Magnetic Field, Zodiacal Light

106.001 On the configuration of interplanetary shock waves produced by powerful chromospheric flares (based on space probes). K. G. Ivanov.
Astron. Zhurn. Akad. Nauk SSSR, Vol. 50, 146 - 149 (1973). In Russian. English translation in Soviet Astron. AJ, Vol. 17, No. 1.
The average head part configuration of shock waves due to powerful chromospheric flares is studied on the base of magnetic and plasma measurements obtained by solar wind sounding and from optical ground observations.

106.002 Further study of the θ component of the interplanetary magnetic field.
R. L. Rosenberg, P. J. Coleman, Jr., N. F. Ness.
Journ. Geophys. Res., Vol. 78, 51 - 58 (1973).
Measurements of the interplanetary magnetic field taken with IMP 3, Pioneer 6, Explorer 34 constitute a large portion of the data available at low and moderate solar activity and provide nearly continuous coverage from mid-1965 through 1966 without radial effects. Study of these observations provides further evidence for the following $B\theta$ effect initially discovered with Mariner 2, 4, and 5.

106.003 Gegenschein observations from Pioneer 10.
M. S. Hanner, J. L. Weinberg.
Sky Telescope, Vol. 45, 217 - 218 (1973).

106.004 Zodiacal light photometry off the ecliptic in quadrature and in opposition with the sun.
R. Dumont, F. Sánchez-Martinez.
Astron. Astrophys., Vol. 22, 321 - 328 (1973). In French.
The following results were obtained from 1964 - 1971 at the Teide Observatory (Tenerife) with a 30 cm-Cassegrain telescope. The zodiacal brightness L_z (in two colours, 4600 and 5020 Å) is expressed in 10th (V) G 2 V stars per square-degree (S_{10}), and was computed from the rough deflection I according to our observational technique (Dumont, 1965 and 1967). We also discuss the possible accuracy of ground-based zodiacal photometry.

106.005 Optical properties of single-component zodiacal light models. R. H. Giese.
Planet. Space Sci., Vol. 21, 513 - 521 (1973).
To provide material for interpretations of forthcoming zodiacal light measurements the characteristics of 468 single-component, in-ecliptic models are summarized in two survey diagrams. The models are based on Mie theory and on a power law for the dependence of the particle number density n on solar distance r and on the size parameter α. The main results are summarized in this paper, especially by two survey diagrams which allow the comparison of the characteristics features simultaneously for the complete field of model parameters.

106.006 Influence du champ magnétique interplanétaire sur les perturbations magnétiques des régions de haute latitude: mise en évidence d'une asymétrie par rapport à la direction terre—soleil. A. Berthelier.
Comptes Rendus Acad. Sci. Paris, Sér. B, Vol. 276, 681 - 684 (1973).
On montre que la composante dans le plan de l'écliptique du champ magnétique interplanétaire responsable des perturbations des régions polaires fait un angle de 10 à 40° Ouest avec la perpendiculaire à la direction terre—soleil. Ce résultat peut être interprété par une asymétrie de la magnétosphère par rapport à la direction terre—soleil.

106.007 Sectorial structure of the interplanetary magnetic field and magnetic disturbances in the polar region.
P. V. Sumaruk, Ya. I. Fel'dshtejn.
Kosmich. Issled., Vol. 11, 155 - 160 (1973). In Russian. Brief information.

106.008 On the fine structure of interplanetary plasma velocities. N. A. Lotova, I. V. Chashej.
Astron. Zhurn. Akad. Nauk SSSR, Vol. 50, 348 - 356 (1973). In Russian. English translation in Soviet Astron. AJ, Vol. 17, No. 2.
The question of the size of diffraction patterns of interplanetary scintillations caused by the spectrum of velocities of non-random inhomogeneities is studied.

106.009 Evidence for an interplanetary source of diffuse HeI, 584 Å radiation.
S. Bowyer, F. Paresce, S. Kumar.
Bull. American Astron. Soc., Vol. 5, 37 - 38 (1973). – Abstr. AAS.

106.010 Earth amidst dust and stones.
B. Yu. Levin, A. N. Simonenko.
Priroda, No. 4.73, p. 7 - 16 (1973). In Russian.

106.011 On the analysis of the observations of interplanetary scintillations obtained with three space receivers.
T. Kakinuma, H. Washimi, M. Kojima.
Publ. Astron. Soc. Japan, Vol. 25, 271 - 280 (1973).
An analytic method of deriving the solar wind velocity from the observations of interplanetary scintillations obtained with three spaced receivers is proposed.

106.012 Structural phenomena in interplanetary medium.
A. V. Bogdanov.
Mezhplanet. sreda i fiz. magnitosfery. Moskva, Nauka, 1972, p. 74 - 88. In Russian. – Abstr. in Referativ. Zhurn. 51. Astron., 4.51.600; 62. Issled. kosmich. prostranstva, 4.62.231 (1973).

106.013 Interaction between the interplanetary medium and the geomagnetosphere. I. V. Kovalevskij.
Probl. kosmich. fiz. Mezhved. nauch. sb., 1972, vyp. (No.) 7, p. 23 - 60. In Russian. – Abstr. in Referativ. Zhurn. 62. Issled. kosmich. prostranstva, 4.62.253 (1973).

106.014 Interplanetary magnetic field and geomagnetic Dst variations. V. L. Patel, U. D. Desai.
Astrophys. Space Sci., Vol. 20, 431 - 437 (1973).
The interplanetary magnetic field has been shown to influence the ring current field represented by Dst. Explorer 28 hourly magnetic field observations have been used with the hourly Dst values. The moderate geomagnetic storms of 60 γ and quiet-time fluctuations of 10−30 γ are correlated with the north to south change of the interplanetary field component perpendicular to the ecliptic. This change in the interplanetary field occurs one to three hours earlier than the corresponding change in the Dst field.

106.015 Magnetotail response to sudden changes in the interplanetary magnetic field.

A. Nishida, N. Nagayama.
Astrophys. Space Sci., Vol. 20, 459 - 472 (1973).

The effect of southward or northward changes in the interplanetary magnetic field is examined statistically in the nightside magnetosphere over the range of 6.6 to 80 R_E from the earth. After southward changes, the deformation of the magnetosphere toward a greater anti-sunward extension of field lines occurs at 6.6 R_E with $\lesssim$ 10 min delay and spreads down the tail to 80 R_E in $\lesssim$ 30 min. Around the onset of the field-line collapse that occurs 1–2 hr later, the southward-directed field is observed briefly in the distant tail. The effect of northward changes could not be recognized in the lobe region of the tail.

106.016 The Pioneer 9 electric field experiment. III. Radial gradients and storm observations.
F. L. Scarf, I. M. Green, J. S. Burgess.
Astrophys. Space Sci., Vol. 20, 499 - 507 (1973).

We present a detailed analysis of the Pioneer 9 VLF electric field observations for 20 selected storm periods covering a heliocentric range extending from 0.754 AU to 0.99 AU. Although data from only two low frequency channels are available, the results of the present study tend to confirm the preliminary speculation by Scarf and Siscoe (1971) that the turbulent E-field spectrum in the disturbed solar wind has a significant radial gradient.

106.017 Alfvén wave refraction by interplanetary inhomogeneities. W. D. Daily.
Journ. Geophys. Res., Vol. 78, 2043 - 2053 (1973).

Pioneer 6 magnetic data reveal that the propagation direction of Alfvén waves in the interplanetary medium is strongly oriented along the ambient field.

106.018 Identification of interplanetary tangential and rotational discontinuities. E. J. Smith.
Journ. Geophys. Res., Vol. 78, 2054 - 2063 (1973).

Discontinuities in the interplanetary magnetic field detected by the Mariner 5 magnetometer were analyzed to determine the magnetic field component normal to the surface of the discontinuity. The existence or absence of a normal field component distinguishes between tangential and rotational discontinuities.

106.019 Observed properties of interplanetary rotational discontinuities. E. J. Smith.
Journ. Geophys. Res., Vol. 78, 2088 - 2093 (1973).

In a prior study, discontinuities were identified as tangential or rotational principally on the basis of the existence or absence of a magnetic field component perpendicular to the plane of the discontinuity. The observed properties of the rotational discontinuities are described in this paper.

106.020 Messung von interplanetarem Staub auf dem HEOS 2 Satelliten. H.-J. Hoffmann, H. Fechtig.
Max-Planck-Inst. Kernphysik Heidelberg, Jahresbericht 1972, p. 234 - 235 (1973).

106.021 The southern component of the interplanetary magnetic field and magnetospheric substorms.
K. G. Ivanov, N. V. Mikerina.
Geomagn. Aeronom., Vol. 13, 482 - 485 (1973). In Russian.

106.022 Diagnostics of the interplanetary magnetic field by ground data on micropulsations of Pc2–4 type.
A. V. Gulielmi, O. V. Bolshakova.
Geomagn. Aeronom., Vol. 13, 535 - 537 (1973). In Russian. Brief information.

106.023 Evidence for waves and/or turbulence in the vicinity of shocks in space. J. K. Chao.

Cosmic plasma physics. Conference 1971, (see 012.016), p. 113 - 122 (1972).

106.024 Fermi acceleration in interplanetary space.
G. Wibberenz, K. P. Beuermann.
Cosmic plasma physics. Conference 1971, (see 012.016), p. 339 - 348 (1972).

106.025 A photometric study of the counterglow from space.
F. E. Roach, B. Carroll, L. H. Aller, J. R. Roach.
Planet. Space Sci., Vol. 21, 1179 - 1184 (1973).

Photometric observations of the region of the counterglow (Gegenschein) made from OSO-6 are examined. A comparison is made between the photometric gradients as measured from the spacecraft and similar gradients deduced from ground-based observations.

106.026 A photometric perturbation of the counterglow.
F. E. Roach, B. Carroll, J. R. Roach, L. H. Aller.
Planet. Space Sci., Vol. 21, 1185 - 1189 (1973).

A weakening of the radiance of the counterglow in the anti-solar direction relative to the regions 5°–15° away is interpreted as evidence for a cloud of scattering material in the general region of the earth–moon system. Further evidence is indicated, by the relative brightening at −180° in the vicinity of the L4 libration point, that the cloud is significantly denser there than in adjacent locations.

106.027 Observations of the zodiacal light at 2.4 μm.
W. Hofmann, D. Lemke, C. Thum, U. Fahrbach.
Nature, Phys. Sci., Vol. 243, 140 - 141 (1973). − Letter.

106.028 Variability of the Z_{SE}-component of the interplanetary magnetic field at the boundary of the sector structure. P. V. Sumaruk, Ya. I. Feldshtein.
Astron. vestn., Vol. 7, 111 - 112 (1973). In Russian.

106.029 On the relation between the pattern and wind velocities in interplanetary scintillations.
J. R. Jokipii, L. C. Lee.
Astrophys. Journ., Vol. 182, 317 - 319 (1973).

The relation between the pattern and wind velocities in interplanetary scintillations is reexamined, and some previous results are corrected. It is demonstrated that there is not a unique, rigid motion of the pattern and that care must be used in defining the velocity.

106.030 Ergebnisse eines Raketenexperimentes zur Messung des Zodiakallichts bei kleinen Elongationen.
C. Leinert, H. Link, E. Pitz, R.-H. Giese.
Mitt. Astron. Ges., No. 32, p. 161 - 162 (1973). − Abstract.

106.031 Mikrowellen-Analogieversuche und Lichtstreuung an kosmischem Staub. R.-H. Giese.
Mitt. Astron. Ges., No. 32, p. 162 (1973). − Abstract.

106.032 Extraterrestrische Ultraviolettstrahlung und die Parameter des sonnennahen H I - Mediums.
H. J. Fahr, G. Lay.
Mitt. Astron. Ges., No. 32, p. 198 - 202 (1973).

106.033 Interplanetary protons and geomagnetic disturbances. O. A. Troshichev, Ya. I. Fel'dshtejn.
Kosmich. Issled., Vol. 11, 486 - 488 (1973). In Russian. Brief information.

106.034 Single-component zodiacal light models.
R. H. Giese.
Dudley Obs., *Albany, N.Y.*, Rep. No. 7, 3 + 52 pp. (1972).

To provide material for interpretations of forthcoming

zodiacal light measurements the characteristics of 468 single-component, in-ecliptic models are presented as a catalog and summarized in two survey diagrams. The models are based on Mie theory (scattering by spherical particles) and on a power law $dn \sim r^{-\nu} \alpha^{-k} \, d\alpha$ for the dependence of the particle number density n on solar distance r and on the size parameter α (circumference/wavelength).

106.035 **Particle acceleration by shock waves in the solar system.** K. A. Anderson.
IV Leningr. mezhdunar. seminar "Edinoobrazie uskoreniya chastits v razlich. masshtabakh kosmosa, 1972". Leningrad, 1972, p. 199 - 211. − Abstr. in Referativ. Zhurn. 62. Issled. kosmich. prostranstva, 7.62.199 (1973).

106.036 **Motion of charged particles in arbitrary magnetic fields in interplanetary space.**
L. I. Dorman, M. E. Kats.
IV Leningr. mezhdunar. seminar "Edinoobrazie uskoreniya chastits v razlich. masshtabakh kosmosa, 1972". Leningrad, 1972, p. 237 - 246. In Russian. − Abstr. in Referativ. Zhurn. 62. Issled. kosmich. prostranstva, 7.62.200 (1973).

Interplanetary medium and physics of the magneto-sphere. See Abstr. 003.135.

Stellar atmospheres and interplanetary plasma. Technical details of radio astronomical reception. See Abstr. 003.138.

The interplanetary medium and physics of the magnetosphere. See Abstr. 003.139.

Propagation and damping of magnetohydrodynamic waves in interplanetary space. See Abstr. 062.068.

Circular polarization by single scattering of unpolarized light from loss-less, non-spherical particles. See Abstr. 063.041.

On the variation of the coronal λ5303 intensity relative to the interplanetary and solar magnetic sector structure, and to geomagnetic activity. See Abstr. 074.021.

Observations of the F corona and inner zodiacal light during the 1972 July 10 total solar eclipse. See Abstr. 074.024.

Test for detection of fine structure of the solar wind velocity. See Abstr. 074.036.

Nonlinear model of high-speed solar wind streams. See Abstr. 074.046.

Power spectrum of density irregularities in the solar wind plasma. See Abstr. 074.047.

Prediction of coronal and interplanetary magnetic fields. See Abstr. 074.049.

The propagation of Alfvén waves and their directional anisotropy in the solar wind. See Abstr. 074.052.

Close connexion between flare-generated coronal and interplanetary shock waves. See Abstr. 074.056.

Anomalously low proton temperatures in the solar wind following interplanetary shock waves − evidence for magnetic bottles? See Abstr. 074.069.

Evolution of large-scale solar wind structures

beyond 1 AU. See Abstr. 074.072.

On the role of fluctuations in the interplanetary magnetic field on heat conduction in the solar wind. See Abstr. 074.093.

Propagation anisotropies of solar flare protons and electrons at low energies in interplanetary space. See Abstr. 078.003.

Pitch angle distribution of solar flare particles in interplanetary space. See Abstr. 078.004.

Numerical studies of the transport of solar protons in interplanetary space. See Abstr. 078.009.

Low-energy protons of solar origin and investigation of the interplanetary medium. See Abstr. 078.039.

Radial variation of magnetic fluctuations and the cosmic-ray diffusion tensor in the solar wind. See Abstr. 078.043.

Scattering of particles in the interplanetary space and properties of solar corpuscular streams. See Abstr. 078.049.

On solar cosmic-ray propagation in interplanetary medium. See Abstr. 078.052.

Enhancement of upper atmospheric sodium from sporadic dust influxes. See Abstr. 082.089.

Semiannual variation of geomagnetic activity. See Abstr. 084.201.

Polar cap magnetic variations and their relationship with the interplanetary magnetic sector structure. See Abstr. 084.252.

Correspondence of solar field sector direction and polar cap geomagnetic field changes for 1965. See Abstr. 084.253.

Periodic variations in geomagnetic activity and sector structure of the interplanetary magnetic field. See Abstr. 084.267.

Magnetic fields near the moon. See Abstr. 094.763.

Lunar microcraters and interplanetary dust. See Abstr. 094.796.

Orientation of interstellar and interplanetary grains. See Abstr. 131.002.

Kosmischer Staub in Astronomie und Weltraumforschung. See Abstr. 131.035.

Interplanetary-scintillation observations of 203 sources identified as radio galaxies or quasars. See Abstr. 141.099.

Analysis of the non-Gaussian spectra of interplanetary scintillations. See Abstr. 141.125.

The Forbush predecrease. See Abstr. 143.013.

Energy losses of galactic cosmic rays in the interplanetary medium. See Abstr. 143.031.

The direct and reverse problems of cosmic ray propagation in interplanetary space.   See Abstr. 143.033.

Interplanetary scintillations of cosmic rays. See Abstr. 143.036.

Cosmic ray electrons from 0.2 to 8 Mev: Pioneer 8 and 9 measurements of their spectrum, time variations, and interplanetary radial gradient.   See Abstr. 143.038.

Interplanetary radial gradients of galactic cosmic ray protons and helium nuclei: Pioneer 8 and 9 measurements from 0.75 to 1.10 AU.   See Abstr. 143.043.

The interplanetary conditions associated with cosmic ray Forbush decreases.   See Abstr. 143.047.

The diurnal effect of cosmic rays and its dependence on the interplanetary magnetic field. See Abstr. 143.048.

Forbush decreases and their relation with solar activity and the parameters of interplanetary matter. See Abstr. 143.078.

Interplanetary shock waves and cosmic rays. See Abstr. 143.088.

107 Cosmogony of the Planetary System

107.001 On the accretion mechanism for the formation of a
        protoplanetary disc.   C. Aust, M. M. Woolfson.
Monthly Notices Roy. Astron. Soc., Vol. 161, 7 - 13 (1973).
    The accretion mechanism for the formation of a protoplanetary disc as proposed by Lyttleton has been examined. It is shown that a cloud of the original dimensions would be gravitationally distorted and eventually disrupted by the approach of the sun. Encounters between the sun and larger and/or denser clouds in which the degree of distortion is greatly reduced have been investigated over ranges of cloud parameter values. It is found that amounts of material satisfying the mass and angular momentum requirements of a protoplanetary disc may still be captured by the accretion process. The similarity between this theory and the capture theory for planetary formation is suggested.

107.002 High-temperature protoplanetary processes. (On
        the formation of metallic cores of the planets).
A. P. Vinogradov.
1-j Mezhdunar. geokhim. kongr., 1971. Tom (Vol.) 1.
Moskva, 1972, p. 11 - 36. In Russian.

107.003 Formation of the outer planets.
        A. G. W. Cameron.
Space Sci. Rev., Vol. 14, 383 - 391 (1973).
    A discussion is given of a number of physical processes which were probably important during the formation of the outer planets if these formed from a gaseous solar nebula in which magnetic effects were not important. Arguments are given that large-scale gravitational instabilities in the solar nebula did not occur. Conditions which may lead to the formation of the regular satellite systems are discussed, and the associated problem of removal of primordial angular momentum from Jupiter, Saturn, and Uranus.

107.004 Fractionation of refractory elements in the solar
        nebula.   L. Grossman.
Meteoritics, Vol. 8, 41 - 42 (1973). – Abstract.

107.005 Heterogeneous accumulation revisited.
        K. K. Turekian, S. P. Clark, A. Davis, L. Grossman.
Meteoritics, Vol. 8, 77 - 78 (1973). – Abstract.

107.006 Numerical models of the primitive solar nebula.

A. G. W. Cameron, M. R. Pine.
Icarus, Vol. 18, 377 - 406 (1973).
    Numerical models have been constructed to represent probable conditions in the primitive solar nebula. A two solar mass fragment of a collapsing interstellar gas cloud has been represented by a uniformly rotating sphere. Two cases have been considered: one in which the internal density of the sphere is uniform and the other in which the density falls linearly from a central value to zero at the surface (the uniform and linear models). These assumptions served to define the distribution of angular momentum per unit mass with mass fraction. The radial pressure gradient of the gas was included in the force balance.

107.007 Accumulation processes in the primitive solar nebula.
        A. G. W. Cameron.
Icarus, Vol. 18, 407 - 450 (1973).
    Particle accumulation processes are discussed for a variety of physical environments, ranging from the collapse phase of an interstellar cloud to the different parts of the models of the primitive solar nebula constructed by Cameron and Pine.

107.008 On the process of accretion in the formation of the
        planets and comets.   J. G. Hills.
Icarus, Vol. 18, 505 - 522 (1973).
    The physically reasonable assumption is made that the seed bodies which initiated the accretion of the individual asteroids, planets, and comets are formed by stochastic processes.

107.009 Evolutionary processes in the solar system.
        G. Colombo.
Phys. Earth Planet. Interiors, Vol. 7, 1 - 8 (1973). – Frontiers lecture given at the First European Earth and Planetary Physics Colloquium, held at Reading University, 1971.

107.010 From cosmic dust to problems of cosmogony.
        F. Singer.
Priroda, No. 4.73, p. 17 - 19 (1973). In Russian.

107.011 Effect of the solar magnetic field and corpuscular
        emission on the evolution of the solar system.
K. P. Butusov.
Trudy Leningr. in-t aviats. priborostr., 1972, vyp. (No.) 75, p.

92 - 96. In Russian. — Abstr. in Referativ. Zhurn. 51. Astron., 4.51.394 (1973).

107.012 L'âge de formation des planètes et des météorites. C. J. Allègre.
Stellar ages. Proc. IAU Colloquium No. 17, (see 012.015), XXXVI, 1 - 5 (1973).

107.013 Structure and evolutionary history of the solar system, III. H. Alfvén, G. Arrhenius.
Astrophys. Space Sci., Vol. 21, 117 - 176 (1973).

The present part (Part III) deals primarily with the plasma processes and the hydromagnetic aspects. A plasma surrounding a rotating central body may attain a state of partial corotation which is determined by the balance between gravitation and the centrifugal force acting on a plasma in a dipole field. Condensation from a partially corotating plasma results in grains orbiting in ellipses with $e = 1/3$ and finally accreting to bodies at $2/3$ of the central distance of the point of condensation. An application of the theory to the Saturnian rings and to the asteroidal belt shows that the fall-down ratio $2/3$ (derived from the geometry of a dipole field) is essential for the understanding of their structure. The structure of the groups of planets and satellites is also discussed but only in a preliminary way.

107.014 Deuterium in the early solar system. D. C. Black.
Icarus, Vol. 19, 154 - 159 (1973).

An attempt is made to construct a self-consistent picture of the deuterium abundance in the early solar system based on the assumption of chemical equilibrium in the solar nebula.

107.015 The end of the iron-core age. R. A. Lyttleton.
The Moon, Vol. 7, 422 - 439 (1973). — Paper dedicated to Professor Harold C. Urey on the occasion of his 80th birthday on 29 April, 1973.

The terrestrial planets aggregated essentially from small particles, to begin as solid cool bodies with the same general compositions, and there is no possibility of an iron-core developing within any of them at any stage. Their differing internal and surface properties receive ready explanation from their different masses which determine whether the pressures within are sufficient to bring about phase-changes.

107.016 Sulla teoria di Hoyle della formazione del sistema solare. N. Dallaporta, L. Secco.
Mem. Soc. Astron. Italiana, Nuova Ser., Vol. 43, 705 - 713 (1973).

A quantitative formulation of the general scheme of Hoyle's theory on the formation of the solar system is attempted. The main point consists in assuming that the original solar magnetic field of the sun is a dipole field. This allows to calculate the total quantity of matter emitted at the solar equator, and the distance reached by the ring thus formed, as functions of the radius of the sun in gravitational contraction along the Hayashi track, and of the initial values of the magnetic field and of the total angular momentum. Preliminary results are presented.

107.017 Comets and the formation of planets. E. J. Öpik.
Astrophys. Space Sci., Vol. 21, 307 - 398 (1973). Paper dedicated to H. C. Urey on the occasion of his 80th birthday on 29 April, 1973.

A morphological study of the physical and dynamical processes of planet formation is presented, with emphasis on the intermediary role of comet nuclei. The items evaluated physically, dynamically, or statistically comprise: (1) the total number mass of comets in Oort's cloud; (2) a re-evaluation of the diameters and masses of cometary nuclei; (3) the processes of nucleation from gravitational and 'Boltzmann' instabilities

of gaseous media to agglomerations of particulate matter as conditioned by inbuilt angular momentum; (4) the statistical-dynamical conditions and time scales of orbital interaction of comets with the planets and the consequences of disintegration. A consistent model proposes the formation of comets and planets in pre-planetary rings of the residual solar nebula, with subsequent ejection, chiefly by Jupiter, of the comets to Oort's sphere.

107.018 Neue Gesichtspunkte zur Entstehung des Planeten-systems. L. Biermann.
Umschau, 73. Jahrgang, p. 375 - 376 (1973).

Some new astronomical developments are described, which seem to be of interest in view of the gap which exists between those informations on the origin of the planetary system derived from astronomical observations and those which are based on research on meteorites and geophysics.

107.019 Cosmogony of planets. B. Yu. Levin.
Physics of the moon and planets, (see 012.024), p. 209 - 219 (1972). In Russian.

107.020 Peculiarities of growth of planetary nuclei. V. S. Safronov, E. V. Zvyagina.
Physics of the moon and planets, (see 012.024), p. 219 - 223 (1972). In Russian.

107.021 On the origin of the solar system. D. ter Haar.
Nederl. Tijdschr. Natuurk., Vol. 39, No. 4, p. 45 - 49 (1973).

The author discusses some of the features which any successful theory of the origin of our solar system should explain and considers some recent theories which have tried to reach such an explanation.

107.022 Contribution to Kant and Laplace's nebular hypothesis. II. A. J. Rutgers.
Proc. Koninkl. Nederl. Akad. Wet., Ser. B, Vol. 76, 35 - 36 (1973).

The transportation of angular momentum in a slowly rotating, contracting sphere of particles obeying Maxwell's distribution law is considered again, this time introducing the resultant particle acceleration towards the centre, equal to $fM/R^2 - v^2/R$. The result is in agreement with the earlier result, a slight outward flow of angular momentum.

107.023 On jetstreams (formation of planetary systems). J. Trulsen.
Phys. Norvegica, Vol. 6, 210 (1972). — See Phys. Abstr., Vol. 76, No. 42124 (1973).

^{247}CM as a short-lived r-process chronometer.
See Abstr. 061.035.

A new cosmological model: Formation of organic molecules, planets, and comets. See Abstr. 065.068.

Potential atmospheric composition of smaller bodies in the solar system and some aspects of planetary evolution. See Abstr. 091.003.

The abundance of CH_3D and the D/H ratio in Jupiter. See Abstr. 099.004.

Ages of the Allende chondrules and inclusions: Do they set a time scale for the formation of the solar system? See Abstr. 105.099.

Extrasolar planetary systems. See Abstr. 117.021.

Stars

111 Stellar Parallaxes

**111.001 First results of the Herstmonceux parallax pro-
gramme.** D. V. Thomas.
Monthly Notices Roy. Astron. Soc., Vol. 161, 335 - 346
(1973).

Parallaxes of 24 stars have been determined from observations made with the 26-inch refractor at Herstmonceux. The
results are in the mean two to five times more accurate, depending on right ascension, than the parallaxes determined
with the same telescope at Greenwich. Twenty of the stars are
in Gliese's catalogue of nearby stars, mostly without trigonometric parallaxes.

**111.002 The parallax and proper motion of the carbon star
X Cancri.** A. R. Upgren.
Astrophys. Journ., (*Letters*), Vol. 179, L121 - L122 (1973).

A new solution for the parallax and proper motion of the
carbon star X Cnc has been made using new measures of the
same plates from which the only existing parallax value was
determined. The former values of $0\rlap{.}''048 \pm 0\rlap{.}''019$ (m.e.) for
the absolute parallax and $0\rlap{.}''436 \pm 0\rlap{.}''027$ (m.e.) for the annual
proper motion are in error and are superseded by values of
$0\rlap{.}''011 \pm 0\rlap{.}''008$ (m.e.) and $0\rlap{.}''026 \pm 0\rlap{.}''014$ (m.e.), respectively.

**111.003 The correlation of parallax errors and the intrinsic
dispersion of the K3–M2 main sequence.**
A. R. Upgren.
Astron. Journ., Vol. 78, 79 - 82 (1973).

The significance of published errors in the trigonometric
parallaxes of K3–M2 dwarfs from several observatories is
evaluated from their correlation with the difference between
the absolute magnitude of a star determined from its parallax,
and that predicted from its color.

**111.004 Parallaxes of 23 stars determined from plates taken
with the McCormick 26-inch refractor.**
L. W. Fredrick, P. A. Ianna.
Astron. Journ., Vol. 78, 93 - 94 (1973).

This 41st list of trigonometric parallaxes by the Leander
McCormick Observatory is a continuation of previous similar
publications in this Journal. Changes in the reduction procedures have been made and are briefly summarized here. Attention is called to the list star number 2134 as an interesting star.
This star has been reported on in detail (Appelbaum 1972).

**111.005 Photographic determinations of the parallaxes of
25 stars with the Thaw refractor.** W. R. Beardsley,
N. E. Wagman, E. Erskine, E. Hubbell, G. Gatewood.
Astron. Journ., Vol. 78, 95 - 96 (1973).

A heterogeneous list of parallax determinations is presented.

**111.006 Trigonometric parallaxes determined with the
Yerkes Observatory 40-inch refractor, II.**
W. F. van Altena, E. U. Vilkki.
Astron. Journ., Vol. 78, 201 - 205 (1973).

Parallaxes and proper motions based on recent observations with the Yerkes Observatory's 40-inch refractor are presented for 13 stars in ten fields.

**111.007 Secular parallaxes of stars and solar velocity in
space derived from absolute proper motions of
14600 stars relative to galaxies.** N. V. Fatchikhin.
Astron. Zhurn. Akad. Nauk SSSR, Vol. 50, 377 - 389 (1973).
In Russian. English translation in Soviet Astron. AJ, Vol. 17,
No. 2.

Solar apex coordinates, secular parallaxes and constants
of galactic rotation for stars with photographic magnitude of
14.6–15.5 have been determined, the large proper motions
being taken into account. The determinations of 14600 stars
were made according to the proper motions relative to galaxies
(the necessary data were obtained at Pulkovo).

111.008 Astrometry and photometry of G95-59.
C. C. Dahn, J. B. Priser.
Astron. Journ., Vol. 78, 253 - 255 (1973).

Plates taken with the 61-inch astrometric reflector yield
a relative parallax of $0\rlap{.}''019 \pm 0\rlap{.}''003$ (m.e.) for the large proper
motion star G95-59. Photometry of this star on both the *UBV*
and *PVI* systems does not confirm the unusually red color
reported previously by Eggen.

111.009 The significance of parallaxes in x and y coordinates.
W. S. Mesrobian, A. R. Upgren.
Bull. American Astron. Soc., Vol. 5, 265 (1973). – Abstr. AAS.

**111.010 Revised parallaxes for ten nearby stars from plates
taken with the Sproul 24-inch refractor.**
S. L. Lippincott.
Astron. Journ., Vol. 78, 426 - 428 (1973).

Observational data and results are given for ten nearby
stars with an average observational interval of three decades.
The probable errors of the parallaxes range from $\pm 0\rlap{.}''002$ to
$\pm 0\rlap{.}''006$. Seven of the stars are being further observed for
evaluation of suspected variable proper motion, or to improve
the parallax.

Astrometric analysis of the field of AC + 65°6955
from plates taken with the Sproul 24-inch refractor.
See Abstr. 041.020.

UBV photometry of large proper motion stars.
See Abstr. 113.057.

**Trigonometric parallax determination for the central
star in the planetary nebula NGC 7293.** See Abstr. 133.027.

112 Proper Motions, Radial Velocities, Space Motions

112.001 Proper motion of BD + 16° 516.
G. D. van Albada.
Astron. Astrophys., Vol. 22, 157 (1973).
A new proper motion is given for the Hyades white dwarf eclipsing binary BD + 16° 516.

112.002 Relative proper motions of faint stars in the Pleiades. B. F. Jones.
Astron. Astrophys., Suppl. Ser., Vol. 9, 313 - 345 = Lick Obs. Bull., No. 629 (1973).
New relative proper motions have been determined for faint stars in the region of the Pleiades, including most suspected members. Membership probabilities have been derived. These have been used to investigate the form of the Pleiades lower main sequence. The results show the Pleiades lower main sequence to be below both the Hyades and the sequence determined by the nearby stars.

112.003 Determination of absolute proper motions of stars relative to galaxies in the SA 32.
A. V. Bolbochanu.
Astron. Zhurn. Akad. Nauk SSSR, Vol. 50, 362 - 370 (1973). In Russian. English translation in Soviet Astron. AJ, Vol. 17, No. 2.
143 galaxies on Pulkovo normal astrograph plates (2° × 2°) including the SA 32 were identified using Zwicky's Catalogue of Galaxies and of Clusters of Galaxies. Absolute proper motions of stars were determined from two plate pairs using 92 reference galaxies ($m_p = 15^m.0$) measured on all the plates. Formulas are derived for comparing the reductions of relative proper motions of stars to absolute proper motions by means of galaxies and reference stars. Statistical corrections for the reduction of relative proper motions to absolute proper motions using the secular parallaxes of Binnendijk and Vasilevskis are computed and the relative proper motions of 17 bright stars compared to the absolute proper motions given in the AGK3.

112.004 The radial velocities of SEG stars and the distance to Scorpius X-1. J. E. Felten, R. M. Humphreys.
Bull. American Astron. Soc., Vol. 5, 24 (1973). − Abstr. AAS.

112.005 Lowell proper motions XVI. Proper Motion Survey of the southern hemisphere with the 13-inch photographic telescope of the Lowell Observatory.
H. L. Giclas, R. Burnham, Jr., N. G. Thomas.
Lowell Obs. Bull., Flagstaff, Arizona, No. 160, Vol. 7, 223 - 300 (1973).

112.006 The radial velocity variations of HD 125823 a Centauri.
A. B. Underhill, D. A. Klinglesmith, H. Frey.
Astron. Astrophys., Vol. 25, 141 - 144 (1973).
Radial velocities are presented from 66 spectrograms of HD 125823 a Centauri obtained on ten consecutive nights. The behaviour of some lines is discussed.

112.007 The radial velocities of SEG stars and the distance to Scorpius X-1. J. E. Felten, R. M. Humphreys.
Astrophys. Journ., Vol. 181, 543 - 546 (1973).
Contrary to an earlier report, the radial velocities of stars SEG 13, 25, and 37 eliminate them as probable members of the Sco-Cen association. The proper-motion argument identifying Sco X-1 as an association member is therefore invalid. A distance $D \approx 0.4-1$ kpc for Sco X-1 seems most acceptable at present.

112.008 Recent determinations of relative proper motions in star clusters. W. F. van Altena.
Stellar ages. Proc. IAU Colloquium No. 17, (see 012.015), VIII, 1 - 5 (1973).

112.009 On determinations of proper motions of stars in ten areas of the sky with open clusters.
A. A. Latypov.
Tsirk. Astron. Inst., Tashkent, No. 34 (381), p. 21 - 28 (1972). In Russian.

112.010 Proper motions of stars in the area of the open cluster NGC 457. A. A. Latypov.
Tsirk. Astron. Inst., Tashkent, No. 37 (384), p. 11 - 28 (1972). In Russian.

112.011 Proper motions of stars in the area of the open cluster NGC 663. A. A. Latypov.
Tsirk. Astron. Inst., Tashkent, No. 39 (386), p. 12 - 37 (1973). In Russian.

112.012 Eigenbewegungen in einem Feld um Alpha Persei.
W. Dieckvoss.
Mitt. Astron. Ges., No. 32, p. 213 (1973). − Abstract.

112.013 Radialgeschwindigkeitsschwankungen des Ap-Sternes HD 224801. W. W. Weiss.
Mitt. Astron. Ges., No. 32, p. 253 - 257 (1973).

112.014 On the possibility of determining stellar radial velocities to 0.01 km s^{-1}. R. and R. Griffin.
Monthly Notices Roy. Astron. Soc., Vol. 162, 243 - 253 (1973).
Dissimilarities in the illumination of spectrographs by star light and by comparison sources, respectively, normally prevent the realization of radial-velocity accuracies anywhere near those which high-resolution spectrographs ought to provide. These difficulies can be entirely circumvented by the use of telluric absorption lines as the stationary comparison source.

112.015 Proper Motion Survey with the forty-eight inch Schmidt telescope. XXXIII. Proper motions for 3478 faint stars. W. J. Luyten.
Separate print Univ. Minnesota, Minneapolis, Minnesota. 32 pp. (1972).
In continuation of No. XXXI of this series the present publication gives data for another 3478 stars. All these data were obtained with the automated-computerized plate scanner and measuring machine. I have included only data for those stars for which no earlier determination of proper motion is available. Of the 3478 stars listed, the motions of 3432 are believed to be new.

Die Eignung des Mikrodensitometers PDS 1000 zum Messen von Radialgeschwindigkeiten. See Abstr. 031.034.

Confronto fra i cataloghi MD e GC ed analisi dei moti propri. See Abstr. 041.017.

Mount Stromlo PZT results. See Abstr. 041.027.

The parallax and proper motion of the carbon star X Cancri. See Abstr. 111.002.

Trigonometric parallaxes determined with the Yer-

kes Observatory 40-inch refractor, II.
See Abstr. 111.006.

On a correlation between the magnitude and the radial velocity of hot stars.   See Abstr. 113.016.

A spectral analysis of κ Piscium, I.
See Abstr. 114.142.

Wide moving pairs among K and M dwarfs.
See Abstr. 117.006.

Improved elements for the Hyades group binary 43
Persei.   See Abstr. 119.006.

Proper motions of nova Per 1901 and the neighbouring stars.   See Abstr. 124.109.

Relative proper motions in the region of the open
cluster NGC 1664.   See Abstr. 153.011.

Another look at the absolute proper motions obtained from the Lick pilot programme.   See Abstr. 155.042.

Luminosity and velocity distribution of high-luminosity red stars near the sun. I. The very young disk population.
See Abstr. 155.099.

113 Stellar Magnitudes, Colors, Photometry

113.001 Infrared photometry of southern Wolf-Rayet stars.
D. A. Allen, F. C. Porter.
Astron. Astrophys., Vol. 22, 159 - 160 (1973).
With the exception of a few WC stars, the nitrogen and carbon sequences of the southern Wolf-Rayet stars can be distinguished by photometry at 1.6 and 2.2μ.

113.002 Three-color photometry of a field in the galactic anticentre section near NGC 1664.
W. Becker, C. Fang.
Astron. Astrophys., Vol. 22, 187 - 194 (1973).
In a field of 0.2 square degree 1743 stars have been measured in the RGU system on Palomar Schmidt plates. The limiting magnitude of the photometry is $G = 18.0$. The color-color diagrams for given intervals of apparent magnitude have been used for the determination of the absolute magnitude and the reddening of main sequence stars. The interstellar reddening is caused by two clouds between 80 pc and 400 pc and between 2.2 kpc and at least 11 kpc. The density gradients for main sequence stars in different luminosity intervals generally are negative. Density gradients are also given for both types of giants separately. The luminosity functions for the distance intervals from 0.5−1.5 kpc and from 1.5−2.0 kpc have lower values, but the same slope as the one for the solar neighbourhood.

113.003 Three-colour photometry in a field in the direction of the galactic anticentre near M 35.
T. Hersperger.
Astron. Astrophys., Vol. 22, 195 - 202 (1973).
2330 stars down to a limiting magnitude of $G = 18.2$ have been measured in a field of 0.27 $\Box°$ in the direction to the galactic anticentre in the RGU system. The interstellar reddening is caused by two absorbing clouds, between 150 and 1300 pc and around 7 kpc, absorbing together 1.78 mag in G. The density functions for the main sequence stars of given intervals in absolute magnitude as well as for the late-type giants are all monotonously decreasing. The luminosity functions of the main sequence stars have been determined for distances between 1.0 and 1.4 kpc and between 1.4 and 2.4 kpc.

113.004 Application of the theory of radiative transfer to stellar images in the photographic emulsion.
W. Seboldt.
Astron. Astrophys., Vol. 22, 217 - 227 (1973). In German.
The theory of radiative transfer is used to derive a theoretical relationship between the brightness of a star and its effective radius in the developed photographic emulsion. The latter is obtained with a variable iris diaphragm photometer. Account is taken of the thickness of the photographic emulsion and the anisotropic multiple scattering of light from the silver halide crystals in the emulsion during the exposure.

113.005 Infra-red photometry of R Coronae Borealis type variables and related objects.
M. W. Feast, I. S. Glass.
Monthly Notices Roy. Astron. Soc., Vol. 161, 293 - 303 (1973).
Infra-red photometry $(J, H, K. L)$ is given for 12 R CrB stars, three HdC stars and two helium stars. N, Q observations are given for six of the R CrB stars. Most of the R CrB stars show infra-red excesses. Although the HdC and helium stars are related spectroscopically to the R CrB stars they do not show IR excesses. Recently proposed models for the R CrB phenomenon are considered. Ejection of particles at minima rather than geometrical eclipses by circumstellar dust blotches

seems required by the observations.

113.006 Recent photometric variability of HD 184279.
C. C. Dahn, H. H. Guetter.
Astrophys. Journ., Vol. 179, 551 - 554 (1973).
Photometric observations made during the fall of 1971 indicate that HD 184279, a UBV standard and a known Be star, was then several tenths of a magnitude fainter than the standard values. Six-color and UBV observations are presented which show the wavelength dependence of this decrease in brightness.

113.007 Near infra-red magnitudes of 248 early-type emission-line stars and related objects. D. A. Allen.
Monthly Notices Roy. Astron. Soc., Vol. 161, 145 - 166 (1973).
Nearly 250 early-type emission-line stars have been observed in the near infra-red; approximately two thirds of these are found to exhibit excess radiation at these wavelengths. Two distinct mechanisms may contribute to the infra-red excesses: thermal radiation from circumstellar dust, and electron bremsstrahlung originating in a shell of ionized gas. It is not currently possible to distinguish between these two alternatives using infra-red photometry alone.

113.008 A rectification of five-colour photometry of LMC supergiants. A. M. van Genderen.
Astron. Astrophys., Vol. 22, 467 - 468 (1973).
It appeared that among the nearly 70 LMC supergiants photo-electrically observed by the author in the Walraven five-colour system, ten non-members, mostly with high $V−B$ indexes, were involved. Upon removing these non-members, the apparent differences between LMC and galactic super-giants largely disappear, although a complete resemblance is not yet proven.

113.009 Population II giants in the DDO photometric system. W. Osborn.
Astron. Astrophys., Vol. 23, 151 - 152 (1973).
Five extremely metal poor population II giants have been observed with the David Dunlap Observatory intermediate-band photometric system. It is shown that such stars can be easily recognized by DDO photometry. The effective temperatures and surface gravities of the stars have been derived from the observed colors.

113.010 Photoelectric UBV photometry in four areas of intermediate-high galactic latitude.
A. Ardeberg, K. Särg, S. Wramdemark.
Astron. Astrophys. Suppl. Ser., Vol. 9, 163 - 181 (1973).
Results of photoelectric UBV photometry for a total of 127 stars are presented. The stars observed are situated in four fields with galactic latitudes between +55° and +59°. Preliminary analysis of the results shows the stars to be predominantly of spectral type F and later. Giant stars and pop. II stars seem to be well represented. The interstellar absorption is generally quite weak.

113.011 Four-color observations of early-type stars. I. Standards and secondary standards.
A. G. D. Philip, K. D. Philip.
Astrophys. Journ., Vol. 179, 855 - 861 (1973).
Four-color observations have been made of 39 standard and 42 secondary standard stars of spectral types O9 to K3. The zero point errors of the indices $b − y$, c_1, and m_1 are + 0.001, − 0.002, and 0.000 mag, respectively, relative to the

published values of Crawford and Barnes. Most of the secondary standards are fainter than $V = 6.0$ mag and are intended for use with large telescopes.

113.012 **The interstellar reddening law in the ultraviolet deduced from filter photometry obtained by the OAO-2 satellite.** M. Laget.
Astrophys. Journ., Vol. 180, 61 - 70 (1973).

Filter photometry has been obtained of 16 B0 stars at 10 effective wavelengths in the range $\lambda\lambda 4250-1430$ Å. The wavelength dependence of the interstellar reddening law, deduced from a least-squares fit of the observed values to the reddening line at each band, is found in satisfactory agreement with that derived by Bless and Savage. Toward the shorter wavelengths the increase of the computed probable error of the slope of the mean reddening line suggests that large fluctuations in the law may occur from star to star.

113.013 **A search for faint blue objects near the north galactic pole.** D. Weistrop.
Astron. Astrophys., Vol. 23, 215 - 219 (1973).

Thirteen faint blue objects have been found near the north galactic pole. Three show significant variations in magnitude over a four-year period, and may be QSO's. Two have been identified as white dwarfs or subdwarfs by other authors, and one is most probably a white dwarf or subdwarf.

113.014 **Error analysis of the Photoelectric Catalogue.** M. P. FitzGerald.
Astron. Astrophys., Suppl. Ser., Vol. 9, 297 - 311 = Contr. Univ. Waterloo Obs., No. 18 (1973).

The data contained in the Photoelectric Catalogue has been analyzed in order to ascertain the internal consistency of the observations, detect transcribing errors, and detect possible variable stars. Tables of the deviations by individual reference number and by spectral class are also shown.

113.015 **RCW 117 and DR 15 observed in the far infrared.** J. P. Emerson, R. E. Jennings, A. F. M. Moorwood.
Nature, Phys. Sci., Vol. 241, 108 - 109 (1973).

Broadband measurements from 40 μm to 350 μm of a number of far infrared sources were made from balloon altitudes during August and September 1972, on flights carried out at the National Center for Atmospheric Research Scientific Balloon Flight Station in Texas. Here we report the discovery of two new far infrared objects associated with the HII regions RCW 117 and DR 15.

113.016 **On a correlation between the magnitude and the radial velocity of hot stars.** J. Lefèvre.
Astron. Astrophys., Vol. 24, 99 - 101 (1973).

The difference between mean apparent magnitudes for B stars with a positive particular radial velocity and for B stars with a negative one is shown to be due to the nearby stars and does not prove the existence of dust in front of hot stars.

113.017 **Faint O−B3 stars in the Centaurus section of the Milky Way.** C. C. McCarthy, E. W. Miller.
Astron. Journ., Vol. 78, 33 - 36, 155 (1973).

Photographic and transmission grating plates have been combined in a search for faint blue stars in Centaurus near $l = 307°.1$, $b = −1°.3$ in an area covering 1350 sq min of arc. Nineteen stars with V magnitudes between 10.0 and 14.4 and with spectral types B3 and earlier have been detected. The region contains no obvious concentrations of OB stars and is probably not located in a major spiral feature.

113.018 **Photoelectric UBV photometry of late-type stars in two regions at high galactic latitude.** J. S. Drilling.
Astron. Journ., Vol. 78, 44 - 46 = Contr. Louisiana State Univ. Obs., *Baton Rouge* (1973).

Photoelectric UBV photometry and objective-prism spectral types have been determined for 62 stars brighter than $V = 15$ in two regions at high galactic latitude. Most of these stars have spectral classes later than A7, and their spectral types and colors are consistent with the view that they are primarily normal F-type stars and G−K giants showing little or no interstellar reddening.

113.019 **$uvby$ and Hβ photometry of early-type stars in a region at high galactic latitude.**
J. S. Drilling, P. Pesch.
Astron. Journ., Vol. 78, 47 - 52, 157 - 158 = Contr. Louisiana State Univ. Obs., *Baton Rouge*, No. 77 (1973).

$uvby$ photometry has been obtained for 85 stars near $l = 179°$, $b = −47°$ which are of spectral class A7 or earlier and brighter than $V = 14.6$ arccording to the finding list of Philip and Drilling. Hβ photometry and slit spectra have been obtained for a number of the brighter stars. The method of quantitative spectral classification developed by Strömgren is applied to the stars for which Hβ photometry was obtained.

113.020 **Two methods for computing monochromatic extinction from BV measurements.** B. J. Taylor.
Astron. Journ., Vol. 78, 61 - 66 (1973).

Two computational methods for obtaining reference wavelengths and monochromatic extinction at those wavelengths from BV observations are presented and discussed, together with calculated parameters necessary for their use. Both methods are applications of the second-order extinction equation derived by King; each is applicable to a particular way of observing for extinction.

113.021 **A finding list of faint UV-bright stars in the galactic plane.** H. H. Lanning.
Publ. Astron. Soc. Pacific, Vol. 85, 70 - 84 (1973).

Eighty-two UV-bright stars have been found on 17 two-color 48-inch Schmidt plates centered on the galactic plane, and on one high-latitude plate. The sources are divided into three categories according to their apparent ($U−B$) values. Some of the most prominent sources are discussed, and finding charts are included for all sources listed.

113.022 **UBV photometry of selected Ap stars.** J. Huchra, S. P. Willner.
Publ. Astron. Soc. Pacific, Vol. 85, 85 - 86 (1973).

UBV photometry is presented for 24 Ap stars which either were previously unmeasured or are suspected to be variable.

113.023 **$uvby$ photometry of Am and Ap stars.** W. H. Warren, Jr.
Astron. Journ., Vol. 78, 192 - 199 = Publ. Goethe Link Obs., Indiana Univ., *Bloomington*, No. 150 (1973).

Four-color observations are reported for about 270 peculiar and metallic-line A stars. Many of the stars on the program are newly discovered Am and Ap stars classified on objective prism spectra by Bond (1970); two Ba II stars in this list have also been observed. The (c_1, m_1)-diagrams for Am and Ap stars are discussed and several prospective magnetic stars are listed on the basis of the diagram for Ap stars.

113.024 **Speckle interferometry: Color-dependent limb darkening evidenced on Alpha and Omicron Ceti.**
D. Bonneau, A. Labeyrie.
Astrophys. Journ., (*Letters*), Vol. 181, L1 - L4 (1973).

Speckle-interferometry images of α Orionis and o Ceti, recorded with a television system on the 200-inch telescope, were studied visually as a preliminary reduction procedure. Both stellar disks have a limb-darkened profile which is appreciably wider in blue than in yellow and red light, suggesting a contribution of scattering phenomena in chromospheric opacity.

113.025 Four-color photometry of blue horizontal-branch stars in NGC 6809. II.   A. G. D. Philip.
Bull. American Astron. Soc., Vol. 5, 14 (1973). – Abstr. AAS.

113.026 Atmospheric extinction – effects of variations.   R. G. Roosen.
Bull. American Astron. Soc., Vol. 5, 25 (1973). – Abstr. AAS.

113.027 An extreme-ultraviolet survey of the northern sky.   M. Lampton, P. Henry, R. Cruddace, F. Paresce, S. Bowyer.
Bull. American Astron. Soc., Vol. 5, 34 (1973). – Abstr. AAS.

113.028 Mariner 9 star photography.   T. E. Thorpe.
Applied Optics, Vol. 12, 359 - 363 (1973).
  Successful photography of stars by the Mariner 9 spacecraft has confirmed both mathematical prediction of point source response by vidicons and preflight calibration results. Camera-B limiting magnitude and integrated image data are presented to provide information relevant to absolute photometric reduction. The effect of image motion on star detection thresholds is also discussed.

113.029 The brightest infrared sources.   G. Neugebauer, E. E. Becklin.
Sci. American, Vol. 228, No. 4, p. 28 - 40 (1973).
  Certain celestial objects radiate thousands of times more energy at infrared wavelengths that the sun does at all wavelengths. They evidently include stars being born and the debris of dying stars.

113.030 Far-ultraviolet photometry of stars in Scorpius and Ophiuchus.   G. R. Carruthers, K. L. Bromberg.
Bull. American Astron. Soc., Vol. 5, 38 - 39 (1973). – Abstr. AAS.

113.031 Ultraviolet photometry of metal rich G dwarf stars.   D. M. Gottlieb.
Bull. American Astron. Soc., Vol. 5, 39 (1973). – Abstr. AAS.

113.032 On the *UBVRI* colours of strong-line G dwarfs and subgiants.   D. Branch, J. B. Alexander.
Monthly Notices Roy. Astron. Soc., Vol. 161, 409 - 420 (1973).
  The broad-band colours of 31 Aql (G8 IV), δ Pav (G8 V), and α Men (G5 V) differ markedly from the colours of Hyades dwarfs of similar temperatures. The colour anomalies of these stars do not appear to be caused by general increases in atomic line strengths. Molecular blanketing has been found to be insufficient and no appropriate source of continuous opacity is known.

113.033 Photoelectric observations of red stars in Cygnus.   K. Voelcker, W. Hofmann, H. Elsässer.
17th Colloquium International Astrophys. Liège 1971, (see 012.003), p. 141 - 143 (1972).

113.034 Infrared emission from stars in the h and χ Persei association.   R. E. Schild.
17th Colloquium International Astrophys. Liège 1971, (see 012.003), p. 295 - 298 (1972).

113.035 Circumstellar shells in the young cluster NGC 2264.   S. E. Strom, K. M. Strom, A. L. Brooke, J. N. Bregman, J. Yost.
17th Colloquium International Astrophys. Liège 1971, (see 012.003), p. 299 - 300 (1972).

113.036 Dreifarbenphotographie mit einer Aufnahme.   S. Rössiger.
Sterne, 49. Jahrgang, p. 117 - 121 (1973).

113.037 Intrinsic color indices $(U-V)°$ and $(B-V)°$ for stars of various spectral types.   G. I. Zaitseva, N. S. Komarov.
Astron. Tsirk., No. 747, p. 5 - 7 (1973). In Russian.

113.038 Calibration of direct photographs using brightness profiles of field stars.   J. Kormendy.
Astron. Journ., Vol. 78, 255 - 262 (1973).
  A method is discussed for calibrating direct photographs to be used for surface photometry. The standards used are the brightness profiles of field stars: their shapes calibrate relative intensities, and their magnitudes provide the zero point.

113.039 UBV photometry of some southern stars. (Second list).   A. W. J. Cousins.
Monthly Notes Astron. Soc. Southern Africa, Vol. 32, 11 - 15 (1973).
  The present list of stars is in continuation of that published 1972 and gives three-colour photometry for 105 stars. This completes the photometry of the early type stars requested by the Radcliffe Observer and of nearly all the brighter stars taken from the 40-inch reflector programme.

113.040 *UBV* photometry of stars in F2.   A. W. J. Cousins, G. M. Harvey.
Monthly Notes Astron. Soc. Southern Africa, Vol. 32, 27 (1973). – Note.

113.041 Photoelectric observations of metallic stars in the Vilnius seven-colour system.   D. Rajkova.
Izv. Sekts. astron. Blg. AN, Vol. 5, 65 - 66 (1972). In Bulgarian. – Abstr. in Referativ. Zhurn. 51. Astron., 4.51.719 (1973).

113.042 Five-colour TV photometry of DI Cep.   P. P. Petrov.
Izv. Krymskoj Astrofiz. Obs., Vol. 46, 25 - 34 (1972). In Russian.
  A special 4-colour photometric system has been used for TV photometry of DI Cep. This system is free from the influence of bright lines in the spectra of T Tau stars. A fifth passband has been used for estimate of Hβ emission. Observations show irregular variations of colour temperature between $9500°K$ and $5000°K$. The position of DI Cep on the theoretical log R vs log T diagram permits to estimate its mass within $1.2 - 1.7 M_\odot$.

113.043 Multicolour observations of stars in the vicinity of the Orion nebula.   M. V. Penston.
Stellar ages. Proc. IAU Colloquium No. 17, (see 012.015), IX, 1 - 6 (1973).

113.044 On the variability of Θ Coronae Borealis.   E. S. Brodskaya.
Inform. Bull. Variable Stars (I. A. U. Commission 27), Konkoly Obs., Budapest, No. 774, 3 pp. (1973).

113.045 On the relations between theoretical and observed colours in the *BVRI* system.   F. Caputo, A. Natta.
Astrophys. Space Sci., Vol. 21, 73 - 77 (1973).
  Transformation relations from theoretical apparent magnitude and $(b-v), (v-r), (v-i)$ colours to the observed m_v and $(B-V), (V-R), (V-I)$ quantities are derived. Black body colours in the *BVRI* system are also computed.

113.046 **Ten-color intermediate-band photometry of stars.**
S. M. Faber.
Astron. Astrophys.,Suppl. Ser., Vol. 10, 201 - 216 (1973).
The present paper reports on observations of 148 stars on a 10-color photometric system designed to study old stellar populations in globular clusters and galaxies. Transformations between the resultant photometric system and data already published are presented. Observations of stars having known abundances are used to study the behavior of certain line indices as a function of abundance variations in G and K dwarfs and giants.

113.047 **Intrinsic ultraviolet colors from OAO-2 Celescope observations for stars on the main sequence.**
K. Haramundanis, C. Payne-Gaposchkin.
Astron. Journ., Vol. 78, 395 - 400 (1973).
Intrinsic ultraviolet colors $(U_l - V)_0$ have been determined with ultraviolet magnitudes derived from several hundred observations taken with the photometers of OAO-2's Celescope experiment. Relationships between intrinsic ultraviolet color, $(B-V)_0$, and spectral class are presented for main-sequence stars from B 0 to G 3. A brief description of the Celescope and its calibration is included, and a comparison of the determined colors is made with theoretical colors.

113.048 *UBV* photometry of the Groningen—Palomar variable star fields.   P. A. Wehinger, B. Hidajat.
Astron. Journ., Vol. 78, 401 - 407, 445 - 452 (1973).
UBV photoelectric sequences have been established for 170 stars in the four Groningen—Palomar variable star fields covering the range 6 to 18 visual magnitude. On the average, three to four stars per magnitude interval were observed in each field. An analysis of the extinction data and of the photometric errors associated with the Kitt Peak observations is presented. Spectral types of field stars brighter than $B \simeq 12.0$ mag have been determined using objective-prism plates.

113.049 **Untersuchungen der Ap-Sterne HD 49976, HD 98088 und HD 203006.**   H. M. Maitzen.
Mitt. Astron. Ges., No. 32, p. 252 (1973). — Abstract.

113.050 **Extrem rote Sterne bei Eta Carinae.**   E. Reichert.
Mitt. Astron. Ges., No. 32, p. 275 - 276 (1973).

113.051 **UBV photometry of some southern stars, (third list).**   A. W. J. Cousins.
Monthly Notes Astron. Soc. Southern Africa, Vol. 32, 43 - 47 (1973).

113.052 **New infrared objects in Orion.** V. S. Shevchenko.
Young stellar groups. Astroclimate, (see 003.012), p. 3 - 7 (1972). In Russian.

113.053 **The blue objects in the region around the globular cluster M92.**
R. G. Mnatsakanian, K. A. Sahakian (*Saakyan*).
Soobshch. Byurakan. Obs., vyp. (No.) 44, p. 43 - 80 (1972). In Russian.
In a region of 16 square degrees around the globular cluster M92 843 blue objects with $B = 11^m.5 - 18^m.5$ and with colours $U-B < + 0^m.2$, $B-V < + 0^m.2$ were detected by blinking the U and B, B and V, U and V pairs of plates. The data for these objects are presented in a supplement of the paper. The probability of discovery of such objects, which depends on their colours and brightnesses, was calculated and presented in a table.

113.054 **A catalog of colorimetric measures of stars on the six-color system of Stebbins and Whitford.**
G. E. Kron, H. H. Guetter, B. Y. Riepe.
Publ. United States Naval Obs., *Washington*, Second Ser., Vol.

20, Part 5, 71 pp. (1972).
This catalog collects in one place previously published six-color photometric data on approximately 800 stars and new data on an additional 500. In all, 2230 photometric observations are presented for 1339 stars. Included in the catalog are many of the brightest and nearest stars, early type stars both reddened and unreddened, degenerate stars, supergiant and super-supergiant stars, helium rich stars, shell stars, and main sequence stars covering a range in metal abundance. Included also are stars in the nearby open clusters and associations, and some of the brighter stars in globular clusters.

113.055 **Comparaison des systèmes photométriques uvby β et de Genève.**   E. Lindemann, B. Hauck.
Bull. Soc. vaudoise Sci. nat., Vol. 71, 201 - 210 (1972) = Publ. Obs. Genève, Sér. A, Fasc. 79/I (1973).
In this study, we compare the properties of the Strömgren and Geneva photometric systems. First for the temperature parameter, then the luminosity and blanketing parameters. For stars with spectral type between A0 and G5 the possibilities of both systems are equivalent.

113.056 **Photometry of metal-deficient A-type stars found on objective-prism plates.**
H. E. Bond, A. G. D. Philip.
Publ. Astron. Soc. Pacific, Vol. 85, 332 - 334 = Contr. Louisiana State Univ. Obs., *Baton Rouge*, No. 82 (1973).
A number of A-type stars suspected to have weak metallic lines were selected on Curtis Schmidt objective-prism plates, and four-color photometry was obtained for seven of these objects. Four of the stars are field horizontal-branch stars (including two RR Lyrae variables), one is an eclipsing binary, and one is probably a λ Bootis-type star. HD 107369 is an unusual star of extraordinarily low surface gravity.

113.057 *UBV* photometry of large proper motion stars.
J. C. Muzzio.
Publ. Astron. Soc. Pacific, Vol. 85, 358 - 361 (1973).
The results of *UBV* photometry for 32 large proper motion stars are presented. The $(B-V)$ vs. $(U-B)$ diagram is used to select probable subluminous stars. A table presents the stars which are nearer than 20 pc according to their photometric parallaxes.

113.058 **Narrow-band infrared photometry of bright stars.**
G. V. Khozov, V. V. Shalberova, L. V. Danilova.
Trudy Astron. Obs., *Leningrad*, Vol. 29 (= Uchenye Zapiski Leningr. Un-ta, No. 363 = Seriya Matem. Nauk, vyp. (No.) 48), p. 80 - 85 (1973). In Russian.
The results of narrow-band infrared photometry of 63 bright stars are presented. The observations for 1.58μ and 1.70μ show the existence of an excess radiation due to a minimum in the H opacity.

Centre de Données Stellaires, Inform. Bull. No. 4. See Abstr. 002.007.

Data reduction techniques for direct astronomical electronography.   See Abstr. 031.030.

Schnellphotometrie veränderlicher Sterne. See Abstr. 031.036.

Reducción de observaciones fotoeléctricas (Reduction of photoelectric observations).   See Abstr. 031.064.

Optimum astronomical photoelectric photometry. Terrestrial operations in the UV-IR band up to 1 μ wavelength. See Abstr. 034.025.

The use of electronographic-type image tubes in

astronomical photometry.　See Abstr. 034.067.

Étude d'astres faibles en lumière totale avec la caméra electronique.　See Abstr. 034.068.

Photometry with the electronic camera. See Abstr. 034.069.

The analysis of direct Spectracon exposures obtained on the Isaac Newton telescope.　See Abstr. 034.070.

A photon-counting detector for stellar spectrophotometry.　See Abstr. 034.077.

Astrometry and photometry of G95-59. See Abstr. 111.008.

Infra-red observations of young stars—III. Nebulous emission-line stars.　See Abstr. 114.004.

Spectroscopic and photometric observations of luminous stars in Carina-Centaurus ($l = 282° - 305°$). See Abstr. 114.012.

Photometry and classification of stars in several H II regions.　See Abstr. 114.026.

Feasibility of UV astronomy by balloon-borne observations. I. Stellar spectrophotometry. See Abstr. 114.032.

Additional observations of supergiants and foreground stars in the direction of the Large Magellanic Cloud. See Abstr. 114.045.

Spectral types and UBV photometry of G−K giants at the North Galactic Pole.　See Abstr. 114.050.

The temperature and continuum of the N stars. See Abstr. 114.055.

Una estrella muy roja en la asociación I Scorpii. See Abstr. 114.097.

The carbon abundance and colours of Delta Pavonis.　　　See Abstr. 114.099.

Low-temperature free-free emission: infrared excesses in Be stars.　See Abstr. 114.120.

Photometric and spectroscopic study of silicon Ap-star HD 193722.　See Abstr. 114.143.

Infrarotobjekte in Aquila und Cygnus. See Abstr. 114.151.

Identifizierung von IRC-Objekten. See Abstr. 114.152.

Variability of T Tauri-like stars in NGC 2264. See Abstr. 114.165.

Luminosity functions for K giant stars derived from the two-micron sky survey.　See Abstr. 115.006.

Photometry of field horizontal-branch stars. See Abstr. 115.014.

Photometric information on globular cluster stars from the Palomar Sky Survey prints.　See Abstr. 115.024.

Ultraviolet photometry from the Orbiting Astronomical Observatory. VII. α^2 Canum Venaticorum. See Abstr. 116.004.

Multicolour observations of UZ Librae. See Abstr. 122.007.

A photometric study of 21 Comae Berenices. See Abstr. 122.009.

Four-color photometry of short-period variable stars.　See Abstr. 122.025.

Some variables of spectral type K. See Abstr. 122.029.

A search for Ap stars with very long periods. See Abstr. 122.074.

Modellrechnungen mit dreidimensional variabler interstellarer Absorption.　See Abstr. 131.172.

On the nature of the infrared point source in the Orion nebula.　See Abstr. 132.028.

Redshift of OQ172.　See Abstr. 141.090.

Periodic light variations in HDE 226868 (Cyg X-1). See Abstr. 142.035.

High-speed UBV photometry of Scorpius X-1 flares. See Abstr. 142.093.

Infrarot - Beobachtungen von Sternen in und nahe der Assoziation VI Cyg O B 2.　See Abstr. 152.010.

The open cluster NGC 2527. See Abstr. 153.005.

Photometric study of the open cluster NGC 2516. See Abstr. 153.006.

Photometric study of the galactic cluster NGC 2335. See Abstr. 153.007.

Three color photometry of the five open clusters NGC 7039, NGC 7062, NGC 7067, NGC 7082, IC 1369. See Abstr. 153.008.

NGC 1893 and NGC 6838, standard stars. See Abstr. 153.035.

Errata

113.901　Erratum: 'Magnitudes, colours and coordinates of 175 ultraviolet excess objects in the field 13^h, +36°' [Astron. Astrophys., Vol. 5, 264 - 279 (1970)]. A. Braccesi, L. Formiggini, E. Gandolfi. Astron. Astrophys., Vol. 23, 159 - 160 (1973).

114 Stellar Spectra, Temperatures, Spectroscopy

114.001 Observations spectroscopiques de HD 50138.
Y. Andrillat, L. Houziaux.
Astrophys. Space Sci., Vol. 18, 324 - 328 (1972).
Observations of the Be star HD 50138 have been extended to the photographic infrared region up to 9500 Å.

114.002 Infra-red observations of young stars—I. Stars in young clusters. M. Cohen.
Monthly Notices Roy. Astron. Soc., Vol. 161, 85 - 95 (1973).
Narrow-band infra-red observations in the 2 to 20 μ region are presented for a selection of stars in the young clusters NGC 2264, IC 5146 and the VI Cygni association. Additionally, infra-red data are given for stars in the grouping of emission-Hα stars associated with NGC 7000, including the peculiar star Lk Hα-190 = V1057 Cyg.

114.003 Infra-red observations of young stars—II. T Tauri stars and the Orion population. M. Cohen.
Monthly Notices Roy. Astron. Soc., Vol. 161, 97 - 104 (1973).
Multifilter infra-red observations in the 2 to 22 μ region are presented for a number of T Tauri variables and early-type stars of the Orion population.

114.004 Infra-red observations of young stars—III. Nebulous emission-line stars. M. Cohen.
Monthly Notices Roy. Astron. Soc., Vol. 161, 105 - 111 (1973).
Multifilter infra-red observations of early-type nebulous stars are presented in the 2 to 22 μ region. Emission features in the 10 μ region are found in the stars AB Aur and V380 Ori. It is found that stars associated with cometary nebulae are characterized by large [11]–[18] indices and this is discussed in the context of a possible picture of these nebulae.

114.005 Lk Hα 101 – not a northern η Carinae?
D. A. Allen.
Monthly Notices Roy. Astron. Soc., Vol. 161, 1P - 2P (1973).
It has been reported that the spectrum of LkHα 101 in the photographic infra-red resembles that of η Carinae. At visible wavelengths the spectrum is found to be much less remarkable.

114.006 Bright infrared sources in M17. D. E. Kleinmann.
Astrophys. Letters, Vol. 13, 49 - 54 (1973).
Observations of four bright infrared sources discovered in M17 are detailed. Two point sources bright at 2.2 μm are identified with the star BD −16°4816 and with a star 1.6ˢE and 77″ N of BD −16°4816; the possible role of these stars in the excitation of the nebula is considered. Two extended, optically thin sources were mapped at 10 μm with a 21-arc sec beam, and 5- to 25-μm photometry of their central regions was obtained. The observations indicate that the 10-μm radiation of these extended sources is thermal emission from dust grains and that these grains are distributed similarly to the ionized gas.

114.007 Carbon stars in the northern Milky Way.
J. Hardorp, K. Lübeck, C. B. Stephenson.
Astron. Astrophys., Vol. 22, 129 - 131 (1973).
59 carbon stars, 4 of which are new, were found in the areas I, III, and V of the catalog 'Luminous Stars in the northern Milky Way' and were subdivided into classes R and N by means of the 4350 Å spectral discontinuity. Equatorial coordinates of about 1″ precision were determined, and for many stars approximate photographic magnitudes were also derived.

114.008 Cyclic variations of the Be star β^1 Monocerotis.
A. Cowley, E. Gugula.
Astron. Astrophys., Vol. 22, 203 - 208 (1973).
Since the mid-1920's the Be star β^1 Mon has undergone regular changes in the shell velocity and intensity ratio of the double hydrogen emission components. A "period" of 12.5 years persisted for three and a half cycles. During the last few years these variations have ceased, and the spectrum appears much as it did between 1905 and 1925 when there were also no significant spectral variations. A qualitative model is presented.

114.009 Spectral variations of 53 Cam (AX Cam).
R. Faraggiana.
Astron. Astrophys., Vol. 22, 265 - 272 (1973).
A study of the variability of the radial velocities and spectral line intensities of 53 Cam has been made on spectra of 9.7 Å/mm and 12.4 Å/mm dispersion taken during the period 1967–1970 at the Haute Provence Observatory.

114.010 Revised chemical abundances of four population-II A-type stars. K. Kodaira.
Astron. Astrophys., Vol. 22, 273 - 279 (1973).
The previous spectral analyses of the population-II A-type stars BD +39°4926 (Kodaira et al., 1970), HD 161817 (Kodaira, 1964), and HD 86986 and HD 109995 (Kodaira et al., 1969) are critically examined to yield revised chemical abundances in accordance with new gf-values for Mg, Si, Fe and Ni. The bearing of the revision on the theory of the element synthesis is discussed.

114.011 Atmospheric abundances in the carbon star HD 156074.
T. F. Greene, J. Perry, T. P. Snow, G. Wallerstein.
Astron. Astrophys., Vol. 22, 293 - 298 (1973).
Atomic and molecular lines have been studied from 6.7 Å/mm coudé spectrograms of HD 156074. From the colors and spectra an effective temperature of 4750°K has been derived with a possible uncertainty of 250°K. The ratio of iron to hydrogen is normal and the ratios of other elements, from sodium to neodymium, to iron are found to deviate from the solar ratio by no more than a factor of three. The CNO abundances are abnormal with ^{13}C enhanced, ^{14}N enhanced and oxygen deficient.

114.012 Spectroscopic and photometric observations of luminous stars in Carina-Centaurus ($l = 282° − 305°$).
R. M. Humphreys.
Astron. Astrophys., Suppl. Ser., Vol. 9, 85 - 96 (1973).
New observational data are presented for 176 luminous stars of all spectral types in the Carina-Centaurus region. The new data consist of radial velocities, MK spectral types and UBV photometry.

114.013 Variation or disappearance of the emission in 8 Be stars. H. Hubert.
Astron. Astrophys., Suppl. Ser., Vol. 9, 133 - 148 (1973).
In French.
We give qualitative results on eight Be stars (HD 50658, HD 168797, HD 175863, HD 177648, HD 210129, HD 214168, HD 217543, HD 212076) which present the same type of variation of the emission: cyclic disappearing or weakening of the emissive shell. For each star we determined the emission variation period; five of them have a period inferior to twenty years. On the other hand we found that all these stars presented some variations of absorption lines (H, HeI, MgII, NaI, CaII).

114.014   Energy distribution in the spectra of six stars.
E. B. Gusev, N. S. Komarov, Ju. A. Medvedev.
Astron. Zhurn. Akad. Nauk SSSR, Vol. 50, 223 - 225 (1973).
In Russian. English translation in Soviet Astron. AJ, Vol. 17,
No. 1. − Short note.

114.015   On the nature of the Sagittarius object IRC−20385.
R. F. Wing, J. W. Warner, M. G. Smith.
Astrophys. Journ., Vol. 179, 135 - 145 (1973).
The infrared source IRC−20385 consists of the extended
object Terzan 5 and a nearby star of type M8. The observations of Terzan 5 are consistent with the hypothesis that it is
a normal globular cluster suffering 9.4 mag visual absorption.
This hypothesis leads to a photometric distance of 8 kpc,
which is consistent with independent arguments that is must
lie in front of the nuclear bulge of our Galaxy.

114.016   The spectrum of Herbig-Haro object No. 1.
K.-H. Böhm, J. F. Perry, R. Schwartz.
Astrophys. Journ., Vol. 179, 149 - 160 (1973).
A spectroscopic and spectrophotometric study of Herbig-
Haro object No. 1 has been carried out using four image-tube
spectra covering the spectral range 3700 Å $\lesssim \lambda \lesssim$ 6850 Å.
New line identifications include additional [Fe II] and [N II]
lines as well as the Balmer lines H10, H11, and H12.

**114.017   Spectroscopic studies of O-type stars. II. Compari-
son with non-LTE models.**   P. S. Conti.
Astrophys. Journ., Vol. 179, 161 - 179 (1973).
Equivalent widths of strong He I and He II lines and Hγ
in a large sample of O stars are compared with predictions
from the plane-parallel non-LTE models of Auer and Mihalas.
There is very good agreement indicating the essential correct-
ness of their models. However, for many stars there is a dra-
matic difference between the predictions and the observations
of λ4686 He II. The runs of equivalent widths of representative lines of C III, N III, O III, and Si IV are also presented
for many O stars.

**114.018   Spectroscopic studies of O-type stars. III. The ef-
fective-temperature scale.**   P. S. Conti.
Astrophys. Journ., Vol. 179, 181 - 188 (1973).
An effective-temperature scale for O-type stars is presented, based upon a comparison of measures of λ4471 He I/
λ4541 He II and predictions from the non-LTE models of
Auer and Mihalas. There is reasonably good agreement between this scale and the effective temperatures inferred from
O stars in H II regions by the Zanstra method. The problems
associated with the low effective temperature derived for
ζ Pup by interferometric methods are also discussed.

114.019   A spectroscopically distinguished class of Be stars.
R. Schild.
Astrophys. Journ., Vol. 179, 221 - 230 (1973).
The pole-on hypothesis to account for a group of Be stars
having sharp He I absorption lines and broad hydrogen absorption wings is critically examined. On the basis that the stars
are not typical Be stars, the hypothesis is rejected and the true
pole-on stars are identified. A relation between emission- and
absorption-line radial velocities is established, and a model of
differential rotation to explain the different line widths is
discussed with respect to the observations.

**114.020   On ultraviolet stellar fluxes. IV. Importance of
bound-free absorption of S I in B to K stars.**
S. P. Tarafdar, M. S. Vardya.
Astrophys. Journ., Vol. 179, 231 - 234 (1973).
The bound-free absorption of S I has been found to be
an important source of opacity in the spectral region $\lambda\lambda$1000−
1197 Å in stars as early as B5.

114.021   A new helium-rich B-type star.   J. S. Drilling.
Astrophys. Journ., (Letters), Vol. 179, L31 - L32 =
Contr. Louisiana State Univ. Obs., Baton Rouge, No. 74
(1973).
Star No. 3378 in the catalog of Stephenson and Sanduleak, which lies near l = 329°, b = +6°, has been identified as
a helium star. A description of the spectrum at 125 Å mm^{-1}
is given along with the radial velocity of the star and its apparent magnitude and colors on the UBV system.

114.022   IRC+10420 − another Eta Carinae?
R. M. Humphreys, D. W. Strecker, T. L. Murdock,
F. J. Low.
Astrophys. Journ., (Letters), Vol. 179, L49 - L52 (1973).
At long wavelengths (1−22μ) the infrared source IRC
+10420 resembles very closely the peculiar object η Car. Pho-
tographic spectra of +10420 show it to be a very high lumi-
nosity star with spectral type about F8 or G0. This is remark-
ably similar to the cF5 absorption-line spectrum reported for
ηCar in 1892.

114.023   The interstellar spectrum of HD 154368.
J. C. Blades, P. D. Bennewith.
Monthly Notices Roy. Astron. Soc., Vol. 161, 213 - 216
(1973).
Strong absorption lines of the interstellar species Ca$^+$,
CH, CH$^+$ and CN have been detected in the spectrum of the
O type star HD 154368. There is evidence that these interstel-
lar species lie in gas situated close to the sun, and also that the
molecular species are more localized in space than is the inter-
stellar Ca$^+$.

**114.024   An abundance analysis of the F-type giant
HD 116745 in the globular cluster Omega Centauri.**
R. J. Dickens, A. L. T. Powell.
Monthly Notices Roy. Astron. Soc., Vol. 161, 249 - 255
(1973).
Spectrograms obtained with a McGee spectracon image
tube at a dispersion of 22.5 Å mm^{-1} have been used for a differential curve-of-growth analysis of the F giant HD 116745 in
ω Cen. The principal result of the analysis of one spectrum
yields a logarithmic iron-to-hydrogen ratio with respect to the
sun of [Fe/H] = -1.2 ± 0.3 (estimated error), as compared to a
value ~ -1.7 expected for cluster giants from ultra-violet excess and Deutsch/Kinman classification. A mass of 0.37 $M_\odot$
$^{+0.37}_{-0.18}$ is derived from the temperature, gravity and luminosity.

**114.025   Observation and interpretation of the infra-red spec-
trum of Eta Carinae.**
G. Robinson, A. R. Hyland, J. A. Thomas.
Monthly Notices Roy. Astron. Soc., Vol. 161, 281 - 292
(1973).
Infra-red spectral scans of Eta Carinae in the wavelength
range 8−13 μm are presented together with new broad-band
photometric measurements. These observations show that the
spectrum of η Carinae exhibits a feature in the 8−13 μm region
characteristic of thermal emission from silicate grains, thus
providing additional evidence that the large infra-red flux is due
to absorption and thermal re-radiation by particles surrounding
a central exciting source. The infra-red spectrum cannot be ade-
quately described by a single dust shell model, but is well represented by a two-shell model.

**114.026   Photometry and classification of stars in several H II
regions.**   P. Mayer, P. Macák.
Bull. Astron. Inst. Czechoslovakia, Vol. 24, 50 - 51, 56c, d
(1973).
MK spectral types and UBV photoelectric photometry are
given for stars in serveral small HII regions.

114.027 **Spectra of the Becklin-Neugebauer point source and the Kleinmann-Low nebula from 2.8 to 13.5 microns.** F. C. Gillett, W. J. Forrest.
Astrophys. Journ., Vol. 179, 483 - 491 (1973).

Spectra of the Becklin-Neugebauer point source and the Kleinmann-Low nebula complex have been obtained from 2.8 to 13.5 μ with $\Delta\lambda/\lambda \sim 0.01$ to 0.02. The spectrum of the point source shows strong absorption features at 3.1 and 10 μ. The Kleinmann-Low nebula also shows evidence of a strong absorption at 10 μ, possibly much stronger than that seen against the point source.

114.028 **Infrared observations of a highly reddened star near NGC 6231.**
R. F. Knacke, K. M. Strom, S. E. Strom.
Astrophys. Journ., Vol. 179, 493 - 494 (1973).

Infrared observations of NGC 6231-92 appear to confirm Herbig's recent suggestion that this object is a highly reddened background star of type F5 Ia.

114.029 **An infrared object probably associated with OH 338.5 + 0.1.** I. S. Glass, M. W. Feast.
Astrophys. Letters, Vol. 13, 81 - 83 (1973).

A bright infrared object has been found near OH 338.5 + 0.1. JHKL photometry is given. The spectral type in the visual region is approximately M3.

114.030 **The infrared spectrum and angular size of Eta Carinae.**
R. D. Gehrz, E. P. Ney, E. E. Becklin, G. Neugebauer.
Astrophys. Letters, Vol. 13, 89 - 93 (1973).

Observations of the spectrum and angular size of η Car favor a model in which short wave energy from a central object of high luminosity is absorbed by grains and thermally reradiated. We observe an inner shell of diameter 2 arc sec at 3 μm and an outer shell probably containing silicates of diameter 6 arc sec at 10 and 20 μm.

114.031 **Results from the Utrecht orbiting spectrophotometer.**
T. M. Kamperman, K. A. van der Hucht, H. J. Lamers, R. Hoekstra.
Sky Telescope, Vol. 45, 85 - 87 (1973).

114.032 **Feasibility of UV astronomy by balloon-borne observations. I. Stellar spectrophotometry.**
C. Navach, M. Lehmann, D. Huguenin.
Astron. Astrophys., Vol. 22, 361 - 370 (1973).

Spectrophotometric measurements in the ultraviolet of 48 stars have been obtained by an objective prism Schmidt camera flying at 40 km. A description of the payload, the calibration technique and the data reduction process are given. The intensity distributions of the stars are presented and briefly discussed.

114.033 **Feasibility of UV astronomy by balloon-borne observations. II. Stellar gradients in the near ultraviolet.** C. Navach.
Astron. Astrophys., Vol. 22, 371 - 379 (1973).

The analysis of 48 spectrophotometric stellar distributions obtained during a balloon-flight is performed by means of two absolute gradients defined shortward of the Balmer discontinuity. It is found that for normal stars of luminosity classes V, IV or III the measured stellar fluxes are lower than the predicted fluxes for wavelengths shorter than 3000 Å and for stars of spectral types later than B5. Rapid rotators show a variation of the ultraviolet gradient which is at least two times greater than that predicted by theory for spectral types in the range B 1 V − B 5 V.

114.034 **The new main-line OH/IR stars.**

W. J. Wilson, K. W. Riegel.
Astron. Astrophys., Vol. 22, 473 - 474 (1973).

The two stars, R LMi and IRC + 20082, were found to have 1667–MHz OH emission. Thirteen other stars were surveyed for 18-cm OH emission with negative results.

114.035 **Study of the abundances of heavy elements in F–G type stars. I. Differential analysis of two metal deficient stars: HR 646 and HR 860.** M. Spite, F. Spite.
Astron. Astrophys., Vol. 23, 63 - 68 (1973).

Differential analysis of two F metal deficient stars HR 646 (η Ari) and HR 860 are presented here, using as a standard of reference the F 8 V star HR 458 (υ And). The study of these two stars is part of a program covering nine F–G metal deficient stars, all slowly rotating.

114.036 **Hα-emission stars in and near NGC 7000.**
G. Welin.
Astron. Astrophys. Suppl. Ser., Vol. 9, 183 - 197 (1973).

A region in Cygnus comprising NGC 7000 and IC 5070 was searched for Hα-emission stars on objective-prism plates obtained with the Uppsala-Kvistaberg Schmidt telescope. Altogether 141 stars of this category were found, 35 of which were previously known. In the catalogue are listed positions (1950.0), V and B magnitudes, Hα intensities and, where known, spectral types. Stars of special interest are discussed in the text.

114.037 **Equivalent widths for γ Cap and the standard 30 L Mi.** P. L. Selvelli.
Astron. Astrophys. Suppl. Ser., Vol. 9, 199 - 207 (1973).

A list of the identifications and of the equivalent widths for the A peculiar star γ Cap and the comparison standard 30 L Mi is presented.

114.038 **Stellar and interstellar K lines: Gamma Pegasi and Iota Herculis.** L. M. Hobbs.
Astrophys. Journ., Vol. 179, 823 - 826 (1973).

High-resolution scans show that the relatively strong (~ 90 mÅ) K lines of Ca II in the early B stars γ Peg and ι Her are almost entirely stellar in origin, although the latter case includes a small interstellar contribution.

114.039 **Premiers résultats du levé spectrophotométrique du ciel dans l'ultraviolet à l'aide du satellite TD-1 A.**
P. Barker, A. Boksenberg, H. E. Butler, S. Gardier, L. Houziaux, C. Humphries, C. Jamar, D. Macau-Hercot, D. Malaise, A. Monfils, K. Nandy, G. I. Thompson, R. Wilson, H. Wroe.
Comptes Rendus Acad. Sci. Paris, Sér. B, Vol. 276, 199 - 202 (1973)

Après avoir décrit succinctement le concept général de l'instrument utilisé à bord du satellite TD-1 A de l'ESRO, les auteurs présentent quelques-uns des résultats obtenus pendant les premières semaines de vol de l'engin. Une brève comparaison est esquissée avec, d'une part, les résultats d'observations similaires, et, d'autre part, avec des modèles théoriques.

114.040 **The infrared sources in M8.**
N. J. Woolf, W. A. Stein, F. C. Gillett, K. M. Merrill, E. E. Becklin, G. Neugebauer, T. J. Pepin.
Astrophys. Journ., (Letters), Vol. 179, L111 - L115 (1973).

The spectral energy distribution and size of the source of infrared radiation from Herschel 36 in M8 have been investigated. Radiation from the diffuse source in the vicinity of the hourglass has been remeasured. It is shown that a significant fraction of the 10-μ flux comes from an unresolved source centered on Herschel 36. Other stars in M8 have also been investigated as potential sources of infrared radiation.

114.041 **Spectral classification of Of stars in VI Cygni (Cygnus OB2).** N. R. Walborn.

Astrophys. Journ., (*Letters*), Vol. 180, L35 - L37 (1973).

Two-dimensional spectral classifications have been obtained for seven O stars in the interesting association VI Cygni. One of them, star No. 7, is of type O3 If$_*$; it is the second known member of this new category, and the first in the northern hemisphere. Its spectrum is illustrated. A distance of 1800 pc to the association is derived.

114.042 Calculation of hyperfine structure of scandium and vanadium for stellar spectral analysis.
M. G. Edmunds.
Astron. Astrophys., Vol. 23, 311 - 316 (1973).

An approximate method is presented for the estimation of the hyperfine splitting constant for atoms obeying LS coupling, which will allow the inclusion of hyperfine structure in the analysis of many more lines than was previously possible. This method is applied to certain low-lying terms of scandium and vanadium, and comparison with published laboratory data indicates that the electron configuration assignment of some of these terms should be revised.

114.043 An analysis of the line spectra of some G and K Ib supergiants. J. van Paradijs.
Astron. Astrophys., Vol. 23, 369 - 379 (1973).

We present the results of an analysis of the line spectra of seven supergiants of type G and K Ib. This analysis is based on observations of line strengths on spectrograms of 1.6 and 6.5 Å/mm, in the wavelength region between 5000 Å and 6650 Å, and on calculations of weak line strengths for a grid of model atmospheres. Curves of growth have been made using solar gf-values. From the condition of minimum scatter in the curve of growth effective temperatures are derived. The gravity of the stars has been obtained from the requirement that two ionisation states of one element should give the same abundances. In order to check the result of Bakos (1971) that there is a correlation between the abundances of heavy elements and the luminosity of the star we have also analyzed five early K type giants, using the equivalent widths of Cayrel and Cayrel (1963) and of Koelbloed (1972). No indication of such an effect is found.

114.044 The weak- helium- line star α Sculptoris. I. The line-spectrum. A. Schmitt.
Astron. Astrophys., Suppl. Ser., Vol. 9, 427 - 436 (1973).

Using spectrograms with dispersions from 7 to 20 Å/mm, the spectrum of the peculiar B star αScl is investigated in the wave-length range 3170–7800 Å. About 400 lines could be assigned to 20 different ions of 18 elements, 26 lines remain unidentified. Equivalent widths of all lines and profiles of the strongest lines are given. From the available material no definite conclusion about spectrum variability can be derived.

114.045 Additional observations of supergiants and foreground stars in the direction of the Large Magellanic Cloud.
J.-P. Brunet, L. Prévot, E. Maurice, G. Muratorio.
Astron. Astrophys., Suppl. Ser., Vol. 9, 447 - 458 (1973).

We present a continuation of a previous work by Ardeberg et al. (1972) concerning new photometric and spectrographic data for 47 supergiants and 40 foreground stars in the direction of the Large Magellanic Cloud. We give V magnitudes, $(B-V)$ and $(U-B)$ colour indices, MK spectral types, radial velocities and remarks concerning positions, spectroscopic features, magnitudes and colours.

114.046 5 GHz observations of the infrared star MWC 349, and the H II condensation W3 (OH).
J. E. Baldwin, C. S. Harris, M. Ryle.
Nature, Vol. 241, 38 - 39 (1973).

We have observed the radio source MWC 349 with the 5-km radio telescope at Cambridge, and find it to have a flux density at 5 GHz of 0.22 ± 0.01 f.u. This source has been compared with the continuum emission from W3 (OH), the compact component associated with OH emission in the H II region W3.

114.047 Spectrum of a dust-embedded Wolf-Rayet star in Cygnus OB2. G. E. Bromage, K. Nandy.
Nature, Phys. Sci., Vol. 241, 30 - 31 (1973).

Herbig and Mendoza (1960) have suggested that three Wolf-Rayet stars may be members of the young association Cyg OB2 (VI Cyg). As part of a wider programme of study of diffuse interstellar spectral features, the brightest of these Wolf-Rayet stars, WR2 (V=11.8), was observed, and the results are presented here.

114.048 A study of Ca II K$_2$ and H$_a$ line widths in late type stars. D. Reimers.
Astron. Astrophys., Vol. 24, 79 - 87 (1973).

In the present study it is tried to relate the line widths of Ca II K$_2$ and H$_a$ to basic quantities such as gravity g, effective temperature T_e, chromospheric turbulent velocities and optical depths in the lines.

114.049 De sterspectrometer S 59.
T. M. Kamperman, K. A. van der Hucht, H. J. Lamers, R. Hoekstra.
Hemel en Dampkring, Vol. 71, 95 - 99 (1973).

114.050 Spectral types and UBV photometry of G–K giants at the North Galactic Pole. R. E. Schild.
Astron. Journ., Vol. 78, 37 - 44 (1973).

New spectral types for 145 late-type stars near the North Galactic Pole are compared with new UBV photometry.

114.051 The helium content of 99 Herculis. M. Spite.
Astrophys. Letters, Vol. 13, 137 - 138 (1973).

We estimate the helium content of the metal-deficient star 99 Herculis, first by use of the mass-luminosity relation, and second from the comparison between the spectroscopic gravity and the gravity deduced from the mass of the star. Both cases yield a helium abundance higher than the sun's.

114.052 Fourier spectroscopy in astronomy.
G. Sęk, A. Starnawski.
Postępy Astron., Vol. 21, 53 - 64 (1973). In Polish.

The principle of work and the construction of Fourier spectroscopy devices of high resolution in the optical and near infrared spectral regions are described. Examples of application to astronomy are given.

114.053 Spectroscopic observations of the Sanduleak-Seggewiss star. L. A. Milone.
Astrophys. Journ., Vol. 180, 631 - 634 (1973).

Spectrograms of the Sanduleak-Seggewiss star taken at the Córdoba Observatory show spectral lines belonging to Fe I multiplets, Ca I, Ca II, Mn I, and H; the G-band is also present. Some peculiarities are seen on the spectrograms.

114.054 Spectroscopy using a 16-inch telescope.
J. C. Castley.
Proc. Astron. Soc. Australia, Vol. 2, 137 - 138 (1972).

114.055 The temperature and continuum of the N stars.
M. S. Bessell, L. Youngbom.
Proc. Astron. Soc. Australia, Vol. 2, 154 - 155 (1972).

114.056 Statistical line blanketing in Arcturus. I. Blanketing-coefficient description of observations.
F. N. Edmonds, Jr.
Publ. Astron. Soc. Pacific, Vol. 85, 24 - 41 (1973).

A statistical description of line blanketing is presented

for Arcturus (α Boo, K2 IIIp, V = +0.06) which is based on equivalent widths and identifications of roughly 10,000 spectral lines within the wavelength region $\lambda\lambda 3600-25,000$ Å. Strong lines are listed individually and the variation or scatter of blanketing within intervals is described. Corrections to this description for limitations in the observations and their analysis are discussed.

114.057 **ζ^1 Scorpii: a variable spectrum supergiant.**
M. Jaschek, C. Jaschek.
Publ. Astron. Soc. Pacific, Vol. 85, 127 - 130 (1973).

Very conspicuous and unexpected variations in the strength of lines belonging to different elements are reported on the basis of 42 Å mm^{-1} spectrograms taken between 1965 and 1970. The variations involve principally He I, N II, and O II.

114.058 **Dust emission nebulae around Orion O and B stars.**
E. P. Ney, D. W. Strecker, R. D. Gehrz.
Astrophys. Journ., Vol. 180, 809 - 816 (1973).

The present study of a number of nearby stars has revealed the existence of extended dust shells around five other O and B stars in the Orion nebula. To determine whether the emission feature is characteristic of the Orion stars or of early stars in general, 20 other early stars were observed but with negative results. In addition, four of the Trapezium stars were observed individually to separate the stellar and nebular emission components of the long-wave radiation.

114.059 **Scanner photometry of weak TiO bands near 1 micron in cool M stars.** G. W. Lockwood.
Astrophys. Journ., Vol. 180, 845 - 855 (1973).

Continuous scans in the region 9660–10,400 Å reveal that overlapping TiO bands depress the continua of cool M stars significantly, shortward of 10,000 Å, and are present, though weaker, from 10,000 to 10,300 Å. The bands become visible at about spectral type M3 and increase in strength throughout the M sequence. Band strengths of TiO and VO, measured in the 1-μ region by scanner observations at discrete wavelengths, are found to be correlated over the spectral range M7–M9.8.

114.060 **Contribution to the study of supermetallicity in late-type giants.**
M.-J. Blanc-Vaziaga, G. Cayrel, R. Cayrel.
Astrophys. Journ., Vol. 180, 871 - 893 = Contr. Lick Obs., No. 377 (1973).

This research is a detailed analysis of ϕ Aur and μ Leo, two of the best examples of "super-metal-rich" stars. The spectra have been interpreted with theoretical line computations using a grid of model atmospheres. The models have been calculated on the assumption of LTE, hydrostatic equilibrium, and radiative equilibrium. Convective transport in the convective zone has been taken into account. Line blanketing has been included using a statistical model of line absorption.

114.061 **The effective temperatures of B type stars.**
M. Kubiak.
Acta Astron., Vol. 23, 23 - 30 (1973).

Continuous energy distributions in the visual range of the spectrum have been measured spectrophotometrically for reddened and unreddened B-type stars, and reduced to absolute fluxes using recent calibrations of Vega. Comparison of the observations with model atmospheres has been carried out.

114.062 **Observations of intense 100-micron objects at 3.5-millimeter wavelength.**
R. L. Brown, J. J. Broderick.
Astrophys. Journ., Vol. 181, 125 - 134 (1973).

Measurements have been made at 85 GHz of 12 objects identified by Hoffman, Frederick, and Emery as sources of 100-μ emission in excess of 10^{-22} W cm^{-2} Hz^{-1}. In an attempt to detect the long-wavelength tail of the far-infrared spectral component, the observed 3.5-mm flux densities for each of the sources are compared with that expected from an extrapolation of the radio spectrum from lower frequencies.

114.063 **On Jaschek and Brandi's line identifications for HD 25354.** S. J. Adelman.
Astron. Astrophys., Vol. 24, 325 - 328 (1973).

Jaschek and Brandi's suggested line identifications for HD 25354 are examined and found to be highly suspect.

114.064 **The H and K absorption features as luminosity indicators for MK classification.**
T. E. Lutz, I. Furenlid, J. H. Lutz.
Bull. American Astron. Soc., Vol. 5, 3 (1973). – Abstr. AAS.

114.065 **Ca II K line reversals in stellar spectra.**
D. R. Hollars, H. A. Beebe.
Bull. American Astron. Soc., Vol. 5, 3 (1973). – Abstr. AAS.

114.066 **Spectrophotometric results from the Copernicus satellite. I. Instrumentation and performance.**
J. B. Rogerson, L. Spitzer, J. F. Drake, K. Dressler, E. B. Jenkins, D. C. Morton, D. G. York.
Bull. American Astron. Soc., Vol. 5, 4 (1973). – Abstr. AAS.

114.067 **Improved time resolution in stellar spectroscopy.**
B. A. Goldberg, G. A. H. Walker, J. W. Glaspey, G. J. Odgers.
Bull. American Astron. Soc., Vol. 5, 5 - 6 (1973). – Abstr. AAS.

114.068 **Observations of some early type supergiants with variable H-alpha profiles.** J. D. Rosendhal.
Bull. American Astron. Soc., Vol. 5, 10 (1973). – Abstr. AAS.

114.069 **Infrared spectra of γ^2 Velorum and ζ Puppis.**
T. G. Barnes, D. L. Lambert, A. E. Potter.
Bull. American Astron. Soc., Vol. 5, 10 (1973). – Abstr. AAS.

114.070 **The C^{12}/C^{13} ratio of the CH stars in ω Centauri.**
R. A. Bell, R. J. Dickens.
Bull. American Astron. Soc., Vol. 5, 11 (1973). – Abstr. AAS.

114.071 **Observations of CO in the 5 micron spectra of NML-Cygnus, IRC + 10216, and VY Canis Majoris.**
T. R. Geballe, E. R. Wollman, D. M. Rank.
Bull. American Astron. Soc., Vol. 5, 31 (1973). – Abstr. AAS.

114.072 **Ultraviolet spectrophotometry of Sirius from spectra obtained on Gemini XII.**
G. G. Spear, Y. Kondo, K. G. Henize.
Bull. American Astron. Soc., Vol. 5, 39 (1973). – Abstr. AAS.

114.073 **OAO-2 observations of the helium spectrum variable a Centauri.** M. R. Molnar.
Bull. American Astron. Soc., Vol. 5, 39 (1973). – Abstr. AAS.

114.074 **A model atmosphere analysis of the super-supergiant HR 5171.** P. R. Warren.
Monthly Notices Roy. Astron. Soc., Vol. 161, 427 - 444 (1973).

A model atmosphere analysis based on five high dispersion spectra of the super-supergiant HR 5171 is presented. The abundances are essentially solar except for sodium which may be overabundant. Shell effects have been considered and, although no evidence of a low-temperature shell has been found from the spectra, the existence of a high temperature, low-density envelope is indicated by the presence of weak nebular [N II] emission.

114.075 **Infrared and microwave spectra of late-type stars.** Introductory report. J.-E. Gaustad.
17th Colloquium International Astrophys. Liège 1971, (see 012.003), p. 87 - 93 (1972).

114.076 **Recent observations of stellar infrared spectra.** M. J. Smyth.
17th Colloquium International Astrophys. Liège 1971, (see 012.003), p. 97 - 100 (1972).

114.077 **Observations of infrared stellar absorption lines.** L. Mertz.
17th Colloquium International Astrophys. Liège 1971, (see 012.003), p. 101 - 106 (1972).

114.078 **Interpretation of the $1-4\,\mu$ spectra of some C stars.** R. I. Thompson, H. W. Schnopper.
17th Colloquium International Astrophys. Liège 1971, (see 012.003), p. 107 - 109 (1972).

114.079 **Two-micron spectra of carbon stars.** J. A. Frogel, A. R. Hyland.
17th Colloquium International Astrophys. Liège 1971, (see 012.003), p. 111 - 120 (1972).

114.080 **Molecular absorption spectra of S-type stars in the one-micron region.** R. F. Wing.
17th Colloquium International Astrophys. Liège 1971, (see 012.003), p. 123 - 140 (1972).

114.081 **Observation of the OH radical in Betelgeuse.** R. Beer, R. H. Norton, R. B. Hutchison, D. L. Lambert, J. V. Martonchik.
17th Colloquium International Astrophys. Liège 1971, (see 012.003), p. 145 (1972).

114.082 **The $^{12}C/^{13}C$ ratio in the atmosphere of Arcturus.** D. L. Lambert, D. S. Dearborn.
17th Colloquium International Astrophys. Liège 1971, (see 012.003), p. 147 - 165 (1972).

114.083 **Infrared observations and atmospheres of cool stars. I. Support mechanisms for circumstellar shell.** K. S. K. Swamy, M. S. Vardya.
17th Colloquium International Astrophys. Liège 1971, (see 012.003), p. 169 - 178 (1972).

114.084 **Recent stellar spectra obtained with a 0.5 cm^{-1} interferometer.** F. F. Forbes.
17th Colloquium International Astrophys. Liège 1971, (see 012.003), p. 229 - 231 (1972).

114.085 **Données de l'observation et de la théorie et prévisions concernant les spectres (IR et ondes ultracourtes) des étoiles normales et particulières de type peu avancé.** Rapport introductif. J.-C. Pecker.
17th Colloquium International Astrophys. Liège 1971, (see 012.003), p. 243 - 292 (1972).

114.086 **A spectroscopic study of Wolf-Rayet stars in the infrared.** F. Bertola, F. Ciatti.
17th Colloquium International Astrophys. Liège 1971, (see 012.003), p. 303 - 310 (1972).

114.087 **Infrared emission in [Ne V].** H. Nussbaumer.
17th Colloquium International Astrophys. Liège 1971, (see 012.003), p. 351 - 356 (1972).

114.088 **Balloon observations of galactic and extragalactic objects at 100 microns.** W. F. Hoffmann.
17th Colloquium International Astrophys. Liège 1971, (see 012.003), p. 357 - 363 (1972).

114.089 **Spectrométrie stellaire par transformation de Fourier de 2 à 5 microns.** J. Laurent, M. Portat.
17th Colloquium International Astrophys. Liège 1971, (see 012.003), p. 497 - 506 (1972).

114.090 **Energy distribution in the stellar spectra of different spectral types and luminosities.** V. Straižys, Z. Sviderskienė.
Bull. Vilnius Astron. Obs., No. 35, 92 pp. (1972). In Russian.
The authors present a catalogue of detailed energy distribution curves of representative stars of different spectral types and luminosities in the interval 3000−10000 Å. It includes 18 main-sequence stars, 3 subgiants, 14 giants and 14 supergiants.

114.091 **Coarse analysis of the peculiar star 53 Aur.** J. Zverko.
Bull. Astron. Inst. Czechoslovakia, Vol. 24, 71 - 75 (1973).
From a curve-of-growth analysis of the Ap star 53 Aur anomalies of the chemical composition of the atmospheric layers result.

114.092 **Some recent aspects of spectroscopy at UV and X-ray wavelengths.** R. J. Speer.
Atoms and molecules in astrophysics. Proc. Scottish Univ. Summer School in Physics 1971, (see 012.005), p. 285 - 310 (1972).

114.093 **Metallicism in border regions of the Am domain. II. Analysis of the Fm stars.** M. A. Smith.
Astrophys. Journ., Suppl. Ser., No. 219, Vol. 25, 277 - 313 = Contr. Lick Obs., No. 374 (1973).
In this study the incidence and abundance anomalies of the Fm stars are considered. This incidence decreases rapidly coolward of a sharp peak at type A6−7 because a rotational velocity "barrier" imposes itself such that only stars with $V_{rot} < V_{crit}\,(T_{eff})$ can maintain metallicism. The V_{crit} decreases very quickly with T_{eff} and eventually, at F2, approaches 0 km sec^{-1}. This effect suggests that rotation and convection act in concert to disrupt the stability of a subsurface radiative zone that would otherwise allow elemental diffusion. Equivalent-width data of several Fm stars are treated by detailed curve-of-growth atmosphere analyses, and the derived abundances are placed on the system of Smith. According to the subsurface diffusion hypothesis, the mergence of the convective zones with T_{eff} should produce different degrees of abundance anomalies in the Am and Fm groups. An extensive qualitative discussion of the variations in anomalies is given.

114.094 **Possible new binaries among the Wolf-Rayet stars.** V. N. de Monteagudo.
Bol. As. Argentina Astron., No. 16, (see 012.007), p. 30 (1971). − Abstract.

114.095 **Peculiar classified Wolf-Rayet stars, HD 90657 and HD 117688.** V. N. de Monteagudo.
Bol. As. Argentina Astron., No. 16, (see 012.007), p. 30 (1971). − Abstract.

114.096 **Variaciones espectrales en estrellas Ap.** E. Brandi.
Bol. As. Argentina Astron., No. 16, (see 012.007), p. 31 (1971). − Abstract.

114.097 **Una estrella muy roja en la asociación I Scorpii.** L. A. Milone.
Bol. As. Argentina Astron., No. 16, (see 012.007), p. 31 (1971).

114.098 Spectroscopic observations of M giant stars at the South Galactic Pole.
D. Crampton, T. Lloyd Evans.
Monthly Notices Roy. Astron. Soc., Vol. 162, 11 - 15 (1973).

Radial velocities and spectral types determined from slit spectra are presented for 30 M giant stars within 30° of the South Galactic Pole. The velocities found here, together with those in the literature giving 67 stars in all, show an increase in velocitiy dispersion from the youngest to the oldest group.

114.099 The carbon abundance and colours of Delta Pavonis. R. A. Bell.
Monthly Notices Roy. Astron. Soc., Vol. 162, 37 - 42 (1973).

Synthetic spectra calculations have been carried out, using different carbon abundances, and the carbon abundance of δ Pav, $[C/H] = +0.45$, obtained by matching observation and calculation. The influence of this enhanced carbon abundance, and the adopted value of the damping, on the UBV and $uvby$ colours of δ Pav has been examined.

114.100 Nature of the light variation of the peculiar A-star HD 221568. K. Kodaira.
Astron. Astrophys., Vol. 25, 93 - 97 (1973).

The light variation of the extraordinary Ap-star HD 221568 is interpreted as the result of the varying line blanketing. The line blocking factors are measured for $\lambda\lambda$ 4100—5900 Å, amounting up to 43% for the scanner pass-band of λ4167 Å at the 'red' phase.

114.101 The equivalent widths of Mg II lines near 2800 Å in the spectra of 31 stars.
H. J. Lamers, K. A. van der Hucht, M. A. J. Snijders, N. Sakhibullin.
Astron. Astrophys., Vol. 25, 105 - 112 (1973).

The total equivalent width of the Mg II lines λ2795.5, λ2802.7 and λ2790.8 λ2798.0 has been measured in the spectrograms of 31 stars, observed by the Orbiting Stellar Spectrophotometer S 59. After corrections for the interstellar components have been applied, the equivalent width in various types of stars are compared with each other. In middle and late-B type stars of class II—V the observed equivalent widths agree with non-LTE predictions, but in early-B stars the observed lines are too strong. Several possible explanations for the origin of this discrepancy are discussed.

114.102 MWC 645 and MWC 819: two stars resembling Eta Carinae. J. P. Swings, D. A. Allen.
Astrophys. Letters, Vol. 14, 65 - 68 (1973).

Two faint stars in the Mount Wilson Catalogue bear strong similarity to Eta Carinae in their spectra and energy distributions.

114.103 On the relative abundance of helium in the envelopes of Wolf—Rayet stars (Study of Wolf—Rayet stars. III). S. V. Rublev.
Astrofiz. Issled., Izv. Spets. Astrofiz. Obs., Vol. 4, 3 - 17 (1972). In Russian.

The abundance of helium and hydrogen in the atmospheres of 5 bright Wolf—Rayet stars (HD 191765, 192103, 192163, 192641, and 193077) is evaluated from ratios of the equivalent widths of bright Balmer lines to those of blended Pickering He II lines by using the method of transition to the spectral series limit. A minimum evaluation of the abundance of helium is obtained without using specific hypotheses concerning the model atmosphere.

114.104 On the spectrophotometric temperatures of Wolf—Rayet stars (Study of Wolf—Rayet stars. IV).
S. V. Rublev.
Astrofiz. Issled., Izv. Spets. Astrofiz. Obs., Vol. 4, 18 - 31 (1972). In Russian.

A critical revision of Kuhi's results has been performed concerning the spectrophotometric temperatures of Wolf—Rayet stars. An analysis of a relationship between the «intrinsic» color temperatures and wavelength carried out under the weakest limiting assumptions has shown the average electron temperatures in the envelopes of Wolf—Rayet stars to be lower than the color temperatures of their «cores» in the photographic region of the spectrum.

114.105 Spectrophotometric study of the magnetic variable star α^2 CVn using high-resolution spectrograms. II. Some features of spectral variability.
R. N. Kumajgorodskaya, I. M. Kopylov.
Astrofiz. Issled., Izv. Spets. Astrofiz. Obs., Vol. 4, 50 - 68 (1972). In Russian.

A study of variations in intensities, central depths, and half-widths is carried out of some lines of a number of chemical elements with phase in the spectrum of α^2 CVn.

114.106 An investigation of the atmospheres of metallic stars. II. A quantitative analysis of the atmospheres of β CrB, 68 Tau, 49 Ari by the curve-of-growth method.
K. I. Kozlova.
Astrofiz. Issled., Izv. Spets. Astrofiz. Obs., Vol. 4, 69 - 80 (1972). In Russian.

Results are given of a curve-of-growth analysis of the atmospheres of β CrB, 68 Tau and 49 Ari. The turbulent velocities, excitation and ionization temperatures, electron densities, and abundances of elements are determined.

114.107 The atmosphere of the supergiant 6 Cas. I. Spectral material and its photometric processing with the aid of an electronic digital computer.
G. I. Abbasov, S. K. Zejnalov, E. L. Chentsov.
Astrofiz. Issled., Izv. Spets. Astrofiz. Obs., Vol. 4, 81 - 90 (1972). In Russian.

A procedure is described and estimates of the accuracy are given of a photometric processing of spectrograms with the aid of a diagram-to-code converter and a computing machine. Central depths, equivalent widths, and half-widths are obtained for 6 Cas in the range 3660—4740 Å on 4 and 14 Å mm^{-1} spectrograms.

114.108 A spectrophotometry of stars in the Pleiades. I. Observations of 19 early-type stars.
A. V. Kharitonov, V. G. Klochkova.
Astrofiz. Issled., Izv. Spets. Astrofiz. Obs., Vol. 4, 91 - 105 (1972). In Russian.

The distribution of energy in continuous spectra of 19 stars in the Pleiades is investigated. The monochromatic illuminations at the boundary of the terrestrial atmosphere and the equivalent widths of the Balmer lines are obtained. Equipment and observational technique are briefly described.

114.109 Scanner observations of hot helium—carbon stars.
T. Faÿ, R. K. Honeycutt, W. H. Warren, Jr.
Astron. Journ., Vol. 78, 246 - 252 = Publ. Goethe Link Obs. Indiana Univ., *Bloomington*, No. 151 (1973).

Photoelectric spectral scans at 20 Å resolution of four hot helium—carbon-rich stars have been reduced to fluxes and are presented in graphical form. Similar flux curves for several normal (hydrogen-rich) stars in the same temperature range are presented for comparison.

114.110 Spectrum variability of the shell star 88 Herculis.
S. N. Svolopoulos.
Bull. Astron. Inst. Czechoslovakia, Vol. 24, 167 - 169 (1973).

Spectrophotometric data from spectra of 88 Her taken in September 1970 and in July 1971 are compared and the noticed differences summarized.

114.111 **C₂ in Eta Aquilae spectrum.**
K. R. Bondal, G. C. Joshi, M. C. Pande.
Bull. Astron. Inst. Czechoslovakia, Vol. 24, 169 - 170 (1973).

The equivalent width of the R_1 (16) line of the 0–0 band of the Swan system of C_2 calculated for the minimum phase of light variation in η Aql comes out to be 27 mÅ.

114.112 **On the observability of CO and CO⁺ in Eta Aquilae.**
M. C. Pande, G. C. Joshi.
Bull. Astron. Inst. Czechoslovakia, Vol. 24, 171 - 172 (1973).

The equivalent widths of the R (29) line belonging to the 0–2 band of CO in η Aql are found to be 0.020 mÅ and 4.17 mÅ and that of R (31) 0.019 mÅ and 3.71 mÅ at maximum and minimum phases respectively.

114.113 **Some tests and consequences of the identification of promethium in HR 465.**
R. Mitalas, J. M. Marlborough.
Astrophys. Journ., Vol. 181, 475 - 480 (1973).

We discuss several ways to confirm the identification of Pm in HR 465. The emission of X-rays and γ-rays resulting from the decay of unstable isotopes of Pm and other rare earths makes it worthwhile to attempt to detect HR 465 as an X-ray or γ-ray source.

114.114 **The infrared spectrum of χ Cygni from 4000 to 6700 cm⁻¹.**
H. L. Johnson, R. I. Thompson, F. F. Forbes, D. L. Steinmetz.
Publ. Astron. Soc. Pacific, Vol. 85, 179 - 186 (1973).

The infrared spectrum of χ Cyg from 4000-6700 cm⁻¹ is presented. Preliminary identification of atomic and molecular features have been made. χ Cyg shows the most intense CO bands observed to date in any star. The bearing of these bands on observed $^{12}C/^{13}C$ ratios is discussed.

114.115 **Technetium in S Sculptoris.** J. G. Cohen.
Publ. Astron. Soc. Pacific, Vol. 85, 187 (1973).

Evidence is given for the presence of technetium in S Scl, a long-period Me variable.

114.116 **New emission-line stars in OH clouds.**
G. L. Grasdalen, L. V. Kuhi, E. A. Harlan.
Publ. Astron. Soc. Pacific, Vol. 85, 193 - 199 (1973).

Eleven dust clouds exhibiting radio OH lines have been surveyed for Hα emission-line stars. Eight new objects were discovered in four such clouds and none in the remaining seven to a limiting magnitude of $m_v \simeq 18$. The objects are most likely T Tauri stars.

114.117 **The helium-rich star HD 184927.**
N. A. Higginbotham, P. Lee.
Publ. Astron. Soc. Pacific, Vol. 85, 215 - 219 = Contr. Louisiana State Univ. Obs., *Baton Rouge*, No. 78 (1973).

A preliminary analysis of HD 184927 yields a helium-to-hydrogen ratio of 2.0, T_{eff} = 22,000° K, and log g = 3.38. A comparison is made with similar type stars.

114.118 **Some magnetic null lines of astrophysical interest.**
S. J. Adelman.
Publ. Astron. Soc. Pacific, Vol. 85, 227 - 229 (1973).

A line list is presented for magnetic null lines found in the Ultraviolet Multiplet Table and in selected spectral analyses for use in detailed studies of magnetic peculiar A star atmospheres.

114.119 **Wavelength shifts and probable isotopic structure of Pt II λ 4046 in mercury stars.**
M. M. Dworetsky, A. H. Vaughan, Jr.
Astrophys. Journ., Vol. 181, 811 - 816 (1973).

Measurements on coudé spectrograms of nine Hg-Pt stars show that Pt II λ4046 is shifted toward longer wavelength by

0.04 to 0.09 Å from the centroid for terrestrial platinum. High-resolution scans made with a Fabry-Perot interferometer show a single exceedingly sharp component in χ Lup A, which probably represents a single heavy isotope. In ι CrB two or more heavy isotopes appear to be present.

114.120 **Low-temperature free-free emission: infrared excesses in Be stars.** R. W. Milkey, H. M. Dyck.
Astrophys. Journ., Vol. 181, 833 - 839 (1973).

We discuss the possibility that the infrared emission observed for certain Ae and Be stars may be produced by low temperature free-free processes in the circumstellar region. The problems associated with H⁻ are discussed. We are left with the fact that a free-free–like opacity fits the observed infrared fluxes but without a corresponding physical picture of the specific mechanism.

114.121 **Spectrophotometric results from the Copernicus satellite. I. Instrumentation and performance.**
J. B. Rogerson, L. Spitzer, J. F. Drake, K. Dressler, E. B. Jenkins, D. C. Morton, D. G. York.
Astrophys. Journ., (*Letters*), Vol. 181, L97 - L102 (1973).

The Princeton telescope-spectrometer on the OAO spacecraft Copernicus scans stellar spectra with a resolution of about 0.05 Å between 950 and 1450 Å, and twice this in first order between 1650 and 3000 Å. The measured photometric precision in the shorter wavelength range is limited only by the statistics of photon counts, with 14-s counts of about 10³ on an unreddened B 1 star, m_V = 5.0, at 1100 Å. In the 1650–3000 Å wavelength range, phototube noise resulting from cosmic rays makes observations difficult on stars fainter than m_V = 3.0.

114.122 **Spectrophotometric results from the Copernicus satellite. II. Composition of interstellar clouds.**
D. C. Morton, J. F. Drake, E. B. Jenkins, J. B. Rogerson, L. Spitzer, D. G. York.
Astrophys. Journ., (*Letters*), Vol. 181, L103 - L109 (1973).

This paper describes a Copernicus survey of selected spectral regions in the reddened stars ξ Per, α Cam, λ Ori, ζ Oph, and γ Ara. The range of $B-V$ color excesses from 0.09 to 0.33 provides a representative sample of paths through interstellar clouds.

114.123 **Spectrophotometric results from the Copernicus satellite. III. Ionization and composition of the intercloud medium.**   J. B. Rogerson, D. G. York, J. F. Drake, E. B. Jenkins, D. C. Morton, L. Spitzer.
Astrophys. Journ., (*Letters*), Vol. 181, L110 - L115 (1973).

Interstellar lines have been studied in the unreddened stars $[E(B-V) \leqslant 0.03]$ λ Sco, υ Sco, α Leo, and α Eri, which are located between 20 and 150 pc from the sun. The lines are on the linear and saturated Doppler portions of the curve of growth. Preliminary analysis shows that processes producing large amounts of highly ionized species are not dominant. Abundances relative to nitrogen are in the cosmic ratios or slightly lower, with some indication of general depletion of the heavier elements. Hydrogen number densities range from a measured value of 0.22 cm⁻³ for λ Sco to a derived value of 0.02 cm⁻³ for α Leo.

114.124 **Spectrophotometric results from the Copernicus satellite. IV. Molecular hydrogen in interstellar space**
L. Spitzer, J. F. Drake, E. B. Jenkins, D. C. Morton, J. B. Rogerson, D. G. York.
Astrophys. Journ., (*Letters*), Vol. 181, L116 - L121 (1973).

Strong H_2 lines are measured in all 11 reddened stars $[E(B-V) > 0.10]$ observed; the fraction f of hydrogen gas in molecular form exceeds 10⁻¹. In eight out of nine unreddened stars $[E(B-V) < 0.05]$ there is no trace of H_2 absorption, with f less than 10⁻⁷. In two stars of intermediate reddening f is between 10^{-5} and 10^{-6}. Measures of two HD lines in nine stars indicate a ratio of HD to H_2 equal to about 10^{-6}; correction for the more rapid disruption of HD molecules, in the absence of effective optical shielding by many other such molecules, indicates that one HD molecule is formed and dissociated for about every 200 H_2 molecules.

114.125 Spectrophotometric results from the Copernicus satellite. V. Abundances of molecules in interstellar clouds. E. B. Jenkins, J. F. Drake, D. C. Morton, J. B. Rogerson, L. Spitzer, D. G. York.
Astrophys. Journ., (*Letters*), Vol. 181, L122 - L127 (1973).

Scans were made at the absorption wavelengths of CO, CO^+, CN^+, CS, C_2, OH, NO, NO^+, NH^+, SiO, MgH^+, and H_2O. Except for CO, no features were detected in the stars observed with $E(B-V)$ ranging up to 0.3. The greatest abundance of CO was found in the direction of ζ Oph.

114.126 Spectrophotometric investigation of κ Cassiopeiae. IV. B. Kovachev.
Izv. Sekts. astron. Blg. AN, Vol. 5, 47 - 63 (1972). In Bulgarian. – Abstr. in Referativ. Zhurn. 51. Astron., 4.51.730 (1973).

114.127 Spectroscopically peculiar stars near the south galactic pole. J. A. Graham, A. Slettebak.
Astron. Journ., Vol. 78, 295 - 301, 347 (1973).

A group of stars with peculiar spectroscopic characteristics near the south galactic pole is discussed. These stars were noted by Slettebak and Brundage in an objective prism survey for stars of early spectral type. In this paper, we present *uvby* photometry for 112 stars and spectroscopic data for 40 stars from the Slettebak—Brundage list.

114.128 A new shell phase in Pleione.
W. W. Morgan, R. A. White, J. W. Tapscott.
Astron. Journ., Vol. 78, 302, 349 - 351 (1973).

The reappearance of the absorption shell spectrum of Pleione has been observed on spectrograms obtained on 9 December 1972 and thereafter.

114.129 A study of B 6 stars. A. B. Underhill.
Astron. Astrophys., Vol. 25, 161 - 174 (1973).

A study of the spectra of ζ Draconis, B 6 III, β Sextantis, B 6 V, and α Leonis, B 7 V, has been made from high dispersion spectrograms which cover the spectral region 3100—6700 Å and from OAO-II spectral scans covering the spectral region 1100—3600 Å. Profiles, equivalent widths and central intensities of many lines in the spectrum of ζ Draconis and of the major lines of β Sextantis are presented. Flux envelopes for ζ Draconis and α Leonis from 1100—6050 Å are derived from published spectrum scans and from new spectrum scans obtained with OAO-II. Fitting these energy distributions to those from reference model atmospheres leads to estimates of the radii and masses of these single stars.

114.130 Spectralphotometry and quantitative analysis of the hydrogen-deficient stars HD 144941 and CPD–69°2698. K. Hunger, J. P. Kaufmann.
Astron. Astrophys., Vol. 25, 261 - 270 (1973).

Wavelengths and equivalent widths are given for absorption lines in the spectra of the hydrogen-deficient stars HD 144941 and CPD–69°2698. The coudé-spectrograms are taken with the ESO 152 cm telescope at La Silla. Fine analyses have been performed by using a grid of flux constant hydrogen-line blanketed models with varying T_{eff}, log g and ϵ_H (number fraction of H). The models are adapted to match the following criteria: H-profiles, He I-equivalent widths, the ionization equilibria of Si II/Si III and Si IV, and also the turbulent velocity.

114.131 On the chemical abundances in the weak-helium-line star α Sculptoris. B. Baschek.
Astron. Astrophys., Vol. 25, 333 - 335 (1973).

A brief discussion and comparison of the recent abundance determinations in α Scl by Schmitt (1972) and Vilhu (1972) are given.

114.132 Lithium as a stellar age indicator. S. Vauclair.
Stellar ages. Proc. IAU Colloquium No. 17, (see 012.015), XXXVIII, 1 - 12 (1973).

114.133 The metallic line A type stars. D. J. Stickland.
Stellar ages. Proc. IAU Colloquium No. 17, (see 012.015), XLVIII, 1 - 9 (1973).

114.134 Chemical composition of the alleged SMR (*super-metal-rich*) stars. G. Cayrel de Strobel.
Stellar ages. Proc. IAU Colloquium No. 17, (see 012.015), LVII, 1 - 9 (1973).

114.135 On the chemical composition and age of the high-velocity giant HD 6497. L. E. Pasinetti.
Stellar ages. Proc. IAU Colloquium No. 17, (see 012.015), LVIII, 1 - 3 (1973).

114.136 CSV 2851: an emission-line, M-dwarf star.
N. Sanduleak, C. B. Stephenson.
Inform. Bull. Variable Stars (I. A. U. Commission 27), Konkoly Obs., Budapest, No. 770 (1973).

114.137 Near-ultraviolet stellar spectra obtained with the experiment S 59 on board the European satellite TD-1A. H. J. Lamers, K. A. van der Hucht.
Bull. American Astron. Soc., Vol. 5, 267 (1973). – Abstr. AAS.

114.138 Anomalous Fe I lines in HD 125823 a Centauri. A. B. Underhill, D. A. Klinglesmith.
Astron. Astrophys., Vol. 25, 405 - 407 (1973).

Evidence is presented for the appearance of broad, shallow lines of Fe I in the spectrum of HD 125823 a Centauri between phases 0.9 and 0.22 of the helium-line variation.

114.139 The unusual red giants in M 5, M 10 and M 92. R. Zinn.
Astron. Astrophys., Vol. 25, 409 - 413 (1973).

The stars that were reported to be unusual red giants in the globular clusters M 5, M 10, and M 92 have been observed spectroscopically at a dispersion of 130 Å/mm. The unusual star in M 5 (IV – 59) and the one in M 10 (1018) have radial velocities that are consistent with cluster membership. New radial velocity data together with the available proper motion and photometric data indicate that the M 92 stars are not cluster members, but are field stars with normal compositions. The spectrograms of IV – 59 and 1018 confirm Osborn's earlier result that these stars have enhanced λ 4216 CN bands.

114.140 Catalogue of Am stars with known spectral types. B. Hauck.
Astron. Astrophys., Suppl. Ser., Vol. 10, 385 - 403 (1973).

We give a catalogue of all Am stars with known spectral types with respect to hydrogen-lines, K-line and metallic lines. This catalogue contains 418 stars.

114.141 Ricerche sulle stelle con regioni attive. C. Blanco, S. Catalano, G. Godoli.
Mem. Soc. Astron. Italiana, Nuova Ser., Vol. 43, 663 - 667 (1973).

Some results of photoelectric observations of stars with a chromospheric activity, carried out or running at Catania with reference to the research program on stellar activity of solar type, are exposed.

114.142 A spectral analysis of κ Piscium, I.
P. Galeotti, E. Lovera.
Mem. Soc. Astron. Italiana, Nuova Ser., Vol. 43, 759 - 765 (1973).

Analyzing 9 spectra of dispersion 34 Å/mm of the magnetic A2p-star κ Piscium, we calculated the spectral variations and the radial velocities. Our results are in agreement with the second of two different photometric periods proposed for this star: P = 0.5853 d and P = 0.5805 d.

114.143 Photometric and spectroscopic study of silicon Ap-star HD 193722. I. A. Aslanov, G. Heildebrandt, V. L. Khokhlova, W. Shöneich (Schöneich).
Astrophys. Space Sci., Vol. 21, 477 - 485 (1973).

The period of light variation $P = 1^d.13316$ has been found for the silicon B9 IVp star HD 193722. Spectroscopic study of this star was based on 35 spectrograms with dispersion 4 Å mm^{-1} well distributed in phase. The measurements of radial velocities of spectral line components for Si II, He I, Eu II, Fe II and Sr II allowed us to localize several regions on the surface of the star with enhanced abundance of these elements.

114.144 On the metallicity of the main-sequence stars in M67.
W. W. Morgan, H. A. Abt.
Astron. Journ., Vol. 78, 386 - 388, 441 (1973).

New determinations of the spectral types of six of the seven stars discussed by Spinrad, Greenstein, Taylor and King (1970) in M67 have been carried out; the same spectrograms used by Spinrad et al., kindly loaned by Sandage, were used for the purpose. The results do not confirm the discovery by Spinrad et al. that there is a pronounced difference in spectral types determined from the hydrogen lines and those from the metallic lines – in the sense that the metallic line types are later. Our redetermination of the spectral types shows that there is no appreciable difference between those from the H lines and those from the metallic lines. New determinations of the color excess from hydrogen and from metallic-line spectral types result in values of $E(B-V) = +0^m.04$ and $+0^m.05$, respectively.

114.145 Low-dispersion spectroscopic classification of the unidentified sources in the Two-Micron Sky Survey.
S. S. Vogt.
Astron. Journ., Vol. 78, 389 - 394, 443 (1973).

The 235 unidentified sources north of −4° declination in the Two-Micron Sky Survey have been classified using low-dispersion near-infrared spectroscopy. They are found to be predominantly late M stars with a few early M stars, nine carbon stars, and several unusual objects. The galactic-latitude distribution of the late M group is shown to be consistent with that of late-type giants. The early M stars lie extremely close to the galactic plane and are believed to be highly reddened supergiants.

114.146 Metalliniensterne. E. Hundt.
SuW, Vol. 12, 171 - 174 (1973).

114.147 Hydrogen lines in A and Ap stars. I. Photoelectric observations. D. F. Gray, J. C. Evans.
Astrophys. Journ., Vol. 182, 147 - 158 (1973).

The authors present photoelectrically measured profiles of Hγ and Hβ in twelve normal A stars and nine peculiar A stars. They find the functional relation between hydrogen line strength and color indices to be the same for these two groups of stars to within a 1 percent uncertainty.

114.148 The G-band anomaly of the asymptotic-branch stars in M92. R. Zinn.
Astrophys. Journ., Vol. 182, 183 - 187 (1973).

An examination of the spectra of 20 giants in M92 revealed that six stars, which are probably asymptotic-branch

stars, have anomalously weak G bands in their spectra. These stars probably have a low abundance of CH. It is suggested that the low abundance of CH is a result of stellar evolution and may be due to the diffusion of carbon during the blue-horizontal-branch phase.

114.149 Infrarotastronomie. M. Grewing.
Mitt. Astron. Ges., No. 32, p. 65 - 90 (1973). − Review paper presented at the assembly of the Astron. Ges., Wien, 1972 Sept.

114.150 Periodische Änderung der Linienprofile im Spektrum von 73 Draconis. K. D. Rakosch.
Mitt. Astron. Ges., No. 32, p. 252 (1973). − Abstract.

114.151 Infrarotobjekte in Aquila und Cygnus.
M. Ü. Akyol, B. Hidajat.
Mitt. Astron. Ges., No. 32, p. 260 (1973). − Abstract.

114.152 Identifizierung von IRC-Objekten.
G. V. Schultz, W. Wiemer.
Mitt. Astron. Ges., No. 32, p. 261 - 262 (1973).

114.153 Das Spektrum des Wolf-Rayet-Sterns HD 152270.
W. Seggewiß.
Mitt. Astron. Ges., No. 32, p. 277 - 278 (1973).

114.154 Observations of diffuse interstellar features in the spectra of dust-embedded and field stars.
G. E. Bromage, K. Nandy.
Astron. Astrophys., Vol. 26, 17 - 32 (1973).

Observations are presented here of twelve interstellar features of the spectra of dust-embedded and field stars in comparable direction in the galactic plane, the wavelength range being 550–700 nm. Equivalent widths, central depths, half widths and central wavelengths of these features have been studied.

114.155 The Mg II lines at 2800 Å in the spectrum of α² Canum Venaticorum.
M. Burger, K. A. van der Hucht, H. J. Lamers.
Astron. Astrophys., Vol. 26, 149 - 153 (1973).

Ultraviolet spectra of α² CVn have been obtained by the Utrecht Orbiting Stellar Spectrophotometer S 59. The observations are compared with LTE-model calculations. Upper limits of the combined Mg-abundance and microturbulent velocity are obtained.

114.156 Observations of some early-type supergiants with variable Hα profiles. J. D. Rosendhal.
Astrophys. Journ., Vol. 182, 523 - 530 (1973).

Study of multiple high-dispersion Hα spectrograms of 20 early-type supergiants indicates that 13 of these stars showed large, visually evident changes in the strength or structure of the Hα as well as in the strength of other prominent lines in the red region of the spectrum. These stars should be suitable candidates for investigating possible coupling between mass loss and turbulence.

114.157 Equivalent width data for supergiants of type G and K Ib. J. van Paradijs.
Astron. Astrophys., Suppl. Ser., Vol. 11, 25 - 40 (1973).

The equivalent widths of lines used in a curve of growth analysis of seven supergiants of type G and K Ib are presented and compared with data from other studies.

114.158 The history of astronomical spectroscopy I. Qualitative chemical analysis and radial velocities.
D. H. Menzel.
Ann. New York Acad. Sci., Vol. 198, (see 012.020), 225 - 234 (1972).

114.159 Spectrophotometry of 5 Wolf-Rayet stars in the photographic region. S. V. Rublev.
Soobshch. Spets. Astrofiz. Obs. AN SSSR, *Zelenchukskaya*, vyp. (No.) 4, 34 pp. (1971). In Russian.

114.160 New shell episode for Pleione. A. F. Gulliver.
IAU Circ., No. 2491 (1973).

114.161 γ^2 Velorum.
S. Jeffers, W. Weller, A. Sanyal.
IAU Circ., No. 2495 (1973).

114.162 γ^2 Velorum. A. Moffat, P. Kjaergaard.
IAU Circ., No. 2508 (1973).

114.163 γ^2 Velorum. S. Jeffers, W. Weller,
A. Sanyal, B. Madore.
IAU Circ., No. 2531 (1973).

114.164 Accurate wavelengths of stellar and telluric absorption lines near $\lambda 7000$ Å. R. and R.Griffin.
Monthly Notices Roy. Astron. Soc., Vol. 162, 255 - 260 (1973).
Wavelengths of 81 stellar and 159 telluric lines in the range $\lambda\lambda 6841 - 7424$ Å have been determined from 103a-U spectrograms of Arcturus and Procyon. Random errors are expected to be only 1 or 2 mÅ.

114.165 Variability of T Tauri-like stars in NGC 2264.
K. Nandy, N. Pratt.
Astrophys. Space Sci., Vol. 19, 219 - 224 (1972).
The variability of the T Tauri-like stars in NGC 2264 in U, B, V, R and I colours has been studied. It is found that the range of variability in amplitude in I is less than in U, B and V, A method of determining relative opacities at these wavelengths from the variability in different colours of these dust embedded stars is also described.

114.166 Spectrophotometry of i Herculis. N. L. Ivanova.
Soobshch. Byurakan. Obs., vyp. (No.) 44, p. 81 - 85 (1972). In Russian.
Two spectra of i Herculis obtained in the coudé focus of the 2-m telescope of the Shemakha Observatory are investigated. The following physical parameters of the star are obtained: the electron density, the electron pressure, the number of atoms in the second quantum state in a column with one sm² cross-section, the Balmer jump, the thickness of the homogeneous atmosphere, and the gravitational acceleration on the star surface.

114.167 Observations of γ Cas in 1968 - 1970.
N. L. Ivanova, N. K. Andreasian.
Soobshch. Byurakan. Obs., vyp. (No.) 44, p. 86 - 90 (1972).
In Russian.
The results of a spectrophotometric investigation of γ Cas in 1968 - 70 are given.

114.168 Spectrophotometric study of weak-line and strong-line stars. C. Casini, L. E. Pasinetti.
Contr. Oss. Astron. Milano-Merate, Nuova Ser., No. 339, 19 pp. (1971).
In the present work we examine some weak-lines and strong-lines stars of spectral type F6–G0, studied by Roman (1950).

114.169 Spectrophotometric study of weak-CN and strong-line stars. C. Casini, L. E. Pasinetti.
Rend. Ist. Lombardo A, Vol. 105, 966 - 973 = Contr. Oss. Astron. Milano-Merate, Nuova Ser., No. 347 (1971/72).
In the present work we have studied some wk-CN stars (Roman, 1952) of G8 III type.

114.170 Rotation and shell spectra among A-type dwarfs.
H. A. Abt, K. I. Moyd.
Astrophys. Journ., Vol. 182, 809 - 816 (1973).
Rotational velocities for 66 metallic-line and 123 normal A5–A9 IV or V stars are given and used to determine the frequency distributions of equatorial rotational velocities. Among the 35 most rapidly rotating normal stars, eight were found to have shell spectra.

114.171 The red star in the open cluster Trumpler 27.
H. Albers.
Astrophys. Journ., Vol. 182, 817 - 819 (1973).
The spectrum of a peculiar red star in the cluster Tr 27 is described and its relationship to other spectral classes is discussed.

114.172 The application of wavelength coincidence statistics to line identification: HR 465 and HR 7575.
M. R. Hartoog, C. R. Cowley, A. P. Cowley.
Astrophys. Journ., Vol. 182, 847 - 858 (1973).
A Monte Carlo technique for the experimental determination of probabilities has been used to supplement standard line-identification procedures in the Ap stars HR 465 and HR 7575. The resulting identifications are divided into three categories ranging from definite to possible identifications. The identification of U II and Pt II in HR 465, the possible identification of Te II, Os II, and Pm II in HR 465, and the possible identification of Se II and Cs II in HR 7575 are discussed in more detail.

114.173 OAO-2 and Mariner 9 ultraviolet observations of δ Persei. M. R. Molnar.
Publ. Astron. Soc. Pacific, Vol. 85, 307 - 308 (1973).
The star δ Per is identified as a variable star from OAO-2 photometry although the data are insufficient to determine the period or the exact nature of the photometric variations. The Mariner 9 spectrometer observations show a strong variable emission feature at $\lambda 1650$, tentatively identified as Ca II (UV (1) multiplet).

114.174 Spectra of some peculiar and luminous bright stars.
A.Cowley.
Publ. Astron. Soc. Pacific, Vol. 85, 314 - 316 (1973).
Spectral classifications on the Yerkes system are given for some newly-recognized peculiar and luminous bright stars. The high frequency of composite spectra with primaries which are bright-giant or supergiant F and G stars is noted.

114.175 Classification of some bright F-type stars with unusual spectra. S. Malaroda.
Publ. Astron. Soc. Pacific, Vol. 85, 328 - 329 (1973).
The spectral classes of a group of F-type stars with unusual spectra are given and a reclassification of F-type supergiants is provided.

114.176 On some peculiarities of the UV-spectra of WR stars.
A. A. Nikitin, T. Kh. Feklistova.
Trudy Astron. Obs., *Leningrad*, Vol. 29 (= Uchenye Zapiski Leningr. Un-ta, No. 363 = Seriya Matem. Nauk, vyp. (No.) 48), p. 39 - 44 (1973). In Russian.
Spectral schemes of C III, N IV and O V as well as their laboratory spectra are used to study the possible structure of the UV spectra of WR stars. Laboratory and stellar IR spectra of C II, C III and other elements enable one to reveal some peculiarities of their UV spectra.

Spectra of rapidly rotating objects.
Sky Telescope, Vol. 45, 226 (1973).

Centre de Données Stellaires, Inform. Bull. No. 4.
See Abstr. 002.007.

Celescope catalog of ultraviolet stellar observations. 5068 objects measured by the Smithsonian experiment aboard the Orbiting Astronomical Observatory (OAO-2). See Abstr. 003.044.

Multiplets in astrophysics.   See Abstr. 022.083.

Recent progress in infrared and microwave techniques of astronomical interest.   See Abstr. 031.014.

Quantitative performance of single- and two-stage image tubes in spectroscopy.   See Abstr. 034.065.

Application of an image isocon and computer to direct digitization of astronomical spectra. See Abstr. 034.075.

Application of new ultraviolet television detectors in an astronomical satellite.   See Abstr. 034.083.

Spectroscopy in the Madrid astronomical observatory.   See Abstr. 034.120.

Stellar astronomy objectives.   See Abstr. 051.012.

On the atmospheric abundances of seven Am SB2 systems.   See Abstr. 064.006.

The metallic-line star 15 UMa and the F 5 V star 5 And.   See Abstr. 064.007.

Analyses of light-ion spectra in stellar atmospheres. III. Nitrogen III in the O stars.   See Abstr. 064.012.

Comparison of Celescope magnitudes with model-atmosphere predictions for A, F, and G supergiants. See Abstr. 064.014.

Wolf-Rayet stars. V. The temperature stratification. See Abstr. 064.026.

The ultraviolet flux envelopes of main-sequence B stars.   See Abstr. 064.057.

Microturbulence in A stars as derived from line profiles.   See Abstr. 064.065.

The atmosphere of the hydrogen-deficient star HD 96446.   See Abstr. 064.073.

A model-atmosphere abundance analysis of the B9V star Nu Capricorni.   See Abstr. 064.074.

The atmosphere of Epsilon Leonis. See Abstr. 064.076.

Molecular dissociation equilibria in SC stars. See Abstr. 064.081.

Ages and chemical compositions of stars. See Abstr. 065.132.

Metallicism and evolution of the Am stars. See Abstr. 065.133.

BD−10°4662 interpreted as a post−T Tauri star. See Abstr. 065.144.

The radial velocity variations of HD 125823 a Centauri.   See Abstr. 112.006.

On the possibility of determining stellar radial velocities to 0.01 km s^{-1}.   See Abstr. 112.014.

Near infra-red magnitudes of 248 early-type emission-line stars and related objects.   See Abstr. 113.007.

Photoelectric UBV photometry of late-type stars in two regions at high galactic latitude.   See Abstr. 113.018.

UBV photometry of selected Ap stars. See Abstr. 113.022.

Speckle interferometry: Color-dependent limb darkening evidenced on Alpha and Omicron Ceti. See Abstr. 113.024.

Photometry of metal-deficient A-type stars found on objective-prism plates.   See Abstr. 113.056.

La classification stellaire BCD: Paramètre caractéristique du type spectral calibration en magnitudes absolues. See Abstr. 115.003.

A C IV emission feature in the near-ultraviolet spectrum of the Wolf-Rayet star γ^2 Velorum. See Abstr. 119.011.

Line strengths in the spectra of seven double-lined Am binary stars.   See Abstr. 119.016.

A search for rapid spectral variations in 10 Lacertae. See Abstr. 122.075.

Interstellar molecular hydrogen observed in the ultraviolet spectrum of Delta Scorpii.   See Abstr. 131.014.

Infrared circular polarization of NML Cygni and VY Canis Majoris.   See Abstr. 131.018.

New H$_2$O sources associated with infrared stars. See Abstr. 131.050.

Interstellar molecular hydrogen detected in the UV spectrum of δ Sco.   See Abstr. 131.082.

Recent polarization observations in the 1.25 to 3.6 micron wavelength range.   See Abstr. 131.092.

Interstellar gas abundances from rocket observations of ultraviolet absorption lines.   See Abstr. 131.137.

Infrared 10 μ emission from condensation nuclei of interstellar grains.   See Abstr. 131.152.

Mariner 9 ultraviolet spectrometer experiment: Interstellar absorption at Lyman alpha in OB stars. See Abstr. 131.165.

Polarization changes of some M type supergiant stars.   See Abstr. 131.195.

The dipole nebula IC 2220, a southern reflection nebula around the variable red giant HD 65750. See Abstr. 132.008.

Ring nebulae and Wolf-Rayet stars; observations of NGC 2359.   See Abstr. 132.025.

The spectrum of HDE 226868 (Cygnus X-1). See Abstr. 142.022.

Spectroscopic observations of the Cygnus X-1 optical candidate.   See Abstr. 142.023.

Spectroscopic observations of the optical candidate for Cygnus X-1.   See Abstr. 142.024.

On X Persei as a stellar X-ray source: Comparison with γ Cassiopeiae.   See Abstr. 142.027.

Galactic X-ray polarimetry and high-resolution X-ray spectroscopy.   See Abstr. 142.066.

HD 153919 and the X-ray source 2 U 1700–37. See Abstr. 142.144.

Some characteristics of the Eta Carinae complex. See Abstr. 152.001.

A young stellar group in the vicinity of R Coronae Austrinae.   See Abstr. 152.002.

Emission-line stars in the Chamaeleon T association. See Abstr. 152.003.

General properties of red giants. See Abstr. 153.021.

A search for Ap stars in open clusters. See Abstr. 153.030.

The brightest stars in the Small Magellanic Cloud. See Abstr. 159.005.

115   Stellar Luminosities, Masses, Diameters, HR-Diagrams and Others

115.001 Note on terminology – specific luminosity.
R. S. Kandel.
Astron. Astrophys., Vol. 22, 155 - 156 (1973).
A new term – specific luminosity – is defined for use in the description of nonisotropic astrophysical sources. Its relation to other quantities of astrophysics and photometry is discussed.

115.002 Luminosity functions of stars in the Praesepe cluster.   N. M. Artiukhina.
Astron. Zhurn. Akad. Nauk. SSSR, Vol. 50, 101 - 106 (1973). In Russian. English translation in Soviet Astron. AJ, Vol. 17, No. 1.
Luminosity functions of stars brighter than 12^m5 pg $(M_{pg} = +6^m5)$ for the nucleus and corona regions of the Praesepe cluster are derived. Photographic magnitudes of the probable members of the cluster are derived. A figure shows luminosity functions of stars brighter than $M_{pg} = +6^m0$ in the regions of nucleus and corona for the α Persei, Pleiades and Praesepe clusters.

115.003 La classification stellaire BCD: Paramètre caractéristique du type spectral calibration en magnitudes absolues.   D. Chalonge, L. Divan.
Astron. Astrophys., Vol. 23, 69 - 79 (1973).
Since the beginning of the new series of spectrophotometrical research described in four papers, the instruments and methods have been improved and several hundreds of stars were observed and classified in the BCD system. From these new data is deduced a more accurate twin-diagram $\lambda_1 D$ for the classification of the stars referred to as "normal". The $\lambda_1 D$ diagram has been calibrated in absolute magnitudes for O, B and the first sub-classes A, from $M = 0$ to $M = -8$.

115.004 Die Durchmesserbestimmung von Sternen mit

interferometrischen Methoden.   P. Buser.
Orion Schaffhausen, 31. Jahrgang, p. 7 - 12 (1973).

115.005 Mean absolute magnitudes and dispersions for selected spectral groups.
S. W. McCuskey, R. S. McMillan.
Astron. Journ., Vol. 78, 73 - 78 (1973).
Mean visual absolute magnitudes and dispersions have been derived from frequency distributions of apparent magnitude, total proper motion and radial speed for the following stellar groups: B8–A3, $5.50 \le m \le 6.50$; A0, $m \le 6.50$; F0-F5, $5.50 \le m \le 6.50$; G8-K3, $4.00 \le m \le 6.00$; and gM0-M4, $4.50 \le m \le 6.50$. Data were taken from the Yale Bright Star Catalogue.

115.006 Luminosity functions for K giant stars derived from the two-micron sky survey.   E. E. Hughes, Jr.
Publ. Astron. Soc. Pacific, Vol. 85, 57 - 67 (1973).
This paper presents a method for determining either the space density or the luminosity function from star counts covering large areas of sky. Space density is assumed to vary only in the direction perpendicular to the galactic plane. Application of the method to a selection of IRC stars dominated by K giants shows that if these stars obey Oort's determination of their normalized space density perpendicular to the galactic plane then the dispersion σ of their 2.2 μ luminosity distribution must be large.

115.007 Possible horizontal-branch stars at high galactic latitudes. VI.   A. G. D. Philip.
Publ. Astron. Soc. Pacific, Vol. 85, 68 - 69 (1973).
Positions, B magnitudes, and spectral types are given for ten new possible field horizontal-branch stars in two regions, the north and south galactic poles.

115.008 **Lunar occultations from Cerro Tololo I. The carbon star, TX Piscium.**
B. M. Lasker, S. B. Bracker, W. E. Kunkel.
Publ. Astron. Soc. Pacific, Vol. 85, 109 - 111 (1973).

Observations from a lunar occultation of the carbon star, TX Psc, give a diameter of 0.″009 ± 0.001 for a uniformly illuminated disk. This diameter and Eggen's (1972) photometric parameters imply a radius of 120 $R_\odot$ or an effective temperature of 3140°K.

115.009 **The angular diameter of Upsilon Capricorni and an occultation of SAO 118655.** D. W. Dunham,
D. S. Evans, E. C. Silverberg, J. R. Wiant.
Astron. Journ., Vol. 78, 199 - 201 (1973).

The angular diameter of υ Cap was determined on 29 June 1972 at occultation using the McDonald 107-inch telescope to be 4.1 arc msec for a uniform disk and 4.7 arc msec for a fully darkened disk. A double observation of SAO 118655 on 24 April at the 107-inch and 36-inch telescopes yields a time difference in good accordance with their relative positions and the lunar limb slope determined for the latter observation.

115.010 **The classification of intrinsic variables. III. Calibration of the luminosities of small amplitude red variables in the old disk population.** O. J. Eggen.
Astrophys. Journ., Vol. 180, 857 - 870 (1973).

Nineteen small-amplitude (less than 0.5 mag in V_E) red variables in the Hyades and old disk population groups, including 14 new variables (μ Gem, HR 4333, ϵ Oct, HD 80567, HR 4184, HR 4463, ϵ Mus, HR 1003, HD 111499, HR 7625, 27 Cnc, HD 115332, HD 123598 and HR 5331) are used to calibrate the (M_{bol}, $R-I$) relation.

115.011 **Infrared luminosity vs. stellar luminosity in H II regions.** H. M. Johnson.
Bull. American Astron. Soc., Vol. 5, 25 (1973). – Abstr. AAS.

115.012 **Hα emission-line stars in the Large Magellanic Cloud: a quick look at the colour-magnitude diagram.**
B. Bohannan.
Bull. American Astron. Soc., Vol. 5, 25 - 26 (1973). – Abstr. AAS.

115.013 **The mean absolute magnitude of the Am stars.**
S. Rocha, C. Jaschek.
Bol. As. Argentina Astron., No. 16, (see 012.007), p. 32 (1971). – Abstract.

115.014 **Photometry of field horizontal-branch stars.**
A. G. D. Philip.
Stellar evolution, (see 012.014), p. 397 - 417 (1972).

115.015 **The Hertzsprung-Russell diagram and stellar ages.**
G. Larsson-Leander.
Stellar ages. Proc. IAU Colloquium No. 17, (see 012.015), II, 1 - 30 (1973). – Review paper.

115.016 **Age from location in the H-R diagram. I. Concluding remarks.** O. J. Eggen.
Stellar ages. Proc. IAU Colloquium No. 17, (see 012.015), X, 1 - 3 (1973).

115.017 **Construction des diagrammes des groupes d'étoiles sur la base de résultats des calculs théoriques.**
O. Dloujnevskaia (*Dluzhnevskaya*).
Stellar ages. Proc. IAU Colloquium No. 17, (see 012.015), XII, 1 - 2 (1973).

115.018 **Horizontal branch morphology.** R. T. Rood.
Stellar ages. Proc. IAU Colloquium No. 17, (see 012.015), XX, 1 - 10 (1973).

115.019 **Photometric detection of red horizontal branch stars.** M. Grenon.
Stellar ages. Proc. IAU Colloquium No. 17, (see 012.015), XXIX, 1 - 6 (1973).

115.020 **Age from location in the H-R diagram. II. Concluding remarks.** J. Delhaye.
Stellar ages. Proc. IAU Colloquium No. 17, (see 012.015), XXXI, 1 - 2 (1973). In French.

115.021 **Age measurements of G type subgiants.**
J. B. Hearnshaw.
Stellar ages. Proc. IAU Colloquium No. 17, (see 012.015), XLI, 1 - 18 (1973).

115.022 **Age des étoiles B à émission.** R. Herman.
Stellar ages. Proc. IAU Colloquium No. 17, (see 012.015), XLIV, 1 - 6 (1973).

115.023 **Measuring the angular diameters of stars.**
R. Hanbury Brown.
Uspekhi fiz. nauk, Vol. 108, 529 - 547 (1972). In Russian.
Abstr. in Referativ. Zhurn. 51. Astron., 5.51.296 (1973).

115.024 **Photometric information on globular cluster stars from the Palomar Sky Survey prints.**
V. Castellani, F. A. D'Antona, R. De Amicis, F. Smriglio.
Astrophys. Space Sci., Vol. 22, 71 - 78 (1973).

Photometric information that can be taken from the Palomar Sky Survey prints can be used to discriminate among the various evolutionary stages of stars that are members of globular clusters, as well as to obtain some idea of the HR diagram for very faint stars. For a test case, it is shown that in a check analysis for M3 the known turn-off luminosity is given to within about 0.5 mag. The globular cluster NGC 5466 is examined and it is concluded that no turn-off occurs before $P \approx 20$ mag.

115.025 **Apparent diameters of 172 B5V–A5V stars of the catalogue of Geneva Observatory.**
M. Fracassini, G. Gilardoni, L. E. Pasinetti.
Astrophys. Space Sci., Vol. 22, 141 - 152 (1973).

The spectrophotometric method for computing the apparent stellar diameters, proposed by Chalonge and Divan (1950) and modified by Fracassini and Pasinetti (1967) has been applied to 172 B5V–A5V stars (single and belonging to galactic clusters) of the catalogue of Geneva Observatory (Rufener, 1971).

115.026 **On the influence of the chemical composition of stars on the position of the initial main sequence.**
Yu. N. Efremov, I. M. Kopylov.
Soobshch. Spets. Astrofiz. Obs. AN SSSR, *Zelenchukskaya*, vyp. (No.) 3, p. 17 - 27 (1971). In Russian.

Unusual objects and high energy astronomy.
See Abstr. 051.016.

Age determination and theoretical luminosity functions for clusters of an intermediate population II.
See Abstr. 065.126.

The correlation of parallax errors and the intrinsic dispersion of the K3–M2 main sequence.
See Abstr. 111.003.

The infrared spectrum and angular size of Eta Carinae. See Abstr. 114.030.

Feasibility of UV astronomy by balloon-borne observations. II. Stellar gradients in the near ultraviolet. See Abstr. 114.033.

Energy distribution in the stellar spectra of different spectral types and luminosities. See Abstr. 114.090.

Metallieniensterne.   See Abstr. 114.146.

Hertzsprung's luminosity-colour relation for visual binaries.   See Abstr. 118.005.

The spectral classification of the β Cephei stars and their location in the theoretical Hertzsprung-Russell diagram. See Abstr. 122.003.

Non-pulsating stars in the RR Lyrae and cepheid instability strips.   See Abstr. 122.040.

The age and evolutionary state of the β Canis Majoris variables.   See Abstr. 122.084.

Blanketing theory and the $G-I$ index. See Abstr. 126.016.

Infrared nebular luminosity versus stellar luminosity in five H II regions.   See Abstr. 131.184.

Luminosity of thermal X-ray sources with a strong magnetic field.   See Abstr. 142.086.

On the use of UBV photometric diagrams for inferring the existence of an open star cluster. See Abstr. 153.023.

Current problems on horizontal branch stars. I: H. B.-topology.   See Abstr. 154.012.

Current problems on horizontal branch stars. II: H. B.-population.   See Abstr. 154.013.

Four-color observations of early-type stars. III. M 4. See Abstr. 154.014.

The effects of a variation in the CNO abundances on the position of initial horizontal-branch models. See Abstr. 154.018.

New M dwarfs in the south galactic cap. See Abstr. 155.007.

Luminosity and velocity distribution of high-luminosity red stars near the sun. I. The very young disk population. See Abstr. 155.099.

The brightest stars in the Small Magellanic Cloud. See Abstr. 159.005.

116   Stellar Magnetic Field, Figure, Rotation

116.001   **Origin of stellar magnetic fields.**
   P. Raychaudhuri.
Astrophys. Space Sci., Vol. 18, 425 - 436 (1972).
   We discuss the present status of the theory of the origin of magnetic fields in stars, magnetic variability of the star and their interpretation from the theoretical point of view, possible origin of the stellar magnetic fields associated with stellar evolution.

116.002   **Photometric investigations of magnetic stars.**
   M. J. Stift.
Astron. Astrophys., Vol. 22, 209 - 215 (1973).
   The first part of this paper deals with UBV photoelectric observations of four peculiar stars: HD 358, HD 32633, HD 215441, and HD 25823. The analysis yields respective periods for the first three stars of 0.9636 days, 6.431 days, and 9.4866 days; no period can be found for the star HD 25823. In the second part, assuming the oblique rotator model, we determine the spatial distribution of the rotational axes of the magnetic stars.

116.003   **HD 215441 and 53 Camelopardalis: Intrinsic polarization of Hβ and the continuum.**

J. C. Kemp, R. D. Wolstencroft.
Astrophys. Journ., (*Letters*), Vol. 179, L33 - L37 (1973).
   Variable linear polarization has been detected in the Hβ absorption line of the magnetic Ap stars HD215441 and 53 Cam, probably caused by the transverse Zeeman effect at least in the latter case. The relationship to the continuum polarization proves that both the Hβ and line polarizations are intrinsic in the two stars.

116.004   **Ultraviolet photometry from the Orbiting Astronomical Observatory. VII. α^2 Canum Venaticorum.**
   M. R. Molnar.
Astrophys. Journ., Vol. 179, 527 - 537 (1973).
   Far-ultraviolet photometry by OAO-2 is presented for α^2 CVn covering the entire 5^d47 period of this magnetic Ap variable. The light curves ranging from $\lambda 1330$ to $\lambda 3320$ indicate that there are two important sources of back warming from far-ultraviolet absorption. Strong line blanketing by the rare earths redistributes flux longward of the Balmer discontinuity, causing the major observed photometric variations. In addition, a second source, which may be due to a combination of continuous opacities and line blanketing from the iron-peak and rare-earth groups below $\lambda 1600$, apparently redistributes flux

into the region of the Balmer discontinuity.

116.005 The projected rotational velocity for 101 southern OB stars.   E. N. Walker.
Observatory, Vol. 93, 75 - 77 (1973).

116.006 Stellar rotation and age determinations.
A. Maeder.
Stellar ages. Proc. IAU Colloquium No. 17, (see 012.015), VII, 1 - 8 (1973).

116.007 Need of planned observations of magnetic stars.
C. Blanco, F. A. Catalano, G. Godoli, S. Vaccari.
Inform. Bull. Variable Stars (I.A.U. Commission 27), Konkoly Obs., Budapest, No. 761, 4 pp. (1973).

116.008 Ricerche sulle stelle magnetiche.
C. Blanco, F. A. Catalano, G. Godoli, S. Vaccari.
Mem. Soc. Astron. Italiana, Nuova Ser., Vol. 43, 655 - 661 (1973).
Photoelectric observations of magnetic stars carried out or running at Catania with reference to the research program on stellar activity of solar type, are reported. Furthermore, preliminary results of a comparative analysis of all of the available magnetic, spectroscopic and photometric observations of magnetic stars, are presented.

116.009 Magnetische Sterne.   K. D. Rakosch.
Mitt. Astron. Ges., No. 32, p. 33 - 54 (1973). − Review paper presented at the assembly of the Astron. Ges., Wien, 1972 Sept.

116.010 Genesis and classification of magnetic stars. I.
E. M. Drobishevsky.
Astrofizika, Vol. 9, 119 - 138 (1973). In Russian. − English translation in Astrophysics, Vol. 9, No.1.
A synthesis of ideas concerning generation of magnetic fields and its relict nature in stars is achieved.

About the influence of a magnetic field on the model atmosphere of a magnetic star.
See Abstr. 064.043.

A post-Newtonian study of differentially rotating polytropes.   See Abstr. 065.014.

Models for rapidly rotating pre-main sequence stars.
See Abstr. 065.015.

Spectra of rapidly rotating objects.
See Abstr. 065.021.

Rapidly rotating stars. VIII. Zero-viscosity polytropic sequences.   See Abstr. 065.022.

On the oscillations and stability of rapidly rotating stellar models. III. Zero-viscosity polytropic sequences.
See Abstr. 065.023.

Rotating magnetosphere: A simple relativistic model.   See Abstr. 065.025.

The adiabatic stability of stars containing magnetic fields − I. Toroidal fields.   See Abstr. 065.062.

A variational principle for magnetoelastic, rotating stars in general relativity.   See Abstr. 065.100.

Stellar magnetism and rotation.
See Abstr. 065.115.

Toroidal and poloidal oscillations of magnetic rotating stars in a decay field.   See Abstr. 065.139.

On the angular momentum of stars and the main sequence stellar models.   See Abstr. 065.140.

Stellar magnetohydrodynamics.
See Abstr. 065.141.

Magnetic fields in rapidly rotating stars.
See Abstr. 065.156.

Pinch instabilities in magnetic stars.
See Abstr. 065.157.

Meridian circulation in differentially rotating stars.
See Abstr. 065.160.

The role of the Coriolis force on the stability of rotating magnetic stars and the origin of convective motions.
See Abstr. 065.162.

Magnetic fields of the sun and stars.
See Abstr. 080.049.

Feasibility of UV astronomy by balloon-borne observations. II. Stellar gradients in the near ultraviolet.
See Abstr. 114.033.

Spectrophotometric study of the magnetic variable star α^2 CVn using high-resolution spectrograms. II. Some features of spectral variability.   See Abstr. 114.105.

Rotation and shell spectra among A-type dwarfs.
See Abstr. 114.170.

A search for Ap stars with very long periods.
See Abstr. 122.074.

The internal magnetic fields of white dwarfs.
See Abstr. 126.031.

Some limits on cosmic-ray heating of HI clouds by magnetic stars.   See Abstr. 131.003.

On cosmic rays and magnetic stars.
See Abstr. 143.028.

117 Binary and Multiple Stars, Theory

117.001 Tidal evolution in close binary systems, II.
    Z. Kopal.
Astrophys. Space Sci., Vol. 18, 287 - 305 = Lunar Sci. Inst., *Houston, Texas,* Contr. No. 100 (1972).

The aim of the present investigation will be to determine the explicit forms of differential equations which govern secular perturbations of the orbital elements of close binary systems in the plane of the orbit, arising from the lag of dynamical tides due to viscosity of stellar material. The results obtained are exact for any value of orbital eccentricity comprised between $0 \leqslant e < 1$; and include the effects produced by the second, third and fourth-harmonic dynamical tides, as well as by axial rotation with arbitrary inclination of the equator to the orbital plane.

117.002 Families of isoenergetic escapes and ejections in the problem of three bodies.   V. Szebehely.
Astron. Astrophys., Vol. 22, 171 - 177 (1973).

A triple stellar system with mass ratios 4:1:1 is subjected to isoenergetic variations and its dynamical behavior is found. The initial positions of the two small masses are fixed and the third body is placed so as to keep the potential energy of the system constant. The behavior of this dynamical system displays a series of continuous transitions between escapes, ejections and interplays. The dynamical characteristics of the ejected or escaping body and of the binary left behind are presented and the escape or ejection parameters are evaluated.

117.003 A modified Lucy model for W Ursae Majoris systems.
    D. L. Moss, J. A. J. Whelan.
Monthly Notices Roy. Astron. Soc., Vol. 161, 239 - 248 (1973).

The model of Lucy for W UMa systems is modified and model contact binary systems are constructed with unequal specific entropies in the adiabatic parts of the convection zones of the components, with energy being allowed to be transferred between the superadiabatic parts of the convection zones. The effect of a difference in mixing length between primary and secondary is also investigated.

117.004 Evolution of contact binaries and star models for cataclysmic binaries.   P. Biermann, H.-C. Thomas.
Astron. Astrophys., Vol. 23, 55 - 61 (1973).

We present calculations of the evolution of contact binary systems. From the observed light curve we can deduce some properties of the common convective envelope. During evolution of the systems mass ratio and period increases. The observed distribution of mass ratios in the color-period diagram is in accord with the picture. In addition we calculated some models for cataclysmic binary systems, consisting of a main sequence star filling its Roche lobe and a white dwarf. We discuss briefly the properties of these systems.

117.005 The peculiar stellar system G 95−57 A/B and G 95−59.   W. F. van Altena.
Astrophys. Journ., Vol. 179, 865 - 867 (1973).

Astrometric observations at the Yerkes Observatory of G 95−59 in conjunction with recomputed proper motions and published parallaxes of G 95−57 A/B, and spectroscopic and photometric observations by Eggen and Greenstein, indicate that (1) the stellar system is not gravitationally bound; (2) the spectroscopic subdwarfs G 95−57 A/B lie 0.90 mag below the Hyades main sequence in the $(M_V, B - V)$-plane; (3) G 95−59 lies 6 mag above and 0.5 mag to the red of the main sequence in the $(M_V, B - V)$-plane, while it lies near the main sequence in the $(M_I, R - I)$-plane.

117.006 Wide moving pairs among K and M dwarfs.
    P. K. Lü, A. R. Upgren.
Astrophys. Journ., Vol. 180, 91 - 97 (1973).

From a sample of 417 K3 V−M2 V stars listed in the McCormick or Michigan catalogs of nearby dwarfs for which space velocities can be determined, the incidence of wide pairs was found to be high and to include about 60 percent of the stars. This incidence is shown to be far higher than that which could be produced by a random distribution of the space velocities.

117.007 Symbiotic star BF Cygni from 1965 to 1970.
    P. Merlin.
Astron. Astrophys., Vol. 23, 363 - 367 (1973). In French.

Sixteen spectra of the symbiotic star BF Cygni covering the period 1965−1970 are investigated. Variations of the line intensities, the Balmer decrements, the color and electronic temperatures are given. Approximate values of the electron density are deduced in order to derive an order of magnitude for the sizes of the emitting regions of BF Cygni. A crude stratification and evolution model is proposed.

117.008 On nova explosions in close binary systems.
    A. V. Tutukov, L. R. Yungelson.
Astrofizika, Vol. 8, 381 - 386 (1972). In Russian. English translation in Astrophysics, Vol. 8, No. 3.

Assuming that explosive phenomena such as those observed for novae occur in binary systems in which a white dwarf accretes mass from an expanded companion, evaluations have been made of the time intervals between successive explosions, the masses of ejected envelopes, and the energy of explosions.

117.009 The distortion of line profiles and velocities in the spectra of contact binaries.   J. B. Hutchings.
Astrophys. Journ., Vol. 180, 501 - 515 (1973).

Use is made of physical models of four contact binary systems, to synthesize line profiles in their spectra. Eclipse and tidal distortion effects displace the profiles at all phases to values which over-estimate the stellar velocities while blending effects, when present, tend to produce an opposing displacement. The four systems are placed on the theoretical H-R and mass-luminosity diagrams.

117.010 Multiplicity of the Sirius system.   I. W. Lindenblad.
    Astron. Journ., Vol. 78, 205 - 207 (1973).

Relative, multiple exposure, photographic positions of the components of Sirius are reported for the years 1970−1972. These positions extend a series of observations which was begun in 1965 (Lindenblad 1970). An analysis of the entire series does not support a suspected short period perturbation and indicates the need for orbit revision.

117.011 On the envelopes of close cool members of some close binaries.   R. H. Koch.
Acta Astron., Vol. 23, 31 - 35 (1973).

A spectral type criterion is developed to test whether cool members of selected close binaries have radiative or convective envelopes. The latter structure, already known by other theoretical and observational evidence, is shown to be correct.

117.012 Orbital motion and parallax of the unresolved binary BD +67°552 from plates taken with the Sproul 24-inch refractor.   S. L. Lippincott.
Bull. American Astron. Soc., Vol. 5, 7 (1973). − Abstr. AAS.

117.013 Astrometric analysis of the triple system BD +66°34.

J. L. Hershey.
Bull. American Astron. Soc., Vol. 5, 7 (1973). — Abstr. AAS.

117.014 Reflection in close binary stars. D. B. Wood.
Bull. American Astron. Soc., Vol. 5, 42 (1973).
Abstr. AAS.

117.015 Black holes in binary systems: observational appearances. N. I. Shakura, R. A. Sunyaev.
IAU Symposium No. 55, (see 012.002), p. 155 - 164 (1973).

117.016 Multiple systems.
Journ. Roy. Astron. Soc. Canada, Vol. 67, 65 - 68 (1973). — Presented at IAU Colloquium No. 18, (see 012.004).

117.017 On the evolutionary stage of V 448 Cygni.
M. Kumsiashvili, L. Yungelson.
Nauchn. Informatsii, vyp. (No.) 23, p. 74 - 81 (1972) In Russian.

On the basis of evolutionary sequences of close binary models with mass exchange, the evolutionary stage of V 448 Cygni and of some other massive binaries is investigated.

117.018 Evolutionary stages of some close binary systems.
T. Galkina, G. Rodionova, L. Yungelson.
Nauchn. Informatsii, vyp. (No.) 23, p. 82 - 84 (1972). In Russian.

The evolutionary stages of close binaries AO Cas, HD 47129, α Vir, HD 190918, HD 193793 are investigated comparing the spectrophotometrical data with evolutionary sequences of mass exchanging close binaries.

117.019 The evolution of a contact binary.
J. Hazlehurst, E. Meyer-Hofmeister.
Astron. Astrophys., Vol. 24, 379 - 392 (1973).

The evolution of a contact binary with a primary of $1.5 M_\odot$ and a secondary of $1.1 M_\odot$ is studied. We find that mass flows from the primary to the secondary component and we estimate that the system would reach equal mass after about 10^8 yr. The relative frequency and other properties of the W UMa systems are discussed in the light of these results.

117.020 Solution of the transfer equation in the atmospheres of binaries. V. P. Merezhin.
Trudy Kazan. Gorod. Astron. Obs., No. 38, p. 3 - 15 (1972). In Russian.

117.021 Extrasolar planetary systems. S.-S. Huang.
Icarus, Vol. 18, 339 - 376 (1973).

The article deals with the occurrence of planetary systems in the universe. In Section I, the terms "planet" and "planet-like objects" are defined. In Section II, the observational search for extrasolar planetary systems is described, as performable by earthbound optical telescopes, by space probes, by long baseline radio interferometry, and finally by inference from the reception of signals sent by intelligent beings in other worlds. In Section III we show that any planetary system must be preceded by a rotating disk of gas and dust around a central mass. In Section IV, a brief review of theories of the formation of the solar system is given along with a proposed scheme for classification of these theories. In Section V, the evidence for magnetic activity in the early stages of stellar evolution is presented, as developed from six independent clues. The magnetic braking theories of solar and stellar rotation are discussed in Section VI, thereby introducing the idea of formation of a rotating disk of gas and dust around stars in Section VII. Section VIII gives an estimate for the frequency of occurrence of planetary systems in the universe.

117.022 Dynamical tides in close binary systems. I. The case of an isothermal envelope. Yu. P. Korovyakovsky.
Astrofiz. Issled., Izv. Spets. Astrofiz. Obs., Vol. 4, 115 - 129 (1972). In Russian.

Tidal phenomena are considered in the envelope of the cooler component of a close binary system whose axial rotation period differs from its period of revolution around the center of mass. A numerical integration is performed of the non-linear equations of the dynamical tides theory. The conditions of matter detachment from the satellite are found for the case when the inner Roche sphere is not filled.

117.023 On the comparison between theoretical and empirical limb-darkening coefficients from the determination of elements of close binary systems.
M. E. Choodnovskij (*Chudnovskij*), A. M. Shulberg.
Astron. Tsirk., No. 754, p. 7 - 8 (1973). In Russian.

117.024 Orbital motion and parallax of the two unresolved astrometric binaries BD +6°398 and BD +67°552 from plates taken with the Sproul 24-inch refractor.
S. L. Lippincott.
Astron. Journ., Vol. 78, 303 - 306 (1973).

Two nearby stars are found each of which has an unseen companion. The components of BD +6°398, Sproul $\pi_{abs} =$ +0".130 ± 0".003 (p.e.), have $\Delta m > 5$; the mass for the unseen component ~ $0.10 M_\odot$. A period of 50 yr satisfies the observations. BD +67°552, Sproul $\pi_{abs} = $ +0".065 ± 0".002 has an unseen companion some two to three magnitudes fainter than the primary with a mass ~ $0.3 M_\odot$; a period of 23 yr satisfies the observations.

117.025 Close binaries. B. Paczyński.
Stellar evolution, (see 012.014), p. 271 - 288 (1972).

117.026 Evolutionary considerations involving the internal density concentration parameter of binary stars.
A. F. Petty.
Astrophys. Space Sci., Vol. 21, 189 - 209 (1973).

The evolutionary changes that occur in the internal density concentration parameter k_2 (called the apsidal constant for brevity) for a star of given mass and initial composition are examined in detail. The purpose is to ascertain whether or not such an approach leads to a reduction in the differences now noted between the theoretically derived values of k_2 and the observed values derived from the secular advance of the periastron in close eclipsing binary systems. A series of stellar models of mass 2.0, 5.0, 10.0 and 20.0 $M_\odot$ were employed, with an initial compositional mixture of $X = 0.739$, $Y = 0.24$ and $Z = 0.021$.

117.027 Alcune considerazioni sull'evoluzione dei sistemi binari. P. Giannone, M. A. Giannuzzi.
Mem. Soc. Astron. Italiana, Nuova Ser., Vol. 43, 767 - 771 (1973).

Some general considerations about the evolution of binary systems are discussed. With regard to the comparison between theory and observations, some open problems are pointed out.

117.028 Equi-density surfaces in synchronously rotating close binaries built on polytropic model $\nu = 3$.
L. C. Green, E. K. Kolchin.
Astrophys. Space Sci., Vol. 21, 285 - 288 (1973). — Presented at the IAU Colloquium No. 16 held at the University of Pennsylvania, Philadelphia, Pa., U.S.A., September 8—11, 1971.

A direct numerical method, involving no perturbational procedures, is described and used for the computation of the surfaces of constant potential, pressure and density in synchronously rotating close binaries with both components built on polytropic model $\nu = 3$.

117.029 **Eftersøgningen af fremmede planetsystemer.**
H. Nielsen.
Astron. Tidssk., Årg. 6, p. 46 - 68 (1973).

117.030 **The radiative interaction and evolution of medium- and low-mass binaries.** R. H. Koch.
Astron. Journ., Vol. 78, 410 - 412 (1973).
The spectral types of the cool components of numerous close binaries are computed on the assumption of convective envelopes for these components. Difficulties of interpreting moderate- and low-mass systems with theoretical cases A and B of binary evolution are described. Possible solutions for these problems are briefly described.

117.031 **Evolution of contact binaries.**
P. Biermann, H.-C. Thomas.
Mitt. Astron. Ges., No. 32, p. 236 - 237 (1973).

117.032 **The W UMa-type systems as contact binaries. I. Two methods of geometrical elements determination. Degree of contact.** S. M. Ruciński.
Acta Astron., Vol. 23, 79 - 120 (1973).
A grid of 48 theoretical contact model light curves for different combinations of the geometrical elements has been computed. The W UMa-type systems form contact common envelopes rather close to the inner critical common equipotential surface; the envelopes of A-type systems reveal slightly stronger contact than that of the W-type systems. Other quite numerous differences between both types of the W UMa-type systems are summarized and discussed. A supplementary method of elements determination making use of the amplitude and half-width of minimum relation is described and applied to V566 Oph and SW Lac, the systems chosen as two extrema in the "solving difficulty" range.

117.033 **Close binaries and their significance to the theory of evolution.** D. Ya. Martynov.
Uspekhi fiz. nauk, Vol. 108, 701 - 732 (1972). In Russian. Abstr. in Referativ. Zhurn. 51. Astron., 5.51.728 (1973).

117.034 **Tidal evolution in close binary systems, III.**
W. E. Haymes.
Astrophys. Space Sci., Vol. 22, 165 - 192 (1973).
The author investigates the energy of close binary systems of constant momentum taking into consideration the first-order effects of rotation and tidal attraction of the components of finite size. The equations for the momentum and the energy of the system are set up and perturbation theory is applied to these equations using the results of Kopal (1972) as initial values. The author compares his results with the initial values and then discusses variations in his constants and the application to various real systems.

117.035 **Fundamental data for contact binaries: RZ Comae Berenices, RZ Tauri, and AW Ursae Majoris.**
R. E. Wilson, E. J. Devinney.
Astrophys. Journ., Vol. 182, 539 - 547 (1973).
Differential corrections analyses of three binaries of the W UMa type show that RZ Tau and AW UMa have common envelopes and the relatively small gravity darkening predicted by Lucy, while RZ Com seems to have a larger gravity effect and is only marginally in contact. Some details of our method for computing contact-binary light curves are given.

117.036 **On the masses of cataclysmic variable stars.**
B. Warner.
Monthly Notices Roy. Astron. Soc., Vol. 162, 189 - 196 (1973).
Masses are derived for the components of 10 short period cataclysmic variable stars. With one exception, the masses of the white dwarf primaries lie in the range $1.2 \pm 0.2 M_\odot$, and are thus approximately three times heavier than isolated white dwarf stars.

117.037 **Roche potentials including radiation effects.**
D. W. Schuerman.
Astrophys. Space Sci., Vol. 19, 351 - 358 (1972).
A modified Roche potential which incorporates the effects of radiation pressure due to one component of a binary system is mathematically explored. In some cases, the resulting potentials do not exhibit the familiar contact surfaces of the classical Roche potential. The concept of a contact surface, which has been fundamental to the investigations of close binary systems, must be used with discretion for close binaries in which one component is very luminous. A convenient criterion for the existence of a contact surface is given.

117.038 **Shock waves in the gaseous streams in close binary systems of dwarf stars.** V. I. Taranov.
Astrofizika, Vol. 8, 567 - 575 (1972). In Russian. – English translation in Astrophysics, Vol. 8, No. 4.
The shock wave in the gaseous stream is formed when it flows into the envelope of the primary star. It is supposed that there is a loss of radiative energy behind the shock-wave front. The distribution of the gasdynamic parameters is found. The thermal instability of stationary currents behind the shock front is examined.

117.039 **Physical relations of triple system components.**
Zh. P. Anosova.
Trudy Glav. Astron. Obs. Pulkovo, Ser. 2, Vol. 77, 126 - 150 (1969). In Russian.
All the triple stars contained in the IDS catalogue were investigated. A comparison of the assumed number of optical triple systems with the observed number of triple stars shows that the triple stars with distance between components of the system $\rho > 100''$ are probably optical; on the other hand, among the triple stars with $\rho > 50''$ there is apparently a considerable number of physical systems. A dynamical criterion was used for distinguishing physical triple systems from optical ones. In the results of application of the dynamical criterion it appears that 235 stars from 1302 triple stars are surely optical systems and 650 stars are possibly physical triple systems.

Binary and multiple systems of stars. (International Series of Monographs in Natural Philosophy, Vol. 51). See Abstr. 003.029.

Black holes in binary systems. Observational appearance. See Abstr. 066.049.

Cyclic variations of the Be star β^1 Monocerotis. See Abstr. 114.008.

Genetical relation between RS CVn binaries and W UMa binaries: a tentative suggestion. See Abstr. 121.018.

On the nature of the luminous central objects in NGC 3603 and 30 Doradus. See Abstr. 131.168.

Radio emission from the close binary b Persei. See Abstr. 141.035.

Wolf-Rayet systems and the origin of massive X-ray binaries. See Abstr. 142.039.

Binary systems as X-ray sources: a review. See Abstr. 142.061.

A research on double stars in T-associations: Preliminary results. See Abstr. 152.007.

118 Visual Binaries

118.001 **Revised elements of ten visual binaries.**
W. D. Heintz.
Astron. Journ., Vol. 78, 208 - 211 (1973).

118.002 **Extended catalogue of visual binaries with known orbits and Barr effect in these systems.**
V. A. Bakanov.
Uch. zap. Gor'kov. gos. ped. in-t, 1972, vyp. (No.) 124, p. 50 - 63. In Russian. – Abstr. in Referativ. Zhurn. 51. Astron., 3.51.673 (1973).

118.003 **The tools to determine orbital parameters.**
W. D. Heintz.
Journ. Roy. Astron. Soc. Canada, Vol. 67, 52 - 55 (1973).
Presented at IAU Colloquium No. 18, (see 012.004).

118.004 **Remarks on orbits and dynamical parallaxes.**
J. Dommanget.
Journ. Roy. Astron. Soc. Canada, Vol. 67, 56 - 57 (1973).
Presented at IAU Colloquium No. 18, (see 012.004).

118.005 **Hertzsprung's luminosity-colour relation for visual binaries.** W. S. Finsen, P. N. J. Wisse.
Journ. Roy. Astron. Soc. Canada, Vol. 67, 58 - 65 (1973).
Presented at IAU Colloquium No. 18, (see 012.004).

118.006 **The photometry of visual binaries.**
R. H. Hardie.
Journ. Roy. Astron. Soc. Canada, Vol. 67, 68 - 72 (1973).
Presented at IAU Colloquium No. 18 (see 012.004).

118.007 **Les étoiles doubles faibles.** P. Couteau.
Journ. Roy. Astron. Soc. Canada, Vol. 67, 77 - 80 (1973). – Presented at IAU Colloquium No. 18, (see 012.004).

118.008 **Observational procedures for visual double-star work.** O. G. Franz.
Journ. Roy. Astron. Soc. Canada, Vol. 67, 81 - 87 (1973).
Presented at IAU Colloquium No. 18, (see 012.004).

118.009 **On the orbit of the visual binary ADS 10561.**
G. A. Starikova.
Astron. Tsirk., No. 746, p. 3 - 4 (1973). In Russian.

118.010 **Orbits of ten visual binaries.** W. D. Heintz.
Astron. Journ., Vol. 78, 307 - 309 (1973).
Elements and dynamical parallaxes have been computed for the double stars ADS 1360, 1796, 2765, 3686, 6989, 10480, 12911, I 658, Cou 14, and ADS 16235 BC.

118.011 **Résultats préliminaires d'une recherche systématique d'étoiles doubles nouvelles entre +60° et le pôle boréal.** P. Muller.
Publ. Obs. Paris, 46 pp. (1973).
La présente publication est le développement d'une suite de listes de couples nouveaux annoncés dans les Circulaires d'Information de la Commission 26 de l 'U.A.I. depuis son numéro 49 (novembre 1969). Le programme avait été présenté brièvement avec la première liste.

118.012 **New double stars (9th series) discovered at Nice.**
P. Couteau.
Astron. Astrophys.,Suppl. Ser., Vol. 10, 273 - 280 (1973).
In French.
The author gives a list of 100 double stars discovered at the 50 and 74 cm refractor.

118.013 **Photometrische Untersuchungen an engen visuellen Doppelsternen.** H. Jenkner.
Mitt. Astron. Ges., No. 32, p. 249 - 252 (1973).

118.014 **Orbit of the visible binary star ADS 6126−AB = Σ 1104.** R. R. De Freitas Mourao.
An. Acad. Brasil. Cienc.,Vol.44, 19 - 23 (1972).
The physical and orbital elements of the binary star ADS 6126−AB = Σ 1104 are determined.

118.015 **Orbites nouvelles.** P. Muller.
Circ. Inform. (U. A. I. Commission des Étoiles Doubles), Obs. Meudon, No. 59 (1973).

118.016 **Étoiles doubles découvertes à Nice** (Lunette de 50 cm). P. Couteau.
Circ. Inform. (U. A. I. Commission des Étoiles Doubles), Obs. Meudon, No. 59 (1973).

118.017 **Orbites nouvelles.** P. Muller.
Circ. Inform. (U. A. I. Commission des Étoiles Doubles), Obs. Meudon, No. 60 (1973).

118.018 **Étoiles doubles découvertes à Nice** (Lunette de 50 cm). P. Muller.
Circ. Inform. (U. A. I. Commission des Étoiles Doubles), Obs. Meudon, No. 60 (1973).

Orbital and physical parameters of double stars.
Proceedings of Colloquium No. 18 of the International Astronomical Union held at Swarthmore College, Pennsylvania, U.S.A., April 12−15, 1972. See Abstr. 012.004.

Multiplicity of the Sirius system.
See Abstr. 117.010.

The binary frequency for Ap stars.
See Abstr. 119.008.

The spectrographic binary HD 180553.
See Abstr. 119.017.

Detection of the secondary spectrum in SX Cassiopeiae. See Abstr. 121.032.

Complanar system of binaries in Aquila.
See Abstr. 155.029.

119 Spectroscopic Binaries

119.001 The least-squares determination of mass ratios of spectroscopic binaries. J. B. Irwin.
Astrophys. Journ., Vol. 179, 241 - 247 = Contr. Louisiana State Univ. Obs., *Baton Rouge*, No. 69 (1973).

The method of O. C. Wilson, as applied to double-lined spectroscopic binaries, has been modified and extended so that definitive determinations of mass ratios and systemic velocities can be made; appropriate weighting of individual velocities is possible. Applications to one visual binary and to six eclipsing binaries are given.

119.002 Spectroscopic binaries with circular orbits. L. B. Lucy, M. A. Sweeney.
Observatory, Vol. 93, 37 - 39 (1973). – Letter.

119.003 Properties and nature of shell stars. 3. Periodic radial-velocity changes of 4 Herculis. P. Harmanec, P. Koubský, J. Krpata.
Astron. Astrophys., Vol. 22, 337 - 341 (1973).

The radial velocity of shell hydrogen lines of 4 Her measured on Ondřejov and Victoria high-dispersion spectrograms of 1969–1972 was found to vary with a period of 46.023^d instead of 0.97625^d established by Heard (1940). The original binary hypothesis is revived.

119.004 A highly evolved, low-mass binary, HZ 22. J. L. Greenstein.
Astron. Astrophys., Vol. 23, 1 - 7 (1973).

The short-period binary, HZ 22 = UX CVn has a small mass function, low total mass, large radial-velocity amplitude and continuous light variation. The invisible secondary is probably a white-dwarf, the primary resembles a normal early B star, with surface gravity near 10^4, $T \approx 28000°K$. The system must have suffered extensive mass exchange, and mass loss. The future life of the B star is very short. The available evolutionary tracks for close binaries suggest how the initially more massive star has become a white-dwarf; the low-mass of the present primary, however, is not explained. Relations with bright blue stars in the halo and globular clusters are discussed.

119.005 The apsidal constant and structure of α Virginis. J. S. Mathis, A. P. Odell.
Astrophys. Journ., Vol. 180, 517 - 529 (1973).

The binary system Spica (α Vir) is analyzed by means of theoretical models. Masses, radii, and orbital elements are taken from the work of Herbison-Evans et al. The primary is assumed to be evolving from the zero-age main sequence, the secondary on or near it. Comparison of the models is made with the observed apsidal constant and with the flux ratio $F(\lambda2905)/F(\lambda5470)$, which has been observed by Bless. It is not possible to fit both of these quantities, even by adjusting the observed quantities within reasonable limits.

119.006 Improved elements for the Hyades group binary 43 Persei. G. Wallerstein.
Publ. Astron. Soc. Pacific, Vol. 85, 115 - 116 (1973).

New orbital elements from 14 high-dispersion spectrograms are derived for this double-line binary of type F5. The orbital elements are combined with the absence of an eclipse to derive minimum masses of $0.94\ M_\odot$ and $0.89\ M_\odot$.

119.007 Spectroscopic binaries – 12th complementary catalogue. A. Pédoussaut, J.-M. Carquillat.
Astron. Astrophys., Suppl. Ser., Vol. 10, 105 - 124 (1973).
In French.

This catalogue is a continuation of the eleven complementary catalogues which began to be published in 1952. In the table are recently investigated orbital elements. In the notes we put forward that the concerned binary is either a new binary or a yet known one. Particularities of each binary and the most important remarks given by authors are mentioned.

119.008 The binary frequency for Ap stars. H. A. Abt, M. S. Snowden.
Astrophys. Journ., Suppl. Ser., No. 215, Vol. 25, 137 - 162 (1973).

A search was made with coudé spectra of the 62 brightest northern Ap stars for spectroscopic binaries. Orbital elements are given for seven newly discovered binaries. The binary frequency of 40 percent for 15 Hg-Mn stars is roughly normal, but the frequency of 20 percent for 45 Si and Sr-Cr-Eu stars is significantly low. Synchronization of rotational and orbital motions occurs for the few systems with orbital periods less than 6.0 days, but rarely for longer periods. The frequency of visual binaries seems to be normal for all three subgroups of Ap stars. The evidence for the occurrence of very low-mass ($\sim 0.01\ M_\odot$) companions or short-term (30-minute) nonrandom radial-velocity fluctuations is very weak.

119.009 Astrometric study of Epsilon Eridani. P. van de Kamp.
Bull. American Astron. Soc., Vol. 5, 6 - 7 (1973). – Abstr. AAS.

119.010 Introductory remarks on spectroscopic parameters. H. A. Abt.
Journ. Roy. Astron. Soc. Canada, Vol. 67, 73 - 76 (1973).
Presented at IAU Colloquium No. 18, (see 012.004).

119.011 A C IV emission feature in the near-ultraviolet spectrum of the Wolf-Rayet star γ^2 Velorum. K. A. van der Hucht, H. J. Lamers.
Astrophys. Journ., Vol. 181, 537 - 542 (1973).

Spectrophotometric satellite observations in the ultraviolet of the binary γ^2 Vel are presented. Spectral features are tentatively identified. An explanation for the strength of the dominating C IV λ 2529.97 emission line is suggested. The diffuse red wing of this line suggests mass transfer from the Wolf-Rayet star to the O-type component.

119.012 The Wolf-Rayet spectroscopic binary HD 92740. V. S. Niemelä.
Publ. Astron. Soc. Pacific, Vol. 85, 220 - 223 (1973).

A radial velocity study of HD 92740, a WN7 star, shows it to be a binary with $P = 10^d04$. Preliminary velocity curves derived from various lines suggest differing amplitudes; and the weak hydrogen absorption lines present in the spectrum share the orbital motion of the WN star. Undetected violet absorption is shown to cause the redward displacement of He II $\lambda4686$. The possibility of an outward accelerating atmosphere is suggested by the presence of a Balmer progression.

119.013 HR 7955 as a spectroscopic binary. F. Spite, M. Spite.
Astron. Astrophys., Vol. 25, 325 - 326 (1973).

The star HR 7955 was suspected to be metal-rich. However, high resolution spectra show double lines. The metal-abundance anomaly, derived without taking into account the binary nature of this star, has to be reconsidered.

119.014 Polarimetry of selected spectroscopic binaries. R. J. Pfeiffer, R. H. Koch.
Inform. Bull. Variable Stars (I. A. U. Commission 27), Konkoly Obs., Budapest, No. 780, 3 pp. (1973).

119.015 Photoelectric observations of β Arietis. H. Ogata.
Inform. Bull. Variable Stars (I.A.U. Commission 27),
Konkoly Obs., Budapest, No. 784 (1973).

**119.016 Line strengths in the spectra of seven double-lined
Am binary stars.** D. J. Stickland.
Roy. Obs. Bull., [Roy. Greenwich Obs., Herstmonceux], No.
177, p. 149 - 187 (1972).
The equivalent width data for both components of seven
double-lined spectroscopic binary (SB2) stars and two metallic-
line A type (Am) stars are presented. The properties of the Am
stars are reviewed and a procedure for the analysis of Am SB2
spectra is outlined.

119.017 The spectrographic binary HD 180553.
D. P. Hube.
Astron. Astrophys.,Suppl. Ser., Vol. 10, 267 - 272 (1973).
Observational material obtained with the 1.88 m reflector
of the David Dunlap Observatory has been used in determining the orbital elements of the short period B-type spectrographic binary HD 180553. The system is also a close visual
binary, a fact which almost certainly has had an adverse effect
on the reliability of the derived elements.

119.018 Short period spectral variations in γ² Velorum.
S. Jeffers, W. Weller, A. Sanyal.
Nature, Phys. Sci., Vol. 243, 109 - 111 (1973).
We report recent observations of γ^2 Velorum in the
4660 Å region made with a rapid scanning spectrometer
having a dispersion of 67 Å mm^{-1} at H_γ in the second order
which permits spectral variations to be studied on a time
scale of 10 s.

119.019 Variable stars among Ca II emission binaries.
W. Herbst.
Astron. Astrophys., Vol. 26, 137 - 140 (1973).
Data on seven strong Ca II emission binaries, all of which
show evidence of light variation, are discussed.

119.020 RY Scuti. V. A. Hughes, A. Woodsworth.
IAU Circ., No. 2488 (1973).

119.021 HD 153919 (2U 1700–37). J. C. Kemp.
IAU Circ., No. 2512 (1973).

119.022 HD 77581. J. B. Hutchings.
IAU Circ., No. 2537 (1973).

119.023 β Coronae Borealis. P. Couteau.
IAU Circ., No. 2547 (1973).

119.024 HD 74375, an interesting spectroscopic binary.
A. van Hoof.
Meded. Konikl. Vlaamse Acad. Wet., Lettere, Schone Kunsten,
(*Belgium*), Vol. 34, No. 4, p. 3 - 13 (1972).

Older Lick and more recent La Silla observations prove
that HD 74375 is a spectroscopic binary with a period of
133$\overset{d}{.}$920 and a very eccentric orbit (e = 0.57).

119.025 Light variations in HD 217312.
K. Madore, J. R. Percy.
Publ. Astron. Soc. Pacific, Vol. 85, 319 - 320 = Commun.
David Dunlap Obs., Univ. Toronto, *Richmond Hill,* No. 351
(1973).
Photometric observations of the massive spectroscopic
binary system HD 217312 have been obtained. These observations, together with more recent observations by Rao, indicate
that HD 217312 undergoes a shallow eclipse. The time of the
eclipse is in good agreement with the time predicted by the
spectroscopic orbital elements.

Orbital and physical parameters of double stars.
Proceedings of Colloquium No. 18 of the International Astro-
nomical Union held at Swarthmore College, Pennsylvania,
U.S.A., April 12–15, 1972. See Abstr. 012.004.

Spectrum variability of the shell star 88 Herculis.
See Abstr. 114.110.

The spectroscopic orbit of HD 90707.
See Abstr. 121.001.

A spectroscopic study of MR Cygni.
See Abstr. 121.007.

Apsidal motion in the binary system V 453 Cygni.
See Abstr. 121.027.

UBV **light variation and orbital elements of HD**
101799. See Abstr. 121.055.

Radio variability of HDE 226868 (Cygnus X-1).
See Abstr. 141.100.

The spectrum of HDE 226868 (Cygnus X-1).
See Abstr. 142.022.

**Spectroscopic observations of the Cygnus X-1 opti-
cal candidate.** See Abstr. 142.023.

**Spectroscopic observations of the optical candidate
for Cygnus X-1.** See Abstr. 142.024.

λ-Sco, a possible source of soft X-rays.
See Abstr. 142.029.

2U 0900–40. See Abstr. 142.130.

Coplanarity in open clusters.
See Abstr. 153.028.

120 Variable Stars: Catalogues, Ephemerides, Miscellanea

120.001 **L'observation visuelle des étoiles variables.**
 M. Dumont, A. Figer.
L'Astronomie, 87ᵉ année, p. 141 - 163 (1973).

120.002 **Remarks concerning period determination of a**
 variable star. I. Todoran.
Stud. Cerc. Astron., Vol. 18, 37 - 46 (1973).
 A peculiar case of period determination is examined, where the existing methods could not be applied with maximal efficiency. The writer proposes a new method which, although it makes use of interpolation in the light curves, can also be applied when the number of observed branches of a light curve is not large. This new method is illustrated by two numerical examples.

120.003 **A new approach to periodogram analyses.**
 D. F. Gray, K. Desikachary.
Astrophys. Journ., Vol. 181, 523 - 530 (1973).
 We present a method for Fourier analyzing quasi-periodic light and radial velocity variations. The scheme gives a particularly clear interpretation to the periodogram, reduces the alias problem, allows detection of very weak components, has potential for greater precision in frequency measurement, and is a shorter computation than the usual method.

120.004 **Provisional ephemerides of 21 variable stars in a**
 field centered at $\alpha = 13^h$, $\delta = -70°$.
R. Deurinck, B. Vissenberg.
Inform. Bull. Variable Stars, (I.A.U. Commission 27), Konkoly Obs., Budapest, No. 793, 31 pp. (1973).

120.005 **Veränderliche Sterne (Variable stars).**
 A. Werner.
Jahrbuch der Schulphysik, [Aulis-Verlag, Köln], Vol. 1, 196 - 200 (1972).
 Survey of the different types of intrinsic variable stars (without novae) and the present status of theoretical interpretation of the pulsation phenomena.

 Special Supplement to the third edition of the General Catalogue of Variable Stars, containing the list of 32731 stars arranged in the order of right ascensions for the equinox 1950.0. See Abstr. 003.003.

 Variable stars.
See Abstr. 003.116.

121 Eclipsing Variables

121.001 The spectroscopic orbit of HD 90707.
T. Lloyd Evans.
Monthly Notices Roy. Astron. Soc., Vol. 161, 15 - 21 (1973).
HD 90707 (B1 III) is an eclipsing binary and a probable member of the open cluster IC 2581 in Carina. The spectroscopic orbit has been determined from 29 observations of the radial velocity of the primary. The assumption that sin $i = 1$ leads to masses of 18.7 and 20.4 $m_\odot$ for the two components, larger than expected from an earlier comparison of the colour-magnitude diagram with theoretical evolutionary tracks.

121.002 On the determination of minimum times of light curves.
R. A. Breinhorst, J. Pfleiderer, M. Reinhardt, M. T. Karimie.
Astron. Astrophys., Vol. 22, 239 - 245 (1973).
The method of Kwee and van Woerden, frequently used to determine minimum epochs of eclipsing binary stars, is discussed regarding its application to asymmetric light curves. Expressions are derived, in second-order approximation, for the difference between Kwee—van Woerden and true minimum epoch as well as the Kwee—van Woerden mean error of the minimum, as functions of the asymmetry. We propose a least-squares fit by a third-degree polynomial as an alternative method of minimum determination. Both methods are applied to artificial and observed light curves. Several new minimum epochs of i Boo and VW Cep are evaluated.

121.003 The synthesis of close-binary light curves. V. The "contact" systems W Ursae Majoris, AM Leonis, V 566 Ophiuchi, and GK Cephei.
J. B. Hutchings, G. Hill.
Astrophys. Journ., Vol. 179, 539 - 549 (1973).
The programs described in earlier papers are adapted to apply to contact systems. In three of four systems investigated heating effects are found to be negligible, and convective gravity darkening to apply. In three systems the small star is the hotter and in all cases the dependence of opacity on local gravity is important. Best-fit models are presented and discussed for all four systems.

121.004 UBV photometry of Zeta Aurigae during the 1971—72 eclipse.
N. B. Sanwal, M. Parthasarathy, K. D. Abhyankar.
Observatory, Vol. 93, 30 - 32 (1973).

121.005 Photoelectric five-color observations of W Serpentis.
A. M. van Genderen.
Astron. Astrophys. Suppl. Ser., Vol. 9, 157 - 161 (1973).
Photoelectric five-color observations of the eclipsing variable W Serpentis made in three nights are presented and discussed. Part of the observations has been made during the international campaign of 1966.

121.006 Possible explanation for nonthermal radio noise from binary stars. T. W. Jones, N. J. Woolf.
Astrophys. Journ., Vol. 179, 869 - 873 (1973).
Radio emission from the binary stars α Sco, β Per, and β Lyr is interpreted as due to the mass exchange between the components. The electrons in the inflowing stream excite plasma waves near the surface of the receptor star, and these waves generate radio emission. Crude predictions of radio frequency, intensity, and timescales of variation seem compatible with current observations.

121.007 A spectroscopic study of MR Cygni.
G. Hill, J. B. Hutchings.
Astron. Astrophys., Vol. 23, 357 - 362 (1973).
New spectrographic observations of MR Cyg have resulted in the determination of a set of spectroscopic orbital elements for the system which are quite different from those previously published. The effects of distortion upon the theoretical velocity curve have been evaluated and we find that, in this system, they are significant. By comparing the derived physical parameters with those predicted by evolutionary models of single stars we find that the components of MR Cyg are slightly evolved "normal" B stars.

121.008 5-μm infrared emission from Algol.
R. F. Jameson, A. J. Longmore, B. Crawford.
Nature, Vol. 242, 107 - 108 (1973).
During September 1972, we made a number of 5-μm observations of the eclipsing binary β Per (Algol) using the 60-inch infrared flux collector at Izana, Tenerife. A figure shows our experimental results.

121.009 Photoelectric light curve of the Algol system TW Andromedae and the interpretation of its distortions by the effects of hot spots. M. Ammann, K. Walter.
Astron. Astrophys., Vol. 24, 131 - 142 = Mitt. Astron. Inst. Univ. Tübingen, No. 127 (1973).
Photoelectric observations in B and V of this semi-detached Algol system obtained in 1965 – 1970 are discussed. As soon as the light curve outside of eclipses had been freed from the effects of reflection and ellipticity, there remained systematic distortions which could be explained by the existence of an additional light source of about 3% of the intensity of the uneclipsed system. This light source is situated on the bright F-component at high latitudes in front of the subgiant. Light from a third star is present.

121.010 The light variation of the eclipsing variable DI Pegasi.
L. Binnendijk.
Astron. Journ., Vol. 78, 97 - 102 (1973).
A total of 587 photoelectric observations in yellow light and a total of 578 observations in blue light of DI Pegasi made during the 1968 and 1969 observing seasons are presented. No reliable orbital elements could be derived using the Russell model.

121.011 The light variation and orbital elements of VW Bootis. L. Binnendijk.
Astron. Journ., Vol. 78, 103 - 106 (1973).
A total of 291 photoelectric observations in yellow light and a total of 290 observations in blue light of VW Bootis were obtained during May 1971. Orbital elements were derived which show that primary eclipse is caused by a transit.

121.012 The eclipsing binary system RU Ursae Minoris.
I.-S. Nha.
Astron. Journ., Vol. 78, 107 - 109 (1973).
Light curves in two wavelength regions for RU UMi are presented, and new orbital elements are derived. The orbital period has been shortening since 1968. From the observed systemic $B-V$ and the eclipse analysis, the spectral types of the large and small components are calculated to be F2 and K4, respectively.

121.013 New photometric elements of the eclipsing binary X Trianguli. J. S. Shaw, K. L. Kusler.
Publ. Astron. Soc. Pacific, Vol. 85, 112 - 114 (1973).
Photographic observations of Jordan (1929) and visual measures by Dugan (1928) of X Tri are reanalyzed using the Russell model and partial limb darkening. Probable sizes and masses of both components are presented. The system is typi-

cal of semidetached systems.

121.014 A minimum each for GL Carinae and SV Centauri.
A. U. Landolt.
Publ. Astron. Soc. Pacific, Vol. 85, 117 - 118 = Contr. Louisiana State Univ. Obs., *Baton Rouge*, No. 75 (1973).

121.015 *UBV* photometry of ζ Aurigae and 31 Cygni.
L. P. Lovell, D. S. Hall.
Publ. Astron. Soc. Pacific, Vol. 85, 131 - 132 (1973).
Differential *UBV* observations of ζ Aur and 31 Cyg, 28 of the former and 36 of the latter, were obtained shortly before and during their recent 1971–72 eclipses.

121.016 Differential *UBV* photometry of β Lyrae, III.
H. J. Landis, L. P. Lovell, D. S. Hall.
Publ. Astron. Soc. Pacific, Vol. 85, 133 - 138 (1973).
As part of a continuing program of systematic photometry of β Lyrae, 151 differential *BV* observations and 130 differential *UBV* observations were obtained in 1971 at three observatories.

121.017 Preliminary results of the study of β-Lyrae during the international campaign of July-August 1971.
M. Hack.
Astron. Astrophys., Vol. 24, 329 - 333 (1973).
High dispersion spectrograms of β Lyrae were taken with the 152 cm coudé telescope of the Haute Provence Observatory during the period July 18 – August 1. Here we report the results of the study of the II A F plates giving radial velocity and intensity of H, He I, Si II and Na I lines.

121.018 Genetical relation between RS CVn binaries and W UMa binaries: a tentative suggestion.
S.-S. Huang.
Bull. American Astron. Soc., Vol. 5, 41 - 42 (1973). – Abstr. AAS.

121.019 The mass-ratio of W UMa. S. P. Worden, J. Whelan.
Bull. American Astron. Soc., Vol. 5, 42 (1973).
Abstr. AAS.

121.020 OAO-2 observations of the 1971 eclipses of 32 Cygni.
L. R. Doherty, A. V. Holm, J. F. McNall.
Bull. American Astron. Soc., Vol. 5, 42 (1973). – Abstr. AAS.

121.021 A detailed model for an eclipsing binary system.
F. G. Van Landingham.
Bull. American Astron. Soc., Vol. 5, 43 (1973). – Abstr. AAS.

121.022 The structure of H α in the spectrum of VV Cephei.
K. O. Wright.
Bull. American Astron. Soc., Vol. 5, 43 (1973). – Abstr. AAS.

121.023 Barr effect and antiapex effect in eclipsing binaries.
V. P. Tolstykh.
Uch. zap. Gor'kov. gos. ped. in-t, 1972, vyp. (No.) 124, p. 70-73. In Russian. – Abstr. in Referativ. Zhurn. 51. Astron., 3.51.678 (1973).

121.024 Photoelectric photometry of TX Herculis.
M. Vetešník, J. Papoušek, J. E. Purkyně.
Bull. Astron. Inst. Czechoslovakia, Vol. 24, 57 - 70 (1973).
1728 photoelectric observations in *B* and 2070 in *V* colour were carried out at the Brno Observatory in the years 1968, 1969, and 1971. Twenty new times of minimum light and new light elements are presented.

121.025 Approximate figures of β Per-type eclipsing variables. The condition of the hydrostatic equilibrium in the

atmospheres of binaries. V. P. Merezhin.
Trudy Kazan. Gorod. Astron. Obs., No. 38, p. 16 - 25 (1972). In Russian.

121.026 The synthesis of close-binary light curves. IV: An application to TX Ursae Majoris and MR Cygni.
G. Hill, J. B. Hutchings.
Astrophys. Space Sci., Vol. 20, 123 - 148 (1973). – Originally presented at the IAU Colloquium No. 16, held at the University of Pennsylvania, Philadelphia, Pennsylvania, U.S.A., September 8-11, 1971.
A full description is given of the theory used in light curve synthesis, for binary stars showing various degrees of interaction. For strongly interacting pairs, an iterative mutual heating calculation is introduced. Detailed models are derived by these methods, and discussed, for the weakly interacting system TX UMa and the strongly interacting system MR Cyg. A critical appraisal is given of this method of light curve synthesis and its results.

121.027 Apsidal motion in the binary system V 453 Cygni.
A. A. Wachmann.
Astron. Astrophys., Vol. 25, 157 - 158 (1973).
About 2700 photographic and 2000 photoelectric observations in *B* and *V* of V 453 Cyg, covering 3000 epochs, show that there is a well established apsidal motion. The orbital eccentricity is 0.020 and the period of apsidal rotation 72 years.

121.028 Has the system XX Cep apsidal motion?
M. I. Lavrov, N. V. Lavrova.
Astron. Tsirk., No. 756, p. 4 - 5 (1973). In Russian.

121.029 Elements of the eclipsing variable star Wr 104.
V. P. Tsesevich, V. G. Karetnikov.
Astron. Tsirk., No. 757, p. 7 - 8 (1973). In Russian.

121.030 Period of SZ Piscium. G. F. G. Knipe.
Monthly Notes Astron. Soc. Southern Africa, Vol. 32, 16 (1973). – Note.

121.031 Eclipsing binary RT Andromedae (I).
A. Dumitrescu.
Stud. Cerc. Astron., Vol. 18, 47 - 62 (1973). In Romanian.

121.032 Detection of the secondary spectrum in SX Cassiopeiae. J. Andersen.
Publ. Astron. Soc. Pacific, Vol. 85, 191 - 192 (1973).
The detection of the secondary spectrum in SX Cas is reported. Lines of Fe I λλ4045, 4063, and 4071, as well as other features, are doubled near phases 0ᴾ25 and 0ᴾ75.

121.033 A computer program for modeling nonspherical eclipsing binary star systems. D. B. Wood.
Publ. Astron. Soc. Pacific, Vol. 85, 253 (1973).
A computer model, written in FORTRAN IV for the IBM 360, is available which accounts for the important geometric and photometric distortions such as rotational and tidal distortion, gravity brightening, and reflection effect. More information may be obtained from D. B. Wood, Code 110, Goddard Space Flight Center, Greenbelt, Maryland 20771.

121.034 A spectrophotometric study of eclipsing variable systems. II. T. M. Rachkovskaya.
Izv. Krymskoj Astrofiz. Obs., Vol. 46, 35 - 46 (1972). In Russian.
The spectra of the eclipsing variable systems EM Cep, Z Vul, IM Mon, λ Tau, RS Vul, V822 Aql, U Sge and RX Her were obtained with a dispersion of 15 and 36 Å/mm near the phases 0ᴾ.00; 0ᴾ.25; 0ᴾ.50 and 0ᴾ.75. Using the spectrograms

obtained spectral types, absolute visual magnitudes, rotational and synchronous velocities have been determined.

121.035 W Ursae Majoris stars: a model for the W-systems.
S. W. Mochnacki, J. A. J. Whelan.
Astron. Astrophys., Vol. 25, 249 - 252 (1973).

A model for the W-type W Ursae Majoris systems is presented in which the energy exchange occurs largely through the superadiabatic part of the common convective envelope and the secondary component is a few hundred degrees hotter than the primary. It is confirmed, by light curve synthesis, that such a model has the observed light curves.

121.036 The variations of the light curve of HZ Herculis in the 35-day cycle. N. Kurochkin.
Inform. Bull. Variable Stars (I.A.U. Commission 27), Konkoly Obs., Budapest, No. 753 (1973).

121.037 Photoelectric observations of 31 Cygni in the 1972 eclipse.
T. Hayasaka, N. Sato, H. Ogata, M. Kitamura.
Inform. Bull. Variable Stars (I.A.U. Commission 27), Konkoly Obs., Budapest, No. 757, 3 pp. (1973).

121.038 Corrected period for MT Her. Z. Pokorný.
Inform. Bull. Variable Stars (I.A.U. Commission 27), Konkoly Obs., Budapest, No. 772 (1973).

121.039 Minima of eclipsing variables. I. Todoran.
Inform. Bull. Variable Stars (I.A.U. Commission 27), Konkoly Obs., Budapest, No. 775 (1973).

121.040 Minima of eclipsing variables. Z. Klimek.
Inform. Bull. Variable Stars (I.A.U. Commission 27), Konkoly Obs., Budapest, No. 779 (1973).

121.041 Photographic observations of eclipsing variables.
P. Ahnert.
Inform. Bull. Variable Stars (I.A.U. Commission 27), Konkoly Obs., Budapest, No. 786 (1973).

121.042 Instantaneous elements of 4 eclipsing stars.
P. Ahnert.
Inform. Bull. Variable Stars (I.A.U. Commission 27), Konkoly Obs., Budapest, No. 786 (1973).

121.043 Minima of 44i Bootis. I. Rudnick.
Inform. Bull. Variable Stars (I.A.U. Commission 27), Konkoly Obs., Budapest, No. 789 = Rosemary Hill Obs., Dep. Phys. Astron., Univ. Florida, Gainesville, Contr. No. 15 (1973).

121.044 UBV and JHKL photometry of "O'Connell effect" binaries. E. F. Milone.
Bull. American Astron. Soc., Vol. 5, 266 - 267 (1973).
Abstr. AAS.

121.045 Correlations among the parameters of the spherical model for eclipsing binaries.
S. Sobieski, J. White.
Astrophys. Space Sci., Vol. 21, 7 - 12 (1973). – Presented at the IAU Colloquium No. 16, held at the University of Pennsylvania, Philadelphia, Pa., U.S.A., September 8 - 11, 1971.

Correlation coefficients have been computed to investigate the parameters used to describe the spherical model of an eclipsing binary system. Regions in parameter hyperspace have been identified where strong correlations exist and, by implication, the solution determinacy is low. The results are presented in tabular form for a large number of system configurations.

121.046 UBV photometry and photometric orbit of UV Lyn.

H. Bossen.
Astron. Astrophys.,Suppl. Ser., Vol. 10, 217 - 230 (1973).

During photoelectric observations the eclipsing binary UV Lyn turned out to be a W UMa star with a period of 0^d415. Orbital elements were calculated and they show that the system consists of a F8 and a G0V star. Calculation of absolute elements shows that the earlier star has most likely evolved and is therefore oversized for its mass.

121.047 Statistics of β Lyr and W UMa type eclipsing binaries. S. Linnaluoto, O. Vilhu.
Astron. Astrophys., Vol. 25, 481 - 484 (1973).

Statistics relating to EB (β Lyr) and EW (W UMa) type eclipsing binaries have been prepared with the purpose of investigating the uniformity of these groups. A gap in the distribution of the differences between the surface brightnesses of the components exists. This points to the detachment of EB- and EW-groups, or to the discontinuous character of the physical process equalising the surface temperatures in components of EW type systems. A parameter might give a quantitative classification criterion to separate the EB and EW groups.

121.048 Rectification of the light curve of an eclipsing system composed of ellipsoidal components.
Y. Hosokawa.
Sci. Rep. Tôhoku Univ.,First Ser.,Vol. 55, 103 - 123 (1972/73).

The rectification process for the light curve of a moderately separated system composed of ellipsoidal components is revised.

121.049 Minima of eclipsing variables (XI).
M. E. Baldwin.
Inform. Bull. Variable Stars, (I.A.U. Commission 27), Konkoly Obs., Budapest, No. 795, 17 pp. (1973).

121.050 The large period variation in SS Cam explained.
C. N. Arnold, D. S. Hall, R. E. Montle.
Inform. Bull. Variable Stars, (I.A.U. Commission 27), Konkoly Obs., Budapest, No. 796 (1973).

121.051 Envelopes in eclipsing binary systems.
S.-S. Huang.
Astrophys. Space Sci., Vol. 21, 263 - 283 (1973). – Presented at the IAU Colloquium No. 16 held at the University of Pennsylvania, Philadelphia, Pa., U.S.A., September 8–11, 1971.

The present paper consists of a discussion of (1) the relevance of envelopes to the study of the light curves of eclipsing binaries, (2) the disk envelope, and (3) the spherical envelope. They are separately considered in three sections, to the fourth section a brief concluding remark is appended.

121.052 Photometric effects of gas streams in Algol systems and their influence on the light curves outside of and within eclipses. K. Walter.
Astrophys. Space Sci., Vol. 21, 289 - 305 (1973). – Presented at the IAU Colloquium No. 16 held at the University of Pennsylvania, Philadelphia, Pa., U.S.A., September 8–11, 1971.

An attempt is made to interpret the distortions of the light curves of semi-detached Algol systems, in terms of the absorption effects of gas streams and of the emission from the impact regions of these streams. Hypothetical models of the distortions of the light curves outside of primary eclipses are given for different locations of the luminous regions. To exhibit features of light curves of this kind, both published and unpublished observations (TW Dra, Y Psc, TV Cas) of Algol systems are used. The existence of luminous regions is also indicated by observations of primary eclipses (SW Cyg, RV Oph, W Del).

121.053 Radiative transfer in atmospheres of Algol-type binaries. Departures from the state of LTE in late-type

components. I. Pustylnik, L. Toomasson.
Astrophys. Space Sci., Vol. 21, 495 - 503 (1973). – Presented at the IAU Colloquium No. 16, held at the University of Pennsylvania, Philadelphia, Pa., U.S.A., September 8–11, 1971.

Departures from the state of LTE in the atmospheres of the secondary components of Algol-type binaries due to anisotropy of irradiation are investigated. Temperature inversion and the abnormal law of limb darkening are the most serious consequences of anisotropy of irradiation if the ratio of the effective temperatures of the components exceeds 2–2.5, providing that collisions can be neglected.

121.054 **Narrow-band photoelectric observations of the eclipsing binary of Wolf-Rayet type V444 Cyg in the continuum ($\lambda\lambda$ 4244–7512 Å).**
A. M. Cherepashchuk, Kh. F. Khaliullin.
Astron. Zhurn. Akad. Nauk SSSR, Vol. 50, 516 - 525 (1973). In Russian. English translation in Soviet Astron. AJ, Vol. 17, No. 3.

About 300 - 450 narrow-band ($\Delta\lambda \sim 75$ –165 Å) observations of V444 Cyg (WN5 + O6) are obtained in each colour ($\lambda\lambda$ 4244, 4789, 5806, 6320, 7512 Å) during 1970–71. It is shown that the main absorption mechanism in the WR envelope is electron scattering.

121.055 *UBV* **light variation and orbital elements of HD 101799.** R. F. Sisteró, M. E. Castore de Sisteró.
Astron. Journ., Vol. 78, 413 - 421 (1973).

Photoelectric *UBV* observations of the W UMa system HD 101799 are presented; they were made at Cerro Tololo Inter-American Observatory (Chile) in April 1971. The period and linear ephemeris were obtained from 14 times of minimum observed in each color. The eclipses show flat minima during an interval of 31 min. Primary minimum is an occultation. Consistent orbital elements were determined for the three light curves based on the Russell model.

121.056 **Bestimmung von Minimumszeiten bei Veränderlichen.**
R. Breinhorst, J. Pfleiderer, M. Reinhardt, T. Karimie.
Mitt. Astron. Ges., No. 32, p. 212 (1973). – Abstract.

121.057 **UBV - Photometrie und Spektroskopie an dem Bedeckungsveränderlichen EI Cephei.**
R. van Rijsbergen.
Mitt. Astron. Ges., No. 32, p. 278 - 279 (1973).

121.058 **The nature of the binary system VW Cygni.**
V. Ureche.
Stud. Univ. Babeş-Bolyai, Ser. Math.- Mech., Anul 13, p. 87 - 93 (1973). In Romanian.

Using the new photometric elements of Lacy (1970) and the spectroscopic elements of Struve (1946), the absolute elements of the binary system VW Cygni are redetermined. On the basis of the mass-luminosity relation obtained by Svetchnikov (1969), the mass ratio has been computed with Kopal's method (1959). The close binary system VW Cygni is semidetached. The paper is concluded with some considerations about the nature and the evolution of the binary system VW Cygni.

121.059 **A computer study of EE Pegasi and CM Lacertae.**
A. P. Linnell.
Astrophys. Space Sci., Vol. 22, 13 - 43 (1973). – Presented at the IAU Colloquium No. 16 held at the University of Pennsylvania, Philadelphia, U.S.A., September 8–11, 1971.

Computer routines permit the solution of eclipsing binary light curves on the Russell model. Application to EE Peg determines values of x_g and x_s even though secondary minimum is only 0.08 mag. deep. First order perturbation theory is used with the Russell model to calculate a final triaxial ellipsoid model. Solution of the CM Lac light curve shows that the data require an occultation eclipse at primary minimum, in contrast to the available nomographic solution.

121.060 **Analysis of eclipsing binary light curves on automatic computers.** E. Budding.
Astrophys. Space Sci., Vol. 22, 87 - 122 (1973).

A direct method of analysis of the light curves of eclipsing binary systems is proposed, which, as an alternative to numerical integration methods, adopts the associated alpha functions of Kopal as the means of description of eclipses. The method has been applied to a number of light curves including those of YZ Cas, WW Aur, YY Gem and SZ Cam. Second order coefficients of limb darkening have been determined for the YZ Cas stars. Finally, a comparative discussion is given of various methods of light curve analysis.

121.061 **Zeta Aurigae: eclipse observations of 1963–64.**
K. C. Gordon.
Astrophys. Space Sci., Vol. 22, 127 - 132 (1973).

Observations of the 1963–64 ζ Aur eclipse through three interference filters are presented. Six-color observations of ζ Aur during totality are appended.

121.062 **An expanding circumstellar cloud of Zeta Aurigae.**
M. Saitō.
Astrophys. Space Sci., Vol. 22, 133 - 140 (1973).

Strong absorption satellite lines of Ca I λ 6572 were found on spectrograms taken on three successive days just after the fourth contact of the 1971–72 eclipse of Zeta Aurigae. The radial velocities of the satellite lines are -88 km s^{-1}, -74 km s^{-1}, and -180 km s^{-1}, respectively, relative to the K-type primary star (K4 Ib). These absorptions should be due to a circumstellar cloud expanding from this component.

121.063 **Die W UMa–Sterne.** M. Fernandes.
BAV Rundbrief, 22. Jahrgang, p. 1 - 5 (1973).

121.064 **Beobachtungen von RW Tauri.** K. Wälke.
BAV Rundbrief, 22. Jahrgang, p. 12 - 13 (1973).

121.065 **More on the effects of the relative shifts of the minima of eclipsing binaries.** L. Frasiński.
Postępy Astron., Vol. 21, 141 - 143 (1973). In Polish.

The paper refers to a paper by T. Z. Dworak (1972). It is shown that the effect of relative shifts of the minima of eclipsing binaries is twice as large as that given by Dworak.

121.066 **V343 Ori – new elements.** R. Szafraniec.
Inform. Bull. Variable Stars (I. A. U. Commission 27), Konkoly Obs., Budapest, No. 800 (1973).

121.067 **HZ Herculis.** S. Goldsmith, J. N. Bahcall, N. A. Bahcall.
IAU Circ., Nos. 2529, 2535 (1973).

121.068 **HZ Herculis.** W. J. Cocke, P. Hintzen, J. S. Scott, S. P. Worden.
IAU Circ., No. 2543 (1973).

121.069 **The case for a black hole in BM Orionis.**
R. E. Wilson.
Astrophys. Space Sci., Vol. 19, 165 - 171 (1972).

Earlier photometric and spectroscopic observations of the binary BM Ori are interpreted in terms of a thin disk model for the object which causes the eclipses. It is shown that the secondary mass, about which the disk particles orbit, has small dimensions and a mass of 3 to 4 $M_\odot$, which suggests that it can only be a collapsed star.

121.070 **Evidence of orbital eccentricity from a new two-**

color light curve of MR Cygni.
P. Battistini, A. Bonifazi, A. Guarnieri.
Astrophys. Space Sci., Vol. 19, 395 - 409 (1972).

The circular elements of eclipsing binary MR Cygni in yellow and blue light are derived from both minima. The durations of eclipses are different and this is interpreted as an effect of orbital eccentricity.

121.071 **ZZ Cygni.** P. Ahnert.
MVS, *Sonneberg,* Vol. 6, 67 - 70 (1973).

Since 1900 the period of ZZ Cygni decreased two times: between 1929 and 1935 by $0^d.0000018$ and between 1955 and 1965 by $0^d.0000019$. Instantaneous and mean elements are given for the years 1900 to 1972.

121.072 **AB Cassiopeiae.** P. Ahnert.
MVS, *Sonneberg,* Vol. 6, 70 - 74 (1973).

AB Cas shows remarkable changes of its period. Between 1931 and 1948 it increases by $0^d.000022$ (nearly 2 seconds), and 1959 once more by $0^d.0000084$. Instantaneous and mean elements are derived for the time 1928 to 1972.

121.073 **AW Vulpeculae.** P. Ahnert.
MVS, *Sonneberg,* Vol. 6, 75 - 76 (1973).

This up to now poorly observed star shows two periods which differ by only $-0^d.0000024$. Probably the change happened about in 1964. The mean and instantaneous elements refer to the time from 1930 to 1972.

121.074 **BO Vulpeculae.** P. Ahnert.
MVS, *Sonneberg,* Vol. 6, 77 - 78 (1973).

BO Vul has been observed systematically not before 1947. 1957 the period changed from $1^d.94591$ to $1^d.945866$ ($-3^s.8$). Therefore mean elements are not suitable for calculating an ephemeris and instantaneous elements are given only.

121.075 **Right ascensions and declinations of eclipsing binary stars.** M. Yu. Volyanskaya.
Catalogues of positions of stars, *Kiev,* (see 041.030), p. 343 - 356 (1970). In Russian.

This catalogue contains the results of observations of 108 eclipsing binaries made with the Odessa Repsold transit circle. The positions (1950.0) are reduced to the FK4 system. Mean errors corresponding to different numbers of observations are given.

121.076 **The mass distribution of eclipsing binaries.**
E. J. Devinney, Jr.
Publ. Astron. Soc. Pacific, Vol. 85, 330 - 331 = Astron. Contr. Univ. South Florida, *Tampa,* No. 65 (1973).

Two sources of data on the masses of binaries are found to show similar deficiencies in the number of stars at about $3.5\,M_\odot$. The gap in the observed mass distribution has a width of about one solar mass. The gap is apparently cosmic and not a result of systematic error.

121.077 **An analysis of the Rosemary Hill observations of 31 Cygni.** R. H. Bloomer, Jr., F. B. Wood.
Publ. Astron. Soc. Pacific, Vol. 85, 348 - 354 = Rosemary Hill Obs., Univ. Florida, *Gainesville,* Contr. No. 35 (1973).

Observations of 31 Cyg are listed. Three spectral ranges were outlined by narrow-band Gyldenkerne filters, and a fourth was in the red near Hα. The atmospheric nature of the eclipse, at least for the two shortest wavelengths observed, seems clearly established. Detailed discussion is postponed until observations made elsewhere using the same filter system are published.

121.078 **A new method for determining the limb darkening coefficients of eclipsing binary systems and its application to an analysis of brightness curves of YZ Cassio-**

peiae. M. I. Lavrov.
Trudy Kazan. Gorod. Astron. Obs., No. 37, p. 3 - 22 (1970). In Russian.

121.079 **GG Cassiopeiae – an eclipsing binary with an unresolved optical companion.** O. S. Shulov.
Trudy Astron. Obs., *Leningrad,* Vol. 29 (= Uchenye Zapiski Leningr. Un-ta, No. 363 = Seriya Matem. Nauk, vyp. (No.) 48), p. 86 - 97 (1973). In Russian.

The results of observations of the eclipsing variable GG Cas are discussed. The results of two-colour polarimetric and four-colour photometric observations are given in tables. These results together with D. M. Popper's spectral data (1956) suggest a composite structure of the variable. The relative and absolute orbital elements are found. An attempt to estimate the intrinsic polarization of the binary's radiation is undertaken. Great uncertainties arise from interstellar polarization.

121.080 **Photoelectric observations of Beta Lyrae in 1967 and 1969.** T. J. Herczeg.
Inform. Bull. Variable Stars (I.A.U. Commission 27), Konkoly Obs., Budapest, No. 805, 3 pp. (1973).

121.081 **Lists of minima of eclipsing binaries.**
Compiled by R. Diethelm, R. Germann, K. Locher, T. Mallama, R. Meier, P. Morger, H. Peter.
BBSAG Bull. No. 7, p. 1 - 3; No. 8, p. 1 - 3; No. 9, p. 1 - 2 (1973). – 40th - 42nd list of Swiss Astronomical Society's Eclipsing Variable Observers.

121.082 **AA Ceti. Translation of the results published in BBSAG Bulletins 2, 5, 6 to Bloomer's new, totally different elements.** K. Locher.
BBSAG Bull. No. 7, p. 4 (1973).

121.083 **Current elements for V346 Aquilae.**
R. Diethelm.
BBSAG Bull. No. 7, p. 4 - 5 (1973).

121.084 **New elements for RU Eridani.** R. Diethelm.
BBSAG Bull. No. 8, p. 4 (1973).

121.085 **Duration and magnitude at totality of SY Hya.**
K. Locher.
BBSAG Bull. No. 8, p. 5 (1973).

121.086 **Evidence for a variable period of VY Hya.**
K. Locher.
BBSAG Bull. No. 8, p. 5 (1973).

121.087 **The totality duration of TU Her.**
H. Peter, K. Locher.
BBSAG Bull. No. 9, p. 3 (1973).

121.088 **The period of RR Draconis.** R. Diethelm.
BBSAG Bull. No. 9, p. 4 - 5 (1973).

121.089 **The minimum brightness of V391 Oph.** K. Locher.
BBSAG Bull. No. 9, p. 5 (1973).

Eclipsing variable stars. See Abstr. 003.149.

Proper motion of BD + 16° 516.
See Abstr. 112.001.

Evolution of contact binaries and star models for cataclysmic binaries. See Abstr. 117.004.

The evolution of a contact binary.
See Abstr. 117.019.

Evolutionary considerations involving the internal density concentration parameter of binary stars. See Abstr. 117.026.

The radiative interaction and evolution of medium- and low-mass binaries. See Abstr. 117.030.

The W UMa-type systems as contact binaries. I. Two methods of geometrical elements determination. Degree of contact. See Abstr. 117.032.

Fundamental data for contact binaries: RZ Comae Berenices, RZ Tauri, and AW Ursae Majoris. See Abstr. 117.035.

HZ Her – a pulsating star of RR Lyrae type? See Abstr. 122.069.

VV 428–479, variable stars in a Cepheus-Lacerta field of the Milky Way. See Abstr. 123.035.

AR Lacertae. See Abstr. 141.112

Optical identification criteria for binary X-ray sources. See Abstr. 142.006.

On the nature of the optical variability of HZ Her = Her X-1 and BD + 34°3815 = Cyg X-1. See Abstr. 142.007.

On the masses of X-ray sources. See Abstr. 142.025.

Astronomy from an X-ray satellite: Measuring the mass of a neutron star. See Abstr. 142.026.

Wolf-Rayet systems and the origin of massive X-ray binaries. See Abstr. 142.039.

Rapidly rotating degenerate dwarfs as X-ray sources in binaries. See Abstr. 142.042.

The spectrum and variability of Hercules X-1 observed by OSO-7. See Abstr. 142.043.

Light curve of HZ Herculis. See Abstr. 142.044.

Evidence for the binary nature of 2U 1700–37. See Abstr. 142.045.

Comments on the nature of Her X-1. See Abstr. 142.046.

The unusual behavior of HZ Herculis (= Hercules X-1) : 1890–1972. See Abstr. 142.048.

Observations of 2U0900-40 from Uhuru. See Abstr. 142.052.

On the optical behaviour of binary X-ray sources. See Abstr. 142.054.

Limits on the optical pulsations of HZ Herculis. See Abstr. 142.079.

Observations of the binary X-ray source SMC X-1 from OSO-7. See Abstr. 142.081.

A low-mass primary for Cygnus X-1? See Abstr. 142.087.

Model for X-ray sources based on magnetic field twisting. See Abstr. 142.088.

HZ Herculis, der Sonneberger Röntgenstern. See Abstr. 142.089.

OAO-2 observations of HD 153919 = 2U 1700−37. See Abstr. 142.091.

The reflection effect in HZ Herculis. See Abstr. 142.092.

X-ray beaming and mass transfer in HZ Her. See Abstr. 142.098.

Optical appearance of binary X-ray sources. See Abstr. 142.099.

The 35-day periodicity of Hercules X-1. See Abstr. 142.100.

The nature of X-ray binaries III. Evolution of massive close binaries with one collapsed component – with a possible application to Cygnus X-3. See Abstr. 142.102.

Speeding up of pulsing X-ray sources. See Abstr. 142.107.

On the explanation of the 36-day period of Her X-1 by free precession. See Abstr. 142.111.

X-ray pulse profile and celestial position of Hercules X-1. See Abstr. 142.114.

On the accretion model for X-ray double stars. See Abstr. 142.117.

Der Ope-Stern X Persei. See Abstr. 142.118.

Optical observations and model for Cygnus X-1. See Abstr. 142.124.

HD 77581 as the optical counterpart of 2U 0900−40. See Abstr. 142.125.

Optical identifications of X-ray sources. See Abstr. 142.131.

Optical pulsations from HZ Herculis. See Abstr. 142.136.

2U 0900-40 a black hole? See Abstr. 142.139.

On the physical parameters for Centaurus X-3 and Hercules X-1. See Abstr. 142.146.

Uhuru observations of the binary X-ray source 2U 0900−40. See Abstr. 142.150.

Optical studies of *Uhuru* sources. IV. The long-term behavior of HZ Herculis = Hercules X-1. See Abstr. 142.151.

Age of Magellanic eclipsing variables and color distribution of 3323 variables in both Magellanic Clouds. See Abstr. 159.006.

Errata

121.901 **Erratum: 'Computer solution of eclipsing-binary light curves by the method of differential corrections'** [Astrophys. Journ., Suppl. Ser., No. 211, Vol. 24, 449 - 478 (1972)].   D. D. Proctor, A. P. Linnell.
Astrophys. Journ., Suppl. Ser., No. 215, Vol. 25, 163 (1973).

121.902 **Erratum: 'The light-curves of 11 eclipsing variables'** [Acta Astron., Vol. 22, 273 - 303 (1972)].
R. Szafraniec.
Acta Astron., Vol. 23, 189 (1973).

121.903 **Erratum: 'Three eclipsing binaries with eccentric orbits'** [Acta Astron., Vol. 22, 411 - 418 (1972)].
H. Bossen, P. Klawitter.
Acta Astron., Vol. 23, 189 (1973).

122 Physical Variables, Flare Stars, Pulsation Theory

122.001 An abundance analysis of the Delta Scuti variable Delta Delphini.   M. Ishikawa.
Publ. Astron. Soc. Japan, Vol. 25, 111 - 127 (1973).

We have carried out a curve-of-growth analysis of the spectra of δ Delphini in the blue region. The atmospheric parameters have been derived. The abundances of twenty elements are determined.

122.002 The spectrum of the beat cepheid U TrA.
A. W. Rodgers, R. A. Gingold.
Monthly Notices Roy. Astron. Soc., Vol. 161, 23 - 26 (1973).

A spectroscopic estimate of the surface gravity of U TrA is obtained which shows that this amplitude modulated variable has a similar gravity to cepheids of the same period.

122.003 The spectral classification of the β Cephei stars and their location in the theoretical Hertzsprung-Russell diagram.   J. R. Lesh, M. L. Aizenman.
Astron. Astrophys., Vol. 22, 229 - 237 (1973).

Spectral types on the MK system are presented for 17 bright β Cephei variables, 7 bright suspected variables and 23 fainter suspected variables. The bright variables are located in the observational HR diagram by means of the photometric parameters Q and β, and their position relative to normal B stars is discussed. A transformation to the (log T_{eff}, M_{bol}) plane enables us to compare the positions of both normal stars and β Cephei variables with evolutionary tracks.

122.004 Short-period variability of B, A and F stars. VI. New Delta Scuti stars in selected regions.   M. Breger.
Astron. Astrophys., Vol. 22, 247 - 249 (1973).

Twelve more short-period Delta Scuti variables have been discovered, bringing the total number to about 70. The new variables are situated in the field as well as in the Praesepe and Hyades clusters. Visual amplitudes are as small as 0.01 mag. The derived periods are typical for the positions of the stars in the color-magnitude diagram.

122.005 Activity in flare stars of the solar neighborhood.
W. E. Kunkel.
Astrophys. Journ., Suppl. Ser., No. 213, Vol. 25, 1 - 36 (1973).
Presented at I. A. U. Colloquium No. 15: 'New directions and new frontiers in variable star research', Bamberg, 1971.

Activity of flare stars in the vicinity of the sun is compared in terms of a set of descriptive parameters that minimize problems of bias and noise in measurement. The results from an examination of U-band observations of about 1000 flares obtained at the McDonald and Cerro Tololo Observatories between 1966 and 1970 are presented.

122.006 Beta Canis Majoris: Period analysis of recent photometry and published radial velocities.
R. R. Shobbrook.
Monthly Notices Roy. Astron. Soc., Vol. 161, 257 - 267 (1973).

Beta Canis Majoris has been observed in yellow light, mainly during the 1971/1972 season. The light variation with the 6-hr period (P_1) has one-quarter of the amplitude of the 6 hr 2 m period (P_2), whereas in the velocities published from 1908 to 1954 and re-examined in this paper, P_1 has a 50 per cent larger variation than P_2. *Both* periods are changing.

122.007 Multicolour observations of UZ Librae.
W. Z. Wisniewski.
Monthly Notices Roy. Astron. Soc., Vol. 161, 331 - 334 (1973).

Multicolour observations of β Lyrae variable, UZ Librae,

are presented. Light curves in *UBVRI* system are given. The colours show an infra-red excess, which is attributed to a third light.

122.008 On the metal abundance of RR Lyrae stars in the globular cluster M 22.
D. Butler, R. P. Kraft, J. S. Miller, L. B. Robinson.
Astrophys. Journ., (*Letters*), Vol. 179, L73 - L78 = Contr. Lick Obs., No. 387 (1973).

The Preston system for measuring metal abundances of RR Lyrae stars has been reestablished from equivalent-width measurements of $H\gamma$, $H\delta$, and Ca II K in spectra obtained with the new image-tube scanner. The RR Lyraes of M 22 are found to be metal poor. This conflicts with the expectation of metal richness as judged from the small slope of the giant branch in the H-R diagram.

122.009 A photometric study of 21 Comae Berenices.
J. R. Percy.
Astron. Astrophys., Vol. 22, 381 - 383 (1973).

New photometric (B and V) observations of the Ap star 21 Com have been obtained at the Kitt Peak and the David Dunlap Observatories. The observations confirm the 30-minute light variation reported in 1957 by Bahner and Mawridis. The range in V varies from 0^m004 to 0^m019 and the range in $(B-V)$ is very small. The cause of the variation is probably pulsation.

122.010 Observations of Hubble-Sandage variables in M 31 and M 33.   L. Rosino, A Bianchini.
Astron. Astrophys., Vol. 22, 453 - 459 (1973).

Eight blue variables of high luminosity (Hubble-Sandage variables), three in M 31 and five in M 33 have been examined on photographs obtained at Asiago for the nova survey after 1953. Light curves are reproduced in this paper. Var. 1 in M 31 is new, the others have been previously studied by Hubble and by Hubble and Sandage. The general characteristics found by these authors are confirmed by the present material. However, Var. B in M 33 has attained in 1963 a maximum of -9.9 B considerably brighter than previously observed. Some considerations follow.

122.011 The "zone of avoidance" in the period-amplitude diagram of the LMC cepheids.
A. M. van Genderen.
Astron. Astrophys., Vol. 23, 153 (1973).

It is shown that also in the period-amplitude diagram of the LMC cepheids of type I a narrow zone exists near log $P = 1$ in which fewer cepheids occur than in the adjacent zones of the same width. This phenomenon was already noticed for the SMC, galactic and M 31 cepheids.

122.012 Five-color observations of the short-period pulsating star δ Del.   A. M. van Genderen.
Astron. Astrophys. Suppl. Ser., Vol. 9, 149 - 156 (1973).

A discussion is presented of photoelectric five-color observations of the δ Scuti star δ Del of which the period is 0^d135. It is likely that the star belongs to the group of short-period variables with two or more periods of oscillations. It was however not possible to determine these periods, because of too few available observations.

122.013 High-frequency optical variables. I. G 61-29.
H. B. Richer, J. R. Auman, B. C. Isherwood, J. P. Steele, T. J. Ulrych.
Astrophys. Journ., Vol. 180, 107 - 114 (1973).

Observations and analysis of the peculiar object G 61-29

are presented. It is shown that when the star is most active a strong optical periodicity near 105 seconds develops, weakens, and shifts to lower frequency, indicating that the mechanism is probably plasma oscillations. A model for G61-29 is derived in which it is suggested that this object should be an easily detected X-ray source.

122.014 High-speed photometry of Z Camelopardalis.
E. L. Robinson.
Astrophys. Journ., Vol. 180, 121 - 141 (1973).

High-speed photometry of Z Cam shows that its light curve bears a strong resemblance to the light curves of U Gem and VV Pup. Strong flickering is always present in Z Cam and has an amplitude correlated with the position of Z Cam in its eruption light cycle. Z Cam does not show eclipses. Low-amplitude sinusoidal periodicities have been found in the light curve during eruptions with periods between 16 and 19 seconds. A model of Z Cam is constructed whose major new feature is a bright spot on the disk of gas around the white dwarf.

122.015 On the nature of radiation of flares of UV Ceti-type stars.   A. A. Korovyakovskaya.
Astrofizika, Vol. 8, 247 - 260 (1972). In Russian. English translation in Astrophysics, Vol. 8, No. 2.

The radiation of an ionized hydrogen gas behind a shock-wave front propagating through a polytropic medium is considered. The theoretical light curves for the gas and those observed for flares of UV Cet-type stars are in satisfactory agreement. Diagrams are presented to show the time variations of the electron and ion temperatures, the degree of ionization and excitation, the number of L_α-quanta behind the shock-wave front for different initial conditions.

122.016 Delta Scuti stars.
A. Baglin, M. Breger, C. Chevalier, B. Hauck, J. M. le Contel, J. P. Sareyan, J. C. Valtier.
Astron. Astrophys., Vol. 23, 221 - 240 (1973).

The extensive observational and theoretical data on δ Scuti variables (published and previously unpublished) are collected and analyzed. Properties of pulsation are discussed. and relationships between δ Scuti stars and other types of pulsators and nonvariable stars are established.

122.017 Second beat period in the light curve of SX Phoenicis.
E. W. Elst.
Astron. Astrophys., Vol. 23, 285 - 290 (1973).

From the photometric observations of SX Phe by Stock and Tapia (1971) a second beat period of $21\overset{d}{.}0$ is discovered. A formula for maximum light of SX Phe is presented.

122.018 Photoelectric observations of the bright RV Tauri stars R Scuti and U Monocerotis.
P. N. J. Wisse, M. Wisse.
Astron. Astrophys., Vol. 23, 463 - 466 (1973).

Photoelectric observations of the RV Tauri stars R Sct and U Mon are presented. An unusually deep minimum of R Sct and other peculiarities in its light and colour curves have been observed. U Mon shows a decreasing mean magnitude. A short description of two spectra of R Sct, taken at the Radcliffe Observatory, is given.

122.019 Optical circular polarization of X Persei.
R. A. Stokes, R. W. Avery, J. J. Michalsky, Jr.
Nature, Phys. Sci., Vol. 241, 5 (1973).

We observed X Per on six nights in September and October 1972, with the Battelle 31-inch reflector and a polarimeter which utilizes a birefringence modulator. We list the Stokes parameter V_s normalized to unit total intensity. Our positive sign refers to circular polarization such that, to an observer facing the source, the rotation of the electric vector

in a fixed plane is clockwise. It is not clear from the results whether the polarization is intrinsic to the star or the result of processes in the interstellar medium, or even a circumstellar cloud.

122.020 X Leonis, 1920–44.   J. E. Isles.
Journ. British Astron. Ass., Vol. 83, 128 - 135 (1973). – Report of Variable Star Section of the British Astron. Ass.

122.021 Variable stars in globular clusters.   H. S. Hogg.
Journ. Roy. Astron. Soc. Canada, Vol. 67, 8 - 18 = Commun. David Dunlap Obs., Univ. Toronto, *Richmond Hill, Ontario,* Canada, No. 346 (1973).–Presidential address of the Canadian Astron. Soc., delivered at the University of Montreal, May 12, 1972.

122.022 The unique variable V725 Sagittarii.   S. Demers.
Journ. Roy. Astron. Soc. Canada, Vol. 67, 19 - 30 (1973).

V725 Sagittarii is the only known variable star which has shown an important change in its period in an interval of a few years. The variation may have a period of about 45 to 50 days. An analysis of magnitude estimates over the last 70 years suggests that V725 Sgr is a single star, probably an RV Tauri or semi-regular variable.

122.023 Progressi nell'interpretazione e studio delle stelle variabili.   L. Rosino.
Coelum, Vol. 41, 1 - 12 (1973).

122.024 Rätselhafte BL-Lacertae-Objekte.   J. Solf.
SuW, Vol. 12, 49 - 50 (1973).

122.025 Four-color photometry of short-period variable stars.   I. Epstein, A. E. Abraham de Epstein.
Astron. Journ., Vol. 78, 83 - 89 (1973).

Observations in the (u, v, b, y) system are reported for 21 additional variable stars, including 12 of the RRab type, seven RRc stars, and two AI Vel variables. The material is added to observations of 38 variables reported previously. The intrinsic parameters, at minimum light of the variables, separated into astrophysically distinct groups by means of their four-color indices, are used in a rediscussion of interstellar effects in the four-color system, and of correlations of the variable star groups with their kinematic properties for four distinct photometric subgroups into which the material separates.

122.026 Changes of period in semiregular variables.
C. H. Lacy.
Astron. Journ., Vol. 78, 90 - 93 (1973).

Fourteen well observed variables currently classified as semiregular are tested for change of mean cycle length by the method of contingency tables. Three variables are found to have a significant change of period during the interval of 67 years covered by the data.

122.027 Dependence of flare decay rates on flare luminosities for UV Cet stars.
R. E. Gershberg, N. I. Shakhovskaya.
Nature, Phys. Sci., Vol. 242, 85 - 86 (1973).

We have analysed observations of 13 UV Cet flare stars made at the Crimean Astrophysical Observatory during 1969–1972 with the 64 cm meniscus telescope, in a photometric system close to B. There exist statistical relations between flare luminosities at maxima and flare decay rates, and the relations cover the range of absolute magnitudes of flare stars from $M_V = 8$ mag up to $M_V = 16$ mag.

122.028´ Rapid irregular variables with spectral characteristics

sdB. F. I. Lukatskaya.
Astrometriya i Astrofizika, *Kiev,* vyp. (No.) 16, (see 003.006), p. 100 - 112 (1972). In Russian.

A statistical analysis is carried out for the light observations of ten rapid irregular variables with sdB spectral characteristics, and for R Aqr and Sco X-1. Both the types and the parameters of stochastic processes which simulate the light variations are determined. Dependences are found between parameters and some physical characteristics of the variables. Sco X-1 is similar to nova-like stars with respect to the results of statistical analyses of light variations.

122.029 Some variables of spectral type K. O. J. Eggen.
Publ. Astron. Soc. Pacific, Vol. 85, 42 - 56 (1973).

(*UBVRI*) photometry for 22 variables of spectral type K are presented, including one new variable (HD 81410) and two previously suspected variables (HD 118238, HD 25966).

122.030 On variable 39 in IC 1613. J. L. Hutchinson.
Publ. Astron. Soc. Pacific, Vol. 85, 119 - 121 (1973).

In a recent study Sandage (1971) presented data obtained by Baade concerning a number of cepheids and other stars in the local-group galaxy IC 1613. One variable in particular shows an unusual light curve. A further study of the variability of this object was therefore undertaken using data in Sandage's paper.

122.031 Photoelectric observations of flare stars. II.
S. Cristaldi, M. Rodonò.
Astron. Astrophys., Suppl. Ser., Vol. 10, 47 - 104 (1973).

The results of a second series of photoelectric observations of UV Cet-type flare stars carried out at the Catania Astrophysical Observatory during 1969 and 1970 are given. The observations refer to the following stars: YZ CMi, UV Cet, BY Dra, EV Lac, AD Leo, PZ Mon, EQ Peg, V 1216 Sgr, BD + 13°2618, BD + 55°1823. With respect to the first series of observations (Cristaldi and Rodonò 1970), the present one includes also simultaneous *UBV* observations.

122.032 Velocity gradients and microturbulence in cepheids.
A. H. Karp.
Astrophys. Journ., Vol. 180, 895 - 900 (1973).

Variations of the microturbulent velocity with phase and height in the atmosphere have been reported in classical cepheids. It is shown that these effects can be understood in terms of variations of the velocity gradient in the atmospheres of these stars.

122.033 Photoelectric light-curves of v UMa.
T. Z. Dworak.
Acta Astron., Vol. 23, 37 - 42 (1973).

The paper contains photoelectric observations of v UMa, a δ Scuti variable, reduced to the *BV* system. Four light-curves were obtained on the basis of these observations. A more accurate period was determined.

122.034 HD 105563 is a star of VV Cephei type.
G. Lyngå.
Astron. Astrophys., Vol. 24, 303 - 304 (1973).

From its spectrum HD 105563 is classified as a star of VV Cephei type. The star is a member of a small group of early type stars so that an absolute magnitude can be derived. Photometric observations show slight indications of variability.

122.035 Mean absolute magnitude of the RR Lyrae stars.
A. Heck.
Astron. Astrophys., Vol. 24, 313 - 316 (1973). In French.

We determine the mean absolute magnitude and the dispersion in magnitude of a sample of RR Lyrae stars from Hemenway's (1971) paper. In "mean light" magnitude, the results are: $\bar{M}_v = 0.51 \pm 0.20$ and $\sigma^2_{M_v} = 0.56 \pm 0.22$ and are co-

herent with those of Heck (1972) and Hemenway (1971).

122.036 Is δ Cep super iron rich?
J. van Paradijs, H. de Ruiter.
Astron. Astrophys., Vol. 24, 317 - 319 (1973).

A reanalysis of the equivalent width data from four spectrograms of δ Cep, using model atmospheres, shows that the conclusion of Kobayashi and Takeuti (1971), namely that δ Cep is "super iron rich", is highly improbable.

122.037 A Q-method for color excesses of classical cepheids.
S. B. Parsons, R. A. Bell.
Bull. American Astron. Soc., Vol. 5, 3 (1973). – Abstr. AAS.

122.038 On the metal abundance of RR Lyrae stars in the globular cluster M22.
D. Butler, R. P. Kraft, J. S. Miller, L. B. Robinson.
Bull. American Astron. Soc., Vol. 5, 14 (1973). – Abstr. AAS.

122.039 Cepheids and helium abundance.
R. S. Tuggle, I. Iben, Jr.
Bull. American Astron. Soc., Vol. 5, 14 (1973). – Abstr. AAS.

122.040 Non-pulsating stars in the RR Lyrae and cepheid instability strips. A. N. Cox, D. S. King, J. E. Tabor.
Bull. American Astron. Soc., Vol. 5, 14 - 15 (1973). – Abstr. AAS.

122.041 Radial velocity curves according to the iterative theory of nonlinear pulsations.
N. R. Simon, V. K. Sastri.
Bull. American Astron. Soc., Vol. 5, 16 (1973). – Abstr. AAS.

122.042 Non-linear pulsation in RR Lyrae star models.
W. H. Spangenberg.
Bull. American Astron. Soc., Vol. 5, 16 (1973). – Abstr. AAS.

122.043 Long-term changes in amplitude for RR Lyrae stars in M5. C. M. Coutts.
Bull. American Astron. Soc., Vol. 5, 16 (1973). – Abstr. AAS.

122.044 Intrinsic B−V colors of galactic cepheids.
T. Kelsall.
Bull. American Astron. Soc., Vol. 5, 16 - 17 (1973). – Abstr. AAS.

122.045 UBV photometry of l Monocerotis. R. L. Millis.
Bull. American Astron. Soc., Vol. 5, 17 (1973). Abstr. AAS.

122.046 The recent outburst of the symbiotic star CI Cygni.
F. M. Stienon.
Bull. American Astron. Soc., Vol. 5, 17 (1973). – Abstr. AAS.

122.047 Polarimetric observations of R Corona Borealis.
G. V. Coyne.
Bull. American Astron. Soc., Vol. 5, 17 (1973). – Abstr. AAS.

122.048 Light curves, spectral type, and absolute magnitude of the carbon star UV Aurigae.
O. G. Franz, N. M. White.
Bull. American Astron. Soc., Vol. 5, 43 (1973). – Abstr. AAS.

122.049 Period changes in β Cephei stars: Comparison of observation with theory. P. P. Eggleton, J. R. Percy.
Monthly Notices Roy. Astron. Soc., Vol. 161, 421 - 425 (1973).

The rate of period change due to evolution has been calculated for a β Cephei star of 16 $M_\odot$ by means of a theoretical evolutionary track. The dependence of this rate on mass has been estimated. This rate has been compared with observed rates of period change in a few β Cephei stars.

122.050 OH emission from long-period variable stars.
Nguyen-Quang-Rieu, R. Fillit, M. Gheudin.
17th Colloquium International Astrophys. Liège 1971, (see 012.003), p. 233 - 237 (1972).

122.051 The Mira variable S Carinae. D. Shinkawa.
Astrophys. Journ., Suppl. Ser., No. 218, Vol. 25, 253 - 276 (1973).
S Carinae has been studied throughout its cycle with high-dispersion coudé spectra and photoelectric spectrum scans. Atmospheric parameters have been derived with the use of coarse analyses and radiative model atmospheres.

122.052 Photometric study of BL Lacertae. II. Brightness variations from March 1969 to January 1971.
C. Bertaud, G. Wlérick, P. Véron, B. Dumortier, J. Bigay, G. Paturel, M. Duruy, P. de Saevsky.
Astron. Astrophys., Vol. 24, 357 - 368 (1973). In French.
BL Lacertae has been monitored during the years 1969 and 1970 with the same observing methods used in 1968. The photoelectric sequence has been improved. We confirm the result of Racine: there is a correlation between the colour index $B-V$ and the V magnitude. The light curve has been constructed with 370 visual measurements and the values obtained from 355 photographic plates for which the r.m.s. error is 0.13 magnitude.

122.053 BD Herculis. M. I. Lavrov, N. V. Lavrova.
Trudy Kazan. Gorod. Astron. Obs., No. 38, p. 46 - 63 (1972). In Russian.

122.054 Fotometría fotoeléctrica UBV de SX Phoenicis.
J. J. Clariá.
Bol. As. Argentina Astron., No. 16, (see 012.007), p. 18 - 22 (1971).

122.055 Most important stars. Yu. N. Efremov.
Zemlya i Vselennaya, 1973, No. 2, p. 46 - 51.
In Russian.

122.056 The Beta Canis Majoris stars 15 CMa and ξ^1 CMa.
R. R. Shobbrook.
Monthly Notices Roy. Astron. Soc., Vol. 162, 25 - 36 (1973).
The Beta Canis Majoris variables 15 CMa and ξ^1 CMa were observed photoelectrically in yellow light during the 1971/72 season. 15 CMa shows random amplitude variations at its known period of $0^d.1845$; a re-discussion of earlier published photometry and radial velocities confirms this and also indicates that the period is variable. ξ^1 CMa, on the other hand, with a period of $0^d.2096$, showed a remarkably stable light curve shape during 1971/72.

122.057 Periods for two variables in M 13. W. Osborn.
Monthly Notices Roy. Astron. Soc., Vol. 162, 91 - 96 (1973).
Periods of 0.381767^d and 0.392665^d have been derived for the M 13 variables 5 and 9, which form a close optical double. New observations are presented for these stars, and it is shown that the observations given by Kollnig-Schattschneider for variable 5 in fact refer to variable 9. Both variables appear to be of RRc type with B amplitudes of 0.6 magnitude.

122.058 Radial velocity, light and colour curves of RZ Cep, an RR Lyrae star. E. A. Epps, J. E. Sinclair.
Observatory, Vol. 93, 78 - 81 (1973).

122.059 Photometric observations of the Delta Scuti star 44 Tauri. J. R. Percy.

Observatory, Vol. 93, 81 - 83 (1973).

122.060 Early visual detection of rapidly fluctuating variable stars. A. D. Thackeray.
Observatory, Vol. 93, 84 - 85 (1973).

122.061 On the line spectrum of SS Cyg.
N. F. Vojkhanskaya.
Astrofiz. Issled., Izv. Spets. Astrofiz. Obs., Vol. 4, 106 - 114 (1972). In Russian.
The line spectrum of SS Cyg has been studied at minimum light and during three flares. An identification of emission and absorption lines is carried out and their equivalent widths are determined.

122.062 Photoelectric photometry of RU Cam in 1971-1972.
G. V. Zaitseva, V. M. Lyutyj, V. M. Kovalenko, O. N. Kovalenko.
Astron. Tsirk., No. 744, p. 1 - 2 (1973). In Russian.

122.063 Observations of R Coronae Borealis in 1971-1972.
A. G. Totochava.
Astron. Tsirk., No. 744, p. 2 - 4 (1973). In Russian.

122.064 Photoelectric UBV magnitudes at maximum and minimum brightness of RZ Lyrae.
Yu. S. Romanov.
Astron. Tsirk., No. 745, p. 3 - 5 (1973). In Russian.

122.065 Variation of physical parameters of pulsation of SX Phoenicis during the Blazhko effect. M. S. Frolov.
Astron. Tsirk., No. 745, p. 7 - 8 (1973). In Russian.

122.066 Possible eruptive stars in the region of M 31 which need to be confirmed.
A. S. Sharov, A. K. Alksnis.
Astron. Tsirk., No. 750, p. 7 - 8 (1973). In Russian.

122.067 Determination of the period of the Blazhko effect with the aid of a computer.
V. M. Grigorevsky, V. D. Motrich.
Astron. Tsirk., No. 751, p. 6 - 8 (1973). In Russian.

122.068 Polarization observations of BL Lac in 1971.
V. A. Dombrovsky.
Astron. Tsirk., No. 753, p. 1 - 2 (1973). In Russian.

122.069 HZ Her – a pulsating star of RR Lyrae type?
G. S. Bisnovatyj-Kogan, B. V. Komberg.
Astron. Tsirk., No. 757, p. 3 - 5 (1973). In Russian.

122.070 Some conclusions on the mean luminosity of RR Lyrae stars in a galactic field. M. S. Frolov.
Astron. Tsirk., No. 759, p. 4 - 6 (1973). In Russian.

122.071 Further photometric observations of V1216 Sagittarii. A. H. Jarrett, J. P. Eksteen.
Monthly Notes Astron. Soc. Southern Africa, Vol. 32, 18 - 20 (1973). – Note.

122.072 Pulsating star UX Ceti. D. Chiş.
Stud. Cerc. Astron., Vol. 18, 63 - 65 (1973).
In Romanian.
Results of photographic observations of the variable star UX Ceti are reported. Photometric elements are derived. UX Ceti seems to be of RRc type.

122.073 A photometric study of the pulsating star BE Eridani. V. Pop, I. Todoran.
Stud. Cerc. Astron., Vol. 18, 67 - 71 (1973). In Romanian.
From 355 photographic observations of BE Eridani 14

epochs of maxima are determined. According to the shape of the light curve BE Eridani belongs to RR Lyrae stars of type RRab.

122.074 A search for Ap stars with very long periods.
S. C. Wolff, N. D. Morrison.
Publ. Astron. Soc. Pacific, Vol. 85, 141 - 149 (1973).

Four-color (*uvby*) photoelectric observations have yielded new periods for three peculiar A-type stars (HD 8441, HD 12288, HD 216533). The period of 12.448 days for HD 24712 is confirmed. The period of HD 18078 and HD 2453 are longer than one year. These results nearly double the number of Ap stars known to have periods longer than 25 days and add weight to the evidence that magnetic stars undergo rotational deceleration during their lifetimes.

122.075 A search for rapid spectral variations in 10 Lacertae.
J. W. Glaspey, G. A. H. Walker.
Publ. Astron. Soc. Pacific, Vol. 85, 188 - 190 (1973).

The spectrum of 10 Lac in the H δ region has been observed using a low light-level Image Isocon system to search for rapid spectrum variations. Two sets of observations covering 31 and 32 minutes, respectively, as well as a number of shorter observations covering an additional 30 minutes, have not revealed any line depth variations greater than ~25 % in either the O II λ4119 or other spectral lines.

122.076 A magnitude sequence for the variable star ST Monocerotis.   A. U. Landolt.
Publ. Astron. Soc. Pacific, Vol. 85, 203 - 204 = Contr. Louisiana State Univ. Obs., *Baton Rouge,* No. 79 (1973).

A *UBV* photoelectric sequence has been established in the neighborhood of the variable star ST Mon. A finding chart is provided.

122.077 On the variability of Case 621 and MSB 57.
J. H. Baumert.
Publ. Astron. Soc. Pacific, Vol. 85, 205 - 206 (1973).

The unusual SC star Case 621 is identified with the variable VX Aql and the variability of the carbon star MSB 57 is reported.

122.078 Two flare stars behind the Pleiades.
W. B. Weaver, S. A. Naftilan.
Publ. Astron. Soc. Pacific, Vol. 85, 213 - 214 (1973).

The flare stars H_{II} 230 and H_{II} 1069, both nonmembers of the Pleiades, were observed at 200 Å mm^{-1}. These stars are well behind the Pleiades and do not show Ca II or hydrogen emission. The spectral type of H_{II} 230 is earlier than that of any other known UV Ceti-type star.

122.079 Long-period variables: Correlation of stellar period with OH radial-velocity pattern.
D. F. Dickinson, E. J. Chaisson.
Astrophys. Journ., *(Letters)*, Vol. 181, L135 - L138 (1973).

The OH radio spectrum seen in variable stars shows two distinct groups of emission features well separated in radial velocity. In this paper we show that the separation is directly proportional to the period.

122.080 SS Aurigae: a Z Cam star?   I. D. Howarth.
Journ. British Astron. Ass., Vol. 83, 179 - 182 (1973).

122.081 X Leonis, 1945-69.   J. E. Isles.
Journ. British Astron. Ass., Vol. 83, 209 - 216 (1973). – Report of Variable Star Section British Astron. Ass.

122.082 The dependence of the flare activity of UV Cet-type stars on their age. I.   P. F. Chugainov.
Izv. Krymskoj Astrofiz. Obs., Vol. 46, 14 - 24 (1972).

In Russian.

Characteristics of the flare activity of 14 red dwarf stars have been determined. Twelve of them are UV Cet-type stars. The most detailed material is used for 4 stars (AD Leo, EV Lac, YZ CMi and UV Cet), for which observations were made during the International Patrol of Flare Stars.

122.083 Some problems of fluorescence of ionized hydrogen optically thick at frequency lines.
R. E. Gershberg.
Izv. Krymskoj Astrofiz. Obs., Vol. 46, 59 - 78, with an appendix by E. E. Shnol', p. 79 - 82 (1972). In Russian.

In order to improve the methods of photometric analysis of the UV Cet-type star flares some problems of fluorescence of hydrogen plasma optically thick at Balmer lines are considered.

122.084 The age and evolutionary state of the β Canis Majoris variables.   J. R. Lesh.
Stellar ages. Proc. IAU Colloquium No. 17, (see 012.015), V, 1 - 11 (1973).

122.085 Ages of δ Scuti and AI Velorum stars.
C. Chevalier.
Stellar ages. Proc. IAU Colloquium No. 17, (see 012.015), VI, 1 - 7 (1973).

122.086 Flares of YZ CMi.   R. C. Kapoor.
Inform. Bull. Variable Stars (I.A.U. Commission 27), Konkoly Obs., Budapest, No. 758, 4 pp. (1973).

122.087 Photoelectric observations of EV Lac during the 1972, September 1 - 15 international patrol.
S. Cristaldi, M. Rodonò.
Inform. Bull. Variable Stars (I.A.U. Commission 27), Konkoly Obs., Budapest, No. 759, 3 pp. (1973).

122.088 Photoelectric observations of the flare star UV Ceti during the 1972 October 1 - 15 international patrol.
S. Cristaldi, M. Rodonò.
Inform. Bull. Variable Stars (I.A.U. Commission 27), Konkoly Obs., Budapest, No. 760, 3 pp. (1973).

122.089 SY Fornacis – no U Geminorum star.
W. Wenzel.
Inform. Bull. Variable Stars (I.A.U. Commission 27), Konkoly Obs., Budapest, No. 763 (1973).

122.090 Development of a new 4-year cycle in the 41-day period of RR Lyrae.   L. Detre, B. Szeidl.
Inform. Bull. Variable Stars (I.A.U. Commission 27), Konkoly Obs. Budapest, No. 764 (1973).

122.091 Photoelectric observations of the flare star YZ CMi during the 1972–73, 30 December - 12 January international patrol.   S. Cristaldi, M. Rodonò.
Inform. Bull. Variable Stars (I.A.U. Commission 27), Konkoly Obs., Budapest, No. 767 (1973).

122.092 On the period luminosity relation of Mira type variables.   K. Ferrari d'Occhieppo.
Inform. Bull. Variable Stars (I.A.U. Commission 27), Konkoly Obs., Budapest, No. 768 (1973).

122.093 New flare stars in the Pleiades.   W. Götz.
Inform. Bull. Variable Stars (I.A.U. Commission 27), Konkoly Obs., Budapest, No. 771 (1973).

122.094 Visual observations of AD Leonis.   J. E. Isles.
Inform. Bull. Variable Stars (I.A.U. Commission 27), Konkoly Obs., Budapest, No. 772 (1973).

122.095 **New flare stars in the Pleiades region.** L. Pigatto.
Inform. Bull. Variable Stars (I. A. U. Commission 27), Konkoly Obs., Budapest, No. 776 (1973).

122.096 **S10760 – a very distant RR-Lyrae-star.**
L. Meinunger.
Inform. Bull. Variable Stars (I. A. U. Commission 27), Konkoly Obs., Budapest, No. 777 (1973).

122.097 **Investigation of two Delta Scuti suspects.**
A. K. Bhatnagar, S. K. Gupta.
Inform. Bull. Variable Stars (I. A. U. Commission 27), Konkoly Obs., Budapest, No. 778 (1973).

122.098 **A flare of "antiflare" star RZ Psc.**
V. G. Karetnikov, A. F. Pugach.
Inform. Bull. Variable Stars (I. A. U. Commission 27), Konkoly Obs., Budapest, No. 783 (1973).

122.099 **New flare stars in the Pleiades region (1972 - 1973).**
G. Haro, E. Chavira.
Inform. Bull. Variable Stars (I. A. U. Commission 27), Konkoly Obs., Budapest, No. 788, 3 pp. (1973).

122.100 **AD Leo.** K. Osawa, K. Ichimura, Y. Shimizu,
T. Okada, K. Okida, M. Yutani, H. Koyano.
Inform. Bull. Variable Stars (I. A. U. Commission 27), Konkoly Obs., Budapest, No. 790 (1973).

122.101 **Continuous photoelectric photometry of AD Leo during the 1973 international patrol.**
B. R. Pettersen, B. N. Andersen.
Inform. Bull. Variable Stars (I. A. U. Commission 27), Konkoly Obs., Budapest, No. 791, 3 pp. (1973).

122.102 **De veranderlijke van de maand: S Ursae Maioris.**
H. Feyth.
Hemel en Dampkring, Vol. 71, 127 - 129 (1973).

122.103 **De veranderlijke van de maand: T Ursae Majoris.**
H. Feijth.
Hemel en Dampkring, Vol. 71, 173 - 176 (1973).

122.104 **Ricerche sulle stelle a brillamento.**
S. Cristaldi, G. Godoli, M. Rodonò.
Mem. Soc. Astron. Italiana, Nuova Ser., Vol. 43, 697 - 703 (1973).
Some results of photoelectric observations of flare stars, carried out or running at Catania, with reference to the program of research activity of solar type, are shown.

122.105 **Variations of the content of chemical elements of classic cepheids of the Galaxy.** N. N. Yakimova.
Astron. Zhurn. Akad. Nauk SSSR, Vol. 50, 526 - 534 (1973). In Russian. English translation in Soviet Astron. AJ, Vol. 17, No. 3.
It is concluded that the higher a cepheid is situated above the galactic plane, the less is the content of heavy elements and helium in a cepheid envelope.

122.106 **Mira-stjerner.** S. Irgens-Jensen.
Astron. Tidssk., Årg. 6, p. 69 - 72 (1973).

122.107 **S Persei, 1920 – 69.** J. E. Isles.
Journ. British Astron. Ass., Vol. 83, 291 - 295 (1973). – Report of the Variable Star Section of the British Astron. Ass.

122.108 **Raumgeschwindigkeiten und Entstehungsorte von klassischen Cepheiden.** R. Wielen.
Mitt. Astron. Ges., No. 32, p. 212 - 213 (1973). – Abstract.

122.109 **The statistics of the β Cephei stars.**
J. R. Lesh, M. L. Aizenman.
Astron. Astrophys., Vol. 26, 1 - 9 (1973).
We discuss not only the incidence of the β Cephei phenomenon among B stars, but also the galactic distribution of the variables and such other statistical data as may give us an indication of their age and evolutionary state.

122.110 **The behaviour of the hydrogen lines in the β Cephei star ν Eridani.** P. G. Laskarides.
Astron. Astrophys., Vol. 26, 91 - 93 (1973).
Noncentral "filling in" of the absorption hydrogen lines by emission from the extended atmosphere may cause the difference in the behaviour of the hydrogen line profiles, compared to the Si III line profiles, as well as the apparent correlation between the phase of the sharper line profile and the wavelength of the line reported earlier.

122.111 **Photometric observations of RS Gruis.**
A. W. J. Cousins, T. G. Hawarden, A. J. Penny.
Monthly Notes Astron. Soc. Southern Africa, Vol. 32, 35 - 40 (1973).

122.112 **A search for Delta Scuti stars.**
J. R. Percy.
Journ. Roy. Astron. Soc. Canada, Vol. 67, 139 - 141 = Commun. David Dunlop Obs., Univ. Toronto, *Richmond Hill, Ontario,* No. 365 (1973).

122.113 **On the existence of a Hertzsprung progression in the halo/old disk cepheids.** R. S. Stobie.
Observatory, Vol. 93, 111 - 114 (1973).

122.114 **Certain relationships between the photometric characteristics of light curves and the colour indices of cepheids in the UBV system. II.** N. Nikolov, G. Momchev.
Izv. Sekts. astron. Blg. AN, Vol. 5, 73 - 88 (1972). In Russian. Abstr. in Referativ. Zhurn. 51. Astron., 5.51.722 (1973).

122.115 **Beobachtungen an R Scuti.** J. Bauer.
BAV Rundbrief, 22. Jahrgang, p. 5 - 12 (1973).

122.116 **Photometric study of BL Lacertae. II bis – B and V magnitudes during the period March 1969 – January 1971 and bibliographic informations.**
C. Bertaud, G. Wlérick, P. Véron, B. Dumortier, M. Duruy, P. de Saevsky.
Astron. Astrophys., Suppl. Ser., Vol. 11, 77 - 92 (1973). In French.
In a recent paper we have shown the light curve of BL Lacertae in B color for the period March 1969 – January 1971 and we have discussed the properties of the flux variations; we publish here tables of the B and V magnitudes used to construct the light curve. To facilitate the study of the photometric variations, we have also prepared a table that includes the publications concerning the light variations of this object in the visible region.

122.117 **Light curves of Orion variables.**
M. G. Nurmukhamedov.
Young stellar groups. Astroclimate, (see 003.012), p. 52 - 83 (1972). In Russian.

122.118 **Photoelectric observations of the flare star AD Leo during the 1973 January 27 - February 9 international patrol.** S. Cristaldi, M. Rodonò.
Inform. Bull. Variable Stars (I. A. U. Commission 27), Konkoly Obs., Budapest, No. 801, 3 pp. (1973).

122.119 **Photoelectric observations of EV Lac during the**

1972 September 1–15 international patrol.
M. Rodonò.
Inform. Bull. Variable Stars (I. A. U. Commission 27), Konkoly
Obs., Budapest, No. 802, 3 pp. (1973).

122.120 New flares in the Pleiades.
    L. G. Balázs, M. Kun, G. Szécsényi-Nagy.
Inform. Bull. Variable Stars (I. A. U. Commission 27), Konkoly
Obs., Budapest, No. 803, 2 pp. (1973).

122.121 The cepheid SU Cygni.
    Yu. V. Borisov.
Byull. in-ta astrofiz. AN Tadzh. SSR, 1972, No. 61, p. 7 - 11.
In Russian.

122.122 SU Tauri.   E. Mayer, W. E. Pennell,
    C. E. Anderson, C. Hurless.
IAU Circ., No. 2496 (1973).

122.123 The period–radius relation for cepheid variable stars.
    R. Woolley, B. Carter.
Monthly Notices Roy. Astron. Soc., Vol. 162, 379 - 400
(1973).
    Eighteen determinations of the radii of cepheid variables
by the Baade–Wesselink method are described. A very close
linear relation between the logarithm of the radii and the logarithm of the period is found. This is compared with radii
found from data given by Sandage and Tammann, using a modification of Wesselink's tables of specific intensity as a function of colour. This modification is discussed in an appendix.
The special position of W Virginis stars in the radius–period
diagram is discussed, and the field stars are compared with data
from globular cluster stars as reported by Dickens and Carey.

122.124 Some results of the cooperative photometric observa-
    tions of the UV Cet-type flare stars in the years
1967–71.   R. E. Gershberg.
Astrophys. Space Sci., Vol. 19, 75 - 92 (1972).
    The list of the cooperative photometric observations of
the UV Cet-type flare stars that have been organized during the
years 1967 to 1971 by the Working Group on Flare Stars of
the IAU Commission 27 is given. A statistical analysis of the
UV Cet, YZ CMi, EV Lac and AD Leo flares observed in the
B-band is carried out. Distributions of flare rise times and of
rates of flare absolute luminosity increase are considered.

122.125 Population II cepheids in globular clusters.
    F. Caputo, V. Castellani.
Astrophys. Space Sci., Vol. 19, 423 - 429 (1972).
    B and V data of population II cepheids in M2, M5, M10,
M13, M80 and ω Cen are combined in order to obtain period-
luminosity and period-colour relations. After correcting the
observed colours for the amplitude effect, period-colour-
amplitude relations are derived for both short and long period
variables.

122.126 An observational study of Mira variables. I. The
    near-infrared photometry.   T. G. Barnes III.
Astrophys. Journ., Suppl. Ser., No. 221, Vol. 25, 369 - 392
(1973).
    This paper presents the results of the first stage in a pro-
gram to determine the absolute magnitudes of Mira variables.
Observations on the *VRI* photometric system are reported for
239 Mira variables at maximum light. The color and phase
characteristics at maximum light are investigated by means of
the best 150 light curves.

122.127 New flare stars in the Pleiades. I.   E. S. Parsamian.
    Soobshch. Byurakan. Obs., vyp. (No.) 44, p. 3 - 13
(1972). In Russian.
    17 new flare stars and 7 repeated flares were found dur-

ing the observations of the Pleiades in 1967–70 with the 21''
and 40'' Schmidt telescopes of the Byurakan Observatory. The
magnitudes of observed flares, dependences between spectral
types and amplitudes of the flare stars in the Pleiades, the
values of the amplitudes for the flare stars of K2-M7 types are
given.

122.128 Observation of the spectrum of the flare of star No.
    205 in the Pleiades.   E. S. Parsamian.
Soobshch. Byurakan. Obs., vyp. (No.) 44, p. 14 - 16 (1972).
In Russian.

122.129 Two-colour observations of the flare of the star No.
    207 in the Pleiades.
E. S. Parsamian, H. S. Chavushian (*O. S. Chavushyan*).
Soobshch. Byurakan. Obs., vyp. (No.) 44, p. 17 - 20 (1972).
In Russian.
    The magnitudes and colour indices during the flare-up
are given. The observations confirm that during the rapid flare
the star becomes more blue.

122.130 Photometric study of new variable stars in NGC
    2264.
H. S. Badalian (*G. S. Badalyan*), L. K. Erastova.
Soobshch. Byurakan. Obs., vyp. (No.) 44, p. 21 - 27 (1972).
In Russian.
    Data on the light variations in three colours of 31 new
variables in NGC 2264 are given. The existence of an ultra-
violet excess for SVS 1538 and SVS 1720 is discovered.

122.131 Photometry of selected flare stars in the Pleiades.
    L. K. Erastova.
Soobshch. Byurakan. Obs., vyp. (No.) 44, p. 28 - 42 (1972).
In Russian.
    The results of the observations of flare stars in the
Pleiades from September 1968 to March 1971 are presented.

122.132 Polarimetric and photometric observations of the
    stars EV Lac and AD Leo.
K. A. Grigorian, M. A. Eritsian.
Soobshch. Byurakan. Obs., vyp. (No.) 44, p. 104 - 107 (1972).
In Russian.
    The results of photometric observations of the flare stars
EV Lac and AD Leo are given. It is shown that, as well during
the flares as outside of them, no polarization exceeding the
observational errors has been observed.

122.133 SU Doradus.   F. M. Bateson.
    Roy. Astron. Soc. New Zealand, Variable Star Sec-
tion, Circ. No. 190, 5 pp. (1973).
    Improved elements for the Mira type variable, SU Dor,
are presented together with a list of observed maxima derived
from visual observations during the interval J.D. 2,435,463 to
2,441,094.

122.134 Continual photoelectric monitoring of flare stars,
    VIII. YZ CMi, AD Leo and UV Cet (1972).
K. Ichimura, Y. Shimizu, E. Watanabe, T. Okada.
Tokyo Astron. Bull., Second Ser., No. 224, p. 2607 - 2612
(1973).
    Flare stars YZ CMi, AD Leo and UV Cet were monitored
photoelectrically at the Okayama Station of Tokyo Astronomical Observatory during 1972. The observations were made
with the simultaneous three-color photometer attached to 91
cm reflector.

122.135 Neue Flare-Sterne in den Plejaden.   W. Götz.
    MVS, *Sonneberg*, Vol. 6, 85 - 86 (1973).
    Some remarks on the discovery and the light changes of
seven new and of two previously known flare stars in the
Pleiades region are given.

122.136   Flare stars in the Pleiades. III.
V. A. Ambartsumian, L. V. Mirzoyan, E. S. Parsamian, H. S. Chavushian (*O. S. Chavushyan*), L. K. Erastova, E. S. Kazarian, G. B. Ohanian (*Oganyan*).
Astrofizika, Vol. 8, 485 - 508 (1972). In Russian. – English translation in Astrophysics, Vol. 8, No. 4.

The results of the observations of stellar flares in the Pleiades region carried out at Byurakan and Asiago, mainly during the seasons 1970–71 and 1971–72, are given. 76 new flare stars and 74 repetitions of flares of known flare stars have been found. The total number of known flare stars in the Pleiades region is 291 and the number of repeated flare-ups is 168. The enriched observational data on the distribution of the known flare stars according to the number of observed flares can be presented as a superposition of two Poisson distributions, indicating the presence of at least two groups of flare stars: a rich group of low frequency of flares and a small group of stars with high frequency of flares. The total number of flare stars has been evaluated by two different methods to be near 1000. A statistical analysis of the flares having photographic amplitudes larger and correspondingly smaller than 2^m has been carried out.

122.137   V1057 Cygni and pre-main-sequence evolution.
G. L. Grasdalen.
Astrophys. Journ., Vol. 182, 781 - 808 (1973).

The authors have analyzed the pre-main-sequence object V1057 Cyg. The observations lead them to conclude that the photometric outburst of the star was due to a fundamental change in the star. Post-outburst observations of V1057 Cyg indicate that the change results in a star ($\sim 8 M_\odot$) which is now close to its equilibrium radiative pre-main-sequence track. This requires to interpret the pre-outburst star as an object in its hydrodynamic phases of evolution.

122.138   Nonlinear cepheid pulsation calculations and comparison with linear theory.
D. S. King, J. P. Cox, D. D. Eilers, W. R. Davey.
Astrophys. Journ., Vol. 182, 859 - 884 (1973).

The overall purpose of the present paper is to attempt a partial comparison of the results of linear and nonlinear calculations (both fully nonadiabatic) of the radial pulsations of a number of purely radiative envelope models of classical cepheids, using the same input physics in both sets of calculations. Also, the models used are as nearly identical as is feasible in view of the very different types of approaches required in the two sets of calculations.

122.139   Pulsational stability of stars in thermal imbalance.
J. P. Cox, C. J. Hansen, W. R. Davey.
Astrophys. Journ., Vol. 182, 885 - 899 (1973).

Integral expressions have been derived for the stability coefficient for a spherically symmetric star in hydrostatic but not necessarily thermal equilibrium, on the basis of a direct linearization of the pulsation equations. Application to some simple cases leads to agreement with previous results (based on the quasi-adiabatic assumption) in one case, but not in others. Reasons are put forth for not expecting close agreement with previous results except possibly in certain special cases.

122.140   Investigation of pulsating variables. II. Radial pulsations of RZ Lyrae.
I. G. Kolesnik, Yu. S. Romanov.
Astrometriya i Astrofizika, *Kiev,* Vyp. (No.) 18, (see 003. 016), p. 59 - 64 (1973). In Russian.

The brightness gradients for RZ Lyr are obtained and their variations with the Blazhko effect are studied. On the basis of the gradients for different phases of the Blazhko effect the relative amplitudes of radial variations are obtained.

122.141   Changes in photometric characteristics with the period of the Blazhko effect in RZ Lyrae.
Yu. S. Romanov.
Astrometriya i Astrofizika, *Kiev,* Vyp. (No.) 18, (see 003. 016), p. 65 - 78 (1973). In Russian.

Three-colour photoelectric observations of RZ Lyrae made during 1968–1970 are analysed. RZ Lyrae has a peculiar Blazhko effect. Behaviour of $U_{max}, B_{max}, V_{max}, (B-V)_{max}, (U-B)_{max}, U_{min}, B_{min}, V_{min}$ and $(B-V)_{min}$ is investigated.

122.142   Polarimetric and photometric studies of variable stars.
V. A. Dombrovskij, T. A. Polyakhova, V. A. Yakovleva.
Trudy Astron. Obs., *Leningrad,* Vol. 29 (= Uchenye Zapiski Leningr. Un-ta, No. 363 = Seriya Matem. Nauk, vyp. (No.) 48), p. 45 - 57 (1973). In Russian.

The results of polarimetric and photometric observations of variable stars μ Cep, V CVn, AK Peg, S CrB, V CrB, U Her, RY Tau, T Tau are given. Time variations of polarization for all these stars are found. The correlation between the variations of polarization, brightness and colours is not the same in all the cases.

122.143   A study of the intrinsic polarization of R Sct.
V. A. Yakovleva.
Trudy Astron. Obs., *Leningrad,* Vol. 29 (= Uchenye Zapiski Leningr. Un-ta, No. 363 = Seriya Matem. Nauk, vyp. (No.) 48), p. 57 - 65 (1973). In Russian.

The results of UBV polarimetric observations of R Sct are discussed. The interstellar component of polarization is subtracted from the observed polarization. The largest amount of intrinsic polarization is observed before the deep light minima of the star.

122.144   An attempt to detect the polarization difference in molecular bands and the continuum in the spectra of high-luminosity red variables.
T. E. Derviz, V. A. Dombrovskij.
Trudy Astron. Obs., *Leningrad,* Vol. 29 (= Uchenye Zapiski Leningr. Un-ta, No. 363 = Seriya Matem. Nauk, vyp. (No.) 48), p. 65 - 71 (1973). In Russian.

A comparison is given of the polarization measurements in molecular bands of TiO and continuum for V CVn, AK Peg, VY CMa and μ Cep. Within the observational accuracy no differences in polarization in bands versus continuum are found.

122.145   Photoelectric observations of the flare star EV Lac.
G. Asteriadis, L. N. Mavridis, D. Stavridis.
Inform. Bull. Variable Stars (I.A.U. Commission 27), Konkoly Obs., Budapest, No. 809, 4 pp. (1973).

Abundance of lithium in the atmospheres of variable M stars of type SR.   See Abstr. 064.053.

Abundance of lithium in the atmospheres of two dM5 stars.   See Abstr. 064.054.

A linear nonadiabatic analysis of radial oscillations in models for Beta Cephei stars.   See Abstr. 065.011.

Supersonic convection and the structure of T Tauri stars.   See Abstr. 065.044.

Non-linear pulsations of upper main sequence stars–I. A perturbation approach.   See Abstr. 065.084.

Non-linear pulsations of upper main sequence stars–II. Direct numerical integrations.
See Abstr. 065.086.

Stellar stability and stellar pulsation.
See Abstr. 065.108.

Variable stars – realistic star models.
See Abstr. 065.109.

The period-age relationship for δ Cephei stars.
See Abstr. 065.127.

Irregular nebular variables and neutrino emission.
See Abstr. 065.159.

Radial pulsations of pre-white-dwarf-stars. II.
Pulsational stability of ^{12}C shell-burning stars.
See Abstr. 065.174.

Infra-red photometry of R Coronae Borealis type
variables and related objects.   See Abstr. 113.005.

Technetium in S Sculptoris.   See Abstr. 114.115.

The classification of intrinsic variables. III. Calibration of the luminosities of small amplitude red variables in the
old disk population.   See Abstr. 115.010.

VV 428–479, variable stars in a Cepheus-Lacerta
field of the Milky Way.   See Abstr. 123.035.

New OH sources discovered at 1612 MHz.
See Abstr. 131.037.

Two puzzling objects: OJ 287 and BL Lacertae.
See Abstr. 141.018.

BL Lac: Strong short-term variability.
See Abstr. 141.031.

Detection of RY Scuti at radiofrequencies.
See Abstr. 141.063.

Variations in the radio structure of BL Lacertae.
See Abstr. 141.109.

ADS 2859 B as an alternative candidate to X Per for
the X-ray source 2 U 0352 + 30.   See Abstr. 142.015.

Limits on the optical pulsations of HZ Herculis.
See Abstr. 142.079.

Places of formation of nearby classical cepheids.
See Abstr. 151.036.

Variable stars in the galactic cluster NGC 6913.
See Abstr. 153.029.

A search for extragalactic objects in the General
Catalog of Variable Stars.   See Abstr. 158.064.

123 Variable Stars: Lists of Observations, Individual Observations

123.001 **A faint maximum of R Andromedae.** J. E. Isles.
Journ. British Astron. Ass., Vol. 83, 135 - 137
(1973). – Report of Variable Star Section of the British
Astron. Ass.

123.002 **Beobachtungsergebnisse der Berliner Arbeitsge-meinschaft für Veränderliche Sterne e.V. (BAV).**
W. Braune, J. Hübscher, E. Mundry.
Astron. Nachr., Vol. 294, 123 - 129 = BAV-Mitt., No. 25
(1973).
In this 9th compilation of BAV results of observations
are given from the years 1969 up to 1971 208 observed minima of 46 eclipsing binaries, 72 maxima of 22 RR Lyrae- and
δ-Cephei-stars, 232 results of 66 Mira stars,72 results of 12 RV
Tauri-stars and 22 results of 3 irregular variables and U Geminorum stars.

123.003 **On seven variable stars.** N. B. Perova.
Astron. Tsirk., No. 743, p. 7 - 8 (1973). In Russian.

123.004 **New observations of GP And.**
G. A. Lange, P. P. Gusev.
Astron. Tsirk., No. 744, p. 7 - 8 (1973). In Russian.

123.005 **On the variability of SVS 1740.** O. E. Mandel.
Astron. Tsirk., No. 747, p. 8 (1973). In Russian.

123.006 **Observations of three variable stars: V459, V617 and V618 Her.** O. A. Chekanikhina.
Astron. Tsirk., No. 748, p. 8 (1973). In Russian.

123.007 **Application of the statistical χ^2 criterion to the study of the light variation of the variable DF Tau.**
E. P. Strelkova.
Astron. Tsirk., No. 755, p. 7 - 8 (1973). In Russian.

123.008 **On the variable RV Cap.** V. P. Tsesevich.
Astron. Tsirk., No. 757, p. 5 - 7 (1973). In Russian.

123.009 **On two unstudied variable stars in Andromeda FI And = S 9498 and FM And = S 9504.** H. Busch.
Inform. Bull. Variable Stars (I.A.U. Commission 27), Konkoly
Obs., Budapest, No. 754 (1973).

123.010 **V 384 Cas.** H. Busch.
Inform. Bull. Variable Stars (I.A.U. Commission 27),
Konkoly Obs., Budapest, No. 754 (1973).

123.011 **MU Cas.** K. Häussler.
Inform. Bull. Variable Stars (I.A.U. Commission 27),
Konkoly Obs., Budapest, No. 755 (1973).

123.012 **UY Mon.** H. Busch.
Inform. Bull. Variable Stars (I.A.U. Commission 27),
Konkoly Obs., Budapest, No. 755 (1973).

123.013 **On the photometric history of V 1329 Cygni = HBV 475.** V. P. Arhipova, O. E. Mandel.
Inform. Bull. Variable Stars (I.A.U. Commission 27), Konkoly
Obs., Budapest, No. 762 (1973).

123.014 **New faint southern variable stars.** R. Knigge.
Inform. Bull. Variable Stars (I.A.U. Commission 27),
Konkoly Obs., Budapest, No. 765, 7 pp. = Veröff. Remeis-
Sternw. Bamberg, Astron. Inst. Univ. Erlangen-Nürnberg, Vol.
10, No. 105 (1973).

123.015 **Suspected long-period variable near NGC 2368.**
S. Wyckoff, P. Wehinger.
Inform. Bull. Variable Stars (I.A.U. Commission 27), Konkoly
Obs., Budapest, No. 766 (1973).

123.016 **The period of variable 7 in M13.**
M. Ibañez, W. Osborn.
Inform. Bull. Variable Stars (I.A.U. Commission 27), Konkoly
Obs., Budapest, No. 769 (1973).

123.017 **A new variable star in the LMC within the error box of the X-ray source LMC X-1.** A. D. Andrews.
Inform. Bull. Variable Stars (I.A.U. Commission 27), Konkoly
Obs., Budapest, No. 773 (1973).

123.018 **Two new variable extragalactic objects.**
L. Meinunger.
Inform. Bull. Variable Stars (I.A.U. Commission 27), Konkoly
Obs., Budapest, No. 777 (1973).

123.019 **First ephemerides of five variable stars in Eridanus and Fornax.** R. Deurinck, M. Goossens.
Inform. Bull. Variable Stars,(I.A.U. Commission 27), Konko-
ly Obs., Budapest, No. 792, 6 pp. (1973).

123.020 **Discussion of 12 variable stars in a region around $\alpha = 17^h, \delta = -70°$.** R. Deurinck, B. Vissenberg.
Inform. Bull. Variable Stars (I.A.U. Commission 27), Konko-
ly Obs., Budapest, No. 794, 16 pp. (1973).

123.021 **A nova-like variable star.** B. S. Whitney.
Inform. Bull. Variable Stars, (I.A.U. Commission
27), Konkoly Obs., Budapest, No. 797 (1973).

123.022 **Concerning a suspected variable star in M13.**
W. Osborn, M. Ibañez.
Inform. Bull. Variable Stars, (I.A.U. Commission 27), Konko-
ly Obs., Budapest, No. 798 (1973).

123.023 **The non-existence of nova Carinae 1970.**
D. J. MacConnell.
Inform. Bull. Variable Stars, (I.A.U. Commission 27), Konko-
ly Obs., Budapest, No. 799 (1973). – Concerning BV 1543.

123.024 **A search for faint variable objects.**
S. van den Bergh, E. Herbst, C. Pritchet.
Astron. Journ., Vol. 78, 375 - 376, 439 (1973).
An area of 6.2 sq deg near M31 has been searched for
variable objects down to the limit of the 48-inch (126 cm)
Schmidt telescope. Thirteen faint variable objects were dis-
covered.

123.025 **Korrekturen zu den Vorhersagen im BAV-Circular 1973.** R. Diethelm.
BAV Rundbrief, 22. Jahrgang, p. 18 - 19 (1973).

123.026 **Observations of W Comae Berenices.**
D. Hoffleit.
Inform. Bull. Variable Stars (I.A.U. Commission 27), Konkoly
Obs., Budapest, No. 800 (1973).

123.027 **Observations of the variable star V 356 Sgr.**
S. Sakuma.
Mem. Japan Astron. Study Ass., No. 19, Vol. 5, 177 - 180
(1972). In Japanese.

123.028 **Observations of variable stars in 1969 by members**

of the Japan Astronomical Study Association.
E. Mochizuki.
Mem. Japan Astron. Study Ass., No. 19, Vol. 5, 197 - 231 (1972).

123.029 **OJ 287 observing campaign.**
J. Pollock, D. Kolpanen, P. Usher.
IAU Circ., Nos. 2493, 2496, 2503 (1973).

123.030 **SU Tauri.** C. E. Scovil, J. E. Bortle.
IAU Circ., No. 2502 (1973).

123.031 **SU Tauri.** K. Locher.
IAU Circ., No. 2505 (1973).

123.032 **OJ 287.** A. Frohlich.
IAU Circ., No. 2525 (1973).

123.033 **R Coronae Borealis variables.** F. M. Bateson.
IAU Circ., No. 2537 (1973).

123.034 **Observations of β Pegasi.** J. Wieczorek.
Urania Kraków, Vol. 44, 20 - 23 (1973). In Polish.

123.035 **VV 428–479, variable stars in a Cepheus-Lacerta field of the Milky Way.**
W. J. Miller, A. A. Wachmann.
Ric. Astron., Specola Vaticana, *Castel Gandolfo,* Vol. 8, (No. 18), 367 - 407 (1973).
Fifty-two variable stars (fourteen of which are already suspected variables whose types and/or light elements were hitherto unknown) were discovered and processed on Castel Gandolfo plates. Nine tables of data on eleven pages, nine pages of identification charts, five pages of light curves, and notes on each variable star summarize the results.

123.036 **Observations of southern variable stars. RA. 06hrs to 12hrs,** 1967 July 1 – 1970 December 31, J.D. 2,439,673 – 2,440,952. F. M. Bateson.
Roy. Astron. Soc. New Zealand, Variable Star Section, Circ. No. 189, 15 pp. (1972). – These observations continue those published in Circular No. 186 and are presented in the same form.

123.037 **Observations of Orion variables.** F. M. Bateson.
Roy. Astron. Soc. New Zealand, Variable Star Section, Circ. No. 191, 5 pp. (1972).

123.038 **RT Apodis.** F. M. Bateson.
Roy. Astron. Soc. New Zealand, Variable Star Section, Circ. No. 192, 3 pp. (1972).
Seventeen maxima of RT Aps are listed, derived from visual observations from J.D. 2,435,925 to 2,440,847.

123.039 **GU Sagittarii.** F. M. Bateson, A. F. Jones.
Roy. Astron. Soc. New Zealand, Variable Star Section, Circ. No. 193, 4 pp. (1972).
GU Sgr is shown to have superimposed on its typical R CrB type variation a semi-regular period of approximately 38 days, with an amplitude that changes from a few tenths of a magnitude to 1.5 magnitudes at certain stages of its main light curve.

123.040 **V442 Centauri.** F. M. Bateson.
Roy. Astron. Soc. New Zealand, Variable Star Section, Circ. No. 194, 2 pp. (1972).

123.041 **RX Microscopii.** F. M. Bateson, A. F. Jones.
Roy. Astron. Soc. New Zealand, Variable Star Section, Circ. No. 195, 2 pp. (1972).
Twenty two maxima are tabulated for RX Mic covering

the interval JD 2,434,836 to 2,440,210.

123.042 **RY Microscopii.** F. M. Bateson, A. F. Jones.
Roy. Astron. Soc. New Zealand, Variable Star Section, Circ. No. 196, 2 pp. (1972).
Twenty one maxima are tabulated for RY Mic covering the interval JD 2,436,100 to 2,440,210.

123.043 **T Microscopii.** F. M. Bateson, A. F. Jones.
Roy. Astron. Soc. New Zealand, Variable Star Section, Circ. No. 197, 2 pp. (1972).
Observed maxima and minima of the semi-regular variable, T Mic, are listed for the interval JD 2,437,410 to 2,440,185.

123.044 **RS Microscopii.** F. M. Bateson, A. F. Jones.
Roy. Astron. Soc. New Zealand, Variable Star Section, Circ. No. 198, 2 pp. (1972).
A period of 228.5 days is found for the Mira type variable, RS Mic during the interval 2,437,401 to 2,440,173.

123.045 **V436 Centauri.** F. M. Bateson.
Roy. Astron. Soc. New Zealand, Variable Star Section, Circ. No. 199 (1972).

123.046 **Ein neuer U-Geminorum-Stern in Ursa Major.**
H. Huth.
MVS, *Sonneberg,* Vol. 6, 63 - 64 (1973).

123.047 **Visuelle Maxima von Mira-Sternen.**
E. Scheller.
MVS, *Sonneberg,* Vol. 6, 64 (1973).

123.048 **Photographische Reihenbeobachtungen.**
P. Ahnert.
MVS, *Sonneberg,* Vol. 6, 65 (1973).

123.049 **Visuelle Beobachtungen von 23 Mira-Sternen, R. Scuti und SS Cygni.** P. Ahnert.
MVS, *Sonneberg,* Vol. 6, 66 (1973).

123.050 **AH Camelopardalis.** W. Zschocke.
MVS, *Sonneberg,* Vol. 6, 79 - 80 (1973).

123.051 **XZ Camelopardalis.** A. Eichhorn.
MVS, *Sonneberg,* Vol. 6, 80 - 81 (1973).

123.052 **Photographische Beobachtungen von Mira-Sternen auf Platten der Sonneberger Himmelsüberwachung.**
E. Splittgerber.
MVS, *Sonneberg,* Vol. 6, 82 - 84 (1973).

123.053 **Visuelle Beobachtungen langperiodischer Veränderlicher.** D. Böhme.
MVS, *Sonneberg,* Vol. 6, 84 (1973).

123.054 **V 1016 Cygni.** H. Geßner.
MVS, *Sonneberg,* Vol. 6, 86 (1973).

123.055 **Bearbeitung von 41 Veränderlichen am Südhimmel. (Feld γ Phoenicis).** I. Meininger.
MVS, *Sonneberg,* Vol. 6, 87 (1973).

123.056 **Photoelektrische Messungen des unregelmäßigen Veränderlichen RZ Piscium.** W. Wenzel.
MVS, *Sonneberg,* Vol. 6, 88 (1973).

123.057 **Observations of variable stars, July – December 1972. Report No. 23.** L. Plaut, H. Feijth.
Nederlandse Vereniging voor Weer- en Sterrenkunde. Kapteyn Astron. Lab., Groningen – Netherlands. 11 pp. (1973).
This report gives 3407 visual observations of 169 variable

stars, 1972 July – December.

123.058 Note on the photometric history of HBV 475 (=
V1329 Cygni).   P. D. Hicks.
Inform. Bull. Variable Stars (I.A.U. Commission 27), Konkoly
Obs., Budapest, No. 804 (1973).

123.059 Photographic observations of V1057 Cyg.
F. Gieseking.
Inform. Bull. Variable Stars (I.A.U. Commission 27), Konkoly
Obs., Budapest, No. 806, 3 pp. (1973).

123.060 Six new variable B-stars.   A. van Hoof.
Inform. Bull. Variable Stars (I.A.U. Commission 27)
Konkoly Obs., Budapest, No. 807, 4 pp. (1973).

123.061 Note on three of G. Hill's stars.   A. van Hoof.
Inform. Bull. Variable Stars (I.A.U. Commission 27).
Konkoly Obs., Budapest, No. 808 (1973).

123.062 Variable star notes.   M. W. Mayall.
Journ. Roy. Astron. Soc. Canada, Vol. 67, 45 - 48,
101 - 104, 157 - 160 (1973).

123.063 Predicted max. in 1973 of long period variables
desired to be observed.
S. Kanda, O. Hasegawa.
Japan Astron. Study Ass. Circ. 280, 2 pp. (1972). In Japanese.

123.064 Observed max. and min. of long period and irregular
variable stars in 1970.   S. Kanda, E. Mochizuki.
Japan Astron. Study Ass. Circ. 281 - 282, 4 pp. (1973).
In Japanese.

123.065 OU Lyr (α 19^{h}07^m 51^s, δ + 32°25'.8, 1900).
O. Hasegawa.
Japan. Astron. Study Ass. Circ. 283 (1973). In Japanese.

123.066 Prediction of minima of RV Tau type variables.
S. Kanda, O. Hasegawa.
Japan Astron. Study Ass. Circ. 283 (1973). In Japanese.

123.067 Observed max. and min. of long period and irregular
variable stars in 1971.
S. Kanda, E. Mochizuki.
Japan. Astron. Study Ass. Circ. 284 - 285, 4 pp. (1973).
In Japanese.

Errata

123.901 Erratum: 'HBV 479 – 495, variables in a field
around SA 18' [Inform. Bull. Variable Stars, (I.A.U.
Commission 27), Konkoly Obs., Budapest, No. 749 (1972)].
A. Wachmann.
Inform. Bull. Variable Stars (I. A. U. Commission 27),
Konkoly Obs., Budapest, No. 770 (1973).

124 Novae

124.001 Formation of coronal lines in the spectra of novae.
II.   V. G. Gorbatskij.
Astron. Zhurn. Akad. Nauk SSSR, Vol. 50, 19 - 26 (1973).
In Russian. English translation in Soviet Astron. AJ, Vol. 17,
No. 1.

The time dependence of intensities of coronal lines in
spectra of novae is calculated and compared to the observational data. The process of temperature relaxation behind the
shock in a circumstellar envelope is considered in detail, ionization and excitation of the helium atoms being taken into account. On this base the mass loss by recurrent novae during
the interval between outbursts is estimated.

124.002 Discovery of five novae in Messier 33 and a super-
nova in a field galaxy.
L. Rosino, A. Bianchini.
Astron. Astrophys., Vol. 22, 461 - 463 (1973).

124.003 Novae in M 31 discovered and observed at Asiago
from 1963 to 1970.   L. Rosino.
Astron. Astrophys., Suppl. Ser., Vol. 9, 347 - 389 (1973).

The systematic search of novae and peculiar objects in
M 31 has led to the discovery of other 44 novae from 1963 to
1970. Positions, magnitudes, identification charts and light
curves are given in this paper. Repeated flare-ups at distance of
years have been found in two objects. The first (No. 48 =
No. 79) is a recurrent nova; the second may be a peculiar object or a foreground U Gem variable of our Galaxy.

124.004 On the form of nova shells.   V. G. Gorbatskii.
Astrofizika, Vol. 8, 369 - 380 (1972). In Russian.
English translation in Astrophysics, Vol. 8, No. 3.

There are observational evidences of continuous mass
loss from close binary systems of dwarf stars such as novae and
recurrent novae. As a result of the outflow of matter a circumstellar envelope develops near the orbital plane of the
system. The main envelope ejected during the nova outburst
interacts with the circumstellar envelope. The main envelope
changes its original form due to the breaking of matter moving near the orbital plane. Some time after the outburst the
visible shell must become elongated in the direction perpendicular to the orbital plane. The form of the shell as function
of time is calculated in this paper. The mass of the envelope
of DQ Her is estimated using the data on the elongation of the
observed shell.

124.005 Hydrodynamic stellar evolution: Studies of the
nova outburst.
S. G. Starrfield, W. M. Sparks, G. S. Kutter, J. W. Truran.
Bull. American Astron. Soc., Vol. 5, 15 (1973). – Abstr. AAS.

124.006 Pulsational instability of a pre-nova model.
N. R. Simon, V. K. Sastri.
Bull. American Astron. Soc., Vol. 5, 15 (1973). – Abstr. AAS.

124.007 Five thousand years of observations of novae.
Yu. P. Pskovsky.

Priroda, No. 3.73, p. 56 - 62 (1973). In Russian.

124.008   Novae.   W. K. Rose.
Stellar evolution, (see 012.014), p. 289 - 305 (1972).

124.009   Considerazioni su alcune caratteristiche delle stelle novae nella fase esplosiva.   L. Rosino.
Mem. Soc. Astron. Italiana, Nuova Ser., Vol. 43, 797 - 803 (1973).
    After examining the common aspects of novae and supernovae in the explosive phase, some characteristics of the novae observed at Asiago are briefly discussed.

124.010   On the IR-radiation of novae.   A. S. Zentsova.
Vestn. Leningr. un-ta, 1973, No. 1, p. 131 - 135. In Russian. – Abstr. in Referativ. Zhurn. 51. Astron., 6.51.609 (1973).

On thermal waves in stars, II.
See Abstr. 065.153.

On nova explosions in close binary systems.
See Abstr. 117.008.

Scorpius X-1 as an old nova.
See Abstr. 142.008.

124.100   Nova Delphini 1967

Nebular stage of nova Delphini 1967, I.
I. Malakpur.
Astron. Astrophys., Vol. 24, 125 - 130 (1973). In French.
    We have studied the nebular stage of nova Delphini 1967 for the period from December 1968 to June 1970. Using the lines of [O III] and [N II], we have calculated the values of N_e and T_e. We have also determined N_e by the method of surface luminosity. Finally, we have calculated a mass of the order of $3.3 \times 10^{-4} M_\odot$ for the envelope.

Ein Atlas der Nova Delphini 1967 und anderer repräsentativer Novaspektren.   W. Seitter.
Mitt. Astron. Ges., No. 32, p. 246 - 249 (1973).

124.101   Nova Serpentis 1970

Ultraviolet spectrophotometry of nova Serpentis 1970.   J. S. Gallagher III, A. D. Code.
Bull. American Astron. Soc., Vol. 5, 17 (1973). – Abstr. AAS.

Preliminary report on the infrared spectrum of nova Serpentis 1970.   F. Ciatti, A. Mammano.
17th Colloquium International Astrophys. Liège 1971, (see 012.003), p. 301 (1972).

A note on the reddening of nova FH Serpentis 1970.
J. B. Hutchings, W. A. Fisher.
Publ. Astron. Soc. Pacific, Vol. 85, 122 - 126 (1973).
    Measurements of the diffuse interstellar lines at $\lambda\lambda 5780, 5796$ in the early postmaximum spectra of nova FH Ser indicate a value of $E_{B-V} \sim 0.90$.

124.102   Nova RR Telescopii

The spectrum of RR Telescopii in 1968.

L. H. Aller, R. S. Polidan, E. J. Rhodes, Jr., G. W. Wares.
Astrophys. Space Sci., Vol. 20, 93 - 110 (1973).
    The nova-like variable RR Telescopii observed at Cerro Tololo Observatory in 1968 displayed an unusually rich emission line spectrum ranging in excitation from Mg I to [Fe VIII]. A list of lines with their suggested identifications and approximate intensities covers the range from $\lambda 3100$ to 6700. Only a semi-quantitative discussion is possible since photographic measurements of line intensities could not be calibrated photoelectrically.

124.103   Nova AH Herculis

High-speed photometry of AH Herculis.
E. L. Robinson.
Astrophys. Journ., Vol. 181, 531 - 536 (1973).
    High-speed photometry of the dwarf nova AH Her is presented. The observations were made during minimum light and at several points on the rising branch of eruptions. Power spectrum analysis revealed no periodicities in any of the light curves, showing that in AH Her, as in CN Ori and Z Cam, the white-dwarf pulsations exist only at maximum light.

124.104   Nova VW Hydri

28-second oscillations in VW Hyi.
B. Warner, J. M. Harwood.
Inform. Bull. Variable Stars (I.A.U. Commission 27), Konkoly Obs., Budapest, No. 756 (1973).

Kurzzeit-Variabilität der Zwerg-Nova VW Hydri im Minimum.   N. Vogt.
Mitt. Astron. Ges., No. 32, p. 249 (1973). – Abstract.

124.105   Nova Doradus 1971a

A photometric study of nova Doradus 1971a in the Large Magellanic Cloud.   A. Ardeberg, M. de Groot.
Astron. Astrophys., Vol. 26, 53 - 64 (1973).
    Photoelectric UBV photometry is presented for nova Doradus 1971a in the Large Magellanic Cloud from observations on 51 nights during a period of 430 days. The light-curve parameters are deduced and discussed as well as the colour curves. Comparisons are made with other novae in the Magellanic Clouds, with novae in M31, and with galactic novae.

124.106   Nova GK Persei

GK Persei.   L. C. Peltier.
IAU Circ., No. 2482 (1973).

GK Persei.   S. Nishimura, K. Osawa, T. Ishihara, M. Kiyokawa, S. Kikuchi.
IAU Circ., No. 2483 (1973).

GK Persei.   P. Tempesti.
IAU Circ., No. 2498 (1973).

GK Per (Nova 1901, $\alpha 3^h 24^m 24^s$, $\delta + 43°33'.7, 1900$).
Japan. Astron. Study Ass. Circ. 283 (1973). In Japanese.

124.107 Nova V1017 Sagittarii

V1017 Sagittarii. A. Jones.
IAU Circ., Nos. 2492, 2496 (1973).

V1017 Sagittarii. A. Jones.
IAU Circ., No. 2498 (1973).

V1017 Sagittarii. N. V. Vidal, D. T. Wickramasinghe, A. W. Rodgers, B. A. Peterson.
IAU Circ., No. 2505 (1973).

124.108 Nova RS Ophiuchi

**The region of formation of the narrow Fe II lines
of RS Ophiuchi.** M. Friedjung.
Astrophys. Space Sci., Vol. 19, 501 - 503 (1972).
The size of this region is found using a method previously
applied by Friedjung and Malakpur (1971) to nova Delphini.
The conclusion of Pottasch (1967) that the lines are formed
in a circumstellar cloud is supported.

124.109 Nova Persei 1901

**Proper motions of nova Per 1901 and the neighbour-
ing stars.** V. A. Sokolova.
Trudy Glav. Astron. Obs. Pulkovo, Ser. 2, Vol. 77, 115 - 125
(1969). In Russian.

124.110 Nova Cephei 1971

**On the light curve and early spectral changes in
nova Cephei 1971.**
J. C. Thomas, A. P. Cowley, D. J. MacConnell, J. Toney.
Publ. Astron. Soc. Pacific, Vol. 85, 309 - 313 (1973).
A light curve for nova Cephei 1971 of photoelectric and
photographic visual magnitudes observed between 28 June
and 24 September 1971 is presented. A description of the
spectral changes between 14 July and 7 August 1971 is also
given. An absolute visual magnitude of -8.3 at maximum is
derived.

125 Supernovae, Supernova Remnants

**125.001 A numerical computation of the dynamical evolu-
tion of a supernova remnant.**
I. Rosenberg, P. A. G. Scheuer.
Monthly Notices Roy. Astron. Soc., Vol. 161, 27 - 45 (1973).
This paper presents a numerical computation of the
dynamical development of a simple model of a supernova
remnant in which an expanding shell of ejected matter interacts with the interstellar gas. The evolution is followed continuously through the stages approximating to an undecelerat-
ed expansion, adiabatic blast wave and the stage dominated by
radiative cooling.

**125.002 A numerical model of the structure and evolution of
young supernova remnants.** S. F. Gull.
Monthly Notices Roy. Astron. Soc., Vol. 161, 47 - 69 (1973).
The evolution of the structure and radio emission of a
young supernova remnant is discussed in terms of a fluid dynamic model of a supernova explosion. The interaction between the interstellar medium and the ejected material leads to
a convective instability, which makes sufficient turbulent
energy available to account for the observed synchrotron radio
emission.

125.003 Type I supernovae. D. Branch, B. Patchett.
Monthly Notices Roy. Astron. Soc., Vol. 161, 71 -
83 (1973).
Wavelength coincidences and comparison of observed
and synthetic spectra support the hypothesis that type I supernova spectra contain blueshifted absorption features (Fe II,
Ca II, Na I, Si II, Mg II) characteristic of spectra of novae near
maximum light. On the assumptions that the light curve is due
to thermal emission from an expanding, optically thick photo-

sphere and that the temperature may be derived from the
normal stellar colour-temperature relation, a modified Baade
method is applied to a composite type I light curve.

**125.004 Outbursts of supernovae and formation of relativis-
tic objects. I.** O. H. Guseinov (*O. Kh. Gusejnov*),
F. K. Kasumov, V. I. Lazarev, A. V. Osipchuk.
Astron. Zhurn. Akad. Nauk SSSR, Vol. 50, 39 - 47 (1973).
In Russian. English translation in Soviet Astron. AJ, Vol. 17,
No. 1.
It is shown that a genetic connection between a pulsar
and a supernova remnant is convincing only if the distance be-
tween them is no more than 30 pc. Only two pairs satisfy this
citerion, 'P 0531 − Crab' and 'P 0833 − Vela'.

125.005 Gamma-ray lines from an expanding supernova shell.
R. T. Brown.
Astrophys. Journ., Vol. 179, 607 - 613 (1973).
Monte Carlo calculations have been made of the transport
of γ-ray lines out of the expanding shell of a type I supernova
produced by silicon burning. Because of the effect of Compton scattering the line spectrum of a young remnant is found
to be superposed on a strong continuum, with no lines resolvable below 0.8 MeV. The predicted contribution to the diffuse
γ-ray background from a large number of such sources emitting over a long period of time is shown to be a continuum
with no measurable line profiles.

125.006 Supernova remnants and gamma-ray sources.
J. A. de Freitas Pacheco.
Astrophys. Letters, Vol. 13, 97 - 101 (1973).
We have shown that, when evolutionary effects are taken

into account, it is very difficult to explain the γ-ray galactic background flux by the pion decay mechanism in supernova remnants, unless we drastically change our current ideas on the energetics of supernovae.

125.007 **The soft X-ray structure of Cassiopeia A.**
A. C. Fabian, J. C. Zarnecki, J. L. Culhane.
Nature, Phys. Sci., Vol. 242, 18 - 20 (1973).
The supernova remnant Cas A is identified as an extended source of soft X-rays. No evidence is found for the presence of a compact X-ray source. There are indications of non-uniformity in the X-ray surface brightness.

125.008 **A theoretical model for type II supernovae.**
S. W. Falk, W. D. Arnett.
Astrophys. Journ., (*Letters*), Vol. 180, L65 - L68 (1973).
Numerical calculations of shock waves in extended circumstellar envelopes, with radiation transport via photon diffusion, have been performed; they suggest a satisfactory explanation for the light curves of (at least) type II supernovae.

125.009 **Polarization of the supernova remnant HB21 at 11-cm wavelength.**
M. R. Kundu, R. H. Becker, T. Velusamy.
Astron. Journ., Vol. 78, 170 - 173 (1973).
Linear polarization at 11-cm wavelength has been detected in the supernova remnant HB21. The degree of polarization varies from about 2% to as much as 20% in some regions. Comparison between the 11- and 6-cm wavelength polarization measurements yields the rotation measure, depolarization, and magnetic field distributions over HB21.

125.010 **Supernova remnants in the Large Magellanic Cloud.**
D. S. Mathewson, J. N. Clarke.
Astrophys. Journ., Vol. 180, 725 - 738 (1973).
Nine supernova remnants (SNRs) have been discovered in the Large Magellanic Cloud (LMC) using a combination of radio and optical techniques. All of the SNR are in extreme Population I regions. The relation between surface brightness Σ and linear diameter D for the SNRs in the LMC is $\Sigma \propto D^{-3}$. An evolutionary trend was discovered in the Hα + [N II] surface brightness, I, such that $I \propto D^{-2.5}$. An integral luminosity function of the type $n(<D) = 1.8 \times 10^{-2} D^{2.5}$ is suggested from which a frequency of SNRs in the LMC of 1 in 500 years is deduced. Two very extended nonthermal radio sources were discovered with linear sizes of approximately 250 pc. They are tentatively identified as sections of the giant shells of 2 super-supernovae which Westerlund and Mathewson suggested to have occurred in the regions of Constellations II and III.

125.011 **Emission-line spectra of supernova remnants and galaxies.** D. E. Osterbrock.
Bull. American Astron. Soc., Vol. 5, 11 - 12 (1973). – Abstr. AAS.

125.012 **Thermal bremsstrahlung from supernova remnants.**
C. J. Lada, W. C. Straka.
Bull. American Astron. Soc., Vol. 5, 12 (1973). – Abstr. AAS.

125.013 **Buoyancy of supernova remnants.** E. M. Jones.
Bull. American Astron. Soc., Vol. 5, 27 - 28 (1973). Abstr. AAS.

125.014 **Absolute magnitude of type I supernovae.**
C. T. Kowal.
Bull. American Astron. Soc., Vol. 5, 28 (1973). – Abstr. AAS.

125.015 **Possible records of the Crab nebula supernova in the Western United States.**
J. C. Brandt, S. P. Maran, R. Williamson, R. S. Harrington,

C. Cochran, M. Kennedy, W. J. Kennedy, V. D. Chamberlain.
Bull. American Astron. Soc., Vol. 5, 29 (1973). – Abstr. AAS.

125.016 **On the possibility of formation of superheavy elements by supernova outbursts and on the probability of detecting them.** É. E. Berlovich.
Third Soviet Gravitational Conference, Erevan, 1972, (see 012.001), p. 294 - 296 (1972). In Russian.

125.017 **Supernova outbursts and pulsars.**
O. Kh. Gusejnov, F. K. Kasumov.
Third Soviet Gravitational Conference, Erevan, 1972, (see 012.001), p. 308 - 312 (1972). In Russian.

125.018 **X-radiation from supernova remnants.**
K. A. Pounds.
IAU Symposium No. 55, (see 012.002), p. 105 - 117 (1973).

125.019 **Low energy X-ray map of Puppis A supernova remnant.** J. C. Zarnecki, J. L. Culhane, A. C. Fabian,
C. G. Rapley, R. Silk, J. H. Parkinson, K. A. Pounds.
Nature, Phys. Sci., Vol. 243, 4 - 5 (1973).
The low energy X-ray emission from the Puppis A supernova remnant has recently been observed with the Mullard Space Science Laboratory grazing incidence telescopes on Copernicus. The observations reported here are from the 0.5 – 1.5 keV detector system. The whole of the radio remnant has been mapped using the 10 arc min field of view. Some additional data from the 6 arc min field of view are also presented.

125.020 **Interpretation of the radio emission from the supernova remnants Cas A and 3C 10.** S. F. Gull.
Monthly Notices Roy. Astron. Soc., Vol. 162, 135 - 142 (1973).
The observations of the young supernova remnants Cas A and 3C 10 are interpreted on the basis of a fluid dynamic model of the evolution of young supernova remnants. The shell structure and radio emission are explained in terms of instabilities arising at the boundary between the material ejected by the explosion and the interstellar medium.

125.021 **The production of deuterium in supernova shocks.**
S. A. Colgate.
Astrophys. Journ., (*Letters*), Vol. 181, L53 - L54 (1973).
It is the purpose of this *Letter* to summarize a more extensive analysis (to be published) of the conditions for shock spallation by a high-temperature shock precursor.

125.022 **High-velocity gas in supernova remnants. II. Shajn 147.** J. Silk, G. Wallerstein.
Astrophys. Journ., Vol. 181, 799 - 804 (1973).
A component of interstellar Ca II at -69 km s^{-1} (with respect to the local standard of rest) has been found in HD 36665, which lies behind the supernova remnant Shajn 147. By comparing the data for Vela XYZ, the Cygnus Loop, and Shajn 147, we show that these three supernova remnants form a sequence with respect to size, age, expansion velocity, excitation, and radio surface brightness. They appear to be similar phenomena observed at different stages of evolution.

125.023 **Early supernova luminosity.**
S. A. Colgate, C. McKee.
Stellar evolution, (see 012.014), p. 307 - 327 (1972). – Reprinted from Astrophys. Journ., Vol. 157, 623 - 643 (1969).
See Abstr. 02.125.006.

125.024 **On the light curve and properties of type I supernovae.** R. Barbon, F. Ciatti, L. Rosino.
Astron. Astrophys., Vol. 25, 241 - 248 (1973).
By the best fitting of all available light curves of type I supernovae an average curve, representative of the class, has

been drawn. Due to the strong similarity of the light curves, the dispersion of the points is relatively small. From the analysis of the average curve some general properties of the SN-I can be derived. The occurrence of type I supernovae in different types of galaxies and the possibility of a further subdivision in two groups are discussed.

125.025 **Old supernova remnants.** S. A. Ilovaisky.
Stellar ages. Proc. IAU Colloquium No. 17, (see 012.015), XXXIX, 1 - 4 (1973).

125.026 **High resolution interferometry of small diameter supernova remnants.** B. R. Hermann, J. R. Dickel.
Bull. American Astron. Soc., Vol. 5, 284 (1973). – Abstr. AAS.

125.027 **A high resolution 21 cm continuum study of the supernova remnants 3C 10 (Tycho's supernova remnant) and 3C 461 (Cas A).** R. G. Strom, R. M. Duin.
Astron. Astrophys., Vol. 25, 351 - 362 (1973).
Synthesis observations made with the Westerbork telescope have been used to produce maps of the supernova remnants 3C 10 and 3C 461. The Stokes parameters I, Q and U have been determined with a resolution of about $25''$.

125.028 **Secular decrease of the flux of supernova remnants Cas A and SN-1572.**
K. S. Stankevich, V. P. Ivanov, V. A. Torkhov.
Astron. Zhurn. Akad. Nauk SSSR, Vol. 50, 645 - 646 (1973).
In Russian. English translation in Soviet Astron. AJ, Vol. 17, No. 3. – Short note.

125.029 **Supernova remnants.** W. Buscombe.
Journ. Astron. Soc. Victoria, Vol. 25, 78 - 80 (1972).

125.030 **An atlas of supernova spectra.**
J. L. Greenstein, R. Minkowski.
Astrophys. Journ., Vol. 182, 225 - 243 (1973).
A selection of supernova spectra is given. Only a brief attempt is made to interpret these spectra. The atlas should serve as a guide to the observation and interpretation of photographic and spectrophotometric studies of supernova evolution. Peculiar types of supernovae are included, with some details of the evolution of spectra of types I, II, III, and V. Rapid changes and the P Cygni character of lines in types II, III, and V are displayed. The figures provide data for identification, wavelengths and velocities, and the relative importance of emission and absorption.

125.031 **A high-sensitivity search for radio emission from young extragalactic supernova remnants at 1415 MHz.** A. G. de Bruyn.
Astron. Astrophys., Vol. 26, 105 - 112 (1973).
A search for radio emission from the remnants of about 35 extragalactic supernovae has been made with the Westerbork Synthesis Radio Telescope. None of the remnants, which had ages from 1 month to 86 years, has been detected. The results have been compared with the empirically determined radio evolution of (much older) galactic supernova remnants (SNR). The results are also compared with current theories about the evolution of young galactic SNR.

125.032 **Observations of soft X-rays: Upper limits on the flux from SN 1972E and measurements of the diffuse background in Centaurus.** T. M. Palmieri, G. A. Burginyon, R. W. Hill, J. K. Scudder, F. D. Seward, A. Toor.
Astrophys. Journ., Vol. 182, 411 - 416 (1973).
An attempt was made to measure soft X-rays from SN 1972E. The resulting upper limit shows that 19 days after discovery, the X-ray luminosity was not greater than ~10 percent of the optical luminosity. Measurements of the background in the surrounding region show that a localized enhancement, ob-

served previously, dominates the diffuse flux in this area. Two maps give the spatial dependence of the diffuse flux in the energy ranges 0.2–0.6 keV and 0.6–1.6 keV.

125.033 **Buoyant supernova remnants.** E. M. Jones.
Astrophys. Journ., Vol. 182, 559 - 568 (1973).
It is proposed that the time scale for buoyant rise may be short enough that many supernova remnants (SNRs) rise and deform before they lose their identity. Six SNRs from the lists of Shaver and Goss appear to be in the process of forming into toroids. The dominant gravitational term causing the rise appears to be fairly local inhomogeneities in the galactic gravitational field. A two-dimensional hydrodynamic calculation of a 4×10^{49} erg supernova event is presented.

125.034 **Supernovae.** P. Wild, W. L. W. Sargent.
IAU Circ., No. 2476 (1973).
Concerning a probable supernova in NGC 2841 and the supernova in NGC 4254 = M 99.

125.035 **Supernova in anonymous galaxy.** C. Kowal.
IAU Circ., No. 2485 (1973).

125.036 **Supernova in anonymous galaxy.**
M. Schmidt, W. L. W. Sargent.
IAU Circ., No. 2487 (1973).

125.037 **Supernova in anonymous galaxy.** F. Zwicky.
IAU Circ., No. 2525 (1973).

125.038 **Polarisation and brightness distributions across IC443 and W44 at 11-cm wavelength.**
J. R. Baker, E. Preuss, J. B. Whiteoak.
Astrophys. Letters, Vol. 14, 123 - 127 (1973).
The distribution of polarisation and total intensity across IC443 and W44 have been observed at a wavelength of 11 cm with a resolution of 4.8 arc min. IC443 shows strong depolarisation over the associated bright nebulosity, but relatively high polarisation where there are no optical features. W44 has polarisation with a brightness distribution similar to that of the total intensity, and with a fairly uniform change in position angle of polarisation across the source.

125.039 **Supernova explosions.** B. Apagyi.
Fiz. Szemle, (*Hungary*), Vol. 22, 295 - 301 (1972).
In Hungarian.
The relationship between pulsars and supernovae is discussed with reference to astronomical observations since 1967. Three models of supernova explosions namely the Fe star model, neutrino star model and double layer red giant model are outlined and are used to explain various concepts on explosions.

125.040 **Fenomenologia delle supernovae.** F. Bonoli.
Coelum, Vol. 41, 93 - 105 (1973).

125.041 **Cas A X-ray spectrum: Evidence for iron line emission.** P. J. Serlemitsos, E. A. Boldt, S. S. Holt, R. Ramaty, A. F. Brisken.
Goddard Space Flight Center, Greenbelt, Maryland, Prepr. X-661-73-79, 1 + 16 pp. (1973).
A sensitive measurement by rocket borne detectors of the X-ray flux from Cas A has revealed a steep continuum and a broad spectral feature in the region where line radiation from iron nuclei would be expected. The presence of broad iron lines is consistent with a model in which ~13 MeV/nucleon iron nuclei charge exchange with surrounding interstellar oxygen and other heavy atoms. The model suggests that a substantial fraction of the energy from the outburst has gone into low energy cosmic rays which produce the observed H II region surrounding the remnant.

125.042 Supernova remnants in the Magellanic Clouds.
D. S. Mathewson, J. N. Clarke.
Astrophys. Journ., Vol. 182, 697 - 698 (1973).

Details are given of a second supernova remnant (SNR) discovered in the Small Magellanic Cloud. This brings the number of known SNRs in the Magellanic Clouds to 14. A relationship between flux density and angular diameter is established which may be used to estimate the distances to galactic SNRs.

125.043 The galactic supernovae of the second millennium A.D. S. van den Bergh.
Publ. Astron. Soc. Pacific, Vol. 85, 335 - 340 (1973).

This paper discusses the five most recent known galactic supernovae and their remnants. The possible importance of circumstellar shells ejected before the explosion of Kepler's supernova and the Cas A supernova is emphasized. Some new observations are presented of the optical remnants of the supernova of 1604 and of Cas A.

A possible explanation for the origin of lithium, beryllium, and boron. See Abstr. 061.063.

Charged particle thermonuclear reactions in nucleosynthesis. See Abstr. 065.001.

Discovery of five novae in Messier 33 and a supernova in a field galaxy. See Abstr. 124.002.

A minimum kinematic distance estimate to the nonthermal source in W 51 based on OH absorption line measurements. See Abstr. 131.142.

Time-dependent radiative cooling of a hot low-density cosmic gas. See Abstr. 131.179.

Evidence for ejection of radio sources from supernova remnants. See Abstr. 141.009.

A low frequency search for compact radio sources in supernova remnants. See Abstr. 141.051.

A low-frequency search for compact radio sources in supernova remnants. See Abstr. 141.078.

Radio halos around old pulsars – ghost supernova remnants. See Abstr. 141.504.

A test of Tsarevsky's pulsar–supernova analysis. See Abstr. 141.535.

Pulsars in supernova remnants. See Abstr. 141.563.

Short-term temporal studies of the X-ray emission from Cassiopeia A, Tycho, and Scorpius X-1. See Abstr. 142.040.

Galactic continuum loops and the diameter-surface brightness relation for supernova remnants. See Abstr. 155.023.

Remarks on the soft X-ray emission from the galactic radio spurs. See Abstr. 155.092.

Errata

125.901 Corrigendum: 'Classification of supernova remnants and H II regions from their recombination line emis- sion' [Australian Journ. Phys., Vol. 25, 539 - 544 (1972)].
J. R. Dickel, D. K. Milne.
Australian Journ. Phys., Vol. 26, 267 (1973).

125.100 Supernova in NGC 5055

Photographic observations of the supernova 1971 in NGC 5055. H. Dürbeck.
Astron. Astrophys., Vol. 22, 317 - 318 (1973).

Photographic magnitudes of the supernova 1971 in NGC 5055 are given. The light curve is discussed briefly.

The light curve of supernova 1971 I.
D. Deming, B. W. Rust, E. C. Olson.
Publ. Astron. Soc. Pacific, Vol. 85, 321 - 327 (1973).

Light and color curves are presented for the 1971 supernova in NGC 5055. The light curves were derived from photoelectric observations, and photographic observations calibrated with a photoelectric sequence. The dates and B, V magnitudes of maximum light are established. The form of the light curve is consistent with previous spectrographic determinations that the supernova was of type I. A $(B-V)$ color curve is presented and is used to derive a reddening of $0^m\!.35$ for the supernova. The consequent visual absorption of $1^m\!.05$ is used to obtain an absolute visual magnitude for the supernova of -18.6.

125.101 Supernova in NGC 5253

The infrared spectrum of the supernova in NGC 5253. F. Ciatti.
Astron. Astrophys., Vol. 22, 465 - 466 (1973).

The infrared spectrum from 6500–11000 Å of the type I supernova in NGC 5253 is here described. An attempt to identify the emission features and a comparison with a type II supernova are outlined.

Spectrophotometry of the supernova in NGC 5253 from 0.33 to 2.2 microns. R. P. Kirshner, S. P. Willner, E. E. Becklin, G. Neugebauer, J. B. Oke.
Astrophys. Journ., (Letters), Vol. 180, L97 - L100 (1973).

Combined infrared and optical measurements are presented of the energy distribution of the type I supernova in NGC 5253 from 1972 May 16 to 1972 July 31. The overall shape from 0.4 to 2.2 μ could be represented by a blackbody which decreased in temperature from about 10,000°K on May 23 to 7500°K on June 5, then remained at about 7000°K until at least July 8.

Infrared emission in the spectrum of SN 1972 in NGC 5253. M. F. McCarthy, G. Arraya.
Bull. American Astron. Soc., Vol. 5, 12 (1973). – Abstr. AAS.

Ultraviolet photometry of the supernova in NGC 5253. A. V. Holm, C. C. Wu, J. J. Caldwell.
Bull. American Astron. Soc., Vol. 5, 28 - 29 (1973). – Abstr. AAS.

Recent supernova in NGC 5253 and the supernova rate.
W. Wamsteker, W. Z. Wiśniewski, T. A. Lee, T. J. Wdowiak.
Nature, Phys. Sci., Vol. 241, 7 - 9, with a correction in Nature, Phys. Sci., Vol. 243, 144 (1973).

We have published photoelectric photometry extending to $\lambda = 2.2$ μm of the bright supernova, SN 1972e, in NGC 5253. Here we discuss some of the interesting aspects of this supernova and present some ideas concerning the supernova rate.

Search for high frequency optical pulsar in supernova NGC 5253. See Abstr. 141.510.

125.102 **Supernova in NGC 5457**

Variable radio emission from the extragalactic supernova 1970g in M101.
W. M. Goss, R. J. Allen, R. D. Ekers, A. G. de Bruyn.
Nature, Phys. Sci., Vol. 243, 42 - 44 (1973).

The variable radio source associated with the supernova 1970g in M101 has been mapped at Westerbork (21 cm and 6 cm) and at Effelsberg (2.8 cm). Although there are as yet insufficient data to provide the basis for a unique theoretical model of the event it is already possible to single out promising lines of attack.

125.103 **Supernova in NGC 4975**

Supernova 1968 in NGC 4975.
R. G. Mnatsakanian.
Inform. Bull. Variable Stars (I.A.U. Commission 27), Konkoly Obs., Budapest, No. 785 (1973).

125.104 **Supernova in NGC 3656**

Supernova in NGC 3656. C. T. Kowal.
IAU Circ., No. 2491 (1973).

Supernova in NGC 3656. R. K. Shakhbazyan.
IAU Circ., No. 2507 (1973).

125.105 **Supernova in NGC 2841**

Supernova in NGC 2841.
W. L. W. Sargent, L. Searle.
IAU Circ., No. 2498 (1973).

125.106 **Supernova in NGC 4944**

Supernova in NGC 4944. L. Kohoutek.
IAU Circ., No. 2521 (1973).

125.107 **Supernova in NGC 4939**

Supernova in NGC 4939. P. Wild.
IAU Circ., No. 2538 (1973).

126 Low-luminosity Stars, Subdwarfs, White Dwarfs

126.001 **Upper limits to the 21 cm continuum radiation from two magnetic white dwarfs.** R. D. Ekers.
Astron. Astrophys., Vol. 22, 309 - 310 (1973).

Observations with the Westerbork Synthesis Radio Telescope place a limit of 2×10^{-29} Wm^{-2} Hz^{-1} on the 21 cm continuum radio emission from the magnetic white dwarfs Grw +70°8247 and G 195−19. A list of the field radio sources found in these observations is included.

126.002 **CPD−31° 1701, an extremely helium-rich, subluminous, O-type star.**
R. F. Garrison, W. A. Hiltner.
Astrophys. Journ., (Letters), Vol. 179, L117 - L120 (1973).

The spectrum of CPD−31°1701 is completely dominated by lines of neutral and ionized helium, most of which are extremely Stark-broadened. The spectral type is O8, and the photometric results are $V = 10.52$, $B-V = -0.31$, and $U-B = -1.17$. The absolute magnitude is fainter than $M_V = 0$ and probably +3.

126.003 **The evolution of a white dwarf by accretion of hydrogen-rich matter. I.** Yu. N. Redkoborody.
Astrofizika, Vol. 8, 261 - 282 (1972). In Russian. English translation in Astrophysics, Vol. 8, No. 2.

An evolutionary sequence was calculated for a white dwarf for which a hydrogen-rich envelope is assumed to increase with time. The accretion of matter was assumed to be quasistatic. Near the point of chemical discontinuity a temperature maximum is shown to be built up, after hydrogen is ignited, a thin shell energy source is formed. The new shell source is thermally unstable. The resulting thermal run-away was followed numerically. The hydrogen burning is shown to become unstable, when the mass of the hydrogen envelope exceeds some critical value.

126.004 **Observations of circular polarization in white dwarfs and in the nuclei of planetary nebulae.**
O. S. Shulov, E. T. Belokon.
Astrofizika, Vol. 8, 343 - 352 (1972). In Russian. English translation in Astrophysics, Vol. 8, No. 3.

The results of circular polarization observations are reported for 10 white dwarfs and 5 nuclei of planetary nebulae.

126.005 **Radial pulsations of a white dwarf in the case of non-uniform rotation.**
M. M. Basko, V. S. Imshennik.
Astrofizika, Vol. 8, 387 - 391 (1972). In Russian. English translation in Astrophysics, Vol. 8, No. 3.

The periods of radial pulsations of non-uniformly rotating white dwarfs near Chandrasekhar's limit are estimated with the help of the energetic method. The effects of neutronization cause a substantial increase of radial pulsation periods, but nevertheless the periods of the rotating star can be 4−5 times less than those of a non-rotating one.

126.006 **Evolution of a white dwarf by accretion of hydrogen-**

rich matter. II. Yu. N. Redkoborody.
Astrofizika, Vol. 8, 393 - 403 (1972). In Russian. English
translation in Astrophysics, Vol. 8, No. 3.

The thermal burst in a hydrogen envelope of a white
dwarf is investigated without calculating stellar models. It is
shown that one can obtain the temperature at the bottom of
the hydrogen envelope from the energy balance in the hydrogen burning shell source. Such an approximation permits to
reveal the role of the screening effect in the development of
the flash.

126.007 **Pulsations and stability of flattened rotating white
dwarfs.** Yu. L. Vartanian.
Astrofizika, Vol. 8, 413 - 418 (1972). In Russian. English
translation in Astrophysics, Vol. 8, No. 3.

The effect of the flattening on equilibrium parameters
and the stability of rotating white dwarfs are considered.

126.008 **On the flare possibility of white dwarfs.**
G. A. Gurzadyan.
Astrofizika, Vol. 8, 479 - 482 (1972). In Russian. English
translation in Astrophysics, Vol. 8, No. 3.

It is shown that the non-thermal bremsstrahlung of fast
electrons may be the cause of the flare of a white dwarf. The
general physical parameters of such a flare are derived by comparison with observations.

126.009 **A search for optical circular polarization in white
dwarfs and late-type stars with circumstellar shells.**
A. Rich, W. L. Williams.
Astrophys. Journ., (Letters), Vol. 180, L123 - L126 (1973).

A search for broad-band optical circular polarization in
the light from white dwarfs has been carried out for 15 stars
previously observed for luminosity variations. We have also
searched for broad band optical circular polarization in six
late-type stars, which exhibit intrinsic plane polarization and
large infrared excess.

126.010 **Cooling time and internal characteristics of hot
white dwarfs.** Yu. L. Vartanian, G. S. Hajian
(Adzhyan), A. S. Harutunian (Arutyunyan).
Astron. Zhurn. Akad. Nauk SSSR, Vol. 50, 305 - 311 (1973).
In Russian. English translation in Soviet Astron. AJ, Vol. 17,
No. 2.

The internal characteristics and cooling time are considered for hot configurations of white dwarfs with masses
$M/M_\odot = 0.182, 0.727, 1.082$, which have exhausted all supplies
of nuclear energy. The distribution of internal temperature for
different values of surface temperature is determined.

126.011 **Helium spectra in white dwarfs with large magnetic
fields.** R. H. Garstang, S. Kemic.
Bull. American Astron. Soc., Vol. 5, 10 - 11 (1973). − Abstr.
AAS.

126.012 **The triply-periodic white dwarf HL Tau-76.**
W. S. Fitch.
Bull. American Astron. Soc., Vol. 5, 17 (1973). − Abstr. AAS.

126.013 **On the theory of hot white dwarfs.**
D. M. Sedrakyan, É. V. Chubaryan.
Third Soviet Gravitational Conference, Erevan, 1972, (see
012.001), p. 356 - 358 (1972). In Russian.

126.014 **The magnetohydrodynamic stability of white
dwarfs and neutron stars.** R. J. Tayler.
Monthly Notices Roy. Astron. Soc., Vol. 162, 17 - 23 (1973).

Vandakurov has recently suggested that an arbitrarily
weak destabilizing magnetic field may lead to the occurrence
of convection in superdense stars. It is pointed out here
that, when thermal corrections to the equation of state are

included, instability is unlikely to occur throughout a large
region of such a star unless the magnetic pressure exceeds
the thermal correction to the ideal degenerate pressure. In
the case of white dwarfs it appears that the necessary magnetic fields for instability are several orders of magnitude
stronger than those which are generally believed to exist in
the stars. In contrast, it seems possible that strong enough
magnetic fields to cause instability could be present in neutron stars.

126.015 **Line profiles and rotation in white dwarfs.**
J. L. Greenstein, D. M. Peterson.
Astron. Astrophys., Vol. 25, 29 - 34 (1973).

Two hydrogen-line white dwarfs, 40 Eri B and Wolf 1346,
show sharp cores in Hα and Hβ on Palomar image-tube coudé
spectra. Sharp cores are predicted from model atmospheres in
which departures from LTE are allowed. Theoretical profiles
have been computed for various rotational velocities. Convoluted with the instrumental profile they are compared with the
observations.

126.016 **Blanketing theory and the $G-I$ index.**
M. N. Perrin.
Astron. Astrophys., Vol. 25, 79 - 83 (1973).

The blanketing theory of Wildey, Burbidge, Sandage and
Burbidge is tested by means of the $G-I$ index (Lick system).
It is found that blanketing theory slightly underestimates the
T_{eff} of extreme subdwarfs. We propose that, for these stars, the
blanketing corrections $\Delta (B-V)$, be diminished by about 0.03.
The corrections $\Delta (B-V)$ obtained through using the $(R-I)_J$
or $(R-I)_K$ indices (Johnson et al., 1968; Taylor, 1970; Eggen,
1971) are discussed.

126.017 **The triply periodic white dwarf HL Tau-76.**
W. S. Fitch.
Astrophys. Journ., (Letters), Vol. 181, L95 - L98 (1973).

Analysis of 1530 differential blue magnitudes of HL
Tau-76 shows that this star is apparently a regular, triply periodic variable with little or no random noise in the extremely
complex light variation.

126.018 **Mass deficiency of white dwarfs and unstable
neutron stars.** É. V. Chubaryan.
Uch. zap. Erevan. un-t. Estestv. n., 1972, No. 2 (120), p. 23 -
26. In Russian. − Abstr. in Referativ. Zhurn. 51. Astron.,
4.51.669 (1973).

126.019 **White dwarfs.** J. P. Ostriker.
Stellar evolution, (see 012.014), p. 211 - 269
(1972).

126.020 **The H and K emission lines in the subdwarf
ζ^1 Reticuli.** R. Foy.
Stellar ages. Proc. IAU Colloquium No. 17, (see 012.015),
XLVI, 1 - 10 (1973).

126.021 **The subdwarf Groombridge 1830.** J. Tomkin.
Stellar ages. Proc. IAU Colloquium No. 17, (see
012.015), L, 1 - 5 (1973).

126.022 **Gas-liquid phase transitions and critical point in hot
white dwarf matter.**
L. De Cesare, A. Forlani, G. Platania.
Astrophys. Space Sci., Vol. 21, 461 - 474 (1973).

We consider the gas-liquid first-order phase transitions and
prove the existence of a critical point in white dwarf matter.
The latent heat released in the liquefaction processes can be
used in the interpretation of the spreading of the white dwarf
sequence in H-R diagram. Some thermodynamic quantities,
e.g., the saturation pressure, the latent heat, etc.,are calculated

along the gas-liquid coexistence curve, and their behaviour near the critical point is studied.

126.023 The wavelength dependence of linear and circular polarized radiation from the magnetic white dwarf Grw+70°8247. K. M. Roussel, R. F. O'Connell.
Astrophys. Journ., Vol. 182, 277 - 282 (1973).

The wavelength dependence of linear polarization from the magnetic white dwarf Grw+70°8247 is investigated. The gray-body magnetoemissive linear polarization is corrected for radiative transfer by the use of a model white-dwarf atmosphere. The wavelength dependence found by this method does not fit the wavelength dependence of the observed linear polarization. Also the wavelength dependence of the circular polarization from Grw+70°8247 is reexamined in the light of new observations.

126.024 Atmosphärenmodelle kühler Weisser Zwerge. R. Wehrse.
Mitt. Astron. Ges., No. 32, p. 232 - 234 (1973).

126.025 Häufigkeiten in Weißen Zwergen mit Kohlenstoffbanden. I. Bues.
Mitt. Astron. Ges., No. 32, p. 234 - 235 (1973).

126.026 Double white dwarf. W. J. Luyten, P. Higgins.
IAU Circ., No. 2542 (1973).

126.027 Cooling sequence of a pure helium white dwarf of 0.15 solar masses. F. D'Antona, G. Magni, I. Mazzitelli.
Astrophys. Space Sci., Vol. 19, 151 - 158 (1972).

A new method of integration has been performed and applied to the computation of the cooling sequence of a $0.15\,M_\odot$ white dwarf consisting of helium. The results are discussed and compared with previous ones obtained by other authors.

126.028 Configurations of hot white dwarfs with nuclear sources of energy. R. M. Avakian.
Soobshch. Byurakan. Obs., vyp. (No.) 44, p. 115 - 119 (1972). In Russian.

The configurations of hot white dwafs with sources of nuclear energy are calculated. The values of luminosity have been calculated and compared with observational data.

126.029 On the theory of white dwarfs. G. S. Sahakian (*Saakyan*), D. M. Sedrakian, E. V. Chubarian.
Astrofizika, Vol. 8, 541 - 556 (1972). In Russian. – English translation in Astrophysics, Vol. 8, No. 4.

126.030 On the existence of subdwarfs in the $(M_{bol}, \log T_e)$-plane. III. O. J. Eggen.
Astrophys. Journ., Vol. 182, 821 - 837 (1973).

A discussion of (R, I) observations of parallax stars indicates that (1) the earlier conclusion that F- and G-type subdwarfs populate the Hyades main sequence in the $(M_{bol}, \log T_e)$-plane resulted from the evolved nature of the subdwarfs and (2) the later-type subdwarfs lie about 1 mag below the old-disk main sequence.

126.031 The internal magnetic fields of white dwarfs. G. Chanmugam, M. Gabriel.
Astrophys. Journ., Vol. 182, 915 - 918 (1973).

It is shown that observable white dwarfs are unlikely to have magnetic fields $\gtrsim 10^{11}$ gauss in their interior.

The role of convection in stellar atmospheres. II. Cool main-sequence stars and metal-deficient subdwarfs. See Abstr. 064.066.

Relavistic stellar stability: an empirical approach. See Abstr. 065.099.

Radial pulsations of pre-white-dwarf stars. II. Pulsational stability of ^{12}C shell-burning stars. See Abstr. 065.174.

On convection and gravitational layering in Jupiter and in stars of low mass. See Abstr. 099.040.

A search for faint blue objects near the north galactic pole. See Abstr. 113.013.

UBV photometry of large proper motion stars. See Abstr. 113.057.

A highly evolved, low-mass binary, HZ 22. See Abstr. 119.004.

Pulsational stability of stars in thermal imbalance. See Abstr. 122.139.

Interstellar Matter, Gaseous Nebulae, Planetary Nebulae

131 Interstellar Space, Interstellar Matter, Polarization of Starlight

131.001 Chemical composition of the interstellar gas: X-ray determinations. R. L. Brown.
Astrophys. Space Sci., Vol. 18, 329 - 333 (1972).
Noting that observations of X-ray attenuation at the K-shell ionization edge for many elements provide, prospectively, the least ambiguous means for establishing the chemical composition of the interstellar gas, we have evaluated the expected spectral discontinuities for all atoms from carbon to sodium and have tabulated the results.

131.002 Orientation of interstellar and interplanetary grains. A. Z. Dolginov.
Astrophys. Space Sci., Vol. 18, 337 - 349 (1972).
The explicit expressions for the orientation distribution function of interstellar and interplanetary dust grains in the anisotropic corpuscular or radiation fluxes, with consideration for the magnetic field influence, are obtained. The orientation of these dust grains is considered. The time required for the orientation is estimated. A possibility of explaining the interstellar polarization and polarization of the cometary radiation is discussed.

131.003 Some limits on cosmic-ray heating of HI clouds by magnetic stars. B. N. G. Guthrie.
Astrophys. Space Sci., Vol. 18, 403 - 407 (1972).
Estimates are made of the low-energy, cosmic-ray power generated by the rotational braking of magnetic Ap stars through interaction with the interstellar medium.

131.004 Evaporation of dirty ice particles surrounding early type stars. IV. Various size distributions. S. Isobe.
Publ. Astron. Soc. Japan, Vol. 25, 101 - 109 (1973).
We calculate the evaporation processes of dirty ice particles with various size distributions surrounding early type stars. If the initial distribution function of the grain radius, a, is given by $n(a) = n(0) \exp(-24a^3)$, where a is in microns, the calculated interstellar absorption, after considering evaporation processes, is consistent with the observed values in the directions of the Orion nebula and the Orion association.

131.005 Radio brightness distribution of Ori A at 4.1 mm wavelength.
N. Kaifu, K. Akabane, M. Morimoto.
Publ. Astron. Soc. Japan, Vol. 25, 129 - 133 (1973).
The continuum brightness distribution of Ori A at 4.1 mm wavelength has been mapped using the 6-m millimeter-wave telescope of the Tokyo Astronomical Observatory. Only the central compact source was detected. The estimated flux density, 182 ± 45 f.u., is in good agreement with measurements at longer wavelengths of the core component of Ori A. The possibility of detecting a new component is discussed.

131.006 Lambda-doubling of the $O^{17}H$ molecule at microwave frequencies. I. E. Valtz, V. A. Soglasnova.
Astrophys. Letters, Vol. 13, 23 - 24 (1973).
The microwave frequencies of the most intense main lines of the molecule $O^{17}H$ are presented.

131.007 The cyanoacetylene cloud in Sagittarius B2.
R. X. McGee, L. M. Newton, R. A. Batchelor, A. R. Kerr.
Astrophys. Letters, Vol. 13, 25 - 32 (1973).
A new 3.4-cm cryogenic receiver operating on the Parkes 64-m radio telescope was used to detect the $F = 1 \to 1$, $F = 2 \to 1$, $F = 0 \to 1$ hyperfine lines of the $J = 1 \to 0$ transition of the cyanoacetylene molecule in Sagittarius B2. On the assumption that the natural line frequency of the principal component $(F = 2 \to 1)$ is 9098.332 MHz, the radial velocity at the central position is $+64.4$ km sec^{-1}. The distribution of cyanoacetylene in Sgr B2 was mapped and compared with that of other molecules. The HCCCN line was not detected in 14 other southern Milky Way sources.

131.008 Observations of dark clouds in IC 1795 (W3) in the formaldehyde line at 4830 MHz.
Y. K. Minn, J. M. Greenberg.
Astrophys. Letters, Vol. 13, 39 - 44 (1973).
A survey of H_2CO absorption in dark clouds in IC 1795 (W3) can be distinctly separated into three groups: the first immediately surrounding W3, has a uniform velocity of about -41 km/sec; the second, associated with the W3(OH) source has a velocity of about -49 km/sec; the third, has a velocity of about -20 km/sec and may be associated with an interarm spur in the foreground.

131.009 Large-scale line splitting in five galactic H II regions.
M. A. Dopita, A. H. Gibbons, J. Meaburn, K. Taylor.
Astrophys. Letters, Vol. 13, 55 - 59 (1973).
Many observations of singly and multiply split [O III] and [N II] lines from large areas of five galactic H II regions are presented. Explanations are sought for these and other unusual motions in H II regions and their similarity to spectral features in some planetary nebulae are pointed out.

131.010 The kinematical distribution of dark clouds surveyed in the 4830 MHz H_2CO line.
Y. K. Minn, J. M. Greenberg.
Astron. Astrophys., Vol. 22, 13 - 25 (1973).
A survey of over 80 dark clouds has been made for formaldehyde absorption using the 140-foot telescope of the National Radio Astronomy Observatory. Most of the clouds have been selected from Lynds' Catalogue of Dark Clouds. Additional ones have been chosen from the Palomar Sky Atlas.

131.011 Interferometer observations of W3 (OH) at 2.695 GHz and 8.085 GHz.
J. E. Wink, W. J. Altenhoff, W. J. Webster, Jr.
Astron. Astrophys., Vol. 22, 251 - 255 (1973).
Observations made at 2.695 GHz and 8.085 GHz with the NRAO interferometer show two continuum sources. At 8.085 GHz a half power width (HPW) of $1.''5$ was measured for the previously known continuum source. The continuum spectrum of this source and its connection with the OH sources in the area is discussed.

131.012 On the problem of large-scale peculiar motion of interstellar gas in the Galaxy. V. I. Ariskin.
Astron. Zhurn. Akad. Nauk SSSR, Vol. 50, 83 - 87 (1973).

In Russian. English translation in Soviet Astron. AJ, Vol. 17, No. 1.

It is indicated the possibility of existence of large-scale regions in the Galaxy of expanding neutral hydrogen towards the arms of Sagittarius and Scutum. These regions are essentially responsible for the thermal background radio emission of the continuous spectrum at 21 cm in this direction. The distance to these regions is $\sim$ 10 - 12 kpc, and the mean dimensions 2 kpc. The possible reason for their formation are supernova explosions of type III which took place in the galactic plane 10^7 years ago.

131.013 Astrophysical masers. II. Polarization properties.
P. Goldreich, D. A. Keeley, J. Y. Kwan.
Astrophys. Journ., Vol. 179, 111 - 134 = Contr. Lick Obs., No. 366 (1973).

The equations governing the transfer of polarized radiation in astrophysical masers are derived. It is found that the magnetic field and the plasma in maser sources play a central role in determining the polarization of the emitted radiation. The character of the polarization depends upon the relative sizes of the decay constant of the maser levels, the stimulated-emission rate, the Zeeman splitting, and the bandwidth of the amplified radiation.

131.014 Interstellar molecular hydrogen observed in the ultraviolet spectrum of Delta Scorpii. A. M. Smith.
Astrophys. Journ., (Letters), Vol. 179, L11 - L15 (1973).

Molecular-hydrogen bands of the Lyman and Werner systems have been observed in the ultraviolet, interstellar spectrum of δ Sco. The average molecular column density is 3.5 (+2.2, −0.9) $\times$ 10^{19} cm^{-2}, and the average temperature of the gas of which the molecules are part is 47°K. The observed column density of hydrogen atoms is $(1.5 \pm 0.5) \times 10^{21}$ cm^{-2}.

131.015 On the differing molecular line widths in dense interstellar clouds. C. Heiles.
Astrophys. Journ., (Letters), Vol. 179, L17 - L19 (1973).

In Sgr B2 and Orion A, radio astronomers observe different line widths for different molecules, or for different transitions within the same molecule. This behavior is consistent with the increasing density and decreasing collapse velocity toward the center found in theoretical models of isothermal collapsing protostars.

131.016 The heating of interstellar clouds by vibrationally excited molecular hydrogen.
T. P. Stecher, D. A. Williams.
Monthly Notices Roy. Astron. Soc., Vol. 161, 305 - 311 (1973).

We discuss the possibility that vibrationally excited H_2 may be collisionally de-excited, so providing a heating mechanism for interstellar clouds which operates by coupling the stellar radiation to the gas. We compare the heating rate obtained in this way with other mechanisms which have been postulated, and present the results of calculations of temperature as a function of depth into clouds of different densities.

131.017 Pressure equilibrium of finite-size clouds in the interstellar medium. R. Graham, W. D. Langer.
Astrophys. Journ., Vol. 179, 469 - 481 (1973).

We consider the pressure equilibrium between the cloud and the intercloud medium for clouds of finite size. Whereas the balance between the cosmic-ray heating and the cooling of the interstellar medium explains the coexistence of cloud and intercloud medium, surface effects due to heat conduction play a role in determining the cloud sizes. In general we find that a mass flow exists across the cloud-intercloud interface and that, for small clouds, this mass flow becomes an important part of the dynamics.

131.018 Infrared circular polarization of NML Cygni and VY Canis Majoris. K. Serkowski.
Astrophys. Journ., (Letters), Vol. 179, L101 - L106 (1973).

Circular polarization with a maximum of 0.6 percent at 1.7μ has been detected for the infrared source NML Cyg. Circular polarization of 0.1 percent at 2.2μ and 0.2 percent at 1.7μ has been found for VY CMa. The linear polarization of several infrared sources and of the planet Mars has also been observed.

131.019 A simple probabilistic theory of fragmentation.
R. B. Larson.
Monthly Notices Roy. Astron. Soc., Vol. 161, 133 - 143 (1973).

A simple model for the fragmentation process in a collapsing interstellar cloud is developed, which is based on the assumption that the successive stages of the fragmentation process can be treated as random events. Some possible ways of accounting for differences in the initial mass spectrum between different stellar systems are also discussed.

131.020 Interstellar scattering and the angular diameters of OH components. L. T. Little.
Astrophys. Letters, Vol. 13, 115 - 118 (1973).

The possibility that the apparent angular diameters of OH source components in W3, W24 and W49 are due to scattering of their radiation by plasma irregularities in the intervening medium is considered.

131.021 Discovery of interstellar methanimine (formaldimine).
P. D. Godfrey, R. D. Brown, B. J. Robinson, M. W. Sinclair.
Astrophys. Letters, Vol. 13, 119 - 121 (1973).

The $1_{10}-1_{11}$ transition of methanimine $H_2C=NH$ has been detected in emission in the spectrum of Sagittarius B2 with the Parkes 64-m telescope. The frequencies of the observed multiplet, near 5.290 GHz, agree well with the multiplet detected in the laboratory, if a radial velocity of 63 ± 2 km sec^{-2} is adopted for the Sgr B2 emission.

131.022 On the presence of H_2 molecules inside neutral globules imbedded in H II regions. G. Stasinska.
Astron. Astrophys., Vol. 22, 355 - 360 (1973).

We study the abundance of the H_2 molecule in neutral globules inside H II regions. We compute the H_2 equilibrium abundances for different values of the incident radiation field, and of the globular mass and density. We show that, if we take into account evolutionary effects, those globules that do not undergo gravitational collapse and remain unionized for a sufficiently long time are definitely molecular during the whole lifetime of the H II region.

131.023 Neutral hydrogen in the unusual dark cloud, Khavtassi 713. S. C. Simonson III.
Astron. Astrophys., Vol. 23, 19 - 23 (1973).

A prominent neutral hydrogen feature has been found to correspond in shape and position with the sharp-edged dark cloud, Khavtassi 713. The feature possesses an unusually strong velocity gradient in latitude, $dV/db = -2.7$ km s^{-1} deg^{-1}. As indicated by stellar data and by its mean radial velocity of 21 km s^{-1}, the distance to Kh 713 is about 1 kpc. Its neutral hydrogen mass is then $\gtrsim 3400\,M_\odot$; the kinetic energy associated with its velocity gradient is $\sim 2 \times 10^{47}$ erg; and its age is $\sim 6 \times 10^6$ y.

131.024 The H_2CO absorption against the Carina nebula.
F. F. Gardner, H. R. Dickel, J. B. Whiteoak.
Astron. Astrophys., Vol. 23, 51 - 54 (1973).

An investigation of the 6 cm H_2CO absorption against the Carina nebula has been made with the Parkes 64-m telescope (4.'2 beam). The optical depths are generally low, but are greatest in two areas where the optical obscuration is high. The formaldehyde results do not contribute significantly to the understanding of the dynamics of the nebula, as revealed by earlier H 109 α observations.

131.025 **On the detection of H$_2$ from interstellar clouds in the wavelength range 4.4 μ to 28.2 μ.** E. Bussoletti. Astron. Astrophys., Vol. 23, 125 - 129 (1973).

The expected fluxes from the regions SgrA, SgrB2, Orion, (the HCN molecular cloud and the Kleinmann and Low nebula) ξ Per and IC 1499, are calculated utilizing radio and UV data to estimate the temperature and H$_2$ densities. The results show that some fluxes are higher than the detection threshold of the present IR detectors (normal N.E.P. $\simeq 4 \times 10^{-14}$ W Hz$^{-1/2}$). It follows that the IR wavelength range 4.4–28.2 μ is a good candidate for future search of molecular hydrogen.

131.026 **Internal motions analysis of the H II region NGC 7635.** A. Maucherat, A. Vuillemin. Astron. Astrophys., Vol. 23, 147 - 150 (1973). In French.

We give radial velocities of the H II regions NGC 7635 and Sharpless 162 measured from two interferograms. There is no significant discontinuity in radial velocities between NGC 7635 and S 162 which seem to belong to the same H II region. We try to explain the form of NGC 7635.

131.027 **Excitation of interstellar OH by the collisional dissociation of water.** W. D. Gwinn, B. E. Turner W. M. Goss, G. L. Blackman. Astrophys. Journ., Vol. 179, 789 - 813 (1973).

Mechanisms are proposed for the collisional pumping of type I OH emission sources (which radiate principally in the 1665 and 1667 MHz lines). The required physical conditions for these processes are shown to exist in protostars, although they may also exist in shock fronts.

131.028 **Time-dependent models for the formation of interstellar clouds.** V. N. Mansfield. Astrophys. Journ., Vol. 179, 815 - 822 (1973).

The time-dependent model for interstellar cloud formation of Schwarz, McCray, and Stein is extended to a three-dimensional treatment, the effect of an interstellar magnetic field is studied, and ultraviolet photoelectric heating from grains is included.

131.029 **Interstellar deuterium: The hyperfine structure of DCN.** R. W. Wilson, A. A. Penzias, K. B. Jefferts, P. M. Solomon. Astrophys. Journ., (Letters), Vol. 179, L107 - L110 (1973).

The hyperfine structure of emission from deuterated hydrocyanic acid, DCN, in the $J = 1 - 0$ line at 72.4 GHz has been observed in the Orion cloud. The observed intensities of the three components are in agreement with theoretical line strengths. A comparison with observations of H^{13}C^{14}N and H^{12}C^{15}N leads to the extraordinary result D/H $= 6 \times 10^{-3}$ in hydrocyanic acid located in the Orion nebula molecular cloud; this cannot be accounted for by any known process of nucleosynthesis. It is suggested that enrichment of the deuterated molecule takes place through chemical fractionation.

131.030 **Cosmic-ray heating and molecular cooling of dense clouds.** A. D. Glassgold, W. D. Langer. Astrophys. Journ., (Letters), Vol. 179, L147 - L151 (1973).

We estimate the heating by low-energy cosmic rays of molecular clouds whose main constituent is H$_2$ to be 19 eV per primary ionization. We determine the steady-state temperature as a function of density where H$_2$ and CO are responsible for cooling for a range of ionization rates and CO abundances.

131.031 **Do interstellar gas clouds exist between spiral arms?** W. J. Quirk.

Astrophys. Journ., Vol. 180, 25 - 30 (1973).

This paper discusses what can happen to clouds as they move from arm regions to interarm regions, through a density-wave shock, and back to arm regions again. We point out that cloud-cloud collisions and other processes may lead to the destruction of interstellar clouds in less time than the 10^8 years it takes clouds to travel across arms.

131.032 **Molecular clouds in W49 and W51.** N. Z. Scoville, P. M. Solomon. Astrophys. Journ., Vol. 180, 31 - 53 (1973).

Radio observations of six molecular lines have been obtained in the W49 and W51 H II region sources as part of an investigation of the physical conditions in molecular clouds and the relationship of these clouds to the H II regions. The principal observations are maps with 4' spacing of the 6-cm formaldehyde (H$_2$CO) absorption and strip maps with 1' spacing of carbon monoxide (CO, $J = 1 \rightarrow 0$) emission at 2.6 mm. A few selected positions were also observed in ^{13}CO and C^{18}O as well as the carbon monosulfide lines (CS, $3 \rightarrow 2$ and $2 \rightarrow 1$) at 2 and 3 mm. Seven distinct clouds are found, and five of these are associated with or near H II regions.

131.033 **Observations of circumstellar circular polarization in four more infrared stars.** J. R. P. Angel, P. G. Martin. Astrophys. Journ., (Letters), Vol. 180, L39 - L41 (1973).

Circular polarization at 0.84 μ has been discovered for four stars with characteristics similar to VY CMa and NML Cyg. This polarization is attributed to multiple scattering in an asymmetric circumstellar dust cloud. Some common properties of the class of stars expected to show circumstellar circular polarization are discussed.

131.034 **Rotational cooling by carbon monoxide in dark clouds.** P. S. Berger, M. Simon. Astrophys. Journ., (Letters), Vol. 180, L43 - L46 (1973).

The role of cooling by the rotational transitions of CO in the dark clouds is discussed. It is shown that the strength of the $J = 1 - 0$ transition of ^{12}C^{16}O and ^{13}C^{16}O observed by Penzias et al. in the ρ Oph cloud is sufficient to permit the cloud to collapse, with the ^{12}C^{16}O line principally cooling the surface of the cloud and the ^{13}C^{16}O line cooling the interior.

131.035 **Kosmischer Staub in Astronomie und Weltraumforschung.** R.-H. Giese. Umschau, 73. Jahrgang, p. 37 - 43 (1973).

The methods to investigate interplanetary dust and cosmic dust outside the solar system have been considerably improved by use of space-borne equipment. Some of the techniques are described and the problems of theoretical interpretations are summarized.

131.036 **Observations of the H II region S101 in Cygnus.** T. Velusamy, M. R. Kundu. Astron. Journ., Vol. 78, 31 - 32, 153 (1973).

Observation of S101 at 11- and 6-cm wavelengths with beams of 5 and 6 arcsec respectively, are presented. The physical parameters of the H II region have been deduced from the radio data. The excitation parameter of the central exciting star of spectral type O7 is 47 pc cm^{-2}.

131.037 **New OH sources discovered at 1612 MHz.** A. Winnberg, W. M. Goss, B. Höglund, L. E. B. Johansson. Astrophys. Letters, Vol. 13, 125 - 131 (1973).

More than 30 new sources of OH emission have been found at 1612 MHz with the 25-m antenna of the Onsala Space Observatory. Most of these sources are type II OH/IR emitters. Many high-velocity emission components are seen.

131.038 **The radio continuum emission of the OH/IR−source ON−4.** J. W. M. Baars, H. J. Wendker.
Astrophys. Letters, Vol. 13, 139 - 141 (1973).

The OH/IR−source ON−4 has been observed in the continuum at 1415 MHz with the Synthesis Radio Telescope at Westerbork. The upper limit to the flux density is 1.5×10^{-29} W m^{-2} Hz^{-1}. A weak radio source of 3.5×10^{-29} W m^{-2} Hz^{-1} is found 1.5 arc min north of ON−4. It coincides with the 10th magnitude star HD 195214.

131.039 **A new correlation among high-velocity clouds in the Galaxy.** J. Silk, R. S. Siluk.
Astrophys. Letters, Vol. 13, 143 - 145 (1973).

An analysis of data for 85 high-velocity clouds with $v_r \leq -39$km/sec in the region $102° \leq l \leq 130°$ and $0 \leq b \leq 25°$ indicates that there is a correlation between the maximum hydrogen column density of each feature and the corresponding radial velocity v_r.

131.040 **Infrared emission from the OH/H_2O sources in W49.** E. E. Becklin, G. Neugebauer, C. G. Wynn-Williams.
Astrophys. Letters, Vol. 13, 147 - 149 (1973).

Infrared sources have been found coincident with both sources of maser emission in W49. The ratios of their 1.35-cm H_2O line to 20-μm continuum emission, however, differ by a factor of 10^3. If the OH and H_2O masers are the result of infrared pumping, there must be some absorption of the infrared energy between the masering region and the sun.

131.041 **The temperature of the turbulent intercloud medium.** P. L. Baker, V. Diadiuk.
Astrophys. Letters, Vol. 13, 199 - 203 (1973).

The equation of state has been computed for interstellar hydrogen heated by cosmic rays and subject to density and temperature inhomogeneity due to turbulence. The results of this computation show that the observed intercloud components of galactic hydrogen-line profiles arise from a thermally stable, high-temperature phase of the interstellar medium.

131.042 **The evolution of interstellar clouds. II. Hydrodynamic treatment of the phase change.** P. Mészáros.
Astrophys. Journ., Vol. 180, 381 - 396 (1973).

We investigate the fate of isolated H I clouds, using the two-phase model of the interstellar medium and a one-dimensional hydrodynamic code, under the assumption that a significant depletion of cooling elements on grains goes on with time.

131.043 **The evolution of interstellar clouds. III. Cloud collisions and statistical theory.** P. Mészáros.
Astrophys. Journ., Vol. 180, 397 - 419 (1973).

The assumption that interstellar clouds become progressively poor in their cooling elements by adsorption on grain surfaces leads to the conclusion that the collision of clouds among themselves must result in their evaporation (change of phase), on a timescale short compared to the time between encounters. This conclusion is applied to the development of a theoretical picture of the cloud population statistics of the Galaxy.

131.044 **Accretion onto black holes: The emergent radiation spectrum.** S. L. Shapiro.
Astrophys. Journ., Vol. 180, 531 - 546 (1973).

The luminosity and frequency spectrum of radiation resulting from interstellar gas accreting onto a black hole are calculated. The model is that of spherically symmetric, steady-state accretion onto a nonrotating black hole at rest in the interstellar medium. A fully relativistic treatment of both the fluid mechanics and radiation processes has been used.

131.045 **Analytic approximation for the saturation behavior** of OH emission regions. R. Lang, P. L. Bender.
Astrophys. Journ., Vol. 180, 647 - 660 (1973).

One of the simplest models for a cosmic maser consists of a homogeneous spherical region in which inverted population is created at a uniform rate. We have investigated the radiation transfer problem including saturation for this case under the following assumptions: (1) two levels only, without magnetic sublevels; (2) a square line shape; (3) excitation by spontaneous emission at a uniform rate through the region, and (4) no scattering in the medium or reflection at the boundary.

131.046 **Interstellar Na I, K I, Ca II, and CH^+ line profiles toward Zeta Ophiuchi.** L. M. Hobbs.
Astrophys. Journ., (*Letters*), Vol. 180, L79 - L82 (1973).

Interferometric, photoelectric scans of interstellar absorption-line profiles of Na I, K I, Ca II, and CH^+ toward ζ Oph are compared.

131.047 **Interstellar deuterium: Chemical fractionation.** P. M. Solomon, N. J. Woolf.
Astrophys. Journ., (*Letters*), Vol. 180, L89 - L92 (1973).

If molecules form or perform exchange reactions under low-temperature conditions, the deuterated forms of some molecules will be relatively overabundant. Observations of DCN and HCN in the Orion cloud are translated into atomic abundances on this hypothesis, and we find that [D]/[H] is between $1:1 \times 10^5$ and $1:8 \times 10^6$.

131.048 **The extinction curve for Cygnus OB2 No. 12.** R. Chaldu, R. K. Honeycutt, M. V. Penston.
Publ. Astron. Soc. Pacific, Vol. 85, 87 - 90 = Publ. Goethe Link Obs., Indiana Univ., *Bloomington*, No. 147 (1973).

We present an extinction curve for the star Cygnus OB2 No. 12, derived from scans of this star and less highly reddened stars of similar spectral types. Possible new diffuse interstellar bands are found.

131.049 **The problem of X-ogen.** D. Buhl, L. E. Snyder.
Astrophys. Journ., Vol. 180, 791 - 800 (1973).

Four new galactic sources exhibiting emission from the unidentified line X-ogen are reported. A rest frequency of 89.189 GHz is derived from the new sources. Several possible molecules with calculated transition frequencies near the X-ogen line are considered; however, identifications of HNC, HCO$^+$, or CCH with X-ogen are difficult. The molecular-hydrogen density necessary to excite the various molecules seen in Orion is derived and related to the angular size of the region observed.

131.050 **New H_2O sources associated with infrared stars.** D. F. Dickinson, K. P. Bechis, A. H. Barrett.
Astrophys. Journ., Vol. 180, 831 - 844 (1973).

We searched for 1.35-cm water-vapor emission from all 138 infrared stars in the Caltech 2-μ survey with a recorded K-magnitude less than zero, and found four new sources. In addition, we searched for microwave H_2O emission in 50 other objects, including all known OH emission sources associated with infrared stars observable from our latitude, several other types of OH emission sources, various galactic objects, and a number of Mira variable stars. Four other new H_2O sources were detected.

131.051 **An anomalous emission line in low-frequency radio spectra toward W49A.** V. Pankonin, A. Parrish, Y. Terzian.
Astrophys. Journ., (*Letters*), Vol. 180, L113 - L116 (1973).

An unidentified emission feature has been detected in the 247, 248α spectra toward W49A. We suggest that it is due to recombination transitions of atomic carbon located in a spiral arm in the line of sight to W49A.

131.052   **Interferometric observations of the $^2\Pi_{3/2}$, $J = 5/2$ state of interstellar OH.**
S. H. Knowles, K. J. Johnston, J. M. Moran, J. A. Ball.
Astrophys. Journ., (*Letters*), Vol. 180, L117 - L121 (1973).

The $^2\Pi_{3/2}$, $J=5/2$, $F=3\rightarrow3$ hyperfine transition of OH at 5 cm wavelength was detected interferometrically with a fringe spacing of 0."05 in W3 OH and NGC 6334N. Both sources were found to consist of two complexes of features separated by about 1".

131.053   **Circular polarization in the B stars X Persei, 5 Persei, 55 Cygni, and 30 Pegasi.**
R. W. Avery, J. J. Michalsky, Jr., R. A. Stokes.
Astrophys. Journ., (*Letters*), Vol. 180, L127 - L128 (1973).

A wavelength-dependent circular polarization component is found in the visible emissions from the suspected supernova remnant X Persei. The circular polarization of 55 Cygni in the *B*-filter is confirmed. The probable circular polarization of 30 Pegasi and 5 Persei in the *B*-filter is reported.

131.054   **Interferometric studies of interstellar CH$^+$ molecules.**   L. M. Hobbs.
Astrophys. Journ., Vol. 181, 79 - 93 (1973).

Interferometric, photoelectric scans of the interstellar λ 4232 line of CH$^+$ have been obtained for 28 stars at a resolving power of 300,000. The limiting detectable equivalent widths are near 1 mÅ for many stars, and are near 0.5 mÅ for a few selected stars, so that a number of new, weak lines are detected. The observations are compared with theoretical predictions concerning interstellar CH$^+$ molecules.

131.055   **The formation of interstellar molecules from negative ions.**   A. Dalgarno, R. A. McCray.
Astrophys. Journ., Vol. 181, 95 - 100 (1973).

The possible contribution of chemical reactions involving negative ions to the formation of interstellar molecules is briefly explored.

131.056   **Electrostatic potential of interstellar grains.**
B. Feuerbacher, R. F. Willis, B. Fitton.
Astrophys. Journ., Vol. 181, 101 - 113 (1973).

The equilibrium electrostatic potential of interstellar grains, subjected to an ultraviolet radiation field in a dilute ambient plasma, is calculated for a variety of radiation fields and plasma parameters, and for two different grain materials. The results of the present calculation suggest that there might be regions where the various components (graphite, silicates) are charged to potentials of opposite sign.

131.057   **Ultraviolet stars and the interstellar gas.**
W. K. Rose, D. G. Wentzel.
Astrophys. Journ., Vol. 181, 115 - 123 (1973).

If luminous red giants evolve to white dwarfs via a very hot stage (ultraviolet star), then these stars may have a significant influence on the interstellar medium. We extend the analysis of Hills by considering the evolution of ultraviolet stars and how these stars affect the dynamics of the resulting H II regions.

131.058   **Interstellar dust and distances to planetary nebulae.**   J. H. Lutz.
Astrophys. Journ., Vol. 181, 135 - 145 (1973).

UBV and Hγ data are used to construct diagrams of color excess versus distance for early-type stars within 1°5 of six planetary nebulae. These diagrams are used in conjunction with the observed extinctions of the planetaries to derive distances for NGC 6741, NGC 6894, NGC 7026, NGC 7354, IC 1747, and IC 289.

131.059   **On the observability of far infrared line emission originating from the interstellar medium.**
S. R. Pottasch.
Astron. Astrophys., Vol. 24, 305 - 307 (1973).

Recent measurements of interstellar absorption lines in the ultraviolet, originating from energy levels above the ground level, can be used to directly compute the expected infrared line radiation. This is done for the λ 156 μ line of C$^+$ in the directions of ζ Oph and δ Sco, and lines of C°, O° and Si$^+$. The computed line strengths are considerable and may be strong enough to be detectable with present techniques.

131.060   **Results of Fabry-Perot Hα spectrometer observations of galactic H II regions.**   V. F. Zhidkov.
Astron. Zhurn. Akad. Nauk SSSR, Vol. 50, 287 - 296 (1973).
In Russian. English translation in Soviet Astron. AJ, Vol. 17, No. 2.

131.061   **Two-component equilibrium model of H I regions of the interstellar medium satisfying the complex of radio observations.**   G. K. Beysekova, N. G. Bochkarev.
Astron. Zhurn. Akad. Nauk SSSR, Vol. 50, 424 - 426 (1973).
In Russian. English translation in Soviet Astron. AJ, Vol. 17, No. 2.

A two-component equilibrium model of H I regions of the interstellar medium has been constructed. The model corresponds to the hydrogen concentration in clouds $n_c \approx 6$ cm^{-3} and to a cloud temperature of $T_c \approx 70-80°$K.

131.062   **Interstellar molecules.**   B. E. Turner.
Sci. American, Vol. 228, No. 3, p. 50 - 62, 67 - 69 (1973).

Twenty-six kinds of molecule have now been discovered in the gas between the stars of our Galaxy. Among them are carbon monoxide, water, ammonia, hydrogen sulfide, formaldehyde and methyl alcohol.

131.063   **Spectrophotometric results from the Copernicus satellite. II. Composition of interstellar clouds.**
D. C. Morton, E. B. Jenkins, J. F. Drake, J. B. Rogerson, L. Spitzer, D. G. York.
Bull. American Astron. Soc., Vol. 5, 4 (1973). – Abstr. AAS.

131.064   **Spectrophotometric results from the Copernicus satellite. III. Ionization and density of the intercloud medium.**   D. G. York, J. B. Rogerson, J. F. Drake, E. B. Jenkins, D. C. Morton, L. Spitzer.
Bull. American Astron. Soc., Vol. 5, 4 (1973). – Abstr. AAS.

131.065   **Spectrophotometric results from the Copernicus satellite. IV. Molecular hydrogen in interstellar space.**   L. Spitzer, J. F. Drake, E. B. Jenkins, D. C. Morton, J. B. Rogerson, D. G. York.
Bull. American Astron. Soc., Vol. 5, 4 (1973). – Abstr. AAS.

131.066   **Spectrophotometric results from the Copernicus satellite. V. Abundances of molecules in interstellar clouds.**   E. B. Jenkins, J. F. Drake, D. C. Morton, J. B. Rogerson, L. Spitzer, D. G. York.
Bull. American Astron. Soc., Vol. 5, 4 - 5 (1973). – Abstr. AAS.

131.067   **Spectrophotometric results from the Copernicus satellite. VI. Extinction by grains at wavelengths between 1200 and 1000 A.**   D. G. York, J. F. Drake, E. B. Jenkins, D. C. Morton, J. B. Rogerson, L. Spitzer.
Bull. American Astron. Soc., Vol. 5, 5 (1973). – Abstr. AAS.

131.068   **A map of galactic absorption.**   J. H. Cahn.
Bull. American Astron. Soc., Vol. 5, 9 (1973).
Abstr. AAS.

131.069   **Intrinsic polarization of ζ Tauri.**
R. W. Capps, G. V. Coyne, H. M. Dyck.

Bull. American Astron. Soc., Vol. 5, 11 (1973). – Abstr. AAS.

131.070 **Time dependent radiative cooling of a hot low-density cosmic gas.** M. C. Kafatos.
Bull. American Astron. Soc., Vol. 5, 21 (1973). – Abstr. AAS.

131.071 **Rotational cooling by carbon monoxide in dark clouds.** P. S. Berger, M. Simon.
Bull. American Astron. Soc., Vol. 5, 21 (1973). – Abstr. AAS.

131.072 **The distribution of galactic carbon monoxide.** P. R. Schwartz, W. J. Wilson, E. E. Epstein.
Bull. American Astron. Soc., Vol. 5, 21 - 22 (1973). – Abstr. AAS.

131.073 **Observations of galactic carbon monoxide.** W. J. Wilson, P. R. Schwartz, E. E. Epstein.
Bull. American Astron. Soc., Vol. 5, 22 (1973). – Abstr. AAS.

131.074 **A correlation study of carbon emission lines and hydroxyl absorption lines toward galactic nebulae.** E. J. Chaisson.
Bull. American Astron. Soc., Vol. 5, 22 (1973). – Abstr. AAS.

131.075 **Carbon recombination line clouds.** A. K. Dupree.
Bull. American Astron. Soc., Vol. 5, 22 - 23 (1973). – Abstr. AAS.

131.076 **Masering of water molecules in a self-exciting proto-stellar gas cloud.** T. de Jong.
Bull. American Astron. Soc., Vol. 5, 23 (1973). – Abstr. AAS.

131.077 **A search for radio frequency transitions from meta-stable states of H_2 and CO.** R. H. Gammon, R. L. Brown, M. A. Gordon.
Bull. American Astron. Soc., Vol. 5, 23 (1973). – Abstr. AAS.

131.078 **Faint optical emission lines from the interstellar medium.** R. J. Reynolds, F. Scherb, F. L. Roesler.
Bull. American Astron. Soc., Vol. 5, 24 (1973). – Abstr. AAS.

131.079 **100 micron observations of the galactic H II region W-3.** P. A. Aannestad, R. J. Emery, W. F. Hoffmann.
Bull. American Astron. Soc., Vol. 5, 31 (1973). – Abstr. AAS.

131.080 **On the excitation of some interstellar molecules.** B. E. Turner, M. Morris, P. Palmer, B. Zuckerman.
Bull. American Astron. Soc., Vol. 5, 31 - 32 (1973). – Abstr. AAS.

131.081 **The nature of molecular clouds.** D. Buhl, L. E. Snyder.
Bull. American Astron. Soc., Vol. 5, 32 (1973). – Abstr. AAS.

131.082 **Interstellar molecular hydrogen detected in the UV spectrum of δ Sco.** A. M. Smith.
Bull. American Astron. Soc., Vol. 5, 32 (1973). – Abstr. AAS.

131.083 **Extinction by ice-coated silicate grains.** W. Fullerton, W. F. Huebner.
Bull. American Astron. Soc., Vol. 5, 34 (1973). – Abstr. AAS.

131.084 **Extinction and polarization in dark clouds.** L. Carrasco, K. M. Strom, S. E. Strom.
Bull. American Astron. Soc., Vol. 5, 34 (1973). – Abstr. AAS.

131.085 **Simulation of interstellar polarization.** F. M. Johnson.

Bull. American Astron. Soc., Vol. 5, 34 - 35 (1973). – Abstr. AAS.

131.086 **Ultraviolet absorption lines in interstellar clouds.** B.-Z. Kozlovsky, M. J. Rees, G. Steigman.
Bull. American Astron. Soc., Vol. 5, 38 (1973). – Abstr. AAS.

131.087 **The opactiy of the interstellar medium to extreme ultraviolet radiation.** R. Cruddace, F. Paresce, C. S. Bowyer, M. Lampton.
Bull. American Astron. Soc., Vol. 5, 38 (1973). – Abstr. AAS.

131.088 **Interaction of the interstellar medium with the solar wind.** W. I. Axford.
Space Sci. Rev., Vol. 14, 582 - 590 (1973).
The possibilities for making observations of the region of interaction between the solar wind and the interstellar medium, and of the interstellar medium itself, are discussed. It is emphasized that missions to the outer planets undertaken in 1977 - 1979 are ideally suited for making such observations.

131.089 **Recent experimental and theoretical investigations on infrared and microwave molecular spectra of astronomical interest. Introductory report.** T. Oka.
17th Colloquium International Astrophys. Liège 1971, (see 012.003), p. 37 - 55 (1972).

131.090 **Condensation nuclei for interstellar grains and the IR emission of late type stars.** T. de Jong, F. Kamijo.
17th Colloquium International Astrophys. Liège 1971, (see 012.003), p. 191 - 196 (1972).

131.091 **The interpretation of continuum and line absorption and radiation by circumstellar dust.** J. M. Greenberg, R. T. Wang.
17th Colloquium International Astrophys. Liège 1971, (see 012.003), p. 197 - 207 (1972).

131.092 **Recent polarization observations in the 1.25 to 3.6 micron wavelength range.** F. F. Forbes.
17th Colloquium International Astrophys. Liège 1971, (see 012.003), p. 217 - 228 (1972).

131.093 **Microwave recombination spectra of molecular ions.** F. Joly, R. McCarroll.
17th Colloquium International Astrophys. Liège 1971, (see 012.003), p. 341 - 344 (1972).

131.094 **The broadening of radio recombination lines emitted by H II regions.** D. Hoang-Binh.
17th Colloquium International Astrophys. Liège 1971, (see 012.003), p. 367 - 370 (1972).

131.095 **Observational, theoretical, and predicted data on the infrared and microwave spectra of interstellar matter. Introductory report.** D. M. Rank.
17th Colloquium International Astrophys. Liège 1971, (see 012.003), p. 377 - 387 (1972).

131.096 **Spectroscopy of tetrabenzporphin molecules and possible astrophysical implications.** F. M. Johnson.
17th Colloquium International Astrophys. Liège 1971, (see 012.003), p. 391 - 407 (1972).

131.097 **Characteristics of the diffuse (tenuous) interstellar medium determined from radio recombination lines.** M. A. Gordon, S. T. Gottesman.
17th Colloquium International Astrophys. Liège 1971, (see 012.003), p. 409 - 416 (1972).

131.098 **Polarizations of the infrared stars and the galactic centre in near infrared.**
T. Maihara, H. Okuda, S. Sato.
17th Colloquium International Astrophys. Liège 1971, (see 012.003), p. 417 - 424 (1972).

131.099 **Ultraviolet effects on the chemical composition and optical properties of interstellar grains.**
J. M. Greenberg, A. J. Yencha, J. W. Corbett, H. L. Frisch.
17th Colloquium International Astrophys. Liège 1971, (see 012.003), p. 425 - 436 (1972).

131.100 **The predicted shape of the infrared extinction curve of silicate grains.** K. Nandy, A. Kelly.
17th Colloquium International Astrophys. Liège 1971, (see 012.003), p. 437 - 442 (1972).

131.101 **Observations of the 6.2 cm formaldehyde line.**
T. L. Wilson.
17th Colloquium International Astrophys. Liège 1971, (see 012.003), p. 447 - 452 (1972).

131.102 **Interferometric observations of formaldehyde absorption in front of strong galactic sources.**
E. B. Fomalont, L. Weliachew.
17th Colloquium International Astrophys. Liège 1971, (see 012.003), p. 453 - 464 (1972).

131.103 **A search for interstellar para-formaldehyde by its 73 GHz transition.** M. Morimoto, N. Kaifu.
17th Colloquium International Astrophys. Liège 1971, (see 012.003), p. 467 (1972).

131.104 **Microwave detection of interstellar formamide.**
R. H. Rubin, R. C. Benson, H. L. Tigelaar, W. H. Flygare.
17th Colloquium International Astrophys. Liège 1971, (see 012.003), p. 471 - 474 (1972).

131.105 **New OH emission sources.**
B. J. Robinson, J. L. Caswell, W. M. Goss.
17th Colloquium International Astrophys. Liège 1971, (see 012.003), p. 475 - 479 (1972).

131.106 **mm-wave lines of organic molecules.**
D. Buhl, L. E. Snyder.
17th Colloquium International Astrophys. Liège 1971, (see 012.003), p. 481 - 486 (1972).

131.107 **Molecules in interstellar clouds.** R. D. Davies.
17th Colloquium International Astrophys. Liège 1971, (see 012.003), p. 489 - 496 (1972).

131.108 **OH excited state emissions and their temporal variations in W3(OH), and W75B.**
O. E. H. Rydbeck, J. Elldér, E. Kollberg, B. Höglund.
17th Colloquium International Astrophys. Liège 1971 (see 012.003), p. 507 - 527 (1972).

131.109 **Interstellar absorption and the bound exciton spectra of SiC.** S. Nikitine.
17th Colloquium International Astrophys. Liège 1971, (see 012.003), p. 529 - 540 (1972).

131.110 **Molecule formation. I. In normal H I clouds.**
P. A. Aannestad.
Astrophys. Journ., Suppl. Ser., No. 217 (I), Vol. 25, 205 - 222 (1973).
We study the formation of simple molecules employing both gas–phase reactions and catalytic surface reactions on interstellar grains. We show that it is likely that interstellar

grains will grow mantles, and we follow the variation in molecular abundances and mantle composition as the cooling elements deplete onto the grains, causing a cloud to heat up and expand.

131.111 **Molecule formation. II. In interstellar shock waves.**
P. A. Aannestad.
Astrophys. Journ., Suppl. Ser., No. 217 (II), Vol. 25, 223 - 252 (1973).
We study the processes of grain destruction and molecule formation due to the sputtering of grain mantles as the grains move relative to the gas in the hot shock wave. It is found that mantles are destroyed completely by this process if the abundance of molecular hydrogen is about 1 percent or more; the corresponding lifetime of a mantle is about 5×10^7 years. We follow the variation of the molecular species through the shock wave using the sputtering process as a source for H_2O, NH_3, and CH_4.

131.112 **Non-equilibrium processes in interstellar molecules.**
M. M. Litvak.
Atoms and molecules in astrophysics. Proc. Scottish Univ. Summer School in Physics 1971, (see 012.005), p. 201 - 276 (1972).

131.113 **Radio recombination lines: An observer's point of view.** E. Churchwell.
Atoms and molecules in astrophysics. Proc. Scottish Univ. Summer School in Physics 1971, (see 012.005), p. 277 - 283 (1972).

131.114 **The formation of H_2 molecules in dark interstellar clouds.** T. de Jong.
Atoms and molecules in astrophysics. Proc. Scottish Univ. Summer School in Physics 1971, (see 012.005), p. 327 - 335 (1972).

131.115 **Emission-line spectra as probes of dust clouds.**
B. E. J. Pagel.
Atoms and molecules in astrophysics. Proc. Scottish Univ. Summer School in Physics 1971, (see 012.005), p. 341 - 345 (1972).

131.116 **Evolution models of the interstellar gas.**
H. Gerola, E. Iglesias, Z. Gamba.
Astron. Astrophys., Vol. 24, 369 - 378 (1973).
A preliminary exploration of the hypothesis that all of the excitation state of the interstellar gas is due to supernova explosions is presented.

131.117 **Detection of interstellar thioformaldehyde.**
M. W. Sinclair, N. Fourikis, J. C. Ribes, B. J. Robinson, R. D. Brown, P. D. Godfrey.
Australian Journ. Phys., Vol. 26, 85 - 91 (1973).
The $2_{11} \leftarrow 2_{12}$ transition of thioformaldehyde (HCHS) has been observed in absorption in the direction of Sagittarius B2. Comparison with the $2_{11} \leftarrow 2_{12}$ absorption of formaldehyde (HCHO) at 15 GHz allows the relative abundance of the two molecular species to be computed as a function of the rotational excitation temperature.

131.118 **Zur Kosmogonie der interstellaren Materie.**
H. Lambrecht.
Veröff. Forschungsbereich Kosm. Phys., Akad. Wiss. DDR, No. 1, p. 27 - 59 (1973).

131.119 **Interaction between the interstellar medium and solar wind plasma.** M. K. Wallis.
Astrophys. Space Sci., Vol. 20, 3 - 18 (1973).
The interaction processes governing the penetration of the interstellar gas into the solar neighbourhood are re-examin-

ed. The termination of the solar wind is rediscussed in the light of both electron heating and the stronger gas/plasma interaction.

131.120 A reappraisal of several extensive phenomena of the high galactic latitudes.
K. H. Elliott, J. Meaburn.
Astrophys. Space Sci., Vol. 20, 111 - 122 (1973).

This is the first of two papers dealing with the extensive high latitude phenomena. The relationships between the diffuse and filamentary gaseous nebulosity, the giant radio loops and the high, medium and low velocity neutral hydrogen are considered in the light of recent observations. The existence of other giant loops is also suggested.

131.121 Correlación entre velocidades radiales de líneas interestelares ópticas y de radio del hemisferio sur.
D. Goniadzki.
Bol. As. Argentina Astron., No. 16, (see 012.007), p. 12 (1971).

131.122 Análisis estadístico de 123 concentraciones de hidrógeno neutro. S. Garzoli, C. Jaschek.
Bol. As. Argentina Astron., No. 16, (see 012.007), p. 13 - 14 (1971).

131.123 La función estructural del material interestelar.
A. Feinstein, H. Marraco.
Bol. As. Argentina Astron., No. 16, (see 012.007), p. 16 - 17 (1971).

131.124 Fabry-Perot interferometric studies on H II regions.
H. H. Hippelein.
Astron. Astrophys., Vol. 25, 59 - 62 (1973).

The radial velocities of several H II regions in Cygnus have been measured with a Fabry-Perot interferometer. The nebula IC 1318 has been studied in detail.

131.125 Thermal and ionization equilibrium in a dense hydrogen cloud. M. Walmsley.
Astron. Astrophys., Vol. 25, 129 - 135 (1973).

The equations of ionization and thermal equilibrium are solved as a function of radius for a dense ($n_H = 1000, 5000$ cm^{-3}) spherical homogeneous interstellar gas cloud. It is found that silicon is the most abundant ionized component in regions where carbon is neutral. Temperatures are in the range $5-15°$ K. The relationship of the models to carbon recombination line observations is discussed.

131.126 Interstellar extinction in the Southern Milky Way.
D. C. B. Whittet, I. G. van Breda, K. Nandy.
Nature, Phys. Sci., Vol. 243, 21 - 23 (1973).

Interstellar extinction curves have been derived from photoelectric scanner observations of twenty-two reddened early-type stars in a region of the Southern Milky Way, including the galactic centre. It has been found that considerable variations exist and that these are not directly related to position in the galactic plane. This indicates that the extinction law for these stars is influenced significantly by the physical conditions prevailing in local dust clouds.

131.127 The temperatures of some interstellar clouds.
M. Milgrom, N. Panagia, E. E. Salpeter.
Astrophys. Letters, Vol. 14, 73 - 75 (1973).

We suggest that dissociation of H_2 molecules by UV photons may be effective in heating the gas in clouds of high-enough density. We examine the case of a cloud of constant density $n = 250$ cm^{-3}.

131.128 On the polarization of the interstellar OH-emission.
L. M. Hall, D. ter Haar.

Monthly Notices Roy. Astron. Soc., Vol. 162, 97 - 108 (1973).

We use density matrix techniques to discuss the behaviour of a saturated OH-maser in which only the four lines corresponding to the ground state $^2\pi_{3/2}$, $J = 3/2$ quartet are taken into account. We write down the equation of motion for the molecular density matrix and discuss its solution. We show that in a single-line model – which should be applicable to the class I OH-sources which show strong circularly polarized 1665 MHz emission – saturated maser effects could explain the observed polarization.

131.129 The interstellar radiation density between 1250 and 4250 Å. A. N. Witt, M. W. Johnson.
Astrophys. Journ., Vol. 181, 363 - 368 (1973).

Using a model distribution of stars and dust, we have calculated the interstellar radiation density in the solar vicinity. We find excellent agreement between our calculations and empirical data determined by Lillie, but we predict a radiation density higher by about a factor of two compared to earlier calculations by Habing for the wavelength range from 1000 to 2000 Å. We also find a very steep rise in the spectrum of the radiation field toward shorter wavelengths in the interval 2000–1500 Å.

131.130 Extinction properties of dust in the ultraviolet and the ionization of helium in galactic H II regions.
E. M. Leibowitz.
Astrophys. Journ., Vol. 181, 369 - 377 (1973).

Some observational evidence is found in NGC 2024 to suggest that the presence of dust grains in an H II region affects the He$^+$/H$^+$ abundance ratio in the region. The evidence is presented by an anticorrelation between the Hα/Hβ line ratio, as measured in different points in the nebula, and the He I λ 5876/Hβ line ratio in these points. No such relationship is found in NGC 6334. The observations of weak radio helium recombination lines in H II regions may thus be explained on the basis of selective absorption of helium ionizing photons by dust grains in those nebulae.

131.131 A search for Hα emission from interstellar clouds.
R. J. Reynolds, F. Scherb, F. L. Roesler.
Astrophys. Journ., (Letters), Vol. 181, L79 - L82 (1973).

Upper limits have been determined for the emission measures of five interstellar 'clouds'. These limits are significantly smaller than the emission measures predicted by a recent model in which the variation in the column-density ratio Na^0/Ca + among interstellar clouds is attributed to differences in the temperature, density, and hydrogen ionization fraction of the clouds.

131.132 Interstellar methylacetylene and isocyanic acid.
L. E. Snyder, D. Buhl.
Nature, Phys. Sci., Vol. 243, 45 - 46 (1973).

We have reported the radio detection of an emission line from the $J_K = 5_0 \rightarrow 4_0$ transition of interstellar methylacetylene (CH_3C_2H) at 85, 457.29 MHz in the direction of the galactic centre source Sgr B2. In April 1972, we obtained new and better methylacetylene data with enough spectral quality to allow the observation of more than one component of the $J = 5-4$ transition; in addition we detected a new transition of interstellar isocyanic acid (HNCO) as well as a weak unidentified line. We observed the CH_3C_2H line again in February 1973, to determine the frequency of the weak unidentified line and here we discuss the results of our accumulated observations.

131.133 Measured extinction efficiency of graphite smoke in the region 1200–6000 Å.
K. L. Day, D. R. Huffman.
Nature, Phys. Sci., Vol. 243, 50 - 51 (1973). – Letter.

131.134 On the gravitational collapse of interstellar magnetic clouds. P. Ingvarson.
School Electrical Engineering, Chalmers Univ. Technology, Göteborg, Sweden, Techn. Rep. No. 31, 7 + 129 pp. (1973).

It is shown that magnetic interstellar clouds are likely to assume disk-like structures. The dynamical behavior of such a disk is studied by means of a computer program. The fragmentation properties of magnetic and rotating infinite disks are derived by a linear analysis, and nonlinear effects are discussed. The results agree well with the computer calculations.

131.135 Evaporation of dirty ice particles surrounding early type stars.V. Variety of interstellar extinction curves. S. Isobe.
Publ. Astron. Soc. Japan, Vol. 25, 253 - 270 = Tokyo Astron. Obs. Repr., No. 434 (1973).

It is shown that mixtures of graphite core-ice mantle grains with large mean size, small mean size, and graphite grains provide good fits to the observed interstellar extinction curves from infrared through far-ultraviolet wavelengths, and how these grains are formed in interstellar space.

131.136 Ionization of the intercloud medium and the central disk regions of spiral galaxies. J. Silk.
Astrophys. Journ., Vol. 181, 707 - 724 (1973).

The two-phase model of the interstellar medium is reconsidered in terms of a hydrogen ionization rate that may vary significantly between cloud and intercloud medium. In view of recent observational and theoretical studies directly pertaining to the ionization of H I clouds, it is proposed that the observed free-free absorption of extragalactic radio sources and nonthermal radio continuum can be produced predominantly in the intercloud medium. Various observational consequences of this hypothesis include interpretations of diffuse Hα and Hβ emission, radio recombination lines from distributed interstellar gas, and line emission from the central disk regions of nearby Sc galaxies. All these observational data can be explained in terms of a simple model of the intercloud medium, which is consistent with recent 21-cm studies of the interstellar medium. Several ionization sources are discussed, and of these, blast-wave emission from supernova remnants and soft X-rays from ultraviolet dwarfs are found to be potentially viable candidates for the role of ionizing the intercloud medium.

131.137 Interstellar gas abundances from rocket observations of ultraviolet absorption lines. E. B. Jenkins.
Astrophys. Journ., Vol. 181, 761 - 779 (1973).

Absorption lines from the ground states of O I, C II, and Si II were observed in the ultraviolet spectra of δ, τ, λ, κ, and υ Sco, ζ Oph, and ζ Per, which were photographed on three Aerobee sounding rocket flights. While the oxygen and silicon absorptions are purely interstellar, some of the C II absorption arises from the star's atmosphere and some presumably from gas in the H II region near the star. Curves of growth for the ultraviolet lines were derived from visual absorption-line profiles, when available. When compared with H I column densities from the Lα absorptions, the amounts of O I and Si II are consistent with the cosmic abundance ratios. C II, on the other hand, appears for various stars to be from 3 to 10 times overabundant.

131.138 Interferometric observations of formaldehyde absorption in front of strong galactic sources.
E. B. Fomalont, L. Weliachew.
Astrophys. Journ., Vol. 181, 781 - 794 (1973).

Observations have been made of formaldehyde absorption at 4830 MHz in front of Sgr A, Sgr B2, W3, NGC 2024, W31, W33, W43, W49A, and W51. The main absorption features show variations of optical depth over the sources, but there is no significant clumping. The previously determined

isotopic ratio $^{12}C/^{13}C$ of 10 for Sgr A and Sgr B2 should be increased to $\sim 25 \pm 5$ for Sgr A and ≥ 20 for Sgr B2. The +40 km s^{-1} cloud associated with Sgr A appears to be rotating as a solid body. The mass of the cloud is calculated to be $\sim 3 \times 10^5\, M_\odot$.

131.139 On the interstellar CO$^+$/CO abundance ratio. L. M. Hobbs.
Astrophys. Journ., Vol. 181, 795 - 798 (1973).

A search at high sensitivity for interstellar absorption in the $\lambda 4250$ line of CO$^+$ yields negative results toward 14 stars, despite the pervasive presence of large amounts of CO in the interstellar gas. This result is in accord with theory, but still lacks by a large factor sufficient sensitivity to test any details of the theory.

131.140 Deuterium in interstellar molecules. W. D. Watson.
Astrophys. Journ., (Letters), Vol. 181, L129 - L133 (1973).

We show that the ratio (atomic D/atomic H) / (total D/total H) approaches $\sqrt{[(\text{total H/total D})]}$ in the more dense regions of clouds, and thus predict a similar enhancement for deuterium in molecules other than HD. For the usual abundance of deuterium the predicted enhancement is $\sim 10^2$, in excellent agreement with the observed (DCN/HCN) ratio.

131.141 A discussion of the distribution of interstellar matter close to the sun. E. Falgarone, J. Lequeux.
Astron. Astrophys., Vol. 25, 253 - 260 (1973).

Using the new data of Radhakrishnan et al. (1972) and of Hughes, Thompson and Colvin (1971), the distribution and kinematics of the interstellar clouds and of the hot intercloud medium are discussed separately.

131.142 A minimum kinematic distance estimate to the nonthermal source in W 51 based on OH absorption line measurements. T. L. Wilson.
Astron. Astrophys., Vol. 25, 329 - 331 (1973).

1667 MHz OH observations of the non-thermal shell source in W 51 show absorption extending to 58.6 km/s. Use of the Schmidt (1965) velocity-distance relation indicates that the kinematic distance to this source is at least 5.4 kpc.

131.143 Large polarization variations in CIT 6. A. Kruszewski.
Inform. Bull. Variable Stars (I. A. U. Commission 27), Konkoly Obs., Budapest, No. 781, 3 pp. (1973).

131.144 On the polarization variability of YY Eri. V. A. Oshchepkov.
Inform. Bull. Variable Stars (I. A. U. Commission 27), Konkoly Obs., Budapest, No. 782 (1973).

131.145 Flux density of OH471 between 1963 and 1970. J. D. Kraus, M. R. Gearhart, J. R. Ehman.
Nature, Phys. Sci., Vol. 243, 94 (1973). – Letter.

131.146 Photographs of neutral atomic hydrogen in the sky. C. Heiles, E. Jenkins.
Bull. American Astron. Soc., Vol. 5, 265 (1973). – Abstr. AAS.

131.147 On UV-star heated models of the interstellar medium. P. Meszaros.
Bull. American Astron. Soc., Vol. 5, 267 (1973). – Abstr. AAS.

131.148 Cold neutral hydrogen in a large region towards the galactic anticenter. P. L. Baker.
Bull. American Astron. Soc., Vol. 5, 283 (1973). – Abstr. AAS.

131.149 The effect of loop III on interstellar neutral hydrogen. I. Fejes, G. L. Verschuur.

Bull. American Astron. Soc., Vol. 5, 284 (1973). – Abstr. AAS.

131.150 Observations of new galactic H II regions and exciting stars.
Y. M. Georgelin, Y. P. Georgelin, S. Roux.
Astron. Astrophys., Vol. 25, 337 - 350 (1973). – In French.

We present *UBV* photometry and spectral types for 45 exciting stars together with radial velocities of 60 new H II regions obtained with Fabry-Perot rings.

131.151 Growth and destruction of interstellar grains in the presence of low-energy cosmic rays.
T. de Jong, F. Kamijo.
Astron. Astrophys., Vol. 25, 363 - 370 (1973).

Using the flux of low-energy cosmic rays postulated to account for the heating and ionization of the interstellar gas we investigate their interaction with grains.

131.152 Infrared 10 μ emission from condensation nuclei of interstellar grains. F. Kamijo, T. de Jong.
Astron. Astrophys., Vol. 25, 371 - 377 (1973).

The particles emitting the 10 μ excess radiation observed in the spectra of many late-type stars, are assumed to provide the condensation cores of interstellar core-mantle grains. The rate at which silicate particles are produced by M supergiants is estimated from the 10 μ excess of α Ori. If M giants, and possibly other late-type stars, also contribute to the total production rate, enough condensation nuclei with a radius of about 100 Å are produced to account for the presently observed interstellar grains. The 10 μ flux observed from the Trapezium nebula in Orion is attributed to radiation of hot silicate cores, left behind after evaporation of the dirty ice mantles. Some implications for the extinction in H II regions and for the dynamical evolution of H II regions are briefly discussed.

131.153 Interstellar matter: an observer's view.
P. G. Mezger.
Interstellar matter, (see 003.011), p. 1 - 205 (1972).

131.154 Interstellar dust. N. C. Wickramasinghe.
Interstellar matter, (see 003.011), p. 207 - 340 (1972).

131.155 Interstellar gas dynamics. F. D. Kahn.
Interstellar matter, (see 003.011), p. 341 - 437 (1972).

131.156 The dynamical behavior of the interstellar gas, field, and cosmic rays. E. N. Parker.
Cosmic plasma physics. Conference 1971, (see 012.016), p. 195 - 202 (1972).

131.157 Infrared pumping for maser OH sources associated with H II regions.
V. V. Burdyuzha, D. A. Varshalovich.
Astron. Zhurn. Akad. Nauk SSSR, Vol. 50, 481 - 490 (1973). In Russian. English translation in Soviet Astron. AJ, Vol. 17, No. 3.

H II regions in which powerful H II OH maser sources have been found are shown by recent observations to be at the same time strong IR sources. A pumping mechanism of OH molecules by far-IR radiation in these sources is considered.

131.158 Interstellarer Raum und interstellare Materie. Globale Eigenschaften in unserem Milchstraßensystem.
P. G. Mezger.
SuW, Vol. 12, 166 - 171 (1973).

131.159 A survey of radio recombination lines toward the galactic center. F. J. Lockman, M. A. Gordon.

Astrophys. Journ., Vol. 182, 25 - 39 (1973).

Observations at 14 positions in the direction of the galactic center region show H159α (1.621 GHz) hydrogen recombination lines over an extended area. The lines do not originate under conditions of LTE. The data is best interpreted as emission from a region with electron temperature of 20° K and an electron density of 3 cm^{-3} lying along the line of sight to the galactic center. This model is also consistent with the observations of diffuse H157α and H166α line emission from the inner regions of the Galaxy.

131.160 The absence of formaldehyde radiation toward cold regions of the galactic plane: Further investigation.
M. A. Gordon, B. Höglund.
Astrophys. Journ., Vol. 182, 41 - 44 (1973).

A survey of three 1° × 1° regions, located on the galactic plane at galactic longitudes of 48°, 70°, and 110°, fails to show the 6-cm line of H_2CO either in absorption or emission to a peak-to-peak noise limit of 0.15° K in antenna temperature. We consider our results in terms of a crude distribution function of dust clouds.

131.161 The radio detection of a giant dust complex in the Perseus arm. B. Höglund, M. A. Gordon.
Astrophys. Journ., Vol. 182, 45 - 54 (1973).

An optically invisible giant dust complex has been detected through observations in H_2CO, OH, and H I. Its distance is about 4400 pc and the dimensions roughtly 38 × 13 pc. Derived column densities are 3.8 × 10^{13} for H_2CO, 1 × 10^{14} for OH, and 2.3 × 10^{19} cm^{-2} for H I. The plausibility of these values is discussed.

131.162 Astrophysical masers. III. Trapped infrared lines and cross-relaxation.
P. Goldreich, D. A. Keeley, J. Y. Kwan.
Astrophys. Journ., Vol. 182, 55 - 66 = Contr. Lick Obs., No. 386 (1973).

The rate of cross-relaxation due to resonance radiation trapping is derived. The effects that this cross-relaxation has on maser source size and polarization are investigated. The influence of trapped resonance radiation on the relative amplification of the individual hyperfine components of the 1.35-cm H_2O line is discussed. The theoretical results are compared with observations.

131.163 Thirteen new H_2O sources associated with OH emission in H II regions.
K. J. Johnston, R. M. Sloanaker, J. M. Bologna.
Astrophys. Journ., Vol. 182, 67 - 75 (1973).

Thirteen new H_2O emission sources associated with OH emission sources in H II regions have been discovered. Taking the great variability of the intensities of H_2O sources into account, the statistics indicate that probably all type I OH emission sources are associated with H_2O sources although the H_2O emission has not been detectable in many cases.

131.164 Interstellar dust in the Rho Ophiuchi dark cloud.
L. Carrasco, S. E. Strom, K. M. Strom.
Astrophys. Journ., Vol. 182, 95 - 109 (1973).

The dark cloud near ρ Oph is shown to be an ideal region for studying several characteristics of the dust component of the interstellar medium. Infrared and optical photometry and measurements of the wavelength dependence of polarization suggest that the grains are larger in the higher-density regions of the ρ Oph cloud.

131.165 Mariner 9 ultraviolet spectrometer experiment: Interstellar absorption at Lyman alpha in OB stars.
R. C. Bohlin.
Astrophys. Journ., Vol. 182, 139 - 145 (1973).

Column densities of neutral hydrogen N_H toward 10

stars are derived from spectra obtained with the ultraviolet spectrometer on the Mariner 9 Mars orbiter. The deduced values of N_H are systematically lower than those derived from OAO-2 data.

131.166 Spectrophotometric results from the Copernicus satellite. VI. Extinction by grains at wavelengths between 1200 and 1000 Å.
D. G. York, J. F. Drake, E. B. Jenkins, D. C. Morton, J. B. Rogerson, L. Spitzer.
Astrophys. Journ., (Letters), Vol. 182, L1 - L6 (1973).

Extinction curves for the interstellar material are extended from 9 to 10 μ^{-1} (1100–1000 Å) for the reddened stars ζ Oph, ζ Per, ξ Per, and α Cam. Good agreement with earlier results is obtained from 7.5 to 9 μ^{-1}. In all cases, the extinction continues to increase in the far-ultraviolet.

131.167 HCN radio emission from the Hourglass region of M8. P. T. Giguere, L. E. Snyder, D. Buhl.
Astrophys. Journ., (Letters), Vol. 182, L11 - L12 (1973).

Radio emission from the $J = 1-0$ line of $H^{12}C^{14}N$ has been detected from the Hourglass region of M8, which includes Herschel 36, an O7 star of known optical properties embedded in dust. The HCN cloud is extended from Herschel 36 throughout the Hourglass region. Implications of this detection are discussed.

131.168 On the nature of the luminous central objects in NGC 3603 and 30 Doradus. N. R. Walborn.
Astrophys. Journ., (Letters), Vol. 182, L21 - L23 (1973).

Classification spectrograms and short-exposure photographs of the central objects in the giant H II regions NGC 3603 and 30 Doradus are illustrated and discussed. The NGC 3603 object is shown to be a very luminous Trapezium-like system including a Wolf-Rayet component. Because of the spectroscopic similarities, it is suggested that the 30 Dor object may also be such a system.

131.169 Polarisation des Sternlichtes zwischen beiden Magellanschen Wolken. T. Schmidt.
Mitt. Astron. Ges., No. 32, p. 113 - 116 (1973).

131.170 Die Temperatur des 'Zwischen-Wolken-Gases' der interstellaren Materie.
O. Hachenberg, U. Mebold.
Mitt. Astron. Ges., No. 32, p. 138 - 143 (1973).

131.171 Interstellar absorption in Norma.
K. Bredow, U. Haug.
Mitt. Astron. Ges., No. 32, p. 263 (1973). – Abstract.

131.172 Modellrechnungen mit dreidimensional variabler interstellarer Absorption.
K. Ferrari d'Occhieppo, H. Jenkner.
Mitt. Astron. Ges., No. 32, p. 264 - 265 (1973).

131.173 Fabry-Perot-Interferometrie an galaktischen H II-Regionen. H. H. Hippelein.
Mitt. Astron. Ges., No. 32, p. 265 - 266 (1973).

131.174 Hydrodynamik interstellarer Wolken.
G. Hertel, H. Röser.
Mitt. Astron. Ges., No. 32, p. 267 - 268 (1973).

131.175 A catalog of data on optically visible H II regions.
P. Angerhofer, E. Churchwell, M. Walmsley.
Mitt. Astron. Ges., No. 32, p. 269 - 272 (1973).

131.176 Interpretation of the carbon recombination line.
M. Walmsley.
Mitt. Astron. Ges., No. 32, p. 274 - 275 (1973).

131.177 Probing the interstellar medium with a point explosion. V. Icke.
Astron. Astrophys., Vol. 26, 45 - 52 (1973).

The interstellar gas density ahead of a shock front caused by a point explosion is calculated, subject to simplifying assumptions. The precision of the method can be estimated by the calculation of the emissivity behind the shock and comparing this to observations. Approximations for the age, mass and thickness of the shock front are also derived. The method is applied to the galactic object NGC 7635.

131.178 Transition radiation from interstellar dust grains.
L. Durand.
Astrophys. Journ., Vol. 182, 417 - 432 (1973).

The spectrum of transition radiation emitted by interstellar dust grains subject to bombardment by relativistic electrons is calculated in detail for grains for realistic sizes. The author concludes that transition radiation is not a significant source of diffuse galactic X-rays, and that it can be important in astrophysical phenomena only under very special conditions.

131.179 Time-dependent radiative cooling of a hot low-density cosmic gas. M. Kafatos.
Astrophys. Journ., Vol. 182, 433 - 447 (1973).

Detailed calculations are presented for the radiative cooling of a hot ($10^4 °K \leqslant T \leqslant 10^6 °K$) interstellar gas. Results are presented in three cases: the first two have initial conditions determined when a 40- or 100-eV photon burst suddenly ionizes the gas, corresponding to a "fossil Strömgren sphere" suddenly formed by an ultraviolet or soft X-ray supernova burst. In the third case the gas is cooling from steady-state ionic abundances at $10^6 °K$ (e.g., a supernova shell that has reached the late radiative-cooling stage).

131.180 Thermal structure and evolution of interstellar gas exposed to a soft X-ray burst. J. Schwarz.
Astrophys. Journ., Vol. 182, 449 - 475 (1973).

The author calculates the temperature and ionization structure of interstellar gas exposed to a burst of monochromatic X-rays. The duration of the burst is assumed short compared to the cooling and hydrogen recombination time scales in the medium. He examines two limiting cases: (1) all photoionizations at a point occur before any photoelectron can thermalize with the atoms and ions; and (2) photoelectrons are created sufficiently slowly to allow electron and heavy-particle gases to remain at a common temperature. He examines the constant-density thermal behavior of both regions subsequent to the photon burst, and discusses implications for time-dependent models for the interstellar medium. He examines as well the effect of the diffuse radiation emitted by the central fully ionized region as it cools.

131.181 Upper limits for interstellar fulvene and nitric acid.
P. T. Giguere, F. O. Clark, L. E. Snyder, D. Buhl, D. R. Johnson, F. J. Lovas.
Astrophys. Journ., Vol. 182, 477 - 479 (1973).

Radio searches for the $1_{01}-0_{00}$ transition of fulvene at 6399 MHz and for the $1_{10}-1_{11}$ transition of nitric acid at 5839 MHz were not successful. Upper limits are reported for fulvene against three galactic sources and for nitric acid against four. Our measurements indicate that nitric acid may be underabundant with respect to cyanoacetylene in the direction of Sgr B2.

131.182 The doublet-ratio method and interstellar abundances. P. Nachman, L. M. Hobbs.
Astrophys. Journ., Vol. 182, 481 - 487 (1973).

The method of doublet ratios, used in obtaining interstellar Na I and Ca II abundances, is generalized to include realistic multiple-cloud cases. Entirely apart from any errors of observation, the simplified velocity distribution used in the method leads to errors in the inferred column densities which are systematic and which can be as large as a factor of ten or more in some practical cases. The D lines of Na I toward ζ Oph illustrate such order-of-magnitude underestimates.

131.183 Observations of the H66α recombination line.
E. B. Waltman, K. J. Johnston.
Astrophys. Journ., Vol. 182, 489 - 496 (1973).
Observations of the H66α recombination line in W3A, Orion A, Orion B, M17, W49, and W51 show that the line intensities are higher than is expected from present non-LTE theory. This may be accounted for by an increase in the density and emission measure of these H II regions.

131.184 Infrared nebular luminosity versus stellar luminosity in five H II regions. H. M. Johnson.
Astrophys. Journ., Vol. 182, 497 - 502 (1973).
The ratio of stellar to nebular luminosity is found in M8, M17, M42, NGC 2024, and NGC 6357. The various observable and theoretical parameters which are brought into the comparison are clearly specified.

131.185 Ammonia in DR 21(OH) and NGC 2264.
C. H. Mayer, J. A. Waak, A. C. Cheung, M. F. Chui.
Astrophys. Journ., (Letters), Vol. 182, L65 - L69 (1973).
NH_3 (1,1) and (2,2) line emission spectra were measured in the DR 21 (OH) region and in a new NH_3 region in NGC 2264. Hyperfine structure of the (1,1) transition was resolved, providing data on optical depth and excitation temperature for calculation of column densities and kinetic temperatures.

131.186 Concerning anomalous formaldehyde hyperfine lines in the dust cloud L 1436.
R. M. Crutcher.
Astrophys. Journ., (Letters), Vol. 182, L71 - L72 (1973).
We have shown that there is no evidence for nonequilibrium hyperfine component intensities of formaldehyde in the dust cloud L 1436.

131.187 Formation of the HD molecule in the interstellar medium. W. D. Watson.
Astrophys. Journ., (Letters), Vol. 182, L73 - L76 (1973).
Ion-molecule, isotope exchange reactions in the interstellar gas that can form HD preferentially in comparison with H_2 are examined. Emphasis is placed upon conditions that are representative of interstellar clouds in which the unexpectedly large HD/H_2 abundance ratio has recently been observed.

131.188 Moleküle im interstellaren Raum. S. Marx.
Astron. in der Schule, 10. Jahrgang, p. 50 - 54 (1973).

131.189 Interstellar molecules.
K. Akabane, M. Morimoto, N. Kaifu.
IAU Circ., No. 2486 (1973).

131.190 Ortho—para transitions in H_2 and the fractionation of HD. A. Dalgarno, J. H. Black, J. C. Weisheit.
Astrophys. Letters, Vol. 14, 77 - 79 (1973).
The otherwise nearly independent forms of H_2, ortho- and para-hydrogen, can be mixed efficiently by rapid proton interchange collisions in interstellar clouds. Similar processes involving deuteron—proton interchange may dominate the fractionation of deuterium in HD and DCN.

131.191 5 GHz observations of small-scale structure in DR21.
S. Harris.
Monthly Notices Roy. Astron. Soc., Vol. 162, 5P - 10P (1973).
The H II region DR21 has been observed with the Cambridge 5-km telescope at 5.00 GHz. It is resolved into four

compact components of high density and small linear dimensions, surrounded by diffuse envelopes of lower density. The condensations may be at a higher electron temperature than the envelopes.

131.192 Radio scintillations due to plasma irregularities with power law spectra: the interstellar medium.
L. T. Little, D. N. Matheson.
Monthly Notices Roy. Astron. Soc., Vol. 162, 329 - 338 (1973).
The scintillation effects which occur when radiation is incident on a medium containing phase changing irregularities which have a spatial frequency spectrum of power law form $P(S) \propto S^{-n}$ are discussed for the case when the rms phase deviation is greater than one radian. An attempt is made to resolve the apparent discrepancies between the works of other authors who have tackled this problem, usually from the standpoint of ray optics. The theory is applied to the interstellar medium.

131.193 Interstellar obscuration in the direction of Maffei 1 and 2. K. Nandy, F. Smriglio.
Monthly Notices Roy. Astron. Soc., Vol. 162, 25P - 29P (1973).
The visual obscuration in the direction of Maffei 1 and 2 is found to be between 4 and 5 magnitude, implying that Maffei 1 is a member of the local group.

131.194 Microwave water vapor emission from galactic sources. W. T. Sullivan III.
Astrophys. Journ., Suppl. Ser., No. 222, Vol. 25, 393 - 432 (1973).
Water vapor emission arising from the 1.35-cm $6_{16} \to 5_{23}$ transition has been observed from the galactic sources W3—OH, Ori A, VY CMa, NGC 6334 (N), Sgr B2, W49, W51, ON 1, and W75(S) in nine distinct periods extending from 1969 January to 1970 June. For each source the author summarizes its properties, describes the H_2O variations, and presents time sequences of the profiles as well as diagrams showing the behavior of individual features with time.

131.195 Polarization changes of some M type supergiant stars. E. D. Arsenievich.
Soobshch. Byurakan. Obs., vyp. (No.) 44, p. 91 - 103 (1972). In Russian.
The results of polarimetric observations of some cold supergiant stars (μ Cep, VV Cep, RW Cep, ST Cep, RW Cyg and TZ Cas) are given. It is shown that in the main all these stars have an intrinsic polarization.

131.196 Interstellar dust. S. Hayakawa, T. Nishimura.
Solid State Phys., (Japan), Vol. 7, No. 9, p. 44 - 49 (1972). In Japanese.
A brief summary is given of observation and astrophysical implications of interstellar dust. Recent development in the studies of optical, magnetic and electric properties of interstellar dust is described.

131.197 Interstellar molecules and the interstellar medium.
D. M. Rank.
Molecular spectroscopy: Modern research, [Academic Press Inc., London], p. 73 - 78 (1972).
Molecules such as OH, NH_3, CN, CH_3OH, H_2CO, HCOOH and H_2 which have been detected in the Galaxy by combining microwave spectroscopy and radio astronomy techniques, are discussed.

131.198 On the energy source for driving the interstellar gas.
B. Basu.
Indian Journ. Phys., Vol. 46, 379 - 386 (1972).

On the basis of the recent observations of high speed mass ejection from the blue giants and supergiants, a quantitative estimate of the available kinetic energy by the process has been made. It is found that the amount of kinetic energy thus produced may be comparable to that available from the ultraviolet radiation of the early type stars.

131.199   Sound vibrations in cosmic magneto-active space.
E. Ya. Gidalevich.
Astrofizika, Vol. 8, 591 - 597 (1972). In Russian. — English translation in Astrophysics, Vol. 8, No. 4.

In a non-relativistic approximation the role of cosmic rays for total elasticity of the interstellar gas is examined. The magnetic field is a means of impulse transfer from the cosmic rays to the "usual" interstellar gas. It is shown that the cosmic rays make an important contribution to the total speed of sound in the presence of a magnetic field.

131.200   A photon rest mass and the propagation of longitudinal electric waves in interstellar and intergalactic space.   R. Burman.
Journ. Phys. A, General Phys., Vol. 6, 434 - 444 (1973).

The author deals with the effect of a non zero photon rest mass m on the propagation of longitudinal electric waves in a plasma for which the electrons are treated as a warm fluid with scalar pressure.

131.201   Origin of interstellar molecules.   M. Shimizu.
Progr. Theor. Phys., (*Japan*), Vol. 49, 153 - 164 (1973).

It is concluded from a thermochemical calculation that the interstellar molecules in compact H II regions may be formed in places of high temperature and high pressure. A possibility of the locations of molecular formation may be the prestellar atmospheres, but the atmospheres of late type stars are other possible candidates. Detectable interstellar molecules suggested from the above calculation are enumerated.

131.202   Studies of interstellar formamide.
C. A. Gottlieb, P. Palmer, L. J. Rickard, B. Zuckerman.
Astrophys. Journ., Vol. 182, 699 - 710 (1973).

The $1_{11} - 1_{10}$ transition of interstellar formamide (NH_2CHO) was detected at 19-cm wavelength in the galactic center sources Sgr A and Sgr B2. Statistical equilibrium calculations suggest an NH_2CHO projected density of $\sim 10^{15}$ cm^{-2} in Sgr B2. Arguments are presented that all of the previously observed K-doublet transitions of formamide, methyl alcohol, formic acid, and acetaldehyde are inverted and amplify the background continuum radiation.

131.203   Radiative transport in interstellar masers.
M. M. Litvak.
Astrophys. Journ., Vol. 182, 711 - 730 (1973).

Calculations are given of the size, intensity, and bandwidth of emission from OH or H_2O maser spheres that are expanding or contracting, and that are homogeneously or inhomogeneously pumped. Amplification of resonant scatter is shown to be less important than that of spontaneous emission in determining the small apparent interferometer sizes of emitters. Appreciable saturation line broadening is calculated for partially or fully saturated spheres, in contradiction to the results of Peters and Allen.

131.204   Heating and ionization of H I regions by discrete soft X-ray and subcosmic-ray sources.
S. M. Lea, J. Silk.
Astrophys. Journ., Vol. 182, 731 - 753 (1973).

The authors calculate the nonuniform ionization of a homogeneous interstellar medium produced by discrete sources of soft X-rays and subcosmic rays. They first solve the ionization equation appropriate to the type of source considered, and then obtain the electron density $n_e(r)$ as a function of radial distance r from the source. They next use this function to evaluate the expected electron density $\langle n_e \rangle$ which would be observed at a random point in the Galaxy, under the assumption that the ionization sources are uniformly distributed. A similar procedure is applied to time-dependent models of the interstellar medium. The authors present a simplified model for the ionization of intercloud matter by bursts of both soft X-rays and subcosmic rays, and calculate the resulting mean and root mean square ionization. They also calculate the ionization and heating rates, and solve the energy equation for a mean temperature. The results are displayed as contours of ionization fraction, ionization rate, and temperature plotted as a function of the source parameters, specifically source strength and mean separation.

131.205   Evidence for a third thermally stable phase of the interstellar gas.   R. Giovanelli, R. L. Brown.
Astrophys. Journ., Vol. 182, 755 - 766 (1973).

Recent neutral-hydrogen observations taken at high spectral resolution of high- and intermediate-velocity clouds have revealed that the velocity dispersions in such H I components divide into two classes; the velocity widths of these classes correspond to gas temperatures of $\sim 1200°$ and $\sim 10,000°$K, respectively. As these observations are not immediately interpretable within the framework of the usual theoretical descriptions of the interstellar gas, the authors have considered modifications of these theories, particularly by changing the chemical abundance of particular species, that may influence the phase equilibrium of the interstellar gas. By uniformly depleting all the coolants or by selectively depleting particular coolants, it is shown that a "cloud" phase at $T \sim 10,000°$K can exist in pressure equilibrium with a thermally stable intercloud medium at $T \simeq 3-5 \times 10^4$ °K.

131.206   Absence of H_2CO 6-centimeter hyperfine anomalies in a dust cloud.   C. Heiles, B. E. Turner.
Astrophys. Journ., (*Letters*), Vol. 182, L121 - L124 (1973).

Apparent hyperfine anomalies inferred earlier by Dieter for a dust cloud are shown to result instead from the presence of a second velocity component.

131.207   The H II region G333.6−0.2, a very powerful 1−20 micron source.   E. E. Becklin, J. A. Frogel, G. Neugebauer, S. E. Persson, C. G. Wynn-Williams.
Astrophys. Journ., (*Letters*), Vol. 182, L125 - L129 (1973).

G333.6−0.2 has been found to be one of the brightest 20-micron sources in the sky, and the most luminous H II region known in the wavelength range $1-25\,\mu$. Despite its high luminosity it consists of only a single component with a diameter of $11''$ (0.2 pc). The source appears to be a dust-filled compact H II region.

131.208   The photochemistry of interstellar molecules.
L. J. Stief.
Molecular Photochem., (*USA*), Vol. 4, 153 - 170 (1972).

131.209   A more accurate formula of accretion by stars.
F. Cernuschi, F. R. Marsicano.
Revista Mat. Fis. Teor., Vol. 21, 217 - 225 = Dep. Astron. Fís., Fac. Humanidades Ciencias, Univ. Montevideo, Publ. No. 42 (1971).

It is analyzed critically the mechanical process assumed in the theory of accretion of mass by a star moving with relative velocity in a cloud. A more accurate formula for the rate of accretion is deduced, the limit of which, when the time necessary to reach the steady state tends to infinite is the formula given by Hoyle and Lyttleton. The new formula gives values for the rate of accretion smaller by a factor of 10^{-7}, when the relative velocity is of the order of 10^5 cms /s , with respect to Hoyle and Lyttleton's predictions.

131.210 **Interstellar absorption of light in the Local System belt.** A. D. Chuadze.
Soobshch. AN GruzSSR, Vol. 68, 577 - 579 (1972). In Russian. – Abstr. in Referativ. Zhurn. 51. Astron., 7.51.726 (1973).

131.211 **Recent work on interstellar grains.**
N. C. Wickramasinghe, K. Nandy.
Rep. Progr. Phys., Vol. 35, 157 - 234 (1972).
   Observational data accumulated over the past decade have considerably narrowed down the choice of proposals for grain models. Mixtures of particles composed of refractory species such as graphite, silicates, quartz and iron appear likely. Such particles may form in a wide range of astronomical situations, for example, cool stars, protostars and supernovae explosions. Dust grains are responsible for infrared radiation in galactic sources, and also play an important role in the synthesis of interstellar molecules.

Light scattering functions for small particles with applications in astronomy. See Abstr. 003.124.

Introduction to molecular spectra. See Abstr. 022.041.

Formation of molecular hydrogen on cold surfaces. See Abstr. 022.043.

Observations of formamide at 6 cm in Sagittarius B2. See Abstr. 022.044.

The effects of Stark broadening in the radio recombination line temperatures. See Abstr. 022.049.

Radio recombination lines from H^0 regions. See Abstr. 022.078.

Extra-galactic astronomy objectives. See Abstr. 051.013.

Thermal instability due to formation of molecular or solid hydrogen. See Abstr. 061.031.

A possible explanation for the origin of lithium, beryllium, and boron. See Abstr. 061.063.

Infrared extinction cross sections of silicate grains. See Abstr. 063.048.

The effect of interstellar medium parameters on the accretion by neutron stars. See Abstr. 065.082.

Interaction of singly charged insterstellar helium ions with the solar wind. See Abstr. 074.006.

Low-intensity Balmer emissions from the interstellar medium and geocorona. See Abstr. 082.008.

The interstellar reddening law in the ultraviolet deduced from filter photometry obtained by the OAO-2 satellite. See Abstr. 113.012.

RCW 117 and DR 15 observed in the far infrared. See Abstr. 113.015.

On a correlation between the magnitude and the radial velocity of hot stars. See Abstr. 113.016.

Bright infrared sources in M 17. See Abstr. 114.006.

The interstellar spectrum of HD 154368. See Abstr. 114.023.

Photometry and classification of stars in several H II regions. See Abstr. 114.026.

Infrared observations of a highly reddened star near NGC 6231. See Abstr. 114.028.

An infrared object probably associated with OH 338.5 + 0.1. See Abstr. 114.029.

Two new main-line OH/IR stars. See Abstr. 114.034.

5 GHz observations of the infrared star MWC 349, and the H II condensation W3 (OH). See Abstr. 114.046.

New emission-line stars in OH clouds. See Abstr. 114.116.

Spectrophotometric results from the Copernicus satellite. II. Composition of interstellar clouds. See Abstr. 114.122.

Spectrophotometric results from the Copernicus satellite. III. Ionization and composition of the intercloud medium. See Abstr. 114.123.

Spectrophotometric results from the Copernicus satellite. IV. Molecular hydrogen in interstellar space. See Abstr. 114.124.

Spectrophotometric results from the Copernicus satellite. V. Abundances of molecules in interstellar clouds. See Abstr. 114.125.

Observations of diffuse interstellar features in the spectra of dust-embedded and field stars. See Abstr. 114.154.

Infrared luminosity vs. stellar luminosity in H II regions. See Abstr. 115.011.

HD 215441 and 53 Camelopardalis: Intrinsic polarization of $H\beta$ and the continuum. See Abstr. 116.003.

Optical circular polarization of X Persei. See Abstr. 122.019.

A search for optical circular polarization in white dwarfs and late-type stars with circumstellar shells. See Abstr. 126.009.

Infrared and microwave emission from nebulae in the Galaxy. See Abstr. 132.018.

Observations of the bright rim of the Horsehead nebula in $H\alpha$ and [N II]. See Abstr. 132.030.

Compact radio source associated with the OH source ON-1 (OH69.5−1.0). See Abstr. 141.087.

Observations of the neutral-hydrogen absorption spectrum of Cygnus X-3. See Abstr. 142.012.

Absorption and production of soft X-rays in the Galaxy. See Abstr. 142.075.

Production of astrophysical X-rays by transition radiation. See Abstr. 142.113.

Transition radiation in astrophysics. See Abstr. 143.035.

On the role of plasma effects in the cosmic ray propagation and isotropization in the Galaxy. See Abstr. 143.041.

Density shock waves driven by star formation – a mechanism. See Abstr. 151.009.

Connection of T-associations with the interstellar medium in the northern part of Monoceros. See Abstr. 152.011.

High-velocity clouds and 'normal' galactic structure. See Abstr. 155.005.

Note on Verschuur's article on high-velocity clouds and 'normal' galactic structure. See Abstr. 155.006.

Molecules and evolution in the Galaxy. See Abstr. 155.010.

H I absorption in the galactic center region and between galactic longitudes 350° and 359°. See Abstr. 155.011.

Survey of molecular lines near the galactic center. III. 6-centimeter formaldehyde absorption at $b = -2'$ from $l = 2°0$ to $l = 4°5$ and at $b = -12'$ from $l = 358°5$ to $l = 2°0$. See Abstr. 155.013.

Underabundance of ionized helium in the galactic centre. See Abstr. 155.019.

The large-scale distribution of low-velocity hydrogen gas at high galactic latitudes. See Abstr. 155.020.

Intermediate-negative-velocity neutral hydrogen at $b \geqq + 15°$. See Abstr. 155.021.

Intermediate-negative-velocity gas around $l = 238°$, $b = + 75°$. See Abstr. 155.022.

On the kinematics of a local component of the inter-

stellar hydrogen gas possibly related to Gould's Belt. See Abstr. 155.027.

Kinematic disturbances in the local neutral hydrogen. See Abstr. 155.032.

Comparison of stellar and neutral hydrogen galactic kinematics. See Abstr. 155.034.

The effect of loop III on interstellar neutral hydrogen. See Abstr. 155.054.

Anomalous helium abundance of the galactic centre H II region G0.5–0.0. See Abstr. 155.058.

Diffuse galactic light and the albedo of interstellar dust in the 1500 Å to 4250 Å region. See Abstr. 155.082.

Analysis of the neutral hydrogen within 1 kpc of the sun. See Abstr. 155.087.

Magneto-gravitational and thermal instability in the galactic disk. See Abstr. 155.090.

A statistical investigation of neutral hydrogen line profiles. See Abstr. 157.002.

The distribution of neutral hydrogen and the velocity field of the galaxy NGC 3109. See Abstr. 158.003.

A neutral hydrogen survey of the galaxy M33. II. Distribution and kinematics of the neutral hydrogen. See Abstr. 158.004.

Aperture synthesis study of neutral hydrogen in the galaxies NGC 6946 and IC 342. See Abstr. 158.005.

Neutral hydrogen in Markarian galaxies. See Abstr. 158.007.

Diameters of H II regions in M31 and comparison with the largest regions in M33. See Abstr. 158.013.

The absorbing material in the Andromeda nebula. See Abstr. 158.038.

A high resolution neutral hydrogen study of the galaxy M 51. See Abstr. 158.039.

H I clouds in clusters of galaxies. See Abstr. 160.016.

132 Emission Nebulae, Reflection Nebulae

132.001 Predictions of exceptionally strong '4430' in the backscattered light from reflection nebulae.
G. E. Bromage.
Astrophys. Space Sci., Vol. 18, 449 - 461 (1972).
    The strength of the '4430' interstellar spectrum feature in the backscattered light from certain reflection nebulae is shown to be a sensitive indicator of the feature's origin. Backscattering '4430' profiles have been computed for several models of silicate, graphite, dirty-ice-coated graphite, and silicon carbide grains with impurities. The effect on these profiles of both non-sphericity of the grains and finite optical thickness of the reflection nebula have been considered.

132.002 Measurements of the electron temperatures in M42 from the profiles of Hα, [N II], Hβ and [O III].
M. A. Dopita, A. H. Gibbons, J. Meaburn.
Astron. Astrophys., Vol. 22, 33 - 39 (1973).
    Two single-etalon, pressure-scanned Fabry-Perot monochromators have been used to obtain the profiles of the Hα, [N II], Hβ and [O III] lines from many positions across M42 (the Orion nebula). The electron temperatures predicted from the profiles of the Hα and [N II] lines from each position have been critically compared to the temperatures from the corresponding profiles of the Hβ and [O III] lines.

132.003 Deuterium in the Orion nebula.
K. B. Jefferts, A. A. Penzias, R. W. Wilson.
Astrophys. Journ., (Letters), Vol. 179, L57 - L59 (1973).
    We have found line emission which we attribute to the $J = 2$ to $J = 1$ transition of DCN in the Orion nebula. This identification has been strengthened by our detection of the corresponding $J = 1$ to $J = 0$ multiplet.

132.004 Observations of optical nebulae at 2695 MHz.
E. Churchwell, C. M. Walmsley.
Astron. Astrophys., Vol. 23, 117 - 124 (1973).
    We report radio continuum observations at 11 cm of seventeen optically identified nebulae from the Sharpless catalog (1959). The spectrum is plotted and a discussion of the physical nature of each source is given. The relationship between excitation parameter and spectral type is rederived on the basis of both LTE and non-LTE model atmosphere results and is used to interpret the observations.

132.005 Comparison of far-infrared, optical, and radiofrequency data of diffuse nebulae. H. M. Johnson.
Astrophys. Journ., (Letters), Vol. 180, L7 - L10 (1973).
    The relation of far-infrared grain emission to bremsstrahlung in gas-and-dust nebulae is discussed. Application is made to diffuse nebulae in a 100-μ catalog which are identified with sources in 6-cm catalogs. Comments are made on some individual objects and on the significance of the observed $S(100\,\mu)/S(6\,\text{cm})$ ratio of flux densities.

132.006 Diffuse nebulae at high galactic latitudes.
G. Grasdalen, J. G. Cohen.
Astrophys. Journ., (Letters), Vol. 180, L11 - L13 (1973).
    The nature of the diffuse nebulae at high galactic latitudes is discussed. Although these nebulae show Hβ in emission, evidence is presented that they are reflection nebulae illuminated by the integrated light of the Galaxy. In the integrated spectra of spiral disks, we find that Hα is in emission.

132.007 The X-ray surface brightness of the Cygnus Loop.
J. C. Stevens, G. P. Garmire.
Astrophys. Journ., (Letters), Vol. 180, L19 - L26 (1973).
    The surface brightness of the Cygnus Loop in X-rays was obtained on a scale of 0.25 deg². The X-ray emission shows strong limb brightening, implying a shell-like source structure. The shell exhibits several regions of enhanced emission. X-ray emission which was observed from a region near the center of the Loop may be from a remnant object.

132.008 The dipole nebula IC 2220, a southern reflection nebula around the variable red giant HD 65750.
J. Dachs, J. Isserstedt.
Astron. Astrophys., Vol. 23, 241 - 245 (1973).
    IC 2220 has been found to be a peculiar butterfly-shaped reflection nebula surrounding the bright variable M 3 III giant HD 65750 = HR 3126. The filaments of the nebula resemble the structure of a magnetic dipole field centered on the red giant. From UBV photometry of the star and the nebula, the $(B-V)$ and $(U-B)$ colours of the nebula are bluer by $0^m.23$ and $0^m.98$, respectively, than those of the illuminating star. A very large closed loop which is visible in IC 2220 is tentatively interpreted as a giant prominence originating in the central star.

132.009 The dynamical effects of stellar mass loss on diffuse nebulae. II. Approximate similarity solutions.
J. E. Dyson.
Astron. Astrophys., Vol. 23, 381 - 385 (1973).
    Approximate similarity solutions are calculated to describe the interaction of a high velocity stellar wind with surrounding ambient gas. Calculations are made for three density distributions in the ambient gas. The distributions have the form $\varrho \propto r^q$, and $q = 0$, -1 and -2.

132.010 The object Fourcade Figueroa, a shred associated with NGC 5128? H. A. Dottori, C. R. Fourcade.
Astron. Astrophys., Vol. 23, 405 - 409 = Publ. Obs. Astron. Cordoba, Tirada Aparte No. 193 (1973).
    The object Fourcade-Figueroa (F–F) ($\alpha_{50} = 13^h32^m5$; $\delta_{50} = -45°25'$) has a total luminosity in the photographic $m_{pg} = 12.14$ and a corrected recession velocity of 830 km/s. The possibility is discussed that it is physically associated with the elliptic galaxy NGC 5128.

132.011 Radio observations of the Gum nebula below 20 MHz. G. R. A. Ellis.
Proc. Astron. Soc. Australia, Vol. 2, 158 - 159 (1972).

132.012 Kinematics of the Huyghenian region of the Orion nebula. D. Fischel, W. A. Feibelman.
Astrophys. Journ., Vol. 180, 801 - 808 (1973).
    The nebular [O II] λ3726 and [O III] λ5007 radial-velocity measures of Wilson et al. are presented in contour map form. The space motions of θ^1 and θ^2 Ori and the nebular measures are discussed.

132.013 Line radiation transfer in extended envelopes. I. Lyman-α radiation in a pure hydrogen nebula.
N. Panagia, M. Ranieri.
Astron. Astrophys., Vol. 24, 219 - 227 (1973).
    The transfer of Ly-α radiation in a spherically symmetric, pure hydrogen nebula has been studied by means of Monte Carlo techniques. The results on the mean number of scatterings and the emergent profiles are extensively discussed. Comparison with other calculations are also made. The dependence of the mean number of scatterings from the nebular parameters, as well as the diffusion effects are examined.

132.014 Internal dust in the Orion nebula.
F. H. Schiffer III, J. S. Mathis.

Bull. American Astron. Soc., Vol. 5, 23 (1973). – Abstr. AAS.

132.015 Internal motion of the Orion cluster.
H. J. Reitsema, W. L. Sanders.
Bull. American Astron. Soc., Vol. 5, 23 (1973). – Abstr. AAS.

132.016 A dynamical model of the Carina nebula.
H. R. Dickel.
Bull. American Astron. Soc., Vol. 5, 24 (1973). – Abstr. AAS.

132.017 A versatile program for modeling low density
gaseous nebulae. R. C. Kirkpatrick.
Bull. American Astron. Soc., Vol. 5, 35 (1973). – Abstr. AAS.

132.018 Infrared and microwave emission from nebulae in
the Galaxy. Introductory report. L. Goldberg.
17th Colloquium International Astrophys. Liège 1971, (see 012.003), p. 315 - 324 (1972).

132.019 The peculiar nebula NGC 7635 and the stars in it.
H. M. Johnson.
17th Colloquium International Astrophys. Liège 1971, (see 012.003), p. 345 - 350 (1972).

132.020 The spectra of gaseous nebulae. M. J. Seaton.
Atoms and molecules in astrophysics. Proc. Scottish Univ. Summer School in Physics 1971, (see 012.005), p. 121 - 153 (1972).

132.021 A reexamination of the Gum nebula.
K. P. Beuermann.
Astrophys. Space Sci., Vol. 20, 27 - 38 (1973).
A detailed analysis of interstellar measurements in the direction of the Gum nebula is carried out. The ionized region is shown to have an angular radius of $\sim 18°$ and appears to be bounded by a shell of neutral gas. The mean electron temperature deduced from radio-frequency absorption measurements is found to be $\lesssim 8500 K$. These parameters suggest that the nebula is the normal and possibly evolved H II region of ζ Pup and γ^2 Vel, rather than the fossil Strömgren sphere of the Vela supernova as suggested by Brandt *et al.* (1971).

132.022 The hydrogen line spectrum of gaseous nebulae.
H. Gerola, M. Salem, N. Panagia.
Bol. As. Argentina Astron., No. 16, (see 012.007), p. 4 (1971). Abstract.

132.023 High radio frequency observations of the Omega
nebula. K. J. Johnston, R. W. Hobbs.
Astron. Journ., Vol. 78, 235 - 238, 279 (1973).
Observations of M17 at 0.955, 1.65, and 2.73 cm yield flux densities of 575 ± 80, 644 ± 64, and 644 ± 64 flux units. We conclude that M17, although having a very complex angular structure over approximately one square degree of the sky, is composed of sources with thermal spectra.

132.024 Infrared linear polarization measures of the Orion
nebula. D. A. Allen, R. W. Capps, H. M. Dyck,
W. J. Forrest, F. C. Gillett, S. J. Loer.
Bull. American Astron. Soc., Vol. 5, 267 - 268 (1973). Abstr. AAS.

132.025 Ring nebulae and Wolf-Rayet stars; observations of
NGC 2359. T. A. Lozinskaya.
Astron. Zhurn. Akad. Nauk SSSR, Vol. 50, 496 - 500 (1973). In Russian. English translation in Soviet Astron. AJ, Vol. 17, No. 3.
Hα interferometric observations of NGC 2359 have been made with a high-contrast Fabry-Perot etalon and contact image converter. The mean radial velocity and the mean half-width of the Hα line are determined. A detailed study of the shell around HD 56925 shows the Hα line being of 150 - 200 km/sec width. An expansion of the shell with a velocity of 55 ± 25 km/sec has been discovered.

132.026 Temperature gradient in gaseous nebulae.
T. B. Pyatunina, V. A. Soglasnova.
Astron. Zhurn. Akad. Nauk SSSR, Vol. 50, 508 - 515 (1973). In Russian. English translation in Soviet Astron. AJ, Vol. 17, No. 3.
A method of computing the electron temperature distribution in gaseous nebulae from radio isophotes at two different frequencies is suggested. From the data at wavelengths 1.95 cm and 75 cm the distribution of the electron temperature averaged along the line of sight is computed for the two nebulae Orion A and Omega. Using the same data an actual electron temperature distribution in the Orion nebula is computed as a function of the distance from the center of the nebula.

132.027 Observations of gas motions in and near the central
cavity of the Rosette nebula. M. G. Smith.
Astrophys. Journ., Vol. 182, 111 - 120 (1973).
Emission-line profiles from the central regions of the Rosette nebula show that from a dynamical point of view the H II region can be split into two distinct regions as predicted, qualitatively at least, by theory. The outer parts of the nebula are once again shown to lack significant large-scale motions. The symmetry of the gas motions about the central cavity and estimates from models by Dyson are consistent with the conclusion that mass loss from the central cluster of stars is the source of energy required to drive the observed supersonic motions.

132.028 On the nature of the infrared point source in the
Orion nebula.
E. E. Becklin, G. Neugebauer, C. G. Wynn-Williams.
Astrophys. Journ., *(Letters)*, Vol. 182, L7 - L9 (1973).
New photometric observations of the BN infrared source in the Orion nebula eliminate the possibility that it is simply a highly reddened star, and undermine the conclusion that there are 80 mag of visual extinction to the source. We suggest that there is a generic relationship between it and the infrared source IRS 5 in the H II region W3.

132.029 The variations of the ratio of the Hα and [N II]
lines throughout the North American (NGC 7000)
and Pelican (IC 5070) nebular complex.
C. Goudis, J. Meaburn.
Astron. Astrophys., Vol. 26, 65 - 70 (1973).
The variations of the ratios of the brightnesses of the Hα (6563 Å) and [N II] (6584 Å) lines were measured across the North American and Pelican nebular complex. Distinct variations were detected. These were converted into variations in the ratios of the volume emissivities of these lines by use of a shell model. The principal conclusions are that the ionization appears to be caused by a star or group of stars behind the dust lane and situated near the centre of the radio contours. Moreover, strong evidence is presented which indicates that these nebulae are part of one H II region in which a huge dust lane is embedded.

132.030 Observations of the bright rim of the Horsehead
nebula in Hα and [N II]. R. Louise, C. Sapin.
Astrophys. Letters, Vol. 14, 119 - 122 (1973).
The 'Horsehead' nebula has been photographed with narrow interference filters centered respectively on Hα and [N II] (λ = 6584 Å). The [N II]/Hα ratio is deduced and its variation towards the bright rim of the Horsehead is studied. It is shown that the maximum value of the temperature is not reached on the rim itself but in a narrow zone located just behind it and ahead of the dark matter.

132.031 Microwave spectroscopic mapping of gaseous nebulae. I. Excited hydrogen, helium, and carbon in Orion B. E. J. Chaisson.
Astrophys. Journ., Vol. 182, 767 - 780 (1973).

This communication reports 7.8-GHz observations of the H94α and the C94α recombination lines at eight locations distributed over the Ori B radio source. He 94α is found at three of these locations, and the results are used to examine critically the apparent underabundance of ionized helium in this nebula.

132.032 Catalogue of nebulae in Crux, Centaurus, Circinus and Norma. B. M. Smith.
Published by Steward Observatory, Tucson, Arizona, 2 + 97 + 22 pp. (1972).

The catalogue presented here is a compilation of published data for all known non-stellar galactic (and several extragalactic) emission sources in the region $302° \leqslant l \leqslant 335°$, $-5° \leqslant b \leqslant +5°$; it may be considered an extension of the "Catalogue of Nebulae in Carina" (1971) of Mrs A. H. Ferguson. As in Mrs. Ferguson's catalogue, the format and the choice of data to be included have been influenced by the catalogue's primary purpose as an aid in the study of galactic structure.

The brightest infrared sources.
See Abstr. 113.029.

New infrared objects in Orion.
See Abstr. 113.052.

Infra-red observations of young stars—II. T Tauri stars and the Orion population. See Abstr. 114.003.

Infra-red observations of young stars—III. Nebulous emission-line stars. See Abstr. 114.004.

Bright infrared sources in M17.
See Abstr. 114.006.

Spectra of the Becklin-Neugebauer point source and the Kleinmann-Low nebula from 2.8 to 13.5 microns.
See Abstr. 114.027.

Hα-emission stars in and near NGC 7000.
See Abstr. 114.036.

The infrared sources in M8.
See Abstr. 114.040.

Dust emission nebulae around Orion O and B stars.
See Abstr. 114.058.

Observations of intense 100-micron objects at 3.5-millimeter wavelength. See Abstr. 114.062.

Light curves of Orion variables.
See Abstr. 122.117.

Observations of Orion variables.
See Abstr. 123.037.

Evaporation of dirty ice particles surrounding early type stars. IV. Various size distributions.
See Abstr. 131.004.

The H_2CO absorption against the Carina nebula.
See Abstr. 131.024.

Internal motions analysis of the H II region NGC 7635. See Abstr. 131.026.

Interstellar deuterium: The hyperfine structure of DCN. See Abstr. 131.029.

Observations of the H II region S101 in Cygnus.
See Abstr. 131.036.

Interstellar deuterium: Chemical fractionation.
See Abstr. 131.047.

The problem of X-ogen. See Abstr. 131.049.

An anomalous emission line in low-frequency radio spectra toward W49A. See Abstr. 131.051.

A correlation study of carbon emission lines and hydroxyl absorption lines toward galactic nebulae.
See Abstr. 131.074.

Extinction properties of dust in the ultraviolet and the ionization of helium in galactic H II regions.
See Abstr. 131.130.

Interferometric observations of formaldehyde absorption in front of strong galactic sources.
See Abstr. 131.138.

The radio detection of a giant dust complex in the Perseus arm. See Abstr. 131.161.

Thirteen new H_2O sources associated with OH emission in H II regions. See Abstr. 131.163.

HCN radio emission from the Hourglass region of M8. See Abstr. 131.167.

On the nature of the luminous central objects in NGC 3603 and 30 Doradus. See Abstr. 131.168.

Observations of the H66α recombination line.
See Abstr. 131.183.

Infrared nebular luminosity versus stellar luminosity in five H II regions. See Abstr. 131.184.

Microwave water vapor emission from galactic sources. See Abstr. 131.194.

Radio emission nebulae surrounding MWC 349 and RY Scuti. See Abstr. 141.040.

Giant loops as fossil Strömgren spheres: Their radio and X-ray emission. See Abstr. 155.089.

133 Planetary Nebulae

133.001 A study of the radio continuum spectra of planetary nebulae. L. A. Higgs.
Monthly Notices Roy. Astron. Soc., Vol. 161, 313 - 330 (1973).
Model radio spectra have been fitted to all observational radio data on planetary nebulae known to the author in 1971. From the two parameters of the fitted spectra, electron temperature and optical depth at 10 GHz, values of emission measure, electron density, nebular radius, distance and extinction at Hα and at Hβ, have been derived for ~140 nebulae. All results of this study are in good agreement with similar results obtained from optical studies of planetary nebulae.

133.002 The structure and evolution of planetary nebulae. E. R. Capriotti.
Astrophys. Journ., Vol. 179, 495 - 516 (1973).
The dynamical evolution of a model of an ideal planetary nebula is studied. In this picture, the H I shell, which surrounds an ionized sphere during the early stages in the evolution of a nebula, is accelerated outward by the thermal and dynamical pressure at its inner boundary or ionization front. The ionization front becomes dynamically unstable when the nebula has expanded and rarefied to the point where recombinations and ionizations become too slow to suppress positive (away from the central star) and negative (toward the central star) displacements, respectively, of the ionization front.

133.003 Visual observations of twenty faint planetary nebulae. D. A. Allen.
Observatory, Vol. 93, 28 - 30 (1973).

133.004 Line intensities and radial velocities for 12 planetary nebulae.
S. D'Odorico, V. C. Rubin, W. K. Ford, Jr.
Astron. Astrophys., Vol. 22, 469 - 472 (1973).
Image tube spectra of twelve planetary nebulae (NGC 2371−72, 2392, 6818, 6905, 7008, 7026, 7048, IC 2149, IC 5217, Anon 3^h50^m, 21^h31^m, 22^h54^m) have been studied to determine relative line strengths (λ 4340−λ6731) and radial velocities.

133.005 New southern planetary nebulae.
S. van den Bergh, R. Racine, S. van Agt, T. Barnes, C. Coutts, B. Madore, A. Skill.
Astrophys. Journ., Vol. 179, 863 (1973).
Seven new planetary nebulae have been found during a two-color survey of the southern Milky Way.

133.006 The influence of dust upon the dynamics and thermal stability of planetary nebulae.
J. H. Hunter, Jr.
Astrophys. Journ., Vol. 180, 99 - 105 (1973).
The objectives of the present study are (a) to determine the consequences of the existence of dust to the dynamics of the nebula within the context of Faulkner's treatment, and (b) to reexamine the thermal stability of the nebular gas including the effects of the dust.

133.007 Photoelectric photometry of NGC 7027. R. D. Schwartz, M. Peimbert.
Astrophys. Letters, Vol. 13, 157 - 160 (1973).
From new infrared observations the reddening of NGC 7027 is obtained. By comparison with observations in the blue and the radio region small differences with respect to the normal reddening law are derived, and possible causes for these differences are discussed. The auroral lines of [N I] are detected for the first time in this object.

133.008 On ionization instability as the cause of planetary nebulae. E. H. Scott III.
Astrophys. Journ., Vol. 180, 487 - 500 (1973).
The probable mass-loss process driven by the ionization instability in the extended envelope of a 1 $M_\odot$ red supergiant was investigated by using sequences of mass-reduced models in hydrostatic equilibrium. Tests for nonadiabatic pulsational instability applied to models with mass not reduced led to a tentative identification of long-period variables as the precursors of the nuclei of planetary nebulae.

133.009 Spectral observations of southern planetary nebulae. Part I. A. E. Ringuelet, R. H. Méndez.
Publ. Astron. Soc. Pacific, Vol. 85, 96 - 98 (1973).
Spectra of eight planetary nebulae are described in the region λ3400−λ7200.

133.010 A note on the planetary 292 + 1°1. A. E. Ringuelet, R. H. Méndez.
Publ. Astron. Soc. Pacific, Vol. 85, 99 (1973).

133.011 The spatial distribution of the 11.7 micron radiation of NGC 7027. R. F. Knacke, A. M. Dressler.
Publ. Astron. Soc. Pacific, Vol. 85, 100 - 102 (1973).
The planetary nebula NGC 7027 was observed at 11.7 μ with a spatial resolution of 4 arc seconds. The infrared radiation is generally correlated with the Hα and radio fluxes indicating that the thermal emission is from hot grains within the ionized regions of the nebula.

133.012 Abundances and ionization distribution in planetary nebulae. E. G. Buerger.
Astrophys. Journ., Vol. 180, 817 - 830 (1973).
Three rather low-excitation nebulae (IC 418, IC 4593, and IC 3568) have been studied with theoretical models, as well as one high-excitation nebula (NGC 7662). The models give the run of electron temperature and ionization with radius, thereby permitting emission-line intensities to be calculated. Matching calculated (and observed) emission-line intensities then gives the abundances of the elements.

133.013 The spectrum of NGC 7027. J. B. Kaler, S. J. Czyzak, L. H. Aller.
Bull. American Astron. Soc., Vol. 5, 12 - 13 (1973). − Abstr. AAS.

133.014 A search for planetary nebulae in globular clusters. A. W. Peterson.
Bull. American Astron. Soc., Vol. 5, 13 (1973). − Abstr. AAS.

133.015 Identification of planetary nebulae in the elliptical galaxies NGC 185, NGC 205, and NGC 221.
H. C. Ford, D. C. Jenner, H. W. Epps.
Bull. American Astron. Soc., Vol. 5, 13 (1973). − Abstr. AAS.

133.016 Chemical abundances in a planetary nebula in the elliptical galaxy NGC 185.
D. C. Jenner, H. C. Ford, H. W. Epps.
Bull. American Astron. Soc., Vol. 5, 13 (1973). − Abstr. AAS.

133.017 Expansion velocities in old planetary nebulae. T. J. Bohuski, M. G. Smith.
Bull. American Astron. Soc., Vol. 5, 13 (1973). − Abstr. AAS.

133.018 High resolution observations of NGC 7027 at 5 GHz. P. F. Scott.
Monthly Notices Roy. Astron. Soc., Vol. 161, 35P - 38P

(1973).

The planetary nebula NGC 7027 has been observed at a frequency of 5.0 GHz with 2.0 arcsec resolution. The observed brightness distribution is consistent with emission from a thick cylindrical shell with its axis at 30° to the line-of-sight and having a uniform electron temperature and density of $(13 \pm 2) \times 10^3$ K and $(5.0 \pm 0.5) \times 10^4$ cm^{-3}, respectively.

133.019 Mean electron density of NGC 7027 inferred from H 85α line data. D. Hoang-Binh.
17th Colloquium International Astrophys. Liège 1971, (see 012.003), p. 365 - 366 (1972).

133.020 Comentarios sobre 272 + 12°1, 292 + 1°1, 328 − 17°1. A. E. Ringuelet, R. H. Méndez.
Bol. As. Argentina Astron., No. 16, (see 012.007), p. 30 (1971). − Abstract.

133.021 Observaciones de nebulosas planetarias australes - I. R. H. Méndez, A. E. Ringuelet.
Bol. As. Argentina Astron., No. 16, (see 012.007), p. 30 (1971). − Abstract.

133.022 A new planetary nebula. D. A. Allen.
Observatory, Vol. 93, 85 (1973). − Letter.

133.023 On the interpretation of IR-radiation of planetary nebulae. V. V. Vityazev.
Astron. Tsirk., No. 746, p. 6 - 8 (1973). In Russian.

133.024 New planetary nebula.
M. A. Kazarian, E. Ya. Oganesyan.
Astron. Tsirk., No. 753, p. 3 - 6 (1973). In Russian.

133.025 Improved optical positions for 153 planetary nebulae. D. K. Milne.
Astron. Journ., Vol. 78, 239 - 242 (1973).

Optical positions have been measured for 153 planetary nebulae within the declination range +27° to −45°. Significant errors have been found in many previously published positions.

133.026 The expansion of NGC 7293. S. Grandi.
Publ. Astron. Soc. Pacific, Vol. 85, 200 - 202 (1973).

An attempt has been made to measure the expansion of NGC 7293 by measuring the filament positions on two 200-inch plates taken nineteen years apart. No evidence for a common expansion or contraction is seen.

133.027 Trigonometric parallax determination for the central star in the planetary nebula NGC 7293.
C. C. Dahn, A. L. Behall, J. W. Christy.
Publ. Astron. Soc. Pacific, Vol. 85, 224 - 226 (1973).

Plates obtained with the 61-inch astrometric reflector yield a relative parallax of 0.″001 ± 0.″005 (m.e.) for the central star in the planetary nebula NGC 7293. This result contradicts the value of 0.″038 ± 0.″015 (m.e.) obtained by van Maanen (1923).

133.028 Spectrophotometric studies of nebulae. XXI. The remarkable planetary NGC 6778.
S. J. Czyzak, L. H. Aller.
Astrophys. Journ., Vol. 181, 817 - 823 (1973).

The relatively faint planetary NGC 6778 is distinguished by an extraordinarily strong spectrum of permitted lines of C II, N II, N III, O II, O III, etc., and pronounced stratification effects. It is suggested that these permitted lines arise from direct excitation from the nucleus rather than by recombination in dense filaments. The [O III] and other forbidden lines seem weaker than in other planetaries.

133.029 Condensations in planetary nebulae.

J. Mottmann.
Astrophys. Journ., Vol. 181, 825 - 831 (1973).

The observed low strengths of hydrogen radio recombination lines in planetary nebulae are investigated. The inclusion of a high degree of clumping in the nebular material is required to reduce the line enhancement predicted by non-LTE theory. Condensations with number densities of about 10^5 cm^{-3} seem to be required. The nature of such condensations is briefly discussed.

133.030 Distance and mass estimates for a few double shell planetaries. P. S. Riherd.
Bull. American Astron. Soc., Vol. 5, 267 (1973). − Abstr. AAS.

133.031 A study of the planetary nebula NGC 1360 and of its central star. V. T. Doroshenko.
Astron. Zhurn. Akad. Nauk SSSR, Vol. 50, 501 - 507 (1973). In Russian. English translation in Soviet Astron. AJ, Vol. 17, No. 3.

Interferometric, spectral and photometric observations of the planetary nebula NGC 1360 carried out during January-February 1972 allowed to obtain the following data: 1. the heliocentric radial velocity, 2. the range of expansion velocity, 3. the distance, 4. the electron density and 5. the energy flux in Hα, OIII + Hβ, λ4686 He II (estimated). For the central star the absolute spectral distribution, the temperature and UBV magnitudes have been found.

133.032 On the presence of a scattered continuum in the planetary nebula BD+30°3639.
S. E. Persson, J. A. Frogel.
Astrophys. Journ., Vol. 182, 177 - 181 (1973).

A measurement of the strength of Hβ relative to that of the nebular continuum in the planetary nebula BD+30°3639 is reported. The equivalent width of Hβ alone, exclusive of the central star, is found to be 1000 ± 200 Å, while the theoretical value at $T_e = 10,000°$K is ~2000 Å. It seems likely that the excess continuum flux is due to starlight that is scattered by particles within the nebula.

133.033 Infrared photometry of planetary nebulae.
S. E. Persson, J. A. Frogel.
Astrophys. Journ., Vol. 182, 503 - 507 (1973).

Photometric observations from 1.6 to 3.5 μ of 28 planetary nebulae are presented. Most of these emit 3.5-μ radiation in excess of that expected from thermal emission by ionized gas.

133.034 Spectrophotometric studies of gaseous nebulae. XXII. The irregular ring nebula NGC 6445.
L. H. Aller, S. J. Czyzak, E. Craine, J. B. Kaler.
Astrophys. Journ., Vol. 182, 509 - 515 (1973).

The relatively high excitation, low surface brightness nebula NGC 6445 shows a spectrum characterized by strong lines of He II λ4685, [Ne III] as well as [O II] λ3727, [S II], [O I], and probably [N II]. The observational data strongly suggest an excess of helium.

133.035 Forbidden lines of neutral carbon in NGC 7027.
I. J. Danziger, L. E. Goad.
Astrophys. Letters, Vol. 14, 115 - 117 (1973).

The identification in NGC 7027 of forbidden emission lines of neutral carbon at 9849.5, 9823.4, and 8727.4 Å is reported. Some implications of their strengths, under the assumption that the metastable levels are populated by electron collisions, are discussed.

133.036 Observation of 9.0-micron line emission from Ar III in NGC 7027 and NGC 6572.
T. R. Geballe, D. M. Rank.
Astrophys. Journ., (Letters), Vol. 182, L113 - L116 (1973).

The fine-structure line of Ar III at 1112 cm^{-1} has been observed in the planetary nebulae NGC 7027 and NGC 6572. The measured line intensities are compared with theoretical estimates for these nebulae.

133.037   Fine structure in NGC 7027.
B. Balick, C. Bignell, Y. Terzian.
Astrophys. Journ., (Letters), Vol. 182, L117 - L120 (1973).

Synthesis observations at 8085 MHz with spatial resolution of ~2″ show the distribution of radio emission from the nebula NGC 7027 to be complex. There is a general, but not detailed, agreement between radio and optical features of the nebula. Consequences of the radio and optical brightness distributions and radio optical depth are discussed.

133.038   Radial velocities of A77 and A72.
S. Brown, P. Lee.
Publ. Astron. Soc. Pacific, Vol. 85, 317 - 318 = Contr. Louisiana State Univ. Obs., Baton Rouge, No. 81 (1973).

New radial velocities are reported for the planetary nebulae A77 and A72 and additional velocities at variance with previously reported velocities are given for NGC 7293, NGC 7094, NGC 2724, and IC 289.

133.039   Model atmospheres of the nuclei of planetary nebulae.   N. A. Sakhibullin.
Trudy Kazan. Gorod. Astron. Obs., No. 37, p. 55 - 65 (1970). In Russian.

Scattering of resonance radiation in a sphere. See Abstr. 063.015.

Models for carbon-rich stars with helium envelopes. See Abstr. 065.088.

Near infra-red magnitudes of 248 early-type emission-line stars and related objects. See Abstr. 113.007.

The spectrum of Herbig-Haro object No. 1. See Abstr. 114.016.

Infrared emission in [Ne V]. See Abstr. 114.087.

Observations of circular polarization in white dwarfs and in the nuclei of planetary nebulae. See Abstr. 126.004.

Large-scale line splitting in five galactic H II regions. See Abstr. 131.009.

Interstellar dust and distances to planetary nebulae. See Abstr. 131.058.

Catalogue of nebulae in Crux, Centaurus, Circinus and Norma. See Abstr. 132.032.

134 Crab Nebula

134.001 Calculation and preliminary interpretation of the observed spectra of gaseous filaments in the Crab nebula. V. V. Golovaty, V. I. Pronik.
Astron. Zhurn. Akad. Nauk SSSR, Vol. 50, 208 - 209 (1973). In Russian. English translation in Soviet Astron. AJ, Vol. 17, No. 1.

The spectra of gaseous filaments are calculated for a number of models. There is good agreement between the theoretical spectrum obtained for one model and the observed spectrum averaged over all the filaments. The attempt was made to interpret the observed differences occurring in the spectra of different filaments.

134.002 Evolution of the Crab nebula.
F. Pacini, M. Salvati.
Astrophys. Letters, Vol. 13, 103 - 104 (1973).

We investigate the time evolution of the magnetic field strength in the Crab nebula. We also show that the Crab nebula cannot contain any energetic particles produced during the early phases.

134.003 Gamma-ray pulses from the Crab nebula.
R. L. Kinzer, G. H. Share, N. Seeman.
Astrophys. Journ., Vol. 180, 547 - 549 (1973).

Evidence is presented for a pulsed flux of γ-rays between 5 and 25 MeV. Possibly significant features in the pulse structure appear at 0.5-ms resolution.

134.004 Reddening of the Crab nebula from observations of [S II] lines. J. S. Miller.
Astrophys. Journ., (Letters), Vol. 180, L83 - L87 = Contr. Lick Obs. No. 370 (1973).

Photoelectric measurements of four infrared and two violet lines of [S II] at two places in the Crab nebula are used to derive a value for the visual interstellar extinction.

134.005 Synchrotron radiation of the Crab nebula.
K. V. Bychkov.
Astron. Zhurn. Akad. Nauk SSSR, Vol. 50, 243 - 252 (1973). In Russian. English translation in Soviet Astron. AJ, Vol. 17, No. 2.

A picture of the distribution of relativistic particles and the magnetic field of the Crab nebula is suggested. This picture helps to explain the general regularities of the synchrotron radiation of the nebula in the whole observed frequency band and peculiarities of the polarization of the synchrotron radiation in the radio and optical frequency bands.

134.006 Detection of pulsed high energy gamma rays from the Crab. M. Campbell, S. E. Ball, Jr., B. McBreen, K. Greisen, D. Koch.
Bull. American Astron. Soc., Vol. 5, 19 (1973). — Abstr. AAS.

134.007 Production of magnetic fields in the Crab nebula and related objects. J. E. Gunn, M. J. Rees.
Bull. American Astron. Soc., Vol. 5, 19 (1973). — Abstr. AAS.

134.008 Abundances in the Crab filaments. K. Davidson.
Bull. American Astron. Soc., Vol. 5, 19 (1973). Abstr. AAS.

134.009 Gamma-ray emission above 20 MeV from the Crab nebula and NP 0532.
B. Parlier, B. Agrinier, M. Forichon, J. P. Leray, G. Boella,

L. Maraschi, R. Buccheri, N. R. Robba, L. Scarsi.
Nature, Phys. Sci., Vol. 242, 117 - 120 (1973).

We have investigated the gamma-ray emission from the Crab nebula in a series of balloon flights of a multiplate spark chamber triggered by a Čerenkov detector-scintillator coincidence system. Here we report data on the continuum and pulsed gamma-ray flux from the Crab for the six successful flights analysed so far.

134.010 On the value of the magnetic field of the Crab nebula. V. V. Usov, Ya. M. Khazan.
Astron. Tsirk., No. 756, p. 1 - 4 (1973). In Russian.

134.011 The excitation mechanism for the filaments in the Crab nebula. K. M. V. Apparao.
Nuovo Cimento Lettere, Ser. 2, Vol. 5, 877 - 878 (1972).

There is growing evidence that the supernova remnant Crab nebula owes its existence to the pulsar NP 0532 at its centre. The pulsar is assumed to be a rotating magnetic neutron star with its magnetic axis inclined to the rotation axis (oblique rotator). These oblique rotators emit low-frequency radiation. The author suggests that the low-frequency radiation from the pulsar supplies energy for the excitation of the filaments of the Crab nebula.

134.012 Observation of linear polarization of the Crab nebula during an occultation by the solar corona.
N. Kawajiri, N. Kawano, Y. Sofue, K. Kawabata.
Journ. Radio Res. Lab., (Japan), Vol. 19, 229 - 247 (1972).

Position angles of linearly polarized radio waves from the Crab nebula are observed at the wavelength, λ = 7.2 cm, when the source is occulted by the southern solar corona in the middle of June, 1971. In order to avoid errors due to the reception of the solar radiation by the side lobes, two independent procedures are used for the observation and reduction of data.

Experimental X-ray astronomy.
See Abstr. 061.064.

Possible records of the Crab nebula supernova in the Western United States. See Abstr. 125.015.

Secular decrease of the flux of supernova remnants Cas A and SN-1572. See Abstr. 125.028.

VLBI observations of the Crab nebula pulsar.
See Abstr. 141.506.

Phenomena expected from the ejection of particles by the pulsar NP 0532 into the Crab nebula. See Abstr. 141.513.

The expected variability of pulsar NP 0532 radiation after spin jumps. See Abstr. 141.524.

On the origin of high-frequency radiation from the Crab nebula. See Abstr. 141.546.

Observations of Taurus X-1 by the 1–60 keV X-ray detector on the OSO-7. See Abstr. 142.009.

Plasma effects and the acceleration of charged particles in pulsar fields. See Abstr. 143.003.

Radio Sources, Quasars, Pulsars, X Ray-, Gamma Ray-Sources, Cosmic Radiation

141 Radio Sources, Quasars, Pulsars

Radio Sources, Quasars

141.001 **On the distance of the galactic radio source W51.**
F. Sato.
Publ. Astron. Soc. Japan, Vol. 25, 135 - 141 (1973).
The purpose of the present study is to obtain the distances to the thermal component No. 1 (G 48.6 + 0.0) of the radio source W 51 and to the nonthermal components Nos. 7 and 9 by examining the neutral hydrogen absorption features of the W 51 region in the Maryland-Green Bank Galactic 21-cm Line Survey, Second Edition (Westerhout 1969).

141.002 **Absorption spectra of quasars.**
C. F. McKee, C. B. Tarter, J. C. Weisheit.
Astrophys. Letters, Vol. 13, 13 - 18 (1973).
The ionization equilibrium of the gas producing the absorption lines in quasar spectra is studied in detail. Lower limits are placed on the density of the absorbing gas, and it is shown that in some cases this gas must be at large distances from the quasar. The results are applied to 4C 05.34.

141.003 **Internal Faraday rotation of Cygnus A.**
S. Mitton.
Astrophys. Letters, Vol. 13, 19 - 22 (1973).
It is shown that the distribution of rotation measures across Cygnus A correlates with the radio source structure. It is argued that the anomalous rotation is intrinsic to the source.

141.004 **Circular polarization studies of selected compact sources at 3240 MHz.**   E. R. Seaquist.
Astron. Astrophys., Vol. 22, 299 - 308 (1973).
This paper is concerned with circular polarization observations of six compact radio sources at 3240 MHz over an extended period of time. The sources observed (PKS 1127−14, BL Lac, CTA 102, 3 C 138, NRAO 140, DA 344) were selected from my original survey (Seaquist, 1969). The observations presented here reveal variable circular polarization in PKS 1127−14, BL Lac, and CTA 102, and for possible circular polarization in 3 C 138 and NRAO 140. The results are compared with those of other observers at different frequencies.

141.005 **The haloes of double radio sources.**
V. N. Kurilchik.
Astron. Zhurn. Akad. Nauk SSSR, Vol. 50, 65 - 71 (1973).
In Russian. English translation in Soviet Astron. AJ, Vol. 17, No. 1.
The long-wave excesses of the radio emission observed in the spectra of many double radio sources are connected with extended haloes of these objects. A close connection of the haloes radio emission spectra with spectra of the main double structures is recognized. This connection shows that double structure and its halo are the result of one event. There is a tendency for the radio galaxies of the clusters and those of the field to have different spectral type of the haloes (L and M spectra respectively). An estimate of the physical parameters of haloes shows that these substructures may be

powerful sources of Compton radiation (inverse Compton effect) at 3° background radiation of the Metagalaxy.

141.006 **Preliminary results of observations of discrete sources and Jupiter at 2 cm in Pulkovo.**
V. M. Bogod, O. A. Golubchina, V. G. Mirovsky, T. B. Pjatunina, N. S. Soboleva, I. A. Strukov, P. A. Fridman.
Astron. Zhurn. Akad. Nauk SSSR, Vol. 50, 72 - 82 (1973).
In Russian. English translation in Soviet Astron. AJ, Vol. 17, No. 1.
Preliminary results of investigations of discrete sources and Jupiter made at λ = 2.04 cm with the large Pulkovo radiotelescope are given. Point sources in the Orion nebula with flux density higher than 3 f. u. are absent. Hence the conclusion is made that compact details are density fluctuations which are in equilibrium with their surroundings. A new source is revealed not far from the optical object in the Omega nebula. Electron density within the region of strong polarization in the Crab nebula is estimated, being smaller than interstellar medium density. The brightness distribution of radio emission over the disk of Jupiter at λ = 2 cm reveals darkening to the edge. Hence the temperature in Jupiter's atmosphere increases with depth.

141.007 **Long-term variations of total and polarized fluxes, absolute energy distribution, and line strength of BL Lacertae and four quasi-stellar sources.**   N. Visvanathan.
Astrophys. Journ., Vol. 179, 1 - 20 (1973).
Observations of blue magnitude, polarization, wavelength dependence of polarization, absolute energy distribution of the continuum, and the strength of emission lines were obtained simultaneously for the variable QSSs 3C 279, 3C 345, 3C 446, 3C 454.3, and the object BL Lacertae on several occasions during the period 1967−1970.

141.008 **Polarization of radio sources. IV. The compact source PKS 2134 + 004.**
A. G. Pacholczyk, T. L. Swihart.
Astrophys. Journ., Vol. 179, 21 - 28 (1973).
Data on flux, circular polarization, linear polarization, and angular size for apparently simple, nonvariable compact radio sources are brought together and analyzed in an attempt to restrict the range of possible models. Three such sources are investigated: PKS 2134 + 004, PKS 0237 − 23, and PKS 1148 − 00.

141.009 **Evidence for ejection of radio sources from supernova remnants.** D. S. Mathewson, J. N. Clarke.
Astrophys. Journ., Vol. 179, 89 - 95 (1973).
Evidence is presented that the radio source 0525−66.0 has been ejected from N49, a supernova remnant in the Large Magellanic Cloud, to a distance of at least 13 times the radius of the shell of N49. A similar process is thought to have occurred in N11 L, another supernova remnant in the Large Cloud.

141.010 **Optical variations of OJ 287, ON 231, and OQ 208.**
E. R. Craine, J. W. Warner.
Astrophys. Journ., (*Letters*), Vol. 179, L53 - L56 (1973).

Optical observations of three Ohio radio sources made during the early part of 1972 are presented. Old plate collections have yielded a 20-year period of observations of OJ 287, with a possible indication of "intraday" variations; OQ 208 is now suspected to be a long-term optical variable.

141.011 **An analysis of the absorption spectrum of PHL 957.**
J. N. Bahcall, P. C. Joss.
Astrophys. Journ., Vol. 179, 381 - 389 (1973).

Five absorption redshifts ranging from $z = 2.67$ to $z = 2.22$ are identified in the observed spectrum of PHL 957, in good agreement with the earlier analysis of Lowrance et al. Our results are to be compared with an average of 1.0 ± 0.9 redshifts found in random-number spectra having the same essential characteristics as the observed spectrum. The distribution of identified absorption redshifts is presented and discussed for five quasars, whose rich absorption spectra have all been studied with the same methods as those used in the present investigation.

141.012 **Generalized angular-velocity formula and kinematical analysis of 3C 279.** E. Alvarez, L. Bel.
Astrophys. Journ., Vol. 179, 391 - 393 (1973).

A figure has been given recently by Whitney for the relative angular velocity of the two radio components of 3C 279. Using the generalized redshift formula obtained previously and the generalized angular-velocity formula for uniform models that we have derived, we analyze the kinematics of the two components of 3C 279 using the figure mentioned before and several simplifying assumptions.

141.013 **Multifrequency polarization observations of eight extragalactic sources.**
F. S. Gauss, S. J. Goldstein, Jr.
Astrophys. Journ., Vol. 179, 439 - 444 (1973).

Observations of linear polarization in forty 5-MHz bands between 1250.4 and 1445.4 MHz show the expected linear relation between position angle of the polarized vector and wavelength squared for eight sources. Seven of the sources have Faraday rotations in agreement with values in the literature obtained at higher frequencies. Upper limits to the difference in Faraday rotation of the two components of five double sources are derived.

141.014 **Quasars as events in the nuclei of galaxies: The evidence from direct photographs.** J. Kristian.
Astrophys. Journ., (Letters), Vol. 179, L61 - L65 (1973).

Similarities between quasars and the nuclei of N galaxies and Seyfert galaxies suggest that quasars may also occur in galaxy nuclei. 200-inch direct photographs of quasars are consistent with this hypothesis. Those quasars which are predicted to show underlying galaxies do so, and those which are predicted not to show underlying galaxies do not.

141.015 **Observations of 3C 129, 3C 129.1 and 3C 83.1 B at 2.7 and 5 GHz.** J. M. Riley.
Monthly Notices Roy. Astron. Soc., Vol. 161, 167 - 180 (1973).

The radio sources 3C 129 and 3C 83.1 B, each of which consists of a long 'tail' extending away from the associated galaxy, have been mapped at 2.7 and 5 GHz with the Cambridge One-Mile telescope. In each case, the double structure previously observed near the galaxy at 1.4 GHz is found to be of considerable complexity. Observations have also been made of 3C 129.1 and reveal four components, two on each side of the associated galaxy. Tables are given summarizing the characteristics of 3C 129, 3C 83.1 B, IC 310 and 5C 4.81, and of the compact radio galaxies with which they may be physically associated.

141.016 **The relation of the redshifts of radio sources to their**
angular diameters. G. M. Richter.
Astrophys. Letters, Vol. 13, 63 - 64 (1973).

The observed angular diameter-redshift distribution for radio sources is shown to be in good agreement with the predictions from the Ryle-Longair concept for the evolution of individual radio sources.

141.017 **Some radio sources with enhanced spectra at centimeter wavelengths.** M. A. Stull, R. D. Carpenter.
Astrophys. Letters, Vol. 13, 73 - 75 (1973).

8000-MHz flux densities are given for 32 radio sources having flat or enhanced high-frequency spectra.

141.018 **Two puzzling objects: OJ 287 and BL Lacertae.**
M. A. Stull.
Sky Telescope, Vol. 45, 224 - 226 (1973).

141.019 **The Hubble diagram for the brightest quasars.**
J. N. Bahcall, R. E. Hills.
Astrophys. Journ., Vol. 179, 699 - 703 (1973).

The slope of the magnitude-redshift relation for the optically most luminous quasars with redshifts ranging from 0.2 to more than 2 is consistent with the value of 5 expected from the expansion of the universe if luminosities are evaluated assuming quasars are at the cosmological distances implied by their redshifts.

141.020 **Variations of the radio source OJ 287 at optical wavelengths.** N. Visvanathan, J. L. Elliot.
Astrophys. Journ., Vol. 179, 721 - 730 (1973).

Accurate photoelectric observations of the radio source OJ 287 and a standard star during 1972 March indicate the possible presence of a periodic component with a period of 39.2 min and an amplitude of 0.0065 mag. The light curve obtained from the available photographic plates, dating from 1894 to the present, exhibits four bursts with time-scales of several months. A slowly varying component rising and falling over a period of 2 years can be seen in the 1971 outburst. Similarities between OJ 287 and BL Lac are discussed.

141.021 **A statistical investigation of the properties of quasars.**
R. Fanti, L. Formiggini, C. Lari, L. Padrielli, J. K. Katgert-Merkelijn, P. Katgert.
Astron. Astrophys., Vol. 23, 161 - 170 (1973).

A discussion is given of most of the observational material concerning QSO's and QSS's. From the $N(m)$-relation and the V/V_{max}-test it is found that the density of both classes of objects increases with increasing distance. An analysis of the bivariate luminosity function of QSS's shows that this function can be conveniently described by postulating a correlation between the radio luminosity function and the optical luminosity such that the radio luminosity function depends on the ratio of radio and optical luminosity. Such a correlation makes it possible to have identical optical luminosity functions for QSO's and QSS's.

141.022 **A first 1415 MHz survey with the Westerbork Synthesis Radio Telescope: An attempt to detect radio emission from quasi-stellar objects.** P. Katgert, J. K. Katgert Merkelijn, R. S. Le Poole, H. van der Laan.
Astron. Astrophys., Vol. 23, 171 - 194 (1973).

A survey has been carried out with the Westerbork Synthesis Radio Telescope in an attempt to detect radio emission from a sample of suspected QSO's. Due to the high sensitivity of the instrument, sources with flux densities down to ≈ 0.007 f.u. can be detected in a single twelve-hour observation; positional accuracy is of the order of an arc second. Out of a total of 99 blue stellar objects, four are detected and another four may be radio emitting near the detection limit. A catalogue of an additional 220 sources found in the survey is presented. An identification programme has been carried out for these sources, yielding 14 possible quasars (including three of the suspected QSO's), 25 galaxies and nine possible galaxies.

141.023 Optical identifications of radio sources from the B 2 catalogue. Quasi-stellar sources. R. Bergamini, A. Braccesi, G. Colla, C. Fanti, R. Fanti, A. Ficarra, L. Formiggini, E. Gandolfi, I. Gioia, C. Lari, B. Marano, L. Padrielli, P. Tomasi, M. Vigotti.
Astron. Astrophys., Vol. 23, 195 - 207 (1973).

This paper presents a homogeneous sample of 70 blue stellar objects, which exhibit the typical ultraviolet excess of quasars. These have been found during the programme of optical identification of radio sources from the B 2 catalogue, using *UBV* plates of the 48″ Schmidt telescope of Mount Palomar. Some of these objects have been observed with the Westerbork instrument, to improve the radio positions and determine the distribution of the spectral indices between 408 and 1415 MHz. Finally we examine the counts as function of the flux of our quasars, on the basis of the M. Schmidt model on the existence of an universal function giving the distribution of the ratios between the radio and the optical flux.

141.024 The distances of the quasars. M. Rowan-Robinson.
Astron. Astrophys., Vol. 23, 331 - 336 (1973).

The implications of the proposal that there are two distinct classes of quasar (one "local", the other "cosmological") for observations of radio, optical and X-ray frequencies are examined, and some possible models to explain the intrinsic redshift component in "local" quasars are considered.

141.025 2.8 cm radio emission from α Orionis, HBV 475 and MWC 349. W. J. Altenhoff, H. J. Wendker.
Nature, Vol. 241, 37 - 38 (1973).

We have used the Max-Planck-Institut 100-m telescope to search for radio emission from stars that are believed to have ejected or to be ejecting matter. This communication concerns the three definite new detections: α Ori, HBV 475 and MWC 349.

141.026 Radio sources identified with stellar objects using precise radio and optical positions.
H. Gent, J. H. Crowther, R. L. Adgie, D. G. Hoskins, H. S. Murdoch, C. Hazard, D. L. Jauncey.
Nature, Vol. 241, 261 - 263, with a correction, Vol. 242, 486 (1973).

It is difficult to establish the identification of a radio source with an object of stellar appearance and neutral colour. We show in this letter that such identifications can be made by comparing radio and optical positions of accuracy better than 1 arc sec.

141.027 Ionization potential—redshift correlations in absorption line QSOs. A. Evans.
Nature, Phys. Sci., Vol. 241, 5 - 7 (1973).

Bahcall (1966) drew attention to an apparent correlation between ionization potential (IP) and redshift in the absorption line spectrum of the QSO 3C 191. The author concludes that the IP-Δz correlation in 3C 191 is not a significant one, and that similar correlations do not exist in the redshift systems of other QSOs.

141.028 Intensity variations in a complete sample of radio sources at 2,300 MHz. G. D. Nicolson.
Nature, Phys. Sci., Vol. 241, 90 - 92 (1973).

Studies of a statistically complete sample of radio sources show that variable sources form a significant component of the strong radio source population above 1,400 MHz.

141.029 Photon mass, quasar redshifts and other abnormal redshifts. J. C. Pecker, W. Tait, J. P. Vigier.
Nature, Vol. 241, 338 - 340 (1973).

We discuss the application to quasars of a mechanism which provides a simple explanation of the essential observational paradoxes which have prevented a coherent understanding of the nature and distance of QSO sources.

141.030 Confirmation of I Zw 1727+50 as a radio source. A. N. Argue, C. M. Kenworthy, M. Ryle, J. R. Shakeshaft.
Nature, Phys. Sci., Vol. 241, 139 (1973). – Letter.

141.031 BL Lac: Strong short-term variability. D. Weistrop.
Nature, Phys. Sci., Vol. 241, 157 - 158, with a correction in Nature, Phys. Sci., Vol. 243, 144 (1973).

As part of an international effort to observe BL Lacertae (VRO 42.22.01) simultaneously at several wavelengths, the author obtained photoelectric magnitudes for this object on December 5, 6 and 7, 1972. The results indicate larger short-term changes in magnitude than have yet been observed in BL Lac.

141.032 Chain of radio sources near 4C 19.15. B. H. Andrew, J. R. Ehman.
Nature, Phys. Sci., Vol. 241, 158 (1973). – Letter.

141.033 Effect of the availability of search lines in determining QSO emission line redshift distribution. D. Basu.
Nature, Phys. Sci., Vol. 241, 159 - 160, with a correction in Nature, Phys. Sci., Vol. 243, 144 (1973).

The present analysis shows that there is an appreciable effect of availability of search lines at various redshifts on the distribution of emission line redshifts of QSOs in the list of De Veny et al. (1971). This supports the suggestion of Karitskaya and Komberg (1970) and of Roeder (1971) that selection effects are present in the emission line redshift distribution of QSOs.

141.034 Radio source counts in cosmology. F. Hoyle, with a reply by M. Schmidt.
Nature, Vol. 242, 108 - 109 (1973). – Letter.

141.035 Radio emission from the close binary b Persei. R. M. Hjellming, C. M. Wade.
Nature, Vol. 242, 250 (1973).

We report the detection of faint intermittent radio emission from the close binary star b Persei. Concurrent observations at 2,695 and 8,085 MHz were made with the NRAO interferometer on spacings of 900, 1,800 and 2,700 m. The total observing time was 14.4 h, mostly in short periods between February 7 and February 15, 1972. Further brief observations were obtained on April 28 and June 21, 1972.

141.036 Frequency dependence of circular polarization in three compact radio sources. E. R. Seaquist, P. C. Gregory, F. Biraud, T. R. Clarke.
Nature, Phys. Sci., Vol. 242, 20 - 23 (1973).

We report simultaneous observations of circular polarization of 3C 138, 3C 279, and BL Lacertae (VRO 42.22.01) at frequencies 1,666, 2,695, 8,085, and 13,500 MHz. In addition we report separate observations of the same sources made at other epochs at 3,240 MHz and 13,500 MHz. The observations are used to examine the frequency dependence of the circular polarization of these three sources.

141.037 First results from the Texas interferometer: Positions of 605 discrete sources. J. N. Douglas, F. N. Bash, F. D. Ghigo, G. F. Moseley, G. W. Torrence.
Astron. Journ., Vol. 78, 1 - 17 (1973).

A new five-element interferometer, whose operation is based on the principle of space-frequency equivalence, is engaged in the determination of 365-MHz positions (accurate to ±1 arcsec), flux densities and fringe visibilities of small diameter (<30 arcsec) discrete sources brighter than 0.25×10^{-26} $Wm^{-2} sec^{-1}$. Positions and flux densities for 605 sources are presented here, together with indications of the progress of a parallel program in optical identification of sources on the basis of precise position coincidence. The weighted rms residual between precise radio and optical positions in the present sample (85 objects) is ± 1.5 arcsec in right ascension, ± 1.0 arcsec in declination.

141.038 Redshift of OH 471.
R. F. Carswell, P. A. Strittmatter.
Nature, Vol. 242, 394 - 395 (1973).

The purpose of this letter is to draw attention to the very high redshift, $z = 3.40$, of the QSO OH 471.

141.039 Cyclicity of optical variations of 3C 273.
V. E. Chertoprud, L. I. Gudzenko, L. M. Ozernoy.
Nature, Phys. Sci., Vol. 242, 70 - 71 (1973).

Wheeler (1972) has discussed our method for the analysis of optical variability of 3C 273 but has come to different conclusions. The aim of this letter is to establish the reasons for this difference.

141.040 Radio emission nebulae surrounding MWC 349 and RY Scuti.
R. M. Hjellming, L. C. Blankenship, B. Balick.
Nature, Phys. Sci.,.Vol. 242, 84 - 85 (1973).

We report observations of MWC 349 and RY Scuti which show that the radio emission from each source is a consequence of resolvable radio nebulosities with thermal radio spectra. The observations were made with the NRAO interferometer at 2,695 and 8,085 MHz.

141.041 Identification of southern quasi-stellar objects – III.
B. A. Peterson, J. G. Bolton.
Astrophys. Letters, Vol. 13, 187 - 192 (1973).

Forty new southern quasi-stellar objects have been found from combined radio and optical observations.

141.042 Circular polarization in inverse Compton scattering of synchrotron radiation.
S. A. Bonometto, A. Saggion.
Astrophys. Letters, Vol. 13, 193 - 197 (1973).

A simple model is used to show that inverse Compton scattering of synchrotron radiation by the very electrons that produced it leads to an emission in which the degree of circular polarization is expected to be much smaller than the degree of circular polarization of the synchrotron radiation itself. The relevance of these results in connection with recent observations of Landstreet and Angel is outlined.

141.043 Magnetic dipole radiation from a supermassive oblique rotator.   L. M. Ozernoy, V. V. Usov.
Astrophys. Letters, Vol. 13, 209 - 214 (1973).

We present the results of a calculation of magnetic dipole radiation from a supermassive oblique rotator (SOR) with a non-negligible contribution of radiation pressure. Consequences of the theory of an SOR are suggested that can serve as observational tests for its validity.

141.044 On the calibration of flux densities and the determination of spectra at radio frequencies.
B. J. Wills.
Astrophys. Journ., Vol. 180, 335 - 350 (1973).

The confidence with which one can define spectra for the majority of catalogued radio sources is limited by the inaccuracy of their flux densities relative to that of Cassiopeia A, and by uncertainty in the spectrum of Cas A itself. These uncertainties are investigated, revising the spectrum for Cas A

and using accurate measurements of relative intensity.

141.045 Observations of variability in OJ 287.
S. Goldsmith, D. Weistrop.
Astrophys. Journ., Vol. 180, 661 - 664 (1973).

Photographic observations in V and B of the radio source OJ 287 have been made in 1972 February – May. The shortest period in which the variations were looked for was 5 minutes. On a timescale of days OJ 287 seems to have periods of intensive activity, variations of 0.3 mag, and periods of inactivity. The data do not indicate variations over a period of minutes.

141.046 A search for radio variations in Virgo A and Cygnus A.   D. S. De Young, D. E. Hogg.
Astrophys. Journ., (Letters), Vol. 180, L61 - L64 (1973).

Results of observations of Virgo A and Cygnus A at 8085 MHz and 2695 MHz which cover a 2-year period are presented. The implications of these results for various theoretical models are briefly discussed.

141.047 Variation in the polarization across bends in the spectra of self-absorbed synchrotron sources.
D. B. Melrose.
Proc. Astron. Soc. Australia, Vol. 2, 140 - 142 (1972).

141.048 31.4-GHz flux density measurements of variable radio sources.   W. A. Dent, R. W. Hobbs.
Astron. Journ., Vol. 78, 163 - 169 = Contr. Five College Obs., Univ. Mass., Amherst, No. 158 (1973).

Measurements of change in flux density with time are presented for 21 sources at 31.4 GHz. The sources, in general, show a higher degree of variability at 31.4 GHz than they do at lower frequencies.

141.049 Simplified photoionization analysis of quasar emission spectra.   K. Davidson.
Astrophys. Journ., Vol. 181, 1 - 14 (1973).

Simplified photoionization calculations, considering quasar emission lines with ultraviolet rest-wavelengths, can be made including only hydrogen, helium, and carbon ions. With plausible assumptions about the helium abundance and ionizing spectral slope, estimates of the (radiation/gas) pressure ratio and carbon abundance are possible.

141.050 A survey of thermal radio sources at 8.2 mm wavelength with high resolution.
I. I. Berulis, R. L. Sorochenko.
Astron. Zhurn. Akad. Nauk SSSR, Vol. 50, 270 - 282 (1973). In Russian. English translation in Soviet Astron. AJ, Vol. 17, No. 2.

A survey of thermal radio sources at 8.2 mm wavelength was carried out with a 22-m radio telescope of resolution 1.9 arc min. Coordinates for 26 selected sources were determined and other physical parameters were measured.

141.051 A low frequency search for compact radio sources in supernova remnants.   D. E. Harris.
Bull. American Astron. Soc., Vol. 5, 28 (1973). – Abstr. AAS.

141.052 Source positions from very long baseline interferometer observations.
T. A. Clark, C. C. Counselman III, H. F. Hinteregger, C. A. Knight, D. S. Robertson, A. E. E. Rogers, I. I. Shapiro, A. R. Whitney.
Bull. American Astron. Soc., Vol. 5, 30 (1973). – Abstr. AAS.

141.053 The source count of weak soruces at 2695 MHz.
E. B. Fomalont.
Bull. American Astron. Soc., Vol. 5, 30 (1973). – Abstr. AAS.

141.054 Observations of Ohio-survey radio sources at 430

MHz. M. A. Stull.
Bull. American Astron. Soc., Vol. 5, 30 - 31 (1973). – Abstr. AAS.

141.055 High resolution observations of Cas-A at 26.3 MHz.
L. K. Hutton, T. A. Clark, W. M. Cronyn.
Bull. American Astron. Soc., Vol. 5, 35 (1973). – Abstr. AAS.

141.056 Redshift-magnitude banding among quasi-stellar sources. W. G. Tifft.
Bull. American Astron. Soc., Vol. 5, 40 (1973). – Abstr. AAS.

141.057 Relativistic motion of ram pressure confinement models for extragalactic radio sources.
W. A. Christiansen.
Bull. American Astron. Soc., Vol. 5, 40 - 41 (1973). – Abstr. AAS.

141.058 Cosmological information from surveys of radio source spectra. B. L. Fanaroff, M. S. Longair.
Monthly Notices Roy. Astron. Soc., Vol. 161, 393 - 407 (1973).

The problem of relating the spectral index distributions of radio sources at different frequencies and flux densities is discussed in terms of evolutionary world models which incorporate strong evolution of the radio source population. It is shown that even in the simplest world models significant changes with flux density in the spectral index distributions are expected in high frequency surveys. The implications of this result for source counts at different frequencies and for the identifications of radio sources are discussed.

141.059 Depolarization in flat-spectra quasars.
A. G. Pacholczyk, S. A. Gregory.
Monthly Notices Roy. Astron. Soc., Vol. 161, 31P - 34P (1973).

It is shown that the wavelength dependence of integrated polarization typical of a composite flat spectrum radio source can be represented as resulting from a superposition of components with polarization spectra similar to those of steep-spectra objects.

141.060 Further observations of the radio star MWC 349.
P. C. Gregory, E. R. Seaquist.
Nature, Phys. Sci., Vol. 242, 101 - 102 (1973).

The authors present observations of MWC 349 at 10.52 and 6.63 GHz obtained during December 1972 and February 1973. Over this period they find no evidence for a change in the flux density of MWC 349 greater than 0.02 f.u. Combining these measurements with the measurements at 1.4 and 5 GHz results in a spectrum consistent with a thermal radiation mechanism.

141.061 A new red-shift mechanism for quasars.
D. K. Ross.
Astron. Astrophys., Vol. 24, 471 - 474 (1973).

A new red-shift mechanism for quasars is proposed which uses the variation of particle rest masses and hence atomic energy levels in a scalar gravitational field.

141.062 Accurate flux densities at 8.87 GHz of 195 radio sources. A. J. Shimmins, J. V. Wall.
Australian Journ. Phys., Vol. 26, 93 - 109 (1973).

Accurate flux densities at 8.87 GHz have been determined with the Parkes 64 m telescope for 195 radio sources, using an on-off integration method. Eighty of the selected sources are identified with QSO's, 40 with galaxies, and one with an H II region, while 74 have not been identified.

141.063 Detection of RY Scuti at radiofrequencies.
V. A. Hughes, A. Woodsworth.

Nature, Phys. Sci., Vol. 242, 116 - 117 (1973). – Letter.

141.064 Culgoora-1 list of radio source measurements at 80 MHz. O. B. Slee, C. S. Higgins.
Australian Journ. Phys., Astrophys. Suppl., No. 27, 43 pp. (1973).

The Culgoora radioheliograph operating at 80 MHz has been used to observe 999 radio sources selected from published catalogues and distributed over the declination range $-48°$ to $+35°$. The Culgoora-1 list contains measurements with $3'.7$ arc resolution of positions, flux densities, and angular sizes of 777 sources, while upper limits to the flux densities of 222 undetected sources are given in a separate list. Success rates for the detection of various classes of radio source at 80 MHz are listed and discussed. Comparisons are made between the Culgoora flux densities and those given in other catalogues.

141.065 On a possible mechanism responsible for the differential energy spectrum of relativistic electrons and non-linear low-frequency spectra of cosmic radio sources.
S. Ya. Braude and E. A. Kaner.
Astrophys. Space Sci., Vol. 20, 59 - 70 (1973).

The paper suggests an explanation of the deviations from the power law which are observed in frequency spectra of discrete radio sources at decametric wavelengths. The distribution function of the relativistic electrons, empirically established in an earlier paper (Braude et al., 1971) has been derived from the kinetic equation. For a number of discrete sources the turbulence energy density and the plasma concentration are deduced with the aid of experimental data on low-frequency radio spectra.

141.066 The velocity of separation of the components of extragalactic radio sources. C. D. Mackay.
Monthly Notices Roy. Astron. Soc., Vol. 162, 1 - 9 (1973).

Observations of a statistically complete sample of radio sources are used to derive limits to the velocity of separation of the components of typical double radio sources. For the most probable rate of luminosity evolution, the components of a typical source are unlikely to be moving away from the parent galaxy with a velocity exceeding 0.08 c.

141.067 Do number counts tell us anything about the distribution of distant sources?
R. F. Carswell.
Monthly Notices Roy. Astron. Soc., Vol. 162, 61 - 72 (1973).

The possibility of obtaining an approximate density evolution using a log N–log S related method with redshift information from a subsample is examined in some artificial cases with the hope that something better than a $(1 + z)^n$ type law may be found. It appears that for realistic models there is little information on the evolutionary behaviour at large redshifts, and a rough criterion for when this is true is given.

141.068 Occultation observations of six radio sources with flat spectra.
V. K. Kapahi, M. N. Joshi, J. Kandaswamy.
Astrophys. Letters, Vol. 14, 31 - 35 (1973).

Accurate radio positions and high-resolution structures derived from lunar occultations observed at 327 MHz, together with accurate optical positions measured from the Sky Survey prints, are presented for six radio sources with flat or peaked spectra. From a total of about 500 radio sources observed in the Ooty occultation survey, nine can be classified in this spectral category. All the nine sources show compact components unresolved with resolutions of 1 or 2 arc sec and are optically identified.

141.069 The radio source Sagittarius A West: thermal or non-

thermal? T. W. Jones.
Astrophys. Letters, Vol. 14, 47 - 50 (1973).

Sgr A West, the radio source recently discovered in the galactic center, is discussed. It is shown that the source may be nonthermal. Since the position of the source agrees closely with that of the $10\,\mu m$ infrared source in the galactic nucleus, the two sources may be associated with nonthermal activity in the nucleus.

141.070 **Luminosity functions of quasars and Seyfert galaxies.** P. Notni, G. M. Richter.
Astron. Nachr., Vol. 294, 95 - 104 (1973).

The optical luminosities of extragalactic objects with broad emission lines, i.e. quasi-stellar radio sources, radio quiet quasi-stellar objects and Seyfert galaxies are compared.

141.071 **Quasares y pulsares.** A. Romañá.
Urania Barcelona, Año 57, No. 275, p. 78 - 122 (1972).

141.072 **Identification of the 5C3 radio sources.**
A. G. Parkes, M. V. Penston.
Monthly Notices Roy. Astron. Soc., Vol. 162, 117 - 126 (1973).

A set of optical identifications for the radio survey 5C3 are presented. Both photometric and positional criteria are used, the latter utilizing the full authority of the radio positions. Identifications of quasars are complete to $U = 19.0$ and of galaxies to $V \simeq 20$.

141.073 **Evidence for luminosity evolution of quasars.**
N. V. Zotov, W. Davidson.
Monthly Notices Roy. Astron. Soc., Vol. 162, 127 - 133 (1973).

Diagrams are presented showing the systematic increase with z of the optical luminosities of 229 confirmed quasars (QSO and QSS). This does not in itself imply, but is consistent with, the presence of powerful luminosity evolution. We then examine two schemes of pure density evolution of quasars recently advanced by Schmidt and compare their predictions at low flux density with comprehensive Parkes data at 1410 MHz and 2700 MHz. The results are in marked disagreement in such a way as to suggest that the postulate of pure density evolution, with neglect of luminosity evolution, is incorrect.

141.074 **Are there two types of quasars?**
B. C. Chiu, P. Morrison, L. Sartori.
Astrophys. Journ., Vol. 181, 295 - 303 (1973).

We show that the hypothesis of two types of quasars is consistent with association and red-shift statistics, maintains the physical unity of categories of objects as well as does the single-type model, and suggests a possible interpretation of such objects as BL Lac and OJ 287.

141.075 **Hubble diagrams for quasars.**
G. Setti, L. Woltjer.
Astrophys. Journ., (Letters), Vol. 181, L61 - L63 (1973).

Quasars are subdivided into several groups. Those whose radio characteristics resemble the strong radio galaxies are found to show a clear Hubble relation.

141.076 **Black holes and absorption redshifts in quasi-stellar objects.** B. Mashhoon.
Astrophys. Journ., (Letters), Vol. 181, L65 - L69 (1973).

On the basis of the possible existence of black holes in galactic nuclei, the jet mechanism proposed by Wheeler is used to explain the absorption redshifts in QSOs.

141.077 **Position of OH 471.** J. H. Crowther.
Nature, Vol. 243, 25 - 26 (1973). – Letter.

141.078 **A low-frequency search for compact radio sources in supernova remnants.** D. E. Harris.
Astron. Journ., Vol. 78, 231 - 234 (1973).

We have made interplanetary-scintillation observations of 25 suggested supernova remnants at frequencies between 53 and 318 MHz to determine if any of these source contain compact radio components similar to that in the Crab nebula. No positive detections resulted.

141.079 **Extended extragalactic radio sources.**
D. S. De Young, G. Burbidge.
Comments Astrophys. Space Phys., Vol. 5, 29 - 36 (1973).

Our purpose is to give a critical discussion of the various proposals which have been made which try to account for the form of the extended sources which we see.

141.080 **A list of quasi-stellar radio sources and quasi-stellar radio-source candidates from the 3C and 4C catalogs between declination $-7°$ and $+40°$.**
D. Agnew, H. Arp.
Publ. Astron. Soc. Pacific, Vol. 85, 162 - 173 (1973).

A list of all 3C and 4C radio sources between declinations $-7°$ and $+40°$ identified by various works has been tabulated. This list of 225 objects provides a large and homogeneous sample of QSS and includes a brief description of their observed characteristics.

141.081 **Effects of Faraday rotation on the degree of polarization in QSOs and Seyfert galaxies.**
M. Fukui.
Publ. Astron. Soc. Japan, Vol. 25, 181 - 189 (1973).

The wavelength dependence of the degree of polarization of QSOs and Seyfert galaxies in the 3–11 cm wavelength region is interpreted in terms of internal Faraday rotation. If this interpretation is accepted, the non-relativistic electron density should be equal to $10^{-1}-10$ cm^{-3}.

141.082 **The absorption lines in quasi-stellar objects.**
J. G. Cohen.
Astrophys. Journ., Vol. 181, 619 - 625 (1973).

A model in which the gas is part of the intergalactic medium or the extreme outer parts of galaxies is discussed. In this case, the gas may be collisionally ionized. For solar metal abundance, it is surprising that Lα is the strongest line in absorption. An attempt to enhance the Lα absorption by adding a cool region in which hydrogen is predominantly neutral fails as this region is photoionized by radiation from the hot region. Therefore, the metal abundance in the absorbing gas must be less than solar if the gas is not photoionized.

141.083 **The helium abundance of 3C 273.**
M. Jura.
Astrophys. Journ., Vol. 181, 627 - 632 (1973).

Models with a time-dependent ionizing flux are sketched for 3C 273. In contrast to models with a steady ionizing flux in which a very low helium abundance is inferred, in time-dependent models the observation of weak He II 4686 is consistent with a solar helium abundance.

141.084 **Observations of Ohio-survey radio sources at 430 MHz.** M. A. Stull.
Astron. Journ., Vol. 78, 285 - 294 (1973).

The 1000-ft radio telescope of the Arecibo Observatory has been used at 430 MHz to measure flux densities and positions of 259 weak sources in the Ohio State 1415-MHz catalog. Comparison of their 430-MHz and 1415-MHz flux densities shows that 87 of the 259 appear to have spectra which are either flat or show a low-frequency cutoff.

141.085 **Polarimetric observations of non-stable stars and extragalactic objects. III. Polarization of quasars**

and galactic nuclei. Yu. S. Efimov, N. M. Shakhovskoy.
Izv. Krymskoj Astrofiz. Obs., Vol. 46, 3 - 13 (1972).
In Russian.

The results of multicolour polarimetric observations of
some quasars, N-galaxies, compact galaxies and nuclei of
Seyfert galaxies, obtained in 1969–1970 with the 2.6-m reflecting telescope, are presented.

**141.086 Origin of broad emission lines from quasistellar
objects.** R. Ptak, R. Stoner.
Nature, Vol. 243, 280 (1973).

The optical spectra of QSOs exhibit very broad high-
excitation emission lines. We have recently proposed a
radially-streaming-proton model that gives an excellent fit to
profiles of the broad $H\alpha$ emission lines from three different
Seyfert galaxy nuclei. We suggest here that a similar idea will
also explain the observed width of QSO emission lines. In the
model, suprathermal protons slow down as they stream radially through a hydrogen gas that is at rest with respect to their
source at the centre.

**141.087 Compact radio source associated with the OH
source ON-1 (OH69.5–1.0).**
A. Winnberg, H. J. Habing, W. M. Goss.
Nature. Phys. Sci., Vol. 243, 78 - 81 (1973).

We report the discovery of an isolated high density H II
region associated with a class 1 OH source; there is no detectable low density H II region surrounding the compact source.
This source may well be the first stage in the evolutionary
process of the H II region.

141.088 Radio stars AR Lacertae and Cygnus X-2.
R. M. Hjellming, L. C. Blankenship.
Nature, Phys. Sci., Vol. 243, 81 - 82 (1973).

We report observations of radio emission from two stars:
AR Lacertae, an Algol-type eclipsing and spectroscopic binary
with a period of 1.98 d; and the peculiar blue star suggested as
the optical counterpart of the X-ray source Cygnus X-2.

**141.089 Observations at 1415 MHz of radio sources in the
field of the double-galaxy system NGC 2798/99.**
R. J. Allen, W. T. Sullivan III.
Astron. Astrophys., Vol. 25, 187 - 190 (1973).

The radio sources in a field of about one square degree
centered near the galaxies NGC 2798 and NGC 2799 have
been observed with the Westerbork Synthesis Radio Telescope
in the continuum at 21-cm wavelength. The purpose of our
observations was to search for further evidence of a physical
interaction between these two galaxies in the form of bridges
or jets of nonthermal radio emission. We find no such evidence
in the present observations.

141.090 Redshift of OQ172. E. J. Wampler,
L. B. Robinson, J. A. Baldwin, E. M. Burbidge.
Nature, Vol. 243, 336 - 337 (1973).

We report the discovery of a second QSO with a redshift
greater than 3. Spectra taken at the 120-inch at Lick Observatory give z = 3.53 for OQ172.

141.091 Optical behaviour of four quasi-stellar objects.
W. Pfau.
Inform. Bull. Variable Stars (I. A. U. Commission 27), Konkoly
Obs., Budapest, No. 787, 3 pp. (1973).

**141.092 Interferometry, scintillation, and minimum angular
diameter.** W. M. Cronyn, M. H. Cohen.
Bull. American Astron. Soc., Vol. 5, 284 (1973). – Abstr. AAS.

**141.093 Structure and apparent motions in compact radio
sources. The quasar patrol.** K. I. Kellermann.
Bull. American Astron. Soc., Vol. 5, 285 (1973). – Abstr. AAS.

**141.094 Flux density measurements of selected radio sources
relative to Cas A at 21.8 GHz.** M. J. Klein.
Bull. American Astron. Soc., Vol. 5, 285 (1973). – Abstr. AAS.

141.095 Compact radio source structure at 2.8 cm.
D. Shaffer.
Bull. American Astron. Soc., Vol. 5, 286 (1973). – Abstr. AAS.

**141.096 A comparison between Compton-synchrotron and
Compton black-body emission in radio sources.**
J. R. Albano, R. Terlevich, J. Frank.
Astrophys. Space Sci., Vol. 21, 177 - 187 (1973).

A simple relation between observational parameters of
radio sources that allows to state the predominance in the
X-ray emission of either the Compton-synchrotron or the
Compton black-body effect is derived. The obtained results
suggest that the Compton-synchrotron contribution is dominant in compact radio sources.

141.097 High-resolution methods in radio astronomy.
G. Silvestro.
Mem. Soc. Astron. Italiana, Nuova Ser,, Vol. 43, 599 - 627
(1973). – Invited paper presented to the 'Giornate di studio
dedicate al Prof. F. Zagar' (see 012.023).

Current methods in high-precision measurements of
radio source position and angular size are reviewed, with regard to the problem of the source structure and evolution.
The present resolution limit is of the order 0.001 second of
arc at the highest radio frequencies, but is still not very satisfactory at low frequencies. A new method is presented, based
on the radio source occultation by the earth, detected outside the ionosphere, that could allow a significant low-frequency resolution improvement, and appears to be suitable
for detecting weak radio sources.

141.098 QSO historical light curves. R. J. Angione.
Astron. Journ., Vol. 78, 353 - 368 (1973).

Optical variability of 20 QSO's and three Seyfert or
Seyfert-like galaxies has been studied based on light curves
obtained from the Harvard historical plate collection. Individual light curves are presented along with discussion of their
interpretation. Each of these 23 objects, selected solely because of apparent brightness, is found to be optically variable.
The variations range from barely detectable ($\sigma > 0.13$ mag) to
over two magnitudes.

**141.099 Interplanetary-scintillation observations of 203
sources identified as radio galaxies or quasars.**
D. E. Harris.
Astron. Journ., Vol. 78, 369 - 375 (1973).

Optically identified radio sources have been observed to
determine the presence of any small-diameter components.
The data, which were obtained at the Arecibo Observatory at
318 and 430 MHz, emphasize the statistical differences between quasars and radio galaxies and demonstrate a correlation between redshift and scintillation visibility for quasars.

141.100 Radio variability of HDE 226868 (Cygnus X-1).
R. M. Hjellming.
Astrophys. Journ., (*Letters*), Vol. 182, L29 - L31 (1973).

The author analyzes all of the observations of Cyg X-1
taken by the NRAO interferometer between 1971 February
23 and 1972 October 21, mostly with simultaneous measurements at 2695 and 8085 MHz. The data show that the initial
"turn-on" of the radio source occurred in 1971 between
March 22 and March 31, just as the X-ray flux was beginning
its change in mean flux level; in addition, another change in
level of mean radio flux occurred sometime between 1972
April 25 and 1972 September 3, and significant variations on
the time scale of days have been seen.

141.101 The evolution of extra-galactic sources reproduced by the radio data. W. Hirth.
Mitt. Astron. Ges., No. 32, p. 116 - 118 (1973).

141.102 Automatic reduction of radio astronomical maps: a map of the W 43 region at 4.5 cm.
L. A. Higgs.
Journ. Roy. Astron. Soc. Canada, Vol. 67, 123 - 138 (1973).

A computer program that produces and analyzes contour maps from radio astronomical data, recorded at the telescope directly onto magnetic tape, is briefly described and the results of a test survey of W 43 at 4.5 cm are presented.

141.103 Scintillations of sources of finite angular dimensions.
V. I. Shishov.
Izv. vyssh. ucheb. zavedenij. Radiofizika, Vol. 15, 1277 - 1285 (1972). In Russian. – Abstr. in Referativ. Zhurn. 51. Astron., 5.51.272 (1973).

141.104 Circular polarization of radiation from cosmic objects. V. N. Sazonov.
Uspekhi fiz. nauk, Vol. 108, 583 - 594 (1972). In Russian. Abstr. in Referativ. Zhurn. 51. Astron., 5.51.278 (1973).

141.105 Angular diameters of quasars of unusual colour.
I. W. A. Browne, R. G. Conway, R. J. Davis, R. E. Spencer, D. Stannard, R. S. Warwick.
Nature, Vol. 244, 19 - 20 (1973). – Letter.

141.106 Radio source depolarization, size and cosmology.
R. G. Strom.
Nature, Phys. Sci., Vol. 244, 2 - 4 (1973).

Data on radio source depolarization and the largest angular size against redshift correlation demonstrate the presence of geometrical effects predicted by relativistic cosmology and suggest the existence of significant amounts of intergalactic gas.

141.107 On the apparent association of quasi-stellar objects with clusters or groups of galaxies with about the same redshift. G. R. Burbidge, S. L. O'Dell.
Astrophys. Journ., (Letters), Vol. 182, L47 - L51 (1973).

The statistical evidence for the association of QSOs with clusters or groups of galaxies of about the same redshift is examined. It is shown that, due to the uncertainty in the number density of cluster centers for all but moderately rich compact clusters, it is premature to conclude that the present evidence for the association of QSOs with such groups or clusters is statistically significant.

141.108 On the character of optical variability of the quasar 3C 273.
V. E. Chertoprud, L. I. Gudzenko, L. M. Ozernoy.
Astrophys. Journ., (Letters), Vol. 182, L53 - L56 (1973).

We discuss the results of recent works referring to 'slow' (~ 10 years) brightness variations of the quasar 3C 273, as well as the pertinent observational data. The validity of our earlier conclusion that the light curve of 3C 273 is incompatible with the model of independent random pulses is confirmed in a new way.

141.109 Variations in the radio structure of BL Lacertae.
B. G. Clark, K. I. Kellermann, M. H. Cohen, D. B. Shaffer, J. J. Broderick, D. L. Jauncey, L. I. Matveyenko, I. G. Moiseev.
Astrophys. Journ., (Letters), Vol. 182, L57 - L60 (1973).

We have observed the structure of the rapid variable radio source BL Lac (VRO 42.22.01) using long baseline interferometer systems with baselines up to 266 million wavelengths. Despite large variations in the total flux and in the overall size of this source, it has maintained an elongated brightness distribution, and the direction of elongation has not changed during the 1.3 years of observation.

141.110 On the spectral index distribution of radio sources selected from the B2 sky survey.
G. Grueff, M. Vigotti.
Astron. Astrophys., Suppl. Ser., Vol. 11, 41 - 66 (1973).

Measurements at 5 GHz of B2 radio sources are presented. The distribution of the spectral index between 408 MHz and 5000 MHz is discussed. The data indicate the existence of a correlation between the average spectral index and the parameter R defined as the ratio between the radio and optical power for quasi-stellar sources. The spectral data are also used to gain information about the nature of sources with no optical counterpart on the Palomar Sky Survey.

141.111 Bubble model of extragalactic radio sources.
S. F. Gull, K. J. E. Northover.
Nature, Vol. 244, 80 - 83 (1973).

The components of radio sources are identified with bubbles of relativistic plasma rising through the hot gas which produces the X-ray emission from clusters of galaxies. Continuous release of energy in the nucleus of a galaxy will lead to the formation of two such bubbles, moving in opposite directions along the axis of rotation.

141.112 AR Lacertae. R. M. Hjellming, L. C. Blankenship.
IAU Circ., No. 2502 (1973).

141.113 Radio emission from V1016 Cygni.
P. A. Feldman, K. A. Marsh, C. R. Purton.
IAU Circ., No. 2543 (1973).

141.114 Radio emission from HD 167362 and VY 2-2.
P. A. Feldman, K. A. Marsh, C. R. Purton.
IAU Circ., No. 2549 (1973).

141.115 V1016 Cygni. W. J. Altenhoff, L. L. E. Braes, H. J. Habing, F. M. Olnon, A. A. Schoenmaker, E. P. J. van den Heuvel, H. J. Wendker.
IAU Circ., No. 2549 (1973).

141.116 The optical spectrum of the radio source B2 1101 + 38. M.-H. Ulrich.
Astrophys. Letters, Vol. 14, 89 - 90 (1973).

Low-dispersion spectra of the optical counterpart of the radio source B2 1101 + 38 show continuum emission only. The absence of discrete features in the optical spectrum suggests that this object may be of the BL Lac type.

141.117 Quasar redshifts and peculiar velocities.
H. Dehnen, O. Obregón.
Astrophys. Letters, Vol. 14, 91 - 97 (1973).

For quasars with high peculiar velocities the functional relations among the cosmological redshift, the peculiar velocity, and the change of the angular separation are given. The results are applied to the quasar 3C 279, assuming three different models for this object. In the special case in which the optical source of this object is co-moving with one of the separating radio components, a drastic reduction of the distance of this object by about 67 per cent would be necessary. This harmonizes with the observations of quasar–galaxy pairs.

141.118 Optical positions for 24 radio sources.
A. N. Argue, C. M. Kenworthy, P. M. Stewart.
Astrophys. Letters, Vol. 14, 99 - 104 (1973).

Positions have been measured photographically for 24 specially selected radio objects, and compared with accurate 2695 MHz positions. Six objects show significant differences which, when combined with other data, suggest the morphology of these objects. Eleven objects that show no significant differences may be suitable as astrometric reference points; the remainder are unsuitable because of finite resolution.

141.119 On the interpretation of the redshift—angular size diagram for quasars. J. C. Jackson.
Monthly Notices Roy. Astron. Soc., Vol. 162, 11P - 13P (1973).

Certain natural assumptions about the evolution of individual quasars allow their redshift–angular size diagram to be interpreted in terms of a quasar population which, with respect to linear dimensions, is not changing with cosmic epoch.

141.120 QSO's near bright galaxies.
I. W. A. Browne, N. J. McEwan.
Monthly Notices Roy. Astron. Soc., Vol. 162, 21P - 24P (1973).

The number of apparent associations between bright galaxies and QSO's from the Parkes 2700 MHz survey is investigated in the light of improved optical identifications.

141.121 Spectroscopic and photometric observations of the quasar 4C 31.63.
K. P. Tritton, S. N. Henbest, M. V. Penston.
Monthly Notices Roy. Astron. Soc., Vol. 162, 31P - 34P (1973).

Spectra have been obtained of Olsen's suggested identifications of 4C 31.63 and 4C 24.6. These show that the former is a quasar but the latter is a foreground star. Photoelectric *UBVRI* magnitudes place the quasar among the 10 brightest known.

141.122 The circular polarization of sources of synchrotron radiation. V. N. Sazonov.
Astrophys. Space Sci., Vol. 19, 3 - 23, 25 - 45 (1972). In Russian and English.

The degree of circular polarization p_c is calculated for two models of a source of synchrotron radiation: (1) a source with an inhomogeneous magnetic field and isotropic angular distribution of the electrons with respect to the magnetic field; (2) a source with a homogeneous magnetic field and anisotropic angular distribution of the electrons in which the anisotropy of angular distribution substantially increases with the electron energy.

141.123 Statistical analysis of multiple absorption spectra in QSO. G. Shaviv, U. Feldman, B.-Z. Koslovsky.
Astrophys. Space Sci., Vol. 19, 159 - 163 (1972).

The purpose of this work is to analyze the level of confidence of the identifications of various z-systems in quasars with multiple red-shift systems. This report discusses the analysis of 4C-0534, where close to 100 absorption lines were found (Lynds, 1971).

141.124 Infrared spectra of quasars and related objects.
J. Dorschner, C. Friedemann, J. Gürtler, H. Oleak, K.-H. Schmidt.
Astrophys. Space Sci., Vol. 19, 263 - 270 (1972).

An attempt is made to explain the infrared radiation observed for several quasars and Seyfert galaxies as thermal radiation of a dust envelope surrounding the cores of these objects. Two kinds of dust particles (graphite and silica) are taken into consideration.

141.125 Analysis of the non-Gaussian spectra of interplanetary scintillations.
S. K. Alurkar, R. P. Sarker, V. Brahmananda Rao.
Astrophys. Space Sci., Vol. 19, 271 - 278 (1972).

The non-Gaussian intensity fluctuation spectra observed by Cohen et al. (1967) are analysed. Computations of the length scales derived from the phase autocorrelation functions using Buckley's method (1971) indicate that for a rms phase deviation of 4 radians or more the diffracting medium behaves as one with its phase structure having 'inner' and 'outer' scales of turbulent blobs or eddies which are present in a turbulent medium.

141.126 Some trends in the red-shift distribution of quasi-stellar objects and related peculiar galaxies.
D. Basu, M. A. Abdu.
Astrophys. Space Sci., Vol. 19, 303 - 308 (1972).

Gaps in the red-shift distribution of quasi-stellar objects and related peculiar galaxies have been studied using 205 sources. The result indicates certain definite trends in the distribution of the gaps but does not suggest any periodicity when the entire sample is considered.

141.127 On the completeness of radio source lists.
R. A. Vardanian, Yu. K. Melik-Alaverdian.
Soobshch. Byurakan. Obs., vyp. (No.) 44, p. 108 - 110 (1972). In Russian.

A new method for determination of the completeness of radio source lists is suggested. It was applied to the radio source list of Fitch et. al. (1969), which was shown to be complete by ~82%.

141.128 Quasars: the hopes confounded. M. Różyczka.
Urania Kraków, Vol. 44, 66 - 72 (1973). In Polish.

141.129 On scintillations of quasars on an inhomogeneous interstellar plasma.
V. V. Vitkevich, V. I. Shishov.
Trudy. fiz. in-ta. AN SSSR, Vol. 62, 42 - 45 (1972). In Russian. – Abstr. in Referativ. Zhurn. 51. Astron., 6.51.777 (1973).

141.130 Preliminary quasar model based on the Yilmaz exponential metric. R. E. Clapp.
Phys. Rev. D, Particles and Fields, Vol. 7, 345 - 355 (1973).

A partially collapsed spherical matter distribution is analyzed with the aid of an integral equation for the gravitational potential, derived from the Yilmaz exponential metric. The quasars are interpreted as gravitationally compacted protogalaxies, with galactic masses and dimensions which are initially stellar but increase as the quasar evolves into a galaxy.

141.131 Recent light changes in three variable radio sources.
G. H. Folsom, A. G. Smith, H. W. Schrader.
Quarterly Journ. Florida Acad. Sci., Vol. 34, 195 - 205 = Rosemary Hill Obs., Univ. Florida, *Gainesville*, Contr. No. 29 (1971).

University of Florida observations in the period 1968–1970 show continued fluctuations in the light of three extragalactic radio sources previously reported as optical variables. The most active of the sources during this period was the N galaxy 3C 371, while the least active was the quasar 3C 454.3 A Seyfert galaxy, 3C 120, displayed an intermediate level of activity.

141.132 The B2 catalogue of radio sources - third part.
G. Colla, C. Fanti, R. Fanti, A. Ficarra, L. Formiggini, E. Gandolfi, I. Gioia, C. Lari, B. Marano, L. Padrielli, P. Tomasi.
Lab. Radioastron. CNR, Ist. Fis., Bologna. Separate print, October 1972, 2 + 52 pp.

The catalogue lists 3227 radio sources observed at 408 MHz with the Bologna Northern Cross Telescope. It covers an area of 0.55 ster. between 34°02′ and 40°18′ down to 0.25 f.u. Results are given for the differential logS–logN relationship and for extended radiosources.

141.133 On the spectral index distribution of radiosources selected from the B2 Sky Survey.

G. Grueff, M. Vigotti.
Lab. Radioastron. CNR, Ist. Fis., Bologna. Separate print,
December 1972, 1 + 21 + 23 pp.

Measurements at 5 GHz of B2 radiosources are presented. The distribution of the spectral index between 408 MHz and 5000 MHz is discussed. The data indicate the existence of a correlation between the average spectral index and the parameter R defined as the ratio between the radio and optical power, for quasi-stellar sources. The spectral data are also used to gain information about the nature of sources with no optical counterpart on the Palomar Sky Survey (Empty Fields).

141.134 The absorption-line spectrum of the bright QSO Markarian 132.

C. F. McKee, W. L. W. Sargent.
Astrophys. Journ., (Letters), Vol. 182, L99 - L101 (1973).

The authors list 24 absorption lines found in the wavelength range $\lambda\lambda 3200-4600$. Eight of these lines are identified with an absorption redshift $z_{abs} = 1.7319$. There is weaker evidence for additional redshifts $z_{abs} = 1.6834$, 1.4721, and 1.2750. Even if these are correct, many absorption lines remain unidentified.

141.135 Photographic photometry of compact extragalactic objects. I.

M. K. Babadzhanyants, V. A. Hagen-Thorn, E. N. Kopatskaya, V. B. Nebelitskij, E. L. Polyanskaya.
Trudy Astron. Obs., Leningrad, Vol. 29 (= Uchenye Zapiski Leningr. Un-ta, No. 363 = Seriya Matem. Nauk, vyp. (No.) 48), p. 72 - 80 (1973). In Russian.

Results are given of a photographic photometry of compact extragalactic objects. The accuracy of the measurements is discussed. Several flares of QSS 3C 345 and N-galaxies 3C 371 and 3C 390.3 were observed in 1968−1970.

141.136 Nuclei of quasars and active galaxies.

L. M. Ozernoj.
Zemlya i Vselennaya, 1973, No. 3, p. 25 - 33. In Russian.

141.137 Aperture synthesis of extragalactic objects.

J. H. Spencer, B. F. Burke.
Mass. Inst. Technol., Res. Lab. Electronics, Quarterly Progr. Rep., No. 107, p. 19 - 23 (1972).

A report on the MIT observations of M31, M33 and M51 at 3.7 and 11.1 cm using the NRAO 3-element interferometer. − DKM

141.138 Kosmische Radioquellen (Cosmic radio sources).

A. Werner.
Jahrbuch der Schulphysik, [Aulis-Verlag, Köln], Vol. 1, 143 - 147 (1972).

Review article concerning the radio frequency radiation of the sun, moon, planets, the Galaxy, and the extragalactic sources.

141.139 Radioastronomische Untersuchungen extragalaktischer Objekte.

A. Witzel.
Thesis, Univ. Münster (Westfalen), Fachber. Phys. 60 pp. (1972).

Radio astrophysics. Non-thermal processes in galactic and extragalactic sources. See Abstr. 003.101.

Relativistische Astrophysik. See Abstr. 011.006.

On the use of radio interferometers with a large base for astrometric work. See Abstr. 041.018.

International Information Bureau on Astronomical Ephemerides. See Abstr. 041.029.

The production of discrete, quantized outflow velocities by radiation pressure in stars, Seyfert nuclei, and quasi-stellar objects. See Abstr. 064.011.

Spectra of rapidly rotating objects. See Abstr. 065.021.

On the temperature of the microwave background radiation at a large redshift. See Abstr. 066.126.

Method and results of an observation of the 3C 144 radio source occultation by the far-off solar corona with the DKR-1000 cross radio telescope of the Physical Institute of the Academy of Sciences. See Abstr. 074.103.

A search for faint blue objects near the north galactic pole. See Abstr. 113.013.

5 GHz observations of the infrared star MWC 349, and the H II condensation W3 (OH). See Abstr. 114.046.

Observations of intense 100-micron objects at 3.5-millimeter wavelength. See Abstr. 114.062.

RY Scuti. See Abstr. 119.020.

Possible explanation for nonthermal radio noise from binary stars. See Abstr. 121.006.

Rätselhafte BL-Lacertae-Objekte. See Abstr. 122.024.

Photometric study of BL Lacertae. II. Brightness variations from March 1969 to January 1971. See Abstr. 122.052.

Photometric study of BL Lacertae. II bis − B and V magnitudes during the period March 1969−January 1971 and bibliographic informations. See Abstr. 122.116.

Radio brightness distribution of Ori A at 4.1mm wavelength. See Abstr. 131.005.

The cyanoacetylene cloud in Sagittarius B2. See Abstr. 131.007.

The radio continuum emission of the OH/IR−source ON−4. See Abstr. 131.038.

Infrared emission from the OH/H_2O sources in W49. See Abstr. 131.040.

Interferometric observations of formaldehyde absorption in front of strong galactic sources. See Abstr. 131.138.

Observations of the H66α recombination line. See Abstr. 131.183.

The H II region G333.6−0.2, a very powerful 1−20 micron source. See Abstr. 131.207.

Recent observations of Cyg X-3 at 365 MHz. See Abstr. 142.031.

21-cm absorption spectrum in front of Cygnus X-3. See Abstr. 142.032.

Particle injection in the Cygnus X-3 radio outburst. See Abstr. 142.036.

Westerbork and Effelsberg observations of Cygnus X-3.   See Abstr. 142.038.

Radio observations of X-ray sources. See Abstr. 142.064.

Radio counterparts of X-ray sources and X-ray counterparts of radio stars.   See Abstr. 142.065.

H I absorption in the galactic center region and between galactic longitudes 350° and 359°. See Abstr. 155.011.

Photoelectric spectrophotometry of Markarian 205 and a nearby suspected radio source.   See Abstr. 158.002.

3C 299: a faint radio galaxy of intermediate redshift.   See Abstr. 158.015.

Redshifts of a BSO and galaxies in the vicinity of the radio source RN 8.   See Abstr. 158.017.

Models for extragalactic objects with very high IR and X-ray luminosity.   See Abstr. 158.024.

Absence of variations in the nucleus of Virgo A. See Abstr. 158.029.

Identification and radio spectra of bright galaxies in the second Bologna Catalogue of radio sources and their radio luminosity function.   See Abstr. 158.040.

Radiospår i galaxhopar.   See Abstr. 158.044.

The redshift-distance relation. IV. The composite nature of N galaxies, their Hubble diagram, and the validity of measured redshifts as distance indicators. See Abstr. 158.055.

The spectrum of the extranuclear regions of Ton 256.   See Abstr. 158.065.

Observational consequences of inverse Compton models for Seyfert galaxies and quasars. See Abstr. 158.081.

Peculiar morphology of the outer regions of NGC 1265 (3C 83.1 B) and NGC 7720 (3C 465). See Abstr. 158.095.

Redshift-magnitude bands, quasi-stellar sources, and systems of redshift.   See Abstr. 158.106.

Faraday depolarization of radio galaxies and quasars with simple spectra.   See Abstr. 158.112.

Redshifts for 51 galaxies identified with radio sources in the 4C catalog.   See Abstr. 158.118.

Infra-red observations of NGC 7552 and NGC 7582 and their identification with PKS radio sources. See Abstr. 158.150.

Clustering effects among clusters of galaxies and quasi-stellar sources.   See Abstr. 160.022.

Pulsars

141.501   **Frequency dependence of pulsar polarization.**
R. N. Manchester, J. H. Taylor, G. R. Huguenin.
Astrophys. Journ., (*Letters*), Vol. 179, L7 - L10 (1973).
Observations show that the fractional linear polarization of integrated pulsar profiles is constant up to some critical frequency, but decreases with increasing frequency above this point. Above the critical frequency the polarization is, in most cases, approximately inversely proportional to frequency.

141.502   **The pulse energy distribution in pulsars.**
F. G. Smith.
Monthly Notices Roy. Astron. Soc., Vol. 161, 9P - 10P (1973).
The histograms of pulse energy for three pulsars show that each histogram is precisely repeatable, but that they differ markedly between different pulsars. The characteristic shapes and repeatability of these histograms show that the coherence of the emission process does not vary randomly.

141.503   **Gamma rays from NP 0532.**
Sky Telescope, Vol. 45, 154 - 155, 163 (1973).

141.504   **Radio halos around old pulsars – ghost supernova remnants.**
R. D. Blandford, J. P. Ostriker, F. Pacini, M. J. Rees.

Astron. Astrophys., Vol. 23, 145 - 146 (1973).
Old pulsars are expected to be surrounded by radio halos caused by relativistic electrons generated by the pulsar and diffusing away at the local Alfvén speed. The non-thermal source observed around CP 1919 may be of this type, since, in addition to the coincidence in position, there is rough quantitative agreement between observed and anticipated source properties. Other possible candidates are considered.

141.505   **Rotation in high-energy astrophysics.**
F. Pacini, M. J. Rees.
Sci. American, Vol. 228, No. 2, p. 98 - 105 (1973).
What is the source of the energy of pulsars, quasars and other strange objects? It may be gravitational energy converted into rotational energy as a large object contracts into a small one.

141.506   **VLBI observations of the Crab nebula pulsar.**
N. R. Vandenberg, T. A. Clark, W. C. Erickson, G. M. Resch, J. J. Broderick, R. R. Payne, S. H. Knowles, A. B. Youmans.
Astrophys. Journ., (*Letters*), Vol. 180, L27 - L29 (1973).
Observations of the Crab nebula pulsar at meter wavelengths using VLBI techniques have been made. The results are presented in detail.

141.507   **Soft X-ray pulsations from PSR 0833−45.**
F. R. Harnden, Jr., P. Gorenstein.

Nature, Vol. 241, 107 - 108 (1973).

We have recently observed the X-ray structure of the Vela and Puppis supernova remnants in the energy range 0.1—1.5 keV with a scanning focusing collector. Here we report only results pertaining to PSR 0833—45.

141.508 Evolution of the pulsar magnetosphere.
V. G. Endean.
Nature, Vol. 241, 184 - 185 (1973).

The author interprets the observational results in the light of the theoretical results in order to obtain more insight into the evolution of the pulsar magnetosphere.

141.509 Relativistic plasma and pulsar emission mechanisms.
S. A. Kaplan, V. N. Tsytovich.
Nature, Phys. Sci., Vol. 241, 122 - 124 (1973).

From the two established observation data of pulsar emission, namely, that the density of the energy flux from 1 cm^2 of the surface of the emitting region is extremely large, and that the maximum effective temperature of radio emission also reaches extremely large values, it necessarily follows that the plasma in the emitting region must be ultrarelativistic. The implications of this consequence are worked out and some numerical estimates are given.

141.510 Search for high frequency optical pulsar in supernova NGC 5253. I. R. Beresford, J. G. Greenhill, P. A. Hamilton, R. D. Watson.
Nature, Phys. Sci., Vol. 241, 126 (1973).

Observations of Kowal's supernova were undertaken with a pulse counting photometer attached to the University of Tasmania's 40-cm telescope. Data were collected on nine nights spread over the period June to September 1972. The records were analysed for periodicities in the frequency range up to 1 kHz using the fast Fourier transform algorithm for real time power spectrum analysis. No significant periodicities were observed in this frequency range.

141.511 Position of PSR 0833—45.
A. E. Vaughan, W. B. McAdam.
Nature, Phys. Sci., Vol. 241, 138 - 139 (1973). — Letter.

141.512 Decametric pulse radio emission from PSR 0809, PSR 1133, and PSR 1919.
Yu. M. Bruck, B. Yu. Ustimenko.
Nature, Phys. Sci., Vol. 242, 58 - 59 (1973).

We have received pulsed signals from three pulsars, PSR 0809 (at approximately 10, 12.6, 20 and 25 MHz), PSR 1133 and PSR 1919 (at 16.7, 20 and 25 MHz). Peak values of the radiation flux reach 20—50 f.u. The measured value of dispersion coincides with the generally accepted magnitudes with relative accuracy 10^{-2}—10^{-3}. A great variety has been observed in the mean amplitude and shape of pulses, with rapid transitions from one shape to another (sometimes occurring over time periods of 2—5 min).

141.513 Phenomena expected from the ejection of particles by the pulsar NP 0532 into the Crab nebula.
L. M. Ozernoy, V. V. Usov.
Astrophys. Letters, Vol. 13, 151 - 156 (1973).

We consider the consequences of the possible existence of a rather dense, quasistationary, magnetosphere around pulsar NP 0532 and of an instability of this magnetosphere appearing with the accumulation of particles.

141.514 Observations at 11 cm of recently discovered pulsars.
D. Graham, G. C. Hunt.
Nature, Phys. Sci., Vol. 242, 86 - 87 (1973).

The 100-m telescope at Effelsberg has been used to determine more accurate positions and periods for seven recently discovered pulsars. The derived positions and other quantities determined are quoted in a table.

141.515 Relativistic turbulent plasma in pulsars.
V. N. Tsytovich, S. A. Kaplan.
Astrofizika, Vol. 8, 441 - 460 (1972). In Russian. English translation in Astrophysics, Vol. 8, No. 3.

Near pulsars the ultra-relativistic plasma is contained in a strong magnetic field. Then the synchrotron losses of transverse particle energy pull out the distribution function along the magnetic field lines. The dispersive properties of such a plasma are studied. The types of oscillations and waves are found.

141.516 Observations of pulsars at 4850 MHz.
J. Crovisier.
Astrophys. Letters, Vol. 13, 221 - 223 (1973).

Eight pulsars have been observed at 4850 MHz with the Nançay radio telescope, and five of these have been detected. The integrated pulse profiles are presented for four of them.

141.517 Mean energies of pulsars at 1420 MHz.
W. Sieber, R. Wielebinski.
Astrophys. Letters, Vol. 13, 225 - 228 (1973).

Mean pulse energies at 1420 MHz are given for 15 pulsars. The results are compared with those of other authors.

141.518 21-cm line absorption profiles of pulsars PSR 0740—28 and PSR 1818—04.
J. Gómez González, M. Guélin, E. Falgarone, P. Encrenaz.
Astrophys. Letters, Vol. 13, 229 - 232 (1973).

21-cm line spectra have been determined in front of the pulsars PSR 0740—28 and 1818—04. Towards PSR 0740—28 absorption is detected up to +19.5 km/sec, and this places the pulsar at a distance of between 1.5 and 2.5 kpc from the sun. Towards PSR 1818—04, absorption is present only near to the local-standard-of-rest velocity. An upper limit of 1.5 kpc is set for its distance.

141.519 The Vela pulsar: member of an association?
W. C. Straka.
Astrophys. Journ., Vol. 180, 907 - 910 (1973).

Brandt et al. (1971) suggest that the pulsar in Vela, PSR 0833-45 may be associated with γ Vel and other stars. Several further pieces of evidence are considered.

141.520 Rotating magnetospheres: an exact 3-D solution.
F. C. Michel.
Astrophys. Journ., (*Letters*), Vol. 180, L133 - L137 (1973).

We derive the basic equations governing the magnetic-field-line configuration and plasma flow about a rotating object having an axisymmetric field in the limit that the plasma inertia can be neglected (strong magnetic fields). We show that the field-line equations can be written in a very simple compact form as a nonlinear second order partial differential equation. We also show that, in the monopole case, this differential equation has an exact solution.

141.521 Report on a search for new optical pulsars.
M. R. Nelson, E. J. Groth.
Astrophys. Journ., Vol. 181, 157 - 159 (1973).

Using on-line digital techniques at the Princeton 36-inch (91-cm) telescope, time series were obtained on pulsar candidates. We examined the data through autocorrelation and power-spectrum analyses for both narrow- and wide-band features. The system was tested by observing the one known optical pulsar. With this exception, no pulsation was detected down to the limits which are reported.

141.522 Pulsar magnetospheres, braking index, polar caps, and period—pulse-width distribution.
D. H. Roberts, P. A. Sturrock.
Astrophys. Journ., Vol. 181, 161 - 180 (1973).

Recent studies indicate that pulsar magnetospheres may contain nonrelativistic material in amounts sufficient to alter drastically the magnetic-field configuration. We have constructed approximate models for the magnetic field structure, considering in detail the aligned and orthogonal cases. Such a magnetospheric structure leads to a braking index $n = 7/3$, in good agreement with the observed braking index of the Crab pulsar. We calculate the polar-cap boundaries, and the resulting period–pulse-width distribution agrees well with observational data. It is suggested that the accumulation of gas and the occurrence of instabilities can explain the timing irregularities ("noise") and glitches observed in the Crab pulsar.

141.523   Notches in the average pulse profile of the pulsar
          PSR 1919+21.   T. H. Hankins.
Astrophys. Journ., (Letters), Vol. 181, L49 - L52 (1973).
     Deep notches in the average pulse profile of PSR 1919+ 21 have been revealed by a predetection dispersion-removal technique, suggesting that the pulsar emission mechanism is distinct from the mechanism that forms the average pulse profile.

141.524   The expected variability of pulsar NP 0532 radiation after spin jumps.   L. M. Ozernoy, V. V. Usov.
Astron. Zhurn. Akad. Nauk SSSR, Vol. 50, 422 - 424 (1973). In Russian. English translation in Soviet Astron. AJ, Vol. 17, No. 2. – Short note.

141.525   The radio spectrum of the short subpulses from
          PSR 0950+08.   B. J. Rickett, T. H. Hankins.
Bull. American Astron. Soc., Vol. 5, 18 (1973). – Abstr. AAS.

141.526   Subpulse time scales in several pulsars.
          J. M. Cordes, T. H. Hankins.
Bull. American Astron. Soc., Vol. 5, 18 (1973). – Abstr. AAS.

141.527   Three station observations of interstellar scintillation patterns of pulsars.
D. C. Backer, A. G. Lyne, G. A. Zeissig.
Bull. American Astron. Soc., Vol. 5, 18 (1973). – Abstr. AAS.

141.528   Simultaneous six frequency observations of pulsar
          fluxes.   D. C. Backer, J. R. Fisher.
Bull. American Astron. Soc., Vol. 5, 18 (1973). – Abstr. AAS.

141.529   Stimulated linear acceleration radiation: a possible pulsar emission mechanism.   W. J. Cocke.
Bull. American Astron. Soc., Vol. 5, 18 - 19 (1973). – Abstr. AAS.

141.530   Pulsar rotation measures and the galactic magnetic
          field.   R. N. Manchester.
Bull. American Astron. Soc., Vol. 5, 35 (1973). – Abstr. AAS.

141.531   An atlas of Stokes parameters based on a synchrotron radiation model.   B. J. Eastlund, B. Miller.
Bull. American Astron. Soc., Vol. 5, 35 (1973). – Abstr. AAS.

141.532   Radio emission from pulsars and surface temperature of neutron stars.   T. N. Rengarajan.
Nature, Phys. Sci., Vol. 242, 102 - 104 (1973).
     The author assumes that the radio emission from pulsars is due to bunches of electrons streaming from the polar caps along the magnetic field lines, and investigates the interaction between these electrons and the blackbody photons emitted from the surface.

141.533   Rotating neutron stars: a model for pulsars.
          M. Grewing, H. Heintzmann.
Zeitschr. Naturforschung, Vol. 28a, 377 - 382 (1973).
     The main properties of neutron star matter and of neu-

tron star models are reviewed with particular emphasis on those aspects that can be directly be related to pulsar observations.

141.534   Pulsars: Three years into a mystery.
          M. C. Hansell, W. Buscombe.
Irish Astron. Journ., Vol. 10, 173 - 189 (1972).

141.535   A test of Tsarevsky's pulsar–supernova analysis.
          V. Combe, M. I. Large.
Astrophys. Letters, Vol. 14, 59 - 60 (1973).
     The correlation between supernova distance and pulsar dispersion measure is attributed to the similar spatial distributions of supernovae and pulsars.

141.536   Pulsars: Properties of PSR 0301 + 19 and PSR
          2020 + 28.   R. E. Schönhardt, W. Sieber.
Astrophys. Letters, Vol. 14, 61 - 64 (1973).
     A power fluctuation spectrum analysis as a function of phase in the pulse window is presented for the pulsars PSR 0301 + 19 and PSR 2020 + 28. For both pulsars periodic power fluctuations have been found, which, in the case of PSR 0301 + 19, can be shown to be due to a drifting subpulse phenomenon.

141.537   Pulsar NP 0532: Variability of dispersion and
          scattering.
J. M. Rankin, C. C. Counselman III.
Astrophys. Journ., Vol. 181, 875 - 889 (1973).
     Observations of the Crab nebula pulsar NP 0532 over a period of 22 months are analyzed for effects of interstellar scattering and dispersion. The two phenomena are distinguished on the basis of a simple model which considers the possible contributions of two scattering regions following Counselman and Rankin. Variations are observed in dispersion and in one of the scattering regions which are imperfectly correlated.

141.538   Slow variations of pulsar intensities.
          G. R. Huguenin, J. H. Taylor, D. J. Helfand.
Astrophys. Journ., (Letters), Vol. 181, L139 - L142 (1973).
     We present observations of the day-to-day intensity fluctuations of five pulsars, extending throughout most of 1971 and 1972. Each source shows a total variation of about a factor of ten over this period, with characteristic fluctuation times of 20 to 70 days.

141.539   Pulsar detections at frequencies of 8.4 and 15.1 GHz.
          G. S. Downs, P. E. Reichley, G. A. Morris.
Astrophys. Journ., (Letters), Vol. 181, L143 - L146 (1973).
     Eleven pulsars were observed and five were detected at 15.1 GHz. Several exhibit strong scintillations at 8.4 GHz. Estimates are made of the spectral indices of five pulsars. Average pulse shapes are presented for the stronger signals.

141.540   A review of theories of pulsars.   H.-Y. Chiu.
          Stellar evolution, (see 012.014), p. 351 - 396
(1972). – Reprinted from Publ. Astron. Soc. Pacific, Vol. 82, 487 - 533 (1970). – See Abstr. 03.141.196.

141.541   Emission mechanism in pulsars.   F. G. Smith.
          Nature, Vol. 243, 207 - 210 (1973).
     In a new model of the emission from pulsars based on the relativistic beaming effect, the source of the radiation is the circular motion of high energy electrons round magnetic field lines. The radio emission at the gyrofrequency is from coherent bunches of electrons, and the optical and X-ray emission is the incoherent synchrotron radiation from the same electrons.

141.542   Continuum radio emission from the vicinity of

pulsars. R. E. Schönhardt.
Nature, Phys. Sci., Vol. 243, 62 - 63 (1973).

Observations aimed at the detection of weak radio emission from the surroundings of pulsars have been performed with the 100-m radio telescope of the Max-Planck-Institut für Radioastronomie, Bonn, between September 1972 and March 1973. An area of about one square degree centred on the relevant pulsar was observed for eighteen pulsars. In addition, a small field around the pulsar PSR0611+22, which is believed to be associated with the supernova remnant IC443, was observed extensively. For almost all the listed pulsars only point sources were detected in the surroundings.

141.543   Wide integrated pulse profiles of pulsars.
D. C. Backer, V. Boriakoff, R. N. Manchester.
Nature, Phys. Sci., Vol. 243, 77 - 78 (1973).

We draw attention to several pulsars which emit over large fractions of their pulse period and consequently represent a fundamental aspect of pulsar radiation which needs to be considered in future pulsar models.

141.544   Slow variations of pulsar intensities.
G. R. Huguenin, J. H. Taylor, D. J. Helfand.
Bull. American Astron. Soc., Vol. 5, 285 (1973). – Abstr. AAS.

141.545   Polarization of individual pulses from pulsars.
R. N. Manchester, J. H. Taylor, G. R. Huguenin.
Bull. American Astron. Soc., Vol. 5, 285 (1973). – Abstr. AAS.

141.546   On the origin of high-frequency radiation from the Crab nebula. A. Ferrari.
Mem. Soc. Astron. Italiana, Nuova Ser., Vol. 43, 715 - 729 (1973).

The oblique magnetic rotator model suggests that pulsars are the primary energy source for the long term activity of the surrounding supernova remnant. They emit e.m. waves of large amplitude and low frequency that are likely to be absorbed by the nebular plasmas. The absorption process is here investigated referring to recent theories of nonlinear plasma physics. A phenomenological model of the Crab nebula is outlined.

141.547   Radio and optical observations of pulsars.
F. D. Drake.
Cosmic plasma physics. Conference 1971, (see 012.016), p. 225 - 231 (1972).

141.548   A three-dimensional relativistic computation for the pulsar magnetosphere.
G. Kuo-Petravic, M. Petravic, K. V. Roberts.
Cosmic plasma physics. Conference 1971, (see 012.016), p. 239 - 247 (1972).

141.549   On the origin of pulsar radiation.
V. V. Zheleznyakov.
Cosmic plasma physics. Conference 1971, (see 012.016), p. 249 - 259 (1972).

141.550   Strong magnetic field effects in the pulsar crusts and atmospheres.
G. Kalman, P. Bakshi, R. Cover.
Cosmic plasma physics. Conference 1971, (see 012.016), p. 261 - 268 (1972).

141.551   Pulsar fluctuation spectra and the generalized drifting-subpulse phenomenon. D. C. Backer.
Astrophys. Journ., Vol. 182, 245 - 276 (1973).

The variation with longitude of the fluctuation spectrum of pulsar radio emission from a fixed longitude is discussed for 13 pulsars. In particular, features in the fluctuation spectra

are quantified. It has been shown that in 1919 + 21 the well-defined feature at 0.23 cycles per pulse period results from a drifting-subpulse phenomenon. In 2016 + 28 a poorly defined feature is caused by similar drifting-subpulse behavior. Thus it is suggested that the similar fluctuation spectrum features of other pulsars arise from the same subpulse phenomenon, generalized sufficiently to include all objects. Three variants of this generalized drifting-subpulse phenomenon are described. A qualitative discussion of the role of intensity fluctuations in various pulsar models is given.

141.552   Magnetic fields of pulsars. G. Chanmugam.
Astrophys. Journ., (*Letters*), Vol. 182, L39 - L41 (1973).

It is shown that one cannot neglect time variations in the magnetic fields of pulsars if one hopes to understand detailed pulsar observations.

141.553   Arrival times of 100 to 400 keV pulses from NP0532.
J. D. Kurfess, G. H. Share.
Nature, Phys. Sci., Vol. 244, 39 (1973).

The authors conclude that the primary peak observed at energies between 100 and 400 keV arrives at the earth within about 0.5 ms of the primary optical peak. This 0.5 ms uncertainty places an upper limit of ~ 150 km for the distance between the regions producing the optical and X-ray emissions.

141.554   Five new pulsars. M. M. Komesaroff, P. A. Hamilton, P. M. McCulloch, J. G. Ables, D. J. Cooke.
IAU Circ., No. 2505 (1973).

141.555   Pulsar X-ray source association. J. G. Ables, P. M. McCulloch, M. M. Komesaroff, P. A. Hamilton, D. J. Cooke.
IAU Circ., No. 2508 (1973).

141.556   Generalized electromagnetic torque on a vacuum pulsar model. S. R. K. Soper.
Astrophys. Space Sci., Vol. 19, 249 - 258 (1972).

The Deutsch solution to the electromagnetic field in a vacuum surrounding a perfectly conducting obliquely rotating sphere with a dipolar distribution of magnetic flux is extended to the oblique rotator with a general axisymmetric surface flux.

141.557   A statistical study of pulsars. E. H. Harutjunian (*Eh. A. Arutyunyan*), Yu. K. Melik-Alaverdian.
Soobshch. Byurakan. Obs., vyp. (No.) 44, p. 111 - 114 (1972).
In Russian.

Statistical investigation of pulsars showed that the distribution of pulsars by galactic longitude substantially differs from that of supernova remnants. A dependence between pulsar periods and radio luminosities is obtained.

141.558   Possible origin of cosmic γ ray flux from pulsars.
S. K. Saha.
Journ. Phys. A, General Phys., Vol. 6, 120 - 124 (1973).

The flux of celestial γ rays has been calculated from neutron stars (pulsars) on the basis of the model of neutron stars as developed by Tsuruta and Cameron and the cooling rate calculated by Raychaudhuri taking into account neutrino emission according to the photon-neutrino coupling theory. The result is found to be in good agreement with the experimental results as obtained in OSO III experiments.

141.559   Remarks on 'A new method for astronomical observation' by B. Kaplan. R. Burman.
Nuovo Cimento Lettere, Ser. 2, Vol. 5, 1054 - 1055 (1972).

Kaplan (see Abstr. 08.141.553) suggested that it might be possible to detect electromagnetic radiation, generated by pulsars, with frequencies near their rotation or pulsation frequencies. It was pointed out that the earth's ionosphere is likely to have an effect on these waves, but no mention was made of effects of the plasma surrounding the pulsar or of the interstellar plasma. The author considers these effects and the possibility of whistler mode propagation.

141.560 Charged particle motion in superstrong electromagnetic fields. M. Grewing, H. Heintzmann.
Phys. Letters A, (*Netherlands*), Vol. 42 A, 325 - 326 (1972).

The motion of charged particles in superstrong electromagnetic fields is studied analytically and numerically. An analytic solution is given for constant fields which also describes satisfactorily the initial part of the motion in the slowly varying fields surrounding a pulsar.

141.561 Pulsar intensity variations as a result of scintillations on inhomogeneous plasma.
V. V. Vitkevich, N. A. Lotova.
Trudy fiz. in-ta. AN SSSR, Vol. 62, 46 - 52 (1972). In Russian. Abstr. in Referativ. Zhurn. 51. Astron., 6.51.613 (1973).

141.562 The magnetic fields of pulsars. D. M. Sedrakian, K. M. Shahabasian (*Shakhabasyan*).
Astrofizika, Vol. 8, 557 - 560 (1972). In Russian. – English translation in Astrophysics, Vol. 8, No. 4.

Two generation mechanisms of the magnetic fields in pulsars are considered. It is shown that if we consider the proton fluid as super-conductive and the electron gas as normal, then in the case of rotation a magnetic momentum is generated. For the pulsar in the Crab nebula it is of the order of 10^{32} gauss cm^3.

141.563 Pulsars in supernova remnants. P. R. Amnuel, O. H. Guseinov (*O. Kh. Gusejnov*), F. K. Kasumov.
Astrofizika, Vol. 8, 561 - 566 (1972). In Russian. – English translation in Astrophysics, Vol. 8, No. 4.

The radio and X-ray spectra of six supernova remnants have been studied. The data on the characteristic break in the spectra indicate the possible presence of a pulsar (neutron star) in SNR Tycho (except the known remnants with pulsars: Crab and Vela X). Apparently, there are no active sources in the supernova remnants Cas A, Pup A and Cyg Loop.

141.564 On a possible role of superbright radiation on pulsar conditions. V. Ya. Eidman (*Ehjdman*).
Astrofizika, Vol. 8, 609 - 612 (1972). In Russian. – English translation in Astrophysics, Vol. 8, No. 4.

The mechanism of superbright synchrotron radiation in vacuum is applied to the interpretation of some peculiarities of pulsar radiation.

141.565 Acceleration of charged particles in the electromagnetic field of pulsars.
W. Fischer, N. Straumann.
Helvetia Phys. Acta, Vol. 45, 1089 - 1093 (1973).

The acceleration of charged particles in the wave zone of the electromagnetic field of pulsars is studied numerically and compared with previous approximate analytic solutions.

141.566 Pulsar radio emission mechanism.
J. Virtamo, P. Jauho.
Astrophys. Journ., Vol. 182, 935 - 949 (1973).

This paper is intended to further develop the radio emission model first proposed by Chiu and Canuto. An assumption made for this particular model is that radiation is due to induced electron bremsstrahlung in an intense magnetic field as the electrons flow, accelerated by an electric field. The bremsstrahlung emission rate used includes an arbitrary direction of photon propagation. The macroscopic radiation pattern is calculated, with the refraction of rays taken into account. The narrow, hollow beam of this radiation which results from this model semiquantitatively explains the double pulse feature of pulsar radiation.

141.567 Magnetosphere structure and radiation mechanisms of pulsars.
D. H. Roberts, P. A. Sturrock, J. S. Turk.
Inst. Plasma Res., Stanford Univ., Stanford, California, SUIPR Rep. No. 503, 28 pp. (1973). – Text of lecture presented by P. A. Sturrock at sixth Texas symposium on relativistic astrophysics, New York, December 20, 1972.

141.568 Absolute timing of twelve pulsars.
D. W. Richards.
Thesis Cornell Univ., Ithaka, New York, 140 pp. (1972).

141.569 System for photo-registration of pulsar impulses.
N. S. Solomin.
Trudy fiz. in-ta. AN SSSR, Vol. 62, 178 - 182 (1972). In Russian. – Abstr. in Referativ. Zhurn. 51. Astron., 7.51.260 (1973).

Energy spectrum of He II in a strong magnetic field and bound-bound transition probabilities.
See Abstr. 022.048.

Non-linear Compton and inverse Compton effect.
See Abstr. 061.001.

Evaluation of astrophysical hypotheses.
See Abstr. 061.044.

On the limiting polarization of radio-waves.
See Abstr. 062.017.

Aligned rotating magnetospheres. I. General analysis.
See Abstr. 062.065.

Formation of neutron star spots and its connection with pulsars. II. Close similarities between radiation from the sun and pulsars. See Abstr. 065.003.

The equilibrium, stability and evolution of a rotating magnetized gaseous disk. See Abstr. 065.158.

Outbursts of supernovae and formation of relativistic objects. I. See Abstr. 125.004.

Supernova outbursts and pulsars.
See Abstr. 125.017.

Gamma-ray emission above 20 MeV from the Crab nebula and NP 0532. See Abstr. 134.009.

Magnetic configuration in the neighborhood of a collapsed star. See Abstr. 142.010.

Short-term temporal studies of the X-ray emission from Cassiopeia A, Tycho, and Scorpius X-1.
See Abstr. 142.040.

Pulsars and X-ray sources. See Abstr. 142.069.

Plasma effects and the acceleration of charged particles in pulsar fields. See Abstr. 143.003.

Path length distribution of cosmic rays from pulsar nebula complexes. See Abstr. 143.068.

142 X Ray-, Gamma Ray-Sources

142.001 A cocoon model for thermal X-ray sources and 'oscillars'. K. M. V. Apparao.
Astrophys. Space Sci., Vol. 18, 334 - 336 (1972).

A gas cocoon surrounding a neutron star can be heated to a high temperature by the low frequency radiation emitted by the neutron star whose rotation axis is inclined to its magnetic axis. This heated gas can emit X-rays and may be identified with thermal X-ray sources. If the neutron star emission shows periodicities larger than the cooling time of the gas, these will be reflected in the emission of X-ray; the recently observed X-ray sources which show oscillations and quasiperiodicities ('oscillars') may be such sources.

142.002 Energy spectrum and time variations of hard X-rays from Cgy X-1. P. C. Agrawal, G. S. Gokhale, V. S. Iyengar, P. K. Kunte, R. K. Manchanda, B. V. Sreekantan.
Astrophys. Space Sci., Vol. 18, 408 - 424 (1972).

Experimental results on the intensity, energy spectrum and time variations in hard X-ray emission from Cyg X-1 based on a balloon observation made on 1971, April 6 from Hyderabad (India) are described. The binary model proposed by Dolan is examined and the difficulties in explaining the observed features of Cyg X-1 by this model are pointed out.

142.003 Diffuse cosmic gamma rays observed at an equatorial balloon altitude.
R. R. Daniel, G. Joseph, P. J. Lavakare.
Astrophys. Space Sci., Vol. 18, 462 - 467 (1972).

A $3'' \times 3''$ NaI(Tl) crystal-photomultiplier assembly with a 4π charged particle anticoincidence shield is used to determine the gamma ray spectrum in the energy region of about 100 keV to 8.5 MeV at a balloon altitude of 4.7 g cm^{-2} over Hyderabad, India. The atmospheric growth curves are used to obtain the contribution of the diffuse cosmic gamma ray flux in the above energy range.

142.004 Time variations of hard X-rays from Sco X-1.
M. Matsuoka, M. Fujii, S. Miyamoto, J. Nishimura, M. Oda, Y. Ogawara, S. Hayakawa, I. Kasahara, F. Makino, Y. Tanaka, P. C. Agrawal, B. V. Sreekantan.
Astrophys. Space Sci., Vol. 18, 472 - 490 (1972).

Simultaneous hard X-ray and optical observations of Sco X-1 were carried out on 1971 May 1 at Hyderabad, India, when Sco X-1 was optically bright. The X-ray intensity observed by balloon-borne counter telescopes increased in coincidence with optical enhancements, while the plasma temperature derived by fitting the X-ray spectrum in the energy range 20—40 keV to the thermal bremsstrahlung spectrum did not appreciably change over the whole period of observation.

142.005 Infall of gas from intergalactic space and soft X-ray background. M. Tosa, T. Kato.
Astrophys. Space Sci., Vol. 18, 504 - 513 (1972).

As the origin of the soft X-ray background, emission of soft X-rays from shocks occurred in the accretion of intergalactic gas onto the Galaxy is studied. Formation of the high velocity cloud by thermal instability in the shocked gas is discussed briefly.

142.006 Optical identification criteria for binary X-ray sources. S. M. Lea, B. Margon.
Astrophys. Letters, Vol. 13, 33 - 37 (1973).

There is now evidence that several galactic X-ray sources are members of binary systems. Although one member of each system is believed to be highly evolved, probably a neutron star or a black hole, the orbits show no observable ellipticity. We propose that the orbits are circularized following the super-

nova, by tidal forces. If this theory is correct, then the rotational period of the primary should be equal to the orbital period of the binary system.

142.007 On the nature of the optical variability of HZ Her = Her X-1 and BD + 34°3815 = Cyg X-1.
V. M. Lyutyj, R. A. Sunyaev, A. M. Cherepashchuk.
Astron. Zhurn. Akad. Nauk SSSR, Vol. 50, 3 - 11 (1973). In Russian. English translation in Soviet Astron. AJ, Vol. 17, No. 1.

The photographic and photoelectric light curves of HZ Her = Her X-1 testify the strong reflection effect ($\sim 1\overset{m}{.}5$) connected with transforming of X-ray radiation in the photosphere of the visible component. The optical variability of the BD + 34°3815 = Cyg X-1 system is caused mainly by the ellipticity of the visible component. The amplitude of variability depends both on the oblateness of the star and on the inclination angle i. The oblateness of the star for the given star mass ratio can be found from the theory of the stars filling the critical Roche lobe. Therefore the data of photoelectric observations allow to find sin i and to determine a lower bound $M_x > 7.8 M_\odot$ for the mass of Cyg X-1.

142.008 Scorpius X-1 as an old nova.
S. N. Shore, W. L. Gebel.
Astrophys. Journ., Vol. 179, 257 - 261 (1973).

We adopt an old-nova model to derive physical parameters for Sco X-1. The model consists of a standing shock, the source of the X-ray continuum, which is formed by mass accretion at the surface of a white dwarf. The source of the matter is taken to be a red companion which fills its Roche lobe. We also comment on the possible origin of the radio sources associated with Sco X-1.

142.009 Observations of Taurus X-1 by the 1—60 keV X-ray detector on the OSO-7. G. W. Clark, H. V. Bradt, W. H. G. Lewin, T. H. Markert, H. W. Schnopper, G. F. Sprott.
Astrophys. Journ., Vol. 179, 263 - 268 (1973).

The 1—60 keV X-ray detector on the OSO-7 satellite is described, and a measurement of the X-ray spectrum of the Crab nebula is reported.

142.010 Magnetic configuration in the neighborhood of a collapsed star. R. H. Cohen, B. Coppi, A. Treves.
Astrophys. Journ., Vol. 179, 269 - 275 (1973).

It is shown that the magnetic configuration in the neighborhood of a collapsed star with parameters appropriate for models of X-ray stars or pulsars is nearly force-free, with $(\nabla \times \mathbf{B})/\mathbf{B}$ nonconstant. In the case where the magnetic axis coincides with the rotation axis, a differential equation for the magnetic surfaces is derived. A proper double-expansion technique is used to obtain a significant asymptotic solution of this equation and to derive explicit expressions for the relevant magnetic-field components.

142.011 Extended X-ray and radio observations of Scorpius X-1. C. R. Canizares, G. W. Clark, W. H. G. Lewin, H. W. Schnopper, G. F. Sprott, R. M. Hjellming, C. M. Wade.
Astrophys. Journ., (*Letters*), Vol. 179, L1 - L5 (1973).

A 23-day observation reveals temporal variations at energies of 3—10 keV similar to those found in previous optical data. However, there is no evidence of more than one preferred intensity level or of an intensity threshold for flares. We rule out X-ray periodicities containing $\geqslant 5$ percent of the mean quiescent-period intensity with periods of from 6 minutes to 4 hours. The X-ray intensity is strongly correlated with hardness but is not correlated with radio data collected over eight days.

142.012 Observations of the neutral-hydrogen absorption spectrum of Cygnus X-3.
K. W. Chu, J. H. Bieging.
Astrophys. Journ., (*Letters*), Vol. 179, L21 - L23 (1973).

We have made interferometric observations of the neutral-hydrogen absorption spectrum of Cygnus X-3. The radio source is more than 10.4 kpc from the sun, and the columnar density of neutral hydrogen along the line of sight is at least $1.7 \times 10^{20} T_s \mathrm{cm}^{-2}$.

142.013 A search for the Cygnus X-3 infra-red candidate at one micron. J. E. Gaustad, B. Margon.
Monthly Notices Roy. Astron. Soc., Vol. 161, 15P - 17P (1973).

An attempt to observe the recently discovered infra-red candidate for Cygnus X-3 at one micron has yielded the limit $I > 16.3$. For a typical O9 star, this implies $A_v > 17.1$, probably excluding membership of Cygnus X-3 in the Cyg OB2 association.

142.014 Neutron-star accretion in a stellar wind: Model for a pulsed X-ray source.
K. Davidson, J. P. Ostriker.
Astrophys. Journ., Vol. 179, 585 - 598 (1973).

Many of the characteristics of the pulsing X-ray sources Cen X-3 and Her X-1 are explicable in terms of a simple model. A rotating magnetized neutron star orbits a more massive slightly evolved star and accretes mass from the stellar wind emanating from that star. An analysis of the tidal disruption problem shows that the period, velocity amplitude, and eclipse duration of the X-ray source permit the neutron star to have a mass of 1 $M_\odot$, determine the inclinations to be 70°–90°, and require separations, masses, radii, and spectral types of the companion stars to be approximately (18 $R_\odot$, 8 $R_\odot$), (20 $M_\odot$, 2 $M_\odot$), (12 $R_\odot$, 4 $R_\odot$), and (B2 III, F5 III), for Cen X-3 and Her X-1, respectively.

142.015 ADS 2859 B as an alternative candidate to X Per for the X-ray source 2 U 0352 + 30.
W. Haupt, A. F. J. Moffat.
Astrophys. Letters, Vol. 13, 77 - 79 (1973).

ADS 2859 B, like X Per (ADS 2859 A), lies very near to the most likely position of the Uhuru X-ray source 2 U 0352 + 30. UBV observation show that ADS 2859 B exhibits extreme T Tauri-like characteristics. Its total luminosity ($\sim 10 L_\odot$) and colours suggest that it is a pre-main-sequence object with a mass $\sim 2 M_\odot$ and an age $\sim 10^6$ yr.

142.016 Polarization of optical radiation and magnetic field of X-ray stars.
A. Z. Dolginov, Yu, N, Gnedin, N. A. Silant'ev.
Astrophys. Letters, Vol. 13, 85 - 87 (1973).

Circular polarization of the optical radiation from X-ray stars has been calculated on the assumption that magneto-active plasma is a source of continuous optical radiation of those stars. The influence of a strong magnetic field on the linear polarization of radiation is briefly discussed.

142.017 The distribution of X-ray sources in our Galaxy.
F. D. Seward.
Sky Telescope, Vol. 45, 220 - 223 (1973).

142.018 HD 154431 and the pulsating X-ray source in Hercules. P. Murdin, A. Savage.
Observatory, Vol. 93, 32 - 33 (1973).

142.019 On the pulsation of X-ray sources.

W. A. Baan, A. Treves.
Astron. Astrophys., Vol. 22, 421 - 424 (1973).

The emission from the X-ray sources Cen X-3 and Her X-1 is considered in the light of a model in which a neutron star is accreting matter transferred from a companion. It is shown that the accretion should occur through a magnetic funnel of which the angular width is estimated. Some difficulties of explaining the beaming of the X-radiation are outlined, and possible solutions are suggested.

142.020 On the isotropy of the X-ray background radiation.
K. Brecher.
Astron. Astrophys., Vol. 23, 105 - 110 (1973).

The spatial fluctuations expected for various models of the X-ray background radiation are examined. The constraints that observations place on the source distribution and radiation mechanisms are then presented. Most models are at present consistent with the lack of observed small scale anisotropy. Future measurements of the beam-to-beam fluctuations could offer a positive test of the inverse Compton explanation for the origin of the X-ray background radiation.

142.021 Observations of spatial structure in the soft X-ray diffuse flux. A. N. Bunner, P. L. Coleman,
W. L. Kraushaar, D. McCammon, F. O. Williamson.
Astrophys. Journ., Vol. 179, 781 - 788 (1973).

Repeated observations of a diffuse source in the region $l^{\mathrm{II}} \sim 200°$, $b^{\mathrm{II}} \sim +10°$ illustrate the complexity of the low-energy X-ray background. The source is not well correlated with known objects or features, and cannot be explained by geometrically simple models of diffuse emission.

142.022 The spectrum of HDE 226868 (Cygnus X-1).
N. R. Walborn.
Astrophys. Journ., (*Letters*), Vol. 179, L123 - L124 (1973).

A refined spectral classification of O9.7 Iab (p-var) has been derived for the supergiant star associated with the X-ray source Cyg X-1. The blue-violet absorption-line spectrum is entirely normal; however, a peculiar, variable emission line is present at He II λ4686. Two spectrograms of the star are illustrated.

142.023 Spectroscopic observations of the Cygnus X-1 optical candidate. H. E. Smith, B. Margon, P. S. Conti.
Astrophys. Journ., (*Letters*), Vol. 179, L125 - L128 (1973).

Data from nine high-dispersion spectra of HDE 226868, the optical candidate for Cyg X-1, are presented. Our data are not compatible with models where the emission line arises from the unseen secondary component of the system. We propose that the emission originates in a gas stream falling toward the secondary; this provides observational evidence that the X-ray source is powered by accretion.

142.024 Spectroscopic observations of the optical candidate for Cygnus X-1. R. Brucato, J. Kristian.
Astrophys. Journ., (*Letters*), Vol. 179, L129 - L133 (1973).

Spectroscopic observations of BD+34°3815 (=HDE 226868) indicate that the mass of the secondary is greater than upper limits for white dwarfs and neutron stars. If it is the X-ray source, it is therefore an interesting candidate for a black hole.

142.025 On the masses of X-ray sources.
R. W. Leach, R. Ruffini.
Astrophys. Journ., (*Letters*), Vol. 180, L15 - L18 (1973).

An analysis of X-ray sources based on the Roche model is here presented. On this basis we can conclude that pulsating sources appear to have systematically smaller masses than nonpulsating sources. We suggest identifying the first objects as neutron stars and the second as totally collapsed objects or black holes. Detailed predictions are presented. Discriminating features between neutron stars and black holes are also given.

142.026 Astronomy from an X-ray satellite: Measuring the mass of a neutron star. W. D. Metz.
Science, Vol. 179, 884 - 885 (1973). — Concerning Hercules X-1.

142.027 On X Persei as a stellar X-ray source: Comparison with γ Cassiopeiae.
A. F. J. Moffat, W. Haupt, T. Schmidt-Kaler.
Astron. Astrophys., Vol. 23, 433 - 439 (1973).

A series of coudé spectra taken September, 1971 reveals that, except for its large light and V/R variations. X Per has the characteristics of a normal Be star. Its spectral type is O9.5 (III–V)e and there is no detectable veiling. γ Cas was scanned by the Uhuru detectors five times and its apparent optical brightness is 25 times greater than that of X Per. Also both stars had similar V/R ratios (> 1) in September, 1971. Thus it is concluded that X Per may not be directly responsible for the X-radiation of the source 2U 0352 + 30; however, our data do not exclude a binary model.

142.028 Observations of periodic variations in the X-ray intensity of Cygnus X-3.
C. R. Canizares, J. E. McClintock, G. W. Clark, W. H. G. Lewin, H. W. Schnopper, G. F. Sprott.
Nature, Phys. Sci., Vol. 241, 28 - 30 (1973).

We present here a study of the time variability of the X-ray intensity of Cygnus X-3 with data collected by the OSO-7 satellite in December 1971 and in January, June and July 1972.

142.029 λ-Sco, a possible source of soft X-rays.
J. A. M. Bleeker, A. J. M. Deerenberg, J. Heise, K. Yamashita, Y. Tanaka.
Nature, Phys. Sci., Vol. 241, 55 - 56 (1973).

An X-ray telescope was flown on May 26, 1971, 1010 UT from Kauai. Here we report the detection of a strong soft X-ray source in a direction near the galactic centre. A bright nearby star, λ Sco, lies close to the edge of the error box.

142.030 Diffuse cosmic gamma rays: Present status of theory and observations. F. W. Stecker.
Nature, Phys. Sci., Vol. 241, 74 - 77 (1973).

The totality of the observations in the 10^{-3} to 100 MeV range follows an E^{-2} trend in the differential isotropic photon spectrum but significant features appear. Possible theoretical interpretations of these features are discussed here. New results on the diffuse flux from the Galaxy substantiate the pion-decay origin hypothesis for gamma-radiation above 100 MeV.

142.031 Recent observations of Cyg X-3 at 365 MHz.
F. N. Bash, F. D. Ghigo.
Nature, Phys. Sci., Vol. 241, 93 - 94 (1973).

We observed Cyg X-3 at 365 MHz using the Broadband-Synthesis-Interferometer at the University of Texas Radio Astronomy Observatory. The observations consist of measurements of the flux density of the object at meridian transit on 20 day from September 6, 1972, to October 10, 1972, inclusive. We have also determined the position of the radio source to an accuracy of approximately 2 arc s.

142.032 21-cm absorption spectrum in front of Cygnus X-3.
R. Lauqué, J. Lequeux, Nguyen-Quang-Rieu.
Nature, Phys. Sci., Vol. 241, 94 - 95 (1973).

We recently obtained good comparison observations of Cyg X-3. Comparing these with the initial observations of September 4 and 5 we find three changes: an optical depth $\tau \simeq 2.5$ in the Cygnus feature near +9 km s⁻¹; an optical depth $\tau \simeq 0.4$ for the hydrogen in the arm at 8 kpc near −45 km s⁻¹;

the definite presence of absorption with an optical depth $\tau \simeq 0.3$ in the farthest arm at −68 km s⁻¹, which lies at a distance of about 11 kpc.

142.033 Determination of the position of GX2 + 5 with Copernicus.
F. J. Hawkins, K. O. Mason, P. W. Sanford.
Nature, Phys. Sci., Vol. 241, 109 - 111 (1973). — Letter.

142.034 Hard X-ray emission from Uhuru sources.
R. K. Manchanda, B. V. Sreekantan.
Nature, Phys. Sci., Vol. 241, 124 - 125, with a correction, Vol. 242, 48 (1973).

In two balloon flights carried out from Hyderabad by the TIFR group, one on April 16, 1969, and the other on May 5, 1970, a large part of the sky defined by RA = 190°–330° and $\delta = -20°$ to +50° was scanned for hard X-ray emission in the 16–150 keV band. The results of the observations are presented.

142.035 Periodic light variations in HDE 226868 (Cyg X-1).
D. F. Lester, I. G. Nolt, J. V. Radostitz.
Nature, Phys. Sci., Vol. 241, 125 - 126 (1973).

We have photometrically monitored the Hα index and the intermediate-band b magnitude of HDE 226868 for several months using the 24-inch telescope of the Pine Mountain Observatory. No significant variations were detected in the photometric Hα index. We have however, observed changes of 0.07 in the b magnitude for which the minima coincide with the spectroscopic phases of conjunction as given by Bolton's radial measurements.

142.036 Particle injection in the Cygnus X-3 radio outburst.
F. W. Peterson.
Nature, Vol. 242, 173 - 177 (1973).

Particle injection played an important role in the recent Cygnus X-3 radio outburst. Calculations on the basis of an improved expanding cloud model show that a maximum mass of $0.76 \times 10^{-8} M_\odot$ was injected over a period of 1.2 days.

142.037 X-ray astronomy (III): Searching for a black hole.
W. D. Metz.
Science, Vol. 179, 1113 - 1115 (1973).

142.038 Westerbork and Effelsberg observations of Cygnus X-3. L. L. E. Braes, G. K. Miley, W. W. Shane, J. W. M. Baars, W. M. Goss.
Nature, Phys. Sci., Vol. 242, 66 - 69 (1973).

Extensive radio observations during September and October 1972 show the behaviour of the Cyg X-3 outburst at 1.4 and 2.7 GHz. Neutral hydrogen absorption measurements place a lower limit of 10 ± 1.5 kpc on the distance of the radio source. No pulsations were detected.

142.039 Wolf-Rayet systems and the origin of massive X-ray binaries. E. P. J. van den Heuvel.
Nature, Phys. Sci., Vol. 242, 71 - 72 (1973).

The author concludes that the system parameters of massive X-ray binaries and of short period WR binaries support the hypothesis that these two types of system represent different stages of evolution of the same kind of objects.

142.040 Short-term temporal studies of the X-ray emission from Cassiopeia A, Tycho, and Scorpius X-1.
S. S. Holt, E. A. Boldt, P. J. Serlemitsos, A. F. Brisken.
Astrophys. Journ., (Letters), Vol. 180, L69 - L74 (1973).

No evidence for stable 2−10 keV periodic emission from Cas A or Tycho in the period range 1 ms to 10 s is found. Upper limits to the pulsed fraction are presented as a function of the assumed light curve, with absolute 99 percent confidence upper limits of 0.089 and 0.195 for Cas A and Tycho, respectively. Previously reported transient 1–10 Hz oscillations from Sco X-1 are not observed.

142.041 **Observation of structure in the X-ray spectrum of Puppis A.**
G. Burginyon, R. Hill, F. Seward, B. Tarter, A. Toor.
Astrophys. Journ., (*Letters*), Vol. 180, L75 - L77 (1973).

X-ray data from two rocket flights which scanned Puppis A show prominent structure in the energy spectrum that cannot be explained by continuum models alone. The spectral observations are well fitted by combining the effects of line radiation, radiative recombination, and bremsstrahlung from the thermal source.

142.042 **Rapidly rotating degenerate dwarfs as X-ray sources in binaries.** K. Brecher, P. Morrison.
Astrophys. Journ., (*Letters*), Vol. 180, L107 - L112 (1973).

We suggest that the recently discovered pulsing binary X-ray sources Her X-1 and Cen X-3 may be rapidly rotating degenerate dwarfs. The model accounts in a natural way for their presence in binary systems, observed luminosity, pulse period, and rate of spin up. No black holes are required even in possibly massive star systems such as Cyg X-1.

142.043 **The spectrum and variability of Hercules X-1 observed by OSO-7.**
M. P. Ulmer, W. A. Baity, W. A. Wheaton, L. E. Peterson.
Astrophys. Journ., (*Letters*), Vol. 181, L33 - L37 (1973).

We report observations of Her X-1 (HZ Her) with the UCSD OSO-7 X-ray telescope 1971 December 8–11 and 1972 May 29–June 1.

142.044 **Light curve of HZ Herculis.**
L. Petro, W. A. Hiltner.
Astrophys. Journ., (*Letters*), Vol. 181, L39 - L42 (1973).

The optical counterpart of Her X-1 was observed for *B*-magnitude on 34 nights between 1972 July 30 and 1972 October 10. The scatter in the light curve is about 0.5 mag and may be related to the 35-day cycle.

142.045 **Evidence for the binary nature of 2U 1700–37.**
C. Jones, W. Forman, H. Tananbaum, E. Schreier, H. Gursky, E. Kellogg, R. Giacconi.
Astrophys. Journ., (*Letters*), Vol. 181, L43 - L48 (1973).

Analysis of *Uhuru* data on the X-ray source 2U 1700–37 has revealed the presence of regular occultations. The period is 3.4 days with an occultation duration of 1.1 days. The source is highly variable on a time scale of minutes. Significant intensity variations have also been observed on a time scale of 0.1 second with a 3σ upper limit of 17 percent on the percentage periodically pulsed. The 6.6-mag O7f star HD 153919 is suggested as the optical candidate.

142.046 **Comments on the nature of Her X-1.**
I. S. Shklovsky.
Astron. Zhurn. Akad. Nauk SSSR, Vol. 50, 233 - 242 (1973). In Russian. English translation in Soviet Astron. AJ, Vol. 17, No. 2.

It is shown that the anisotropic mechanism (of synchrotron or cyclotron type) with knife-like diagram can naturally explain the complete picture of X-emission of the source. Some empirical arguments are given that Her X-1 is a real neutron star and its X-ray emission is connected with the accretion of gas flowing out from the optical component. The precession of the rotation axis of the neutron star is explained by the dynamical effect of the infalling gaseous stream.

142.047 **X-ray astronomy (II): A new breed of pulsars.**
W. D. Metz.
Science, Vol. 179, 986 - 988 (1973).

142.048 **The unusual behavior of HZ Herculis (= Hercules X-1): 1890-1972.**
C. A. Jones, W. Forman, W. Liller.
Bull. American Astron. Soc., Vol. 5, 32 (1973). – Abstr. AAS.

142.049 **Hard X-ray dip in Her X-1.**
W. A. Mahoney, K. A. Anderson.
Bull. American Astron. Soc., Vol. 5, 32 - 33 (1973). – Abstr. AAS.

142.050 **Spectrum of the X-ray source at M 87.**
R. C. Catura, L. W. Acton, H. M. Johnson, W. T. Zaumen, P. C. Fisher.
Bull. American Astron. Soc., Vol. 5, 33 (1973). – Abstr. AAS.

142.051 **Balloon observation of high-energy X-ray sources from the southern sky.** W. H. G. Lewin, G. R. Ricker, J. E. McClintock, M. Gerassimenko, S. G. Ryckman.
Bull. American Astron. Soc., Vol. 5, 33 (1973). – Abstr. AAS.

142.052 **Observations of 2U0900-40 from Uhuru.**
W. Forman, C. Jones, H. Tananbaum, E. Kellogg, H. Gursky, R. Giacconi.
Bull. American Astron. Soc., Vol. 5, 33 (1973). – Abstr. AAS.

142.053 **Accretion models for galactic X-ray sources.**
J. Buff, R. McCray.
Bull. American Astron. Soc., Vol. 5, 33 - 34 (1973). – Abstr. AAS.

142.054 **On the optical behaviour of binary X-ray sources.**
J. E. Pringle, M. J. Rees.
Bull. American Astron. Soc., Vol. 5, 42 (1973). – Abstr. AAS.

142.055 **Gamma-ray astronomy and cosmic rays.**
V. L. Ginzburg.
Uspekhi fiz. nauk, Vol. 108, 273 - 283 (1972). In Russian. Abstr. in Referativ. Zhurn. 51. Astron., 3.51.657 (1973).

142.056 **Gamma-astronomy and cosmic rays.**
V. L. Ginzburg.
Zemlya i Vselennaya, 1973, No. 1, p. 2 - 6. In Russian.

142.057 **Discrete sources of gamma-quanta.**
A. M. Gal'per, V. G. Kirillov-Ugryumov, B. I. Luchkov.
Zemlya i Vselennaya, 1973, No. 1, p. 6 - 11. In Russian.

142.058 **X-ray and gamma-ray astronomy. Introductory remarks.** B. B. Rossi.
IAU Symposium No. 55, (see 012.002), p. 1 - 6 (1973).

142.059 **UHURU results on galactic X-ray sources.**
H. D. Tananbaum.
IAU Symposium No. 55, (see 012.002), p. 9 - 28 (1973).

142.060 **Observations of cosmic X-ray sources by the MIT instrument on the OSO-7.** G. W. Clark.
IAU Symposium No. 55, (see 012.002), p. 29 - 35 (1973).

142.061 **Binary systems as X-ray sources: a review.**
R. P. Kraft.
IAU Symposium No. 55, (see 012.002), p. 36 - 50 = Contr. Lick Obs., No. 368 (1973).

142.062 **Hard cosmic X-ray sources.** L. E. Peterson.
IAU Symposium No. 55, (see 012.002), p. 51 - 73 (1973).

142.063 **Simultaneous X-ray, optical and radio observations of galactic X-ray sources.** W. A. Hiltner.

IAU Symposium No. 55, (see 012.002), p. 74 - 85 (1973).

142.064   Radio observations of X-ray sources.
L. L. E. Braes, G. K. Miley.
IAU Symposium No. 55, (see 012.002), p. 86 - 97 (1973).

142.065   Radio counterparts of X-ray sources and X-ray counterparts of radio stars.   R. M. Hjellming.
IAU Symposium No. 55, (see 012.002), p. 98 - 104 (1973).

142.066   Galactic X-ray polarimetry and high-resolution X-ray spectroscopy.   R. Novick.
IAU Symposium No. 55, (see 012.002), p. 118 - 131 (1973).

142.067   Models for compact X-ray sources.
E. E. Salpeter.
IAU Symposium No. 55, (see 012.002), p. 135 - 142 (1973).

142.068   Models for compact pulsing X-ray sources.
J. P. Ostriker, K. Davidson.
IAU Symposium No. 55, (see 012.002), p. 143 - 154 (1973).

142.069   Pulsars and X-ray sources.   F. Pacini.
IAU Symposium No. 55, (see 012.002), p. 165 - 167 (1973).

142.070   UHURU results on extragalactic X-ray sources.
E. M. Kellogg.
IAU Symposium No. 55, (see 012.002), p. 171 - 183 (1973).

142.071   The properties of extragalactic X-ray sources from visible light observations.   W. L. W. Sargent.
IAU Symposium No. 55, (see 012.002), p. 184 - 198 (1973).

142.072   Extragalactic X-ray sources.   G. R. Burbidge.
IAU Symposium No. 55, (see 012.002), p. 199 - 207 (1973).

142.073   Extragalactic X-ray sources and their contribution to the diffuse background.
G. Setti, L. Woltjer.
IAU Symposium No. 55, (see 012.002), p. 208 - 211 (1973).
Invited paper.

142.074   The soft X-ray background.
H. Friedman, G. Fritz, S. D. Shulman, R. C. Henry.
IAU Symposium No. 55, (see 012.002), p. 215 - 234 (1973).

142.075   Absorption and production of soft X-rays in the Galaxy.   S. Hayakawa.
IAU Symposium No. 55, (see 012.002), p. 235 - 249 (1973).

142.076   'Evolutionary' theories of the X-ray background.
M. J. Rees.
IAU Symposium No. 55, (see 012.002), p. 250 - 257 (1973).

142.077   Diffuse background of energetic X-rays.   Y. Pal.
IAU Symposium No. 55, (see 012.002), p. 279 - 302 (1973).

142.078   High-energy discrete sources.   G. G. Fazio.
IAU Symposium No. 55, (see 012.002), p. 303 - 323 (1973).

142.079   Limits on the optical pulsations of HZ Herculis.
A. Frohlich.
Astrophys. Letters, Vol. 13, 233 - 235 (1973).
Two sets of observations were made to detect the optical counterpart in HZ Herculis of the 1.2-sec X-ray pulsation of Her X-1. The first set of measurements placed an upper limit of 1 per cent of the optical power of HZ Her in the pulsed mode. The second set of observations placed upper limits between 0.2 to 0.8 per cent of the optical power of HZ Her with pulse widths of at least 0.2 sec.

142.080   Periodicity of Cyg X-3.   A. Treves.
Nature, Phys. Sci., Vol. 242, 121 (1973).
The author examines a different explanation for the periodicity P = 4.8 h of Cyg X-3, namely that P is the period of precession of a rapidly rotating neutron star.

142.081   Observations of the binary X-ray source SMC X-1 from OSO-7.
M. P. Ulmer, W. A. Baity, W. A. Wheaton, L. E. Peterson.
Nature, Phys. Sci., Vol. 242, 121 - 123 (1973).
We report here the results of 40 days of coverage, April 15, 1972, to May 25, 1972, of the binary X-ray source SMC X-1 with the UCSD OSO-7 X-ray telescope. These observations above 7 keV complement the 2-6 keV Uhuru satellite results reported by Schreier et al. and permit an extension of the spectrum to at least 35 keV. Combined with other observations, they could help determine long term periodic variations, if any exist.

142.082   Extragalactic X-ray sources and the X-ray background.   A. C. Fabian.
Nature, Phys. Sci., Vol. 242, 134 (1973). - Letter.

142.083   X-ray cluster of galaxies identified in error box 2U2358-29.   L. A. Thompson.
Nature, Phys. Sci., Vol. 243, 5 - 6 (1973). - Letter.

142.084   Comments on the paper by Mack and Robbins relating to a recent theory on the origin of the universal X-ray background. [Astrophys. Space Sci., Vol. 16, 336 - 337 (1972)].   P. Raychaudhuri, P. Bandyopadhyay.
Astrophys. Space Sci., Vol. 20, 43 - 44 (1973). - Research note.

142.085   On a recent theory of the origin of the X-ray background—further arguments: Comments on the preceding note.   J. E. Mack, D. E. Robbins.
Astrophys. Space Sci., Vol. 20, 45 - 46 (1973). - Research note.

142.086   Luminosity of thermal X-ray sources with a strong magnetic field.
Yu. N. Gnedin, R. A. Sunyaev.
Monthly Notices Roy. Astron. Soc., Vol. 162, 53 - 59 (1973).
The luminosity of thermal X-ray sources with a strong magnetic field and large optical depth with respect to Thomson scattering $r_T > (m_e c^2 / \kappa T_e)^{1/2}$ is shown to be many times that of an optically thin source having the same emission measure but radiating due to bremsstrahlung. This increase in luminosity is caused by the Compton process of energy transfer from electrons to photons. A solution of the problem of Compton energy exchange between radiation and magnetoactive plasma is presented.

142.087   A low-mass primary for Cygnus X-1?
V. Trimble, W. K. Rose, J. Weber.
Monthly Notices Roy. Astron. Soc., Vol. 162, 1P - 3P (1973).
If the primary of HDE 226868 (Cyg X-1) is a low mass (0.3 - 0.5 $M_\odot$), low surface gravity B star of the type of which the primary of HZ 22 is the prototype, then the secondary falls well within the mass range of stable neutron stars (and white dwarfs), and need not be a black hole.

142.088   Model for X-ray sources based on magnetic field twisting.
J. N. Bahcall, M. N. Rosenbluth, R. M. Kulsrud.
Nature, Phys. Sci., Vol. 243, 27 - 28 (1973).

Several X-ray sources have been identified with binary stellar systems. We propose a model for the production of X-rays based on magnetic field twisting. We assume that the binary system consists of two ordinary, but magnetic, rotating stars. We assume that magnetic flux links these two stars and that they are not corotating. The idea of the model is that the linked lines of force will be twisted up, increasing the magnetic energy until instabilities set in and release the magnetic energy in the form of heat which is radiated as X-rays.

142.089 HZ Herculis, der Sonneberger Röntgenstern.
R. Kippenhahn.
SuW, Vol. 12, 133 - 136 (1973).

142.090 On Compton models of the isotropic X-ray background. K. Brecher.
Astrophys. Journ., Vol. 181, 255 - 259 (1973).
It is shown that the spectral shape of the X-ray flux arising from Compton scattering of fast electrons on the microwave background radiation, contrary to a recent assertion, is not inconsistent with present observations of the isotropic X-ray background.

142.091 OAO-2 observations of HD 153919 = 2U 1700−37.
S. R. Heap.
Astrophys. Journ., (Letters), Vol. 181, L71 - L73 (1973).
HD 153919, an O7f star recently proposed as the optical identification of the X-ray source 2U 1700−37, was monitored with the Wisconsin experiment on OAO-2 in 1972 September. These observations indicate that HD 153919 is optically a variable.

142.092 The reflection effect in HZ Herculis.
R. E. Wilson.
Astrophys. Journ., (Letters), Vol. 181, L75 - L77 (1973).
The theoretical phase law for the reflection effect is not in agreement with the photometric observations of HZ Her. The discrepancy may be explained by transfer of energy from the irradiated side of the optical component to the side facing away from the X-ray source.

142.093 High-speed UBV photometry of Scorpius X-1 flares.
T. J. Moffett, G. Grupsmith, P. A. Vanden Bout.
Publ. Astron. Soc. Pacific, Vol. 85, 177 - 178 (1973).
High-speed UBV observations of Sco X-1 during flare activity reveal that the color indices $(B-V)$ and $(U-B)$ change little if at all during a flare. This indicates that the mechanism responsible for flaring is essentially the same as that governing more gradual intensity changes.

142.094 Distance determination of variable X-ray sources.
J. Trümper, V. Schönfelder.
Max-Planck-Inst. Phys. Astrophys., Inst. Extraterr. Phys., München, MPI - PAE/Extraterr. 82, 3 + 20 pp. (1973). − To be published in Astron. Astrophys.
A new method is described for the direct determination of the geometric distance of X-ray sources. The method is applicable to variable X-ray sources and implies the existence of an X-ray halo, which is formed by scattering on interstellar dust. Time structures are delayed and smeared out in the halo. Since the time scales involved depend linearly on the source distance the latter can be determined by measuring damping effects across the halo. Also, information can be obtained about the dust density distribution along the line of sight.

142.095 The cosmic γ-ray spectrum between 0.3 and 27 MeV measured on Apollo 15.
J. I. Trombka, A. E. Metzger, J. R. Arnold, J. L. Matteson, R. C. Reedy, L. E. Peterson.
Astrophys. Journ., Vol. 181, 737 - 746 (1973).
The spectrum of the total (diffuse and discrete sources)

cosmic γ-ray background over the 0.3−27 MeV range has been measured with an uncollimated NaI(Tl) scintillation counter 7.0 cm in diameter × 7.0 cm long located on a boom 7.6 m from the Apollo 15 service module. Data on cosmic γ-rays were taken during trans-earth coast at various boom extensions, detector gains, and with the plastic anticoincidence scintillator enabled and disabled.

142.096 On the distances to transient X-ray sources.
J. Silk.
Astrophys. Journ., Vol. 181, 747 - 751 (1973).
From a discussion of the X-ray data, it is argued that the transient X-ray sources must be galactic, and must possess peak luminosities comparable with the strongest known galactic X-ray sources. The distribution of transient sources is predicted to be that of a halo population, and their frequency of occurrence is estimated as a function of flux level.

142.097 The number-intensity distribution of X-ray sources observed by Uhuru.
T. Matilsky, H. Gursky, E. Kellogg, H. Tananbaum, S. Murray, R. Giacconi.
Astrophys. Journ., Vol. 181, 753 - 759 (1973).
The Uhuru catalog of X-ray sources is used to analyze the number versus apparent-intensity relation of X-ray objects, by constructing (log N, log S)-plots similar to those used in radio astronomy. Two distinct distributions for objects at low $(|b| < 20°)$ and high $(|b| > 20°)$ galactic latitude are discussed.

142.098 X-ray beaming and mass transfer in HZ Her.
P. A. Strittmatter, J. Scott, J. Whelan, D. T. Wickramasinghe, N. J. Woolf.
Astron. Astrophys., Vol. 25, 275 - 284 (1973).
The optical and X-ray properties of HZ Her are analyzed within the framework of a model in which the primary is illuminated by X-ray radiation from a degenerate secondary companion. System parameters are derived on the basis of a simplified model.

142.099 Optical appearance of binary X-ray sources.
J. E. Pringle.
Nature, Phys. Sci., Vol. 243, 90 - 94 (1973).
The expected optical behaviour of binary X-ray sources is considered with reference to their optical identification. The system Hercules X-1 is studied in some detail, in particular the 35 d period and optical pulses. A model is proposed for transient X-ray sources.

142.100 The 35-day periodicity of Hercules X-1.
R. McCray.
Nature, Phys. Sci., Vol. 243, 94 - 96 (1973).
The author suggests as an alternative hypothesis that the 35 d cycle is caused by a nearly periodic regulation of the accretion gas flow which is presumed to be the source of the X-ray luminosity of Hercules X-1. The model explains not only the mechanism for the 35 d periodicity, but also the phase-locking of the 35 d cycle to the 1.7 d orbital period and some details of the profile of the cycle and of X-ray spectral changes within the cycle.

142.101 Balloon observations of Sco X-1 in the energy interval 17−106 keV.
A. K. Jain, U. B. Jayanthi, K. Kasturirangan, U. R. Rao.
Astrophys. Space Sci., Vol. 21, 107 - 116 (1973).
The paper presents the intensity and spectral nature of the X-ray emission from Sco X-1 in the energy interval 17−106 keV based on the observations made by a balloon borne scintillation telescope system flown on November 15, 1971 from Hyderabad, India. Comparing the present data with those obtained elsewhere, the temporal characteristics of the X-ray

emission from Sco X-1 are discussed.

142.102 The nature of X-ray binaries III. Evolution of massive close binaries with one collapsed component — with a possible application to Cygnus X-3.
E. P. J. van den Heuvel, C. De Loore.
Astron. Astrophys., Vol. 25, 387 - 395 (1973).

The evolution is computed for close binaries consisting of a massive primary ($15\,M_\odot$ or $21\,M_\odot$) together with a neutron star secondary ($M_\odot$ or $2\,M_\odot$) with orbital periods between 2 and 6 days. Two cases are considered, viz. (1) the neutron star is able to accrete the $10^{-3}\,M_\odot$/yr lost by the primary after this star has started to overflow its Roche lobe, and (2) accretion onto the neutron star is limited to $\lesssim 10^{-7}\,M_\odot$/yr, due to the critical luminosity. It is suggested that the final systems may be identified with X-ray sources such as Cygnus X-3 ($P_x = 4\overset{h}{.}8$).

142.103 Distance determination of variable X-ray sources.
J. Trümper, V. Schönfelder.
Astron. Astrophys., Vol. 25, 445 - 450 (1973).

A new method is described for the direct determination of the geometric distance of X-ray sources. The method is applicable to variable X-ray sources and implies the existence of an X-ray halo, which is formed by scattering on interstellar dust. Time structures are delayed and smeared out in the halo.

142.104 On the size and nature of the X-ray source in Cen A.
G. C. Perola, M. Tarenghi.
Astron. Astrophys., Vol. 25, 461 - 465 (1973).

On the assumption that the low energy cut-off observed in the X-ray spectrum of NGC 5128 (Cen A) is due to photoelectric absorption, and that the X-ray source is concentric with the galaxy, the radio data in the continuum and the 21 cm line on the inner radio source of Cen A are used to set an upper limit on the size of the X-ray source of 1/2 kpc and 20 pc respectively if the absorbing gas is neutral or ionized. The nature of the source assumed coincident with the nucleus of the galaxy is then examined in the hypothesis of a synchrotron or inverse Compton origin of the X-rays. It is concluded that in both cases electrons of thousands of GeV or hundreds of MeV respectively need to be injected continuously, to compete with radiative losses and escape.

142.105 Alcuni problemi dell'astronomia X. L. Gratton.
Mem. Soc. Astron. Italiana, Nuova Ser., Vol. 43, 585 - 598 (1973). — Invited paper presented to the 'Giornate di studio dedicate al Prof. F. Zagar' (see 012.023).

142.106 Plasma turbulent heating and thermal X-ray sources.
B. Coppi, A. Treves.
Cosmic plasma physics. Conference 1971, (see 012.016), p. 215 - 223 (1972).

142.107 Speeding up of pulsing X-ray sources.
W. Y. Chau.
Nature, Phys. Sci., Vol. 243, 133 - 134 (1973).

One of the most intriguing results emerging from extended observations of X-ray sources is that in at least two cases the period of the rapid variation is decreasing. With Henriksen and Feldman, the author has discussed (1972) the outline of a plausible model based on accretion on a rapidly rotating, magnetic white dwarf evolving along the so-called "Jacobi sequence" where it has the peculiar property of speeding up while losing angular momentum. He now discusses the spin-up aspect of the model further. The idea involves the equilibrium configuration of a uniformly rotating homogeneous fluid mass in the classical theory of hydrodynamics.

142.108 Possible correlation between the soft X-ray flux and features in the radio continuum in the anticentre.
E. M. Berkhuijsen.
Nature, Phys. Sci., Vol. 243, 135 - 136 (1973).

Bunner et al. (1973) have reported the observation of extended soft X-ray emission in the region $190° < l < 220°$, $-5° < b < +35°$ in three energy bands. They showed that the high emission features found are not of solar or terrestrial origin. The author proposes a possible correlation with features in the galactic continuum radiation.

142.109 Photographic photometry of the proposed optical candidate for SMC X-1.
C. J. Butler, P. B. Byrne.
Nature, Phys. Sci., Vol. 243, 136 - 138 (1973).

The authors present independent photographic material in three colours which indicate that Sanduleak 160 is, indeed, SMC X-1.

142.110 On the variability of high-energy gamma-ray sources.
B. M. Vladimirsky, A. A. Stepanian, V. P. Fomin.
Astron. Zhurn. Akad. Nauk SSSR, Vol. 50, 449 - 452 (1973). In Russian. English translation in Soviet Astron. AJ, Vol. 17, No. 3.

Some results of observations of the variable source of high-energy gamma-rays in the Cassiopeia region are presented. The problem of the variability of such a type of sources is discussed.

142.111 On the explanation of the 36-day period of Her X-1 by free precession. I. D. Novikov.
Astron. Zhurn. Akad. Nauk SSSR, Vol. 50, 459 - 461 (1973). In Russian. English translation in Soviet Astron. AJ., Vol. 17, No. 3.

The theory explaining the 36-day period of Her X-1 by free precession of a slightly asymmetric neutron star is advocated. It is shown that in a binary system accretion of matter on the magnetic poles of the neutron star can lead to asymmetry and free precession of this star.

142.112 X-ray background intensity fluctuations.
Ya. M. Khazan.
Astron. Zhurn. Akad. Nauk SSSR, Vol. 50, 469 - 474 (1973). In Russian. English translation in Soviet Astron. AJ, Vol. 17, No. 3.

X-ray background intensity fluctuations are calculated in the case, when the background results from radiation of non-resolved discrete sources.

142.113 Production of astrophysical X-rays by transition radiation. W. D. Watson.
Astrophys. Journ., Vol. 182, 17 - 24 (1973).

Results of detailed calculations are presented for the spectrum of transition radiation produced when relativistic charged particles traverse astrophysical dust grains. When competing processes and upper limits to the galactic electron flux are considered, neither the low-energy nor the higher-energy X-rays associated with the galactic disk can be explained by transition radiation from grains.

142.114 X-ray pulse profile and celestial position of Hercules X-1. R. Doxsey, H. V. Bradt, A. Levine, G. T. Murthy, S. Rappaport, G. Spada.
Astrophys. Journ., (*Letters*), Vol. 182, L25 - L28 (1973).

The celestial position of the binary X-ray source Her X-1 has been measured with a precision of 30" and, within the uncertainties, agrees with the position of the optical variable HZ Herculis. An average light curve for the 1.24-s periodicity has been obtained with ~10 ms resolution. The average pulse is a double-peaked structure which shows significant intensity changes in time scales down to 30 ms.

142.115 Zeeman effect in the X-ray star candidates HD

77581 and θ^2 Orionis.
J. C. Kemp, R. D. Wolstencroft.
Astrophys. Journ., (*Letters*), Vol. 182, L43 - L46 (1973).

The discovery of Zeeman effects is reported in HD 77581 and θ^2 Orionis, optical candidates for the X-ray sources Vela XR-1 and 2U 0525−06, respectively. The maximum longitudinal magnetic fields recorded were −10,000 gauss in HD 77581 and +1500 gauss in θ^2 Ori. Various polarimetric data are also given, including evidence for a variable linear polarization in HD 77581.

142.116 Kosmische Quellen von Röntgen- und Gamma-strahlung. J. Trümper.
Mitt. Astron. Ges., No. 32, p. 91 - 106 (1973). − Review paper presented at the assembly of the Astron. Ges., Wien, 1972 Sept.

142.117 On the accretion model for X-ray double stars.
G. Börner, F. Meyer, H. U. Schmidt, H.-C. Thomas.
Mitt. Astron. Ges., No. 32, p. 237 - 239 (1973).

142.118 Der Ope-Stern X Persei.
A. F. J. Moffat, W. Haupt, T. Schmidt-Kaler.
Mitt. Astron. Ges., No. 32, p. 242 - 243 (1973). − Abstract.

142.119 A method of determining polarization and energetic spectrum of cosmic γ-quanta.
A. M. Gal'per, S. R. Kel'ner, Yu. D. Kotov, V. M. Logunov.
Izv. AN SSSR, Ser. fiz., Vol. 36, 2354 - 2358 (1972). In Russian. − Abstr. in Referativ. Zhurn. 51. Astron., 5.51.703 (1973).

142.120 Rocket observations of the X-ray absorption measure of Sco X-1.
A. J. M. Deerenberg, J. A. M. Bleeker, P. A. J. de Korte, K. Yamashita, Y. Tanaka, S. Hayakawa.
Nature, Phys. Sci., Vol. 244, 4 - 6 (1973).

The authors report two rocket observations of Sco X-1 with similar instruments. The first experiment, LEINAX I, was launched on May 26, 1971, and the second experiment, LEINAX II, on May 22, 1972.

142.121 Possible candidate for LMC X-1.
P. B. Byrne, C. J. Butler.
Nature, Phys. Sci., Vol. 244, 6 - 7 (1973). − Letter.

142.122 Possible low ionosphere response to very hard X-rays from Cygnus X-3 bursts in September 1972.
P. Kaufmann, L. R. Piazza, S. Ananthakrishnan.
Astrophys. Space Sci., Vol. 22, 67 - 70 (1973).

Coincidence of VLF propagation anomalies in the low terrestrial ionosphere with the flaring period of Cyg X-3 tentatively suggest that this source has shown an output in very hard X-rays. Difficulties in interpretation are discussed.

142.123 Description of small-scale fluctuations in the diffuse X-ray background.
A. Cavaliere, A. Friedland, H. Gursky, G. Spada.
Astrophys. Journ., Vol. 182, 405 - 410 (1973).

An analytical study of the fluctuations on a small angular scale expected in the diffuse X-ray background in the presence of unresolved sources is presented. The source population is described by a function $N(S)$, giving the number of sources per unit solid angle and unit apparent flux S. The distribution of observed flux, s, in each angular resolution element of a complete sky survey is represented by a function $Q(s)$. The analytical relation between the successive, higher-order moments of $N(S)$ and $Q(S)$ is described.

142.124 Optical observations and model for Cygnus X-1.
J. B. Hutchings, D. Crampton, J. Glaspey, G. A. H. Walker.
Astrophys. Journ., Vol. 182, 549 - 557 (1973).

Spectrophotometric observations have been made with an image isocon camera of the λ4686 region of the spectrum of HDE 226868. The authors present the results and an analysis of the light variations, both of which suggest a model and mass ratio for the system.

142.125 HD 77581 as the optical counterpart of 2U 0900−40.
N.V. Vidal, D. T. Wickramasinghe, B. A. Peterson.
Astrophys. Journ., (*Letters*), Vol. 182, L77 - L79 (1973).

Photoelectric observations of the light curve of HD 77581 made in 1973 between January 22 and March 6 are found to be consistent with the 8.95-day X-ray period of 2U 0900−40.

142.126 Observations of gamma-ray bursts of cosmic origin.
R. W. Klebesadel, I. B. Strong, R. A. Olson.
Astrophys. Journ., (*Letters*), Vol. 182, L85 - L88 (1973).

Sixteen short bursts of photons in the energy range 0.2−1.5 MeV have been observed between 1969 July and 1972 July using widely separated spacecraft. Burst durations ranged from less than 0.1 s to ~30 s, and time-integrated flux densities from ~10^{-5} ergs cm^{-2} to ~2×10^{-4} ergs cm^{-2} in the energy range given. Significant time structure within bursts was observed.

142.127 Highly variable X-ray source. T. H. Markert, G. W. Clark, W. H. G. Lewin, H. W. Schnopper, G. F. Sprott.
IAU Circ., No. 2483 (1973).

142.128 2U 1700−37. E. N. Walker, A. D. Thackeray.
IAU Circ., No. 2493 (1973).

142.129 Cepheus X-4. M. P. Ulmer, W. A. Baity, L. E. Peterson, W. A. Wheaton.
IAU Circ., No. 2493 (1973).

142.130 2U 0900−40. W. A. Hiltner.
IAU Circ., Nos. 2502, 2515 (1973).

142.131 Optical identifications of X-ray sources.
N. V. Vidal, D. T. Wickramasinghe, B. A. Peterson, C. Jones, W. Liller.
IAU Circ., No. 2503 (1973).

142.132 Cygnus X-2.
R. M. Hjellming, L. C. Blankenship.
IAU Circ., No. 2509 (1973).

142.133 Cepheus X-4. T. H. Markert, G. W. Clark, D. R. Hearn, W. H. G. Lewin, H. W. Schnopper, G. F. Sprott.
IAU Circ., No. 2512 (1973).

142.134 Tentative identification of Centaurus X-3.
W. Liller, W. Forman.
IAU Circ., No. 2518 (1973).

142.135 WRA 795. N. V. Vidal, D. T. Wickramasinghe, B. A. Peterson, M. S. Bessell, M. E. Perry.
IAU Circ., No. 2521 (1973).

142.136 Optical pulsations from HZ Herculis.
E. J. Groth, S. Z. Yeung, C. Papaliolios, C. R. Pennypacker, G. Spada, J. Middleditch.
IAU Circ., No. 2523 (1973).

142.137 Scheduled observations of optically identified X-ray sources.
S. P. Maran, L. E. Peterson, G. W. Clark.
IAU Circ., No. 2523 (1973).

142.138　**Centaurus X-3.**　E. Schreier, R. Giacconi, H. Gursky, E. Kellogg, R. Levinson, H. Tananbaum, J. H. Parkinson, K. A. Pounds, P. W. Sanford, F. Hawkins, K. Mason.
IAU Circ., No. 2524 (1973).

142.139　**2U 0900-40 a black hole?**　D. T. Wickramasinghe, N. V. Vidal, B. A. Peterson, M. S. Bessell, M. E. Perry.
IAU Circ., No. 2525 (1973).

142.140　**HD 153919 (2U 1700-37).**　E. P. J. van den Heuvel.
IAU Circ., No. 2526 (1973).

142.141　**Scheduled observations of optically identified X-ray sources.**　R. Thomas.
IAU Circ., No. 2534 (1973).

142.142　**New X-ray source in Perseus.**　C. Heinz, H. Bradt, G. Clark, W. Lewin, G. Sprott.
IAU Circ., No. 2540 (1973).

142.143　**Cygnus X-1.**　E. N. Walker.
IAU Circ., No. 2551 (1973).

142.144　**HD 153919 and the X-ray source 2 U 1700−37.**　E. N. Walker.
Monthly Notices Roy. Astron. Soc., Vol. 162, 15P - 20P (1973).

HD 153919 is the prime optical candidate for the X-ray source 2 U 1700−37. Radcliffe spectra show the star to be an O5.5f star with a strong line of C III. The radial velocities indicate that the star has an expanding atmosphere. The star is both a radial velocity and spectrum variable.

142.145　**Production of metagalactic X-rays by relativistic dust grains.**　S. Hayakawa.
Astrophys. Space Sci., Vol. 19, 173 - 179 (1972).

The relativistic dust grains which may be responsible for ultra-high energy cosmic rays, as suggested by the present author, interact with the cosmic black-body radiation. This results in the energy loss of the relativistic dust grains, so that their energy spectrum is cut-off at the Lorentz factor as large as $2 \times 10^3 (0.1 \mu/a)$, where a is the grain radius. The black-body radiation is scattered and absorbed by the dust grains. The photons scattered and reemitted contribute to metagalactic X-rays. The X-ray intensity estimated is comparable to the observed one in the soft X-ray region.

142.146　**On the physical parameters for Centaurus X-3 and Hercules X-1.**　G. E. McCluskey, Jr., Y. Kondo.
Astrophys. Space Sci., Vol. 19, 279 - 284 (1972).

Upper and lower limits on the physical parameters for Cen X-3 and Her X-1 have been computed from a simple assumption involving the mass function.

142.147　**The shape of the diffuse cosmic X-ray spectrum.**　C. S. Dyer, A. R. Engel, J. J. Quenby.
Astrophys. Space Sci., Vol. 19, 359 - 367 (1972).

Recent work by Dyer and Morfill has shown that satellite measurements of the diffuse cosmic X-ray spectrum made with crystal scintillators may include errors due to radioactive spallation products formed in the detector by inner belt and cosmic ray protons. An estimate is made of the magnitude of this source of background for the various experimental situations. A review is made of experiments covering the range 1 keV− 100 MeV in order to ascertain whether a single exponent spectrum is capable of fitting the experimental results. The astrophysical implications of such a spectrum are briefly considered. Suggestions are made for the location and correction for background of future experiments.

142.148　**Collimator corrections to the measured diffuse X-ray background.**　A. Dumas, H. Horstman, E. Horstman-Moretti.
Astrophys. Space Sci., Vol. 19, 495 - 500 (1972).

The contribution to the main detector counting rate from photons scattered and reemitted from the collimator walls is considered in two particular cases, that of our group's old S-11 rocket data and OSO-III. We find that the S-11 results are not changed appreciably while our consideration of the OSO-III results has lead us to the belief that these results should be strongly modified.

142.149　**Circular polarization observations of Sco X-1.**　O. S. Shulov, E. N. Kopatskaya.
Astrofizika, Vol. 8, 621 - 624 (1972). In Russian. – English translation in Astrophysics, Vol. 8, No. 4.

New observations carried out in July 1972 confirm the existence of variable circular polarization in the red spectral region of Sco X-1.

142.150　*Uhuru* **observations of the binary X-ray source 2U 0900−40.**　W. Forman, C. Jones, H. Tananbaum, H. Gursky, E. Kellogg, R. Giacconi.
Astrophys. Journ., (*Letters*), Vol. 182, L103 - L107 (1973).

Uhuru observations show that 2U 0900−40 is an eclipsing binary with a period of 8.95 ± 0.02 days, confirming the report by Ulmer et. al. The X-ray spectrum is flat and shows a large amount of low-energy absorption. The 2−6 keV intensity of the source is highly variable. The identification of 2U 0900−40 with HD 77581 is discussed.

142.151　**Optical studies of** *Uhuru* **sources. IV. The long-term behavior of HZ Herculis = Hercules X-1.**　C. A. Jones, W. Forman, W. Liller.
Astrophys. Journ., (*Letters*), Vol. 182, L109 - L112 (1973).

Harvard photographs of HZ Herculis taken during the period 1890−1972 reveal the following: (1) The orbital period has remained constant at 1.70017 days to within two parts in 10^5 since 1900. (2) HZ Her does not always display the 1.7-day light variation. (3) When the 1.7-day variation is not detectable, one observes a second brightness fluctuation with an amplitude of $\Delta m_{pg} = 0.28 \pm 0.06$ and a period half that of the orbital motion.

Relativistische Astrophysik.　See Abstr. 011.006.

The High Energy Astronomical Observatory.　See Abstr. 054.021.

X-ray astronomy – results and instruments.　See Abstr. 061.045.

Experimental X-ray astronomy.　See Abstr. 061.064.

Possible production of the soft X-ray background by the coronae of red giants.　See Abstr. 064.042.

The thermal radiation spectra of supermassive stars and X-ray sources.　See Abstr. 064.079.

The beaming of radiation from an accreting magnetic neutron star and the X-ray pulsars.　See Abstr. 065.117.

Electric field around an accreting star.　See Abstr. 065.161.

Black holes in binary systems. Observational appearance.　See Abstr. 066.049.

Lunar composition from Apollo orbital measurements. See Abstr. 094.301.

GX 5-1. See Abstr. 096.024.

The radial velocities of SEG stars and the distance to Scorpius X-1. See Abstr. 112.004.

The radial velocities of SEG stars and the distance to Scorpius X-1. See Abstr. 112.007.

Some tests and consequences of the identification of promethium in HR 465. See Abstr. 114.113.

HD 77581. See Abstr. 119.022.

HZ Herculis. See Abstr. 121.067.

HZ Herculis. See Abstr. 121.068.

Optical circular polarization of X Persei. See Abstr. 122.019.

A new variable star in the LMC within the error box of the X-ray source LMC X-1. See Abstr. 123.017.

Supernova remnants and gamma-ray sources. See Abstr. 125.006.

The soft X-ray structure of Cassiopeia A. See Abstr. 125.007.

Low energy X-ray map of Puppis A supernova remnant. See Abstr. 125.019.

Observations of soft X-rays: Upper limits on the flux from SN 1972E and measurements of the diffuse background in Centaurus. See Abstr. 125.032.

Thermal structure and evolution of interstellar gas exposed to a soft X-ray burst. See Abstr. 131.180.

Heating and ionization of H I regions by discrete soft X-ray and subcosmic-ray sources. See Abstr. 131.204.

The X-ray surface brightness of the Cygnus Loop. See Abstr. 132.007.

Radio stars AR Lacertae and Cygnus X-2. See Abstr. 141.088.

Radio variability of HDE 226868 (Cygnus X-1). See Abstr. 141.100.

Gamma rays from NP 0532. See Abstr. 141.503.

Soft X-ray pulsations from PSR 0833−45. See Abstr. 141.507.

Pulsar X-ray source association. See Abstr. 141.555.

Possible origin of cosmic γ ray flux from pulsars. See Abstr. 141.558.

Transition radiation in astrophysics. See Abstr. 143.035.

An estimate of the energy spectrum of gamma rays from the central region of the Galaxy and some implications. See Abstr. 155.001.

A soft X-ray survey from the galactic center to Cassiopeia. See Abstr. 155.008.

On the gamma ray line from the galactic centre. See Abstr. 155.017.

'Local' theories of the X-ray background. See Abstr. 155.043.

Accretion by neutron stars at the galactic center. See Abstr. 155.055.

Giant loops as fossil Strömgren spheres: Their radio and X-ray emission. See Abstr. 155.089.

Remarks on the soft X-ray emission from the galactic radio spurs. See Abstr. 155.092.

Models for extragalactic objects with very high IR and X-ray luminosity. See Abstr. 158.024.

X-ray observations of NGC 5128 (Centaurus A) from *UHURU*. See Abstr. 158.057.

Clusters of galaxies are the possible source of a background X-ray emission. See Abstr. 160.023.

Errata

142.901 Erratum: 'The interaction of Sco X-1 with its environment' [Astron. Astrophys., Vol. 20, 287 - 291 (1972)].
J. Silk, D. W. Goldsmith, G. B. Field, L. Carrasco.
Astron. Astrophys., Vol. 23, 321 (1973).

143 Cosmic Radiation

143.001 Cosmic ray electrons of $E > 1$ Gev—Some new measurements and interpretations.
W. R. Webber, J. M. Rockstroh.
Journ. Geophys. Res., Vol. 78, 1 - 11 (1973).

In the summer of 1971 we conducted new measurements of the primary electron spectrum from 1 to 20 Gev with our electron spectrometer telescope at Fort Churchill. Our measured primary electron spectrum is uncomfortably close to that calculated for secondary electrons produced by the interaction of cosmic ray nuclei with interstellar hydrogen in the Galaxy, and therefore the question of the origin of the high-energy cosmic ray electrons is reopened. The problem of the age of cosmic ray electrons must also be reexamined if our measured spectrum is found to extend to still higher energies.

143.002 Energy dependent time lag in the long-term modulation of cosmic rays.
J. J. Burger, B. N. Swanenburg.
Journ. Geophys. Res., Vol. 78, 292 - 305 (1973).

The intensity variations in the cosmic ray electron spectrum above 500 Mev observed outside the radiation belts from March 1968 to August 1971 are reported. The data show the existence of a large energy dependent hysteresis in the long-term modulation and two different types of steplike changes in the modulation parameter at rigidities below those observed with neutron monitors.

143.003 Plasma effects and the acceleration of charged particles in pulsar fields. W. H. Kegel.
Astron. Astrophys., Vol. 22, 475 - 477 (1973).

We consider the acceleration of cosmic ray particles in strong electromagnetic fields of the kind expected in the vicinity of pulsars. The results are applied to the Crab pulsar and Crab nebula.

143.004 Loi générale de production multiple de 10 à 10^{12} GeV.
J.-N. Capdevielle, J. Dupuy, A. Cachon.
Comptes Rendus Acad. Sci. Paris Sér. B, Vol. 276, 219 - 222 (1973).

De l'étude par notre méthode de simulation des propriétés moyennes des grandes gerbes du rayonnement cosmique, se dégage, entre 10^{14} et 3×10^{17} eV, la loi de multiplicité des particules secondaires créées. Le domaine de validité de cette loi, compatible avec les résultats obtenus avec les accélérateurs à partir de 10 GeV, semble pouvoir être étendu jusqu'aux plus hautes énergies connues.

143.005 Cosmic rays in a random magnetic field: Breakdown of the quasilinear derivation of the kinetic equation.
T. B. Kaiser, F. C. Jones, T. J. Birmingham.
Astrophys. Journ., Vol. 180, 239 - 245 (1973).

We consider the problem of deriving a kinetic equation for the cosmic-ray distribution function in a random magnetic field. A model is adopted which is mathematically simple but which contains the essential physics. We investigate the perturbation expansion upon which the quasilinear treatment employed by previous authors is based. As pointed out by Klimas and Sandri, the existence of resonant particles causes the breakdown of the adiabatic approximation. We find further that resonant particles cause a general secular growth of higher-order terms in the expansion which invalidates the entire perturbative approach.

143.006 A phenomenological study of cosmic ray propagation. I. The nuclei component.
F. Le Guet, J. A. de Freitas Pacheco.
Astron. Astrophys., Vol. 23, 337 - 346 (1973).

We have considered a phenomenological model for cosmic ray propagation through the Galaxy. The diffusion equation has been solved for several nuclei species. The actual data are consistent with a flat disk diffusion region which has a radius of about 14 kpc and a half-height of about 2 kpc.

143.007 Charge dependence of the energy spectra of cosmic rays. J. F. Ormes, V. K. Balasubrahmanyan.
Nature, Phys. Sci., Vol. 241, 95 - 96 (1973).

We report new experimental results in the energy range 3 to 50 GeV nucleon^{-1} which indicate that the well known enrichment of heavy nuclei in cosmic rays is increasing with increasing energy. The measurements were made with a balloon-borne ionization spectrometer.

143.008 Evidence for differences in the energy spectra of cosmic ray nuclei.
W. R. Webber, J. A. Lezniak, J. C. Kish, S. V. Damle.
Nature, Phys. Sci., Vol. 241, 96 - 98 (1973).

In a table we show data on the intensities of He-, (C+O)-, ($Z=17-25$)-, (Fe+Ni)-nuclei measured at several energies from ~ 1 to 50 GeV nucleon^{-1}. The data in this table show a very interesting behaviour. The fraction of $Z=17-25$ to Fe+Ni nuclei changes from $\gtrsim 1$ at energies $\lesssim 1$ GeV nucleon^{-1} to values $\lesssim 0.2$ at the highest energies measured.

143.009 Mean path length of high energy galactic cosmic rays in the galactic disk. J. Audouze, C. J. Cesarsky.
Nature, Phys. Sci., Vol. 241, 98 - 100 (1973).

The leakage term which describes the probability that galactic cosmic rays leave the Galaxy must be energy dependent. Here we present a quantitative estimation of this dependence and discuss the implications of such a dependence.

143.010 Energy dependence of primary cosmic ray nuclei abundance ratios. M. Meneguzzi.
Nature, Phys. Sci., Vol. 241, 100 - 101 (1973).

Recent measurements of cosmic ray chemical composition between 1 and 100 GeV nucleon^{-1} show a decrease of the abundance ratios of secondary to parent nuclei above $\simeq 10$ GeV/N. The same observations also show an energy dependence of the primary nuclei abundance ratios. The presently observed variations of these ratios seem to be consistent with (and give more support to) the conclusion drawn from secondary to parent nuclei ratios, without any need for additional hypothesis, such as charge dependence of the mean path length or charge dependence of the source spectral index.

143.011 Tracks from extinct radioactivity, ancient cosmic rays, and calibration ions.
R. L. Fleischer, H. R. Hart, Jr.
Nature, Vol. 242, 104 - 105 (1973).

We present here an alternative explanation of the fossil tracks in lunar minerals that have been attributed to extinct radioactivity of superheavy elements.

143.012 A novel type of interaction near 10^4 GeV?
J. E. F. Baruch, G. Brooke, E. W. Kellermann.
Nature, Phys. Sci., Vol. 242, 6 - 7 (1973). – Letter.

143.013 The Forbush predecrease.
R. E. Gold, D. S. Peacock.
Journ. Geophys. Res., Vol. 78, 577 - 587 (1973).

Forbush decreases observed by superneutron monitors are frequently preceded by an almost equally large predecrease. Five such events are studied here, and there are two main observational results: (1) the predecrease represents a cosmic ray depletion in a narrow range of pitch angles (±30°) around the mean interplanetary field direction and (2) the predecrease has an unusual rigidity dependence inasmuch as its amplitude increases with rigidity, going something like $R^{0.5 \pm 1.5}$ up to a limit of about 80 Gv.

143.014 The heliocentric radial gradient in cosmic ray density and the 'Swinson' sidereal time variation.
D. M. Thomson.
Planet. Space Sci., Vol. 21, 133 - 143 (1973).

The sidereal time variation reported by Swinson depends on the existence of a heliocentric radial gradient of cosmic ray density in the rigidity range ≲ 100 GV and appears because of the inclination of the axis of rotation of the earth to the normal to the ecliptic plane. In this paper results obtained at Makerere on the earth's surface and at Kilembe in Uganda, East Africa, are compared with data previously reported by Swinson from Chacaltaya and Embudo.

143.015 The upper boundary of antinuclei content in primary cosmic radiation.
N. S. Ivanova, V. N. Kulikov, E. A. Yakubovskij.
Kosmich. Issled., Vol. 11, 163 - 165 (1973). In Russian. Brief information.

143.016 Manganese-54 and the lifetime of relativistic cosmic rays. M. Cassé.
Astrophys. Journ., Vol. 180, 623 - 629 (1973).

The influence of a possible ^{54}Mn decay on predicted Mn/Fe and Mn/(Ti + V + Cr) ratios is discussed in the light of present observations.

143.017 Electron streams with energies ≳ 80 MeV at the equator from data of measurements aboard Cosmos 490. R. N. Basilova, N. L. Grigorov, L. F. Kalinkin, G. I. Pugacheva, E. A. Pryakhin, I. A. Savenko.
Kosmich. Issled., Vol. 11, 160 - 161 (1973). In Russian. Brief information.

143.018 A search for cosmic rays streaming out of the Galaxy. A. G. Fenton, K. B. Fenton.
Proc. Astron. Soc. Australia, Vol. 2, 139 - 140 (1972).

143.019 A rigorous cosmic-ray transport equation with no restrictions on particle energy.
A. J. Klimas, G. Sandri.
Astrophys. Journ., Vol. 180, 937 - 954 (1973).

A new transport equation for the cosmic-ray omnidirectional intensity is obtained. This equation follows exactly from the coupled pair of differential moment equations we presented earlier. It can be characterized as a nonlocal convection-diffusion equation in which the usual transport coefficients are replaced by time integral operators. The large- and small-gyroradius transport theories due originally to Jokipii are regained. The validity of these theories as asymptotic limits and as approximate theories in the interplanetary magnetic field is discussed.

143.020 A measurement of cosmic-ray rigidity spectra above 5 GV/c of elements from hydrogen to iron.
L. H. Smith, A. Buffington, G. F. Smoot, L. W. Alvarez, M. A. Wahlig.
Astrophys. Journ., Vol. 180, 987 - 1010 (1973).

This paper presents measurements of the differential rigidity spectra of primary cosmic-ray nuclei between 5 GV/c and 100 GV/c. These measurements were performed with a balloon-borne superconducting magnetic spectrometer containing scintillation detectors, optical spark chambers, and associated electronics. We present a rigidity spectrum for each element from hydrogen through oxygen and for groups of elements through the iron group.

143.021 Investigation of resonance integrals occurring in cosmic-ray diffusion theory.
F. C. Jones, T. J. Birmingham, T. B. Kaiser.
Astrophys. Journ., (Letters), Vol. 180, L139 - L142 (1973).

Using an analytic expression for the power spectrum we have performed, by contour integration, certain integrals that arise in the theory of cosmic rays propagating in a static, random magnetic field. We find that as $t \to \infty$ these integrals relax to a "resonance" form employed by previous authors. We also find, however, that the time scale over which this relaxation occurs can depend critically on the correlation length of the random field and on μ, the cosine of the particle's pitch angle with respect to the background field, and in particular for $\mu = 0$ this relaxation does not occur.

143.022 Preliminary Pioneer-10 intensity gradients of galactic cosmic rays. R. B. McKibben, J. J. O'Gallagher, J. A. Simpson, A. J. Tuzzolino.
Astrophys. Journ., (Letters), Vol. 181, L9 - L13 (1973).

Selected data from the University of Chicago charged-particle telescope on the Pioneer-10 spacecraft bound for Jupiter have been examined and compared with data from the University of Chicago charged-particle telescopes on the earth satellites IMP-5 and IMP-6 at 1 a.u. to derive a preliminary integral intensity gradient for relativistic galactic protons and helium nuclei. Preliminary differential gradients also have been obtained for the energy range 29−67 MeV per nucleon.

143.023 Cosmic rays in the outer solar system.
E. N. Parker.
Space Sci. Rev., Vol. 14, 576 - 581 (1973).

A space mission to Jupiter and Saturn, and beyond, provides an opportunity to explore the low energy galactic cosmic rays, which are largely excluded from the inner solar system by the outward sweep of the magnetic fields in the solar wind. The nuclear abundances, and in particular the presence or absence of high Z nuclei, will give critical information on the proximity of cosmic ray sources.

143.024 On the theory of cosmic ray transfer with anisotropic scattering of particles. V. F. Zakharchenko.
Geomagn. Aeronom., Vol. 13, 14 - 19 (1973). In Russian.

143.025 Modulation of galactic cosmic rays by the solar wind, asymmetrically with regard to the heliolatitude.
L. I. Dorman, Z. Kobilinsky, T. S. Khadakhanova.
Geomagn. Aeronom., Vol. 13, 20 - 25 (1973). In Russian.

143.026 Solar modulation of galactic cosmic rays. 3. Implications of the Compton-Getting coefficient.
L. A. Fisk, M. A. Forman, W. I. Axford.
Journ. Geophys. Res., Vol. 78, 995 - 1006 (1973).

We discuss the implications of the behavior of the Compton-Getting coefficient with energy, and, in particular, we consider conditions in the interplanetary medium that can lead to a near-zero Compton-Getting coefficient at energies below ~200 MeV/nucleon.

143.027 The origin of cosmic rays — new interest in an old question. M. Grewing, H. Heintzmann.
Zeitschr. Naturforschung, Vol. 28a, 369 - 376 (1973).

After briefly reviewing the experimental facts that seem of direct relevance for the question of the origin of cosmic rays, and after quickly going through the "classical" theories of cosmic ray origin, we discuss in some detail those theories that have been put forward over the past few years since the discovery of pulsars.

143.028 On cosmic rays and magnetic stars. O. Havnes.
Astron. Astrophys., Vol. 24, 435 - 440 (1973).

In an earlier paper (Havnes, 1971), rotating magnetic stars which accelerate ionized interstellar gas particles were suggested as possible cosmic ray sources. The relative cosmic ray abundance from such sources should depend on the ionization state of the interstellar gas as it encountered the rotating magnetic field of the star. In this paper we have computed this ionization state for a number of stellar temperatures representative of the magnetic stars.

143.029 Composition of galactic cosmic rays with 30<E<10 MeV/nucleon.
D. O'Sullivan, A. Thompson, P. B. Price.
Nature, Phys. Sci., Vol. 243, 8 - 9 (1973). – Letter.

143.030 Ultrahigh energy photons, electrons, and neutrinos, the microwave background, and the universal cosmic-ray hypothesis.   F. W. Stecker.
Astrophys. Space Sci., Vol. 20, 47 - 57 (1973).

The production of ultrahigh energy photons, electrons and neutrinos as the decay products of pions produced in photomeson interactions between cosmic-ray nucleons and the blackbody microwave background is discussed in terms of the resultant energy spectra of these particles. Simple asymptotic formulas are given for calculating the ultrahigh energy photon spectrum predicted for the universal cosmic-ray hypothesis and the resulting spectra are compared with those obtained previously by numerical means using a different propagation equation for the photons. Approximate analytic solutions for the photon spectra are given in terms of simple power-law energy functions and slowly varying logarithmic functions. The generic relation between the various secondary components is discussed in terms of their astrophysical implications.

143.031 Energy losses of galactic cosmic rays in the interplanetary medium.   I. H. Urch, L. J. Gleeson.
Astrophys. Space Sci., Vol. 20, 177 - 185 (1973).

Using realistic models of cosmic-ray propagation in interplanetary space we present, for electrons, protons and helium nuclei of a given energy near earth, calculations of their distribution in energy before entering the solar cavity and their mean energy loss. Interplanetary conditions appropriate for the epochs 1965 and 1969 have been used. Cosmic-ray energies in the range of 20 MeV/nucleon to 1000 MeV/nucleon have been considered.

143.032 High-energy cosmic rays.
A. F. Titenkov, L. A. Vedeshin.
Zemlya i Vselennaya, 1973, No. 2, p. 31 - 35. In Russian.

143.033 The direct and reverse problems of cosmic ray propagation in interplanetary space.   I. N. Toptygin.
Geomagn. Aeronom., Vol. 13, 212 - 218 (1973). In Russian.

143.034 Diffusion of charged particles in a random magnetic field.   J. A. Earl.
Astrophys. Journ., Vol. 180, 227 - 238 (1973).

When charged particles move in a random magnetic field superposed upon a relatively large constant field, their pitch-angle distribution can be calculated to any desired precision by an iterative approximation procedure. Improved knowledge of the pitch-angle distribution and of the characteristic time for relaxation of anisotropy leads to an accurate expression for the coefficient of diffusion parallel to the mean field.

143.035 Transition radiation in astrophysics.
G. B. Yodh, X. Artru, R. Ramaty.
Astrophys. Journ., Vol. 181, 725 - 736 (1973).

Transition radiation produced by relativistic electrons traversing cosmic grains is investigated as a possible source of celestial X-rays. The detailed theory of transition radiation including the formation-zone effect is used to calculate the X-ray emissivity in interstellar space. It is found that the largest contribution of transition radiation to the observed X-ray emission from the galactic disk is at about 2 keV where less than about 0.3 percent of the observed emissivity is due to transition radiation.

143.036 Interplanetary scintillations of cosmic rays.
A. J. Owens, J. R. Jokipii.
Astrophys. Journ., (Letters), Vol. 181, L147 - L150 (1973).

Statistically significant broad-band fluctuations or "scintillations" in the high-energy ($T \sim 1$ GeV) cosmic-ray flux observed by neutron monitors are interpreted. A mechanism by which interplanetary magnetic-field fluctuations induce cosmic-ray scintillations is presented and shown to be in quantitative agreement with observation.

143.037 Electrons in cosmic rays.   S. V. Bulanov.
Trudy XVI Nauch. konf. Mosk. fiz.-tekhn. in-t, 1970, Ser. "Obshch. i priklad. fiz.", "Molekulyar. i khim. fiz.", Moskva, 1972, p. 32 - 42. In Russian. – Abstr. in Referativ. Zhurn. 51. Astron., 4.51.813 (1973).

143.038 Cosmic ray electrons from 0.2 to 8 Mev: Pioneer 8 and 9 measurements of their spectrum, time variations, and interplanetary radial gradient.
W. R. Webber, J. A. Lezniak, S. V. Damle.
Journ. Geophys. Res., Vol. 78, 1487 - 1501 (1973).

143.039 Solar and geomagnetic modulation of low-energy secondary cosmic ray electrons.
J. G. Luhmann, J. A. Earl.
Journ. Geophys. Res., Vol. 78, 1502 - 1514 (1973).

This paper presents experimental results on electrons in two energy ranges, 15–65 Mev and 45–150 Mev, obtained throughout the solar cycle during an extensive series of balloon flights at high and intermediate latitudes. In the discussion, a theoretical treatment of return albedo is presented and is used to interpret some of these results.

143.040 On a possibility of recording primary cosmic electrons from synchrotron radiation in the geomagnetic field.   O. F. Prilutskij.
Pis'ma v ZhurnEhTF, Vol. 16, 452 - 454 (1972). In Russian. Abstr. in Referativ. Zhurn. 62. Issled. kosmich. prostranstva, 4.62.143 (1973).

143.041 On the role of plasma effects in the cosmic ray propagation and isotropization in the Galaxy.
V. L. Ginzburg, V. S. Ptuskin, V. N. Tsytovich.
Astrophys. Space Sci., Vol. 21, 13 - 38 (1973).

The purpose of this paper is to give a systematic answer to all questions concerning the excitation by cosmic rays of waves of different kinds (magnetohydrodynamic, high-frequency plasma waves, i.e. Langmuir waves etc.), damping of these waves in the interstellar medium and their role in scattering and isotropization of cosmic rays.

143.042 Composition of low energy cosmic radiation from silicon to nickel.
K. Söderström, S. Lindstam, S. Behrnetz, K. Kristiansson.
Astrophys. Space Sci., Vol. 21, 211 - 222 (1973).

The charge composition of the cosmic radiation for the elements with $Z \geqslant 14$ has been studied in a stack of nuclear emulsions exposed at Fort Churchill in July 1967. Particles stopping in the stack have been measured with a nuclear track photometer to determine the charge. Relative abundances are given for the elements silicon to nickel and comparison is made with other investigations.

143.043 Interplanetary radial gradients of galactic cosmic ray protons and helium nuclei: Pioneer 8 and 9 measure-

ments from 0.75 to 1.10 AU. W. R. Webber, J. A. Lezniak. Journ. Geophys. Res., Vol. 78, 1979 - 2000 (1973).

We report measurements of the radial interplanetary cosmic ray intensity gradient made with cosmic ray instruments aboard the Pioneer 8 and 9 spacecraft. The principal time period covered in these studies is from January 1968 to June 1969. During this time the radial distance of these two spacecraft varied from 0.75 to 1.10 AU from the sun, and the radial separation of the individual spacecraft was as large as 0.30 AU. Our results show that the determination of a radial cosmic ray gradient requires the simultaneous determination of the energy dependence of the solar modulation effects on protons and helium nuclei over the same energy range for which the gradients are measured.

143.044 Model of a moving magnetic mirror and the Forbush effect in cosmic rays.
A. V. Belov, L. I. Dorman, B. A. Shakhov.
Geomagn. Aeronom., Vol. 13, 399 - 405 (1973). In Russian.

143.045 Cosmic ray spectrum and plasma turbulence.
V. N. Tsytovich.
Cosmic plasma physics. Conference 1971, (see 012.016), p. 269 - 271 (1972).

143.046 The galactic cosmic ray diurnal variation as a streaming plasma interaction between galactic and solar corpuscular radiation. V. J. Kisselbach.
Cosmic plasma physics. Conference 1971, (see 012.016), p. 349 - 350 (1972).

143.047 The interplanetary conditions associated with cosmic ray Forbush decreases. L. R. Barnden.
Cosmic plasma physics. Conference 1971, (see 012.016), p. 351 - 358 (1972).

143.048 The diurnal effect of cosmic rays and its dependence on the interplanetary magnetic field.
E. Bussoletti, N. Iucci.
Cosmic plasma physics. Conference 1971, (see 012.016), p. 359 - 364 (1972).

143.049 On the hypothesis of dust grain origin of cosmic rays air showers. V. S. Berezinsky, O. F. Prilutsky.
Astrophys. Space Sci., Vol. 21, 475 - 476 (1973). − Research note.

143.050 Periodic variations of the cosmic radiation − III. The 27-day variation. W. Messerschmidt.
Planet. Space Sci., Vol. 21, 1141 - 1150 (1973).

It is the aim of the present paper to investigate the existence of a persistent 27-day wave for the following time and the relation between the cosmic radiation, the terrestrial magnetism and the solar activity.

143.051 Influence of energy-dependent escape from the Galaxy on cosmic electrons.
R. F. Silverberg, R. Ramaty.
Nature, Phys. Sci., Vol. 243, 134 - 135 (1973).

The escape of cosmic rays from the Galaxy may be energy dependent. The authors point out that the consequence of such a variation for cosmic electrons at high energies is that the effects of synchrotron and Compton losses on the electron spectrum are significantly diminished by comparison with the expected effects for a constant lifetime.

143.052 Cosmic ray sources: Evidence for two acceleration mechanisms.
R. Ramaty, V. K. Balasubrahmanyan, J. F. Ormes.
Science, Vol. 180, 731 - 733 (1973).

The difference between the energy spectra of iron and other cosmic rays is interpreted in terms of two source mechanisms. One mechanism, possibly acceleration at neutron star surfaces, produces the iron, and another is responsible for the rest of the primary nuclei. Within this model, observations of high-energy cosmic rays could determine whether secondary nuclei are produced in the sources or in the interstellar medium.

143.053 The fluorine abundance in the galactic cosmic radiation.
B. G. Cartwright, M. Garcia-Munoz, J. A. Simpson.
Astrophys. Journ., Vol. 182, 9 - 15 (1973).

The abundance of fluorine relative to oxygen, $\Gamma(F/O)$, has been measured in the energy ranges 100−200 MeV per nucleon and greater than or equal to 1000 MeV per nucleon on the satellite IMP-5 in 1969−1970 and compared with earlier satellite measurements. Production of fluorine by spallation in the interstellar medium has been calculated with account taken of all elements with nuclear charge number $8 \leq Z \leq 16$, and iron−including secondary, tertiary, and higher-order production−for several models of galactic cosmic-ray propagation.

143.054 Investigation of cosmic radiation in the surroundings of the moon on Luna 10, 11, 12.
N. L. Grigorov, V. G. Kurt, V. N. Lutsenko, V. L. Maduev, N. F. Pisarenko, I. A. Savenko.
Mezhplanet. sreda i fiz. magnitosfery. Nauka, Moskva, 1972, p. 109 - 126. In Russian. − Abstr. in Referativ. Zhurn. 62. Issled. kosmich. prostranstva, 5.62.275 (1973).

143.055 Relativistic heavy cosmic rays.
R. A. Mewaldt, J. I. Fernandez, M. H. Israel, J. Klarmann, W. R. Binns.
Astrophys. Space Sci., Vol. 22, 45 - 65 (1973).

During three balloon flights of a 1 m² sr ionization chamber Čerenkov counter detector system, the authors have measured the atmospheric attenuation, flux, and charge composition of cosmic-ray nuclei with $16 \leqslant Z \leqslant 30$ and rigidity greater than 4.5 GV.

143.056 The cosmic rays of superhigh energies.
L. E. Gurevich, A. A. Rumyantsev.
Astrophys. Space Sci., Vol. 22, 79 - 85 (1973).

The paper considers the generation mechanism of the relativistic particles of superhigh energies ($\gtrsim 10^{18}$ eV) in a plasma where the supersonic turbulence and the hydrodynamic shock waves occur. It is found that the conditions necessary for the formation of this turbulence are realized in supernovae shells during the period of the outburst.

143.057 A new test for solar modulation theory: the 1972 May−July low-energy galactic cosmic-ray proton and helium spectra.
M. Garcia-Munoz, G. M. Mason, J. A. Simpson.
Astrophys. Journ., (Letters), Vol. 182, L81 - L84 (1973).

For a three-month period in 1972 we have measured the low-energy galactic cosmic-ray proton and helium spectra using the University of Chicago IMP-5 satellite experiment. Although the proton spectrum has a characteristic rising slope in the interval 30−100 MeV, the helium spectrum shows a previously unobserved feature: an intensity essentially independent of kinetic energy below 80 MeV per nucleon. Indeed, below ~ 35 MeV per nucleon, the helium flux exceeds the proton flux.

143.058 High energy particle astronomy.
A. Buffington, R. A. Muller, L. H. Smith, G. F. Smoot.
Publ. American Astronaut. Soc., Sci. Techn. Ser., Vol. 28, (see 012.018), p. 289 - 298 (1972).

143.059 Spuren der schweren Ionen der kosmischen Strah-

lung in lunaren und meteoritischen Mineralen: Beschleunigerexperimente und Stabilitätsuntersuchungen zur Teilchenidentifizierung.
T. Plieninger, W. Krätschmer, W. Gentner.
Max-Planck-Inst. Kernphysik Heidelberg, Jahresbericht 1972, p. 232 - 233 (1973).

143.060 Theory of cosmic-ray variations. L. I. Dorman.
Cosmic rays No. 13, (see 003.013), p. 5 - 66 (1972). In Russian.

143.061 Energy spectrum of galactic electrons in the 2×10^{10} - 2×10^{11} eV energy range.
V. I. Zatsepin, V. I. Rubtsov, N. S. Svirzhevsky.
Cosmic rays No. 13, (see 003.013), p. 67 - 68 (1972). In Russian.

143.062 Solar activity and cosmic rays in 1963–1965.
A. G. Zusmanovich, E. V. Kolomeets, R. A. Chumbalova, Yu. A. Shakhova.
Cosmic rays No. 13, (see 003.013), p. 73 - 77 (1972). In Russian.

143.063 Calculation of cosmic-ray modulation in cosmic space. L. I. Dorman, Z. Kobilinski.
Cosmic rays No. 13, (see 003.013), p. 108 - 134 (1972). In Russian.

143.064 Cosmic-ray study by means of the artificial lunar satellites Luna 11 and Luna 12.
N. L. Grigorov, V. G. Kurt, V. N. Lutsenko, V. L. Maduev, N. F. Pisarenko, I. A. Savenko.
Cosmic rays No. 13, (see 003.013), p. 135 - 148 (1972). In Russian.

143.065 Study of cosmic rays on the Proton series of artificial earth satellites. N. L. Grigorov, I. A. Savenko, R. N. Basilova, N. N. Volodichev, S. I. Voropaev, L. F. Kalinkin, G. P. Kakhidze, V. A. Labutin, A. S. Melioransky, E. A. Pryakhin, I. D. Rapoport.
Cosmic rays No. 13, (see 003.013), p. 149 - 161 (1972). In Russian.

143.066 Study of cosmic-ray motion in the cosmic space near the earth. V. K. Budilov, V. I. Ivanov, L. V. Kozak, L. A. Mirkin, I. G. Tsukerman.
Cosmic rays No. 13, (see 003.013), p. 232 - 235 (1972). In Russian.

143.067 On the secondary production of galactic cosmic ray electrons. L. M. Choate, J. R. Wayland.
Astrophys. Space Sci., Vol. 19, 195 - 200 (1972).
The differential energy spectrum and charge ratio of primary cosmic ray electrons produced by collisions of primary cosmic ray particles with the interstellar medium is calculated by means of the two temperature statistical model of high-energy interactions. Two realistic models for the primary cosmic ray flux are considered. The distribution of matter in the Galaxy and energy loss of produced secondaries and electrons are considered. The results are compared to recent experimental data.

143.068 Path length distribution of cosmic rays from pulsar nebula complexes.
K. M. V. Apparao, T. N. Rengarajan.
Astrophys. Space Sci., Vol. 19, 293 - 296 (1972).
Particles accelerated in the vicinity of pulsars have to traverse the nebular matter surrounding pulsars. Using the variation of pulsar luminosity and nebular expansion, the path length distribution for the particle radiation is deduced and compared with that obtained from experimental observation.

143.069 Peculiarities of the influence of the solar wind on cosmic-ray intensity in 1969.
A. V. Belov, L. I. Dorman, L. Kh. Shatashvili.
Kosmich. Issled., Vol. 11, 418 - 430 (1973). In Russian.

143.070 Calculated cosmic ray neutron monitor response to solar modulation of galactic cosmic rays.
K. O'Brien, G. de P. Burke.
Journ. Geophys. Res., Vol. 78, 3013 - 3019 (1973).
We have modulated galactic cosmic ray spectra by means of the Ehmert potentials, and then by means of a nucleonic transport code (O'Brien, 1971) determined the counting rate of several NM-64 cosmic ray neutron monitors.

143.071 A model for the break in the primary cosmic-ray energy spectrum at $\sim 10^{15}$ eV. S. Naranan.
Nuovo Cimento Lettere, Ser. 2, Vol. 5, 817 - 823 (1972).
The author examines the possibility that the change in the primary cosmic-ray spectrum at $\sim 10^{15}$ eV arises from a change in the hadron interaction characteristics of high-energy cosmic rays with matter in the volume where they are accelerated by a Fermi-type mechanism.

143.072 Unorthodox ideas concerning the origin of cosmic radiation. K. Sitte.
Nuovo Cimento Lettere, Ser. 2, Vol. 5, 1033 - 1037 (1972).
The author examines two unconventional views on cosmic-ray origin: the assumption of extragalactic origin of the entire cosmic radiation and the 'cosmic ray big bang' theory. He considers that neither theory can be considered as definitely disqualified.

143.073 The generation of the highest cosmic ray energies. M. Grewing, H. Heintzmann.
Phys. Letters A, (*Netherlands*), Vol. 42 A, 345 - 346 (1973).
The acceleration of charged particles in superstrong electromagnetic fields to energies $> 10^{21}$ has been studied numerically, taking into account radiation reaction effects.

143.074 Expected gradients of the density of cosmic radiation for various models of galactic cosmic-ray propagation in interplanetary space. L. I. Dorman, Z. Kobilinski.
Izv. AN SSSR. Ser. fiz., Vol. 36, 2346 - 2353 (1972). In Russian. – Abstr. in Referativ. Zhurn. 51. Astron., 6.51.477 (1973).

143.075 Frequency spectrum of intensity variations of cosmic rays and solar activity.
E. V. Kolomeets, R. A. Chumbalova, Yu. A. Shakhova, Ya. E. Shvartsman, B. N. Shigaev.
Izv. AN SSSR, Ser. fiz., Vol. 36, 2405 - 2410 (1972). In Russian. – Abstr. in Referativ. Zhurn. 51. Astron., 6.51.515 (1973).

143.076 Issues of the linear and non-linear theory of cosmic-ray modulation.
A. V. Belov, I. V. Dorman, L. I. Dorman, B. A. Shakhov.
Izv. AN SSSR. Ser. fiz., Vol. 36, 2324 - 2331 (1972). In Russian. – Abstr. in Referativ. Zhurn. 51. Astron., 6.51.521 (1973).

143.077 On the relation between the 27-day cosmic-ray variations and various indices of solar activity during 1957 - 1970.
G. A. Bazilevskaya, V. P. Okhlopkov, T. N. Charakhch'yan.
Izv. AN SSSR. Ser. fiz., Vol. 36, 2376 - 2382 (1972). In Russian. – Abstr. in Referativ. Zhurn. 51. Astron., 6.51.522 (1973).

143.078 Forbush decreases and their relation with solar activity and the parameters of interplanetary matter.

L. I. Dorman, N. S. Kaminer, A. E. Kuz'micheva.
Izv. AN SSSR. Ser. fiz., Vol. 36, 2391 - 2395 (1972). In Russian. – Abstr. in Referativ. Zhurn. 51. Astron., 6.51.523 (1973).

143.079 **Analysis of cosmic-ray variations of magnetospheric and interplanetary origin by data of a spectrograph.**
V. M. Dvornikov, L. I. Dorman, A. A. Luzov, A. V. Sergeev, A. L. Yanchukovskij.
Izv. AN SSSR. Ser. fiz., Vol. 36, 2427 - 2434 (1972). In Russian. – Abstr. in Referativ. Zhurn. 51. Astron., 6.51.524 (1973).

143.080 **Some informations about heavy primary nuclei ($Z \geqslant 33$) outside the terrestrial magnetosphere.**
N. S. Ivanova, V. V. Varyukhin, V. N. Kulikov, E. A. Yakubovskij.
Izv. AN SSSR. Ser. fiz., Vol. 36, 2359 - 2362 (1972). In Russian. – Abstr. in Referativ. Zhurn. 51. Astron., 6.51.656 (1973).

143.081 **Propagation of cosmic rays in the Galaxy.**
S. V. Bulanov, V. A. Dogel', V. S. Ptuskin, S. I. Syrovatskij.
Izv. AN SSSR. Ser. fiz., Vol. 36, 2292 - 2296 (1972). In Russian. – Abstr. in Referativ. Zhurn. 51. Astron., 6.51.739 (1973).

143.082 **Investigation of the propagation of cosmic radiation in the interplanetary magnetic field on the basis of the kinetic equation.** L. I. Dorman, M. E. Kats.
Izv. AN SSSR. Ser. fiz., Vol. 36, 2271 - 2277 (1972). In Russian. – Abstr. in Referativ. Zhurn. 62. Issled. kosmich. prostranstva, 6.62.221 (1973).

143.083 **Anisotropy of cosmic rays in the disturbed period from 25.X. - 10.XI. 1968.** Ya. L. Blokh,
L. I. Dorman, E. A. Eroshenko, O. I. Inozemtseva.
Izv. AN SSSR. Ser. fiz., Vol. 36, 2383 - 2386 (1972). In Russian. – Abstr. in Referativ. Zhurn. 62. Issled. kosmich. prostranstva. 6.62.238 (1973).

143.084 **Polar coupling factors and generalization of the spectrographic method for investigating variations of cosmic radiation of magnetospheric and interplanetary origin.** L. I. Dorman. G. Sh. Shkhalakhov.
Izv. AN SSSR. Ser. fiz., Vol. 36, 2420 - 2426 (1972). In Russian. – Abstr. in Referativ. Zhurn. 62. Issled. kosmich. prostranstva, 6.62.239 (1973).

143.085 **Capture of primary cosmic radiation in the upper atmosphere as a source of excess radiation.**
D. Kh. Morozov.
Izv. AN SSSR. Ser. fiz., Vol. 36, 2447 - 2450 (1972). In Russian. – Abstr. in Referativ. Zhurn. 62. Issled. kosmich. prostranstva, 6.62.241 (1973).

143.086 **The possibility of registering primary cosmic electrons by means of synchrotron radiation in the geomagnetic field.** O. F. Prilutskii.
JETP Letters, (USA), Vol. 16, 320 - 321 (1972).
Considers a possible method of registering primary cosmic electrons of high energy (>100 GeV) which makes it possible, in principle, to overcome a) the small fluxes of primary cosmic electrons and b) the strong background of nuclear-active particles.

143.087 **A possible cosmic ray primary particle energy spectrum above 10^4 GeV and its astrophysical implications.** A. Subramanian.
Proc. Indian Acad. Sci., Ser. A, Vol. 76, 121 - 128 (1972).

A primary cosmic ray particle energy spectrum of galactic origin with a sharp cut-off at an energy per nucleon of 3×10^4 GeV for protons and 5×10^6 GeV for heavier particles and extending only up to energies $\sim 10^{15}$ eV is deduced in part from the observed cosmic ray phenomena at the highest energies. It appears that the cut-off is not due to the magnetic rigidity of the particles in the Galaxy but due to a cut-off in or near the sources themselves.

143.088 **Interplanetary shock waves and cosmic rays.**
L. I. Dorman.
Comments Astrophys. Space Phys., Vol. 5, 67 - 74 (1973).
The cosmic ray intensity increases before sudden magnetic storms. This effect begins many hours before shock wave arrival at the earth, increases gradually and reaches peak amplitude (1–2%) by the moment the storm begins. The effect is fairly fine and difficult to isolate on the background of various fluctuations.

143.089 **Long-term variation in the cosmic-ray diurnal anisotropy.** N. Iucci, M. Storini.
Nuovo Cimento B, Ser. 11, Vol. 13B, 361 - 378 (1973).
The long-term modulation of the diurnal variation of the cosmic-ray intensity, observed in quiet times during the ascending phase of solar cycle number 20 is fully explained by the change in the interplanetary sector pattern.

143.090 **Relevance of cosmic ray data above 10^{12} eV to models of high energy interactions.**
J. Wdowczyk, A. W. Wolfendale.
Journ. Phys. A, General Phys., Vol. 6, L48 - L51 (1973).
Recent experiments on proton-proton collisions with the intersecting storage ring facility have shown a number of features which may be explained by the high energy 'scaling' model of Feynman. The authors have examined cosmic ray data in some detail to 10^{18} eV and find that they do not support the model being valid, at least without serious modification, to these much higher energies.

143.091 **Contribution of pion production by primary cosmic-ray nucleons to the interstellar electron-positron flux.** J. Dooher.
Phys. Rev. D, Particles and Fields, Vol. 7, 1406 - 1411 (1973).
The secondary electron-positron component of high-energy cosmic radiation is calculated.

143.092 **The aeonic flux of ultra-energetic cosmic rays.**
L. G. Van Loon.
Nuovo Cimento B, Ser. 11, Vol. 14B, 267 - 283 (1973).
Recently it has been found that cosmic-ray interactions of ultra-high energy leave visible marks of more than several micron in lunar-rock minerals. In the present work the method for determining the energy of these interactions is given and the cosmic-ray flux in past aeons is determined.

143.093 **Galactic cosmic rays in interplanetary space.**
E. Leer.
Phys. Norvegica, Vol. 6, 193 - 198 (1972). – See Phys. Abstr., Vol. 76, No. 42096 (1973).

143.094 **Null correlation method for estimation of the primary energies of cosmic ray jets.**
R. E. Streitmatter.
Canadian Journ. Phys., Vol. 51, 804 - 813 (1973).
A new method for estimating the center of mass of high energy cosmic ray interactions is introduced and tested with a simple Monte Carlo model and a small number of cloud chamber jets.

143.095 **Differential energy spectra of low energy (<8.5 MeV per nucleon) heavy cosmic rays during solar**

quiet times.
D. Hovestadt, O. Vollmer, G. Gloeckler, C. Y. Fan.
Max-Planck-Inst. Phys. Astrophys., Inst. Extraterr. Phys.,
München, MPI–PAE/Extraterr. 84, 12 pp. (1973).

Carbon, oxygen and heavier nuclei have been observed below 8.5 MeV per nucleon during solar quiet times. The authors find that the C/O abundance ratio is 0.50 ± 0.15, and the differential energy spectra below 1 MeV have the form $KE^{-4.9 \pm 0.3}$. They infer from this ratio that most of these particles are likely to be of solar origin.

143.096 On stochastic mechanisms of acceleration under cosmic conditions.
Yu. N. Gnedin, A. Z. Dolginov, V. N. Fedorenko.
IV Leningr. mezhdunar. seminar "Edinoobrazie uskoreniya chastits v razlich. masshtabakh kosmosa, 1972". Leningrad, 1972, p. 27 - 46. In Russian. – Abstr. in Referativ. Zhurn. 51. Astron., 7.51.648 (1973).

143.097 On rocket measurements of corpuscular radiation intensities on the Heiss island in 1968–1969.
V. F. Tulinov, G. F. Tulinov, V. V. Tulyakov.
Trudy Tsentr. aerhol. observ., 1972, vyp. (No.) 111, p. 24 - 28. In Russian. – Abstr. in Referativ. Zhurn. 51. Astron., 7.51.650 (1973).

Cosmic rays in the earth's magnetic field.
See Abstr. 003.048.

Partial cross-sections in high-energy nuclear reactions, and astrophysical applications. I. Targets with $Z \leqslant 28$.
See Abstr. 022.045.

Partial cross-sections in high-energy nuclear reactions, and astrophysical applications. II. Targets heavier than nickel. See Abstr. 022.046.

Gamma astronomy and cosmic rays. II.
See Abstr. 061.002.

Charged particle acceleration in strong dipole fields.
See Abstr. 061.038.

Annual and three-monthly variations in solar activity and cosmic ray intensity. See Abstr. 072.030.

Record-breaking cosmic ray storm stemming from solar activity in August 1972.
See Abstr. 078.006.

Comparative characteristics of the soft component of solar and galactic cosmic radiation according to rocket and stratospheric measurements on the island with the coordinates $71°.2$ N and $155°.0$ E and in Apatity. See Abstr. 078.058.

Calculations of neutron flux spectra induced in the earth's atmosphere by galactic cosmic rays.
See Abstr. 082.111.

Lunar composition from Apollo orbital measurements. See Abstr. 094.301.

High resolution time averaged (millions of years) energy spectrum and chemical composition of iron-group cosmic ray nuclei at 1 A.U. based on fossil tracks in Apollo samples. See Abstr. 094.509.

Ultra-heavy cosmic rays in the moon.
See Abstr. 094.510.

Search for stable, fractionally charged particles (quarks) in lunar material. See Abstr. 094.514.

Solar flare and galactic cosmic ray studies of Apollo 14 and 15 samples. See Abstr. 094.806.

Charge assignment to cosmic ray heavy ion tracks in lunar pyroxenes. See Abstr. 094.807.

Spatial variations of cosmic rays derived from the radio activity of meteorites with known orbits.
See Abstr. 105.145.

Cosmic-ray heating and molecular cooling of dense clouds. See Abstr. 131.030.

The dynamical behavior of the interstellar gas, field, and cosmic rays. See Abstr. 131.156.

Sound vibrations in cosmic magneto-active space.
See Abstr. 131.199.

Heating and ionization of H I regions by discrete soft X-ray and subcosmic-ray sources. See Abstr. 131.204.

Gamma-ray astronomy and cosmic rays.
See Abstr. 142.055.

Gamma-astronomy and cosmic rays.
See Abstr. 142.056.

Galactic magnetic field irregularities and their effect on cosmic ray propagation at energies above 10^{17} eV.
See Abstr. 156.003.

Stellar Systems

151 Kinematics and Dynamics of Stellar Systems

151.001 **An exact solution for a collisionless flat galactic model.** S. Aoki.
Publ. Astron. Soc. Japan, Vol. 25, 35 - 50 (1973).

A method is established for the exact solution of a stationary collisionless self-gravitating frequency function of a flat galaxy with rotationally symmetric density distribution. The method is to determine a pertinent function of the energy and angular momentum integrals to match with a given density distribution in the plane.

151.002 **Tidal interaction of galaxies.**
T. M. Eneev, N. N. Kozlov, R. A. Sunyaev.
Astron. Astrophys., Vol. 22, 41 - 60 (1973).

A qualitative analysis and numerical calculations allow to determine the main effects associated with the passage of a massive body in a number of characteristic hyperbolic orbits with respect to the galaxy. The behaviour of 800–2000 non-interacting point satellites initially travelling in circular Keplerian orbits around central regions of the galaxy and perturbed by the close passage of a massive body is analysed.

151.003 **On 'fast' and 'slow' density waves in spiral galaxies.**
A. B. Mikhailovsky, A. M. Fridman.
Astron. Zhurn. Akad. Nauk SSSR, Vol. 50, 88 - 96 (1973). In Russian. English translation in Soviet Astron. AJ, Vol. 17, No. 1.

The attempt is made to explain the known observational fact – 'splitting' of spiral arms of galaxies into arms of red, yellow and blue stars – on the basis of the hypothesis on the displacement of a spiral density wave in the gaseous disk plane. The waves moving in the direction of rotation are called 'fast', the waves moving in the opposite direction 'slow'. As a classical example of an axisymmetric system a rotating homogeneous cylinder is considered. It is shown that in it only slow density waves can develop. The condition of formation of slow density waves in a plane rotating disk with arbitrary dependence on the radius of the angular rotation velocity and surface density is obtained.

151.004 **Stability of gravitating systems with a quadratic potential. I. Methods of investigation of stability of systems with a restricted phase volume. A spectrum of oscillations of Maclaurin's stellar disk.**
V. L. Poljachenko, I. G. Shukhman.
Astron. Zhurn. Akad. Nauk SSSR, Vol. 50, 97 - 100 (1973). In Russian. English translation in Soviet Astron. AJ, Vol. 17, No. 1.

151.005 **The stability of a self-gravitating, nonrotating gas layer with stellar, magnetic, and cosmic-ray components. II.** S. A. Kellman.
Astrophys. Journ., Vol. 179, 103 - 109 (1973).

A time-independent linear stability analysis is performed on a self-gravitating, plane-parallel, isothermal layer of nonrotating gas with magnetic and cosmic-ray components. The gas layer is immersed in a rigid plane-stratified isothermal layer of stars which supply a self-consistent gravitational field. The stability analysis is confined to disturbances which propagate along the magnetic field.

151.006 **Nonaxisymmetric kinematics in galaxies with axisymmetric mass distributions.** C. L. Berry.
Astrophys. Journ., Vol. 179, 395 - 415 (1973).

We investigate the dynamics of a galaxy with an axisymmetric, time-independent mass distribution but a nonaxisymmetric, time-dependent distribution of stars in phase space. The system considered is of infinitesimal thickness, and the motion of a star thus has two degrees of freedom. Associated with this motion is a third isolating integral in addition to the energy and the angular momentum. An elementary model showing nonaxisymmetric differential motions and a deviation of the vertex is presented.

151.007 **Tidal origin of elliptical galaxies of high surface brightness.** S. M. Faber.
Astrophys. Journ., Vol. 179, 423 - 426 = Contr. Lick Obs., No. 378 (1973).

Three peculiar elliptical galaxies having high surface brightness and low luminosity are discussed. Several lines of evidence indicate that they are the tightly bound cores of normal elliptical galaxies whose outer regions have been stripped away in tidal interactions with more massive companions.

151.008 **Star migration studies have not yet revealed the presence of a spiral density wave.** A. J. Kalnajs.
Observatory, Vol. 93, 39 - 42 (1973). – Letter.

151.009 **Density shock waves driven by star formation – a mechanism.** P. Biermann.
Astron. Astrophys., Vol. 22, 407 - 412 (1973).

Here a mechanism is proposed in which the energy input from H II regions of newly formed stars and energy dissipation from colliding clouds drive a density shock wave that exists exclusively in the gas flow. Star and cloud formation is thought to be initiated by the shock. The mechanism is discussed in form of a simple numerical model. Some basic properties of the wave, such as length scales and energy balance, relate well with properties of interstellar matter.

151.010 **Numerical study of a four-dimensional mapping. II.**
C. Froeschlé, J.-P. Scheidecker.
Astron. Astrophys., Vol. 22, 431 - 436 (1973).

We use here a new numerical method, which enables us to study the variation with n of the two largest eigen-values – in absolute magnitude – of the linear tangential mapping T^{n*} of the mapping T^n. This variation appears to be a very sensitive indicator of stochasticity. We show that these dynamical systems seem to follow the general behaviour of C-systems in the ergodic zone. We confirm in particular earlier results concerning the effects of coupling and of the variation of the initial conditions.

151.011 **The density-wave theory of galactic spirals.**
J. H. Piddington.
Astrophys. Journ., Vol. 179, 755 - 770 (1973).

Two questionable assumptions made two decades ago have persisted; that because there is no dust between the spiral arms of the Andromeda galaxy there is no gas, and that the elongated features of cool gas (H I) seen in our Galaxy are parts of twin spiral arms. In the present paper these assumptions and their extensions have been tested in our Galaxy and some others. We confirm that the spiral tracers which delineate spiral forms occur in regions of low H I density yet are absent from nearby regions of high density. The density-wave theory must now include an X-factor of star formation, perhaps a shock wave as already suggested.

151.012 Random force in gravitational systems.
A. Ahmad, L. Cohen.
Astrophys. Journ., Vol. 179, 885 - 896 (1973).

Numerical experiments have been performed to study the probability distribution of the random force in gravitational systems. The results are in good agreement with the theory of Chandrasekhar and von Neumann. Also a derivation of the Holtsmark distribution for a finite number of particles is given and plotted.

151.013 The verification of Minardi's instability criterion for nonhomogeneous self-gravitating equilibria.
S. Cuperman, I. Tzur.
Astrophys. Journ., Vol. 180, 181 - 193 (1973).

Results of numerical integrations of the time-dependent Vlasov and Poisson equations are presented in support of Minardi's linear instability criterion for inhomogeneous non-Maxwellian Vlasov equilibria.

151.014 Phase mixing of the second kind in stellar systems. II. L. P. Osipkov.
Astrofizika, Vol. 8, 295 - 304 (1972). In Russian. English translation in Astrophysics, Vol. 8, No. 2.

Phase-mixing of the second kind is studied for the case when the phase-space is represented as a set of tori enclosed into each other. The characteristic time of mixing is estimated on the basis of geometric considerations. For systems admitting Lindblad-Oort's third integral mixing is absent.

151.015 Velocity variation of a star as a purely discontinuous random process. III. Stars with different masses in an open cluster. V. S. Kaliberda, I. V. Petrovskaya.
Astrofizika, Vol. 8, 305 - 313 (1972). In Russian. English translation in Astrophysics, Vol. 8, No. 2.

The evolution of the velocity distribution function of a group of stars in an open cluster is considered as a purely discontinuous random process. The mass of the star under consideration is supposed to be half the average mass of a cluster star. Using the second Kolmogorov–Feller equation, the velocity distribution function of the 2 considered stars, the escape rate and the amount of energy taken away by the dissipated stars are found for different moments of time. The results for the quasi-stationary state are compared with the solution of the Fokker-Planck equation by Spitzer and Härm (a continuous random process).

151.016 On the dynamics of gravitating systems on the neutrino background of the universe. T. B. Omarov.
Astrofizika, Vol. 8, 315 - 323 (1972). In Russian. English translation in Astrophysics, Vol. 8, No. 2.

The gravitational influence of a homogeneous neutrino sea of the expanding universe on the dynamics of a pair of quasi-punctiform bodies separating on its background is investigated.

151.017 The dynamics of dense stellar systems.
W. C. Saslaw.
Publ. Astron. Soc. Pacific, Vol. 85, 5 - 23 (1973).

This is a review of the main physical processes which occur in dense stellar systems: gravitational relaxation, evaporation of stars, antiequipartition, growth of dense cores, stellar coalescence, stellar disruption, star-gas interactions, and star

formation. These processes are combined to sketch theories of the formation, intermediate evolution, and ultimate fate of galactic nuclei.

151.018 Galactic orbits and tidal radii of the clusters M67, NGC 188 and ω Centauri.
D. W. Keenan, K. A. Innanen, F. C. House.
Astron. Journ., Vol. 78, 173 - 179 (1973).

The galactic orbits of the clusters ω Cen, M67, and NGC 188 were calculated in the mass models of Schmidt (1956, 1965) and Innanen (1966) using an advanced numerical method. Tidal radii of the clusters were calculated using the parameters of the orbits and a formula of King (1962). Comparisons were made with the observed limiting radii of the clusters. Stars with opposite angular momentum are unaffected by the galactic perturbing field out to much greater distances from the cluster. Simple numerical experiments and the recent results of Toomre and Toomre (1972) and Wright (1972) confirm this effect.

151.019 Primordial random motions and angular momenta of galaxies and galaxy clusters.
J. Silk, S. Lea.
Astrophys. Journ., Vol. 180, 669 - 686 (1973).

We study the decay of primordial random motions of galaxies and galaxy clusters in an expanding universe by solving a kinetic equation for the relaxation of differential energy spectra $N(E, t)$. Systematic dissipative energy losses are included, involving gravitational drag by, and accretion of, intergalactic matter, as well as the effect of collisions with other systems. Formal and numerical solutions are described for two distinct modes of galaxy formation in a turbulent medium, corresponding to formation at a distinct epoch and to continuous formation of galaxies.

151.020 Resonant stellar orbits in spiral galaxies.
P. O. Vandervoort.
Astrophys. Journ., Vol. 180, 739 - 758 (1973).

This paper describes an epicyclic theory of the orbits of stars near the Lindblad resonances of a spiral galaxy. The spiral component of the gravitational field is assumed to be stationary, in a uniformly rotating frame of reference, and tightly wound. The equations of motion are solved by the method of harmonic balance of Bogoliubov and Mitropolsky. The solution represents families of periodic orbits, tube orbits, and nonresonant orbits. Integrals of the motion are obtained in a representation suitable for the extension of the theory of density waves into the regions of the Lindblad resonances.

151.021 On the "thermodynamics" of self-gravitating n-body systems. R. H. Miller.
Astrophys. Journ., Vol. 180, 759 - 782 (1973).

The experiments reported in this paper represent a tentative step toward seeing how "laboratory" stellar systems would respond to being subjected to "thermodynamic experiments". The results are so far removed from a "thermodynamic" behavior that it is out of the question to interpret them on thermodynamic terms. These experiments do not confirm the "gravothermal catastrophe". An attempt to understand why this might be so leads to a rather lengthy examination of some of the premises underlying some attempts to apply thermodynamic ideas to stellar dynamics. The reasons why the experimental results behaved as they did becomes clear as a result of these examinations.

151.022 The perpendicular oscillations of a homogeneous slab of stars. A. J. Kalnajs.
Astrophys. Journ., Vol. 180, 1023 - 1034 (1973).

A complete set of normal modes, their frequencies, and their energies have been found for the above one-dimensional, self-gravitating, collisionless system. The modes are purely oscillatory, and the system is stable. A finite-amplitude homologous pulsation mode is also described.

151.023 Numerical experiments on the stability of spherical stellar systems. M. Hénon.
Astron. Astrophys., Vol. 24, 229 - 238 (1973).

The concentric shell model is used to investigate numerically the stability of spherical steady-state stellar systems with respect to spherical disturbances. Polytropic models with an isotropic velocity distribution are found to be stable almost down to the limiting index value $n = 1/2$. "Generalized polytropes", with a distribution function depending on energy and angular momentum, show instability when n is low and the velocity distribution is radially elongated.

151.024 Gradient instability in a system of gravitating point masses.
G. S. Bisnovaty-Kogan, A. B. Mikhailovskij.
Astron. Zhurn. Akad. Nauk SSSR, Vol. 50, 312 - 319 (1973). In Russian. English translation in Soviet Astron. AJ, Vol. 17, No. 2.

The existence of a gradient (drift) instability is shown which is connected with the dependence of the longitudinal temperature on the radius of a rotating, infinitely long Jeans stable cylinder of finite radius in equilibrium. The stability of the cylinder with radial dependence of density is investigated; the possibility of stabilization of temperature gradient instability by the density gradient is shown, and the boundary of the stability region is found.

151.025 Computer simulation of dynamic systems.
W. J. Quirk.
Bull. American Astron. Soc., Vol. 5, 9 (1973). – Abstr. AAS.

151.026 A numerical study of the collision of two star clusters. T. Arny.
Bull. American Astron. Soc., Vol. 5, 24 (1973). – Abstr. AAS.

151.027 Equilibrium theories of rotating figures of self-gravitating masses in the presence of a magnetic field.
R. S. Oganesyan, M. G. Abraamyan.
Third Soviet Gravitational Conference, Erevan, 1972, (see 012.001), p. 339 - 340 (1972). In Russian.

151.028 Theories of galactic spiral structure. Comparisons with observations. J. H. Piddington.
Monthly Notices Roy. Astron. Soc., Vol. 162, 73 - 89 (1973).

The patterns of star creation which provide the optical spiral structure of galaxies have been attributed either to a gravitational (density-shock) or hydromagnetic origin; in the latter model gas clumping results from varying tilt of an oblique magnetic field. The predictions of the two theories are compared with observational data; in each theory the dust, young-star and older-star arms are separated as observed. The major differences lie in the surface density and velocity distributions of cool gas.

151.029 Contribución al estudio de la dinámica de los sistemas estelares a simetria cilindrica.
A. Catalá Poch.
Urania Barcelona, Año 57, No. 275, p. 3 - 41 (1972).

151.030 A third integral of motion in a system with a potential of the fourth degree. II. Resonance case 1 : 2.
P. Andrle.
Bull. Astron. Inst. Czechoslovakia, Vol. 24, 161 - 164 (1973).

Assuming that a steady-state galaxy has a plane and an axis of symmetry and that its potential is a polynomial of the fourth degree, the first approximation to a formal third integral was found in terms of Weierstrassian elliptic functions (Andrle, 1966). Higher approximations can only be derived

for some resonances of the periods of the two relevant elliptic functions in a simple form. The case leading to the resonance 1 : 2 in real periods and 1 : 1 in imaginary periods is treated in greater detail.

151.031 Cosmic turbulence and the origin of galaxies.
B. J. T. Jones.
Astrophys. Journ., Vol. 181, 269 - 294 (1973).

It is shown that, contrary to previous assertions, turbulence in the universe cannot be supported against viscous decay after the epoch t_{eq} when the energy densities of matter and radiation are equal. The problem of galaxy formation in cosmological models with density parameter $\Omega h^2 < \Omega_m h^2$ is also considered. An alternative theory of galaxy formation, based on the dissipation of strong turbulence of t_{eq}, is presented.

151.032 Excitation of spiral density waves by gas flow in a star-gas disk. S. Kato.
Publ. Astron. Soc. Japan, Vol. 25, 231 - 242 (1973).

The excitation of spiral density waves in a rotating star-gas disk by a kind of two-stream instability due to gases flowing relative to a rotating stellar disk is examined. Brief discussions are made in order to apply the results to actual spirals.

151.033 The particle resonance in spiral galaxies. Nonlinear effects. G. Contopoulos.
Astrophys. Journ., Vol. 181, 657 - 684 (1973).

A theory is developed to account for the nonlinear effects, near the particle resonance, found by numerical integration. There are four equilibrium points in the rotating frame of reference: two of them unstable, at the minima of potential (L_1, L_2), and two stable, at the maxima of potential (L_4, L_5). Many particles are trapped in librating orbits around L_4, L_5. Using the lowest-order terms of a "third integral" of motion near L_4, L_5, we find the behavior of the trapped orbits and compare the approximate theoretical results with the orbits found numerically by computer. An estimate of the trapped mass is given. Some effects due to these mass concentrations are found numerically and theoretically.

151.034 Patterns of waves in galactic disks.
C. Hunter.
Astrophys. Journ., Vol. 181, 685 - 705 (1973).

The ray methods of geometrical optics are used to calculate both steady and unsteady patterns of waves produced by localized sources in galactic disks. Caustics or ray envelopes are shown to form in a number of instances, and are important because they mark the most prominent features of wave patterns. Unsteady spiral wave patterns in a stellar disk can have caustics that propagate both toward and away from the galactic center. The study of these unsteady waves in a specific model shows that one-armed spiral waves have the widest range of propagation, can propagate into the galactic center and be reflected there, and allow the development of leading as well as trailing spiral wave patterns.

151.035 Stellar coalescence. A. G. W. Cameron.
Stellar evolution, (see 012.014), p. 807 - 812 (1972).

151.036 Places of formation of nearby classical cepheids.
R. Wielen.
Astron. Astrophys., Vol. 25, 285 - 297 (1973).

We determine and discuss the places of formation of 19 nearby classical cepheids. The ages of the cepheids are derived from the theoretical period-age relation which we check by cluster cepheids. The galactic orbits of the cepheids are numerically integrated backwards in time using the present space velocities and positions derived and discussed elsewhere. The derived birthplaces of most nearby cepheids are in agreement with the predictions of Lin's theory. Most of the nearby cepheids probably originated in the Sagittarius arm and in the Perseus arm.

151.037 **The origin of rotation of galaxies.**
A. G. Doroshkevich.
Astrophys. Letters, Vol. 14, 11 - 13 (1973).
It is proposed that galactic rotation originates as vorticity that is generated by shock waves arising in accordance with the nonlinear theory of gravitational instability.

151.038 **Small perturbations in flat galaxies. II. Time-dependent azimuthal perturbations.**   M. Clutton-Brock.
Astrophys. Space Sci., Vol. 21, 79 - 106 (1973).
This paper describes methods of calculating the response of a flat galaxy of stars to a perturbation which can depend on time and on angle. The starting point is the response to a pulse: the response to any other time dependence can be found by convolution. A single orbit responds with growing oscillations at the resonant frequencies, so the Laplace transform of the orbital response has a set of double poles along the real frequency axis. A simple expression for the orbital response is found in terms of the derivatives of Hankel-Laguerre functions with respect to action-angle variables. The Laplace transform of the system response is expanded in a series of simple basis functions.

151.039 **Peculiar velocities of galaxies and signs of the local cluster.**   B. I. Fesenko.
Astron. Zhurn. Akad. Nauk SSSR, Vol. 50, 491 - 495 (1973). In Russian. English translation in Soviet Astron. AJ, Vol. 17, No. 3.
The method of angular diameters is considered for the determination of the peculiar velocity of the Galaxy and of the peculiar velocity variance of galaxies.

151.040 **On the star dissipation in open clusters. II.**
V. M. Danilov.
Astron. Zhurn. Akad. Nauk SSSR, Vol. 50, 541 - 548 (1973). In Russian. English translation in Soviet Astron. AJ, Vol. 17, No. 3.
A new method of calculation of the dissipation of star clusters with non-uniform stellar mass composition is presented. The method is applied to a uniform spherical model of a quasi-stationary cluster consisting of 100 stars with radius of 3 pc. It was found that the rate of cluster dissipation is equal to -8.58×10^{-8} star per year, that corresponds to the full disintegration of the cluster in the course of 10^9 years.

151.041 **On the theory of stability of gravitating systems relative to surface perturbations.**   I. G. Shukhman.
Astron. Zhurn. Akad. Nauk SSSR, Vol. 50, 651 - 653 (1973). In Russian. English translation in Soviet Astron. AJ, Vol. 17, No. 3. – Short note.

151.042 **Motion of fourteen stars in the Orion nebula cluster.**
E. G. Vaerewyck, W. R. Beardsley.
Astrophys. Journ., Vol. 182, 121 - 127 (1973).
Photographs of the Orion nebula cluster obtained with the Thaw refractor of the Allegheny Observatory have been utilized to determine relative motions among a number of bright blue stars near the center of the association Id Ori. No cluster expansion was detected for these stars. Results indicate instead that these massive stars may be moving inward toward the center.

151.043 **Gravothermal catastrophe in rotating galaxies.**
M. Clutton-Brock.
Astrophys. Space Sci., Vol. 19, 201 - 206 (1972).
Violent relaxation in a rotating galaxy can lead to a core-halo system. In order to understand this, a very crude model

is studied in which all details are ignored, and only the dimensions of physical quantities are retained.

151.044 **Small perturbations in flat galaxies. I. Equilibrium models and adiabatic perturbations.**
M. Clutton-Brock.
Astrophys. Space Sci., Vol. 19, 225 - 247 (1972).
The aim of this series of papers is to develop straightforward methods of computing the response of flat galaxies to small perturbations. This Paper I considers steady state problems. The general approach is to study the dynamics of each individual orbit. The response to a small perturbation is found by seeking the response of each orbit. When the perturbations are axisymmetric and slowly varying, the response can be easily found using adiabatic invariants. An analytic approximation to the response matrix is derived, and applied to estimate the eccentricity needed for stability against local perturbations.

151.045 **Energy conservation for a combined system of gas and stars.**   H. Niimi.
Journ. Phys. Soc. Japan, Vol. 33, 1183 (1972).
Considers a combined system of gas and collisionless stars under mutual interaction only through the gravitational field. This paper obtains the energy conservation equation for the combined system with collective gravitational interaction of interstellar gas with collisionless stars.

151.046 **On the instability of equilibrium figures consisting of a few uniformly rotating parts.**
V. A. Antonov.
Vestn. Leningr. un-ta, 1973, No. 1, p. 27 - 130. In Russian. Abstr. in Referativ. Zhurn. 51. Astron., 6.51.729 (1973).

151.047 **Resonance damping of oscillations in a model of a spherical star cluster.**
V. S. Synakh, A. M. Fridman, I. G. Shukhman.
Astrofizika, Vol. 8, 577 - 585 (1972). In Russian. – English translation in Astrophysics, Vol. 8, No. 4.
The stability of a spherically symmetric system of rotating masses is considered. The density of the system is assumed to be decreasing with the radius as r^{-2} in accordance with observations. It is shown that aperiodical perturbations as well as neutral oscillations are absent and resonance damping takes place in such systems.

151.048 **Numerical experiments concerning isolated and non-isolated stellar systems.**   P. Bouvier.
Comput. Phys. Commun., (*Netherlands*), Vol. 4, 345 - 346 (1972).
A brief review is given of the principal methods used to deal with the dynamics of stellar systems, mostly star clusters, which are nearly all, at least partly, relaxed with respect to stellar encounters; particular stress is laid upon the direct numerical integration of the N-body problem.

151.049 **Computing experiments on stellar systems.**
P. Bouvier.
Arch. Sci. Genève, Vol. 25, 1 - 10 (1972) = Publ. Obs. Genève, Sér. A, Fasc 79/II (1973). – Review paper presented at the 2nd Conference of the European Physical Society (joint session of the Physics in Astronomy and Computational Physics Divisions) at Wiesbaden, October 1972.

151.050 **Liouville's theorem and the third integral of motion for steady-state stellar systems. II. The integrals determining a "monotach" stream.**   L. P. Osipkov.
Trudy Astron. Obs., *Leningrad*, Vol. 29 (= Uchenye Zapiski Leningr. Un-ta, No. 363 = Seriya Matem. Nauk, vyp. (No.) 48), p. 106 - 114 (1973). In Russian.
A relation connecting the integrals which are independent of the velocities and determine a monotach stream of

stars is found using Liouville's theorem and a hydrodynamic relation.

151.051 On the theory of spherical stellar systems.
M. A. Belozerova.

Trudy Astron. Obs., *Leningrad,* Vol. 29 (= Uchenye Zapiski Leningr. Un-ta, No. 363 = Seriya Matem. Nauk, vyp. (No.) 48), p. 114 - 122 (1973). In Russian.

The equations of motion are derived for a stationary spherical stellar system corresponding to some phase density of integrals of motion. The equation of motion for the simplest spherical one-phase systems is solved numerically.

151.052 On the stability of an infinite cylinder with aniso-tropic velocity distribution.
V. A. Antonov, E. M. Nezhinskij.

Trudy Astron. Obs., *Leningrad,* Vol. 29 (= Uchenye Zapiski Leningr. Un-ta, No. 363 = Seriya Matem. Nauk, vyp. (No.) 48), p. 122 - 140 (1973). In Russian.

The stability is studied with respect to longitudal perturbations of an infinite cylinder homogeneous along the z axis. The problem is reduced to a one-dimensional one by proper choice of the model. It is shown that the cylinder is unstable for any dispersion of the velocities (σ^2). An asymptotic estimate of the critical wavelength is found for several density distributions. It shows that the critical wavelength increases so that it can hardly play any role in physical applications.

151.053 Construction of a model of a rotating stellar system by a numerical experiment. II.
S. P. Yakimov.

Trudy Astron. Obs., *Leningrad,* Vol. 29 (= Uchenye Zapiski Leningr. Un-ta, No. 363 = Seriya Matem. Nauk, vyp. (No.) 48), p. 141 - 149 (1973). In Russian.

A model of a quasistationary rotating stellar system is constructed by numerical integration of the equations of motion for five bodies. The initial ratio of rotation energy to total kinetic energy is taken equal to 0.8. The distributions of density, centroid velocity and dispersions of peculiar velocities are obtained. The mean flattening of the model is 0.25. The rotation curve has several maxima. A similar fact is observed for a number of spiral galaxies.

151.054 Structure des amas à symétrie sphérique dans la phase de relaxation.
Y. Talpaert.

Acad. Roy. Belgique, Bull. Cl. Sci., Vol. 58, 229 - 244 (1972).

151.055 Discrete Newtonian gravitation and the N-body problem.
D. Greenspan.

Utilitas Mathematica, Vol. 2, 105 - 126 (1972).

The author develops a discrete theory of Newtonian gravitation which, unlike other difference formulations, will be energy conserving. The ease with which the theory can be applied to resolve nondegenerate n-body problems is illustrated by several examples related to perturbations of orbital motion. It is shown that the perihelion motion of the planet Mercury contains significant negative perturbations in addition to the well-known positive ones.

151.056 A forcing mechanism for spiral density waves in galaxies.
S. I. Feldman, C. C. Lin.

Studies Applied Math., [Mass. Inst. Techn., Cambridge, Mass.], Vol. 52, 1 - 20 (1973).

The authors consider gas dynamical models of spiral galaxies in which there is a rigidly rotating, weakly barlike structure in the central regions. It is found that, in the neighborhood of the corotation circle, this barlike structure forces a trailing spiral wave. Such a driven wave could then propagate inwards to the bar and complete a feedback loop to maintain the spiral structure.

On the impossibility of free precession of a liquid mass approaching the state of relative equilibrium.
See Abstr. 042.051.

Dynamical contraction of rotating gaseous spheroids.
See Abstr. 061.061.

Star formation and evolution in spiral galaxies.
See Abstr. 065.010.

Interaction of proto-stars in a collapsing cluster.
See Abstr. 065.103.

A simple probabilistic theory of fragmentation.
See Abstr. 131.019.

Do interstellar gas clouds exist between spiral arms?
See Abstr. 131.031.

On star dissipation in open clusters.
See Abstr. 153.002.

On the use of *UBV* photometric diagrams for inferring the existence of an open star cluster.
See Abstr. 153.023.

Structure and dynamics of the galactic system. A report.
See Abstr. 155.084.

Magneto-gravitational and thermal instability in the galactic disk.
See Abstr. 155.090.

Structure of the Galaxy.
See Abstr. 155.100.

On the structure of peculiar galaxies.
See Abstr. 158.102.

Spiral structure and nuclear activity in galaxies.
See Abstr. 158.116.

On the problem of the origin of spiral structure.
See Abstr. 158.120.

Photoionization by massive stars in protogalaxies.
See Abstr. 161.003.

152 Stellar Associations

152.001 **Some characteristics of the Eta Carinae complex.**
N. R. Walborn.
Astrophys. Journ., Vol. 179, 517 - 525 (1973).
   The nature and spatial relationship of Trumpler 14 and Trumpler 16 have been clarified; Trumpler 14 is shown to be a more distant, exceedingly young cluster whose brightest member is a possible pre- or post-Wolf-Rayet star. The interpretation of the Wolf-Rayet stars in this region is discussed. Finally a comment is made concerning the excitation of the H II region NGC 3372 associated with the two clusters.

152.002 **A young stellar group in the vicinity of R Coronae Austrinae.** R. F. Knacke, K. M. Strom, S. E. Strom, E. Young, W. Kunkel.
Astrophys. Journ., Vol. 179, 847 - 854 (1973).
   Infrared and optical observations are presented for 11 stars in the dark cloud located near the emission-line variable R CrA. Many of these objects are found to have significant infrared excesses suggesting the presence of circumstellar envelopes. The age of the group is probably less than 10^6 years. Most of these stars are T Tauri-like objects with a range of masses between 1 and 4 $M_\odot$.

152.003 **Emission-line stars in the Chamaeleon T association.**
K. G. Henize, E. E. Mendoza V.
Astrophys. Journ., Vol. 180, 115 - 119 (1973).
   Objective-prism spectrum surveys conducted in 1962 and in 1970 reveal 32 stars showing $H\alpha$ emission in the region of the Chamaeleon T association. Twelve of these stars show variations in either continuum intensity or $H\alpha$ intensity. The second brightest emission-line star, CD$-76°486$, shows significant spectrum variations between 1962 and 1972.

152.004 **Internal motions in the association Cep OB3.**
C. D. Garmany.
Astron. Journ., Vol. 78, 185 - 191 (1973).
   Cep OB3 has been studied astrometrically and spectroscopically in order to learn more about the internal motions of its members. Relative proper motions of 77 O- and B-type stars were determined from two sets of plates with a 47-year interval between them, by using a reference frame of approximately 100 field stars per plate pair. Radial velocities were obtained for half of these stars at Kitt Peak National Observatory.

152.005 **Parámetros cinemáticos de grupos estelares.**
A. E. Gómez.
Bol. As. Argentina Astron., No. 16, (see 012.007), p. 36 (1971). – Abstract.

152.006 **The reddening effect and dust shells of hot stars in associations.** V. I. Kardopolov.
Astron. Tsirk., No. 754, p. 5 - 6 (1973). In Russian.

152.007 **A research on double stars in T-associations: Preliminary results.** M. M. Zakirov.
Astron. Tsirk., No. 757, p. 1 - 2 (1973). In Russian.

152.008 **The nuclear and kinematic ages of stellar associations.** A. Maeder.
Stellar ages. Proc. IAU Colloquium No. 17, (see 012.015), XXIV, 1 - 13 (1973).

152.009 **Formation and ages of the associations.** R. Sancisi.
Stellar ages. Proc. IAU Colloquium No. 17, (see 012.015), XXV, 1 - 5 (1973).

152.010 **Infrarot-Beobachtungen von Sternen in und nahe der Assoziation VI Cyg OB 2.**
K. Voelcker, H. Elsässer.
Mitt. Astron. Ges., No. 32, p. 258 - 260 (1973).

152.011 **Connection of T-associations with the interstellar medium in the northern part of Monoceros.**
V. E. Slutskij.
Young stellar groups. Astroclimate, (see 003.012), p. 29 - 46 (1972). In Russian.

152.012 **On the distribution of Be stars.** V. I. Kardopolov.
Young stellar groups. Astroclimate, (see 003.012), p. 47 - 51 (1972). In Russian.

152.013 **On the dependence of the total mass of stars in a unit volume on the time of their evolution.**
P. Ye. Zakharova, M. A. Svechnikov.
Astrofizika, Vol. 9, 143 - 147 (1973). In Russian. – English translation in Astrophysics, Vol. 9, No. 1.
   The dependence of the distribution of the total mass of stars in a unit volume on their main sequence lifetimes makes it possible to estimate the times of the outset and the termination of star formation in stellar aggregates. It is found, that the age of the galactic cluster NGC 6866 is about 1.3×10^9 years; the star formation lasted nearly 10^9 years.

   Spectral classification of Of stars in VI Cygni (Cygnus OB2). See Abstr. 114.041.

   Spectrum of a dust-embedded Wolf-Rayet star in Cygnus OB2. See Abstr. 114.047.

   The Vela pulsar: member of an association? See Abstr. 141.519.

   OB stars in young star clusters connected with nebulae. See Abstr. 153.001.

   Helium abundances in NGC 2264, II Scorpii, and I Lacertae. See Abstr. 153.014.

   The open cluster Collinder 107 and a stellar ring in Monoceros. See Abstr. 153.033.

   Cold stars in O-clusters. See Abstr. 153.041.

153 Galactic Clusters

153.001 OB stars in young star clusters connected with nebulae. V. I. Kardopolov.
Young stellar groups. Astroclimate, (see 003.012), p. 8 - 28 (1972). In Russian.

153.002 On star dissipation in open clusters.
V. M. Danilov.
Astron. Zhurn. Akad. Nauk SSSR, Vol. 50, 217 - 220 (1973). In Russian. English translation in Soviet Astron. AJ, Vol. 17, No. 1.

A new method of calculation of stellar clusters dissipation with non-uniform stellar composition with regard to mass is presented. The method is applied to a uniform spherical model of a quasi-stationary cluster consisting of 100 stars with radius of 3 pc. Previous results on the existence of two active regions of dissipation in such clusters are confirmed. It was found that the rate of cluster dissipation is equal to − 8.58 x 10^8 stars per year; that correponds to the complete disintegration of the cluster in the course of 10^9 years.

153.003 Membership of the open cluster NGC 6633.
W. L. Sanders.
Astron. Astrophys. Suppl. Ser., Vol. 9, 213 - 220 (1973).

Probabilities of membership based on relative proper motions for 497 stars in the field of NGC 6633 are given. The cluster proper motion dispersion ($m.e.$) of $0''.0007$ yields 113 probable members.

153.004 Membership of the open cluster NGC 6913 (M 29).
W. L. Sanders.
Astron. Astrophys. Suppl. Ser., Vol. 9, 221 - 227 (1973).

Probabilities of membership, based on relative proper motions, for 228 stars in the field of NGC 6913 are given. The cluster proper motion dispersion ($m.e.$) of $0''.0025$ yields 92 probable members.

153.005 The open cluster NGC 2527. U. Lindoff.
Astron. Astrophys. Suppl. Ser., Vol. 9, 229 - 232 (1973).

Photographic magnitudes on the UBV system based on a photoelectric sequence have been determined for stars in NGC 2527, designed as an open cluster. Spectral classes have been determined for the brightest stars from slit spectra and objective prism plates. NGC 2527 seems to consist of about 25 stars, all of them brighter than $V \approx 13^m$. The brightest stars show spectral classes A2–A8, and NGC 2527 is likely to be a real cluster. The color excess, E_{B-V}, in front of the cluster is about 0^m1 and the distance 550 pc.

153.006 Photometric study of the open cluster NGC 2516.
A. Feinstein, H. G. Marraco, I. Mirabel.
Astron. Astrophys. Suppl. Ser., Vol. 9, 233 - 250 (1973).

104 stars were observed in the UBV system, of which 28 and 23 were also in the RI and $H\beta$ system, respectively. It was found that the dispersion of the observed color excesses is almost completely due to the intrinsic color dispersion rather than by that of the true excess. A corrected modulus of $8^m00 \pm 0^m20$ and an age of 6×10^7 years were obtained. All the peculiar stars observed by Abt and Morgan (1969) were found to be members.

153.007 Photometric study of the galactic cluster NGC 2335.
J. J. Clariá.
Astron. Astrophys. Suppl. Ser., Vol. 9, 251 - 260 (1973).

Photoelectric results (UBV) are presented for 60 stars down to $V = 14^m0$. A mean color excess $E(B-V) = +0.40$, corresponding to visual absorption of 1^m20 was found. A mini-

mum membership of 19 stars, including two possible giant members was confirmed. The true distance modulus and the distance were found to be 10^m05 and 1020 pc respectively. An age of 1.5×10^8 years from the turn off point on the main sequence was found. It is suggested that the open clusters NGC 2335 and NGC 2343 are members of a double system connected with the H II region S 296 and the Canis Majoris OB 1 association.

153.008 Three color photometry of the five open clusters NGC 7039, NGC 7062, NGC 7067, NGC 7082, IC 1369. S. M. Hassan.
Astron. Astrophys. Suppl. Ser., Vol. 9, 261 - 287 (1973).

The application of the method of three color UBV-photometry to the five open clusters lead to new distance determinations of these objects. Considerations on age and membership as well as other cluster parameters are discussed. NGC 7067 is confirmed to be a young open cluster located at the +I spiral arm. The other clusters have been found to be too old for spiral arm indicators.

153.009 The young galactic cluster NGC 3766.
W. Winnenburg.
Astron. Astrophys., Vol. 24, 157 - 159 (1973).

The southern galactic cluster NGC 3766 has been studied by means of the UBV photometry using plates taken with the ADH telescope of the Boyden Observatory. Apparent diameter, colour excess, apparent distance modulus, true distance modulus have been derived. By using different methods the age is found to be less than 3×10^7 years. These results fit the Becker diagram of spiral arms very well.

153.010 Junge offene Sternhaufen. Ihre Natur und ihre Bedeutung für die Milchstraßenforschung.
N. Vogt, A. F. J. Moffat.
SuW, Vol. 12, 74 - 78 (1973).

153.011 Relative proper motions in the region of the open cluster NGC 1664.
S. J. Kerridge, R. M. Nelson, W. S. Mesrobian.
Astron. Journ., Vol. 78, 53 - 60 (1973).

Relative proper motions of 222 stars in the region of NGC 1664 have been determined from two plate pairs taken with the 30-inch Thaw refractor of the Allegheny Observatory. The epoch difference of 40 yr yielded centennial proper motions with an average mean error of $0''.13$ in x and $0''.12$ in y coordinate. The solutions for probability of cluster membership imply that there are 57 cluster members brighter than $B = 13.8$ within $7'.5$ of the adopted cluster center.

153.012 A new age indicator for galactic clusters.
L. G. Taff, J. E. Littleton.
Astrophys. Letters, Vol. 13, 133 - 135 (1973).

We have investigated the correlation of the color index of the second bluest star in galactic clusters with two quantitative and one qualitative method for determining relative ages of galactic clusters. The results support the conclusion that this color index is a monotonic function of the age of the cluster.

153.013 Four-color observations of early-type stars. II. The new open cluster in line of sight with the Large Magellanic Cloud. A. G. D. Philip.
Astrophys. Journ., Vol. 180, 421 - 424 (1973).

Analysis of four-color and $H\beta$ photometry of A and F stars in the new open cluster indicates that it resembles the Hyades. Effective temperatures and $\log g$'s are calculated for the 19 stars confirmed as cluster members by Murray, Dickens, and Walker.

153.014 **Helium abundances in NGC 2264, II Scorpii, and I Lacertae.** D. M. Peterson, H. L. Shipman.
Astrophys. Journ., Vol. 180, 635 - 645 (1973).

We determine the helium abundances of three clusters and associations: NGC 2264, II Sco, and I Lac. The very young cluster NGC 2264 has an abundance of $N(He) = 0.08$ while the two associations have abundances of $N(He) = 0.10$. The difference is significant at the 3 σ level.

153.015 **A study of the open cluster NGC 1778.**
R. Barbon, S. M. Hassan.
Astron. Astrophys., Suppl. Ser., Vol. 10, 1 - 10 (1973).

Magnitudes and colours in the UBV system, based on a combination of photoelectric and photographic photometry, have been determined for 114 stars in the open cluster NGC 1778. The colour excess is $0^m.33$ and the distance to the cluster is 1670 pc. The age is estimated to 1.5×10^8 years. Ten bright stars in the region of the cluster have been observed spectroscopically, among which three could be identified as Be stars.

153.016 **A list of suspected clusters in the southern Milky Way.** L. O. Lodén.
Astron. Astrophys., Suppl. Ser., Vol. 10, 125 - 133 (1973).

A list with identification maps is presented of 44 stellar clusterings in the Carina–Centaurus region, which are presumably of physical nature. UBV photometry has been performed on a few of the conspicuous members.

153.017 **Intermediate-band photometry of M67.**
K. A. Janes.
Bull. American Astron. Soc., Vol. 5, 2 - 3 (1973). – Abstr. AAS.

153.018 **The absolute proper motion of the Pleiades cluster.**
W. F. van Altena, B. F. Jones.
Bull. American Astron. Soc., Vol. 5, 7 (1973). – Abstr. AAS.

153.019 **Membership of the open cluster IC 4756.**
A. D. Herzog, W. L. Sanders.
Bull. American Astron. Soc., Vol. 5, 7 (1973). – Abstr. AAS.

153.020 **Evolution of integral parameters of open clusters.**
A. E. Piskunov.
Nauchn. Informatsii, vyp. (No.) 23, p. 97 - 102 (1972). In Russian.

On the basis of evolutionary tracks computed by Paczynski (1970), and Salpeter's mass function (1955), models of open clusters of different ages have been calculated and evolutionary changes of integral magnitudes, and colours of open cluster models have been derived.

153.021 **General properties of red giants.**
A. E. Vasilevsky.
Nauchn. Informatsii, vyp. (No.) 23, p. 103 - 133 (1972). In Russian.

A detailed review and analysis of modern data on photometric and spectroscopic properties, peculiarities of chemical composition, space distribution and velocities of red giants is presented. Special attention is paid to properties of red giants in open clusters, such as distribution on colour–absolute magnitude diagrams of open clusters, and abundances of heavy elements of these objects.

153.022 **Fotometría fotoeléctrica UBV del cúmulo abierto NGC 5460.** J. J. Clariá.
Bol. As. Argentina Astron., No. 16, (see 012.007), p. 25 - 29 (1971).

153.023 **On the use of UBV photometric diagrams for inferring the existence of an open star cluster.**
G. Burki, A. Maeder.
Astron. Astrophys., Vol. 25, 71 - 77 (1973).

In order to discuss the arguments which may be drawn from UBV photometric data on the existence or non-existence of an open star cluster, we have made simulations of UBV measurements for field stars under very classical assumptions on interstellar reddening, luminosity function, distribution of stars in space and spread on the main sequence. The simulations clearly show that field stars may give well-behaved 'sequences' in the UBV diagrams. The reasons why field stars give such features are shown. Some characteristics are defined, allowing to discriminate 'sequences' of field stars from those of real clusters.

153.024 **Über den Aufbau von 11 offenen Sternhaufen.**
W. Lohmann.
Astron. Nachr., Vol. 294, 105 - 111 = Astron. Rechen-Inst. Heidelberg, Mitt. Ser. B (1973).

The strip functions, circular velocity functions and various characteristic parameters of NGC 659, 1027, 1245, 1502, 1528, 1907, 2420, 6830, 6866, 7062 and IC 1848 are derived. From these functions the relative density distribution in the "mean" cluster defined by the 11 clusters is determined.

153.025 **Junge offene Sternhaufen. Ihre Natur und ihre Bedeutung für die Milchstraßenforschung (2. Teil).**
N. Vogt, A. F. J. Moffat.
SuW, Vol. 12, 109 - 112 (1973).

153.026 **Southern open star clusters. II. UBV–Hβ photometry of 11 clusters between galactic longitudes 259° and 280°.** N. Vogt, A. F. J. Moffat.
Astron. Astrophys., Suppl. Ser., Vol. 9, 97 - 131 (1973).

Of the eleven clusters investigated four have spectral types younger than B3. Their distribution suggests the presence of a spiral feature at $l \sim 270°$ in agreement with recent observations of other types of spiral arm tracers. Suspected supergiants of luminosity class II have been found in two of the clusters.

153.027 **The luminosity function of the cluster NGC 752.**
O. A. Chekanikhina.
Astron. Tsirk., No. 746, p. 1 - 3 (1973). In Russian.

153.028 **Coplanarity in open clusters.**
O. Ferrer, C. Jaschek.
Publ. Astron. Soc. Pacific, Vol. 85, 207 - 212 (1973).

A new procedure is given for deriving the distribution of the orbital planes of binaries in clusters, based upon the statistics of the observed mass functions. Tables are provided for applications of the method. The procedure is applied to IC 4665 and an average inclination of 45° and a small dispersion of the inclinations is found.

153.029 **Variable stars in the galactic cluster NGC 6913.**
G. A. Bakos.
Bull. Astron. Inst. Czechoslovakia, Vol. 24, 164 - 167 (1973).

Photoelectric observations of eight bright stars of the galactic cluster NGC 6913 made during 1968–1972 are presented and the light variations of a number of stars discussed.

153.030 **A search for Ap stars in open clusters.**
A. Young, A. E. Martin.
Astrophys. Journ., Vol. 181, 805 - 810 (1973).

Low-dispersion spectrograms of 62 stars in 13 open clusters have been observed in the intrinsic color range $-0.13 < B - V < +0.06$ in order to search for Ap stars. The purpose is to contribute to the statistics of the occurrence of Ap stars in homogeneous age groups, which is necessary as a critical test for some theories regarding the nature of Ap stars. Three new Ap stars are positively identified (all silicon-enhanced), and six other marginal candidates are noted.

153.031 Southern open star clusters III. UBV–Hβ photometry of 28 clusters between galactic longitudes 297° and 353°.   A. F. J. Moffat, N. Vogt.
Astron. Astrophys.,Suppl. Ser., Vol. 10, 135 - 193 (1973).
   The photoelectric photometry yields 22 certain clusters. The parameters of each cluster are summarized in a table. Seven clusters have spectral types earlier than B3: two of these fall in Becker and Fenkart's (1970) inner arm − II; the other five are located in the inner arm −I and yield no change in the structure obtained from previous studies of young open clusters. Suspected supergiants have been found in 10 of the clusters.

153.032 Chemical composition and the location of the Hyades main sequence in the color-magnitude and mass-luminosity diagrams.   D. Koester, V. Weidemann.
Astron. Astrophys., Vol. 25, 437 - 443 (1973).
   A method is developed which serves to answer the question how far an observed metal overabundance as measured by the atmospheric parameter [Fe/H] can be explained by lower original hydrogen content, X, or must be traced back to true metal enrichment, Z. Relative displacements of zero-age sequences of stars with changing X and Z in color-magnitude diagrams are calculated with atmospheric blanketing effects taken into account. Comparison with observation in the Hyades case, which is thoroughly rediscussed, yields a slight reduction of the original hydrogen content ($X = 0.6$) and a metal enrichment ($Z = 0.035$) as compared to a standard composition ($X = 0.70$, $Z = 0.02$). The mass-luminosity discrepancy is considerably reduced.

153.033 The open cluster Collinder 107 and a stellar ring in Monoceros.   J. Isserstedt, T. Schmidt-Kaler.
Astron. Astrophys., Suppl. Ser., Vol. 10, 365 - 383 (1973). In German.
   A star field of 60′ diameter within the OB aggregate Mon OB1, centered on the inconspicuous open cluster Collinder 107, and containing a stellar ring, is investigated by photographic UBV photometry and low dispersion spectral classification.

153.034 On a model of globular clusters with continuously distributed stellar mass.   V. M. Bagin.
Astron. Zhurn. Akad. Nauk SSSR, Vol. 50, 653 - 657 (1973). In Russian. English translation in Soviet Astron. AJ, Vol. 17, No. 3. − Short note.

153.035 NGC 1893 and NGC 6838, standard stars.   J. Cuffey.
Astron. Journ., Vol. 78, 408 - 409 (1973).
   Photoelectric magnitudes and colors in the UBV system for 56 stars in NGC 1893 and for 13 stars in NGC 6838 are given in tables with identification charts in figures.

153.036 Photographische UBV-Photometrie des galaktischen Sternhaufens NGC 6819.   G. Auner.
Mitt. Astron. Ges., No. 32, p. 138 (1973). − Abstract.

153.037 Der Sternhaufen NGC 7142.   E. S. Pendl.
Mitt. Astron. Ges., No. 32, p. 213 - 214 (1973). − Abstract.

153.038 On the brightness function of the star cluster M67.   M. Popova.
Izv. Sekts. astron. Blg. AN, Vol. 5, 67 - 71 (1972). In Bulgarian. − Abstr. in Referativ. Zhurn. 51. Astron., 5.51.756 (1973).

153.039 Photographic UBV photometry of ten open star clusters in a galactic field at $l = 135°$.   A. F. J. Moffat, N. Vogt.
Astron. Astrophys., Suppl. Ser., Vol. 11, 3 - 23 (1973) In German.
   In a field 5° × 5° just NE of h and χ Persei photographic UBV photometry ($V \leqq 16.5$) has been carried out in ten open clusters for which no previous UBV data exist. Three of these are well defined young clusters suitable as spiral arm tracers.

153.040 Photographic photometry of the open cluster M35.   N. V. Vidal.
Astron. Astrophys., Suppl. Ser., Vol. 11, 93 - 106 (1973).
   UBV photographic photometry of 436 stars in the region of M35 is presented. Using a recent proper motion survey, C-M and C-C diagrams for probable cluster members are drawn. It is found that M35 is as young as the Pleiades. Salpeter's initial luminosity function seems to fit best the luminosity function for cluster members.

153.041 Cold stars in O-clusters.   R. A. Vardanian, N. G. Kchatchatrian (Khachatryan).
Astrofizika, Vol. 8, 613 - 619 (1972). In Russian. − English translation in Astrophysics, Vol. 8, No. 4.
   It is shown by a statistical method that some of M type stars from the CIT catalogue belong to O-clusters and O-associations while their connection with B-clusters is week or does not exist at all.

153.042 On the estimate of the age of the cluster α Per.   P. Ye. Zakharova, M. A. Svechnikov.
Astrofizika, Vol. 9, 147 - 150 (1973). In Russian. − English translation in Astrophysics, Vol. 9, No. 1.
   The times of the beginning and the end of star formation in the young galactic cluster α Per are determined by an investigation of the total máss of the main sequence stars per unit interval in log $M/M_\odot$.

Centre de Données Stellaires, Inform. Bull. No. 4. See Abstr. 002.007.

Formation of stars in a rotating cloud with magnetic field. See Abstr. 065.004.

Core-helium-burning stars in extremely young clusters. See Abstr. 065.041.

Models of population I clump giants. See Abstr. 065.042.

Solar neutrinos, Martian rivers, and Praesepe. See Abstr. 080.037.

Relative proper motions of faint stars in the Pleiades. See Abstr. 112.002.

Recent determinations of relative proper motions in star clusters. See Abstr. 112.008.

On determinations of proper motions of stars in ten areas of the sky with open clusters. See Abstr. 112.009.

Proper motions of stars in the area of the open cluster NGC 457. See Abstr. 112.010.

Proper motions of stars in the area of the open cluster NGC 663. See Abstr. 112.011.

Infrared emission from stars in the h and χ Persei association. See Abstr. 113.034.

Circumstellar shells in the young cluster NGC 2264.
See Abstr. 113.035.

Infra-red observations of young stars—I. Stars in young clusters.   See Abstr. 114.002.

A spectrophotometry of stars in the Pleiades. I. Observations of 19 early-type stars.   See Abstr. 114.108.

On the metallicity of the main-sequence stars in M67.
See Abstr. 114.144.

Variability of T Tauri-like stars in NGC 2264.
See Abstr. 114.165.

The red star in the open cluster Trumpler 27.
See Abstr. 114.171.

Luminosity functions of stars in the Praesepe cluster.   See Abstr. 115.002.

The classification of intrinsic variables. III. Calibration of the luminosities of small amplitude red variables in the old disk population.   See Abstr. 115.010.

Construction des diagrammes des groupes d'étoiles sur la base de résultats des calculs théoriques.
See Abstr. 115.017.

Age des étoiles B à émission.
See Abstr. 115.022.

The spectroscopic orbit of HD 90707.
See Abstr. 121.001.

New flares in the Pleiades.   See Abstr. 122.120.

New flare stars in the Pleiades. I.
See Abstr. 122.127.

Observation of the spectrum of the flare of star No. 205 in the Pleiades.   See Abstr. 122.128.

Two-colour observations of the flare of the star No. 207 in the Pleiades.   See Abstr. 122.129.

Photometry of selected flare stars in the Pleiades.
See Abstr. 122.131.

Neue Flare-Sterne in den Plejaden.
See Abstr. 122.135.

Flare stars in the Pleiades. III.
See Abstr. 122.136.

New flare stars in the Pleiades.
See Abstr. 122.093.

New flare stars in the Pleiades region.
See Abstr. 122.095.

New flare stars in the Pleiades region (1972 - 1973).
See Abstr. 122.099.

Ammonia in DR 21(OH) and NGC 2264.
See Abstr. 131.185.

Velocity variation of a star as a purely discontinuous random process. III. Stars with different masses in an open cluster.   See Abstr. 151.015.

On the star dissipation in open clusters. II.
See Abstr. 151.040.

Motion of fourteen stars in the Orion nebula cluster.
See Abstr. 151.042.

On the dependence of the total mass of stars in a unit volume on the time of their evolution.
See Abstr. 152.013.

An up-to-date picture of galactic spiral features based on young open star clusters.   See Abstr. 155.015.

A photometric study of the integrated light of clusters in the Magellanic Clouds and the Fornax dwarf galaxy.
See Abstr. 154.010.

Luminosity and velocity distribution of high-luminosity red stars near the sun. I. The very young disk population.
See Abstr. 155.099.

154   Globular Clusters

154.001   Chemical composition of globular cluster stars and morphological properties of their horizontal branches.   A. V. Mironov.
Astron. Zhurn. Akad. Nauk SSSR, Vol. 50, 27 - 38 (1973). In Russian. English translation in Soviet Astron. AJ, Vol. 17, No. 1.

The dependence of the form of the horizontal branch of globular clusters on the degree of mass loss happening during stellar evolution and on the helium abundance Y in the star envelopes are considered. A correlation between the helium abundance and the richness of W Vir variables in the clusters is found. It is discovered that there exist two types of globular clusters distinguished by features of dependence of the shape of the horizontal branch on the chemical composition. A connection between the most probable degree of mass loss by the cluster stars and their chemical composition is found.

154.002   The observed deficiency of ionized gas in globular clusters and the companions of M 31.
J. G. Hills, M. J. Klein.
Astrophys. Letters, Vol. 13, 65 - 68 (1973).

An ultra-sensitive search at 3.8 cm for free-free emission from ionized gas in the globular clusters M 3, M 5, M 13, M 75 and M 92 and the two galactic companions of M 31 reveals no radiation above the experimental error which is typically $4 \times 10^{-29} \mathrm{Wm}^{-2} \mathrm{Hz}^{-1}$. The upper limit on the amount of gas in most of these systems is about an order of magnitude below that computed on the assumption that the mass shed by evolving stars has been retained in the systems.

154.003   Masses of old LMC globular clusters.
K. C. Freeman, C. Munsuk.
Proc. Astron. Soc. Australia, Vol. 2, 151 - 152 (1972).

154.004   Intrinsic colors of globular clusters in the UBV system.   R. Racine.
Astron. Journ., Vol. 78, 180 - 184 (1973).

It is shown that the color excess ratio $E(U-B)/E(B-V)$ for globular clusters increases with advancing integrated spectral types, that the spectral types are closely correlated with the intrinsic color index $(B-V)_0$, and that the intrinsic colors $(U-B)_0$ and $(B-V)_0$ of galactic globular clusters define a dispersion-free relation in the two color diagram. New reddening values are determined for 86 clusters, and the advantages and limitations of broadband UBV photometry of the integrated light of globular clusters are discussed.

154.005   Probably new globular clusters of the Andromeda nebula.   A. S. Sharov.
Astron. Zhurn. Akad. Nauk SSSR, Vol. 50, 263 - 269 (1973). In Russian. English translation in Soviet Astron. AJ, Vol. 17, No. 2.

Twenty five diffuse objects in the region of the Andromeda nebula having magnitudes $V = 15^{\mathrm{m}}6 - 18^{\mathrm{m}}2$ and colours $B-V < 1^{\mathrm{m}}0$ are discovered on plates obtained with the Schmidt telescope of the radioastrophysical observatory of the Latvian Academy of Sciences (Baldone). The objects concentrate clearly to the Andromeda nebula, and it is necessary to consider them as probably new globular clusters of this galaxy.

154.006   Neutral hydrogen observations of eight globular clusters.   G. R. Knapp, F. J. Kerr, W. K. Rose.
Bull. American Astron. Soc., Vol. 5, 24 (1973). – Abstr. AAS.

154.007   On the structure of the horizontal branch and on the age of the globular clusters.   A. M. Eigenson.
Astron. Tsirk., No. 743, p. 1 - 3 (1973). In Russian.

154.008   Brightness distribution in globular clusters and generalized isochronic models.   G. G. Kuzmin, Ü.-I. K. Veltmann, Yu. A. Vennik, O. A. Tõeleid, E. V. Tago.
Astron. Tsirk., No. 745, p. 1 - 3 (1973). In Russian.

Comparison of models with observations by G. E. Kron and N. U. Mayall has shown that the concentration classes a, b, c, which they introduce, can be described very well by the Schuster model, the isochronic model and the limiting model, respectively. It seems possible that the degree of the central concentration correlates with the metal abundance in the globular clusters. Some original measurements of the globular clusters M3, M5, M13, M92, M15, M2 are also discussed.

154.009   The globular cluster NGC 6934.
W. E. Harris, R. Racine.
Astron. Journ., Vol. 78, 242 - 245, 281 - 282 (1973).

New photoelectric and photographic photometry in B and V is presented for the globular cluster NGC 6934. The color-magnitude diagram is determined down to $V \cong 18$ and is basically similar to that for M3. The reddening and the distance modulus are derived.

154.010   A photometric study of the integrated light of clusters in the Magellanic Clouds and the Fornax dwarf galaxy.   I. J. Danziger.
Astrophys. Journ., Vol. 181, 641 - 655 (1973).

The results of an 11-color photometric study of 20 globular clusters and eight open clusters in the Magellanic Clouds and three globulars in Fornax are reported. Because the photometric system is designed to measure absorption-line features in late-type spectra the important results relate to the metal content of globular clusters or old evolved systems. The special nature of the unusual globular clusters in the Clouds studied by Gascoigne and observed here is discussed, but strong conclusions are avoided. A brief discussion of the results on open clusters is given.

154.011   Masses of red giants on the asymptotic branch of globular clusters.   R. M. Rusev.
Astron. Zhurn. Akad. Nauk SSSR, Vol. 50, 535 - 540 (1973). In Russian. English translation in Soviet Astron. AJ, Vol. 17, No. 3.

On the basis of the synthesis of the globular clusters NGC 5466 and NGC 6397 the mean masses of the red giants (RG) and of the giants on the asymptotic branch (AGB) are compared. A comparison of the density ranges permitted us to conclude that $\bar{M}^{\mathrm{AGB}} \lesssim \bar{M}^{\mathrm{RG}}$.

154.012   Current problems on horizontal branch stars. I: H. B.-topology.
F. Caputo, A. Natta, V. Castellani.
Astrophys. Space Sci., Vol. 22, 199 - 211 (1973).

After a short historical survey, present available information on the location of H. B. stars in the colour-magnitude diagram are collected. A general agreement is found between observations and theoretical predictions as deduced from the available B.C.- and $(B-V)$-log T_e relations. A comparison among presently available evaluations of the B.C.-log T_e and $(B-V)$-log T_e relations is made in order to derive informations about the range of reliability, as well as in order to emphasize some possible peculiarities.

154.013   Current problems on horizontal branch stars. II: H. B.-population.
F. Caputo, A. Natta, V. Castellani.

Astrophys. Space Sci., Vol. 22, 213 - 225 (1973).

In the light of our present knowledge on stellar evolution, it is shown that red ZAHB will not develop for very low metal globular clusters if Y is appreciably greater than 0.1. By comparing evolutionary paths with the observed populations of the horizontal branches the authors find evidence for mass loss in blue H.B., and evidence for mass dispersion in red H.B. not depopulated in RR Lyrae. The expected correlations between mass loss and spatial distribution of H.B. stars are briefly discussed.

154.014 Four-color observations of early-type stars. III. M4.
A. G. D. Philip.
Astrophys. Journ., Vol. 182, 517 - 521 (1973).

Four-color measures have been made of ten blue-horizontal-branch stars in the globular cluster M4. The quantities θ_e and log g have been calculated for stars with 0.00 mag $\leqslant$ $(b - y)_0 \leqslant 0.12$ mag, and these parameters are in accord with the (log g, θ_e)-relation found for NGC 6397 and other clusters by Newell.

154.015 Photometry of southern globular clusters–I. Bright stars in ω Centauri. R. D. Cannon, R. S. Stobie.
Monthly Notices Roy. Astron. Soc., Vol. 162, 207 - 225 (1973).

The techniques used for observing globular cluster stars are described, and several sources of error are discussed. Photoelectric UBV data are presented for 126 stars brighter than $V = 16.5$ in ω Cen. These data are compared with those of previous photoelectric and photographic studies.

154.016 Photometry of southern globular clusters–II. Bright stars in NGC 6752. R. D. Cannon, R. S. Stobie.
Monthly Notices Roy. Astron. Soc., Vol. 162, 227 - 234 (1973).

Photoelectric UBV data are presented for nearly 100 stars with $V \leq 16$ in NGC 6752. The cluster has reddening $E(B-V)=$ 0.05 mag and distance modulus $(m-M) \approx 13.5$. One new small amplitude red variable has been detected.

154.017 Structure of the horizontal branch and age of globular clusters. A. M. Eigenson (*Ehjgenson*).
Astrofizika, Vol. 9, 107 - 118 (1973). In Russian. – English translation in Astrophysics, Vol. 9, No. 1.

The ratio of the number of stars from the blue side of the gap of variable stars and that of the total number of non-variable horizontal branch stars is compared with other characteristics of globular clusters.

154.018 The effects of a variation in the CNO abundances on the position of initial horizontal-branch models.
F. D. A. Hartwick, D. A. Vanden Berg.
Publ. Astron. Soc. Pacific, Vol. 85, 355 - 357 (1973).

From the results of horizontal-branch model computations in which the nitrogen abundance was varied from 0.1 to 10 times normal we conclude, following the suggestion of Hartwick and McClure, that a nitrogen overabundance of $\lesssim 10$ times normal can explain the anomalous red horizontal branch in the globular cluster NGC 7006.

The blue objects in the region around the globular cluster M92. See Abstr. 113.053.

On the nature of the Sagittarius object IRC–20385. See Abstr. 114.015.

An abundance analysis of the F-type giant HD 116745 in the globular cluster Omega Centauri. See Abstr. 114.024.

The C^{12}/C^{13} ratio of the CH stars in ω Centauri. See Abstr. 114.070.

The unusual red giants in M 5, M 10 and M 92. See Abstr. 114.139.

The G-band anomaly of the asymptotic-branch stars in M92. See Abstr. 114.148.

Horizontal branch morphology. See Abstr. 115.018.

Photometric information on globular cluster stars from the Palomar Sky Survey prints. See Abstr. 115.024.

On the metal abundance of RR Lyrae stars in the globular cluster M22. See Abstr. 122.008.

Variable stars in globular clusters. See Abstr. 122.021.

On the metal abundance of RR Lyrae stars in the globular cluster M22. See Abstr. 122.038.

Long-term changes in amplitude for RR Lyrae stars in M5. See Abstr. 122.043.

Population II cepheids in globular clusters. See Abstr. 122.125.

A search for planetary nebulae in globular clusters. See Abstr. 133.014.

Galactic orbits and tidal radii of the clusters M67, NGC 188 and ω Centauri. See Abstr. 151.018.

Variations in spectral-energy distributions and absorption-line strengths among elliptical galaxies. See Abstr. 158.027.

155 Structure and Evolution of the Galaxy

155.001 An estimate of the energy spectrum of gamma rays from the central region of the Galaxy and some implications. K. C. Anand, S. A. Stephens.
Astrophys. Space Sci., Vol. 18, 387 - 402 (1972).

High resolution surveys of the galactic centre suggest the existence of an extended non-thermal source (bulge) with an intensity much larger than the total background radiation in that direction. In this paper, we have first evaluated the physical conditions existing in this restricted region of space from an analysis of the radio spectrum. The gamma ray spectra from the bulge arising from interactions of cosmic rays with matter and radiation are then calculated in detail. A comparison has been made with the estimated background gamma ray spectra from the disk.

155.002 Statistical significance of some optical evidence for the bending of the galactic plane. D. Sher.
Astrophys. Space Sci., Vol. 18, 468 - 471 (1972).

It is argued on the basis of an analysis of variance test that the distribution of Wolf-Rayet stars in the Galaxy departs systematically from the plane in the same sense as the neutral hydrogen.

155.003 Galactic structure from observations of interstellar calcium lines. I. Analysis of the radial velocities.
M. Chu-Kit.
Astron. Astrophys., Vol. 22, 69 - 74 (1973). In French.

As part of a galactic survey the distribution and kinematics of the interstellar calcium are investigated from $l = 285°$, through the galactic center to $l = 29°$. 122 radial velocities of interstellar calcium lines have been measured in 68 O−B stars with a dispersion of 12.3 Å mm^{-1} in La Silla Observatory (ESO). The kinematical distances are calculated from Schmidt's model.

155.004 Galactic differential rotation derived from the radial velocities of some population I objects. M. Crézé.
Astron. Astrophys., Vol. 22, 85 - 89 (1973).

Solution for Oort's constant A of galactic differential rotation and for the components of the solar motion are derived from the radial velocities of three population I samples (population I cepheids, H II regions, open clusters).

155.005 High-velocity clouds and 'normal' galactic structure. G. L. Verschuur.
Astron. Astrophys., Vol. 22, 139 - 151 (1973).

Most of the high-velocity clouds of neutral hydrogen which lie at intermediate and high galactic latitudes, as well as the highest-velocity clouds which lie near the plane are probably part of distant spiral arms. Since the arms are extremely non-uniform and non-planar they have not here-to-fore been recognized in galactic plane surveys.

155.006 Note on Verschuur's article on high-velocity clouds and 'normal' galactic structure.
A. N. M. Hulsbosch, J. H. Oort.
Astron. Astrophys., Vol. 22, 153 - 154 (1973).

Verschuur's model which explains most of the high-velocity clouds as being parts of distant spiral arms is discussed.

155.007 New M dwarfs in the south galactic cap.
D. H. P. Jones.
Monthly Notices Roy. Astron. Soc., Vol. 161, 19P - 24P (1973).

Twenty-two new dwarf M stars are reported. If the M dwarfs are distributed homogeneously in the solar vicinity then the spatial mass density of the present sample is too small to explain the 'missing mass' required by stellar dynamics. There is a large uncertainty arising from the small sample size.

155.008 A soft X-ray survey from the galactic center to Cassiopeia. G. Burginyon, R. Hill, T. Palmieri, J. Scudder, F. Seward, J. Stoering, A. Toor.
Astrophys. Journ., Vol. 179, 615 - 625 (1973).

Results are presented of a survey of the galactic plane between $l^{II} = 0°$ and $l^{II} = 145°$. The detector was sensitive to X-rays with energy from 0.2 to 18 keV and had an effective area of 640 cm^2. Source locations within 0.02 deg^2 and spectral parameters have been derived for Cyg 1, Cyg 2, the Cygnus Loop, Ser 1, and the supernova remnant Cas A. Spectral parameters have also been derived for GX 5−1, GX 9+1, GX 13+1, and GX 17+2; however, these sources were located in only one dimension. Some spectral information and locations in one dimension have been obtained for Tycho's supernova remnant, Cyg 3, and seven other sources.

155.009 Observations of the galactic nucleus at 350 microns.
D. Y. Gezari, R. R. Joyce, M. Simon.
Astrophys. Journ., (*Letters*), Vol. 179, L67 - L70 (1973).

We have observed the regions in the galactic center in the far-infrared and detected the flux $(3.91 ± 0.76) × 10^{-11}$ W m^{-2} from Sgr B and obtained an upper limit $1.5 × 10^{-11}$ W m^{-2} from Sgr A, corresponding to flux densities at 350μ of $(4.30 ± 0.86) × 10^{-22}$ and $1.6 × 10^{-22}$ W m^{-2} Hz^{-1}, respectively. Scanning observations of Sgr B in right ascension indicate that the width of the source is $5'-10'$. A close association between the far-infrared source and the molecular region in Sgr B is indicated, and an optically thin thermal model for the far-infrared radiation from dust grains is discussed.

155.010 Molecules and evolution in the Galaxy.
D. Buhl.
Sky Telescope, Vol. 45, 156 - 158 (1973).

155.011 H I absorption in the galactic center region and between galactic longitudes 350° and 359°.
I. Kazès, D. Aubry.
Astron. Astrophys., Vol. 22, 413 - 420 (1973).

Thirteen absorption profiles along the galactic center region as well as five in the region $l^{II} = 350°$ to 359° are presented. Distance estimates have been derived for radio sources in these regions. In particular, in the regions of the galactic center, relative abundances among H I, OH and H$_2$CO are discussed in the light of the detected neutral gas in different molecular clouds.

155.012 Statistical principles of galactic optical astronomy. Part I. H. Eelsalu.
Tartu Astron. Obs. Teated, No. 41, 80 pp. (1973). In Russian.

155.013 Survey of molecular lines near the galactic center. III. 6-centimeter formaldehyde absorption at $b = -2'$ from $l = 2°0$ to $l = 4°5$ and at $b = -12'$ from $l = 358°5$ to $l = 2°0$. N. Z. Scoville, P. M. Solomon.
Astrophys. Journ., Vol. 180, 55 - 59 (1973).

New observations suggest that the group of dense, low-positive-velocity clouds does not continue beyond $l = 2°$ and that the negative-velocity clouds are part of an arm structure within 300 pc of the galactic center and having a total mass of $\sim 10^6 M_\odot$. We also note the existence of molecular gas at $l = 2°3$ with a velocity of 240 km s^{-1}.

155.014 327-MHz observations of the galactic center: Possible detection of a deuterium absorption line.

D. A. Cesarsky, A. T. Moffet, J. M. Pasachoff.
Astrophys. Journ., (*Letters*), Vol. 180, L1 - L6 (1973).

We have observed the spectrum of radiation from the galactic center in the vicinity of the deuterium ground-state hyperfine transition. With ~ 100 hours of observing time the spectrum shows rms fluctuations ~7 × 10^{-5} of the on-source power level. An absorption feature at 327.38837±0.00001 MHz (corrected to the local standard of rest) has a depth of 2 × 10^{-4} of the continuum level. This feature is probably the deuterium line at $v_{LSR} = -3.7$ km s^{-1}.

155.015   **An up-to-date picture of galactic spiral features based on young open star clusters.**
A. F. J. Moffat, N. Vogt.
Astron. Astrophys., Vol. 23, 317 - 320 (1973).

The number of young open star clusters previously studied with three-colour photometry (Becker and Fenkart, 1971:88) now stands at 110. Revisions for four clusters are given, of which Pişmiş 20 is especially interesting. This cluster coincides with the supernova remnant G 320.4−1.2 and the X-ray source 2 U 1509−58. The distribution of young clusters is compared with that of five other young spiral tracers out to ~4 kpc from the sun.

155.016   **Lunar occultations of the galactic center region in H I, OH and H$_2$CO lines. I. Observations and contour maps.**   A. Sandqvist.
Astron. Astrophys., Suppl. Ser., Vol. 9, 391 - 425 (1973).

In the period from January, 1968 to October, 1970, eleven lunar occultations of the galactic center region have been observed with the 140 foot radio telescope at the National Radio Astronomy Observatory, Green Bank, West Virginia, using a 100 channel autocorrelator for the first two occultations, and a 413 channel autocorrelator for the remaining nine. Eight events have been observed in the 1667 MHz OH line, six in the 1665 MHz OH line, one in the 1720 MHz OH line, six in the 1420 MHz H I line, and two in the 4830 MHz H$_2$CO line. Baseline-corrected antenna temperature contour maps on grid of velocity versus sidereal time have been produced for all the observed lines. Opacity contour maps have also been produced, as well as differentiated contour maps of selected parts of the occultations.

155.017   **On the gamma ray line from the galactic centre.**
P. Guthrie, E. Tademaru.
Nature, Phys. Sci., Vol. 241, 77 - 79 (1973).

Johnson, Harnden and Haymes (1972) have published a spectrum of low energy gamma rays from the direction of the galactic centre. In addition, they find a spectral feature at 473±30 keV. A physical explanation of the feature, if it is a real line at 473±30 keV, is presented.

155.018   **Radio pulses from the direction of the galactic centre.**   V. A. Hughes, D. S. Retallack.
Nature, Vol. 242, 105 - 107 (1973).

We conclude that in the direction of the galactic centre there are sources which emit pulses of radio waves of duration less than 1 s and a source which seems to contribute about 24 % of the events, with pulses of consistently greater amplitude than the average of all events. The question how these events relate to known radio, optical, infrared, X-ray and γ-ray sources is discussed.

155.019   **Underabundance of ionized helium in the galactic centre.**   E. Churchwell, P. G. Mezger.
Nature, Vol. 242, 319 - 320 (1973).

Most cosmological models predict a helium abundance of approximately 10 % by number. In an effort to determine whether this value holds throughout our Galaxy we have observed hydrogen and helium radio recombination lines in 40 galactic H II regions. A brief summary of the primary results is given.

155.020   **The large-scale distribution of low-velocity hydrogen gas at high galactic latitudes.**
I. Fejes, P. R. Wesselius.
Astron. Astrophys., Vol. 24, 1 - 13 (1973).

The distribution of low-velocity neutral hydrogen gas ($-21 < V < +15$ km/s) at latitudes $|b| \geqq 15°$ is discussed on the basis of the Groningen High Latitude Survey (Tolbert, 1971). A number of strong neutral hydrogen ridges are described, some of which have not been noticed before.

155.021   **Intermediate-negative-velocity neutral hydrogen at $b \geqq +15°$.**   P. R. Wesselius, I. Fejes.
Astron. Astrophys., Vol. 24, 15 - 34 (1973).

The properties of intermediate-negative-velocity (INV) neutral hydrogen are described using the Groningen High Latitude Survey (Tolbert, 1971). The distribution of INV gas is compared to those of low-velocity and high-negative-velocity gas. Distance estimates for some of the INV complexes and one intermediate-positive-velocity complex are derived by comparing intermediate-velocity features of the K line of ionized calcium and the 21-cm line of neutral hydrogen. Two interpretations for the occurrence of such large INV complexes at high latitudes are discussed: a nearby supernova remnant and the interaction of a big gas complex with the galactic layer. The latter model is preferred.

155.022   **Intermediate-negative-velocity gas around $l = 238°$, $b = +75°$.**   P. R. Wesselius.
Astron. Astrophys., Vol. 24, 35 - 39 (1973).

The neutral hydrogen distribution, especially for intermediate-negative-velocities (INV, $-91 < V < -21$ km/s), has been studied in a region of about 10° × 10° centred at $l = 238°$, $b = +75°$. In most of the region INV gas is present. A number of subconcentrations can be distinguished. All the sub-concentrations are probably parts of one large concentration. There appears to be a connection between this INV concentration and the low-velocity gas, indicating recent interaction between the two.

155.023   **Galactic continuum loops and the diameter−surface brightness relation for supernova remnants.**
E. M. Berkhuijsen.
Astron. Astrophys., Vol. 24, 143 - 147 (1973).

The linear diameter and average surface brightness at 1 GHz of 5 galactic continuum loops have been estimated in order to find their positions in the $D - \Sigma$ diagram for supernova remnants. The loops appear to fit the relation derived by Ilovaisky and Lequeux (1972) reasonably well, which confirms the hypothesis that they are supernova remnants of great age (~ 10^6 years). A new $D - \Sigma$ relation has been computed for two cases.

155.024   **Galactic loops as supernova remnants in the local galactic magnetic field.**   T. A. T. Spoelstra.
Astron. Astrophys., Vol. 24, 149 - 155 (1973).

Results from model calculations based on Van der Laan's model for a shell expanding in a magnetoionic medium are presented for Loop IV, the Lupus Loop, Monoceros Loop and Origem Loop. It is indicated that the spatial orientation of the loops contains information on the direction of the magnetic field of the undisturbed medium outside the shell.

155.025   **Gamma-ray emission from the region of the galactic center.**
G. H. Dahlbacka, P. S. Freier, C. J. Waddington.
Astrophys. Journ., Vol. 180, 371 - 379 (1973).

A combination nuclear-emulsion-spark-chamber γ-ray ($E > 100$ MeV) telescope has been used to study the region of sky that includes the galactic center.

155.026 **The Perseus spiral arm at 21-cm.**
G. L. Verschuur.
Astron. Astrophys., Vol. 24, 193 - 200 (1973).
It is shown that the neutral hydrogen emission apparently associated with the Perseus spiral arm at latitude zero may be due to matter in a more distant spiral arm. When this effect is recognized it is possible to account for the apparent difference between the neutral hydrogen velocities and the stellar velocities in the Perseus arm without invoking the presence of shock waves or other streaming motions in the Perseus arm.

155.027 **On the kinematics of a local component of the interstellar hydrogen gas possibly related to Gould's Belt.**
P. O. Lindblad, K. Grape, A. Sandqvist, J. Schober.
Astron. Astrophys., Vol. 24, 309 - 312 (1973).
It is shown that a local component of the interstellar neutral hydrogen displays motions characteristic for a shell or cloud expanding in the field of differential rotation. Some evidence is presented that this feature is related to the Gould Belt system of early type stars.

155.028 **On the problem of background radio emission of the Galaxy.** V. I. Ariskin.
Astron. Zhurn. Akad. Nauk SSSR, Vol. 50, 283 - 286 (1973).
In Russian. English translation in Soviet Astron. AJ, Vol. 17, No. 2.
An analysis of the distribution of the Galaxy background radio emission in direction of the longitudes $l^{II} = 20°.3 - 27°.3$ in the wavelength range from 3.5 m to 3.75 cm has been made.

155.029 **Complanar system of binaries in Aquila.**
V. V. Radzievsky, E. V. Radzievskaya.
Astron. Zhurn. Akad. Nauk SSSR, Vol. 50, 371 - 376 (1973).
In Russian. English translation in Soviet Astron. AJ, Vol. 17, No. 2.
An area on the celestial sphere with the centre in Aquila limited by $28°30' < l < 71°$ and $-9° < b < 18°30'$, containing 36 visual binaries with determined orbits including 32 stars with retrograde motion, was detected. The pole of complanarity of the system ($l = 55°; b = 7°$), the coefficient of complanarity (0.70) and the probability of a chance of this distribution are determined. It is shown on the basis of the field and kinematic features that the solar system belongs to the detected stellar cluster. The ensuing problems are formulated.

155.030 **The abundances and ages of F and G stars in the solar neighborhood.** R. E. S. Clegg, R. A. Bell.
Bull. American Astron. Soc., Vol. 5, 2 (1973). – Abstr. AAS.

155.031 **The kinematics of galactic spiral structure.**
W. B. Burton.
Bull. American Astron. Soc., Vol. 5, 7 - 8 (1973). – Abstr. AAS.

155.032 **Kinematic disturbances in the local neutral hydrogen.** P. L. Baker.
Bull. American Astron. Soc., Vol. 5, 8 (1973). – Abstr. AAS.

155.033 **Galactic spiral structure from 21-cm data: A new map.** G. L. Verschuur.
Bull. American Astron. Soc., Vol. 5, 8 (1973). – Abstr. AAS.

155.034 **Comparison of stellar and neutral hydrogen galactic kinematics.** T. Bania, W. B. Burton.
Bull. American Astron. Soc., Vol. 5, 8 (1973). – Abstr. AAS.

155.035 **A density-wave map of the galactic spiral structure.**

S. C. Simmons III.
Bull. American Astron. Soc., Vol. 5, 8 (1973). – Abstr. AAS.

155.036 **One-armed spiral waves.** C. Hunter.
Bull. American Astron. Soc., Vol. 5, 8 - 9 (1973).
Abstr. AAS.

155.037 **Fine structure in the galactic center.**
B. Balick, R. M. Hjellming.
Bull. American Astron. Soc., Vol. 5, 31 (1973). – Abstr. AAS.

155.038 **Observations of the galactic nucleus at 350 microns.**
D. Y. Gezari, R. R. Joyce, M. Simon.
Bull. American Astron. Soc., Vol. 5, 31 (1973). – Abstr. AAS.

155.039 **100 micron survey of southern Milky Way and Magellanic Clouds.**
W. F. Hoffmann, C. L. Frederick, R. J. Emery.
Bull. American Astron. Soc., Vol. 5, 31 (1973). – Abstr. AAS.

155.040 **Gamma radiation from the galactic center region.**
R. L. Kinzer, G. H. Share, N. Seeman.
Bull. American Astron. Soc., Vol. 5, 33 (1973). – Abstr. AAS.

155.041 **High velocity gas motion near the galactic center.**
R. H. Sanders, G. T. Wrixon.
Bull. American Astron. Soc., Vol. 5, 41 (1973). – Abstr. AAS.

155.042 **Another look at the absolute proper motions obtained from the Lick pilot programme.**
S. V. M. Clube.
Monthly Notices Roy. Astron. Soc., Vol. 161, 445 - 463 = Contr. Lick Obs., No. 379 (1973).
A general model describing first-order variations in the velocity field with respect to distance in the wide solar neighbourhood is applied to the analysis of absolute proper motions of faint stars. The results are interpreted as implying a distance to the galactic centre of 7 kpc in good agreement with other determinations. The velocity field also indicates a local expansion of the solar neighbourhood away from the galactic centre. The observations of H I and OH absorption lines at the galactic nucleus are reinterpreted as providing very strong support for this discovery. Some problems arising from this discovery are mentioned. The velocity field of the nearby stars is also examined, and this clearly gives some physical reality to the stellar drifts once conjectured as an explanation of stellar kinematics in the solar neighbourhood. The relationship with stellar groups is also indicated.

155.043 **'Local' theories of the X-ray background.**
J. E. Felten.
IAU Symposium No. 55, (see 012.002), p. 258 - 275 (1973).

155.044 **Infrared line radiation from the neighbourhood of the galactic center.** S. R. Pottasch.
17th Colloquium International Astrophys. Liège 1971, (see 012.003), p. 327 - 333 (1972).

155.045 **High resolution radio recombination line observations in the direction of the galactic center.**
E. Churchwell, P. G. Mezger.
17th Colloquium International Astrophys. Liège 1971, (see 012.003), p. 335 (1972). – Abstract.

155.046 **Far infrared observations of the galactic center region.** R. J. Van Duinen, T. J. Helmerhorst, H. Olthof, J. Slingerland, J. J. Wijnbergen.
17th Colloquium International Astrophys. Liège 1971, (see 012.003), p. 597 - 600 (1972).

155.047 **Space distribution of stars in the southern Milky**

Way: II. A region in Vela. W. H. Wooden II.
Publ. Warner and Swasey Obs., Cleveland, Ohio, Vol. 1, No. 2, 83 pp. (1971).

V magnitudes, B−V colors, and MK spectral types have been determined for 6195 stars in the region LF 13, an area of 17.1 square degrees centered at $l = 281°$, $b = +3°.9$. A catalogue of the data together with identification charts is given in two appendices A and B. The magnitude limit is approximately V = 12.5. Space densities of the stars in LF 13 have been computed as a function of spectral type and distance from the sun. General luminosity functions evaluated at 100, 200, 400, and 600 parsecs do not differ substantially from the standard van Rhijn function.

155.048 The determination of the mass of the Galaxy using the density wave model of the spiral arms.
B. Basu, A. K. Roy.
Bull. Astron. Inst. Czechoslovakia, Vol. 24, 76 - 79 (1973).

The mass of the Galaxy has been calculated from a simple relation obtained from the study of the density wave model of the spiral arms of the Galaxy. The resulting mass is $2.1 \times 10^{11} M_\odot$.

155.049 On the kinematical and spatial coincidence of optical and radio spiral arms in our Galaxy.
Y. K. Minn, J. M. Greenberg.
Astron. Astrophys., Vol. 24, 393 - 404 (1973).

Radial velocity distributions of such spiral tracers as H I, H II regions, open clusters, and O associations in the outer part of our Galaxy are compared. It is found that the kinematics of the stellar objects is discrepant with that of the H I and H II regions. Rotation curves for gas and stars within 4 kpc of the sun are compared with the Schmidt rotation curve; whereas the H II gas follows the Schmidt curve, the stellar objects diverge. A plausible theoretical explanation for the velocity differences between stellar objects and their associated H II regions is given on the basis of the density-wave theory of spiral structure.

155.050 Search for infrared anomalies associated with gravitational events at the galactic centre.
R. E. Slusher, J. A. Tyson.
Nature, Vol. 243, 25 (1973).

Weber has reported bursts of gravitational radiation apparently arriving from the direction of our galactic centre. However, radio pulses noncoincident with Weber's pulses have recently been reported. Here we report the results of our search for bursts of infrared radiation from the galactic centre region.

155.051 Limit to pulses of radiofrequency emission from the galactic centre. E. Ó'Mongáin.
Nature, Phys. Sci., Vol. 242, 136 - 137 (1973).

The nucleus of our galaxy, at the position of the Sgr A radio source, has been searched for pulses of radio emission in the frequency range 1,650 to 1,720 MHz. A lack of electromagnetic pulses associated with gravitational pulses is no evidence against gravitational pulses from this or other regions; however, the limits to pulses, other than gravitational, from this and other work place interesting constraints on the rate of catastrophic events or emission from them.

155.052 Gamma-ray observations of the galactic centre.
H. F. Helmken, J. A. Hoffman.
Nature, Phys. Sci., Vol. 243, 6 - 8 (1973).

To within statistical accuracy, the present results require the presence of a Compton-scattered γ-ray flux from the galactic centre in addition to pion-decay γ-rays.

155.053 Una nueva interpretación de los recuentos estelares promedios de Seares, van Rhijn, Joyner y Richmond

1925 Ap J 62. H. Wilkens.
Bol. As. Argentina Astron., No. 16, (see 012.007), p. 32 - 34 (1971).

155.054 The effect of loop III on interstellar neutral hydrogen. I. Fejes, G. L. Verschuur.
Astron. Astrophys., Vol. 25, 85 - 91 (1973).

Observations of neutral hydrogen along loop III in the galactic region $152° < l < 167°$, $+21° < b < +31°$ are presented. An analysis of the velocity distribution of the gas shows that loop III strongly effects the motion of the gas along its continuum ridges, but not in a way which could be explained by a supernova phenomenon.

155.055 Accretion by neutron stars at the galactic center.
L. Maraschi, A. Treves, M. Tarenghi.
Astron. Astrophys., Vol. 25, 153 - 155 (1973).

The central region of the Galaxy is considered and the density of stars and interstellar matter is estimated. The power owed to accretion by neutron stars is calculated for an r.m.s. velocity of ~ 160 km/s as from the virial theorem. An upper limit for the accretion luminosity at X-ray wavelengths is proposed.

155.056 Galactic latitude dependence of the diffuse far-ultraviolet background.
C. S. Weller, G. R. Carruthers, R. C. Henry.
Bull. American Astron. Soc., Vol. 5, 38 (1973). − Abstr. AAS.

155.057 The diffuse far-ultraviolet background at 1440−1620 Å. S. Shulman, R. C. Henry.
Bull. American Astron. Soc., Vol. 5, 38 (1973). − Abstr. AAS.

155.058 Anomalous helium abundance of the galactic centre H II region G0.5−0.0.
W. K. Huchtmeier, R. A. Batchelor.
Nature, Vol. 243, 155 - 156 (1973).

We report here a search for the He 109α line from the galactic centre H II region G0.5−0.0. Our failure to detect the line sets an upper limit to the abundance of helium which is about 10% of the abundance found in many H II regions.

155.059 Neutral hydrogen in the Perseus arm of the Galaxy.
N. V. Bystrova.
Astrofiz. Issled., Izv. Spets. Astrofiz. Obs., Vol. 4, 130 - 135 (1972). In Russian.

On the basis of observations with the large Pulkovo radio telescope having a receiver band of 90 kHz the antenna temperature isophotes are constructed of neutral hydrogen in the Perseus arm of the Galaxy. The region of the sky between galactic longitudes $l = 165°$ and 240°, and latitudes $b = -10°$ and +10° was investigated.

155.060 The fine structure of the neutral hydrogen distribution in the Perseus arm of the Galaxy.
I. V. Gosachinsky.
Astrofiz. Issled., Izv. Spets. Astrofiz. Obs., Vol. 4, 136 - 142 (1972). In Russian.

It is found that in the areas investigated the Perseus arm of the Galaxy consists of a diffused background and numerous concentrations having an elongated shape along the arm.

155.061 Distance of the sun from the galactic center.
K. A. Barkhatova, T. P. Gerasimenko, D. G. Ryvkin.
Astron. Tsirk., No. 743, p. 3 - 4 (1973). In Russian.

155.062 Improvement of the galactic rotation parameters for a system of open clusters.
K. A. Barkhatova, V. M. Danilov.
Astron. Tsirk., No. 743, p. 5 - 6 (1973). In Russian.

155.063 A simple model of the Galaxy with a logarithmic density law. B. Basu, G. Saha.
Bull. Astron. Inst. Czechoslovakia, Vol. 24, 157 - 161 (1973).

A model of the Galaxy based on a logarithmic density law has been constructed. The potential and attraction in the plane of symmetry have been calculated, and the circular velocities deduced at different distances from the centre. The volume and projected densities have been obtained at different distances and the total mass has been evaluated.

155.064 Evidence for a shock associated with the local spiral arm. W. J. Quirk, R. M. Crutcher.
Astrophys. Journ., Vol. 181, 359 - 362 (1973).

The "cold cloud" of Riegel and Crutcher is probably a region of high gas density caused by a shock induced by the gravitational field of the local spiral arm, at an inclination angle of 5°–8°.

155.065 A new programme to determine the distance to the centre of the Galaxy. R. Woolley.
Monthly Notes Astron. Soc. Southern Africa, Vol. 32, 17 - 18 (1973).

An observing programme, to be carried out with the 20-inch and 40-inch reflectors at the new Observatory at Sutherland, has been initiated to determine the distance to the centre of the Galaxy by means of W Virginis stars.

155.066 Galactic three-dimensional shock waves and its effect on the formation of stars. M. Tosa.
Publ. Astron. Soc. Japan, Vol. 25, 191 - 205 (1973).

Steady flows of the interstellar gas through the spiral arms of the density wave in the Schmidt (1965) model of the Galaxy are investigated under the assumption that the thickness of the gaseous disk may change as the gas flows. Numerical calculations confirm the existence of two periodically-located shock waves propagating along and within the imposed two-armed background spiral pattern. The hydrostatic equilibrium of the gaseous disk in the z-direction and its implications to star formation are discussed and a mechanism of negative feedback to control both the equilibrium and the star formation is suggested. The breakdown of the hydrostatic equilibrium and the effect of the cosmic rays in shock waves are also discussed briefly.

155.067 Radio spurs and spiral structure of the Galaxy. I. Optical and radio brightness distributions in the Milky Way. Y. Sofue.
Publ. Astron. Soc. Japan, Vol. 25, 207 - 229 (1973).

A clear correlation is found between the positions of galactic radio spurs, optically obscured regions (dark patches) in the Milky Way, and tangential directions of the spiral arms. This correlation suggests that the spurs are physically connected with dense interstellar gas along the spiral arms of the Galaxy.

155.068 On the anisotropy in the brightness of B-type stars. V. A. Arshinova.
Uch. zap. Gor'kov. gos. ped. in-t, 1972, vyp. (No.) 124, p. 44 - 49. In Russian. – Abstr. in Referativ. Zhurn. 51. Astron., 4.51.861 (1973).

155.069 The fine-structure of the next inner spiral arm. T. Schmidt-Kaler, W. Schlosser.
Astron. Astrophys., Vol. 25, 191 - 202 (1973). In German.

Fine-structure perpendicular to the galactic plane is revealed by wide-angle photographs and photoelectric surface photometry of the Southern Milky Way in the ultraviolet. Three main filaments with an average length of around 40° and an inclination to the galactic equator of about 12° are seen. The pattern may thus be described as "shingle-like".

155.070 Updating galactic spiral structure. B. J. Bok.
American Scient., Vol. 60, 708 - 722 (1972).

155.071 Age and kinematic properties. A. Blaauw.
Stellar ages. Proc. IAU Colloquium No. 17, (see 012.015), XXII, 1 - 12 (1973).

155.072 The kinematic age of the Gould Belt. J. R. Lesh.
Stellar ages. Proc. IAU Colloquium No. 17, (see 012.015), XXIII, 1 - 7 (1973).

155.073 The velocity dispersion of young stars. A. Gomez.
Stellar ages. Proc. IAU Colloquium No. 17, (see 012.015), XXVI, 1 - 2 (1973).

155.074 Dépendance de la déviation du vertex avec l'âge des étoiles. M. Mayor.
Stellar ages. Proc. IAU Colloquium No. 17, (see 012.015), XXVII, 1 - 8 (1973).

155.075 The kinematics of K-giant stars. K. A. Janes, R. D. McClure.
Stellar ages. Proc. IAU Colloquium No. 17, (see 012.015), XXVIII, 1 - 13 (1973).

155.076 Ages of the 'SMR' (super-metal-rich) stars. P. M. Williams.
Stellar ages. Proc. IAU Colloquium No. 17, (see 012.015), LIII, 1 - 11 (1973).

155.077 Observational evidence for chemical and kinematical evolution in our Galaxy. J. B. Hearnshaw.
Stellar ages. Proc. IAU Colloquium No. 17, (see 012.015), LVI, 1 - 13 (1973).

155.078 Gaussian components of stellar velocity distributions from star counts at high galactic latitude. A. R. Upgren, J. C. Titter, R. L. Dessureau.
Bull. American Astron. Soc., Vol. 5, 267 (1973). – Abstr. AAS.

155.079 OB star distribution in Ara. R. J. Havlen.
Bull. American Astron. Soc., Vol. 5, 267 (1973). Abstr. AAS.

155.080 Galactic background spectrum between 130 kHz and 2600 kHz. L. W. Brown.
Bull. American Astron. Soc., Vol. 5, 283 - 284 (1973). Abstr. AAS.

155.081 327 MHz observations of the galactic center: Possible detection of a deuterium absorption line. D. A. Cesarsky, A. T. Moffet, J. M. Pasachoff.
Bull. American Astron. Soc., Vol. 5, 284 (1973). – Abstr. AAS.

155.082 Diffuse galactic light and the albedo of interstellar dust in the 1500 Å to 4250 Å region. A. N. Witt, C. F. Lillie.
Astron. Astrophys., Vol. 25, 397 - 404 (1973).

The surface brightness of the night sky has been measured with stellar photometers of the Orbiting Astronomical Observatory OAO–2 in 29 of Kapteyn's Selected Areas at ten wavelengths between 1500 Å and 4250 Å. The residuals are positive and show a pronounced increase in magnitude with decreasing galactic latitude and are interpreted as diffuse galactic light. A comparison of the results with existing models for the radiative transfer of diffuse galactic light suggests a wavelength dependent albedo of the interstellar grains with a pronounced minimum around 2200 Å and a rapid increase towards values near unity at wavelengths shortward of 2000 Å.

155.083 A new method for distance estimates in the galactic plane. J. H. Cahn.
Astron. Astrophys., Vol. 25, 477 - 479 (1973).

A procedure for estimating distances near the galactic plane is proposed, following the original work of Crézé (1972). The distances obtained depend upon independently determined extinctions and distances of planetary nebulae. For a given object, color excesses and celestial coordinates must be observed to estimate distance. More accurate distance can be inferred if the distance moduli, $m_v - M_v$, are observed.

155.084 Structure and dynamics of the galactic system. A report. S. W. McCuskey.
Warner and Swasey Obs., Case Western Reserve Univ., East Cleveland, Ohio, U.S.A. 161 pp. (1973). — This report of IAU Commission 33 is a compilation of information from several authors. The topical order follows closely the order in the shorter version of the report which will appear in IAU Trans., Vol. 15 A. Contents: I. Introduction; II. Basic data and calibration problems; III. Local galactic structure; IV. Overall structure of the Galaxy; V. Kinematics; VI. Dynamics.

155.085 Le popolazioni stellari e l'evoluzione della Galassia. V. Castellani.
Mem. Soc. Astron. Italiana, Nuova Ser., Vol. 43, 687 - 690 (1973).

It is shown that observational behaviours of high-velocity stars as well as supermetallicity of some old cluster contradict the classical hypothesis on the formation of the halo population. A picture is proposed in which high velocity-stars, disk and spiral arms form a family historically connected. In this context globular clusters are shown to be a decoupled family of objects, likely originating from the galactic nucleus, via explosive processes.

155.086 HNCO in the galactic centre. D. Buhl, L. E. Snyder, P. R. Schwartz, J. Edrich.
Nature, Vol. 243, 513 - 514 (1973).

The $1_{01} - 0_{00}$ transition of the HNCO molecule at 21.9817 GHz has been detected using a 1.4 cm parametric amplifier at the 36 foot telescope of the US National Radio Astronomy Observatory (Buhl et al., 1972). The present observations were made with the same receiver and the 85-foot Maryland Point telescope of the US Naval Research Laboratory. A contour map of the region around the galactic centre source Sgr B2 was made.

155.087 Analysis of the neutral hydrogen within 1 kpc of the sun. A. P. Henderson.
Astron. Journ., Vol. 78, 381 - 386 (1973).

Neutral hydrogen about the LSR (Local Standard of Rest) is assumed to approximate a plane-parallel layer. Line profiles $-10° \geqslant l \geqslant 250°$) are analyzed using differential motion equations and small nonuniform circular effects are found. Eight latitude positions, symmetric with respect to the plane, yield small velocities (< 3 km/sec) which are found to be functions of z-position relative to the plane. The result is a "shearing" effect whereby the gas above and below the plane has a slightly different velocity from the gas closer to the plane.

155.088 Resultate einer photographischen Flächenphotometrie der südlichen Milchstraße.
W. Schlosser, T. Schmidt-Kaler.
Mitt. Astron. Ges., No. 32, p. 263 - 264 (1973). — Abstract.

155.089 Giant loops as fossil Strömgren spheres: Their radio and X-ray emission. M. C. Kafatos, P. Morrison.
Astron. Astrophys., Vol. 26, 71 - 77 (1973).

The loops are examined as radio and soft X-ray sources under the assumption that they are objects related to the Gum nebula but much older than it. The loop diameters are

likely on the scale of 100 pc or more, as reported recently. This is naturally understood with the fossil Strömgren sphere model.

155.090 Magneto-gravitational and thermal instability in the galactic disk. S. Ames.
Astrophys. Journ., Vol. 182, 387 - 404 (1973).

The stability of a system containing thermal gas, cosmic rays, a magnetic field, and a gravitational field, which is used as a model for the disk of the Galaxy, is investigated. The ionizing and heating effects of the cosmic rays on the thermal gas are included, as is the thermal response of the gas, thus allowing the possibility of thermal instability assisting in the gas concentrations which can develop by an adiabatic magneto-gravitational instability.

155.091 The development of our insight into the structure of the Galaxy between 1920 and 1940. J. H. Oort.
Ann. New York Acad. Sci., Vol. 198, (see 012.020), 255 - 266 (1972).

155.092 Remarks on the soft X-ray emission from the galactic radio spurs. P. Raychaudhuri.
Astrophys. Space Sci., Vol. 19, 145 - 149 (1972).

It is pointed out that all old supernova remnants are not in general sources of soft X-ray emission. Again it is pointed out that the galactic radio spur (Cetus arc) may be an old supernova remnant, but it has already ceased to be a source of X-ray emission. Finally the X-ray flux from Vela is estimated from the cooling rate of neutron star by neutrino emission.

155.093 On the spiral structure of our Galaxy.
L. S. Marochnik, Yu. N. Mishurov. A. A. Suchkov.
Astrophys. Space Sci., Vol. 19, 285 - 292 (1972).

It is shown that properties of the spiral wave in the Galaxy are determined by the mass distribution of its flat subsystem rather than by the full mass distribution. The hypothesis is suggested which associates the generating mechanism of spiral waves with the rotating bar of old stars in the center of the Galaxy.

155.094 The galactic wind. M. Sroczyńska.
Urania Kraków, Vol. 44, 138 - 144 (1973). In Polish.

155.095 A model for peaking of galactic gravitational radiation. G. A. Campbell, R. A. Matzner.
Journ. Math. Phys., New York, Vol. 14, 1 - 6 (1973).

Geometrical optics is used to calculate the radiation pattern from a source in orbit in a strong gravitational field. No specific mechanism is postulated for the radiation itself, and only the field's effect on the radiation enters. The model proposes a 'black hole' at the galactic center.

155.096 The spiral structure of the Milky Way. B. Balazs.
Fiz. Szemle, (Hungary), Vol. 22, 372 - 377 (1972). In Hungarian.

155.097 Searches for pulses of electromagnetic radiation from the galactic centre. J. V. Jelley.
General Relativity and Gravitation, (GB), Vol. 4, 23 - 27 (1973).

155.098 The puzzle of galactic deuterium. S. Mitton.
New Scient., (GB), Vol. 57, 537 - 538 (1973).

Within the past two months radio astronomers have discovered substantial concentrations of deuterium in the Milky Way. Since this second lightest element is destroyed inside stars its unexpected presence poses a number of intriguing problems.

155.099 Luminosity and velocity distribution of high-luminosity red stars near the sun. I. The very young disk
population.   O. J. Eggen.
Publ. Astron. Soc. Pacific, Vol. 85, 289 - 306 (1973).
    The luminosities and space motions of the very young
disk-population red stars near the sun are discussed on the
basis of *UBVRI* photometry and accurate apparent motions.
The sample includes all stars brighter than $V_E = 5^m0$ as well
as some objects apparently fainter because of interstellar reddening. The FK4 system of proper motions, with corrections
to Newcomb's precessional constant, is found to be preferred
to the N 30 system. Very young disk stars are defined by the
velocity distribution of the early B-type stars and about one-
third of the sample of luminous red stars in this population
belong to the Pleiades group.

155.100 Structure of the Galaxy.
    O. J. Eggen, K. C. Freeman, A. W. Rodgers.
Rep. Prog. Phys., Vol. 36, 625 - 694 (1973).
    We have summarized the present knowledge of the stellar
content of our Galaxy and discussed at some length the use of
this knowledge in creating a dynamical framework that can be
used in the attempt to understand galaxies in general.

    Radio recombination lines from H° regions.
See Abstr. 022.078.

    On the origin of deuterium.   See Abstr. 061.012.

    ^{187}Re, recycling *r*-process elements through stars,
and the age of the Galaxy.   See Abstr. 065.076.

    On metal rich and SMR (*super-metal-rich*) stars.
See Abstr. 065.135.

    Secular parallaxes of stars and solar velocity in
space derived from absolute proper motions of 14600 stars
relative to galaxies.   See Abstr. 111.007.

    Three-colour photometry of a field in the galactic
anticentre section near NGC 1664.   See Abstr. 113.002.

    Three-colour photometry in a field in the direction
of the galactic anticentre near M 35.   See Abstr. 113.003.

    Faint O−B3 stars in the Centaurus section of the
Milky Way.   See Abstr. 113.017.

    Balloon observations of galactic and extragalactic
objects at 100 microns.   See Abstr. 114.088.

    Infrarotastronomie.
See Abstr. 114.149.

    Luminosity functions for K giant stars derived from
the two-micron sky survey.   See Abstr. 115.006.

    Wide moving paris among K and M dwarfs.
See Abstr. 117.006.

    On the problem of large-scale peculiar motion of
interstellar gas in the Galaxy.   See Abstr. 131.012.

    A new correlation among high-velocity clouds in the
Galaxy.   See Abstr. 131.039.

    Results of Fabry-Perot Hα spectrometer observa-
tions of galactic H II regions.   See Abstr. 131.060.

    Polarizations of the infrared stars and the galactic
centre in near infrared.   See Abstr. 131.098.

    Interstellar extinction in the Southern Milky Way.
See Abstr. 131.126.

    A discussion of the distribution of interstellar mat-
ter close to the sun.   See Abstr. 131.141.

    Cold neutral hydrogen in a large region towards the
galactic anticenter.   See Abstr. 131.148.

    Observations of new galactic H II regions and ex-
citing stars.   See Abstr. 131.150.

    Interstellarer Raum und interstellare Materie. Glo-
bale Eigenschaften in unserem Milchstraßensystem.
See Abstr. 131.158.

    The absence of formaldehyde radiation toward cold
regions of the galactic plane: Further investigation.
See Abstr. 131.160.

    Interstellar absorption of light in the Local System
belt.   See Abstr. 131.210.

    Catalogue of nebulae in Crux, Centaurus, Circinus
and Norma.   See Abstr. 132.032.

    The distribution of X-ray sources in our Galaxy.
See Abstr. 142.017.

    Diffuse cosmic gamma rays: Present status of theory
and observations.   See Abstr. 142.030.

    The number-intensity distribution of X-ray sources
observed by *Uhuru*.   See Abstr. 142.097.

    Possible correlation between the soft X-ray flux and
features in the radio continuum in the anticentre.
See Abstr. 142.108.

    Description of small-scale fluctuations in the diffuse
X-ray background.   See Abstr. 142.123.

    The shape of the diffuse cosmic X-ray spectrum.
See Abstr. 142.147.

    Collimator corrections to the measured diffuse
X-ray background.   See Abstr. 142.148.

    Mean path length of high energy galactic cosmic
rays in the galactic disk.   See Abstr. 143.009.

    Influence of energy-dependent escape from the
Galaxy on cosmic electrons.   See Abstr. 143.051.

    The density-wave theory of galactic spirals.
See Abstr. 151.011.

    The young galactic cluster NGC 3766.
See Abstr. 153.009.

    A statistical investigation of neutral hydrogen line
profiles.   See Abstr. 157.002.

    A 5-GHz survey of the galactic center region with
a resolution of 4 arc min.   See Abstr. 157.004.

    An atlas of galactic neutral hydrogen for the region
$270° \leq l \leq 310°; -7° \leq b \leq 2°$.   See Abstr. 157.008.

    The Berkeley low-latitude survey of neutral hydro-
gen. Part I. Profiles.   See Abstr. 157.009.

Continuum radio structure of the galactic disk.
See Abstr. 157.012.

Die Spiralstruktur der Sternsysteme. I.
See Abstr. 158.100.

155.901 **Erratum: 'A survey of linear polarization at 1415 MHz. IV. Discussion of the results for the galactic spurs '** [Astron. Astrophys., Vol. 21, 61 - 84 (1972)].
T. A. T. Spoelstra.
Astron. Astrophys., Vol. 24, 161 (1973).

156 Galactic Magnetic Field

156.001 **A stochastic model of the galactic magnetic field.**
T. B. Kaiser.
Astrophys. Journ., Vol. 181, 349 - 357 (1973).
    Existing stochastic models of the galactic magnetic field are considered and found to suffer certain defects. A new model is proposed which overcomes faults of the previous theories while retaining their strengths.

156.002 **A photon rest mass and electric currents in the Galaxy.**   J. C. Byrne, R. R. Burman.
Journ. Phys. A, General Phys., Vol. 6, L10 - L14 (1973).
    Following an idea of Goldhaber and Nieto, conditions for the existence and stability of the currents which are required to support the observed galactic magnetic field are used to obtain an upper limit for the photon rest mass.

156.003 **Galactic magnetic field irregularities and their effect** on cosmic ray propagation at energies above 10^{17} eV.
J. L. Osborne, E. Roberts, A. W. Wolfendale.
Journ. Phys. A, General Phys., Vol. 6, 421 - 433 (1973).
    Astronomical data have been examined with regard to the characteristics of the irregular component of the magnetic field in the Galaxy. A model has been formulated for the irregular field, and cosmic ray trajectories have been traced through the aggregate field.

Pulsar rotation measures and the galactic magnetic field.   See Abstr. 141.530.

Galactic loops as supernova remnants in the local galactic magnetic field.   See Abstr. 155.024.

Galactic magnetic fields: cellular or filamentary structure?   See Abstr. 158.028.

157 Galactic Radio Radiation

157.001 A survey of the radio background at 38 MHz.
J. Milogradov-Turin, F. G. Smith.
Monthly Notices Roy. Astron. Soc., Vol. 161, 269 - 279 (1973).

The Jodrell Bank Mark I radio telescope has been used to survey the northern sky at a frequency of 38 MHz, with a beamwidth of 7.5°. The map is intended for comparisons of the spectral indices of various features of galactic radio emission.

157.002 A statistical investigation of neutral hydrogen line profiles. P. L. Baker.
Astron. Astrophys., Vol. 23, 81 - 92 (1973).

Statistical parameters for two ensembles of observed hydrogen line profiles have been derived and used to describe the turbulent state of the interstellar medium. We find that the scale size of the turbulence is about 7 parsecs. Thermal and turbulent broadening can be separated; the derived gas temperature is 5300° in the broad local component and 590° in the anticenter direction at $b = -24°$. The Mach number of the turbulence is of the order of unity. At $l = 87.5°$, $b = 24.5°$, where the line profile is asymmetric, the turbulence is anisotropic; the larger scale sizes move predominantly inward indicating that the turbulence may arise in a flow or streaming motion.

157.003 Calibration profiles for observations in the 21 cm line. W. G. L. Pöppel, E. R. Vieira.
Astron. Astrophys., Suppl. Ser., Vol. 9, 289 - 295 = Argentine—Carnegie Radio Astron. Station Inst. Argentino Radioastron., Contr. No. 32 (1973).

For a set of 5 calibration points on the sky, a large number of hydrogen-line observations have been obtained. Since the statistical noise of the resulting average profiles is relatively low, they seem adequate to be used by other observers working with similar resolutions in angle and frequency.

157.004 A 5-GHz survey of the galactic center region with a resolution of 4 arc min.
J. B. Whiteoak, F. F. Gardner.
Astrophys. Letters, Vol. 13, 205 - 207 (1973).

A 5-GHz survey with a beamwidth of 4.1 arc min has been made of the galactic center region between RA $17^h 37^m$ and $17^h 52^m$, and Dec $-30°$ and $-27° 30'$. H 110α recombination-line observations were obtained at the eight peak positions listed previously by Downes and Maxwell. No line emission was detected at three of the positions, including Sgr A.

157.005 The galactic radio spectrum between 130 and 2600 kHz. L. W. Brown.
Astrophys. Journ., Vol. 180, 359 - 370 (1973).

The IMP-6 radio astronomy experiment has provided new measurements of the galactic background spectrum at 22 frequencies between 130 and 2600 kHz. A highly accurate spectrum is presented which corresponds to the minimum galactic radiation observed with a short dipole antenna. The estimated maximum spectrum is presented also.

157.006 Galactic plane radiation at 2.7 GHz. N. J. Keen.
Astron. Astrophys., Vol. 24, 299 - 301 (1973).

Stable receiver and atmospheric conditions have permitted measurements of the unresolved component of the galactic plane at 2.7 GHz. A value of $\beta = 2.85$ is obtained for the spectral index of the brightness temperatures between 820 MHz and 2.7 GHz.

157.007 Observaciones de la Galaxia a bajas latitudes en la
línea de 21-cm. W. G. L. Pöppel, E. R. Vieira.
Bol. As. Argentina Astron., No. 16, (see 012.007), p. 14 - 15 (1971).

157.008 An atlas of galactic neutral hydrogen for the region $270° \leq l \leq 310°$; $-7° \leq b \leq 2°$. S. L. Garzoli.
Carnegie Instn. Washington Publ. No. 629, 123 pp. Price $ 4.00 (1972). In Spanish and English.

We present in this work diagrams of the distribution of neutral hydrogen obtained in the region $270° \leq l \leq 310°$; $-7° \leq b \leq 2°$, complementing the existing observations. We also show superpositions of profiles from various surveys for comparison. The method of obtaining these diagrams, as well as the reduction techniques used for the observations, details of the characteristics of the equipment used, and a comparison with those employed at other observatories, is explained.

157.009 The Berkeley low-latitude survey of neutral hydrogen. Part I. Profiles.
H. Weaver, D. R. W. Williams.
Astron. Astrophys., Suppl. Ser., Vol. 8, 1 - 503 (1973).

We here present 38,961 HI profiles covering the galactic zone $-10°$ to $+10°$ between the longitude limits $l = 10°$ to 250°. We discuss in some detail the observing program, the accuracy of telescope pointing and receiver calibration, and the principles of data reduction.

157.010 Studies of four regions for use as standards in 21-cm observations. D. R. W. Williams.
Astron. Astrophys., Suppl. Ser., Vol. 8, 505 - 516 (1973).

Four 21-cm regions including three I.A.U. recommended standards have been studied for suitability as standard regions. Maps are presented of the regions of both peak temperatures and integrated intensity. Correction curves are given to convert the measured antenna temperature into the more meaningful brightness temperature. These curves have been derived using detailed measurements of the 85-foot telescope antenna pattern and assuming source brightness distributions of simple form.

157.011 136-MHz/400-MHz radio-sky maps.
R. E. Taylor.
Proc. IEEE, Vol. 61, 469 - 472 (1973).

Detailed sky maps for the 136-and 400-MHz space research (space-to-earth) satellite frequency bands have been generated, using a computer, by NASA.

157.012 Continuum radio structure of the galactic disk.
R. M. Price.
Bull. American Astron. Soc., Vol. 5, 285 (1973). — Abstr. AAS.

157.013 A survey of neutral hydrogen in low galactic latitudes between $l = 356°$ and $24°$. II. Atlas of 21 cm hydrogen-line profiles and contour maps.
S. C. Simonson III, R. Sancisi.
Astron. Astrophys., Suppl. Ser., Vol. 10, 283 - 364 (1973).

Observations in the 21-cm line of neutral hydrogen were made with the Dwingeloo 25-m radio telescope at a resolution of 16 kHz on a 0°.5 grid over the region $l = 356°$ to $2°$, $b = 0°$ to 5°; $l = 2°$ to 10°; $b = 0°$; and $l = 10°$ to 24°, $b = -8°$ or $-6°$ to $+5°$. The observations are presented individually as hydrogen-line profiles (part IX of the "Dwingeloo Atlas of 21 cm profiles") and collectively as contour maps of brightness temperature in (l, V) and (b, V) coordinates.

A reappraisal of several extensive phenomena of the high galactic latitudes.   See Abstr. 131.120.

Kosmische Radioquellen (Cosmic radio sources). See Abstr. 141.138.

Possible correlation between the soft X-ray flux and features in the radio continuum in the anticentre. See Abstr. 142.108.

On the problem of background radio emission of the Galaxy.   See Abstr. 155.028.

Radio spurs and spiral structure of the Galaxy. I.

Optical and radio brightness distributions in the Milky Way. See Abstr. 155.067.

Errata

157.901   Errata: 'A longitude survey of radio recombination lines from the diffuse interstellar medium' [Astrophys. Journ., Vol. 176, 587 - 596 (1972)]. M. A. Gordon, T. Cato. Astrophys. Journ., Vol. 182, 649 (1973).

157.902   Errata: 'The latitude extent of diffuse ionization in the Galaxy' [Astrophys. Journ., Vol. 178, 119 - 124 (1972)].   M. A. Gordon, R. L. Brown, S. T. Gottesman. Astrophys. Journ., Vol. 182, 649 (1973).

158 Single und Multiple Galaxies

158.001 Aperture synthesis observations of Maffei 2 at λ 21 cm. M. C. H. Wright, G. A. Seielstad.
Astrophys. Letters, Vol. 13, 1 - 7 (1973).

Maffei 2 is established as rotating early-type spiral galaxy with a total mass exceeding $2.5 \times 10^9 D[\text{Mpc}] M_\odot$ and a fractional hydrogen content less than $0.015 D$, where D is the distance to the galaxy. The continuum emission has nuclear, disk, and halo components.

158.002 Photoelectric spectrophotometry of Markarian 205 and a nearby suspected radio source.
J. B. Oke, H. M. Tovmassian.
Astrophys. Letters, Vol. 13, 9 - 11 (1973).

A star-like object at the position of a radio source near Markarian 205 is found to be a late-type star. Markarian 205 is a very luminous Seyfert galaxy which has a spectrum almost identical with that of the Markarian galaxy 79.

158.003 The distribution of neutral hydrogen and the velocity field of the galaxy NGC 3109. W. Huchtmeier.
Astron. Astrophys., Vol. 22, 27 - 31 (1973).

The irregular-type galaxy NGC 3109 has been observed in the 21 cm hydrogen line with the Nançay radio telescope. The H I content is found to be $22 \times 10^9 \, M_\odot$, or 20% of the total mass of $1.1 \times 10^{10} \, M_\odot$. The H I has a flat-topped distribution, being reasonably well represented by a gaussian function. Parameters of the H I distribution and the rotation curve are deduced by a model-fitting procedure.

158.004 A neutral hydrogen survey of the galaxy M33. II. Distribution and kinematics of the neutral hydrogen.
W. Huchtmeier.
Astron. Astrophys., Vol. 22, 91 - 109 (1973).

The nearby Sc-type galaxy NGC 598 (M33) has been observed within the 21-cm line of neutral atomic hydrogen with the 200-m Nançay transit telescope of Meudon Observatory. The distribution of neutral hydrogen and the kinematical properties of the galaxy have been derived.

158.005 Aperture synthesis study of neutral hydrogen in the galaxies NGC 6946 and IC 342.
D. H. Rogstad, G. S. Shostak, A. H. Rots.
Astron. Astrophys., Vol. 22, 111 - 119 (1973).

Synthesis observations of neutral hydrogen emission in the two spiral galaxies NGC 6946 and IC 342 are presented. Results are presented in the form of contour maps of the integral and mean velocity of the HI profile at each position in the galaxy, as well as a line profile for each galaxy as a whole.

158.006 The distribution of the mass-to-luminosity ratio of spiral and irregular galaxies.
R. Y. Chiao, M. Reinhardt.
Astron. Astrophys., Vol. 22, 257 - 264 (1973).

The ratio of total mass to photographic luminosity M/L of spiral and irregular galaxies contained in the compilation of Roberts (1972) is analyzed statistically. The properties of the distribution of M/L are investigated in detail, and an analytic fit is given.

158.007 Neutral hydrogen in Markarian galaxies.
L. Bottinelli, L. Gouguenheim, J. Heidmann.
Astron. Astrophys., Vol. 22, 281 - 287 (1973).

The 21-cm neutral hydrogen line has been measured for the first time in eleven non-Seyfert Markarian galaxies with the Nançay radio telescope. The data (radial velocity, hydrogen flux and velocity dispersion) are presented. The integral properties determined from the 21-cm line together with the optical properties are given and discussed. Individual galaxies are commented upon.

158.008 Velocity dispersions in galaxies. II. The ellipticals NGC 1889, 3115, 4473, and 4494.
D. C. Morton, R. A. Chevalier.
Astrophys. Journ., Vol. 179, 55 - 68 (1973).

Coudé spectra with 0.9 Å resolution between 4185 and 4415 Å were obtained from the galaxies NGC 1889 (E0), 3115 (E7/S0), 4473 (E5), and 4494 (E1) with an integrating television sensor. Comparison with spectra of G5 III, G8 III, and K0 III stars broadened by various Gaussian velocity profiles showed that these galaxies have velocity dispersions in their nuclei of $\sigma = 110 \pm 20$, 215 ± 20, 160 ± 30, and 160 ± 20, respectively. Rotation curves were derived for the inner parts of NGC 3115 and 4473.

158.009 Hα fluorescence in the filaments of M82.
D. Van Blerkom, J. I. Castor, L. H. Auer.
Astrophys. Journ., Vol. 179, 85 - 88 (1973) = Contr. Five College Obs., Univ. Mass., *Amherst*, No. 145 (1973).

Polarization of Hα radiation from the filaments of M82 has previously been attributed to scattering off dust of light emitted by a central source. It is difficult, however, to find a physically plausible velocity distribution of the scattering material to account for the observed line displacements in the filaments. The simplest interpretation, that of matter ejected by a single explosive event, requires the Hα emission to be intrinsic to the filaments. We propose that Hα is produced in the filaments by fluorescence.

158.010 A reassessment of the structure of the Seyfert galaxy NGC 4151. R. D. Davies.
Monthly Notices Roy. Astron. Soc., Vol. 161, 25P - 30P (1973).

Twenty-one centimetre neutral hydrogen measurements of NGC 4151 show that the position angle of its major axis is $26° \pm 3°$ rather than the generally adopted value of $\sim 130°$ derived from the brighter isophotes. Neutral hydrogen extends $\sim 9'$ arc along the major axis and has a normal surface density for a Sab galaxy in the regions where the outer optical spiral arms are extremely faint. Some implications of these results are discussed.

158.011 A search for hard X-rays from extragalactic objects.
J. G. Laros, J. L. Matteson, R. M. Pelling.
Astrophys. Journ., Vol. 179, 375 - 380 (1973).

Three known extragalactic sources — M87, M31, and 3C 273 — were studied in the 20–300 keV range from balloons in 1970. With the possible exception of M87, a null result was obtained in each case. To reconcile the M87 measurement of $(4 \pm 2) \times 10^{-10}$ ergs cm^{-2} s^{-1} between 19 and 50 keV with the other X-ray data requires time variability or a two-component emission model.

158.012 Observations of mass motions in Markarian 78.
T. F. Adams.
Astrophys. Journ., Vol. 179, 417 - 422 (1973).

Spectroscopic observations are presented showing the presence of emitting clouds in the inner regions of Markarian 78. These clouds are similar to the "nuclear clouds" found by Walker in the Seyfert galaxies NGC 1068 and NGC 4151. Interference-filter photographs are also presented showing Markarian 78 to be enveloped in a network of filaments similar to those seen in "exploding" galaxies. An attempt is made to understand the outlying filaments as remnants of a previous explosion ionized by nonthermal radiation from the nucleus.

158.013 **Diameters of H II regions in M31 and comparison with the largest regions in M33.**
H. Arp, F. Brueckel.
Astrophys. Journ., Vol. 179, 445 - 451 (1973).

In order to obtain diameters of H II regions which are comparable to those that are measured in more distant Sb spirals, small-scale plates in Hα wavelengths have been obtained for M31. It is found that H II regions with low and medium surface brightness mostly range from less than 7 to about 40 arc sec in diameter at the distance of M31. High-surface-brightness H II regions show a wider distribution of sizes, ranging from less than 7 to about 70 arc sec in apparent diameter. Observations carried out with exactly the same plate-filter and exposure combinations on M33 show differences in the structure of the average H II region in M31 and M33.

158.014 **Neutral hydrogen at two Holmberg radii from M33.**
M. C. H. Wright.
Astrophys. Journ., Vol. 179, 453 - 460 (1973).

Neutral hydrogen emission has been detected at two Holmberg radii from M33 in the region of a continuum radio source 4C 29.3. Interferometric measurements show that there is no absorption to an rms limit of 3 percent of the flux of the source. The emission is probably an extension of the H I in the plane of M33.

158.015 **3C 299: a faint radio galaxy of intermediate redshift.** H. Spinrad, H. E. Smith.
Astrophys. Journ., (*Letters*), Vol. 179, L71 - L72 (1973).

3C 299 is optically identified with a double-nucleus, 18.4 (red) magnitude radio galaxy. It has a high excitation emission line spectrum which yields $z = 0.367$.

158.016 **Spectrophotometry of Ton 524a, b.**
L. B. Robinson, E. J. Wampler.
Astrophys. Journ., (*Letters*), Vol. 179, L79 - L82 = Contr. Lick Obs., No. 388 (1973).

Spectra of Ton 524a, b show that these two compact emission-line galaxies have nearly identical redshifts. This is the first pair of Seyfert-like galaxies that appear to form a physical pair.

158.017 **Redshifts of a BSO and galaxies in the vicinity of the radio source RN 8.**
J. S. Miller, L. B. Robinson, E. J. Wampler.
Astrophys. Journ., (*Letters*), Vol. 179, L83 - L87 = Contr. Lick Obs., No. 389 (1973).

Spectrograms of a blue stellar object and five nearby galaxies in the vicinity of the radio source RN 8 have been obtained with the 120-inch Lick reflector and image tube scanner. The BSO has a spectrum similar to Seyfert nuclei, and the measurement of five emission lines yielded a redshift $z = 0.184$. Three of the galaxies measured have redshifts very similar to that of the BSO. A fourth galaxy appears to have a much lower redshift, while the fifth galaxy observed did not show any features suitable for a redshift determination.

158.018 **The broad component of Hα in the Seyfert galaxy NGC 5548.** R. L. Ptak, R. E. Stoner.
Astrophys. Journ., (*Letters*), Vol. 179, L89 - L92 (1973).

The broad component of the Hα emission from NGC 5548 is in excellent agreement with a profile calculated from a new model for the source of this radiation. In the model, protons are ejected with energy greater than 200 keV from a small source and subsequently stream outward through uniformly dense, partially ionized hydrogen gas.

158.019 **Optical polarization in the nuclei of E galaxies.**
D. S. Heeschen.
Astrophys. Journ., (*Letters*), Vol. 179, L93 - L96 (1973).

Observations have been made of optical polarization in the nuclei of several giant elliptical galaxies.

158.020 **Optical polarization in the nucleus of M87.**
T. D. Kinman.
Astrophys. Journ., (*Letters*), Vol. 179, L97 - L99 (1973).

Heeschen's discovery of significant linear polarization in the ultraviolet radiation of the nucleus of M87 (NGC 4486) is confirmed. There is evidence that this polarization is variable.

158.021 **Continuum radio emission from NGC 4656/7 and NGC 891 at 408 MHz.**
J. E. Baldwin, G. G. Pooley.
Monthly Notices Roy. Astron. Soc., Vol. 161, 127 - 132 (1973).

New radio continuum observations of NGC 4656/7 and NGC 891 are described and the results compared with the radio discs of other spiral galaxies.

158.022 **A ring in a galaxy.** A. J. Penny, A. P. Fairall.
Observatory, Vol. 93, 27 - 28 (1973).

The purpose of this note is to draw attention to a compact object which contains a ring. It is VII Zw 8 [α = $03^h 32^m 4$, δ = $+72° 24'(1950)$].

158.023 **A discussion of the new variations observed in the nucleus of the Seyfert galaxy NGC 3516.**
S. Collin-Souffrin, D. Alloin, Y. Andrillat.
Astron. Astrophys., Vol. 22, 343 - 354 (1973).

By comparing new spectroscopic observations (1970–1971) with previous observations (1967) performed in the same conditions, we essentially deduce: (1) an enhancement of the Balmer lines (the intensity is about four times larger); (2) a decrease of the intensity of the [O III] line λ 4363; (3) no variations of the continuum in the range 4000–6000 Å. An attempt is made to explain such variations over a period of three years at most.

158.024 **Models for extragalactic objects with very high IR and X-ray luminosity.** J. Bergeron, E. E. Salpeter.
Astron. Astrophys., Vol. 22, 385 - 406 (1973).

We consider cases where the observed X-ray luminosity $L_{X, ob}$ is higher than (or about equal to) the (thermal) optical luminosity of the compact region (or the galaxy). We investigate simple models which attempt to fit the observed data on both the X-ray and IR emission. In one model the IR represents synchrotron radiation of relativistic electrons in a magnetic field and in other model thermal radiation by dust grains.

158.025 **A neutral hydrogen study of the Scd-galaxy NGC 4244.** W. Huchtmeier.
Astron. Astrophys., Vol. 23, 93 - 96 (1973).

The late-type (Scd) galaxy NGC 4244 has been observed in the 21 cm hydrogen line with the Nançay radio telescope using frequency resolutions of 330 and 60 kHz. Six per cent $(7 \times 10^9 M_\odot)$ of the total mass of $1.2 \times 10^{11} M_\odot$ is in form of neutral hydrogen. An asymmetry in the mass distribution of H I was found; there is 20 % more H I to the east of the optical center. No deviations from circular motions were noted.

158.026 **Neutral hydrogen study of 21 small angular diameter galaxies.**
C. Balkowski, L. Bottinelli, L. Gouguenheim, J. Heidmann.
Astron. Astrophys., Vol. 23, 139 - 143 (1973).

Neutral hydrogen contents, systemic radial velocities and indicative total masses of 21 galaxies with relatively small angular diameters from Sb to irregular types have been measured with the Nançay radiotelescope. Determination of systemic radial velocities have been made for 5 galaxies with previously unknown velocities; their group membership is confirmed, except for one of them. The effective neutral hydrogen diameters are determined for 6 galaxies.

158.027 Variations in spectral-energy distributions and absorption-line strengths among elliptical galaxies.
S. M. Faber.
Astrophys. Journ., Vol. 179, 731 - 754 (1973).

33 elliptical galaxies and 10 globular clusters have been observed on a 10-color intermediate-band photometric system designed for the study of old stellar populations. To within the accuracy of the present data, the integrated colors of elliptical galaxies can be specified with only two independent parameters which in turn may be utilized to determine both the intrinsic colors and the reddening between the galaxy and the observer. The intrinsic colors and line strengths are closely correlated with absolute magnitude at all luminosities.

158.028 Galactic magnetic fields: cellular or filamentary structure? F. C. Michel, A. Yahil.
Astrophys. Journ., Vol. 179, 771 - 780 (1973).

We have examined the possibility that the galactic field is composed of cells or filaments elongated by plasma motions or by the differential rotation of the galactic disk. Such cells could result from special events, and each would then contain its own independent magnetic-field structure. Our analysis of available data indicates that there is a significant correlation between the signs of the observed rotation measure between different radio sources as a function of their angular separation out to angles of about 40°. The anticorrelation expected for a uniform field component is absent. We discuss these results in terms of a cellular field model.

158.029 Absence of variations in the nucleus of Virgo A.
K. I. Kellermann, B. G. Clark, M. H. Cohen, D. B. Shaffer, J. J. Broderick, D. L. Jauncey.
Astrophys. Journ., (*Letters*), Vol. 179, L141 - L144 (1973).

Repeated observations of M87 at 3.8 cm, between 1971 February and 1972 August, show no change in intensity or size of the compact radio nucleus.

158.030 Galaxies with ultraviolet continuum. V.
B. E. Markarian, V. A. Lipovetsky.
Astrofizika, Vol. 8, 155 - 164 (1972). In Russian. English translation in Astrophysics, Vol. 8, No. 2.

The fifth list of galaxies with ultraviolet continuum is presented. The list contains observational data for 106 new objects.

158.031 On the nature of galaxies with ultraviolet continuum. I. Principal spectral and colour characteristics.
B. E. Markarian.
Astrofizika, Vol. 8, 165 - 176 (1972). In Russian. English translation in Astrophysics, Vol. 8, No. 2.

The results of spectral and photoelectric observations of galaxies with UV excess are considered. On the slit spectrograms of almost all galaxies having a strong UV continuum emission lines are present. The results of photoelectric observations show that according to the character of energy distribution in their spectra, there exists a similarity between the considered galaxies and QSOs. The overwhelming majority of the s-sd type galaxies and a small part of those of the d-ds type have condensed stellar-like nuclei. Each fourth of these objects has the Seyfert characteristics.

158.032 The spectra of Markarian galaxies. V.
M. A. Arakelian, E. A. Dibay, V. F. Yesipov (*Esipov*).
Astrofizika, Vol. 8, 177 - 186 (1972). In Russian. English translation in Astrophysics, Vol. 8, No. 2.

The results of spectral observations of 76 objects from the fourth list of galaxies with ultraviolet continuum are presented. Emission lines are detected in the spectra of 61 objects. Six objects show the spectral peculiarities of the nuclei of Seyfert galaxies.

158.033 Hydrogen lines in the spectrum of the Markarian 6 galaxy during its activity.
V. I. Pronik, K. K. Chuvaev.
Astrofizika, Vol. 8, 187 - 196 (1972). In Russian. English translation in Astrophysics, Vol. 8, No. 2.

Spectra of the Markarian 6 (IC 450) galaxy were obtained in 1970–1971 with dispersion of 335 and 100 Å/mm using an image-tube spectrograph at the 2.6 m telescope of the Crimean Astrophysical Observatory. During one year of observations the H_β flux decreased by nearly two times and the intensity of the continuum by more than three times. A strong correlation between these values has been found. On all spectra the hydrogen lines have very broad wings. The complex structure of these lines indicates the presence of separate gas clouds with different velocities.

158.034 H I observations of compact galaxies.
R. Lauqué.
Astron. Astrophys., Vol. 23, 253 - 257 (1973).

35 galaxies with velocities up to 6800 km/s, taken from Zwicky's seven lists of compact galaxies have now been observed at Nançay. From the 21-cm neutral hydrogen emission line flux detected in eight of these objects, it is shown that a high proportion of these compact galaxies with sharp emission line optical spectra are indeed hydrogen rich, though similar to late type galaxies. A possible instrumental selective effect is discussed.

158.035 The manifold of galaxies. Galaxies with known dynamical parameters. P. Brosche.
Astron. Astrophys., Vol. 23, 259 - 268 (1973).

For the sample of galaxies with known rotation curves the following independent integral properties were compiled: morphological type, photometric radius, radius and velocity of the maximum of the rotation curve, absolute luminosity, color, mass of neutral hydrogen. The main result of a component analysis is the presence of two significant dimensions in the sample. The relationships between the different variables indicate that the star formation rate is not a function of interstellar density alone.

158.036 Properties of the radio continuum emission from interacting galaxies.
R. J. Allen, R. D. Ekers, B. F. Burke, G. K. Miley.
Nature, Vol. 241, 260 - 261 (1973).

Several interacting galaxies have been observed recently in different programmes using the Westerbork Synthesis Radio Telescope at 21 cm wavelength. We give here a compilation of the data in order to examine the morphology, strength, and the frequency of occurrence of the radio emission from these interacting systems.

158.037 Optical amplitudes and radio spectra of variable galaxies. D. L. Hall, P. D. Usher.
Nature. Phys. Sci., Vol. 241, 31 - 32 (1973).

We have examined the optical record of I Zw 187 from the Harvard plate collection and published data, and find a total optical amplitude of at least 2.1 blue photographic magnitudes, which places the object within the range of variability of the other proposed members of the class (BL Lac). There is also a suggestion of an empirical relationship between optical amplitude and radio spectral index for members of the class, in the sense that more inverted (centimetre excess) radio spectra are associated with larger total optical amplitudes.

158.038 The absorbing material in the Andromeda nebula.
A. M. van Genderen.
Astron. Astrophys., Vol. 24, 47 - 51 (1973).

A discussion is presented of the distribution of the absorbing material in the Andromeda nebula, obtained with the aid of population I cepheids in Baade's four variable-star fields. The results are interpreted as indicating that in the regions investigated the dust is concentrated in the optical arms themselves rather than near their inner edges as is observed in some other spiral galaxies. The dust-to-gas ratio is found to be nearly equal to that in the Galaxy.

158.039 A high resolution neutral hydrogen study of the galaxy M 51. L. Weliachew, S. T. Gottesman.
Astron. Astrophys., Vol. 24, 59 - 67 (1973).

A 2 arc min synthesis study of the spiral galaxy M 51 (NGC 5194) has been made in the λ 21 cm emission line of neutral hydrogen. A morphological H I symmetry axis is found at a position angle of 22° while the dynamically determined major axis is found at a position angle of − 8°. The velocities in the north-east outer spiral structure are found to be strongly noncircular. A possible continuation in neutral hydrogen of the outermost south-west spiral arm has been detected. M 51 also exhibits a strong deficiency in H I surface density at its center.

158.040 Identification and radio spectra of bright galaxies in the second Bologna Catalogue of radio sources and their radio luminosity function.
R. Fanti, I. Gioia, C. Lari, J. Lequeux, R. Lucas.
Astron. Astrophys., Vol. 24, 69 - 78 (1973).

We have identified radio sources in the second Bologna Catalogue (Colla et al., 1970, 1972) with galaxies in the Reference Catalogue of Bright Galaxies (G. and A. de Vaucouleurs, 1964) and in Vol. 2 and 3 of the Catalogue of Galaxies and of Clusters of Galaxies (Zwicky and Herzog, 1963, 1966); the two samples contain respectively 25 and 40 identified galaxies. These objects have been observed at 1415 MHz at Nançay to obtain their spectral indices between 408 and 1415 MHz. Using various criteria,we confirm that there exists two classes of elliptical galaxies on the basis of their radio properties, and we derive the radio luminosity function for the galaxies with steep radio spectra.

158.041 Dependence of the integrated background light on cosmology, galactic spectra, and galactic evolution.
B. M. Tinsley.
Astron. Astrophys., Vol. 24, 89 - 98 (1973).

The theoretical spectrum and intensity of the integrated light of distant galaxies is studied as a function of uncertain parameters entering its calculation. These include the cosmological model, epochs of formation of galaxies, evolution of the magnitudes and colors of galaxies of different types, the ultraviolet radiation of galaxies, and the local luminosity density of the universe.

158.042 On the black-hole model of galactic nuclei.
C. A. Norman, D. ter Haar.
Astron. Astrophys., Vol. 24, 121 - 124 (1973).

We show that the observed large infrared emission from some galactic nuclei finds a natural explanation, if one takes plasma turbulence into account in Lynden-Bell and Rees' black-hole model of galactic nuclei.

158.043 L'Observatoire de Byurakan et les galaxies de Markarian. J. Heidmann.
L'Astronomie, 87e année, p. 101 - 103 (1973).

158.044 Radiospår i galaxhopar. L. Bååth.
Astron. Tidssk., Årg. 6, p. 18 - 24 (1973).

158.045 High-frequency radio observations of normal galaxies. W. H. McCutcheon.
Astron. Journ., Vol. 78, 18 - 25 (1973).

Observations have been made of 22 normal galaxies at a wavelength of 4.5 cm. Of the 17 detected, four were observed at 2.8 cm. Five of the galaxies have spectra which steepen

near either 1400 or 2695 MHz. Seven galaxies have straight spectra over the range 178–6630 MHz. The average low-frequency spectral index is $\bar{\alpha}_{low}$ = 0.71 ± 0.07. The average high-frequency spectral index $\bar{\alpha}_{high}$ = 0.88 ± 0.05. The irregular I and spiral galaxies show a correlation between total radio luminosity and total photographic luminosity.

158.046 The instability of the double system of galaxies NGC 7752–53. F. Bertola, S. D'Odorico.
Astrophys. Letters, Vol. 13, 161 - 164 (1973).

The rotation curve of NGC 7753, the main component of an M51-type system of galaxies, has been derived. A comparison of the radial velocity of the companion, NGC 7752, with the velocity field in NGC 7753 leads to the conclusion that the system is not bound.

158.047 Rapid variations of Hα intensity in the nuclei of Seyfert galaxies NGC 4151, 3516, 1068.
A. M. Cherepashchuk, V. M. Lyutyi.
Astrophys. Letters, Vol. 13, 165 - 168 (1973).

Rapid variations of the intensity of the emission lines Hα + [N II] in the spectra of the nuclei of Seyfert galaxies NGC 1068, 3516 and 4151 have been discovered. The time scale of these rapid variations is ~ 5–15 days and the amplitude is about 10–35 per cent.

158.048 The radio spectrum of M31.
E. M. Berkhuijsen, R. Wielebinski.
Astrophys. Letters, Vol. 13, 169 - 172 (1973).

Observations have been made of the Andromeda nebula, M31, at 2695 MHz, beamwidth 4.8 arc min, with the 100-m radio telescope of the Max-Planck-Institut für Radioastronomie. The results show that the spectra of the spiral arms and of the nucleus of M31 are non-thermal between 408 and 2695 MHz.

158.049 The spectra of Markarian galaxies, VI.
M. A. Arakelian, E. A. Dibay, V. F. Yesipov (Esipov).
Astrofizika, Vol. 8, 329 - 335 (1972). In Russian. English translation in Astrophysics, Vol. 8, No. 3.

The results of spectral observations of twenty-nine objects from the lists of galaxies with ultraviolet continuum are presented. Emission lines are detected in the spectra of twenty-five objects.

158.050 Preliminary data on the optical variability of the NGC 4486 jet.
V. I. Pronik, A. G. Scherbakov (Sherbakov).
Astrofizika, Vol. 8, 337 - 342 (1972). In Russian. English translation in Astrophysics, Vol. 8, No. 3.

Relative changes of brightness for some jet condensations of NGC 4486 were found when comparing photographs taken with the 2.6-meter telescope, and those taken by several authors during 1934–1956.

158.051 Estimates of the brightness of selected Markarian galaxies. II.
M. A. Arakelian, E. A. Dibay, V. M. Lyutiy (Lyutyj).
Astrofizika, Vol. 8, 473 - 476 (1972). In Russian. English translation in Astrophysics, Vol. 8, No. 3.

The results of UBV observations of galaxies with ultraviolet continuum from 3 Markarian lists are presented.

158.052 Optical and near-infrared observations of the nearby spiral galaxy Maffei 2. H. Spinrad, J. Bahcall, E. E. Becklin, J. E. Gunn, J. Kristian, G. Neugebauer, W. L. W. Sargent, H. Smith.
Astrophys. Journ., Vol. 180, 351 - 358 (1973).

Spectra, photographs, and photometric measurements have been used to show that Maffei 2 has a distance of 5 ± 2 Mpc and that it has a morphological type near Sbc II − in agreement with similar conclusions made earlier by radio observers. We discuss the possible relationships between Maffei 2 and the elliptical galaxy Maffei 1; there are serious inconsistencies in the existing data which bear on this question.

158.053 Anomalies in the low-frequency radio spectra of some bright galaxies.   O. B. Slee.
Proc. Astron. Soc. Australia, Vol. 2, 159 - 161 (1972).

158.054 A list of galaxies with peculiar nuclei.   J. L. Sérsic.
Publ. Astron. Soc. Pacific, Vol. 85, 103 (1973).
The purpose of this letter is to make a available to observers a list of galaxies with peculiar nuclei which the author found in the Hubble Plate Collection at the Hale Observatories in 1966.

158.055 The redshift-distance relation. IV. The composite nature of N galaxies, their Hubble diagram, and the validity of measured redshifts as distance indicators.
A. Sandage.
Astrophys. Journ., Vol. 180, 687 - 697 (1973).
N galaxies exhibit color and magnitude gradients with changing aperture size that require the presence of two components. A centrally peaked blue source superposed on an extended red component is indicated by new photometry. Calculations by two methods show the central source to have colors of a quasar, and the distended source to have colors and the radial intensity distribution of a giant E galaxy. N systems are redder than quasars but bluer than normal E galaxies. The Hubble diagram for the galaxy component alone in 12 N systems has the same slope, scatter, and zero point as the diagram for radio galaxies, requiring that redshifts of N galaxies have no component, Δz, other than the expansion redshift to within the accuracy of the test. The Hubble diagram for the companion E galaxies to 3C 303, 3C 371, and 3C 390.3, using the redshifts of their associated N galaxies, leads to the same conclusion.

158.056 Velocity dispersions in galaxies. III. The nucleus of M31.   D. C. Morton, T. X. Thuan.
Astrophys. Journ., Vol. 180, 705 - 714 (1973).
We have measured the dispersion in the velocity distribution of the stars in the nucleus of M31 using a coudé spectrum obtained with the integrating television camera. We also have investigated the nuclear rotation curve at a position angle of 38° from the major axis.

158.057 X-ray observations of NGC 5128 (Centaurus A) from UHURU.   W. Tucker, E. Kellogg,
H. Gursky, R. Giacconi, H. Tananbaum.
Astrophys. Journ., Vol. 180, 715 - 724 (1973).
We have previously reported on the existence of an X-ray source identified with the galaxy NGC 5128. We have now extended our work on NGC 5128 with the analysis of several days of production data. The most significant feature of the new observational results is the X-ray spectrum which we find to have a 3.4 keV low-energy cutoff and a slope of −0.7 for the energy index of a power-law fit. We also have a refined source location box, a 2 σ upper limit of 15' on the angular size of the X-ray source, and an improved upper limit on the X-ray emission from the extended radio lobes of Cen A. We consider implications from the identification of the X-ray source with the core of NGC 5128.

158.058 Spectroscopy of outlying faint galaxies in the region of the Coma cluster.   W. G. Tifft, S. A. Gregory.
Astrophys. Journ., Vol. 181, 15 - 17 (1973).
Redshifts are reported for 27 outlying galaxies in Coma. Twenty-four have redshifts typical for Coma galaxies, and 11 galaxies show emission lines.

158.059 Electrographic UBV photometry of the jet in M87.   H. D. Ables, G. E. Kron.
Astrophys. Journ., Vol. 181, 19 - 25 (1973).
UBV magnitudes of the jet in M87 have been measured from electronic camera exposures obtained with the 40-inch Ritchey-Chrétien telescope. The observed magnitude and color indices are $V = 15.73$, $B - V = 0.06$, and $U - B = -0.44$. These color indices are not consistent with a single synchrotron spectrum.

158.060 A tidally truncated elliptical galaxy.
I. R. King, J. Kiser.
Astrophys. Journ., Vol. 181, 27 - 30 (1973).
Surface photometry of the elliptical companion of NGC 5846 shows that it lacks an extended envelope but instead has a profile resembling that of a low-concentration star cluster. It appears to be tidally limited by its giant neighbor.

158.061 Southern galaxies. VI. Luminosity distribution in the Seyfert galaxy NGC 1566.
G. de Vaucouleurs.
Astrophys. Journ., Vol. 181, 31 - 50 (1973).
Isophotes, luminosity profiles, and photometric parameters of the brightest southern Seyfert galaxy NGC 1566, revised type SAB(s)bc, are derived from long- and short-exposure photographs taken with the Mount Stromlo 75-cm reflector and the Cordoba 152-cm reflector.

158.062 Gas motions in the nucleus of the Seyfert galaxy NGC 4151.   M.-H. Ulrich.
Astrophys. Journ., Vol. 181, 51 - 59 (1973).
This is a report of some spectrographic observations of the nucleus of the Seyfert galaxy NGC 4151 and a qualitative description of the properties of the nuclear clouds of ionized gas as can be deduced from the structure of the emission lines. The observing procedure is described, the geometry and the velocity field in the cloud complex are analyzed, and the results are discussed.

158.063 Stellar motions near the nucleus of M31.
V. C. Rubin, W. K. Ford, Jr., C. K. Kumar.
Astrophys. Journ., Vol. 181, 61 - 77 (1973).
Velocities of stars along the major and minor axes in the nuclear bulge of M31 have been measured from the λ 5269 Fraunhofer E absorption line (Fe I + Ca I). The mean uncertainty of each velocity is estimated to be 25 km s^{-1}. We can offer no realistic explanation of the complex stellar motions, or of their relation to the presumably younger gas.

158.064 A search for extragalactic objects in the General Catalog of Variable Stars.   H. E. Bond.
Astrophys. Journ., (Letters), Vol. 181, L23 - L24 = Contr. Louisiana State Univ. Obs., Baton Rouge, No. 80 (1973).
As part of a search for optically variable extragalactic objects, some 700 variables listed in the General Catalog of Variable Stars were examined on the Palomar Sky Atlas prints. Two objects, X Comae and V1102 Cygni, have the appearance of compact galaxies and, together with such previously discoverd objects as V395 Herculis and Zw 0039.5+4003, define a class of radio-quiet, optically variable galaxies.

158.065 The spectrum of the extranuclear regions of Ton 256.   J. Silk, H. E. Smith, H. Spinrad, G. B. Field.
Astrophys. Journ., (Letters), Vol. 181, L25 - L26 (1973).
Scans of the nuclear and extranuclear regions of Ton 256 are consistent with a two-component model consisting of a central quasar-like nucleus embedded in a giant elliptical galaxy. The integrated visual magnitudes of the two components are comparable both to one another and to the magnitude of the brightest galaxy in a cluster at the same redshift.

158.066 Infrared and radio observations of the nucleus of NGC 253.
E. E. Becklin, E. B. Fomalont, G. Neugebauer.
Astrophys. Journ., (*Letters*), Vol. 181, L27 - L31 (1973).
An extended nuclear core to NGC 253 has been mapped in both the infrared and radio wavelengths. The 10″ core shows a large 10- and 20-μ flux and a relatively flat radio spectrum. The infrared is conjectured to come from thermal reradiation from dust grains.

158.067 A survey of elliptical galaxies at 6 cm.
R. D. Ekers, J. A. Ekers.
Astron. Astrophys., Vol. 24, 247 - 253 (1973).
This paper reports the results of a 6 cm radio continuum survey of 191 E and S0 galaxies with known redshifts. Nineteen galaxies were detected. These were reobserved at higher resolution and could be divided into well separated classes of compact and extended sources. We discuss the possibility that the S0 galaxies detected are in fact misclassified E or D galaxies.

158.068 Photometry and some features of spiral Seyfert galaxies beyond the nucleus.
A. V. Zasov, V. M. Lyutyj.
Astron. Zhurn. Akad. Nauk SSSR, Vol. 50, 253 - 262 (1973). In Russian. English translation in Soviet Astron. AJ, Vol. 17, No. 2.
The results of electrophotometry of ten Seyfert galaxies: NGC 1068, 1275, 3227, 3516, 4051, 4151, 5548, 7469, Mrk 10 and Mrk 79 in the UBV system are analysed.

158.069 Resolution of a dwarf spheroidal companion to the Andromeda nebula. S. van den Bergh.
Bull. American Astron. Soc., Vol. 5, 5 (1973). — Abstr. AAS.

158.070 Spectral characteristics of the "red" and "blue" spiral arms in NGC 5194.
S. M. Simkin, M. L. West.
Bull. American Astron. Soc., Vol. 5, 9 (1973). — Abstr. AAS.

158.071 Integrated spectral energy distributions of galaxies.
D. C. Wells.
Bull. American Astron. Soc., Vol. 5, 26 (1973). — Abstr. AAS.

158.072 Radio observations of nearby spiral galaxies.
J. H. Spencer, B. F. Burke.
Bull. American Astron. Soc., Vol. 5, 29 (1973). — Abstr. AAS.

158.073 High-resolution radio continuum observations of bright spiral galaxies at 1415 MHz.
P. C. van der Kruit.
Bull. American Astron. Soc., Vol. 5, 30 (1973). — Abstr. AAS.

158.074 Luminosity distribution in the Seyfert galaxy NGC 1566. G. de Vaucouleurs.
Bull. American Astron. Soc., Vol. 5, 39 - 40 (1973). — Abstr. AAS.

158.075 A two-component photometric model of Seyfert galaxies.
G. de Vaucouleurs, A. de Vaucouleurs, H. G. Corwin, Jr.
Bull. American Astron. Soc., Vol. 5, 40 (1973). — Abstr. AAS.

158.076 Is NGC 1068 an infrared variable?
D. Morrison, T. Simon.
Bull. American Astron. Soc., Vol. 5, 40 (1973). — Abstr. AAS.

158.077 The interaction of gas and stars in dense galactic nuclei. D. S. De Young.
Bull. American Astron. Soc., Vol. 5, 41 (1973). — Abstr. AAS.

158.078 Données d'observation et interprétation des spectres infrarouges et microondes des galaxies et de la matière intergalactique. Rapport introductif. Y. Andrillat.
17th Colloquium International Astrophys. Liège 1971, (see 012.003), p. 547 - 569 (1972).

158.079 Le tri des atomes et des grains de poussière au voisinage d'une galaxie, l'évolution de cette galaxie, et l'observation, dans les domaines IR et radio, de son rayonnement. J.-C. Pecker.
17th Colloquium International Astrophys. Liège 1971, (see 012.003), p. 573 - 574 (1972).

158.080 A synchrotron radiation model of the infrared radiation from the nucleus of NGC 1068.
W. G. Fogarty, A. G. Pacholczyk.
17th Colloquium International Astrophys. Liège 1971, (see 012.003), p. 575 - 581 (1972).

158.081 Observational consequences of inverse Compton models for Seyfert galaxies and quasars.
J. Bergeron, E. E. Salpeter.
17th Colloquium International Astrophys. Liège 1971, (see 012.003), p. 583 (1972).

158.082 Observations of interacting galaxies at 2.2 and 4.6 cms. A. E. Wright, C. R. Purton.
17th Colloquium International Astrophys. Liège 1971, (see 012.003), p. 585 - 588 (1972).

158.083 Is the far-infrared radiation from galactic nuclei due to molecular masers?
P. M. Solomon, M. J. Rees.
17th Colloquium International Astrophys. Liège 1971, (see 012.003), p. 591 - 595 (1972).

158.084 Dust models for infrared galaxies.
N. C. Wickramasinghe.
17th Colloquium International Astrophys. Liège 1971, (see 012.003), p. 601 - 606 (1972).

158.085 Aperture synthesis study of neutral hydrogen in NGC 2403 and NGC 4236. I. Observations.
G. S. Shostak, D. H. Rogstad.
Astron. Astrophys., Vol. 24, 405 - 410 (1973).
Observational results of an aperture synthesis study of neutral hydrogen in the nearby galaxies NGC 2403 and NGC 4236 are presented. Maps of the integral of the line profile at each position on the galaxies are given, as well as maps showing lines of constant radial velocity determined from the peaks of the profiles. In addition, line profiles for the galaxies as a whole and maps of the 1420 MHz continuum radiation from each object are displayed.

158.086 Aperture synthesis study of neutral hydrogen in NGC 2403 and NGC 4236. II. Discussion.
G. S. Shostak.
Astron. Astrophys., Vol. 24, 411 - 419 (1973).
The H I synthesis results presented in the previous paper are discussed.

158.087 On the diameter-luminosity relation of galaxies.
I. S. Balinskaja, I. D. Karachentsev.
Astrometriya i Astrofizika, *Kiev*, vyp. (No.) 16, (see 003.006), p. 112 - 120 (1972). In Russian.
The regressions of M on γ and of γ on M are examined for the distributions of galaxies on absolute magnitude M and log-diameter γ.

158.088 Un estudio sobre la irregular magallánica NGC 7764.
M. G. Pastoriza, E. L. Agüero.

Bol. As. Argentina Astron., No. 16, (see 012.007), p. 3 - 4 (1971).

158.089 Búsqueda de grupos de galaxias y de galaxias peculiares.
H. A. Dottori, A. G. Samuel, J. L. Sérsic.
Bol. As. Argentina Astron., No. 16, (see 012.007), p. 4 - 6 (1971).

158.090 El objeto Fourcade-Figueroa. C. R. Fourcade.
Bol. As. Argentina Astron., No. 16, (see 012.007), p. 10 - 11 (1971).

158.091 Fotometría superficial de galaxias a dos colores.
E. L. Agüero.
Bol. As. Argentina Astron., No. 16, (see 012.007), p. 17 - 18 (1971).

158.092 Fotometría fotoeléctrica UBV de galaxias con núcleos peculiares.
M. G. Pastoriza, J. J. Clariá.
Bol. As. Argentina Astron., No. 16, (see 012.007), p. 22 - 24 (1971).

158.093 Espectro colisional de galaxias con líneas de emisión.
L. Coscia, H. Gerola.
Bol. As. Argentina Astron., No. 16, (see 012.007), p. 34 - 36 (1971).

158.094 The stellar content of bright galactic nuclei.
J. R. Baldwin, I. J. Danziger, J. A. Frogel, S. E. Persson.
Astrophys. Letters, Vol. 14, 1 - 6 (1973).

The strengths of absorption bands due to H_2O and CO at 2.1 and 2.3 μm have been measured for the nuclear regions of M31, M81, and NGC 5195. When compared with models of the stellar content of these galaxies, the observations seem to rule out the possibility of a dwarf-enriched sequence. Optical observations of M81 and NGC 4594 indicate that the stellar population and metal abundance of the nuclear regions of these galaxies are similar to those of giant ellipticals. NGC 5195, on the other hand, seems to be a metal-poor system which is reddened by dust.

158.095 Peculiar morphology of the outer regions of NGC 1265 (3C 83.1 B) and NGC 7720 (3C 465).
F. Bertola, G. C. Perola.
Astrophys. Letters, Vol. 14, 7 - 10 (1973).

Deep photographs of NGC 1265 and 7720 reveal an asymmetry in the outer regions of these two galaxies, which appears to be correlated with the peculiar geometry of the associated radio sources, 3C 83.1 B and 3C 465. The possible nature of this asymmetry, as due to stars or to diffuse matter, and its origin are briefly discussed.

158.096 The expulsion of dust from galaxies.
R. Y. Chiao, N. C. Wickramasinghe.
Astrophys. Letters, Vol. 14, 19 - 23 (1973).

We examine the conditions for the expulsion by radiation pressure of dust particles from several types of galaxy. Particles composed of graphite, iron, silicate, or ice, with sizes typical of interstellar grains, may be expelled from spiral and irregular galaxies, but not from elliptical galaxies.

158.097 The 49-cm linear polarization distribution in 3C 327 and the density of intergalactic gas.
P. P. Kronberg, J. M. Crelinsten.
Astrophys. Letters, Vol. 14, 25 - 30 (1973).

Strip distributions at wavelengths of 49.1 and 21.1 cm of linearly polarized and total radiation are presented for the radio galaxy 3C 327.

158.098 A colorimetric and spectrophotometric investigation of the irregular galaxy M 82.
B. P. Artamonov, L. S. Nazarova.
Astrofiz. Issled., Izv. Spets. Astrofiz. Obs., Vol. 4, 143 - 153 (1972). In Russian.

Photometric sections are obtained of the surface brightness of M 82 on negatives taken through narrow-band filters with reference to out-of-focus images of stars with known energy distribution in the spectrum. The energy distribution in the continuous spectrum for different regions of the galaxy is found. The power flux is calculated in the Hα line for the emission region of the galaxy. The observed energy distribution in the continuous spectrum for three regions of M 82 is compared with a distribution calculated under the assumption that the continuous spectrum is due to the radiation from reddened A0—F0 stars.

158.099 On the origin of the rapid-moving gas system in NGC 1275. G. A. Pirog.
Astrofiz. Issled., Izv. Spets. Astrofiz. Obs., Vol. 4, 154 - 156 (1972). In Russian.

A new estimate is obtained of the age of the rapid-moving gas system in NGC 1275 (of the order of 10^8 years) on the basis of the explosion hypothesis. It is suggested that there should be stellar-type sources of ionization within the system.

158.100 Die Spiralstruktur der Sternsysteme. I.
H. -E. Fröhlich.
Sterne, 49. Jahrgang, p. 65 - 72 (1973).

158.101 Short-period brightness variations of the nuclei of the N-galaxies 3C 371 and 3C 390.3.
L. A. Urasin, N. V. Lavrova.
Astron. Tsirk., No. 744, p. 4 - 5 (1973). In Russian.

158.102 On the structure of peculiar galaxies. J. L. Sérsic.
Bull. Astron. Inst. Czechoslovakia, Vol. 24, 150 - 157 (1973).

The dynamical problem of the motion of a test particle in a gravitational field of variable mass is studied to explain some properties of galaxies. A general potential is reduced to one with axial symmetry and the properties of singular regions and points of equipotentials are studied. The theoretical results are used to give an interpretation of some appearances of peculiar objects (ejection of massive bodies from galaxies, the existence of annular structure, etc.).

158.103 Spectral observations of some Markarian galaxies with broad hydrogen emission lines from the fifth and sixth lists.
I. M. Kopylov, V. A. Lipovetsky, V. I. Pronik, K. K. Chuvaev.
Astron. Tsirk., No. 755, p. 1 - 3 (1973). In Russian.

158.104 Spectral observations of Markarian galaxies. IV.
E. K. Denisyuk.
Astron. Tsirk., No. 759, p. 6 - 8 (1973). In Russian.

158.105 New observations of two compact galaxies.
M. V. Penston, M. J. Penston.
Monthly Notices Roy. Astron. Soc., Vol. 162, 109 - 116 (1973).

Very deep integration prints are presented of the compact galaxies BL Lacertae and 3C 390.3. That of BL Lac shows a faint extended object (possibly a galaxy) about 1 arcmin south of BL Lac. The other shows that 3C 390.3 lies in the direction of several galaxies. Scanner observations of 3C 390.3 demonstrate that the light from an annulus centred on the nucleus is probably emitted by stars. Observations of the two other galaxies close to 3C 390.3 on the sky show that one is a foreground object, whereas the other has a very similar redshift to that of 3C 390.3.

158.106 Redshift-magnitude bands, quasi-stellar sources, and systems of redshift. W. G. Tifft.
Astrophys. Journ., Vol. 181, 305 - 326 (1973).

The visibility and character of redshift-magnitude bands in a diagram of general field galaxies compared to specific clusters depends upon the cosmological model. Band structure is, however, shown to exist among quasi-stellar sources and to show the identical slope and band spacing characteristics as do the galaxies in the Coma cluster. At least 14 bands can be identified, forming a convergent band series among QSS emission-line objects. The band phenomenon is briefly discussed in terms of a model invoking multiple states of matter, rapid galaxy evolution, and possible time evolution of matter observed as a function of look-back time in a singular-origin universe.

158.107 I4329A: an extreme Seyfert galaxy. M. J. Disney.
Astrophys. Journ., (Letters), Vol. 181, L55 - L60 (1973).

The southern spiral I4329A is found to be the most extreme Seyfert yet discovered. $H\alpha$ is 13,000 km s^{-1} wide, and $f(H\beta)$ is 7×10^{42} ergs s^{-1}, which approaches that from the QSOs even if they are at cosmological distances.

158.108 Surface photometry of galaxies: Comparison of the luminosity profiles and photometric parameters of southern galaxies measured at Cordoba and Mount Stromlo G. de Vaucouleurs, E. Agüero.
Publ. Astron. Soc. Pacific, Vol. 85, 150 - 161 (1973).

The equivalent luminosity profiles and main photometric parameters of 14 galaxies in common between the Cordoba Atlas and Mount Stromlo Survey are compared. Systematic and accidental errors are discussed. Total photographic magnitudes based on provisional zero points are transformed to the B system. The relations between photometric effective (half-power) diameters and apparent photographic diameters are derived. The relations between photometric concentration indices and morphological type are illustrated.

158.109 Hot-spot nucleus galaxy NGC 2782. K. Sakka, S. Oka, K. Wakamatsu.
Publ. Astron. Soc. Japan, Vol. 25, 153 - 173 (1973).

The results of photographic, spectroscopic, and photoelectric observations of NGC 2782 are presented. This galaxy, having four discrete clouds in the nuclear region, is found to be a hot-spot nucleus galaxy in Sérsic and Pastoriza's (1965) nomenclature. Comparison between the nuclear clouds of NGC 2782 and those of NGC 1068 and M82 is made with respect to the nuclear activity of the former.

158.110 The energy distribution of the jet in Messier 87. N. Kaneko, M. Nishimura, K. Toyama.
Publ. Astron. Soc. Japan, Vol. 25, 175 - 180 (1973).

The blue, visible, and infrared fluxes of the jet in M 87 are found to be 1.64, 3.03, and 4.88×10^{-29} Wm^{-2} Hz^{-1} respectively. The spectral index of 1.74 is derived between the blue and infrared fluxes.

158.111 Spectrophotometry of three emission-line galaxies. W. T. Forrester.
Astrophys. Journ., Vol. 181, 633 - 640 (1973).

Analysis of spectrophotometric data concerning three blue emission-line galaxies indicates a level of excitation similar to that of low-or moderate-excitation galactic H II regions. The relative emission-line intensities and the shapes of the nuclear continua over the blue and near-ultraviolet wavelengths indicate that a population of hot stars is responsible for both the gas excitation and the blue continua.

158.112 Faraday depolarization of radio galaxies and quasars with simple spectra. R. G. Strom.
Astron. Astrophys., Vol. 25, 303 - 312 (1973).

Measurements are presented of the linear polarization of a number of radio galaxies at a wavelength of 49 cm; these along with previous measurements made at 49 cm and 73 cm have been used individually, and in combination with observations made at shorter wavelengths, to investigate Faraday depolarization. Evidence is given that for sources with simple radio spectra, depolarization increases with decreasing component separation. The effects of galactic depolarization, source luminosity and brightness are also discussed and found to be relatively insignificant.

158.113 Distance of two galaxies in Stephan's quintet and possible non-velocity redshifts. C. Balkowski, L. Bottinelli, P. Chamaraux, L. Gouguenheim, J. Heidmann.
Astron. Astrophys., Vol. 25, 319 - 324 (1973).

Twenty one-cm line observations of two galaxies in Stephan's quintet have been carried out with the Nançay radio-telescope: NGC 7319 and NGC 7320. From the measured H I fluxes and line widths, combined with optical data on magnitudes and photometric diameters, we deduce distances for these galaxies using various distance criteria worked out in previous studies. It is possible that at least part of the redshift of NGC 7319 is anomalous.

158.114 Upper limits to the H I content in several elliptical galaxies. L. Bottinelli, L. Gouguenheim, J. Heidmann.
Astron. Astrophys., Vol. 25, 451 - 454 (1973).

Six bright elliptical galaxies have been searched for neutral hydrogen 21-cm radiation with the Nançay radiotelescope. Our observations show that there is no 21-cm line whose width would be smaller than our spectral range of 900 km s^{-1}. Upper limits for neutral hydrogen content as small as a few times $10^8 M_\odot$ are obtained, in particular for NGC 4472.

158.115 Kompakte Galaxien. N. Richter.
Astron. in der Schule, 10. Jahrgang, p. 4 - 7 (1973).

158.116 Spiral structure and nuclear activity in galaxies. P. C. van der Kruit.
Nature, Phys. Sci., Vol. 243, 127 - 130 (1973).

Spiral galaxies with strong density wave compression have a lower specific angular momentum than weak compression galaxies of the same mass. Such galaxies also have a larger degree of central mass concentration and generally better developed and possibly more active nuclei.

158.117 New redshifts of circumpolar southern galaxies. G. de Vaucouleurs, A. de Vaucouleurs.
Astron. Journ., Vol. 78, 377 - 381 (1973).

Redshifts of nine galaxies south of $\delta = -60°$ have been determined from ten spectrograms obtained with the Page—Carnegie image-tube spectrograph attached at the Newtonian focus of the Cordoba Observatory 154-cm reflector. The reciprocal dispersions are 150 Å/mm at $H\alpha$ in the first order and 78 Å/mm at $\lambda 3727$ in the second order. The derived velocities have mean errors of ~ 40 km sec^{-1}. The velocities are used to discuss group membership.

158.118 Redshifts for 51 galaxies identified with radio sources in the 4C catalog. W. L. W. Sargent.
Astrophys. Journ., (Letters), L13 - L15 (1973).

Redshifts are given for 51 galaxies identified by Olsen and by Hazard and Jauncey with radio sources in the 4C catalog between $+20°$ and $+40°$ declination.

158.119 Radio observations of chains and groups of galaxies. J. W. Sulentic, M. A. Kaftan-Kassim.
Astrophys. Journ., (Letters), Vol. 182, L17 - L19 (1973).

Several chains and groups of galaxies have been observed at low frequencies using the Arecibo radio telescope. No excess radiation over normal galaxies was detected.

158.120 On the problem of the origin of spiral structure.
J. H. Oort.
Mitt. Astron. Ges., No. 32, p. 15 - 31 (1973). — Karl Schwarzschild lecture presented at the assembly of the Astron. Ges., Wien, 1972 Sept.

158.121 A model for exploding galaxies.
I. Appenzeller, K. Fricke.
Mitt. Astron. Ges., No. 32, p. 107 (1973). — Abstract.

158.122 Die Mannigfaltigkeit der Galaxien.   P. Brosche.
Mitt. Astron. Ges., No. 32, p. 107 - 108 (1973).
Abstract.

158.123 Radio luminosity function of galaxies.
J. Pfleiderer.
Mitt. Astron. Ges., No. 32, p. 108 -111 (1973).

158.124 Die Verteilung des Masse-Leuchtkraft-Verhältnisses von irregulären und Spiralgalaxien.
R. Y. Chiao, M. Reinhardt.
Mitt. Astron. Ges., No. 32, p. 119 (1973). — Abstract.

158.125 Redshifts of companion galaxies.
L. Bottinelli, L. Gouguenheim.
Astron. Astrophys., Vol. 26, 85 - 89 (1973).

The study of the differential velocities of companion galaxies with respect to their main galaxy made by Arp is extended to the nearby groups of galaxies in which the main galaxy is at least 50% more luminous than each of the others. An effect similar to the one found by Arp is brought to light. The mean value of the radial velocity difference between the companion and the main galaxies is equal to 90 km s^{-1}. Three possible observational biases are considered which do not appear to be responsible of the observed effect.

158.126 Study of the central region of M31. I. Spectrophotometric observational results.
M. Joly, Y. Andrillat.
Astron. Astrophys., Vol. 26, 95 - 103 (1973).

We try to obtain some information about the stellar content of M31's central bulge. We describe the spectrophotometric observational results obtained for a region of 3 kpc diameter, centered on the nucleus and divided into sections 250 pc long.

158.127 Measurements of diameters of galaxies on the Palomar Observatory Sky Survey.
M. Kalinkov, K. Stavrev, D. Karadjov, N. Mihnevsky.
Izv. Sekts. astron. Blg. AN, Vol. 5, 97 - 147 (1972).
Abstr. in Referativ. Zhurn. 51. Astron., 5.51.787 (1973).

158.128 Surface distribution of interacting and normal galaxies around the north galactic pole.
M. Kalinkov, N. Spasova.
Izv. Sekts. astron. Blg. AN, Vol. 5, 149 - 164 (1972). In Bulgarian. — Abstr. in Referativ. Zhurn. 51. Astron., 5.51.788 (1973).

158.129 Time variation of metal abundance in galaxies. Super-metal-rich stage.
S. Ikeuchi, H. Sato, T. Sato, H. Takeda.
Progr. Theor. Phys., (*Japan*), Vol. 48, 1885 - 1898 (1972).

Time variation of metal abundance Z in the interstellar gas is studied and it is found that the abundance Z does not necessarily increase monotonously with time. Super-metal-rich stars are explained as the stars formed during the peak stage of Z, which appeared at $(1 \pm 0.5) \times 10^9$ years after the start of star formation in galaxies.

158.130 On the variability of the optical radiation of the nuclei of Seyfert galaxies.   M. K. Babadzhanjants, V. A. Hagen-Thorn, V. M. Ljutiy (*Lyutyj*).
Astrofizika, Vol. 8, 509 - 528 (1972). In Russian. — English translation in Astrophysics, Vol. 8, No. 4.

The data of spectral, photometric and polarimetric observations of the nuclei of Seyfert galaxies lead to the conclusion that in the nuclei of these galaxies there are sources of synchrotron radiation of small sizes. They are responsible for the observed variability of the optical radiation and its polarization.

158.131 Spectral investigation of the galaxy Markarian 8.
E. Ye. Khachikian.
Astrofizika, Vol. 8, 529 - 539 (1972). In Russian. — English translation in Astrophysics, Vol. 8, No. 4.

The results of a detailed spectral investigation of the galaxy Markarian 8 are presented. The galaxy has a complicated morphological and dynamical structure. It contains five superassociations with different redshifts. The unusual velocity field in the galaxy is discussed.

158.132 The relation between the gradient of surface brightness and other properties of Seyfert galaxies.
M. A. Arakelian.
Astrofizika, Vol. 8, 624 - 627 (1972). In Russian. — English translation in Astrophysics, Vol. 8, No. 4.

It is shown that the gradient of surface brightness is in rather close correlation with some other properties of Seyfert galaxies—the luminosity of the nucleus and that of the galaxy itself, the powers of radio emission, infrared emission and that of hydrogen-line emission.

158.133 The dynamics of the Andromeda nebula.
V. C. Rubin.
Sci. American, Vol. 228, No. 6, p. 30 - 36 (1973).

The stars, dust and gas of this spiral galaxy are all in motion. Spectrographic observations show that they do not simply wheel around the galactic center but move in a quite complex pattern.

158.134 On the nature of galaxies with ultraviolet continuum. II. Objects with broad emission lines.
B. E. Markarian.
Astrofizika, Vol. 9, 5 - 19 (1973). In Russian. — English translation in Astrophysics, Vol. 9, No. 1.

46 objects with broad emission lines detected by a number of investigators are considered. Some data including the Doppler widths of permitted and forbidden lines of these objects are given in a table. The objects having broad emission lines as well as Seyfert galaxies are separated into types I and II, depending on the character of the energy distribution and widths of line profiles in their spectra.

158.135 Detailed UBV photometry of the spiral galaxy NGC 2276.   R. K. Shahbasian (*Shakhbazyan*).
Astrofizika, Vol. 9, 21 - 37 (1973). In Russian. — English translation in Astrophysics, Vol. 9, No. 1.

The results of a detailed three-color photometry of the late type spiral galaxy NGC 2276 are presented. A chart of the distribution of surface brightness and color-indexes over the galaxy is given. Integral magnitudes, U−B, B−V colors along the axes of the galaxy are given. The results of UBV photometry of 16 superassociations in NGC 2276 are presented. The U−B, B−V diagram is constructed. The results of detailed colorimetry are used to investigate the dependence of relative intensities and mean surface brightness on the color.

158.136 Spectroscopic observations of the galaxy Markarian 6.
P. Notni, E. Ye. Khachikian, M. M. Butslov, G. T. Gevorkian.
Astrofizika, Vol. 9, 39 - 50 (1973). In Russian. – English translation in Astrophysics, Vol. 9, No. 1.
    The results of the spectral observations of Markarian 6 obtained with the 2-m universal telescope of the Tautenburg Observatory are presented.

158.137 On the central condensations in E- and S0-galaxies.
K. A. Sahakian (*Saakyan*).
Astrofizika, Vol. 9, 51 - 55 (1973). In Russian. – English translation in Astrophysics, Vol. 9, No. 1.
    The results of the classification of central condensations of 31 elliptical and lenticular galaxies are presented. Most of the observed galaxies are estimated to class 3. Starlike nuclei are observed very seldom.

158.138 On the classification of the central parts of some bright galaxies. R. G. Mnatsakanian.
Astrofizika, Vol. 9, 57 - 62 (1973). In Russian. – English translation in Astrophysics, Vol. 9, No. 1.
    The data of five-mark Byurakan classification of the central parts of 87 bright galaxies of different morphological types are presented.

158.139 Long baseline interferometry of the Seyfert galaxy 3C84. T. H. Legg, N. W. Broten, D. N. Fort,
J. L. Yen, F. V. Bale, P. C. Barber, M. J. S. Quigley.
Nature, Vol. 244, 18 - 19 (1973).
    As an initial result the visibility curve for 3C84, taken in November 1972, is presented here, with two possible interpretations.

158.140 Possible systematic redshifts in a chain of galaxies.
S. A. Gregory, L. P. Connolly.
Astrophys. Journ., Vol. 182, 351 - 355 (1973).
    Redshifts are reported for two groups of galaxies. Although these two groups have previously been assigned to the same cluster, their redshifts indicate that they are not associated. A systematic variation of redshifts is found across the chain of one of these groups, Abell 2247, and several interpretations are discussed.

158.141 An upper limit to the angular diameter of the nucleus of NGC 4151. M. Schwarzschild.
Astrophys. Journ., Vol. 182, 357 - 361 (1973).
    During the seventh flight of Stratoscope II, a 36-inch balloon-borne telescope, five photographs of the nucleus of the Seyfert galaxy NGC 4151 as well as four equivalent photographs of a comparison star were obtained. From a densitometric analysis a safe upper limit to the half-intensity diameter of the nucleus of $0\overset{''}{.}08$ is derived.

158.142 On the emission lines of NGC 1068.
J. A. Eilek, J. R. Auman, T. J. Ulrych, G. A. H. Walker, L. V. Kuhi.
Astrophys. Journ., Vol. 182, 363 - 368 (1973).
    The Balmer lines and the forbidden lines of oxygen and nitrogen have been observed in the spectrum of NGC 1068 with an image isocon at 4 Å and 1.3 Å resolutions.

158.143 The internal motions and nuclear mass of the Seyfert galaxy NGC 7469. K. S. Anderson.
Astrophys. Journ., Vol. 182, 369 - 380 (1973).
    Image-tube spectrograms of high resolution have been utilized to determine the velocity distribution within the central regions of NGC 7469. The observed velocity curves largely reflect the rotation of the galaxy. A component of much smaller amplitude, attributed to radial motions, is also present. If distributed uniformly, the mass interior to 400 pc

is about $10^9 M_\odot$. The significance of these mass determinations in relation to models for Seyfert nuclei is discussed.

158.144 A note on the stellar content of NGC 5195.
H. Spinrad.
Astrophys. Journ., Vol. 182, 381 - 385 (1973).
    The stellar content of NGC 5195, investigated and synthesized by scanner spectroscopy, is remarkably similar to that found for M 82 by O'Connell.

158.145 Far-infrared observations of galactic nuclei.
D. A. Harper, Jr., F. J. Low.
Astrophys. Journ., (*Letters*), Vol. 182, L89 - L93 (1973).
    M82 (NGC 3034) and NGC 253 have been detected at wavelengths between 27 and 125 μ. Their respective 3–3000 μ luminosities are $5 \times 10^{10} L_\odot$ and $3 \times 10^{10} L_\odot$. A 2 σ_m upper limit of 760×10^{-26} W m^{-2} Hz^{-1} has been established for the 110-μ flux from NGC 1068.

158.146 Radio outburst of NGC 1275. E. Epstein.
IAU Circ., No. 2519 (1973).

158.147 Flare-up of activity in the nucleus of NGC 5548.
G. de Vaucouleurs, A. de Vaucouleurs.
IAU Circ., No. 2529 (1973).

158.148 An attempt to detect radio emission from Markarian galaxies at 408 MHz.
P. Thomasson, V. H. Malumian.
Monthly Notices Roy. Astron. Soc., Vol. 162, 295 - 297 (1973).
    Upper limits to the radio emission and luminosities at 408 MHz from 21 Markarian galaxies having either quasi-stellar and Seyfert characteristics or stellar-type spectra are reported.

158.149 The $V-K$ colours of the nuclei of bright galaxies.
M. V. Penston.
Monthly Notices Roy. Astron. Soc., Vol. 162, 359 - 366 (1973).
    Photometric observations of the nuclei of the galaxies M32, M33, M51, NGC 5195 and M101 are reported. These give $U-B$, $B-V$, $H-K$ and $V-K$ colours for each object and the $K-L$ colour for M32.

158.150 Infra-red observations of NGC 7552 and NGC 7582 and their identification with PKS radio sources.
I. S. Glass.
Monthly Notices Roy. Astron. Soc., Vol. 162, 35P - 37P (1973).
    JHKL fluxes are given for NGC 7552 and 7582 and upper limits on the *JHK* fluxes are obtained for NGC 7590. The identification of the first two of these with PKS 2313–428 and PKS 2315–426 is suggested.

158.151 Chain of galaxies in Centaurus.
J. L. Sérsic, E. L. Agüero.
Astrophys. Space Sci., Vol. 19, 387 - 394 (1972).
    Further photometric and spectroscopic observations of a chain of galaxies in Centaurus are presented and discussed.

158.152 On the nature of NGC 5253.
J. L. Sérsic, G. Carranza, M. Pastoriza.
Astrophys. Space Sci., Vol. 19, 469 - 494 (1972).
    It is proposed that NGC 5253 has been the place of a violent event like M82. The presence of a long jet and the characteristics of its nuclear complex points to this interpretation.

158.153 Search for dwarf spheroidal galaxies.
N. Sh. Mailian.
Astrofizika, Vol. 9, 63 - 69 (1973). In Russian. – English

translation in Astrophysics, Vol. 9, No. 1.

104 dwarf spheroidal galaxies are discovered on the Palomar Sky Survey charts in the region with $\alpha > 56°$ and $b > 20°$. The space density of these objects is estimated to $10^{-1} Mpc^{-3}$. The distribution of the angular diameters of the objects under consideration is presented.

158.154 A possible evidence for the recent origin of Markarian galaxies. J. Heidmann, A. T. Kalloghlian.
Astrofizika, Vol. 9, 71 - 77 (1973).

A list of pairs of Markarian galaxies with strong ultraviolet continuum is compiled. The statistical considerations show that most of them are physical. However, the observed differences of radial velocities of components of several pairs strongly suggest that a large proportion of these physical systems have positive energy. If so, the kinematical ages of these systems are of the order of 10^8 years, which serves as an argument in favour of the recent formation of Markarian galaxies.

158.155 On the physical peculiarities of the nucleus of the galaxy Markarian 6. E. Ye. Khachikian.
Astrofizika, Vol. 9, 139 - 142 (1973). In Russian. — English translation in Astrophysics, Vol. 9, No. 1.

For the explanation of broad wings of hydrogen lines $H\alpha$ and $H\beta$, which recently appeared in the spectra of Markarian 6 simultaneously with their new violet components, it is suggested that from the nucleus of this galaxy two hydrogen clouds have been ejected in opposite directions.

158.156 Observations of the bar of NGC 4314.
B. T. Lynds, I. Furenlid, J. Rubin.
Astrophys. Journ., Vol. 182, 659 - 669 (1973).

Image-tube photographs in blue light and at wavelengths 6563 and 6650 Å are used to obtain colors of the bar of NGC 4314. A nuclear region of about 5″ in size is found to have a strong red continuum. This nucleus is surrounded by a region of extensive hydrogen emission with discrete H II regions identified. The color of the bar is relatively uniform and bluer than the central region. Dust lanes in the bar are found to have extinctions of at least 1.0 mag in the blue.

158.157 The structure and content of NGC 205.
P. W. Hodge.
Astrophys. Journ., Vol. 182, 671 - 695 (1973).

The structure and the content of NGC 205, an elliptical galaxy companion to M31, is studied by means of a series of photoelectric and photographic measurements. Isophotometry of the galaxy provides information on its detailed shape, and was carried out in three colors on the UBV system. Luminosity profiles based on direct photoelectric measurements from the center along the north-south and east-west axes are presented. From these, and using isophotes, the major- and minor-axis profiles are obtained and are compared with measures made previously by others. The galaxy is found to have an anomalous color distribution for an elliptical galaxy, with the center being decidedly bluer than the outer regions. Examination of Hale Observatory plates in the blue and ultraviolet provides information on the number and distribution of the resolved O- and B-type stars in the central area of NGC 205. Twelve dark nebulae are identified, and their distribution is found to be similar to the distribution of OB stars, with the exception that the largest and most conspicuous dust clouds lie outside of the area inhabited by these stars. Lick Observatory 120-inch telescope photoelectric measurements of the brightness and colors of all of the eight globular star clusters known for this system provide a luminosity function for these objects that is compared with the luminosity function for the clusters in M31 and found to lie approximately 2 mag fainter than the latter.

158.158 Photoelectric photometry of some galaxies in the region of the Virgo cluster. W. G. Tifft.

Publ. Astron. Soc. Pacific, Vol. 85, 283 - 285 (1973).

Four-color photometry of 26 galaxies, mostly in the region of the Virgo cluster, is presented.

158.159 On the $(U-B)$ color indices and hydrogen line strengths of elliptical galaxies.
P. W. Hodge.
Publ. Astron. Soc. Pacific, Vol. 85, 286 - 288 (1973).

The presence of a contaminating population of young stars in nearby elliptical galaxies can probably explain the anomalous hydrogen line strengths observed in at least one of them (NGC 205) and can partly explain the correlation between $(U-B)$ color index and total absolute magnitude for elliptical galaxies.

158.160 Statistical investigation of the fluctuation of the number of observed galaxies.
G. V. Andreev, R. G. Lazarev.
Materialy 3-j Nauch. konf. Tomsk. un-ta po mat. i mekh. Vyp. (No.) 2. Tomsk, Tomsk. un-t, 1973, p. 85 - 86. In Russian. — Abstr. in Referativ. Zhurn. 51. Astron., 7.51.732 (1973).

Adriaan van Maanen's influence on the island universe theory: Part 2. See Abstr. 004.050.

Extra-galactic astronomy objectives.
See Abstr. 051.013.

The production of discrete, quantized outflow velocities by radiation pressure in stars, Seyfert nuclei, and quasi-stellar objects. See Abstr. 064.011.

Condensation of stars and formation of a magnetic field in protogalaxies. See Abstr. 065.009.

The history of star formation and the colors of late-type galaxies. See Abstr. 065.016.

Some very tentative evidence for recent star formation in elliptical galaxies. See Abstr. 065.129.

Star formation and the chemical history of galaxies.
See Abstr. 065.134.

Observations of Hubble-Sandage variables in M31 and M33. See Abstr. 122.010.

Novae in M 31 discovered and observed at Asiago from 1963 to 1970. See Abstr. 124.003.

Emission-line spectra of supernova remnants and galaxies. See Abstr. 125.011.

Ionization of the intercloud medium and the central disk regions of spiral galaxies. See Abstr. 131.136.

Interstellar obscuration in the direction of Maffei 1 and 2. See Abstr. 131.193.

Identification of planetary nebulae in the elliptical galaxies NGC 185, NGC 205, and NGC 221.
See Abstr. 133.015.

Chemical abundances in a planetary nebula in the elliptical galaxy NGC 185. See Abstr. 133.016.

Quasars as events in the nuclei of galaxies: The evidence from direct photographs. See Abstr. 141.014.

Observations of 3C 129, 3C 129.1 and 3C 83.1 B at 2.7 and 5 GHz.   See Abstr. 141.015.

Confirmation of I Zw 1727+50 as a radio source. See Abstr. 141.030.

Circular polarization in inverse Compton scattering of synchrotron radiation.   See Abstr. 141.042.

Magnetic dipole radiation from a supermassive oblique rotator.   See Abstr. 141.043.

Luminosity functions of quasars and Seyfert galaxies.   See Abstr. 141.070.

Effects of Faraday rotation on the degree of polarization in QSOs and Seyfert galaxies.   See Abstr. 141.081.

Polarimetric observations of non-stable stars and extragalactic objects. III. Polarization of quasars and galactic nuclei.   See Abstr. 141.085.

Observations at 1415 MHz of radio sources in the field of the double-galaxy system NGC 2798/99. See Abstr. 141.089.

QSO historical light curves. See Abstr. 141.098.

Interplanetary-scintillation observations of 203 sources identified as radio galaxies or quasars. See Abstr. 141.099.

QSO's near bright galaxies.   See Abstr. 141.120.

Some trends in the red-shift distribution of quasi-stellar objects and related peculiar galaxies. See Abstr. 141.126.

Recent light changes in three variable radio sources. See Abstr. 141.131.

Photographic photometry of compact extragalactic objects. I.   See Abstr. 141.135.

Nuclei of quasars and active galaxies. See Abstr. 141.136.

Aperture synthesis of extragalactic objects. See Abstr. 141.137.

Rotation in high-energy astrophysics. See Abstr. 141.505.

Transition radiation in astrophysics. See Abstr. 143.035.

Theories of galactic spiral structure. Comparisons with observations.   See Abstr. 151.028.

The observed deficiency of ionized gas in globular clusters and the companions of M 31.   See Abstr. 154.002.

Probably new globular clusters of the Andromeda nebula.   See Abstr. 154.005.

A photometric study of the integrated light of clusters in the Magellanic Clouds and the Fornax dwarf galaxy. See Abstr. 154.010.

Structure and dynamics of the galactic system. A report.   See Abstr. 155.084.

Low-dispersion spectra of galaxies III. Abell No. 194.   See Abstr. 160.014.

Errata

158.901   Erratum: 'The radio emission of NGC 4258 and the possible origin of spiral structure' [Astron. Astrophys., Vol. 21, 169 - 184 (1972)].
P. C. van der Kruit, J. H. Oort, D. S. Mathewson.
Astron. Astrophys., Vol. 22, 479 (1973).

159 Magellanic Clouds

159.001 Dust in the core of the Small Magellanic Cloud.
W. L. Martin, A. D. Thackeray.
Monthly Notices Roy. Astron. Soc., Vol. 161, 5P - 8P (1973).

Of 11 stars in the SMC core selected from UBV photometry as possibly reddened early-type stars, four appear to be members, one a Be star. The result is consistent with a normal dust content of the Cloud, but further photometry is desirable.

159.002 Kinematic properties of 30 H II regions in the Small Magellanic Cloud. M. G. Smith, D. W. Weedman.
Astrophys. Journ., Vol. 179, 461 - 467 (1973).

Emission-line profiles and radial velocities for 30 H II regions in the Small Magellanic Cloud have been observed with a pressure-scanned Fabry-Perot interferometer having a spectral resolution of 11 km s^{-1}. The most probable velocities of internal gas motions vary little among the H II regions and average 14.2 km s^{-1}. Radial velocities accurate to 2.5 km s^{-1} show a steep velocity gradient across the SMC but do not prove that the SMC is rotating.

159.003 A search for neutral hydrogen remnants of strong tidal disruption of the Small Magellanic Cloud.
I. F. Mirabel, K. C. Turner.
Astron. Astrophys., Vol. 22, 437 - 440 (1973).

A region of the sky in the south galactic hemisphere has been observed in the 21 cm hydrogen line following a suggestion by A. Toomre that a "tail" of neutral hydrogen might exist, removed, by gravitational interaction with our Galaxy, from the Small Magellanic Cloud during an earlier close encounter. No such regularly distributed material was observed, and an upper limit on the amount of such gas may be put as of the order of 1% of the amount of material in the bridge region of the Magellanic Clouds.

159.004 Is the core of the Large Magellanic Cloud placed in NGC 1910? Yu. N. Efremov.
Astron. Tsirk., No. 749, p. 7 - 8 (1973). In Russian.

159.005 The brightest stars in the Small Magellanic Cloud.
P. S. Osmer.
Astrophys. Journ., Vol. 181, 327 - 348 (1973).

Photometric observations on the $uvby$ system of the 169 luminous stars in Sanduleak's finding list for the SMC have been made along with $H\beta$ measures of all his stars brighter than $V = 12$ and some representative fainter ones. This paper reports the results of an analysis of these observations.

159.006 Age of Magellanic eclipsing variables and color distribution of 3323 variables in both Magellanic Clouds. S. Gaposhkin.
Stellar ages. Proc. IAU Colloquium No. 17, (see 012.015), IV, 1 - 4 (1973).

159.007 Large Magellanic Cloud. 2nd list of L.M.C. members and list of galactic stars.
C. Fehrenbach, M. Duflot.
Astron. Astrophys.,Suppl. Ser., Vol. 10, 231 - 265 (1973).
In French.

We publish two new lists of stars in the direction of the Large Magellanic Cloud. 88 of the first list are Large Magellanic Cloud members and 253 of the second list are galactic stars.

Time table of star formation in the Large Magellanic Cloud. See Abstr. 065.120.

A rectification of five-colour photometry of LMC supergiants. See Abstr. 113.008.

Additional observations of supergiants and foreground stars in the direction of the Large Magellanic Cloud. See Abstr. 114.045.

Hα emission-line stars in the Large Magellanic Cloud: a quick look at the colour-magnitude diagram. See Abstr. 115.012.

The "zone of avoidance" in the period-amplitude diagram of the LMC cepheids. See Abstr. 122.011.

A photometric study of nova Doradus 1971a in the Large Magellanic Cloud. See Abstr. 124.105.

Supernova remnants in the Large Magellanic Cloud. See Abstr. 125.010.

Supernova remnants in the Magellanic Clouds. See Abstr. 125.042.

Polarisation des Sternlichtes zwischen beiden Magellanschen Wolken. See Abstr. 131.169.

Evidence for ejection of radio sources from supernova remnants. See Abstr. 141.009.

Masses of old LMC globular clusters. See Abstr. 154.003.

A photometric study of the integrated light of clusters in the Magellanic Clouds and the Fornax dwarf galaxy. See Abstr. 154.010.

100 micron survey of southern Milky Way and Magellanic Clouds. See Abstr. 155.039.

160 Clusters of Galaxies

160.001 Properties of the redshift-magnitude bands in the Coma cluster. W. G. Tifft.
Astrophys. Journ., Vol. 179, 29 - 44 (1973).
Redshifts of additional faint galaxies in the Coma cluster are given. A total of 108 galaxies with redshift and magnitude data is now available and is discussed with regard to both nuclear-region and total magnitude.

160.002 Stability of clusters of galaxies with mass loss to gravitational radiation. D. S. Dearborn.
Astrophys. Journ., Vol. 179, 45 - 53 (1973).
The effects of various rates of assumed mass loss to gravitational radiation on a cluster of galaxies are studied using a N-body problem solver. The distributions of the theoretical clusters are compared to the observed distributions of the spiral and elliptical clouds in the Virgo cluster.

160.003 Physics of the X-radiation from clusters of galaxies. S. Sofia.
Astrophys. Journ., (*Letters*), Vol. 179, L25 - L29 (1973).
By examining the data relating to the Perseus, Virgo, and Coma clusters of galaxies, it is shown that the X-ray luminosity in the 2–10 keV range is directly proportional to the average virial mass per galaxy in the cluster. The X-radiation is optically thin bremsstrahlung radiation produced by the (not necessarily isothermal) hot gas. The consequences of this model are examined, and it is shown that it leads to values for the cluster parameters in good agreement with the observations.

160.004 Luminosity functions of clusters of galaxies. H. J. Rood, G. O. Abell.
Astrophys. Letters, Vol. 13, 69 - 72 (1973).
Magnitudes determined independently by Abell and by Rood for galaxies in a 0.32 sq deg core region of the Coma cluster are found to be in good agreement with each other. At the 100 per cent completeness limit of Rood's sample, Abell's data are 91 per cent complete by number and 97.5 per cent complete by luminostiy. Cluster luminosity functions display a prominent plateau near the bright end.

160.005 On the reality of the velocity dispersions in groups of galaxies. J. C. Jackson.
Observatory, Vol. 93, 19 - 23 (1973).
The reality of the large velocity dispersions in small groups of galaxies has been investigated by calculating the fictitious dispersion V_f generated in each group by random errors in measured radial velocities, using published weights of these velocities.

160.006 Clusters of galaxies and the cosmic light. S. A. Shectman.
Astrophys. Journ., Vol. 179, 681 - 698 (1973).
Because galaxies cluster in space, the cosmic background light due to galaxies is not smoothly distributed on the night sky. Based on an explicit description of clusters of galaxies, spatial power spectra for the cosmic-light fluctuations are predicted. These spectra depend strongly on the luminosity density of the universe and the covariance structure of the galaxy distribution. They also depend somewhat on cosmological model.

160.007 Shakhbazian I: a distant cluster of compact galaxies. L. B. Robinson, E. J. Wampler.
Astrophys. Journ., (*Letters*), Vol. 179, L135 - L139 = Contr. Lick Obs., No. 390 (1973).
Slit spectra, obtained with the Lick Observatory image-tube scanner showed that a distant compact cluster of red compact galaxies in Ursa Major has a velocity dispersion of 62 km s^{-1}. The calculated mass-to-light ratio is 2, a value that may be too high owing to uncertainties in the measurements of radial velocities of individual members.

160.008 Gravity of neutrinos of nonzero mass in astrophysics. R. Cowsik, J. McClelland.
Astrophys. Journ., Vol. 180, 7 - 10 (1973).
If neutrinos have a rest mass of a few eV/c^2, then they would dominate the gravitational dynamics of the large clusters of galaxies and of the universe. A simple model to understand the virial mass discrepancy in the Coma cluster on this basis is outlined.

160.009 The cluster of galaxies Abell 2670. A. Oemler, Jr.
Astrophys. Journ., Vol. 180, 11 - 23 (1973).
Abell 2670 is a rich, compact cluster of galaxies dominated by a very large central cD galaxy. A new method of galaxy photometry has been used to obtain its luminosity function, which resembles that of Coma and other clusters. The cD galaxy has an elliptical-galaxy core and a diffuse envelope which has been traced to almost 1 Mpc and which probably fills the cluster.

160.010 The Sculptor group. B. M. Lewis, B. J. Robinson.
Astron. Astrophys., Vol. 23, 295 - 301 (1973).
Results from high resolution H I studies of five of the seven members of the Sculptor group are discussed.

160.011 List of clusters of galaxies with published redshifts. T. W. Noonan.
Astron. Journ., Vol. 78, 26 - 31 (1973).
This paper lists the 138 clusters with published redshifts. The Humason–Mayall–Sandage designation is extended to all clusters in the list, and the Abell-catalog and Zwicky-catalog numbers are given where possible.

160.012 The dependence of Compton X-ray emission from clusters of galaxies on the velocity dispersion of the cluster. R. L. Brown.
Astrophys. Journ., (*Letters*), Vol. 180, L49 - L53 (1973).
In considering the expansion of radio sources in a cluster of galaxies it can be shown that X-ray emission associated with such discrete sources arising from Compton interactions of the synchrotron electrons with the microwave background radiation depends on the parameters of the intracluster medium. The implications of this result are discussed.

160.013 Upper limits on an ionized intracluster medium in the Coma cluster. J. Holberg, S. Bowyer, M. Lampton.
Astrophys. Journ., (*Letters*), Vol. 180, L55 - L59 (1973).
An upper limit of 200 photons cm^{-2} s^{-1} is given for Lα emission from the Coma cluster of galaxies. This, in conjunction with previous measurements at X-ray and radio wavelengths, effectively rules out models in which the cluster is bound by an ionized intracluster gas.

160.014 Low-dispersion spectra of galaxies III. Abell No. 194. A. G. D. Philip, J. W. Sulentic.
Publ. Astron. Soc. Pacific, Vol. 85, 104 - 108 (1973).
Low-dispersion spectra ($\simeq$ 10,000 Å mm^{-1}) have been obtained of the brighter members of the cluster of galaxies,

Abell 194. The reddest and bluest galaxies in the cluster are identified and a color-magnitude diagram is presented.

160.015 **Core radii of clusters of galaxies at different redshifts.** N. A. Bahcall.
Astrophys. Journ., Vol. 180, 699 - 704 (1973).

Number counts of galaxies for five rich clusters of galaxies with redshifts ranging from $z = 0.022$ to $z = 0.38$ are fitted to isothermal gas-sphere models. A model with the same cutoff fits all five clusters. Core radii are obtained for each of the clusters.

160.016 **H I clouds in clusters of galaxies.** N. C. Smart.
Astron. Astrophys., Vol. 24, 171 - 180 (1973).

It is proposed that the Coma cluster of galaxies is bound by clouds of neutral hydrogen gas. These clouds will be optically thick to 21 cm line radiation so that their total emission will be less than the observed upper limit from the cluster. They must also be rotationally supported against collapse. These clouds also provide a mechanism for heating a diffuse gas between the clouds and galaxies to $\sim 7 \times 10^{7} {}^{\circ}$K in agreement with the X-ray observations.

160.017 **Mariner 9 ultraviolet spectrometer experiment: upper limits on the Lyman-alpha flux from clusters of galaxies.** R. C. Bohlin, R. C. Henry, J. R. Swandic.
Bull. American Astron. Soc., Vol. 5, 29 (1973). — Abstr. AAS.

160.018 **Gas in clusters of galaxies.**
A. Yahil, J. P. Ostriker.
Bull. American Astron. Soc., Vol. 5, 29 (1973). — Abstr. AAS.

160.019 **Sobre una cadena de galaxias en Centauro.**
J. L. Sérsic, E. L. Agüero.
Bol. As. Argentina Astron., No. 16, (see 012.007), p. 6 - 10 (1971).

160.020 **Fotometría de cúmulos de galaxias. Experiencias con efecto Sabatier. III.** H. A. Dottori.
Bol. As. Argentina Astron., No. 16, (see 012.007), p. 24 - 25 (1971).

160.021 **Sizes of clusters of galaxies in the Zwicky catalog.**
T. W. Noonan.
Astron. Journ., Vol. 78, 227 - 230 (1973).

Counts of the number of clusters in the Zwicky catalog show the following: (1) Galactic extinction increases with decreasing galactic latitude. (2) There is an rms dispersion of order 0.2 in the $\log_{10}$ of the intrinsic cluster sizes for clusters at the faint limit. (3) If the spatial density of clusters is uniform, then the correlation of angular catalog diameter with distance for the nearer clusters is different for the different types of clusters in the catalog.

160.022 **Clustering effects among clusters of galaxies and quasi-stellar sources.**
R. S. Bogart, R. V. Wagoner.
Astrophys. Journ., Vol. 181, 609 - 618 (1973).

A nearest-neighbor statistical test for angular correlations among sets of extragalactic objects is described. Using this test, it is found that within their respective distance groups, distant rich clusters of galaxies from Abell's catalog are significantly clustered. On the other hand, there is (somewhat

contradictory) evidence for an anticorrelation between the less distant and more distant clusters.

160.023 **Clusters of galaxies are the possible source of a background X-ray emission.**
O. F. Prilutsky, I. L. Rozental.
Astron. Zhurn. Akad. Nauk SSSR, Vol. 50, 462 - 468 (1973). In Russian. English translation in Soviet Astron. AJ, Vol. 17, No. 3.

A model in which heated gas radiation in clusters of galaxies is the main source of a background emission in the X-ray range is proposed. Intensity and energetic spectrum of the background emission are calculated within the limits of this supposition.

160.024 **Mariner 9 ultraviolet spectrometer experiment: Upper limits on the Lyman-alpha flux from clusters of galaxies.** R. C. Bohlin, R. C. Henry, J. R. Swandic.
Astrophys. Journ., Vol. 182, 1 - 7 (1973).

Data from the ultraviolet spectrometer on the Mariner 9 Mars orbiter have been used to set upper limits on the redshifted $L\alpha$ flux from the Perseus and Pegasus I (NGC 7619) clusters of galaxies.

160.025 **Die Verteilung der Rotverschiebungen im Virgo-Haufen.** G. A. Tammann.
Mitt. Astron. Ges., No. 32, p. 111 - 113 (1973).

160.026 **On the existence of second-order clusters of galaxies.**
M. Kalinkov.
Izv. Sekts. astron. Blg. AN, Vol. 5, 165 - 177 (1972). In Bulgarian. — Abstr. in Referativ. Zhurn. 51. Astron., 5.51.802 (1973).

160.027 **The redshifts of ten distant clusters of galaxies.**
W. L. W. Sargent.
Publ. Astron. Soc. Pacific, Vol. 85, 281 - 282 (1973).

Redshifts are given for the clusters Abell 98, 274, 655, 665, 2029, 2224, and 2670 and for the Zwicky clusters Zw Cl 0659 + 63, Zw Cl 1710 + 64, and Zw Cl 1101 + 70.

On the apparent association of quasi-stellar objects with clusters or groups of galaxies with about the same redshift. See Abstr. 141.107.

X-ray cluster of galaxies identified in error box 2U2358–29. See Abstr. 142.083.

Radiospår i galaxhopar. See Abstr. 158.044.

Spectroscopy of outlying faint galaxies in the region of the Coma cluster. See Abstr. 158.058.

Redshift-magnitude bands, quasi-stellar sources, and systems of redshift. See Abstr. 158.106.

Photoelectric photometry of some galaxies in the region of the Virgo cluster. See Abstr. 158.158.

Comment on the Noerdlinger hypothesis. See Abstr. 161.002.

Ultraviolet background radiation. See Abstr. 162.005.

161 Intergalactic Matter

161.001 **About the heating of metagalactic gas.**
  O. F. Prilutsky, I. L. Rosental.
Astron. Zhurn. Akad. Nauk SSSR, Vol. 50, 216 - 217 (1973).
In Russian. English translation in Soviet Astron. AJ, Vol. 17,
No. 1.
  A possible mechanism of heating of metagalactic gas in
the neighbourhood of a cluster of galaxies is considered. The
heating of gas is caused by electron thermal conductivity.

161.002 **Comment on the Noerdlinger hypothesis.**
  J. K. Hill.
Astrophys. Journ., Vol. 179, 371 - 373 (1973).
  Diffuse X-radiation resulting from intergalactic gas in
clusters of galaxies is computed, on the assumption that all
galaxies originally formed as members of compact clusters
which contained enough gas to bind them gravitationally. This
hypothesis is shown to be untenable if thermal energy is transferred to the gas in order to expand or dissolve the clusters.

161.003 **Photoionization by massive stars in protogalaxies.**
  B. M. Tinsley.
Astrophys. Letters, Vol. 14, 15 - 17 (1973).
  It is postulated that the abundance of heavy elements

found in the halo populations of galaxies is attributable to a
short period of formation of massive stars during the collapse
of the protogalaxies. The ultraviolet radiation from these stars
fully ionizes the protogalaxy if its radius exceeds about 20
kpc. The combined effect of all protogalaxies is sufficient to
ionize the intergalactic medium if their smoothed density
exceeds about 0.06 that of intergalactic gas.

  The absorption lines in quasi-stellar objects.
See Abstr. 141.082.

  **Infall of gas from intergalactic space and soft X-ray
background.**   See Abstr. 142.005.

  **Données d'observation et interprétation des spec-
tres infrarouges et microondes des galaxies et de la matière
intergalactique.**   See Abstr. 158.078.

  **The 49-cm linear polarization distribution in 3C 327
and the density of intergalactic gas.**   See Abstr. 158.097.

  **Distance-redshift relations for universes with some
intergalactic medium.**   See Abstr. 162.007.

162 Structure and Evolution of the Universe, Cosmology

162.001 **On the generation of density fluctuations due to turbulence in self-gravitating media.** T. Sasao.
Publ. Astron. Soc. Japan, Vol. 25, 1 - 33 (1973).
  The spectra of density fluctuations formed by turbulence with given energy spectra and their time variations are studied in the framework of the theory of homogeneous, isotropic, and compressible turbulence. The theoretical results are applied to the formation of density fluctuations in the expanding universe and in contracting protogalaxies.

162.002 **The thermal future of the universe.**
P. C. W. Davies.
Monthly Notices Roy. Astron. Soc., Vol. 161, 1 - 5 (1973).
  The thermal behaviour of matter and electromagnetic radiation in the late stages of ever expanding cosmologies is examined. The cooling rate of ions is calculated, and the absorption of radiation discussed, leading to a condition for complete opaqueness.

162.003 **The case for a chaotic cosmogony.** J. Silk.
Comments Astrophys. Space Phys., Vol. 5, 9 - 14 -(1973).

162.004 **Nucleosynthesis in the symmetric universe.**
E. Schatzman.
Comments Astrophys. Space Phys., Vol. 5, 23 - 28 (1973).

162.005 **Ultraviolet background radiation.** R. C. Henry.
Astrophys. Journ., Vol. 179, 97 - 102 (1973).
  The high-galactic-latitude ultraviolet background flux has been measured and is 1900(+0, −950) photons (cm²s sterad Å)⁻¹, at 1450 Å, as seen with a 10° field-of-view detector. This is in good agreement with a value obtained by Lillie with the Wisconsin experiment on OAO-2, which had a much smaller field of view.

162.006 **The role of neutrinos in the evolution of primeval adiabatic perturbations.** P. J. E. Peebles.
Astrophys. Journ., Vol. 180, 1 - 5 (1973).
  According to a recent suggestion, neutrinos may significantly hinder the growth of density irregularities in an expanding, radiation-dominated cosmological model. The effect is evaluated by direct numerical integration of the perturbation equations. Contrary to the suggestion, the effect is shown to be quite small.

162.007 **Distance-redshift relations for universes with some intergalactic medium.** C. C. Dyer, R. C. Roeder.
Astrophys. Journ., (*Letters*), Vol. 180, L31 - L34 (1973).
  The distance-redshift relations for universes with a constant fraction α of the mean density in the form of intergalactic matter are shown to be solutions of a hypergeometric equation. Solutions can be given in closed form for $\alpha = 0, 2/3$, and 1.

162.008 **A limit on the redshift due to interaction with electromagnetic radiation.**
R. L. Cohen, G. K. Wertheim.
Nature, Vol. 241, 109 (1973). − Letter.

162.009 **Faraday rotation and antimatter in the universe.**
A. H. Nelson.
Nature, Vol. 241, 185 - 186 (1973).
  Data on Faraday rotation of polarized radio waves from extragalactic sources have led to various interpretations. For symmetric ambiplasma, its lepton residue, and therefore the rotation rate, vanishes. The author considers the consequences

of this for the interpretation of the astronomical data.

162.010 **CPT conservation in the oscillating model of the universe.** M. G. Albrow.
Nature, Phys. Sci., Vol. 241, 56 - 57 (1973).
  Davies (1972) has proposed an oscillating model of the universe in which the direction of time flow reverses between alternate cycles. It seems sensible to ask whether we may expect any conservation laws to hold. In the speculative spirit of the model, I make the hypothesis that CTP is rigorously conserved, where C=charge conjugation, P=parity and T=time reversal.

162.011 **Difficulties concerning a finite photon rest mass.**
J. F. Woodward, W. Yourgrau.
Nature, Vol. 241, 338 (1973).
  Pecker, Roberts and Vigier have suggested that much, if not all, of the cosmological redshift might be attributable to inelastic photon–photon scattering—if the photon possesses a finite rest-mass. Pecker and Vigier have attempted to apply Vigier's ideas to the electromagnetic propagation anomalies observed by Sadeh and others. We agree in part with the ideas of Vigier and his co-workers, but their hypothesis, as regards these phenomena, cannot be completely correct as stated.

162.012 **Multiple universes.** T. Gold.
Nature, Vol. 242, 24 - 25 (1973).
  The author considers the properties of a closed or nearly closed universe in which there are substantial variations of mean density on a large scale, so that close "sub-universes" could exist within it.

162.013 **Produktion von Antimaterie.** Yu. D. Prokoshkin.
Umschau, 73. Jahrgang, p. 153 - 154 (1973).

162.014 **Modelli cosmologici.** G. Romano.
Coelum, Vol. 41, 16 - 23 (1973).

162.015 **Why is the universe isotropic?**
C. B. Collins, S. W. Hawking.
Astrophys. Journ., Vol. 180, 317 - 334 (1973).
  We examine the question of whether the present isotropic state of the universe could have resulted from initial conditions which were "chaotic", in the sense of being arbitrary, any anisotropy dying away as the universe expanded. We show that the set of spatially homogeneous cosmological models which approach isotropy at infinite times is of measure zero in the space of all spatially homogeneous models.

162.016 **New limit on small-scale irregularities of "blackbody" radiation.** Y. N. Parijskij.
Astrophys. Journ., (*Letters*), Vol. 180, L47 - L48 (1973).
  There are no fluctuations of blackbody radiation above the 0.8×10^{-4}°K rms level at 2.8 cm wavelength on the scales $3'-1°$. This new limit rejects Ree's model of the universe filled by long-wavelength gravitational radiation and restricts some hypotheses of the origin of galaxies as well as clusters of galaxies.

162.017 **A search for isolated dispersed radio-frequency pulses.** G. R. Huguenin, E. L. Moore.
Bull. American Astron. Soc., Vol. 5, 6 (1973). − Abstr. AAS.

162.018 **Hydrodynamics of blast waves in an expanding universe.** J. Schwarz, J. P. Ostriker, A. Yahil.
Bull. American Astron. Soc., Vol. 5, 41 (1973). − Abstr. AAS.

162.019 **The mechanical aspect of the expansion of the universe.** O. Onicescu.
Prog. şti., Vol. 8, No. 5, p. 209 - 212 (1972). In Romanian. Abstr. in Referativ. Zhurn. 51. Astron., 3.51.780 (1973).

162.020 **Compatibility of the de Sitter model with Mach's principle.** B. L. Al'tshuler.
Third Soviet Gravitational Conference, Erevan, 1972, (see 012.001), p. 6 - 11 (1972). In Russian.

162.021 **A ball of dust in empty space.** M. P. Korkina.
Third Soviet Gravitational Conference, Erevan, 1972, (see 012.001), p. 78 - 82 (1972). In Russian.

162.022 **Gravitational fields with 3-parametric groups of motions on two-dimensional invariant manifolds and the problem of isotropization of inhomogeneous models.** I. S. Shikin.
Third Soviet Gravitational Conference, Erevan, 1972, (see 012.001), p. 178 - 181 (1972). In Russian.

162.023 **On the quantization of a closed isotropic model with cosmological term.**
M. I. Kalinin, V. N. Mel'nikov.
Third Soviet Gravitational Conference, Erevan, 1972, (see 012.001), p. 237 - 240 (1972). In Russian.

162.024 **On cosmological initial conditions.** A. V. Byalko.
Third Soviet Gravitational Conference, Erevan, 1972, (see 012.001), p. 297 - 299 (1972). In Russian.

162.025 **The cause of expansion and turbulence of the universe in the scalar-tensor theory of gravitation and scalar gravitational radiation.**
L. É. Gurevich, S. D. Dynkin, A. M. Finkel'shtejn.
Third Soviet Gravitational Conference, Erevan, 1972, (see 012.001), p. 307 - 308 (1972). In Russian.

162.026 **Structural features of galaxies as a consequence of screening of the Newtonian gravitational potential.**
V. P. Kolpakov, Ya. P. Terletskij.
Third Soviet Gravitational Conference, Erevan, 1972, (see 012.001), p. 316 - 320 (1972). In Russian.

162.027 **Simple numerical relations between the fundamental constants in cosmology and microphysics.**
P. N. Kropotkin.
Third Soviet Gravitational Conference, Erevan, 1972, (see 012.001), p. 324 - 327 (1972). In Russian.

162.028 **Kinetic theory of small perturbations of the Newtonian universe.**
V. B. Magalinskij, V. V. Seliverstov.
Third Soviet Gravitational Conference, Erevan, 1972, (see 012.001), p. 334 - 337 (1972). In Russian.

162.029 **On the cosmological influence on island systems.** A. V. Mandzhos.
Third Soviet Gravitational Conference, Erevan, 1972, (see 012.001), p. 337 - 339 (1972). In Russian.

162.030 **Inhomogeneous cosmological models with planar and pseudospherical symmetries.** V. A. Ruban.
Third Soviet Gravitational Conference, Erevan, 1972, (see 012.001), p. 348 - 351 (1972). In Russian.

162.031 **Locally inhomogeneous cosmological models.** É. Saar.
Third Soviet Gravitational Conference, Erevan, 1972, (see 012.001), p. 353 - 356 (1972). In Russian.

162.032 **On physical limitations of topological model universes.** D. D. Sokolov.
Third Soviet Gravitational Conference, Erevan, 1972, (see 012.001), p. 361 - 362 (1972). In Russian.

162.033 **Friedmann universe in a centrally symmetric reference frame and the general covariance principle.**
K. P. Stanyukovich, O. Sh. Sharshekeev.
Third Soviet Gravitational Conference, Erevan, 1972, (see 012.001), p. 362 - 365 (1972). In Russian.

162.034 **Generalized dependence "redshift – stellar magnitude" in Friedmann's cosmology.** V. E. Yakimov.
Third Soviet Gravitational Conference, Erevan, 1972, (see 012.001), p. 368 - 371 (1972). In Russian.

162.035 **On the free motion in Friedmann space with "dusty" matter.** V. N. Yakovlev.
Third Soviet Gravitational Conference, Erevan, 1972, (see 012.001), p. 372 - 374 (1972). In Russian.

162.036 **Derivation of the differential equation for the simpler cosmological models.** P. T. Landsberg.
Nature, Phys. Sci., Vol. 242, 104 (1973). – Letter.

162.037 **Die Geburtsstunde des Universums.** H.-U. Keller.
Orion Schaffhausen, 31. Jahrgang, p. 45 - 49 (1973).

162.038 **Singularity and matter creation in cosmological models.** J. V. Narlikar.
Nature, Phys. Sci., Vol. 242, 135 - 136 (1973).
The author suggests that the two drawbacks in the classical cosmological discussion may be connected and that a proper consideration of matter creation should resolve the problem of singularity.

162.039 **Cosmological hadronic fireball.** R. F. Sisteró.
Astrophys. Space Sci., Vol. 20, 19 - 25 (1973).
Relativistic non-zero pressure cosmology describes a hadronic fireball in the primordial stage of the universe near the singularity. It is phenomenologically matched to Hagedorn's equation of state based on thermodynamics of strong interactions.

162.040 **Imperfect fluid Friedmannian cosmology.**
M. Heller, Z. Klimek, L. Suszycki.
Astrophys. Space Sci., Vol. 20, 205 - 212 (1973).
The bulk viscosity is introduced into the frame of ordinary Friedmannian cosmology (under highly idealized assumption of the constant coefficient of bulk viscosity). Explicit solutions are given for the viscous flat universe filled with the dust-substratum and for the viscous radiative universe. The problem, how does the introduction of viscosity affect the appearance of singularity, is briefly discussed.

162.041 **Galaxienbewegung und Frühphase des Kosmos.**
H. Oleak, H.-J. Treder.
Astron. Nachr., Vol. 294, 89 - 90 (1973).
The peculiar motions of the galaxies prevent the existence of a radiation universe in the past, if the present mean mass density is larger than the critical value $10^{-30} g/cm^3$.

162.042 **The definition of rest in empty universes.**
S. von Hoerner.
Astrophys. Journ., Vol. 181, 261 - 268 (1973).
Using two test particles, a definition of rest is given which holds for any type of curved 3-space, empty or not. A two-rocket experiment is described which establishes the absolute rest frame without regard to any matter or background radiation. A cosmic twin paradox is mentioned.

162.043 **An empty bubble in nonstatic cosmological models.**
P. Sengupta.
Stud. Univ. Babeş-Bolyai, Ser. Phys., Anul 18, Fasc. 1, p. 13 - 23 (1973).

162.044 **The problem of extrapolation of knowledge in cosmology.** A. Tursunov.
Izv. AN TadzhSSR. Otd. fiz.-mat. i geol.-khim. n., 1972, No. 3, p. 14 - 21. In Russian. – Abstr. in Referativ. Zhurn. 51. Astron., 4.51.1 (1973).

162.045 **L'hypothèse cosmologique de Gamov est-elle confirmée?** A. Polikarov.
Izv. Sekts. astron. Blg. AN, Vol. 5, 89 - 95 (1972). – Abstr. in Referativ. Zhurn. 51. Astron., 4.51.935 (1973).s

162.046 **Study of gravitational fields in an anisotropic model with matter and neutrinos.** I. S. Shikin.
Zhurn. ehksperim. i teor. fiz., Vol. 63, 1529 - 1537 (1972). In Russian. – Abstr. in Referativ. Zhurn. 51. Astron., 4.51.939 (1973).

162.047 **On the construction of spinor field equations in cosmological space.**
D. F. Kurdgelaidze, L. N. Plyusnin.
Izv. vyssh. ucheb. zavedenij. Fizika, 1972, No. 10, p. 20 - 25. In Russian. – Abstr. in Referativ. Zhurn. 51. Astron., 4.51.948 (1973).

162.048 **The applications of relativistic kinetic theory to cosmological models – some observational consequences.** R. Hakim, C. Vilain.
Astron. Astrophys., Vol. 25, 211 - 217 (1973).

In this paper, a gas consisting of clusters of galaxies is studied using kinetic theory. It is assumed to have been in equilibrium at some arbitrarily earlier epoch, and to have the density observed at the present time. The model is also assigned a finite pressure. We abstract an equation of state together with a number of observables: these are used to establish a comparison with other Friedmann models. A comparison with observation leads to the conclusion that when clusters of galaxies are treated as test particles, their random velocities do not exceed some value close to 4500 km/s. We discuss the significance of this result, and also some other applications of the same model.

162.049 **The age of the universe and its expansion.**
J. Heidmann.
Stellar ages. Proc. IAU Colloquium No. 17, (see 012.015), XXXVII, 1 - 22 (1973).

162.050 **The age of the universe.** S. van den Bergh.
Stellar ages. Proc. IAU Colloquium No. 17, (see 012.015), XL, 1 - 13 (1973).

162.051 **The Copernican character of Einstein's cosmology.**
A. Harder.
Ann. Sci., Vol. 29, 339 - 347 (1972).

162.052 **Kinematics of the metagalaxy based on the theory of special relativity.** K. Hara.
Sci. Rep. Tôhoku Univ., First Ser., Vol. 55, 169 - 191 (1972/73).

The apparent magnitude–spectral shift relation for an observer located far from the metagalactic center is derived special–relativistically. Two models are considered as the metagalaxy.

162.053 **Modelli cosmologici (3. - 5.).** G. Romano.
Coelum, Vol. 41, 58 - 63 (1973).

162.054 **Closed time and absorber theory.** D. T. Pegg.
Nature, Phys. Sci., Vol. 243, 143 - 144 (1973).
Letter.

162.055 **New limit of fluctuations of relict emission of the universe.** Yu. N. Parijsky.
Astron. Zhurn. Akad. Nauk SSSR, Vol. 50, 453 - 458 (1973). In Russian. English translation in Soviet Astron. AJ, Vol. 17, No. 3.

It is found that the fluctuations of the brightness of the metagalaxy averaged on the scale of $12' \times 40'$ at 4 cm wavelength are about 1.3×10^{-4} °K. It seems the large part of these fluctuations is due to weak background sources. It gives us an upper limit $< 4 \times 10^{-5}$ °K to the relative fluctuation of "black body" radiation. It limits the energy of large-scale gravitational waves in the universe and excludes some variants of the "fluctuating" and "curled" models of the origin of clusters and galaxies.

162.056 **Olbers' paradoks.** Ø. Grøn.
Astron. Tidssk., Årg. 6, p. 73 - 77 (1973).

162.057 **Dark skies and fixed stars.** M. Hoskin.
Journ. British Astron. Ass., Vol. 83, 254 - 262 (1973). – Christmas lecture 1972 given at a meeting of the British Astron. Ass.

162.058 **The role of the electron–neutrino interaction in the primordial gas.** H. F. Hecht.
Astron. Astrophys., Vol. 26, 123 - 125 (1973).

We report the results of some calculations concerning the importance of the electron–neutrino interaction in the primordial gas. Since the total strength of this interaction is not very well established, we allow deviations from the currently accepted value and compute the ramifications in the early stages of cosmological evolution.

162.059 **Lobachevskij's geometrical ideas and some problems of modern cosmology.** N. A. Litsis.
Filos. vopr. estestvoznaniya. Riga, "Zinatne", 1972, p. 96 - 123. In Russian. – Abstr. in Referativ. Zhurn. 51. Astron., 5.51.1 (1973).

162.060 **A critique of Hoyle and Narlikar's new cosmology.** J. M. Barnothy, B. M. Tinsley.
Astrophys. Journ., Vol. 182, 343 - 349 (1973).

The magnitudes of spiral and elliptical galaxies, and colors of distant ellipticals, predicted by Hoyle and Narlikar's new cosmology, are derived on the assumption that the variation of the gravitational constant G affects only the stellar temperatures and luminosities. It appears very difficult to reconcile this theory with observations of galaxies unless it can be shown that a strong increase in G results in fundamental changes in stellar structure which counteract the effects considered here.

162.061 **The rotation and distortion of the universe.** C. B. Collins, S. W. Hawking.
Monthly Notices Roy. Astron. Soc., Vol. 162, 307 - 320 (1973).

The isotropy of the microwave background is used to place upper limits on the large-scale anisotropies of the universe. This is done by considering the behaviour of the microwave background in all types of spatially homogeneous models that could reasonably represent our universe. The significance of the results for galaxy formation is briefly discussed.

162.062 **The problem of primeval stellar matter.** L. W. Mirzojan.
Urania Kraków, Vol. 44, 2 - 9, 39 - 46 (1973). In Polish.

162.063 **Primordial irregularities in the early universe.** K. Tomita.

Progr. Theor. Phys. (*Japan*), Vol. 48, 1503 - 1516 (1972).

In the early universe, there is an 'anti-Newtonian stage' when the dimension of an irregularity exceeds the horizon ct, far before the decoupling epoch. The evolution of an irregularity at this stage is clarified, in which general relativistic non-linearity plays an essential role.

162.064 Nonlinear gravitational growth in expanding gases.
M. Sasaki, K. Kusukawa.
Journ. Phys. Soc. Japan, Vol. 33, 1182 (1972).

The problem of nonlinear gravitational growth in an expanding universe based on relativistic theory has been discussed by Kihara (1967, 1968). In the present paper, the same results are obtained by the local nonlinear analysis of the Newtonian cosmological model.

162.065 The principle of reciprocity. I. Khan.
International Journ. Theor. Phys., (*GB*), Vol. 6, 383 - 397 (1972). – See Phys. Abstr., Vol. 76, No. 16118 (1973).

162.066 A redshift magnitude relation for radiation univers-es. S. E. Kaufman.
Nuovo Cimento B, Ser. 11, Vol. 13 B, 91 - 100 (1973).

A closed formula for the relation between distance and redshift in an expanding radiation-filled universe is derived for the case of positive cosmological constant and positive space curvature. This formula is then generalized so that it is valid for all values of the cosmological constant and the three possible space curvatures. From this distance redshift formula, a generalized redshift magnitude relation is found.

162.067 Origin of magnetic fields in the early universe.
E. R. Harrison.
Phys. Rev. Letters, Vol. 30, 188 -190 (1973).

Primordial turbulence in the radiation era generates a weak seed magnetic field on all scales of the turbulence. Field generation ceases at the epoch of equal radiation and matter densities, when the field strength is of the order of 1G. It is estimated that the present intergalactic magnetic field has an intensity of $\sim 10^{-8}$G and a scale-length of ~ 1 Mparsec.

162.068 An hypothesis connecting the entropy with the inhomogeneity of the universe.
Ya. B. Zel'dovich.
Zhurn. ehksperim. i teor. fiz., Vol. 64, 58 - 66 (1973). In Russian. – Abstr. in Referativ. Zhurn. 51. Astron., 6.51.781 (1973).

162.069 Metrics for the static plane-symmetric de Sitter space-time. B. V. Prepelitsa.
Izv. vyssh. ucheb. zavedenij. Fizika, 1973, No. 1, p. 152 - 153. In Russian. – Abstr. in Referativ. Zhurn. 51. Astron., 6.51.789 (1973).

162.070 On physical and mathematical models of space, time and motion. B. G. Lepekhin.
Trudy Mosk. in-ta ehlektron. mashinostr., 1972, vyp. (No.) 28, p. 120 - 167. In Russian. – Abstr. in Referativ. Zhurn. 51. Astron., 6.51.808 (1973).

162.071 On the absorption of super-high energy photons in the universe. S. G. Matinian.
Astrofizika, Vol. 8, 587 - 590 (1972). In Russian. – English translation in Astrophysics, Vol. 8, No. 4.

Probabilities of γ-quanta hadronic absorption in a photon gas from universal thermal radiation, extragalactic radiation, optical radiation of galaxies and X-ray photons are calculated. It follows from the obtained results that the universe is opaque for γ-quanta with $E_\gamma \geqslant 10^{14}$ eV.

162.072 On the estimate of mean density of matter in the metagalaxy. M. A. Arakelian.
Astrofizika, Vol. 9, 151 - 153 (1973). In Russian. – English translation in Astrophysics, Vol. 9, No. 1.

Evidence is presented that the space density of dwarf galaxies ($M_{Pg} > -16$) in the metagalactic field is close to the value provided by the members of the Local Group. Then the mean density of matter due to such galaxies is at least several times greater than the value due to galaxies of higher luminosities.

162.073 The Copernican character of Einstein's cosmology.
A. Harder.
Ann. Sci., (*GB*), Vol. 29, 339 - 347 (1972).

162.074 An anisotropic universe with torsion.
W. Kopczynski.
Phys. Letters A, (*Netherlands*), Vol. 43A, 63 - 64 (1973).

The problem of a spatially flat and homogeneous gravitational field produced by spinning matter is considered within the framework of the Einstein-Cartan theory. If the effect of torsion is greater than that of shear, solutions of the field equations have no cosmological singularity.

162.075 Densities of baryons and neutrinos in the universe from an analysis of big-bang nucleosynthesis.
H. Reeves.
Phys. Rev. D, Particles and Fields, Vol. 6, 3363 - 3368 (1972).

The contribution from the 'big-bang' to the observed 'cosmic' abundances of D, ^{3}He, ^{4}He, and ^{7}Li can, in principle, be used to determine and, in practice, give limits to the baryonic number B and the electron leptonic number L_e of the universe.

162.076 High-frequency sound waves to eliminate a horizon in the mixmaster universe. D. M. Chitre.
Phys. Rev. D, Particles and Fields, Vol. 6, 3390 - 3396 (1972).

From the linear wave equation for small-amplitude sound waves in a curved space-time, there is derived a geodesiclike differential equation for sound rays to describe the motion of wave packets. These equations are applied in the generic, non-rotating, homogeneous closed-model universe (the 'mixmaster universe', Bianchi type IX).

162.077 Gravitational instability of regular model-universes in a modified theory of general relativity.
H. Nariai.
Progr. Theor. Phys., (*Japan*), Vol. 49, 165 - 180 (1973).

Any regular model-universe must be non-singular for an infinitesimal perturbation of the metric if it exists at all. To examine this problem, the gravitational instability of regular isotropic model-universes in a modified theory of general relativity is investigated.

162.078 Uranium and helium and the determination of cosmic ages.
R. Gallino, A. Ferrari, A. Masani.
Quaderni Ric. Sci., No. 64, p. 3 [96] - 45 [138] = Contr. Oss. Astron. Milano-Merate, Nuova Ser., No. 333 (1970).

162.079 Consequences of field quantization in de Sitter type cosmological models. E. A. Tagirov.
Ann. Physics, Vol. 76, 561 - 579 (1973).

162.080 Construction of a general cosmological solution of the Einstein equation with a time singularity.
V. A. Belinskii, E. M. Lifshitz, I. M. Khalatnikov.
Soviet Phys. JETP, (*USA*), Vol. 35, 838 - 841 (1972).

It is shown how to describe the process of alternation of Kasner epochs in the oscillatory regime of approach to a singularity of the general (nonhomogeneous) solution to the

Einstein equation.

162.081 Problems of gravitation and relativity theory in modern cosmology. H.-J. Treder.
Wiss. Zeitschr. Techn. Hochschule Karl-Marx-Stadt, No. 3, p. 347 - 350 (1972). In German.

Concretizes the saying that everything compatible with the laws of nature happens in the cosmos, to propose a conception of cosmology on the basis of two main hypotheses: 'smoothing' (averaging) and 'relativity'.

162.082 Dynamics of the Friedmann universe using Regge calculus. P. A. Collins, R. M. Williams.
Phys. Rev. D, Particles and Fields, Vol. 7, 965 - 971 (1973).

Models for the Friedmann universe are constructed from 5, 16, or 600 dust-filled tetrahedrons connected so as to form a closed space. Using the techniques of Regge calculus the time development of these model universes is determined and the results compared with the standard analytic solution for an isotropic dust-filled universe.

162.083 Observational constraints imposed by Brans-Dicke cosmologies. R. E. Morganstern.
Phys. Rev. D, Particles and Fields, Vol. 7, 1570 - 1579 (1973).

The flat-space Brans-Dicke (BD) Friedmann cosmologies previously found are analyzed in more detail.

162.084 A universe filled with magnetofluid. T. H. Date.
Current Sci., (*India*), Vol. 42, No. 1, p. 15 - 16 (1973).

The field equations for relativistic magnetohydrodynamics proposed by Lichnerowicz (1967) have been solved. From this solution, expressions for the total pressure, total density and total internal energy density were evaluated.

162.085 The Brans-Dicke theory and anisotropic cosmologies. R. A. Matzner, M. P. Ryan, E. T. Toton.
Nuovo Cimento B, Ser. 11, Vol. 14B, 161 - 172 (1973).

The Hamiltonian method developed to study homogeneous, anisotropic cosmological models can be extended to the equivalent models in the Brans-Dicke theory. Universes of Bianchi types I and IX are studied and diagrammatic solutions are obtained in both cases.

162.086 Limits on the cosmological constant and space curvature in Friedmann cosmologies.
P. T. Landsberg, B. M. Brown. ⸱
Astrophys. Journ., Vol. 182, 653 - 658 (1973).

The half-plane defined by the deceleration parameter q and the density parameter σ is divided into seven regions. Big-bang zero-pressure Friedmann models can lie in six of these. For each of these six regions algebraic and numerical inequalities are established for the cosmological constant Λ and for the space curvature k/R^2. The shapes of the seven regions are determined by new inequalities.

162.087 Primordial irregularities in the early universe. K. Tomita.
Res. Inst. Theor. Phys. Hiroshima Univ., (*Japan*), No. 11, p. 1 - 24 (1972).

162.088 On the origin of galactic rotation. K. Tomita.
Res. Inst. Theor. Phys. Hiroshima Univ., (*Japan*), No. 14, p. 1 - 7 (1972).

162.089 Gravitational instability of regular model-universes in a modified theory of general relativity.
H. Nariai.
Res. Inst. Theor. Phys. Hiroshima Univ., (*Japan*), No. 15, p. 1 - 28 (1972).

L'univers relativiste.   See Abstr. 003.082.

Über die Richtung der Zeit.
See Abstr. 022.031.

Big-bang nucleosynthesis revisited.
See Abstr. 061.005.

Equations for a plasma consisting of matter and antimatter. See Abstr. 062.036.

Origin of the microwave background.
See Abstr. 066.004.

Black holes in an expanding universe.
See Abstr. 066.008.

Gravitation, cosmology, cosmogony (general relativity in the physical description of the world).
See Abstr. 066.022.

Fine-scale anisotropy of the microwave background: an upper limit at $\lambda = 3.5$ millimeters.   See Abstr. 066.051.

On singularities in general relativity and cosmology.
See Abstr. 066.056.

Search for small-scale anisotropy in the 2.7° K cosmic background radiation at a wavelength of 3.56 centimeters.
See Abstr. 066.076.

Singularities in general relativity and cosmology.
See Abstr. 066.077.

The implications for geophysics of modern cosmologies in which G is variable.   See Abstr. 081.010.

Photon mass, quasar redshifts and other abnormal redshifts.   See Abstr. 141.029.

Cosmological information from surveys of radio source spectra.   See Abstr. 141.058.

Radio source depolarization, size and cosmology.
See Abstr. 141.106.

On the dynamics of gravitating systems on the neutrino background of the universe.   See Abstr. 151.016.

Primordial random motions and angular momenta of galaxies and galaxy clusters.   See Abstr. 151.019.

Cosmic turbulence and the origin of galaxies.
See Abstr. 151.031.

Dependence of the integrated background light on cosmology, galactic spectra, and galactic evolution.
See Abstr. 158.041.

Clusters of galaxies and the cosmic light.
See Abstr. 160.006.

Errata

162.901 Erratum: 'Cosmology and information' [Publ. Astron. Soc. Pacific, Vol. 84, 818 - 822 (1972)].
D. H. Gudehus.
Publ. Astron. Soc. Pacific, Vol. 85, 254 (1973).

Author Index

BELL, B.
074.058
BELL, E. E.
031.020
BELL, M. B.
033.001
BELL, P. M.
094.072 .074 .174 .638
BELL, P. R.
094.381 .432
BELL, R. A.
076.027
114.070 .099
122.037
155.030
BELL, R. J.
003.032
BELOKON, E. T.
126.004
BELON, A. E.
084.013
BELOROSSOVA, T. S.
031.055
BELOUS, L. M.
103.131
BELOV, A. V.
143.044 .069 .076
BELOV, I. F.
077.025
BELOV, R.
094.876
BELOZEROVA, M. A.
151.051
BELTON, M. J. S.
099.064
101.003
BELVEDERE, G.
080.032
BELY, O.
022.010
BELYAEV, N. A.
011.032
103.136
BELYAEV, V. A.
073.112
BELYAEVA, E. E.
064.039 .040
BELY-DUBAU, F.
022.070
BEM, J.
032.008
103.112
BENADA, J.
094.394
105.003
BENCE, A. E.
094.076 .169 .204 .216
.344 .552 .632 .647
.914
BENDEL, W. L.
080.047
BENDER, C. F.
022.008
BENDER, P. L.
094.927
131.045
BENES, K.
094.843 .945
097.102
BENKOVA, N. P.
081.004

BEN-MENAHEM, A.
081.009
BENNETT, D. J.
084.237
BENNETT, G.
084.015
BENNETT, L.
094.054
BENNEWITH, P. D.
114.023
BENSON, R. C.
131.104
BENUKH, V. V.
SEE BENYUKH, V. V.
BENYUKH, V. V.
104.019 .036
BENZ, A. O.
077.010
BENZI, V.
065.170
BERDICHEVSKY, M. N.
084.233
BERDOT, J. L.
094.256 .269 .803 .808
BERENDZEN, R.
004.050
005.002
012.020
014.024 .032 .036 .044
BERESFORD, I. R.
141.510
BEREZINSKY, V. S.
143.049
BERGAMINI, R.
141.023
BERGE, G. L.
093.066
BERGER, P. S.
131.034 .071
BERGERON, J.
022.052
158.024 .081
BERGH, C. DE
093.004
BERGH, S. VAN DEN
065.129
123.024
125.043
133.005
158.069
162.050
BERGSTRALH, J. T.
093.006
099.058
100.020
BERKHUIJSEN, E. M.
142.108
155.023
158.048
BERKING, B.
094.175
BERKO, F. W.
084.036
BERKOVICH, L. M.
042.048
BERKOVSKIJ, B. M.
062.015
BERLOVICH, E. E.
125.016
BERNAS, R.
094.410

BERNHARD, H.
014.008
BERNOT, M.
075.016
BERNSTEIN, J.
003.033
BERRY, C. L.
151.006
BERRY, H.
094.411 .705
BERTAUD, C.
122.052 .116
BERTAUX, J. L.
082.003
103.102
BERTHEL, R. O.
022.007
BERTHELIER, A.
084.229
106.006
BERTHELIER, J.-J.
084.205
BERTIAU, F. C.
082.121
BERTOLA, F.
114.086
158.046 .095
BERTOLETTI, M.-J.
094.388
BERTOTTI, B.
084.259
091.039
BERTSCH, D. L.
078.012 .031 .032
BERULIS, I. I.
141.050
BESSELL, M. S.
114.055
142.135 .139
BESSEY, R.
074.062
BEST, G. T.
022.002
BEST, J. B.
094.176 .646 .663
BETANCOURT, O.
032.028
BEUERMANN, K. P.
106.024
132.021
BEUGLASS, L. K.
045.029
BEWICK, A.
084.401
BEYER, R. L.
094.376
BEYER, R. R.
034.085
BEYSEKOVA, G. K.
131.061
BHANDARI, N.
094.257 .507 .508 .509
.800 .808
BHARGAVA, B. N.
084.213 .263
BHAT, S.
094.507 .508 .509
BHATIA, P. K.
062.053
BHATNAGAR, A. K.
122.097

BHATNAGAR, P. L.
014.027
BHATTACHARJI, J. C.
046.002
BIANCHI, G.
009.017
079.101
BIANCHINI, A.
122.010
124.002
BIANCHINI, G.
045.015
BIBRON, R.
094.723
BICAK, J.
065.165
BIDELMAN, W. P.
008.033
BIEGING, J. H.
142.012
BIEL, J.
031.065
BIELICKA, K.
035.001
BIEMANN, K.
094.443 .444 .955
BIERMANN, L.
102.016
107.018
BIERMANN, P.
061.025 .041
117.004 .031
151.009
BIETKOWSKI, H.
003.034
BIGAY, J.
122.052
BIGGAR, G. M.
094.194 .348 .619
BIGNAMI, G. F.
031.045
BIGNELL, C.
133.037
BILD, R.
094.139
105.059
BILLAUD, G.
044.002
BILLINGSLEY, F. C.
094.483
BILSON, E.
094.827
BINDER, A. B.
099.061
BINNENDIJK, L.
121.010 .011
BINNS, W. R.
143.055
BINZ, C. M.
105.095
BIRAUD, F.
141.036
BIRD, M. L.
094.350
BIRK, J.-L.
094.924
BIRKEBAK, R. C.
094.010 .474 .484 .838
BIRKLE, K.
082.098

BIRMINGHAM, T. J.
062.041 .044
143.005 .021
BIRN, J.
091.017 .067
BIRULYA, T. A.
036.003
BIRYUKOV, YU. L.
093.084
BISCHOFF, M.
002.007
BISHOFF, K.
083.055
BISHOP, G. R.
061.053
BISHOP, R. H.
083.076
BISKUP, M.
003.130
BISNOVATYI-KOGAN, G. S.
065.158
BISNOVATYJ-KOGAN, G. S.
065.009 .143
122.069
BISNOVATY-KOGAN, G. S.
151.024
BISWAS, B. N.
033.058
BISWAS, S.
078.001 .031 .032
BITTENCOURT, J.
082.042
BJORDAL, J.
084.011
BJORKHOLM, P.
094.060 .301 .571 .587
.755 .756
BLAAUW, A.
155.071
BLACK, D. C.
107.014
BLACK, J. H.
131.190
BLACK, L. P.
094.416
BLACK, M. S.
094.692
BLACKMAN, G. L.
131.027
BLADES, J. C.
114.023
BLAIR, A. G.
082.015
BLAIR, I. M.
094.286 .287 .809 .810
BLAIR, P. M.
094.521
BLAKE, J. B.
061.022 .035
065.018
084.405
BLAMONT, J. E.
076.011
082.003
103.102
BLANCHARD, D. P.
094.086
BLANCHARD, M. B.
104.011
BLANCO, B. M.
046.020

BLANCO, B. M.
103.117
BLANCO, C.
114.141
116.007 .008
BLANCO, V. M.
046.020
BLANC-VAZIAGA, M.-J.
114.060
BLANDER, M.
094.468
105.060
BLANDFORD, R. D.
141.504
BLANK, D.
003.121
BLANKENSHIP, L. C.
141.040 .088 .112
142.132
BLANTON, J. N.
042.039
BLATTNER, W.
082.076
BLAU, P. J.
094.084
105.041
BLEEKER, J. A. M.
142.029 .120
BLERKOM, D. VAN
158.009
BLEVINS, B. A.
033.021
BLINNIKOV, S. I.
065.143 .158
BLINOV, N. S.
041.012 .018
BLOCH, M. R.
094.512 .799
105.127
BLOCK, L. P.
083.901
BLODGET, H.
094.587 .755
BLOKH, YA. L.
143.083
BLOOMER JR., R. H.
121.077
BLUEHDORN, J.
003.014
BLUM, P. W.
082.030
BLUMSACK, S. L.
097.013
BOBROV, M. S.
100.055
BOCCHIO, F.
094.588
BOCEK, J.
104.003
BOCHKAREV, N. G.
131.061
BOCHSLER, P.
094.437
BOCSA, G.
098.011 .012
103.001
BODE, H. J.
010.041
BODENHEIMER, P.
065.022 .023

BOEHM, K. H.
114.016
BOEHM, S.
075.011
BOEHME, A.
075.011
BOEHME, D.
123.053
BOEHME, S.
002.034
BOEHM-VITENSE, E.
064.047 .063
065.073
BOELLA, G.
134.009
BOENKOVA, N. M.
083.067
BOERNER, G.
142.117
BOESE, R. W.
093.009 .026
BOGARD, D. D.
094.140 .258 .729
105.011 .012 .056 .070
BOGART, R. S.
160.022
BOGDANOV, A. V.
097.006 .126
106.012
BOGDANOV, M. B.
104.035
BOGOD, V. M.
141.006
BOGOMOLOV, E. A.
034.095
BOHACHEVSKY, I. O.
093.036
BOHANNAN, B.
115.012
BOHLIN, R. C.
131.165
160.017 .024
BOHUSKI, T. J.
133.017
BOIGEY, F.
042.052
BOISCHOT, A.
091.071
BOJKO, P. N.
034.021
BOK, B. J.
155.070
BOK, I. I.
013.009
BOKHAN, N. A.
103.109
BOKSENBERG, A.
034.076 .083
114.039
BOLAND, B. C.
074.004
BOLBOCHANU, A. V.
112.003
BOLDT, E. A.
125.041
142.040
BOLELLI, C.
103.117
BOLKVADZE, O. R.
099.090

BOLLIN, E. M.
105.037
BOLOGNA, J. M.
131.163
BOLSHAKOVA, O. V.
106.022
BOLTENKOV, B. S.
078.047
BOLTON, A. J. C.
065.072
BOLTON, J. G.
141.041
BOLTOVSKIJ, V. G.
045.019
BONANOMI, J.
008.082
BONAZZOLA, S.
066.075
BOND, H. E.
113.056
158.064
BONDAL, K. R.
114.111
BONDAR', L. N.
094.876
BONEV, N.
093.060
094.590
097.116
101.018
BONIFAZI, A.
121.070
BONNEAU, D.
113.024
BONNELLE, C.
076.029
BONNER, J.
094.442 .904
BONNET, R. M.
054.024
BONNOR, W.
003.035
BONOLI, F.
125.040
BONOMETTO, S.
061.007
BONOMETTO, S. A.
141.042
BONOV, A.
072.068 .069
073.096
BOOTS, J. N.
041.027
045.021
BOOZER, A. H.
065.087
BORDOVITSYNA, T. V.
099.102
BOREMAN, J. A.
094.359
BORG, J.
094.459
BORGHT, R. VAN DER
061.016
BORIAKOFF, V.
141.543
BORISOV, YU. V.
122.121
BORN, G. H.
097.018 .082

BORN, W.
094.720
BORNATICI, M.
062.044
BORNER, G.
065.097
BORODIN, N. F.
093.037
097.006
BORODINA, E. G.
022.003
BORODITSKIJ, I. M.
041.025
BOROVIK, V. N.
077.017 .031
BORTLE, J.
103.100 .114 .116 .119
.124
BORTLE, J. E.
103.106 .108
123.030
BORUKHOV, M. YU.
094.860 .861
BOSCHI, E.
093.082
BOSSEN, H.
121.046 .903
BOSTICK, W. H.
073.084
BOTTEMA, M.
032.027
BOTTINELLI, L.
158.007 .026 .113 .114
.125
BOTTINGA, Y.
094.334
BOTTINO, M. L.
094.412 .689
BOTTON, C.
099.019
BOTVINOVA, V. V.
094.920
BOUCHET, M.
094.388
BOUDETTE, E. L.
094.055
BOUGERET, J. L.
077.011
BOUGHNER, R. E.
063.026
BOURGEOIS, J.
096.006
BOURY, A.
065.054
BOUSKA, J.
103.007 .008
BOUSKA, V.
105.003
BOUT, P. A. VANDEN
142.093
BOUVIER, J.-L.
094.019 .227
BOUVIER, P.
151.048 .049
BOVA, B.
003.036
BOWELL, E.
094.821 .862
BOWEN, I. S.
034.045

BROWN, B. M.
162.086
BROWN, D. R.
073.053
BROWN, G. E.
065.171
094.308
BROWN, G. M.
094.067 .124 .177 .306
.346 .614
BROWN, J. C.
073.090
076.003
BROWN, L. W.
084.408
155.080
157.005
BROWN, R. D.
022.044
131.021 .117
BROWN, R. HANBURY
115.023
BROWN, R. L.
114.062
131.001 .077 .205
157.902
160.012
BROWN, R. R.
084.006
BROWN, R. T.
125.005
BROWN, R. W.
094.099 .212 .615 .628
.913
BROWN, S.
133.038
BROWN, W.
094.569
BROWNE, I. W. A.
141.105 .120
BROWNING, R.
082.033 .902
BROWNLEE, D.
094.525
BROWNLEE, D. E.
094.141 .275
BRUBAK, H.
096.014
BRUCATO, R.
142.024
BRUCK, YU. M.
141.512
BRUECK, H. A.
008.041
BRUECKEL, F.
158.013
BRUECKNER, G. E.
076.028
BRUECKNER, W.
072.013
BRUENN, S. W.
066.020
BRUIN, F.
075.010
BRUN, A.
005.026
BRUNER JR., E. C.
073.054 .094
BRUNET, J.-P.
114.045

BRUNFELT, A. O.
094.217 .393 .679
BRUWER, J. A.
098.058
BRYAN, W. B.
094.362
BRYANT, D. A.
084.015
BRZOSTKIEWICZ, S. R.
005.032
097.098
BUCCHERI, R.
134.009
BUCHELE, D. R.
034.033
BUCHER, W.
094.525
BUCHHEIT, R. D.
094.944
BUCHHOLZ, V. L.
034.075
BUCHWALD, V. F.
105.063 .064
BUCK, R. M.
084.228 .278 .279
BUDDING, E.
121.060
BUDILOV, V. K.
143.066
BUDINE, P. W.
099.049 .091
BUECHTEMANN, W.
082.052
BUEREN, H. G. VAN
064.018
BUERGEL, W. G.
094.933
BUERGER, E. G.
133.012
BUES, I.
126.025
BUFF, J.
142.053
BUFFINGTON, A.
143.020 .058
BUFFONI, L.
103.110
BUGAENKO, L. A.
099.074
103.102
BUGAENKO, O. I.
103.102
BUHL, D.
061.013
131.049 .081 .106 .132
.167 .181
155.010 .086
BUHLER, F.
084.050
BUJ, J.
097.119
BULANOV, S.
061.068
BULANOV, S. V.
143.037 .081
BULEKOV, V. P.
094.603
BULGAKOV, B. L.
046.006
BULLEN, K. E.
091.038 .060

BULLER, A. T.
097.125
BUMBA, V.
072.006
BUNAKOVA, A. M.
091.022
097.037
BUNCH, T.
094.352
BUNCH, T. E.
094.058 .097 .142 .306
.329
105.065
BUNNENBERG, E.
094.441
BUNNER, A. N.
034.116
142.021
BURBIDGE, E. M.
141.090
BURBIDGE, G.
141.079
BURBIDGE, G. R.
141.107
142.072
BURCH, J. L.
084.227
BURDO, R. A.
094.382
BURDYUZHA, V. V.
131.157
BURGER, J. J.
143.002
BURGER, M.
114.155
BURGESS, D. E.
034.076
BURGESS, J. S.
106.016
BURGINYON, G.
142.041
155.008
BURGINYON, G. A.
125.032
BURINSKIJ, A. YA.
066.029
BURKE, B. F.
141.137
158.036 .072
BURKE, G. DE P.
143.070
BURKE, P. G.
022.038
BURKE, W. J.
094.789
BURKE-GAFFNEY, M. W.
007.000
BURKI, G.
153.023
BURLAGA, L. F.
074.071 .075
103.102
BURLINGAME, A. L.
094.102 .245 .254 .444
.752 .901
BURMAN, R.
131.200
141.559
BURMAN, R. R.
156.002

COYNE, G. V.
122.047
131.069
CRAIG, I. J. D.
073.004
CRAINE, E.
133.034
CRAINE, E. R.
141.010
CRAM, L. E.
073.020
CRAMPTON, D.
022.084
114.098
142.124
CRANNELL, C. J.
022.087
CRANNELL, H.
022.087
CRAWFORD, B.
121.008
CRAWFORD, D. L.
032.009
CRAWFORD, M. L.
094.080
CRELINSTEN, J. M.
158.097
CREMERS, C. J.
094.027 .474 .484 .786
CRESSY, P. J.
105.070
CRESSY JR., P. J.
105.011 .024
CREW, E. W.
082.013
CREZE, M.
155.004
CRIPE, J.
094.746
CRISTALDI, S.
122.031 .087 .088 .091
    .104 .118
CRISTESCU, C.
011.011 .014
098.011 .012
101.010
103.001
CRISWELL, D. R.
012.022
094.052 .792
CRISWELL, S. J.
094.570 .916
CROCE, V.
075.022
CROFT, T. A.
074.092
CROMWELL, R. H.
034.066
CRONYN, W. M.
033.023
100.013 .018
141.055 .092
CROOM, D. L.
077.013
CROSBIE, A. L.
063.014
CROSS, D. J.
004.004
CROVISIER, J.
141.516

CROWTHER, J. H.
141.026 .077
CROZAZ, G.
094.081 .255 .713 .806
CRUDDACE, R.
113.027
131.087
CRUIKSHANK, D. P.
009.023
093.062
094.584 .867
098.016
099.011 .066
101.001
CRUMP, P. C.
097.078
CRUTCHER, R. M.
131.186
155.064
CUFFEY, J.
082.035
153.035
CUKIERMAN, M.
094.182 .787
CULHANE, J. L.
125.007 .019
CULLUM, M. J.
031.030
034.070 .077
CUNNINGHAM, B. E.
094.523
CUNNINGHAM, G. G.
104.011
CUPERMAN, S.
074.050 .054 .093 .099
151.013
CUPP, R. E.
063.027
CURRAN, R. J.
097.075
CURRIE, D. G.
094.927
CURRIE, R. G.
072.049
084.260
CURTIS, A. C.
010.012
CURTIS, G. H.
072.067
CURTIS, N. A.
034.061
CUTTITTA, F.
094.218 .220 .385 .634
    .684
CUTTS, J. A.
097.008 .059 .069
CZANK, M.
094.643
CZECHOWSKY, P.
084.051
CZYZAK, S. J.
133.013 .028 .034

DACHS, J.
132.008
D'ADDARIO, L. R.
033.043
DAHLBACKA, G. H.
155.025

DAHLEN, F. A.
045.003
DAHMS, R. G.
094.523
DAHN, C. C.
111.008
113.006
133.027
DAILY, W. D.
106.017
DAINTY, A.
094.779
DAINTY, J. C.
031.050
DALGARNO, A.
008.026
061.065
092.001
131.055 .190
DALLAPORTA, N.
065.124
107.016
DALL'OGLIO, G.
071.052
D'AMATO, R.
033.002
D'AMICO, J.
094.439 .724
DAMLE, S. V.
143.008 .038
DAMNITZ, B.
003.108
DAMON, M. P.
033.026
DANCHIN, R. V.
094.379
DANDEKAR, B. S.
034.044
D'ANGELO, N.
074.078
102.017
DANIEL, R. R.
142.003
DANIELL JR., R. E.
091.014
093.018
DANIELSON, R. E.
032.026
100.040
101.013 .014
DANIELSSON, L.
062.038
DANILIN, V. A.
093.072
DANILKIN, N. P.
083.048
DANILOV, A. M.
051.009
DANILOV, V. M.
151.040
153.002
155.062
DANILOV, YU. A.
004.083
DANILOVA, L. V.
113.058
DANKANITS, A.
004.044
DANTEL, M.
074.028 .029

DERMENDZHIEV, V.
073.096
DERMOTT, S. F.
091.070
DERR, V. E.
063.027
DERVIZ, T. E.
122.144
DE SAEVSKY, P.
122.052 .116
DESAI, U. D.
106.014
DESBOROUGH, G. A.
094.359
DESER, S.
066.071
DESIKACHARY, K.
120.003
DE SILVA, H. A. B. M.
065.156
DESMARAIS, D. J.
094.247
DESSUREAU, R. L.
155.078
DETRE, L.
122.090
DEUBNER, F.-L.
071.062
080.040
DEUBNER, J.
094.531 .732
DEURINCK, R.
120.004
123.019 .020
DEUTSCHMAN, W. A.
003.044
034.084
DEVANEY, J. R.
105.037 .058
DEVINNEY, E. J.
117.035
DEVINNEY JR., E. J.
121.076
DEVORKIN, D.
014.044
DEWITT, H. E.
062.021 .022
DE YOUNG, D. S.
SEE YOUNG, D. S. DE
DIADIUK, V.
131.041
DIAMOND, H.
094.143 .262 .418 .714
DIBAY, E. A.
158.032 .049 .051
DICK, K. A.
099.084
DICK, M. L.
097.025
DICKE, R. H.
080.005
094.927
DICKEL, H. R.
131.024
132.016
DICKEL, J. R.
125.026 .901
DICKENS, R. J.
114.024 .070
DICKEY JR., J. S.
094.351

DICKINSON, D. F.
122.079
131.050
DICKINSON, R. E.
082.004
091.080
093.046
DICKS, L. A.
031.038
DICKSON, J.
034.079
DIECKVOSS, M.
098.028 .030
DIECKVOSS, W.
112.012
DIEGO Q., F.
079.003
096.020
DIERENFELDT, K. E.
093.025
DIETHELM, R.
121.081 .083 .084 .088
123.025
DIETRICH, W. F.
078.013
DIETZ, D.
003.047
DIETZ, R. S.
105.071 .072
DIETZEL, H.
105.136
DINESCU, A.
011.015
DINGLE, L. A.
072.054
DIRIKIS, M.
047.006
DISNEY, M. J.
158.107
DISTASIO, A. J.
034.003
DIVAN, L.
115.003
DIVARI, N. B.
082.062
DIXON, D.
094.619
DIXON, R. S.
041.014
DLOUJNEVSKAIA, O.
SEE DLUZHNEVSKAYA, O.
DLUZHNEVSKAYA, O.
115.017
DMITRENKO, D. A.
077.025
DMITRIEVA, I. D.
083.055
DOAN, A. S.
094.330
DOBACZEWSKA, W.
010.014
DOBROWOLNY, M.
102.017
DOBRZYCKI, J.
003.130
004.053
DODD, R. J.
101.004
D'ODORICO, S.
133.004
158.046

DODSON, H. W.
072.047
DOE, B. R.
094.271 .415 .708 .739
DOELL, R. R.
094.500
DOGEL', V. A.
143.081
DOHERTY, L. R.
051.013
121.020
DOKUCHAEVA, O. D.
036.003
DOLAN, W. W.
004.040
DOLES, J. H.
083.072
DOLGINOV, A. Z.
062.062
131.002
142.016
143.096
DOLGINOV, SH. SH.
093.055
DOLLASE, W. A.
094.314
DOLLFUS, A.
091.073
094.482 .821 .853 .862
097.015 .101
100.053
DOMBROVSKIJ, V. A.
122.142 .144
DOMBROVSKY, V. A.
122.068
DOMINA, G.
075.009
DOMINGO, V.
084.240
DOMKE, H.
063.005
DOMMANGET, J.
118.004
DONANGELO, R.
022.086
DONN, B.
103.102
DONNELLY, R. F.
073.039
076.018
DOOHER, J.
143.091
DOORNINK, D. G.
063.013
DOPITA, M. A.
034.001
131.009
132.002
D'ORAZI, R.
065.170
DORETY, N.
094.282 .498 .771
DORMAN, I. V.
143.076
DORMAN, J.
094.583 .778 .779
DORMAN, L. I.
003.013 .048
077.040
078.054 .066
106.036

DORMAN, L. I.
143.025 .044 .060 .063
.069 .074 .076 .078
.079 .082 .083 .084
.088
DOROSHENKO, V. T.
082.064
133.031
DOROSHKEVICH, A. G.
066.033 .061
151.037
DORSCHNER, J.
141.124
DOSCHEK, G. A.
073.012 .040 .083
DOSSIN, F.
103.100
DOSTOVALOV, S. B.
093.053
DOTTORI, H. A.
132.010
158.089
160.020
DOUGLAS, A. V.
005.008 .029
DOUGLAS, B. C.
054.002
055.008
DOUGLAS, J. A. V.
094.306 .327
DOUGLAS, J. N.
141.037
DOUGLASS, D. H.
045.014
DOWNING, H. D.
022.019
DOWNS, G. S.
097.002
141.539
DOWTY, E.
094.058 .075 .097 .142
.184 .312 .634
DOXSEY, R.
142.114
DOYLE, F. J.
094.929
DOYLE, R. J.
080.046
094.712
DRAGON, J. C.
094.710
DRAGUNOVA, A. V.
064.046
DRAKE, E. M.
094.543
DRAKE, F. D.
008.006
092.005
141.547
DRAKE, J. C.
094.185 .326 .676
DRAKE, J. F.
114.066 .121 .122 .123
.124 .125
131.063 .064 .065 .066
.067 .166
DRAKE, M. J.
094.105 .162
105.160
DRAKE, S.
004.036

DRAMBA, C.
004.042
005.025
DRAN, J. C.
094.804
DRAPER, C. S.
010.037
DRAVINS, D.
034.014
DRAVSKIKH, A. F.
077.018 .030
DRAVSKIKH, Z. V.
077.030
DRAWIN, H. W.
062.032
DREIBUS, G.
094.686
105.031 .073
DRESSLER, A. M.
133.011
DRESSLER, K.
114.066 .121
DREVER, H. I.
094.616
DREWRY, J. W.
042.001
DRILLING, J. S.
113.018 .019
114.021
DROBDZEV, V. I.
SEE DROBZHEV, V. I.
DROBISHEVSKY, E. M.
116.010
DROBOVA, L. V.
094.857
DROBZHEV, V. I.
083.051
DROZD, R.
094.255 .713 .806
DROZD, R. J.
094.005
DROZYNER, A.
052.029
DRYER, M.
003.009
062.048
074.050
DRYUCHENKO, D. D.
094.603
DUBA, A.
094.031 .581
DUBE, A.
105.077 .078 .141
DUBININ, E. M.
084.255 .268
DUBININ, EH. M.
084.244
DUBINSKIJ, B. A.
011.017
DUBINSKY, B. A.
033.008
DUBISCH, R.
097.062 .063
DUCATI, H.
094.530
DUCROCQ, A.
097.122
DUDINOV, V. N.
091.077
DUDKIN, V. E.
082.110

DUENNEBIER, F.
094.583 .778 .779
DUERBECK, H.
125.100
DUERR, H. P.
022.074
DUFLOT, M.
159.007
DUFOUR, L.
015.004
DUGGAL, S. P.
078.004 .006
DUIN, R. M.
125.027
DUINEN, R. J. VAN
155.046
DUKAS, H.
003.062
DUKE, C. M.
094.055
DUKE, M. B.
094.054 .350
DUL'KIN, L. Z.
051.009
DUMA, D. P.
094.131
DUMANSKIJ, Z. O.
063.021
DUMAS, A.
142.148
DUMITRESCU, A.
121.031
DUMONT, M.
015.002
120.001
DUMONT, R.
098.020
106.004
DUMONT, S.
071.008
DUMORTIER, B.
122.052 .116
DUNCAN, A. R.
094.376 .683
DUNCOMBE, J. S.
094.300
DUNCOMBE, R. L.
041.048
DUNHAM, A. C.
094.331
DUNHAM, D. W.
096.002 .022 .025
115.009
DUNHAM, J. B.
094.299
DUNLAP, J. R.
034.019 .072
094.118
DUNN, J. R.
094.284 .497 .768 .772
DUNN, P. J.
045.030
DUNN, R. B.
034.050
071.034
073.102
DUNPHY, P. P.
076.008
DUPREE, A. K.
071.045
073.057

DUPREE, A. K.
080.025
131.075
DUPUY, D. L.
009.019
DUPUY, J.
143.004
DURAND, L.
131.178
DURAUD, J. P.
094.804
DURNEY, B. R.
073.058
074.102
DURRANI, S. A.
094.260 .286 .287 .810
105.055
DURRIEU, L.
094.459 .804
DURUSSEL, R.
031.027
DURUY, M.
122.052 .116
DUXBURY, T. C.
097.082 .083
DVORNIKOV, V. M.
143.079
DWORAK, T. Z.
004.073
122.033
DWORETSKY, M. M.
114.119
DWORNIK, E.
094.495
DWORNIK, E. J.
094.218 .220 .385 .684
.774
DYAL, P.
094.493 .764
DYCE, R. B.
092.003
DYCK, H. M.
114.120
131.069
132.024
DYER, C. C.
162.007
DYER, C. S.
142.147
DYMANUS, A.
022.025
DYNKIN, S. D.
162.025
DYRING, E.
084.266
DYSON, J. E.
132.009
DYVIG, R. R.
034.066
DZHAPIASHVILI, V. P.
094.863
DZHORDZHIO, N. V.
084.043
DZHUCHENKO, YU. M.
SEE ZHUCHENKO, YU. M.
DZYAMAN, D. D.
033.010 .011
DZYUBENKO, N. I.
084.023

EADES JR., J. B.
042.001
EARDLEY, D. M.
066.116
EARL, J. A.
143.034 .039
EAST, G.
079.002
EASTLUND, B. J.
141.531
EATHER, R. H.
084.004
EATON, A. L.
094.054
EBERHARDT, P.
084.050
094.437 .730
EBERLEIN, D.
064.027
ECCLES, D.
083.014
ECKHARDT, D. H.
094.014
ECKLUND, W. L.
084.037
ECONOMOU, T. E.
094.517
097.033
EDBERG, S.
072.052
EDBERG, S. J.
073.002
EDDY, J. A.
071.032
073.059
079.106
EDELSON, S.
077.062 .068
EDEN, H. F.
097.090
EDGINGTON, J. A.
094.286 .287 .809 .810
EDLEN, B.
022.073
EDMONDS JR., F. N.
114.056
EDMONDSON, F. K.
008.018
014.034
EDMUNDS, M. G.
114.042
EDRICH, J.
155.086
EDWARDS, P. J.
008.039
EELSALU, H.
155.012
EFANOV, V. A.
093.065
EFENDIEVA, S. A.
072.010 .078
EFIMENKO, G. G.
052.028
EFIMOV, YU. S.
141.085
EFREMOV, YU. N.
003.003
115.026
122.055
159.004

EGGEN, O. J.
115.010 .016
122.029
126.030
155.099 .100
EGGLETON, P. P.
065.029 .072
122.049
EGGLETON, R.
094.569
EGGLETON, R. E.
094.055
EGIDI, A.
084.240
EGLINTON, G.
094.065 .440 .748 .900
EGOROV, V. A.
052.027
EHJDMAN, V. YA.
141.564
EHL'YASBERG, P. E.
082.034
EHMAN, J. R.
131.145
141.032
EHMANN, W. D.
094.089 .144 .222 .387
.680
105.030
EHRLICH, E. N.
094.890
EHSHMATOV, M. R.
096.011
EICHHORN, A.
123.051
EICHHORN, G.
053.015
EIDMAN, V. YA.
SEE EHJDMAN, V. YA.
EIGENSON, A. M.
154.007 .017
EILEK, J. A.
158.142
EILERS, D. D.
122.138
EISENTRAUT, K. J.
094.409 .692
EKBERG, J. O.
022.036 .037
EKERS, J. A.
158.067
EKERS, R. D.
125.102
126.001
158.036 .067
EKONOMOV, A. P.
093.083
EKSTEEN, J. P.
122.071
EL-BAZ, F.
094.608 .609 .610 .934
ELDRIDGE, J. S.
094.230 .233 .261 .381
.432 .716 .717
EL GORESY, A.
094.113 .145 .146 .306
.323 .534 .624 .626
105.074 .125
ELLDER, J.
131.108

ELLER, E.
094.587 .755
ELLIOT, J.
099.053
ELLIOT, J. L.
141.020
ELLIOTT, K. H.
131.120
ELLIS, D.
084.047
ELLIS, G. F. R.
003.058
ELLIS, G. R. A.
033.006 .025
099.018 .026
132.011
ELLIS, W.
094.450
ELLIS, W. L.
094.520
ELMABSOUT, B.
042.012
EL-RAEY, M.
077.066
ELSAESSER, H.
009.024
082.098
113.033
152.010
ELSAESSER, K.
062.043
ELST, E. W.
063.046
122.017
ELSTE, G.
071.021
ELSTON, D. P.
094.055
ELSTON, W. E.
097.100
ELTERMAN, L.
082.054
ELWIN, S. J.
099.041
ELZIE, J. L.
094.375
EMELEUS, C. H.
094.177 .346 .614
EMEL'YANOV, I. A.
033.036
EMERSON, G.
098.035
EMERSON, J. P.
113.015
EMERY, R. J.
131.079
155.039
EMETS, A. I.
081.028
ENALSKY, V. A.
021.003
ENCRENAZ, P.
141.518
ENCRENAZ, T.
091.043
099.082
ENDEAN, V. G.
141.508
ENDIKOV, G. I.
085.015

ENEEV, T. M.
151.002
ENGEL, A. E. J.
094.336
ENGEL, A. R.
142.147
ENGEL, C. G.
094.336
ENGELHARDT, W. VON
094.211 .361 .653
ENGIBARYAN, N. B.
063.012 .045
ENGLAND, A. W.
094.055
ENGSTROM, S. F. T.
074.004
ENGVOLD, O.
054.006
071.003
072.015
ENSLIN, H.
045.020
ENSOR, D. S.
082.051
ENSSLIN, N.
080.047
EOLL, J. G.
064.028
EPIKHIN, E. N.
066.042
EPPS, E. A.
122.058
EPPS, H. W.
133.015 .016
EPSTEIN, A. E.
ABRAHAM DE
122.025
EPSTEIN, E.
158.146
EPSTEIN, E. E.
131.072 .073
EPSTEIN, G. L.
072.055
073.023 .060
EPSTEIN, I.
122.025
EPSTEIN, L.
022.058
031.051
EPSTEIN, S.
094.106 .407 .700
ERASTOVA, L. K.
122.130 .131 .136
ERGMA, E.
064.037
ERICKSON, E.
093.021
ERICKSON, W. C.
033.024
141.506
ERITSIAN, M. A.
034.096
122.132
ERLANK, A. J.
094.239 .379 .687
105.044
ERMOLAYEV, G. G.
041.031
ERNEST, O.
036.003

ERNSBERGER, F. M.
094.743
ERNSBERGER, K.
054.004
EROSHENKO, E. A.
143.083
EROSHENKO, E. G.
093.055
EROSS, B.
097.062 .063 .064
ERPYLEV, N. P.
011.019
ERSHKOVICH, A. I.
084.223 .224 .241 .243
.261
102.007
ERSKINE, E.
111.005
ESEPKINA, N. A.
033.014
ESIPOV, V. F.
158.032 .049
ESPOSITO, P. B.
097.018
ESSEX, E. A.
083.022 .023
ESSEX, J. D.
082.054
ESTEP, P. A.
094.290 .469 .817
ESTERKIN, V.
077.037
EUGSTER, O.
094.410
EVANGELISTI, F.
076.013
EVANS, A.
141.027
EVANS, D. E.
094.758 .760
EVANS, D. S.
115.009
EVANS, J. C.
094.434
114.147
EVANS, J. V.
083.013 .063
093.056
EVANS, J. W.
072.046
EVANS, L. G.
077.052
EVANS, T. LLOYD
114.098
121.001
EVANS, W. F. J.
082.007
EVDOKIMOV, I. Y.
103.126
EVDOKIMOV, I. YU.
103.126
EVDOKIMOV, Y. V.
103.126
EVDOKIMOV, YU. V.
103.105 .126
104.010
EVENSEN, N. M.
094.706
EVERHART, E.
042.032
102.012

FROLOV, M. S.
003.003
122.065 .070
FRONDEL, C.
094.326 .353
FRUCHTER, J. S.
094.434 .683 .722 .909
FRY, D. J. I.
098.035
FRYAR, J.
082.033 .902
FRYDER, V.
031.015
FUBARA, D. M. J.
046.014
FUCHS, L.
094.740
FUCHS, L. H.
105.042 .079 .106
FUCHS, W. R.
003.051
FUDALI, R. F.
105.080
FUENMAYOR SUAREZ, F. J.
065.181
FUERST, E.
077.002 .009
FUERSTENBERG, F.
075.011
077.065
FUJII, M.
142.004
FUJII, N.
094.029 .781
FUJIMOTO, M.
065.003
FUJIMOTO, M.-K.
065.095
FUJITA, H.
094.319
FUKAO, S.
062.049
FUKAYA, R.
082.048 .049
FUKUI, M.
141.081
FULCHIGNONI, M.
094.170 .367 .675
FULIGNI DI GRANDE, M. T.
076.013
FULLAGAR, P. D.
094.412 .689
FULLER, M.
094.768
FULLER, M. D.
094.284 .497 .772
FULLER JR., E. L.
094.457 .554
FULLERTON, W.
131.083
FUNICIELLO, R.
094.170 .219 .367 .675
FUNK, H.
094.737
105.142
FUNKHOUSER, J.
094.403 .709
FURENLID, I.
114.064
158.156

FUSSMANN, G.
022.033

GABRIEL, A. H.
022.042
GABRIEL, M.
065.053 .118
126.031
GAFFEY, M. J.
098.013
GAGARIN, YU. F.
034.086
GAIL, H.-P.
061.070
GAJDUK, A. R.
097.087 .120
GAJNOVA, L. E.
078.040
GALEOTTI, P.
114.142
GALINDO, V.
003.020
GALKIN, L. S.
099.074
103.102
GALKIN, V. D.
022.082
GALKINA, T.
117.018
GALLAGHER III, J. S.
124.101
GALLINO, R.
061.032
162.078
GAL'PER, A.
061.020
GAL'PER, A. M.
142.057 .119
GALPERIN, B. O.
095.006
GAL'PERIN, YU. I.
051.008
083.032 .056 .057
GAMBA, Z.
131.116
GAMBURG, S. S.
099.078
GAMMAGE, R. B.
094.457 .462 .554
GAMMELIN, P.
105.136
GAMMON, R. H.
131.077
GANAPATHY, R.
094.023 .094 .154 .228
.374 .380 .594 .696
.697
105.007 .008 .045 .053
.093
GANCARZ, A. J.
094.004 .088 .173
GANDOLFI, E.
071.052
113.901
141.023 .132
GANEA, I.-M.
066.052
GANEKO, Y.
096.028

GANGADHARAM, E. V.
094.382
GANZ, R.
008.008 .088
GAPOSHKIN, S.
159.006
GARCIA-MUNOZ, M.
143.053 .057
GARDIER, S.
114.039
GARDINER, J. G.
033.051
GARDNER, F. F.
131.024
157.004
GARFINKEL, B.
042.009
GARLICK, G. F. J.
094.480 .481 .793 .811
GARMANY, C. D.
152.004
GARMIRE, G. P.
132.007
GARNER, E. L.
094.702
GARRISON, R. E.
094.364
GARRISON, R. F.
126.002
GARSTANG, R. H.
074.098
126.011
GARTMANOV, V. N.
078.047
GARY, B.
099.006
GARZOLI, S.
131.122
GARZOLI, S. L.
157.008
GAST, P. W.
094.054 .121 .268 .373
.707 .981
GATEWOOD, G.
111.005
GATLAND, K. W.
015.006
GAULT, D. E.
094.008 .030 .147 .275
.795 .796
105.081 .116
GAUR, V. P.
072.037
GAUSS, F. S.
141.013
GAUSTAD, J. E.
114.075
142.013
GAUTIER, D.
099.082
GAVRILENKO, V. G.
061.067
GAVRILOV, I. V.
094.018 .134 .882
GAVRYUK, M. I.
046.008
GAY, J.
034.011 .118
GAY, P.
094.186 .332 .627

GDALEVICH, G. L.
  083.053 .054 .055
GEAKE, J. E.
  094.480 .481 .482 .793
    .811 .821
GEARHART, M. R.
  131.145
GEBALLE, T. R.
  114.071
  133.036
GEBBIE, H. A.
  082.073
GEBBIE, K. B.
  073.075
  080.018
GEBEL, W. L.
  142.008
GEBHARDT, W.
  014.016
GEHRELS, T.
  099.063
  103.118 .121
GEHRKE, C. W.
  094.751 .949 .952
GEHRZ, R. D.
  114.030 .058
GEISS, J.
  084.050
  094.437 .730
GEL'FGAT, B. E.
  042.048
GELFREJKH, G. B.
  033.016
  077.016 .018 .020 .033
GELINAS, A.
  065.032
GELLES, R.
  031.052
GELTMAN, S.
  022.069
GENDEREN, A. M. VAN
  113.008
  121.005
  122.011 .012
  158.038
GENNER, R.
  033.059
GENT, H.
  141.026
GENTNER, W.
  094.113 .512 .535 .807
  105.082 .131
  143.059
GENTRY, R. V.
  094.318
GEORGELIN, Y. M.
  131.150
GEORGELIN, Y. P.
  131.150
GEORGEVIC, R. M.
  052.020
GEORGIEV, N.
  041.005
GERARD, E.
  100.012
GERARD, J.
  094.301 .587 .755
GERARD, J. T.
  094.382
GERASIMENKO, S. I.
  103.127

GERASIMENKO, T. P.
  155.061
GERASSIMENKO, M.
  142.051
GERLING, E. K.
  094.120
GERMANN, R.
  121.081
GEROLA, H.
  131.116
  132.022
  158.093
GERSHBERG, R. E.
  122.027 .083 .124
GERSHENGORN, G. I.
  083.037
GERTSENSHTEJN, M. E.
  066.031
GESSNER, H.
  123.054
GETMANTSEV, G. G.
  083.059
GETZELEV, I. V.
  078.023
GEVORKIAN, G. T.
  158.136
GEZARI, D. Y.
  071.036
  155.009 .038
GHETU, I.
  098.011 .012
  103.001
GHEUDIN, M.
  122.050
GHIGO, F. D.
  141.037
  142.031
GHOSE, S.
  094.082 .636
GIACAGLIA, G. E. O.
  042.028
  081.034
GIACCONI, R.
  012.002
  142.045 .052 .097 .138
    .150
  158.057
GIANNONE, P.
  117.027
GIANNUZZI, M. A.
  117.027
GIBB, F. G. F.
  094.331 .616
GIBB, T. C.
  094.775
GIBBINS, C. J.
  082.073
GIBBONS, A. H.
  131.009
  132.002
GIBERT, J. M.
  094.446 .749
GIBSON, E. K.
  094.054
GIBSON, J.
  098.023 .027 .039 .041
    .053 .056 .058 .061
  103.114 .117
GIBSON, U.
  098.023 .027 .041 .053
    .056 .058 .061

GIBSON, U.
  103.114 .117
GIBSON JR., E. K.
  094.056 .148 .250 .252
    .400 .401 .741 .744
    .746 .903
  105.012
GICLAS, H. L.
  098.025
  103.100 .114 .118 .124
    .127 .132
  112.005
GIDALEVICH, E. YA.
  131.199
GIERASCH, P.
  097.057 .062 .063
GIERASCH, P. J.
  097.013
GIESE, R. H.
  106.005 .030 .031 .034
  131.035
GIESEKING, F.
  123.059
GIFFEN, R.
  098.002 .018
GIGUERE, P. T.
  131.167 .181
GILARDONI, G.
  115.025
GILBERT, J. C.
  099.081
GILES, H.
  094.219
GILES, H. N.
  094.331
GILES, J.
  099.054
GILLETT, F. C.
  099.016
  114.027 .040
  132.024
GILLINGHAM, P. R.
  032.004
GILLUM, D. E.
  094.089 .144 .222 .680
  105.030
GILMAN, R. C.
  064.001
GILMORE, A. C.
  103.103 .133
GILVARG, A. B.
  097.022
GINGOLD, R. A.
  065.046
  122.002
GINZBURG, A. S.
  093.078
  097.019
GINZBURG, V. L.
  003.052
  022.076
  061.002 .026
  066.056
  142.055 .056
  143.041
GIOIA, I.
  141.023 .132
  158.040
GIOVANELLI, R.
  131.205

GOREL, L. F.
  041.038 .040
GORELIK, G. E.
  066.089
GORENSTEIN, P.
  034.121
  094.060 .301 .571 .587
    .755 .756
  141.507
GORGOLEWSKI, S.
  009.021
GORIN, V. D.
  105.145
GORODETSKIJ, D. E.
  103.016
GORONINA, K. A.
  094.876
GOROSHANKIN, B. N.
  083.053
GORSKIJ, YA.
  066.032
GORTON, M.
  094.237
GOSACHINSKY, I. V.
  155.060
GOSE, W. A.
  094.051 .280 .769 .773
GOSLING, J. T.
  074.051 .069
GOSS, W. M.
  125.102
  131.027 .037 .105
  141.087
  142.038
GOSWAMI, J. N.
  094.257 .800 .808
GOTO, T.
  045.011 .012
GOTO, Y.
  032.016
GOTTESMAN, S. T.
  131.097
  157.902
  158.039
GOTTLIEB, B.
  082.084
GOTTLIEB, C. A.
  131.202
GOTTLIEB, D. M.
  113.031
GOUDIS, C.
  034.023
  132.029
GOUDY, A. J.
  105.020
GOUGH, P. T.
  034.002
GOUGUENHEIM, L.
  158.007 .026 .113 .114
    .125
GOUVEIA, H.
  082.042
GOWER, A. C.
  034.075
GOZHIJ, A. V.
  032.029
GRABOSKE, H. C.
  062.021 .022
  065.024 .061
GRACHEV, S. I.
  062.046

GRACHEV, YU. A.
  077.025
GRAEVE, E. DE
  082.121
GRAF, H.
  094.713 .730
GRAF, L. EH.
  094.603
GRAF, W.
  077.008 .015
GRAHAM, A.
  094.377
GRAHAM, D.
  141.514
GRAHAM, J. A.
  114.127
GRAHAM, R.
  131.017
GRAMLICH, J. W.
  094.702
GRANDI, S.
  133.026
GRANGEL, J. C.
  022.085
GRANT, R. W.
  094.468 .673
GRAPE, K.
  155.027
GRARD, R. J. L.
  094.790
GRASDALEN, G.
  132.006
GRASDALEN, G. L.
  114.116
  122.137
GRASSERBAUER, M.
  105.156
GRATTON, L.
  142.105
GRAY, C. M.
  094.705
GRAY, D. F.
  114.147
  120.003
GRAYSON, H. W.
  066.013
GRAYSON, M. A.
  105.010
GREATREX, R.
  094.775
GREATRIX, G. R.
  072.067
GREBENIKOV, E. A.
  003.055
GREBENKEMPER, C. J.
  072.018
GRECHISHCHEVA, I. M.
  093.012 .016
GREELEY, R.
  094.030
GREEN, A. E. S.
  099.087
GREEN, D. H.
  094.038 .189 .347 .551
    .618
GREEN, I. M.
  106.016
GREEN, L. C.
  117.028
GREEN, R.
  103.132

GREEN, R. H.
  094.519
GREEN, R. R.
  097.002
GREEN, S.
  022.008
GREEN, W. B.
  097.008
GREENBERG, J. M.
  063.044
  131.008 .010 .091 .099
  155.049
GREENBERG, R.
  091.042
GREENBERG, R. J.
  100.028
GREENE, A. E.
  064.081
GREENE, C. H.
  094.461
GREENE, T. F.
  099.067
  114.011
GREENHILL, J. G.
  141.510
GREENLAND, L. P.
  094.385 .684
GREENMAN, N. N.
  094.477 .812
GREENSPAN, D.
  151.055
GREENSTEIN, J. L.
  119.004
  125.030
  126.015
GREENWALD, R. A.
  084.037
GREENWOOD, N. N.
  094.775
GREENWOOD, W. R.
  094.936
GREGORY, P. C.
  141.036 .060
GREGORY, S. A.
  141.059
  158.058 .140
GREISEN, K.
  134.006
GRENON, M.
  065.135
  115.019
GREVE, A.
  031.024
  032.002
  076.007 .026
GREVESSE, N.
  012.003
GREWING, M.
  061.038
  114.149
  141.533 .560
  143.027 .073
GREYBER, H. D.
  012.018
GRIB, A. A.
  066.030
GRIBBIN, J.
  044.008
  091.068
GRIBOVAL, D.
  036.007

GURNEY, J. J.
094.239 .379 .687
GURSHTEJN, A. A.
005.013
GURSKY, H.
061.045
094.587 .755
142.045 .052 .097 .123
.138 .150
158.057
GURTOVENKO, E. A.
071.023 .024
GURZADYAN, G. A.
126.008
GUSEINOV, O. H.
SEE GUSEJNOV, O. KH.
GUSEINOV, R. E.
074.016
GUSEJNOV, O. KH.
065.038 .082
125.004 .017
141.563
GUSEV, E. B.
114.014
GUSEV, P. P.
123.004
GUSSENHOVEN, M. S.
074.044
GUSTAFSSON, B.
064.062
GUTHRIE, B. N. G.
131.003
GUTHRIE, P.
155.017
GUTIERREZ, J. A.
075.012
GUTKIN, A. M.
094.892
GUTMAN, I. I.
066.024
GUTZWILLER, M. C.
008.128
042.046
GVOZDEVA, L. G.
105.138
GWINN, W. D.
131.027

HAAR, D. TER
062.017
107.021
131.128
158.042
HABING, H. J.
012.010
141.087 .115
HACHENBERG, O.
077.009
131.170
HACK, M.
064.060
121.017
HACKNEY, R.
014.010
HADDOCK, F. T.
074.043
077.047 .069
HAEKLI, T. A.
105.023

HAEMEEN-ANTTILA, K. A.
100.015
HAERENDEL, G.
083.075
HAEUSSLER, K.
123.011
HAFNER, S. S.
094.310 .635 .644
HAGAN, L.
008.122
HAGEN, J. P.
077.036
HAGEN-THORN, V. A.
141.135
158.130
HAGFORS, T.
008.006
093.033 .056
094.877
HAGGERTY, S. E.
094.085 .135 .190 .191
.192 .625
HAIT, M. H.
094.055 .607
HAJDUK, A.
104.002
HAJIAN, G. S.
SEE ADZHYAN, G. S.
HAKIM, R.
162.048
HALL, D. L.
158.037
HALL, D. N. B.
071.002 .051 .055
080.006
HALL, D. S.
121.015 .016 .050
HALL, F. G.
094.574
HALL, J. E.
083.026
HALL, J. S.
008.044
100.026
HALL, L. A.
076.018
HALL, L. M.
131.128
HALL, R. C.
010.037
HALL, R. G.
041.026
HALL, T. A.
094.055
HALLAM, M.
094.151
HALLIDAY, I.
104.012
HALPERN, B.
094.441
HALVORSEN, H. D.
071.003
HAMANO, Y.
094.486 .781
HAMILTON, P. A.
141.510 .554 .555
HAMILTON, P. B.
094.069 .251 .907
HAMMERTON, M.
101.016

HAMMOND, A. L.
094.129
097.024
HAMMOND, D. A.
094.742
HAN, S. M.
073.104
HANBURY BROWN, R.
115.023
HANEL, R. A.
097.075
HANEMAN, D.
094.503
HANKINS, T. H.
141.523 .525 .526
HANNER, M. S.
106.003
HANSELL, M. C.
141.534
HANSEN, C. J.
065.030
122.139
HANSEN, O. L.
099.012
HANSON, J. N.
042.050
HANSON, W. B.
083.008
HANSSON, N.
007.000
HANTZSCHE, E.
003.057
HANZAL, J.
005.019
HAPKE, B.
094.580 .856
HARA, H.
031.018
HARA, K.
162.052
HARA, T.
035.003
044.035
061.061
HARADA, K.
094.248 .750 .906
HARADA, Y.
096.028
HARAMUNDANIS, K.
002.007
113.047
HARAMUNDANIS, K. L.
003.044
HARDCASTLE, K.
094.405
HARDCASTLE, K. G.
094.249
HARDER, A.
162.051 .073
HARDIE, R. H.
118.006
HARDING, G. A.
008.027
HARDORP, J.
114.007
HARDY, J.
033.056
HARE, P. E.
094.248 .750 .906 .950
HARGRAVES, R. B.
094.282 .343 .498 .771

HARLAN, E. A.
114.116
HARMANEC, P.
119.003
HARMON, R. S.
094.025 .628
HARNDEN JR., F. R.
141.507
HARNIK, A. B.
094.643
HARO, G.
122.099
HARPER JR., D. A.
158.145
HARRINGTON, J. P.
063.023
HARRINGTON, R. S.
046.020
125.015
HARRIS, B.
034.121
094.587 .755
HARRIS, C. S.
114.046
HARRIS, D. E.
141.051 .078 .099
HARRIS, I.
082.030
HARRIS, S.
131.191
HARRIS, W. E.
154.009
HARRISON, E. F.
053.020
HARRISON, E. R.
162.067
HARSTAD, K. G.
063.008
HART, R.
004.050
005.002
HART JR., H. R.
073.098
094.040 .045 .114 .264
    .265 .505 .506 .801
    .808
143.011
HARTLE, R. E.
074.006
097.096
HARTLEY, K. F.
031.029
HARTMANN, W. K.
094.550 .584
097.082
HARTOOG, M.
071.021
HARTOOG, M. R.
114.172
HARTUNG, J. B.
094.147 .275 .511 .535
    .795 .796
HARTWICK, F. D. A.
065.058
154.018
HARUTJUNIAN, E. H.
 SEE ARUTYUNYAN, EH. A.
HARUTUNIAN, A. S.
 SEE ARUTYUNYAN, A. S.

HARUTYUNIAN, G. G.
 SEE ARUTYUNYAN, G. G.
HARVEY, C. C.
077.005
HARVEY, G. M.
113.040
HARVEY, J.
072.004
HARVEY, J. W.
031.054
071.001 .037
072.002
080.020
HARVEY, K.
072.004
HARWIT, M.
034.031
097.085
HARWOOD, J. M.
124.104
HASEGAWA, O.
123.063 .065 .066
HASKELL, G. P.
084.401
HASKIN, L. A.
094.086 .223 .396 .688
HASSAN, S. M.
153.008 .015
HATANAKA, H.
103.114
HAUCK, B.
002.007
065.133
113.055
114.140
122.016
HAUCK, J.
094.341 .617
HAUER, F.
007.000
HAUG, U.
131.171
HAUGE, O.
071.054
HAUPT, H.
074.088
079.004 .102
HAUPT, W.
097.026
142.015 .027 .118
HAUSER, J. S.
094.444
HAUSKA, H.
084.266 .267
HAVAS, P.
042.018
HAVLEN, R. J.
155.079
HAVNES, O.
143.028
HAWARDEN, T. G.
122.111
HAWKING, S. W.
003.058
162.015 .061
HAWKINS, F.
142.138
HAWKINS, F. J.
142.033
HAWKINS, J. W.
094.372

HAYAKAWA, M.
084.212
HAYAKAWA, S.
084.412
131.196
142.004 .075 .120 .145
HAYASAKA, T.
121.037
HAYASHI, C.
065.163
HAYASHI, K.
066.119
HAYASHI, N.
071.062
HAYES, J. M.
094.247 .897
HAYMES, W. E.
117.034
HAYNES, N. R.
097.007
HAYS, J. F.
094.200 .613 .656
HAYS, P. B.
082.027 .029 .041 .115
084.017
HAZARD, C.
141.026
HAZLEHURST, J.
117.019
HEAD, J.
094.054
HEAD, J. W.
094.055 .130 .483 .605
    .822 .940
HEALY, L. G.
034.062
HEAP, S. R.
142.091
HEARN, A. G.
064.009 .019
HEARN, D. R.
142.133
HEARNSHAW, J. B.
115.021
155.077
HEASLEY, J. N.
064.082
073.068
HEATH, D. F.
034.122
076.033
HECHT. H. F.
162.058
HECK, A.
008.055
103.100
122.035
HECKMANN, O.
005.021
HEDEMAN, E. R.
072.047
HEDGE, C. E.
094.271 .708
HEDGECOCK, P. C.
084.240
HEDIN, A. E.
082.071
HEERAN, M. P.
084.019
HEESCHEN, D. S.
008.031 .050 .116

HEYMANN, D.
094.731 .738
HEYWOOD, H.
094.455
HIBBERSON, W. O.
094.347 .618
HIBINO, A.
094.311
HICKS, P. D.
123.058
HICKSON, P. J.
094.542
HIDAJAT, B.
113.048
114.151
HIEBERT, R. D.
082.015
HIGASHI, K.
022.051
HIGBIE, P. R.
076.008
HIGGINBOTHAM, N. A.
114.117
HIGGINS, C. S.
141.064
HIGGINS, G. H.
081.005
HIGGINS, P.
126.026
HIGGS, _. A.
133.001
141.102
HIGH, R. W.
094.524
HILEMAN, F. D.
094.692
HILL, G.
121.003 .007 .026
HILL, J. K.
161.002
HILL, R.
142.041
155.008
HILL, R. W.
125.032
HILLER, H.
081.901
HILLS, H. K.
094.466 .759
HILLS, J. G.
064.042
107.008
154.002
HILLS, R. E.
093.008
141.019
HILTNER, W. A.
008.004
126.002
142.044 .063 .130
HILTON, H. H.
084.203
HIMMEL, G.
022.033 .063
HINNOV, E.
062.010
HINTENBERGER, H.
094.422
105.149
HINTEREGGER, H. E.
034.114

HINTEREGGER, H. F.
141.052
HINTHORNE, J. R.
094.073 .354
HINTZE, K.-H.
004.030
HINTZEN, P.
121.068
HIPPELEIN, H. H.
131.124 .173
HJRABAYASHI, H.
021.005
HIRAO, K.
099.034
HIRAYAMA, T.
021.004
073.001
103.127
HIRNER, A.
032.034
HIRSHBERG, J.
071.016
072.048
074.061
HIRTH, W.
077.009
141.101
HIRUMA, T.
034.063
HISLOP, J. S.
094.389
HIVELY, R.
066.004
HJALMARSON, A.
062.067
HJELLMING, R. M.
141.035 .040 .088 .100
.112
142.011 .065 .132
155.037
HLAVA, P. F.
094.028
HOANG-BINH, D.
131.094
133.019
HOBBS, B. A.
094.003
HOBBS, L. M.
114.038
131.046 .054 .139 .182
HOBBS, R. W.
072.055 .059
073.023 .060
077.043
132.023
141.048
HOCH, R. J.
084.025
HODGE, P.
094.525
HODGE, P. W.
082.067
094.141 .586
158.157 .159
HODGES, C. A.
094.055
HODGES JR., R. R.
094.758
HODGSON, G. W.
094.441 .905

HODGSON, R. G.
098.079
HOEBEL, P.
031.065
HOEG, E.
031.033
HOEGLUND, B.
131.037 .108 .160 .161
HOEGNER, W.
102.011
HOEHN, D. H.
082.052
HOEKSTRA, R.
114.031 .049
HOEPPNER, W.
065.079 .150
HOERNER, S. VON
033.007
162.042
HOERZ, F.
094.141 .147 .275 .511
.795 .796
105.001
HOFF, D.
014.045
HOFFLEIT, D.
034.101
123.026
HOFFMAN, H. S.
022.002
HOFFMAN, J.
084.031
HOFFMAN, J. A.
096.024
155.052
HOFFMAN, J. H.
034.119
094.758
HOFFMAN, K. A.
094.278
HOFFMAN, M.
074.063
HOFFMAN, M. M.
079.101
HOFFMAN, W. P.
094.620
HOFFMANN, B.
003.062
HOFFMANN, H.-J.
105.135
106.020
HOFFMANN, W. F.
051.014
114.088
131.079
155.039
HOFMANN, W.
034.009
082.099
106.027
113.033
HOFMEYR, P. K.
094.379
HOGAN, P. A.
066.124
HOGG, D. E.
141.046
HOGG, H. S.
007.000
122.021

HOHENBERG, C. M.
094.005 .152 .255 .713
.806
HOHMANN, W.
003.153
HOLBERG, J.
160.013
HOLDSWORTH, E.
105.051 .085
HOLDSWORTH, E. F.
105.016 .066
HOLLAND, J. G.
094.124 .177 .346 .614
HOLLAND, P. T.
094.752 .901
HOLLANDSKIJ, O. P.
064.055
HOLLARS, D. R.
114.065
HOLLEBEKE, M. A. I. VAN
078.033
HOLLISTER, L. S.
094.343 .553 .633
HOLLWEG, J. V.
074.040
HOLM, A. V.
121.020
125.101
HOLMES, H. F.
094.457 .554
HOLT, H. E.
094.055
HOLT, J. N.
064.019
071.018
HOLT, S. S.
125.041
142.040
HOLWEGER, H.
071.061
HOLZER, E.
074.094
HONDA, M.
094.722
HONES JR., E. W.
084.202
HONEYCUTT, R. K.
114.109
131.048
HOOF, A. VAN
119.024
123.060 .061
HOPKINS, H. D.
083.015
HOPKINS, J.
081.030
HOPPE, J.
004.011
HOPPER, R. W.
094.788
HORAI, K.-I.
094.546
HORAK, H.
082.076
HORAK, H. G.
063.019
074.024
HORAK, Z.
066.050
HORD, C. W.
097.009 .025 .042 .055

HOREDT, G.
061.009
HORIAI, K.
035.003
HORN, P.
094.161 .527 .732
105.124
HORNE, D. F.
003.063
HORNSTEIN, S.
022.008
HORSTMAN, H.
076.014
142.148
HORSTMAN-MORETTI, E.
076.014
142.148
HORZ, F.
094.054
HOSHI, R.
065.163
HOSKIN, M.
162.057
HOSKINS, D. G.
141.026
HOSOKAWA, Y.
121.048
HOUCK, J. R.
097.041
HOURANI, H.
075.010
HOUSE, F. C.
151.018
HOUSLEY, R. M.
094.288 .468 .488 .673
.782
HOUSTON, W. N.
094.452 .832 .833
HOUZIAUX, L.
114.001 .039
HOVANESIAN, R. S.
SEE OGANESYAN, R. S.
HOVESTADT, D.
143.095
HOVLAND, H. J.
094.016 .833
HOWARD, C. J.
022.005
HOWARD, K. A.
094.605 .931 .939
HOWARD, R.
073.002
075.016
080.026
HOWARD, R. F.
080.029
HOWARTH, I. D.
122.080
HOWELL, F. J.
098.035
HOWELL, J. A.
098.026
HOYLE, F.
003.064
061.012
141.034
HOYT JR., H. P.
094.479 .806 .813
HSIEH, T.
094.794

HUANG, C. K.
094.196 .197 .649
HUANG, S.-S.
117.021
121.018 .051
HUANG, W. H.
094.370
HUANG, Y. H.
084.222
HUANG, Y.-N.
083.006
HUBATSCH, W.
003.014
HUBBARD, E. C.
091.069
HUBBARD, N. J.
094.054 .121 .333 .373
.707 .741 .981
HUBBARD, W. B.
051.002
062.008
091.029 .041 .066
099.079
HUBBELL, E.
111.005
HUBE, D. P.
119.017
HUBER, M. C. E.
071.045
HUBERT, H.
114.013
HUCHRA, J.
113.022
HUCHRA, J. P.
103.132
HUCHT, K. A. VAN DER
114.031 .049 .101 .137
.155
119.011
HUCHTMEIER, W.
158.003 .004 .025
HUCHTMEIER, W. K.
155.058
HUDSON, H. S.
073.061 .109
HUEBNER, J. S.
094.312
HUEBNER, W.
094.529
HUEBNER, W. F.
131.083
HUEBSCHER, J.
123.002
HUECKEL, P.
100.025
HUEY, J. M.
094.416
105.151
HUFFMAN, D. R.
131.133
HUFFMAN, G. P.
094.283 .826
HUG, K.
094.929
HUGGINS, F. E.
094.637
HUGHES, A. R. W.
083.005
HUGHES, D. W.
104.005

HUGHES, T. C.
094.398
HUGHES, V. A.
119.020
141.063
155.018
HUGHES JR., E. E.
115.006
HUGUENIN, D.
114.032
HUGUENIN, G. R.
141.501 .538 .544 .545
162.017
HUGUENIN, R.
097.061
HULSBOSCH, A. N. M.
155.006
HUMMER, D. G.
064.012 .032
HUMPHREYS, R. M.
112.004 .007
114.012 .022
HUMPHREYS JR., W. C.
003.065
HUMPHRIES, C.
114.039
HUMPHRIES, D. J.
094.194 .348 .619
HUNDHAUSEN, A. J.
074.046 .072
HUNDT, E.
114.146
HUNEKE, J. C.
094.088 .426 .711
105.014 .109
HUNG, R. J.
074.001 .011 .012 .023
HUNGER, K.
064.068
114.130
HUNSUCKER, R. D.
084.013
HUNT, G. C.
141.514
HUNT, G. E.
091.019
093.007 .010
099.005 .044 .045 .058
HUNT, G. R.
094.291 .818
097.039 .040
HUNT, R. H.
022.019
HUNTEN, D. M.
051.002
091.048
097.004 .103
099.030 .099
HUNTER, C.
151.034
155.036
HUNTER JR., J. H.
133.006
HURD, J. M.
094.701
HURLESS, C.
122.122
HURUKAWA, K.
103.127
HUSAIN, D.
022.017

HUSAIN, L.
094.266 .709 .935
HUSEYNOV, O. H.
065.038
HUSS, G. I
105.018 .086
HUTCHEON, I. D.
094.276 .518 .802 .805
.808
HUTCHINGS, J. B.
008.118
117.009
119.022
121.003 .007 .026
124.101
142.124
HUTCHINSON, J. L.
122.030
HUTCHISON, R.
105.087 .088
HUTCHISON, R. B.
114.081
HUTH, H.
123.046
HUTTON, L. K.
141.055
HYLAND, A. R.
065.046
114.025 .079
HYNDS, R. J.
084.401
HYNEK, J. A.
008.043 .065
034.019 .072
094.118

IACOB, C.
004.045
IANNA, P. A.
111.004
IBANEZ, A. L.
077.058
IBANEZ, M.
123.016 .022
IBEN JR., I.
065.104 .119 .121
122.039
IBRAGIMOV, N. B.
099.073
ICHIMARU, S.
062.056
ICHIMURA, K.
122.100 .134
ICKE, V.
131.177
IDEL'SON, N. I.
004.061
IFEDILI, S. O.
078.059
IGLESIAS, E.
131.116
IGNAT'EV, P. P.
093.080
IHOCHI, H.
094.416
IKE, K.
103.124
IKEDA, Y.
094.612

IKEUCHI, S.
065.163
158.129
IKHSANOV, R. N.
072.024 .025 .026
IKHSANOVA, V. N.
077.017
IKUTA, K.
062.036
ILENCIK, J.
078.022
ILJASOV, U. I.
071.027
073.036
ILK, K. H.
052.043
ILLARIONOV, A. F.
064.079
ILOVAISKY, S. A.
125.025
ILYASOV, YU. P.
033.031
074.103
IMAMURA, M.
094.434 .722
IMSHENNIK, V. S.
065.102
126.005
INCIONG, S. V.
047.016 .017 .018
INGALLS, R. P.
092.003
093.003 .056
INGERSON, P. G.
033.070
INGVARSON, P.
131.134
INNANEN, K. A.
151.018
INNANEN, W. G.
078.018
INOZEMTSEVA, O. I.
143.083
IONESCU-GULIAN, C.
004.043
IOSHPA, B. A.
071.030
080.036
IP, W.-H.
098.006
IPAT'EV, V. I.
078.045
IPSER, J. R.
065.052
IRGENS-JENSEN, S.
122.106
IROSHNIKOV, R. S.
099.037 .038
IRVINE, W. M.
008.003
091.052
094.297 .565
097.107
100.005
IRWIN, G.
082.100
IRWIN, J. B.
119.001
ISAKA, H.
082.010

ISARD, J. O.
094.456
ISAVNINA, I. V.
094.956 .959
ISENHOUR, T. L.
094.409
ISHERWOOD, B. C.
034.075
122.013
ISHIHARA, T.
124.106
ISHII, H.
031.018
045.013
082.049
ISHIKAWA, M.
122.001
ISHIMARU, A.
091.081
ISKHAKOV, I. A.
077.061
ISLES, J. E.
010.012
122.020 .081 .094 .107
123.001
ISLIKER, W.
009.005
ISOBE, S.
131.004 .135
ISRAEL, M.
081.009
ISRAEL, M. H.
143.055
ISRAEL, W.
066.099
ISSERSTEDT, J.
132.008
153.033
ITO, J.
094.353
ITOH, N.
065.176
IUCCI, N.
143.048 .089
IVANENKO, D. D.
066.025
IVANITSKAYA, O. S.
066.065
IVANOV, A. N.
033.031 .033
IVANOV, K. G.
062.016
073.025
106.001 .021
IVANOV, N. M.
053.005
IVANOV, S. N.
033.031 .033
074.103
IVANOV, V. I.
082.065 .129
143.066
IVANOV, V. P.
083.059
125.028
IVANOV, YU. G.
083.042
IVANOVA, N. L.
114.166 .167
IVANOVA, N. S.
034.086

IVANOVA, N. S.
143.015 .080
IVANOVA, T. A.
078.016
IVANOV-KHOLODNY, G. S.
083.038 .047
IVANOV-KHOLODNYJ, G. S.
076.036
IVELSKAYA, M. K.
083.047
IWADATE, K.
034.040
035.003
IWANISZEWSKA, C.
003.066
IWANOWSKA, W.
004.064
005.009
IYENGAR, V. S.
142.002
IZVEKOVA, V. A.
033.030

JACCHIA, L. G.
082.120
JACKA, F.
084.044
JACKISCH, G.
004.031
JACKSON, E. D.
094.055 .556
JACKSON, E. K.
094.054
JACKSON, J. C.
066.010
141.119
160.005
JACKSON, J. J.
034.119
JACKSON, P.
094.569
JACKSON, P. F. S.
094.235 .694
JACKSON, R. F.
094.102
JACKSON, R. J.
094.254
JACKSON, T. J.
094.447
JACKSON IV, A. A.
066.011
JACOBS, J. A.
081.017
JACOBS, J. W.
094.086
JACOBSON, I. D.
042.039
JACOBY, J.
094.657
JAERVI, P.
055.004
JAFFE, L. D.
094.064 .855
JAGER, C. DE
003.067
010.017
076.005
JAGGI, R. K.
084.272

JAGODA, N.
034.111
JAGODZINSKI, H.
094.175 .639
JAGOUTZ, E.
094.383 .674
JAHELKA, E. D.
097.008
JAHN, BOR-MING
094.706
JAHN, R. A.
094.286 .809
JAIN, A. K.
142.101
JAIN, A. V.
105.039
JAKES, P.
094.025 .628
JAKI, S. L.
003.068
JAMAR, C.
114.039
JAMES, G. L.
033.071
JAMES, O. B.
094.925
JAMESON, R. F.
121.008
JAMIESON, H. D.
094.852
JAMIESON, W. D.
094.369 .671
JANES, K. A.
098.031
153.017
155.075
JANGHORBANI, M.
094.089 .144 .222
105.030
JANSSEN, M. A.
093.008 .035
097.045
JANSSENS, T. J.
034.052
JAROSEWICH, E.
094.655 .943
105.021 .105 .165 .166
JARRETT, A. H.
009.009
122.071
JASCHEK, C.
114.057
115.013
131.122
153.028
JASCHEK, M.
114.057
JASCHEK, W.
034.110
JASTRZEBSKI, T.
103.112
JAUHO, P.
141.566
JAUNCEY, D. L.
100.006
141.026 .109
158.029
JAYANTHI, U. B.
142.101
JEDWAB, J.
094.195 .366

JEDWAB, J.
105.025
JEFFERIES, J. T.
008.058
071.039
076.020
JEFFERS, S.
114.161 .163
119.018
JEFFERTS, K. B.
131.029
132.003
JEFFERYS, W. H.
042.049
JELLEY, J. V.
034.005
155.097
JENKEN, M. E. J.
033.048
JENKINS, E.
131.146
JENKINS, E. B.
114.066 .121 .122 .123
   .124 .125
131.063 .064 .065 .066
   .067 .137 .166
JENKINS, E. F.
008.119
JENKINS, R. W.
078.059
JENKINS, V.
033.052
JENKNER, H.
118.013
131.172
JENNER, D. C.
133.015 .016
JENNINGS, R. E.
113.015
JENNISON, R. C.
094.797
JENSEN, P.
041.001
JENSEN, V. O.
074.078
JEROME, D. Y.
094.090
105.017 .089
JESSBERGER, E.
094.403
JETTNER, F. C.
014.041
JETZER, F.
100.010
JEUKEN, M. E. J.
033.060
JEZKOVA, M.
104.003
JOCKERS, K.
103.102
JOERGENSEN, B. G.
098.021
103.005
JOERGENSEN, H. E.
065.125
JOGLEKAR, P. J.
085.005
JOHAN, Z.
094.078
JOHANSEN, R.
061.062

JOHANSEN, R. A.
097.008
JOHANSSON, L. E. B.
131.037
JOHN, T. L.
064.088
JOHN, W.
094.420
JOHNSON, D. R.
131.181
JOHNSON, F. M.
065.068
131.085 .096
JOHNSON, F. S.
082.084
094.760
JOHNSON, H. L.
114.114
JOHNSON, H. M.
115.011
131.184
132.005 .019
142.050
JOHNSON, H. R.
064.015 .050
JOHNSON, L. B.
011.026
JOHNSON, L. C.
062.010
JOHNSON, M. W.
131.129
JOHNSON, P. H.
094.450
JOHNSON, S. M.
094.401
JOHNSON, S. W.
094.453 .830
JOHNSON, T. V.
091.045
094.840
098.004 .015
JOHNSON JR., G. G.
094.458 .828
JOHNSTON, D. A.
094.628
JOHNSTON, K. J.
131.052 .163 .183
132.023
JOHNSTON, R.
094.616
JOKIPII, J. R.
078.030 .043 .046
106.029
143.036
JOLY, F.
131.093
JOLY, M.
158.126
JONES, A.
124.107
JONES, A. F.
123.039 .041 .042 .043
   .044
JONES, B. B.
074.004
JONES, B. F.
112.002
153.018
JONES, B. J. T.
151.031

JONES, B. R.
094.240 .761 .918
JONES, B. W.
094.780
JONES, C.
094.439
142.045 .052 .131 .150
JONES, C. A.
142.048 .151
JONES, D. H. P.
155.007
JONES, E. M.
125.013 .033
JONES, F. C.
143.005 .021
JONES, H. P.
073.060
080.027
JONES, K. L.
083.020
097.066
JONES, P. B.
031.009
JONES, R. A.
084.016 .017 .046
JONES, R. V.
009.012
JONES, T. J. L.
034.083
JONES, T. W.
121.006
141.069
JONES JR., E. C.
080.047
JONG, T. DE
131.076 .090 .114 .151
   .152
JORDAN, J. A.
094.055
JORDAN, J. F.
097.018
JORDAN, J. L.
094.242
JORDAN, J. R.
033.055
JORDAN, P.
003.014
JORDAN, S. D.
003.069
072.059
077.043
JORGIO, N. V.
SEE DZHORDZHIO, N. V.
JOSEPH, G.
142.003
JOSHI, G. C.
114.111 .112
JOSHI, M. N.
141.068
JOSS, P. C.
065.087 .088 .089
066.126
102.013 .015
141.011
JOSTIES, F. J.
098.026
JOURET, C.
094.459 .804
JOURNET, A.
034.011 .118

JOVANOVIC, S.
  094.098 .156 .231 .298
       .391 .715 .740
  105.090
JOYCE, R. R.
  071.036
  155.009 .038
JOZKOVICH, B.
  033.008
JUAN, V. C.
  094.196 .197 .649
JUDGE, D. L.
  022.079
  082.056
JUERSS, F.
  004.024
JULL, E. V.
  033.072
JUNG, J.
  002.007
  009.013
JUNGE, C.
  082.091
JUNKES, J.
  008.028
JUNKINS, J. L.
  042.039
JUPP, A. H.
  042.004 .019 .901
JURA, M.
  141.083
JURTCHENKO, B. N.
  SEE YURCHENKO, B. N.

KACPEREK, A.
  094.286
KAEHLER, H.
  022.034
  061.039
  065.148
KAFATOS, M.
  131.179
KAFATOS, M. C.
  131.070
  155.089
KAFTAN-KASSIM, M. A.
  158.119
KAGANOVSKIJ, G. M.
  032.020
  045.018
KAGIWADA, H.
  063.010
KAHAN, E.
  012.017
  036.006
KAHN, F. D.
  003.011
  131.155
KAI, K.
  021.005
  077.060
KAIFU, N.
  131.005 .103 .189
KAISER, M. L.
  084.408
  099.033 .050
KAISER, T. B.
  143.005 .021
  156.001

KAISER, W. A.
  094.423
KAJNAROVA, YA.
  083.054
KAKHIDZE, G. P.
  143.065
KAKINUMA, T.
  106.011
KAKKURI, J.
  034.108
KAKUTA, C.
  045.010
KALABA, R.
  063.010
KALACHEV, P. D.
  033.035 .036 .037 .038
KALACHEV, V. L.
  054.017
KALACHNIKOV, A. A.
  091.079
KALBITZER, S.
  094.530
KALER, J. B.
  133.013 .034
KALIBERDA, V. S.
  151.015
KALININ, A. P.
  022.064
KALININ, M. I.
  162.023
KALINKIN, L. F.
  143.017 .065
KALINKINA, O. M.
  093.012 .016
KALINKOV, M.
  158.127 .128
  160.026
KALINYAK, A. A.
  071.029
KALKOFEN, W.
  080.003
KALLARAKAL, V. V.
  098.026
KALLIOMAEKI, K.
  034.108
KALLOGHLIAN, A. T.
  158.154
KALMAN, G.
  141.550
KALNAJS, A. J.
  151.008 .022
KALRA, G. L.
  062.064
KAMEL, A. A.
  042.043
KAMIJO, F.
  131.090 .151 .152
KAMIJO, I.
  031.018
KAMINER, N. S.
  078.066
  143.078
KAMMER, H.
  105.126
KAMP, L. W.
  064.023
KAMP, P. VAN DE
  008.109
  119.009
KAMPERMAN, T. M.
  114.031 .049

KANAMORI, H.
  094.486
KANDA, S.
  004.071
  123.063 .064 .066 .067
KANDA, T.
  032.031
  034.024
KANDASWAMY, J.
  141.068
KANDEL, R. S.
  115.001
KANE, J.
  097.093
KANE, S. R.
  076.019
KANE, W. T.
  094.659
KANEKO, N.
  158.110
KANEOKA, I.
  094.732
KANER, E. A.
  141.065
KANGAS, J.
  072.060
  080.043
  084.008 .009
KANTZ, M. L.
  098.025
  103.100 .114 .118 .124
       .127 .132
KAPAHI, V. K.
  141.068
KAPLAN, G.
  094.299 .388
KAPLAN, I. R.
  094.246 .253 .404 .902
KAPLAN, M. H.
  094.911
KAPLAN, S. A.
  022.089
  061.034 .069
  141.509 .515
KAPOOR, R. C.
  122.086
KARACHENTSEV, I. D.
  158.087
KARADJOV, D.
  158.127
KARAKADKO, V. K.
  034.095
KARAS, R. H.
  084.006
KARDOPOLOV, V. I.
  152.006 .012
  153.001
KARDOS, J. L.
  094.479
KARETNIKOV, V. G.
  064.002
  121.029
  122.098
KAREV, V. I.
  076.031
KARIMIE, M. T.
  121.002
KARIMIE, T.
  121.056
KARP, A. H.
  122.032

KARP, D.
093.056
KARPMAN, V. I.
083.001
KARPOVA, L. M.
004.079
KARPOWICZ, M.
003.070
KARR JR., C.
094.290 .469 .817
KARTASHOV, V. F.
094.895
KASAHARA, I.
142.004
KASATKIN, A. M.
097.020 .023
KASIMOV, U.
084.413
KASINSKIJ, V. V.
072.077
073.108 .111
KASOLD, J.
097.093
KASTEL', G. R.
103.100
KASTELEIN, W.
004.018
KASTURIRANGAN, K.
142.101
KASUMOV, F. K.
065.038
125.004 .017
141.563
KATASEV, L. A.
104.016
KATGERT, P.
141.021 .022
KATGERT-MERKELIJN, J. K.
141.021 .022
KATH, J.
094.937
KATO, S.
151.032
KATO, T.
084.412
142.005
KATS, M. E.
106.036
143.082
KATSUBE, T. J.
094.491
KATTAWAR, G. W.
063.007 .024
KATTERFELD, G. N.
094.945
KATZ, J. I.
065.088
KATZ, M. E.
077.040
KAUFFELDT, A.
004.029
KAUFMAN, S. E.
162.066
KAUFMANN, J. P.
064.068
114.130
KAUFMANN, P.
033.002
077.058
142.122

KAULA, W. M.
091.082
094.757 .927
KAULBACH, F.
003.014
KAWABATA, K.
134.012
KAWAJIRI, N.
134.012
KAWANO, N.
134.012
KAWATA, Y.
100.029
KAY, H. F.
094.461
KAYE, M.
094.377 .685 .705
KAYE, M. J.
094.411
KAZANASMAS, M. S.
003.150
KAZANTSEV, A. N.
093.072 .073 .074
KAZARIAN, E. S.
122.136
KAZARIAN, M. A.
133.024
KAZES, I.
100.012
155.011
KAZIMIRCHAK-POLONSKAYA,
E. I.
104.026
KCHATCHATRIAN, N. G.
SEE KHACHATRYAN, N. G.
KEARNEY, P. D.
078.008
KEATING, G. M.
082.072
KEAY, C. S. L.
099.062
KEAYS, R. R.
094.374
KEELEY, D. A.
131.013 .162
KEEN, N. J.
033.047
157.006
KEENAN, D. W.
151.018
KEENEY, J.
094.600
KEGEL, W. H.
143.003
KEIHM, S.
094.544
KEIL, K.
094.028 .058 .075 .097
.142 .184 .306 .329
.650
105.021 .060 .076 .091
KEILHACKER, M.
062.045
KEITH, J. E.
094.054 .718
KELLENBENZ, H.
003.014
KELLER, C. F.
074.030
KELLER, H.-U.
102.003

KELLER, H.-U.
103.102
162.037
KELLER, W. D.
094.370
KELLERMANN, E. W.
143.012
KELLERMANN, K. I.
141.093 .109
158.029
KELLMAN, S. A.
151.005
KELLOGG, E.
142.045 .052 .097 .138
.150
158.057
KELLOGG, E. M.
142.070
KELLY, A.
131.100
KELLY, A. N.
077.044
KELLY, D. C.
062.005
KELLY, W. R.
094.746
105.092
KEL'NER, S. R.
142.119
KELSALL, T.
122.044
KEMIC, S.
126.011
KEMP, J. C.
063.041
100.033
116.003
119.021
142.115
KENKNIGHT, C. E.
099.063
KENNEDY, G. C.
081.005
KENNEDY, J. R.
099.028
KENNEDY, M.
125.015
KENNEDY, W. A. G.
033.001
KENNEDY, W. J.
125.015
KENNEL, C. F.
051.002
084.257 .271 .280
091.034
KENWORTHY, C. M.
141.030 .118
KENYON, W. J.
094.712
KERDEMELIDIS, V.
033.071
KERIMBEKOV, M. B.
071.047
KERR, A. R.
033.045
131.007
KERR, F. J.
014.028
033.024
154.006

KERRIDGE, S. J.
153.011
KERRIGAN, T. C.
105.040
KERZHANOVICH, V. V.
093.037
KESSLER, D. J.
094.524
KESTEN, H.
003.071
KHABIBULLIN, SH. T.
094.921
KHACHATRYAN, N. G.
153.041
KHACHATUR'YANTS, L. S.
051.003
KHACHIKIAN, E. YE.
158.131 .136 .155
KHADAKHANOVA, T. S.
143.025
KHAIMOV, I. M.
034.099
KHALATNIKOV, I. M.
162.080
KHALIL, H. K.
063.014
KHALIULLIN, KH. F.
121.054
KHALTAR, D.
082.127
KHAN, H. A.
094.260
KHAN, I.
162.065
KHAN, M. A.
081.026 .031
KHANTADZE, A. G.
062.047
KHARE, B. N.
063.022
KHARITONOV, A. V.
003.079
114.108
KHARITONOVA, G. A.
099.093
100.004 .016
101.017
KHARKAR, D. P.
094.395
KHATIPOV, A. E.-A.
004.081
KHAZAN, YA. M.
134.010
142.112
KHEJFETS, S. A.
091.064
KHENTOV, A. A.
054.018
KHERSONSKY, V. K.
022.071
KHETSELIUS, V. G.
082.104 .105
KHO, T. H.
063.001 .034 .036
KHOKHLOVA, V. L.
114.143
KHOLOPOV, P. N.
003.003
KHOLSHEVNIKOV, K. V.
042.056
052.039

KHOR'KOV, V. D.
078.055
KHOSKOVICH, B.
033.008
KHOZOV, G. V.
113.058
KHRAMOV, A. N.
081.004
KHRAPKO, R. I.
066.047 .086
KHRISTICH, V. G.
031.059
KHRULEV, V. V.
033.018
KHRUNOV, E. V.
051.003
KIANG, T.
103.103
KIEFFABER, L. M.
082.018 .074 .108
KIEFFER, H. H.
091.055
100.030
KIELKOPF, J. F.
031.021
KIESL, W.
105.155
KIEWIET DE JONGE, J.
008.093
KIKO, J.
094.530 .732
104.003
105.123
KIKUCHI, H.
084.258
KIKUCHI, N.
044.011
045.011
KIKUCHI, S.
124.106
KILADZE, R. I.
100.054
KILFOYLE, B. P.
084.044
KILMARTIN, P. M.
103.133
KIM, C. K.
094.356
KIM, I. S.
073.009
KIM, J. H.
094.356
KIM, Y. K.
094.356
KIMBALL, D. S.
084.020
KIMBERLIN, J.
094.375
KIMURA, S.
096.021
KING, D. A.
004.051
KING, D. S.
122.040 .138
KING, I. R.
158.060
KING, J. W.
082.088
083.014
KING JR., E. A.
012.022

KING JR., E. A.
094.178 .355 .648
KING-HELE, D. G.
054.014
081.901
KINMAN, T. D.
158.020
KINOSHITA, H.
042.007 .026
052.011
KINZER, R. L.
134.003
155.040
KIPPENHAHN, R.
061.025 .041
142.089
KIRCHGRABER, U.
042.044
KIRICHUK, V. V.
082.119
KIRILLOV, I. V.
081.020
KIRILLOV-UGRYUMOV, V. G.
142.057
KIRK, D. B.
053.011
KIRKPATRICK, R. C.
132.017
KIRNOZOV, F. F.
093.002
KIRSCH, E.
078.002
KIRSHNER, R. P.
125.101
KIRSTEN, T.
094.161 .425 .527 .528
      .529 .530 .732
104.003
105.082 .123 .131
KISELEVA, T. P.
093.014
097.049
099.039
KISER, J.
158.060
KISH, J. C.
143.008
KISJUN, L. N.
094.131
KISLIUK, V. S.
094.018 .132 .133
KISLYAKOV, A. G.
093.065
094.872
KISS, E.
094.347
KISSEL, J.
105.136
KISSELBACH, V. J.
143.046
KISSELL, K. E.
034.065
KITAMURA, M.
121.037
KITAYAMA, K.
094.340
KITCHIN, C. R.
064.033 .034
KIVELSON, M.
084.411

KOLCHIN, E. K.
117.028
KOLENKIEWICZ, R.
045.030
KOLESNIK, I. G.
065.177
122.140
KOLESNIKOV, S. M.
066.043
KOLESOV, A. K.
063.047 .049
KOLESOV, YU. I.
077.061
KOLLBERG, E.
131.108
KOLOMEETS, E. V.
072.030
078.014 .040
143.062 .075
KOLOPUS, J. L.
094.501
KOLOSNITSYN, N. I.
066.040
KOLPAKOV, V. P.
162.026
KOLPANEN, D.
123.029
KOMAROV, N. S.
113.037
114.014
KOMBERG, B. V.
122.069
KOMESAROFF, M. M.
099.071
141.554 .555
KOMKOVA, T. G.
082.037
KOMPANEETS, A. S.
061.066
KOMRAKOV, G. P.
083.059
KONDO, Y.
114.072
142.146
KONDRAT'EV, K. YA.
097.037
KONDRAT'EV, N. YA.
052.009
KONDRAT'EVA, E. D.
102.023
104.010
KONDRATIEV, K. J.
091.004
KONENKO, A. F.
085.015
KONG, J. A.
094.848
KONOPLEVA, V. P.
103.102
KONTOR, N. N.
078.052
093.080
KONYAKHINA, S. S.
078.010 .028 .064
084.409
KONYUKOV, M. V.
064.080
074.104
KOPAL, Z.
002.036
003.073

KOPAL, Z.
006.000
007.000
094.841
117.001
KOPATSKAYA, E. N.
141.135
142.149
KOPCZYNSKI, W.
162.074
KOPECKY, M.
064.005
072.034
KOPELEVICH, YU. KH.
005.023
KOPPEL, A. A.
066.026
KOPVILLEM, U. KH.
066.044 .045 .060
KOPYLETS, K. N.
079.103
KOPYLOV, I. M.
114.105
115.026
158.103
KORCHAK, A.
066.053
KORDA, E. J.
094.659
KOREKAWA, M.
094.175 .639
KORFF, S. A.
082.112 .113 .114
KORKINA, M. P.
066.027
162.021
KORMENDY, J.
113.038
KORNEEV, V. V.
076.031
KORNHERR, M.
062.045
KOROL', A. N.
094.863
KOROLEV, F. A.
097.022
KOROLEV, O. S.
077.034
KOROTEV, R. L.
094.396 .688
KOROTKIKH, T. N.
064.002
KOROTTSEV, O.
005.030
KOROVKINA, T. L.
014.047
KOROVYAKOVSKAYA, A. A.
122.015
KOROVYAKOVSKY, YU. P.
117.022
KORPUSOV, V. N.
104.018
KORSUN, A. A.
044.018
KORZHAVIN, A. N.
077.020 .033
KOSAI, H.
103.124 .127
KOSIN, G. S.
041.039 .045 .046

KOSLOVSKY, B.-Z.
141.123
KOSTIK, R. I.
071.056
KOSTYLEV, K. V.
104.022
KOTHARI, B. K.
094.243 .745
KOTOV, V. A.
072.045
KOTOV, YU. D.
142.119
KOTRC, P.
064.005
KOUBSKY, P.
051.010
119.003
KOUNS, C. W.
094.689
KOURGANOFF, V.
003.074
KOUTCHMY, S.
034.026
KOVACH, J. J.
094.290 .469 .817
KOVACH, R. L.
094.044 .932
KOVACHEV, B.
114.126
KOVADLO, P. G.
082.065 .129
KOVAL', I. K.
097.110
KOVAL, V. I.
105.139
KOVALENKO, O.
103.105
KOVALENKO, O. N.
103.100 .114
122.062
KOVALENKO, V.
103.105
KOVALENKO, V. A.
046.017
KOVALENKO, V. M.
103.100 .114
122.062
KOVALEVSKIJ, I. V.
106.013
KOVALEVSKY, J.
007.000
KOVAL'SKAYA, I. YA.
094.883
KOVNER, M. S.
084.225 .230 .231 .262
KOWAL, C.
125.035
KOWAL, C. T.
098.036 .038 .055
103.100
125.014 .104
KOYANO, H.
122.100
KOZAI, Y.
032.031
042.007
052.011 .032
055.001
KOZAK, L. V.
143.066

KOZHEVNIKOV, A. A.
083.034
KOZHEVNIKOV, N. I.
072.063
080.017
KOZIEL, K.
094.880
KOZLOV, N. N.
151.002
KOZLOVA, K. I.
114.106
KOZLOVSKAYA, S. V.
092.007
KOZLOVSKY, B.-Z.
131.086
KOZLOWSKI, M.
065.152
KRAATZ, P.
094.465
KRAEHENBUEHL, U.
094.094 .154 .228 .696
.697 .730
105.053 .093
KRAETSCHMER, W.
094.807
143.059
KRAFT, R. P.
008.105
122.008 .038
142.061
KRAHENBUHL, U.
094.594
KRAHN, D.
002.034
KRALL, N. A.
003.075
KRAMER, D.
066.035 .103
KRANJC, A.
052.033
KRANZER, W.
014.003
KRASHENINNIKOVA, G. V.
094.861
KRASIKOV, V. A.
094.959
KRASNOPOL'SKIJ, V. A.
094.869
KRASOVSKII, G. N.
097.020
KRASSOVSKY, V. I.
022.021
KRAUS, J. D.
008.035
131.145
KRAUS, K.
022.031
KRAUSCHE, D. S.
099.028
KRAUSE, F.
061.043
062.025
KRAUSHAAR, W. L.
034.116
142.021
KRAVTSOV, F. I.
103.112
KRAY, W. C.
094.445
KREJNIN, E. I.
041.019

KREMPEC, J.
066.018
KREMSER, G.
084.008 .009
KREPLIN, R. W.
073.083
KREPSKI, C.
003.046
KRIDELBAUGH, S. J.
094.057 .091 .198
KRIEGER, A. S.
074.083
KRIMIGIS, S. M.
078.036
KRISHAN, S.
074.090
KRISTIAN, J.
141.014
142.024
158.052
KRISTIANSSON, K.
143.042
KRIUK, V. I.
094.128
KRIVSKY, L.
072.071
073.026 .047
078.022 .057
KROGDAHL, W. S.
015.001
KRON, G. E.
034.069
113.054
158.059
KRONBERG, P. P.
158.097
KROPOTKIN, P. N.
162.027
KROPP, K.
094.450
KROTIKOV, V. L.
094.875
KRPATA, J.
119.003
KRUEGER, A.
077.048 .065
KRUG, E.
005.015
KRUGER, R. A.
034.036
KRUGOV, V. D.
100.056
KRUIT, P. C. VAN DER
158.073 .116 .901
KRUPENIO, N. N.
093.068
KRUSE, H.
094.383 .686
KRUSZEWSKI, A.
131.143
KRUTOV, V. V.
076.031
KRYGIER, B.
066.018
KRYMSKIJ, G. F.
078.051
KRYMSKY, G. F.
078.045
KSANFOMALITI, L. V.
034.017
094.863

KSANFOMALITI, L. V.
097.020 .023 .036
KUBIAK, M.
114.061
KUBIERSCHKY, K.
034.111
KUCHINA, T. M.
066.036
KUCHOWICZ, B.
066.092
KUDRIN, V. B.
051.005
KUDRYAVTSEV, M. I.
076.025
KUEHNER, E. C.
094.702
KUENDIG, H.
091.012
KUENZEL, H.
072.072
KUGAENKO, B. V.
082.034
KUHI, L. V.
064.026 .059
114.116
158.142
KUIMOV, K. V.
072.019
KUIPER, T. B. H.
077.003
KUJI, S.
034.039
KUKARKIN, B. V.
003.003
KUKARKINA, N. P.
003.003
KUKHARSKAYA, N. F.
093.077
KUKIN, L. M.
094.873
KUKLIN, G. V.
072.079
073.112
KULCIZKY, A. P.
034.047
KULICK, C. G.
094.343 .642
KULIKOV, V. N.
034.086
143.015 .080
KULIKOVA, N. V.
104.016
KULKARNI, P. V.
082.022
KULLERUD, G.
094.362
KULSRUD, R. M.
142.088
KULUS, H.
094.735
KUMAJGORODSKAYA, R. N.
114.105
KUMAR, C. K.
158.063
KUMAR, S.
082.002 .040
106.009
KUMAZAWA, M.
094.489
KUMSIASHVILI, M.
117.017

MARCUS, A. H.
094.549
MARECHAL, A.
034.012
MAREK, J.
022.075
MAREK, K.-H.
055.005
MARGOLIS, J. S.
091.019 .049
099.044
MARGOLIS, S. V.
094.365 .742
MARGON, B.
142.006 .013 .023
MARIN, M.
036.007
MARIS, G.
077.045
MARKARIAN, B. E.
158.030 .031 .134
MARKEEV, A. P.
042.002
MARKERT, T. H.
142.009 .127 .133
MARKINA, O. T.
041.011 .037 .043
MARKOV, A. V.
094.865
MARKOV, M. S.
094.892
MARKOVA, L.
103.105
MARKOWITZ, W.
045.022
055.010
MARKS, G. H.
080.007
MARKUSHEVICH, A.
004.097
MARLBOROUGH, J. M.
114.113
MAROCHNIK, L. S.
155.093
MAROV, M.
093.005
MAROV, M. YA.
093.011 .037 .050 .075
.083 .901
097.099
MARPEGAN, J.
008.024
MARQUARDT, C. L.
094.281 .770
MARRACO, H.
131.123
MARRACO, H. G.
153.006
MARSDEN, B. G.
098.025 .033 .034 .041
.042 .045 .047 .054
.056 .057 .060
102.008
103.100 .109 .114 .117
.118 .121 .122 .124
.126 .127 .130 .133
.134
MARSH, J. G.
054.002
055.008

MARSH, K. A.
141.113 .114
MARSHALL, M. P.
065.174
MARSICANO, F. R.
131.209
MARSTON, A. C.
094.475
MARTI, K.
094.372 .421 .733
105.099
MARTIN, A. E.
153.030
MARTIN, C. F.
081.035 .037
MARTIN, L. J.
097.051 .067 .078
MARTIN, M. R.
094.376
MARTIN, P. G.
131.033
MARTIN, R. T.
094.831
MARTIN, S. F.
034.053
073.046
MARTIN, T. V.
081.035
MARTIN, T. Z.
009.023
MARTIN, W. L.
159.001
MARTINEZ, J.
036.007
MARTINI, L.
076.034
MARTINSEN, P.
061.037
MARTJANOV, S. A.
084.261
MARTONCHIK, J. V.
114.081
MARTRES, M.-J.
073.041
075.030
MARTSVALADZE, N. M.
082.039
MARTYANOV, S. A.
062.019
084.223 .224 .242
MARTYNENKO, V. V.
104.017 .020 .025
MARTYNJUK, A. I.
095.006
MARTYNOV, A. I.
053.005
MARTYNOV, D. YA.
012.024
117.033
MARTYNOVA, A. I.
042.055
MARTYNOVA, N. F.
052.018
MARUSSI, A.
043.002
MARVIN, U. B.
094.105 .351 .669
105.033
MARX, S.
036.005
041.050

MARX, S.
131.188
MASAJTIS, V. L.
105.049 .158
MASANI, A.
012.023
061.032
162.078
MASCART, P.
082.010
MASEVICH, A. G.
003.005
MASHHOON, B.
141.076
MASINI, R.
003.080
MASON, B.
003.081
094.201 .202 .338 .655
.943
105.057 .121 .157 .164
MASON, C. C.
094.851 .942
097.125
MASON, G. M.
143.057
MASON, H. P.
097.108
MASON, K.
142.138
MASON, K. O.
142.033
MASSEY, H.
061.018
MASSON, C. R.
094.369 .671
MASUDA, A.
094.690 .926
105.005
MATEO, J.
044.028
MATERNI, A.
100.010
MATESHVILI, YU. D.
082.062
MATHER, J. C.
082.044
MATHER, R. S.
046.015 .024
081.025
MATHESON, D. N.
131.192
MATHEWS, T.
078.007
MATHEWSON, D. S.
125.010 .042
141.009
158.901
MATHIS, J. S.
119.005
132.014
MATILSKY, T.
142.097
MATINIAN, S. G.
162.071
MATROSOV, V. M.
051.009
MATSCHINSKI, M.
022.022
MATSON, D. L.
098.015

MIRONOV, A. V.
154.001
MIROSHNICHENKO, L. I.
078.015
MIROVSKY, V. G.
141.006
MIRZOJAN, L. W.
162.062
MIRZOYAN, L. V.
122.136
MISHCHENKO, M. P.
044.031
MISHUROV, YU. N.
155.093
MISKOVIC, V. V.
003.088
MISRA, M.
066.122
MISRA, R. K.
083.074
MITALAS, R.
080.028
114.113
MITANI, T.
081.022
MITCHELL, F. J.
094.520
MITCHELL, J. K.
094.016 .452 .832 .833
MITCHELL, J. M.
094.447
MITRA, R. K.
073.097
085.004
MITROFANOVA, L. A.
022.082
MITROPOLSKAYA, O. N.
071.019
MITSKEVIC, N. V.
066.042
MITSKEVICH, N. V.
080.013
MITTER, J.
007.000
MITTON, S.
011.001
141.003
155.098
MIYAJIMA, M.
094.479
MIYAKE, G.
105.101
MIYAMOTO
103.114
MIYAMOTO, S.
094.889
097.113
142.004
MIYASHITA, A.
082.050
MIYAUCHI, N.
031.018
MIZERA, P. F.
084.405
MIZUTANI, H.
094.486 .781
MNATSAKANIAN, R. G.
113.053
125.103
158.138

MOCHIZUKI, E.
123.028 .064 .067
MOCHNACKI, S. W.
121.035
MODALI, S. B.
073.023
MODEL, F.
002.009
MODZELESKI, J. E.
094.069 .251
MODZELESKI, V. E.
094.069 .251 .947
MOE, K.
082.068
MOERGELI, M.
094.730
MOFFAT, A.
114.162
MOFFAT, A. F. J.
142.015 .027 .118
153.010 .025 .026 .031
.039
155.015
MOFFET, A. T.
155.014 .081
MOFFETT, R. J.
083.008 .025
MOFFETT, T. J.
142.093
MOGILEVSKIJ, E. I.
077.034
MOGILEVSKY, E. I.
073.048
MOHAMMED, M. A. J.
094.069 .947
MOHR, J. M.
005.035
MOISEEV, I. G.
093.065
141.109
MOLER, R. B.
094.467
MOLES, M.
022.047
MOL'KOVA, L. A.
094.875
MOLNAR, M. R.
114.073 .173
116.004
MOLODENSKY, M. M.
074.017 .041
MOLTON, P.
015.005
099.081
MOLTON, P. M.
015.008 .012
094.109
MOMCHEV, G.
122.114
MONFILS, A.
114.039
MONIN, A.
093.041
MONNIN, M.
094.414 .713
MONTEAGUDO, V. N. DE
114.094 .095
MONTGOMERY, M. D.
051.002
074.031 .070 .074
084.274

MONTLE, R. E.
121.050
MOODY, J. R.
094.702
MOORE, C. B.
094.054 .252 .400 .746
.903
105.018 .084 .085 .092
.097
MOORE, D. R.
094.054
105.012
MOORE, E. L.
162.017
MOORE, G. W.
094.056 .148 .250 .744
MOORE, H. J.
094.798
MOORE, J.
053.007
MOORE, L. J.
094.702
MOORE, P.
003.089 .090 .091 .092
.093
010.012
094.302
MOORE, R. L.
072.016
MOORE-SITTERLY, C.
022.035 .083
MOORWOOD, A. F. M.
113.015
MOOS, H. W.
082.005
099.054
MORALES CABRERA, C. G.
031.064
MORAN, J. M.
131.052
MORAN, P. E.
042.006 .011
MORANDO, B.
004.008
079.101
MORANZINO, C.
044.024
MORAWSKI, A.
094.203
MOREL, J. A.
094.938
MORELL, M.
099.009
MORENO, G.
074.101
MORFILL, G.
084.214 .401
MORGAN, B. L.
034.059 .070 .077
MORGAN, J. W.
051.006
094.023 .094 .137 .154
.228 .374 .380 .387
.594 .680 .696 .697
105.007 .053 .093 .143
MORGAN, R. G.
094.523
MORGAN, W. W.
114.128 .144
MORGANSTERN, R. E.
162.083

MORGENTHALER, G. W.
012.018
MORGER, P.
121.081
MORI, T.
096.028
MORIMOTO, M.
131.005 .103 .189
MORITA, I.
082.049
MORITZ, H.
046.016
MORIYAMA, S.
097.121
MORKOWSKA, B.
098.007
MORODER, E.
082.012
MOROZ, V. I.
091.076
097.020 .022 .023 .036
MOROZHENKO, A. V.
092.004
097.112
099.003 .043 .074
103.102
MOROZOV, D. KH.
143.085
MOROZOVA, I. M.
094.120
MORRIS, C. S.
103.106 .107
MORRIS, G. A.
097.002
141.539
MORRIS, M.
131.080
MORRISON, D.
002.020
009.023
091.040 .053 .057
098.016
099.011
100.036
101.001
158.076
MORRISON, D. A.
094.054 .333 .357 .652
.668 .798 .936
MORRISON, F.
052.041
MORRISON, G. H.
094.382
MORRISON, J. A.
094.489
MORRISON, L. V.
044.003
096.015 .023 .024
MORRISON, N. D.
002.020
122.074
MORRISON, P.
141.074
142.042
155.089
MORRISON, S. L.
032.033
MORRO, A.
066.095
MORROW, W. E.
093.056

MORTON, D. C.
114.066 .121 .122 .123
.124 .125
131.063 .064 .065 .066
.067 .166
158.008 .056
MORTON, J. M.
034.102
MORZENTI, S. P.
094.678
MOSELEY, G. F.
141.037
MOSHKIN, B. E.
093.083
MOSIER, S. R.
084.408
MOSKALENKO, A. M.
062.018
MOSKOVKINA, L. A.
042.054
MOSLEN, M. T.
014.024
MOSS, D. C.
094.454
MOSS, D. L.
065.015
117.003
MOSTEPANENKO, V. M.
066.030
MOTOVILOV, EH. A.
094.603
MOTRICH, V. D.
122.067
MOTTA, S.
073.082
080.032
MOTTMANN, J.
133.029
MOTZ, H.
062.066
MOTZ, L.
066.079
MOUNT, G. H.
071.040
MOUROT, S.
004.004
MOUTSOULAS, M.
002.036
MOYD, K. I.
114.170
MOZER, F. S.
084.204
MRKOS, A.
098.050
103.009 .013 .100 .114
.115 .116 .123 .124
.127
MROZOWSKI, S.
010.021
MUAN, A.
094.341 .617
MUCKE, H.
004.089
MUEHLBERGER, W. R.
094.054 .055 .940
MUEHLNER, D.
066.102
MUELLER, E. A.
012.020
014.025

MUELLER, G.
094.155 .460
MUELLER, H.
044.013
MUELLER, H. W.
094.530 .532
MUELLER, O.
094.107 .113 .229 .403
.533 .589 .747
105.127 .132
MUELLER, W. F.
094.361 .640
MUENCH, G.
051.002
099.030
100.041
MUENOW, D. W.
094.742
MUHLEMAN, D. O.
094.539
MUIR, I. D.
094.186 .332 .627
MUIR, P.
094.237 .377 .685
MUIR JR., A. H.
094.468
MUIRDEN, J.
074.087
MUKHERJEE, N. R.
094.470 .762
MUKHERJEE, P. K.
091.015
MULHOLLAND, J. D.
011.008
022.088
042.021
094.166 .570 .927
MULLAN, D. J.
072.011 .017 .073 .074
MULLEN, E. G.
084.002
MULLER, P.
118.011 .015 .017 .018
MULLER, R.
072.039
MULLER, R. A.
143.058
MULLER ZUM HAGEN, H.
066.121
MULLINS, L. D.
042.041
MULLINS, N. E.
054.002
MUNDRY, E.
123.002
MUNIZ BARRETO, L.
044.027
MUNN, M. W.
065.100
MUNSUK, C.
154.003
MURADOV, A.
083.066
MURAI, T.
065.003 .163
MURAKAMI, G.
044.035
MURAKAMI, T.
084.412
MURATORIO, G.
114.045

PLAKHOV, YU. V.
042.014 .036
PLANNER, H. N.
105.060
PLANT, A. G.
094.306 .327 .629
PLASS, G. N.
063.024
PLATANIA, G.
126.022
PLATONOV, I. N.
103.126
PLATOV, YU. V.
073.011 .107
PLAUT, L.
123.057
PLECHKOV, V. M.
094.871
PLIENINGER, T.
094.807
143.059
PLOCIENIAK, S.
071.044
PLOTKIN, H. H.
094.927
PLYASOVA-BAKUNINA, T. A.
084.236
PLYUSNIN, L. N.
162.047
PLYUSNINA, L. A.
072.077
PNEUMAN, G. W.
074.003 .066
PODGORNY, I. M.
084.255 .268
PODGORNYJ, I. M.
062.031
084.244
PODOSEK, F. A.
094.088 .426 .711
105.014 .109 .147
POEPPEL, W. G. L.
157.003 .007
POGNON, E.
009.006
POKORNY, Z.
084.027
099.031
121.038
POKROVSKY, G. I.
051.004
POLENOV, B. V.
097.006
POLEZHAEV, V. I.
093.075
POLIDAN, R. S.
124.102
POLIEVKTOV-NIKOLADZE,
N. M.
066.046
POLIKAROV, A.
162.045
POLISHCHUK, R. F.
066.134
POLJACHENKO, V. L.
061.036
151.004
POLKOWSKI, G.
094.352
105.116

POLLACK, H. N.
081.016
POLLACK, J.
097.062 .063
POLLACK, J. B.
091.058
093.021
097.041 .082
100.038 .044 .048
POLLOCK, J.
123.029
POLNAREV, A. G.
066.019 .061
POLNITZKY, G.
041.021
POLOZHENTSEV, D. D.
041.003 .047
POLUPAN, P. N.
073.028
POLYAKHOVA, E. N.
052.038
POLYAKHOVA, T. A.
122.142
POLYAKOV, V. M.
083.037
POLYANSKAYA, E. L.
141.135
POMA, A.
041.017
POMERANTZ, M. A.
078.004 .006
PONNAMPERUMA, C.
094.441 .751 .899 .905
.952
105.009
PONOMAREV, V. N.
083.032 .056 .057
POOLEY, G. G.
158.021
POP, V.
122.073
POPKOV, I. V.
083.059
POPOV, G. M.
031.026
POPOVA, M.
153.038
POPOVA, M. P.
094.295
POPOVICI, C.
005.024
011.013
013.005
POPPEN, R. F.
092.002
PORCH, W. M.
082.051
PORFIRJEVA, G. A.
071.005
POROSHIN, A. P.
096.012
PORTAT, M.
114.089
PORTENIER, W. R.
094.054
PORTEOUS, H. L.
004.002
PORTER, F. C.
113.001
PORUBCAN, V.
104.001

POSPERGELIS, M. M.
094.864
POSS, H. L.
079.100
POSTPISCHL, D.
046.012
POTAPOV, A. S.
084.234
POTTASCH, S. R.
131.059
155.044
POTTER, A. E.
114.069
POTTER, N. M.
094.382
POUNDS, K. A.
125.018 .019
142.138
POUPEAU, G.
094.161 .256 .269 .803
.808
POVENMIRE, H.
096.026
POWELL, A. L. T.
114.024
POWELL, B. A.
022.055
POWELL, B. N.
094.351 .658
POWELL, J. R.
034.061
POWERS, W. T.
034.072
POYNTER, R. L.
099.051
POZO, E.
033.008
PRACHYABRUED, W.
094.287 .810
PRADERIE, F.
064.008 .029
PRAETORIUS, H. M.
084.218
PRASAD, B.
077.056
PRATAP, R.
084.250
PRATT, N.
114.165
PRENDERGAST, K. H.
063.028
PRENTICE, A. J. R.
065.044
PREPELITSA, B. V.
162.069
PRESS, F.
094.583 .778 .779
PRESS, W. H.
065.092
PRESTON, G. W.
051.012
PRETI, G.
094.443
PREUSS, E.
125.038
PREVOT, L.
114.045
PREWITT, C. T.
094.308
PRIBOEVA, N. V.
099.095

REHSE, H.
046.025
REICHERT, E.
113.050
REICHLEY, P. E.
097.002
141.539
REID, A. F.
094.306
REID, A. M.
094.025 .054 .099 .212
.328 .628 .913
105.111
REID, G. C.
083.061
REID, M. J.
094.004
REID JR., J. B.
094.206 .351 .669
REIGBER, C.
052.042
REIMERS, D.
064.067
114.048
REINBOLD, S. J.
043.001
REINES, F.
080.002
REINHARDT, M.
066.048
121.002 .056
158.006 .124
REIPURTH, B.
098.021
103.005
REITER, R.
085.001
REITMEYER, W. L.
008.065
013.001
REITSEMA, H.
093.031
098.031
REITSEMA, H. J.
132.015
REMY-BATTIAU, L.
012.003
RENGARAJAN, T. N.
141.532
143.068
RENNILSON, J. J.
094.055
RENSE, W. A.
076.012
REQUIEME, Y.
034.008
RESCH, G. M.
141.506
RESKIN, G.
103.122
RETALLACK, D. S.
155.018
REY, P.
094.397 .727
REYNOLDS, J. H.
094.428 .728
REYNOLDS, M. A.
094.444
105.056
REYNOLDS, R. J.
082.008

REYNOLDS, R. J.
131.078 .131
REYNOLDS, R. T.
094.548 .845
REYSS, J. L.
094.274 .723
RHO, J. H.
094.442 .753 .904
RHODES, J. M.
094.054 .232 .981
RHODES JR., E. J.
124.102
RIBES, J. C.
022.044
131.117
RICE, C. M.
094.317
RICH, A.
126.009
RICHARDS, D. W.
141.568
RICHARDS, P. L.
082.044
RICHARDSON, D. L.
042.040
RICHARDSON, K. A.
094.718
RICHARDSON, M. F.
094.409
RICHER, H. B.
122.013
RICHTER, A. K.
074.053
RICHTER, G. M.
141.016 .070
RICHTER, J.
022.065 .075
RICHTER, K. R.
093.081
RICHTER, N.
102.011
158.115
RICKARD, L. J.
131.202
RICKER, G. R.
142.051
RICKETT, B. J.
074.047
141.525
RICORT, G.
071.009
RIDDLE, A. C.
077.014
RIDGWAY, S. T.
099.065
RIDLEY, E. C.
091.080
RIDLEY, W. I.
094.025 .054 .099 .212
.615 .628 .631 .913
RIDPATH, I.
033.042
RIEDER, R.
094.383
RIEDLER, W.
084.008 .009
RIEGEL, K. W.
082.045
114.034
RIEKE, G. H.
099.062

RIEPE, B. Y.
113.054
RIETSCHEL-KLUGE, R.
003.154
RIGAUD, P.
082.011
RIGBY, B. J.
084.221
RIGHINI, A.
082.012
RIGUTTI, M.
073.014
RIHERD, P. S.
133.030
RIJSBERGEN, R. VAN
121.057
RIKITAKE, T.
084.207
RILEY, D. L.
094.467
RILEY, J. M.
066.003
141.015
RILEY, L. A.
100.026
RING, J.
034.070
RINGUELET, A. E.
133.009 .010 .020 .021
RINGWOOD, A. E.
094.031 .036 .189 .347
.551 .581 .618
RISHBETH, H.
083.003
RISSE, H.
014.014
ROACH, F. E.
106.025 .026
ROACH, J. R.
106.025 .026
ROBBA, N. R.
134.009
ROBBINS, D. E.
142.085
ROBBINS, M. K.
094.054
ROBERT, M.
034.118
ROBERTS, D. H.
141.522 .567
ROBERTS, E.
156.003
ROBERTS, J. R.
022.059 .060
ROBERTS, K. V.
141.548
ROBERTS, M. J.
012.005
ROBERTS, P. H.
061.043
062.025
ROBERTS, R.
034.123
ROBERTS, W. O.
080.014
ROBERTSON, D. S.
141.052
ROBERTSON, J. W.
065.041
ROBIE, R. A.
094.490

SAGAN, C.
097.064 .082 .094
099.053
100.021
SAGGION, A.
061.007
141.042
SAGNIER, J.-L.
099.032
SAHA, G.
155.063
SAHA, S. K.
141.558
SAHADE, J.
012.019
SAHAKIAN
SEE SAAKYAN
SAHAL-BRECHOT, S.
073.015
SAITO, K.
079.001
093.013
097.038
SAITO, M.
121.062
SAKAI, H.
094.246 .253
SAKHAROV, V. I.
021.002
045.006
SAKHIBULLIN, N.
114.101
SAKHIBULLIN, N. A.
133.039
SAKKA, K.
158.109
SAKUMA, S.
123.027
SAKURAI, K.
065.162
077.012 .042 .057
078.024
SALABUN, J.
011.030
SALEM, M.
132.022
SALIE, H.
003.017
SALISBURY, J. W.
002.015 .036
094.291 .818
097.016 .040
098.014
SALMON, B.
097.104
SALPETER, E. E.
065.087 .088 .089
080.009
099.040
131.127
142.067
158.024 .081
SALVATI, M.
134.002
SAMARDZHIEV, D. T.
083.053
SAMARTSEV, V. V.
066.044
SAMSONOV, I. S.
078.045

SAMUEL, A. G.
158.089
SANCHEZ-MARTINEZ, F.
106.004
SANCISI, R.
152.009
157.013
SANDAGE, A.
158.055
SANDEL, B. R.
084.012
SANDER, M. J.
097.008
SANDERS, R. H.
155.041
SANDERS, W. L.
031.011
132.015
153.003 .004 .019
SANDERS, W. M.
074.063
SANDERSON, R. B.
031.020
SANDFORD II, M. T.
063.020
064.013
074.024
SANDOMIRSKIJ, A. B.
081.015
SANDQVIST, A.
155.016 .027
SANDRI, G.
143.019
SANDULEAK, N.
114.136
SANFORD, P. W.
142.033 .138
SANOVICH, A. N.
094.115 .864
SANTANGELO, N.
076.014
SANTIN, P.
077.067
SANWAL, N. B.
121.004
SANYAL, A.
114.161 .163
119.018
SAPIENZA, G.
075.009
SAPIN, C.
132.030
SAPPENFIELD, K. M.
094.702
SARABHAI, V.
084.250
SARANGI, S.
091.007
SAREYAN, J. P.
122.016
SARGENT, W. L. W.
065.016
103.100
125.034 .036 .105
141.134
142.071
158.052 .118
160.027
SARKAR, S. K.
073.097
085.004

SARKER, R. P.
141.125
SARRIS, E. T.
073.013
SARTORI, L.
141.074
SARYCHEV, V. A.
052.030
SASAKI, M.
162.064
SASAO, T.
162.001
SASLAW, W. C.
151.017
SASTRI, V. K.
122.041
124.006
SASTRY, C. V.
077.004
SATAROVA, L. M.
105.169
SATHER, R.
103.121
SATO, F.
141.001
SATO, H.
032.031
158.129
SATO, J.
094.274 .723
SATO, N.
121.037
SATO, S.
131.098
SATO, T.
066.076
158.129
SAUNDERS, R. S.
097.054 .058
SAUSE, G.
103.100
SAVAGE, A.
142.018
SAVENKO, I. A.
076.025
078.048
143.017 .054 .064 .065
SAVUN, O. I.
084.402
SAWYER, R. F.
066.114
SAZONOV, A. Z.
046.021
SAZONOV, V. N.
141.104 .122
SCALISE JR., E.
077.058
SCANLAN, M. J. B.
033.062
SCARF, F. L.
084.280 .411
100.001
106.016
SCARGLE, J. D.
064.011
SCARSI, L.
134.009
SCHAACK. D.
097.041
SCHAAL, R. E.
077.058

SCHABER, G.
094.569
SCHABER, G. G.
094.055 .304 .305
SCHAEFFER, O. A.
094.709 .732 .935
SCHAFER, J. P.
094.055
SCHAIFERS, K.
003.114
010.010
012.021
SCHAIRER, J. F.
094.341
SCHAMEL, H.
062.043
SCHAPER, P. W.
082.102
SCHARLEMANN, E. T.
062.065
SCHATTEN, K. H.
074.049
SCHATZMAN, E.
065.137
162.004
SCHATZMAN, E. L.
014.030
SCHAUDY, R.
094.375
105.059
SCHAYES, G.
082.126
SCHEER. M. L.
053.012
SCHEFFER, U.
002.034
SCHEIDECKER, J.-P.
151.010
SCHELLER, E.
123.047
SCHERB, F.
082.008
131.078 .131
SCHERBAKOV, A. G.
158.050
SCHERER, G.
094.788
SCHERER, M.
074.055
083.027
SCHERRER, P.
077.066
SCHERRER, P. H.
080.014 .029
SCHEUER, P. A. G.
125.001
SCHIDLOWSKI, M.
082.091
SCHIFF, D.
065.142
SCHIFFER, F. H.
064.030
SCHIFFER, J. P.
094.514
SCHIFFER III, F. H.
132.014
SCHILD, R.
114.019
SCHILD, R. E.
113.034
114.050

SCHINDLER, K.
012.016
084.249 .259
SCHINDLER, R. A.
082.102
SCHINDLER, S. M.
078.008
SCHLEGEL, K.
083.011
SCHLEGEL, R.
066.012
SCHLOSSER, W.
097.026
155.069 .088
SCHMADEBECK, R.
094.587 .755
SCHMADEL, L. D.
031.035
SCHMEIDLER, F.
003.014
004.091 .092 .093
005.006
SCHMID, R.
094.175 .643
SCHMID-BURGK, J.
061.070
063.029
SCHMIDT, D. S.
042.029 .053
SCHMIDT, H. H.
094.940
SCHMIDT, H. U.
072.065
080.041
142.117
SCHMIDT, K.-H.
141.124
SCHMIDT, M.
125.036
141.034
SCHMIDT, R.
094.453 .830
SCHMIDT, T.
131.169
SCHMIDT, T. E.
105.112
SCHMIDT, W.
102.006
SCHMIDT-BLEEK, F.
105.002
SCHMIDT-KALER, T.
142.027 .118
153.033
155.069 .088
SCHMIDTKE, G.
071.015
SCHMIED, L.
075.007
SCHMITT, A.
114.044
SCHMITT, G. A.
082.079
SCHMITT, R. A.
094.092 .224 .225 .226
.386 .397 .682
105.061 .069
SCHMUTZER, E.
061.060
SCHNEEWEISS, A. B.
SEE SHNEJVAJS, A. B.

SCHNEIDER, E.
094.277 .512 .535 .799
SCHNEIDER, H.
094.211 .653
SCHNEIDER, K.
032.034
SCHNEIDER, M.
021.006
SCHNEIDER, W. C.
054.019
SCHNEIDMILLER, R. F.
094.465
SCHNETZLER, C. C.
094.157 .378 .412 .689
SCHNOPPER, H. W.
114.078
142.009 .011 .028 .127
.133
SCHNUR, G.
082.098
SCHOBER, H. J.
034.087
SCHOBER, J.
155.027
SCHOENBERNER, D.
064.069
SCHOENEICH, W.
114.143
SCHOENFELDER, V.
142.094 .103
SCHOENHARDT, R. E.
141.536 .542
SCHOENMAKER, A. A.
141.115
SCHOLL, H.
100.019 .045
SCHOLZ, C.
094.489
SCHOLZ, D.
075.011
SCHOLZ, E.
007.000
SCHOLZ, M.
064.027
SCHONFELD, E.
094.054 .233 .333 .381
.432 .698 .717
SCHONFELDER, V.
032.034
SCHOOLMAN, S. A.
034.126
073.072
SCHOPF, J. W.
094.448
SCHORN, R. A.
093.027 .028
SCHRADER, H. W.
141.131
SCHRAMM, D. N.
061.008 .015 .022 .035
065.018 .060
SCHREIBER, E.
094.489
SCHREIER, E.
142.045 .138
SCHREURS, J. W. H.
094.659
SCHROEDER, W.
003.152
SCHROETER, E. H.
080.038

SUNYAEV, R. A.
117.015
142.007 .086
151.002
SURI, A. N.
076.008
SURKOV, IU. A.
093.002
SURKOV, YU. A.
093.012 .016 .048
094.854
SURMELIAN, G. L.
022.006 .048
SURNIN, YU. V.
052.008
SUROWIECKI, J.
003.043
SUSZYCKI, L.
162.040
SUTTER, J.
094.709
SUTTON, A. L.
094.336
SUTTON, G.
094.583 .778 .779
SUTTON, R. L.
094.055 .304 .607
SUZUKI, K.
103.100 .114 .116 .132
SUZUKI, Y.
034.063
SVALGAARD, L.
072.057
080.014
084.252
SVECHNIKOV, M. A.
152.013
153.042
SVESTKA, Z.
073.089
078.029
SVETASHKOVA, N. T.
104.028
SVIDERSKIENE, Z.
114.090
SVIRZHEVSKY, N. S.
143.061
SVOLOPOULOS, S. N.
008.059
114.110
SWAMY, K. S. K.
114.083
SWANDIC, J. R.
160.017 .024
SWANENBURG, B. N.
143.002
SWANN, G. A.
094.055 .305 .605 .607
SWANSON, P. N.
077.036
SWARTZ, W. E.
082.006
SWARZTRAUBER, P.
074.057
SWEDLUND, J. B.
100.027
SWEENEY, M. A.
119.002
SWEIGART, A. V.
065.074

SWIERKOWSKA, S.
103.102
SWIHART, T. L.
003.113
014.020
141.008
SWINDELL, W.
099.063
SWINGS, J. P.
114.102
SYKES, M. J.
097.077
SYKORA, J.
071.050
SYMES, R. F.
105.088
SYNAKH, V. S.
151.047
SYROVATSKII, S. I.
073.107
SYROVATSKIJ, S. I.
061.068
143.081
SYTINSKII, A. D.
085.002
SZAFRANIEC, R.
121.066 .902
SZEBEHELY, V.
117.002
SZECSENYI-NAGY, G.
122.120
SZEIDL, B.
122.090
SZYMANSKI, W.
072.070

TABOR, J. E.
122.040
TABORDA, J.
042.027
TABORKO, I. M.
094.884
TACKETT, S. L.
105.020
TADDEUCCI, A.
094.170 .219 .367 .675
TADEMARU, E.
155.017
TAFF, L. G.
153.012
TAGIROV, E. A.
162.079
TAGO, E. V.
154.008
TAIT, W.
141.029
TAKAGI, K.
022.061
TAKAGI, S.
044.010 .035
045.008
TAKAHASHI, H.
082.042
TAKAHASHI, K.
052.036 .037
061.052
TAKAKURA, T.
077.001
TAKAOKA, N.
061.058

TAKAOKA, N.
094.422
TAKAYANAGI, A.
077.060
TAKECHI, A.
082.050
TAKEDA, H.
094.328 .333 .615 .631
105.111
158.129
TAKENS, R. J.
064.022
TALBOT JR., R. J.
065.076
TALPAERT, Y.
151.054
TALWAR, S. P.
064.038
091.015
TAMBOVSKI, G. A.
034.097
TAMHANE, A. S.
094.507 .508 .509 .800
TAMMANN, G. A.
006.000
160.025
TANABE, H.
082.050
TANAKA, H.
077.048
TANAKA, K.
073.077 .078 .081
076.017
TANAKA, T.
094.690 .926
099.034
105.005
TANAKA, Y.
084.412
142.004 .029 .120
TANANBAUM, H.
142.045 .052 .097 .138
.150
158.057
TANANBAUM, H. D.
142.059
TANDBERG-HANSSEN, E.
072.051
073.065 .101
TANDON, J. N.
062.064
065.007
TANNER, R. W.
004.017
032.032
041.028
TANSKANEN, P.
084.008 .009
TAPLEY, B. D.
042.003
TAPPING, K. F.
075.028
TAPSCOTT, J. W.
114.128
TARAFDAR, S. P.
064.077
114.020
TARANOV, V. I.
117.038
TARANOVA, O. G.
093.015

THROWER, P. A.
094.620
THUAN, T. X.
158.056
THUM, C.
106.027
THYSSEN-BORNEMISZA, S.
066.080
TIDWELL, E. D.
022.016
TIDY, E.
094.082
TIFFT, W. G.
141.056
158.058 .106 .158
160.001
TIGELAAR, H. L.
131.104
TIKHONOV, T. V.
094.888
TIKHONOVA, T. V.
094.874
TIMOFEEV, B. V.
077.025
TIMOFEEVA, P. M.
079.103
TIMOSHKOVA, E. I.
042.030
052.039
TIMOTHY, A. F.
074.083
TIMOTHY, J. G.
034.034 .115
TINBERGEN, J.
034.006
TINSLEY, B. A.
082.032 .042
TINSLEY, B. M.
065.010
158.041
161.003
162.060
TIOMNO, J.
066.081 .108 .112
TISHCHENKO, A. P.
094.959
TITARCHUK, L. G.
082.057
091.016
093.084
TITENKOV, A. F.
143.032
TITHERIDGE, J. E.
083.021
TITLE, A. M.
034.055
073.071
TITTER, J. C.
155.078
TITTMANN, B. R.
094.288 .488 .782 .917
TITULAER, C.
094.482
TJIN A DJIE, H. R. E.
064.022
TKACHENKO, V. I.
078.023
TKACHEV, G. N.
083.042
TLAMICHA, A.
077.048

TOBAILEM, J.
105.036
TODD, T.
094.487 .783
TODORAN, I.
120.002
121.039
122.073
TOELEID, O. A.
154.008
TOJO, A.
079.001
TOKSOEZ, M. N.
094.168 .575 .779
TOKSOEZ, N.
094.583 .778
TOLBERT, C. R.
008.031
TOLLAND, H. G.
081.019
TOLSTYKH, V. P.
121.023
TOMA, E.
041.013
TOMANOV, V. P.
102.019 .020
TOMASI, P.
141.023 .132
TOMASKO, M. G.
099.055 .063
TOMELLERI, V.
045.017
TOMITA, K.
032.031
034.024
162.063 .087 .088
TOMKIN, J.
126.021
TOMOSCHEIT, D.
031.065
TOMOZOV, V. M.
062.004
073.033
TONEY, J.
071.007
124.110
TOOLIN, R. B.
082.054
TOOMASSON, L.
121.053
TOON, O. B.
097.057
TOOR, A.
125.032
142.041
155.008
TOPTYGIN, I. N.
062.033 .068
078.037· .049 .065
143.033
TORBIN, S. I.
082.038
TORELLI, M.
075.022
TORII, Y.
032.031
TORKHOV, V. A.
125.028
TORNBERG, N. E.
094.819

TORRAO, T.
041.049
046.023
TORRENCE, G. W.
141.037
TORRES, C.
103.117 .128 .129
TOSA, M.
142.005
155.066
TOTH, R. A.
082.102
TOTOCHAVA, A. G.
122.063
TOTOMANOV, I.
041.005
TOTON, E. T.
162.085
TOTUBALINA, M. G.
034.095
TOVMASSIAN, H. M.
158.002
TOYAMA, K.
158.110
TRAFTON, L.
091.006
099.088
100.031 .039 .047
101.011
TRAFTON, L. M.
100.042
TRAILL, R. J.
094.327
TRAINOR, J. B.
010.008
011.027
103.004
TRAINOR, J. H.
051.002
TRAJMAR, S.
084.041
TRAKHTENGERTS, V. YU.
061.069
TRANSKIJ, I. A.
078.051
TRAN-ZUJ-TKHOAN
041.024
TRASCO, J. D.
099.023
TRAUB, W. A.
066.021
093.022
099.057
TRAUGER, J. T.
099.057
TRAUTMAN, A.
066.014
TRAVESI, A.
094.392
TRAVING, G.
003.114
064.027
TRAVIS, L. D.
064.048 .066
080.008
TREANOR, P. J.
008.028
082.101 .121
TREDER, H.-J.
003.010
004.010 .013

YENCHA, A. J.
131.099
YENGIBARIAN, N. B.
063.012
YENGIBARIAN, N. B.
SEE ENGIBARYAN, N. B.
YEOMANS, D. K.
102.008
103.127 .131
YERBURY, M.
094.827
YERBURY, M. J.
100.006
YESEPKINA, N. A.
SEE ESEPKINA, N. A.
YESIPOV, V. F.
158.032 .049
YEUNG, S. C.
097.050
YEUNG, S. Z.
142.136
YIN, L.
094.587 .755
YIN, L. I.
094.301
YIOU, F.
022.077
YODH, G. B.
143.035
YOKOI, K.
065.164
YOKOYAMA, K.
044.035
YOKOYAMA, Y.
094.274 .723
105.034
YONGE, C. J.
094.123
YORK, D.
094.712
YORK, D. G.
114.066 .121 .122 .123
.124 .125
131.063 .064 .065 .066
.067 .166
YOSHIDA, J.
042.902
YOSHIOKA, O.
105.133
YOST, E.
094.483
YOST, J.
113.035
YOUMANS, A. B.
141.506
YOUNG, A.
153.030
YOUNG, A. T.
080.037
093.001 .006 .017 .019
097.008
YOUNG, D. S. DE
141.046 .079
158.077
YOUNG, E.
152.002
YOUNG, J. W.
093.006
094.055
YOUNG, L. D. G.
093.001

YOUNG, L. G.
093.006 .019 .029
YOUNG, R. S.
094.946
YOUNGBOM, L.
114.055
YOURGRAU, W.
162.011
YOUSEF, S.
076.006
YU, G.
062.027
YUDIN, O. I.
077.025
YUDIN, V. M.
066.033
YUDOVICH, L. A.
084.254
YUDOVICH, V. L.
084.246
YUEN, P.
083.006
YUGAJ, M. A.
079.103
YUHAS, D.
094.255 .414 .806
YUHAS, D. E.
094.808
YUKHIMUK, A. K.
077.040
YUMI, S.
045.023 .024
YUN, H. S.
072.074
YUNGELSON, L.
117.017 .018
YUNGELSON, L. R.
117.008
YURCHENKO, B. N.
082.093
YUROVSKAYA, L. I.
033.008
077.022 .050
YUROVSKIJ, YU. F.
077.021 .022
YUROVSKY, YU. F.
077.049 .050
YUSUPOV, U.
095.005
YUTANI, M.
122.100

ZABELINA, I. A.
082.128
ZABUSKY, N. J.
083.072
ZADUNAISKY, P. E.
102.010
ZAEHRINGER, J.
094.403 .425
105.046
ZAGREBIN, D. V.
081.001
ZAITSEVA, G. I.
113.037
ZAITSEVA, G. V.
122.062
ZAJDLER, L.
007.000

ZAJTSEV, N. A.
066.043
ZAKHARCHENKO, V. F.
078.050
143.024
ZAKHAR'EV, B. V.
094.603
ZAKHAROV, V. D.
003.132
ZAKHAROVA, G. A.
073.032
ZAKHAROVA, P. YE.
152.013
153.042
ZAKIROV, M. M.
152.007
ZAMPIROLLO, P.
045.015
ZAPPALA, R. A.
073.082
075.009
ZAPPALA, V.
103.114 .116
ZARNECKI, J. C.
125.007 .019
ZARNITSYNA, I. G.
093.076
ZASOV, A. V.
158.068
ZATSEPIN, V. I.
143.061
ZAUMEN, W. T.
076.024
142.050
ZAVARZIN, YU. M.
034.021
ZAVELEVICH, F. S.
093.075
ZAVRIEV, A. V.
085.009
ZAYARNAYA, E. S.
083.045
ZDUNKEVICH, M. D.
091.020
ZECH, G.
002.034
ZEILIK, M.
092.001
ZEINALOV, R. A.
052.001
ZEISSIG, G. A.
141.527
ZEJNALOV, S. K.
114.107
ZEL'DOVICH, S. A.
066.044
ZEL'DOVICH, YA. B.
066.034
162.068
ZELENKOVA, L. V.
083.050
ZELENY, L. M.
084.232
ZELIKMAN, M. A.
062.062
ZELLNER, B.
100.023
ZELMANOV, A. L.
066.074
ZEMLYAKOV, A. S.
051.009

Subject Index